THE GEOLOGY OF IRELAND

The Giant's Causeway, where the visitor can walk down to the sea across a pavement of upturned basalt columns (the colonnade). Ponded lava at the bottom of an old river valley offered ideal conditions for slow and uniform cooling, conditions that generated an almost perfect hexagonal array of fractures growing at right angles to the cooling surface at the base of the flow. Each basalt column then contracted independently along its length breaking into short segments terminated by convex or concave joints, as in the foreground. A dolerite dyke separates the Grand Causeway (pictured) from the Middle Causeway (farther west) and in the gap rests a fallen block of basalt from the curvi-columnar zone (the entablature) of a higher flow.

Photograph by C. H. Holland; caption by J. Preston

THE GEOLOGY OF IRELAND

Second Edition

Edited by

Charles Hepworth Holland
and
Ian S. Sanders

DUNEDIN

Published by
Dunedin Academic Press ltd
Hudson House
8 Albany Street
Edinburgh EH1 3QB
Scotland

www.dunedinacademicpress.co.uk

First Edition published 2001

Second Edition ISBN:
9781903765715 (Cloth)
9781903765722 (Paperback)

British Library Cataloguing in Publication data
A catalogue record for this book is available from the British Library

Design and typesetting by Makar Publishing Production, Edinburgh
Printed by Hussar Books in Poland

Contents

Contributing Authors

D. M. Chew	Department of Geology, Trinity College, Dublin
P. Coxon	Department of Geography, Trinity College, Dublin
J. S. Daly	UCD School of Geological Sciences, University College, Dublin
J. R. Graham	Department of Geology, Trinity College, Dublin
G. L. Herries Davies	Department of Geography, Trinity College, Dublin
C. H. Holland	Department of Geology, Trinity College, Dublin
S. G. McCarron	Department of Geography, National University of Ireland, Maynooth
F. J. G. Mitchell	Department of Botany, Trinity College, Dublin
D. Naylor	Department of Geology, Trinity College, Dublin
J. Preston	Formerly: Department of Geology, Queen's University, Belfast
I. S. Sanders	Department of Geology, Trinity College, Dublin
G. D. Sevastopulo	Department of Geology, Trinity College, Dublin
P. M. Shannon	UCD School of Geological Sciences, University College, Dublin
M. J. Simms	Department of Geology, Ulster Museum, Belfast
C. J. Stillman	Department of Geology, Trinity College, Dublin
P. N. Wyse Jackson	Department of Geology, Trinity College, Dublin

Preface

The first edition of *The Geology of Ireland* was published by Dunedin Academic Press in 2001. Although it was based on *A Geology of Ireland* (1981), it was a fully altered book with very many references. It took account of the considerable amount of work on Irish geology undertaken during the preceding two decades. It has rather quickly gone out of print.

The present Second Edition is fully revised. There are significant changes in authorship. In the past few years there has been considerable additional research, naturally unevenly distributed. There are many additional references to the extensive literature. Diagrams and illustrations are now largely in colour.

As was the intention in the previous book, we hope that this edition will be of value to students of Irish and British geology, of all ages, professional and amateur, and also to those many earth scientists internationally who must take interest in this remarkable island, poised on the edge of Europe and yet linked in so many ways with North America. Our treatment again is in terms of historical geology, but our new arrangement of chapters takes better account of palaeogeographical evolution. Again we include a short survey of geophysical evidence, an account of the geology of offshore Ireland and a review of the fascinating history of Irish geology itself.

This book, as was the case with its predecessor, depends upon the work of many people, some of whom are mentioned in the text and others whom it has not been possible to acknowledge. Sources of illustrations are given and acknowledged as appropriate. Elaine Cullen of the Department of Geology, Trinity College has, with her usual flair, undertaken the formidable task of drafting the diagrams. In this evolving enterprise we have continued to enjoy the confidence of those working with Dunedin Academic Press, especially Dr Douglas Grant. Their sustained advice and help in the preparation of this new edition is much appreciated.

Charles Hepworth Holland
Ian S. Sanders
Trinity College, Dublin, January 2009

Under bare Ben Bulben's head
In Drumcliff churchyard Yeats is laid.
An ancestor was rector there
Long years ago, a church stands near,
By the road an ancient cross.
No marble, no conventional phrase;
On limestone quarried near the spot
By his command these words are cut:

Cast a cold eye
On life, on death.
Horseman, pass by!

W. B. Yeats
Under Ben Bulben 1938 *

1

Introduction

Charles Hepworth Holland and Ian S. Sanders

This book is about the island of Ireland as a physical whole. Studies of marine geology and geophysics, stimulated by the search for petroleum, have shown that the Irish coastline, like that of mainland Britain, does not, as was once believed, capriciously cut off from observation a solid geology continuous in character from that of the land. The configuration of the coastline in places follows the margins of submarine subsided basins of thick Mesozoic and Tertiary rocks. Thus, in this book, we do not confine ourselves to the onshore, but provide also an account of Ireland's economically important offshore geology (Chapter 17).

The land of Ireland is largely of Palaeozoic rocks, amply covered in places by the glacial deposits of the very latest episodes of geological history. Clearly there were earlier connections between the environments represented by the Irish Lower Palaeozoic rocks and those recorded in rocks of the same age in mainland Britain and even in the New World. Evidence of these connections is to some extent concealed by the younger rocks of the marginal basins and has been made more elusive by the effects of ocean floor spreading, episodes of subduction, and the displacement of terranes. Certainly, as always in geology, the international context must be kept in mind.

The details of the Irish landscape, rural and urban; the distribution of bog and forest; the nature of the present day vegetation; all depend upon the activities of man during his short occupancy of this island. The scenery of both upland and lowland Ireland (Fig. 1.1) is also much affected by the climatic extremes of the Quaternary Period discussed in Chapter 15. We have but small areas of those Mesozoic rocks which William Smith, 'The Father of English Geology', saw in succession in southern and eastern England, where they are widely displayed, inclined to the south-east 'like slices of bread and butter on a plate'. Except for the relatively small area of the Lough Neagh basin we lack, also, the Tertiary basins of London and Hampshire. The exceedingly varied and sometimes very beautiful landscape of Ireland (Whittow, 1974; Holland, 2003) is controlled, regionally at least, by a very varied but largely Precambrian and Palaeozoic solid geology (Fig. 1.2), the skeleton of its anatomy. These rocks form the mountainous rim of Ireland, and the mountainous 'islands' which arise from the central plains are largely of Old Red Sandstone or of Lower Palaeozoic rocks. Within the rim the Lower Carboniferous widely provides the concealed floor to the Quaternary blanket. In the north and west, Precambrian scenery, particularly that of its quartzite mountains, is beautifully depicted in the paintings of Paul Henry in Connemara and Derek Hill in Donegal. Farther south the splendid coastline of the Burren in County Clare is carved in Lower Carboniferous limestones; the famed Cliffs of Moher in Upper Carboniferous sedimentary rocks; the peninsulas of Kerry and Cork in Old Red Sandstone; and so on. An exception is County Antrim with its Giant's Causeway in columnar basalts of Tertiary age (Frontispiece) and where white Chalk and dark igneous rocks can be seen juxtaposed in striking contrast.

In what follows we trace the geological history of what is now Ireland from Precambrian times to the present. The period of time (Table 1.1) is immense. The basis of historical geology has itself oscillated over the years between uniformitarianism, catastrophism, and a kind of all-embracing cyclicity of varied physical and organic processes. A modern view should recognise large-scale cyclicity in the grand pattern of crustal evolution. It should also take account of the many events, some regional, some global, which have punctuated the stratigraphical record. An example would be widespread sea level changes related to the waxing and waning of glaciations.

In consideration of Mesozoic and later times, the advent of the science of palaeomagnetism, followed by the accumulated evidence for ocean floor spreading, has lent much support to the idea that there has been continental drifting apart of what are now regarded as Eurasian and American plates from a spreading axis along the Mid-Atlantic ridge. The Antrim igneous rocks described in

Fig. 1.1 Twelve Bens, Connemara, County Galway, across blanket bog.
Reproduced by permission of the Department of the Environment, Heritage and Local Government, Ireland.

Chapter 14 are but part of a much larger North Atlantic province associated with this process of break up and separation. This closely understood latest episode of continental drift now provides a background to the understanding of such matters as palaeobiogeography.

A contrasting situation exists if we return to the Palaeozoic, where the results of palaeontology, stratigraphy, and structural geology must themselves be used as evidence in suggesting the original disposition of continents and oceans. Palaeogeographical maps for the Early Palaeozoic, such as those of Cocks and Fortey (1982) and Cocks and Torsvik (2002), show north-western Ireland attached to a North American plate and the remainder of the island associated at an earlier stage with Gondwana and, by later Ordovician times, with the Baltic region. Thus the Iapetus Ocean is seen to have separated what are now two parts of Ireland. Evidence from Irish Lower Palaeozoic rocks is consistent with the picture of a gradual closing of Iapetus to be completed by Late Silurian times, with a final shunt, locking the opposing continents together during the Early Devonian. We refer to this drawn-out saga of the shrinking ocean, which began in Cambrian times, as the Caledonian orogenic cycle.

Terminology relating to the demise of Iapetus can be confusing. The last, late shove in the Early Devonian is known as the Acadian Orogeny, and is described in Chapter 8. It was associated with the injection of copious volumes of granitic magma, which now form the batholiths of Donegal, Galway, and Leinster. However, the term 'Late Caledonian' is preferred for these intrusions, since the Acadian has been defined elsewhere as a brief episode of tectonism within the protracted period over which the individual granite bodies were emplaced.

The closure of Iapetus was a complex process: in Early Ordovician times, subduction led to the convergence of a volcanic island arc, built up on the ocean floor, with the northern continental landmass (known as Laurentia). Chapter 4 outlines how the resulting collision led to thickening, folding, and heating of the continental margin in a 'fast and furious' event called the Grampian Orogeny.

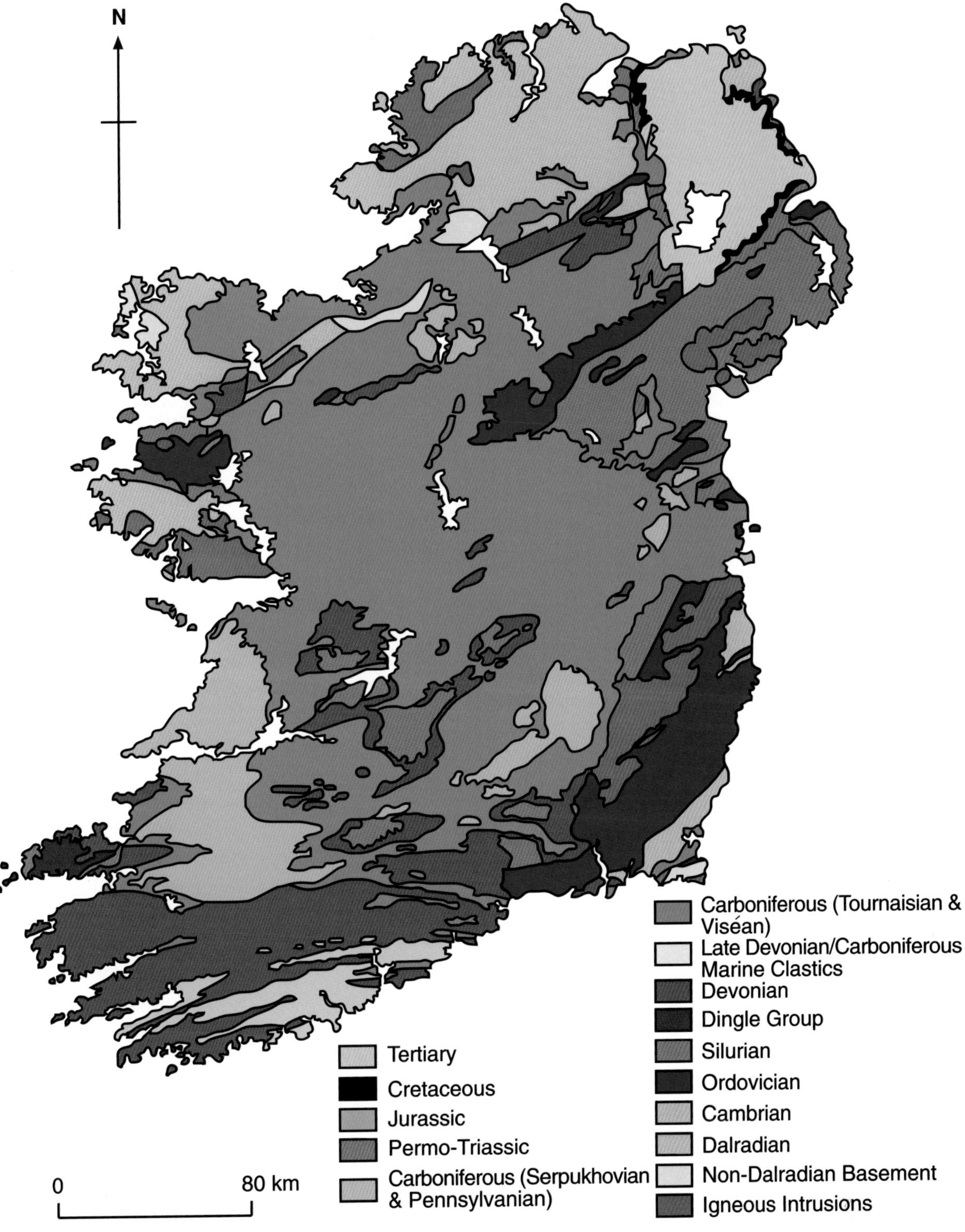

Fig. 1.2 Geological map of Ireland (compiled from many sources). The width of the outcrop of Cretaceous rocks is exaggerated.

Table 1.1 Geochronometric scale (extracted largely from a chart compiled by Gradstein *et al.*, 2004).

Eras	Periods			Approximate age in millions of years (Ma)
Cenozoic	Quaternary			2.65
	Tertiary	Neogene	Pliocene	5.3
			Miocene	23
		Palaeogene	Oligocene	34
			Eocene	56
			Palaeocene	65
Mesozoic			Cretaceous	146
			Jurassic	200
			Triassic	251
Palaeozoic		Upper	Permian	299
			Carboniferous	359
			Devonian	416
		Lower	Silurian	444
			Ordovician	488
			Cambrian	542
Precambrian eras	Proterozoic			2,500
	Archaean			4,567
origin of the Earth				4,567

Dispassionate consideration remains essential. If we are to understand this Palaeozoic past, palaeontologists, sedimentologists, and structural geologists must be mutually supportive. It seems that in making palaeogeographical models it is no longer safe glibly to accept orthogonal reconstructions. The recognition of laterally displaced terranes (see Hutton, 1987, for a useful treatment) is important, but there must be real evidence for these. It is unhelpful to regard every strip of fault-bounded country as a terrane.

During the Late Palaeozoic, Ireland as a whole formed part of a Laurentian plate and our splendid display, particularly of Lower Carboniferous rocks, can be clearly linked with its counterparts elsewhere in Europe. These rocks continue to be much studied. They are important in international stratigraphical classification. They are of economic importance.

If order is to be maintained in the study of this long geological history, use of a reasonably consistent and agreed stratigraphical nomenclature is needed. Geologists in Europe, where so much of the early stratigraphical work was done, have the problem of operating under an umbrella of long-established but very varied terminology. They must steer a course between the maintenance of well known and well tried terms, which do not fit into modern rules, and the provision of a stratigraphical nomenclature which is understandable from one worker to another and, above all, is understandable internationally. The old and the new are intermingled in the stratigraphy that follows. A case in point is the subdivision of the Cenozoic Era (Table 1.1) where we retain the long-established names Tertiary and Quaternary, even though the International Commission for Statigraphy recommends that their stratigraphical status is no longer valid. We have been as consistent as possible. These matters of principle are discussed in Holland (1978, 1986).

The scope of the present book is a full account of the geology – the actual rock record – in Ireland, with less emphasis on reconstructing the plate motions and other underlying mechanisms that produced the geology. For a broad synthesis of the latter, one should consult the book on the geological history of Britain and Ireland by Woodcock and Strachan (2004). A general guide to the geology of Northern Ireland is that by Mitchell (2004a) for the Geological Survey. The *Special Reports* of the Geological Society of London contain correlation charts for the various systems throughout the British Isles. The elaborate series of palaeogeographical and lithofacies maps in the Atlas published by the Geological Society (Cope *et al.*, 1992a) covers the whole of the British Isles. A pleasing addition to the literature is Wyse Jackson's (1993) walking guide to *The Building Stones of Dublin*.

A useful geological map of Ireland is the 1:500,000 sheet published in 2006 by the Geological Survey of Ireland in conjunction with the Geological Survey of Northern Ireland. Regional compilation maps covering the entire island (1:100,000 in the South and 1:50,000 in the North) are also available from the respective geological surveys. Several larger-scale geological maps of parts of Ireland, privately produced in recent years, are referred to at appropriate points in the text that follows.

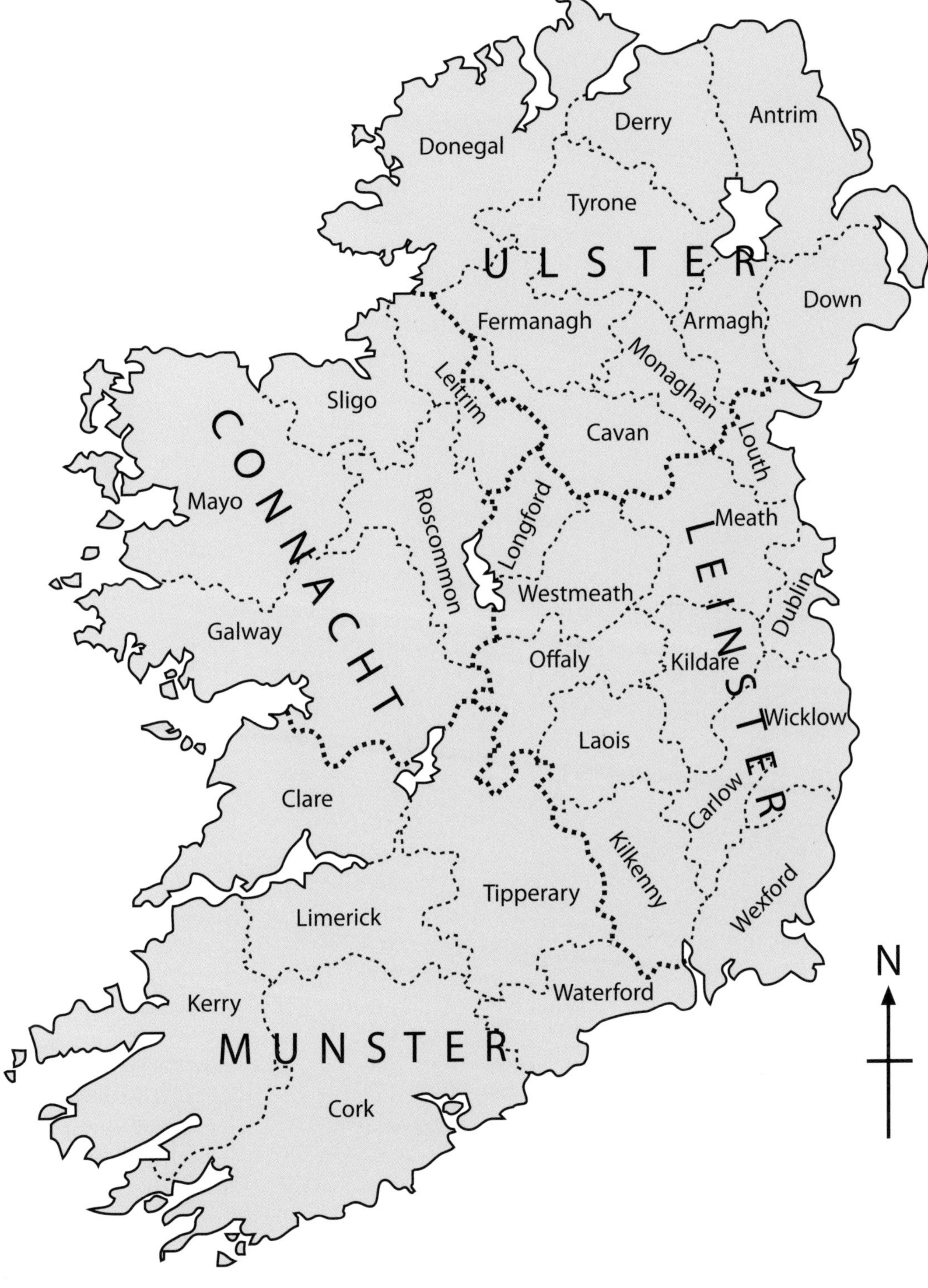

Fig. 1.3 County map of Ireland. Provinces also are indicated. The name Connacht shown on the map and based upon the Irish spelling is widely used, though the version Connaught is often found in the literature, for instance of the coalfields.

2

Precambrian

J. S. Daly

Precambrian rocks make up about 10% of the surface area of the island of Ireland (Fig. 2.1). The oldest known Precambrian rocks are syenitic orthogneisses (i.e. gneisses with igneous protoliths) of the Rhinns Complex, exposed on Inishtrahull (Fig. 2.1) off Malin Head, County Donegal. The rocks of Inishtrahull are dated at 1779 ± 3 Ma (Fig. 2.2, Daly *et al.*, 1991) and thus belong to the Palaeoproterozoic (see Table 2.1 for Precambrian stratigraphic nomenclature). This era of Earth history is also represented in Ireland by the somewhat younger Palaeoproterozoic rocks of the Annagh Gneiss Complex of NW Mayo (Fig. 2.1, Fig. 2.10), where Mesoproterozoic orthogneisses also occur. Altogether orthogneisses comprise only a small proportion of the total exposed Precambrian. By far the greater proportion of Precambrian rocks by area is made up of supracrustal rocks, i.e. rocks with sedimentary and volcanic protoliths. Almost all of these belong to the Dalradian Supergroup, a thick succession of rocks which accumulated during the Neoproterozoic and which are exposed in a series of inliers in Connacht and Ulster (Fig. 2.1). Although Dalradian deposition probably continued at least into the Middle Cambrian in Ireland, the Dalradian sequence is treated as a whole in this chapter. Other Precambrian supracrustal units shown in Fig. 2.1 include the metasediments of the Rosslare Complex in SE Ireland, the enigmatic Slishwood Division, and the Tyrone Central Inlier, which has recently been shown to be part of the Dalradian (Chew *et al.*, 2008).

The Precambrian rocks of Ireland have all been affected by Caledonian events. As discussed further in Chapters 4 and 8, the Caledonian orogenic cycle refers to the sum of tectonic events surrounding the opening and closing of the Iapetus Ocean (Bingen *et al.*, 1998; Kamo *et al.*, 1989; Svenningsen, 2001), which separated the North American continent (Laurentia) in north-western Ireland from the Avalonian continent in south-eastern Ireland. The Caledonian orogenic cycle involved two main collisional orogenic episodes. The second of these was the collision associated with the terminal closure of the Iapetus Ocean in the Silurian (Chapter 8), and is referred to here as the Late Caledonian Orogeny. This tectonic event was fundamentally important in laying the foundations of the modern Irish crust. During the Late Caledonian Orogeny the bulk of the Irish crust was assembled from a series of terranes (Murphy *et al.*, 1991) generally striking NE–SW or E–W. Thus many if not all of the Precambrian outcrops are allochthonous. As a result, their original tectonic setting is open to interpretation, giving rise to much debate and ongoing research.

Earlier in the Caledonian orogenic cycle, at about 470 Ma (early Middle Ordovician) the Precambrian rocks in north-western Ireland were deformed and metamorphosed in an event known as the Grampian Orogeny (Chapter 4; Lambert and McKerrow, 1976; Lindsay *et al.*, 1989). It occurred while the ocean was still quite wide, and subduction led to the collision of an island arc with the northern (Laurentian) continental margin. All the Precambrian supracrustals and the older basement rocks in this region were affected. The rocks most prominently associated with the Grampian Orogeny are the supracrustal rocks of the Dalradian Supergroup, mentioned above. Dalradian rocks form the backbone of much of the spectacular scenery of the western seaboard and its adjacent mountains, and make up approximately ten percent of the land area of Ireland. In Scotland, the Dalradian rocks were intruded, possibly after some deformation, by the Ben Vuirich granite at *c.*590 Ma (Rogers *et al.*, 1989; Tanner and Leslie, 1994; Tanner, 1996). This, together with the ion microprobe U–Pb zircon age of 601 ± 4 Ma for a felsic tuff within the Tayvallich volcanics (Dempster *et al.*, 2002), a unit high within the Dalradian stratigraphy, demonstrates a Precambrian age for the deposition of most of the Dalradian Supergroup.

Precambrian rocks that are not Dalradian are exposed in only a few places in north-western Ireland, and their

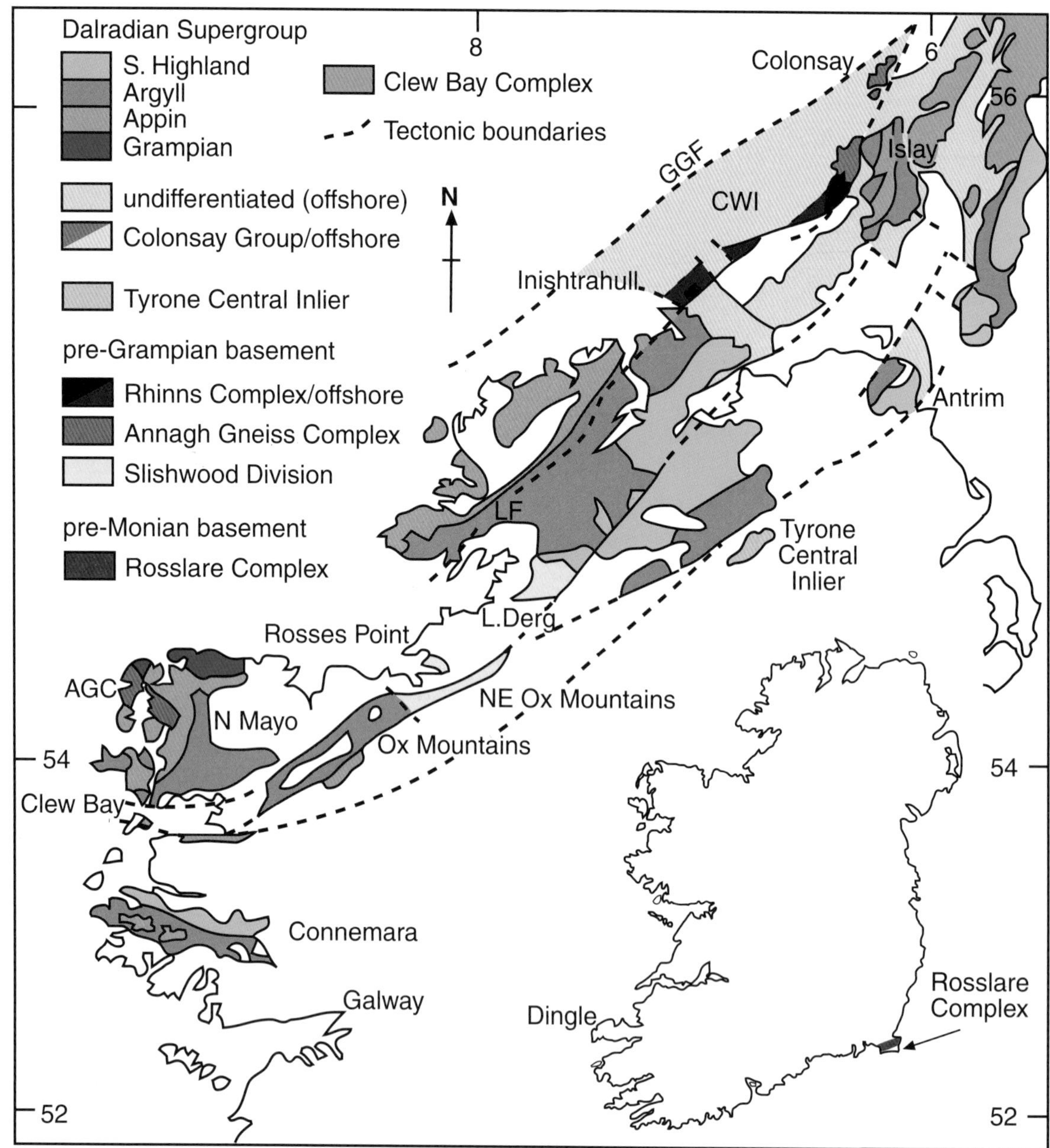

Fig. 2.1 Sketch map showing the location of selected Precambrian rocks in Ireland and south-western Scotland and the main stratigraphical subdivisions of the Dalradian Supergroup. The distribution of the Rhinns Complex and Colonsay group is after Muir *et al.* (1994b). Abbreviations: AGC = Annagh Gneiss Complex, CWI = Colonsay West Islay Block, GGF = Great Glen Fault, LF = Leannan Fault.

total outcrop comprises about a half of one percent of the land area. These Precambrian outcrops have a more complex history before the Grampian Orogeny than the Dalradian, and have been designated as *pre-Grampian* basement using two distinct criteria – Precambrian igneous protolith ages and evidence of pre-Grampian metamorphism. Pre-Grampian orthogneisses outcrop in two areas in Ireland (Fig. 2.1), as already mentioned. These are the Rhinns Complex (or Colonsay–West Islay block) on Inishtrahull off Malin Head in Donegal, and the Annagh Gneiss Complex (AGC) in north Mayo. Two other occurrences traditionally considered to be pre-Grampian basement comprise supracrustal rocks (paragneisses). These are the Slishwood Division (Max and Long, 1985) in the neighbouring NE Ox Mountains, Lough Derg and Rosses Point inliers in counties Donegal, Fermanagh and Sligo, and the Central Inlier of Tyrone (Fig. 2.1). Chew *et al.* (2008) have suggested

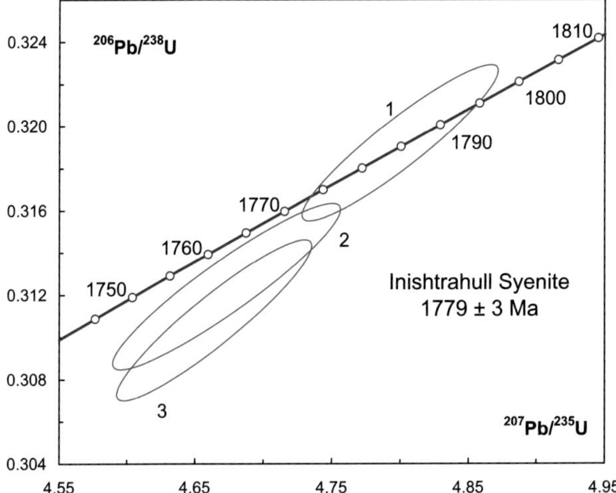

Fig. 2.2 Concordia diagram showing U–Pb zircon data for the Inishtrahull Syenite, Rhinns Complex, Inishtrahull (Daly, 2001) and its calculated age of 1779 ± 3 Ma (Daly *et al.*, 1991). Error ellipses show data from different hand-picked zircon fractions: 1 = prisms, 2 = best prisms (both abraded) and 3 = detached tips. The latter were analysed separately to see if there were age differences between different parts of the zircon grains. Lack of an age difference between the tips and cores of the grains and the concordant or near-concordant ages indicate a lack of zircon inheritance, consistent with the juvenile character of the Rhinns Complex indicated by Sm–Nd data (see text).

Table 2.1 Subdivision of Precambrian time after Hofmann (1992).

Eon		Ma	Era
Cryptozoic	Proterozoic	542	Neoproterozoic
		1000	Mesoproterozoic
		1600	Palaeoproterozoic
	Archaean	2500	Neoarchaean
		2800	Mesoarchaean
		3200	Palaeoarchaean
		3600	Eoarchaean
		4000	

that the Central Inlier may correlate with the Dalradian, thus casting doubt on its pre-Grampian status.

Seismic refraction studies (Lowe, 1988; Lowe and Jacob, 1989; Jacob *et al.*, 1985; Chapter 18) indicate the presence of rocks with velocities greater than 6 km s^{-1} at a relatively shallow depth (<10 km) beneath the Palaeozoic cover in central Ireland. These high velocity rocks have generally been assumed to be crystalline Precambrian basement. However, investigation of high-grade crustal xenoliths transported in Lower Carboniferous volcanics at localities in the Irish midlands (Westmeath and Offaly; Strogen, 1974b; Van den Berg *et al.*, 2005) indicate a likely absence of Precambrian rocks. The xenoliths yield PT values compatible with an origin in the lower crust (Van den Berg, 2005) and their experimentally determined seismic velocities match present-day lower crustal values (Lowe and Jacob, 1989) derived from the seismic refraction data (Van den Berg *et al.*, 2005). U–Pb zircon dating indicates that the xenoliths comprise sedimentary, volcanic and granitic protoliths of Palaeozoic age (Daly and Van den Berg, 2004) and thus represent high metamorphic grade Caledonian rocks, as suggested by Max and Long (1985), rather than Precambrian basement. The deep crustal xenoliths are further discussed in Chapter 18.

Recognition that the two halves of the island belonged to different tectonic plates in Caledonian times has had a profound influence on geological thinking. Rocks of the Avalonian continental margin, exposed in south-eastern Ireland (Chapters 6, 7 and 8) preserve evidence of at least three orogenic events, the last of which relates to the Late Caledonian collision. The earliest of the three is the Avalonian Orogeny, which took place in the Neoproterozoic, and was followed by the Monian Orogeny in the early Ordovician (e.g. Tietzsch-Tyler, 1996). Pre-Monian (Avalonian) rocks occur in the Rosslare Complex in County Wexford in the south-eastern corner of Ireland. These rocks may continue westwards as far as the west coast, as suggested by the occurrence of southerly-derived detritus of Rosslare Complex aspect in Devonian rocks in the Dingle area (Fig. 2.1; Todd, 2000).

Coloured large-scale maps (1:100,000 or larger) of all of the Precambrian regions are available: Connemara (Leake *et al.*, 1981; Long and McConnell, 1995); North Mayo (Max *et al.*, 1992); north-west Donegal (Pitcher and Berger, 1972); North Donegal (McConnell and Long, 1997); South Donegal (Long *et al.*, 1999); Donegal, Derry, Fermanagh, Tyrone and Antrim (Geological

Survey of Northern Ireland, 1997); Mayo–Sligo–Leitrim (Harney *et al.*, 1996) and Wexford (Tietzsch-Tyler and Sleeman, 1994a).

In this chapter the various groups of Precambrian rocks are treated under three headings. The first covers the pre-Grampian basement in the north-west, which includes the orthogneisses of the Rinns Complex and the Annagh Gneiss Complex, and the younger paragneisses of the Slishwood Division and the Tyrone Central Inlier. Moving next to the south-east corner of Ireland, the second heading covers the pre-Monian rocks of the Rosslare Complex. Finally, returning to the north-west, the Dalradian Supergoup is described and discussed.

Before starting, however, given the importance of geochronology and isotopic tracers in the investigation of Precambrian rocks, it is worth pausing to discuss the divison of Precambrian time and to review some of the isotopic dating methods that will be referred to in this and later chapters.

Precambrian time and stratigraphy

Precambrian rocks, unlike those of the Phanerozoic, contain few fossils. It is therefore necessary to rely on field relationships, lithostratigraphy and isotopic dating techniques, to establish the ages of geological events affecting them. In this chapter and throughout the book, ages are reported either in units of millions (10^6) of years before present (Ma) or billions (10^9) of years before present (Ga).

A younger limit for the Precambrian is taken at 542 Ma – the age of the base of the Cambrian (cf. Gradstein *et al.*, 2004b). A scheme to subdivide Precambrian time (Plumb, 1991), based on the work of the Subcommission on Precambrian Stratigraphy of the International Commission on Stratigraphy, has been ratified by the International Union of Geological Sciences. Larger scale subdivisons into eons and eras have been widely accepted and are now in general use. These include the division of the Precambrian into Archaean and Proterozoic eons with a boundary between them at 2500 Ma. The Proterozoic has been further divided into three eras named Palaeoproterozoic, Mesoproterozoic and Neoproterozoic with boundaries at 2500, 1600 and 1000 Ma respectively. These divisions are adopted in this chapter (Table 2.1). The Subcommission on Precambrian Stratigraphy has recommended subdivision of the Palaeoproterozoic, Mesoproterozoic and Neoproterozoic into eras (Gradstein *et al.*, 2004b) but, as yet, only the tripartite division of the Neoproterozoic into Tonian, Cryogenian and Ediacaran has gone into widespread use. Because of the lack of definitive ages for the relevant stratigraphic sequences in Ireland, use of these terms in this chapter is considered premature. Sadly, the attractively simple 100 Ma interval 'geon' scale, proposed by Hofmann (1990, 1999) has not met with wide acceptance.

Radiogenic isotope geochemistry and isotopic dating

Isotopic dating of orthogneisses and minor intrusions, integrated with structural observations and metamorphic petrology, are essential techniques for deciphering the geological history of Precambrian terrains. Although many primary features may survive, e.g. sedimentary structures and igneous intrusive contacts, in some cases the protoliths may be identifiable only through geochemistry. Tectonic modelling is often inhibited by the lack of clear pointers to the tectonic environment. As in modern orogens, determining the relative contribution of juvenile crust and reworked older material in ancient orogenic belts is of particular interest (e.g. Daly *et al.*, 2001, 2006). Methods of assessing the relative contributions of older and juvenile crust include Sm–Nd isotopic analysis of whole-rocks whose crystallisation ages have been obtained by U–Pb dating, typically of zircon, as well as Lu–Hf isotopic analysis of the same mineral. These techniques are among the variety of methods available (Table 2.2) to date magmatic, metamorphic, deformational and cooling events.

In this chapter the results of U–Pb, Rb–Sr, Sm–Nd, Lu–Hf, and Ar–Ar dating are used. The U–Pb decay scheme is one of the most powerful because it may be applied directly to fairly common accessory minerals such as zircon and monazite, which have high parent/daughter (i.e. U/Pb ratios). The age calculation depends only on solving the decay equation – the correction for initial lead incorporated when the mineral crystallised is very small. Similarly individual mineral species may be analysed to obtain ages based on the K–Ar decay scheme, because Ar, being a gas, is usually expelled from the mineral lattice at high temperatures (see discussion below).

In contrast, determination of Rb–Sr and Sm–Nd ages generally depends on the isochron method because the parent/daughter ratios are relatively low in most materials. This requires additional analysis of either a second mineral or the whole-rock in order to determine and correct for the isotopic composition of the daughter isotope originally present in the mineral (the initial isotopic composition). This relies on the assumption that the two minerals (or rocks) were in isotopic equilibrium

– i.e. had the same isotopic composition at the time determined by the isochron age. One exception is muscovite (white mica), which often has a sufficiently high Rb/Sr ratio such that the initial Sr content of the mineral is very small in comparison to Sr produced by radioactive decay. Thus muscovite Rb–Sr ages may sometimes be obtained without the need to analyse a second mineral, and the initial Sr may be ignored in the age calculations.

In general, all isotopic ages are subject to leakage of the radiogenic daughter isotope from the host mineral. To that extent all isotopic ages record the time when minerals quantitatively retain the accumulating daughter isotopes. This occurs when a mineral cools to or crystallises below a temperature known as the 'blocking' or 'closure' temperature. The method chosen for dating depends on the information required – different methods may provide different ages that date different events in the rock's history. Minerals with different closure temperatures can in principle provide a range of ages that date different points in the rock's thermal history (e.g. Sanders *et al.*, 1987). In reality closure temperatures are actually rather poorly known and are very difficult to calibrate. Moreover, other factors, especially deformation and the style of diffusion involved (intracrystalline versus grain boundary), are probably important factors in the interpretation of mineral ages in metamorphic rocks. However, as a general rule the sequence of ages is usually U–Pb zircon > Lu–Hf garnet > Sm–Nd garnet > U–Pb monazite > U–Pb titanite > Ar–Ar hornblende > Rb–Sr muscovite > Rb–Sr biotite, reflecting the relative closure temperatures of these mineral systems (Table 2.2).

Readers are referred to Faure (1986), Heaman and Ludden (1991), Gill (1997), Hanchar and Hoskin (2003), Dickin (2005) and Allègre (2008) for a detailed account of the various isotopic dating methods. Application of isotopic dating techniques to metamorphic problems are reviewed by Cliff (1985), Mezger (1990), Vance *et al.* (2003) and Vernon and Clarke (2008). Field geological techniques, geochronological applications and sampling techniques employed in Precambrian high-grade terrains are discussed by Passchier *et al.* (1990).

U–Pb dating

The U–Pb dating method (Table 2.2) is based on two independent decay schemes, ^{235}U–^{207}Pb and ^{238}U–^{206}Pb, differences in whose half-lives permit the construction of the Concordia curve – the locus of points with identical ^{235}U–^{207}Pb and ^{238}U–^{206}Pb ages plotted using coordinates of $^{207}Pb/^{235}U$ and $^{206}Pb/^{238}U$ (Fig. 2.2). Samples plotting on Concordia have identical ^{235}U–^{207}Pb and ^{238}U-^{206}Pb ages. Thus the method has an internal check on accuracy. The Tera–Wasserburg Concordia diagram is a useful alternative that depicts the data in co-ordinates of $^{207}Pb/^{206}Pb$ and $^{238}U/^{206}Pb$ (Fig. 2.3). In particular, discordance due to lead loss will cause affected data points to be displaced from Concordia along a horizontal line, i.e. with constant $^{207}Pb/^{206}Pb$ but increased $^{238}U/^{206}Pb$ ratios (Fig. 2.3). Alternatively if common (i.e. unradiogenic) lead is inadvertently included in the analysis (e.g. when an ion microprobe spot strays over the edge of a grain into the surrounding resin), the datum will be displaced towards increasing $^{207}Pb/^{206}Pb$ but decreasing $^{238}U/^{206}Pb$

Table 2.2 Summary of isotopic dating methods used in this chapter (after Heaman and Parrish, 1991; Mezger, 1990; Scherer *et al.*, 2001 and references therein).

Parent-Daughter	Half life (Ma)	Decay Constant (yr^{-1})	Material commonly dated	Mineral closure temperature
^{235}U–^{207}Pb ^{238}U–^{206}Pb	704 4,468	9.8485×10^{-10} 1.55125×10^{-10}	zircon baddeleyite monazite titanite	>800°C 650°C 600°C
^{40}K–^{40}Ar possibly measured as ^{39}Ar–^{40}Ar	1,250	5.543×10^{-10}	biotite muscovite hornblende Kfeldspar	300°C 350°C 530°C <250°C
^{87}Rb–^{87}Sr	48,800	1.42×10^{-11}	biotite muscovite hornblende igneous rocks	350°C ? 500°C 500°C
^{147}Sm-^{143}Nd	106,300	6.54×10^{-12}	garnet igneous and clastic sedimentary rocks (t_{DM} ages)	? *c.*700°C
^{176}Lu-^{176}Hf	37,200	1.865×10^{-11}	garnet zircon (model ages)	> 700°C

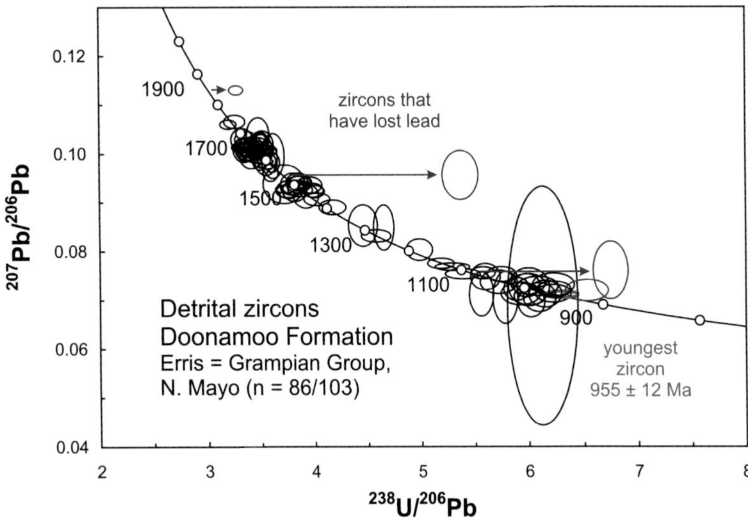

Fig. 2.3 Tera-Wasserburg Concordia diagram showing U–Pb analyses of detrital zircon (McAteer *et al.*, 2008) from a sample of psammite from the Doonamoo Formation, Grampian Group Dalradian on the Mullet peninsula (Fig. 2.10). 86 of 103 analyses are concordant and are plotted as black 2σ error ellipses that intersect Concordia. Discordant data points are shown in blue. Arrows indicate the direction in which these may have moved off Concordia as a result of lead loss. The youngest detrital zircon is shown as a red ellipse together with its U–Pb age, 955 ± 12 Ma, which provides a maximum age limit for the deposition of the Doonamoo Formation.

ratios in the direction of modern common lead, which has coordinates of 0.83 and zero, respectively (Stacey and Kramers, 1975).

Major advances have taken place in recent years in the application of the U–Pb dating technique by isotope dilution thermal ionisation mass spectrometry (IDTIMS, reviewed by Parrish and Noble, 2003) especially on minerals such as zircon, based mainly on the work of Krogh (1973, 1982) and Mattinson (2005).

Mineral grains, especially zircon, may be air-abraded to remove the outer altered parts of the grains or 'chemically abraded' by heating to high temperature (800–1100°C) and selectively leaching the thermally annealed grains to remove domains that have suffered lead loss, thereby achieving concordant analyses of the residue. Data points that fall off Concordia were traditionally interpreted in terms of lead loss, sometimes with misleading consequences (cf. Rogers and Pankhurst, 1993). Improvements in mass-spectrometry and the reduction of lead blank levels in the laboratory mean that very small fractions of zircon, including single grains or mechanically selected parts of grains, may be analysed by IDTIMS. This means that complex populations can be deconvoluted, e.g. mixtures of two or more mineral components whose ages differ, or magmatic cores and metamorphic overgrowths. Thus several generations of zircon growth can be dated to provide a detailed calibration of the rock's history. Moreover, inherited grains (unmelted grains from the protolith of a granite or contaminants from the wallrocks) may be dated, providing constraints on igneous petrogenesis.

The complexity of many zircon crystals (possibly containing inherited cores, exhibiting several growth stages of both magmatic and metamorphic origin and including metamict domains likely to have lost some or all of their radiogenic lead) has driven the need for *in situ* analytical methods. This approach was led by Compston (e.g. Compston *et al.*, 1984), who developed the Sensitive High Resolution Ion Microprobe (SHRIMP) at the Australian National University. This instrument (and others such as the Cameca IMS 1270 and 1280) employs secondary ion mass spectrometry (SIMS) to measure the U/Pb ratio and Pb isotopic composition *in situ* on single grains of zircon, which have been cut and polished to reveal the zircon interiors. In this way ages can be obtained from shallow (2 μm deep) spots *c.*20 μm in diameter. Although the resulting ages are less precise than the best obtainable by IDTIMS, the ion microprobe has the advantage that selected areas within a single grain may be dated while preserving the petrographic relationship between them. Thus overgrowths and cores may be identified and dated – avoiding many of the problems inherent in interpreting complex zircon populations.

An alternative approach to *in situ* U–Pb geochronology employs laser-ablation sampling and analysis by inductively coupled plasma mass spectrometry (LA-ICPMS – see Košler and Sylvester, 2003, for a review). Compared with the SIMS method, this technique offers rapid analysis (*c.*4 minutes per analysis compared with about 20 minutes by SIMS) but somewhat poorer precision, mainly due to fractionation of the U/Pb ratio during laser ablation. Typical laser ablation pits (50 μm across and 30 μm deep) involve much larger sample volumes than SIMS analysis, which limits the spatial resolution. However, the method is widely used, especially in detrital zircon studies (see below).

Prior to employing either of the above techniques of *in situ* analysis, careful imaging by cathodoluminescence or electron backscattering is essential to select target regions for dating and to avoid inclusions, fractures and metamict regions. In addition to U–Pb dating of discrete components of zircon grains, *in situ* analysis of Hf (see below) and oxygen isotopes is proving to be a powerful geochemical tracer of continental evolution (e.g. Hawkesworth and Kemp, 2006).

U–Pb dating of detrital minerals as a sedimentary provenance tool

Zircon is a major component of the heavy mineral suite in many siliciclastic deposits and its U–Pb age distribution has been used in the analysis of the sedimentary record (Fedo *et al.*, 2003) since the pioneering work of Gaudette *et al.* (1981). Early U–Pb dating studies relied on TIMS dating of single grains or parts of grains (e.g. Drewery *et al.*, 1987; Cliff *et al.*, 1991) and were thus limited in scope, due to the relatively small number of grains that could be practically analysed and the inherent risk of sample bias in selecting grains likely to yield concordant ages.

Ion microprobe and especially laser ablation ICPMS methods, coupled with careful selection of analytical points, permits rapid analysis of large numbers of grains (e.g. Figs. 2.3, 2.4) sufficient to ensure representative sampling (Dodson *et al.*, 1988). These methods have been widely applied in the investigation of Precambrian sedimentary sequences in the north Atlantic region (e.g. Cawood and Nemchin, 2001; Rainbird *et al.*, 2001; Cawood *et al.*, 2003, 2004; Kinnaird *et al.*, 2007; Chew *et al.*, 2008; Kirkland *et al.*, 2008). Careful selection of analytical points using cathodoluminescence or back-scattered electron imaging ensures that only the appropriate parts of grains are analysed. For example, post-depositional metamorphic overgrowths can be excluded from the detrital population and metamict regions and fractures can be avoided, thus reducing the risk of a discordant analysis. While the age distribution of the detrital zircon population can be compared with putative source regions, the data also permit rigorous statistical comparison of zircon populations (e.g. Kirkland *et al.*, 2008) thereby providing a fresh correlation tool and a maximum depositional age for a sedimentary sequence through the age of the youngest detrital grain. Occasionally, dating of detrital zircon populations can reveal a cryptic volcaniclastic zircon component and thus provide a direct estimate of the depositional age (e.g. Kirkland *et al.*, 2005).

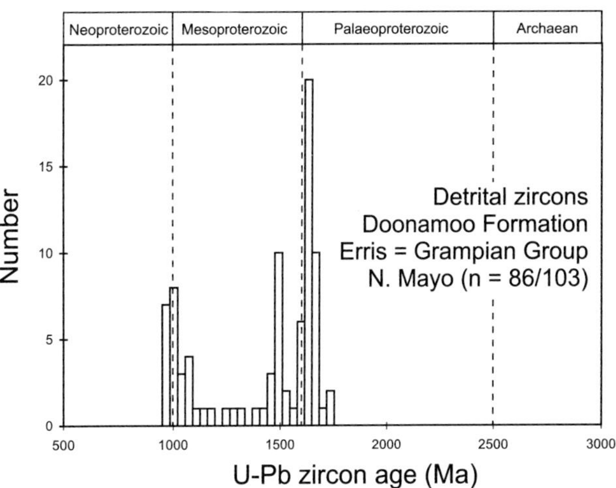

Fig. 2.4 Histogram of concordant U-Pb ages for the sample of Dalradian psammite shown in Fig. 2.3. The data show that the source rocks have contributed predominantly late Palaeoproterozoic to early Neoproterozoic zircons. Early Palaeoproterozoic and Archaean zircons are not present.

Rb–Sr dating

The Rb-Sr method (Table 2.2) is based on the decay of ^{87}Rb to ^{87}Sr. For many years, Rb–Sr whole-rock dating was the main method applied to the geochronology of Precambrian terrains – particularly in attempting to date igneous protolith ages of metamorphosed felsic orthogneisses. Whole-rock isochron ages are based on analyses of several samples with a range of Rb/Sr ratios. Rocks of the same age that are in isotopic equilibrium display a linear correlation (an isochron) between the Sr isotopic composition and Rb/Sr ratio. Solving this linear equation provides both the age and the isotopic composition of Sr at the time isotopic equilibrium was achieved between the samples. The precision of the Rb–Sr whole-rock method depends on the range of Rb/Sr ratios, which is intrinsically rather limited. Hence the precision of Rb–Sr whole-rock ages is generally poor compared with, e.g. the U–Pb zircon method. Whole-rock ages are open to interpretation because a variety of processes may have controlled equilibration of Sr isotopes between different samples. Thus it may be uncertain whether a Rb–Sr whole-rock age is dating a metamorphic or a magmatic event. Inheritance from the protoliths of igneous rocks is also possible, and in some cases a spurious linear relationship between isotopic composition and Rb/Sr ratio may reflect binary mixing between different components, e.g. in magmatic systems and in certain cases in sedimentary basins. Because of the poor precision and questionable accuracy, the Rb–Sr whole-rock isochron method has largely fallen into disuse.

Rb–Sr mineral ages are, however, being increasingly applied to the investigation of metamorphic processes and the thermal history of orogenic belts (Cliff *et al.*, 1996; Flowerdew *et al.*, 2000; Dallmeyer *et al.*, 2001; Glodny *et al.*, 2002; Chew *et al.*, 2003; Bröcker and Franz, 2006; Flowerdew *et al.*, 2007; Kirkland *et al.*, 2007; Glodny *et al.*, 2008). Suitable minerals include muscovite, biotite and hornblende. In high grade metamorphic rocks these minerals record cooling ages that may provide valuable information on the cooling history of metamorphic belts. At lower metamorphic temperatures, below the closure temperature, mineral ages may date actual events (Cliff, 1985). It has recently become possible to date very small amounts of sample, e.g. fine-grained seams of white mica defining specific foliations, potentially providing insight into the timing of specific deformational phases. In the micas, as previously mentioned, the Rb/Sr ratios may be sufficently high that the age is independent of the initial isotopic composition of Sr (van Breemen *et al.*, 1978). Otherwise correction for initial Sr is usually made by analysing co-existing plagioclase.

Sm–Nd dating

The Sm–Nd mineral isochron dating method (Fig. 2.5) has provided important results from the Irish Precambrian (e.g. Sanders *et al.*, 1987; Flowerdew and Daly, 2005) and so is also described briefly here. Dating of magmatic minerals in gabbros and dolerites (e.g. Flowerdew and Daly, 2005) yields potentially accurate but imprecise ages, the major source of uncertainty being the restricted difference in Sm/Nd ratio between magmatic clinopyroxene and plagioclase. Sm–Nd dating of metamorphic garnet yields precise ages because garnet usually has a very high Sm/Nd ratio. However, the interpretation of Sm–Nd garnet ages is controversial. Retention of radiogenic Nd in a mineral (e.g. garnet) that would naturally prefer to exclude Nd in favour of the heavier rare earth elements such as Sm is probably temperature-dependent, and there is some disagreement as to the appropriate 'closure temperature'. A high closure temperature for Nd in garnet (e.g. as advocated by Hensen and Zhou, 1995) would mean that high-grade metamorphic events could be dated, while a low closure temperature would provide only a minimum estimate of the metamorphic age and instead provide a point on a temperature–time path for the rock (e.g. Mezger, 1990). An additional problem lies in the common presence in garnet of mineral inclusions such as monazite, whose light rare-earth element enrichment has the effect of lowering the apparent Sm/Nd ratio of the garnet and compromising the age. Significant improvements in both precision and accuracy have, however, been achieved by selectively leaching garnet using sulphuric acid, which preferentially dissolves the monazite (Anczkiewicz and Thirlwall, 2003).

Sm–Nd model ages

Apart from garnet, most minerals, and therefore whole-rocks, have low Sm/Nd ratios and display little variation in the ratio. For this reason the Sm–Nd whole-rock isochron method is rarely used. However, Sm–Nd analysis of individual whole-rock samples may be used to calculate model ages (Fig. 2.6). Sm–Nd model ages are the time when, projecting back in time from the present day, the neodymium isotopic composition of a sample coincides with that of the Earth's mantle or some other reference material (e.g. the bulk Earth, which is the same as the Chondrite Uniform Reservoir, or CHUR). Relative to the mantle the crust is enriched in light rare-earth

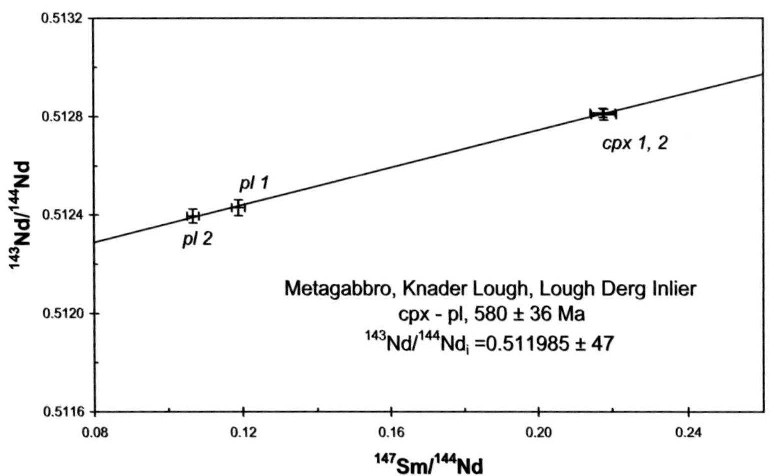

Fig. 2.5 Sm–Nd isochron for relict magmatic minerals in a metagabbro from Knader Lough, Co. Donegal from Flowerdew and Daly (2005). pl1, pl2 = plagioclase fractions, cpx1, cpx2 = clinopyroxene fractions. *Reproduced by kind permission of the Royal Irish Academy.*

elements, i.e. crustal rocks have a low Sm/Nd ratio. As a consequence of partial melting in the mantle and transfer of the resulting magma to the crust, the mantle has a high Sm/Nd ratio and thus Nd in the mantle has more radiogenic isotopic ratios (higher $^{143}Nd/^{144}Nd$) than the crust at any time. Usually the model ages are calculated relative to this 'depleted' mantle (DePaolo, 1981) assuming that the isotopic composition of the mantle has steadily increased over geological time owing to extraction of continental crust. Such ages are known as depleted mantle model ages (t_{DM}). These 'ages' need not date specific geological events.

However, in the simplest case, t_{DM} ages of, e.g. granitoid gneisses, date 'crustal extraction', 'crust formation' or 'crustal residence'. Such ages estimate the time of fractionation of the Sm/Nd ratio during the formation of juvenile continental crust from depleted mantle. In general, model ages must be interpreted cautiously (cf. Arndt and Goldstein, 1987) because granitoids may form by mixing components from two or more sources. Although Sm–Nd data alone can sometimes indicate the time of crust formation, the data are much more powerful when the crystallisation age of the rocks is known or can be constrained independently (Fig. 2.7). In these circumstances, the $^{143}Nd/^{144}Nd$ isotopic ratio at the time of crystallisation may be calculated and expressed as ε_{Nd}. Positive initial ε_{Nd} values serve to identify juvenile crust, i.e. crust formed with minimal contributions from older material. In this case there will be close agreement between the t_{DM} age and the crystallisation age. In contrast, ancient crust and its derivatives have negative ε_{Nd} values and t_{DM} ages will be older than the crystallisation age. In crust of mixed origin, it may be possible to

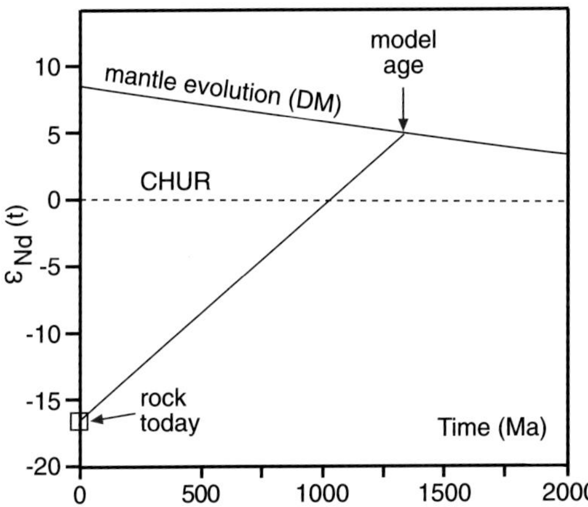

Fig. 2.6 Sm–Nd isotopic evolution diagram in coordinates of ε_{Nd} and time. ε_{Nd} is the fractional deviation in parts per thousand of the $^{143}Nd/^{144}Nd$ isotopic ratio of a sample from that of the bulk earth at the same time. By definition, the ε_{Nd} value of the bulk earth (or Chondrite Uniform Reservoir, CHUR) is zero and remains constant over time. The $^{143}Nd/^{144}Nd$ ratio of all materials increases over time due to radioactive decay of ^{147}Sm. However, certain geochemical domains (reservoirs) within the earth – especially the mantle – undergo changes in their Sm/Nd ratio over time. Thus the curve labelled 'mantle evolution (DM)' depicts the changing rate of accumulation of ^{143}Nd in the mantle due to the extraction of crust. This happens because crust formation depletes the mantle in large ion lithophile elements such as Nd, i.e. because the crust prefers Nd to Sm, the Sm/Nd ratio of the mantle increases. Hence the mantle becomes progressively more radiogenic (more positive ε_{Nd} values) over time compared with CHUR or the crust (DePaolo, 1981). A neodymium evolution line is shown for a typical continental crustal rock, which today has a negative ε_{Nd} value. The model age for this rock is calculated as the time when its Nd evolution line (calculated from the Sm–Nd decay equation) intersects the model mantle curve. Model ages do not necessarily date geological events but they can provide an indication of major crust formation events.

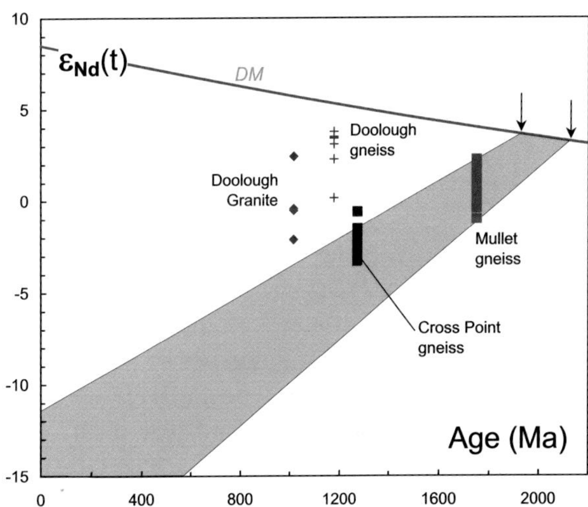

Fig. 2.7 Sm-Nd isotopic evolution diagram for selected rocks from the Annagh Gneiss Complex (Menuge and Daly 1990; Fitzgerald, Daly and Menuge, unpublished). Calculated ε_{Nd} values for the Mullet, Cross Point and Doolough gneisses and the Doolough Granite are plotted at their time of formation (U-Pb zircon ages; Table 2.4). The range of Nd evolution lines (magenta) is shown for the *c.*1750 Ma old Mullet gneisses. This intersects the depleted mantle curve (DM) yielding a range of depleted mantle model ages (tDM) from *c.*1.92 to 2.11 Ga (arrowed). At *c.*1270 Ma, εNd values of most of the Cross Point samples overlap the range of the Mullet gneisses while some have more positive εNd values. From this it can be deduced that the Cross Point gneisses could have formed by melting the Mullet gneisses with the addition of some mantle-derived material, presumably of more basic composition. In contrast, at *c.*1177 Ma, the Doolough gneisses have positive εNd values close to the mantle value indicating that they represent juvenile crust derived mainly from the mantle but possibly contaminated by continental crust.

calculate the proportions of the contributing components (cf. Patchett and Arndt, 1986).

Sm–Nd isotopes in provenance studies

While the Sm–Nd system is most directly applicable to igneous protoliths, it can also provide valuable constraints on the crustal residence ages of rocks parental to sediments and metasediments (e.g. Morton *et al.*, 1991). Clastic sediments can be regarded as composites derived from subaerially exposed crust, sampling a much wider area than intrusive rocks (Frost, 1993). Since the results are representative of the upper crust, Sm–Nd studies of metasediments complement those based on middle or lower crustal plutonic rocks. The Sm–Nd depleted mantle model age (t_{DM} age) of clastic sediments or metasediments may be interpreted as the mean crustal residence age of the Nd in the detritus. In general this will reflect mixing of several source materials of differing crustal extraction ages. The t_{DM} age of the mixture will then give a minimum crustal residence age for the oldest component and a maximum crustal residence age for the youngest component. The frequency distribution for a large number of samples may permit the identification of components of intermediate crustal residence age and allow more accurate estimates of the crustal residence ages of the source rocks contributing detritus. Fine-grained clastics such as shales, or metashales, provide the best estimate of the average crustal residence age of the source, since they are well mixed and usually show no evidence for sedimentary fractionation of the Sm/Nd ratio. Sandstones, and their metamorphic equivalents, are more likely to yield information on a more local scale and to afford a means of discrimination between different source regions. Thus some degree of reconstruction of the provenance history may be possible.

Lu–Hf

Although yet to be applied to dating events affecting Irish rocks, the Lu–Hf dating method has been widely applied to the dating of metamorphic garnet (Scherer *et al.*, 2000; Anczkiewicz *et al.*, 2004, 2007; Lagos *et al.*, 2007) and offers great potential in dating metamorphic events. High precision can be achieved because garnets can have very high Lu/Hf ratios ($^{176}Lu/^{177}Hf$ ratios up to 8, compared with 0.001 to 0.1 for many whole-rocks) and the shorter halflife of ^{176}Lu compared with ^{147}Sm (Table 2.2) offers the possibility of greater precision than the Sm–Nd method. Moreover, the closure temperature is probably higher (other factors being equal) than that of the Sm–Nd system (Scherer *et al.*, 2000) so that crystallisation ages can be determined. Unfortunately the method is compromised by the possible presence of submicroscopic Hf-rich inclusions, such as zircon, which may be impossible to remove by hand-picking. The presence of zircon inclusions has the effect of reducing the apparent Lu/Hf ratio of the garnet and hence the precision (and accuracy) of the calculated age.

The high Hf content of zircon coupled with its very low Lu/Hf ratio (<0.001) makes it a useful geochemical tracer since zircon thus preserves its initial $^{176}Hf/^{177}Hf$ ratio, which is a record of the Hf isotopic composition of the magma from which it formed. In a manner analogous to that employed for the Sm–Nd isotopic system, the Hf isotopic composition of magmatic or detrital zircons can

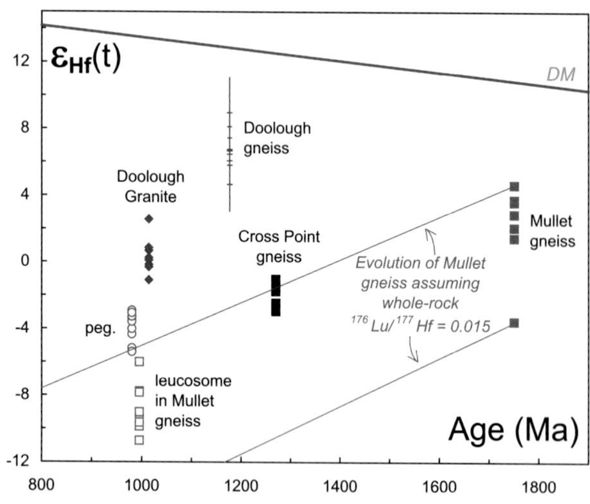

Fig. 2.8 Lu-Hf evolution diagram for the Annagh Gneiss Complex (Daly *et al.*, 2008a) showing laser-ablation analyses of zircon grains of known U-Pb age. Zircons in *c.*995 Ma migmatitic leucosomes within the Mullet gneisses have identical Hf isotopic compositions to those of the Mullet gneisses at 995 Ma (evolving between the red lines, assuming a typical crustal Lu/Hf ratio), suggesting that the leucosomes derived their Hf from their host by in situ melting. Circa 1270 Ma old Cross Point gneisses probably also formed by melting the *c.*1750 Ma Mullet gneisses, perhaps with the addition of more radiogenic Hf (higher ε_{Hf}). In contrast, zircons from the *c.*1177 Ma Doolough gneisses (shown as blue error crosses) contain much more radiogenic Hf, requiring a juvenile (mantle) source. These rocks probably represent but a small fragment of a significant volume of juvenile crust, produced by arc magmatism. The Doolough gneisses were tectonically juxtaposed with the rest of the Annagh Gneiss Complex during the Grenville Orogeny. *c.*980 Ma pegmatites (peg.) cutting the Mullet gneiss also require an additional input of radiogenic Hf, which could have come from tectonically juxtaposed Doolough gneiss. DM = depleted mantle.

be compared with those of the contemporaneous mantle to assess the crustal residence time and hence to distinguish juvenile from reworked crustal sources (Fig. 2.8). Hawkesworth and Kemp (2006) review the methodology and applications to crustal evolution studies.

K–Ar and Ar–Ar dating

Because argon is a gas, the application of the K–Ar decay scheme (Table 2.2) to dating geological events is complicated by the rock's thermal history. In many cases radiogenic argon is retained in K-bearing minerals, such as micas and amphiboles, only below a relatively low temperature and the measured ages are thus 'cooling ages' rather than crystallisation ages. Since most of the Precambrian in Ireland has been metamorphosed during the Grampian Orogeny, relatively few minerals record Precambrian ages. Those that do so occur where Grampian heating or deformational effects have been minimal, e.g. within parts of the Rhinns Complex offshore Donegal where hornblendes record Mesoproterozoic ages (Roddick and Max, 1983; see below).

Acquisition and interpretation of argon isotopic dates has been greatly improved by the development of the ^{40}Ar–^{39}Ar step-heating method and through the use of laser-ablation methods of intragrain sampling (e.g. Chew *et al.*, 2003). Samples used in ^{40}Ar–^{39}Ar dating are first irradiated to convert ^{39}K to ^{39}Ar by nuclear reactions. This has the advantage that the relative amounts of parent and daughter isotopes can be measured by the same technique and on the same aliquot of the sample. Incremental heating of the sample allows a series of age determinations to be made by measuring the amount and isotopic composition of Ar released from the sample at different temperatures. The integrity of the results may be assessed in several ways. Acceptable ages require replication of at least three successive steps involving at least 50% of the argon evolved from the sample. These are known as 'plateau ages' (Fig. 2.9). The presence of extraneous argon ('excess argon' – i.e. additonal to that produced within the sample by decay of potassium) will result in steps with anomalously old ages (Fig. 2.9). In such cases useful age information may be unobtainable.

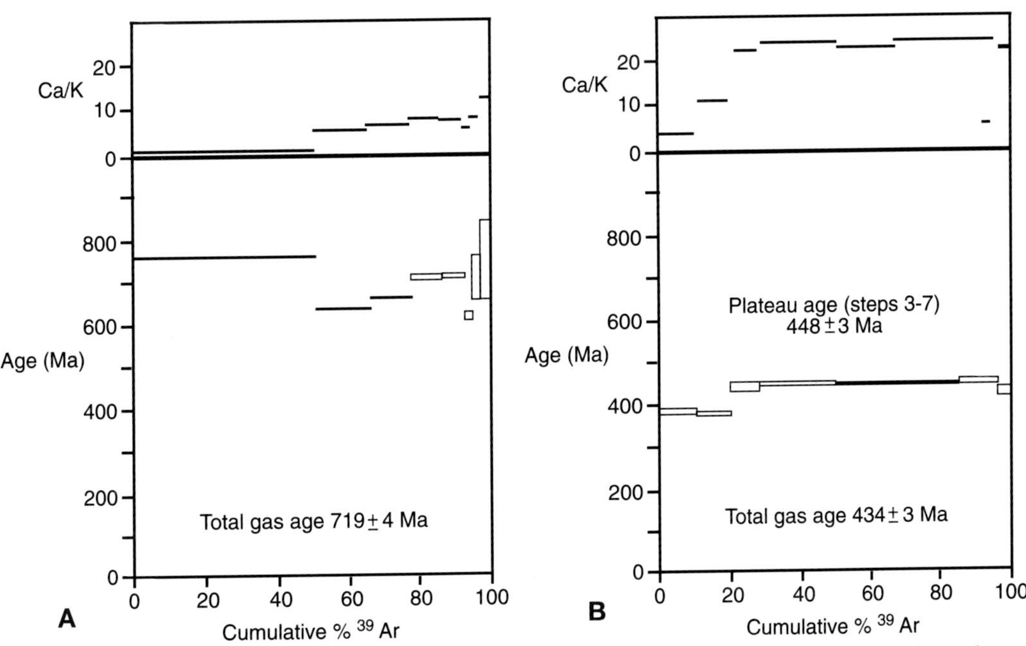

Fig. 2.9 Ar–Ar step heating spectra for metamorphic hornblendes (based on Daly and Flowerdew, 2005; Flowerdew *et al.*, 2000) to illustrate the use of Ca/K ratios in identifying impurities and explaining anomalously young ages in the low temperature steps. **A)** Pre-Grenville hornblende from the Annagh Gneiss Complex showing a low temperature step with excess argon possibly contained in secondary biotite as suggested by the very low Ca/K ratio of this step. The excess argon may have been derived from the enclosing potassium-rich Mesoproterozoic rocks. The staircase pattern of increasing apparent age in the later steps is impossible to interpret. No useful age information can be obtained from spectra of this complexity. **B)** Grampian hornblende from Argyll Group volcanics in the Central Ox Mountains Inlier showing a good plateau age of 448 ± 3 Ma. The material yielding these gas increments has a constant Ca/K ratio consistent with pure hornblende. The lower temperature increments display younger ages and correlated lower Ca/K ratios consistent with the presence of small amounts of biotite probably as inclusions that were invisible during sample preparation.

Differences in closure temperature between minerals may result in mixed ages when minerals are intergrown on a scale too small to be recognised during conventional mineral separation. Such mixed ages may be assessed by observing variations in apparent Ca/K ratio, measured by proxy as the $^{37}Ar/^{39}Ar$ ratio, at different temperatures (Fig. 2.9). Flat age spectra or plateaux with constant Ca/K ratios are regarded as optimal, implying homogeneity in terms of age and mineral composition and indicating a lack of excess argon.

Pre-Grampian basement

As has been stated already, all of the Precambrian rocks in north-western Ireland have been affected by Grampian deformation and metamorphism (Johnston, 1995; Roddick and Max, 1983; Sanders *et al.*, 1987), i.e. by tectonic events in the Ordovician. The Dalradian Supergroup was metamorphosed for the first time in the Grampian Orogeny, but other rocks show evidence of an earlier metamorphic history. Such rocks are here grouped as pre-Grampian basement, and as outlined in the introduction, they include the Rhinns Complex exposed on Inistrahull, the Annagh Gneiss Complex in north-west Mayo, the Slishwood Division in Sligo, Fermanagh and Donegal, and the Central Inlier of Tyrone (though the last has recently been shown to be probable Dalradian).

The existence and nature of pre-Grampian events is the subject of debate and active research (Hutton and Alsop, 2004; Tanner, 2005; Daly and Flowerdew, 2005). Grenvillian (*c.*1.0 Ga) deformation and metamorphism has been demonstrated in north-west Mayo, (e.g. Aftalion and Max, 1987; van Breemen *et al.*, 1978; Daly, 1996) but there is as yet no definitive demonstration of the effects of later Neoproterozoic orogenic events, such as the 'Knoydartian' Orogeny recognised in Scotland (Vance *et al.*, 1998). The Rhinns Complex has provided evidence for both Laxfordian (*c.*1.7 Ga, Roddick and Max, 1983; Loewy *et al.*, 2003) and Grampian or Late Caledonian (pre-*c.*0.37 Ga, Macintyre *et al.*, 1975b) metamorphic events but no record of the Grenvillian or Knoydartian having affected these rocks. The Slishwood Division experienced high-grade metamorphism while the Dalradian was being deposited, so it cannot be pre-Dalradian basement. This enigma will be discussed below.

Sm–Nd model ages of exposed pre-Grampian rocks range from *c.*2.0 Ga to *c.*1.3 Ga (Fitzgerald *et al.*, 1993; Daly *et al.*, 1991; Menuge and Daly, 1990; Sanders *et al.*, 1987 and unpublished data) indicating that all of the basement units in Ireland represent crust formed during the Proterozoic. However, Lu–Hf isotopic data from zircons in granitic rocks intruding the Tyrone Central Inlier indicate the presence of Archaean crust at depth (Flowerdew *et al.*, 2008) consistent with the suggestion of Long and McConnell (1997), based on consideration of deep seismic data, that Archaean lithosphere might be present at depth under NW Ireland.

Rhinns Complex

The Rhinns Complex (Muir *et al.*, 1992; 1994a, b) is one of two important occurrences of Palaeoproterozoic orthogneisses within the pre-Grampian basement of north-western Ireland, the other being the Annagh Gneiss Complex. It occurs within the fault-bounded Colonsay–West Islay block (Fig. 2.1; Bentley *et al.*, 1988; Fitches *et al.*, 1990; Marcantonio *et al.*, 1988), which extends from Colonsay in the north-east to the Inishtrahull platform (Max and Long, 1985) in the south-west. Its Irish exposure is limited to the island of Inishtrahull.

Geological and geochemical investigations of the Rhinns Complex (Fitches *et al.*, 1990; Muir *et al.*, 1994a) have shown that the protolith was an alkaline igneous association of syenite and subordinate mafic material, intruded by gabbro sheets. The rocks are everywhere deformed, often gneissose and generally exhibit amphibolite-facies metamorphic assemblages except where retrogressed. To the north-east, on Colonsay and Islay, the Rhinns Complex is unconformably overlain by a thick sequence of Neoproterozoic, low-grade cover rocks, termed the Colonsay Group (e.g. Bentley, 1988) which may extend over a large area off the north coast of Donegal (Fig. 2.1). The Colonsay Group may correlate with the Dalradian (cf. Muir *et al.*, 1997; Rock, 1985; McAteer *et al.*, 2006). Exposures on the Garvan Islands, 1 km north-east of Malin Head, show that the Rhinns Complex underlies the Dalradian rocks of Donegal along the Malin Head Thrust (Max, 1981; Max and Long, 1985). South-east of Inishtrahull, lower amphibolite-facies metasediments which are of higher metamorphic grade than the adjacent Dalradian have been recovered by submarine sampling. These rocks may correspond to the Moine Supergroup of Scotland (Holdsworth *et al.*, 1994; Strachan *et al.*, 2002; Max and Long, 1985; Evans, 1983).

The major lithology on Inishtrahull, comprising *c.*90% of the exposed rock, is a deformed coarse-grained pink syenitic gneiss. This consists predominantly of alkali feldspar and hornblende (after clinopyroxene), with minor plagioclase (An20–An36), biotite, chlorite and quartz.

Accessory minerals include zircon, titanite, apatite, epidote and oxides. Minor amounts of fine-grained amphibolite are also present, locally brecciated by syenite forming an agmatite (see Bowes and Hopgood, 1975; Plate VIII D). The dominant foliation is defined by millimetre-scale mafic/felsic compositional banding. This foliation is composite, representing several generations of transposition. Locally, intrafolial folds of the banding are preserved.

Sheets of coarse-grained gabbro, up to 15 m thick and traceable for up to 200 m, make up about 5% of the outcrop. The gabbro sheets are generally concordant with the foliation in the enclosing syenite, but several exhibit cross-cutting relationships and chilled margins, demonstrating emplacement after the development of a foliation in the syenite host (Table 2.3). Similar observations were made by Roddick and Max (1983) on the Tor Rocks, *c.*2 km north-west of Inishtrahull. The gabbros have a primary igneous mineralogy comprising mainly clinopyroxene and plagioclase (An52). After emplacement, the gabbros were metamorphosed under amphibolite-facies conditions, as shown by the growth of green-brown hornblende at the expense of pyroxene. Both the gabbros and the syenite are cut by thin post-tectonic quartzo-feldspathic pegmatite veins.

Metasediments are probably absent (Daly *et al.*, 1991). Previously reported calcareous metasediments (Bowes and Hopgood, 1975) are thought to be intensely sheared lamprophyre dykes. These dykes are similar to Caledonian lamprophyres found within the basement rocks on Islay (Bentley, 1988).

A summary of the geochronology and geological history of the Rhinns Complex in Ireland and Scotland is given in Table 2.3 (Muir *et al.*, 1994a; Loewy *et al.*, 2003). The Inishtrahull syenitic gneiss has yielded a U–Pb zircon age of 1779 ± 3 Ma, interpreted as dating the igneous crystallisation of the gneiss's protolith (Fig. 2.2, Daly *et al.*, 1991). As said at the start of the chapter, this is the oldest known igneous crystallisation age in Ireland. Sm–Nd depleted mantle model ages range from 1912 Ma to 1978 Ma. The similarity of the model ages and the crystallisation age indicates that the protoliths were derived from a depleted mantle source essentially uncontaminated by older crust. These data confirm previous tentative correlations of the Inishtrahull rocks with those of Islay in SW Scotland where Marcantonio *et al.* (1988) were the first to demonstrate that the Rhinns Complex is Palaeoproterozoic in age and that it comprises juvenile mantle-derived material. They reported a U–Pb zircon age of 1782 ± 5 Ma, which they interpreted as dating the igneous crystallisation of the protolith of the Islay gneiss. Importantly, their Sm–Nd and Pb isotopic data showed that very little old crust could have participated in the genesis of the Islay basement. Metamorphic zircons from a Rhinns Complex metagabbro at Lossit Bay, Islay yielded U–Pb ages of 1725–1729 Ma (Loewy *et al.*, 2003).

Table 2.3 Evolution of the Rhinns Complex (after Loewy *et al.*, 2003; Muir *et al.*, 1994a, 1997 and references therein).

Lithological additions	Tectonic events	Thermal conditions and metamorphism	Age (Ma)	Method
	cooling	c.300°C	c.370	K–Ar biotite
appinite intrusion into Colonsay Group (on Colonsay)			439 ± 9	U–Pb zircon SHRIMP
	Grampian deformation			
deposition of Colonsay Group (on Islay)	uplift and exhumation	cooling		
		amphibolite facies	c.1710	Ar–Ar hornblende
			c.1730	U–Pb met. zircon
gabbro and minor mafic and felsic intrusions				
	deformation, gneiss formation, possibly syn-plutonic			
syenite protolith to Inishtrahull syenitic gneiss			1779 ± 3	U–Pb zircon
			<1970	Sm–Nd model age

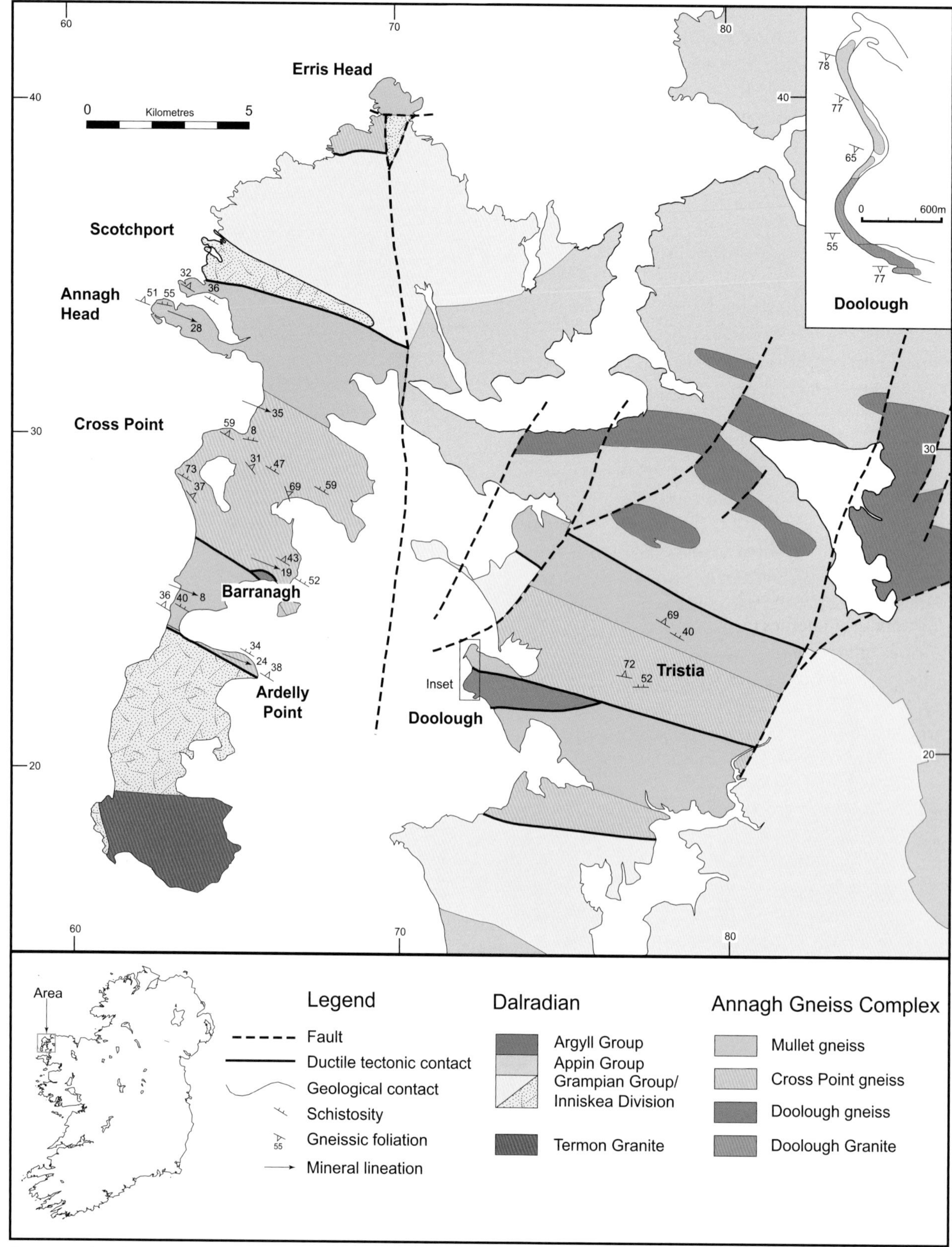

Fig. 2.10 Sketch geological map of part of the North Mayo Inlier on the Mullet peninsula and adjacent mainland. The geological subdivision of the Annagh Gneiss Complex is based on U–Pb geochronology, Sm–Nd and geochemical data, and is based in part on work by Rose Fitzgerald (written communication, 1996).

This is consistent with a minimum age of 1710 Ma based on $^{40}Ar/^{39}Ar$ step-heating of an amphibole concentrate from a discordant metagabbro on Tor Beg, located 2 km north-west of Inishtrahull (Roddick and Max, 1983). Rb–Sr and K–Ar ages in the range of 520 Ma to 370 Ma indicate strong Grampian to Late Caledonian overprinting (Macintyre *et al.*, 1975b).

Previous correlations of tectonic and isotopic events in the Rhinns Complex on Inishtrahull with Badcallian (late Archaean) events in the Lewisian Complex of NW Scotland (Hull *et al.*, 1890; McCallien, 1930; Bowes and Hopgood, 1975; Macintyre *et al.*, 1975b) are now ruled out because the Rhinns Complex is exclusively Palaeoproterozoic in age. Although some distance away, possible correlation with Laxfordian events affecting the Lewisian Complex is suggested by the similar metamorphic age, *c.*1740 Ma, of the latter (Corfu *et al.*, 1994).

Annagh Gneiss Complex (AGC)

The Annagh Gneiss Complex (Menuge and Daly, 1994) is the second example of Palaeoproterozoic orthogneiss in north-western Ireland, and it makes up the pre-Grampian basement of the North Mayo Inlier (Figs. 2.1, 2.10). It structurally underlies the Dalradian and is exposed in the cores of two south-east-plunging anticlines on the Mullet peninsula and the adjacent 'mainland' eastwards from Doolough (Figs. 2.10, 4.13). The Dalradian rocks were deformed and metamorphosed under medium pressure amphibolite-facies conditions (Yardley, 1980) during the Grampian Orogeny. However, it is well established that the basement gneisses reached a somewhat higher metamorphic grade with widespread migmatisation in Grenville times (van Breemen *et al.*, 1978).

The AGC has previously been grouped with metasediments of the Inishkea Division, jointly comprising the Erris Complex (Max and Long, 1985). Here, however, the Inishkea Division (also known as the Scotchport schists) is interpreted as Dalradian (mainly Grampian Group) in line with arguments by Kennedy and Menuge (1992). Like the Dalradian, it tectonically overlies the Annagh Gneiss Complex, and its previous interpretation as reconstituted gneisses or as pre-Dalradian sediments compared with the Moine Supergroup (Holdsworth *et al.*, 1994) of Scotland (e.g. Max and Long, 1985) is unwarranted.

The AGC is generally well exposed along the coast of the Mullet peninsula, but exposures are poor inland especially in the east on the mainland between Doolough and Tristia (Fig. 2.10). Early investigations established that the AGC comprises intermediate to acid and partially migmatitic orthogneisses. In many areas the gneisses are banded on a scale of 1–10 cm, with alternating layers rich and poor in mafic minerals. The overall composition is granodioritic to granitic (Winchester and Max, 1984). Early amphibolitised basic bodies are concordant with the main gneissose foliation and may represent dykes or sills. Several generations of granitoid (and pegmatite) sheets up to a few metres wide range from syn- to post-tectonic with respect to the main foliation. Generally the gneisses exhibit a strong NW–SE trending gneissose fabric, which tends to obscure the relationships between the various lithologies because they all tend to be aligned parallel to it.

At Annagh Head (Figs. 2.10, 2.11) and Doolough (Figs. 2.10, 2.12), cross-cutting relationships make it possible to work out a geological history (e.g. Sutton, 1972; Phillips, 1981a; Max and Long, 1985; Johnston, 1995) including separating the effects of the Grenville Orogeny into at least three deformational events, D_G1–D_G3 (Daly, 1996; Daly and Flowerdew, 2005). The relationships are particularly clear at Doolough, where late Grenville granite pegmatites cross-cut folded Doolough gneisses, themselves of early Grenville age (Fig. 2.12, see below). Among the important features to be seen at Annagh Head is a late cross-cutting suite of metabasite dykes (Fig. 2.11). These are now unmigmatised but foliated metadolerites. They have been correlated with metabasites that cut the overlying Dalradian rocks (R. C. Fitzgerald *et al.*, 1994). Because their correlatives share all deformational events that affect the Dalradian, the late metabasite suite cutting the AGC is considered to have been deformed only in the Grampian. The metabasites may thus be used to distinguish between pre-Grampian and Grampian events.

Crustal evolution of the Annagh Gneiss Complex

U–Pb zircon dating (Table 2.4) has yielded a chronology (Daly, 1996) that, together with geological mapping, borehole information, and geochemical and isotopic data, result in a threefold subdivision of the AGC (Fig. 2.10) into *c.*1.75 Ga Mullet gneisses, *c.*1.28 Ga Cross Point gneisses and *c.*1.18 Ga Doolough gneisses (Daly, 1996; Fitzgerald *et al.*, 1996). Unfortunately none of the geological boundaries between the three sets of gneisses is exposed, but it is thought likely that the original contacts between the Cross Point and Mullet gneisses were intrusive.

Most of the AGC has Sm–Nd model ages (t_{DM}, DePaolo, 1981) in the *c.*1.8–2.0 Ga range (Menuge and

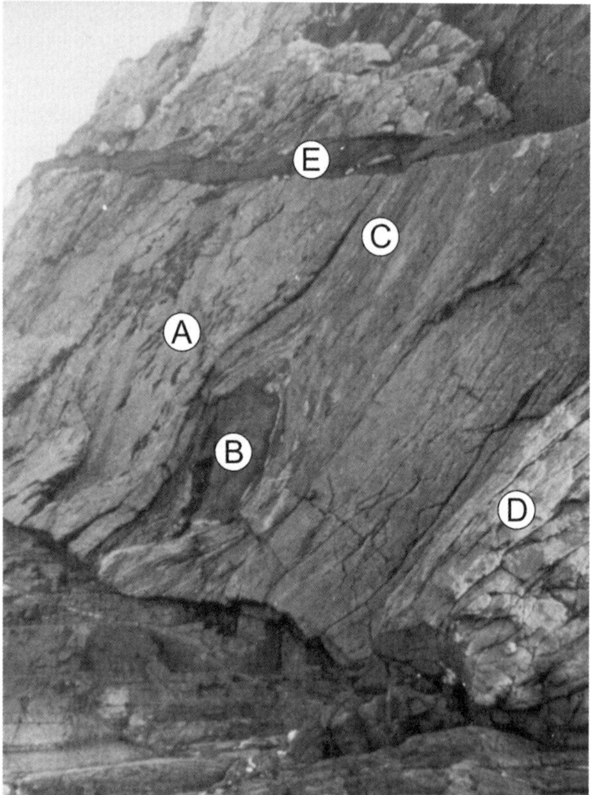

Fig. 2.11 Photograph of the Mullet gneisses on a vertical cliff face at Annagh Head (Fig. 2.9) looking east. The vertical scale is about 20 m; the large amphibolite boudin is about 1 m across. Two samples of Mullet gneiss (A) at this locality have yielded a U–Pb zircon age of 1753 ± 3 Ma. The main foliation in the gneiss wraps around boudins of early amphibolite (B). The gneiss and amphibolite are affected by migmatitic leucosomes (e.g. at C), one of which has a U–Pb zircon age of 995 ± 6 Ma. A deformed granite pegmatite (D) cuts the foliation and migmatitic leucosomes in the gneisses and has a U–Pb zircon age of 980 +38/−12 Ma. Grampian strain is low at this locality because the metadolerite (E), which cuts the 980 Ma pegmatite, is only weakly deformed. However, just north and south of the area in the photograph, the metadolerite, together with the other rocks, is strongly deformed in a north-north-westerly trending Grampian shear zone. U–Pb ages are from Daly (1996).

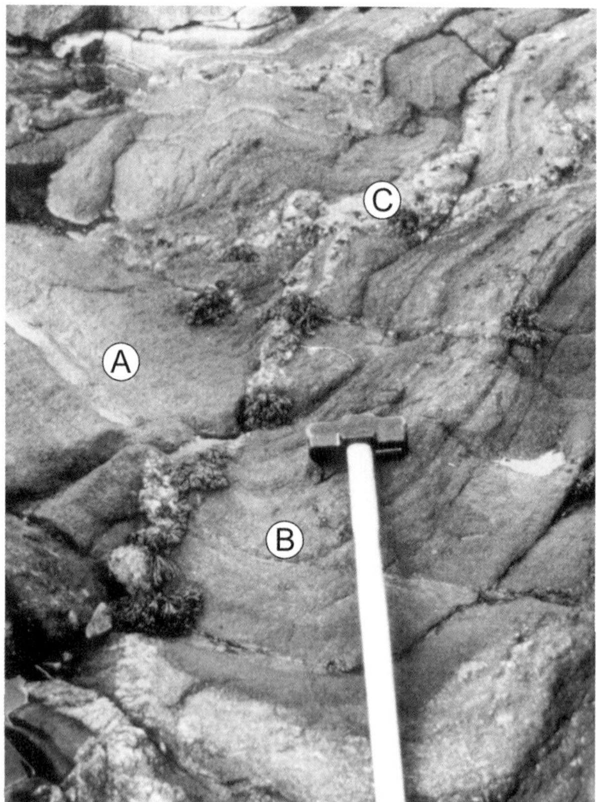

Fig. 2.12 Photograph of deformed Doolough gneiss at Doolough (Fig. 2.10). The Doolough gneiss at this locality (A) has a U–Pb zircon age of 1177 ± 4 Ma. An early deformation event (D_G1) produced a strong foliation and produced or accentuated the compositional banding. The banding was subsequently folded into spectacular sheath-like F_G2 folds, here with a vertical plunge and steep north–south axial plane (B). Both events predate 1015 ± 4 Ma, the U–Pb zircon age obtained from the Doolough Granite (Daly 1996). The F_G2 fold is cut by an undeformed pegmatite (C) with a U–Pb zircon age of 995 ± 6 Ma. U–Pb ages from Daly (1996).

Daly, 1990; Fitzgerald *et al.*, 1993; 1996). The Mullet gneisses and Cross Point gneisses are just distinguishable in terms of Sm–Nd model ages with t_{DM} values of 1.9–2.0 Ga and 1.7–1.8 Ga respectively (Fitzgerald *et al.*, 1996). Correlation of the Annagh Gneiss Complex with the Lewisian of north-west Scotland is precluded, given that the latter consists principally of crust formed during the Archaean. Early studies using Sm–Nd data suggested that the Annagh Gneiss Complex developed in two distinct episodes with crustal extraction ages of *c*.1.30–1.40 Ga and *c*.1.71–1.99 Ga, termed 'Doolough' and 'Mullet' gneisses, respectively (Menuge and Daly, 1990; 1994). Further Sm–Nd model age dating (Fitzgerald *et al.*, 1993) using samples from a wider area has shown that the Doolough gneisses (Fig. 2.10) are much less extensive than depicted by Menuge and Daly (1990). The Doolough gneisses are actually restricted to small areas around Doolough and Barranagh (Fig. 2.10).

Mullet gneiss

The Mullet gneisses occur in two NW–SE trending bands extending from the central part of the Mullet peninsula and apparently continuing onto the mainland north and south of Tristia (Fig. 2.10). They also outcrop at Erris Head where they form the sparse host rocks to

Table 2.4 Pre-Grampian history of the Annagh Gneiss Complex based on Daly (1996).

Age (Ma)	Dated sample	Magmatic, metamorphic and structural events
1753 ± 3	Mullet gneiss at Annagh Head	Crust formation from juvenile magma with mantle-like Nd and Hf (evidently part of a major crust-forming event seen also at Islay, Inishtrahull and Rockall).
1271 ± 6 1287 + 38/-35	Cross Point gneiss at Tristia Cross Point gneiss at Cross Point	Anorogenic magmatism.
1177 ± 4	Doolough gneiss at Doolough	Addition to the crust of juvenile magma. Early Grenville deformation. Formation of foliation (D_G1) in the Doolough gneiss and later folding (D_G2). Possible tectonic juxtaposition of Doolough gneiss with the older Cross Point gneiss and Mullet gneiss.
1015 ± 4	Doolough Granite at Doolough	Peralkaline granite intrusion.
995 ± 6 999 + 29/-9 993 ± 6 984 ± 6 980 + 38/-12	Leucosome at Annagh Head Pegmatite at Tristia Pegmatite at Doolough Pegmatite at Cross Point Pegmatite at Annagh Head	Migmatisation and injection of discordant pegmatites.
958 + 17/-19	Metamorphic zircon in Cross Point gneiss	Late Grenville deformation (D_G3). Discordant pegmatites foliated. Further metamorphism.
c.590*	(based on data from outside Annagh Gneiss Complex)	Regional crustal extension. Deposition of Dalradian sediments.

All ages are U–Pb zircon ages from Daly (1996) apart from the age for the Mam sill in Donegal (Daly *et al.*, unpublished data*).

an extensive injection of late (presumably Grenville) granitic pegmatites. The Mullet gneisses are best displayed on the Mullet peninsula near Annagh Head (Fig. 2.10, 2.11). Formation of the Mullet gneiss protoliths at Annagh Head (Fig. 2.11) has been dated at 1753 ± 3 Ma. These rocks have juvenile initial ε_{Nd}values (Fitzgerald *et al.*, 1996) similar to those obtained from the Rhinns Complex of Islay and Inishtrahull (Marcantonio *et al.*, 1988; Daly *et al.*, 1991).

The main lithology is a migmatised quartzo-feldspathic orthogneiss (i.e. originally igneous) locally with more micaceous darker bands. Compositionally the Mullet gneisses are I-type (*sensu* Chappell and White, 1974) metaluminous quartz monzodiorites, granodiorites and granites with granodiorite predominating (Daly, 2001). In a few places, e.g. at Annagh Head and north of Doolough, the gneisses contain K-feldspar megacrysts which probably originated as phenocrysts. Although alkali-calcic, the trace element signatures of the Mullet gneisses are compatible with a volcanic arc (subduction-related) setting (Fitzgerald, *pers. comm.*, 1995). The Mullet gneisses enclose boudins of coarse amphibolite, now generally concordant but originally emplaced as dykes because they are locally cross-cutting. The age of these dykes is unknown but they are succeeded by several episodes of Grenville deformation (see below). Generally the Mullet gneisses exhibit a strong NW–SE trending gneissose fabric. In places this fabric is a composite result of Grenville and later (probably Grampian) strain (cf. Johnston, 1995), but as may be seen at Annagh Head (Fig. 2.11) it is predominantly a Grenville feature (see below). Both before and after the emplacement of the early basic dykes, the Mullet gneisses were migmatised. The later migmatitic leucosomes are sometimes pinkish and often pegmatitic. Some discrete pegmatite sheets (of Grenville age, see below) appear to post-date the migmatisation but the ages of the two events are generally not distinguishable by the U–Pb zircon method.

Cross Point gneiss

The main belt of Cross Point gneiss extends in a broad NW–SE trending belt from Cross Point on the Mullet Peninsula to Tristia on the mainland lying between bands of Mullet gneiss (Fig. 2.10). Smaller outcrops also occur south of Erris Head and along the southern margin of the AGC in fault contact with the Dalradian. The Cross Point gneisses have been dated at two localities (Daly, 1996). At Tristia (Fig. 2.10) a near concordant age of 1271 ± 6 Ma was obtained while at Cross Point (Figs. 2.10, 2.13) severe disturbance during Grenville metamorphism resulted in an upper discordia intercept age of 1287 +36/-35 Ma. These ages are indistinguishable within error. However, the more precise *c*.1270 Ma age is taken as dating the Cross Point gneiss protoliths.

Fig. 2.13 Grenville pegmatite (parallel to hammer) at Cross Point (Fig. 2.10), dated at 984 ± 6 Ma, cuts Cross Point gneisses dated at *c.*1270 Ma, and the foliation (D_G1 or D_G2) affecting them. The pegmatite is later deformed by D_G3 (foliation parallel to hammer), which also affects the gneisses (e.g. foliation parallel to hammer but discordant to layering). Mafic and felsic banding (visible in bottom right and top left of photograph) within the Cross Point gneiss is in part original, reflecting the mixed nature of these rocks – partly granitoid melts from the Mullet gneiss and partly a mantle-derived basic component (see also Figs. 2.7 and 2.8).

The Cross Point gneisses are syenitic to granitic in composition and have A-type within-plate geochemical characteristics (Fitzgerald *et al.*, 1996). They are commonly banded on various scales depending in part on their state of strain and in part on the development of migmatitic leucosomes. Near Cross Point the felsic gneisses are interbanded on a decimetric to centimetric scale with foliated amphibolites. Larger layered mafic bodies also occur but it is uncertain if these share a common origin with the amphibolite bands. Occasionally the felsic gneisses contain mafic clots or tabular mafic enclaves suggestive of mixing between felsic and mafic magmas. In places, e.g. at Tristia, the Cross Point gneiss is relatively homogenous, resembling a foliated granitoid intrusion. This rock is shown on Griffith's (1839) map with an ornament similar to that employed for the more familiar Caledonian granites, which it superficially resembles. Generally the Cross Point gneisses are strongly foliated or lineated. The history of deformation and pegmatite injection is similar to that seen in the Mullet gneisses.

Geochronology and geochemical modelling (Fitzgerald *et al.*, 1996; Fitzgerald, *pers. comm.* 1996) suggests that the Cross Point gneisses may be intrusive into, and in part derived from, the Mullet gneisses, probably under anorogenic conditions. This is consistent with their younger age, their similar initial ε_{Nd}(Fig. 2.7) and ε_{Hf} (Fig. 2.8) values compared with the Mullet gneisses, and their A-type chemistry. At 1270 Ma, the Cross Point gneisses actually have slightly more positive ε_{Nd} values (Fig. 2.7) than the Mullet gneisses, suggesting an additional mantle component in their genesis (Fitzgerald *et al.*, 1996). A possible intrusive relationship between the Cross Point and Mullet gneisses is consistent with the map pattern even though the contacts between the two units are not exposed. Strain (probably of Grampian age) intensifies in both units astride the inferred position of the contact north of Cross Point so that the original relationships may be impossible to decipher.

Doolough gneisses and Doolough granite

The Doolough gneisses (Figs. 2.10, 2.12) are mainly sub-alkaline granites and trondhjemites with subordinate amounts of amphibolite. Mafic and felsic varieties of Doolough gneiss are inter-banded on a decimetric scale. These resemble straight gneisses and suggest that the Doolough gneisses have experienced high strain. Formation of the Doolough gneiss protolith, constrained to be less than *c*.1.3–1.4 Ga based on Nd isotopic data (Menuge and Daly, 1990) has been dated at 1177 ± 4 Ma. The Doolough gneisses are a small component of the Annagh Gneiss Complex and unless originally intruded into the AGC, were probably juxtaposed with the rest of the AGC (see below) during early Grenville deformation (D_G1) when they acquired their high strain fabric.

Migmatisation is sporadic in the Doolough gneisses but at least two generations of granitoid intrusives may be distinguished – the peralkaline Doolough Granite (Winchester and Max, 1987a; Aftalion and Max, 1987), which postdates two deformational events in the Doolough gneisses and a later suite of granite pegmatites (see below).

The Doolough Granite appears to stitch a tectonic contact between the Doolough gneisses to the south and the Mullet gneisses to the north (Fig. 2.10). The gneiss is cut by the Doolough Granite and also occurs within it as partially digested xenoliths. Sm–Nd model ages from the Doolough Granite vary from 1450 Ma to 1758 Ma and tend to be older towards the northern contact, consistent with contamination of the Doolough Granite by the different wall rocks on either side. Magmatic zircons from the Doolough Granite have yielded a U–Pb age of 1015 ± 4 Ma (Daly, 1996). A second population of zircons, with a poorly defined age of *c*.1220 Ma has likely been inherited from the *c*.1177 Ma old Doolough gneiss. This is supported by the Sm–Nd data (Fig. 2.7) and by Hf isotopic analysis of the zircons (Fig. 2.8). The *c*.1015 Ma old magmatic zircons (Fig. 2.8) display resorbed cores, too small to analyse, but their Hf isotopic composition is consistent with a source stored in grains equivalent to those of the Doolough gneiss, which are envisaged to have dissolved and regrown during crystallisation of the Doolough Granite.

Aftalion and Max (1987) dated large unabraded zircon fractions from the Doolough Granite, which yielded a discordia intercept age of 1093 ± 8 Ma, which probably reflects a mixture of 1015 Ma and older zircons, affected by recent lead loss. By air-abrading the zircons (Krogh, 1982) it was possible to remove the effects of lead loss (Daly, 1996). Moreover by analysing small fractions and single grains it has been possible to deconvolute the mixture, thus revealing the correct age.

Grenville deformation and metamorphism

Evidence for at least two Grenville deformational events can be deduced in the field at many localities. U–Pb dating suggests that at least three events have occurred, denoted D_G1, D_G2 and D_G3 (Daly, 1996). However it is not easy to correlate these events throughout the AGC.

At Doolough (Figs. 2.10, 2.12) all three Grenville deformation events may be seen in the field. D_G2 folding of a D_G1 high strain foliation in the Doolough gneisses must predate the 1015 Ma Doolough Granite which contains xenoliths of foliated Doolough gneiss, displaying F_G2 folds. Also at Doolough an undeformed pegmatite cutting folded, foliated Doolough gneiss (Fig. 2.12) was dated at 993 ± 6 Ma. This result confirms a minimum age for D_G2. Locally these pegmatites are deformed either by D_G3 or perhaps by Grampian movements. The generally weak deformation seen in the Doolough granite and in the pegmatites at Doolough suggests that D_G3 was less important here than elsewhere in the AGC.

At Cross Point (Figs. 2.10, 2.13) the Grenville deformation history is also quite well constrained. Since the Cross Point gneisses are *c*.1270 Ma old, all deformation must be Grenville or younger. Two phases of Grenville deformation can be recognised and although it is not possible to distinguish between D_G1 and D_G2 at Cross Point, the later deformation there is probably the D_G3 event. A pegmatite cutting the Cross Point gneisses and an early (D_G1 or D_G2) foliation affecting them has yielded a zircon age of 984 ± 6 Ma. The D_G3 foliation affects this pegmatite (Fig. 2.13) and also transects the layering and the early foliation in the gneisses at a high angle. This locality is unusual in that the foliation throughout the AGC is generally parallel to compositional banding. One group of zircons from the *c*.1270 Ma old Cross Point gneiss, interpreted to be metamorphic, are apparently aligned parallel to the D_G3 foliation and yield an age of 958 +17/-19 Ma. This age is (just) younger than that obtained from the cross-cutting pegmatite. These metamorphic zircons grew during development of the D_G3 foliation possibly as the result of Zr release from amphibole. Post-tectonic titanite from the same locality yields a concordant U–Pb age of *c*.943 Ma (Daly and Flowerdew, 2005).

At Tristia (Fig. 2.10) a post-D_G1 or -D_G2 pegmatite cutting foliated 1270 Ma old Cross Point gneiss yielded a U–Pb zircon age of 999 +29/-9 Ma. The pegmatite is

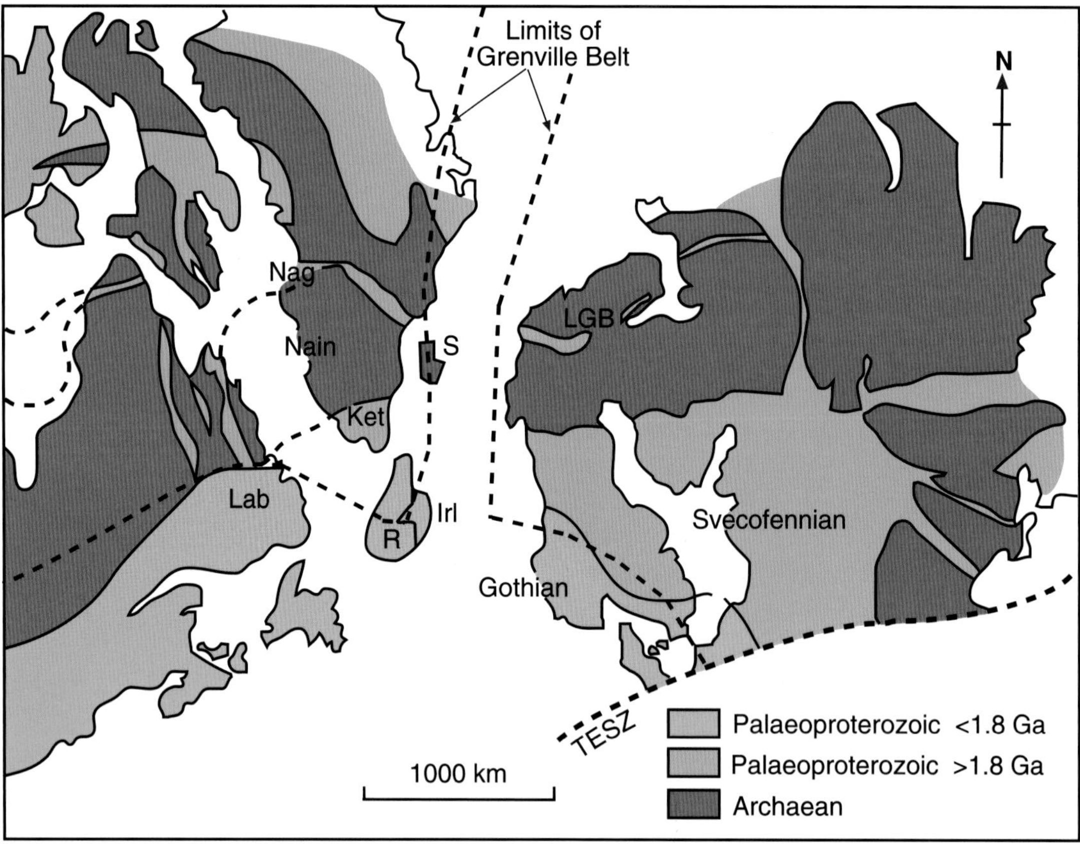

Fig. 2.14 Generalised map of distribution of Precambrian units in the North Atlantic province (after Gorbatschev and Bogdanova, 1993) showing regions underlain by Archaean and Palaeoproterozoic crust and the possible extent of the Grenville belt. Ket = Ketilidian; Lab = Labradorian; LGB = Lapland Granulite Belt; Nag = Nagssuqtoqidian; Irl = Ireland, north of the Iapetus suture; R = Rockall; S = Scotland; TESZ = Trans-European suture zone.

itself deformed by D_G3, which is thus constrained to be younger than *c.*999 Ma in keeping with the results from Cross Point and Doolough. On Annagh Head (Fig. 2.10), zircons from a migmatitic leucosome cutting the 1753 Ma Mullet gneiss have yielded an age of 995 ± 6 Ma. At the same locality a deformed pegmatite cutting the main foliation in the Mullet gneiss has a U–Pb zircon age of 980 +38/-12 Ma. This pegmatite and the foliation affecting it are cut by a weakly deformed metadolerite. If it is assumed that the metadolerite deformation is Grampian, then some pre-Grampian, presumably Grenville, deformation must have occurred at Annagh Head after *c.*980 Ma. This is presumably the D_G3 deformation, which is also present at Cross Point.

In several places on the Mullet Peninsula, kinematic indicators such as oblique extensional crenulation and asymmetric feldspar porphyroclasts indicate sinistral shear in a west-north-westerly direction before the emplacement of pegmatitic granite sheets dated between 993 and 984 Ma (Table 2.4).

Metamorphic titanite from several locations within the AGC probably grew at *c.*980 Ma (Daly and Flowerdew, 2005). The U–Pb data are slightly discordant with some lead loss apparently occurring during the Caledonian. However, these data suggest that prolonged Caledonian heating is unlikely to have exceeded *c.*600°C, the closure temperature for Pb diffusion in titanite (Heaman and Parrish, 1991).

Grenville events may be summarised as follows: D_G1 produced the main foliation in the Doolough gneisses, which were folded by D_G2. These events occurred after 1177 Ma and before 1015 Ma. Very high strains affected the Doolough gneisses and may also have been responsible for the non-cylindrical D_G2 folding seen at Doolough. The Doolough gneisses may have been juxtaposed with the rest of the AGC during D_G1 or D_G2. Further tectonothermal activity resulted in migmatisation and the intrusion of discordant pegmatites (at Annagh Head, Tristia and Cross Point) in the interval 995–980 Ma. D_G3 probably took place at *c.*960 Ma. This event produced a foliation in the pegmatites and locally developed a foliation in the gneisses (e.g. at Cross Point). Late titanite grew at or before *c.*940 Ma at Cross Point (Daly and Flowerdew, 2005).

The evolution of the Annagh Gneiss Complex may be summarised as follows. Much of the Annagh Gneiss Complex originated as juvenile Palaeoproterozoic crust represented by the 1753 ± 3 Ma old calc-alkaline Mullet gneisses. Late Mesoproterozoic Cross Point gneisses with A-type geochemistry and Palaeoproterozoic t_{DM} ages were emplaced as anorogenic granitoids at 1271 ± 6 Ma, probably by melting of the pre-existing Palaeoproterozoic Mullet gneisses with the addition of a mantle-derived mafic component. The Doolough gneisses comprise a small volume of juvenile granitoids and associated basic rocks, which formed at 1177 ± 4 Ma. The Doolough gneisses occur as a tectonic sliver, possibly incorporated along a Grenville high strain zone. Their origin is unclear. They may be a further manifestation of anorogenic magmatism or they may represent a much larger volume of juvenile subduction-related material. Grenville deformation occurred in two stages, between 1177–1015 Ma and from 995–960 Ma. These events were separated by the intrusion of the Doolough peralkaline granite at 1015 ± 4 Ma and by migmatisation and pegmatite emplacement between 995 and 980 Ma. Post-Grenville metadolerites help to distinguish Caledonian strain from earlier deformation.

Correlation of AGC with Precambrian rocks further afield
Available geothermological and isotopic data from the Annagh Gneiss Complex and the Rinns Complex suggest that north-western Ireland and south-western Scotland are underlain by juvenile crust formed *c.*1.75–1.8 Ga ago. This Palaeoproterozoic crust may be correlated westwards with rocks of similar age and geochemistry on the Rockall Plateau (Daly *et al.*, 1995), in the Ketilidian terrain of south Greenland, and with large areas of central North America. The Proterozoic of NW Ireland and NW Scotland may have formed part of a continuous belt along the accreting continental margin of Laurentia in the Palaeoproterozoic (Fig. 2.14, Muir *et al.*, 1992).

Anorogenic magmatism, similar in age to the *c.*1.27 Ga protoliths of the Cross Point gneisses of the AGC, is widely recognised in Laurentia and Baltica (cf. Romer, 1996). Rifting or attempted rifting appears to have occurred semi-continuously between *c.*1450 Ma and 1230 Ma in Laurentia (e.g. Gower, 1996). Regionally extensive mafic dyke swarms including the 1267 Ma Mackenzie dykes (Le Cheminant and Heaman, 1989), the 1268 Ma Nutak and 1273 Ma Harp dykes in Labrador (Cadman *et al.*, 1993) testify to a major episode of continental extension. At the same time, bimodal volcanism developed in rift sequences such as the Wakeham Supergroup of southern Quebec (Martignole *et al.*, 1994). Anorogenic plutonic magmatism formed the *c.*1350 Ma–1290 Ma Nain Plutonic Suite of Labrador and continued in the eastern Grenville Province until *c.*1225 Ma (Gower and Krogh, 2002).

The 1177 ± 4 Ma age for the Doolough gneisses confirms the juvenile origin for these rocks suggested by their positive initial ε_{Nd} (Fig. 2.7, Menuge and Daly, 1990; 1994) and ε_{Hf} values (Fig. 2.8). However their tectonic significance is uncertain. Both a subduction-related and an anorogenic origin are compatible with their geochemistry. They are highly deformed and the small outcrops at Doolough and Barranagh (Fig. 2.10) may coincide with a major Grenville shear zone along which the Doolough gneisses were juxtaposed with the rest of the AGC. If this is true the Doolough gneiss outcrops might only represent a small fraction of their original extent (Menuge *et al.*, 1995).

Alternatively, if the Doolough gneisses are an expression of extensional anorogenic magmatism, they may have intruded into the AGC. If they are anorogenic, correlatives of the protoliths of the Doolough gneisses in Laurentia would include widespread anorthosite magmatism associated with granitoids derived by crustal melting (the anorthosite-mangerite-charnockite-granite, or AMCG suite) under relatively stable mildly extensional conditions. AMCG magmatism extended throughout the Grenville Province from the Adirondacks (McLelland *et al.*, 1996) to Labrador and beyond (Emslie and Hunt, 1990). In Baltica, anorogenic magmatism is also known from approximately the same time, especially in southern Norway in the interval between 1190 and 1130 Ma (Heaman and Smalley, 1994).

Grenvillian deformation affecting the AGC appears to have taken place in two stages – D_G1 and D_G2 before 1015 Ma and D_G3 between 980 and 940 Ma. In a broad sense these events correlate with the Grenville event in Laurentia and with the Sveconorwegian in Baltica. The early deformation may correspond to the *c.*1050 Ma Ottawan Orogeny (cf. McLelland *et al.*, 1996) in the western Grenville Province, and may also correlate with events in NW Scotland and Baltica. In Scotland, metamorphic recrystallisation or subsequent cooling is recorded by Sm–Nd mineral ages of 1082 and 1010 Ma from eclogites of the Glenelg inlier (Sanders *et al.*, 1984) and U-Pb zircon dating of retrogression at 1000 Ma (Brewer *et al.*, 2003) biotite ages indicate post-metamorphic cooling 1024–1122 Ma ago in the Lewisian of South Harris (Cliff and Rex, 1989). In western Sweden, Sveconorwegian deformation and metamorphism are imprecisely dated at 1090 Ma (Daly *et al.*, 1983).

The post-D_G2 peralkaline Doolough Granite has a chemistry suggestive of anorogenic magmatism, but whether this is intra-orogenic or whether it separates the Grenville Orogeny into two distinct phases is uncertain.

Later (D_G3) deformation in the AGC superficially corresponds to the main Grenville deformation in eastern Labrador, which occurred between 1010 and 990 Ma (Gower, 1996). However, this event in eastern Labrador is succeeded by late- to post-tectonic granites between 966 and 956 Ma (Gower, 1996) while in the AGC deformation may have continued until *c.*960 Ma. Both the granitoid magmatism and the late Grenville deformation appear to be diachronous – generally earlier in the west and later in the east. The granitoid/pegmatite ages range from 1090–1076 in the southwest Grenville Province to 966–956 Ma in eastern Labrador to 930–920 Ma in western Baltica (cf. Romer, 1996). The relevant data are still lacking in large parts of the Sveconorwegian Province, but late- to post-tectonic uplift and cooling in the eastern part of the Sveconorwegian Province in southern Sweden occurred *c.*1009–965 Ma with the possibility of a later event at *c.*930–905 Ma based on $^{40}Ar/^{39}Ar$ dating of hornblende and white mica respectively (Page *et al.*, 1996).

Slishwood Division

The north-east Ox Mountains and Lough Derg inliers and the eastern end of the Rosses Point Inlier (Fig. 2.1), collectively known as the Slishwood Division (Max and Long, 1985), share similar lithologies and metamorphic histories. Gravity (Young, 1974) and magnetic data (Max *et al.*, 1983a) suggest that they are all part of one basement block. The Slishwood Division has long been regarded as pre-Grampian basement (Max and Long, 1985; Sanders *et al.*, 1987; Phillips, 1981a; Daly, 2001). Arguments for its pre-Grampian age include the exceptionally high metamorphic grade, the complex metamorphic and structural history compared with the Dalradian, as well as geochronological evidence. The contacts between the Slishwood Division and the Dalradian are everywhere tectonic. In south Donegal, metasediments belonging to a high stratigraphic level in the Dalradian were first thrust onto the Slishwood Division and later extended as a result of gravitational collapse along the Lough Derg Slide (Alsop, 1991; see also Chapter 4). At the SW end of the NE Ox Mountains Inlier, the contact with the Dalradian rocks (North Ox Mountains Slide, Fig. 2.15) involves SE-directed compressional shearing, which reworks the high-grade fabrics in the Slishwood Division and interleaves the Slishwood Division rocks with the Dalradian in a series of tectonic slices (Flowerdew 1998/9).

The Slishwood Division is named after the Slishwood area, south of Lough Gill in the NE Ox Mountains Inlier (Figs. 2.1, 2.15), a narrow, north-east trending ridge of metamorphic rocks extending from Ladies Brae to Manorhamilton (Fig. 2.15). The rocks, described by Lemon (1966, 1971), Phillips *et al.* (1975), Sanders (1979), Molloy and Sanders (1983) and Flowerdew and Daly (2005), are predominantly migmatitic psammitic gneisses with minor pelites, semipelites, calc-silicates, metabasites and serpentinites, the last originally juxtaposed with the other lithologies as peridotite bodies. With the exception of the serpentinites, the Lough Derg (Anderson, 1948; Church, 1969; Unitt, 1997) and Rosses Point (Max, 1984) inliers display similar field and petrographic characteristics. All three inliers are cut by a suite of granitic pegmatites (Flowerdew *et al.*, 2000), while the NE Ox Mountains Inlier is also cut by several tonalite and granite bodies, possibly subduction-related, that intruded between 471 and 467 Ma (Fig. 2.15; Flowerdew *et al.*, 2005).

At Slishwood, high-pressure granulite-facies assemblages are well preserved, and from this area a protracted history of deformation and metamorphism has been determined (Table 2.5, Sanders *et al.*, 1987; Sanders, 1991, 1994; Flowerdew and Daly, 2005; Daly *et al.*, 2008b). Here, Sanders *et al.* (1987) recognised an early eclogite-facies metamorphism with metamorphic pressures estimated in the range 12–14 kbar. High-pressure granulite facies assemblages developed at 11 kbar and *c.*860°C in response to isothermal decompression, probably associated with crustal extension, when original omphacitic clinopyroxene was replaced by sieve-textured plagioclase-augite intergrowths. Subsequently, slow isobaric cooling took place at depth with kyanite replacing sillimanite (Sanders *et al.*, 1987). At Slishwood and elsewhere, deformed garnet-clinopyroxene-plagioclase-bearing metabasite lenses and garnet-kyanite-mesoperthite assemblages in metasediments indicate pressures in excess of 10 kbar. Metamorphic studies by Unitt (1997) and Flowerdew and Daly (2005) in the Lough Derg Inlier and near Manorhamilton in the NE Ox Inlier (Fig. 4.8)

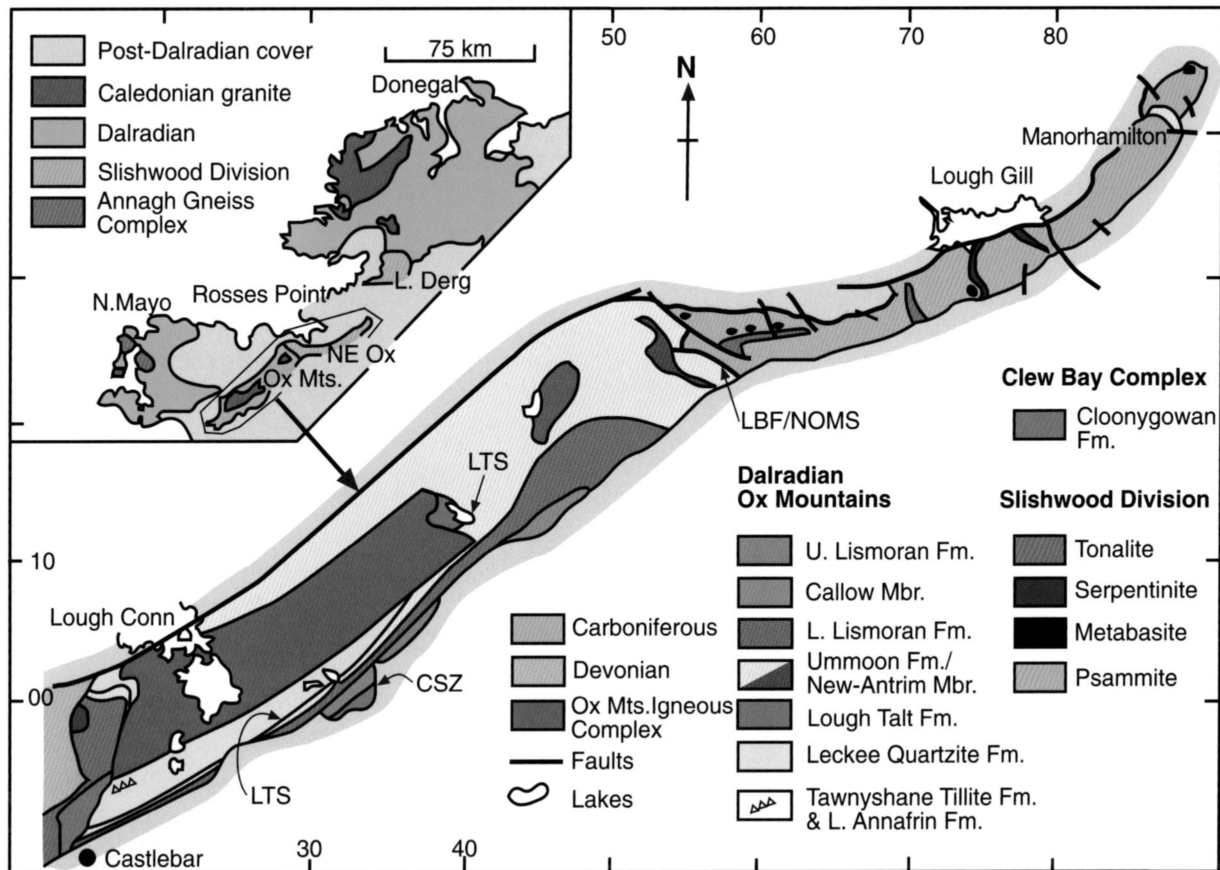

Fig. 2.15 Geological sketch map of the Ox Mountains region based on Alsop and Jones (1991) and Sanders *et al.* (1987). LBF/ NOMS = Ladies Brae Fault/ North Ox Mountains Slide; LTS = Lough Talt Slide; CSZ = Callow Shear Zone. The term 'slide' here means syn-metamorphic ductile fault (Hutton 1979). Inset shows major geological divisions in north-western Ireland together with location of main map.

Table 2.5 Structural history and geochronology of the Slishwood Division, from Sanders (1994), Flowerdew and Daly (2005), Flowerdew *et al.* (2000; 2005), Daly *et al.* (2008a) and unpublished data.

Lithological additions	Tectonic events	Thermal conditions and metamorphism	Age	Method
deposition of Viséan sediments	Final uplift and erosion	cooling	< 350 Ma	stratigraphic
	foliation, mylonites	cooling epidote–amphibolite facies	≥ 455 Ma	Rb–Sr muscovite
granitic pegmatites tonalite and migmatisation	Sinistral shearing and extension	granulite facies	*c.*470 Ma	U–Pb zircon
	open folds	slow cooling at depth		
migmatisation?	extension	decompression to high-pressure granulite facies	605–540 Ma	Sm–Nd garnet
	shearing at high T			
		eclogite facies		
	folding and boudinage			
basic bodies			580 ± 36 Ma	Sm–Nd mineral–whole-rock model age
deposition of arkose and minor pelite and limestone			*c.*940 Ma	Youngest U–Pb detrital zircon age

indicate a broadly similar PTt evolution to that in the Slishwood area (Sanders *et al.*, 1987) with PT conditions of *c.*14–16 kbar and 750–810°C (Flowerdew and Daly, 2005).

The age of the high-grade metamorphism is still poorly known but appears to be pre-Grampian. Sanders *et al.* (1987) and Flowerdew and Daly (2005) obtained Sm–Nd mineral isochrons from various combinations of garnet, clinopyroxene and plagioclase from partially equilibrated high pressure granulite- or partially decompressed eclogite-facies assemblages that range from 605 ± 37 Ma to 539 ± 11 Ma (Fig. 4.8). Interpretion of these isochrons is unclear. Firstly, perfectly preserved eclogite-facies assemblages have not survived decompression. Secondly, as shown in the classic study of Mørk and Mearns (1986), dry metamorphism of gabbroic rocks may fail to achieve isotopic equilibrium because the metamorphic minerals may inherit variations in Nd isotopic composition from their pre-existing igneous precursors. Thus, if garnet inherited relatively unradiogenic Nd from pre-existing plagioclase, which has a low Sm/Nd ratio, the resulting isochron could be systematically too young (Flowerdew and Daly, 1997; 2005). Thirdly, REE-rich inclusions in some of the mineral separates and disequilibrium between some of the phases exacerbate the difficulty of interpreting the isochrons. Nevertheless, it is likely, given the high temperatures, that some or all of these ages are dating cooling as opposed to crystallisation of the dated minerals, and that the eclogite-facies metamorphism occurred before the Grampian Orogeny (Flowerdew and Daly, 2005).

However, two pieces of evidence indicate that the age of the high-grade metamorphism cannot be much older than the Grampian. Firstly, U–Pb detrital zircon ages (Daly *et al.*, 2004) demonstrate a post-Grenvillian age of deposition of the protolith. This rules out, incidentally, a correlation of the high-grade metamorphism with, for instance, that of the Glenelg inlier in Scotland (Sanders *et al.*, 1984; Brewer *et al.*, 2003), though correlation of the Slishwood Division metasediments with the Grampian Group of the Dalradian (Sanders, 1994) is still a possibility, as is the traditional comparison with the post-Grenvillian Moine Supergroup (e.g. Anderson, 1948; Powell, 1965). Secondly, in the Lough Derg Inlier some of the metabasite bodies preserve original gabbroic textures. Flowerdew and Daly (2005) dated one of these basic bodies, from Knader Lough, at 580 ± 36 Ma. This Sm–Nd mineral isochron age of the relict magmatic mineral assemblage (clinopyroxene and plagioclase) in a metagabbro (Figs. 2.5, 4.8) is presumed to date igneous crystallisation, and thus provides a maximum age for the high-grade metamorphism. This age also suggests that gabbroic magmatism may have been related to extension associated with the opening of the Iapetus Ocean (cf. Bingen *et al.*, 1998).

Extreme metamorphism between 540 and 580 Ma suggests that the Slishwood Division was not part of Laurentia at this time, consistent with its interpretation as an exotic microcontinental indenter (Flowerdew and Daly, 1999). Migmatitic leucosomes that developed during sinistral extension (Fig. 2.16) under granulite facies conditions have yielded U–Pb zircon ages about 470 Ma (Daly *et al.*, 2008b) indicating that re-equilibration took place during the Grampian collision.

Throughout the Slishwood Division, the high-grade fabrics are cut by small bodies of granitic pegmatite. These minor intrusions clearly post-date the high-grade metamorphism and also cut structures associated with imbrication of the basement with Dalradian rocks, particularly at the south-western end of the NE Ox Mountains Inlier (Flowerdew *et al.*, 2000, 2005). Muscovite-bearing granite pegmatites in the NE Ox Mountains, Lough Derg and Rosses Point inliers have average Rb–Sr muscovite cooling ages clustering closely

Fig. 2.16 The latest of three generations of migmatitic leucosome (parallel to the Swiss Army knife) developed at *c.*470 Ma (U–Pb zircon age, Daly *et al.*, 2008b) in Slishwood Division paragneisses along late sinistral extensional shears south of Killerry Mountain, NE Ox Mountains Inlier.

around 455 Ma (van Breemen *et al.*, 1978, Flowerdew *et al.*, 2000). These pegmatites may have been generated as a result of extension following underthrusting by low-grade metasediments. Within the Lough Derg Slide Zone the pegmatites have been subject to intense extensional strain (Alsop, 1991).

Tyrone Central Inlier

The Central Inlier of Tyrone (Fig. 2.1) is poorly exposed and is thus the least well known of the Precambrian outcrops in Ireland. It comprises psammitic and pelitic paragneisses described by Hartley (1933b) and Cole (1897), cut by granitic pegmatites. The inlier is referred to as the Corvanaghan Formation on the 1:250 000 geological map of Northern Ireland (Geological Survey of Northern Ireland, 1997; Cooper and Johnston, 2004). Deformation and amphibolite-facies metamorphism of the Corvanaghan Formation pre-date the emplacement of the Tyrone Igneous Complex (Phillips, 1978), the basic plutonic part of which has been identified as an Arenig ophiolite complex (Hutton *et al.*, 1985), emplaced soon after its formation. Reports of a Precambrian age for the metamorphism in the Central Inlier based on unpublished Rb–Sr mica ages from granitic pegmatites have not been substantiated by recent studies (Chew *et al.*, 2008). One Central Inlier pegmatite yielded a U–Pb zircon age of 467 ± 12 Ma while the same sample and one other yielded identical Rb–Sr muscovite-feldspar ages of 457 ± 4 Ma (Chew *et al.*, 2008). All of the published evidence is compatible with a Grampian age for the metamorphism of the Central Inlier. Moreover, U–Pb detrital zircon ages and whole-rock Sm–Nd model ages permit correlation of the metasediments with the Argyll or Southern Highland groups of the Dalradian (see below). Deposition of the Corvanaghan Formation took place after 999 ± 23 Ma – the U–Pb Concordia age of the youngest dated detrital zircon. On this basis, correlation with the Slishwood Division remains possible. However, as discussed above, the metamorphic history of the Central Inlier is quite unlike that of the Slishwood Division in having reached only sillimanite zone (670 °C, 6.8 kbar; Chew *et al.*, 2008) and reportedly containing cordierite (Hartley, 1933b). In addition, the Central Inlier is lithologically distinct in lacking metabasites, calc-silicates and serpentinite.

Pre-Monian basement

Neoproterozoic rocks of the Rosslare Complex outcropping in County Wexford (Figs. 2.1, 2.17) form the basement of the Avalonian continent, which lay on the south-eastern side of the Iapetus Ocean. In what may be a mirror image of events in north-western Ireland, the Rosslare Complex, together with overlying early Palaeozoic rocks, has been affected by the early Ordovician Monian Orogeny, possibly similar in age to the Grampian. The Monian Orogeny may have affected the south-eastern margin of the Caledonian belt over large distances along strike (Tietzsch-Tyler, 1996). The Rosslare Complex also preserves evidence of the Neoproterozoic Avalonian Orogeny.

Rosslare Complex

The pre-Monian Precambrian rocks of the Rosslare Complex, south-east County Wexford, have been reviewed by Tietzsch-Tyler (1996). The Rosslare Complex consists of variably deformed amphibolites and metasedimentary paragneisses cut by several generations of minor intrusions (Table 2.6). Granodioritic rocks also occur but are relatively rare. They appear to be either the products of migmatisation of the paragneisses or to represent minor intrusive sheets of the Late Caledonian Carnsore and Saltees granites.

An angular unconformity is inferred between a cover of Cambrian sediments (Cahore Group) and the Rosslare Complex basement (Tietzsch-Tyler and Sleeman, 1994a; Tietzsch-Tyler, 1996; Fig. 2.17). Following Bennett *et al.* (1989), Sleeman and Tietszch-Tyler (1988), and Tietszch-Tyler (1996) the Cahore Group now includes the Cullenstown Formation, previously considered to be Precambrian in age and part of the Rosslare Complex (F. C. Murphy, 1990).

The Rosslare Complex is divided into two major units – metasediments of the Kilmore Quay Group to the south and amphibolites of the Greenore Point Group to the north. The north-western margin of the complex is a major greenschist facies sinistral mylonite zone (Baker, 1970; Max and Dhonau, 1971; F. C. Murphy, 1990) known as the Ballycogley mylonites. The south-eastern boundary is the Kilmore–Wilkeen Mylonite Zone (F. C. Murphy, 1990) into which the Late Caledonian Saltees and Carnsore granites have been emplaced (Max *et al.*, 1979; O'Connor *et al.*, 1988).

The protolith ages of the Rosslare Complex are not known. Two granodiorite samples have yielded Sm–Nd model ages of 1821 and 1759 Ma (recalculated from Davies *et al.*, 1985) but it is not clear if these data have any direct geological significance. ^{39}Ar–^{40}Ar hornblende ages of 626 ± 6 Ma from the Greenore Point Group

Table 2.6 Summary of geological events in the Rosslare Complex based on Max and Roddick (1989) and Tietzsch-Tyler (1996).

Age	Geological event	Metamorphic conditions
Mesozoic–Tertiary	fault reactivation	
Carboniferous–Permo-Trias	fault reactivation	
*c.*430 Ma (Rb–Sr WR)[1]	**Caledonian**	
	Carnsore granite intruded	
	D5 shear zones	greenschist facies
	intrusion of younger basic dykes	
*c.*437 Ma (Rb–Sr WR)[2]	Saltees granite intruded	
*c.*485 Ma (Rb–Sr WR)[3]	**Monian**	
	D4 deformation and retrograde metamorphism (local?)	low amphibolite to upper greenschist facies
	intrusion of older basic dykes	
	Avalonian	
	D3 deformation	
	intrusion of granodiorites	
*c.*620 Ma (^{40}Ar–^{39}Ar)[3]	metamorphism	
	syn-D2 intrusion of St Helen's Gabbro	amphibolite facies
	migmatisation and D1 deformation	high amphibolite facies
	intrusion of Greenore Diorite	
	deposition of greywackes and formation of amphibolites (main protoliths of Rosslare Complex)	

1: O'Connor *et al.* (1988); 2: Max *et al.* (1979); 3: Max and Roddick (1989).

and 618 ± 5 Ma from the St Helen's Gabbro provide a minimum age for the protoliths of the complex (Max and Roddick, 1989) and also show that the metamorphism affecting it is Neoproterozoic in age. It has been suggested that rocks similar to the Rosslare Complex may form the sub-Palaeozoic basement in south-eastern Ireland, based on isotopic similarities between these granites and the Palaeozoic cover rocks (O'Keeffe, 1986; O'Connor *et al.*, 1988; Kennan *et al.*, 1979).

The Kilmore Quay Group (Fig. 2.17) comprises highly strained paragneisses including semipelitic mica schists, sometimes migmatised and banded paler-coloured psammites interbedded with occasional calcsilicates and minor quartzites. Fine-grained foliated amphibolites are also present in minor amounts. Original bedding is suggested by layering on a 0.5 to 1 m scale, which is usually parallel to the foliations, usually defined by micas and elongated quartz. The chemistry of the metasediments suggests a non-calcareous greywacke protolith with a mainly igneous provenance (Winchester and Max, 1982). The paragneiss mineral assemblages are made up of variable proportions of plagioclase, biotite, quartz, muscovite and garnet with lesser amounts of epidote, titanite, zircon, rutile, opaque oxides, tourmaline, and rare kyanite.

The Greenore Point Group (Fig. 2.17, Tietzsch-Tyler, 1996), formerly known as the St Helen's Amphibolite (Baker, 1970), is made up of a thick sequence of amphibolites probably of igneous origin (Baker, 1970; Thorpe, 1974; Barber *et al.*, 1981; Winchester and Max, 1982; Max and Long, 1985) with both plutonic (e.g. Winchester and Max, 1982) and possibly volcanic components (Max and Roddick, 1989). The chemistry of the amphibolites is similar to that of continental tholeiites (Winchester and Max, 1982; Gibbons and Horak, 1996). Mineral assemblages are entirely metamorphic and comprise hornblende, plagioclase and minor garnet, quartz, epidote, biotite, titanite and ilmenite (Baker, 1970; Max and Dhonau, 1971).

The relative age of the Kilmore Quay and Greenore Point Groups is unknown, but both appear to have been intruded by the St Helen's Gabbro. Max (1975) described a sharp southern contact between the gabbro and the metasediments of the Kilmore Quay Group, which is unfortunately now covered by beach sands. Baker (1970) suggested that the northern contact may be a mylonite zone. He suggested that the amphibolites of the Greenore Point Group might be derived by deformation of a large gabbro body with the relatively undeformed St Helen's

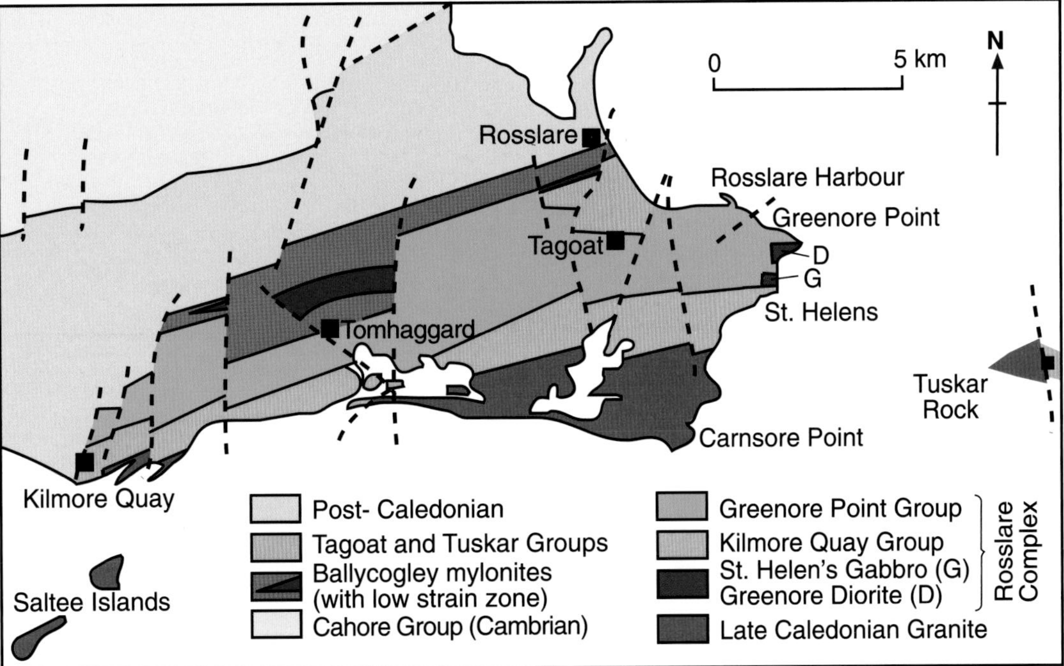

Fig. 2.17 Geological map of the Rosslare–Kilmore Quay area in southeast Ireland (based on Tietzsch-Tyler, 1996).

body representing a low-strain region. Max (1975) regarded both margins as intrusive with the southern margin possibly representing the base of an original sill.

Max (1975) recognised four orogenic cycles within the Rosslare Complex, based on superimposed fold sequences and sequential igneous intrusions, later interpreted as five events by Tietzsch-Tyler (1996). Max (1975) recognised early syntectonic migmatites and several episodes of igneous intrusion, including two generations of basic dykes separated by deformation and the syn-kinematic St Helen's Gabbro (Table 2.6). Subsequent work has demonstrated that only the early part of the history is Precambrian in age. The age of the earliest deformation and metamorphism is not known. The second deformation probably occurred close to *c.*620 Ma (Max and Roddick, 1989) when the St Helen's Gabbro was intruded. The D1 and D2 structures (Avalonian Orogeny) are overprinted by further folding and granitic intrusion followed by basic and intermediate dykes. D3 is probably also Precambrian (Avalonian) but its age is not well constrained (Tietzsch-Tyler, 1996).

Intrusion of a second set of basic dykes marks the onset of Monian events. The dykes were affected by further deformation (D4, 'Monian') that also affects the Cambrian Cahore Group sediments. The Monian deformation took place at or before *c.*480 Ma as recorded by ^{39}Ar–^{40}Ar hornblende ages (Max and Roddick, 1989). Monian deformation is also constrained to be earlier than the Upper Arenig to Llanvirn Tagoat Group (Fig. 2.17, Brenchley *et al.*, 1967; Tietzsch-Tyler and Sleeman, 1994a), which rests unconformably on the Rosslare Complex and contains clasts of the Ballycogley mylonites. Rb–Sr and U–Pb dating of the Saltees and Carnsore granitic intrusions (Max *et al.*, 1979; O'Connor *et al.*, 1988) shows that they and the intervening D5 deformation (Table 2.6) are of Silurian age.

Traditionally, comparisons are made between the Rosslare Complex and the basement of the eastern Appalachians and Avalonia (Murphy, 1990) rather than with the Precambrian of north-western Ireland. Tietzsch-Tyler (1996) has suggested that the metasedimentary clasts in the southerly-derived Devonian Inch Conglomerate Formation (Todd, 2000) within the Dingle Basin of SW Ireland may have been sourced in a south-westward continuation of the Rosslare Complex. On a wider scale Tietzsch-Tyler (1996) also correlated the Rosslare Complex with the Coedana and Sarn complexes in North Wales and with the Neoproterozoic Avalonian rocks of Newfoundland. However, clasts of schist, mylonite, and gneiss from the Inch Conglomerate have yielded Rb–Sr muscovite ages between *c.*435 and 398 Ma consistent with unroofing of a metamorphic belt affected by early Silurian deformation. These display a range of Sm–Nd model ages from 1.38 to 3.34 Ga (Todd, 2000).

While the Mesoproterozoic and Palaeoproterozoic t_{DM} ages are typical of the presumed Avalonian basement south-east of the Iapetus suture, the one gneiss clast yielding an Archaean age is intriguing and indicates that our knowledge of the basement rocks in southern Ireland is far from complete.

Dalradian Supergroup

The Dalradian Supergroup (Figs. 2.1, 2.18) is a thick sequence of deformed and metamorphosed clastic sediments, limestones and basic volcanics deposited on the Laurentian margin of the Rodinia supercontinent (Cawood *et al.*, 2003; 2007a, b). In areal extent, the Dalradian is by far the largest development of Precambrian rocks in Ireland. It outcrops in three major inliers (Fig. 2.1) – in Donegal-Tyrone (Fig. 2.19), in north Mayo and in Connemara. Smaller inliers (Fig. 2.1) occur in north-east Antrim, in Lack (Tyrone), in the Ox Mountains (Alsop and Jones, 1991), at the western end of the Rosses Point inlier (Max, 1984) and at Manorhamilton (Long and Yardley, 1979; Fig. 2.15). The term Dalradian, first used by Geikie (1891) and redefined in Harris and Pitcher (1975), is an anglicisation of *Dalriadan* referring to the *Dalriada*, the early rulers of Ireland and adjacent parts of Scotland. The stratigraphy of the Dalradian rocks in Ireland is described and discussed by Harris and Pitcher (1975), Harris *et al.* (1978), Harris *et al.* (1994), Cooper and Johnston (2004), Hutton and Alsop (2004), and McCay *et al.* (2006). The Dalradian Supergroup in Scotland has been reviewed by Strachan *et al.* (2002). The equivalent rocks along strike to the west are the Fleur de Lys Supergroup of Newfoundland (Kennedy, 1979).

The Laurentian basement that extended to accommodate Dalradian sedimentation is made up of at least two pre-Grampian rock units described earlier in this chapter. In Ireland, these rocks occur in tectonic contact with the Dalradian (Fig. 2.1) in north Mayo (Annagh Gneiss Complex, Fig. 2.8) and off the Donegal coast on Inishtrahull (Rhinns Complex) where a contact is inferred. The unconformity is not preserved, but an original depositional contact can be inferred in North Mayo (cf. R.C. Fitzgerald *et al.*, 1994; McAteer *et al.*, 2008). Three units previously considered to be pre-Grampian 'basement' have been shown to be part of the Dalradian. These are the metasediments of the central Ox Mountains (Alsop and Jones, 1991), the Inishkea Division of the North Mayo Inlier (Kennedy and Menuge, 1992), and the Tyrone Central Inlier (Chew *et al.*, 2008).

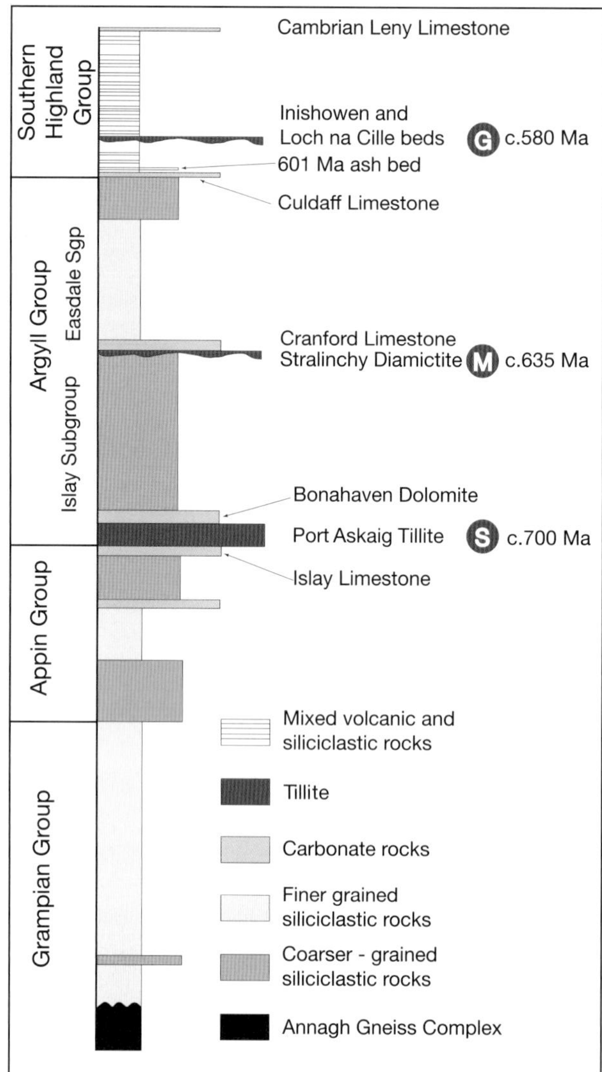

Fig. 2.18 Schematic stratigraphy of the Dalradian Supergroup (after McCay *et al.*, 2006) highlighting the three main episodes of glaciation recorded by the Dalradian sequence. S = Sturtian; M = Marinoan; G = Gaskiers. Note that the base of the Stralinchy Diamictite corresponds to the proposed orogenic unconformity identified by Hutton and Alsop (2004).

Deposition of the Dalradian Supergroup (between *c.* 920 and 550 Ma, see below) spans a critical period in Earth history involving the destruction of the Rodinia supercontinent (Young, 1995; Karlstrom *et al.*, 2000), climatic extremes (Kaufman *et al.*, 1997; Hoffman *et al.*, 1998) and the eventual development of the first hard-bodied organisms (Grotzinger *et al.*, 1995). There is thus great interest in the palaeogeography and tectonic history of this time interval (C.W. Thomas *et al.*, 2004; Rainbird *et al.*, 2001; Lund *et al.*, 2003; Cawood *et al.*, 2003, 2004; Hartz and Torsvik, 2002). However, actualistic

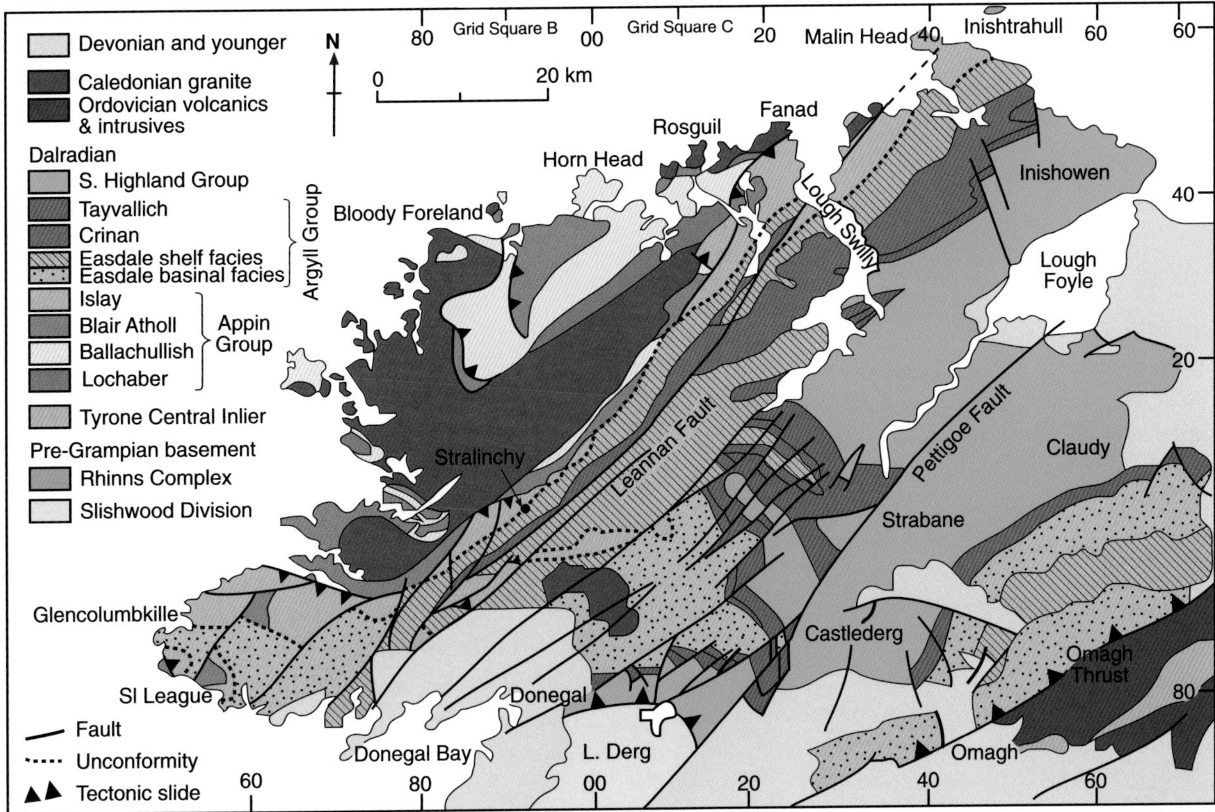

Fig. 2.19 Map of the geology of the western part of the Ulster Dalradian, excluding Antrim, based on Hutton and Alsop (1996), showing the unconformity proposed by Hutton and Alsop (2004).

palaeogeographic reconstructions are hampered by the allochthonous nature of much of the rock record due to major plate reorganisation in the Palaeozoic, e.g. associated within the Caledonian-Appalachian Orogen.

The Dalradian Supergroup is characterised by a distinctive stratigraphy, conventionally divided into four major groups – Grampian, Appin, Argyll, and Southern Highland (Fig. 2.18). The Dalradian was deposited initially in an ensialic rift basin (Harris *et al.*, 1978), which started to form in the late Proterozoic on the Laurentian continental margin of Rodinia with extreme attenuation of the subjacent continental basement, resulting in the exposure and protrusion of sub-continental lithospheric mantle (Chew, 2001) and the development of basaltic volcanism associated with the possible development of oceanic crust at the final stages. The initial rifting is thought to be accompanied by the deposition of the Grampian Group, the oldest exposed rocks of the succession, which outcrop in north-west Mayo (Fig. 2.1). These comprise a repetitive sequence of quartzo-feldspathic meta-arkoses. Later as active rifting waned (Soper and England, 1995) a more lithologically diverse association of shallow water, probably marine sediments were deposited in an upwards-shallowing sequence in response to thermal subsidence, forming the Appin Group. The succeeding Argyll Group is a response to renewed rifting and was deposited initially in confined basins, and later in more laterally continuous sequences as the basins deepened. Eventually, volcanics and turbidites filled the basin near the top of the sequence, which may prograde onto oceanic crust (Anderton, 1985; Soper, 1994a). As discussed below, the site of the Dalradian basin may be confidently placed on the continental margin of Laurentia. Arguments in favour of a location on other continents (e.g. Bluck and Dempster, 1991; Dalziel, 1994) have been made redundant by the changing interpretation both of the age and tectonic significance of igneous rocks cutting the Dalradian and of the structures affecting it.

While the Dalradian has been considered to represent a coherent stratigraphy, there are several problems arising. Firstly the duration of sedimentation, though poorly constrained, is at least 170 Ma, which is rather long for a continually operating depositional system (Dempster *et al.*, 2002). Several authors have suggested that there may be important breaks in deposition (e.g. Dempster *et al.*, 2002) and Hutton and Alsop (2004) have proposed the existence of an orogenic unconformity in County Donegal (Fig. 2.19, and see below). Of critical importance is the existence and significance of possible Neoproterozoic deformation, which may have affected the Grampian and possibly also the early Appin Group rocks (see below).

In a radical re-evaluation of the Dalradian stratigraphy, McCay *et al.* (2006) have proposed that the Dalradian sequence includes three glacial horizons (Fig. 2.18). The widely recognised and hitherto critical marker horizon for the base of the Argyll Group – the Port Askaig Tillite – was suggested by McCay *et al.* (2006) to be equivalent to the *c.*700 Ma Sturtian glaciation. In addition, an entirely new glacial episode was proposed for the Argyll Group rocks overlying the orogenic unconformity in Donegal proposed by Hutton and Alsop (2004). The carbon isotopic characteristics of the dolostone and limestone overlying a proposed glacial diamictite within the Cranford Limestone Formation were argued to represent the *c.*635 Ma Marinoan glaciation (McCay *et al.*, 2006). Finally, the youngest manifestation of glaciation – the ice-rafted debris horizon (Condon and Prave, 2000) within the Fahan Grits of the Southern Highland Group in Inishowen (equivalent to the Loch na Cile Boulder Bed in Scotland) was correlated with the *c.*580 Ma Gaskiers glaciation (McCay *et al.*, 2006).

Dalradian stratigraphy

For many years, the standard Dalradian stratigraphic succession established in Scotland has been correlated into Ireland (Harris *et al.*, 1994) although some of the major units are not widely developed, e.g. the Grampian Group is restricted to the North Mayo inlier. Interpretation of the depositional history is of course complicated by the polyphase deformational and metamorphic events that affected the Dalradian during the Caledonian Orogeny. Several large-scale recumbent folds have been recognised in both areas, modified to varying degrees by later structures resulting in a complicated map pattern (e.g. Leake *et al.*, 1981). Regional variations in strain and metamorphic grade, and the cutting out of parts of the section across slides (i.e. ductile faults – see Hutton, 1979), all complicate stratigraphic analysis and correlation and discourage or inhibit rigorous basin analysis studies. Unfortunately the lack of fossils means that only lithological correlations can be used. The development of turbidites towards the top of the Dalradian succession makes correlation particularly difficult (cf. Leake and Tanner, 1994) in the upper part of the Argyll Group and in the Southern Highland Group. The stratigraphic nomenclature adopted here is that of Harris *et al.* (1994).

Two marker horizons, mappable over hundreds of kilometres, have long been recognised within the Dalradian succession – the Port Askaig Boulder Bed (see above), and the Loch Tay/Tayvallich/Culdaff Limestone, a calcite marble, deposited as a shallow shelf to basin carbonate horizon. Both units have been interpreted as representing time planes, though similar time correlation between other lithological units is probably unsafe on at least two counts. Firstly, in well-exposed ground in Scotland, abundant normal growth faults have been demonstrated (Anderton, 1988; Soper and Anderton, 1984) and so lateral continuity of stratigraphic units is unlikely on a kilometric scale. Secondly, the original geometry of the Dalradian basin is by no means certain. If it developed as a series of continental margin basins, then similar sequences could have filled spatially and temporally separate depocentres as they evolved. Correlations based on trace element geochemistry, e.g. of metabasites (Winchester *et al.*, 1987) may be misleading because chemically similar basic magmas could be generated independently at different times. Nonetheless, stratigraphic variations in the chemistry of Dalradian metasediments do seem to exist on various scales (Winchester *et al.*, 1996; Daly and Menuge, 1989). However, these are probably of less practical value for correlation than standard stratigraphic techniques.

Grampian Group

The Erris Group of North Mayo is generally considered to be equivalent to the Grampian Group of the standard Scottish succession (Harris *et al.*, 1994). In this scheme, the Grampian Group is the lowest unit of the Dalradian (Lindsay *et al.*, 1989; Harris *et al.*, 1978). In Ireland, the Grampian Group is exposed only in the North Mayo inlier in a series of tectonic slices. It is made up of banded, quartz-plagioclase-mica, locally pebbly, psammites up to a total of 4 km in thickness, with abundant cross-stratification, heavy mineral (haematite and magnetite) bands and thin calc-silicate lenses, probably laid down in fluviatile to marginal marine environments. On the Mullet Peninsula, Grampian Group psammites of the Doonamoo Formation pass stratigraphically downwards into the Inishkea metasediments (known locally as the Scotchport schists). These rocks comprise strongly foliated semipelitic, feldspathic schists and psammitic gneisses with very minor heavy mineral seams and calcareous layers belonging to the Inishkea Division (Fig. 2.10;

Sutton, 1972; Winchester and Max, 1987b). Kennedy and Menuge (1992) pointed out that rocks of the Inishkea Division from different places probably occupy different positions within the Dalradian stratigraphy. While those on the Mullet Peninsula (Scotchport schists) and on the Inishkea islands probably belong to the Grampian Group, those at Kinrovar (Kinrovar Schist of Crow and Max, 1976) may belong to the Appin Group.

Appin Group

The base of the Appin Group is defined at the base of the stratigraphically lowest major quartzite. Such a boundary occurs in Achill Island (Slievemore Quartzitic Formation: Kennedy, 1969 a, b), overlain by 4 km of cross-bedded psammites and semipelitic bands with quartzites increasing in abundance upwards. These crop out extensively in both the southern and northern parts of the North Mayo inlier. Above this lies about 1 km of calcareous pelite and semipelite with local bands of marble and quartzite (Ashleam Bay Pelitic and Calcareous formations), which are correlated with the lithologically similar Creeslough Formation of Donegal. This whole 5 km sequence, equivalent to the Lochaber Subgroup of Scotland (Harris and Pitcher, 1975), represents a transition from the marginal marine depositional environment of the Grampian Group to more stable marine shelf conditions. It is succeeded by a quartzite, limestone and graphitic pelite sequence named the Ballachulish Subgroup in Scotland. In Donegal its equivalents are the Altan Limestone, the Ards Black Schist and Quartzite and Sessiagh-Clonmass formations. Similar lithologies occur in north-west Mayo. Overlying this are correlatives of the Scottish Blair Atholl Subgroup comprising approximately 1500 m of graphitic and pyritic schists, limestone and semipelitic bands, well exposed in Donegal as the Falcarragh and Loughros formations (Harris and Pitcher, 1975; Pitcher and Berger, 1972). Grey semipelitic schists with quartzite and calc-silicate layers occur in the Inishkea division (Max and Long, 1985; Winchester and Max, 1987b), and in parts of mainland north-west Mayo (the Inver Group: Sutton, 1971; Crow *et al.*, 1971). In the central axis of Connemara (Tanner and Shackleton, 1979; Leake *et al.*, 1981; Leake and Tanner, 1994), the Appin Group includes the famous green serpentine-bearing Connemara marble, which is widely used as a decorative stone. Based on facies variations within the Appin Group in northwest Donegal, Hutton and Alsop (1996) have suggested that the change from a broad shallow shelf in the north-east to deeper water sediments to the south-west coincides with a major strike swing (see below) controlled by the geometry of the Laurentian continental margin.

Argyll Group

The Argyll Group occupies a major part of all of the Irish Dalradian outcrops. Its base is marked by the Port Askaig Boulder Bed and as such is more or less isochronous above the Appin Group basins. This boulder bed can be traced from Scotland, through Donegal and north-west Mayo and into Connemara. It is a complex diamictite up to 525 m thick with a calcareous matrix. There is a well-developed stratigraphy with the lowest unit sitting in sharp contact with an underlying dolomite. Within the lowest unit the dominant clast type is dolomite. In contrast, the middle unit comprises an arenaceous boulder bed with granitic clasts, whereas the upper unit, in addition to isolated granitic clasts, also contains minor amounts of extra-basinal psammitic and semipelitic gneiss fragments. The upper boulder bed has a quartzitic matrix, which passes gradationally up into the overlying quartzite. A glacial origin for these deposits is inferred, both on their sedimentology (Eyles, 1988) and their widespread distribution. Individual members within the unit can be traced for up to 160 km, indicating the scale of individual basins at the time of its formation. A glacio-marine setting is suggested by evidence of tidal influence (Eyles, 1988; Eyles and Eyles, 1983), though some have argued that the deposit represents a terrestrial grounded ice deposit including some subaqueous deposition (Spencer, 1971; Howarth, 1971; Fairchild, 1985).

The well-washed quartzite overlying the Port Askaig diamictite contains heavy mineral bands and cross-bedding, sometimes with a dolomite and dolomitic pelite near its base. It can be traced from north-east Scotland some 700 km along strike to the Bennabeola Quartzite of Connemara, which forms the rugged backbone of the 'Twelve Bens' of Connemara (Fig. 2.1). There is substantial thickness variation along strike. Some of this is probably tectonic, but much is likely to be original, with facies variation in the overlying units. There is substantial thickness change within the Slieve Tooey Quartzite in the Glencolumbkille region of Donegal, in a region where there are low-angle extensional structures. Anderton (1988) has suggested that on the islands of Jura and Islay the same unit changes its thickness because of growth faults. The quartzite may be the product of winnowing of glacial and peri-glacial deposits, with sands being deposited in a series of individual basins. In the lower part of the quartzite on Islay there is evidence for intertidal depositional conditions with deeper water to the north-west and current flow from the south. Similar flow patterns are recorded by cross-beds in the Glencolumbkille area of Donegal (Fig. 2.19).

Hutton and Alsop (2004) have proposed that the base of the Easdale Subgroup in Donegal marks the position of a major unconformity above the quartzites of the Islay Subgroup, to which they ascribe orogenic status (see also below). At the type locality at Stralinchy (Fig. 2.19), the Cranford Limestone Formation has been interpreted by McCay *et al.* (2006) as a cap carbonate, overlying a glacial diamictite (Stralinchy Conglomerate of Hutton and Alsop, 2004). Carbon isotopic variations within the basal dolostone at Stralinchy and within the Cranford Limestone at Cranford display the characteristic features documented for the widespread Marinoan glaciation (Kennedy *et al.*, 1998; Hoffman and Schrag, 2002). Along strike, the Easdale Subgroup is marked by the appearance of graphitic pelites and semipelites and is characterised by major variations in facies both on a regional and local scale. North-west of the Leannan Fault a considerable unconformity between the Cranford Limestone and the underlying Slieve Tooey Quartzite has long been recognised. Traced southwestwards, the intermittent limestones and evaporites of the Cranford Limestone give way to a turbiditic facies in the Slieve League peninsula. South-east of the Leannan Fault in Donegal (Fig. 2.19) the Easdale subgroup defines a major sub-basin that extends eastwards to the Sperrins and southwards to the Ox Mountains. A major change in facies from predominantly shallow water ripple-dominated pelites to predominantly basinal psammitic grits and turbidites takes place south-westwards, coinciding with the hinge of a regional strike swing. Hutton and Alsop (1996) suggested that these features were inherited from the underlying basement (see below). On Achill Island the Easdale subgroup includes coarse conglomerates, while in Connemara the unit is pelitic and semipelitic with interbeds of pebbly quartzite and limestone. Max *et al.* (1992), however, placed the Achill conglomerates and associated rocks (Keem Conglomerate Formation, etc.) in the Appin Group (Harris *et al.*, 1994). Such differences of opinion reflect the difficulties in correlating polyphase-deformed rocks transected by tectonic discontinuities. The occurrence of Ba-mineralisation associated with the Lakes Marble in Connemara (Reynolds *et al.*, 1990) is similar to that in the Easdale sub-group in Perthshire, Scotland and affords an additional means of stratigraphic correlation.

Near the top of the Easdale Subgroup (Chew, 2001) or within the overlying Crinan Subgroup (Chew, 2003) a series of serpentinitic mélanges occur, for instance, in the Lack inlier in Tyrone, in north-west Mayo (Kennedy, 1980; Chew, 2001), and in north Connemara (Chew, 2003), which have been interpreted as protrusions of sub-continental lithospheric mantle through extremely attenuated continental crust (Chew, 2001; 2003).

Sedimentation in the Crinan Subgroup, overlying the Easdale Subgroup, is characterised by pebbly psammites, essentially turbiditic in character, becoming progressively more distal upwards. This unit crops out across the country as the Streamstown (Connemara), Achill Head (north-west Mayo), Crana (Donegal/Derry), and Murlough Bay (Antrim) formations. In contrast to the Easdale Subgroup, the Crinan Subgroup pinches out southwards in South Donegal and Tyrone (Hutton and Alsop, 1996) consistent with a major depocentre to the north-west.

The Loch Tay Limestone is an extensive shallow-shelf carbonate containing interbedded basic volcanics which provide a useful, probably isochronous marker horizon (Gower, 1973) within the Tayvallich Subgroup at the top of the Argyll Group. It has been mapped out over hundreds of kilometres throughout Scotland and thence to the Torr Head Limestone of Antrim, the Dungiven Limestone of Tyrone (Arthurs, 1976a, b) and the Culdaff Limestone of Donegal. In general it shows a facies change from shelf in the north-west to more basinal facies to the south-east (Gower, 1973). Basic volcanics are extensively developed in Tyrone, the Nephin Beg Ranges of north-west Mayo and the south-west end of the Ox Mountains. Some of these may belong to the Easdale Subgroup, but they all certainly belong to either the upper part of the Argyll Group or the base of the Southern Highland Group. These may continue into the Dalradian of the central Ox Mountains (Long and Max, 1977), much of which contains lithological similarities to the Argyll Group as a whole (Alsop and Jones, 1991; cf. Andrews *et al.*, 1978).

Southern Highland Group

The succession is essentially one of turbidites with varying degrees of proximality, together with local graphitic pelites and basic volcanics. Within the Ben Levy Grit Formation (Leake and Tanner, 1994) of Connemara (e.g. near Tully Cross, Leake *et al.*, 1981), and in the southern part of the Lack Inlier of south-west Tyrone, both here taken as belonging to the Southern Highland Group (cf. Harris *et al.*, 1994), serpentinite bodies up to hundreds of metres long in a graphitic pelite matrix occur as either sedimentary clasts or protrusions onto the seafloor (cf. Chew, 2001, 2003). In Achill, Inishowen and Tyrone, epidote-rich horizons also occur near the base of the succession and in north-east Tyrone, two horizons up to 100 m thick of basic pillow lavas lie near the base of the group (Arthurs, 1976b). Compositions of turbidite clasts are predominantly quartz, blue quartz, quartzite, and feldspar. In northern Inishowen, proximal turbidites with granitic gneiss clasts up to 3 cm in diameter within the Fahan Grits have been interpreted as ice-rafted debris by Condon and Prave (2000). The source of the blue quartz is unclear, but it must have been derived from a high-temperature source rock.

Age of the Dalradian Supergroup and stratigraphic problems

The depositional age of the Dalradian succession is difficult to establish because of the general lack of datable fossils. From a field geological perspective only very broad constraints may be placed on the timing of Dalradian deposition. A minimum age is provided by the unconformable relationship with Upper Llandovery sandstones and probable Upper Llandovery volcanics (Leake and Tanner, 1994) in east Connemara. Thus Dalradian deposition must be older than *c.*430 Ma (see also Chapter 3). Further indirect evidence for a pre-Middle Ordovician age is provided by the presence of Dalradian clasts with metamorphic muscovite Rb–Sr ages of 462 Ma and 471 Ma in the Derryveeney Formation in South Mayo (Graham *et al.*, 1991).

Direct evidence for an upper limit on the timing of Dalradian deposition relies on its stratigraphic relationship with the Clew Bay Complex (Fig. 2.1). Much controversy has surrounded the relationship between the Dalradian and the metasediments of the Highland Border Complex in Scotland and its equivalent in Ireland, the Clew Bay Complex (Williams *et al.*, 1994, 1996; Johnston and Phillips, 1995; Harkin *et al.*, 1996; Chew, 2003; Chew *et al.*, 2003) and the Cloonygowan Formation of the Central Ox Mountains (Fig. 2.15). Until the reported discovery of trilete miospores interpreted to be of Silurian age (Williams *et al.*, 1994, 1996) the clastic metasediments of the Clew Bay Complex had been included within the Southern Highland Group. The exclusion of these rocks from the Dalradian gained general acceptance (e.g. Daly, 2001) and instead they were correlated with the Highland Border Complex of Scotland, which at the time was also excluded from the Dalradian. However, in marked conflict with the fossil evidence for a Silurian age (Williams *et al.*, 1994, 1996), mineral geochronology (Chew *et al.*, 2003) and field evidence for a deformational history shared with the Dalradian (Chew, 2003) have shown that the Clew Bay Complex was deformed and metamorphosed in the earliest Middle Ordovician. Tanner and Sutherland (2007) have reviewed the micropalaeontological evidence from Scotland and concluded that the Highland Border Complex includes no rocks older than Arenig. This, together with compelling field evidence for stratigraphic continuity (Tanner, 1995) between the Highland Group and fossiliferous units of the Highland Border Complex such as the Lower Cambrian Leny Limestone (now termed the Trossachs Group), would seem to resolve the controversy. Dalradian deposition thus continues into the Early Ordovician. Thus the record of *Protospongia hicksi* (Rushton and Phillips, 1973) within the Clew Bay Complex is restored as the only direct biostratigraphic constraint on the age of the Irish Dalradian. The Clew Bay Complex is further discussed in Chapter 4.

Field evidence placing an older limit on Dalradian sedimentation is even less precise because all contacts with basement rocks are tectonic. However, R.C. Fitzgerald *et al.* (1994) and McAteer *et al.* (2008) have suggested that the Annagh Gneiss Complex in north-west Mayo may have been a depositional basement for the Dalradian. They suggest that post-Grenville metadolerites cutting the basement are equivalent to pre-tectonic metadolerites cutting the Dalradian. Thus a post-Grenville, i.e. post *c.*940 Ma (Daly, 1996; Daly and Flowerdew, 2005) age for Dalradian deposition is indicated. U–Pb dating of granitoid clasts and detrital zircons in the Erris Group (= Grampian Group) metasediments in North Mayo indicates a maximum age of 955 ± 12 Ma (McAteer *et al.*, 2008; Fig. 2.3), based on the age of the youngest concordant detrital zircon. This is within error of the maximum age obtained from dating detrital zircons in the Scottish Grampian Group (Cawood *et al.*, 2003).

Geochronological investigations in Scotland have been critical in constraining Dalradian deposition to be mainly late Precambrian (Neoproterozoic). The Tayvallich Volcanics of Scotland, near the top of the Argyll Group, have yielded zircon U–Pb ages of 595 ± 4 Ma (Halliday *et al.*, 1989) and 601 ± 4 Ma (Dempster *et al.*, 2002) interpreted to date their eruption. In addition the Ben Vuirich Granite, which cuts Appin and Argyll Group metasediments in Perthshire, has yielded a high precision conventional U–Pb age of 590 ± 2 Ma U–Pb (Rogers *et al.*, 1989) and a SHRIMP ion microprobe U–Pb age of 597 ± 11 Ma (Pidgeon and Compston, 1992). Early granitoid intrusions have not been recognised within the Irish Dalradian in spite of an extensive search (Cliff *et al.*, 1996).

Dating of the Mam sill, a metadolerite with a well-preserved igneous mineralogy and texture that cuts Appin Group metasediments near Horn Head in Donegal (Fig. 2.19), confirms the Precambrian age for at least the Appin Group in Ireland. This rock has yielded a well-fitted Sm–Nd mineral isochron of 625 ± 47 Ma (Kirwan *et al.*, 1989) for plagioclase and clinopyroxene mineral separates together with the whole-rock. Even allowing for the very large uncertainty associated with this age, which is an unavoidable consequence of the limited variation in Sm/Nd ratio in such rocks, the age is clearly Precambrian. The age is supported by a SIMS U–Pb zircon age of *c.*590 Ma from a gabbro pegmatite within a correlative sill at Dooros Point (Daly, Whitehouse and Kirwan, unpublished data). Kirwan *et al.* (1989) regard the Mam sill as pre-deformational, intruding into sediments of the lower part of the Appin Group in Donegal, which they interpret to have been semi-consolidated at the time of intrusion. Geochemical investigations of the Donegal metadolerites (Kirwan written communications, 1988, 1996) are consistent with a rift origin for the magmatism, and the age of the Mam sill is similar to other age estimates for magmatism related to the rifting of Iapetus (Bingen *et al.*, 1998; Kamo *et al.*, 1989). The most precise of these ages, from the Tayvallich volcanics, suggest that Dalradian deposition up to the Tayvallich subgroup was completed by *c.*601 Ma (Dempster *et al.*, 2002).

Several lines of evidence support the controversial view that the older parts of the Dalradian have been deformed in an orogenic event before the extensional events associated with the eventual opening of the Iapetus ocean.

Firstly, syntectonic pegmatites within Grampian Group metasediments in Scotland (Piasecki and Van Breemen, 1979) have yielded *c.*710 Ma Rb–Sr muscovite ages and monazites from pegmatite veins and a phyllonitic mylonite within the Grampian Shear Zone – a complex zone of repeated ductile thrusting in which the Grampian Group and possible basement rocks, collectively known as the Glen Banchor and Dava successions (formerly known as the Grampian Division or Central Highland Migmatite Complex) – have been imbricated. Pegmatites cutting some of these rocks, supposedly including Grampian Group metasediments, have yielded U–Pb ages of *c.*806 Ma (Noble *et al.*, 1996). The fact that a sedimentary transition from Grampian Group to Appin Group occurs in the Scottish Highlands (Glover *et al.*, 1995) led Noble *et al.* (1996) to suggest that there must be an intra-Dalradian unconformity and a previously unrecognised structural break. Since then even older ages for the metamorphism have been reported (Highton *et al.*, 1999) and it is now widely accepted that the Glen Banchor and Dava successions were affected by deformation and metamorphism between *c.*840 and *c.*800 Ma, which can be correlated with the Knoydartian event affecting the Moine Supergroup. The critical issue is whether or not unequivocal Grampian Group rocks were affected. According to Strachan *et al.* (2002), the isotopic evidence for the Knoydartian event in the Central Highlands is restricted to the Glen Banchor and Dava successions, though Dempster *et al.* (2002) maintain that the possibility of orogenic thickening of the Dalradian rocks during the Knoydartian 'can not be entirely discounted'.

Secondly, disputed evidence regarding the field relations of the *c.*590 Ma Ben Vuirich Granite in Scotland leave open the possibility of early (pre-590 Ma) deformation affecting the Appin and Argyll group rocks that the granite cuts (Dempster *et al.*, 2002; Tanner and Leslie, 1994; Tanner, 1996). In particular, at issue is the extent to which the contact metamorphism affecting the country rocks overprints a pre-existing fabric or not (Tanner, 2006). Dempster *et al.* (2002) point out that the pressure estimates for the contact metamorphism around the Ben Vuirich Granite may be significantly underestimated and imply depths up to 10 km. Since these depths are argued to be incompatible with the available stratigraphic load, a tectonic thickening event is required (Dempster *et al.*, 2002), raising the possibility of an orogenic break within the Dalradian sequence.

Thirdly, in Donegal, Hutton and Alsop (2004) have argued that the Appin and lower Argyll rocks stratigraphically beneath the Stralinchy Conglomerate have experienced compressional deformation – hence their 'orogenic unconformity'. In contrast to Kirwan *et al.* (1989), Hutton and Alsop (1995) consider the Mam sill to truncate compressional F2 folds. Given the *c.*590 Ma age for the sill (see above), if their interpretation of the field relations is correct, then a significant part of the compressional deformation usually ascribed to the Grampian (Chapter 4) must be of Precambrian age.

Apart from these cases, geochronology elsewhere in Ireland (e.g. in north-west Mayo, Daly and Flowerdew, 2005) indicates that the first regional metamorphic events to affect the Dalradian took place in the Middle Ordovician during the Grampian Orogeny (Chapter 4).

On maps published by the Geological Survey of Ireland (Long and McConnell, 1995; Harney *et al.*, 1996) several units generally accepted as belonging to the Dalradian have been designated as 'dubiously Dalradian' and possibly of Cambro-Ordovician age (Long *et al.*, 1995). These include the Ben Levy and Lough Kilbride formations of Connemara, which are separated from the rest of the Dalradian by the Renvyle–Bofin slide (Leake and Tanner, 1994) and the Callow Succession in the Ox Mountains, said to be separated from the Dalradian proper by a terrane boundary (Glenawoo slide, Winchester *et al.*, 1987), whose terrane-bounding status has been disputed by Jones and Leat (1988). Until further evidence is available, most workers continue to regard these rocks as Dalradian (cf. Harris *et al.*, 1994).

Dalradian provenance and the palaeogeography of the Dalradian on the margin of Laurentia

Several lines of evidence indicate that Dalradian deposition took place on the Laurentian continental margin (Cawood *et al.*, 2003, 2007a, 2007b). The possibility that the succession may have evolved along the Gondwana margin (Bluck and Dempster, 1991) has already been mentioned and discounted because the original grounds for this argument now have little support. Palaeocurrent evidence generally suggests a northwesterly derivation in present-day coordinates (Anderton, 1985) although standard methods of palaeocurrent analysis are difficult to apply in view of the polyphase folding as well as the possibility of large-scale strike-slip movements (Johnston and Phillips, 1995). Daly and Menuge (1989) and Menuge *et al.* (2004) found that the mean Sm–Nd t_{DM} age of the Dalradian increases stratigraphically upwards from 1.72 Ga in the Grampian Group (n = 5 samples), to 1.81 Ga in the Appin Group (n = 14), to 1.76 Ga in the Islay Subgroup (n = 9), to 1.84 Ga in the Easdale Subgroup (n = 3) to 2.20 Ga in the Crinan Subgroup (n = 9), similar to 2.06 Ga in the Southern Highland Group (n = 9). These Sm–Nd model ages reflect a change in the average age of the sourcelands contributing detritus to the Dalradian basin, possibly reflecting the gross palaeogeographic pattern of Proterozoic and Archaean belts in Laurentia (Daly and Menuge, 1989). However, the trend is not smooth and Menuge *et al.* (2004) drew attention to the apparently abrupt change at the level of the Crinan Subgroup, which follows the orogenic unconformity proposed by Hutton and Alsop (2004) at Stralinchy, County Donegal and the glacigenic horizon at the same locality, correlated with the *c.*635 Ma Marinoan episode by McCay *et al.* (2006).

Several investigations of Dalradian sedimentary provenance have been based on detrital zircon dating (Cawood *et al.*, 2003, 2007a; McAteer *et al.*, 2008). These studies have shown that the Grampian Group is characterised by a major contribution from late Palaeoproterozoic and late Mesoproterozoic to early Neoproterozoic (so-called 'Grenvillian') zircons while Archaean grains are absent (Fig. 2.4, McAteer *et al.*, 2008) or rare (Cawood *et al.*, 2003, 2007a), occurring in just one Scottish sample from Glen Buck. The Appin Group was supplied by similar sources with the addition of an important late Archaean component that persists throughout the rest of the succession. In Argyll Group times, the Palaeoproterozoic contribution becomes older, especially within the Port Askaig tillite (Cawood *et al.*, 2003), and a wider range of Mesoproterozoic zircon and a younger component of Neoproterozoic zircon is present. By Southern Highland Group times, the Palaeoproterozoic contribution is relatively subdued and the population is dominated by a broad range of late Archaean and Mesoproterozoic ages.

The appearance of older detritus at higher stratigraphic levels might be the result of progressive headward capture as the drainage system feeding the Dalradian matured, thereby reaching sources in the continental interior. Dalradian deposition took place during continental extension in late Neoproterozoic times, overlapping the time interval when the Iapetus Ocean was initiated (Soper and England, 1995). Soper (1994a) has suggested that Dalradian deposition took place close to the Greenland–Labrador promontory between the Appalachian and Greenland continental margins facing the NE–SW and the N–S trending arms of the Iapetus Ocean. Deposition of the Dalradian close to this promontory may explain the great thickness of Dalradian sediments compared with the thinner rift sequences elsewhere along the Laurentian margin. The convex aspect and large scale of the promontory may be reflected in the offshore deep seismic Flannan reflector (Snyder *et al.*, 1996). Hutton and Alsop (1996) have argued that the Greenland–Labrador promontory controlled Dalradian stratigraphy and sedimentation patterns in Donegal and had a major influence on subsequent tectonic events discussed further in Chapter 4. Hutton and Alsop (1996) have identified a N12° E trending 'Donegal Lineament' as the expression of a steep pre-Grampian basement structure, which may have been exploited during the rifting of Iapetus.

A south-easterly provenance for at least part of the Dalradian – particularly the Portaskaig tillite in earliest

Argyll Group times – has been suggested (Anderton, 1982). However, recognition of Palaeoproterozoic basement in the Rhinns Complex and on the Rockall Plateau removes the objection to a northerly provenance for the tillite clasts whose southerly derivation seemed necessary because the clasts did not resemble the Lewisian Complex to the northwest. A small number of clasts have been dated. Fitches *et al.* (1996) obtained an average t_{DM} age of *c.*1.8 Ga for granitoid tillite clasts from the Garvellach islands and Port Askaig in Scotland. Three of the clasts yielded near-concordant U–Pb ages of *c.*1803 Ma and 1876 Ma (Evans *et al.*, 1998; Loewy *et al.*, 2003). These results are consistent with both local and far-travelled sources in both Laurentia and Baltica, whose Proterozoic histories have much in common (Kirkland *et al.*, 2007a) and along whose common margin of Rodinia, at least the early part of the Dalradian sequence was probably deposited.

3

Ordovician of the North

J. R. Graham

In contrast to the limited outcrop of Cambrian rocks, Ordovician rocks are widely distributed in Ireland (Fig. 3.1). Perhaps more than those of any other system, Ordovician rocks display a complex interplay of volcanic activity, sedimentation and tectonics that was associated with the margins of the Iapetus Ocean. Ordovician sedimentary and volcanic rocks were originally formed in belts roughly parallel to the Iapetus continental margins, which were aligned approximately north-east to south-west (present co-ordinates). Later deformation has commonly reutilised earlier faults and has thus emphasised original across-strike differences. This has resulted in a series of geological terranes with internally consistent histories, which, in many cases, have uncertain relationships to adjacent, across-strike terranes.

This chapter deals with rocks generally accepted to have formed along the Laurentian margin of Iapetus, whereas Chapter 6 deals with definitive Avalonian terranes and those of uncertain origin now in the Iapetus suture zone. Biostratigraphical subdivision of Ordovician rocks has its earliest history in Britain where many of the original series names originated. However, it has long been recognised that many faunas from NW Ireland were not easily referable to the British biozonation but rather to that developed in America and Australia. Recent work on global correlation of the Ordovician System has seen the adoption of new international stage names. Biostratigraphical definition and currently adopted correlation of these new stages with the original British subdivisions are shown in Figure 3.2.

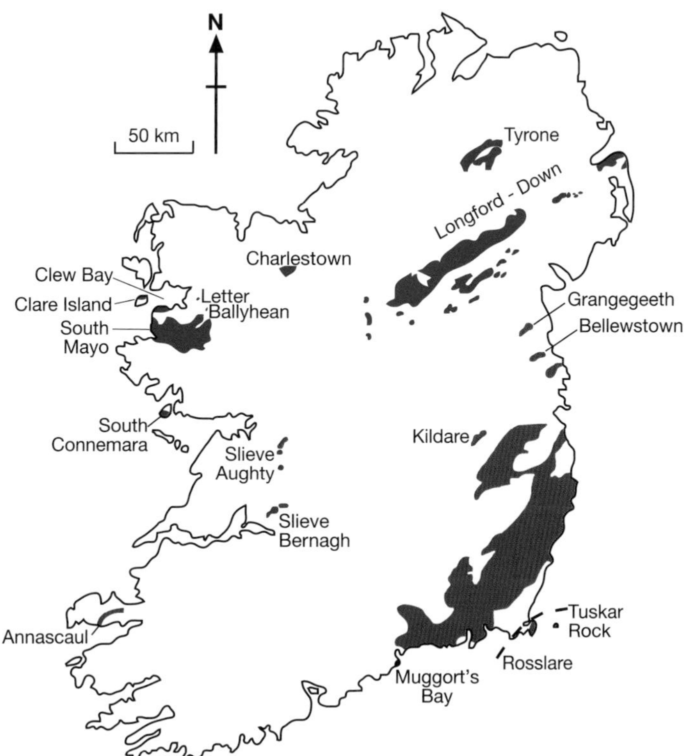

Fig. 3.1 Distribution of Ordovician rocks in Ireland: the rocks discussed in this chapter are shown in purple.

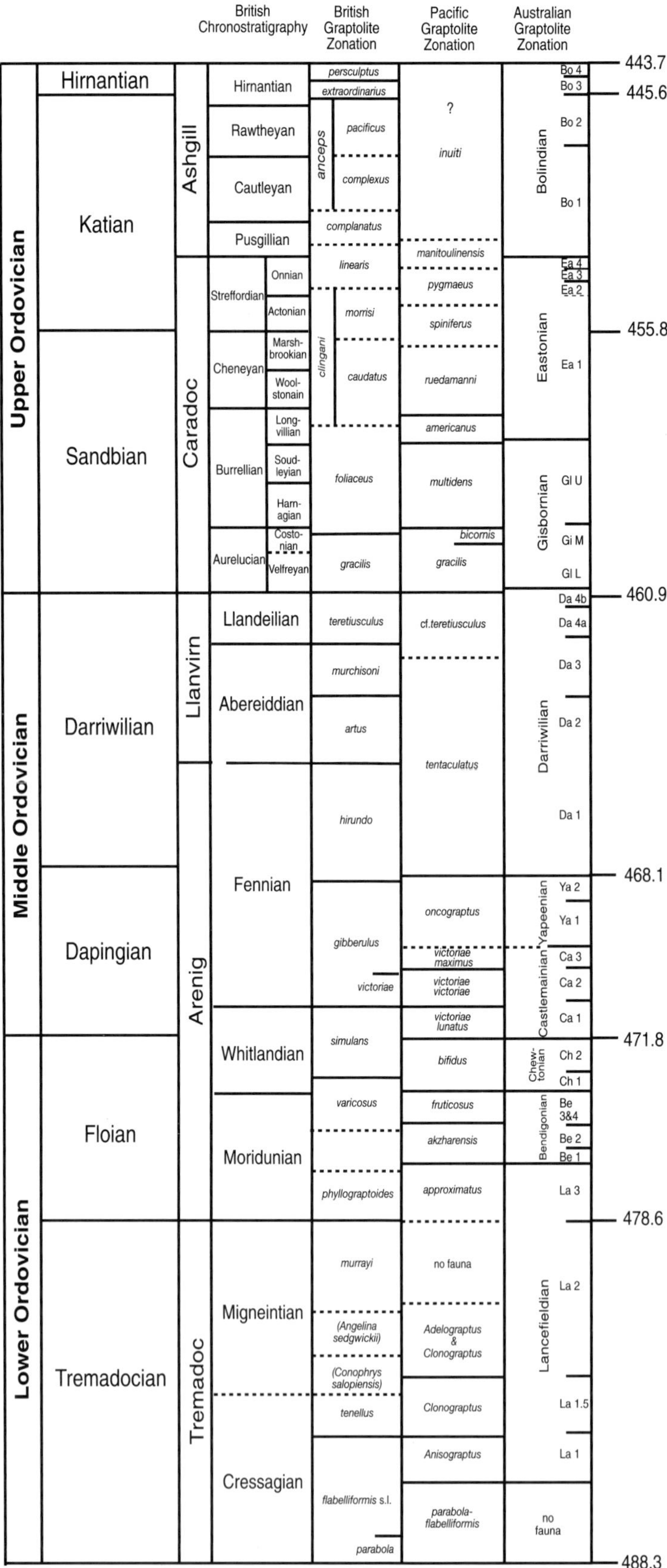

Fig. 3.2 International and British chronostratigraphy of the Ordovician System with graptolite biozonation and best estimates of the numerical ages of stage boundaries indicated (*after* Bergstrom and Xu 2007; Gradstein *et al.*, 2004).

Clew Bay

Clew Bay, County Mayo, exposes an area of enigmatic, low grade metamorphic rocks bounded to the north by rocks of the Dalradian Supergroup (Chapter 2) and to the south by Silurian rocks (Fig. 4.4). Early workers correlated all of these rocks with the Dalradian, although there were problems with the presence of Lower Palaeozoic fossils in the Clew Bay rocks (Rushton and Phillips, 1973; Harper *et al.*, 1989). Rocks within this area comprise two lithological packages which are in tectonic contact with each other.

Deer Park Complex

The Deer Park Complex (Phillips, 1973; Max and Long, 1985) comprises pelitic, psammitic, and amphibolitic schists. These are tectonically interleaved with serpentinites and associated basic igneous rocks which have been shown to have a MORB composition and which have been interpreted as forming an ophiolitic mélange (Ryan, Sawal, and Rowlands, 1983). The neodymium isotopic compositions and rare earth element profiles of metadolerites from the Deer Park Complex support a juvenile MORB origin (Chew *et al.*, 2007). The metasediments are interpreted as deepwater Laurentian margin sediment caught up by obduction of the ophiolite (Chew *et al.*, 2007). The age of the igneous rocks cannot be demonstrated with certainty, but they have been shown to have undergone early Ordovician metamorphism (Chapter 4). By inference these rocks have been widely regarded as early Ordovician.

Clew Bay Complex

The second unit consists of low grade, mainly siliciclastic, metasediments, the Clew Bay Complex (Chew, 2003; Williams *et al.*, 1994). Previously these rocks have also been termed the Clew Bay Cambro-Ordovician (Graham *et al.*, 1985; 1989) or the Clew Bay Group (Max, 1989). Two parts of this complex have now yielded fossils of Ordovician age. The paucity of fossils and the complex nature of the geology have resulted in some dispute in detail as to which rocks really belong to the Clew Bay Complex. The suggestion of Harper *et al.* (1989) that rocks on the southern part of Achill Island are part of this group is at variance with the major contrasts noted across the Achill Beg Fault and with most other interpretations (Chapters 2, 4; Chew, 2003; Max, 1989; Graham *et al.*, 1989). The Clew Bay Group of Max (1989) comprised the Ballytoohy Formation of Clare Island and the Killadangan Formation of the South Clew Bay imbricate zone. The rocks of the South Achill Beg Formation (Chew, 2003) have strong lithological similarities with the Ballytoohy and Killadangan formations and are regarded here as part of this complex. The metasedimentary rocks near Westport (Westport Grit Formation, Fig. 3.3) (McManus, 1972; Graham *et al.*, 1985, 1989; Max, 1989) also belong to this complex. It is possible that all of these represent a single dismembered sedimentary unit.

The Ballytoohy Formation of north-west Clare Island has yielded a single specimen of *Protospongia cf. hicksi* (Rushton and Phillips, 1973) from black and green cherts. This confirmed a Cambro-Ordovician age, with a mid-Cambrian age cautiously suggested as most likely. The recovery of coniform euconodonts (Harper *et al.*, 1989) from the same horizon suggests that an early to mid Ordovician age is more probable. Other fragmental fossils recovered from an impure argillaceous limestone were not age diagnostic. Lithologies in the Ballytoohy Formation include graphitic mudrocks, spilites, greywackes, coarse graded psammites containing abundant blue quartz, and micro-conglomerates with clasts of vein quartz, schist, gneiss, and granite (Phillips, 1973). The deformed appearance at outcrop scale is in many cases of sedimentary or tectono-sedimentary origin (Max, 1989). Such sedimentary disruption renders difficult elucidation of the subsequent tectonic history. Williams *et al.* (1994) considered the whole of the Clew Bay Complex to be a mélange unit. Mudstone rip-up clasts from sandstones within the Ballytoohy Formation, presumably considered to be a mega-clast, have yielded palynomorphs that Williams, Harkin & Higgs (1996) consider to indicate a Wenlock age. Debates concerning the stratigraphy and structure of the Ballytoohy Formation are summarised in Graham (2001b) and also in Chapter 4.

The Killadangan Formation (Graham *et al.*, 1985) as redefined by Max (1989) (Fig. 4.4) contains coarse psammitic greywackes, conglomerates, graphitic mudrocks, and locally cherts and volcanic rocks. An S-element of the conodont *Periodon* has been recovered from a chert block (Williams *et al.*, 1994). Sandstones containing little or no internal fabric occur as discontinuous lenses mantled by a highly cleaved mudrock matrix, suggesting high shearing strains at low metamorphic grade (Max, 1983). These numerous slumped beds were interpreted by Max (1989) as forming at the base of a continental slope. Williams *et al.* (1994) reported simple trilete spores from what they considered to be mélange matrix of the Killadangan Formation and suggested that they indicate an age no older than Wenlock. They claim the contact between the Killadangan Formation and the Silurian Cregganbaun Formation is a fault. However, Maguire and Graham (1996) concurred with earlier workers on the basic unconformable nature of this contact (Bickle *et al.*, 1972; Phillips, 1983; Max, 1989; Johnston and Phillips, 1995) and suggested that the apparent conflict may be due to poorly constrained age ranges for the described palynomorphs. Dewey and Mange (1999) described sandstone clasts from within the Letterbrock Formation of the main Murrisk succession (probably late Arenig in age) which are not only lithologically identical to the Killadangan Formation in hand sample and thin section, but also contain a similar heavy mineral assemblage. Mange *et al.* (2003) also present evidence from the Killadangan Formation that precludes acceptance of a Silurian age. The most likely explanation of the observations is that the Killadangan Formation is part of an early Ordovician succession and that the apparent biostratigraphical conflict is due to poorly constrained age ranges for the palynomorphs.

The South Achill Beg Formation (Chew, 2003) comprises psammites, semi-pelites and calcareous pelites that have, to date, not yielded any fossils. Nevertheless, their strong similarity to the Ballytoohy and Killadangan formations merits inclusion within the Clew Bay Complex.

The coarse-grained sediments of the Clew Bay Complex indicate a provenance from quartzo-feldspathic continental crust. The heavy mineral assemblages of the South Achill Beg and the Killadangan formations are directly comparable (Chew, 2003; Dewey and Mange, 1999; Mange *et al.*, 2003) and they do not contain any definitive 'Grampian' detritus. Detrital zircon analysis from South Achill Beg shows no detritus younger than Grenville (1 Ga) (Chew, personal communication 2008). Neodymium model (T_{DM}) ages from the Killadangan, Ballytoohy, and South Achill Beg formations are directly comparable (2.2–2.65 Ga) and slightly older than those of the Argyll Group (1.65–2.39 Ga) and Southern Highland Group (1.79–2.36 Ga) of the Dalradian Supergroup. It has been demonstrated by Chew *et al.* (2003) that the rocks either side of the Achill Beg Fault share a similar structural history. Ages of *c.*460 Ma from muscovites that grew during formation of the S2 fabric in the Clew Bay Complex also help to disprove the proposed Silurian age for these rocks. Provisional dates of 482 ± 1 Ma from ^{40}Ar–^{39}Ar dating of the metamorphic fabrics in the Deer Park Complex (Chapter 4) suggest that these rocks may record an earlier (Tremadocian) event that could be related to ophiolite obduction. The tectonic setting of the Clew Bay Complex and the Deer Park Complex is discussed in detail in Chapter 4.

The general similarity between the succession in the Clew Bay area and that in the Highland Border Complex of Scotland has received considerable attention (Phillips, 1973; Curry *et al.*, 1984; Graham, 1987; Graham *et al.*, 1989; Harper *et al.*, 1989; Max, 1989; Chew, 2003; Tanner and Sutherland, 2007). A general correlation is likely as both occupy a similar position south of Dalradian rocks within the Highland Boundary Fault Zone. There is a clear similarity between the twofold subdivision recognised in Scotland by Tanner and Sutherland (2007). The Deer Park Complex bears close comparison to the ophiolitic portion of the Highland Border Complex (the Garron Point Group of Tanner and Sutherland, 2007). The lower grade metasedimentary rocks of the Clew Bay Complex are strikingly similar to the Trossachs Group of Tanner and Sutherland (2007). Tanner and Sutherland (2007) interpret the Trossachs Group as having a transitional basal contact with the Southern Highland Group of the Dalradian. In Scotland, Tanner and Sutherland (2007) have shown that there is no unequivocal evidence for rocks younger than Arenig within the Highland Border Complex and, at present, the same is true in Ireland for the Clew Bay Complex.

South Mayo

In South Mayo rocks of known Ordovician age are represented by three different successions, namely the Lough Nafooey, Tourmakeady, and main Murrisk successions (Graham *et al.*, 1985, 1989) (Figs. 3.3, 3.4), which have limited overlap in time. Other rocks which are likely to be Ordovician in age but which have not been dated with certainty are the Bohaun Formation, the rocks of the Ballyhean inlier, and the Farnacht Formation (Fig. 3.3).

Lough Nafooey Group

The Lough Nafooey Group comprises the oldest dated Ordovician rocks in the region and crops out in an elongate east–west belt that extends from Derry Bay on Lough Mask to the western end of Lough Nafooey. This group, which is dominated by the products of basic and intermediate volcanism, was initially described in detail by Gardiner and Reynolds (1912, 1914). More recently Ryan *et al.* (1980) recognised four formations within this group, namely the Bencorragh, Finny, Knock Kilbride, and Derry Bay formations (Fig. 3.5). Graptolites recovered from the Knock Kilbride Formation were compared

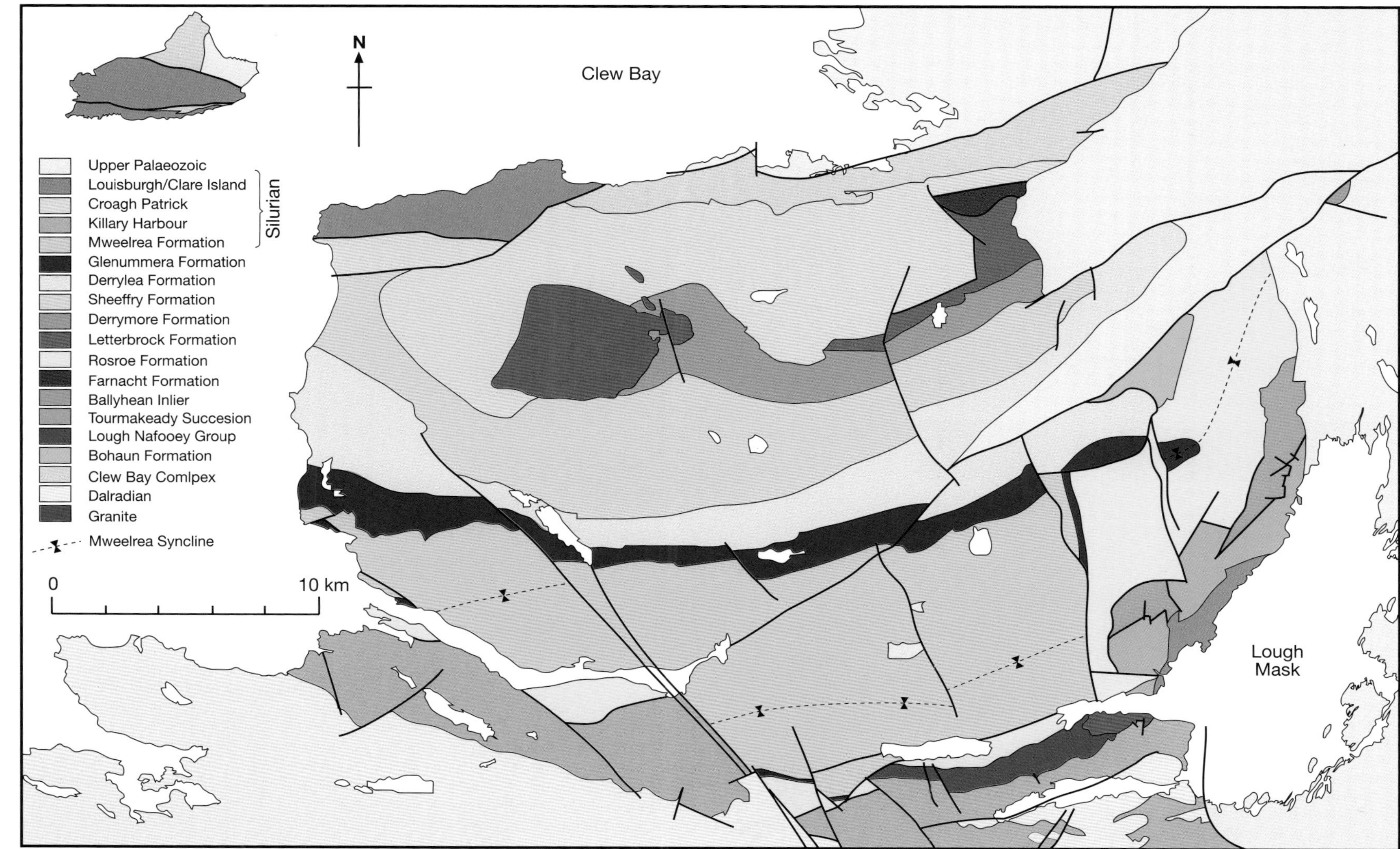

Fig. 3.3 Geological map of South Mayo (after Graham *et al.*, 1985).

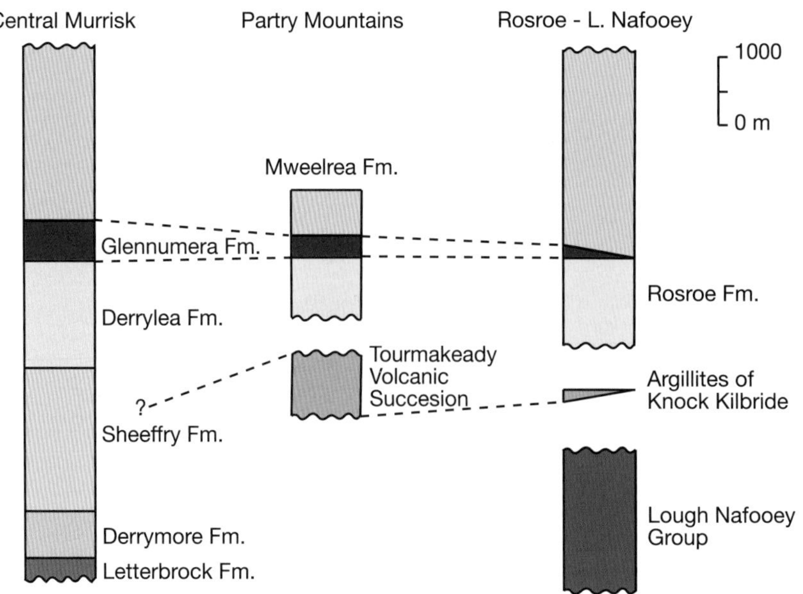

Fig. 3.4

A) Comparison of known thickness of Ordovician successions in South Mayo. The base of the regionally extensive Glenummera Formation is used as a datum.

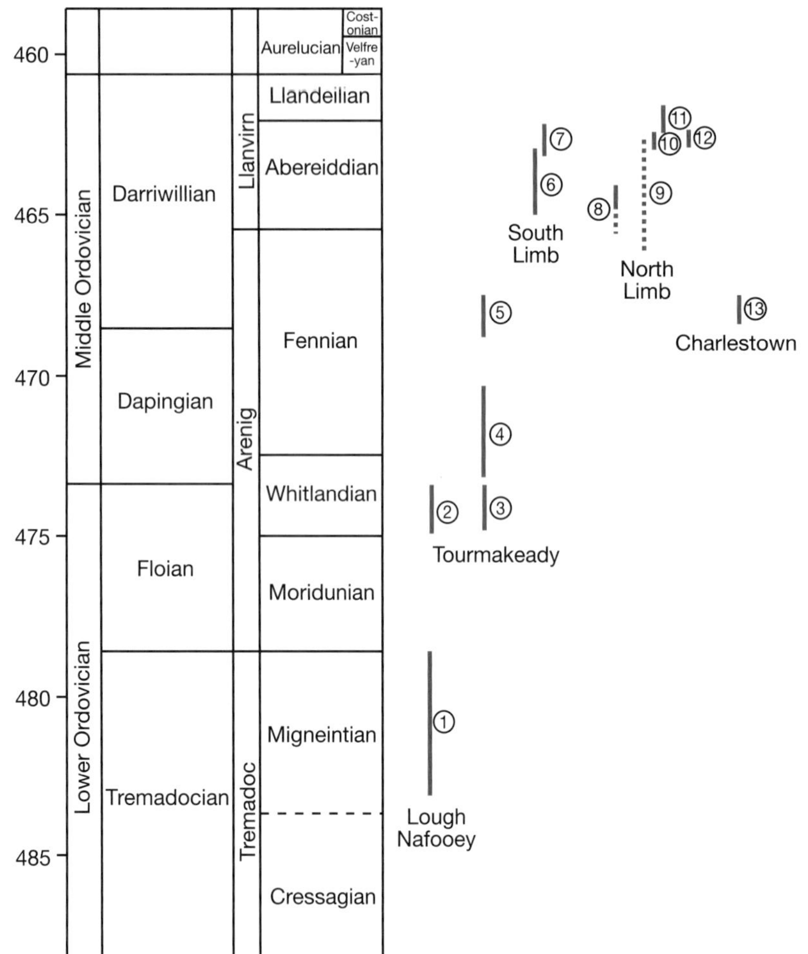

B) Distribution in time of the Ordovician successions of South Mayo. The red bars represent biostratigraphical determinations with associated errors. Note that many biostratigraphical determinations overlap but they are consistent with the succession established from field mapping.

1. Knock Kilbride Formation, Lough Nafooey Group (Williams and Harper, 1994).
2. Knock Kilbride argillites (Dewey *et al.*, 1970).
3. Mount Partry Formation, Tourmakeady Volcanic Succession (Dewey *et al.*, 1970).
4. Glensaul School, Tourmakeady Volcanic Succession (Dewey *et al.*, 1970).
5. Garranagerra, Tourmakeady Volcanic Succession (Dewey *et al.*, 1970).
6. Bencraff, Rosroe Formation (Dewey *et al.*, 1970).
7. Rosroe Formation, Killary Bay Little (Skevington, 1971).
8. Sheeffry Formation, Derrygarve (Pudsey, 1984).
9. Derrylea Formation, Doolough (Pudsey, 1984).
10. Mweelrea Formation, Slate Band 1 (Pudsey, 1984).
11. Mweelrea Formation, Slate Band 2 (Pudsey, 1984; Williams, 1972).
12. Glenummera Formation, Lough Shee (Harper *et al.*, 1988).
13. Charlestown (Dewey *et al.*, 1970).

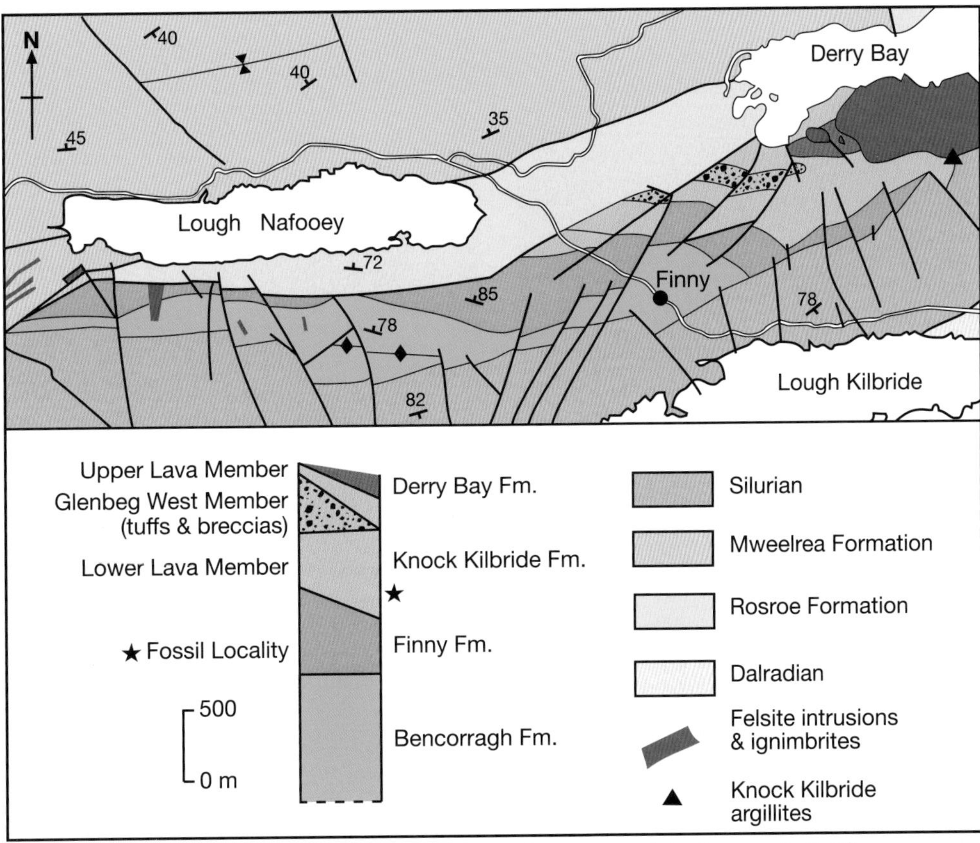

Fig. 3.5 Geological map of the Lough Nafooey Group (*after* Ryan *et al.*, 1980).

with the standard Australian succession (Ryan *et al.*, 1980) and an age within the uppermost part of the Lancefield Zone (La3) which is probably equivalent to the *Tetragraptus approximatus* Biozone (lowermost Arenig) of the standard British succession (Fig. 3.2), was suggested. New collections from this locality by Williams and Harper (1994) suggest a slightly older age (late Tremadoc – La2). Chew *et al.* (2007) have obtained zircon ages of 489.9±3.1Ma and 487.8±2.3Ma from plagiogranite boulders now present in the basal Silurian rocks, but derived from the Lough Nafooey Group. These data suggest both that the magmatism commenced in the late Cambrian and also, on the basis of Nd isotopes, that there had been assimilation of old continental crust, probably subducting Laurentian margin sediments, by 490Ma. This is somewhat earlier than the latter part of the Lough Nafooey Group, which was suggested to record the onset of subduction of Laurentian margins based on Nd isotopic data by Draut *et al.* (2004), and also earlier than the 478Ma for sediment recycling as suggested by Dewey (2005).

The Bencorragh Formation (>850m) is composed of meta-basalts (containing relic igneous pyroxenes) and spilites (containing only low grade minerals) which often exhibit good pillow form (Fig. 3.6A). The base of the Finny Formation (390–590m) is defined by the lowest keratophyric (meta-andesitic) eruptive products. There is considerable lateral variation within this formation from coarse andesitic breccias (Fig. 3.6C) and conglomerates (Fig. 3.6D) to intercalated tuffs, pillow lavas, and green and red (jasper) chert bands. The well-rounded clasts present in the conglomerates imply working by wave action, and thus that at least the upper parts of these volcanoes were in shallow water. As would be expected, sandstones from the Finny Formation plot in the undisected arc field of QFL discriminant diagrams (Dewey and Mange, 1999; Fig. 3.11). The Knock Kilbride Formation (480–1150m) consists of lower (Fig. 3.6B) and upper lava members separated by a unit of epiclastic breccias and conglomerates. Rare graptolitic black shales occur interbedded with pillow lavas, hyaloclastite breccias and cherts in the lava members. The base of the Derry Bay Formation (>150m) is defined by Ryan *et al.* (1980) as a 0.2m thin pyritous shale containing indeterminate graptolite fragments. This is overlain by silts and arkosic arenites (20m) and immature granule grade

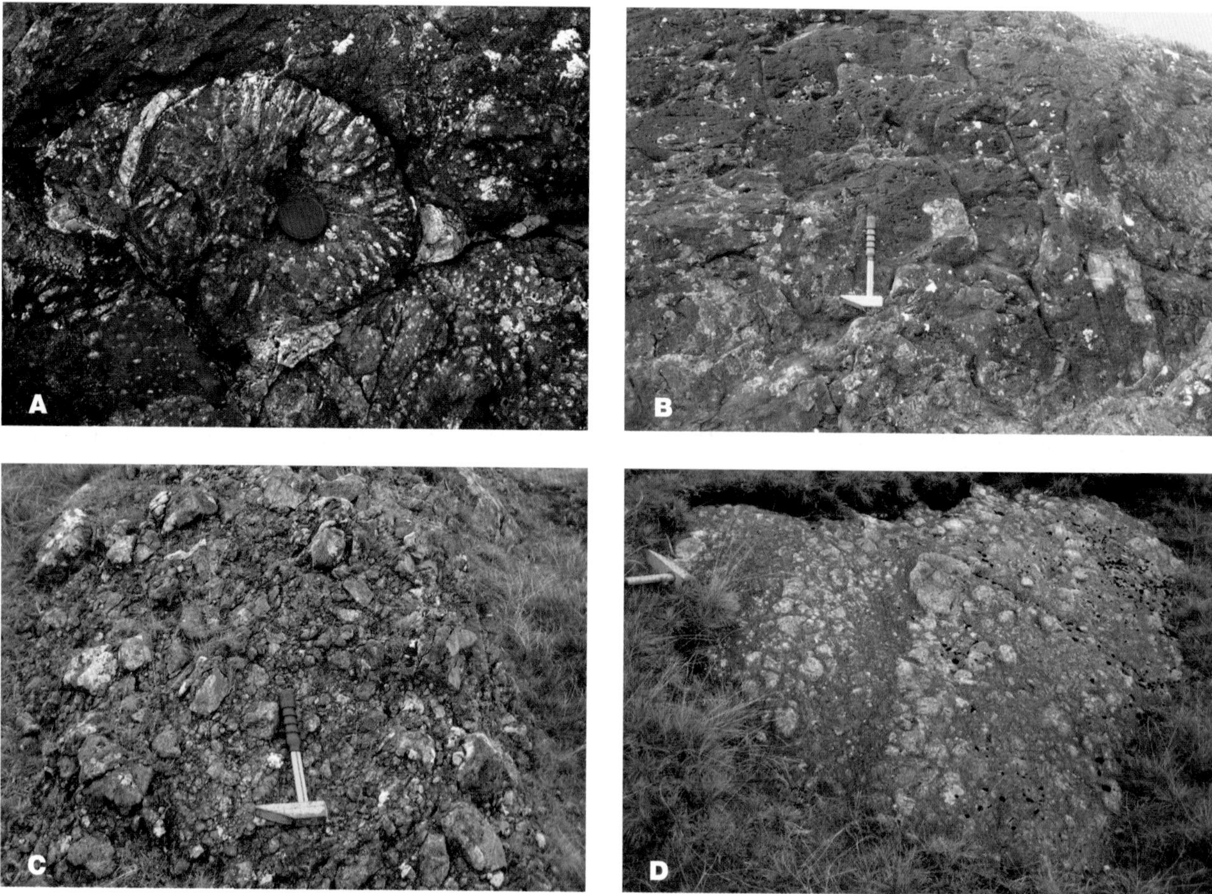

Fig. 3.6 **A)** Vesicular pillow lavas, Bencorragh Formation, Bencorragh. **B)** Pillow lavas, Knock Kilbride Formation, Finny. **C)** Volcanic breccias, Finny Formation, near Finny. **D)** Bedded epiclastic conglomerates, Finny Formation, near Finny.

volcaniclastic rocks (minimum 130m). These contain infrequent dispersed texture conglomerates with clasts of silicic volcanics, granite and spilite. The presence of these lithologies is important as they indicate considerable geochemical evolution of the Lough Nafooey arc.

On Knock Kilbride the Lough Nafooey Group is overlain by argillites lacking volcanic detritus (Gardiner and Reynolds, 1912) that contain graptolites of Arenig age. A local disconformity can be inferred but not proven. These argillites are baked by a pre-Silurian felsite sheet that intrudes the Lough Nafooey Group, the Derry Bay Felsite.

All lavas in the Lough Nafooey Group exhibit low-grade hydrous assemblages with some apparent variation of metamorphic grade with stratigraphical height (Ryan *et al.*, 1980). Lavas at the base of the Bencorragh Formation contain actinolite + chlorite + epidote + albite + quartz +/- magnetite assemblages, whilst higher in the formation rocks of similar basic composition contain albite + chlorite + quartz +/- calcite, pumpellyite, prehnite, sericite, and magnetite. These assemblages are recorded throughout the overlying Finny Formation and in the lower parts of the Knock Kilbride Formation. However, laumontite replaces albite, haematite replaces magnetite, and chalcedony replaces quartz in the Upper Lava Member of the Knock Kilbride Formation. This stratigraphically controlled zonation, from apparent zeolite to greenschist facies assemblages, in steeply dipping rocks is likely to be due to burial metamorphism which predates the mid to late Ordovician folding. The preservation of these low-grade assemblages indicates that subsequent regional metamorphic events associated with Ordovician and Silurian deformational events on the south limb of the Mweelrea Syncline were of very low grade.

The rocks preserved in the Lough Nafooey Group have been interpreted as being produced in an oceanic island arc (Ryan *et al.*, 1980). The position of this arc and its relationship to adjacent units has figured prominently in many palaeotectonic reconstructions of the Caledonides (see Chapter 4 for summary).

Bohaun Volcanic Formation

Volcanic rocks crop out in the region around Bohaun and Sraheen in the Erriff Valley (Fig. 3.3). Pillow lavas were first identified by Geikie (1897) but the formation was not described in any detail until McManus (1972). A summary of unpublished work by P.D. Ryan and J.B. Archer is given in Graham *et al.* (1989). The sequence is dominated by spilitic pillow lavas that have an apparent post-tectonic thickness of at least 1700m, the base being nowhere exposed. However, the rocks are highly deformed and exhibit a strong cleavage fabric, such that the original thickness and actual succession are uncertain.

The age of this succession is not known with great certainty. McManus (1972) proposed that it was laterally equivalent to and interfingered with the Derrylea Formation, but more recent mapping has shown that the contact referred to by McManus (1972) between the volcanic rocks and the Derrylea sedimentary rocks is in fact an early thrust contact that is folded in the Bohaun Shear Zone (McCaffrey and Graham, unpublished). The Bohaun Volcanic Formation can be shown to be unconformably overlain by the conglomerates of the Rosroe Formation of Llanvirn age (Graham, 1987). There is good lithological correlation between the Bohaun Volcanic Formation and the Knock Kilbride Formation of the Lough Nafooey Group (Clift and Ryan, 1994). The greater depletion of incompatible trace elements in the Bohaun Volcanic Formation compared to the Lough Nafooey Group was interpreted as due to a more trenchward position of the formation. A similarity in age to the Lough Nafooey Group is implied. Draut *et al.* (2004) showed that the Bohaun Volcanic Formation is more primitive than even the Bencorragh Formation, based on Nd isotopic data.

Tourmakeady volcanic succession

This classic area along the north-west shores of Lough Mask contains a sequence of Arenig volcanic and associated rocks. Although the fauna of the area has been the subject of several papers (Williams and Curry, 1985; Taylor and Curry, 1985; Donovan, 1986), the lithologies and stratigraphy have not been described in detail since the works of Gardiner and Reynolds (1909, 1910), although a summary using the informal stratigraphical units described below is given in Graham *et al.* (1989) (Fig. 3.7).

The rocks comprising the lower part of the Mount Partry Formation are marked by a sequence of graded volcaniclastic sandstones, siltrocks, and graptolitic mudrocks that are overlain by thin bands of red and green cherts interbedded with angular volcanic breccias containing andesitic to rhyolitic debris. These are lithologically comparable and have yielded a similar Chewtonian graptolite fauna to the argillites on Knock Kilbride (Dewey *et al.*, 1970; Fig. 3.4), that appear to rest on or be faulted against the Lough Nafooey Group. Higher levels of this formation contain a sequence of water-lain tuffs that range from fine ash bands to breccias containing coarse angular blocks of silicic and intermediate lavas which are often scoriaceous. Coarse to medium-grained tuffs with abundant volcanic quartz comprise the highest levels of this formation.

A unit of 'extrusive' rhyolites occurs just above or at the top of the Mount Partry Formation and can be traced from the Glensaul Valley in the south-west to north of Srah in the north-east. Gardiner and Reynolds (1909) recognised the extrusive nature of this pinkish-weathering quartz porphyry in the Tourmakeady area,

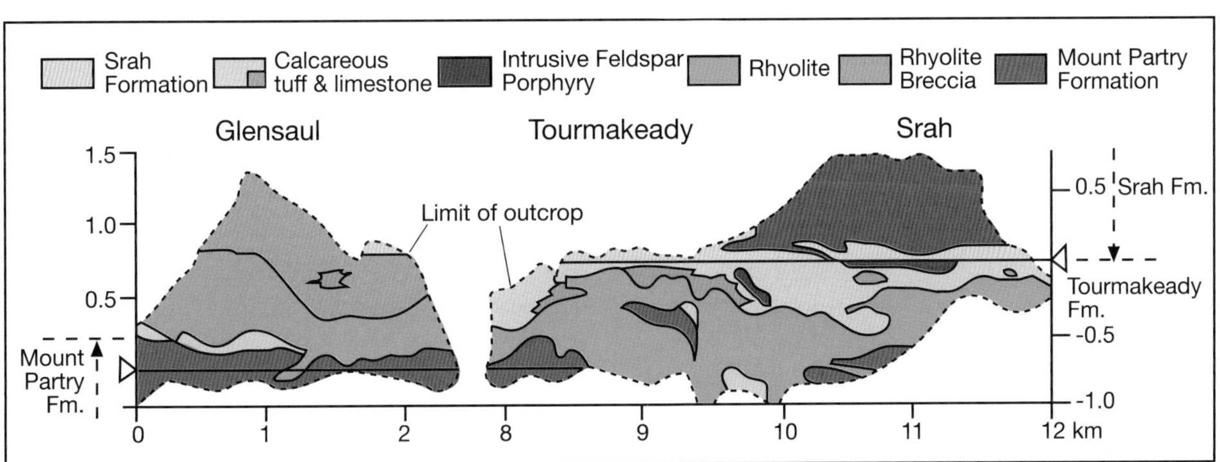

Fig. 3.7 Downdip projection of the Tourmakeady Volcanic Succession emphasising lateral variation (*after* Graham *et al.*, 1989).

but believed it was intrusive in the Glensaul Valley (Gardiner and Reynolds, 1910). This unit is described as 'extrusive' because it occurs at a near constant stratigraphical level and is conformably overlain by limestones. Volcanic breccias contain blocks of similar rhyolites. In spite of the large volume of magma, no contact metamorphic effects occur. Flow banding parallel to bedding and columnar jointing perpendicular to bedding occur locally. This unit probably evolved as a rhyolite dome formed on the sea bed. The overlying rhyolite breccias with their marine fossils represent the brecciated carapace of the dome. A down-dip projection of this rhyolite mass shows that these lavas are thickest and cut through the base of the stratigraphy in the region near Tourmakeady, suggesting that this was a local centre of eruption (Graham *et al.,* 1989; Fig. 3.7). Palaeontological evidence constrains this eruptive event within the late Arenig (Williams and Curry, 1985; Fig. 3.4). The rhyolite unit is up to 500m thick and the prolific shelly faunas could have developed on the uplifted surface of the dome that formed in water depths of several hundreds metres. This would explain why the shelly faunas are uniquely associated with the rhyolites, whilst deeper water planktonic faunas are recorded elsewhere in the Arenig rocks of South Mayo.

The fossiliferous Tourmakeady Formation contains two main facies, a sequence of limestones and limestone breccias and a sequence of volcanic breccias, tuffs, and clastic sediments. Gardiner and Reynolds (1909) termed the former unit the Tourmakeady Beds and the latter the Shangort Beds, which they described together because of the rapid lateral transitions between them. The calcareous horizons which often rest directly upon the extrusive rhyolites are most common at low levels in this formation. They contain massive grey crystalline limestones, partially brecciated limestones, and limestone breccias containing angular blocks of silicic volcanic rocks and cherts. In the north, the breccias often pass laterally into ashy sandstones. These limestone breccias and associated calcareous tuffs have yielded prolific, often silicified, shelly faunas (Williams and Curry, 1985; Taylor and Curry, 1985; Donovan, 1986) which have North American provincial affinities and indicate a late Arenig age (equivalent to the *D. hirundo* Biozone). The volcanic and sedimentary components of this formation comprise calcareous tuffs locally intercalated with rhyolitic breccias and mudrocks in the north, and a sequence of coarse breccias and tuffs in the south containing sediment rafts up to 30m x 5m.

The youngest Srah Formation comprises conglomerates, granule-grade quartzose sandstones, cherts, and mudrocks. At Glensaul, a latest Arenig fauna (Dewey *et al.,* 1970) is recorded in graptolitic mudrocks interbedded with cherts and ashy sandstones which are attributed to this formation (Fig. 3.4).

Geochemically the Tourmakeady volcanic rocks are clearly more evolved than those of the Lough Nafooey Group. Although final physical juxtaposition can only be confirmed by the common overstep of the Rosroe Formation in the early Llanvirn, most authors have treated the two volcanic successions as part of the continued evolution of a single magmatic arc (e.g. Graham *et al.,* 1989; Clift and Ryan, 1994; Dewey and Mange, 1999; Draut and Clift, 2001; Draut *et al.,* 2004; Dewey, 2005), although there is a marked change in most geochemical parameters, e.g. REE geochemistry, Nd isotopic composition, across the two successions (Fig. 4.6).

Ballyhean inlier

A small inlier of pre-Carboniferous rocks exists to the south of Ballyhean (Fig. 3.3) with its northern margin defined by the post-Carboniferous Erriff Valley Fault. The brief description below is based on a reconnaissance examination given in Graham *et al.* (1989). The inlier exposes volcaniclastic conglomerates and sandstones. The sandstones and conglomerates contain both quartzose metamorphic detritus and fresh volcanic components. Other volcaniclastic rocks contain almost exclusively igneous clasts including devitrified glass. The igneous material is mainly acidic to intermediate in composition and bears obvious comparison to the Tourmakeady Volcanic Succession. All the rocks seen in this inlier show a strong fabric comparable to that seen along strike at Bohaun.

Farnacht Formation

This formation crops out a few kilometres south of Westport (Max *et al.,* 1988). It consists of schistose dacitic–andesitic lavas of volcanic arc affinity. It was originally regarded by Kelly and Max (1979) as the lowest part of the Letterbrock Formation of the Murrisk Group. However, Max *et al.* (1988) suggested that these volcanic rocks are a separate entity which is tectonically isolated from the Letterbrock Formation. At present the age of these volcanic rocks is largely unconstrained, although likely to be pre-Wenlock on the basis of argon step-heating data (Max *et al.,* 1988). As most of the volcanic rocks in north-west Ireland that have been dated are assigned to the early to mid Ordovician, this age is considered likely.

Delaney Dome Formation

The Delaney Dome Formation crops out only in a structural window beneath the Mannin Thrust on which the highly deformed Dalradian rocks of Connemara have been transported southward (Leake *et al.*, 1983; Fig. 4.16). Although the Delaney Dome Formation consists exclusively of mylonites, it has been shown that the protoliths were acid igneous rocks, probably volcanic rocks (Leake and Singh, 1986; Tanner *et al.*, 1989). The age of thrusting has been interpreted as 447±4Ma based on Rb–Sr mica ages, and a Rb–Sr whole rock isochron gives a minimum age for the protolith of 472±12 (Tanner *et al.*, 1989). Draut and Clift (2002) presented U–Pb zircon lower intercept ages of 474.6±5.5Ma, which they interpreted as reflecting magmatic crystallisation. They note similarities in geochemistry, as well as in age, with the Tourmakeady Volcanic Succession.

Murrisk Group

Ordovician rocks of the Murrisk Group crop out over a large part of South Mayo (Fig. 3.3) exposing an apparently continuous succession, over 6km thick, on the northern limb of a major open, east–west syncline, the Mweelrea Syncline. The present lithostratigraphy is based on work by Theokritoff (1951), Stanton (1960), Dewey (1963) and Pudsey (1984a). Extension of this succession and its correlatives into eastern Murrisk was described by McManus (1967, 1972) and Graham (1987) and onto the southern limb of the Mweelrea Syncline by Archer (1977).

The lowest formation, whose base is never seen, is the Letterbrock Formation (>250m) which crops out south and west of Westport. It consists of turbiditic sandstones, slates and conglomerates, collectively interpreted as deepwater deposits (Pudsey, 1984a). The clasts in the conglomerates consist of dolerite, gabbro, low-grade metasandstones, basic and silicic volcanic rocks, chert and quartzite (J.P. Wrafter, *pers. comm.*). Dewey and Mange (1999) demonstrated that the clasts of low-grade metasandstones are indistinguishable from sandstones present in the Killadangan Formation (Fig. 3.11). Pudsey (1984a) described some slump folds overturned to the south-east or east in support of a northerly derivation.

The base of the overlying Derrymore Formation (700m) is taken as the lowest massive, convolute, fine sandstone above the conglomerates of the Letterbrock Formation (Pudsey, 1984a). It consists mainly of thin, massive and graded sandstones, interpreted as turbidites, and dark green mudrocks with a general increase in fine material westwards. There are also minor, thin acid tuffs, and fine-grained conglomerates. Palaeocurrent data are sparse but suggestive of both lateral (from north) and axial (from east) transport (see Graham *et al.*, 1989 for summary).

The base of the succeeding Sheeffry Formation is taken at the incoming of black laminae in mudrocks (Pudsey, 1984a) and the formation is characterised by grey and green-grey mudrocks, graded lithic sandstones, and numerous pyroclastic rocks. A small graptolite fauna was recovered from Derrygarve by Stanton (1960) and has been further examined and added to by Pudsey (1984a). The fauna most likely belongs to the *Etherigei* Biozone of North America, which would equate to the earliest Llanvirn of the British subdivisions (Fig. 3.4). Dewey (1963) recognised six subdivisions of the Sheeffry Formation in central Murrisk, based largely on amounts and type of volcanic material. These appear to be only locally recognisable, in part because the volcanic debris is most prominent in the central area studied by Dewey. The tuff horizons are all acidic in composition, most are graded and many are reworked. Almost all the sedimentary rocks show features indicating deposition as turbidites with some black laminae in the mudrocks indicating hemipelagic deposits. There is no obvious sequential organisation in these turbidites (Pudsey, 1984a; McManus, 1972) and a basin plain/fan fringe environment is suggested by Pudsey (1984a). A slight westerly fining of the formation is interpreted as a proximal–distal variation.

The base of the Derrylea Formation (c.2000m) is defined by a prominent tuff band or by the first appearance of white-weathering sandstones. The formation appears to both thicken and coarsen eastwards. It is sandstone-dominant (Fig. 3.10B) with several of the sandstones being granule-bearing, and contains conglomerates in the east in the Erriff valley (McManus 1972). The sandstones are separated by grey mudrocks with black laminae in the west and by paler green mudrocks in the east. There are also interbedded tuff bands. Graptolites have been recovered from Doolough which indicate an age low in the Llanvirn (Pudsey 1984a; Fig. 3.4).

Dewey (1962, 1963) considered this unit to be characterised by two different types of turbiditic sandstones, being axially (from the east) and laterally (from the south) derived. This assertion is at some variance with the palaeocurrent data presented by Pudsey (1984a), which suggests westerly and southerly-directed flows. In eastern Murrisk, McManus (1972) reported mainly axial turbidites (from the north-east) replaced by lateral (from the south-east) turbidites in the Bohaun–Erriff area. These apparently contradictory datasets can be reconciled by simple sediment mixing as initially recognised by Dewey (1961, 1962, 1963). The northerly-derived sediment carries an ophiolitic signature (Wrafter and Graham, 1989; Dewey and Mange, 1999; Fig. 4.5) whereas the southerly- and easterly-derived sediment represents the generally northward progradation of the Rosroe system discussed below. An inlier of Ordovician rocks east of Clew Bay, named the Letter Formation, comprises greywackes, acid pyroclastic volcanics, and mudrocks. This has been dated as Ordovician by chitinozoa (Graham and Smith, 1981) and is correlated with the upper part of the Derrylea Formation by Mange *et al.* (2003) on the basis of its heavy mineral assemblage.

Cropping out beneath the Glenummera Formation between Killary Harbour and Lough Nafooey on the south limb of the Mweelrea Syncline, and thus laterally equivalent to the Derrylea Formation (Figs. 3.3; 3.4), are >1350m of sandstones and conglomerates with minor mudrocks termed the Rosroe Formation by Archer (1977). In the western part of the outcrop these rocks are thrust southwards over the Wenlock rocks of the Salrock Formation. This thrust dies out eastwards, and south-west of Lough Nafooey the Rosroe Formation is unconformably overlain by the Llandovery Lough Mask Formation. The contact between the Rosroe Formation and the Lough Nafooey Group south of Lough Nafooey has been interpreted as a major fault, the Lough Nafooey Fault (Ryan and Archer, 1978). However, the general parallelism of strikes in the Rosroe Formation to this contact, which can also be traced eastwards to the Kilbride Peninsula, suggests that it may represent modification of an original unconformity. Thus the Rosroe Formation represents the oldest part of the main Murrisk succession seen on the southern limb of the Mweelrea Syncline.

The lowest part of this formation is seen in a stream section south-west of Leenane where a mud-dominant member, the Bencraff Member, underlies the coarser-grained lithologies typical of the bulk of the formation. A similar muddy unit is present near Currarevagh at the western end of Lough Nafooey. The contact between the Bencraff Member and the coarser part of the Rosroe Formation was thought to be an unconformity by Theokritoff (1951) but was convincingly shown to be gradational by Archer (1977). The Bencraff Member consists of mudrocks, some of which have yielded early Llanvirn graptolites (Fig. 3.4; Dewey *et al.*, 1970), cherts, and thin sandstones. Theokritoff (1951) also noted a bed of limestone containing ossicles of cystids or crinoids.

The main part of the Rosroe Formation consists of conglomerates (Figs. 3.8; 3.9), coarse-grained sandstones, andesitic pyroclastic rocks, and minor limestone breccias. The conglomerates and many of the sandstones show evidence for mass flow, and many of the thinner sandstones are turbidites. Soft sediment deformation is very common (Archer, 1984). Mudrocks, which are a minor component of the formation, show many graded silt beds and also contain some carbonaceous laminae, which have locally yielded graptolites. The depositional environment is that of a submarine fan complex with transport towards the north-west, based on imbrication in the conglomerates and on sole marks from the sandstones (Archer, 1977; Ryan and Archer, 1977). These flow directions were refuted by Clift *et al.* (2003) although it is difficult to interpret their data, as poorly-defined 'pebble orientations' are mixed with those of cross-beds of uncertain scale and have been converted to azimuth data by interpretation.

Petrographically the Rosroe sandstones are all volcanic lithic arenites (Archer, 1977). The conglomerates are characterised by well-rounded clasts of granite (Fig. 3.8B, D), rhyolite, and psammite with a tendency for metamorphic clasts to increase in proportion upwards. Dewey and Mange (1999) noted a major petrographic change from their heavy mineral analysis part-way through the formation, both on the south limb of the Mweelrea Syncline and in the Partry Mountains, which they equate to that seen in the Derrylea Formation. Some data on trace element geochemistry of granite clasts from the Rosroe Formation are given by Clift *et al.* (2003). They suggested that the granites were produced during a collisional event and were intruded into deformed crust of the Laurentian continental margin. They noted close comparisons to the Oughterard Granite but significant differences from the Connemara quartz diorites.

A series of lithologically identical conglomerates and associated sandstones in the Partry Mountains has been termed the Maumtrasna Formation (Dewey 1963; McManus 1967, 1972). These rocks were regarded as the highest unit in the Murrisk succession, being assigned an age ranging up to the late Llanvirn (Llandeilian) (Williams *et al.,* 1972) or Caradoc (Dewey, 1963; Williams, 1980). It was even suggested by Max *et al.* (1978a) that these rocks could be Devonian in age. Later Williams (1984) suggested that they were lateral equivalents of the Mweelrea Formation. In a re-assessment of

Fig. 3.8 **A)** Argillites at Knock Kilbride. Contact metamorphism from the Derry Bay Felsite has enhanced the fine scale bedding. **B)** Conglomerate rich in granitic clasts, Rosroe Formation, north-east of Lough Nafooey. **C)** Interbedded conglomerates and sandstones in the Rosroe Formation east of Lough Nafooey. **D)** Coarse grained conglomerate dominated by granitic clasts in a coarse sandstone matrix, Rosroe.

the geology of the Partry Mountains, Graham (1987) has shown that these Maumtrasna rocks (800–900m) rest unconformably on the Tourmakeady Volcanic Succession in the east and on the Bohaun Volcanic Formation in the west. Draut *et al.* (2004) claim that this latter contact is faulted, but produce no evidence to support their assertion. The Maumtrasna conglomerates have also been shown to underlie the Glenummera Formation at Lough Shee (Graham, 1987; Harper *et al.*, 1988) and in Glenmask, and are thus laterally equivalent to the Rosroe Formation to the south-west and the Derrylea Formation to the north and west. Both in the Partry Mountains and near Lough Nafooey there is a clear coarsening-upward sequence and a predominance of granite and acid porphyry clasts with a subordinate but important metamorphic component (Fig. 3.9). The major difference in the Partry Mountains is the proposed alluvial fan setting (McManus, 1967; Williams, 1984; Graham, 1987) compared to the submarine fan setting in the Killary Harbour area. The lateral transition was probably via fan deltas, which would also have existed between the Partry Mountains and the Derrylea Formation of the northern limb of the Mweelrea Syncline. The rocks that were termed Derrylea Formation in the Erriff Valley contain coarse-grained sandstones with some conglomerates ranging up to boulder size (McManus, 1972) which are rich in clasts of granite and porphyry (Graham, 1987) and could easily be assigned to the Rosroe Formation. Thus the term Rosroe Formation is used here to describe all of the sub-Glenummera package (Fig. 3.3) in which conglomerates are common, and it is recommended that the term 'Maumtrasna Formation' should be discontinued.

The Glenummera Formation (up to 600m) which crops out on both limbs of the Mweelrea Syncline forms a distinctive lithostratigraphical marker. This formation has a relatively sharp base defined by the incoming of

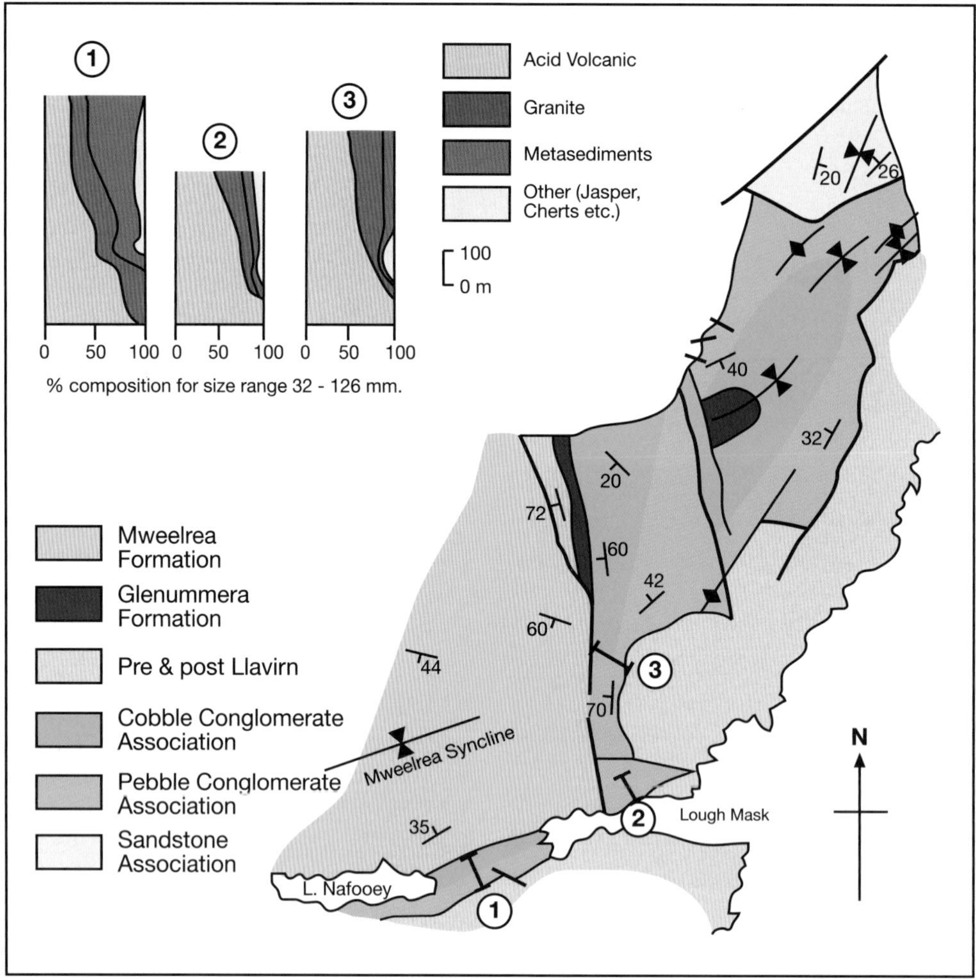

Fig. 3.9 Coarsening-upward sequence and change in clast composition in the Rosroe Formation in the Lough Nafooey–Partry Mountains area (after Graham *et al.*, 1989; clast composition data supplied by J.P. Wrafter).

green-grey slates without black laminae. The unit is mud-dominant with intervals of fine, laminated sandstones and lenses of conglomerate, particularly in the upper part. A graded tuff is reported from the southern limb of the Mweelrea Syncline by Stanton (1960). Although many of the coarser beds are graded, they have a much more lenticular geometry than the turbidites of the underlying formations, and some of the ripple foresets show opposed directions (Pudsey, 1984b). There is common bioturbation and also, locally, a shallow marine shelly fauna of Llanvirn age (Stanton, 1960; Pudsey, 1984a; Harper *et al.*, 1988). The formation shows a well-developed coarsening upward sequence in the upper part (Pudsey, 1984b; McManus, 1972) and thins towards the south-east.

The rocks have been interpreted as the deposits of a slope and a prograding, aggrading shelf (Pudsey, 1984b). The sea was thought to have low to moderate wave energy. Clasts in the conglomerate lenses in the upper part are petrographically identical to those in the overlying Mweelrea Formation.

The transitional base to the Mweelrea Formation (>2100m) is taken at the incoming of trough cross-bedded sandstones at the top of a coarsening-upward sequence that characterises the upper part of the Glenummera Formation. Coarse-grained trough and planar cross-bedded sandstones with locally abundant conglomerates form the bulk of this formation (Fig. 3.10). Five major, laterally extensive tuff bands have been used as time planes to elucidate lateral variation in the Mweelrea Formation (Stanton,1960; Dewey, 1963; Pudsey, 1984b). Some of these tuff bands appear to have formed as ignimbrites. On the basis of unwelded bases, Stanton (1960) suggested that the ignimbrites in western Murrisk were deposited in very shallow water with the tops showing subaerial erosion. Dewey (1963) reported that the tuff beds were thinner farther east in central Murrisk and that

Fig. 3.10 **A)** Gently dipping coarse-grained sandstones and pebble conglomerates of the Mweelrea Formation, Mweelrea. The prominent pale horizon dipping gently left from the col is one of the ignimbrite horizons. **B)** Typical exposure of parallel bedded sandy turbidites of the Derrylea Formation, Glenummera. **C)** Trough cross-bedded pebbly sandstones, Mweelrea Formation, Mweelrea. **D)** Gilbert delta in the Mweelrea Formation, near Allaran Point, north of Killary Harbour. White line indicates topset deposits which show large wave ripples on bedding surfaces. The prominent left inclined surfaces represent the westward advance of the delta foresets. **E)** The Maumtrasna Plateau, thought to be an exhumed Carboniferous peneplain, is cut into the Ordovician Mweelrea Formation. The foreground rocks are upward and northerly younging pillow lavas of the Bencorragh Formation, Lough Nafooey Group.

only one of these beds had an unwelded base, suggesting a palaeoslope down to the north-west. The tuff horizons are generally succeeded by a coarse conglomerate containing abundant tuff clasts. The associated sandstones are often rich in magnetite, which is thought to be derived from the tuff.

Three prominent slate members were recognised within the Mweelrea Formation by Stanton (1960), who noted that they became thinner to the east along strike and southwards across the Mweelrea Syncline. These slates are all sharp-based and show coarsening-upward sequences. The lower two have a shallow marine fauna of Llanvirn age (Stanton, 1960; Williams, 1972; Pudsey, 1984a; Fig. 3.4) and all show evidence of bioturbation and wave activity. It has been suggested by Harper and Parkes (2000b) that the highest slate band may extend into the Caradoc, although the data on which this is based have yet to be published. These slate members are thus basically similar to the upper part of the Glenummera Formation, although of lesser areal extent, and a similar environmental interpretation can be applied. The map patterns thus suggest strongly that the fossiliferous mudrocks at Lough Shee (Harper *et al.*, 1988) are better assigned to the Glenummera Formation rather than any of the members in the Mweelrea Formation as suggested by Harper and Parkes (2000a).

The suggestion of increasing proximality to the south and east is supported by maximum grain size of the conglomerates and by the palaeocurrent data. Williams (1984) demonstrated the presence of a conglomerate member in the Bunnacunneen area, west of Lough Nafooey and showed palaeocurrents from the south-east. Dewey and Mange (1999) noted that these conglomerates are rich in both garnet and euhedral zircon. McConnell and Riggs (*pers. comm.*, 2007) have analysed detrital zircons and granitoid boulders from this conglomerate. Their data show a probability maximum at 466Ma, comparable to ages of acidic magmatism from the Connemara magmatic arc, and they suggest that Connemara was close to its present position in Mweelrea times. They also note differences in granite geochemistry between boulders from the Bunnacunneen conglomerate and those from the Rosroe Formation. Farther west in the Killary Harbour area the palaeocurrents in the Mweelrea Formation show a more east to west trend (Pudsey, 1984b). The palaeogeography was one of a fluvial plain dominated by coarse-grained bedload rivers introducing their load via fan deltas (Fig. 3.10D) directly into a sea with low to moderate wave energy such that limited reworking occurred.

The clasts in the Mweelrea conglomerates are psammitic schist, acid volcanic rocks, granite, vein quartz, jasper, and chert. There appears to be a tendency for metamorphic detritus to form an increasing proportion of the clasts upwards, which is a continuation of the pattern seen in the underlying Rosroe Formation. This pattern is also evident in the heavy mineral analyses of Dewey and Mange (1999). Plots of sandstone composition on QFL diagrams demonstrate the increasingly quartzose nature of the sediment towards the top of the formation (Dewey and Mange, 1999; Fig. 3.11).

On the south limb of the Mweelrea Syncline the base of the Murrisk succession is never seen, but the persistent presence of coarse fans in the Rosroe Formation is suggestive of a proximal position and continued synsedimentary fault activity. The limestone breccias (Archer, 1977; Graham *et al.*, 1989) can be interpreted as slumped remnants of a narrow southern shelf. The evidence from the palaeocurrents and the outcrops in the Partry Mountains suggest that this southern margin swings to a more north-east to south-west orientation in the east. Palaeomagnetic work (MacNiocaill *et al.*, 1998) suggests that this strike swing is likely to be primary rather than due to later oroclinal bending. Such basement orientation may well be responsible for the nature of the eastern termination of the Mweelrea Syncline, where a relatively simple pattern of east–west strikes and gentle to moderate dips is replaced by north–south strikes and steep dips and associated faulting. However, this geometry could also be produced by later block-faulting. The indication of a northern source for much of the north limb succession supports the notion of a small westward-opening basin (present coordinates) and thus the term South Mayo Trough employed by Dewey (1963) is appropriate. In general terms the mudrocks of the Glenummera Formation interrupt a simple westerly and/or northerly prograding, clastic sequence in which flows were predominantly axial. It is not yet clear whether these mudrocks reflect a regional subsidence event or a major eustatic event.

The Murrisk Group also shows major spatial and temporal petrographic changes. These changes are conveniently shown by whole-rock geochemistry of the sandstones, especially in its matrix-rich sandstones, where the original detrital mineralogy is only partially preserved (Wrafter and Graham, 1989; Fig. 4.5). The north limb sequence is best interpreted as the unroofing of an ophiolite that had been obducted onto continental crust. The south limb sequence can be interpreted as the erosion of an acidic magmatic arc, including its subvolcanic plutons.

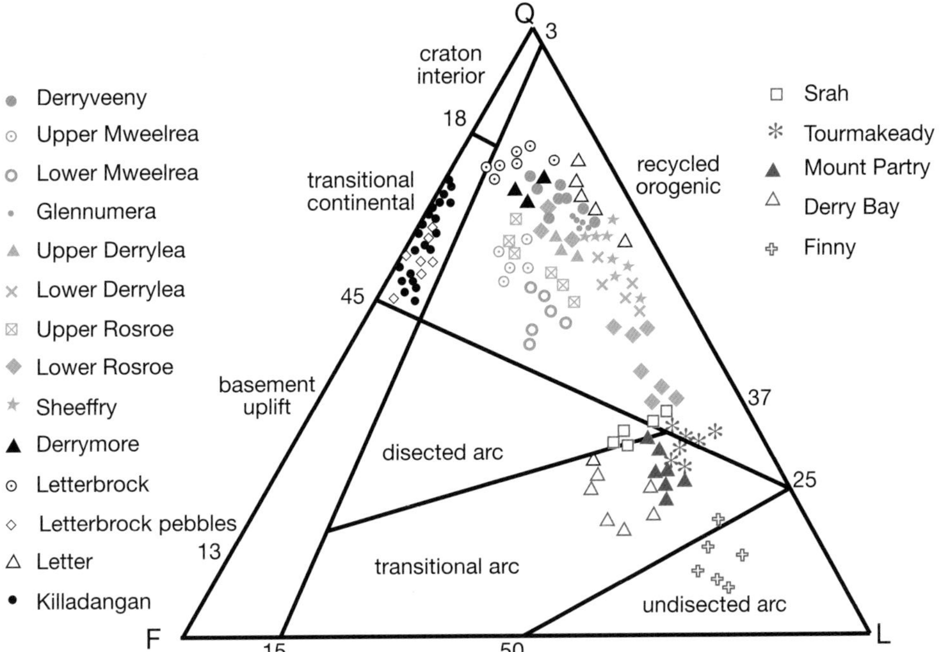

Fig. 3.11 Mineralogical data for sandstones of the South Mayo Ordovician plotted on a QFL diagram (from Dewey and Mange, 1999). *Reproduced by kind permission of the Geological Society of London.*

These data and the overall interpretation are strongly supported by the detailed heavy mineral analysis of Dewey and Mange (1999). These authors detected a horizon within both the Rosroe and the Derrylea formations at which there is a marked influx of metamorphic detritus, and they correlated this horizon across the Mweelrea Syncline. They rejected the possibility that this metamorphic influx reflects the unroofing of southerly Dalradian basement mainly on the basis of the proposed tectonic models for the position of Connemara. However, they and Dewey (2005) also noted that the arrival of metamorphic detritus in the Derrylea Formation is present in darker, green axial turbidites rather than in southerly-derived pale turbidites that contain only arc-derived debris.

It is clear that the Murrisk succession was deposited relatively rapidly in the mid Ordovician (Fig. 3.4). This fact, along with the paucity of biostratigraphical data, means that the detailed correlation of formations by biostratigraphy is not always possible. Although more radiometric age data are now available, the rapid sedimentation rates mean that alternative interpretations within the error limits of the data remain possible. The potential geotectonic scenarios in which accommodation for this sediment could be produced are discussed in more detail in Chapter 4.

Derryveeny Formation

This formation consists of boulder conglomerates and sandstones cropping out west of Lough Mask (Graham *et al.* 1985, 1989, 1991). These were originally thought to underlie the Tourmakeady volcanic rocks (Gardiner and Reynolds, 1909) but more recent investigation has shown that the contacts are faulted (Graham *et al.*, 1989, 1991). The general parallelism of strike in the Derryveeny Formation to its western contact against the Arenig rocks of the Tourmakeady Volcanic Succession suggests that the faults modify an original unconformity. Where the Derryveeny conglomerates crop out close to Silurian rocks on the north shore of Derry Bay they strike south-east, and therefore run directly into the north–south trending base of the Silurian Lough Mask Formation. Excavation of this contact has proven that red sandstones of the Lough Mask Formation lie unconformably on the Derryveeny Formation, and thus that the Derryveeny Formation is pre-upper Llandovery in age (Graham *et al.*, 1991).

Most of the outcrop of the Derryveeny Formation is rather poorly exposed but good sections exist on the northern shore of Derry Bay, in the Treanlaur and Glensaul rivers. A lower conglomerate member (c.220m) passes gradationally upwards into a sandstone member (>400m). Imbrication is common in the conglomerate

Fig. 3.12 **A**) Glaciated surface subparallel to bedding showing the typical texture of the coarse conglomerates in the Derryveeny Formation, Derry Bay. **B**) Interbedded imbricated conglomerates and coarse sandstones, Derryveeny Formation, Derry Bay. **C**) Large boulders of migmatitic gneiss (centre) and granite (lower right), Derryveeny Formation, Derry Bay. The dark areas in the migmatite are restite patches rich in sillimanite. **D**) Large migmatite clast, Derryveeny Formation, Derry Bay. Hammer in all photos is 40 cm long.

member, suggesting derivation from the east or south-east (Graham *et al.,* 1991). Maximum clast sizes commonly exceed 0.7m with some individual clasts exceeding 1m in length. The clasts are predominantly of metamorphic rocks, including migmatitic gneisses, with granites and porphyries being locally very common (Fig. 3.12). Of particular interest are sillimanite-bearing migmatite clasts which offer close comparisons, both petrographically and isotopically, with the Lough Kilbride Formation of the Connemara massif. Graham *et al.* (1991) interpreted the clasts as being derived from this unit, thus establishing a clear Ordovician linkage between South Mayo and the Connemara massif. Rb–Sr dates (muscovite–whole rock pairs) from two schist clasts suggest a maximum age of deposition of 462±7Ma. This is consistent with derivation from Connemara but is not a very precise constraint, given the known rapid sequence of orogenic events (Dewey, 2005). Trace element analyses of granite boulders from the Derryveeny Formation plot, not surprisingly, in the 'within plate' granite field, and, according to Clift *et al.* (2003), are geochemically indistinguishable from those of the Rosroe Formation.

Charlestown

In this Lower Palaeozoic inlier, which lies along the projected strike of South Mayo (Fig. 3.1), the Ordovician rocks are mainly volcanic with subordinate black mudrocks. Although 'double graptolites' were reported by Kilroe (1898), the Arenig age of the succession was established from new collections by Cummins (1954). A re-examination of this fauna by Dewey *et al.* (1970) further refined the age to late Arenig (equivalent to *D. hirundo* Biozone) (Fig. 3.4). These rocks were also described by Charlesworth (1960a) with more detailed accounts incorporating borehole data given by O'Connor (1987) and O'Connor and Poustie (1986).

O'Connor (1987) recognised about 2000m of lava flows, intrusive rocks, tuffs, volcanic breccias, resedimented tuffs, and mudrocks. The graptolitic mudrocks sampled by Cummins (1954) are situated about 150m above the lowest exposed strata. There is a gradual change upward through the succession from a variety of tuff types at the base accompanied by tholeiitic spilites and basalts, to calc-alkaline andesitic tuffs and resedimented tuffs, to more felsic tuffs accompanied by dacite and rhyodacite intrusions at the top of the succession. The environment was marine throughout and the major, minor, and trace element chemistry of both the intrusive and extrusive igneous rocks is consistent with a developing oceanic volcanic arc (O'Connor, 1987).

Tyrone

Some 180km NE of Clew Bay, in the Tyrone area, Lower Palaeozoic rocks crop out immediately adjacent to Dalradian metasediments of the Sperrin Mountains (Fig. 3.13). Despite a position analogous to that of the Clew Bay Complex and South Mayo, the sequence at Tyrone shows important differences from these areas. It may be significant that the geophysical signature of the Highland Boundary Fault passes north of the surface contact (Max and Riddihough, 1975) suggesting southward overthrusting (Chapter 4) of the Dalradian and thus possible omission of some Lower Palaeozoic sequences.

In Tyrone an ophiolite slab (the Tyrone Plutonic Group) has been thrust northwards over a metasedimentary succession of probable Dalradian Supergroup affinity, the Tyrone Central Inlier (Chapters 1, 4). A tonalite (the Craigballyharky Pluton, Fig. 3.13), which is inferred to be intruded into 'hot' ophiolite, gives a discordant lower intercept U–Pb zircon age of 471+2/-4Ma (Hutton *et al.*, 1985). This would indicate an Arenig age using the Gradstein *et al.* (2004) timescale, the age interpreted by Hutton *et al.* (1985). Lying to the north-west of this ophiolite is an extensive series of high-level intrusive rocks and lavas of arc type, the Tyrone Volcanic Group (Hartley, 1933b; Cobbing *et al.*, 1965; Hutton *et al.*, 1985). Black shales associated with these volcanic rocks on Slieve Gallion (Fig. 3.13) have yielded fragments of graptolites originally interpreted as late Llanvirn (Llandeilan) to Caradoc in age (Hartley, 1936). Further collections from Hartley's locality were interpreted as Arenig or Llanvirn in age by Hutton and Holland (1992), strengthening lithological comparisons that had been made to the Ballantrae area of south-west Scotland. Re-collection from the Slieve Gallion locality and re-examination of earlier material has allowed Cooper *et al.* (2008) to give a more precise age of early Castlemainian (Ca 1) of the Pacific graptolite stratigraphy (Fig. 3.2). U–Pb dating of high level intrusive rhyolites in the Tyrone Volcanic Group from Formil Hill (Fig. 3.13), a horizon interpreted as being stratigraphically equivalent to the graptolite locality, has produced a concordant age of 473±0.8Ma (Cooper *et al.*, 2008).

Near Pomeroy (Fig.3.13) a thin, shallow water Upper Ordovician succession unconformably overlies the ophiolite (Fearnsides *et al.*, 1907; Mitchell, W.J., 1977;

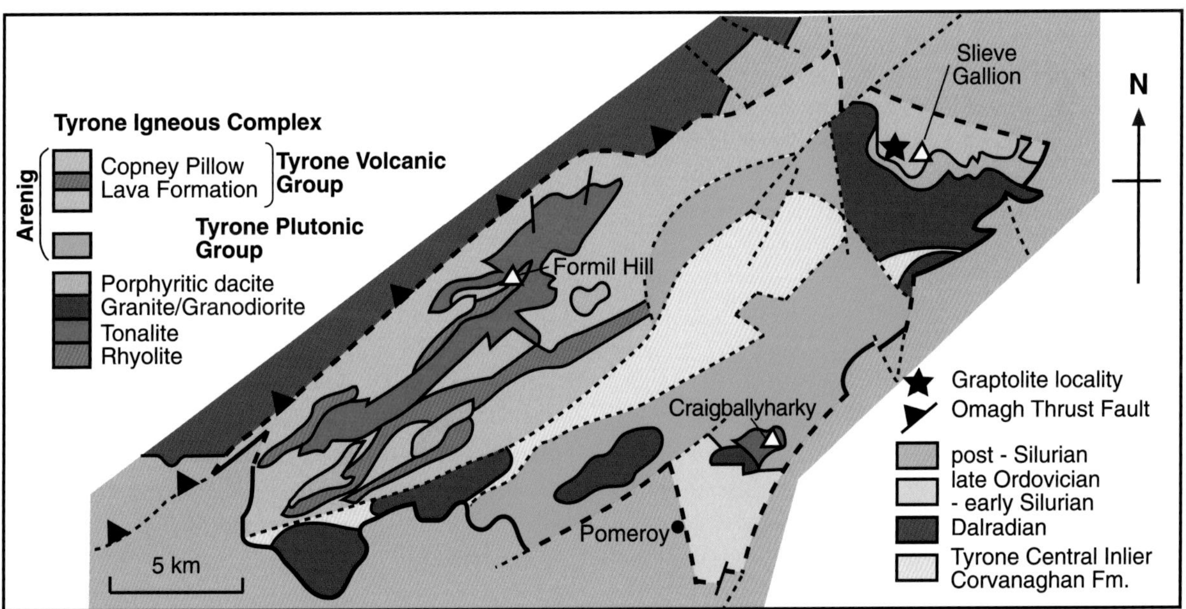

Fig. 3.13 Summary geology of the Tyrone inlier (after Cooper *et al.*, 2008).

Tunnicliff, 1982; Harper *et al.*, 1994). The Bardahessiagh Formation has yielded faunas of Burrellian (Caradoc) age that indicate deposition below storm wave base (Candela, 2002; 2006). The Killey Bridge Formation has yielded faunas of Rawtheyan (Ashgill) age and the overlying Tirnaskea Formation of Hirnantian (latest Ashgill) age. The latter indicate shallow water environments that have been interpreted as having been influenced by the glacially controlled Hirnantian regression (Candela, 2006).

The unconformable contact between the Caradocian mudrocks and the early to mid Ordovician ophiolitic rocks implies intra-Ordovician deformation. The claim that fragments from the plutons associated with the ophiolite can be recognised in the Caradoc sediments (Institute of Geological Sciences, 1979) also indicates significant intra-Ordovician uplift and erosion. Overall the Tyrone succession bears close comparison to that exposed in the Girvan/Ballantrae area in the Midland Valley of Scotland.

Longford-Down

The Longford-Down terrane is only part of a much larger geological entity encompassing the Southern Uplands of Scotland. Its westward continuation beneath the later cover can also be inferred with re-emergence as the Slieve Aughty inlier and probably the South Connemara Group that is preserved as a roof pendant in the *c.*400Ma Galway Granite (Fig. 3.1). This whole region is characterised by deep-water sandstones, conglomerates, pelagic mudrocks, and volcanic rocks. These are arranged in a series of strike-parallel tracts of remarkable lateral continuity distinguished by differences in age and petrography. These tracts are bounded by steep, north-west dipping faults and internally tend to young to the north-west, although on a larger scale the tracts become younger south-eastwards (Chapter 8). Attempts to explain the disposition of strata and their tectono-sedimentary origin have produced divergent interpretations. Various proposals, including a continental margin accretionary prism (Legget *et al.*, 1979; Anderson, 2001; Stone and Merriman, 2004; Mange *et al.*, 2005; Fig. 8.4); a fore-arc basin (Armstrong *et al.*, 1996; Bluck *et al.*, 2006); a back-arc basin (Morris, 1987); and an imbricate thrust stack developed largely in a successor basin (Hutton and Murphy, 1987), have been advanced, with possibly large transcurrent movements proposed on at least one of the major tract-bounding strike faults (Anderson and Oliver, 1986; Hutton, 1987). These differing tectonic scenarios are discussed in more detail in Chapter 8.

The subdivision in Scotland into Northern, Central, and Southern belts can also be applied in Ireland. On a more detailed scale the Irish sequences can also be correlated with the fault-bounded tracts of Leggett *et al.* (1979) defined in Scotland. Ordovician rocks dominate the Northern Belt but also occur, mainly as mudrocks ('Moffat Shale'), in the Central Belt. Biostratigraphical correlation between areas is based mainly on graptolites, which are locally abundant in the black, pelagic shale facies.

Northern Belt

Rocks within the Northern Belt are entirely of Ordovician age with the exception of a small, problematic outlier of mudrocks that have been assigned an upper Wenlock or Ludlow age (Smith, 1979) at Ballinalee. It has been suggested that this outlier may be in thrust contact with the adjacent Ordovician rocks (Morris, 1979, 1987). However, it must be noted that this outlier is dated only by palynology and does not have independent dating by graptolites to confirm the mid-Silurian age. Greywackes are the principal rock type in the northern belt but interbedded mudrocks, cherts and spilitic volcanic rocks are also present (Craig, 1984; Morris *et al.* 1986; Fig. 3.14). In Ireland these volcanic–argillite associations are mainly of late Llanvirn or Caradoc age with possible early Llanvirn rocks recorded in the Arva sector at the western end of the Longford-Down (Corn Hill Formation of Morris, 1983). In Scotland this association is reported as extending down into the Arenig (Leggett *et al.*, 1979) but Armstrong *et al.* (2002) suggested that there may be a break between the Arenig mudrocks and cherts and those of late Llanvirn age.

In Scotland the Northern Belt has been described in terms of three fault-bounded tracts (I to III of Leggett *et al.*, 1979). Their tract I, which lies north of the Glen App Fault, is either not present or not exposed in Ireland. The fault separating tracts II and III is traceable from Scotland into Ireland and has been termed the Northern Belt Median Fault by Morris (1987) (Fig. 3.14). This fault separates two very distinct greywacke 'petrofacies'; a metaclast petrofacies to the north-west (tract II) and a basic clast petrofacies to the south-east (tract III), although there are other minor petrofacies in each tract.

The metaclast petrofacies, which characterises tract II, is represented by the Arva and Strokestown sectors at the western end of the Longford-Down terrane (Morris,

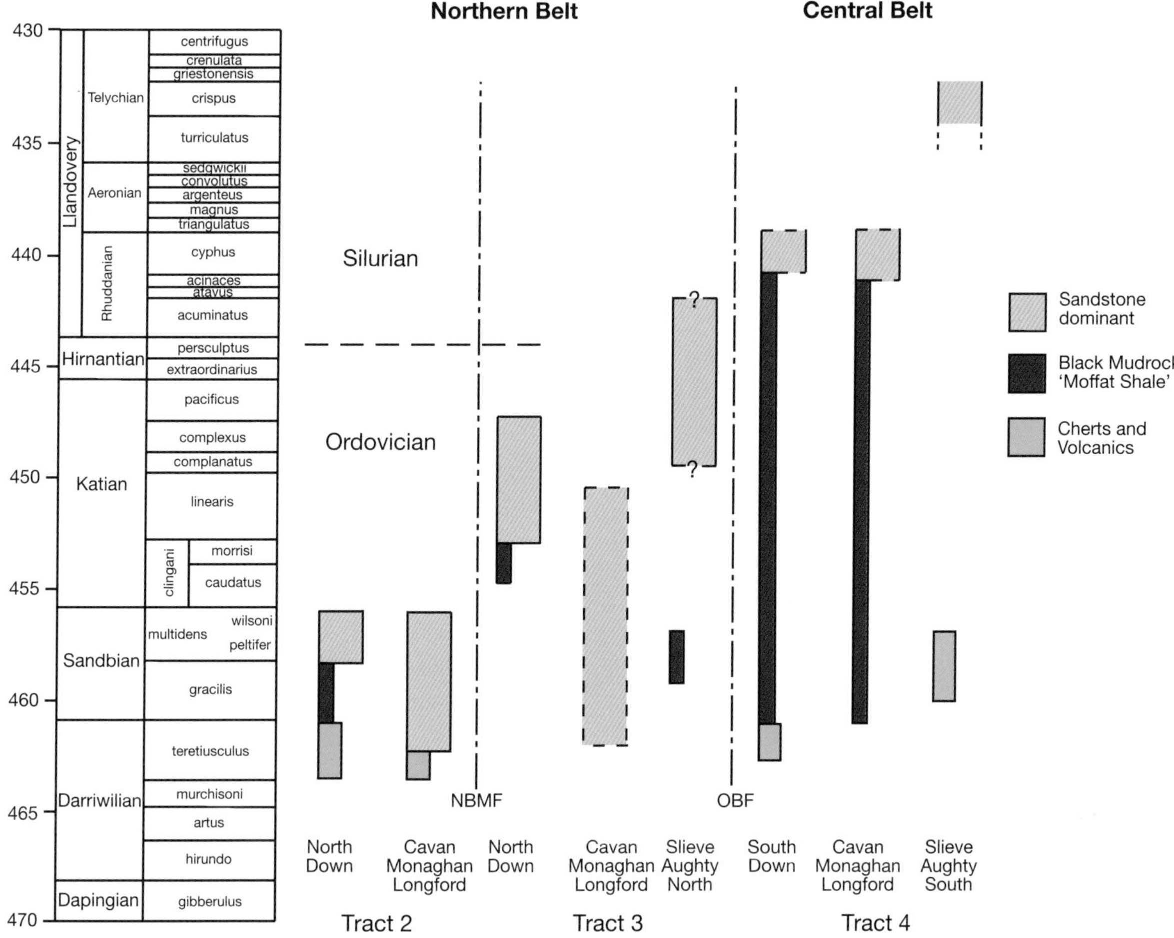

Fig. 3.14 Summary of late Ordovician and early Silurian successions of the Longford-Down belt.

1983) and by the Carrickanoran Formation of the Clontibret area (Morris *et al.*, 1986). This petrofacies plots as 'recycled orogen' on sandstone-discriminant diagrams (Fig. 3.15) and a mixed metamorphic and felsic igneous continental source terrain is indicated. It is considered to be north-westerly derived although palaeocurrent data are scarce. Preliminary detrital mineral ages from this petrofacies in Scotland (Kelley and Bluck, 1989) suggest that the source was a metamorphic block undergoing uplift and cooling during the Ordovician. Bluck (2000a) and Bluck *et al.* (2006) argued that this block has a different uplift history from that of the Dalradian and may be represented by a basement block now largely concealed beneath the Midland Valley of Scotland (the 'Midland Valley Block'). Rocks belonging to this petrofacies are mainly 'classical turbidites' with subordinate thick-bedded, coarse-grained channelised sandstones. Intercalations of pelagic mudrocks, where present, may be tens of metres thick. Deposition on the outer and mid-parts of prograding submarine fans is suggested (Morris, 1979).

The basic clast petrofacies of tract III of Morris (1987) is of magmatic arc provenance (Fig. 3.15) and the volcanogenic detritus suggests derivation from tuffaceous deposits rather than lavas (Kelling, 1961, 1962; Sanders and Morris, 1978; Stone *et al.*, 1987), and this in turn suggests a Plinian type of volcanic regime. Pyroxene mineral chemistry studies (Sanders and Morris, 1978; Morris *et al.*, 1988) indicate a calc-alkaline affinity for this detritus, which in view of its regional distribution, suggests derivation from a regionally extensive calc-alkaline volcanic arc. An exhumed subduction complex and sialic crustal component must also have been present in this source terrane to provide the distinctive blueschist and other metamorphic detritus described by Sanders and Morris (1978) and Kelling (1961). In Scotland sedimentary structures indicate south to north palaeocurrents that suggest that the source area for this material lay to the south, considerably

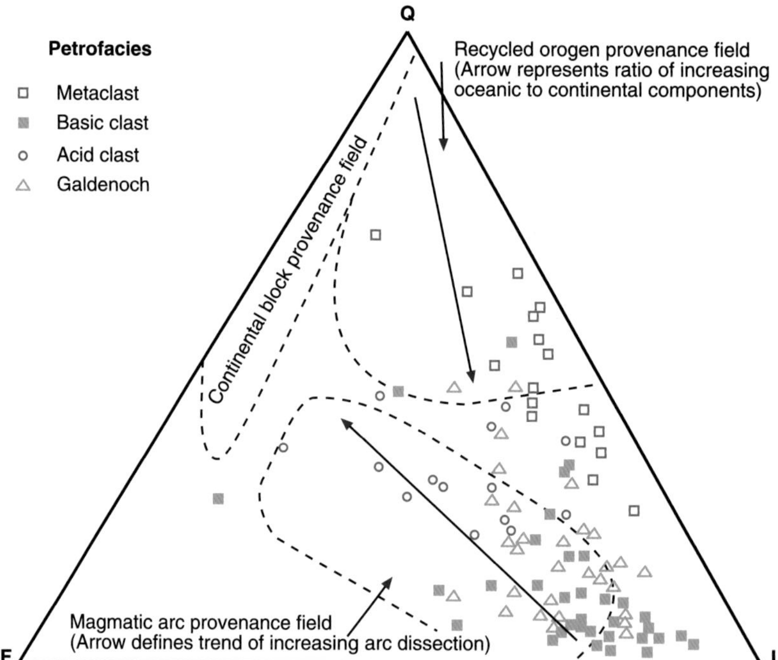

Fig. 3.15 Sandstone compositions and petrofacies for the Northern Belt, Longford-Down (from Morris, 1987). *Reproduced by kind permission of the Geological Society of London.*

outboard of the Laurentian margin. Palaeocurrent data are not available for the Irish sector, but the source is inferred to lie to the south by Morris (1979, 1987) as temporally equivalent greywackes to the north do not contain volcanic arc debris. Rocks of the basic clast petrofacies commonly display amalgamated beds and massive, pebbly sandstones, locally organised in channelised sequences (Morris, 1979; Craig, 1984) as well as more classical thick/medium bedded turbidite sequences. Deposition in more proximal settings than those of the metaclast petrofacies is indicated (Morris, 1987).

A gradation in character between the basic and the acid clast petrofacies, which also characterises tract III, suggests that the latter also had its source in the same general terrane. The freshness of the detritus has been used to infer a proximal and probably contemporaneous volcanic arc (Morris, 1987; Stone *et al.*, 1987). However, dating of the 'fresh' andesitic detritus in Scotland has suggested a source area with U–Pb detrital zircon dates as young as 583±38Ma (Phillips *et al.*, 2003), supporting earlier data from detrital micas (Kelley and Bluck, 1989) indicating an older source. Phillips *et al.* (2003) interpreted these data as indicating a source block derived from the Avalonian margin that may now be buried beneath later strata, a block that was termed 'Novantia' by Armstrong and Owen (2001). The presence of southerly-derived detritus compositionally similar to the basic clast petrofacies, the Galdenoch petrofacies, in tract II in Scotland suggests limited transcurrent displacement on the Northern Belt Median Fault. This is supported by structural indicators of mainly dip-slip displacement in the Longford-Down area (Morris, 1987). This Galdenoch petrofacies is also recognised in rocks of the Gowna sector (Tract III) of the Longford Down (Morris, 1979; 1987).

Diachronous trends are evident in the first arrivals of northern belt clastic suites. The arrival of the metaclast petrofacies progressively youngs along strike to the north-east from Longford to Abington in the Southern Uplands as well as younging across strike to the SE (Morris, 1987). The basic clast petrofacies also defines a north-east younging trend along strike, although with greater complexity. Lack of continuity of minor petrofacies along strike probably indicates lateral impersistence of source areas.

Spilitic volcanic rocks in the Longford-Down area range from probable late Llanvirn on the east coast at Ballygrot (Sharpe, 1970; Craig, 1984) to possibly Caradoc age in Longford (Morris, 1983) and geochemical studies indicate an affinity with tholeitic ocean-floor basalts (Morris, 1979; Craig, 1984). Stratiform, manganiferous, iron-rich mudrocks are spatially associated with some of these spilites, e.g. Coronea Formation (Morris, 1984) and appear to range in age from Arenig (in Scotland) to Caradoc. There is a suggestion from

Scotland that there may be an hiatus between mudrocks and cherts of Arenig age and those of late Llanvirn or earliest Caradoc age that are present beneath the typical 'Moffat Shale' facies (Armstrong *et al.*, 2002), but there is insufficient exposure to test whether a similar situation exists in Ireland.

Central Belt

The northern and central belts are separated by an important fault zone which has been traced throughout the Longford-Down–Southern Uplands belt (the Northern–Central Belts Boundary Fault of Morris, 1987; the Orlock Bridge Fault of Anderson and Oliver, 1986). The zone of fault-associated deformation varies from a few metres to >1km at Slieve Glah, County Cavan. The fault zone can be traced westwards through the Slieve Aughty inlier (Pracht *et al.*, 2004) where it separates Upper Ordovician mudrocks and greywackes (Toberlatan and Gortnagleav formations) in the north from Upper Ordovician mudrocks and bipartite mafic–intermediate volcanic rocks (Caher Hill Formation) and Silurian greywackes to the south (Emo and Smith, 1978). Undated, deformed volcanogenic turbidites penetrated in a borehole at Tynagh, just north of the putative surface trace of this fault, also compare well with northern belt basic clast petrofacies. Moreover, undated spilites intersected beneath the Carboniferous at Craughwell also suggest a comparison with Northern Belt volcanic rocks (J.H. Morris and G.J.H. Oliver *pers. comm.*, 1989). The fault differs from other tract defining faults in its clear evidence of ductile, quasi-plastic deformation at depth (Anderson and Oliver, 1986) and in its uniquely large displacement.

In contrast to the important coarse-grained clastic sequences in the northern belt, Ordovician rocks in the central belt are mainly pelagic or hemi-pelagic mudrocks (Fig. 3.14). Such sequences, termed the Moffat Shales in Scotland, are relatively thin and usually graptolitic, giving good biostratigraphical control. The oldest strata recorded to date in Ireland are Llanvirn mudrocks and cherts near Acton, County Armagh (Institute of Geological Sciences, 1975). The succeeding coarse clastic rocks are mainly of Llandovery age at their base, and become progressively younger across strike to the south-east. An exception may be some of the greywackes in the Monaghan area, which on field relations appear to underlie *acuminatus* Biozone black shales, and thus could extend down into the Ordovician (Morris *et al.*, 1986). These central belt greywackes are characterised by an abundance of granitic detritus and in the earliest 'pyroxenous petrofacies' (Morris, 1987) by the addition of andesitic debris. This latter detritus is readily comparable with the basic clast petrofacies of the northern belt, but in contrast is interpreted as having a north-westerly source. The provenance bipolarity across the Northern–Central Belts Boundary Fault strongly suggests that at this time a volcanic arc source was situated between the northern and central belts (Morris, 1987; Murphy and Hutton, 1986; Fig. 3.16). The age difference in these arc-derived clastic aprons could be explained either by primary distribution

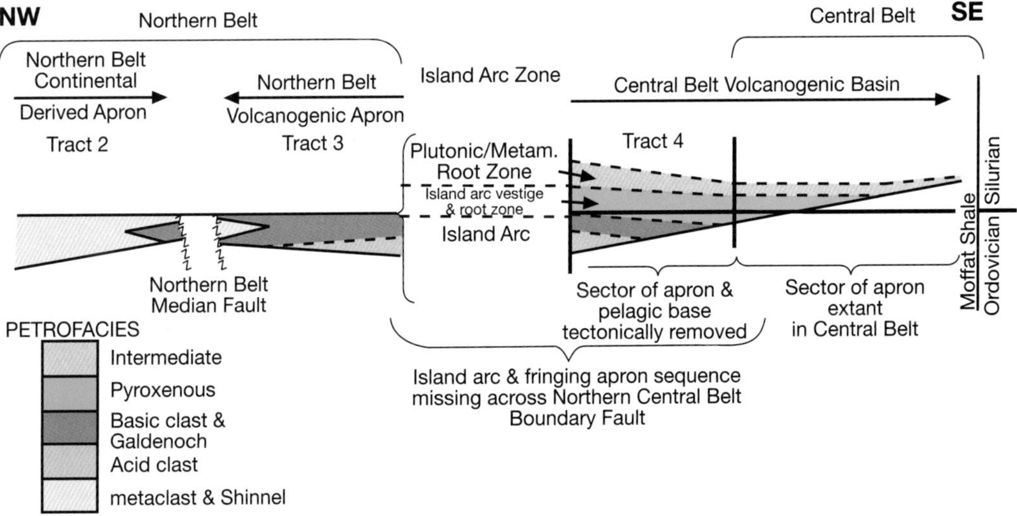

Fig. 3.16 Interpretation of the petrofacies distributions and provenance directions that suggests the former presence of an arc between the North and Central belts.

patterns, or by tectonic removal of part of the southern apron, the preferred interpretation in Fig. 3.16. Evidence from farther south in the Central Belt (Anderson and Rickards, 2000) suggests that pelagic Moffat Shale facies continued throughout the late Ordovician and into the Llandovery (*crispus* Biozone) before the arrival of turbiditic sandstones.

South Connemara

A sequence of volcanic and sedimentary rocks is exposed south of the Dalradian rocks of Connemara and separated from them by the Skird Rocks Fault (Fig. 3.17). Many workers have suggested that this fault represents a continuation of the Southern Uplands Fault (e.g. Leake, 1963; Max and Ryan, 1975; McKerrow, 1986; Hutton, 1987). These rocks have generally been assumed to be Ordovician in age although they were initially only constrained by their presence as a large roof pendant in the c.400Ma Galway granite. Palaeontological data, whilst still imprecise, has now confirmed an Ordovician age (Williams *et al.*, 1988). Thus this South Connemara Group has special importance in constraining any possible continuation of major geological lineaments from Scotland and eastern Ireland to the west coast.

Ffrench and Williams (1984) erected a stratigraphy comprising five formations and totalling >3km in thickness, with extrusive basaltic rocks forming the lowest and highest units. Immobile trace element data from the basalts of both formations suggest that they are ocean floor basalts, possibly formed in a marginal basin (Ryan, Max, and Kelly, 1983). This stratigraphy has more recently been reinterpreted by Ryan and Dewey (2004) who recognised a simpler threefold stratigraphy with many tectonic contacts producing structural repetition. The two stratigraphies are compared in Figure 3.17. As well as recognising three formations, Ryan and Dewey (2004) recognised numerous mélange zones, and it is from one of these that the productive microfossil sample of Williams *et al.* (1988) was obtained. However, the lithology is clearly comparable with the Golam Chert Formation (Fig. 3.17). Age assignation is based largely on a smooth chitinozoan cf. *Lagenochitina esthinica* Eisenack (Williams *et al.*, 1988; Harper and Parkes, 2000a), which is taken to indicate a Middle Ordovician (Arenig-Llanvirn) age. The lower parts of the *in situ* cherts

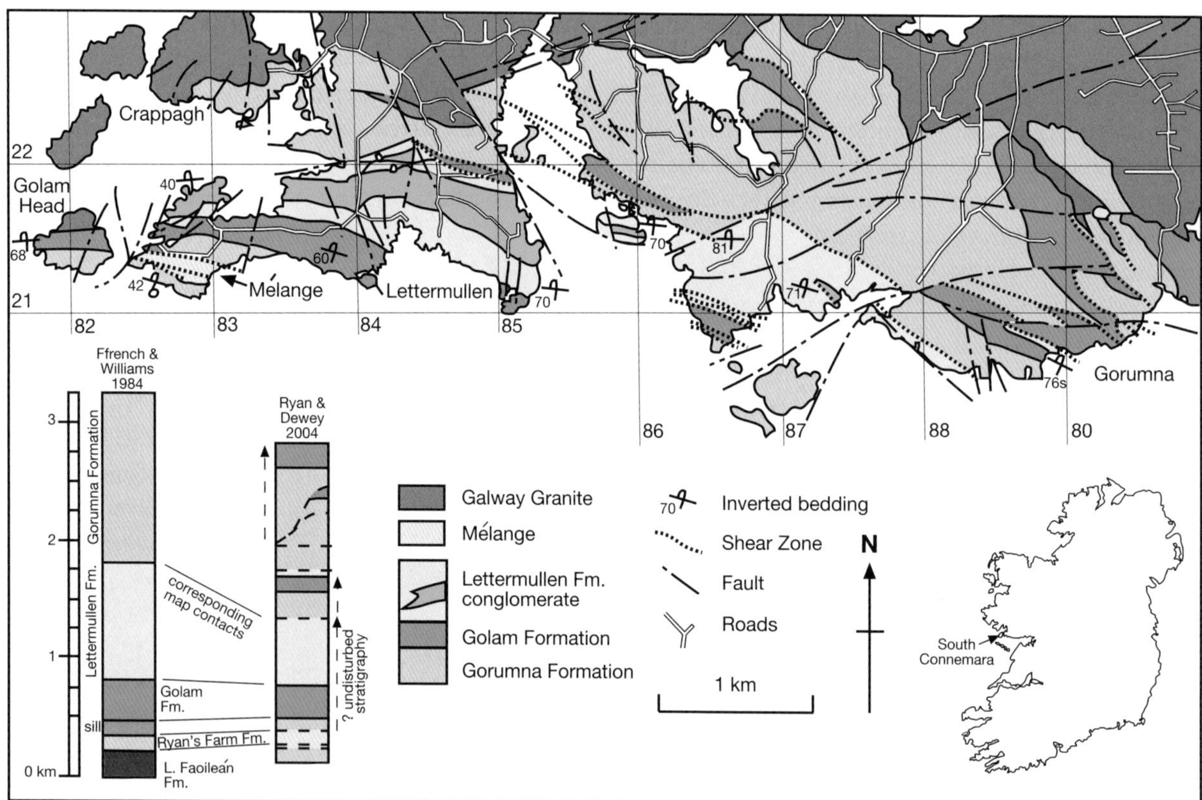

Fig. 3.17 Geological map of the South Connemara Group after Ryan and Dewey (2004) and a comparison to the earlier stratigraphy of Ffrench and Williams (1984).

of the Golam Formation that rest directly on pillow lavas are highly manganiferous, and are interpreted as deep sea, pelagic deposits by Ryan and Dewey (2004). These authors emphasise that the cherts may represent a considerable interval of time, making it difficult to ascertain the age of both the volcanic rocks and the overlying clastic sediments.

The younger, coarser-grained Lettermullen Formation that overlies the cherts comprises boulder, cobble, and pebble conglomerates with subordinate coarse sandstones and finer-grained lithologies. The maximum clast size recorded is 1.8m (Ffrench and Williams, 1984) and sedimentary structures indicate deposition from sediment gravity flows. These coarser lithologies are thought to represent deposition in, or proximal to, submarine channels. The predominance of acid igneous and metamorphic clasts in the conglomerates indicates derivation from a continental source area. The association of foliated granitoids, migmatites and metagabbros, particularly in the lower part of the Lettermullen Formation, has been used by Ryan and Dewey (2004) to suggest derivation from Connemara to the north and, by inference, an age younger than the Grampian orogeny (47–463Ma, Friedrich *et al.*, 1999). This orogeny has been interpreted as very short and accompanied by rapid exhumation to bring these rocks to the surface (Dewey, 2005; Chapter 4).

Correlation of the South Connemara Group has generally been made with northern belt rocks of the Longford-Down area and specifically with parts of the Strokestown–Arva tract (Tract II) (Phillips *et al.*, 1976; Morris, 1983; Ffrench and Williams, 1984; Williams *et al.*, 1988; Ryan and Dewey, 2004). If the preliminary age determinations are confirmed and this correlation accepted, then the diachronous pattern of first arrival of coarse clastic debris noted for the Longford-Down–Southern Uplands would continue westwards (Williams *et al.*, 1988). It would also imply that Tract II is the northernmost tract in western Ireland, as these rocks abut directly against the Skird Rocks Fault, a probable equivalent of the Southern Uplands Fault. It would also provide a clear linkage between Northern Belt rocks and the Dalradian. In Scotland Bluck (2000) has argued that derivation from the Midland Valley block seems more reasonable for proximal Northern Belt conglomerates, although Oliver *et al.* (2000) present a garnet/whole-rock date of 468.7±1.5Ma from detrital garnets in the Northern Belt that implies derivation from the Grampian terrane.

Palaeogeography

Attempting to reconstruct the palaeogeography of the Ordovician of Ireland is possible only in first approximation. There is clear evidence for both obduction and subduction of oceanic crust that has resulted in major, often unquantifiable, changes in spatial arrangements of the rocks. It has been noted above that graptolites from north-west Ireland are more easily placed in biozones developed from successions in Australia and America than in those developed largely in Wales (Skevington, 1974). In contrast, graptolites from south-east Ireland are more easily referred to the British biozonal scheme. Contrasts in shelly faunas also exist (Williams, 1969; 1973) which have been summarised by Fortey and Cocks (1988), Harper and Parkes (1989), Harper (1992), Parkes and Harper (1996), and Harper *et al.* (1996). Lower Ordovician shelly faunas from South Mayo are closely comparable to the Toquima–Table Head faunas found on the north-west shelf of the Iapetus Ocean in North America. These are low latitude warm water faunas and published palaeomagnetic data also indicate low palaeolatitudes (11–19°S) (MacNiocaill *et al.*, 1997).

There is now abundant evidence for an extensive arc-marginal basin complex situated near the Laurentian margin of the Iapetus Ocean in the early Ordovician. Palaeomagnetic data give palaeolatitudes whose range overlaps with those of the Laurentian craton (MacNiocaill *et al.*, 1997). The Ordovician rocks of South Mayo and Tyrone are part of this complex, which can be readily traced into Newfoundland along strike (S.H. Williams *et al.*, 1996). It is thought that the Grampian/Taconic orogeny resulted from the collision of this arc with Laurentia and these events are discussed in detail in Chapter 4. Questions such as what distances were involved in the volcanic arc–trench systems rely largely on comparison with modern analogues. Additionally, to what extent the present disposition of Irish Ordovician rock units reflects their original geometry and to what extent it has been modified by strike-slip displacements is also highly debatable. In some cases structural data on the sense of lateral displacements across faults exist, but generally there is little constraint on the magnitude of these displacements.

Whilst the deepwater nature of the rocks preserved in the Longford-Down terrane is not in doubt, it has been noted above that several models have been advanced to explain their development. These models must also take into account their structural history, and thus are discussed further in Chapter 8.

4

Grampian Orogeny

D. M. Chew

Introduction

The Caledonian orogenic cycle

During the latest Precambrian a wide ocean (termed the Iapetus Ocean) separated the Laurentian continent (Laurentia), which included north-west Ireland, from the Avalonian microcontinent (Avalonia), which included south-east Ireland. The early history of the Iapetus Ocean is discussed in detail in Chapter 2. During the Cambrian, this ocean started to close. A series of tectonic events, ranging in time from the Cambrian to the Devonian, were associated with the ocean's closure. Combined, this series of tectonic events has been termed the Caledonian Orogeny (McKerrow *et al.*, 2000), though here the term *Caledonian orogenic cycle* is used instead.

In Ireland, the Caledonian orogenic cycle comprises two main phases. These are an Early to Middle Ordovician (475–460 Ma) phase, termed the Grampian Orogeny, and an Early Devonian (405 Ma) phase, termed the Acadian Orogeny. The Grampian Orogeny was caused by the collision of the Laurentian continental margin of north-west Ireland with an oceanic arc terrane (Fig. 4.1a). This arc was intra-oceanic, and was produced by the subduction of oceanic crust within the Iapetus Ocean. Following the collision, subduction of oceanic crust continued. The Laurentian margin of north-west Ireland progressively approached the Avalonian microcontinent in south-east Ireland (Fig. 4.1b). By the Late Silurian, all the intervening oceanic crust had been subducted, and a continent–continent collision ensued (Fig. 4.1c). The boundary between Laurentia and Avalonia is termed the Iapetus Suture. The Acadian Orogeny is the final (Early Devonian) stage of this collision. It is punctuated in Ireland by post-orogenic clastic rocks of Middle Devonian age (or younger) resting unconformably on folded cleaved rocks which range in age from Cambrian to Silurian. The Acadian Orogeny is considered in detail in Chapter 8 (Late Caledonian tectonism and magmatism).

The Grampian Orogeny

Lambert and McKerrow (1976) recognised that the Dalradian sequences of the Scottish Highlands had undergone polyphase deformation and metamorphism during the Ordovician. This phase of orogenic activity clearly predated the post-Silurian deformation, also called 'Caledonian', seen elsewhere in Britain and Ireland. They coined the term 'Grampian Orogeny' to distinguish the Ordovician tectonic event.

The Grampian Orogeny is now widely regarded as having resulted from the collision of Laurentia with an oceanic arc during the Arenig (Dewey and Shackleton, 1984). It is broadly equivalent to the Taconic Orogeny of the Appalachians and the Humberian Orogeny of western Newfoundland (McKerrow *et al.*, 2000; Dewey and Mange, 1999; Dewey and Shackleton, 1984).

Figure 4.2 illustrates the rocks that are affected by the Grampian Orogeny in Ireland. They are restricted to the north-western parts of the country, i.e. north-west of the Iapetus Suture. They include rocks that were deposited on the passive margin of Laurentia, in particular the Neoproterozoic to Cambrian Dalradian Supergroup (Chapter 2), and rocks associated with the colliding Early Ordovician arc, such as the volcanics of South Mayo and the volcanics and plutons of the Tyrone Igneous Complex (Chapter 3). Also caught up in the collision (i.e. interleaved between the Dalradian sediments on the Laurentian margin and the colliding arc) were a series of Precambrian metasedimentary rocks including those of the Slishwood Division and Tyrone Central Inlier (Chapter 2) and the Early Ordovician turbidites and volcanics of the Clew Bay Complex (Chapter 3). The boundary between the rocks on the Laurentian margin and those of the colliding arc terrane is marked by the Fair Head–Clew Bay Line (Fig. 4.2). It is the continuation of the Highland Boundary Fault of Scotland into Ireland, and coincides with a strong magnetic lineation extending from Fair Head to the northern coast of Clew Bay (Max and Riddihough, 1975). The surface expression

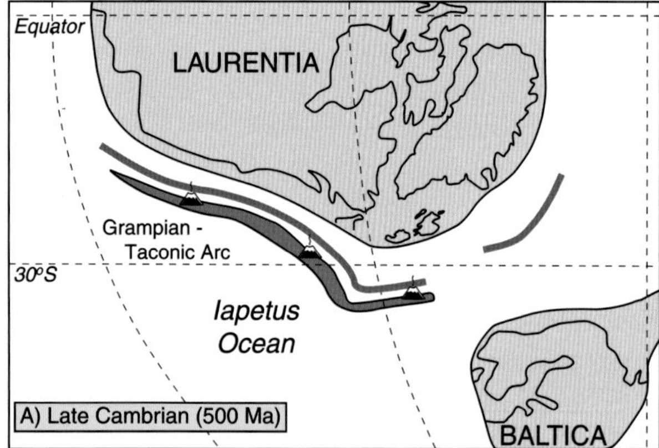

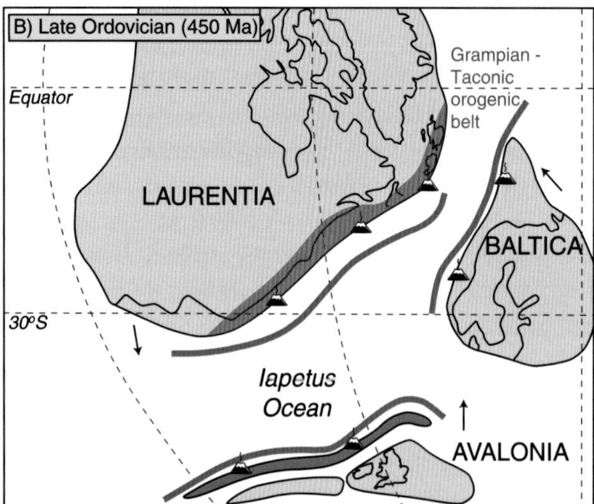

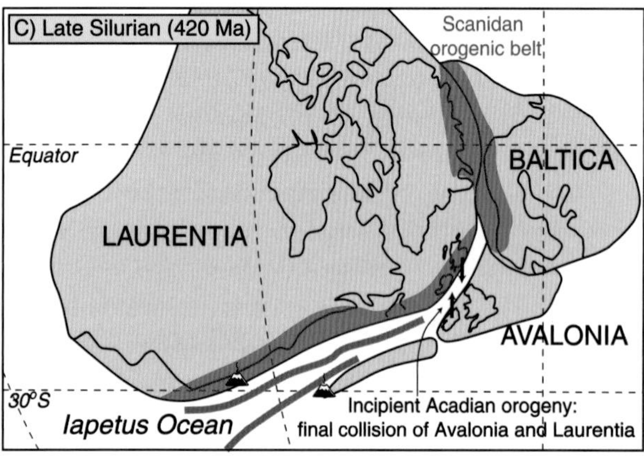

Fig. 4.1 Schematic tectonic evolution of the Caledonian orogenic cycle (the closure of the Iapetus Ocean), showing major orogenic events (e.g. the Grampian and Acadian). Volcanic arcs are shown in green; trenches are shown in blue and indicate the polarity of subduction; collisional orogens are shown in red. A) south-directed subduction creates a volcanic arc within the Iapetus Ocean outboard of Laurentia. B) This arc collides with Laurentia causing the Grampian Orogeny, and north-directed subduction under Laurentia begins, contemporaneous with south-directed subduction beneath Avalonia. C) The Iapetus Ocean has nearly closed. The 'head-on' collision of Baltica and Laurentia causes the Scandian Orogeny, while the highly oblique collision between Laurentia and Avalonia causes the Acadian Orogeny.

of this lineament is commonly a fault, as is particularly evident in the Clew Bay region (Fig. 4.2).

This review of the Grampian Orogeny in Ireland considers the structural, metamorphic and magmatic evolution of each of the constituent elements listed above, with the Grampian evolution of the Dalradian initially presented in a condensed form. There then follows a review of the evidence for the timing of the Grampian Orogeny, and a summary of the tectonic evolution of the Irish segment of the Laurentian margin. At the end of the chapter, supplemental information explains the shorthand terminology (D1, S2, F3 etc) used in describing the structural and metamorphic history, and a detailed synthesis of each separate area of Dalradian rocks is presented in turn.

Grampian evolution of the Dalradian Supergroup

Structure

The Dalradian rocks of Ireland and Scotland experienced multiple episodes of deformation during the Grampian Orogeny. Each deformation event produced its own characteristic folds, schistosities and lineations, resulting in a complex cumulative set of structures. In Ireland, the Dalradian outcrop is fragmented into five inliers: NW Mayo, Donegal and the Sperrin Mountains, the Central Ox Mountains, Connemara, and NE Antrim. In view of the difficulties in correlating deformation events from inlier to inlier, deformational and metamorphic chronologies presented in this synthesis are local; i.e. they relate exclusively to the inlier being discussed and no regional correlation is implied.

The structural history of the Scottish Dalradian was comprehensively studied during the last century and, because of the more complete exposure in Scotland, it provides an extremely useful template with which to compare the Irish Dalradian. Four main deformational episodes (D1 to D4) have been identified in the Scottish Dalradian (e.g. Harris *et al.*, 1976). The structure is dominated by a large, recumbent, south-eastwards-vergent antiform, the D2 Tay Nappe. In general, this flat-lying structure has been folded downwards adjacent to the Highland Boundary Fault (Fig. 4.3a) in a D4 monoform termed the Highland Border Downbend. The core of the

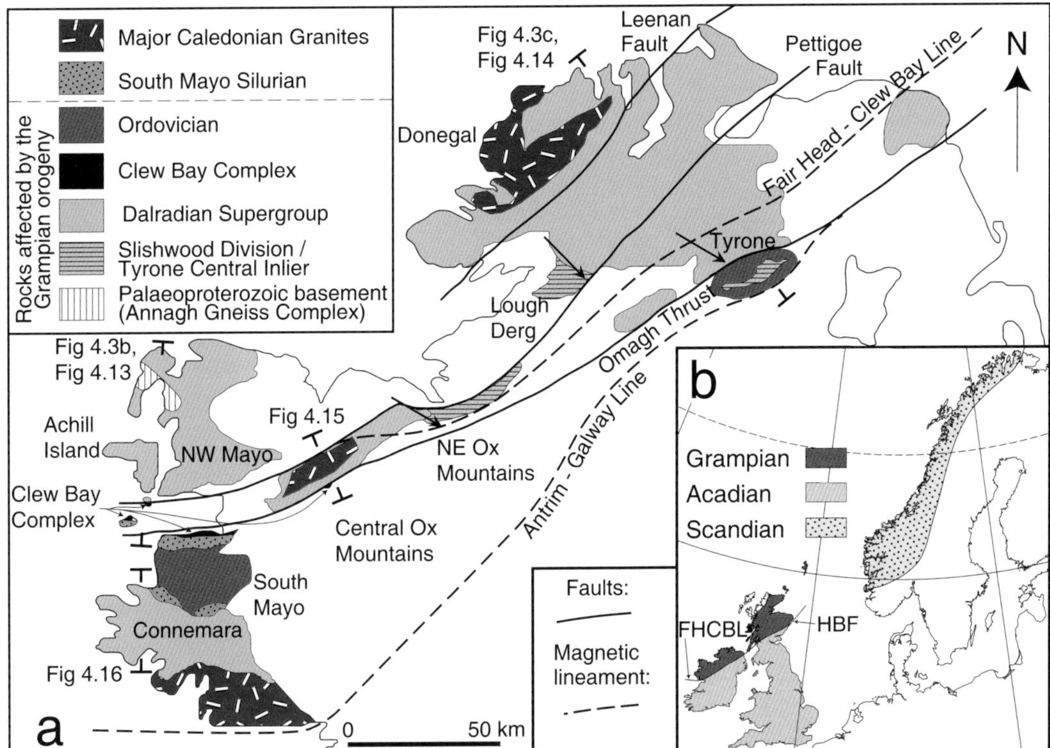

Above: Fig. 4.2 Geological map of NW Ireland showing the major structural features and Precambrian and Lower Palaeozoic inliers discussed in the text. Arrows indicate the transport direction of major nappes close to the Fair Head–Clew Bay Line.

Right: Fig. 4.3 Schematic structural sections through the Grampian Belt in Scotland, NW Mayo and Donegal. Adapted from Strachan (2000a) and Chew (2003).

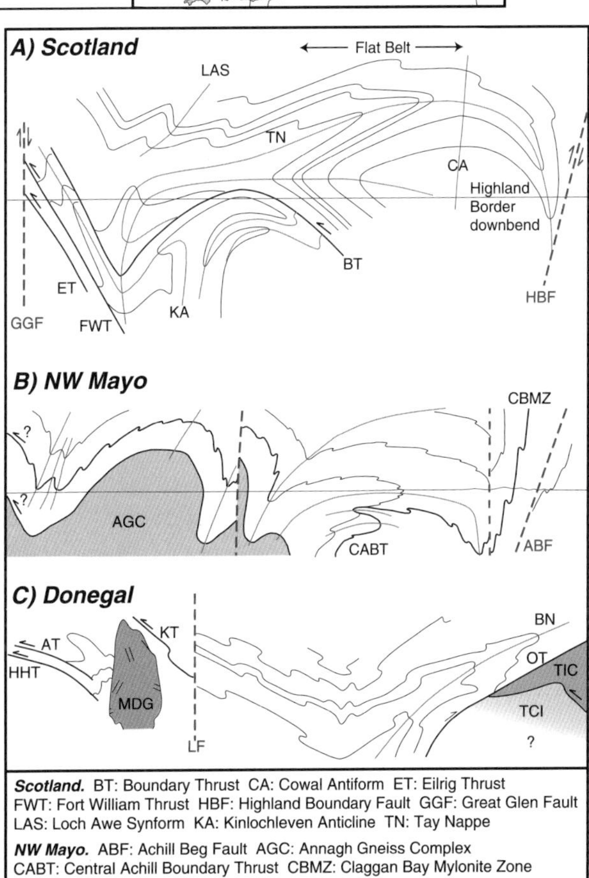

Tay Nappe is exposed solely in the Loch Awe Syncline of the SW Highlands (Fig. 4.3a). Across this steep belt there is a zone of primary facing divergence, with the Islay Anticline verging to the north-west and the Tay Nappe verging to the south-east. This has been interpreted as a root zone to the major nappe structures (Roberts and Treagus, 1977). D4 deformation, which caused the downbend (Johnson, 1991), resulted in upright folds and an associated strong crenulation cleavage close to the Highland Boundary Fault.

The two Dalradian inliers in Ireland that offer the longest transects orthogonal to orogenic strike are the NW Mayo and Donegal inliers. The broad structure of the NW Mayo Inlier (Fig. 4.3b) is similar to that of the SW Highlands of Scotland. There is a series of early bedding-parallel shear zones (commonly referred to as 'slides' in the literature). These are folded by later (F2) folds. There is a primary facing divergence of the F2 folds on either side of a basement core, the Annagh Gneiss Complex. To the north of this 'root zone' shallowly inclined F2 folds face north; to the south, recumbent F2 folds face south (Chew, 2003). Approaching the Achill Beg Fault, the south-facing F2 antiform is rotated into a downward-facing orientation (Fig. 4.3b), analogous to the Highland Border Downbend.

The Dalradian rocks of Donegal show evidence of three phases of Grampian deformation. F1 folds are best seen on the Inishowen peninsula, where there is a primary facing divergence across the Inishowen Syncline. The Inishowen syncline is correlated with the Loch Awe Syncline of Scotland (Hutton and Alsop, 1996). D2 in north Donegal is associated with the development of north-west-directed thrust nappes (Fig. 4.3c). These thrust nappes are separated from each other by D2 slides. Within each nappe, recumbent F2 folds face north-west (Hutton and Alsop, 1995). Further south, to the south-east of the Leannan Fault, isoclinal, recumbent F2 folds, such as the Sperrin Nappe, were transported towards the south-east (Alsop, 1996). Major recumbent F3 nappes (such as the Ballybofey Nappe) show a consistent vergence and sense of movement to the south-east (Alsop, 1996). During this D3 phase of south-east-directed shearing, the locally-inverted Dalradian succession was thrust over the colliding arc terrane represented by the Slishwood Division of the Lough Derg inlier (Alsop, 1991), and the Ordovician Volcanics of the Tyrone Inlier (Alsop and Hutton, 1993).

In summary, the Grampian structural evolution of the NW Mayo and Donegal inliers shares a similar structural evolution to that of the Scottish Dalradian (Fig. 4.3). Early structures, such as thrust nappes and 'slides' commonly face to the north-west (i.e. away from the Fair Head–Clew Bay Line) and also appear to have been transported in this direction. This trend is particularly evident in the northern portions of the NW Mayo and Donegal inliers. Passing southwards towards the Fair Head–Clew Bay Line, the structures become more upright and the facing of the main fold nappes changes. Close to the Fair Head–Clew Bay Line, the facing directions of the main phase of folding is towards the south-east in both inliers, and there is also clear evidence of nappe transport in this direction, over the colliding arc terrane.

Metamorphism

The Dalradian outcrop in Scotland is one of the classic areas for the study of regional metamorphism. Barrow (1893), working in the SE Highlands of Scotland, was the first to show that differing mineral assemblages in pelitic rocks reflect different conditions of metamorphism. In addition to the Barrovian (medium-pressure regional metamorphism) zonal scheme resulting from Barrow's work, Read (1923) described the Buchan zonal scheme in the NE Highlands of Scotland, which represents conditions of low-pressure, high temperature metamorphism.

The metamorphic evolution of the Irish Dalradian is dominated by the development of Barrovian metamorphic assemblages, although the metamorphic grade varies from place to place significantly. In the NW Mayo Inlier, the metamorphic grade is highest closest to the basement core (the Annagh Gneiss Complex) where it locally reaches the sillimanite zone (Max *et al.*, 1983) and post-dates the development of the main folds (locally F2 in age). The metamorphic grade decreases to the south towards Clew Bay, where the timing of porphyroblast growth is earlier, post-dating the development of F1 folds. The metamorphic evolution of the Dalradian of the Central Ox Mountains Inlier is similar. Here the metamorphic grade decreases from the staurolite–kyanite zone in the NW to greenschist facies in the SE, and post-dates the development of the main folds (which are F3 in age in the Central Ox Mountains Inlier). The metamorphism in Donegal reached the garnet zone over most of the inlier (Pitcher and Berger, 1972) and usually post-dates the development of F2 folds. The metamorphic grade increases approaching the Lough Derg inlier to the south, as progressively deeper structural levels are exposed through the inverted lower limb of the Ballybofey Nappe (Alsop, 1991). Here the metamorphism is in the staurolite–kyanite zone and is structurally later, post-dating the development of F3 folds.

However, there are two places in the Dalradian of Ireland where Barrovian metamorphic conditions are not encountered. Firstly, adjacent to the voluminous basic and intermediate Ordovician intrusions of south Connemara, migmatitic melting of metasediments has taken place under low-pressure/high temperature conditions, analogous to the Buchan metamorphism of Scotland. This metamorphic event (which postdates the third phase of deformation in Connemara) overprints an earlier phase of regional metamorphism, which was probably in the staurolite–kyanite zone over much of the Connemara Inlier (Yardley, 1976; 1980). The second exception to the prevalence of Barrovian metamorphic conditions is encountered in the southern portion of the NW Mayo Inlier. On southern Achill Island adjacent to the Fair Head–Clew Bay Line, blueschist-facies metamorphism (indicative of high pressure, low temperature metamorphism) occurred (Gray and Yardley, 1979). The blueschist-facies assemblages developed contemporaneously with Barrovian metamorphic assemblages in the Dalradian further to the north (Chew *et al.*, 2003). The tectonic significance of the Buchan metamorphism of Connemara and the blueschist-facies metamorphism of southern Achill Island will be discussed later.

Magmatism

Magmatism associated with the Grampian Orogeny (475–460 Ma) in the Dalradian of Ireland is apparently restricted to Connemara. All intrusions in NW Mayo, Donegal and the Central Ox Mountains either pre-date or post-date the Grampian Orogeny. Examples include the pre-Grampian (*c.*600 Ma) rift-related basic magmatism in the Dalradian, associated with continental breakup and the formation of the Iapetus Ocean (Chapter 2), and the post-Grampian suite of Late Caledonian granites (Chapter 8).

In Connemara and NE Scotland large volumes of basic and intermediate magma, which were injected into the Dalradian rocks during the Grampian Orogeny, are interpreted as representing the roots of a continental volcanic arc (e.g. Yardley and Senior, 1982; Tanner, 1990). The earliest syn-orogenic intrusions of Connemara are basic (gabbroic), and there is a clear progression with time to more acid magma types (Leake, 1958). The timing of emplacement of a small basic suite in north Connemara is thought to be syn-D2, but immediately prior to D3 (Wellings, 1998), although Tanner (1990) argues for an early D3 emplacement for all the gabbros of Connemara. U–Pb single grain zircon analyses of the Cashel–Lough Whelan gabbro give an age of 470.1 ± 1.4 Ma (Friedrich *et al.*, 1999a). A mafic pegmatite of the Currywongaun intrusion in north Connemara gives a U–Pb zircon age of 474.5 ± 1 Ma (Friedrich *et al.*, 1999a).

Discriminating between the later intermediate to acid magma types and the adjacent migmatised metasedimentary country rock in the field is problematical (e.g. Cliff *et al.*, 1996). The intermediate to acid magma types are generally regarded as being late D3 in age (Leake and Tanner, 1994). Quartz-diorite gneisses in the migmatite zone in south Connemara yield U–Pb zircon ages of 463 ± 4 Ma (Cliff *et al.*, 1996) and 467 ± 2 Ma (Friedrich *et al.*, 1999b). Coeval anatexis (melting) is suggested by *c.*467–468 Ma pegmatite and leucosome formation in the migmatites (Friedrich *et al.*, 1999b). The post-D4 Oughterard granite in Connemara is seen to cut minor F4 folds and the F4 Connemara Antiform, and has yielded a 462.5 ± 1.2 Ma U–Pb xenotime age (Friedrich *et al.*, 1999a). The Oughterard granite postdates all Grampian deformation in Connemara and is hence an important time constraint in the evolution of the Grampian Orogeny in Ireland. Metamorphic mineral ages from the Dalradian metasediments will be discussed in a later section discussing the timing of orogeny.

Grampian evolution of the Clew Bay Complex

The Clew Bay Complex comprises a series of low-grade turbiditic metasediments, and higher grade metamorphosed basic and ultrabasic rocks. The turbiditic portions of the Clew Bay Complex include the Ballytoohy and Portruckagh Formations on Clare Island, the South Achill Beg Formation on the island of Achill Beg and the Killadangan Formation on the south coast of Clew Bay (Fig. 4.4). The Westport Grit Formation (Fig. 4.4) and Cloonygowan Formation in the Ox Mountains (indicated by an arrow in Fig. 4.2) are potential correlatives. The higher-grade serpentinite, amphibolite and metasediment of the Deerpark Complex (Phillips, 1973) which outcrops in the Kill Inlier on Clare Island and on the south coast of Clew Bay (Fig. 4.4) are also regarded as part of the Clew Bay Complex (D.M. Williams *et al.*, 1994, 1996). The Clew Bay Complex as a whole shares many similarities with the Highland Border Complex of Scotland. Stratigraphic and lithological descriptions of the various units are found in Chapter 3.

The two components of the Clew Bay Complex, the low-grade turbiditic metasediments and the higher grade metamorphosed basic and ultrabasic rocks, are always found in tectonic contact, and the contact between the Clew Bay Complex and the Dalradian is also tectonic (the Achill Beg Fault of Fig. 4.4). Hence the original relationship between these units is somewhat problematic. However, the ophiolitic affinity of the high grade metabasites and serpentinites (Ryan *et al.*, 1983b) means that the Clew Bay Complex figures prominently in tectonic models of the Grampian Orogeny (e.g. Dewey and Mange, 1999), with the low-grade turbidites interpreted as an accretionary complex and the high-grade metabasites and serpentinites representing a supra-subduction ophiolite (e.g. Fig. 4.11b).

Structure

Phillips (1973) recognised five distinct deformation phases affecting the Ballytoohy Formation on north

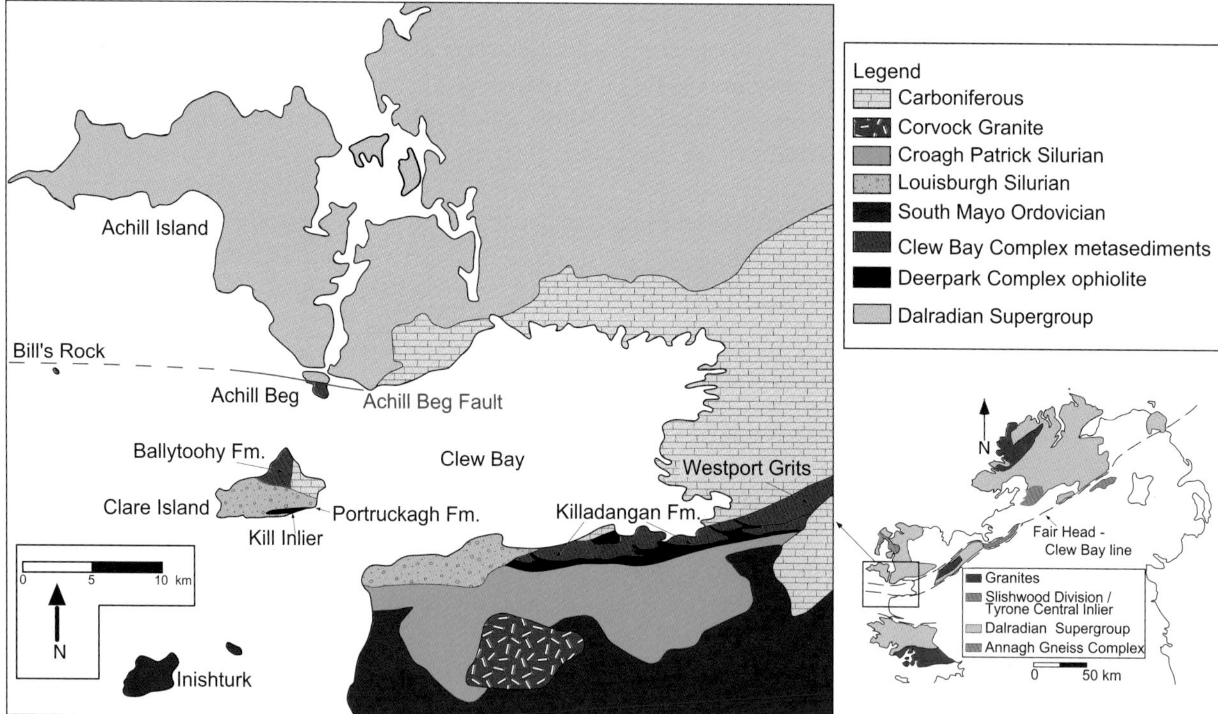

Fig. 4.4 Geological Map of the Clew Bay region illustrating the components of the Clew Bay Complex.

Clare Island (Fig. 4.4). The large-scale distribution of the succession is determined by recumbent, isoclinal F2 folds which fold an earlier S1 cleavage. These are modified by steeply inclined F4 folds; D3 and D5 effects are relatively minor. Phillips (1973) considered that the Ballytoohy Formation experienced the full deformation sequence that affected the North Mayo Dalradian, excepting the minor D5 structures. Jones (1989) concluded that the Ox Mountains Dalradian and the adjacent Cloonygowan Formation had shared the same deformation history. S1 in both units is very poorly preserved, occurring in the hinges of rare F2 folds. These F2 folds are isoclinal with a strong penetrative fabric; quartz pebbles are strongly flattened along the S2 foliation. Abundant F3 folds are open to isoclinal and are associated with an intense fabric. Jones (1989) also regarded the structural history of the Ardvarney and Westport Grit formations as closely resembling that of the Cloonygowan Formation. Chew (2003) concluded that the Dalradian and the Clew Bay Complex on the island of Achill Beg (Fig. 4.4) share the same structural history across the Achill Beg Fault. A D1 high-strain event is common to both units, and is associated with the development of tectonic slides. The D2 event is responsible for the formation of crustal-scale nappes, and in both units, beds are consistently downward-facing on the S2 foliation. The above structural studies all suggest contemporaneous deformation of the Dalradian and the Clew Bay Complex during a mid-Ordovician Grampian orogenic event. Tanner and Sutherland (2007), in a revised model for the tectonic evolution of Highland Border Complex in Scotland, proposed that the majority of the units in the Highland Border and Clew Bay complexes should be regarded as the youngest portions of the Dalradian Supergroup. The only truly allochthonous units in these two complexes in the model of Tanner and Sutherland (2007) are the ophiolitic rocks of the Highland Border ophiolite and the Deerpark Complex.

An alternative interpretation of the structure of parts of the Clew Bay Complex is given by Max (1989), who reinterpreted the disposition of units within the Ballytoohy Formation to be largely tectono-sedimentary in origin, with blocks of greywacke up to 8 m across floating in a black mudstone matrix. Two tectonic fabrics superimposed on 'these often chaotic mélanges' were inferred. Williams *et al.* (1994) interpreted the whole of the Ballytoohy Formation as a mélange. The large, inland exposures of greywacke were regarded as mega-blocks (up to 500 m long) enclosed within the mélange matrix, and it was suggested that these constituted the stratigraphy of Phillips (1973). The Killadangan Formation on the south side of Clew Bay was also reinterpreted as a mélange. The mélange matrix has yielded trilete miospores, interpreted to be of Silurian (Wenlock) age (Williams *et al.*, 1994; D.M. Williams *et al.*, 1996). These microfossil discoveries have proved contentious as they preclude the involvement of the Clew Bay Complex in a mid-Ordovician Grampian Orogeny. A review of the contrasting structural models proposed for the Clew Bay Complex on Clare Island is found in Graham (2001b).

Metamorphism

There has been relatively little published work on the metamorphic rocks of the Clew Bay Complex. Phillips (1973) describes amphibolite-facies (hornblende, oligoclase ± epidote) metabasites tectonically interleaved with garnet-grade semipelitic schists (garnet, biotite, muscovite, and oligoclase) in the Deerpark Complex of Clare Island. Ryan *et al.* (1983b) report identical assemblages for similar lithologies on the south side of Clew Bay. Chew *et al.* (2006) infer P–T conditions of *c.*550°C, 3.3 kbar for the garnet-grade assemblages within the semipelitic schists on the south side of Clew Bay. Petrographic evidence demonstrates that these semipelitic schists have probably experienced the same metamorphic event as the amphibolite-facies metabasites. The lower-grade deep marine sediments of the Clew Bay Complex have experienced low greenschist-facies metamorphism. Phillips (1973) describes growth of fine-grained biotite and muscovite in pelitic rocks of the Ballytoohy Formation. The basic volcanics within the formation (the Tonaltatarrive Spilite Member) have experienced greenschist-facies metamorphism, with development of an assemblage of actinolite, chlorite and albite (Phillips, 1973). Pelitic lithologies in the Clew Bay Complex of South Achill Beg typically contain phengitic muscovite + chlorite + albite + quartz. Estimating peak metamorphic temperature in low greenschist facies assemblages such as these is problematical due to the difficulty in ensuring that the metamorphic phases present are in chemical equilibrium. Thermobarometric estimates utilising local equilibria of chlorite and phengite pairs (Chew *et al.*, 2003) yield P–T conditions *c.*325–400 °C, 10 kbar. Although the thermobarometric dataset is currently extremely limited, P–T estimates from the Clew Bay Complex support models (e.g. Dewey and Mange, 1999) where the low-grade turbidites are an accretionary complex (which would be expected to have experienced high-pressure, low temperature metamorphic conditions), and the high-grade metabasites and serpentinites are a supra-subduction ophiolite (which would have experienced high-temperature, low-pressure metamorphism within the metamorphic sole). Metamorphic mineral ages from the Clew Bay Complex will be discussed in the section on the timing of orogeny.

Magmatism

The meta-igneous rocks of the Deerpark Complex south of Clew Bay were interpreted by Ryan *et al.* (1983b) as a dismembered ophiolite. Ryan *et al.* (1983b) report metagabbro and metadolerite preserving ophitic texture with a primary igneous mineralogy of augite and plagioclase, with one outcrop of metadolerite even displaying the chiral chilled margin pattern typical of ophiolitic sheeted dykes. The chemical composition of the metadolerites suggests they are altered tholeiitic basalts, with a MORB-like chemistry. The CIPW-normative mineralogy of the serpentinites suggests they were originally dunites or harzburgites. Initial Nd isotopic compositions of the metadolerites confirm their juvenile character, with $\varepsilon_{Nd(480)}$ values of +6 (Chew *et al.*, 2007). Gabbroic rocks have also been identified within the Ballytoohy Formation on Clare Island, along with MORB lavas (with some geochemical evidence of an arc influence), umbers, pyrite-rich shales and deep water cherts (D.M. Williams *et al.*, 1997), which together are interpreted as indicating the presence of a fossilised spreading centre.

Grampian evolution of South Mayo

The South Mayo Trough (Chapter 3) is interpreted as a fore-arc basin which developed between the accretionary complex rocks of the Clew Bay Complex and the South Mayo volcanic arc (e.g. Dewey and Mange, 1999). Although many of the basal contacts are either not exposed or are faulted, the sediments in the basin were presumably deposited on the Early Ordovician arc volcanics of South Mayo. These include the Late Tremadoc Lough Nafooey Group, the Tourmakeady Volcanic Succession and the Bohaun Volcanic Succession of unknown age (Chapter 3). The basin fill of the South Mayo Trough (formally known as the Murrisk Group) consists of a series of predominantly deep marine volcaniclastic rocks and tuffs that shallows upwards (Chapter 3). The South Mayo Trough is of critical importance in understanding the evolution of the Grampian Orogeny. This is because the basin was not inverted during the arc–continent collision event, but was merely buckled into a large fold, commonly known as the Mweelrea Syncline. It preserves sediment received from the deforming and unroofing orogen, and also volcanic detritus from the South Mayo arc. Therefore both the detrital record of the basin fill and the chemistry of the arc volcanic rocks can be used to reconstruct the evolution of the orogeny.

Detrital record of the basin fill

The Murrisk Group shows several major changes in petrography, both with time and location (Chapter 3). Figure 4.5 illustrates how the heavy detrital minerals and bulk whole-rock geochemistry of sandstones of the northern limb of the South Mayo Trough changes with time. Whole-rock geochemistry demonstrates that the lower portions of the northern limb are derived from a source enriched in Mg, Cr, and Ni, indicative of an ultramafic (ophiolitic) source region (Wrafter and Graham, 1989). This prominent ultramafic signature decreases up sequence. The abundance of detrital chrome spinel follows a similar pattern (Dewey and Mange, 1999). The drop in detrital chrome spinel abundance coincides with a sudden influx of metamorphic detritus (garnet, staurolite, sillimanite and muscovite) near the top of the Lower Llanvirn Derrylea Formation (Fig. 4.5, Dewey and Mange, 1999). These data suggest the progressive unroofing of an ophiolite complex followed by the exhumation of the Grampian metamorphic belt (Wrafter and Graham, 1989; Dewey and Mange, 1999).

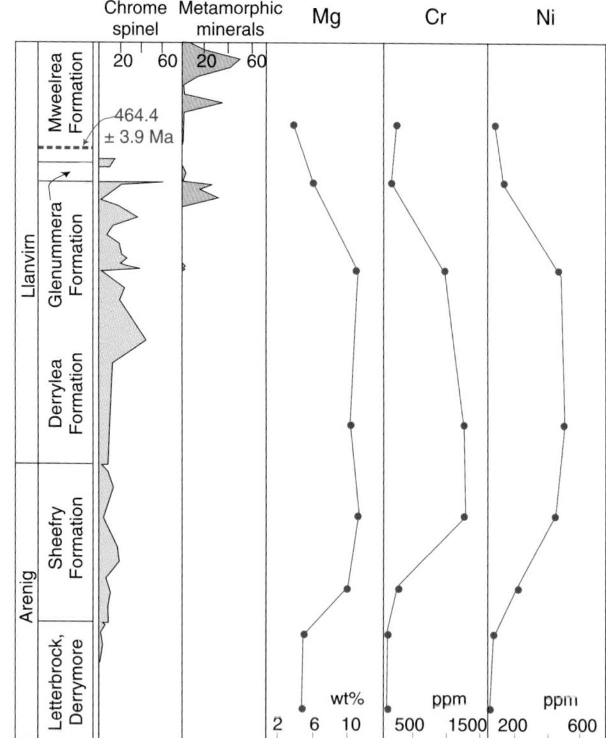

Fig. 4.5 Temporal evolution of detrital heavy mineral assemblages of sandstones from the northern limb of the South Mayo Trough. The percentage of each component (e.g. chrome spinel or metamorphic detritus) is illustrated in the histogram. Also shown is the whole rock geochemistry (Wrafter and Graham, 1989) and a U–Pb zircon age from an ignimbrite (Dewey and Mange, 1999).

Chemistry of the arc volcanics

The South Mayo arc volcanism spans the Grampian arc–continent collision event. Hence the chemistry of the arc volcanics can be used to constrain the onset of collision, much as the detrital heavy mineral record preserved in the basin fill was used above. The volcanics are described in Chapter 3. Figure 4.6 (*from* Draut *et al.*, 2004) illustrates how the chemistry of the arc volcanics changes with time. The oldest (pre-collisional) arc volcanics include the lower portions of the Late Tremadoc Lough Nafooey Group and the Bohaun Volcanic Succession (Fig. 4.6).

The age of basaltic lavas that comprise the Bohaun Volcanic Succession is not precisely known, but their boninitic chemistry (Clift and Ryan, 1994) and strongly positive $\varepsilon_{Nd(t)}$ values could indicate earliest-stage formation above a young subduction zone. The light rare earth element (LREE) depletion and strongly positive $\varepsilon_{Nd(t)}$ values of the tholeiitic basalts at the base of the Lough Nafooey Group and the lack of continental detritus in the oldest sediments of that group also suggest an origin far from the Laurentian margin (Ryan *et al.*, 1980), while younger volcanic units exhibit a trend toward higher-silica, higher-K compositions with increasing LREE enrichment (Ryan *et al.*, 1980) and lower $\varepsilon_{Nd(t)}$ values. Plagiogranite boulders in the basal Silurian succession (the Killary Harbour–Joyce's Country Silurian, Chapter 7) are unequivocally derived from the underlying Lough Nafooey Group and two clasts have yielded U–Pb SIMS zircon ages of 489.9 ± 3.1 Ma and 487.8 ± 2.3 Ma (Chew *et al.*, 2007). Nd isotopic evidence ($\varepsilon_{Nd(490)}$~0) demonstrates that the plagiogranites had assimilated some old continental crust, and therefore the arc volcanics were tapping subducting Laurentian margin sediments by 490 Ma.

The Arenig Tourmakeady Volcanic Group contains andesitic and rhyolitic tuffs and volcaniclastic sediments (Graham *et al.*, 1989). These volcanics are LREE-enriched, and have strongly negative $\varepsilon_{Nd(t)}$ values (Fig. 4.6) indicating substantial assimilation of old continental material. Therefore the Tourmakeady Volcanic Group is believed to span 'hard' arc–continent collision (i.e. orogeny and regional deformation: Draut and Clift, 2001), and its eruption was synchronous with the peak

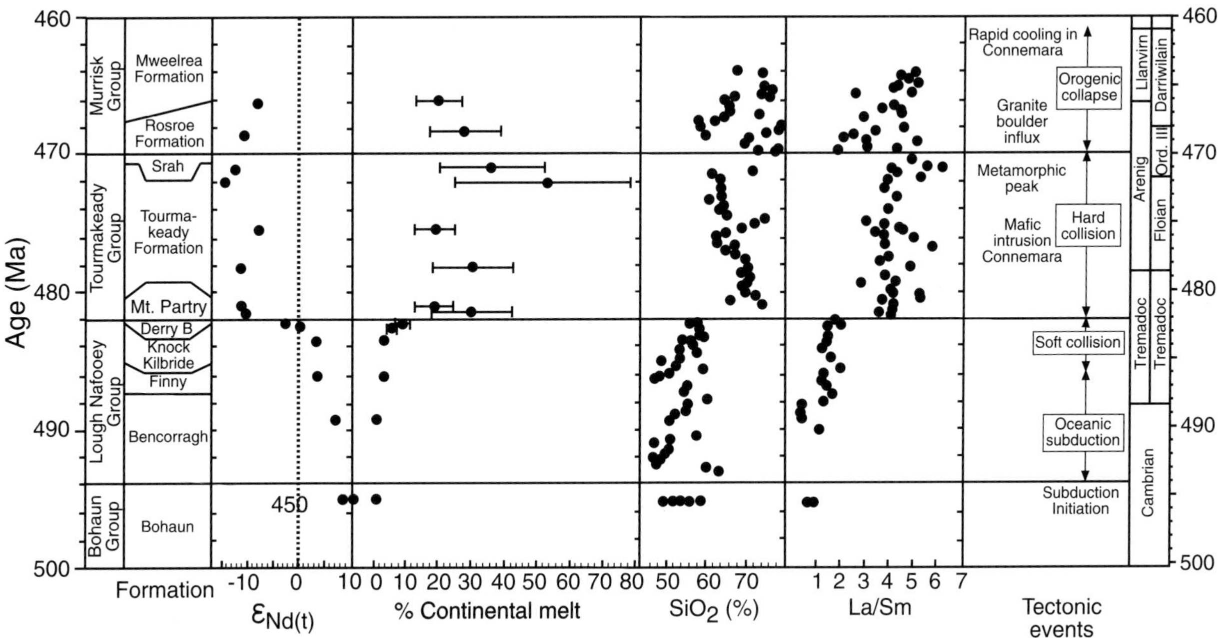

Fig. 4.6 Temporal evolution of some geochemical parameters of the South Mayo volcanic arc and its proposed link with orogenic evolution (Draut *et al.*, 2004). *© 2004, Cambridge University Press.*

regional metamorphism of the Laurentian margin as dated by Friedrich *et al.* (1999a) in Connemara. Volcanic horizons within the Murrisk Group include andesitic tuffs and ignimbrites of the Rosroe Formation and ignimbritic tuffs of the Llanvirn Mweelrea Formation. These are interpreted as syn- to post-collisional arc volcanics, and are LREE-enriched, with strongly negative $\varepsilon_{Nd(t)}$ values (Fig. 4.6). The lowermost ignimbrite horizon within the Mweelrea Formation has yielded a U–Pb zircon age of 464.4 ± 3.9 Ma (Dewey and Mange, 1999).

Structure and metamorphism

Despite their important role in the arc–continent collision that produced the Grampian orogeny, the arc volcanics and fore-arc sediments of South Mayo are relatively undeformed. The Lough Nafooey Group, which is thought to lie underneath the 6 km of sediments which fill the South Mayo Trough, has experienced very low-grade (prehnite-pumpellyite facies) metamorphism (Ryan *et al.*, 1980). This suggests that South Mayo did not undergo substantial tectonic burial during the Grampian Orogeny. In particular, it is very improbable that the allochthonous position of the Connemara terrane with respect to the rest of the Laurentian margin (Fig. 4.1) can be explained by simple southwards thrusting of this terrane over the very low-grade arc and fore-arc basins of South Mayo.

The fold axial plane of the Mweelrea Syncline is vertical and strikes E–W with a horizontal fold axis (Graham *et al.*, 1989). The northern limb exhibits isoclinal folding and penetrative cleavage which are also seen in the Silurian rocks unconformably overlying the Murrisk Group. However, there is evidence that much of the deformation was Ordovician in age and was subsequently enhanced by later Silurian tectonism (Kelly and Max, 1979). In contrast the cleavage in the southern limb is only developed in mudrocks and small-scale folding is absent. Early Caradoc brachiopod faunas have been recovered from the upper part of the Mweelrea Formation (Harper and Parkes, 2000) while Late Llandovery (Telychian) rocks of the Killary Harbour–Joyce's Country Silurian unconformably overlie the Murrisk Group. The development of the Mweelrea Syncline is, therefore, constrained to post-Early Caradoc, pre-Late Llandovery in age (455.8–436 Ma, Gradstein *et al.*, 2004).

Grampian evolution of the Tyrone Igneous Complex and the Tyrone Central Inlier

The Tyrone Igneous Complex is described in Chapter 3. It is composed of three units (Fig. 4.7): an ophiolitic block (the Tyrone Plutonic Group), an arc volcanic sequence (the Tyrone Volcanic Group), and a series of broadly granitic plutons. The ophiolitic block forms a slab that

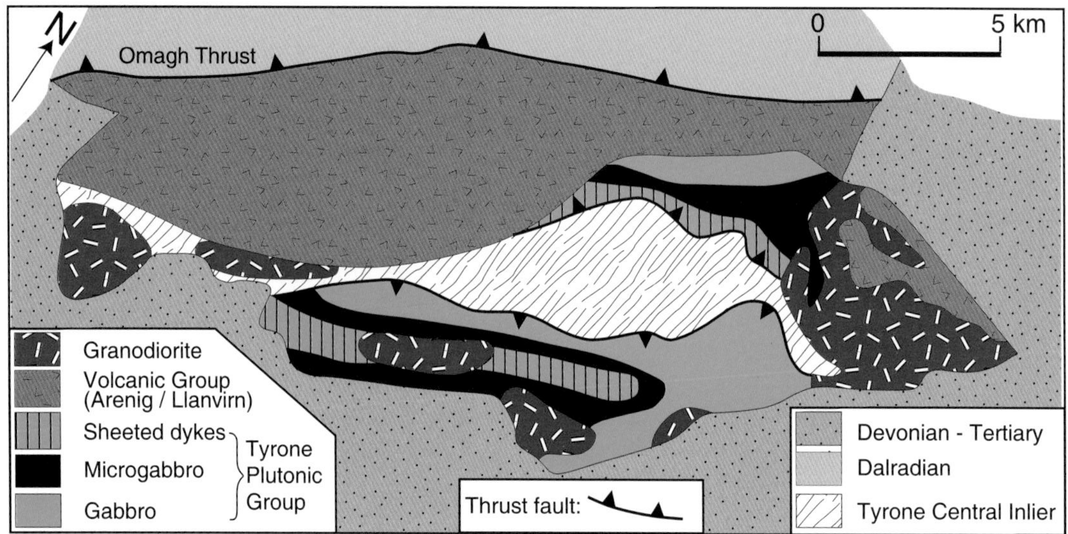

Fig. 4.7 Geological map of the Tyrone Igneous Complex and the Tyrone Central Inlier (after Hutton *et al.*, 1985).

has been thrust over high-grade basement paragneisses of the Tyrone Central Inlier (Hutton *et al.*, 1985). The juxtaposed ophiolite–metasedimentary basement package is unconformably overlain by Arenig–Llanvirn arc volcanics and black shales of the Tyrone Volcanic Group (Hutton and Holland, 1992) (Chapter 3, Fig. 4.7). A series of 470–465 Ma tonalite and granodiorite plutons cut the Igneous Complex and the Central Inlier, while higher-level equivalents cut the Volcanic Group (Cooper and Mitchell, 2004; Fig. 4.7). Low-grade (greenschist facies) metasediments of the Sperrin Mountains Dalradian were then subsequently thrust from the NW over the Tyrone Volcanic Series during regional SE-directed D3 thrusting (Alsop and Hutton, 1993). This complicated sequence of events is potentially of great importance in understanding the evolution of the Grampian Orogeny in Ireland.

The Tyrone Igneous Complex

The ophiolitic rocks (the Tyrone Plutonic Group) consist of a unit of gabbros and dolerites. The gabbros exhibit cumulate layering and are overlain by a sheeted dolerite dyke complex which exhibits classic one-sided chilled margins (Hutton *et al.*, 1985). Of crucial importance is the relationship between the ophiolitic rocks and the *c.*470 Ma tonalite and granodiorite suite. Hutton *et al.* (1985) infer that one such tonalite (the Craigballyharky tonalite) exhibits magma-mixing relationships with the ophiolitic rocks, implying that they were both hot at the time of intrusion of the tonalite, and that the high temperature of the basic rocks was a relic of ocean floor metamorphism. Therefore the U–Pb zircon age of the Craigballyharky tonalite ($471^{+2}/_{-4}$Ma) approximates the age of obduction (Hutton *et al.*, 1985). The upper intercept of $2030^{+630}/_{-500}$Ma shows that the tonalite has inherited zircon from a Palaeoproterozoic source, presumably the basement paragneisses of the Tyrone Central Inlier.

The Tyrone Central Inlier

The Tyrone Central Inlier is a high-grade metasedimentary terrane which is structurally overlain by the ophiolite complex (Hutton *et al.*, 1985). The paragneisses of the inlier have undergone polyphase deformation and metamorphism with a primary assemblage in pelitic lithologies of biotite + plagioclase + sillimanite + quartz ± garnet (i.e. sillimanite zone, below the second sillimanite isograd). Associated with this metamorphic event are abundant leucosomes. The high grade assemblages and leucosomes are cut by post-tectonic pegmatites which post-date at least two deformation fabrics. The leucosomes have yielded a weighted average ^{207}Pb / ^{206}Pb zircon age of 467 ± 12 Ma while the pegmatites have yielded 457 ± 7 Ma and 458 ± 7 Ma Rb–Sr muscovite–feldspar ages (Chew *et al.*, 2008). Biotite from the main fabric in the Tyrone Central Inlier yields a ^{40}Ar–^{39}Ar age of 468 ± 1.4 Ma, while muscovite from the same pegmatites that have yielded *c.*458 Ma Rb–Sr ages yields ^{40}Ar–^{39}Ar ages of 466 ± 1 Ma and 468 ± 1 Ma (Chew *et al.*, 2008). Palaeoproterozoic Nd model ages have also been obtained from the paragneisses of the Tyrone Central Inlier. These overlap with Nd model ages obtained from both the Argyll and Southern Highland Groups (Daly and Menuge, 1989). U–Pb detrital zircon analyses from a psammitic gneiss yield age populations at 1.05–1.2, 1.5, 1.8, 2.7 and 3.1 Ga (Chew *et al.*, 2008) which are ages typical of the Laurentian craton (e.g. Cawood *et al.*, 2003).

Combined, these data suggest that the Tyrone Central Inlier is a high-grade metasedimentary terrane of Laurentian (Dalradian?) affinity which has experienced high-grade metamorphism during the Grampian Orogeny, possibly in the roots of a deforming arc. The ophiolite was juxtaposed with the Central Inlier at this time, and the two units were then intruded by a series of stitching tonalitic–granodioritic plutons at 470–465 Ma and accompanied by the extrusion of arc lavas. The geological history of the Tyrone Igneous Complex and the Tyrone Central Inlier is substantially different to that of the adjacent Sperrin Mountains Dalradian which has experienced greenschist-facies metamorphism and no Grampian magmatism. It is probable that the Tyrone Igneous Complex and the Tyrone Central Inlier evolved outboard of the Laurentian margin during the Grampian Orogeny, and were finally juxtaposed with the margin when the Dalradian was thrust over the Tyrone Inlier during regional SE-directed D3 thrusting (Alsop and Hutton, 1993).

Grampian evolution of the Slishwood Division

The Slishwood Division (Fig. 4.8) is a metasedimentary unit that displays an early metamorphic history that has not been observed in the Dalradian rocks of north-west Ireland. It records pre-Grampian high-pressure granulite- and earlier eclogite-facies metamorphic events (Sanders *et al.*, 1987). These high-pressure metamorphic events are described in detail in Chapter 2 and are briefly outlined here. Sm–Nd garnet-plagioclase whole-rock isochrons from the granulite-facies assemblages yield ages of 605 ± 37 Ma (Sanders *et al.*, 1987) and 544 ± 52 Ma, 539 ± 11 Ma, 596 ± 68 Ma and 540 ± 50 Ma (Flowerdew and Daly, 2005). Pressure–temperature estimates for the granulite-facies assemblages are *c.*15 kbar, 800°C (Flowerdew and Daly, 2005). The Sm–Nd ages and pressure-temperature estimates are illustrated in Figure 4.8.

The Grampian (i.e. *c.*475–465 Ma) histories of the Tyrone Central Inlier and Slishwood Division are very similar. Both have undergone leucosome generation and subsequent intrusion of pegmatites which cut the high-grade fabrics. The high-grade fabrics in the Slishwood Division are cut by early Grampian tonalite intrusions (Fig. 4.8) which are restricted in extent to the Slishwood Division outcrop (Flowerdew *et al.*, 2005). These tonalites

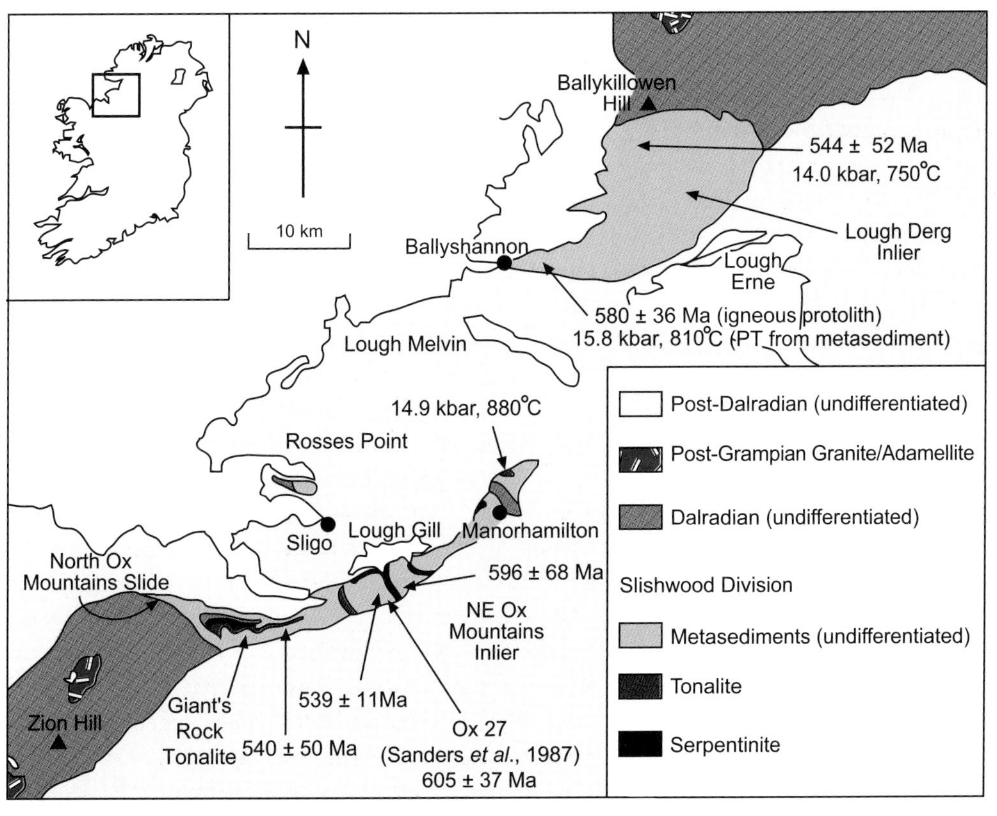

Fig. 4.8 Geological map of the Slishwood Division after Flowerdew and Daly (2005). Pressure–temperature estimates and Sm–Nd garnet ages for the pre-Grampian granulite facies event in the Slishwood Division from Flowerdew and Daly (2005) and Sanders *et al.* (1987) are also illustrated.

have yielded U–Pb SIMS zircon ages of 472 ± 6 Ma and 467 ± 6 Ma (Flowerdew *et al.*, 2005). Rb–Sr muscovite ages from the pegmatite suite in the Slishwood Division cluster at around *c.*460–455 Ma (Table 4.1, Flowerdew *et al.*, 2000). Rb–Sr muscovite-plagioclase ages from the main foliation in Slishwood Division metasediments record 460–450 Ma cooling ages, while retrograde hornblende from metabasite pods within the metasediments record ^{40}Ar–^{39}Ar ages ranging from *c.*480–450 Ma (Table 4.1, Flowerdew *et al.*, 2000). Final imbrication of the Slishwood Division with the Central Ox Mountains Dalradian occurred during regional SE-directed D3 shearing (Flowerdew *et al.*, 2000).

Evidence for the timing of the Grampian Orogeny in Ireland

The cause of the Grampian Orogeny is now widely regarded as the collision of Laurentia with an oceanic arc during the Arenig (Dewey and Shackleton, 1984). However, the timing of this orogeny has proved controversial in the past. In particular, the possible existence of an early Neoproterozoic orogenic phase has been rigorously debated.

It is now accepted that much of the Moine Supergroup and the basal portion of the Grampian Group Dalradian (the Glen Banchor Succession) in Scotland have experienced a Neoproterozoic tectonothermal event, loosely constrained to *c.*800 Ma and commonly referred to as the Knoydartian Orogeny. Whether this event represents a true compressional orogeny (e.g. Piasecki and Van Breemen, 1983; Noble *et al.*, 1996; Vance *et al.*, 1998; Tanner and Evans, 2003) or extension-related tectonism prior to the formation of the Iapetus Ocean (e.g. Dewey and Mange 1999; Soper *et al.*, 1999; Dalziel and Soper 2001) is still debated. Probably the most compelling evidence for a *c.*800 Ma compressional Knoydartian orogeny in Scotland is reported by Vance *et al.* (1998) who present *c.*820–790 Ma Sm–Nd garnet–whole rock ages from the Moine Supergroup. As these garnets grew under P–T conditions of at least 11 kbar and *c.*700 °C (Vance *et al.*, 1998), they are almost certainly associated with crustal thickening.

The evidence that younger horizons of the Dalradian have experienced Neoproterozoic deformation is more speculative. The 590 ± 2 Ma (Rogers *et al.*, 1989) Ben Vuirich granite cuts Appin Group sediments of the Dalradian Supergroup in Scotland. It was originally regarded as a post-D2, pre-D3 intrusion (Bradbury *et al.*, 1976) and hence D1 and D2 deformation must have been Precambrian in age. However, the field relationships of Bradbury *et al.* (1976) are probably erroneous. The Ben Vuirich granite is now considered to have experienced the regional D2 deformation (Tanner and Leslie, 1994) and more recent studies suggest that the only fabric it cuts is a contact hornfels fabric mimetic after bedding and produced by intrusion-related deformation (Tanner, 1996).

Until recently, evidence in Ireland for Neoproterozoic orogeny in the Grampian belt of Ireland was restricted to the 600–550 Ma Sm–Nd garnet ages from granulites of the NE Ox Mountains Inlier (Sanders *et al.*, 1987; Flowerdew and Daly, 2005). However these rocks also have an Ordovician Grampian history and are encountered on the margin of the Grampian belt. It is therefore possible that they experienced Neoproterozoic deformation exotic to Laurentia and were then accreted to Laurentia during the Grampian Orogeny. Recent evidence suggests the possible presence of an orogenic unconformity within the Easdale subgroup of the Argyll Group of Donegal (Hutton and Alsop, 2004). Reworked clasts that contain a pre-existing tectono-metamorphic history are identified within a conglomerate that lies along the Easdale unconformity surface. The underlying sequence also exhibits pre-existing deformational fabrics that display erosional truncation at the base of the conglomerates. The regional extent and significance of this event are still being debated (e.g. Tanner, 2005). No Knoydartian events have been documented in Ireland, despite comprehensive investigation. In particular, a mineral geochronology study across the Dalradian–Annagh Gneiss Complex basement contact in NW Mayo has not identified Knoydartian deformation in either the Annagh Gneiss Complex or in the immediately adjacent Dalradian rocks (Daly and Flowerdew, 2005).

Direct evidence: biostratigraphical

The timing of the Grampian Orogeny is bracketed by the age of the youngest fossiliferous horizon in the Dalradian sequence, and the oldest unconformable strata resting upon it. Diagnostic faunas have yet to be recovered from the Dalradian of Ireland. The youngest reliable biostratigraphical age from the Dalradian of Scotland is the late Early Cambrian trilobite fauna of the Leny Limestone; while a minimum age for the Grampian orogeny is given by the unconformable relationship of the Connemara Dalradian with Upper Llandovery sandstones (Leake and Tanner, 1994). If it is accepted that the Clew Bay

Complex has experienced the entire Grampian deformation cycle, then its biostratigraphic constraints can be used to constrain the timing of orogeny. The most reliable constraint is probably the Llanvirn euconodont fauna of Harper *et al.* (1989).

A maximum age for Grampian D3 deformation in Ulster is provided by the Ordovician volcanics of the Tyrone Inlier. The D3 Omagh Thrust has placed the inverted Dalradian succession over the Tyrone Volcanic Group (Alsop and Hutton, 1993). Black shales in the volcanics have yielded Arenig-Llanvirn graptolites (Hutton and Holland, 1992).

Direct evidence: geochronology of igneous intrusions

Timing of the Grampian Orogeny can also be bracketed by the dating of pre-, syn- and post-orogenic igneous intrusives. The Maam Sill, a pre-deformational metadolerite which cuts Appin Group sediments in Donegal, has yielded an imprecise but well-fitted 625 ± 47 Sm–Nd clinopyroxene–whole rock–plagioclase isochron (Kirwan *et al.*, 1989).

The syn-orogenic igneous intrusives of Connemara were briefly described earlier. They consist of large volumes of basic and intermediate magma that were intruded during the Grampian Orogeny, and are interpreted as representing the roots of a volcanic arc (e.g. Yardley and Senior, 1982; Tanner, 1990). The syn-D2 to early D3 basic intrusions have yielded U–Pb zircon ages of 470.1 ± 1.4 Ma and 474.5 ± 1 Ma (Friedrich *et al.*, 1999a). Late D3 Quartz-diorite gneisses in the migmatite zone in south Connemara yield U–Pb zircon ages of 463 ± 4 Ma (Cliff *et al.*, 1996) and 467 ± 2 Ma (Friedrich *et al.*, 1999b). The post-tectonic Oughterard granite in Connemara has yielded a 462.5 ± 1.2 Ma U–Pb xenotime age (Friedrich *et al.*, 1999a).

In the Tyrone Central Inlier, syn-tectonic leucosomes have yielded a $^{207}Pb/^{206}Pb$ zircon age of 467 ± 12 Ma (Chew *et al.*, 2008). ^{40}Ar–^{39}Ar muscovite ages from post-tectonic pegmatites yield ages of 466 ± 1 Ma and 468 ± 1 Ma (Chew *et al.*, 2008). In the NE Ox Mountains, early Grampian tonalites which cut high-grade (granulite-facies) assemblages in the Slishwood Division have yielded U–Pb SIMS zircon ages of 474 ± 5 Ma, 472 ± 6 Ma, 471 ± 5 Ma and 467 ± 6 Ma (Flowerdew *et al.*, 2005).

Indirect evidence: detrital and volcanic record of adjacent basins

The detrital and volcanic record of the South Mayo Trough, a Grampian fore-arc basin, was reviewed earlier. The detrital record (Figure 4.5) suggests the progressive unroofing of an ophiolite complex during the Arenig followed by the exhumation of the Grampian metamorphic belt (Wrafter and Graham, 1989, Dewey and Mange, 1999). The incoming of metamorphic detritus occurs during the Llanvirn. The LREE depletion and strongly positive $\varepsilon_{Nd(t)}$ values (Figure 4.6) at the base of the arc sequence (Late Tremadoc) suggest minimal assimilation of continental material and an origin far from the Laurentian margin. Younger volcanic units exhibit a trend toward higher-silica, higher-K compositions with increasing LREE enrichment and lower $\varepsilon_{Nd(t)}$ values (Figure 4.6) indicating substantial assimilation of old continental material during arc–continent collision (Draut and Clift, 2001). The transition from juvenile magmatism to magmatism with substantial assimilation of Laurentian material occurs at the base of the Arenig Tourmakeady Volcanic succession (Draut and Clift, 2001).

The docking of the Connemara terrane with the South Mayo Trough is recorded by the presence of high-grade metamorphic clasts within conglomerates of the post-Llanvirn, pre-Llandovery Derryveeney Formation, which sits unconformably on the Murrisk Group of the South Mayo Trough (Graham *et al.*, 1991). These sillimanite-bearing migmatite clasts can be matched to the neighbouring Lough Kilbride Formation of the Connemara Dalradian by a combination of palaeocurrent and Nd isotopic data (Graham *et al.*, 1991). In addition, two metamorphic clasts from within the Derryveeny Formation yield 471 ± 8 Ma and 462 ± 7 Ma Rb–Sr muscovite–whole rock ages (Graham *et al.*, 1991).

Indirect evidence: geochronological

Fabric-forming minerals such as muscovite, biotite and hornblende can easily be related by field studies to discrete deformational events. Hence dating these fabric-forming minerals using the Rb–Sr, ^{40}Ar–^{39}Ar or K–Ar techniques could potentially date a specific phase of deformation. However, these methods only date deformation at relatively low metamorphic grades. At higher grades they record the post-metamorphic cooling history of an orogenic belt (Cliff, 1985). The principles behind mineral geochronology are described in Chapter 2.

Metamorphic crystallisation ages

Estimates for the closure temperature of the Sm–Nd system in garnet range from c.600 °C (Mezger *et al.*, 1992) to *c.*850 °C (Cohen *et al.*, 1988). As Dalradian metamorphism rarely exceeds upper amphibolite facies, the Sm–Nd system in garnet will record garnet growth rather than cooling, particularly if the higher estimates for the closure temperature are accepted.

There are only a limited amount of Sm–Nd metamorphic garnet data from the Grampian belt in Ireland. Peak MP3 metamorphism of Dalradian rocks adjacent to the Slishwood Division in NW Ireland has been dated by two Sm–Nd garnet–whole rock isochrons of 457 ± 16 Ma and 465 ± 28 Ma (Flowerdew *et al.*, 2000). Peak MP2 metamorphism in the Dalradian adjacent to the Annagh Gneiss Complex has been dated at 467 ± 14 Ma by the same method (Flowerdew, 2000). The 600–550 Ma Sm–Nd garnet ages from granulites of the NE Ox Mountains Inlier (Sanders *et al.*, 1987; Flowerdew and Daly, 2005) discussed earlier may have been exotic to Laurentia and therefore would not be part of the Grampian orogenic cycle.

Metamorphic cooling ages

Recent geochronological studies in Connemara, NW Mayo and the Ox Mountains focused on dating discrete deformation events (i.e. fabrics). These studies are consistent with a *c*.470 Ma Grampian Orogeny (Friedrich, 1998, 1999b; Flowerdew *et al.*, 2000; Flowerdew, 2000; Chew *et al.*, 2003; Daly and Flowerdew, 2005). Data are summarised in Tables 4.1–4.4 and Figure 4.9.

The oldest mineral cooling age is a 475 ± 4 Ma hornblende age from the Annagh Gneiss Complex (Daly and Flowerdew, 2005, Table 4.1, Figure 4.9B). The data from the Ox Mountains and NW Mayo Inlier (Flowerdew *et al.*, 2000; Chew *et al.*, 2003; Daly and Flowerdew, 2005) imply a *c*.470–460 Ma age for the main nappe fabric (S3 in the Ox Mountains, S2 in NW Mayo). The *c*.470–460 Ma D3 interleaving of the Dalradian and the Slishwood Division in the NE Ox Mountains is consistent with an Arenig-Llanvirn age for the D3 Omagh Thrust in Tyrone. In addition, D3 is further constrained in the NE Ox Mountains by the presence of *c*.480 Ma retrograde hornblende poikiloblasts in the Slishwood Division, which are demonstrably pre-D3 in age, and by post-D3 *c*.455 Ma pegmatites related to extensional collapse (Flowerdew *et al.*, 2000). The abundance of 460–450 Ma mineral ages (Figure 4.9A) is similar to the pattern of metamorphic mica ages from Scotland (Dempster, 1985; Dempster *et al.*, 1995; Table 2.2) and is likely to record a period of rapid uplift and exhumation (Flowerdew *et al.*, 2000).

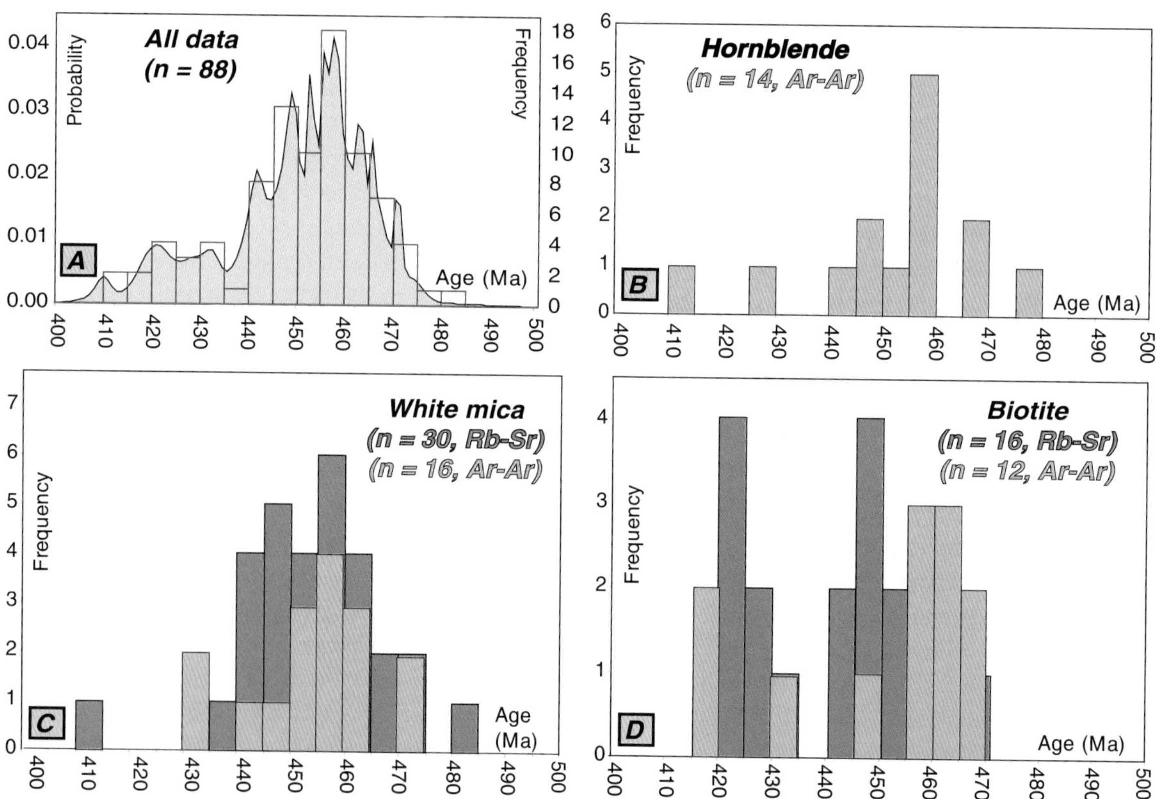

Fig. 4.9 Histograms illustrating the age distribution of mineral cooling ages in the Dalradian rocks from Ireland.

Table 4.1 Summary of metamorphic cooling age data from the Ox Mountains fromFlowerdew *et al.* (2000).

Method and locality	Age (Ma)							
Ar-Ar hornblende (Slishwood Division)	459 ± 3	450 ± 2	457 ± 2					
Rb-Sr muscovite-plag (Slishwood Division)	459 ± 7	449 ± 7						
Ar-Ar hornblende (NE Ox Dalradian)	467 ± 3	457 ± 3	448 ± 3	442 ± 3	429 ± 5	410 ± 3		
Rb-Sr muscovite-plag (NE Ox Dalradian)	472 ± 8	470 ± 7	467 ± 7	464 ± 9	456 ± 12	446 ± 7	446 ± 8	410 ± 8

All ages relate to the main fabric in the inlier; S3 in the Slishwood Division and the NE Ox Dalradian

Table 4.2 Summary of metamorphic cooling age data from NW Mayo from Daly and Flowerdew (2005), Chew *et al.* (2003), and Chew *et al.* (Unpublished data).

Method and locality	Age (Ma)					
Ar-Ar hornblende (Annagh Gneiss Complex)	475 ± 4	459 ± 5				
Rb-Sr hbl-plag (Annagh Gneiss Complex)	481 ± 40					
Ar-Ar hornblende (NW Mayo Dalradian)	458 ± 5	446 ± 9	467 ± 5			
Rb-Sr biotite-plag (NW Mayo Dalradian)	425 ± 7	430 ± 7	421 ± 7	429 ± 6	423 ± 7	421 ± 6
Rb-Sr muscovite-plag (NW Mayo Dalradian)	464 ± 9	450 ± 8	441 ± 7	448 ± 7	451 ± 8	441 ± 7
Ar-Ar biotite (NW Mayo Dalradian)	458 ±1					
Ar-Ar muscovite (S. Achill Dalradian)	463 ± 4	457 ±4	459 ± 1	460 ± 5	462 ± 4	472 ± 1
Rb-Sr muscovite-plag (S. Achill Dalradian)	458 ± 7	460 ± 7	459 ± 7	458 ± 7	442 ± 7	

All ages are of the main S2 fabric in the inlier.

Table 4.3 Summary of metamorphic cooling age data from Connemara from Friedrich (1998), Friedrich *et al.* (1999b) and Elias (1988).

Method and locality	Age (Ma)							
Ar-Ar biotite (North Connemara)	463 ± 1	460 ± 3	464 ± 1	466 ± 1	449 ± 2	456 ± 1	457 ± 3	430 ± 4
Ar-Ar biotite (South Connemara)	418 ± 7	419 ± 4	467 ± 3					
Ar-Ar muscovite (North Connemara)	471 ± 1	453 ± 1	453 ± 1	449 ± 2	458 ± 4			
Ar-Ar muscovite (South Connemara)	454 ± 1	456 ± 2	433 ± 5	433 ± 3	442 ± 2			
Rb-Sr muscovite (North Connemara)	480 ± 19	454 ± 8	455 ± 6	440 ± 7				
Rb-Sr muscovite (South Connemara)	449 ± 5	463 ± 7	439 ± 5	466 ± 6	453 ± 6			
Rb-Sr biotite (North Connemara)	443 ± 5	446 ± 5	422 ± 4	449 ± 5	466 ± 5	441 ± 4		
Rb-Sr biotite (South Connemara)	446 ± 5	449 ± 5	452 ± 5	452 ± 5				

All ages are of the main fabric in the inlier.

Table 4.4 Summary of metamorphic cooling age data from the Clew Bay Complex, Chew *et al.* (2003), Chew *et al.* (2006) and Chew *et al.* (Unpublished data).

Method and locality	Age (Ma)		
Rb-Sr muscovite-plag (S. Achill Beg)	461 ± 7	460 ± 7	
Ar-Ar muscovite (S. Achill Beg)	462 ± 1	490 ± 1*	462 ± 1
Rb-Sr muscovite-plag (Deerpark Complex)	483 ± 7		
Ar-Ar muscovite (Deerpark Complex)	482 ± 1	460 ± 1	

All ages are of the main fabric in the inlier. *Interpreted as due to excess argon as the same sample has yielded a Rb-Sr muscovite age of 461 ± 7.

The cooling history of Connemara differs slightly from that of NW Ireland and Scotland (Friedrich *et al.*, 1999b). Rapid cooling during the period 468–460 Ma seems necessary in order to preserve muscovite and biotite ^{40}Ar–^{39}Ar cooling ages as old as *c.*470 Ma in northern Connemara (Table 4.3). The slightly younger ages in southern Connemara are attributed to thermal resetting adjacent to the gabbro and orthogneiss complexes (Friedrich *et al.*, 1999b).

The only exception to the general trend of *c.*470 Ma and younger mineral cooling ages is in the Clew Bay Complex. Whereas mineral ages from the Clew Bay Complex adjacent to the Dalradian on South Achill Beg cluster around 460 Ma (Table 4.4, Figure 4.9), muscovite from high-grade metasedimentary slivers within the ophiolitic Deerpark Complex has yielded a 483 ± 7 Rb–Sr muscovite age and a 482 ± 1 ^{40}Ar–^{39}Ar age (Table 4.4, Chew *et al.*, 2006).

Summary

Figure 4.10 summarises the numerous timing constraints described above. One of the chief difficulties in constructing such a diagram is the uncertainty in comparing events constrained by biostratigraphy (e.g. the detrital and volcanic record of South Mayo) with those constrained by geochronology. What are clearly needed are more absolute ages for the volcanism in South Mayo, in particular U–Pb zircon dating of tuff horizons. This figure should be consulted in conjunction with the following section, which reviews models for the Grampian Orogeny with specific reference to Ireland.

Tectonic models for the Grampian Orogeny

The Grampian–Taconic Orogen

Closure of the Iapetus Ocean is thought to have commenced with the 'southward' subduction of Iapetus oceanic crust belonging to the Laurentian plate under a chain of primitive, continent-facing oceanic arcs during the Late Cambrian/Tremadoc (Dewey and Mange, 1999; Van Staal *et al.*, 1998). Good evidence for this view comes from the Laurentian margin in North America where the Taconic Orogeny is the direct correlative of the Grampian Orogeny this side of the Atlantic. It has been suggested that the primitive oceanic arcs may have originally nucleated on oceanic transform faults (Karson and Dewey, 1978) during the Middle Cambrian (Dewey and Mange, 1999). With the onset of subduction, these mafic, infant arcs evolved into Early Ordovician intermediate arcs with associated suprasubduction ophiolites (Dewey and Mange, 1999). Collision of the arc with the Laurentian margin, and the consequent obduction of the ophiolite, is well constrained in the Baie Verte Oceanic Tract of the Notre Dame Subzone in Newfoundland. Upper Cambrian–Middle Tremadoc suprasubduction ophiolites and juvenile volcanic–plutonic complexes were obducted onto the Laurentian margin, which is thought (based on seismic reflection data) to structurally underlie the entire Notre Dame Subzone (Keen *et al.*, 1986). The arc/ophiolite allochthon and the underlying ophiolitic mélange are cut by arc-related plutons as old as Early Arenig (Van Staal *et al.*, 1998 and references therein). The isotope geochemistry of these tonalitic–granitic stitching plutons suggests they have ascended through continental crust (Whalen *et al.*, 1997). Hence, if the Baie Verte Oceanic Tract does indeed structurally overlie Laurentian crust, then slab break-off and a subsequent polarity reversal in the direction of subduction (from southwards to northwards) is implied. It has been suggested that this change in subduction polarity, inferred in North America, may also have occurred in Ireland and Scotland. The voluminous basic intrusions in the Dalradian of Connemara and NE Scotland have been interpreted as the roots of a volcanic arc (Yardley *et al.*, 1982; Yardley and Senior, 1982), generated by 'northward' subduction underneath the Laurentian margin.

Several models for the Grampian and Taconic orogens (e.g. Dewey and Mange, 1999; Van Staal *et al.*, 1998; Dewey and Shackleton, 1984) attribute the bulk of the deformation and metamorphism to the obduction of the fore-arc ophiolite onto the Laurentian margin. In such models, the ophiolite obduction event stacks the Dalradian nappe pile and accretes the arc to the margin, with the collisional suture represented by the Baie Verte–Clew Bay–Highland Border Line. This collisional suture incorporates a diverse package of accreted material 'swept up' by the oceanic arc (Dewey and Mange, 1999). Subsequent slab break-off is then envisaged, and the resultant switch in subduction polarity would then have produced syn-collisional, arc-related magmatism on the Laurentian margin (Dewey and Mange, 1999; Van Staal *et al.*, 1998; Dewey and Ryan, 1990). Subduction of Iapetus under the Laurentian margin is believed to have continued on into the Silurian (Van Staal *et al.*, 1998), with large amounts of material being shed from the uplifting orogen into thick accretionary prisms such as the Southern Upland and Longford-Down belts (Hutchison and Oliver, 1998).

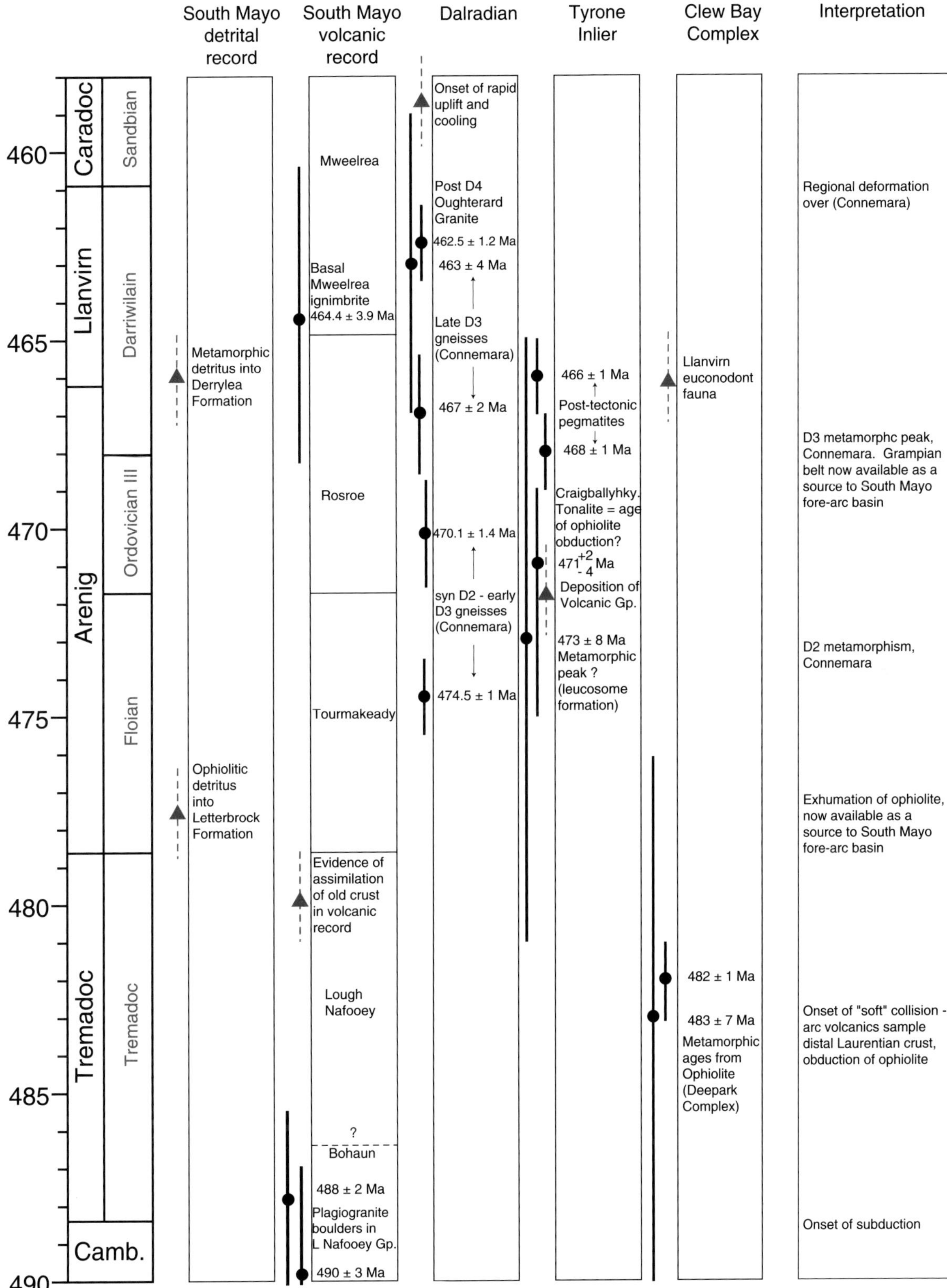

Fig. 4.10 Diagram summarising the timing constraints on the Grampian Orogeny in Ireland. Timescale from Gradstein *et al.* (2004). Black circles indicate geochronological ages; error bars are 2σ. Red triangles indicate events constrained by biostratigraphy; schematic error bars are illustrated.

The Grampian Orogen in Ireland

The development of the Grampian orogen in Ireland is schematically illustrated in Figure 4.11, which is based largely on information derived from the coastal section in western Mayo. This is probably the most complete section through the Grampian orogenic belt. Here all the major components of the orogen are present: Laurentian basement (Annagh Gneiss Complex); Laurentian cover (Dalradian Supergroup); accretionary complex (Clew Bay Complex); supra-subduction ophiolite (Deerpark Complex); and arc volcanics and fore-arc basin (South Mayo). Additionally, information derived from other localities on the Irish segment of the deforming Laurentian margin is projected onto the line of section. This includes the late-stage imbrication of the Slishwood Division into the Dalradian block (Flowerdew, 2000).

Figures 4.11a and 4.11b illustrate the tectonic setting immediately prior to orogeny, and are adapted after Dewey and Mange (1999). The infant oceanic arc initiates in the Tremadoc (Fig. 4.10, Fig. 4.11a). Shortly afterwards, obduction of the Clew Bay Complex ophiolite (the Deerpark Complex) occurs (Figure 4.11b). The timing of this obduction event is constrained by the presence of Arenig ophiolitic detritus in the South Mayo Trough and the *c.*480 Ma metamorphic ages obtained from the ophiolitic sole (Figure 4.10). The change in arc chemistry at this time is probably a result of introducing continental material into the source region of melt production beneath the arc by subduction of distal Laurentian slope sediment (Draut *et al.*, 2004). Although the absolute age of the blueschists in the Dalradian rocks of South Achill Island is still uncertain, the high P, low T metamorphism is texturally early and may have happened at this time.

Figure 4.11c is approximately contemporaneous with the metamorphic peak. Although establishing the kinematics of early structures in a polyphase metamorphic belt is difficult, it appears that many of the structures were directed towards the Laurentian landmass. Thrusting in such a direction may also explain the exhumation of the blueschist-facies metamorphic rocks. If the blueschist-facies metamorphism resulted from subduction of distal Laurentian sediment (e.g. Fig. 4.11b), then a simple and efficient exhumation pathway would be reversal of the subduction path by transferring the high P, low T rocks to the hanging wall plate. Subsequent collisional thickening and exhumation would then have resulted in large amounts of metamorphic detritus being transported into the South Mayo Trough during the Llanvirn (Fig. 4.10).

Although there is no evidence for a micro-continental indenter such as the Slishwood Division or Tyrone Inlier in western Ireland, there is evidence for late stage nappe transport away from the Laurentian continent (Figure 4.11d). Late-stage south-directed nappe transport is well documented in Donegal, the Ox Mountains and Tyrone, where it is associated with the overthrusting of Dalradian rocks over an outboard continental block. By analogy, this is inferred at depth in western Ireland.

Supplemental information on the Grampian orogenic history of individual Dalradian inliers

Structural analysis of polyphase-deformed rocks

This section is intended to give a brief introduction to the terminology used in describing metamorphic rocks which have undergone several episodes of ductile deformation (polyphase deformation).

The term 'phase' itself may be used to describe either an orogeny (e.g. 'the Grampian phase of the Caledonian orogenic cycle') or more commonly a deformation event of regional extent in an orogenic belt (e.g. 'the D1 deformation phase in the NW Mayo inlier').

In an orogenic belt, deformation events (or phases) usually result in large-scale structures such as folds or bedding-parallel ductile shear zones ('slides') with associated small-scale structures such as minor folds and fabrics (foliations, lineations etc.). For example, the D2 deformation event in the NW Mayo Inlier is the second phase of deformation in this region. It led to large-scale F2 folds which are associated with an S2 schistosity, and an L2 lineation. In general, when referring to a specific deformation event ('Dn'), we refer to the folds produced by this event as 'Fn folds'. Correspondingly, the same deformation event will produce a 'Ln lineation' and a 'Sn schistosity'. Metamorphic minerals which grow *during* a Dn event are 'Mn' or 'MSn' in terms of the textural age of the grains, while metamorphic minerals which grow after this event are referred to as 'MPn' in age. For example, again using the NW Mayo Inlier as an example, garnet growth in the blueschist-facies rocks in the southern portion of the inlier is MP1 in age, implying the garnet grew after the first deformation event (and thus overgrows the S1 schistosity) and it grew prior to the second deformation event (it is wrapped by the S2 schistosity).

Constructing a deformation history is undertaken by regional mapping. Commonly one set of structures (e.g. folds) or fabrics (e.g. schistosities) is particularly well-developed and it can therefore be mapped throughout

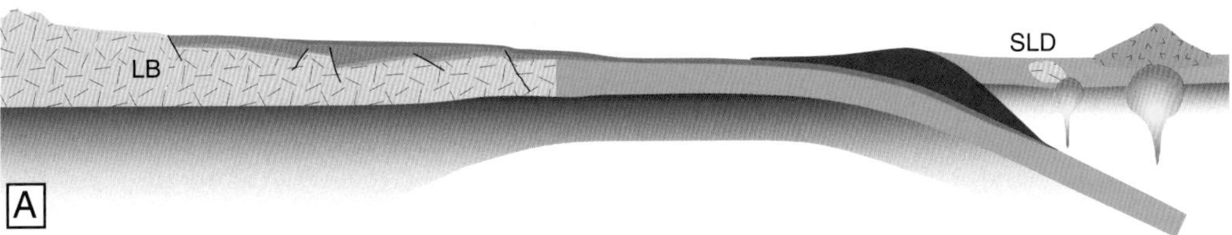

Schematic reconstruction of the Laurentian margin of western Ireland at c. 485 Ma (Tremadoc). The Laurentian margin (Dalradian) is subducted under the infant Lough Nafooey arc. Modified after Dewey & Mange (1999).

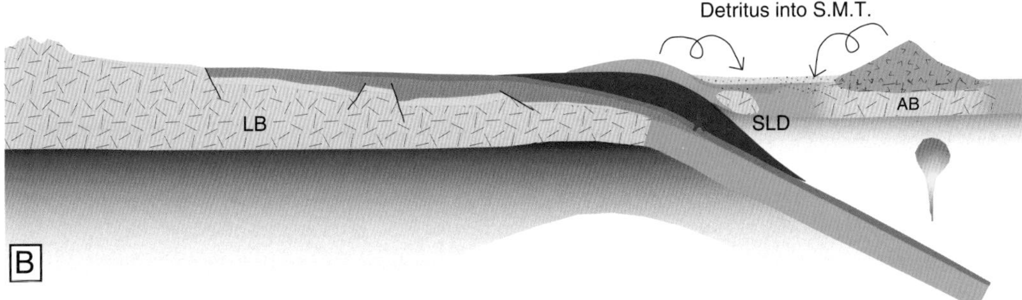

Schematic reconstruction of the Laurentian margin of western Ireland at c. 475 Ma (mid Arenig). Incipient development of blueschist-facies metamorphism (asterisk) in distal Dalradian sediments and the Clew Bay Complex accretionary wedge. Onset of foreland-directed thrusting and ophiolite obduction. Arc volcanics assimilate Laurentian crustal material.

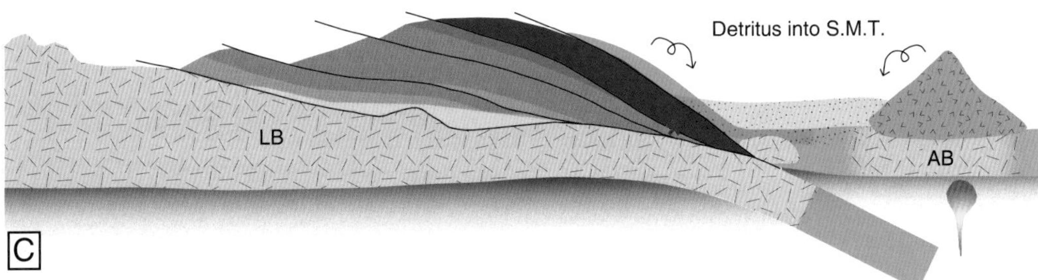

Schematic reconstruction at c. 470 Ma (Early Llanvirn). Collisional thickening produced by foreland-directed thrusting. Blueschist-facies rocks are transferred to the overiding plate and thrust back to the foreland where the deforming Dalradian rocks are experiencing the regional metamorphic peak. Metamorphic detritus shed into South Mayo Trough.

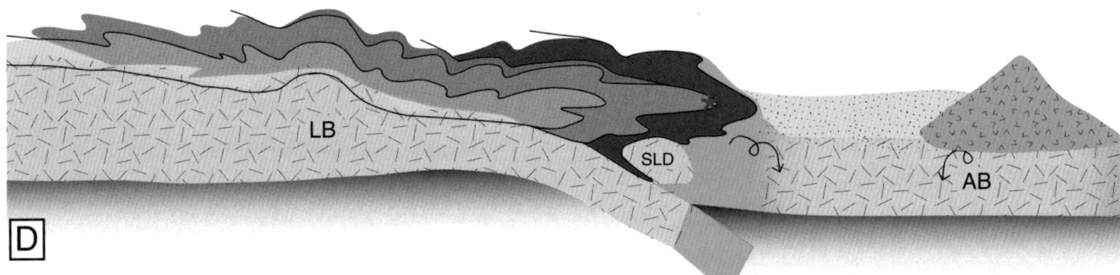

Schematic reconstruction at c. 465-460 Ma (Upper Llanvirn - Llandeilo). Underthrusting of a landmass associated with the arc terrane (inferred to be the Slishwood Division or possibly arc basement) produces south-directed nappes adjacent to the Laurentian margin.

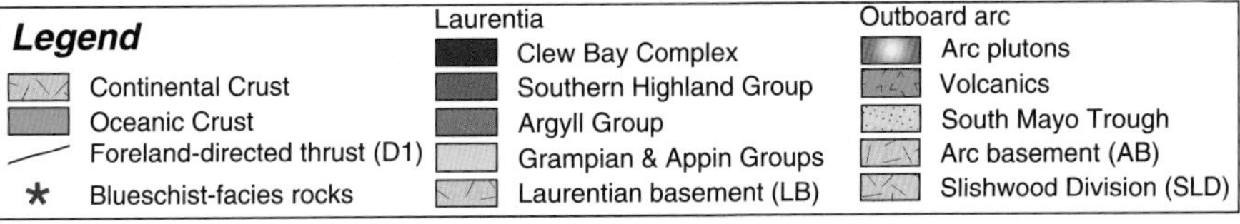

Fig. 4.11 Schematic vertical sections showing inferred tectonic development of the Grampian Orogeny in western Ireland, looking east. The Connemara terrane and its arc plutons produced by north-directed subduction would be located to the south of the South Mayo arc in Figs. 4.11c and 4.11d.

a region with some degree of confidence. Where these folds or fabrics are affected by later deformation phases, refolded folds or overprinting schistosities are encountered. Conversely, evidence of earlier deformation phases may be preserved in porphyroblasts (visible in thin section) or in low-strain zones in the field. Figure 4.12a illustrates a refolded fold from the Dalradian rocks of the NW Mayo Inlier. The most commonly encountered folds are F2 folds with an upright axial planar S2 foliation (Fig. 4.12b). However, these must be second-generation folds because occasionally they are observed to fold earlier F1 folds which have an associated S1 cleavage (Fig. 4.12b). In thin section the second foliation (S2) can be seen to crenulate the earlier S1 foliation. Garnet clearly overgrows the S1 foliation, but is wrapped by the S2 foliation. Hence the garnet growth is texturally MP1 (post D1, pre-D2) in age.

Although a deformation history may be constructed for a given outcrop, it is unwise to assume that each deformation event represents a region-wide 'pulse' of deformation. It is possible (particularly in zones of high non-coaxial strain) to develop multiple refolding in a single, progressive deformation event. Hence correlation of deformation phases should be undertaken with care. In particular it is highly probable that the deformation history of each of the various Irish Dalradian inliers is different, and correlating specific deformation events from inlier to inlier would be unwise.

Grampian evolution of the NW Mayo Inlier

The broad structure of the NW Mayo Inlier is similar to that of the SW Highlands of Scotland. There is a primary facing divergence either side of a basement core, the Annagh Gneiss Complex. To the north of this 'root zone' F2 folds face north; to the south, recumbent F2 folds face south (Chew, 2003; Johnston, 1995; Johnston and Phillips, 1995; Kennedy, 1980). Approaching the Achill Beg Fault, the south-facing F2 antiform is rotated into a downward-facing orientation (Fig. 4.13), analogous to the Highland Border Downbend.

D1 deformation produced a series of tectonic slides (Kennedy, 1969b; 1980). Associated with these high strain zones are minor F1 folds, which are in places sheath folds, and prolate pebble-stretching with the maximum extension direction (x) parallel to the hinges of F1 folds. The disposition of the major units in the inlier is mainly the result of D2 folding (Winchester and Max, 1996; Kennedy, 1980). Johnston (1995), Johnston and Phillips (1995) and Harris (1995, 1993a, 1993b) recognise an east–west trending L2 stretching lineation, despite the apparent movement of the Dalradian nappe to the south (Fig. 4.13). Johnston (1995) argued that D2 structures represent north-westward-vergent thrusts over a crustal-scale ramp zone. Harris (1995) favoured a westward direction of transport for the recumbent D2 nappes, which is contemporaneous with E–W dextral transpression in the steep zone adjacent to the Achill Beg Fault. Alternatively, Sanderson *et al.* (1980), Chew (2003), and Chew *et al.* (2004) consider the strike-swing and generation of downward-facing folds to be the result of later modification by an E–W dextral shear zone along the north margin of Clew Bay.

The northern part of the inlier displays the highest metamorphic grades. Kyanite and staurolite occur sporadically, due to the paucity of pelitic outcrop. P–T estimates of the staurolite–kyanite zone are 8 ± 2 kbar and 620 ± 30 °C (Yardley *et al.*, 1987), and *c.*9–11 kbar and *c.*640 °C (Flowerdew, 2000). There is a general decrease in metamorphic grade to the south. Much of the rest of the inlier is garnet zone, with the lowest grade rocks restricted to South Achill and the Corraun peninsula (Yardley *et al.*, 1987). The metamorphic peak in the staurolite–kyanite zone is MP2 in age (Max, 1973; Yardley *et al.*, 1987), while peak metamorphism in lower grade rocks to the south is predominantly MP1 in age (Chew, 2003; Kennedy, 1969a; Morley, 1966; Crow, 1973). The lower grade rocks (particularly on Achill Island) also show substantial development of both prograde and retrograde porphyroblastic albite. The southern portion of the inlier, adjacent to the Achill Beg Fault is characterised by the development of early (MP1) blueschist-facies assemblages (Chew *et al.*, 2004; Gray and Yardley, 1979). Textural evidence suggests that the blueschist-facies metamorphism was contemporaneous with Barrovian metamorphism in the Dalradian further north in central Achill. Pressure–temperature estimates for the Barrovian metamorphism in central Achill are 525 ± 45°C, 6.5 ± 1.5 kbar (1σ), and 460 ± 45°C, 10.5 ± 1.5 kbar (1σ) for the blueschist-facies metamorphism in south Achill (Chew *et al.*, 2003).

Grampian evolution of Donegal and the Sperrin Mountains

Donegal, in contrast to NW Mayo, the Ox Mountains and Connemara, shows evidence of three major phases of Grampian deformation. F1 folds are best seen on the Inishowen peninsula, where there is a primary facing divergence across the Inishowen Syncline, which is correlated with the Loch Awe Syncline of Scotland (Hutton

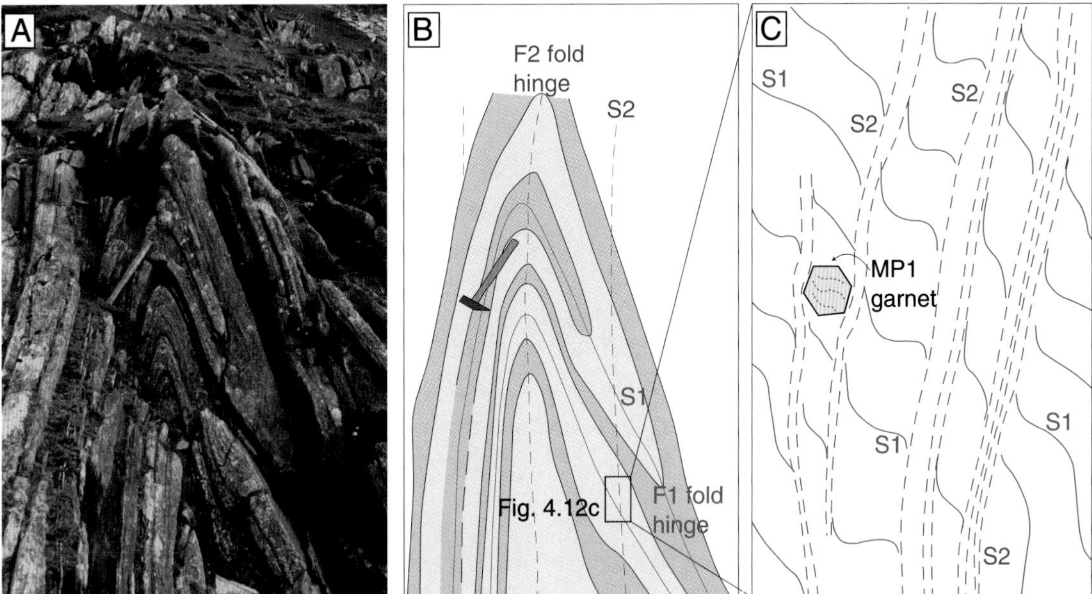

Fig. 4.12 Field photograph and illustrations used to convey the meaning of the terminology employed in describing polyphase-deformed rocks. *See* text.

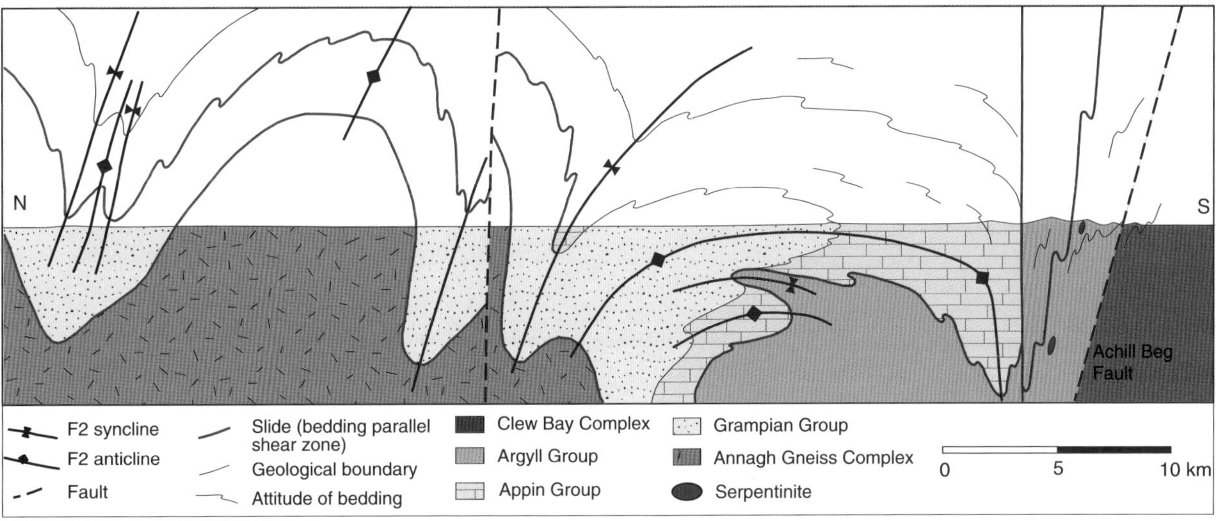

Fig. 4.13 N–S cross section through the NW Mayo inlier after Chew (2003).

and Alsop, 1996). In north Inishowen, bedding faces up to the north-west on S1; to the south, bedding faces up to the south-east on S1. D2 in north Donegal is associated with north-west-directed thrust nappes (Fig. 4.14a). These thrust nappes are separated from each other by D2 slides such as the Horn Head and Knockateen slides. Within each slice recumbent F2 folds face north-west (Hutton and Alsop, 1995). The timing of the development of the F2 folds relative to the D2 slides is somewhat complex due to the 'progressive' nature of the deformation. To the south-east of the Leannan Fault, isoclinal, recumbent F2 folds, such as the Sperrin Nappe (Fig. 4.14b) are inferred to have resulted from south-east-directed overshear (Alsop, 1996).

Major recumbent F3 nappes and shear zones show a consistent vergence and sense of movement to the south-east, and are best developed south-east of the Leannan Fault (Alsop, 1996). The largest of these is the crustal-scale Ballybofey Nappe (Alsop, 1996). During the F3 phase of south-east-directed shearing, the locally-inverted Dalradian succession was thrust over both the Slishwood Division of the Lough Derg Inlier (Alsop, 1991), and the

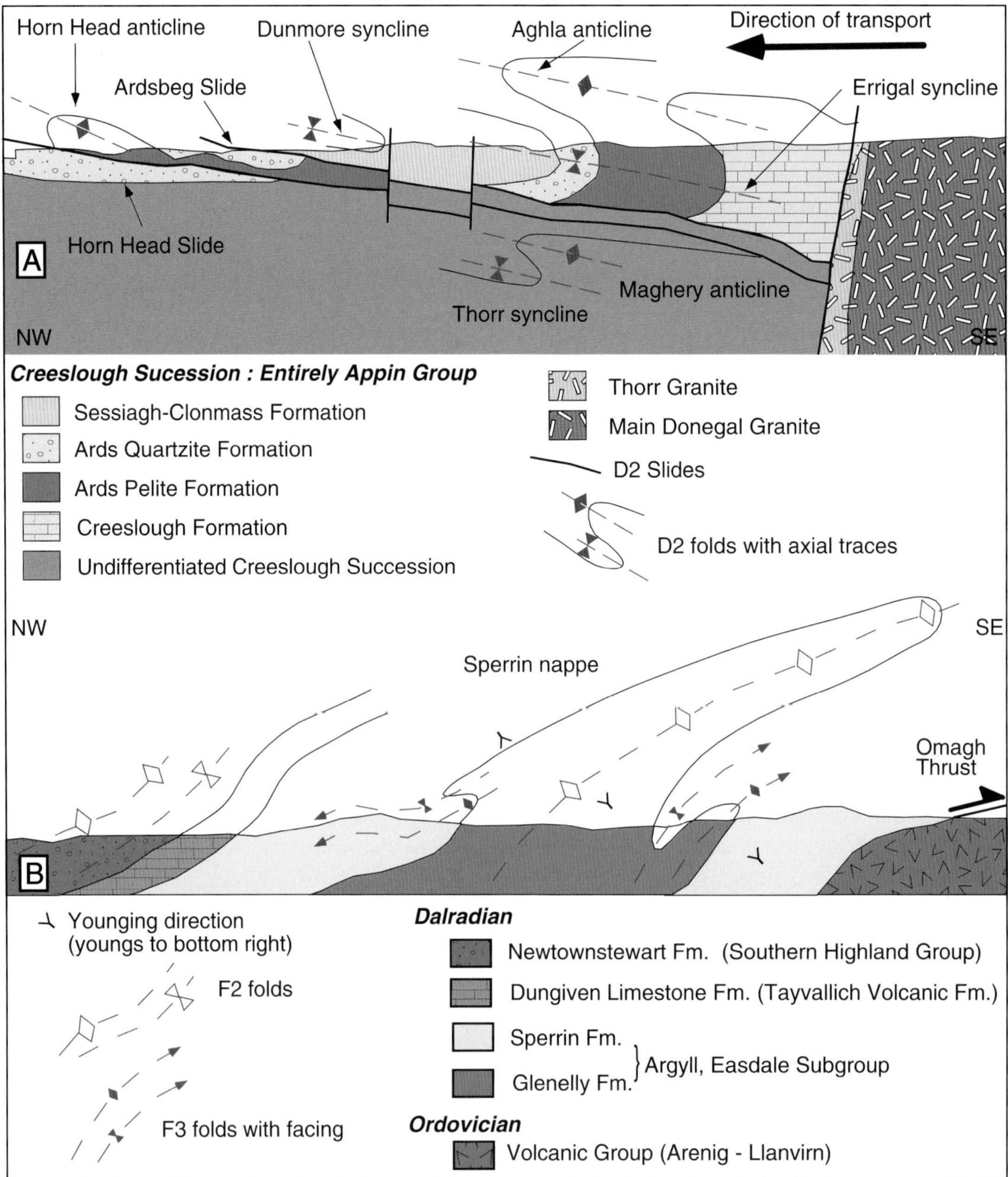

Fig. 4.14 (a) Northwest–southeast cross-section of northern Donegal showing D2 structural features, adapted from Long and McConnell (1997). (b) Northwest-southeast cross-section through mid-Ulster showing major D2 and D3 structures, adapted from Alsop and Hutton (1993).

Ordovician volcanics of the Tyrone Inlier (Alsop and Hutton, 1993; Fig. 4.14b). This south-east-directed D3 thrusting in southern Donegal is very similar to the D3 interleaving of the Dalradian and the Slishwood Division in the NE Ox Mountains Inlier (Flowerdew, 2000).

The metamorphism in Donegal was generally MP2 in age, and reached the garnet zone over most of the inlier (Pitcher and Berger, 1972). The metamorphic grade increases approaching the Lough Derg inlier, as progressively deeper structural levels are exposed through the inverted lower limb of the Ballybofey Nappe (Alsop, 1991). Metamorphic grade in this region was as high as the staurolite–kyanite zone and is MP3 in age (Alsop, 1991).

Grampian evolution of the Central Ox Mountains Inlier

The structural history (and the stratigraphical affinities) of the Ox Mountains Inlier have in the past proved controversial (e.g. Phillips *et al.*, 1975; Long and Max, 1977). The generally accepted chronology of events is that of MacDermot *et al.* (1996) which is adapted slightly from that of Jones (1989). F1 folds are not seen; S1 is preserved as inclusion trails in porphyroblasts or in the hinges of rare F2 folds. Only one major F2 fold has been identified – the isoclinal, upward-facing Lough Anaffrin anticline of Long and Max (1977). A major D3 structure, the Lough Talt antiform, is responsible for the disposition of the major units (Fig. 4.15), although the hinge of this fold has been nearly completely obscured by the later intrusion of the Ox Mountains Granodiorite (Fig. 4.15).

In the NE Ox Mountains (Fig. 4.1) an enigmatic unit composed predominantly of psammitic gneisses known as the Slishwood Division (Max and Long, 1985), was interleaved with Ox Mountains Dalradian during D3 deformation (Flowerdew, 2000). The Slishwood Division contains latest Precambrian granulite-facies assemblages (Sanders *et al.*, 1987) and early Grampian tonalite intrusions (Flowerdew, 1998/9; Flowerdew *et al.*, 2005). Interleaving of the Dalradian and the Slishwood Division was caused by top-to-the- ESE shearing along the flat-lying D3 North Ox Mountains Slide (Flowerdew, 2000).

The Ox Mountains Granodiorite was intruded syn-kinematically during D4 sinistral transpression (McCaffrey, 1992). D4 strain partitioning in the Dalradian is localised as tectonic slides (MacDermot *et al.*, 1996). These D4 high strain zones include the Callow Shear Zone, the Glennawoo Slide and the Lough Talt Slide (Jones, 1989) and are likely to have had an earlier history of movement (MacDermot *et al.*, 1996, Flowerdew, 2000). The age of

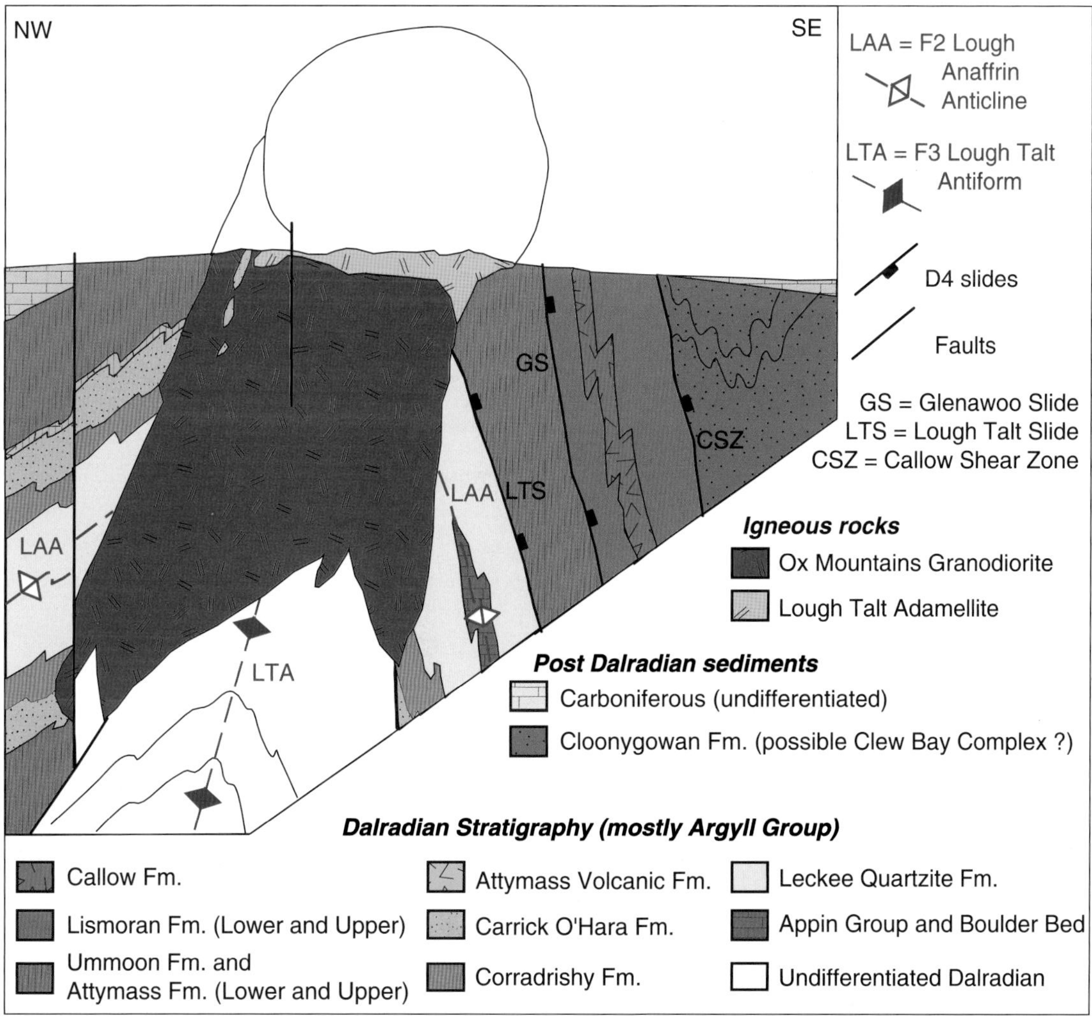

Fig. 4.15 Schematic cross section through the Central Ox Mountains after MacDermot *et al.* (1996). The continuation of the Ox Mountains Granodiorite and the Lough Talt Adamellite above the ground surface is indicated by solid lines.

the D4 event is constrained by the 412.3 ± 0.8 Ma crystallisation age (Chew and Schaltegger, 2005) for the Ox Mountains Granodiorite (Chapter 8) and hence D4 is post-Grampian.

The regional metamorphism in the Ox Mountains Dalradian is MP3 in age (MacDermot *et al.*, 1996). A transect across the inlier shows that metamorphic grade decreases from the staurolite–kyanite zone in the NW to greenschist facies in the SE. Yardley *et al.* (1987) estimate that the PT conditions of the MP3 metamorphism in the staurolite–kyanite zone are similar to those of the staurolite-kyanite zone in NW Mayo at 8 ± 2 kbar and 620 ± 30°C. PT estimates of the slightly earlier D3 interleaving of the Dalradian with the Slishwood Division are *c.*9–10 kbar and 550–600°C (Flowerdew and Daly, 2005).

Grampian evolution of Connemara

The structural development of the allochthonous Connemara terrane closely follows that of the Ox Mountains. Evidence for a D1 structural event is tenuous, with S1 preserved solely as inclusion trails in garnet and plagioclase porphyroblasts, and no unequivocal F1 folds have been found (Leake and Tanner, 1994). A major, isoclinal F2 fold, the Derryclare anticline (Tanner and Shackleton, 1979; Fig. 4.16), is refolded by several tight F3 folds. The fact that only one major D2 fold has been identified in Connemara is somewhat intriguing, especially when one considers that the axial plane is defined by the Dalradian tillite (referred to in Connemara as the Cleggan Boulder Bed). This is presumably at least partly a function of the rather unusual lithologies present in the boulder bed, which greatly aids identification of the stratigraphic repetition caused by the F2 folding. The tight F3 fold complex is north-facing (Leake, 1989). The F3 folds were probably flat-lying nappes (Long *et al.*, 1995) prior to D4.

F4 folds, such as the Connemara Antiform, have a close to open style with subvertical axial planes, and have modified the stacked nappe pile into flat and steep belts (Fig. 4.16). During or just prior to D4, the Connemara terrane was thrust southwards over the silicic volcanics of the Delaney Dome Formation (Leake *et al.*, 1983). This shear zone, known as the Mannin Thrust, is interpreted

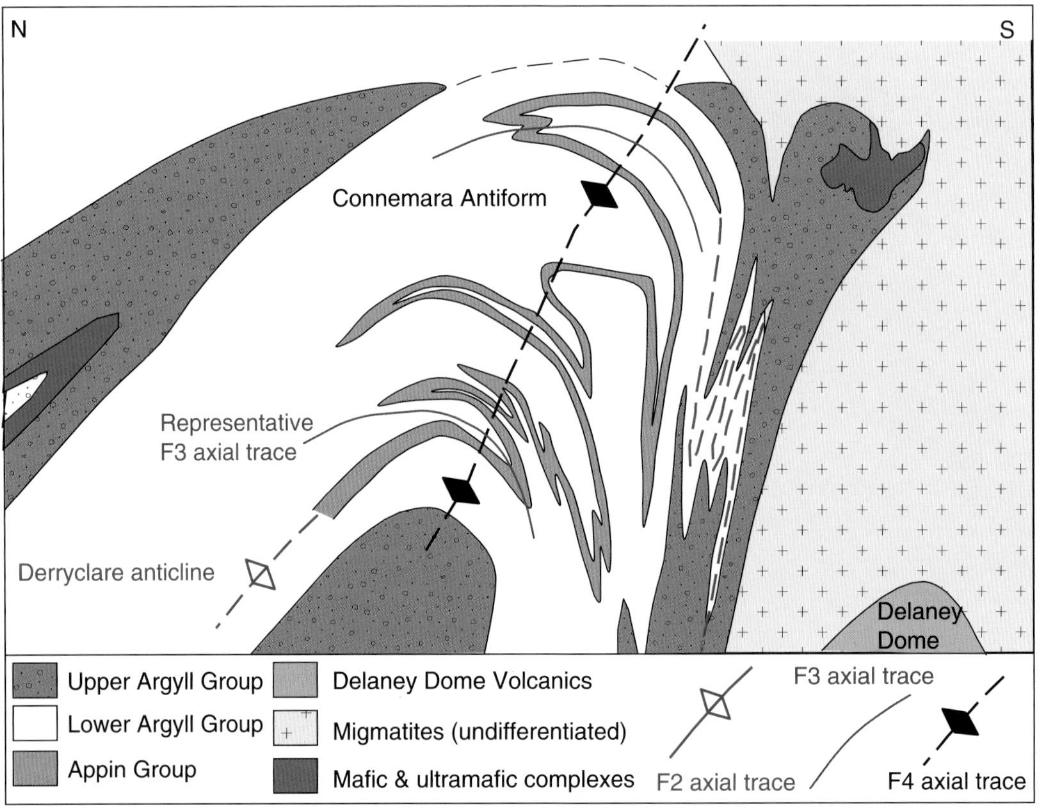

Fig. 4.16 Schematic north–south cross section through Connemara showing the F2 and F4 axial traces of the Derryclare anticline and Connemara antiform. Most F3 traces are omitted for clarity. Adapted from Tanner and Shackleton (1979).

as a 'retrocarriage phase' marking the end of orogenic growth (Dewey and Mange, 1999) and a consequent flip in subduction polarity.

Metamorphic grade in Connemara increases southwards across a series of well-defined east–west isograds, from the garnet zone on the islands of Inishshark and Inishbofin to the sillimanite-K-feldspar zone in the south (Leake and Tanner, 1994). These isograds cut the hinges of F3 folds (Leake and Tanner, 1994). Adjacent to the voluminous basic and intermediate intrusions of south Connemara, migmatic melting of metasediments of appropriate composition has taken place under low-pressure/high temperature conditions, analogous to the Buchan metamorphism of Scotland. Regional metamorphism is generally regarded as staurolite–kyanite zone over most of the Dalradian succession, and MS2–MP2 in age (e.g. Yardley, 1976; 1980), with later overprinting by MS3–MP3 low-pressure/high temperature assemblages.

5

Cambrian of Leinster

C. H. Holland

In mainland Britain, which includes the historical type area for the Cambrian System, these rocks are restricted in outcrop compared with those of the Precambrian, Ordovician, and Silurian. In Ireland, Cambrian rocks are not only even more restricted but have raised problems in stratigraphical dating. Outcrops in south-eastern Ireland are confined largely to Howth Head and Ireland's Eye to the north of Dublin Bay, to the Bray Head district to the south of Dublin, and to south-eastern County Wexford. (Fig. 5.1). Additionally, there is a very small outcrop at the south-eastern end of the Slievenamon inlier (Fig. 7.8) in County Kilkenny.

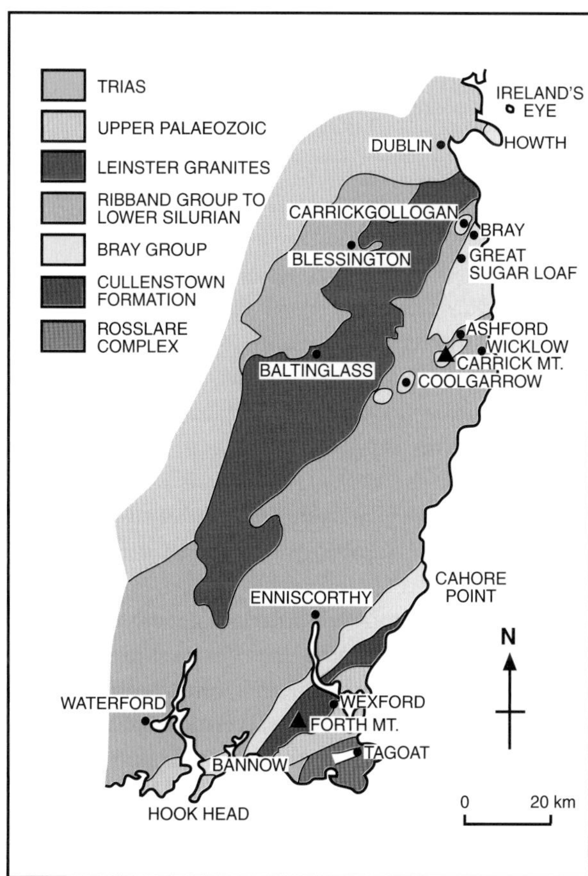

Fig. 5.1 Geological sketch-map of south-eastern Ireland.

In more recent years, microfossils have proved to be useful in Irish Cambrian stratigraphy (for references see below). Body macrofossils have scarcely been found in these rocks, and the characteristic trace fossil *Oldhamia* has itself been the subject of biostratigraphical debate – apart from the enigma of its relationship within the classification of living things. Two forms are known: *Oldhamia radiata*, consisting of ridges radiating from a centre, and *Oldhamia antiqua* with semicircular patterns of radiating ridges that may occur in linear series (Fig. 5.2). The radiating structures of both species may be seen to penetrate several superimposed sedimentary laminae. *Oldhamia* was first recorded from Bray Head by Oldham (1844) and described by Forbes (1848), who suggested that it might be a bryozoan. Since then it has been ascribed to various animal and plant groups, but is now generally regarded as a trace fossil composed of radiating burrows. Other records of its species in Europe and North America are of uncertain age (Dhonau and Holland, 1974), but there are cases of its association with known Cambrian fossils, as in New York State, Boston, and Norway. Kjerulf (1880) described it from the Oslo graben in association with *Paradoxides*. In New York State, Dale (1904) recorded it from the Nassau Formation, now known to be Lower Cambrian. In the Boston area, *Oldhamia* is associated with Cambrian fossils (Howell, 1922; Palmer, 1971).

The Bray Head district

Rocks of the Bray Group (Gardiner and Vanguestaine, 1971; Dhonau and Holland 1974; Brück *et al.*, 1979) are well exposed along the coast at Bray Head and to the south of it, and their outcrop extends over a considerable inland area to the south-west. Brück *et al.* (1974) recorded poorly preserved acritarchs of late Early Cambrian age from the Bray Head Formation (see below) in Rocky Valley, north of the Sugar Loaf mountains. *Oldhamia* is commonly found in the thick (more than 4,500 m) succession of the Bray Group as a whole, particularly in purple and green slates and laminated siltstones at the

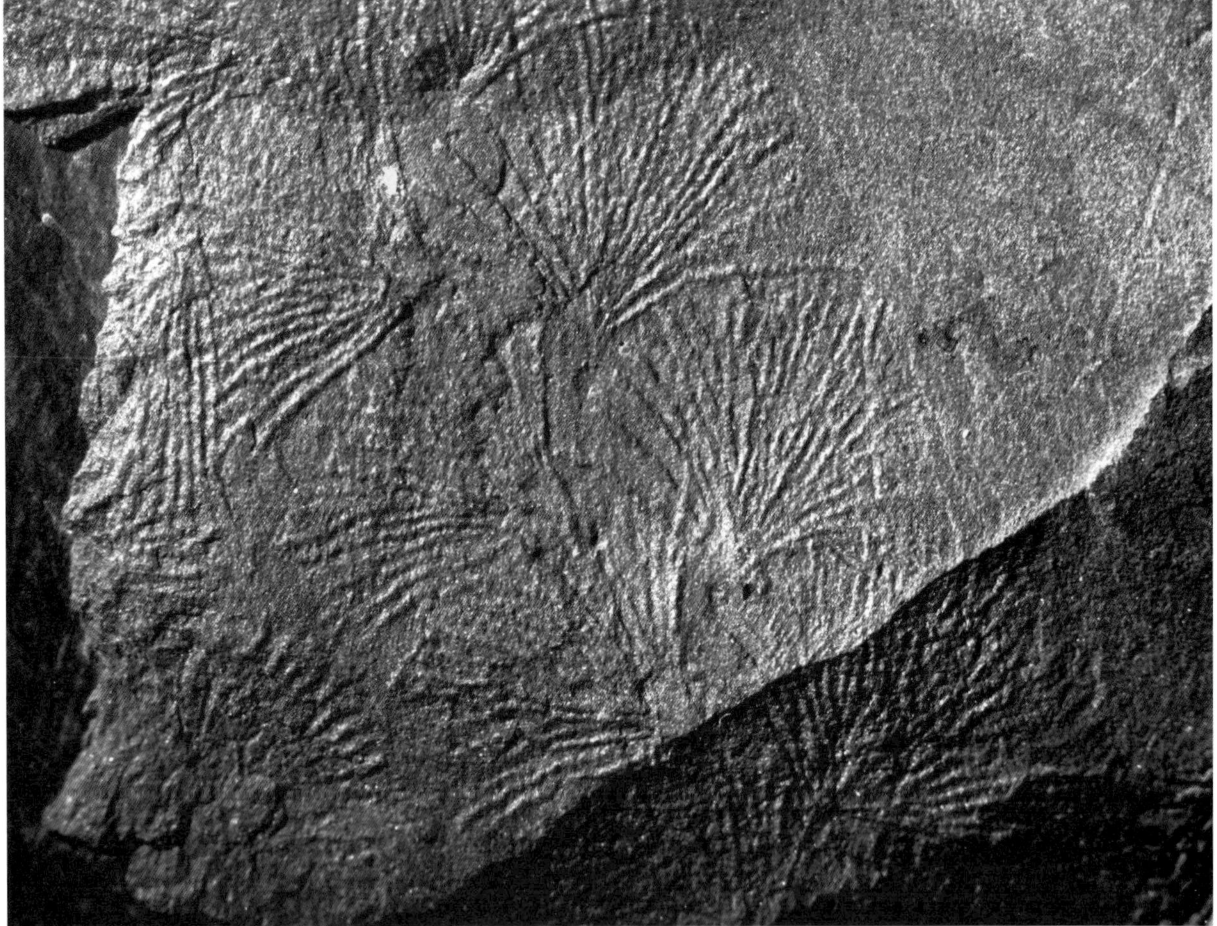

Fig. 5.2 *Oldhamia antiqua* Forbes (National Museum of Ireland specimen NMI G2:1969) Coll.'Mr J.R.Joly' (× 3.8). Part of a specimen figured by Joly (1887).

tops of turbiditic units. Other trace fossils are found, of which the worm burrows *Arenicolites* and *Skolithos* are probably indicative of relatively shallow water. Brindley *et al.* (1973) suggested that the supposed trace fossil 'Histioderma' actually represents small sand volcanoes.

The succession at Bray Head consists of predominantly northward-dipping and younging green, purple, red, and grey slates and interbedded (commonly feldspathic) greywacke sandstones and siltstones, frequently well laminated. Very thickly bedded white to pink quartzites up to at least 100 m but laterally variable in thickness form prominent coastal and inland features, such as the two Sugar Loaf mountains (Fig. 5.3). In some of the inland ridges repeated displacement of the quartzites by cross-faulting is well displayed (Fig. 5.4). Sedimentary structures, such as graded bedding, cross-lamination, convolute bedding, and flute moulds are well seen, and many of the greywackes display the Bouma sequence. They are regarded as proximal turbidites, and the presence of associated quartzites and slump structures is appropriate to this designation. Sedimentary structures are less easily recognised in the quartzites themselves, but some are seen to be slumped, and basal flute moulds are also known. Load moulds and flame structures are found in the greywacke turbidites as well as at the bases of the quartzites.

Brindley *et al.* (1973) summarised directional evidence available in the area neighbouring Bray Head. Brück and Reeves (1976) made additional observations in the wider area of outcrop. Flute moulds indicate a north-westerly derivation for the proximal turbidites, but there is evidence from ripple drift of more distal material having arrived from the north-east. In places, on the other hand, a generally northerly origin is indicated. The quartzites were deposited as slump sheets or grain flows from the east.

Brück and Reeves (1976) recognised a lower, Devil's Glen Formation (2000 m) cropping out anticlinally

Fig. 5.3 *The Great Sugar Loaf* (about 7 km south-west of Bray), County Wicklow. *Photograph: C.H. Holland.*

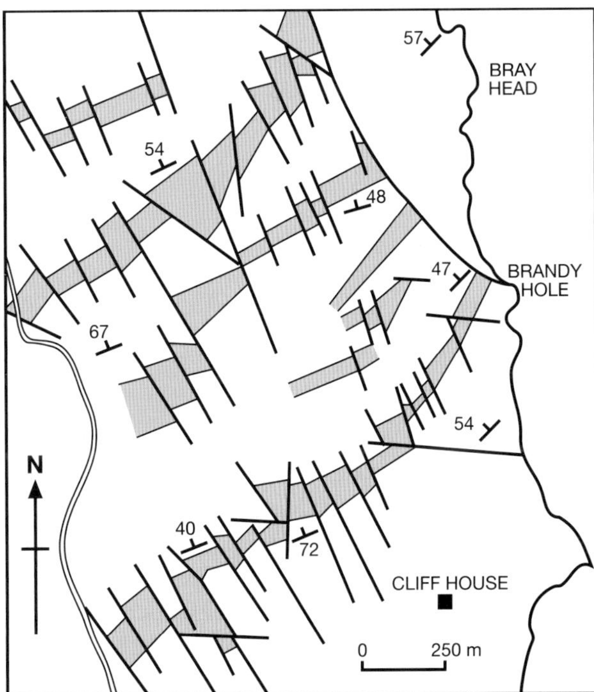

Fig. 5.4 Faulting in quartzite units at Bray Head (modified from Brück and Reeves 1976).

in the west from south of Enniskerry, running east of Roundwood towards Ashford, its type locality in the Vartry River. Its greywackes and slates are thickly bedded and quartzites are very rare. A narrow strip of country west of this is of the succeeding Bray Head Formation (at least 2500 m), as is most of the whole Cambrian outcrop from Bray Head, through the Sugar Loaf mountains, Glen of the Downs, and Newtown Mount Kennedy. McConnell, Philcox, and others (1994) based their account of the Bray Group on work by C. O'Byrne. The transition from the Devil's Glen Formation to the Bray Head Formation was thus taken to be a thick unit of red shales seen in the Powerscourt River and at the base of the Bray Head section. This was interpreted as a hemipelagic deposit, representing a discrete interval within the normal turbiditic conditions.

Brück and Reeves (1976) suggested that an early pre-cleavage phase, which may be largely soft sediment deformation, accounts for major folds in the district as well as many minor ones. An early cleavage is rare. The main east-north-easterly cleavage of the area relates to the very numerous tectonic folds of a few metres in scale.

Significant plunge on the folds is confined to a southern belt. This cleavage may be seen to cut some of the first phase folds, as in the northerly striking syncline of the Sugar Loaf mountains. A third phase of minor angular folds and kink bands is rarely developed. Apart from the many cross-faults cutting the quartzites, some of larger extent are found throughout the area, as for instance the Rocky Valley fault extending eastwards for more then 6 km.

Thrusting is important in defining the outcrop of the Bray Group. An isolated mass of these rocks has been thrust to the north at Carrickgollogan north-west of Bray, the hill being clear even in the profile of the Dublin mountains seen from the north side of Dublin Bay. A faulted, and possibly thrust, contact has been suggested also at Carrick Mountain in the south. A small area of Bray Group, again possibly in thrust contact with the Ribband Group, is found to the south-west of the main outcrop at and near Coolgarrow. The boundaries of the main outcrop of the Bray Group have not proved to be easy to define. To the north and west again there appears to be a thrust boundary with rocks of the Bray Group overlying the younger Ribband Group. Here, to the east of the Leinster Granite, Brück *et al.* (1974) referred to three formations which locally make up the Ribband Group in the area between the granite and the outcrop of the Bray Group. The oldest, Maulin Formation (900 m), adjacent to the granite, is of grey slates and thin quartzites, though there are two thicker lenses of quartzite. It probably extends down into the Cambrian, as similar ('Clara Series') rocks in Wexford have yielded acritarchs of this age. A tentative acritarch dating of the third and youngest part of the Ribband Group (the Glencullen River Formation), adjacent to the outcrop of the Bray Group and separated from it by a thrust, is of early Ordovician age.

The southerly contact of the Bray Group has been described as faulted, though to the south-west Tremlett (1959) recognised a sequence comprising 'Knockrath Series' (over 1,300 m of dark grey and black slates and quartzitic siltstones), 'Bray Series' (purple mudstones followed by grey and green greywackes and massive quartzites more than 300 m in total thickness), and 'Clara Series' (dark slates). The succession here does not (in fact) appear to show any substantial breaks, and Downie and Tremlett (1968) recorded acritarchs from the Clara Series, which, together with evidence of associated brachiopods, indicated a Cambrian age. Brück and Reeves (1976) referred both Knockrath and Clara series to the Ribband Group.

To the south of Roundwood, in the south-western part of the main outcrop of the Bray Group, Brindley and Millan (1973) described variolitic (basic) lavas, some with deformed pillows. Just over 5 km to the south-east of this, there are similar rocks at two horizons, evidently within the Devil's Glen Formation referred to above.

Howth and Ireland's Eye

The Howth Peninsula to the north of Dublin Bay was an island in early post-glacial time and is still linked to the mainland by only a tombolo of sand and gravel (Holland, 1981). A small area of Carboniferous dolomitised limestones is present at the landward end of the peninsula and is faulted against the rocks of the Bray Group, ferruginous material being prominent at the junction. The contact itself is seen clearly in the cliffs at Balscadden Bay on the northern side of the peninsula, where a fault breccia is developed. On the southern, Sutton shore at some conditions of the beach an area of ferruginous staining appears to mark the position of a fault. Inland, drift tends to conceal the outcrop, though topographical features and very limited sub-surface information suggest that there is not one simple fault line.

When a new edition of *The Geology of Dublin* (Lamplugh *et al.*, 1903) was published, the solid geology was taken largely from the earlier memoir by Kinahan and his colleagues (1875), but important notes were added concerning new views and additional information. The section relating to the Howth peninsula (pages 68 to 75) still provides a useful detailed account of the district. These Cambrian rocks are sufficiently similar to those at Bray Head as to allow assignment to the same lithostratigraphical group. On the northern side of the peninsula, *Oldhamia* is known from Puck's Rocks and from cliffs north of the Nose of Howth. Other trace fossils are to be found. Of these Crimes (1976) recorded *Arenicolites, Granularia, Planolites, Skolithos,* and *Teichichnus*, seemingly from beds below those yielding *Oldhamia*, which has usually been found with relatively deep water but proximal turbidites, whereas the lower assemblage indicates shallower water.

Lithologically the most striking feature is the presence of bands and isolated masses of quartzite, relatively resistant to marine erosion and also forming prominent inland ridges and areas of high ground, such as those of Shielmartin Hill and of Black Linn at the summit of the peninsula. These rocks are characteristically ochreously stained, and this so-called 'Howth Stone' is attractively seen in old walls (Fig. 5.5) and buildings, though less so

Fig. 5.5 Howth stone wall, Howth peninsula. *Photograph: C.H. Holland.*

in some of the garnishments of modern houses in the Dublin area. The quartzites are repeated by faulting, but they appear also to be stratigraphically numerous. Other sedimentary rocks are well seen in the cliff sections. There are greenish, red, and grey shales, siltstones, and greywacke sandstones. Bentonitic clays are common. Sedimentary structures are especially well displayed on the bare rock surfaces washed by the sea. They include parallel lamination, graded bedding, and flute moulds. Slump structures of various kinds are present at many localities, for instances at Kilrock, Hippy Hole, and Sheep Hole (Fig. 5.6) on the south side of the peninsula. There are conglomerates in which laminated quartzite clasts set in a matrix show primary lamination disturbed by submarine sliding and subsequently cleaved with the matrix. A greywacke facies with much evidence of soft sediment deformation and re-sedimentation contrasts with the prominent quartzites. In some cases the latter may be simply interbedded with the turbiditic greywackes, but in others they are thought to represent exotic blocks, which may be as much as 700 m in size. In many cases the precise origin of the quartzites is unclear.

Cleavage is widely developed in the finer rocks, and more than one episode may be seen to be present. The quartzites are well jointed and there is much faulting, though not necessarily on a substantial scale. Van Lunsen and Max (1975) mapped five formations at Howth and recognised a sequence from sedimentary, through tectono-sedimentary, to true tectonic structures. The effects of soft sediment deformation appear to increase as the succession is ascended; slump folding and slumped masses of quartzite in the lower beds giving way to 'chaotic sedimentary breccias' at higher levels. All the formational boundaries were thought to involve slumping. The main structure of the area was regarded as a syncline plunging steeply to the east, but there is a profusion of smaller-scale structures. The succession is cut by basic intrusions, mostly in a rotten condition. Some resemble the Ordovician Lambay Porphyry of Lambay Island some 8 km north of Howth.

Ireland's Eye, the small island about one kilometre from Howth Harbour, has quartzites at its northern and southern ends and finer red or variegated rocks, partly covered at high water, in between. The northern quartzite

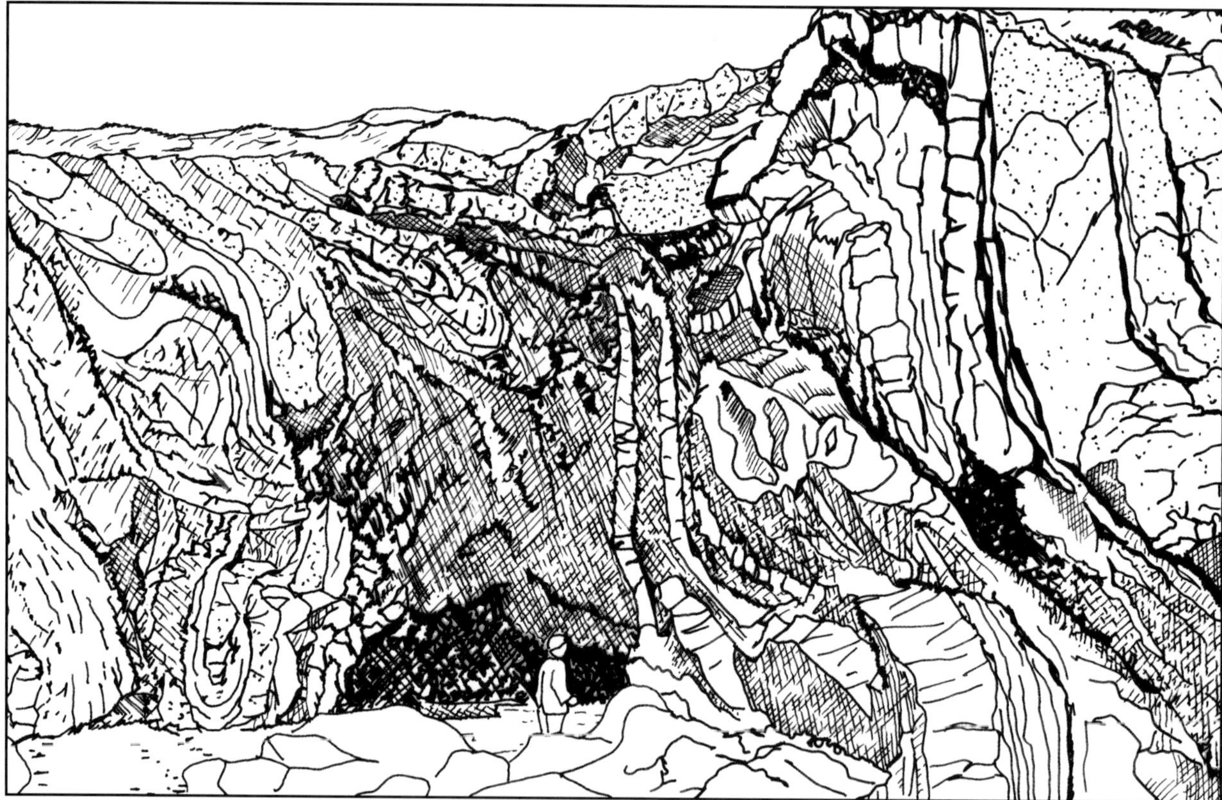

Fig. 5.6 Large-scale slumping at Sheep Hole (approximately 2 km west of Baily Lighthouse), Howth peninsula (after Van Lunsen and Max, 1975).

forms the main mass of the island and its striking profile is seen from Howth. 'A huge rock on its eastern extremity…appears to have been driven under by some convulsion of nature…' (Bartlett, 1842). The succession has been compared with the lower part of that at Howth. The island must represent part of an isolated anticlinal culmination or be separated from the mainland by faulting. In the former case its surface profile may approximate to an original post-Carboniferous erosion surface. Gardiner and Robinson (1970) provided a preliminary account of the geology of the island, awarding it three Cambrian formations. Van Lunsen and Max (1975) considered these to young southwards rather than northwards.

Gardiner and Vanguestaine (1971) described acritarch assemblages from two samples from the Bray Group on Ireland's Eye. By comparison with material from eastern Europe, the assemblages were regarded as Early to Middle Cambrian in age. Smith (1977) extracted an assemblage from Howth, which, by reference to material from Britain as well as the Baltic region, he showed to be of middle Early Cambrian age.

County Wexford

In the extreme south-west of County Wexford, rocks of the Booley Bay Formation to the south and east of Duncannon, at the landward end of the Hook Peninsula, must, largely at least, be assigned to the Cambrian. The formation comprises some 2500 m of dark cleaved mudstones, siltstones, thinly bedded greywacke sandstones, and some conglomerates. It is of distal turbidites and provides an excellent display of sedimentary structures such as flute moulds and groove moulds (Gardiner, 1978). Tietzsch-Tyler and Sleeman (1994a), in their geological description to accompany the South Wexford Sheet of the Geological Survey of Ireland's 1:100,000 Map Series, extended the outcrop of the same facies, though much divided by faulting, north-eastwards to the River Slaney to join rocks which Shannon (1978a) had mapped as the Polldarrig Formation (900 m) at the base of his Ribband Series. Here there are dark grey shales and slates with some greenish greywackes. Thin quartzites are prominent in a middle member.

Moczydlowska and Crimes (1995) obtained acritarchs of Late Cambrian age from various levels in the Booley

Bay Formation. In an adjacent paper, Crimes *et al.* (1995) described and illustrated what they referred to as an 'Ediacaran biota' from Booley Bay with the two genera *Ediacaria* and *Nimbia*. Jensen *et al.* (2002) have since reinterpreted *Nimbia* as 'swing marks'. MacGabhann *et al.* (2007) provide a very full discussion of the supposed *Ediacaria booleyi*, based on study of over 200 specimens. They conclude that these structures should not be so assigned. They are tentatively considered to be, at least in part, biogenic, but the possibility of a wholly inorganic origin is not discounted. Thus an Ediacaran biota is no longer to be recognised in these Cambrian rocks.

A narrow strip of Cambrian rocks runs from the south coast east of Bannow, north-eastwards, eventually to reach the coast again at Cahore Point. The name 'Cahore Group' was applied by Tietzsch-Tyler and Sleeman (1994a) to the whole of this strip and also to the Booley Bay Formation already mentioned. The Cahore Group is best regarded as equivalent to the Bray Group.

On the coast east of Bannow, Dhonau (1973) defined various formations, one of which, the Cross Lake Formation in the east near Blackhall, is in faulted contact with a small outcrop of Caradoc rocks. The succession along this faulted strip of coastal exposures is now understood to begin with the Kiln Bay Formation, which is followed by the Cross Lake Formation and then the Ardenagh Formation. All three yield *Oldhamia*. In contrast to the Booley Bay Formation, all these are dominated by greyish-green greywacke sandstones and variously coloured mudstones. White quartzites occur in the Cross Lake Formation with evidence of slumping and sliding related to the south-eastern margin of the basin of deposition.

The north-eastern coastal section of the Cahore Group was divided by Crimes and Crossley (1968) into three formations: the Glasscarrig Formation (600 m), the Pollduff Formation (more than 750 m), and the Roney Formation (more than 600 m). The second of these contains *Oldhamia* and the third *Oldhamia* and *Arenicolites*. All three consist dominantly of feldspathic sandstones, siltstones, and red and green shales, with quartzites becoming conspicuous in the third formation. Graded bedding, parallel lamination, groove moulds, and load structures are present. A proximal origin has been suggested for the coarser components.

Tectonic units in south Wexford have been compared with those recognised across the Irish Sea in Anglesey (Tietzsch-Tyler, 1989;Tietzsch-Tyler and Sleeman, 1994a). A postulated second unit (within the Cahore Group) to the south-east of the Cambrian rocks mentioned so far is of the Cullenstown Formation (Max and Dhonau, 1974). Its outcrop is a poorly exposed, much faulted north-easterly trending belt from Cullenstown Strand on the south Wexford coast, through Forth Mountain, to the coast at Wexford Town. The chlorite grade metasediments of the Cullenstown Formation include quartzites, greywackes, and pale green and red albite-chlorite-schists. Max and Dhonau (1974) regarded their polyphase deformation as late Precambrian in age. However, Brück and Vanguestaine (2004) have described late Middle Cambrian to early Late Cambrian acritarchs from the formation, comparable with records from eastern Newfoundland.

Brück *et al.* (1979) extended the outcrop of the Cullenstown Formation and thus included the ground that Shannon (1978a) had mapped as Bray Group. In this respect they were following the work of Max and Dhonau (1974). Shannon, in his description of the area to the north-west of Wexford Town, had recognised three formations within his Bray Group, the first dominantly of quartzites and the other two dominantly of greywackes, the whole over 2000 m thick The third formation was said to be followed conformably by the Polldarrig Formation already mentioned. The succession youngs consistently north-westwards. Shannon (1978b) concluded that the origin of the matrix of the greywackes in south-eastern Ireland is best explained as being the result of diagenesis influenced primarily by the effects of overburden. He regarded the subsequent low-grade regional metamorphism as having little effect on the diagenetic features.

Bennett *et al.* (1989) later suggested that the tripartite division north-west of Wexford Town, ascribed to the Bray Group by Shannon (1978a), can be recognised also in the Cullenstown Formation at Cullenstown Strand. They found that the Cullenstown strata are all inverted. They suggested that their slightly higher grade of metamorphism compared with that of the Bray Group elsewhere can be explained by their somewhat deeper burial on the recumbent limb of a major F1 anticline, the normal limb of which may be represented by the rocks described by Shannon (1978a).

To the south-east of the outcrop of the Cullenstown Formation, and across an area of Carboniferous strata, are the older rocks of a third tectonic unit (F.C. Murphy, 1990), the Rosslare Complex of Precambrian gneisses and amphibolites. An outcrop of mylonitised clastic rocks (the Ballycogly mylonite zone of F.C. Murphy, 1990) is present to the north-east beween the Rosslare Complex and the outcrop of the Carboniferous. Tietzsch-Tyler and Sleeman (1994a) assigned the Ballycogly Formation (better than 'Group' as in their account) to the Cambrian. It is certainly older than the Arenig rocks of the Tagoat Formation, but there is no proof of a Cambrian age.

Slievenamon

A very small area in the south-eastern corner of the Lower Palaeozoic inlier of Slievenamon, County Tipperary (Fig. 7.8) was assigned by Colthurst (1974) to the Carricktriss Formation (800+ m). It is in faulted contact with the Ahenny Formation, a fault breccia being seen in a stream section. Much of the succession consists of volcanoclastic rocks: feldspathic tuffs, fine-grained green tuffs, and green breccias. There are associated siltstones and black shales. Rare bands of tuff near the base of the Ahenny Formation suggest that there was continuity of deposition between the two formations. A sample from siltstones with convoluted lamination obtained from the local base of the Carricktriss Formation yielded an acritarch assemblage of Late Cambrian age (Colthurst and Smith, 1977), whereas another sample from the upper part of the formation was dominated by acritarchs giving a Tremadoc to Arenig age. Smith (1979) later mentioned the possibility of reworking into the Silurian in these upper beds.

Palaeogeography

The slumped beds, including conglomerates, in the Cambrian of Wexford indicate proximity to what has long been referred to as the Irish Sea Landmass (Irish Sea Horst Complex of Brasier *et al.*, 1992). The Welsh Cambrian rocks were deposited in a basin to the south-east of this, and the Bray Group was evidently laid down in a corresponding subsiding basin to the north-west. Indications are that this basin was relatively narrow with active, possibly fault-controlled, margins. The Cambrian rocks of the Durness Group of north-western Scotland, with their faunas of North American affinity, are not represented in Ireland. Evidence from the meagrely fossiliferous Cambrian rocks of Ireland, taken alone, demands no more than a subsiding basin, reasonably referred to as the Leinster Trough, and a more active south-eastern margin. The lack of records of trilobites from the Bray Group remains tantalising; but then the Lower Cambrian of Wales is scarcely better endowed.

6

Ordovician of the South

J. R. Graham and C. J. Stillman

This chapter deals with rocks that formed either on the southern side of the Iapetus Ocean, attached to Avalonia, or formed within the main oceanic area and are now present in the Iapetus suture zone. There is little evidence of original continuity between the current, geographically separated, outcrops, most of which have been assigned to separate terranes (Murphy *et al.*, 1991; Harper and Parkes, 1989), partly on the basis of faunal differences. They are considered here in three main groupings; the East-Central Ireland inliers, the main Leinster outcrop, and the Rosslare Block.

East-central and south-west Ireland inliers

A series of small inliers of Ordovician rocks occur between the Southern Uplands–Longford Down tract and the extensive outcrop of the Leinster zone to the south-east. These all show significant contrasts from the sequences in the Longford-Down but varying degrees of similarity to the main Leinster sequence. The position of these inliers is indicated on Figures 6.1 and 6.2 and summary correlations are given in Figure 6.3. It is likely that the unexposed boundaries between adjacent inliers are faults, and thus original geographical relationships are highly debatable.

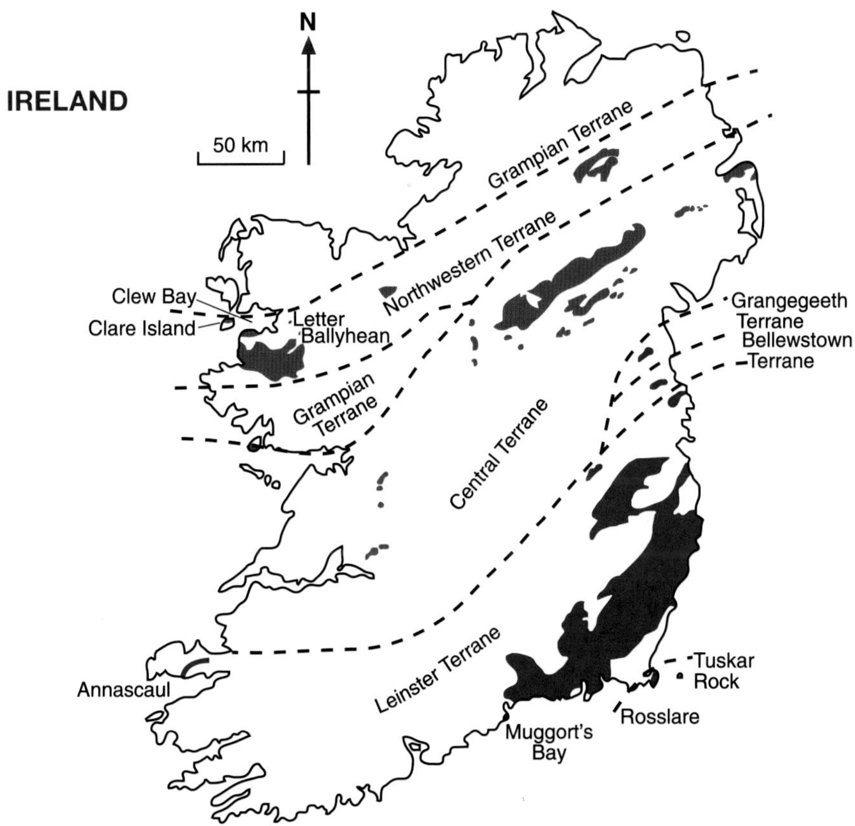

Fig. 6.1 Distribution of Ordovician rocks in Ireland. Those discussed in this chapter are shown in purple.

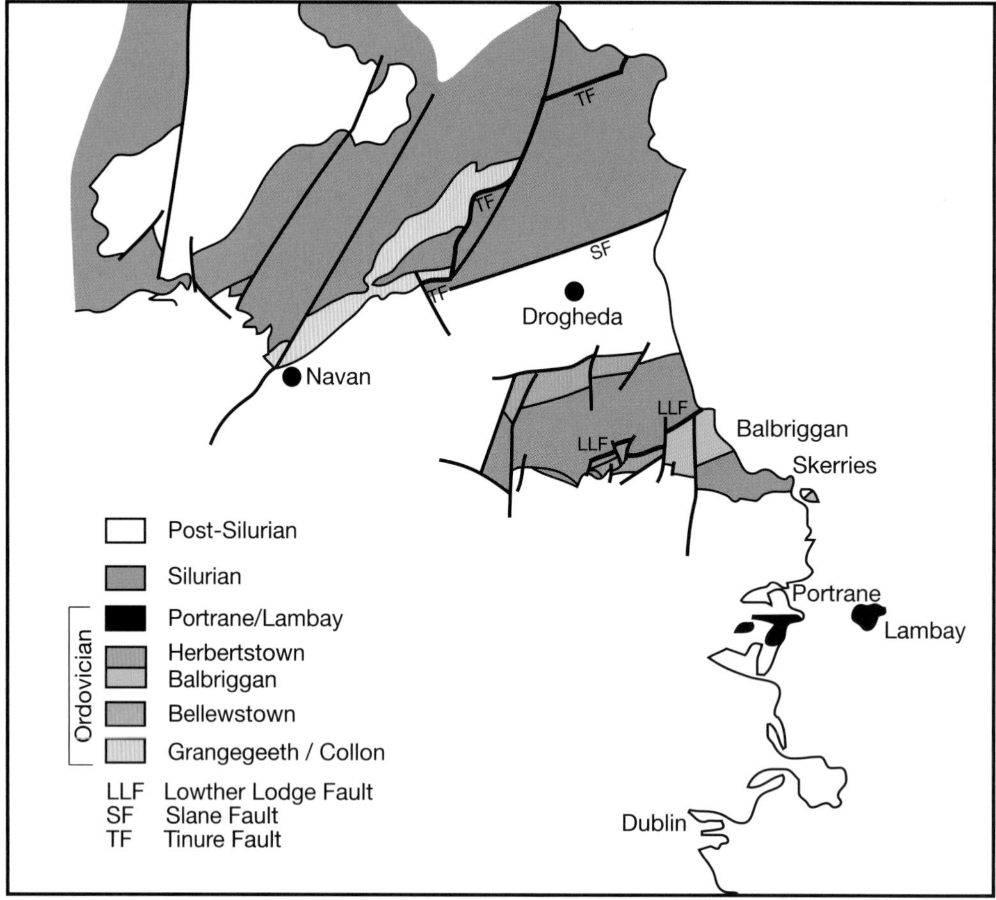

Fig. 6.2 Distribution of the Ordovician inliers in east-central Ireland. Successions are discussed in the text and summarised in Fig. 6.3.

Slane – Grangegeeth – Collon

This inlier is largely surrounded by Silurian rocks. Details of the succession are given in Harper (1948, 1952), Manistre (1952), Brenchley *et al.* (1967b), Harper and Romano (1967), Romano (1970), Brenchley *et al.* (1977), Romano (1980a), Vaughan (1991), and Owen *et al.* (1992) (Fig. 6.2). The lowest Slane Group (>1270 m) is a dominantly volcaniclastic sequence with interbedded mudrocks. It is possible that this group is all of Llanvirn age with the *D. artus* Biozone being confidently recognised in the Hill of Slane Formation, and faunas from the White Island Bridge Formation probably indicating the *D. murchisoni* Biozone. These graptolites have peri-Gondwanan affinities (Owen *et al.*, 1992). The volcanic rocks are predominantly crystal and lapilli tuffs, with some layers possibly resedimented by turbidity currents (Romano, 1980a). Newly discovered exposures confirm the south-westward extent of the Slane Group to near Navan, where the group also occurs beneath the Carboniferous in the subsurface (Vaughan, 1991).

The Grangegeeth Group represents a transgressive sequence that is inferred to unconformably overlie the Slane Group (Romano, 1980a). The oldest (Collon) formation consists of volcanogenic conglomerates interpreted as being produced in high-energy shallow water conditions by erosion of contemporaneous volcanic rocks, which are also evidenced by tuffs. The succeeding formations (Knockerk and Fieldstown) are dominated by sediments, commonly tuffaceous and/or calcareous, which have yielded rich shelly and graptolite faunas of Caradoc (Longvillian) age. The trilobite fauna analysed by Romano and Owen (1993) is interpreted to have Scoto-Appalachian affinities, and they suggested that the final Iapetus suture must lie south of Grangegeeth.

The youngest Mellifont Abbey Group is presumed by Romano (1980a) to overlie the Grangegeeth Group unconformably, although there could only be a limited time gap. The group consists of pelagic mudrocks and cherts that have been assigned an upper Caradoc and Ashgill age by Romano (1980a). Poorly located graptolites from

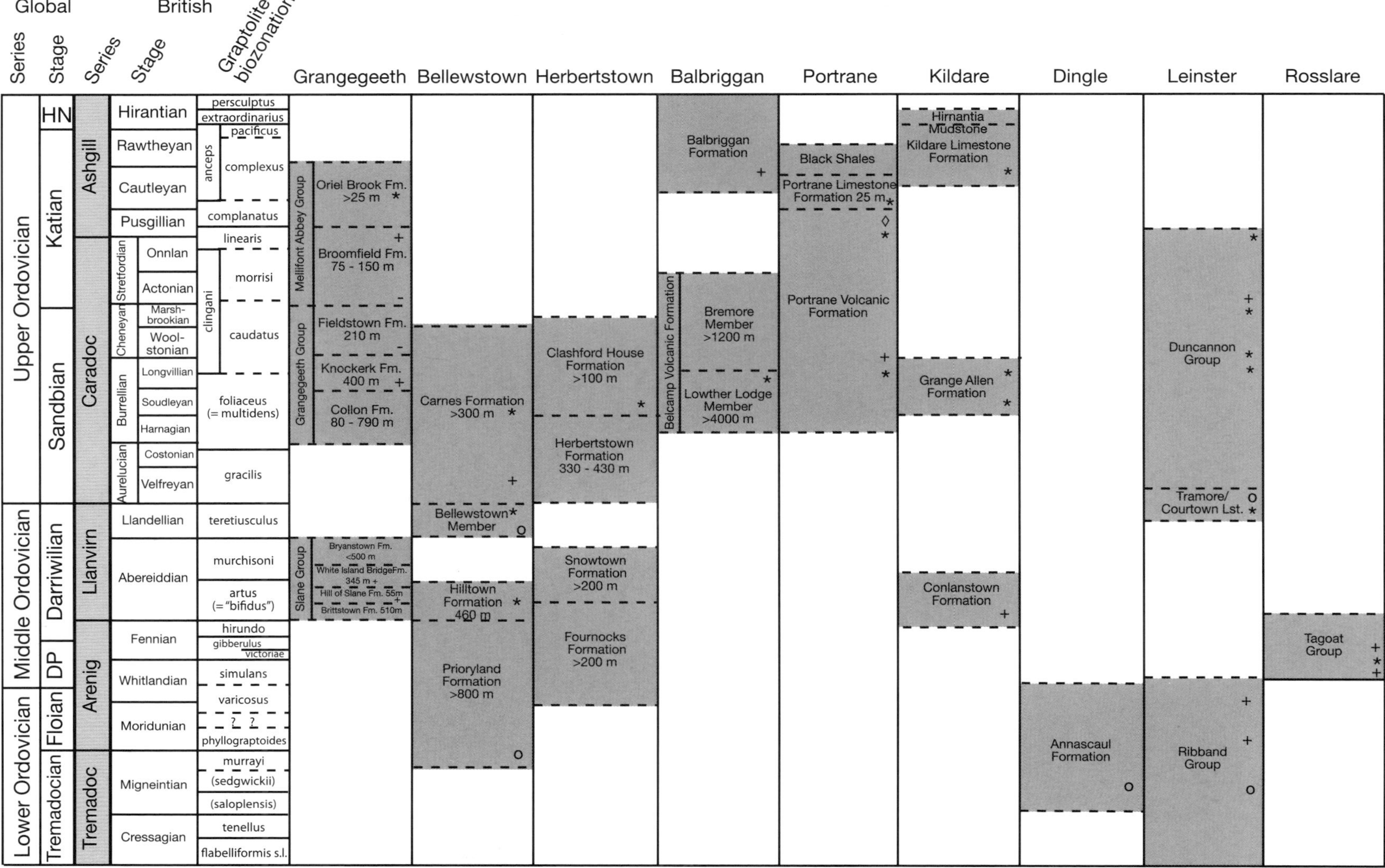

Fig. 6.3 Biostratigraphical correlation of the Ordovician successions of south and east Ireland.
°Microfossil data + Graptolites * Shelly fossils For global stages HN = Hirnantian; DP = Dapingian

the Broomfield Formation have been assigned to the *clingani* Biozone by Rushton and Zalasiewicz (1999) on the basis of *Normalograptus pollex,* which they claim has an age restricted to the lower part of the *clingani* Biozone. Furthermore, these authors suggest that *N. pollex* may be restricted to the southern margin of Iapetus. Although volcanic rocks are less evident than in the underlying groups, ashy mudrocks were recognised by Harper and Mitchell (1982) who interpreted these rocks as the products of a submarine slope adjacent to centres of volcanic activity. The youngest formation in the Mellifont Abbey Group, the Oriel Brook Formation, comprises deep-water mudrocks that have yielded trilobites (Romano, 1980a) and a *Foliomena* brachiopod fauna that indicates a Cautleyan age (Harper and Mitchell, 1982). It has been suggested that there may be a conformable passage into Silurian rocks in an inverted borehole sequence just west of Navan (Lenz and Vaughan, 1994).

In addition to the three groups described above, a series of quartz-phyric lavas crop out about 3 km west of Slane. These rocks are in fault contact with the Slane Group and have generally been assumed to be Ordovician in age (Gillot, 1951). However, their structural style suggests that they may be considerably younger and possibly even Devonian. A newly identified feeder of these lavas near Slane cross-cuts lamprophyres and thus is likely to be post–late Silurian (Vaughan, 1996).

Bellewstown

At Bellewstown >1.5 km of Ordovician rocks crop out which are unconformably overlain by Silurian rocks to the south (Murphy, 1987a), and which are faulted against Carboniferous rocks to the north. These have been described by Harper and Rast (1964), Romano (1970), McKee (1976), and Brenchley *et al.* (1977), and their stratigraphy and structure have been revised by Murphy (1984, 1985, 1987a).

A lower mudrock formation (Prioryland Fm. – >800 m) includes matrix-supported breccias of mass flow origin and contains detritus indicative of a quartzo-feldspathic source. It was assigned to the Arenig by Harper and Rast (1964) on the basis that the overlying formation was of Llanvirn age, but McKee (1976) recorded acritarchs of Tremadoc age from near the base of the formation. The overlying Hilltown Formation (460 m) consists mainly of acid volcanic rocks and interbedded graptolitic mudrocks assigned to the *D. artus* Biozone of the Llanvirn. Ash-flow ignimbrites are interpreted as the product of acid Plinian eruptions of rhyodacite composition and the presence of shelly fossil debris in the associated lahars indicates an at least partially submarine origin. The brachiopod faunas have been referred to the Celtic biogeographical province (Harper *et al.*, 1990). The succeeding Carnes Formation (>300 m) contains a basal, lenticular Bellewstown Member of calcareous sandstones and bioclastic limestones which appear to be late Llanvirn in age (Murphy, 1987a). Overlying graptolitic mudrocks are referred to the *N. gracilis* Biozone and the shelly faunas from the upper parts of the formation suggest an early–mid Caradoc (Harnagian or Soudleyan) age (Brenchley *et al.*, 1977). The bulk of the formation consists of pyroclastic rocks with some mudflows. A non-sequence or unconformity at the base of the calcareous Bellewstown Member is inferred by Murphy (1987a). The limestones may represent a condensed sequence, based on data from conodont faunas (Bergstrom and Orchard, 1985), or there may also be a further gap below the *N. gracilis* mudrocks (Fig. 6.3).

Balbriggan

An Ordovician succession, locally >1.6 km thick, crops out near Balbriggan, some 6 km south of the Bellewstown inlier. In the western, inland part of the inlier near Herbertstown are >930 m of sedimentary and volcanic rocks. The lowest two sedimentary formations (Fournocks and Snowtown formations) comprise unfossiliferous mudrocks that had previously been assigned to the Silurian on lithological grounds (Romano, 1980b; Harper *et al.*, 1985). However, Murphy (1987a) claims that they structurally underlie the tuffaceous, andesitic volcanic rocks and mudrocks of the Herbertstown Formation. The conformably overlying mudrocks of the Clashford House Formation have yielded shelly faunas of mid-Caradoc (probably Soudleyan) age (Harper *et al.*, 1985; Parkes, 1994).

At the eastern, coastal end of the inlier the Belcamp Formation (Murphy, 1987a) comprises mainly andesitic volcanic rocks with subordinate sedimentary levels of Caradoc (probably Longvillian) age (Parkes and Harper, 1996). Volcanism had virtually ceased during the Caradoc *D. clingani* Biozone prior to overstep by Ashgill graptolitic mudrocks of the *D. anceps* Biozone. The andesitic volcanic rocks are probably all submarine deposits and comprise pillow lavas and breccias, hyaloclastite, and crystal tuffs and also massive sheet-like extrusive and intrusive flows with associated breccias produced by intrusion into wet sediment. It has been suggested (Stillman and Sevastopulo, 2005) that the volcanic sequence represents a submarine volcanic seamount. Volcanic rocks at Shenicks Island, Skerries, are separated from the Belcamp Formation by 8 km of Silurian outcrop but comprise similar lithologies, although of more distal facies (Brück and Kennan, 1970; Murphy, 1987a).

Lambay – Portrane

Some 8 km south of the Balbriggan inlier and about 18 km north of Dublin, a further Ordovician inlier at Portrane and Lambay Island exposes mainly volcanic rocks with some associated limestones and mudrocks. On Lambay Island, lavas and volcaniclastic rocks occur in roughly equal amounts (Gardiner and Reynolds, 1898; Stillman *et al.*, 1974). The lavas are petrographically similar to those at Balbriggan, though there is a higher proportion of the more basic pyroxene-phyric basaltic andesites.

The volcanic rocks demonstrate the growth of a submarine volcanic seamount and its emergence as a subaerial island. (Stillman and Sevastopulo, 2005). Outcrops of mudrocks are seen within the volcanic succession. A graptolite fauna from these mudrocks is indicative of a Caradoc age, possibly about *wilsoni* Biozone (within *foliaceus*) (Soudleyan) (B. Rickards, *pers. comm.*). New collection and examination of old survey collections by Matthew Parkes suggest a mid-Caradoc age from the brachiopod faunas (Parkes and Harper, 1996). Overlying the submarine volcanic rocks at Kiln Point is a series of breccias rich in limestone clasts and ashy material, and also thin-bedded limestones. An Upper Ordovician fauna was recorded by Gardiner and Reynolds (1898). Examination of trilobites from the collections of the Geological Survey of Ireland has provisionally suggested a Cautleyan (mid-Ashgill) age (D. Siveter, *pers. comm.*). These limestones may represent a fringing reef around an extinct volcanic island. The breccias occur in slumped beds, which are best explained as volcanically or seismically triggered slides. Above the breccias and mudstones at Kiln Point is a succession of subaerial lavas and minor intrusions, which include a porphyritic andesite (Lambay Porphyry) originally described in 1878 by Von Lasaulx and subsequently recorded in many books on igneous petrography. This porphyry was widely used in Neolithic stone axes.

On the mainland at Portrane are exposed similar andesitic lavas and intrusions, pyroclastic rocks, and breccias containing ash and limestone clasts. There are also thin limestones, calcareous tuffs, and some black mudrocks. Graptolites from these black mudrocks and brachiopods from an overlying nodular limestone indicate a mid-Caradoc age (Parkes and Harper, 1996). Trilobites such as *Tretaspis portrainensis* (Reed) (Lamont, 1941) reported from sediments deposited after most of the volcanic activity are unlikely to be older than late Caradoc (Onnian). In faulted contact with the rocks described above are thicker bedded limestones sandwiched between packets of thinner bedded limestones with complex folds, probably of soft sediment origin (Fig. 6.4). These are collectively termed the Portrane Limestone and, although their insoluble residues show some fragments of pumice which may be contemporaneous (Gardiner and Reynolds, 1897), they appear to post-date the main volcanic episode.

Thus, a time gap has been suggested between the main development of the volcanic rocks and these well-bedded limestones (Murphy, 1987a), and this is consistent with the available data. The Portrane Limestone is of particular interest due to a rich silicified fauna of brachiopods, trilobites, corals, bryozoans and ostracodes (A.D. Wright, 1963, 1964; Kaljo and Klaamann, 1965; Orr, 1987; Price, 1981;Ross, 1966), which allows determination of a precise high Cautleyan (mid-Ashgill) age, thus inviting comparison with the less accessible but similar development on Lambay Island. Brenchley and Newall (1984) suggested that the limestone breccias were rapidly deposited mass flows in channel fill sequences associated with a sea-level low during the late Ashgill (Hirnantian). A palaeokarst development indicating post-Cautleyan emergence has been described by Parkes (1993b), and linked to the Hirnantian global sea level fall (Harper and Parkes, 2000a). The stratigraphical position of black shales above this palaeokarst is not yet fixed biostratigraphically, although they are ascribed a pre-Hirnantian age by Harper and Parkes (2000a). Contacts in the critical area are complicated by later thrusting that has exploited this black shale (Fig. 6.5), although the Silurian can be seen to rest unconformably on the Ordovician in the southern part of the inlier.

Kildare

Lower Palaeozoic rocks surrounded by Carboniferous strata are found in a narrow inlier known as the Chair of Kildare. Ordovician rocks form the eastern part of this inlier, being faulted against Silurian rocks lying to the west. The stratigraphy was originally outlined by Reynolds and Gardiner (1896) and has been subsequently modified during numerous palaeontologically based studies (e.g. Dean, 1971–8; A.D. Wright, 1967, 1968, 1970; Parkes and Palmer, 1994).

Fig. 6.4 Thin- and thick-bedded limestones of the Portrane Limestone Formation demonstrating folds of probable soft sediment origin near northern Martello Tower, Portrane.

Fig. 6.5 Silurian strata (upper and left portion of cliff) thrust over Portrane Limestone Formation, Priest's Chamber, Portrane. (Photograph Alison Graham.)

The central part of the inlier, known as Grange Hill and the Chair, exposes an ESE dipping sequence of sedimentary and volcanic rocks. The oldest strata are poorly described green siltstones (Conlanstown Formation of Parkes and Palmer, 1994), which are almost certainly the source of an old record of *D. bifidus* (Elles and Wood, 1910–1918, p.43) indicating an early Llanvirn age. The overlying ashes and ashy sediments have yielded a mid-Caradoc (Soudleyan) shelly fauna described by Parkes (1994) and these are overlain by andesitic lavas, also seen on the Hill of Allen at the easternmost end of the inlier. Succeeding ashy siltstones have yielded a mid-Caradoc (Longvillian) fauna, thus giving a precise age for the igneous rocks. These Caradoc fossil assemblages suggest very shallow water conditions possibly controlled by a local volcanic edifice (Parkes and Palmer, 1994). Although these Caradoc rocks appear to be structurally conformable with the underlying siltstones, a disconformity omitting the upper parts of the Llanvirn and Caradoc is indicated.

The succeeding Chair of Kildare Limestone has yielded rich faunas of trilobites, corals, brachiopods, bryozoans, and conodonts. Detailed work on the trilobites has permitted the establishment of a late Ashgill (Rawtheyan) age, slightly younger than the Portrane Limestone. Mudstones interbedded with the upper parts of this limestone have yielded an upper Ashgill 'Hirnantia' fauna (A.D. Wright, 1968). However, this does not necessarily imply a Hirnantian (latest Ashgill) age, as this fauna is partly facies controlled (Rong and Harper, 1988). The apparent lack of late Caradoc and early Ashgill rocks suggests a further non-sequence, probably an angular unconformity (Parkes *pers. comm.* 1997), between the Chair of Kildare Limestone and the underlying volcanic rocks, similar to that suggested at Portrane.

Dingle Peninsula

Todd *et al.* (2000) have described palynomorphs from the Annascaul Formation of the Dingle Peninsula that are of early Ordovician (Tremadoc to early Arenig) age and interpret these as *in situ* rather than as reworked into the Silurian, to which the Annascaul rocks were previously assigned (and are assigned in the next chapter). They also present structural evidence to suggest that the Annascaul Formation is older than the adjacent Ballynane Formation that has yielded Wenlock fossils. The Annascaul Formation is at least 500 m thick and is dominated by mudrocks with subordinate quartz wacke sandstones, tuffaceous conglomerates and mélanges. To date fine-grained limestone clasts in some of the mélanges have failed to yield any fossils.

Leinster

Despite being the largest area occupied by Ordovician rocks in Ireland, this remains one of the poorest known, in part due to very limited inland exposure and in part due to a dearth of biostratigraphical control. The major subdivisions presently recognised through the main outcrop north-west of the Rosslare Block are the Ribband Group and the Duncannon Group (Fig. 6.6). Most workers have suggested that reasonable correlations exist between disjunct outcrops of these groups. However, Max *et al.* (1990) have emphasised the importance of major NE–SW faults and shear zones and consider the area to represent a series of separate sub-terranes juxtaposed during Silurian sinistral transpressional deformation.

Ribband Group

This lithostratigraphical group, first erected by Egan *et al.* (1899) and formalised by Gardiner and Vanguestaine (1971) is still rather poorly known. It has also suffered certain amounts of redefinition. The usage here follows that of the recent GSI maps of Leinster (McConnell *et al.*, 1994; Tietzsch-Tyler *et al.*, 1994a, b). Its essential feature is that it is mud rich, commonly grey in colour but including red and green coloured units. A fine-scale interbedding of thin siltrocks and sandstones with the finer-grained mudrocks is a characteristic feature responsible for the name of this group. The group is thought to contain both Cambrian and Ordovician strata as mentioned in Chapter 5, with the inferred conformable contact with the Bray Group (Shannon, 1978; Brück and Stillman, 1978) implying a Middle Cambrian age for the lower parts. In the area just south of Dublin, where the Ribband Group occurs both east and west of the Leinster Granite, poorly preserved palynomorphs indicate a probable Lower Ordovician age (Brück, 1972; Brück *et al.*, 1974).

The outcrop of the Ribband Group is traversed by a series of NE-SW faults and shear zones (Clegg and Holdsworth, 2005; McConnell *et al.*, 1999). Correlation across these structures is by no means certain, and those shown in Fig. 6.7 lack detailed biostratigraphical confirmation. The Aghafarrell Formation west of the granite and the lower part of the Maulin Formation east of the granite have been suggested as distal equivalents of the Bray Formation, and by implication of Cambrian

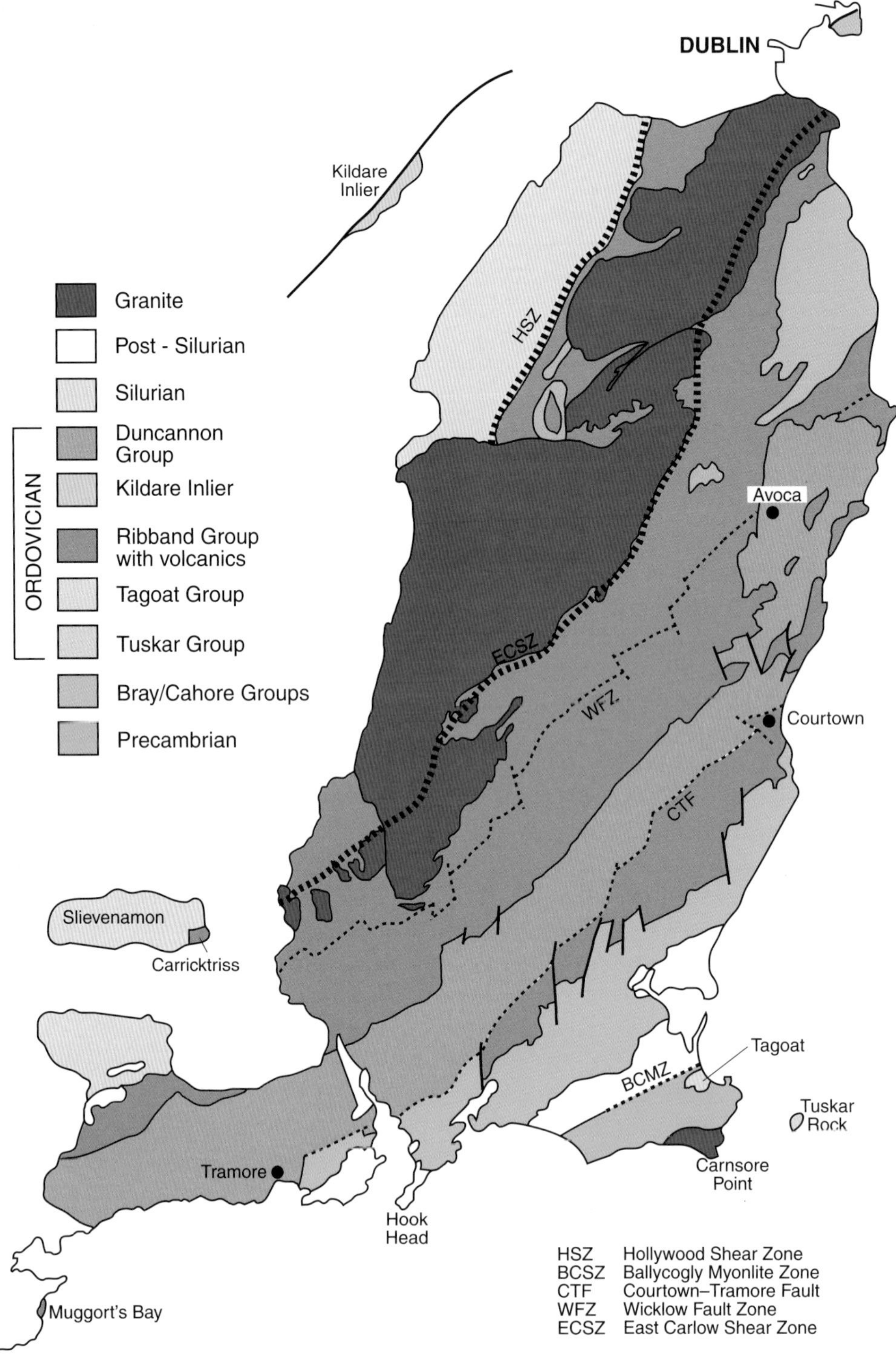

Fig. 6.6 Distribution of Ordovician rocks in Leinster showing main stratigraphical groupings and fault zones.

age, although they are mapped as Lower Ordovician (McConnell *et al.*, 1994). The Aghafarrell Formation contains extrusive tholeiitic basaltic rocks and andesites. From the trace element geochemistry McConnell and Morris (1997) suggest the presence of active subduction. Some volcanic rocks, mainly andesitic, also occur in the overlying Butter Mountain Formation. The presence of coticule rocks, beds rich in Mn garnets, in the Butter Mountain Formation to the west and the Maulin Formation to the east of the granite, is used as a basis

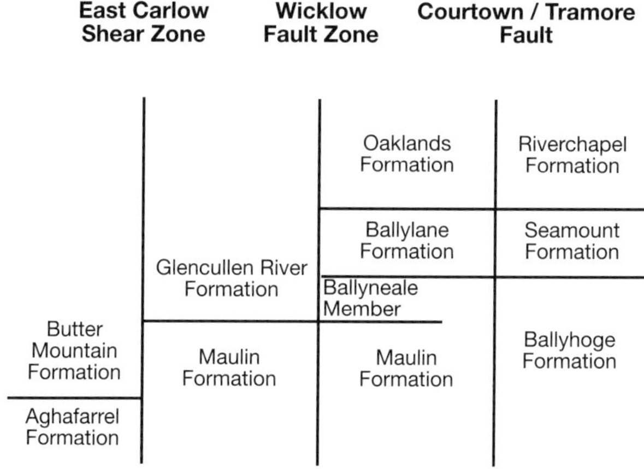

Fig. 6.7 Correlation of formations within the Ribband Group; see text for details.

of correlation. These coticules have been interpreted as alteration products of Mn-rich sea floor sediments associated with black smokers. East of the Leinster Granite the Glencullen River Formation (Brück *et al.*, 1974) overlies the Maulin Formation and consists of tuffs and siltrocks.

More reliable Lower Ordovician ages have been obtained from the Courtown area on the east coast (Crimes and Crossley, 1968; Brenchley *et al.*, 1967a; Brenchley and Treagus, 1970). Here Arenig *D. extensus* graptolite faunas (approximately *varicosus* Biozone) have been obtained from the Riverchapel Formation (Brenchley *et al.*, 1967a). A graptolite fauna from the Oaklands Formation at Kiltrea, near Enniscorthy, was considered by Brenchley *et al.* (1967a) to indicate a late Arenig (*D. hirundo*) or possibly early Llanvirn (*D. artus*) age. However, recent re-examination of this material suggests an early Arenig age is more appropriate (Rushton, 1996). Thus the limited age data on both sides of the Leinster Granite suggests that the Ribband Group extends only up into the Arenig.

Acritarchs recovered from the Blackhall Formation east of Bannow Bay suggest an age no older than late Tremadoc, and some species suggest a late Arenig age (Brück and Vanguestaine, 2004). Definitive Ordovician acritarchs have also been recovered from the Muggort's Bay inlier in SW Waterford where there is clear evidence of reworking of Cambrian assemblages (Brück and Vanguestaine, 2005). The Carricktriss Formation, which crops out in the eastern part of the Slievenamon inlier, contains numerous volcaniclastic rocks interbedded with black slates and siltstones. Palynomorphs recovered from the Carricktriss Formation have also been interpreted as indicating early–mid Ordovician ages (Maziane-Serraj *et al.*, 2000). The undated Ballyneale Volcanic Member in east Kilkenny (Brück *et al.*, 1979; Tietzsch-Tyler *et al.*, 1994a), of probable Tremadoc age, also contains acid tuffs (Shannon, 1978, 1979a).

It is difficult to give an accurate estimate of thicknesses attained by the Ribband Group throughout its outcrop. In the better quality coastal exposures near Courtown an estimate of *c.*1.7 km of strata is possible by tracing through numerous folds. A thickness of nearly 5 km was estimated by Shannon (1978) near New Ross, but this may be an over-estimate if significant repetition is concealed in poorly exposed ground. There are few published data on the depositional environments represented by the Ribband Group. Many of the finely alternating siltrock–mudrock sequences have been interpreted as distal turbidites. However, trace fossil assemblages reported by Crimes and Crossley (1968) from the Courtown area imply rather shallow water conditions. Volcanic rocks in the Ribband Group are volumetrically minor in comparison to the extensive late Llanvirn to Caradoc vulcanism of the Duncannon Group and are of basic to intermediate composition.

Duncannon Group

This group is characterised by the abundance of intermediate to acidic volcanic rocks and rests with marked unconformity on rocks of the Ribband Group. The main outcrop is a north-east–south-west trending belt from near Arklow to Waterford (Fig. 6.6), although the prominent north-east–south-west elongation is partly due to later deformation. The volcanic horizons are not continuous. There were almost certainly a number of eruptive centres, which probably operated at slightly different times and attained different stages of evolution. Significant outcrops also occur in synclinal cores between Arklow and Rathdrum, including the mineralised sequence at Avoca (McConnell *et al.*, 1991) where stratiform ore bodies occur. West of Waterford there is a wider outcrop belt where the structural pattern changes to open, gentle folds broken up by extensive faulting. The overall pattern is that the older rocks are basaltic to andesitic and the younger ones are dominantly rhyolitic.

In places the lowest strata are shallow-water limestones such as the Courtown Limestone Formation (Crimes and Crossley, 1968; Brenchley and Treagus, 1970; Mitchell *et al.*, 1972) and the Tramore Limestone Formation (Mitchell *et al.*, 1972; Carlisle, 1979; Key *et al.*, 2005). Early Caradoc (Costonian) brachiopod–trilobite faunas are known from Courtown, whereas at Tramore the limestone formation is thought to represent much of the Llandeilian (late Llanvirn) and possibly also the Costonian (early Caradoc). At Tramore there is considerable debate concerning the age of the mudrocks underlying the Tramore Limestone, which have variously been considered as Cambrian (Cahore Group), early Ordovician (Ribband Group), or part of the later Ordovician Duncannon Group. The presence of irregularly shaped syn-sedimentary minor intrusions of andesite and rhyolite that appear to have penetrated both the underlying mudrocks and the base of the limestone while both were partly consolidated (Stillman and Sevastopulo, 2005) would suggest that a Duncannon Group age is most likely. There is much lateral variation of the Tramore Limestone Formation (Fig. 6.8) in general showing thin shelf sequences at Tramore in the east and thicker basinal sequences farther west (Carlisle, 1979; Wyse Jackson *et al.*, 2002). The junction between these two facies is inferred to have been controlled by syn-sedimentary north-east–south-west faults. Syn-sedimentary faulting has also been inferred to control the location of the shelf edge on which the Courtown Limestone was deposited before it foundered (Tietzsch-Tyler *et al.*, 1994a). The basinal limestones overlie sediments associated with basaltic and andesitic volcanic rocks near Bunmahon which indicate an age equivalence of these volcanics to the lower parts of the Tramore Limestone (Fig. 6.8). Thus the Bunmahon volcanics are best considered as the oldest volcanic rocks of the Duncannon Group rather than as part of the Ribband Group as had previously been suggested (Carlisle, 1979; Brück *et al.*, 1979).

The limestones and their age equivalents are overlain by dark mudrocks bearing *N. gracilis* Biozone faunas (Carrighalia Formation of Carlisle, 1979; Ballinatray Formation of Brenchley and Treagus, 1970 and Mitchell *et al.*, 1972). These represent quiet euxinic conditions but not necessarily very deep water. In all areas the main volcanic episodes follow on from these mudrocks and are characterised by rapid lateral variation in acid volcaniclastic rocks and shallow intrusions (Schiener, 1974; Boland, 1983; Swennen, 1984). Faunas from these horizons indicate the Burrellian Stage of the Caradoc (Carlisle, 1979; Mitchell *et al.*, 1972; Brenchley *et al.*, 1977; Parkes, 1994). The youngest rocks documented to date are from one locality in the Campile Formation of County Wexford, where Ashgill trilobites have been confirmed (Owen and Parkes, 1996).

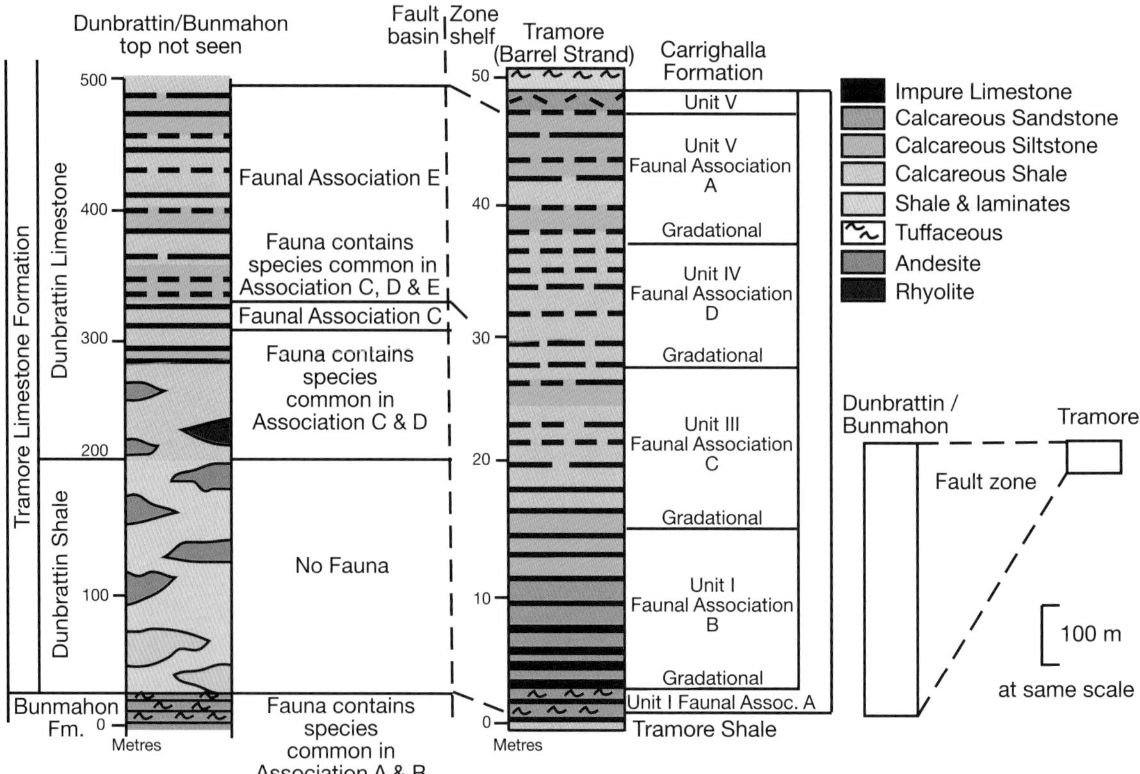

Fig. 6.8 Lateral variation in the Tramore Limestone Formation (*after* Carlisle, 1979).

The northern end of the belt, from Wicklow and Rathdrum to Arklow, contains some of the most intensely studied sequences, due mainly to the presence of the Avoca copper deposit. The Avoca ore body is a volcanogenic massive sulphide deposit of syngenetic exhalative origin. Massive sulphides are associated with largely subaqueous volcaniclastics: rhyolitic, dacitic, and andesitic tuffs with interbedded clastic sediments. The volcanic material is often reworked and apparently deposited from turbidity currents. Proximal units are slumped volcanic mudflows containing abundant brecciated rhyolite and ignimbrite. Intrusive rhyolite bodies are extensively brecciated and associated with hydrothermal mineralisation. At Avoca the ore deposits were mined for copper and iron pyrite for over two centuries (Platt, 1977; Sheppard, 1980).

Elsewhere throughout the Duncannon Group the magmatic liquids were from time to time sulphide-rich; primary pyrite is a common feature of the lavas and juvenile pyroclastics and has in places accumulated to produce sedimentary bodies of massive pyrite. Sericitisation and carbonation of volcaniclastics appear to have resulted from the action of entrapped and heated sea water. Hydrothermal activity associated with rhyolite intrusion and brecciation has been responsible for extensive vein formation, often of quartz-felspar, or quartz-sericite, with some sphalerite-galena-chalcopyrite mineralisation.

At Arklow Head, Stillman and Maytham (1973) described a small subaerially accumulated pile of rhyolitic pumice some 40 to 50 m thick built up on black mudstone and overlain by more extensive vitric and crystal tuffs with some accretionary lapilli. The core of the pile is heavily silicified with what appears to be fumarolic silica. The tuffs pass laterally into well-bedded water-lain tuffs indicating that the fringes of the volcano were below water. The whole structure has subsequently been intruded by a thick sheet of diorite.

Southward from Arklow Head to Waterford Harbour are strung out a number of lenses of volcanics and minor intrusives, petrographically similar to the Avoca sequences, though McConnell (1987) has defined several magma batches on the basis of their geochemistry. It seems likely that they represent a line of volcanoes, strung out along the arc, probably mostly submarine, but occasionally emerging as islands. The volcanoes developed at slightly different times and from magmas which had different sources and degrees of evolution. This line of volcanic centres appears to have been sited close to the south-eastern margin of a sedimentary basin which extended along the edge of a crustal block that was undergoing extensional rifting. The volcanics show a bimodal subdivision into basaltic and rhyolitic compositions. The basalts have characteristics of magmas from both subduction zone and enriched mantle sources and are interpreted as the products of the rifting of a volcanic arc. The rhyolites were originally peralkaline and the trace elements indicate a fractional crystallisation relationship between the basaltic and rhyolitic end members of the series. Thus the setting is similar to that envisaged for the contemporaneous Welsh marginal basin.

At the southern end of the arc in County Waterford, west of Waterford Harbour, the volcanic belt swings to an east–west orientation and the structural pattern changes to open gentle folds broken by extensive faulting. Some of the faults appear to have been active during and immediately following the period of volcanicity, but many are of Variscan age. For more than 20 km the Waterford coastline presents a succession of small bays and sea cliffs 30–40 m high, providing almost continuous exposure of a succession of volcanics and intrusives, which Sir Archibold Geike (1897) described as 'perhaps the most wonderful series of volcanic vents within the British Isles'. In recognition of this outstanding volcanic geology as well as a very significant history of mainly 19th Century copper mining, the region, dubbed 'the Copper Coast', was awarded the designation of 'European Geopark' in 2001, and 'Global Geopark' in 2004, both endorsed by UNESCO.

The volcanics appear to represent the emplacement and growth of just two submarine volcanic centres. (Stillman and Sevastopulo, 2005). The western centre, informally named the Bunmahon Volcano, began to erupt in late Llanvirn (Llandeilian) times. The more extensive eastern centre, named the Kilfarrasy Volcano, is slightly younger and began to erupt in the Caradoc. The two centres overlap just to the east of Bunmahon. Both volcanoes erupted into and onto ocean-floor muds, which continued to be deposited throughout the lifetimes of the volcanoes. The distribution of the volcanic rock units is controlled by faults, and inland the volcanic sequences are faulted against older Ribband Group mudrocks. All were deformed both during the Caledonian (Acadian) orogeny and again in the Variscan orogeny, the latter being responsible for much of the faulting. Despite the considerable deformation, the grade of metamorphism remained extremely low, barely extending above the anchizone. The difference in lithological rheology resulted in considerable differences in reponse to the folding; most of the movement being taken up by the mudrocks, which are strongly cleaved, tightly folded, and locally thrust and overturned, whereas the more massive volcanic rocks appear to be hardly affected.

The Bunmahon Volcano, whose coastal outcrops extend westward from Bunmahon to Ballyvoyle Head, is essentially a basalt-andesite volcano, poor in volcanic gases. It seems to have erupted frequently with the production of small batches of magma, which, rising to the surface, often failed to emerge onto the seabed; instead coming to a halt in the water-saturated seafloor deposits of muds and hyaloclastites, producing peperites and locally heating the mudrocks to raise their metamorphic grade almost to greenschist facies. Eruptions that were large enough and persisted long enough did extrude onto the seabed, forming sheet lavas and sometimes pillow lavas. However, most were shattered by contact with the cold water and produced hyaloclastites. Where slopes had developed these volcaniclastics commonly slipped downslope to generate volcanic mudflows (lahars). These submarine basalts display two types of *in-situ* low-grade metamorphism. Where exposed on the seabed, the cold water affected the crystallisation of the lava, so that they became spilitised. This meant that the normal basaltic calcium-plagioclase feldspars were converted to sodium-plagioclase feldspars, whereas the pyroxenes retained much of their original composition. The occurrence of the minerals prehnite or pumpellyite show that the subsequent orogenic metamorphism did not reheat them significantly. On the other hand, where the magma failed to reach the surface but spread into sills at shallow depth in the waterlogged seafloor deposits, their heat was retained and with the abundant sediment pore-water generated the hydrothermal mineral assemblages of albite-epidote-chlorite. These features are well represented in excellent outcrops at Bunmahon Head, Ballydowane Bay, and Stradbally Cove. Small gabbro and dolerite intrusions penetrated through these deposits, giving cohesion to the core of the volcano, assisting its growth to a seamount form. Later rhyolitic phases of eruption produced complex inter-relationships which are seen in Ballyvooney Bay.

The Kilfarrasey Volcano, whose coastal outcrops are seen from Tramore to Bunmahon, is larger and rather more complex. It appears to have had a number of eruptive centres with two contemporaneous magmas available. Although rhyolite is the more abundant, sometimes both rhyolite and basalt-andesite magmas appear to have erupted at the same time. Activity started rather later than in the Bunmahon Volcano, and shortly before the first eruptions it seems that the sea bed might have risen up; deepwater black shales are replaced for a time by shallow water fossiliferous limestones of the Tramore Limestone Formation, and very shortly after this, volcanic activity seems to have begun (Stillman and Sevastopulo, 2005). This uplift may well have been related to the uprising of the magma chamber that was feeding the Bunmahon volcano, but was short-lived, as the limestones were rapidly succeeded by dark mudrocks, into and through which the bulk of the Kilfarrasey Volcano magmatic rocks were emplaced.

The eruptions of the Kilfarrasey Volcano presented a very different scenario from that of the Bunmahon Volcano. The rising magma was of andesitic or rhyolitic composition, much richer in silica and volcanic gases, and the eruptions occurred in much shallower water, where the load-pressure of the seawater was not sufficient to dampen the violent escape of gases; hence the eruptions were of the much more explosive 'Plinian' type. Most of the rhyolite eruptions appear to have been submarine, occurring at several small underwater centres. At these there is no evidence for the proximity of land and little evidence that many of the centres ever emerged above sea level, though the occasional presence of accretionary lapilli suggests that columns of ash were ejected out of the water into the air. The products were mainly of pyroclastic material which erupted both in vertical columns which fell back as spreads of ash-fall, and as ground-hugging surges which spread laterally from the vent to generate ash-flows (ignimbrites). These pyroclastics lithified to form the tuffs that form the bulk of the Caradoc volcanic rocks in the Duncannon Group. However, sometimes the explosive process degassed the magma and the pyrocastics were followed by more viscous lavas which generated thick flows and domes, with contorted flow banding and abundant small silica nodules (lithophysae). In at least one case, at Ballyscanlon, there is evidence that such a dome formed on the seabed, its outer skin cooling and thermally shattering to a crackle breccia. This was shed onto the surrounding seafloor in many places as proximal coarse volcaniclastic mass flows, and distally as finer volcanic turbidites.

At the 'Metal Man' on Tramore's Great Newtown Head, a rhyolitic pyroclastic flow (ignimbrite) is emplaced on black mudstone. (Fritz and Stillman, 1996). Its base is a 10 m thick layer of angular lumps of black mudstone, pumice and rhyolite in a matrix of ash. The flow has clearly ripped up and baked lumps of soft mud and dragged along fragments of the volcanic vent. Above this layer is a zone of pumice and ash with streaky texture defined by fiamme of pumice flattened while hot and plastic. Above this is a zone with even more pronounced streaky texture in which flattened pumice and glass shards have been welded together. There is also columnar jointing typical of the cooling of ignimbrite sheets. The top of the tuff is almost entirely of very fine-grained ash, again typical of the ash cloud which rises above a pyroclastic flow, but which here contains marine fossil brachiopods. The ignimbrite seems to have retained sufficient heat to weld its pumice whilst flowing under water. At least one vent succeeded in ejecting its column of ash into the air from a series of underwater explosive eruptions. At Dunhill Castle (Stillman, 1971) pumices and accretionary lapilli are found in a series of five upwardly-fining beds of rhyolitic tuff. Accretionary lapilli are produced when fine ash coagulates around a raindrop in the atmosphere and falls back to the ground, with the outer layer drying to form a hard shell which commonly cracks like an eggshell on landing. The tops of the beds demonstrate aqueous re-sorting and are also marked by larger rhyolite clasts which have indented the surface onto which they have fallen. The tuff beds are poorly sorted; even the coarsest contain more than 50% of fine ash. There is a complex repetition of grading which, together with the poor sorting, suggests that each bed is the product of a successive ash fall, in each of which the particles were sorted and graded by size and weight, firstly in the vertically erupted ash column, then partially re-sorted as they sank back through the sea water.

The copper ore-body found in the Duncannon Group at Bunmahon on the Waterford Coast is very different from that seen at Avoca. It is an epigenetic lode or vein-type copper mineralisation formed much later than the Ordovician host rock. Here a large number of copper-bearing lodes (or veins) trend north-west to south-east across the strike of the host rocks. Chalcopyrite is the main copper mineral; it occurs in, and marginal to, large, steeply dipping quartz veins which are up to 20 m thick and often contain fragments of brecciated wall rock, suggesting that they were emplaced during active deformation (Sleeman and McConnell, 1995). Though a pre-Devonian age is suggested by the presence of an unmineralised basal Old Red Sandstone conglomerate that unconformably overlies one of the principal copper lodes, many of the productive lodes are situated on north–south faults that cut the Old Red Sandstone. This suggests that the mineralisation may relate to the presumed Variscan age of the faults, or these Variscan faults have simply reactivated previously mineralised structures. The primary source of the copper is unknown, though the abundance of sulphides in the Duncannon Group mudstones may be contributory.

A significant time-gap exists between the youngest known parts of the Ribband Group and the oldest dated parts of the Duncannon Group. Moreover, there is a marked change in structural style across the boundary between these two groups. The Ribband Group contains recumbent F1 folds with local downward facing, which are absent in the Duncannon Group. A period of significant deformation during the Llanvirn is indicated. This suggests a possible lack of continuity between the Leinster domain and the Central Ireland inliers during the earlier part of the Ordovician when they may have been separate terranes (Murphy *et al.*, 1991).

Rosslare

Rosslare exposes a complex of basement rocks traversed by ENE trending shear zones. Resting, or appearing to rest, on this basement are rocks of known and possible Ordovician age known as the Tagoat Group and the Tuskar Group.

Tagoat Group

This group comprises a basal, unfossiliferous purple conglomerate, sandstone, siltstone, and grey shale overlain by fossiliferous siltstone and turbiditic sandstones. The detritus in the conglomerate includes clasts of quartz mylonite and mica schist with inferred derivation from the Rosslare Complex. An Arenig age was deduced from dendroid graptolites collected by Baker (1966). Re-examination of the fossiliferous localities by Brenchley *et al.* (1967a) yielded graptolites, trilobites, brachiopods and machaeridians. These mainly suggested an Arenig (*D. extensus*) (approximately *varicosus* Biozone) age with one locality suggesting a late Arenig (*D. hirundo* Biozone) age and a new locality at Maytown suggesting a possible early Llanvirn age. Many of the brachiopods are endemic and show close similarities to late Arenig faunas from Anglesey (Williams, 1973; Neuman and Bates, 1978). These faunas are part of the peri-Gondwanan Celtic province (Neuman and Harper, 1992).

Along strike from the Tagoat outcrop to the south-west lies a petrographically similar sequence of sedimentary rocks termed the Silverspring Beds by Baker (1966). These are rather more strained than the Tagoat rocks and locally show effects of contact metamorphism. They may be lateral equivalents of the Tagoat sequence but could also be older (F.C. Murphy, 1990).

Tuskar Group

This succession, described by Max and Ryan (1986), crops out on some offshore rocks and also as xenoliths in the Carnsore granite. It consists of basic volcanic rocks, including pillow lavas and hyaloclastites, tuffaceous sediments, black mudrocks, and medium- to fine-grained greywackes. The volcanic rocks, of tholeiitic to transitional alkaline affinity, possibly developed on a rifted continental basement. A Cambrian age was suggested by Max and Ryan (1986) but this was largely due to constraints imposed by the then current isotopic dates on the Carnsore granite which hornfelses these rocks. More precise dating of this intrusion (c. 428 Ma) and of the Saltee granite (*c.*436 Ma) now allows an Ordovician age for these rocks which might be more in keeping with regional Caledonide geology. However, their age remains uncertain.

Palaeogeography

It has been noted in Chapter 3 that the Ordovician rocks with which we are dealing are but a fragmental record of an ancient fold belt which resulted from the closure of an ocean. Thus critical evidence contained in the ocean floor has been lost during subduction, a process that has also resulted in major, often unquantifiable, changes in spatial arrangements of the rocks. Palaeogeographical reconstructions can be made using evidence from sedimentary facies, faunal provinces, palaeomagnetism, igneous provinces, and provenance links.

Many controls exist on the formation of biogeographical provinces. The presence of migration barriers such as large extents of deep water would be expected to produce differences in the shelf faunas on opposite sides of an ocean. However, other factors are also important such as relative sea level (transgressions and regressions), palaeolatitude, shelf width and ocean currents. Moreover, dispersal of faunas varies not only with time but also with type of organism and additionally with life habits within types (Fortey, 1984; Fortey and Cocks, 1988; Cocks and Torsvik, 2002). In addition, oceanic islands were probably important as faunal dispersal points (Neuman, 1984, 1988).

It has been noted in Chapter 3 that graptolites from north-west Ireland are more easily placed in biozones developed from successions in Australia and America than in biozones developed largely in Wales (Skevington, 1974). In contrast, graptolites from south-east Ireland are more easily referred to the British biozonal scheme. In view of their life habit in surface waters, these differences are most easily interpreted as indicating different temperatures and, by inference, different latitudes. Contrasts in shelly faunas also exist (Williams, 1969, 1973) which have been summarised by Fortey and Cocks (1988), Harper and Parkes (1989), Harper (1992), Parkes and Harper (1996), Harper *et al.* (1996), Williams *et al.* (1996), and Owen and Parkes (2000).

In contrast to the low latitude warm water faunas found in north-west Ireland, the early Ordovician faunas of south-eastern Ireland belong to the Celtic province typical of southern Britain and France. Although the existence of a distinctive Celtic province has been questioned (Cocks and McKerrow, 1993), the statistical tests on differences in the brachiopod faunas (Harper, 1992; Neuman and Harper, 1992; Harper *et al.*, 1996) appear to be convincing. Moreover, published palaeomagnetic data associated with Celtic faunas indicate mid to high latitudes (50°–60° S) (Torsvik and Trench, 1991; MacNiocaill *et al.*, 1997). During the latter part of the Ordovician there is less clear-cut biogeographical differentiation amongst the marginal faunas (Harper and Parkes, 1989; Owen and Parkes, 2000).

In general the interpretation of Ordovician faunal data suggests that three continental areas were involved in the development of the Iapetus Ocean. These were Laurentia, which included north-west Ireland, Avalonia, including south-east Ireland, initially attached to Gondwana but rifting from it pre-Arenig, and Baltica (Fortey and Cocks, 1988). The steady increase in warmer water faunas and facies of Avalonia suggest that it moved northwards towards a nearly static Laurentia. Strong similarities of Baltic and Avalonian faunas suggest possible collision of these blocks by the late Ordovician. This picture of northward movement of Avalonia is supported by the palaeomagnetic data (MacNiocaill and Smethurst, 1994; Torsvik *et al.*, 1992; Torsvik *et al.*, 1993; Bachtadse and Briden, 1990; MacNiocaill *et al.*, 1997; Cocks and Torsvik, 2002), which suggest an ocean no greater than 3000 km wide by the mid–late Ordovician. Baltica displays similar palaeolatitudes to Avalonia in the early Ordovician and longitudinal separation, by Tornquist's Sea, is largely inferred from faunal evidence.

Analogies with modern ocean basins suggest that it is unlikely that the Iapetus Ocean between Laurentia and its marginal arcs and Avalonia was unbroken ocean floor.

The likelihood of oceanic islands that may have acted as migratory staging posts was proposed by Neuman (1984). It is possible that some of the central Ireland inliers are related to this. Areas such as Grangegeeth show Celtic faunas in the early Ordovician and Scoto-Appalachian faunas in the late Ordovician, consistent with rapid northward drift to within migratory reach of the Laurentian margin (Harper *et al.*, 1996). A peri-Avalonian arc, represented in Ireland by the Duncannon Group volcanic rocks, can also be recognised in Britain (Lake District) and Newfoundland and New Brunswick along strike, thus indicating considerable strike extent. There is general agreement between the data from inferred faunal provinces and those from palaeomagnetic investigations (Cocks and Torsvik, 2002).

Reconstruction of ancient oceans and their changing widths relies also on interpretation of igneous rocks that may indicate subduction or extensional regimes. There is evidence in the Ribband Group for volcanic rocks with a subduction signature, a similar situation to that existing in Wales (McConnell, 2000). Later in the Ordovician the volcanic rocks of the Balbriggan–Lambay–Kildare areas, termed the Lambay Belt by McConnell (2000), represent volcanism above a southerly-dipping subduction zone, a similar situation to that envisaged for the English Lake District (Kokelaar, 1988). The Duncannon Group with its peralkaline rhyolites suggests an extensional setting. This is analogous to successions in the Welsh Basin, although it is likely that these areas were separated by the Irish Sea Landmass.

Superimposed on these shifting continental positions during Ordovician times are some major events that affected the whole Iapetus realm. The early Caradoc is represented by a very large transgression and the effects of the late Ordovician glaciation in Gondwana are seen as late Ordovician regression and some distinctive faunal associations such as the Foliomena and Hirnantia faunas (Harper and Parkes, 1989; Rong and Harper, 1988; Parkes and Harper, 1996; Fortey and Cocks, 2005). The general similarity of palaeomagnetic poles from all Silurian successions (MacNiocaill *et al.*, 1997) suggests that the main assembly of continental blocks took place during the Ordovician, perhaps the period of most dramatic plate tectonic activity in the history of Ireland.

Figure 6.9 represents a reasoned palaeogeographical reconstruction based on a wide variety of faunal, palaeomagnetic and lithological criteria that is consistent with the information preserved in the rocks of Ireland.

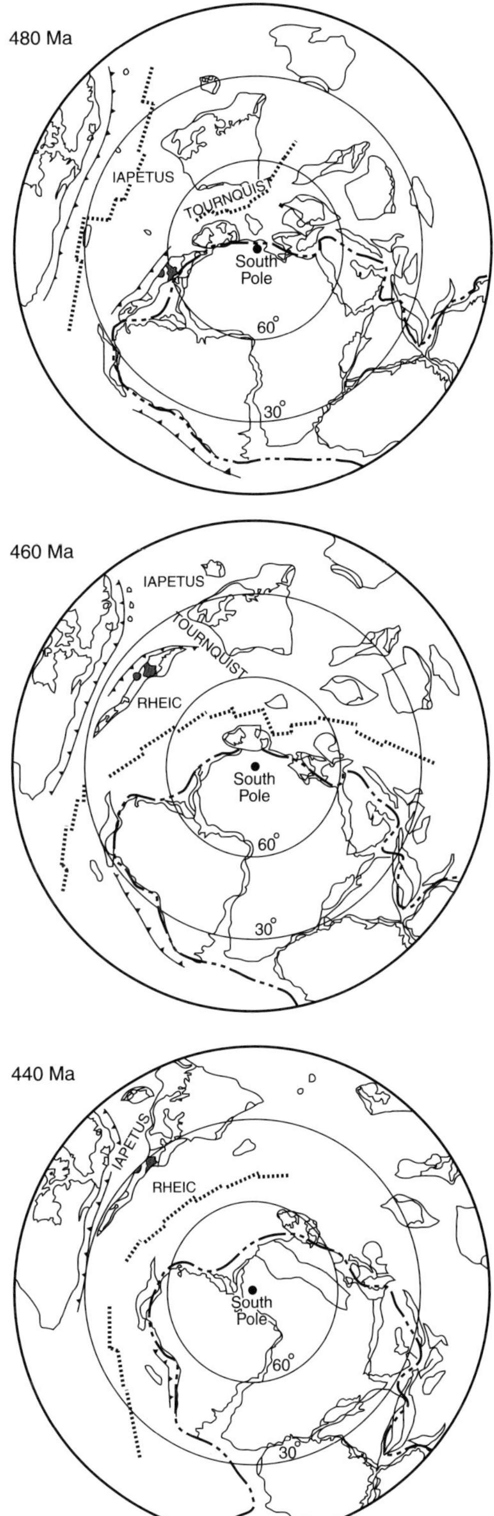

Fig. 6.9 Palaeogeographic reconstructions for the Ordovician (*after* Cocks and Torsvik, 2002). Southern Britain and southern Ireland are shown in red.

Spreading centres — Subduction zones

Approximate limits of Gondwana

7

Silurian

C. H. Holland

In 1948 when J.C. Harper summarised knowledge of the Lower Palaeozoic rocks of Ireland relatively little was known of its Silurian strata. At the time of writing of the Geological Society of London's Silurian correlation paper (Cocks *et al.*, 1971) additions had been few, though of some significance, especially in the north-west, where research workers from Imperial College London, the University of Oxford, and Trinity College Dublin had been active in investigation and publication. During the following decade a Silurian research project at Trinity College Dublin achieved much. Various contributions followed, particularly in County Mayo, in eastern Ireland, and in the Dingle Peninsula. A revised correlation paper was published by the Geological Society (Cocks *et al.*, 1992). Since the publication of the First Edition of the present book a number of the Geological Survey of Ireland's 'Descriptions' to accompany their 1:100,000 bedrock map series have involved Silurian rocks. There have been few other publications on Ireland's Silurian geology.

Mayo and Galway

The Silurian rocks of County Mayo and a small part of north Galway are seen in four separate areas (Fig. 7.1), respectively Louisburgh and Clare Island, Croagh Patrick, Killary Harbour to Joyce's Country, and Charlestown.

Louisburgh – Clare Island

Silurian rocks form the southern part of Clare Island and the southern margin of Clew Bay from Roonah Quay, eastwards through Louisburgh to Old Head (Fig. 3.3). They were described by Graham *et al.* (1989) in a memoir to accompany their map of the Geology of South Mayo (1:63,360), published by the University of Glasgow in 1985. Later, Volume 2 of the Royal Irish Academy's *New Survey of Clare Island*, edited by J.R. Graham and published in 2001, was devoted to geology. On Clare Island the Silurian rests unconformably on the Deer Park Complex. The outcrop on the mainland is terminated to the south by the Emlagh Thrust. Detailed structural analysis recorded in Palmer *et al.* (1989) constrains post-Llandovery strike-slip movement on this fault to less than 25 km. Correlation between the Silurian rocks of the mainland and those of Clare Island was demonstrated by Phillips *et al.* (1970) and Phillips (1974). The nomenclature developed by Phillips (1974) on Clare Island is

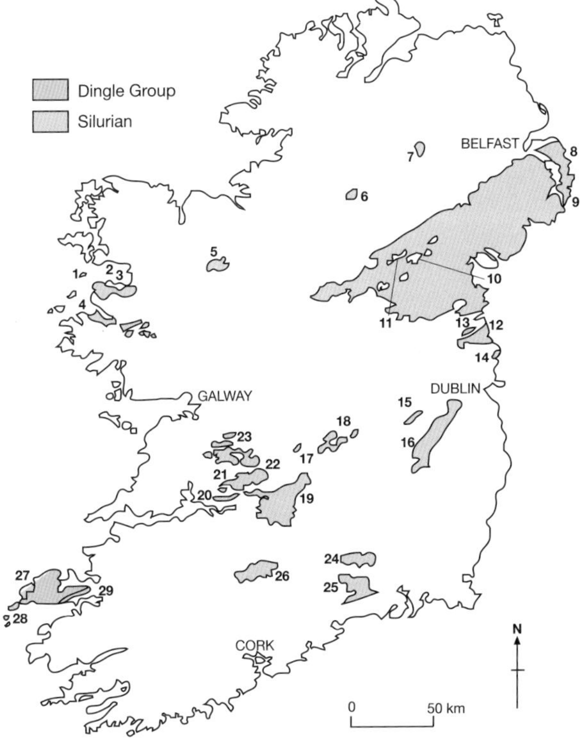

Fig. 7.1 Map showing distribution of Silurian rocks in Ireland. 1. Clare Island. 2. Louisburgh. 3. Croagh Patrick. 4. Killary Harbour. 5. Charlestown. 6. Lisbellaw. 7. Pomeroy. 8. Coalpit Bay. 9. Tieveshilly. 10. Shercock. 11. Lough Acanon. 12. Balbriggan. 13. Bellewstown. 14. Portrane. 15. Kildare. 16. Leinster inlier. 17. Knockshigowna. 18. Slieve Bloom. 19. Slieve Phelim-Silvermines-Devilsbit. 20. Cratloe Hills. 21. Slieve Bernagh. 22. Arra Mountains. 23. Slieve Aughty. 24. Slievenamon. 25. Comeraghs. 26. Galty Mountains. 27. Dunquin. 28. Inishvickillane. 29. Annascaul.

applicable also to the Louisburgh district. The succession is as follows (figures for thickness are as on Clare Island, modified as in Graham, 2001a):

(5) Glen Pebbly Arkose Formation	c.500+ m
(4) Bunnamohaun Siltstone Formation	c.250 m
(3) Knockmore Sandstone Formation	c.450 m
(2) Strake Banded Formation	c.280 m
(1) Kill Sandstone Formation	c.18–30 m

The *Kill Sandstone Formation,* which is found only on Clare Island, begins with a basal conglomerate (6 m) containing well-rounded cobbles of vein quartz, psammite, biotite schist, amphibolite, and red chert. Much of this assemblage can be matched to the underlying basement. Above are cream-coloured cross-stratified pebbly sandstones derived from the north and occurring in what have been interpreted as fluvial channels.

The *Strake Banded Formation* is of laminated red and grey siltstones with subordinate sandstones and graded, pale-coloured acid tuffs. Some of the sandstones, especially those associated with the tops of the tuffs, are themselves tuffaceous; they were derived from the west. Other sandstones are feldspathic. They show large-scale cross-stratification with derivation from the east. There are horizons of pumice blocks up to 10 cm in diameter. In the finer beds, soft sediment deformation on a small scale is common, as are ripple marks and desiccation cracks. Maguire and Graham (1996) suggested deposition of the formation by sheet floods on extensive non-marine mudflats.

On the mainland the contact of the *Knockmore Sandstone Formation* with the underlying Strake Banded Formation, which is well exposed at several localities, is sharp and the basal bed is a coarse, locally pebbly, calcareous sandstone. Most of the Knockmore Sandstone Formation is of buff or green sandstones, typically with erosive bases and with intra-formational breccias. The predominant sedimentary structures are flat lamination and low-angle cross-stratification. Current flow on this sheetflood system was to the west. The interbedded mudrocks, some of which are red, are characterised by desiccation cracks. Smith (1979a) obtained a few simple trilete spores from a single sample collected 'towards the top' of the formation.

The *Bunnamohaun Siltstone Formation* is of laminated red and green siltstones with some lenticular sandstones that are identical to those of the Knockmore Sandstone Formation. Wave ripples, desiccation cracks, and vertical burrows are common. Maguire and Graham (1996) suggested the presence of a non-marine mudflat complex with the sandstones representing fluvial incursions and some grey or brownish finely laminated siltstones having been formed in small temporary lakes. On the mainland there is differentiation of the formation into a dominantly green lower part and a dominantly red upper part.

In the past recognition of the age of the Clare Island and Louisburgh successions depended upon their structural context and, less satisfactorily, on lithological similarities to the Croagh Patrick sequence. Palmer *et al.* (1989) recorded the important discovery of an identifiable fauna of active pelagic agnathans in the mud-silt laminites at the base of the Bunnamohaun Siltstone Formation. It includes the anaspid agnathan *Birkenia* cf. *elegans* (Gilmore, 1992), some thelodont scales, and spiral coprolites. In addition, higher in the same formation on Clare Island there are many specimens of a small frond-shaped fossil originally misidentified as the bryozoan *Glauconome,* which previously had been known also from the Silurian of the Midland Valley of Scotland. Evans and Graham (2001) have described a now substantial collection of the fossil and suggested a possible charophyte relationship under the name *Peltoclados clarus.* Palmer and his colleagues suggested the implication of a 500 km elongate aquatic connection with the similar non-marine rocks with the same fossil assemblage in the Midland Valley of Scotland, which are dated as Wenlock.

The *Glen Pebbly Arkose Formation* is characterised by coarse pebbly sandstones, polymict conglomerates, and subordinate red mudrocks. On the mainland the last commonly show desiccation cracks. The clasts include fine-grained volcanic rocks, spilite, serpentinite, mudrocks, calcareous siltstones with some derived Silurian corals and gastropods, quartzite, schist, and chert. Many of the clasts can be related to the Deer Park Complex. The coarse clastics of the formation occur in channels (well seen, for example, at Roonah Quay on the mainland), cross-stratification indicating a northerly origin. This is different from the easterly or westerly sources indicated lower in the succession. A fluviatile environment is suggested, dominated by bedload rivers.

The north-western part of Clare Island is occupied by the Ballytoohy Formation, over which there has been much debate. Higgs and Williams (2001) have described a Silurian microflora, from which, despite high thermal maturity and poor preservation, they were able to suggest a Silurian age for the formation. This assignment has been refuted by various workers on various grounds. Graham (2001b) reviewed the arguments. Chew (2003) has demonstrated contemporaneous deformation of the Clew Bay Complex, which includes the Ballytoohy Formation, and the Dalradian of North Mayo during a middle Ordovician Grampian orogenic event.

Croagh Patrick

A belt of Silurian rocks of muscovite and chlorite greenschist facies lies south of the Emlagh Thrust and extends eastwards through Croagh Patrick and to the south of Westport. There is some confusion of nomenclature here (Anderson, 1960; Dewey, 1963; Bickle *et al.*, 1972; Kelly and Max, 1979), which Graham *et al.* (1989) attempted to solve by simplification.

A small outcrop to the south-west of the Corvock Granite is of cross-stratified marbles and calcareous siltstones (*Derryheagh Formation*) containing a Lower Silurian shelly fauna. The more widely developed *Cregganbaun Formation*, regarded as of nearshore marine origin, is dominantly of cross-stratified quartzites with a variable basal boulder bed. The deformed clasts of the latter are of quartzite and psammite with some red jaspers. The overlying quartzites form the famous mountain of Croagh Patrick (Fig. 7.2). Its angular scree is familiar to the thousands of pilgrims who, every year, make the ascent to be rewarded, if the mist clears, by one of the most splendid panoramas in the whole of Ireland. The overlying *Borris Formation* of Graham *et al.* (1989) is variable in lithology, including greenish siltstones and some impure limestones. A shelly fauna, particularly of brachiopods and corals, indicates a Lower Silurian age. Subsequently, Williams and Harper (1991) recorded *Pentamerus oblongus* from the Borris Formation 'within a set of Llandovery assemblages'. The highest, synclinally disposed *Lough Nacorra Formation* is mainly of greenish cross-stratified sandstones with finer inter-beds. There are numerous slump structures.

Killary Harbour – Joyce's Country

The Silurian rocks of this southern belt (Fig. 3.3) are separated from the Croagh Patrick succession by the tract of Ordovician rocks described in Chapter 3. There were various earlier contributions on these rocks, but the modern nomenclature was established by Laird and McKerrow (1970) and Piper (1972). The succession in formational terms is as follows:

Fig. 7.2 Croagh Patrick, County Mayo, from the west-north-west. *Photograph*: C.H. Holland.

(6) Salrock Formation	815 m
(5) Lough Muck Formation	200–280 m
(4) Glencraff Formation	65 m
(3) Lettergesh Foirmation	*c.*1500 m
(2) Kilbride Formation	up to 415 m
(1) Lough Mask Formation	0–170 m

The Telychian Stage of the Llandovery Series (Fig. 7.3) is well represented in this southern tract by a clastic succession up to 600 m thick, which rests on an irregular surface of Connemara Schists or of Ordovician rocks. The basal beds of the *Lough Mask Formation* are a thin (0–20 m) locally derived breccia or, to the east in Joyce's Country, a vesicular and pillowed albite trachyte lava which thickens eastwards to 115 m. In places there are sedimentary rocks beneath the lava. The rest of the formation comprises coarse red sandstones, commonly cross-stratified, with interbedded thin mudrocks and some cobble conglomerates. It is probably a braided fluviatile deposit transported from the north (Piper, 1972).

The *Kilbride Formation* is of green and grey sandstones, quartzites with vertical burrows (*Skolithos*), and some quartz microconglomerates and siltstones. There are brachiopod faunas including *Eocoelia curtisi* of middle Telychian age in its lower part (the Annelid Grit of earlier nomenclature), as a nearshore marine environment becomes represented. Higher (Finney School) beds show a deeper water *Clorinda* community. Doyle (1994) gave a detailed facies analysis of the storm-dominated sediments.

The upper part of the Kilbride Formation is locally dominated by bright purple and red mudrocks referred to as the Tonalee Member (Graham *et al.*, 1989). Its maximum thickness is 75 metres. This coloration is widely indicative of the *Monoclimacis crenulata* Biozone, the top graptolite biozone of the Llandovery Series (Fig. 7.3). Ziegler and McKerrow (1975) suggested that such colours might have arisen from oxidised soils in the absence of plant cover, and, in conditions of transgression, might be preserved in marine sediments offshore and in deeper seas in the absence of the biological reworking that would characterise nearshore regimes. Some green feldspathic tuffs are also present. Tabulate coral colonies are seen to have been halted in growth by volcanic debris that swept across the sea floor (Harper *et al.*, 1995).

Where it is conglomeratic the basal part of the *Lettergesh Formation* has been assigned to the Gowlaun Member (Piper, 1967, 1972). To the south of Loch Fee this rests directly on the Connemara Schists. Eastwards it is seen to pass laterally into the Lough Mask and Kilbride formations. Farther east it may rest directly on the latter or be absent. It comprises clast-supported conglomerates, sandstones, and mudrocks. The clasts (up to 0.8 m in size) are of reddish sandstones, metamorphic rocks, and basic to intermediate volcanics. The source of this material is difficult to determine, but fabric studies by Williams and

GLOBAL STANDARD STRATIGRAPHY				GRAPTOLITE BIOZONES
SILURIAN SYSTEM	UPPER SILURIAN	PŘÍDOLÍ		(no biozones recorded in British Isles)
		LUDLOW	LUDFORDIAN STAGE	*bohemicus*
				leintwardinensis
			GORSTIAN STAGE	*incipiens*
				scanicus
				nilssoni
		WENLOCK	HOMERIAN STAGE	*ludensis*
				nassa
				lundgreni
			SHEINWOODIAN STAGE	*ellesae*
				linnarssoni
				rigidus
				riccartonensis
				murchisoni
				centrifugus
	LOWER SILURIAN	LLANDOVERY SERIES	TELYCHIAN STAGE	*crenulata*
				griestoniensis
				crispus
				turriculatus
			AERONIAN STAGE	*sedgwickii*
				convolutus
				argenteus
				magnus
				triangulatus
			RHUDDANIAN STAGE	*cyphus*
				acinaces
				atavus
				acuminatus

Fig. 7.3 Silurian global standard stratigraphy and graptolite biozones.

O'Connor (1987) suggested derivation from the north-east. They postulated a deltaic fan building southwards.

The remainder of the Lettergesh Formation is of poorly sorted and frequently graded sandstones inter-bedded with mudrocks. The sandstones are composed of angular grains of quartz and feldspar and of rock fragments (including jasper and volcanic material), set in a fine matrix which has been chloritised. They can be described as arkosic greywackes and their common grading, sharp bases, and lateral continuity all suggest a turbiditic origin.

Rickards (1973) discovered an abundant graptolite fauna some 30 m above the base of the Lettergesh Formation in the Doon Rock area of the Kilbride peninsula. Several species suggest a Llandovery age and *'Diversograptus' inexpectatus* occurs in the *crenulata* Biozone in the Czech Republic. The presence of common *Monograptus priodon* confirms a very high Llandovery age. Rickards and Smith (1968) recognised both *Cyrtograptus murchisoni* and *Monograptus riccartonensis* biozones (Wenlock Series, Sheinwoodian Stage) from the Clonbur area between Lough Mask and Lough Corrib. A somewhat younger Wenlock graptolite fauna including *Monoclimacis flumendosae* and *Monograptus flemingii* was collected from a locality near Owenduff Bridge. Finally, D.G. Smith (1975) obtained plant spores, mostly occurring in tetrads, from a locality at the north-eastern extremity of the Kilbride peninsula at the south-western end of Lough Mask. Rickards identified graptolites from the same sample as *Monoclimacis* sp., probably of *Cyrtograptus centrifugus* Biozone (basal Wenlock) age.

Formations above the Lettergesh Formation are seen only to the west in the area to the south of Killary Harbour. The *Glencraff Formation* is similar to the rocks below but is more thinly bedded and of finer grain size. Its top is marked by a cream-weathering tuff. Nealon (1989) provided evidence that the formation comprises mainly offshore, storm-influenced deposits.

The *Lough Muck Formation* reflects a return to shallow water conditions (Nealon and Williams, 1988). The succession is well displayed north of Lough Muck and on the southern slope of Knockraff. Mudstones and siltstones are associated with sandstones, arkosic in composition but better sorted than those of the earlier part of the succession. There are frequent channels. Low in the formation a shelly fauna is recorded from Knockraff, which includes brachiopods and gastropods as main components, associated with bryozoans, corals, crinoids, and trilobites. The shells have been transported but certainly are dominated by the species *Eocoelia sulcata*. Higher in the formation *Monoclimacis flumendosae* again is present and there are further shelly assemblages associated with the abundant cross-stratified sandstones.

The succession is closed by the *Salrock Formation,* dominantly of red mudrocks. There are also sandstones, many of which are not red. The beds are well exposed around Killary Bay Little. To the east, greenish beds are more common. The direction of derivation is from the west or north-west. *Lingula* and poorly preserved gastropods are the lingering marine or quasi-marine fauna.

Charlestown

The small, ill-exposed inlier of Lower Palaeozoic rocks which is usually referred to as Charlestown (Fig. 7.1) begins about a kilometre south of that town and forms a south-western end of the Curlew Mountains anticline with its long narrow outcrop of Old Red Sandstone. The Silurian follows unconformably upon the steeply folded Ordovician and itself dips at high angles to the south-west. The upward sequence comprises purple sandstones, limestones, fine sandstones, and grey, reddish, or green sandstones (Charlesworth, 1960a). Parkes (1993a) gave these four divisions formational names as follows: *Glen School Formation* (*c.*300 m), *Uggool Limestone Member* included in the *Cloonnamna Formation* (*c.*300 m), *Cloonierin Formation* (*c.*1500 m). The first of these is fluviatile in origin and resembles the Lough Mask Formation. The thin, poorly exposed, shallow marine Uggool Member yields a rich coral and stromatoporoid assemblage. The Cloonnamna Formation has a more diverse shelly fauna of brachiopods, bivalves, trilobites, etc, though with fewer corals. The tabulate corals are here the best biostratigraphical indicators (Scrutton and Parkes, 1992). The Uggool Limestone is of early *crenulata* Biozone age, or possibly *griestoniensis* Biozone. The *Palaeocyclus porpita–Favosites multipora* fauna of the Cloonnamna Formation is highly characteristic of the *crenulata* Biozone of the uppermost Telychian.

Northern Inliers: Lisbellaw and Pomeroy

The northern inliers of Lisbellaw and Pomeroy (Fig. 7.1) provide evidence of a more basinward Llandovery (Aeronian) succession. At *Lisbellaw,* County Fermanagh slumped conglomerates with clasts from the Dalradian and the Tyrone Igneous Complex are sandwiched within graptolitic shales (Harper and Hartley, 1938). Older rocks are not seen. Graptolites are recorded also from rocks interbedded with the conglomerates. The Llandovery biozones of *triangulatus* to *argenteus* appear to be present. There are a few shelly fossils including bivalves and cephalopods. The village of Lisbellaw itself is built upon one of the features made by the conglomerates, in which some of the clasts reach 0.6 m in diameter.

The folded Ordovician and Silurian rocks around *Pomeroy*, County Tyrone were described in some detail by Fearnsides *et al.* (1907). Following directly upon the Ashgill, the relatively thin Llandovery sequence is richly graptolitic. Plentiful collections were made through the 'Little River Group'(40 m), comprising the first five of six named divisions.

Much of the Rhuddanian and all of the Aeronian Stage appear to be present. A complete sequence was recognised in the Little River from the *Akidograptus acuminatus* Biozone to that of *Monograptus sedgwicki*. The *Monograptus sedgwicki* Biozone is know from Limehill. The top, presumably Telychian division, referred to as the 'Corrycroar Group' (30–46 m) yielded only a few fragments of graptolites. It was described as of 'green and purple mudstones, shales, flags, and grits of Gala type'. Tunnicliffe (1983) described the oldest known nowakiid (Tentaculitoidea) from the type locality of *Monograptus sedgwicki* at Pomeroy.

Longford-Down Inlier

The most extensive area of Silurian rocks in Ireland is that of the Longford-Down inlier, which appears on the geological map as such an obvious continuation of the Southern Uplands of Scotland. The richly graptolitic sections in places along the north-east coast have long been known (Swanston and Lapworth, 1877). Charles Lapworth, reflecting on the close comparison between the Coalpit Bay (Fig. 7.1) succession and that in the Southern Uplands, wrote that 'The Coalpit Bay Division of the County Down Silurians has yielded all the Graptolites of the Birkhill Shales [Rhuddanian and Aeronian], with the exception of one special group, viz., that of the *Rastrites maximus* zone [equivalent to the lower part of the *Monograptus turriculatus* Biozone, basal Telychian], which lies at the very summit of the Moffat series'. Toghill (1968) described a new species from the lower part of the *cyphus* Biozone here and listed the associated graptolites. These graptolitic shales pass up into thickly bedded greywackes (Fig. 7.4). Northwards, beyond a fault gap, older greywackes are laterally equivalent to them.

Coastal exposure in County Down provides an almost complete section across the intensely folded Ordovician and Silurian turbidite belt, which is segmented by various strike faults (Anderson and Cameron, 1979). The pattern is as has been described for the Southern Uplands of Scotland in that the fold envelope in each segment descends northwards but the strike faults throw up progressively younger rocks to the south. At the old locality of Tieveshilly (Fig. 7.1), at the southern end of the Ards Peninsula, old slate pits and other small exposures reveal grey or black graptolitic shales interbedded with the local sandstones. *Monograptus riccartonensis*, a Wenlock (Sheinwoodian) biozonal index, is commonly present in the uppermost beds (Swanston and Lapworth, 1877).

Cameron and Anderson (1980) positively identified and described the common pale green and grey bands in the Silurian of County Down as metabentonites, which occur in both graptolitic shales and turbidites. They concluded that the volcanic eruptions that were the source of the bentonites began not later than *Monograptus gregarius* Biozone times and persisted at least until the end of the Llandovery.

A thoroughly detailed investigation by E.A. O'Connor (1975) of the Shercock–Aghnamullen district (Fig. 7.5), Counties Cavan and Monaghan, is probably representative of the central part of the Longford-Down inlier with its continuation of the Central Belt of the Southern Uplands of Scotland. Here, as elsewhere within the inlier, drift cover is substantial and in places drumlins dot the landscape. The local succession is as follows:

Shercock Formation (*c.*300 m) Lough Avaghan Formation (*c.*2800 m)
Silurian
Taghart Mountain Formation (500–1700 m)
Ordovician
Kehernaghkilly Formation (black shales) (*c.*100 m)

The *Taghart Mountain Formation* is of quartz mudrocks with convolute bedding and sole markings, alternating with bands, also at least several centimetres thick, of soft silty shales. The beds mostly young north-westwards, though the dip is generally to the south-east.

In the *Shercock Formation* medium to thickly bedded greywackes are interbedded with black shales and greenish or greyish silty shales. Many of the greywackes are graded, and sole markings are common. Plentiful graptolites in scattered localities indicate the presence of the

Fig. 7.4 Loaded flute moulds on inverted Silurian greywacke, near Galloways Burn, Donaghadee, County Down. *Photograph: IPR/109-39CT British Geological Survey.*

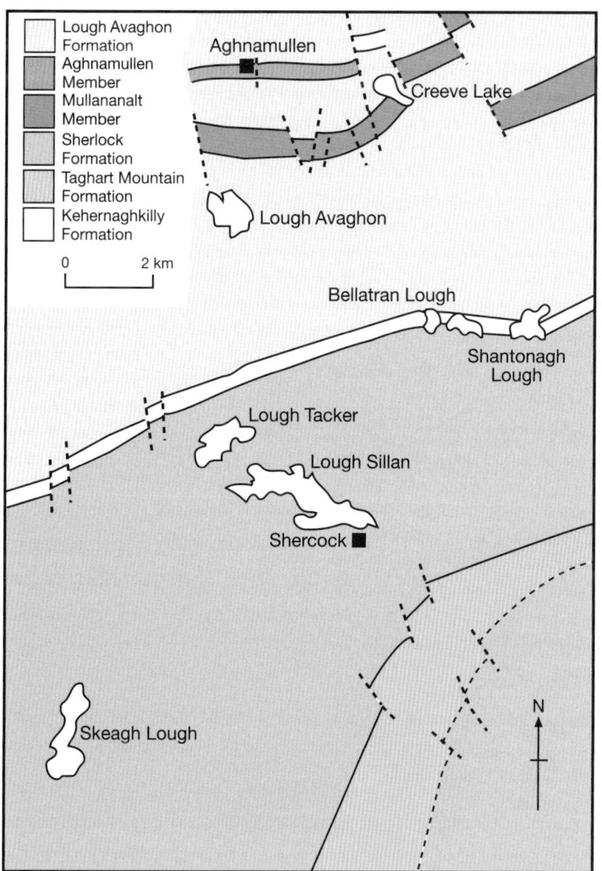

Fig. 7.5 Geological map of the Shercock–Aghnamullen district, Counties Cavan and Monaghan (simplified from E.A. O'Connor, 1975).

Rhuddanian and probably the Aeronian stages from the *Atavograptus atavus* Biozone upwards. As in the older formation, the beds tend to young north-westwards and dip south-eastwards. There are two outcrops, of which the south-eastern is faulted against the Taghart Mountain Formation. In the north-western outcrop, however, the inverted Shercock Formation appears to continue the north-westerly younging seen in the probably older Taghart Mountain Formation.

The *Lough Avaghan Formation* is lithologically similar to the Shercock Formation, though generally somewhat coarser in grain size. Quartz pebble greywackes are more common. O'Connor recognised two finer-grained members which are indicated in Figure 7.5, and farther south, discontinuous thin linear outcrops of an agglomerate which he referred to as the Carrickatee Member. Palaeontological evidence for the age of the formation is sparse, an approximate level of diagnosis being achieved only for the *Monograptus cyphus* Biozone, represented in its lower part.

Proximal and distal turbidites are present in the Silurian succession. A northerly source is indicated, probably from the Tyrone Igneous complex. All the finer beds are prominently cleaved and the beds face downwards on the cleavage and, in places, upwards. Thus pre-cleavage folding and perhaps thrust slicing are involved, as are cross-folds following upon the main phase of folding and its associated cleavage. It is possible that soft-sediment deformation is involved. Elsewhere in the Longford-Down inlier certain thick structureless quartzites appear to be grain-flow deposits, and there are slumped conglomerates.

In general, the appearance of thickly bedded greywackes in the Irish Lower Palaeozoic sequence appears to become younger southwards. Thus they are already present in the Ordovician of the northern belt of the Longford-Down inlier. In the area near Lough Acanon (Fig. 7.1) (Phillips and Skevington, 1968) there appears to be continuity from Ordovician to Silurian graptolitic shales, with greywackes present only in the lower part of the Llandovery, as is the case in the Shercock–Aghnamullen district some 8 km to the east. From a borehole through the Carboniferous south-west of Navan, Lenz and Vaughan (1994) described a faulted and mostly inverted sequence with richly graptolitic pelagic and hemipelagic sedimentary rocks of late Ordovician to middle Wenlock age, followed by greywackes.

In their description to accompany the Monaghan–Carlingford Sheet of the 1:100,000 Geological Survey bedrock map, Geraghty *et al.* (1997)include a substantial part of the Longford-Down inlier. Four numbered tracts are employed within the Central Belt. The same scheme has been used for adjacent sheets. The tracts are strips with considerable stratigraphical and tectonic continuity along strike and with boundaries formed by strike faults. This follows practice in Scotland, though the numbering is different. The formations referred to above fall within the four tracts 4 to 7, but additional lithostratigraphical units are incorporated.

Eastern inliers: Balbriggan, Portrane, and Kildare

Balbriggan

The graptolitic Silurian rocks south of Balbriggan (Fig. 7.1) are well exposed in the coastal section described by Rickards *et al.* (1973). The basal Silurian *Parakidograptus acuminatus* Biozone is here faulted against Ordovician rocks. Interrupted only by minor faults, the succession continues to the *Monograptus turriculatus* Biozone. Above this, the Telychian Stage is largely of greywackes. Graptolitic mudstones of Wenlock age then continue into the *Monograptus riccartonensis* Biozone. There are then associated greywackes, but the graptolitic evidence persists into the *Cyrtograptus lundgreni* Biozone. A development of banded bluish mudstones in the *Cyrtograptus murchisoni* Biozone recalls the Brathay Flags lithology, and indeed the whole Silurian sequence south of Balbriggan is directly comparable with that in the English Lake District.

The completeness of the Balbriggan succession is probably diminished only by the effects of minor faulting. Indeed, the Silurian graptolitic biostratigraphical record, with some fifty species recorded, is certainly the best available in Ireland. Above the graptolitic sequence of the Balbriggan Formation (greater than 560 m), which is also taken to include a thin development of black shales of the Ordovician *Dicellograptus anceps* Biozone, are sparsely fossiliferous greywacke sandstones assigned to the Skerries Formation (more than 350 m), which have yielded fragmentary graptolites of at least *Cyrtograptus lundgreni* Biozone age.

Inland exposures near Balbriggan are much poorer, but the late Veronica Burns continued for some twenty years to accumulate graptolitic evidence from this unpromising ground. The Lowther Lodge Fault is taken in Murphy *et al.* (1991) to separate Bellewstown and Leinster terranes, though by Silurian times the distinction between these may be insubstantial. The several hundred graptolites eventually collected from inland localities came from the Bellewstown terrane (Burns and Rickards, 1993). There is here a west-south-westerly plunging syncline. The lowest beds seen are of *Monograptus riccartonensis* Biozone age. The succession reaches the Ludlow Series, represented by the *Neodiversograptus nilssoni* Biozone.

Burns and Rickards were able to make comparisons across the Lowther Lodge Fault only at some Wenlock horizons. At the *riccartonensis* level they found the beds to the north 'much more ribbon banded (with silt and sand laminae)' and thus perhaps closer to successions in North Wales or the Longford-Down inlier. McConnell *et al.* (2001) provided a more elaborate map of the inland area than that of Rickards and Burns and assigned additional formation names for the Wenlock and Ludlow rocks in this poorly exposed ground.

Portrane

At Portrane (Fig. 7.1), originally described by Gardiner and Reynolds (1897), between Balbriggan and Dublin, a small area of greywacke sandstones, siltstones, and mudstones at the northern end of its outcrop is certainly in tectonic contact with the Ordovician. The coastal section is very well exposed but there is little information inland. Assignment to the Silurian has depended upon lithological comparison (e.g. with the Silurian of Balbriggan). Parkes (1993b) suggested that the rather obscure south-western contact of the Silurian with the Ordovician Portrane Limestone is actually an unconformity.

Kildare

The Kildare inlier (Fig. 7.1), approximately along strike from Portrane and 45 km south-west of Dublin, was described originally by Gardiner and Reynolds (1896). The narrow, broken, 9 km long ridge made by the Lower Palaeozoic rocks is conspicuous in the topography. Parkes and Palmer (1994) provided a modern map and detailed description. The Silurian forms Dunmurry Hill in the southern part of the inlier. It is faulted against the Ordovician succession of Grange Hill, though some of the top formation is present also in the latter area. The sequence comprises the *Guidenstown* (about 50 m), *Rahilla* (110 m), and *Dunmurry* formations. The first of these is of black shales with fragmentary graptolites indicating a *sedgwicki* Biozone (Aeronian) age. The easily recognisable red mudstones of the Rahilla Formation have not yielded fossils. Such a colour is commonly characteristic of the highest part of the Telychian Stage. It may therefore be that the greenish greywacke siltstones, sandstones, and shales of the Dunmurry Formation are Wenlock in age.

Leinster Massif

Knowledge of the Lower Palaeozoic rocks in south-eastern Ireland was much advanced by P.M.Brück's studies of outcrops in the northern part of the Leinster Massif (Brück 1970, 1971a, 1972, 1973) (Fig. 7.1). Small-scale cross-stratification indicates that the steeply dipping succession to the north-west of the granite youngs westwards. This block of country is largely of Silurian rocks referred to by Brück *et al.* (1979) as the Kilcullen Group. McConnell and Philcox *et al.* (1994) have summarised the structural arrangements here, with a shear zone forming the eastern boundary against the Ordovician and the unconformity with the Upper Palaeozoic rocks to the west.

Brück recognised five formations here. The *Pollaphuca Formation* of coarse-graded grey greywacks and dark shales yielded poorly preserved lower Ordovician palynomorphs (Brück *et al.*, 1974); but McConnell and Philcox *et al.* suggested that there may have been reworking into the Silurian. The succeeding *Slate Quarries Formation* (40 m) is based on a locality north of Blessington, for which Brück *et al.* recorded palynological evidence for an upper Ordovician to Lower Silurian age. The *Glen Ding Formation* (430 m) yielded diagnostic assemblages of chitinozoans indicating a Silurian age. The *Titterkevin Formation* (110 m) gave a specifically Llandovery age, one of the samples coming from within 4 metres of its top. Crinoid ossicles and rare bryozoan and brachiopod fragments were recorded from it and the succeeding *Carrighill Formation*.

The greywackes become generally finer upwards. In the Glen Ding Formation their matrix is largely of sericite and chlorite. In the Tipperkevin Formationit it is largely illite, whereas the Carrighill Formation has an unusual carbonate matrix mainly of iron-rich dolomite. This overprints some chlorite and sericite, which can themselves be seen to corrode grains, showing recrystallisation post-dating stresses related to the formation of the regional (S1) cleavage. Graded bedding, convolute bedding, flute moulds, and other characteristic sedimentary structures are commonly present. The increasingly varied sequences of Bouma intervals present in individual graded units as the succession is ascended contribute to evidence that proximal turbidites gave way in time to those of more distal origin. Derivation was from the eastern quadrant throughout.

Central inliers

Knockshigowna Hill

The smallest of the Silurian inliers of central Ireland is that of Knockshigowna Hill, where these rocks form a ridge less than 3 km in extent. It lies within the horseshoe of Lower Palaeozoic inliers of Slieve Bloom, Slieve Phelim–Silvermines–Devilsbit, Cratloe Hills, Broadford Mountains–Slieve Bernagh–Arra Mountains, and Slieve Aughty (Fig. 7.1). The succession described by Prendergast (1972) comprises 500 m of greywacke sandstones, siltstones, and shales followed in the axis of a syncline by some 150 m of conglomerates. Shales associated with the greywackes yield *Monograptus riccartonensis* and *Monclimacis vomerina* indicative of a Wenlock (Sheinwoodian) age The conglomerate contains clasts of quartz, jasper, and chert from 1 to 10 cm in size. In sandy pockets within it an upper Wenlock (Homerian) shelly fauna is found, dominated by brachiopods but with bivalves, gastropods, trilobites, and other forms. The brachiopods include *Eoplectodonta duvalii*, *Isorthis clivosa*, *Leptaena depressa*, and *Meristina obtusa*. The assemblage appears to have been derived from a low intertidal to subtidal environment, though its state of preservation is not indicative of prolonged transportation.

Slieve Bloom

Some sixteen individual inliers of Silurian rocks appear within the high, rolling, and much wooded Old Red Sandstone country of the Slieve Bloom Mountains, Counties Laois and Offaly (Feehan, 1982a). Somewhat separate from the other inliers is the Capard ridge in the north-east described by Holland and Smith (1979). All these Silurian rocks have been assigned to the single *Capard Formation*, which in the Capard area is about 1500 m thick. Grey greywacke sandstones, siltstones and mudstones dominate the succession. The siltstones occur characteristically as banded coarse to fine units, similar to those in the Slieve Phelim inlier. A localised conglomeratic unit resembles that of Knockshigowna Hill. The structure of the area is relatively simple, with folding of Caledonoid orientation and a single variably developed cleavage, which is frequently not axial planar.

Loughlin (1976) recorded graptolites from the north-western part of the area which indicate a *Monograptus riccartonensis* Biozone age, and further material was collected by Feehan (1982a). The fauna includes *Monograptus priodon*, *Monoclimacis vomerina*, *Pristiograptus meneghini*, and others resembling *M. riccartonensis*. Holland and Smith (1979) recorded palynomorphs from the Capard inlier including acritarchs, chitinozoans, and abundant trilete spores, the assemblage indicating a Wenlock age. G.D. Sevastopulo (*in* Feehan, 1982a) commented that crinoids from Slieve Bloom are almost certainly assignable to *Scyphocrinites*, a form commonly associated with basinal facies.

Slieve Phelim–Silvermines–Devilsbit

The relatively large area of Silurian rocks forming the generally elevated country of the Slieve Phelim–Silvermines–Devilsbit inlier is very largely confined to the *Cyrtograptus lundgreni* Biozone (Wenlock, low Homerian), though this is probably more than 1000 m thick. In the southern part of the inlier, Doran (1974) gave detailed description of the repetitive sequence of greywackes, laminated siltstones, and mudstones, assigning the whole to a single *Hollyford Formation*. Archer (1981) used the same formational nomenclature in mapping similar rocks in the Keeper Hill district in the north-western part of the inlier. Doran noted that the Silurian rocks are in places stained red to a depth of at least 240 m below the unconformable Upper Old Red Sandstone. The style of deformation throughout the inlier is of numerous upright folds of various orientations in different structural blocks and with one dominant cleavage that may or may not be penetrative. A few large quarries reveal the presence of numerous minor faults.

Graptolites are found at many localities in the coarser, brown-weathering, 'biscuity' laminae of the laminated siltstones. *Pristiograptus dubius*, *Monograptus flemingii*, *P. pseudodubius*, *Cyrtograptus lundgreni*, *C. hamatus*, and *Paraplectograptus eiseli* are recorded, those listed first being the more common. Evidently a quiet water regime was interrupted sporadically by distal turbidity currents, forming the graded beds now seen as greywackes. The dominant flow was southwards, though farther north in the inlier there are indications of an easterly or north-easterly derivation.

Palmer (1970) showed that the supposedly Ludlow rocks of the north-eastern corner of the inlier (Devilsbit mountain district) are in fact of *Pristiograptus ludensis* Biozone age, at that time the first recognition of this highest Wenlock graptolite biozone in Ireland. *P. ludensis* is associated with *Monograptus auctus*.

Apart from graptolites and orthoconic nautiloid cephalopods, other fossils are relatively rare throughout the whole inlier. Cope (1959) analysed those present, recognising that the greywackes carry derived faunas of thick-shelled brachiopods and crinoids; and certain micaceous siltstones show a terrestrially derived association of plant fragments and a phyllocarid; whereas possibly benthonic thin-shelled brachiopods, small bivalves, and crinoids occur in the more common laminated siltstones.

A most important find of macroplants including the erect fertile land plant *Cooksonia*, preserved as a highly coalified impression, has been recorded from the marine siltstones of the *Monograptus ludensis* Biozone in the Devilsbit district (Edwards *et al.*, 1983). Palynomorphs also are present.

Cratloe Hills

The name Cratloe Hills has been employed since the time of the original geological survey (Kinahan *et al.*, 1860) for the long, very narrow strip of Silurian rocks separated from the Broadford Mountains inlier by a tract of Upper Old Red Sandstone and Carboniferous rocks. The Silurian and Old Red Sandstone form a prominent ridge well seen from the Limerick to Ennis road. The northern boundary fault of the inlier, though it has been regarded as the extension of a supposedly important Navan–Silvermines line, appears to be of small throw (Holland *et al.*, 1988). To the south the faulted boundary of the inlier is interrupted by areas where the unconformity between the Old Red Sandstone and the Silurian is seen. The Silurian rocks of the inlier have undergone brittle deformation and there are many small-scale folds of various axial trends and numerous small fractures. Cleavage is rarely seen.

The Silurian here comprises grey to greenish grey, more or less laminated siltstones and sandstones with many nodules up to 0.5 m in size. A variety of secondary colours have resulted from weathering and staining. The whole succession has been assigned to the *Cratloes Formation* (approximately 300 m). The rocks are sparsely graptolitic but sufficiently so as to indicate an upper Wenlock (Homerian) age (Holland *et al.*, 1988). A locally developed pebbly sandstone member contains a Homerian shelly fauna of rugose corals, tabulates, trilobites, and brachiopods such as *Dicoelosia biloba*, *Eoplectodonta duvalii*, *Leangella segmenta*, and *Meristina tumida*. Weir, who originally described this area in 1962, later recognised the Wenlock age of the shelly assemblage and of one from a thin conglomerate in the Arra Mountains (Weir, 1975).

Broadford Mountains–Slieve Bernagh –Arra Mountains

The extensive Silurian inlier of the Broadford Mountains, Slieve Bernagh, and the Arra Mountains (Fig. 7.1) is cut by the southern end of Lough Derg and the River Shannon in an area of outstanding natural beauty. The presence of Llandovery black shales, the first such beds to become known in the southern half of Ireland, was established by Rickards and Archer (1969) in a very small inlier near Tomgraney, County Clare, where, together with Ordovician graptolitic shales, they are faulted against the younger Silurian rocks of the Slieve Bernagh inlier and are overlain unconformably by Upper Old Red Sandstone. Dr Aubrey Flegg made a detailed investigation of the intricate geology of the eastern part of the Slieve Bernagh inlier and the whole of the Arra Mountains to the north-east. He kindly provides a geological map and notes (Fig. 7.6). The similarly thick basinal succession of the Broadford Mountains anticline to the south-west of Slieve Bernagh was mapped by Weir (1962).

Slieve Aughty

Projecting the strike of the Lower Palaeozoic rocks of the Longford-Down inlier south-westwards, we meet the inliers of Slieve Aughty (Fig. 7.7). Here the regional strike is seen to have swung to a more westerly direction. The area is crossed by a series of strike faults, some at least of which were re-activated in post-Carboniferous times. G.T. Emo (see Emo and Smith, 1978) defined three formations of Silurian or partly Silurian age. The *Gortnagleav Formation* (more than 650 m) is of unfossiliferous greenish or mottled mudstones, siltstones, and greywacke sandstones, with some thin lavas, faulted against the Ordovician in the north of the area. Lithological comparisons suggested that the lower part of the formation is probably Ordovician and the upper part Llandovery. In particular, there are some coarser turbiditic sandstones near the top of the Gortnagleav Formation which may represent a prelude to those found in the Derryfadda Formation.

The much thicker (more than 2500 m) *Derryfadda Formation* is of coarser, green, proximal turbidites interbedded with siltstones and mudstones. There are microconglomerates in the channelled bases of some of the coarser turbidites. The beds young northwards but dip to the south and are crossed by a number of strike faults. One locality with banded siltstones yielded graptolites of the *Monoclimacis griestoniensis* Biozone (Telychian). Emo and Smith (1978) obtained confirmatory palynological evidence.

A wide extent of the southern part of the area is occupied by the *Killanena Formation* (more than 3000 m). In places the unconformity below the Old Red Sandstone is beautifully displayed. The formation is of rocks thinner bedded and finer grained, though grey-greenish greywackes again alternate with laminated and cross-laminated fine sandstones, siltstones, and mudstones. Again the beds young northwards, dip to the south, and are affected by strike faults. Both distal and proximal turbidites are present and slump structures are common. This formation contains more grey siltstones or mudstones suitable for palynological investigation. Miospores and acritarchs both indicate an upper Llandovery to lower Wenlock age. This formation is partly at least equivalent in age to the Derryfadda Formation. One palynological assemblage is of Tremadoc to Arenig aspect and represents that reworking of early Ordovician acritarchs that is so widespread in the British Isles.

Attempting to recognise general relationships in all the Silurian inliers of central Ireland, Harper and

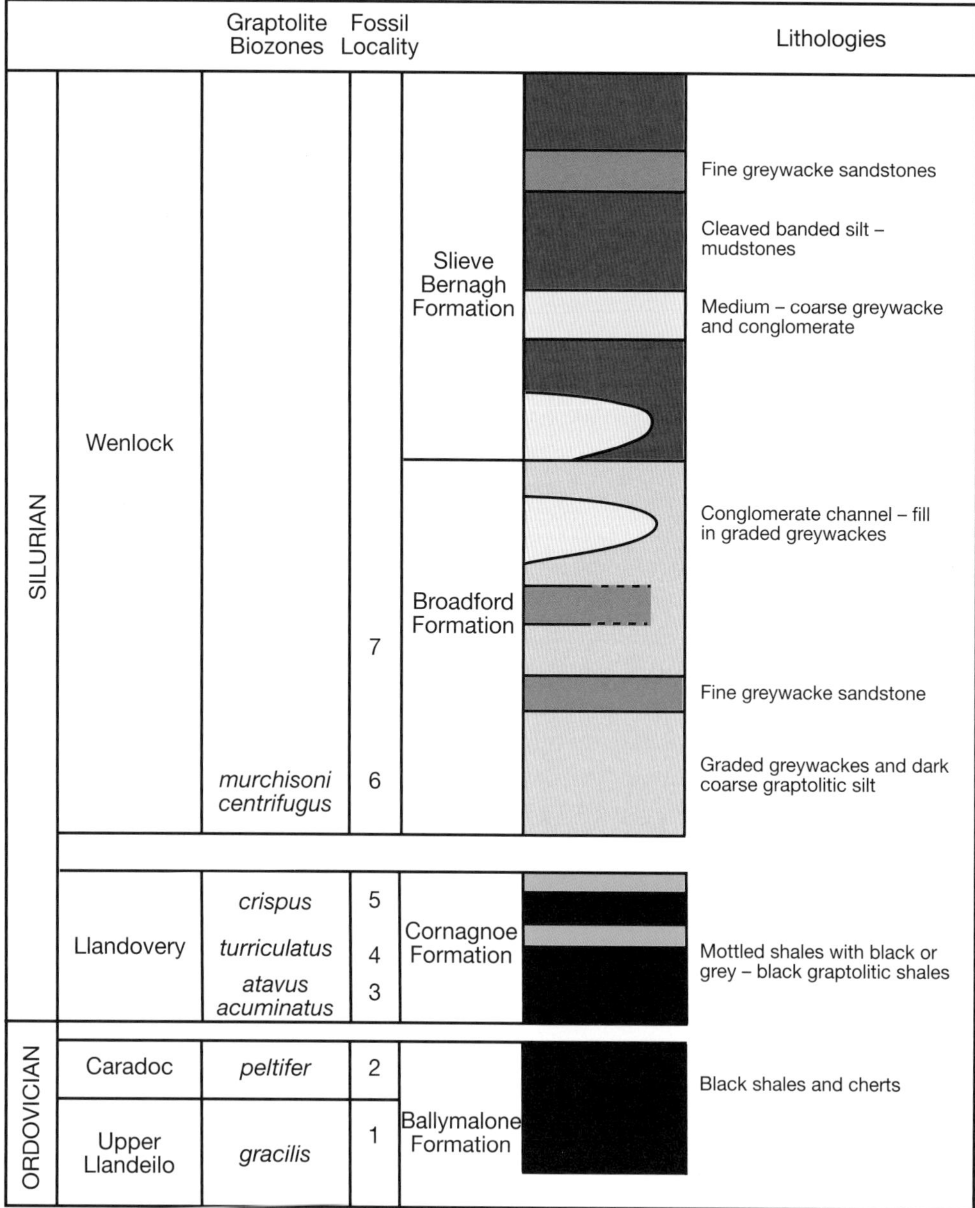

Brenchley (1972) distinguished between the north-western inliers of Slieve Aughty and Slieve Bernagh–Arra Mountains and those of the Cratloe Hills, Slieve Phelim, Knochshigowna, and Slieve Bloom. Shelly faunas (though certainly rare) are present along with graptolites in the latter group, whereas the north-western inliers tend to be devoid of macrofossils. The carrying of a shelly fauna into the Arra Mountains succession has been mentioned already. A north-westerly facing palaeoslope was thus postulated. Weir (1975) subsequently wrote of a Limerick–Tipperary shelf in contrast to the Silurian basin of Slieve Aughty and Slieve Bernagh.

Southern Inliers

Slievenamon

The characteristic topographical expression of the various Silurian inliers of central and southern Ireland is well seen in that of Slievenamon (Fig. 7.1), where the Old Red Sandstone forms a high rim to the inlier, culminating in the mountain of Slievenamon itself at its western end. Details of the roughly rectangular area of Lower Palaeozoic rocks of the Ninemilehouse Tableland (Jukes and du Noyer, 1858) were unknown until Colthurst (1974) established the pattern of north-easterly strike (Fig. 7.8), the sequence younging north-westwards

Opposite and below Fig. 7.6 *Eastern end of the Slieve Bernagh – Ana Mountain Inlier.* Immediately east of the Ordovician fossil localities (1, 2) of the Ballymalone Formation (Archer *et al.*, 1996) is a stream cutting pale grey mudstones with inter-bedded graptolitic mudstones yielding upper Llandovery *turriculatus* Biozone graptolites (4a). Upstream of a small fault, dark graptolitic mudstones yield lower Llandovery graptolites (3) successively from *acuminatus* and *atavus* biozones (Rickards and Archer, 1969). Subsequent field mapping (Aubrey Flegg) revealed a further graptolite locality (4b) in black shales associated with mottled silts, but separated from the Ordovician shales by a fault sliver of fine-grained sandstone. These are of *turriculatus* Biozone age (Hutt, *pers. comm.*). These, together with the extensive occurrences of mottled shale that occur to the west, in the Cornagnoe Valley, are referred to the Cornagnoe Formation. Black shales associated with these have yielded a *crispus* Biozone fauna (also kindly identified by J. Hutt).

In the Broadford Formation four main greywacke lithologies dominate: fine laminated silt/mudstones (60% of outcrop), fine and medium greywacke sandstones (30% and 8%), and coarse greywackes and conglomerates (2%). Alternations are rapid, so it has been possible to represent only the thicker units on the map. Graded bedding is common and well developed. A strong slaty cleavage has enabled slate extraction from some silt/mudstone bands. Two fossil localities give somewhat conflicting evidence. West of Killaloe, black to dark grey cherty silts have yielded graptolites (6) attributable to the *murchisoni* or *centrifugus* biozones of the lower Wenlock (J. Hutt). On the other side of the Shannon, just north of Ballina, Weir (1975) recorded an upper Wenlock, shelly fauna.

In the Slieve Bernagh Formation coarse greywacke sandstones and conglomerates carrying proximally derived mudstone flakes fill broad north–south orientated channels. In contrast, thick sequences of laminated silt/mudstones represent long periods of quiescence. Fine greywacke sandstones are well winnowed and often lack silty tops. Sedimentation appears to be from two sources: a distal source to the east with a proximal source, possibly to the north, for the coarse sediments. The coarse greywackes include igneous and metamorphic fragments. Grains of jasper occur commonly only in the north.

The major structure of the inlier is the Slieve Bernagh Syncline. This is a major Fl fold with a gentle westerly plunge. Superimposed on this are minor, often monoclinal folds, with plunges of between 8° and 25° west. Minor folding is more intense in the fine silt/mudstones, which show an intense slaty cleavage. On the southern limb of the anticline, where the steep limbs of the monoclines overturn and dip south, steep reverse strike-faults represent a final tightening of the structure. To the north of the Portroe fault, the cleavage flattens out to the extent that occasional downward-facing structures are encountered. Later near-vertical folding has induced strike deflections.

In addition to the early strike faulting, normal north-south-west wrench faults occur in pairs. Curved, usually sinistral, wrench faults follow, with late re-activation of the earlier strike faults.

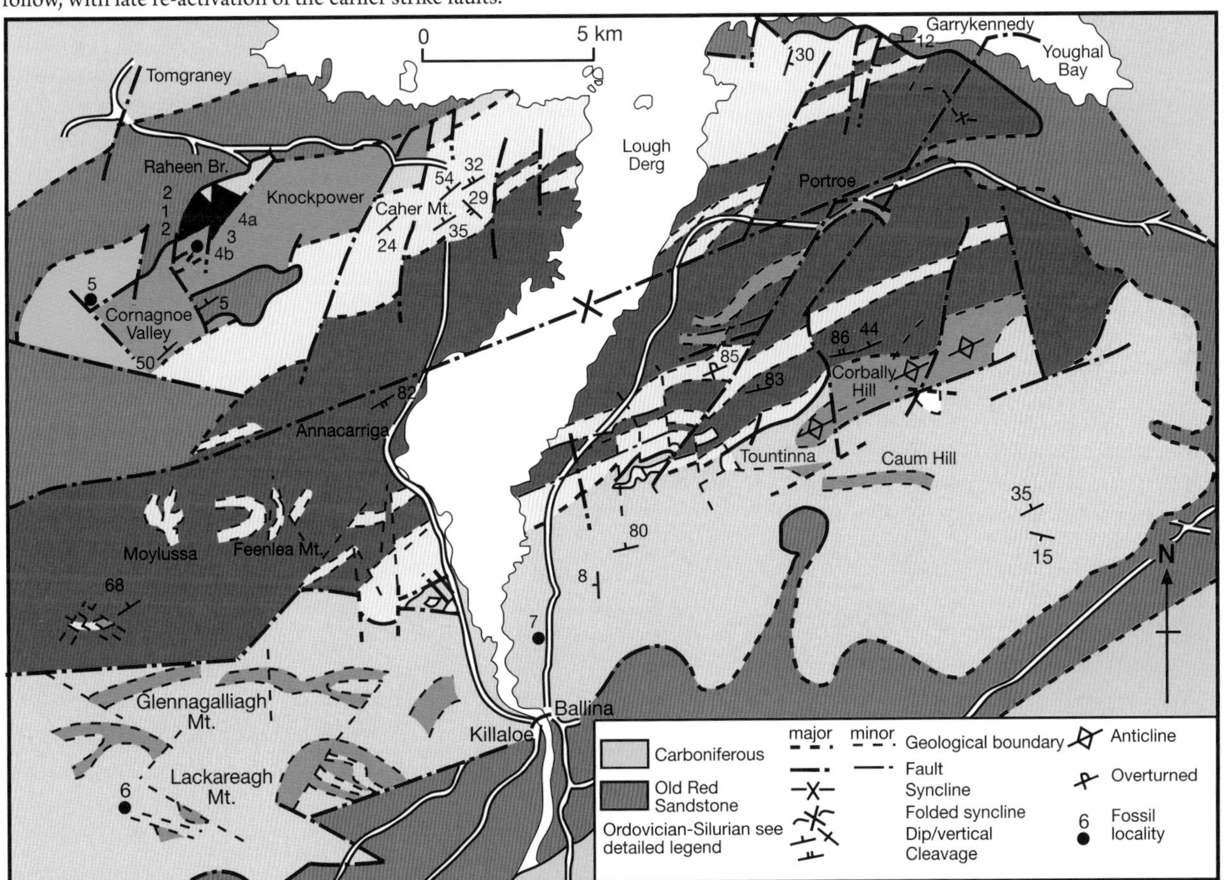

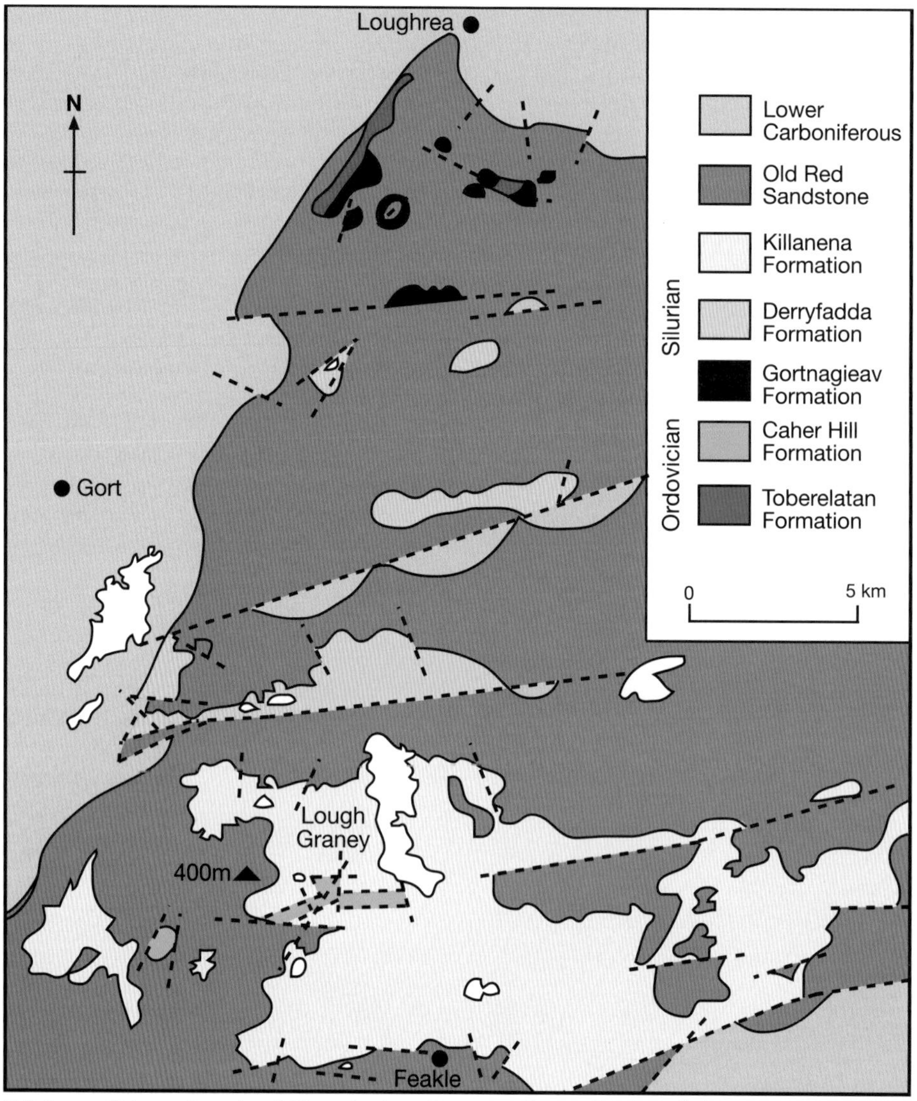

Fig. 7.7 Lower Palaeozoic rocks of Slieve Aughty, Counties Clare and Galway (*after* Emo and Smith, 1978).

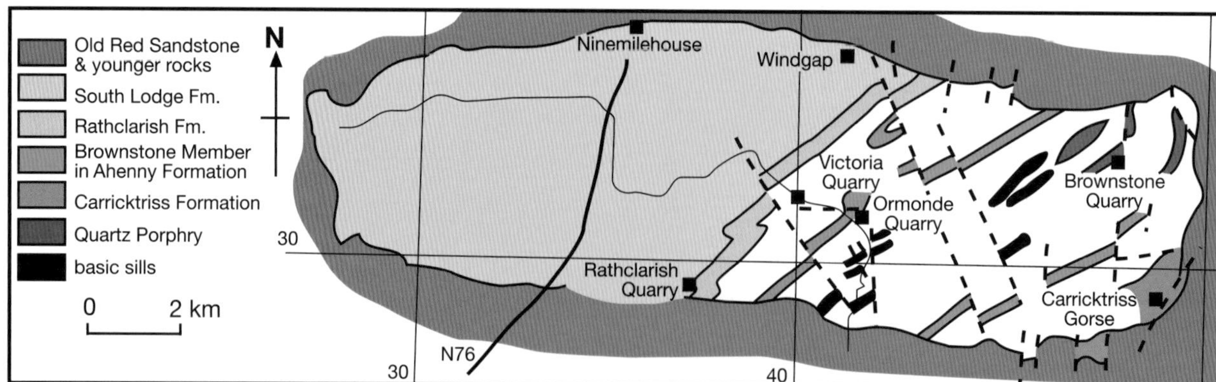

Fig. 7.8 Geological map of the Slievenamon inlier, County Tipperary (*after* Colthurst, 1974).

except for the effects of folding of relatively small amplitude. The succession comprises:

(4) South Lodge Formation (possibly more than 3400 m)
(3) Rathclarish Formation (340 m)
(2) Ahenny Formation (*c.*3000 m)
(1) Carricktriss Formation (more than 800 m)

The *Carricktriss Formation* (Ribband Group according to Brück *et al.*, 1979) and its faulted contact with the *Ahenny Formation* have been referred to in Chapter 5. Grey banded slates and blue-black slates are characteristic of the *Ahenny Form*ation and are well seen in old slate quarries. About the middle of the formation is the distinctive *Brownstown Member* (230 m) of graded and imbricate conglomerates with clasts of varied composition , up to 2 cm in length, in a muddy matrix. The overlying *Rathclarish Formation* is of well developed, graded, grey greywackes, siltstones, and mudstones typically turbiditic in character. The *South Lodge Formation* forms a little more than half of the inlier and is characterised by greenish greywackes and slates. Most of the greywackes have a calcareous matrix and weather easily to a brownish material. These beds are seen in road cuttings along the Callan to Clonmel road (N76).

As in the Galty Mountains area, referred to below, sole markings are not easily seen. The upper three formations are best interpreted as turbidites, those of the Ahenny Formation being distal in character, the others proximal. Colthurst and Smith (1977) and subsequently Smith (1979b) made effective use of micropalaeontology in the biostratigraphy of this succession, in which the only available macrofaunal evidence was of a few unidentified graptolites from the South Lodge Formation. Acritarchs, chitinozoans, scolecodonts, and spores were recovered variously from the samples, though they are usually rare.

The Ahenny and Rathclarish formations certainly contain early Ordovician assemblages that have been reworked. Silurian, probably upper Llandovery acritarchs were obtained from near the base of the Ahenny Formation and the Rathclatrish Formation also contains trilete spores indicative of the Silurian. The South Lodge Formation was shown to be Wenlock by the presence high in the sequence of abundant trilete spores, almost entirely of one type. The samples lack the diversity of Ludlow assemblages.

The Silurian rocks of the Slievenamon inlier were referred to by Brück *et al.* (1979) as the Kilcullen Group, a name which they applied to all the post-Ribband Group non-volcanic strata north-west of the Leinster Granite and in both Slievenamon and the Comeraghs (see below). In Slievenamon, as indicated above, the group must be separated from the local Ribband Group by a substantial stratigraphical break.

Comeragh Mountains

The Lower Palaeozoic rocks of north County Waterford (Fig. 7.9) are separated from those of the Slievenamon inlier by the narrow lower Carboniferous and Old Red Sandstone tract of the Suir Valley. The area of Silurian rocks here has been referred to loosely as the Comeraghs. The Silurian actually forms a low plateau to the west of which is the impressive range of the Comeragh Mountains, with its spectacular corrie of Coumshingaun cut into the Old Red Sandstone. The northern and eastern boundaries of the plateau are formed by a narrower ridge of thinner Old Red Sandstone. To the south are Ordovician rocks of the Waterford volcanic belt. The inlier is almost split into two by the Old Red Sandstone feature of Croaghaun Hill (Fig. 7.9) and the string of very small outliers that extends north-eastwards from it along a fault.

Penney (1980a) divided the succession into three formations. Meagre graptolitic evidence is sufficient to allow assignation of the first of these (the Ross Formation) to the Caradoc. The succeeding *Kilmacthomas Formation* (1100 m) occurs south of Croaghaun Hill and narrowly to the north of it. The lithology is of interbedded laminated green slates and homogeneous or laminated purple and green slates. Thin acid tuffs are common, particularly in the south, but these may well be Ordovician. The northern half of the inlier is occupied by the *Ballindysert Formation* (1900 m) of dark grey slates with numerous thinly bedded greenish-grey greywackes and siltstones. The Ballyhest Member (50–350 m) within the Ballyndysert Formation, but repeated by folding, is of conglomerates and greywacke sandstones. The main phase of folding is on various scales and there is a dominant cleavage, not precisely axial planar.

There is very little biostratigraphical evidence for the age of these rocks. A single specimen of *Orthograptus* recovered from the lower part of the Ballindysert Formation suggests an age older than upper Llandovery. Smith (1979a) obtained sparse acritarch assemblages from this formation indicating a Llandovery to early Wenlock (Sheinwoodian) age. The Kilmacthomas Formation yielded only reworked Ordovician acritarchs. It is important to note that these are *Arenig* acritarchs. Trace fossils are common in parts of this formation, mostly as small horizontal burrows.

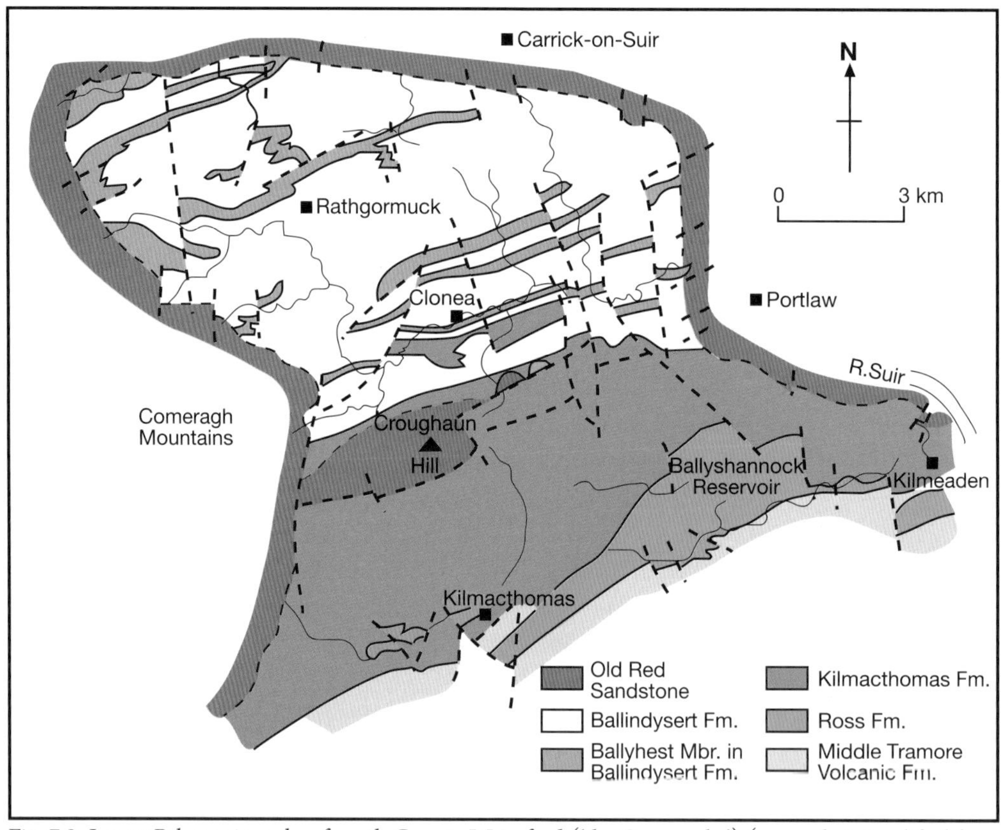

Fig. 7.9 Lower Palaeozoic rocks of north County Waterford ('the Comeraghs') (somewhat simplified from Penney, 1980a).

Penney (1980a) regarded the slates of the Ballindysert Formation as identical to those of the Ahenny Formation in the Slievenamon inlier. The coarse greywackes of the Brownstown Member there are correlateable with the Ballyhest Member. Micro-conglomeratic horizons in the Ballyhest Member also recall the upper part of the Inchacoomb Formation in the Galtys (see below). Penney interpreted the environment of deposition of the Silurian rocks of north Waterford and the Slievenamon inlier as an extensive basinal plain subject to turbidity currents. A large deep-sea fan (producing the Ballyhest Member) spread north-westwards, bearing Waterford volcanic material from the margin of the Irish Sea Landmass. The Brownstown Member in the Slievenamon inlier can be seen as representing the more distal part of this fan as coarse conglomerates and evidence of surface channels are absent here.

Galty Mountains

In the Galty Mountains area (Jackson, 1978) thickly bedded grey and greenish-grey greywackes, siltstones, and mudstones of the *Inchacomb Formation* (550–650 m) are followed by dark grey to black siltier greywackes and shales of the *Ballygeana Formation* (1275 m). The former consists of two members, the older of which forms the core of an anticline crossing the area from east to west (Fig. 7.10). The middle of this lower member has purple mudstones and siltstones interbedded with the normal grey and greenish beds. The upper member contains a high proportion of calcareous siltstones and mudstones interbedded with dark green to grey mudstones, micaceous silty sandstones, and minor greywackes.

The darker grey to black shales of the *Ballygeana Formation* yield graptolite faunas, the lowest recorded being that of a single diplogratid (associated with trilobite fragments) of presumed Llandovery age, at a level 188 m above the base of the formation. In higher beds Jackson (1978) recognised three successive graptolite assemblages. The first, typified by the presence of *Cyrtograptus centrifugus*, *Monograptus priodon*, *Monoclimacis vomerina*, and *Pristiograptus*, indicates the presence of the *C. centrifugus* Biozone. The second yields *Monograptus riccartonensis*, *M. firmus*, *M. firmus sedberghensis*, *M. priodon*, and *Plectograptus*, indicating the *M. riccartonensis*

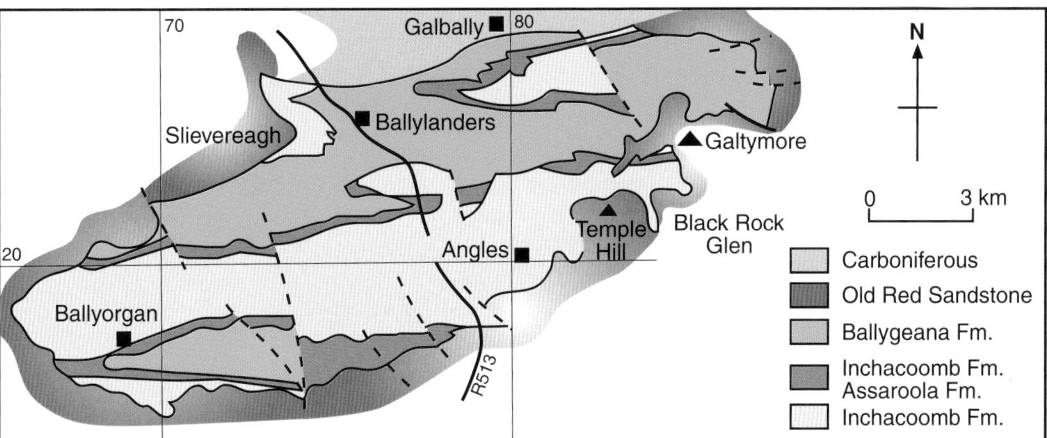

Fig. 7.10 Geological map of the Galty Mountains inlier (*after* Jackson,1978).

Biozone. The third fauna is of *Monograptus flexilis flexilis, Monoclimacis flumendosae, Monograptus flemingii, Pristiograptus dubius, P. menaghini*, and *Cyrtograptus*, the assemblage indicating the *Cyrtograptus linnarssoni* Biozone. Thus there is good representation of the Sheinwoodian Stage of the Wenlock Series.

Within this inlier the Silurian rocks form a relatively subdued topography compared with the magnificent range of the Galty Mountains themselves, which are of Upper Old Red Sandstone, and the lesser hills of the same rocks to the south and east. The best exposures of the Silurian are in streams draining the Galty Mountains, though the relatively poorly exposed lower lying areas have useful old roadstone quarries. The nature of the exposures, together with the effects of cleavage and bedding plane slip, have tended to obscure evidence of sole markings. The modal development of each formation is a 'middle absent' Bouma sequence, though within each there is an upward change in relative proportions, 'top absent' sequences giving way to 'bottom absent' sequences as the formation is ascended. The turbidites are neither clearly distal nor clearly proximal in character. Some conglomeratic beds are present within certain graded units or as channel infillings. Rhyolites and orthoquartzites are conspicuous amongst the clasts.

Dingle Peninsula

There remains for consideration the Dingle Peninsula, County Kerry (Fig. 7.1), an area relatively remote from the inliers discussed so far and one, in various ways, unique in Irish Silurian geology. The area is one of splendid scenery in which the topography clearly reflects the geology. Fossiliferous Silurian rocks of the *Dunquin Group* (Holland, 1969) are present in the Dunquin inlier at the western extremity of the peninsula and in the long narrow strip of the Annascaul inlier (Fig. 7.11). The beautiful mountain Caherconree lies at the eastern end of this inlier and from its summit, itself capped with Upper Old Red Sandstone, the Annascaul inlier is seen as a topographical trough within the higher ground made by the Old Red Sandstone to the south-east and the *Dingle Group* to the north-west. On the north-eastern side of the mountain, Derrymore Glen cuts deeply through the Old Red Sandstone to provide another small inlier of Silurian rocks. At its south-western end the Annascaul inlier reaches the sea in the bay to the west of Minard Head and there is a small disconnected inlier to the north of Bull's Head. Finally, fossiliferous Silurian rocks are seen again in the Blasket Islands, especially Inishvickillane, the most remote of these.

The Dingle Peninsula provides a second area of proved Ludlow strata in Ireland and its Wenlock succession is characterised by rich assemblages of shelly fossils and a splendid display of volcanic rocks. The fossiliferous Silurian rocks of the Dunquin Group in the Dunquin inlier pass up into a thick purplish fluviatile sequence, the Dingle Group, ranging from middle Ludlow, through the Přídolí Series, and into the Lower Devonian–thus providing the only Irish record of this interval.

Dunquin inlier

The succession in the Dunquin inlier (Holland, 1987,1988) is very well displayed in almost continuous coastal sections (Figs. 7.12, 7.13). Its arrangement in an overfold (Fig. 7.14) with a faulted middle limb has long

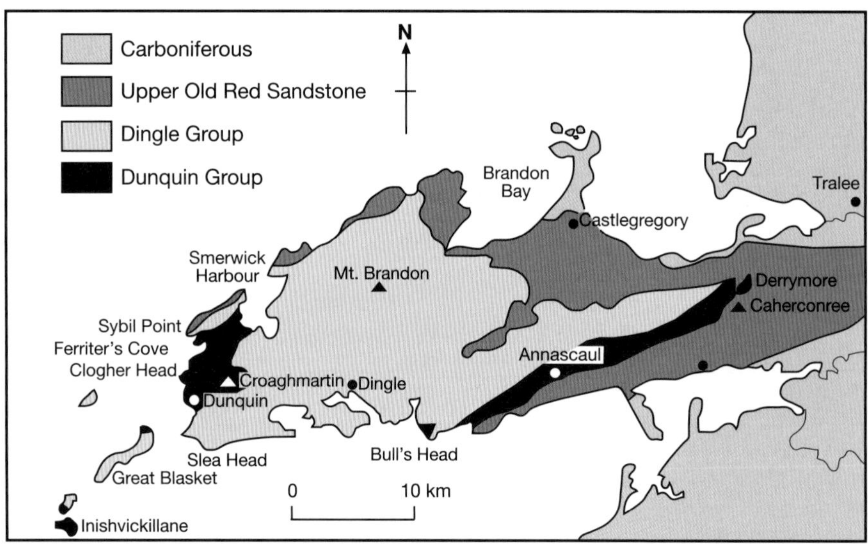

Fig. 7.11 Geological map of the Dingle Peninsula, County Kerry.

Upper Old Red Sandstone
Dingle Group
Croaghmarhin Formation
Drom Point Formation
Mill Cove Formation
Clogher Head Formation
Ferriters Cove Formation
fault
stratigraphical boundary
Coosgorrib
unexposed
Smerwick Harbour
Coogglass
Ferriters Cove
Doon Point
Ballyaglisha
Ballyferriter East
Ballyferriter West
Drom
Trabaneclogher
Clogher Head
Croaghmarhin
Minnaunmore Rock
Marhinmore
Owen
Mill Cove
Dunquin
Cooshaun
N
0
1 km

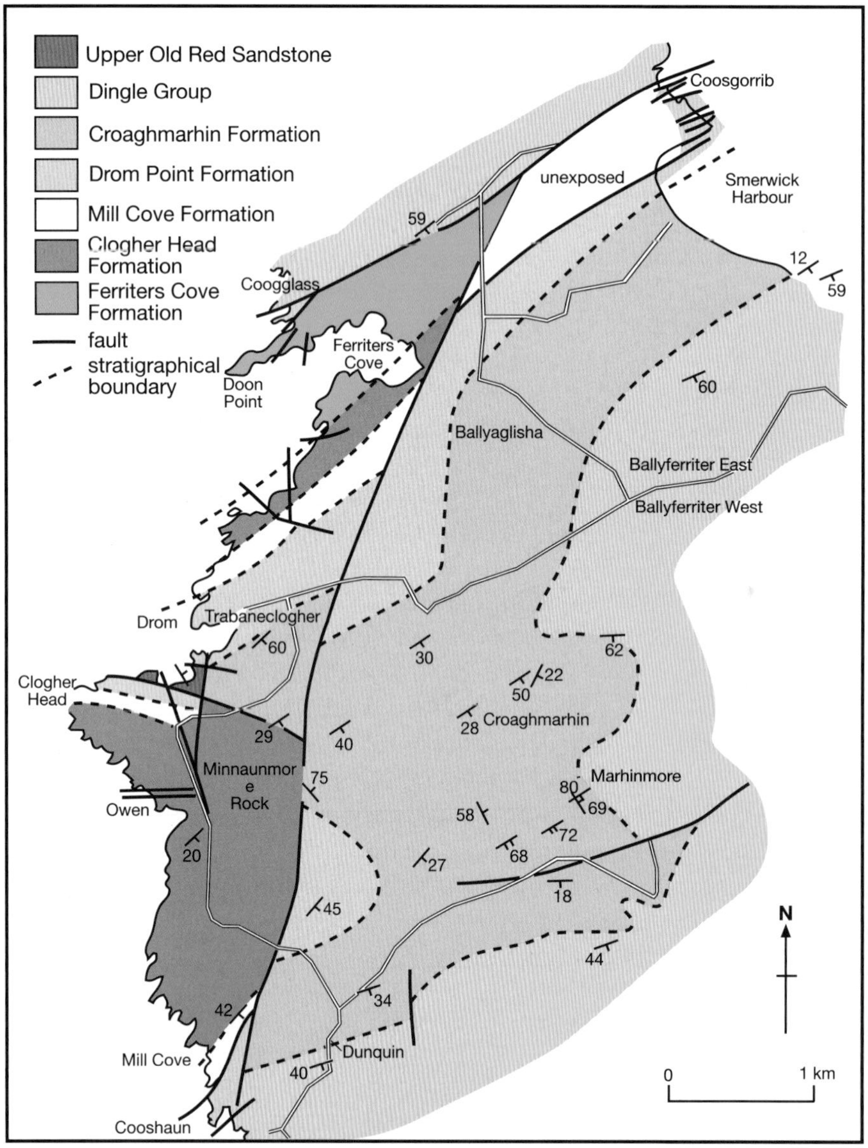

Fig. 7.12 Geological map of the Dunquin inlier, Dingle Peninsula, County Kerry (*after* Holland, 1988). Some authors have proposed unconformities within the Dingle Group, thus justifying the elevation of some of its formations to group status.

Opposite: Fig. 7.13 View from Clogher Head north-eastwards across the northern part of the Dunquin inlier. In the distance the summit of Mount Brandon (Dingle Group) is in cloud. Along the Atlantic coast (back left) the serrated ridge marks the narrow strip of Upper Old Red Sandstone, which dips steeply westwards and rests unconformably on the Dingle Group. The latter forms the slope between Smerwick Harbour (the long inlet in the background) and Ferriters Cove. Across a fault to the south-east the Dunquin Group is reached. Within it the splendidly exposed ascending succession of the Ferriters Cove, Mill Cove, Clogher Head, and Drom Point formations is clearly seen in the low cliffs. Bedding planes of the Drom Point Formation are seen dipping towards the sea in Trabane Clogher, the bay to the right. The photograph is taken from the north face of Clogher Head. The beds here are inverted, and the purple rocks of the Mill Cove Formation and the greyish Drom Point Formation are encountered in the descent to the sea. *Photograph: C. H. Holland.*

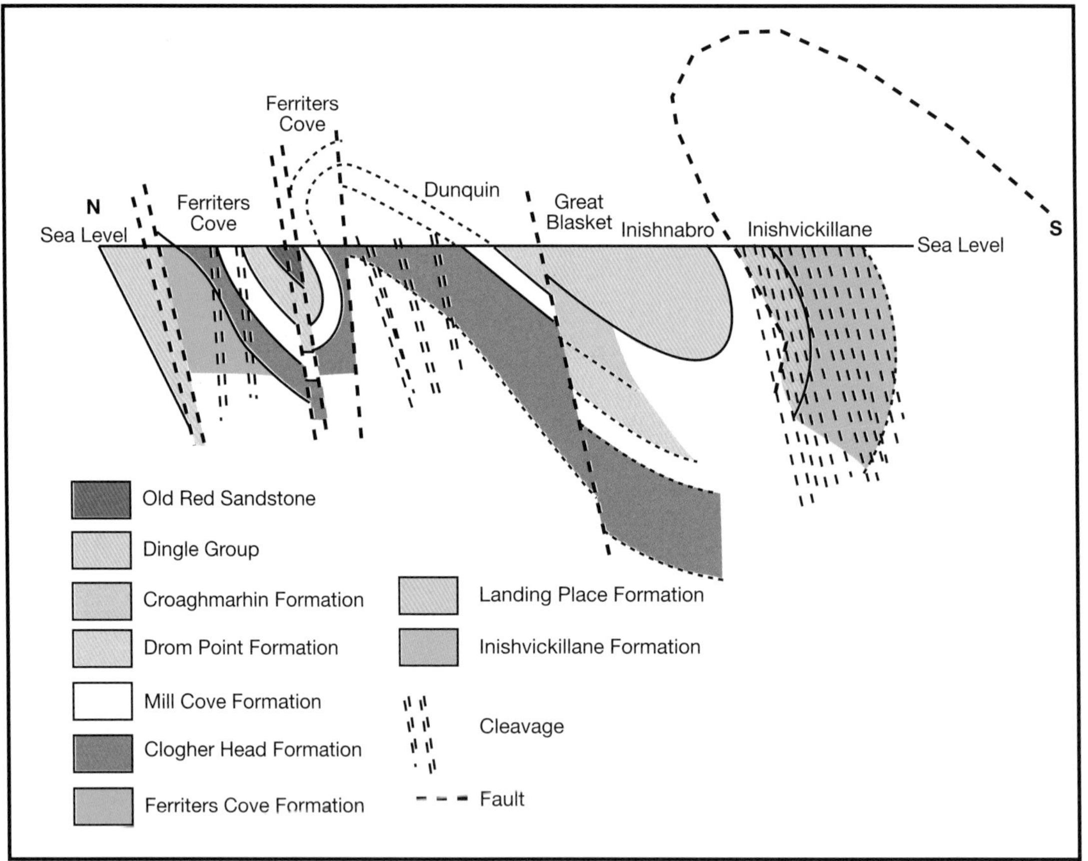

Fig. 7.14 Diagrammatic section across the Dunquin inlier and Inishvickillane, Dingle Peninsula, County Kerry.

been known through the Geological Survey memoir by Jukes and Du Noyer (1863) and through the detailed stratigraphical study by Gardiner and Reynolds (1902). The sequence is as follows:

(5) Croaghmarin Formation	400 m
(4) Drom Point Formation	300 m
(3) Mill Cove Formation	100 m
(2) Clogher Head Formation	200–600 m
(1) Ferriters Cove Formation	600 m

The *Ferriters Cove Formation* is confined to the area close to the bay of the same name. It is characteristically of greenish and yellowish siltstones with subordinate sandstones, conglomerates, and tuffs. There is a rich shelly fauna, much of it in the form of comminuted fragments. There is much evidence of bioturbation. Brachiopods are conspicuous, as are tabulate corals and crinoid ossicles. There are also gastropods, trilobites, and bryozoans. The tabulate corals may be concentrated in layers of broken and disturbed material. The upper Wenlock (Homerian) brachiopod fauna includes *Atrypa reticularis, Hesperorthis davidsoni, Holcospirifer bigugosus, Leptaena depressa, Meristina obtusa,* and *Rhychotreta* cf. *cuneata*. Watkins (1978) saw these beds as corresponding to settling of mud and silt below wave base offshore.

The main volcanic development in the area is represented by the *Clogher Head Formation* with its flow-banded and other rhyolites, ignimbrites, and pyroclastics. These are seen, repeated by the folding, at three places in the coastal sections, but are best developed in the middle inverted limb, where they form substantial cliffs at the . seaward end of Clogher Head and to the south of it. The typical rhyolitic crags extend inland to Minnaunmore Rock, seen to the east of the Dunquin to Ballyferriter road. Farther east they are cut off at a north-south fault. There are some good examples of volcanic mudflows. Wright *et al.* (2003) gave detailed description of a polygenetic palaeosol capping a basalt at the top of the formation at a locality north-east of Clogher Head. It is thought to represent hydrothermal alteration of the basalt with subsequent subaerial weathering. Some of the tuffs are graded and may contain shelly fossils such as gastropods. Associated, more or less calcareous, greenish to yellowish brown siltstones yield a rich shelly fauna similar to that of the Ferriters Cove Formation, but less diverse. Amongst the opportunistic rhychonellid brachiopods, *Ferganella* cf. *borealis* and *Rhynchotreta* cf. *cuneata* are common. Again there are numerous crinoid ossicles. The seemingly endemic brachiopod *Holcospirifer bigugosus* (Bassett *et al.*, 1976) is conspicuous on some bedding planes. A greenish andesitic lava occurs at the top of the formation.

The succeeding *Mill Cove Formation* is not fossiliferous, but its red to purple siltstones and sandstones contrast with the greenish or yellowish rocks below and above (Fig. 7.13). It probably represents a non-marine interval in the succession.

The *Drom Point Formation* is of greenish grey, yellow weathering, more or less laminated calcareous siltstones and sandstones. There are some changes in the Wenlock brachiopod fauna. *Atrypa reticularis, Howellella, Meristina obtusa, Protochonetes,* and *Sphaerirhynchia wilsoni* are common. There are also common bryozoans, crinoid ossicles, and tabulate corals. The shells in this formation tend to be concentrated in lenses. Its most remarkable characteristic is the profusion of the trace fossil *Chondrites* on many bedding planes (Simpson, 1957). Faunal diversity has increased again. A shallow shelf environment seems to have been subject to more rapid sedimentation during storms.

The coastal sections go no higher in the Drom Point Formation, but a small faulted area of fossiliferous Silurian beds at the northern end of the Great Blasket Island shows the same beds passing upwards into a level with rich banks of the pentamerid brachiopod *Rhipidium hibernicum*, a species also evidently endemic to the area (Bassett *et al.*, 1976). Fortunately, the same banks can be identified in the less well exposed inland area of the mainland along the Dunquin River.

Above the *Rhipidum* occurrence the succession can be traced into the thick, irregularly bedded, somewhat calcareous greenish grey siltstones of the succeeding *Croaghmarhin Formation*, which form the conspicuous conical hill of that name. They appear to span the uppermost Wenlock and the lower and middle parts of the Ludlow Series. These conquinoid siltstones, deposited farther from the shore, contain brachiopods such as *Atrypa reticularis, Howellella,* and *Strophonella euglypha*. With volcanic conditions now over, corals flourished again as they did during the time of deposition of the Ferriters Cove Formation. There are characteristic layers rich in tabulates: favositids, halysitids, and heliolitids in the form of branching and broken, hollow, yellow-coated moulds.

The upper part of the formation is of thinner, more evenly bedded, less calcareous siltstones in which the shelly fauna closely resembles that of the Lower Leintwardine Formation (lower Ludfordian Stage) of the Welsh Borderland (Holland *et al.*, 1963). Conspicuous are *Amphistrophia funiculata, Microsphaeridiorhynchus nucula, Chonetes lepismus, Dayia navicula, Isorthis orbicularis, Shaleria ornatella,* and *Sphaerirhynchia wilsoni*. Pterineid bivalves and crinoid ossicles are quite common. There is a very rare record of a graptolite comparable with *Monograptus leintwardinensis* var *incipiens*. With continued silting up of the area the fauna declined until, at the top of the formation, a few small rhynchonellids and scattered crinoid ossicles remain. The formation is capped by a resistant micaceous, light greenish siltstone which in a very few exposures can be seen to be followed directly upwards by typical purple and greenish beds of the Dingle Group.

Blasket Islands

Parkin (1974) described the geology of Inishvickillane. The beds dip steeply to the south-south-east but are all inverted. The *Inishvickillane Formation* (probably *c.*1140 m) is of intermediate lavas and tuffs. There are purple scoriaceous horizons and a thick development of flow-brecciated lava. The *Landing Place Formation* (*c.*250 m) includes fossiliferous grey tuffaceous siltstones with tuff bands. Shelly faunas from its lower part can be correlated with those of the upper part of the Drom Point Formation. They include *Amphistrophia funiculata, Atrypa reticularis, Dicoelosia biloba, Leangella segmentum, Leptaena depressa,* and *Meristina obtusa*. Parkin recorded two specimens of *Rhipidium* sp. Large tabulate coral colonies in living position are a conspicuous feature. *Chondrites* occurs but with thinner tubes and less impressive development than on the mainland. As in the Dunquin inlier, a Hercynian cleavage postdating the main folding is sporadically developed.

S.P. Todd (1991) discovered a sparse Silurian shelly fauna in the southern part of the adjacent island of Inishnabro. The shallow marine sandstones and siltstones are followed conformably by the Dingle Group.

A century ago John Joly (1857–1933), Professor of Geology and Mineralogy at Trinity College Dublin, observed lavas similar to those of Inishvickillane making the Foze Rocks some 5 km from the island outwards into the Atlantic, and hence the most westerly occurrence of Silurian rocks in Europe. He travelled, hazardously it would now seem, in an open four-oared boat with two companions and three local boatmen. During his geological investigations one of his companions remained on board, evidently to ensure the continued loyalty of the crew. It is fair to add that they had come through fog, guided only by a pocket compass, their approach heralded by the weird cries of seals at first unseen (Joly, 1920; Parkin, 1976a, Wyse Jackson, 2006).

Annascaul, Bull's Head, Derrymore Glen

The *Annascaul Formation* (*c.*1160 m) (Parkin, 1976b) occupies the Bull's Head inlier and the Annascaul inlier from Minard Bay (where it is severely crushed and sheared) eastwards to the lower slopes of Caherconree. Its upper Wenlock (Homerian Stage) fauna is mostly confined to the Ballynane Member (55–82 m) found in the Bull's Head inlier, at Ballynane in the centre of the Annascaul strip, and on the slopes of Caherconree. Brachiopods are dominant and include: *Amphistrophia funiculata, Atrypa reticularis, Dicoelosia biloba, Eoplectodonta duvalii, Leangella segmentum, Leptaena depressa,* and *Meristina obtusa*. The fauna varies from place to place. Crinoid ossicles and bryozoans are common. The fossils are characteristically preserved as randomly orientated moulds in volcanic tuff.

Thin bands and nodular masses of limestone from the Ballynane Member on the flanks of Caherconree have yielded an abundant and excellently preserved trilobite fauna, comprising aulacopleurids, calymenids, cheirurids, lichids, odontopleurids, phacopids, proetids, and styginids (Siveter, 1989). The odontopleurids dominate the fauna, with *Odontopleura ovata* as by far the commonest species. One locality has also provided conodonts (Aldridge, 1980). Additionally, Ferretti *et al.* (1993) described problematic microfossils from here.

The *Caherconree Formation* (of uncertain thickness) is seen in stream sections and crags on the mountainside, where it is affected by minor faults and by plunging folds of small amplitude. The rocks are laminated siltstones seen also in Derrymore Glen. The graptolite fauna preserved here includes *Saetograptus* cf. *chimaera semispinosus, Cucullograptus scanicus,* and *Pristiograptus dubius*. There are also orthoconic nautiloid cephalopods, cardiolid bivalves, and other rare shelly fossils. The graptolites are indicative of the *C. scanicus* Biozone of the Ludlow Series, and these siltstones are faulted against the Wenlock.

The younger *Derrymore Formation* (thickness again uncertain) is seen only in Derrymore Glen. It displays a shelly fauna. The trilobite *Calymene puellaris* found here is characteristic of the Upper Leintwardine Formation (Ludfordian Stage) in various British areas. There are appropriate ostracodes and a brachiopod assemblage including *Microsphaeridiorhynchus nucula, Chonetes minimus, Dayia navicula, Howellella elegans, Isorthis orbicularis, Leptostrophia filosa, Protochonetes ludloviensis, Salopina lunata,* and *Sphaerirhynchus wilsoni*.

Parkin (1976b) drew together a local palaeogeographical picture with the volcanic centres of the Blasket Islands and Clogher Head giving way to turbiditic sandstones with lava clasts and volcanic mudflows seen in the equivalent rocks of Annascaul. A palaeoslope dipping approximately east-north-eastwards is thus suggested, and would also account for the development in *C. scanicus* Biozone times of a graptolitic facies in the region of Caherconree. Bassett *et al.* (1976) suggested that the two endemic brachiopod species well seen in the Dunquin inlier may have occupied isolated areas related to volcanic islands. Wenlock volcanicity from Inishvickillane to the Dunquin inlier varies from intermediate to acid with a minor basic component.

Dingle Group

In the coastal sections of the Dunquin inlier at Coosgorrib, Coosglas, and Cooshaun (Fig. 7.12), as along the Annascaul inlier, the Dingle Group is faulted against the fossiliferous Wenlock rocks. The same situation obtains at the northern end of the Great Blasket Island. Fortunately there are inland exposures in the Dunquin inlier where the transition into the continental facies can be seen (Holland, 1987).There is also now additional supporting evidence from Inishnabro. The youngest Silurian fossils presently known from the Dingle Peninsula are not younger than middle Ludlow (early Ludfordian age). The Dingle Group includes the upper part of the Ludlow Series, the Přídolí Series, and part at least of the Lower Devonian. Description of the Dingle Group is given in the next chapter of this book, along with the Devonian Old Red Sandstone.

Silurian palaeogeography

In summary, Llandovery rocks are well represented in Ireland. Indications of the variable history and character of the north-western margin of the marine basin are seen in the west in the rocks and fossils of Galway and Mayo, at Charlestown, and at Lisbellaw. The Llandovery turbidites in Leinster were derived from a Silurian 'Irish Sea Landmass' to the east. Elsewhere the Llandovery record is of turbidite formations of greywackes and siltstones associated with shales or slates representing the background sedimentation of the basin. Llandovery graptolite faunas are well seen in the Longford-Down inlier, in the Balbriggan inlier, and to a limited extent in Slieve Bernagh. Other areas of Llandovery rocks such as those of Slieve Aughty and Leinster are strikingly lacking in macrofossils. On the whole, the advent of greywackes came later in more southerly areas.

The model of an accretionary prism which, rightly or wrongly, has been developed for the Lower Palaeozoic rocks of the Southern Uplands of Scotland (Leggett *et al.*, 1979, etc.) is not applicable in Ireland (Holland, 1986). E.A. O'Connor (1975) drew attention to the pattern of outcrops in the Longford-Down inlier, where belts of black shales, some certainly of Ordovician age, are brought up within the broad expanse of largely Silurian greywackes. A similar pattern obtains in Slieve Aughty (Fig. 7.7). It appears that all these occurrences of black shales are brought up by strike faulting and subsequently revealed in the Hercynian , generally anticlinal structures, which affect Old Red Sandstone and Lower Palaeozoic rocks alike. Weir (1974) saw the Ordovician and Llandovery shales of Slieve Aughty and Slieve Bernagh in the form of 'diapiric anticlines' characterised by closely spaced faults and appearing within the Silurian distal turbidites.

The Wenlock is much the best represented of the four Silurian series in Ireland. The suggested palaeogeography given in Figure 7.15 overleaf can be related to the varied evidence detailed elsewhere in this chapter. The Wenlock evidence does not itself demand that the Iapetus Ocean had been, or was being, eliminated by the coming together of plates, though clearly it is not inconsistent with the earlier completion of such a process. Irish Silurian rocks provide no evidence for an Iapetus suture. There have been suggestions as to the position of such a line, but the least assailable solution is to place it where there are no Silurian outcrops. The presence of Bohemian elements in British and Irish Silurian rocks requires a marine connection into what is now Central Europe.

Marine Ludlow rocks are so far known in Ireland only from the Balbriggan inlier and the Dingle Peninsula. By upper Ludlow (middle and upper Ludfordian) times the record is confined to the fluviatile sedimentary rocks of the Dingle Group, which continued to accumulate during the time of the Přídolí Series. The picture is very different in mainland Britain where a full Ludlow marine sequence in Wales, the Welsh Borderland, and northern England is followed by the rather varied quasi- marine rocks of the Downton Group.

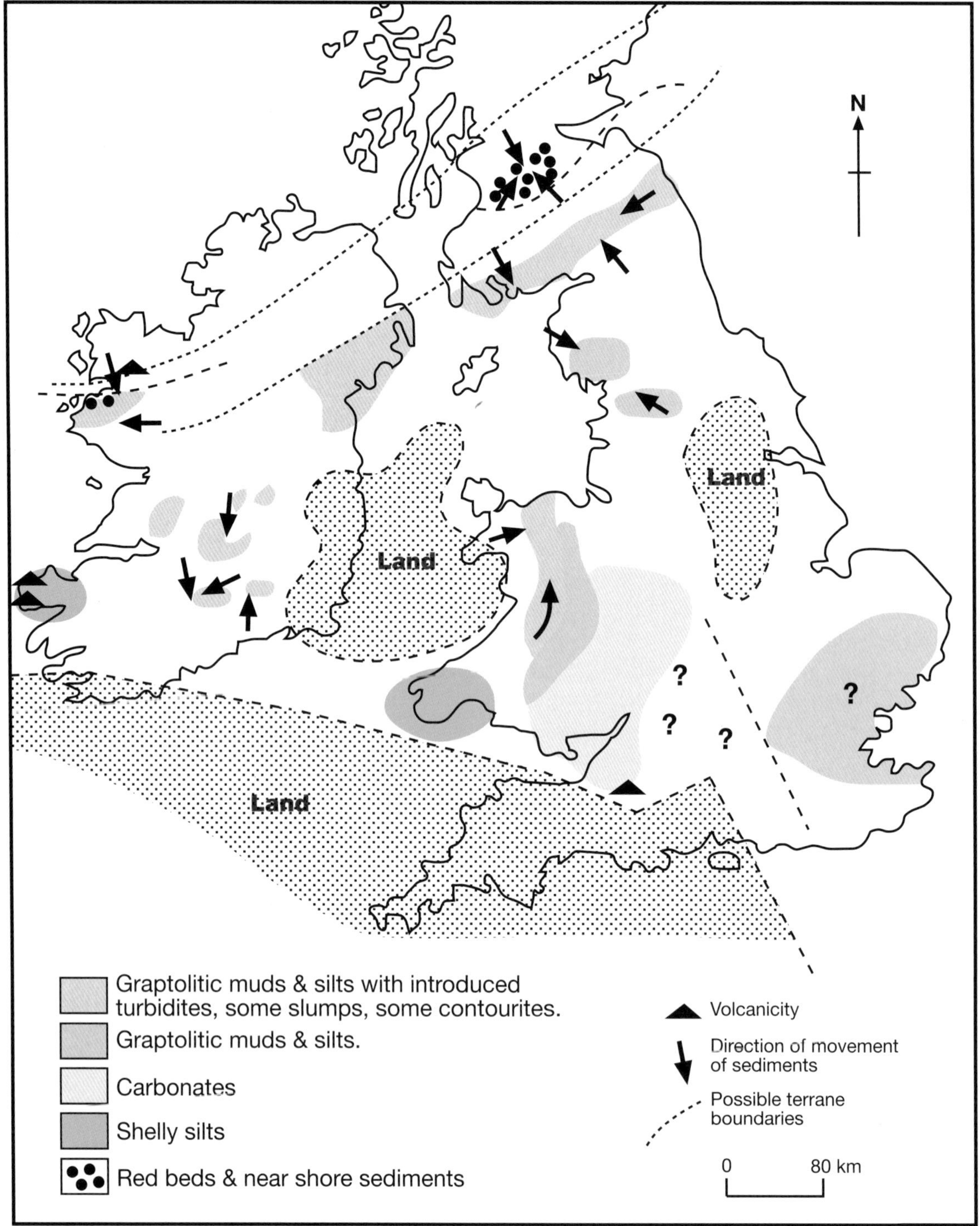

Fig. 7.15 Generalised Wenlock palaeogeography of the British Isles. North arrow and scale relate to present geography.

8

Late Caledonian orogeny and magmatism

D. M. Chew and C. J. Stillman

In Ireland, two main orogenic phases were associated with the closure of the Iapetus Ocean: an Early–Middle Ordovician (475–460 Ma) phase, termed the Grampian Orogeny, and an Early Devonian (400 Ma) phase, termed the Acadian Orogeny. Together these two orogenies are the main components of the so-called Caledonian orogenic cycle (see Chapter 4). The Grampian Orogeny was caused when the Laurentian continental margin in north-west Ireland collided with, and was partly subducted beneath, an oceanic arc terrane (Fig. 4.1a). Following the collision, there was continued subduction of the Iapetus ocean floor, and the Avalonian continent (to which south-east Ireland belongs) progressively approached the deformed and metamorphosed margin of Laurentia, now welded to the arc terrane (Fig. 4.1b). By the Late Silurian, the intervening oceanic crust had been subducted, and the opposing shorelines became juxtaposed along the so-called Iapetus Suture (Fig. 4.1c). The last stage of this collisional tectonic event, the Acadian Orogeny, did not occur until Early Devonian times. It is named after Acadia, the francophone region of maritime Canada, which includes Nova Scotia, where it has long been recognised.

Although this chapter covers the Acadian Orogeny in Ireland, the term *Late Caledonian* was chosen for the chapter's title to permit inclusion here of other igneous and tectonic events that are not strictly Acadian (i.e. Early Devonian), but which nevertheless belong to the later stages of the overall Caledonian orogenic cycle (defined in Chapter 4).

It should be noted that the tectonic cause of Acadian deformation in Ireland remains somewhat uncertain. As it post-dates final (Late Silurian) closure of the Iapetus Ocean by about 25 million years, it has been regarded as 'proto-Variscan' rather than 'Late Caledonian' by Woodcock *et al.* (2007), possibly caused by a renewed 'push from the south' involving subduction of the Rheic Ocean.

Late Caledonian (i.e. Late Silurian–Early Devonian) deformation affected four principal pre-Devonian crustal blocks in Ireland, informally referred to in this chapter as the Grampian terrane, the Midland Valley terrane, the Longford-Down terrane and the Leinster terrane (Fig. 8.1 inset), and its effects in these four areas are considered in turn in the first part of the chapter. Following this, Late Caledonian granites and related intrusive rocks are described. Detailed syntheses of the Late Caledonian tectonism and igneous petrogenesis are presented at the end of the relevant sections in this chapter. Late Caledonian *extrusive* igneous rocks are dealt with in the Silurian and Devonian chapters (Chapters 7 and 9).

Tectonic evolution

The most obvious expression of Late Caledonian orogenesis in Ireland is of clastic rocks of Devonian or Early Carboniferous age resting unconformably on folded, cleaved rocks that range in age from Cambrian to Silurian. The folding and cleavage are in many places demonstrably Early Devonian in age, and resulted from the Acadian Orogeny. In general, the metamorphic grade associated with the Acadian Orogeny is quite low – typically rocks in the Silurian inliers of central Ireland (Fig. 8.1) have experienced sub-greenschist facies metamorphism, in contrast to the high-grade metamorphic rocks commonly produced by the Grampian Orogeny. This is compatible with structural evidence that the collision of Laurentia with Avalonia was strongly oblique (Soper *et al.*, 1992; Dewey and Strachan, 2003), with sinistral transpression and without substantial crustal thickening.

Timing

The timing of the Acadian Orogeny in Ireland is reviewed in Phillips (2001). In most places, deformation is temporally constrained to have occurred between the Silurian (late Wenlock) and Early Devonian.

Near Pomeroy in County Tyrone (Fig. 8.1), Middle Old Red Sandstone sedimentary rocks lie unconformably

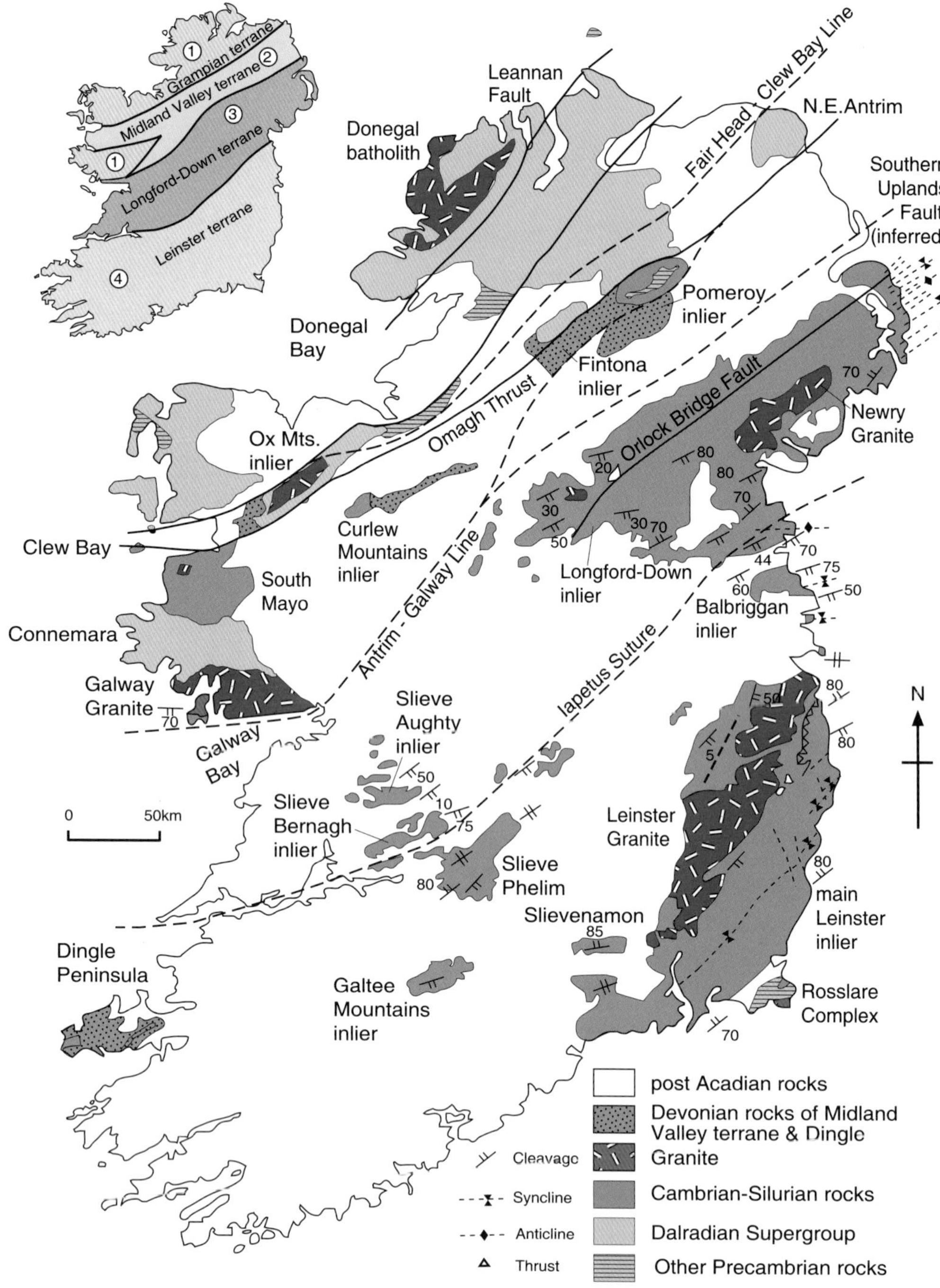

Fig. 8.1 Map of Ireland showing location of Acadian and other Late Caledonian tectonic features and granite plutons mentioned in the text. Inset figure shows the four informally termed tectonic zones or terranes.

on folded Upper Ordovician and Lower Silurian rocks. In south County Mayo, F1 folds in Wenlock rocks are cut by the Middle Devonian Corvock Granite (387 ± 12 Ma, O'Connor, 1989; Table 8.1; Fig. 8.2). To the south in County Galway, Wenlock sedimentary rocks south of Killary Harbour were folded prior to the intrusion of microgranite sills which are probably related to the Lower Devonian (*c.*400 Ma, Table 8.1) Galway granite suite. In County Down, the Lower Devonian Newry Granodiorite (399 ± 3 Ma, Meighan and Neeson, 1979, Table 8.1) cuts F1 folds in Lower Silurian (Llandovery) greywackes. The Lower Devonian Leinster Granite, (405 ± 2 Ma,

Table 8.1 Age data for Irish Caledonian Intrusives (recalculated with $^{87}Rb = 1.42 \times 10^{-11} y^{-1}$)

Rock Suite	Method	Age*	$^{87}Sr/^{86}Sr$ Initial	MSWD	Reference
TYRONE					
Craigballyharky tonalite	U/Pb zircon	471 +2/-4 Ma			Hutton *et al.* (1985)
ANTRIM					
Cushleake	Rb/Sr w.r.	496 ± 128 Ma	0.7042 ± 5		Harrison & Wilson (1978)
CONNEMARA					
Cashel - L. Wheelaun gabbro	U/Pb zircon	477 +25/-6 Ma			Jagger *et al.* (1988)
	U/Pb zircon	470.1 ± 1.5 Ma		0.53	Friedrich *et al.* (1999a)
Currywongaun gabbro	U/Pb zircon	474.5 ± 1.0 Ma		0.11	Friedrich *et al.* (1999a)
Quartz diorite gneisses	U/Pb zircon	463 ± 4 Ma			Cliff *et al.* (1996)
	U/Pb zircon	467.9 ± 1.2 Ma			Friedrich *et al.* (1999a)
Oughterard	Rb/Sr w.r.	407 ± 23 Ma	0.7076 ± 1		Kennan *et al.* (1987)
	U/Pb xenotime	462.5 ± 1.2 Ma		1.16	Friedrich *et al.* (1999a)
Galway: Errisbeg townland & Murvey	Rb/Sr w.r.	399 ± 7 Ma	0.7044 ± 1		Leggo *et al.* (1966)
Galway: Main	Rb/Sr w.r.	398 ± 12 Ma	0.704		Max *et al.* (1978b)
Galway: Megacrystic granite (G1)	U/Pb zircon	394.4 ± 2.2 Ma			Feely *et al.* (2003)
Galway: Megacrystic granite (G2)	U/Pb zircon	c. 402 Ma			Feely *et al.* (2003)
Galway: Enclave (G3)	U/Pb zircon	397.7 ± 1.1 Ma			Feely *et al.* (2003)
Galway: Megacrystic granite (G4)	U/Pb zircon	399.5 ± 0.8 Ma			Feely *et al.* (2003)
Galway: Costelloe Murvey (G5)	U/Pb zircon	380.1 ± 5.5 Ma			Feely *et al.* (2003)
Galway: Carna Granite	U/Pb zircon (bulk, unabraded)	412 ± 15 Ma			Pidgeon (1969)
Galway: Carna Granite	Re-Os molybdenite	407.3 ± 1.0 Ma#			Selby *et al.* (2004)
Galway: Murvey Granite	Re-Os molybdenite	410.7 ± 1.0 Ma#			Selby *et al.* (2004)
Inish	Rb/Sr w.r.	418 ± 8 Ma	0.704 ± 3		Leggo *et al.* (1966)
Omey	Rb/Sr w.r.	402 ± 18 Ma	0.7057 ± 55		Leggo *et al.* (1966)
	Re-Os molybdenite	422.4 ± 1.7 Ma			Feely & Selby (unpub.)
Roundstone	Rb/Sr w.r.	409 ± 83 Ma			Leggo *et al.* (1966)
DONEGAL					
Thorr	Rb/Sr w.r.	418 ± 26 Ma	0.7055 ± 4	74.2	O'Connor *et al.* (1982a)
Main	Rb/Sr w.r. min.	388 ± 4 Ma	0.70654 ± 5	1.69	Halliday *et al.* (1980)
	Rb/Sr w.r. min.	386 ± 33 Ma	0.70604 ± 8	67.8	Halliday *et al.* (1980)
	Rb/Sr w.r.	407 ± 23 Ma	0.7063 ± 5	35.10	O'Connor *et al.* (1982a)
	207Pb-206Pb uraninite	407 ± 4 Ma			O'Connor *et al.* (1984)
Ardara	Rb/Sr w.r.	405 ± 5 Ma	0.7065 ± 2	9.6	Halliday *et al.* (1980)
Rosses	Rb/Sr w.r.	404 ± 3 Ma	0.7062 ± 3	1.57	Halliday *et al.* (1980)
Trawenagh Bay	Rb/Sr w.r.	405 ± 3 Ma	0.7052 ± 4	1.15	Halliday *et al.* (1980)
	U/Pb monazite	411 ± 2 Ma			Halliday *et al.* (1980)
Fanad	Rb/Sr w.r.	402 ± 10 Ma	0.7050 ± 1	1.58	O'Connor *et al.* (1987)
Barnsmore	Rb/Sr w.r.	397 ± 7 Ma	0.7063 ± 5	10.2	O'Connor *et al.* (1987)
MAYO					
Blacksod	Rb/Sr w.r.	381 ± 10 Ma	0.739		Leutwein (1970)
Corvock	Rb/Sr w.r.	381 ± 19 Ma			Leake (1978)
	Rb/Sr w.r.	387 ± 12 Ma	0.708 ± 3	7.41	O'Connor (1989)
Ox Mountains Granodiorite	Rb/Sr w.r.	489 ± 18 Ma	0.7073 ± 2		Max *et al.* (1976)
Ox Mountains Granodiorite	Rb/Sr w.r.	477 ± 6 Ma	0.7055 ± 1	2.4	Pankhurst *et al.* (1976)
Ox Mountains Granodiorite	U/Pb zircon	412.3 ± 0.8 Ma		1.8	Chew & Schaltegger (2005)
Lough Talt and Easky adamellites	Rb/Sr w.r.	401 ± 33 Ma	0.7066 ± 5	8.8	Long *et al.* (1984)
DOWN					
Newry Granodiorite	Rb/Sr-min	399 ± 3	0.70546 ± 14		Meighan & Neeson (1979)
Newry Granodiorite	U/Pb zircon	387 Ma¶			Meighan & Neeson (1979)
Newry, intermediate facies	Rb/Sr w.r. min	403 ± 3	0.70507 ± 6		Meighan & Neeson (1979)
Newry Granodiorite	Rb/Sr w.r.	391 ± 21	0.70595 ± 18	0.21	P.J. O'Connor (1975)
LEINSTER					
N. Pluton	U/Pb monazite	405 ± 2 Ma			O'Connor *et al.* (1989)
N. Units	Rb/Sr w.r.	404 ± 24	0.708 ± 2	35.9	O'Connor & Bruck (1978)
Carrigmore	Rb/Sr w.r.	409 ± 17	0.7055 ± 1	2.97	O'Connor & Reeves (1980)
Camsore	207Pb-206Pb zircon	432 ± 3			O'Connor *et al.* (1988)
Camsore	Rb/Sr w.r.	428 ± 11	0.7068 ± 3	0.65	O'Connor *et al.* (1988)
Saltees	Rb/Sr w.r.	436 ± 7	0.7099 ± 4	1.8	Max *et al.* (1979)
Drogheda	Rb/Sr w.r.	379 ± 18	0.7066 ± 3		McConnell & Kennan (2002)

* Rb-Sr isochrons calculated with $\lambda^{87}Rb = 1.42 \times 10^{-11} yr^{-1}$

§ Initial ratio quoted in the source reference (i.e. will differ if $\lambda^{87}Rb$ used $\neq 1.42 \times 10^{-11} yr^{-1}$)

\# Weighted mean of two replicate analyses

¶ Minimum ^{238}U-^{206}Pb age with no error quoted

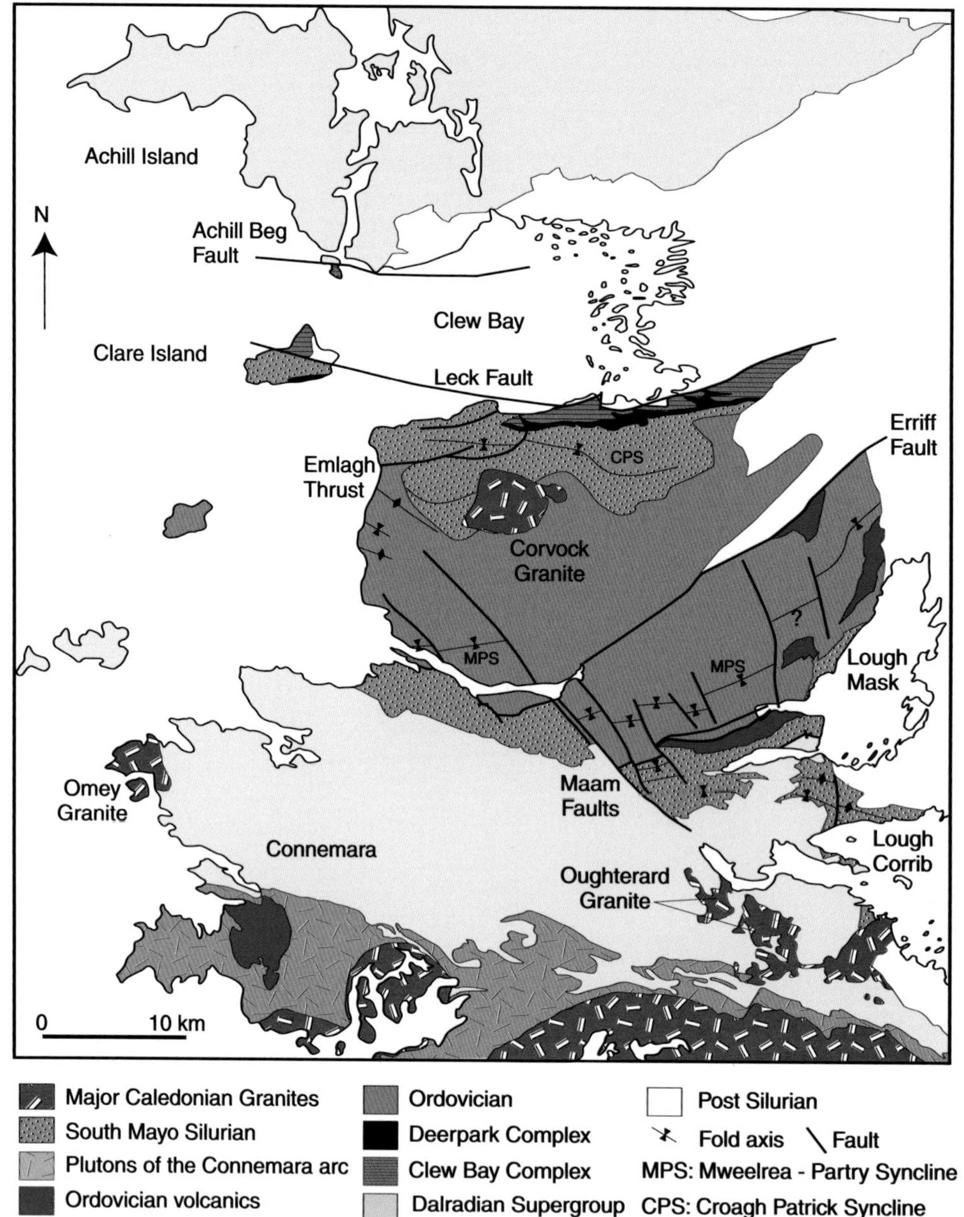

Fig. 8.2 Geology of south County Mayo and northwest County Galway.

O'Connor *et al.*, 1989, Table 8.1) was intruded during D2 deformation of the adjacent Ordovician and Silurian country rocks.

There is also local evidence of post-Acadian, Middle–Late Devonian tectonism. In the Dingle Peninsula, Upper Old Red Sandstone sediments rest unconformably on rocks that range in age from Silurian (Wenlock) to Early Devonian (the Dunquin and Dingle Groups, Chapter 7). In the south-west Ox Mountains, folding and thrusting of Lower and Middle Old Red Sandstone conglomerates and sandstones (Chapter 9) preceded deposition of Lower Carboniferous clastic rocks, while in County Tyrone, deformed Middle–Upper Devonian sedimentary rocks of the Fintona inlier are unconformably overlain by Lower Carboniferous rocks.

Structural style

The location of the four principal pre-Devonian crustal blocks, or terranes, in Ireland that experienced Late Caledonian deformation are shown on the inset map in Figure 8.1. Each crustal block differs in its structural style. The Grampian terrane consists of rocks of the Dalradian

Supergroup and its older gneissic basement (Chapters 2 and 4), which were metamorphosed during the Grampian Orogeny. These rocks behaved in a relatively rigid fashion during Late Caledonian orogenesis. The principal form of deformation was NE–SW trending sinistral faulting and shearing, which facilitated the intrusion of several Late Caledonian granites, such as the Main Donegal Granite (Hutton, 1982) and the Ox Mountains granodiorite (McCaffrey, 1992). The Midland Valley terrane experienced more ductile regional strain in the form of upright folds and cleavage, with local evidence for polyphase deformation. The Longford-Down terrane is typified by numerous steeply-dipping to inverted stratigraphic packages which are bounded by reverse faults. The internal younging within each thrust slice is towards the north-west, yet the gross disposition of units suggests an overall younging towards the south-east. Polyphase folds and cleavages intensify towards the Iapetus Suture in the southern part of the zone. The Leinster terrane is dominated by upright folds with associated cleavage and thrust faulting, and later large-scale NE–SW trending shear zones associated with the emplacement of the Leinster Granite. A detailed account of the deformation history of each of the four terranes follows.

Grampian terrane

Donegal

During the Late Silurian–Early Devonian, the *c.*428–400 Ma Donegal batholith (Halliday *et al.*, 1980; O'Connor *et al.*, 1982a; 1984; Condon *et al.*, 2004; Figs. 8.7, 8.8; Table 8.1) was emplaced into a NE–SW trending ductile sinistral shear zone within Dalradian metasediments (Hutton, 1982). The structural history of the emplacement of the batholith is considered here, while petrological aspects of the various plutons are treated later in the chapter.

Three phases of ductile deformation (local D1–D3) which pre-date granite emplacement are recognised in the Dalradian rocks of Donegal. (Note that the shorthand terminology, D1–D3 etc., is explained in Chapter 4). D4 to D9 are temporally and spatially associated with the emplacement of the batholith. D4 and D5 appear to be related in time to the intrusion of the Ardara and Thorr plutons (Fig. 8.7), while D6 to D9 are closely associated with the emplacement of the Main Donegal Granite into a NE–SW trending sinistral shear zone (Hutton, 1982) whose main cleavage is D6. This is a crenulation cleavage associated with gentle to open, upright, gently-plunging folds. It is weakly developed in the NW margin of the aureole and has an E–W strike. Moving SE towards the granite, the cleavage becomes an intense schistosity coeval with thermal metamorphism. The folds tighten and, with the cleavage, rotate anticlockwise into parallelism with the granite contact. Deformed pebbles in the Dalradian Ards Quartzite show that the D6 strain increases towards the granite, with X/Z values of up to ~25. This pattern of cleavage swing and apparent increase in strain is seen throughout the NW aureole and is typical of one half of a sinistral shear zone (Hutton, 1982). The counterpart half of this shear zone, with strain northwards increasing towards the granite contact, occurs in the aureole on the SE side of the granite.

Ox Mountains

The Central Ox Mountains inlier (Figs. 8.1, 8.9) consists of a sequence of Dalradian metasediments (Long and Max, 1977; Alsop and Jones, 1991) intruded by the 412.3 ± 0.8 Ma Ox Mountains granodiorite (Chew and Schaltegger, 2005; Table 8.1). Grampian deformation of the Dalradian produced three phases of ductile deformation (local D1–D3). The Ox Mountains granodiorite intrudes the core of a major upright D3 antiform that trends NE–SW, subparallel to the length of the inlier (Chapter 4). Its deformation history is reported by Hutton and Dewey (1986), Hutton (1987) and McCaffrey (1992; 1994). High strain zones (the Lough Talt Slide, the Glennawoo Slide and the Callow Shear Zone) are well developed on the limbs of the D3 antiform, and are parallel to the vertical, axial planar S3 fabric. Kinematic indicators such as rotated porphyroblasts and extensional crenulation cleavages display abundant evidence for sinistral shear. In the Ox Mountains granodiorite, the main solid-state foliation is sub-vertical, it strikes NE–SW, and is accompanied by a stretching lineation which plunges gently to the north-east or south-west. NNE trending sinistral S–C fabrics are commonly well-developed subparallel to the asymmetrical extensional crenulation cleavages in the country rock.

Although the Ox Mountains granodiorite was once thought to have been emplaced during D3 (Grampian) deformation in the country rock, it is now thought that sinistral shear-related deformation is Late Caledonian in age (MacDermot *et al.*, 1996; Chew *et al.*, 2004; Chew and Schaltegger, 2005). It therefore intrudes a discrete, later (local D4) sinistral shear zone superimposed on earlier Grampian (local D1–D3) structures, similar to the shear zone intruded by the Main Donegal Granite.

The role of crustal lineaments in the Grampian terrane

The metamorphic rocks of the Grampian terrane behaved in a relatively rigid fashion during the Late Caledonian orogenesis. Deformation was strongly partitioned into major NE–SW trending sinistral faults and shear zones that facilitated the intrusion of granite, as described above. These structures include the Leannan Fault and Fair Head–Clew Bay Line, and can be traced for hundreds of kilometres to the north-east into Scotland.

The Leannan Fault in Donegal (Figs. 8.1, 8.7) is a major splay of the Great Glen Fault of Scotland. It consists of a number of faults, with the main fault cropping out on the south-eastern margin of the fault zone. The main fault is 60–100 m wide, and consists of a zone of predominantly brecciated rock with occasional smooth surfaces with slickensides (Pitcher *et al.*, 1964). The centre of the main fault zone exhibits a steeply-dipping mylonitic fabric with a sub-horizontal lineation. A downthrow of several kilometres has been estimated (Pitcher *et al.*, 1964), while horizontal (sinistral) displacement, estimated at 34 km (Alsop, 1992), juxtaposes rocks of contrasting lithology and metamorphic grade. On the north-western side, chlorite-zone Dalradian metasediments (the Kilmacrenan Succession) run parallel to the fault zone for over 100 km with little change in lithology, structure or metamorphic grade. On the south-eastern side, Grampian (local D1-D3) fold traces trend obliquely to, and are truncated by, the fault. There are prominent lithological facies changes, and the metamorphic grade is higher and increases progressively to the south-west, where it reaches the kyanite zone. Movement on the Leannan Fault clearly occurred during the Devonian because Lower Old Red Sandstone conglomerates of the Ballymastocker inlier are bounded to the south-east by the fault, which itself is overlain by late Tournaisian clastic sediments near Killybegs. The Lower Devonian Barnesmore Granite (397 ± 7 Ma) is sinistrally offset by 3.5 km by a fault parallel to the Leannan Fault (see Figure 8.7).

The Fair Head–Clew Bay Line (Fig. 4.2, Fig. 8.1) forms the south-eastern limit to the Dalradian outcrop in Ireland (with the exception of the allochthonous Connemara block of the Grampian terrane). It is believed to represent the Grampian collisional suture between the deformed and metamorphosed Laurentian margin and the outboard oceanic arc (Dewey and Shackleton, 1984; Chapter 4). It is defined by a conspicuous magnetic lineament (Max and Riddihough, 1975) from north-eastern Ireland to Clew Bay in the west. The major surface expression of the Fair Head–Clew Bay Line is a fault zone that includes the Omagh Thrust and in general lies about 10 km to the south of the magnetic lineament (Fig. 4.2, Fig. 8.1). Both lineament and fault zone have been collectively referred to as the Fair Head–Clew Bay Line (Ryan *et al.*, 1995). Sinistral strike-slip motion along this line in the Central Ox Mountains, described earlier, may continue as a zone of sinistral transpressive shear into the Croagh Patrick Syncline on the southern margin of Clew Bay (Hutton, 1987; Dewey and Strachan, 2003).

Large-scale Late Caledonian brittle deformation along the Fair Head–Clew Bay Line is evident from strike-slip movement on the Leck Fault (Fig. 8.1, Fig. 8.2). On Clare Island, the fault plane is vertical with sub-horizontal slickensides. Further east the Leck Fault curves into a north-east trend in the south-west Ox Mountains, where inter-montane basins of Lower and Middle Devonian sediments (Chapter 9) have been attributed to minor extensional basins formed on a releasing bend during a phase of sinistral strike-slip movement (Phillips, 2001).

Midland Valley terrane

Rocks of the Midland Valley terrane (Fig. 8.1, Fig. 8.2) have experienced widespread regional deformation in the form of upright folds and cleavage, which are locally polyphase. The Silurian rocks of the Louisburgh Succession (Chapter 7) exhibit upright, sub-horizontally plunging, east–west buckle folds with a wavelength of up to 1 km (Phillips *et al.*, 1970). F2 folds are upright and are relatively minor. Subsequent D3 thrusts (such as the Emlagh Thrust) displaced these weakly-deformed rocks south-eastwards over the northern limb of the Croagh Patrick Syncline.

The Silurian rocks of the Croagh Patrick Syncline (Chapter 7) have undergone a complex history of deformation dominated by sinistral transpression (Hutton and Dewey, 1986; Dewey and McManus, 1964). The Croagh Patrick Syncline (Fig. 8.2) is an upright to north-dipping major D1 fold associated with gently to moderately-plunging fold hinges and coaxial pebble and mineral stretching lineations (Hutton and Dewey, 1986). Pelitic rocks in the syncline display a penetrative S1 cleavage defined by muscovite, chlorite and local biotite (Anderson, 1960). Pressure shadows around deformed cobbles in the Silurian conglomerates of Croagh Patrick show evidence of sinistral shear during D1, along with sinistral tension gash arrays and spaced sinistral extensional shears in these rocks. In the serpentinites of the Deer Park Complex to the north, an intense steeply inclined mylonitic fabric with a sub-horizontal stretching lineation is affected pervasively by a steeply inclined sinistral extensional crenulation cleavage (Hutton and Dewey, 1986). This is best seen on the lower reaches of the pilgrimage path to the summit of Croagh Patrick. Hutton and Dewey (1986) interpret the Croagh Patrick Syncline as a ductile flower structure that was produced by a sinistral transpressional shear zone with its axis in Clew Bay. D2 in the Silurian rocks produced upright folds with steep hinges and a steep cleavage that contains a weak to moderately-developed sub-horizontal stretching lineation. S2 consistently cuts the axial surface of the Croagh Patrick Syncline in a clockwise sense, and the fabric intensifies to the south to become the dominant cleavage in the Silurian and Ordovician rocks of the South Mayo Trough. The southward increase in intensity and consistent clockwise obliquity of S2 across D1 structures suggests a large-scale Ramsay and Graham-type (1970) shear zone (Hutton and Dewey, 1986). D3 deformation produced upright folds and cleavages as a result of a late localised strike-parallel sinistral shear zone (Dewey and McManus, 1964). Hutton and Dewey (1986) interpret D1, D2, and D3 of the Silurian rocks as part of a continuum of sinistral transpressive deformation in the South Mayo Trough that initiated in the region of the southern shore of Clew Bay but whose locus switched to the southern bounding faults at a later stage. It is likely that this zone of sinistral transpressive shear continues north-east to the Central Ox Mountains inlier where there is abundant evidence for Late Silurian–Early Devonian sinistral strike-slip movement (see previous section).

Phillips (2001) has reviewed the timing of Late Caledonian deformation in the other inliers of the Midland Valley terrane (Fig. 8.1), where it is predominantly Acadian in age. In the Curlew Mountains inlier, Acadian deformation resulted in steeply-dipping to vertical Silurian rocks, and a north-east trending syncline in probable Lower Old Red Sandstone rocks. Pre–Late Devonian folds and a south-east dipping thrust fault are seen in the Late Ordovician and Silurian rocks of the Pomeroy inlier. Open, north-east trending buckle folds are also seen in the Lower Old Red Sandstone rocks of Cushendall, County Antrim.

In western Ireland, in both the Grampian and Midland Valley terranes, there is a pronounced strike swing from NE–SW to E–W as the Atlantic Ocean is approached. This is likely to be an Early Ordovician (or significantly older) feature. Palaeo-poles from the Middle Ordovician (Llanvirn) Mweelrea ignimbrites of the South Mayo Trough are in agreement with established middle Ordovician reference poles for Laurentia (MacNiocaill *et al.*, 1998). This implies there has been no significant rotation of the South Mayo Trough relative to Laurentia, and that the E–W trend of the trough and the curvature of this segment of the orogen is therefore primary. In contrast, palaeomagnetic evidence to the south of the South Mayo Trough (from both the Connemara block of the Grampian terrane and its Silurian cover sequences) suggests there has been a substantial clockwise rotation of this region relative to Laurentia during the Late Silurian (post-Wenlock) to Devonian (Morris, 1976; Smethurst and Briden, 1988; Briden *et al.*, 1989; Smethurst *et al.*, 1994). This clockwise rotation has been attributed to the superposition of the pre-existing (i.e. E–W) structural grain on the Connemara terrane and its Silurian cover during Late Silurian to Early Devonian sinistral transpression (MacNiocaill *et al.*, 1998).

Longford-Down terrane

The Longford-Down terrane occurs to the south-east of the Grampian and Midland Valley terranes (Fig. 8.1 inset). It consists of deep marine rocks deposited on the NW margin of the Iapetus Ocean during the Lower Palaeozoic (Chapters 3, 7). The Longford-Down terrane is best exposed in the Longford-Down inlier, but it is also visible further west in the Slieve Aughty inlier and other smaller inliers. The Longford-Down terrane continues NE into Scotland as the Southern Uplands terrane; its northern boundary in Scotland is the Southern Uplands Fault, but in Ireland the continuation of this fault is obscured by overlying Lower Carboniferous strata. Ryan *et al.* (1995) used magnetic data from north-west Ireland and Ryan and Dewey (2004) employed structural and provenance arguments to suggest that the Southern Uplands Fault continued across Ireland to Galway Bay, where Ordovician rocks of tract 2 of the northern belt of the Longford-Down terrane are faulted against the Lower Devonian Galway Granite along the Skird Rocks Fault (Fig. 8.1).

In Scotland, Peach and Horne (1899) recognised three belts in the Southern Uplands: the Northern Belt, which consists of Lower Ordovician black shales, cherts and lavas overlain by thick greywacke sequences; the Central Belt, comprising Upper Ordovician and Lower Silurian graptolitic shales and overlying greywackes; and the Southern Belt consisting of a thick sequence of entirely Silurian graptolitic shales and greywackes. A similar pattern exists in the Longford-Down inlier. Although the distinction of Peach and Horne (1899) is somewhat over-simplified, it has proved fundamental to subsequent research, and importantly highlights the gross, large-scale younging to the south-east of the rocks of the Southern Uplands and Longford-Down inliers.

The Ordovician and Silurian rocks of the Longford-Down inlier contain nine (or more) distinct stratigraphic sequences which differ in character and age over distances of a few kilometres across strike, but which can be traced for up to 100 km, perhaps more, along strike. These distinct stratigraphic sequences are bounded by major, steep strike-parallel reverse faults, and are commonly referred to as *tracts*. Although the large-scale younging of the Northern, Central and Southern belts is to the south-east, within each fault-bounded tract the succession generally youngs to the north-west.

Northern Belt

In counties Cavan and Longford, the Northern Belt is up to 30 km wide and comprises two fault-bounded tracts (Morris, 1987). These tracts correspond to tracts 2 and 3 using the numbering scheme of Leggett *et al.* (1979). In the northern tract, typified by greywackes containing clasts of metamorphic rock, strata are flat-lying and inverted. They define a major (10 km wide) recumbent fold which contains a number of isoclinal, frequently hingeless folds. The inverted strata and associated minor folds are cross-cut by the regional folds (F1) and their associated axial planar cleavage (S1). Morris (1987) has suggested that the early inversion represents a major orogenic event at the end of the Ordovician. However, pre-D1 inversion of strata over a distance of 15 km across strike is also found in Silurian rocks of the Central Belt (P.J. O'Connor, 1975; Phillips and Sevastopulo, 1986), and Phillips (2001) argues for large-scale slumping as a more likely explanation for these pre-D1 structures.

Rocks of the Northern Belt continue westwards into the northern part of the Slieve Aughty inlier (Fig. 8.1) and along the south shore of Galway Bay, and thus there is a major strike swing between counties Cavan and Galway. North-east in County Down, the Northern Belt is less than 5 km wide. Craig (1982) recognised five strike-fault bounded blocks, all of which have been correlated with tract 3 by Morris (1987). South-east verging F1 folds have axial surfaces which dip steeply to the south-east and hinges which plunge about 30° north-east. There is a penetrative S1 cleavage that is usually parallel to the axial planes of F1 folds, but which locally transects the axial planes in a clockwise sense. The southern boundary of the Northern Belt is marked by a major sinistral shear zone, the Orlock Bridge Fault (Fig. 8.1) and its lateral continuation to the south-west, the Slieve Glah Shear Zone (Anderson and Oliver, 1986). The fault lies within a zone of ductile sinistral deformation up to 1 km wide that is characterised by well-developed sinistral S–C fabrics and local mylonitisation. Anderson and Oliver (1986) infer a sinistral displacement in excess of 400 km along the fault, which can be traced westwards through the northern part of the Slieve Aughty inlier.

Central Belt

The Central Belt of the Longford Down inlier lies to the south of the Orlock Bridge Fault. It consists of Upper Ordovician–Llandovery pelagic black shales and cherts overlain by a thick sequence of Silurian turbidites. In Scotland, the correlative of the pelagic shale unit is termed the Moffat Shale, while the overlying turbiditic sequence is subdivided into an older Gala Group to the north-west and a younger Hawick Group to the south-east. In general, there is a southward decrease in the age of the Moffat Shale and the base of the overlying turbidites.

In the north-eastern part of the inlier, nine fault-bounded tracts have been recognised on the Ards Peninsula of County Down (Barnes *et al.*, 1987). The more northerly tracts contain the proximal turbidite facies of the Gala Group, while tracts in the south contain distal turbidites of the Hawick Group. The dominant D1 deformation phase imposed a system of upright, south-east verging folds which plunge gently to the north-east. These folds are associated with an S1 slaty cleavage, and the strike-parallel reverse faults are also D1 in age. In the three northernmost tracts of the Central Belt on the Ards Peninsula, the cleavage is parallel to the F1 fold axial planes and the L1 stretching lineation is perpendicular to the F1 fold hinges. South of the these tracts, the E–W trending cleavage transects the NE–SW trending F1 fold axial planes in a clockwise sense, and the L1 stretching lineation is parallel to the F1 fold hinges. Anderson (1987) has shown that this is the result of D1 sinistral transpression, probably during the Llandovery. Open, south-verging F2 folds and north-verging F3 folds and their associated crenulation cleavages are best developed in the south of the Central Belt but do not have a significant influence on the regional map pattern (Cameron, 1981).

To the southwest in counties Louth, Monaghan, Cavan, and Longford, a lack of detailed regional geological mapping and high-resolution biostratigraphy precludes direct comparison with the more detailed tract nomenclature system established for the Southern Uplands (e.g. Floyd, 2001). Only four tracts have been identified in the Central Belt (Geraghty, 1997; Morris *et al.*, 2003). In County Monaghan, F1 folds are upright or steeply inclined, and are tight to isoclinal with wavelengths of several hundred metres. The S1 cleavage is penetrative and transects F1 folds in a clockwise sense, reflecting sinistral shear as described previously in County Down. In the Shercock area of County Cavan, the cleavage dips more steeply than bedding to the south-east. The bedding is predominantly inverted and generally faces downwards on the cleavage. Downward-facing zones up to 15 km across continue westwards across County Cavan and may be related to large-scale slumping. In the northernmost tracts of the Central Belt in this region greywackes of mafic composition show widespread development of syn-D1 prehnite-pumpellyite facies metamorphism (Oliver, 1978). The Central Belt continues westwards into the southern part of the Slieve Aughty inlier (Fig. 8.1), where a fault-bounded unit equivalent to the Moffat Shale is juxtaposed against Upper Llandovery–Lower Wenlock greywackes. Bedding dips south, but generally faces up to the north on the S1 cleavage (Emo, 1978). Transected F1 folds are visible in the southern portion of the inlier. The S1 cleavage locally cuts across earlier folds, which may be either slump-related or early tectonic features (Pracht *et al.*, 2004).

Southern Belt

In the south-eastern part of the Longford-Down inlier in County Louth, the surface expression of the Iapetus Suture zone (the Tinure Fault) is exposed. Vaughan (1991) and Vaughan and Johnston (1992) have shown that to the north of the Tinure Fault, Llandovery–Wenlock strata correlatable with the Central and Southern Belts of the Southern Uplands are present. Llandovery–Wenlock lithologies of Vaughan's Clontail tract closely resemble Hawick Group rocks of the Central Belt, while to the south, greywackes of the Salterstown and Rathkenny tracts correlate with the Southern Belt (the Wenlock Riccarton Group) of the Southern Uplands (Fig. 8.3). The structural evolution of this zone (and the similar Arra Mountain–Slieve Bernagh inlier of counties Tipperary and Clare, Fig. 8.1) is considered in more detail in the section concerning the Iapetus Suture.

Structural synthesis

The most complete sections through the Longford-Down terrane are encountered on the coast of County Down, and this summary is largely derived from field data from this region (e.g. Barnes *et al.*, 1987; Anderson, 1987). The earliest deformation phase is represented by pre-D1 folding which can be on a large-scale, and which was probably caused by slump folding of unconsolidated or partially-lithified sediment. The D1 deformation phase is characterised by SE-directed thrusting and SE-verging folding. These reverse faults are now commonly sub-vertical because of subsequent rotation and steepening. The F1 folds are associated with a cleavage, which developed under low-grade metamorphic conditions (prehnite-pumpellyite facies). In the Northern Belt and northernmost tracts of the Central Belt the cleavage is axial planar, and the stretching lineation is perpendicular to the fold hinges. This is consistent with coaxial (pure shear) deformation. Further to the south in the Central Belt, the cleavage commonly transects the F1 axial planes in a clockwise sense and is accompanied by a sub-horizontal stretching lineation, indicative of sinistral transpression. Late-stage deformation consists primarily of northwest-directed thrusting and north-west verging F2 folds. It is probable that the F1 folds, thrusts and cleavages record progressive deformation in an accretionary prism (see below) during Late Ordovician–Silurian time, and thus are not Acadian structures. It is also probable that the change from deformation dominated by pure-shear in the Northern Belt to general shear with a strong sinistral simple shear component in the southern part of the Central Belt records the change from an orthogonal to an oblique plate collision during Late Ordovician–Silurian time. Later northwest-directed D2 structures affect the earlier structures which formed within the accretionary prism and may be Acadian in age. They are best developed near the southern margin of the inlier close to the Iapetus Suture zone and are discussed later.

The Longford-Down and Southern Upland terranes: an accretionary prism?

The tectonic setting of the Longford-Down and Southern Upland terranes has been the subject of much debate (Kelling, 2001). Any model for it must account for the following observations (paraphrased from Strachan, 2000b):

a the large volume of deep marine sediments (submarine fans and pelagic sediment);
b the location of these terranes between the Iapetus Suture and the Laurentian landmass, and the presence of an active volcanic arc in the Midland Valley terrane

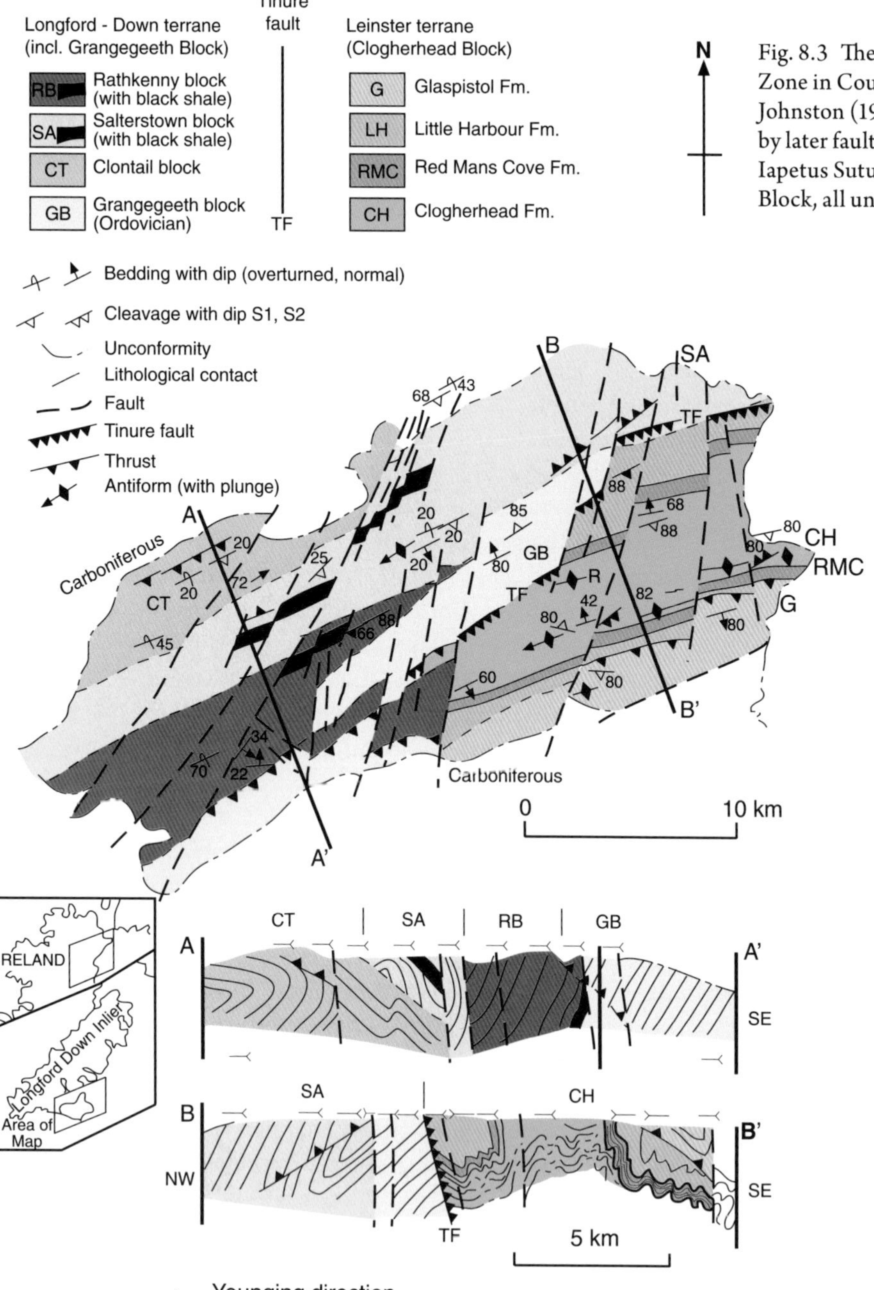

Fig. 8.3 The geology of the Iapetus Suture Zone in County Louth, *after* Vaughan and Johnston (1992). The Tinure Fault, dissected by later faulting is the proposed trace of the Iapetus Suture. Apart from the Grangegeeth Block, all units are Silurian.

(e.g. the Middle–Upper Ordovician volcanism in the South Mayo Trough and the Llandovery–Wenlock volcanism in the north Galway Silurian succession);

c the occurrence of numerous major reverse faults, spaced several kilometres apart and with internal stratigraphic younging to the north-west; and

d the occurrence to the south-east of progressively younger rocks, and in particular the diachronous south-eastward spread of submarine fan deposition.

These observations closely accord with modern accretionary prisms in convergent margin settings. In an accretionary prism, deep sediments on the ocean floor are not subducted, but rather are scraped off the subducting oceanic plate, bulldozer-fashion, and progressively accreted to the front of the overlying plate to form a prism by under-thrusting (Fig. 8.4). As the accretion process continues at the toe of the aggrading prism, early-formed gently-dipping thrusts are steepened by

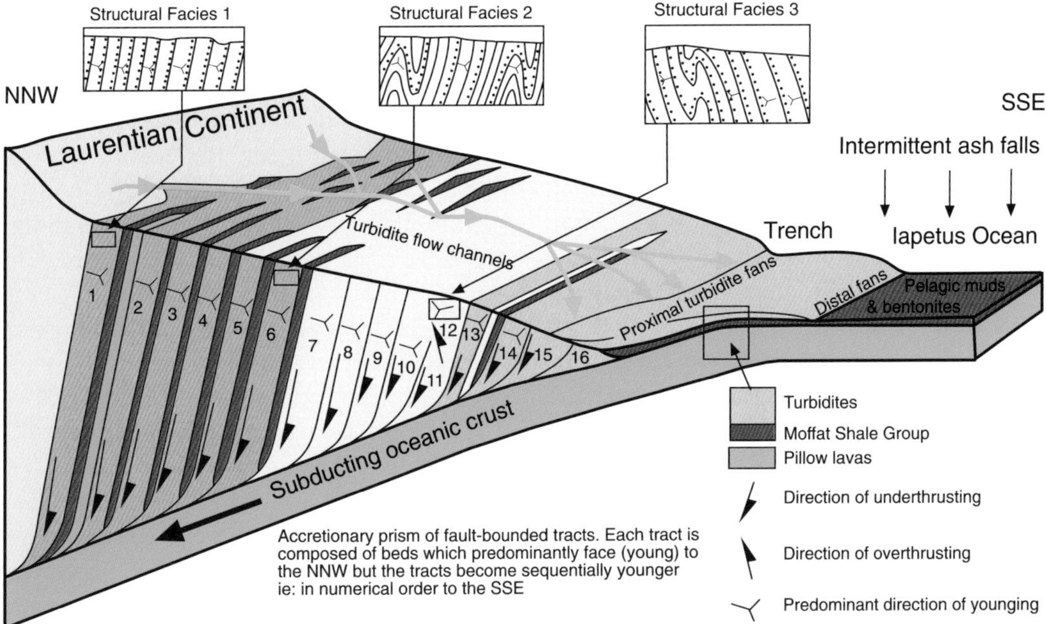

Fig. 8.4 The accretionary prism model for the formation of the Longford-Down–Southern Uplands Terrane, *after* Anderson (2004).

back-rotation. Accretionary prisms commonly produce 'seaward-verging folds and reverse faults that separate landward-dipping packages of accreted sediments' (Byrne and Hibbard, 1987). Such a model was first applied to the Southern Upland–Longford-Down terranes by Mitchell and McKerrow (1975), and gained wide acceptance shortly afterwards. However, various authors (e.g. Murphy and Hutton, 1986) subsequently pointed out that the structural and stratigraphic features observed in the Southern Upland–Longford-Down terranes are essentially those of an evolving imbricate thrust stack, which does not necessarily imply an accretionary prism setting. Structural arguments alone are therefore not diagnostic, and the sedimentary affinities of the sequences of the Southern Upland–Longford-Down terranes must also inform the model. A detailed discussion of this controversy is beyond the scope of this chapter, particularly as it currently remains unresolved, and the reader is encouraged to consult Anderson (2001) and references therein for further information.

Two models, those of Murphy and Hutton (1986), and of Stone *et al.* (1987) and Morris (1987), differ from the standard Mitchell and McKerrow (1975) accretionary prism model, and will be described briefly. The first favours an accretionary prism setting for the Northern Belt but proposes that subduction had essentially ceased by the time the Upper Ordovician–Silurian Central and Southern Belts were deposited. The latter belts are interpreted as a series of fans which prograded to the south-east in a successor basin, and which were subsequently deformed by an oblique sinistral thrust stack. In the second model, Stone *et al.* (1987) and Morris (1987) favour the Northern Belt as a back-arc basin located to the north of a north-dipping subduction zone. This model resolves the problem of volcaniclastic turbidites in the Northern Belt, which contain detritus derived from a volcanic chain to the south. Stone *et al.* (1987) proposed that the subsequent closure of the Iapetus Ocean deformed the back-arc basin (the Northern Belt), creating a SE-propagating thrust stack that buried the southern volcanic source. In this model, the Central and Southern belts represent a southward-migrating foreland basin formed ahead of the SE-propagating thrust stack.

The Iapetus Suture

The Iapetus Suture in Ireland and Britain is the line (or rather zone) that separates the former Laurentian and Avalonian continents. In Ireland it separates the Southern Belt of the Longford-Down terrane from the Leinster terrane. The precise trace of the suture remains somewhat uncertain. This issue is reviewed by Todd *et al.* (1991) who employed four different datasets to constrain the suture's location. These are faunal provinciality, deep crustal evidence (based on geophysical studies and analysis of xenolith suites), isotope geochemistry, and structural evidence.

Faunal provinciality

In their classic study McKerrow and Cocks (1976) demonstrated that during the Lower Palaeozoic there was a gradual increase in the similarity of the faunas between Laurentia and Avalonia. They invoked a wide Iapetus Ocean during Cambrian times, with a progressive migration of the more mobile components of the faunas as the ocean closed. Pelagic animals crossed first, followed during the Late Ordovician by animals (trilobites and brachiopods) with pelagic larval stages. Animals without a pelagic larval stage (such as benthic ostracodes) and freshwater fish were not able to cross until the ocean had closed in the Silurian. Faunal provinciality can thus be used to locate the surface trace of the suture, but it is most applicable to Cambrian and Early Ordovician faunas when provinciality was at a maximum.

The Iapetus suture zone in Ireland is relatively well exposed in County Louth. Two small 'suspect' terranes with Ordovician sediments, the Bellewstown and Grangegeeth terranes (Chapter 3), are caught up between the Longford-Down and Leinster terranes. Faunal contrasts between these terranes support their distinction from each other and from their neighbouring terranes on either side of the suture zone (Harper and Parkes, 1989). Mid-Llanvirn rocks in the Bellewstown terrane of the Balbriggan inlier (Fig. 8.1) contain brachiopods of the Celtic province, a peri-insular Avalonian fauna (Harper *et al.*, 1990). 10 km to the northwest of Bellewstown, the Ordovician rocks of the Grangegeeth terrane (Fig. 8.3) contain Llanvirn Atlantic province graptolites of peri-Gondwanan affinity (Owen *et al.*, 1992). However, the overlying Caradoc rocks contain benthic shelly faunas, which although showing some overlap with Gondwanan faunas show a strong bias at generic and species level towards a Scoto-Appalachian (i.e. northern Laurentian) fauna (Owen *et al.*, 1992). The Grangegeeth terrane is envisaged by Todd *et al.* (1992) as an oceanic island which originated at southern latitudes (i.e. close to Gondwana/Avalonia) during the Early Ordovician, and which drifted north during the Llanvirn, accumulating a more equatorial Laurentian-type fauna en route. It finally docked against the Irish margin of Laurentia (the Longford-Down terrane) during the Silurian. Todd *et al.* (1991) place the Iapetus Suture between the Grangegeeth and Bellewstown terranes.

Deep crustal structure based on geophysical evidence and xenolith suites

Based on seismic refraction data, the Iapetus Suture zone appears to be characterised by a thickening of the mid-crustal layer in two-dimensional P-wave velocity models in both the COOLE (Lowe and Jacob, 1989) and VARNET (Landes *et al.*, 2000) profiles. Both velocity models envisage the suture zone as approximately 60 km wide at mid-crustal levels. Additionally, in both profiles higher velocities are inferred north of the suture zone compared to the south, similar to the N–S LISPB model in northern Britain (Faber and Bamford, 1979).

The BIRPS deep seismic reflection profiles (Klemperer *et al.*, 1991) below the Irish Sea between Ireland and Britain (WINCH 2) and below the Atlantic Ocean to the west of Ireland (WIRE 1) image the Iapetus Suture zone as a stack of inclined reflectors in the crust, approximately 50 km wide, dipping north at 30° between depths of 5 km and about 35 km. In the upper crust, between depths of 5 km and 20 km, the suture zone appears to have been thrust northwards along a south-dipping plane.

Brown and Whelan (1995), using data from a magnetotelluric profile recognised a highly conductive zone (resistivity less than 10 Ohm.m) dipping south-eastwards at an angle of 8° from a depth of *c.*5.5 km to *c.*10 km within the suture region of central Ireland. They interpreted this as meta-sedimentary rocks of the Longford-Down terrane underthrust beneath Avalonian crystalline basement rocks of the Leinster zone. This is in contrast to the seismic reflection data which images north-dipping structures and the evidence from xenolith suites (see below) which suggests the Leinster terrane underlies the Longford-Down terrane.

In the Central Belt of the Longford-Down terrane on the Ards Peninsula in County Down, a lamprophyre dyke intruding Upper Llandovery greywackes after the local D1 deformation has been dated by K–Ar hornblende at 415 ± 12 Ma (Ludlow) by Anderson and Oliver (1996). Two such dykes contain xenoliths of schistose and mylonitic calc-alkaline basalt, similar to basalts of the Borrowdale Volcanics of the English Lake District (equivalent to the Leinster terrane). Anderson and Oliver suggest that they represent fragments of the Lake District Volcanic Terrane, sheared and thrust under the Longford-Down inlier to a depth of 20–30 km during closure of the Iapetus Ocean.

Xenoliths found in Lower Carboniferous volcanic pipes are samples of the lower crust in central Ireland (Strogen, 1974b). The volcanic pipes intrude the Longford-Down terrane above the suture, but the xenoliths appear compositionally to be derived from Avalonian lithosphere (i.e. the Leinster terrane) at depth (Van den Berg, 2005). Based on thermobarometry, metapelitic xenoliths were entrained from depths of *c*.20–25 ± 3.5 km and rare mafic granulites from depths of 31–33 ± 3.4 km (Van den Berg *et al.*, 2005). Experimentally determined seismic velocity (Vp) values for the xenoliths correspond well with model velocities from the ICSSP, COOLE I and VARNET seismic refraction lines when corrected for present-day lower crustal temperature (Van den Berg *et al.*, 2005).

Combining the available geological and geophysical data, the Iapetus Suture zone appears to be up to 50 km wide at depth. North-dipping deep seismic reflectors and limited data available from xenolith suites support a north-dipping suture with subduction of the Leinster terrane beneath the Longford-Down terrane.

Isotopic contrasts across the suture zone

Attempts to delimit the trace of the Iapetus Suture using isotopic data from sediments or granites have so far failed to resolve its precise location (Todd *et al.*, 1991). Nd model ages (T_{DM}) (Chapter 2) of Lower Palaeozoic rocks from the Longford-Down and Leinster terranes either side of the suture exhibit significant overlap (Todd *et al.*, 1991). The Sr isotopic systematics of Caledonian granites could potentially provide an indication of the trace of the suture. Granites north of the suture (i.e. in the Longford-Down and Grampian terranes) contain relatively less radiogenic ^{87}Sr and have lower Rb/Sr ratios compared to granites in the Leinster terrane (Todd *et al.*, 1991). However there is a lack of granite exposure in the central parts of Ireland adjacent to the suture zone. Potentially, modern sedimentary provenance techniques applied to the Lower Palaeozoic sequences either side of the Iapetus Suture zone could refine the trace of the suture. Single grain detrital zircon U–Pb dating (e.g. Košler *et al.*, 2002) or common Pb isotope analysis of detrital K-feldspars (e.g. Tyrrell *et al.*, 2006) could potentially distinguish between Laurentian and Avalonian sediment sources.

Structural evidence

The south-eastern part of the Longford-Down inlier in County Louth is one of the few places where Lower Palaeozoic rocks are exposed adjacent to the putative trace of the Iapetus Suture zone (Vaughan and Johnston, 1992). The Tinure Fault, whose trace has been offset by later faulting (Fig. 8.3), is believed to be the surface expression of the suture. To the north of the Tinure Fault, Silurian strata have been correlated with the Central Belt (the Clontail tract) and the Southern Belt (the Salterstown and Rathkenny tracts) of the Longford-Down terrane (Fig. 8.3). Interposed between the Silurian rocks and the Tinure Fault is the Grangegeeth terrane, a thin fault-bounded sliver of Ordovician volcaniclastics overlain by mudrocks and turbidites, which was discussed earlier in the context of faunal provinciality. South of the Tinure Fault a sequence of Silurian greywackes and mudrocks comprise the Clogherhead block. If the Tinure Fault is the suture, then the Clogherhead block belongs to the Leinster terrane, even though it is exposed within the Longford-Down *inlier*.

Bedding in the Longford-Down and Grangegeeth terranes generally youngs to the north-west (Fig. 8.3). Tight, upright F1 folds trend north-northeast and *verge southwards*. S1 is a weak slaty cleavage which transects F1 fold hinges by up to 16° in a clockwise sense, indicative of sinistral shear (Vaughan and Johnston, 1992). South of the Tinure Fault, D1 structures in the Clogherhead block *verge northwards*, and are southerly-inclined, upright folds also associated with sinistral transpression. There is therefore a fundamental change in the early (D1) structural style across the Tinure Fault. Structures on both sides of the Tinure Fault were modified by D2 northwest-directed thrusting and north-verging folding with a component of dextral transpression.

Evidence has already been presented that the Grangegeeth terrane represents a Middle Ordovician oceanic island, which originated close to Avalonia and must have then drifted north towards Laurentia during the Late Ordovician as it acquired progressively more Laurentian faunas. Therefore on faunal grounds, the Iapetus Suture is believed to pass between its southern boundary (the Tinure Fault) and the northern margin of the Bellewstown terrane 10 km to the south in the Balbriggan inlier (Fig. 8.1). Vaughan and Johnston (1992) argue that the change in the sense of F1 vergence across the Tinure Fault suggests that this fault represents the surface expression of the suture.

In the Arra Mountain–Slieve Bernagh inlier of counties Tipperary and Clare, the structure is similar to that of the SE margin of the Longford-Down terrane in County Louth. F1 folds verge to the south-east, and the S1 cleavage transects the F1 folds in a clockwise sense. In the south of the inlier, the axial planes of F1 folds and S1 are vertical. To the north, these structures become progressively overturned to dip at low angles to the south, comparable to the D2 northward thrusting in County Louth. Using the structural criteria applied previously (the sense of vergence of F1 folds), the Iapetus Suture would be expected to pass south of the Arra Mountain–Slieve Bernagh inlier.

However, in the inlier immediately to the south of the Arra Mountain–Slieve Bernagh inlier (the Slieve Phelim–Silvermines–Devilsbit inlier, Fig. 8.1), the sense of vergence of F1 folds is problematical. Upright F1 folds trend between 020° and 080° and are associated with a steep axial-planar cleavage. In the north-western part of the inlier, Archer (1981) recognised southward-verging upright F1 folds with clockwise transection by the S1 cleavage. Archer (1981) and Todd *et al.* (1991) therefore argue on structural grounds that the suture passes south of the Slieve Phelim-Silvermines–Devilsbit inlier. Phillips (2001) argues that the large-scale F1 folds have neutral vergence and that the Iapetus Suture passes between the Arra Mountains–Slieve Bernagh inlier and the Slieve Phelim–Silvermines–Devilsbit inlier. The latter configuration is illustrated in Fig. 8.1.

Leinster Zone

Phillips (2001) reviews the structure of the various Lower Palaeozoic inliers of the Leinster terrane close to the Iapetus Suture zone. The structure of the Slieve Phelim–Silvermines–Devilsbit inlier is reviewed in the previous section. Further south in the Galtee Mountains inlier, Late Caledonian deformation produced east-west trending, upright, gently-plunging F1 folds which verge to the south. The S1 axial plane cleavage is locally crenulated by north-west trending minor F2 folds.

The Balbriggan inlier on the east coast is bisected by the Lowther Lodge Fault, which divides it into northern and southern sectors, both of which contain Ordovician rocks overlain by Silurian turbidites. The major structure in the northern part of the inlier is an ENE trending F1 syncline. A steep south-dipping S1 cleavage transects minor F1 folds in a clockwise sense on the steep southern limb. Murphy (1985) has shown that this is the result of sinistral transpressive shear intensifying into a major shear zone boundary marked by the Lowther Lodge Fault. On the northern limb, the cleavage transects minor Fl folds in an anticlockwise sense, reflecting transpressive dextral shear along the faulted boundary with Ordovician volcanic rocks of the Bellewstown area. D2 deformation is restricted to the southern part of the southern sector where it is characterised by south-verging F2 folds with an associated crenulation cleavage. Further south, the major structures of the Portrane and Howth inliers are east-west trending upright synclines.

Structure of the main Leinster inlier

Regional deformation in the Leinster inlier is thought to be Acadian in age, based on the apparent broad contemporaneity with the intrusion of the Lower Devonian Leinster Granite (405 ± 2 Ma, O'Connor *et al.* 1989, Table 8.1). The granite was intruded during D2 deformation of the Ordovician and Silurian country rocks. The inlier is cut by a number of regional-scale sinistral shear zones and faults including the Hollywood Shear Zone, Wicklow Fault Zone and East Carlow Deformation Zone (Fig. 8.5, McArdle and Kennedy, 1985; Max *et al.* 1990; McConnell *et al.*, 1999; McConnell, 2000).

D1 regional deformation produced tight to isoclinal folds and a strong penetrative cleavage in pelitic rocks (Sanderson *et al.*, 1980). To the north-west of the Leinster Granite there is pronounced flattening on the S1 cleavage. This cleavage is axial planar to a south-east verging syncline which faces up to the north-west. To the south-east of the Leinster Granite, there are a series of upright F1 folds with an axial planar S1 cleavage. This cleavage locally contains flattened strain markers. The hinges of minor F1 folds often plunge south-eastwards, down the dip of the cleavage. In the southern part of the inlier in counties Wexford and Waterford, major F1 folds appear to fan away from a central upright anticline coincident with the main axis of the granite pluton to the north-east. In the northern part of the inlier, Cambrian rocks of the Bray Group show evidence of slump-related megascopic pre-D1 folding (Brück and Reeves, 1976). They have been thrust north-westwards over the aureole rocks of the Leinster Granite during D1. Refolding of the thrust plane by F2 folds and tightening of F1 folds approaching the thrust plane provides evidence for a D1 age for thrusting (McConnell *et al.*, 1994).

Intrusion of the Leinster Granite post-dates the D1 deformation phase as it cuts a suite of dolerite dykes which possess an S1 cleavage (Brindley, 1956). The granite was deformed during the second (D2) deformation while it was still hot. A broad zone of D2 deformation (the East Carlow Deformation Zone, McArdle and Kennedy, 1985) extends several kilometres out from the south-east margin of the pluton in County Carlow. It is characterised by strain intensification and tightening of earlier F1 folds. A WNW trending schistosity is developed along with minor F2 folds and a strong L2 mineral lineation. The thermal aureole extends up to 2 km from the contact of the pluton, and the growth of porphyroblasts indicative of contact metamorphism (staurolite, andalusite, biotite, and garnet) occurred after the formation of the

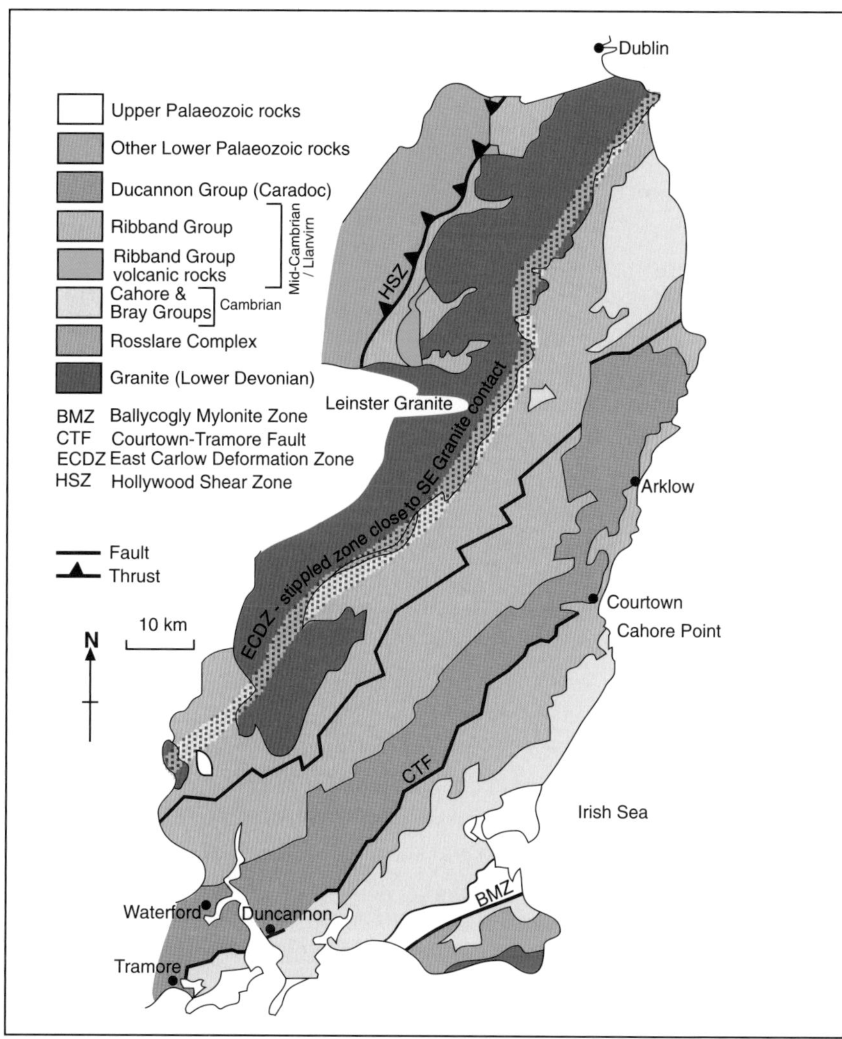

Fig. 8.5 The geology of the Leinster inlier illustrating the major stratigraphic units and major faults. Modified after Clegg and Holdsworth (2005), McConnell and Philcox (1994) and Tietzsch-Tyler *et al.* (1994).

S1 cleavage, during D2 (McArdle and Kennedy, 1985). To the south-east, in the aureole of the Blackstairs Unit of the Leinster Granite, Shannon (1979b) describes a structural history analogous to the East Carlow Deformation Zone with D2 structures increasing in intensity towards the granite contact, and a syn-D2 intrusion age based on the timing of growth of porphyroblastic andalusite. Sanderson *et al.* (1980) document zones of D2 high strain in the country rock adjacent to the pluton contact and a strong stretching lineation (with X/Y ratios exceeding 4). West of the pluton, the lineation plunges obliquely towards the south-west, while east of the pluton the extension lineation plunges obliquely north-east, parallel to the mineral lineation in the granite. The role of D2 shear zones in facilitating granite intrusion has become increasingly apparent. O'Mahony (2001) envisages that granite ascent was facilitated by a multiple sheeting mechanism within an active transtensional shear zone.

Several Acadian NE-SW trending high strain zones have now been identified in the Leinster inlier (Fig. 8.5), and are not just confined to the pluton margins. Clegg and Holdsworth (2005) describe a heterogeneously developed, braided network of ENE- to NE-trending, subvertical sinistral shear zones and fault arrays of Acadian age adjacent to the Courtown-Tramore Fault in County Wexford. The geometric and kinematic characteristics of the shear zone elements suggest components of NW–SE shortening, top-to-the-SW sinistral shear and a small component of top-to-the-SE thrusting. These are kinematically partitioned into zones characterised by contraction- and strike-slip-dominated deformation respectively.

The Dingle Peninsula

The east–west striking rocks of the Dingle Group (Chapter 7) are deformed with the conformably underlying

Silurian succession (the Dunquin Group, Chapter 7). In the north-west of the area, the Smerwick Group (originally considered to be Mid- to Late Devonian in age, Todd *et al.*, 1988; Todd, 1989b) rests unconformably on the Dingle Group. The Acadian event is characterised by regional folding and faulting of Silurian basement, Dingle Group and Smerwick Group sedimentary rocks (Richmond and Williams, 2000). This early stage of folding is characterised by tight, overturned WNW-trending structures (Meere and Mulchrone, 2006).

Following the Acadian Orogeny, the Eifelian (Early Middle Devonian) Pointagare Group was deposited unconformably on the deformed Smerwick and Dingle Groups (Richmond and Williams, 2000). Richmond and Williams therefore infer a Late Silurian–Early Devonian age for the Smerwick Group on structural grounds. Finite strain analysis (reduction spot data and R_f/φ clast analysis) of deformed sedimentary rocks below and above the Acadian unconformity surface demonstrates that this boundary represents a significant bulk strain discontinuity (Meere and Mulchrone, 2006). Acadian deformation resulted in cleavage development and high strain in rocks below the unconformity surface (Meere and Mulchrone, 2006), in contrast to earlier studies (e.g. Parkin, 1976b), which attributed this high strain deformation to later Variscan tectonism. Instead, Variscan effects in this region are weak, restricted to the development of a weak and localised disjunctive cleavage (Meere and Mulchrone, 2006).

Meere and Mulchrone (2006) recognise that the S1 (Acadian) cleavage transects F1 fold axes in an anticlockwise sense, implying dextral closure of the Dingle Basin during the Acadian Orogeny. This is in contrast to the sinistral transpression elsewhere south-east of the Iapetus Suture zone in SE Ireland, the Isle of Man, Wales and the Lake District, inferred from consistent clockwise transection of fold axes by cleavage fabrics (Murphy, 1985; Soper *et al.*, 1987; Woodcock *et al.*, 1988; Clegg and Holdsworth, 2005). However, the structural trend of the basement fabric in the Dingle Peninsula and SW Ireland is ENE–WSW, compared to NE–SW (the typical Caledonide trend) in the zones that exhibit sinistral transpression. Superimposing an Acadian convergence vector of 10–15° west of north (Soper *et al.*, 1987) on a pre-existing basement fabric that strikes ENE–WSW in the Dingle area yields the component of dextral shear observed (Meere and Mulchrone, 2006).

Summary of Late Caledonian tectonism

The Caledonian orogenic cycle is a series of tectonic events associated with the closure of the Iapetus Ocean (McKerrow *et al.*, 2000a). Continuous subduction of the Iapetus Ocean during the Ordovician and Silurian resulted in the Laurentian continent of north-west Ireland progressively approaching the Avalonian continent of south-east Ireland. By the Late Silurian, virtually all the intervening oceanic crust had been subducted, and the two continents had met. This Late Caledonian collision was 'soft', in that it did not result in large-scale crustal thickening, but instead was strongly oblique, dominated by sinistral transpression (Soper *et al.*, 1992; Dewey and Strachan, 2003). In Ireland, the major Late Caledonian orogenic event, the Acadian Orogeny, occurred in the Early Devonian, slightly after the final closure of the Iapetus Ocean.

In general, the metamorphic grade associated with the Late Caledonian tectonism was quite low – typically rocks in the Silurian inliers of central Ireland experienced sub-greenschist facies metamorphism. Late Caledonian deformation varies in structural style across Ireland. The Grampian terrane behaved in a relatively rigid fashion, with deformation partitioned into NE–SW trending sinistral faults and shear zones which facilitated the intrusion of several Late Caledonian granites, such as the Main Donegal Granite (Hutton, 1982). The Midland Valley terrane experienced more ductile regional strain in the form of upright folds and cleavage, with local evidence for polyphase deformation. The Longford-Down terrane is typified by numerous steeply-dipping to inverted stratigraphic packages which are bounded by reverse faults. The internal younging within each thrust slice is towards the north-west, yet the gross disposition of units suggests an overall younging towards the south-east. These observations closely accord with modern accretionary prisms in convergent margin settings. If such a tectonic scenario is applicable to the Longford-Down terrane, then some of the structures observed must be diachronous and significantly pre-date the Acadian Orogeny, and are probably Late Ordovician or Early Silurian in age. This is because in an accretionary prism, deep sediments on the ocean floor are continuously scraped off the subducting oceanic plate, and progressively accreted to the front of the accretionary prism by under-thrusting. The structural style of the Leinster terrane is dominated by upright folds with associated cleavage and thrust faulting, and by later large-scale NE–SW trending Acadian shear zones associated with the emplacement of the Leinster Granite.

The Iapetus Suture in Ireland and Britain is the line or zone that separates the former Laurentian and Avalonian continents. Despite its fundamental importance in the Caledonian Orogeny, its precise trace remains somewhat uncertain. In Ireland, the surface trace of this collisional suture separates the Longford-Down and Leinster terranes. The location of the suture trace in Ireland has been constrained by faunal provinciality, deep crustal evidence (based on geophysical studies and analysis of xenolith suites), structural evidence, and isotope geochemistry. The Iapetus suture zone in Ireland is relatively well exposed in County Louth where it passes between the Grangegeeth and Bellewstown terranes, two small 'suspect' terranes sandwiched between the Longford-Down and Leinster terranes. Faunal contrasts between these two small terranes support their distinction from each other and from their neighbouring terranes. There is also a fundamental change in the early (D1) structural style across the suture zone. To the north of the suture, F1 folds verge southwards. South of the suture, D1 structures in general verge northwards.

Late Caledonian granites and related intrusions

Granite plutons intruded during the Caledonian orogenic cycle in Ireland and Britain have been divided into two suites: the 'Older Granites' and 'Newer Granites' (Read, 1961). Table 8.1 lists all geochronological data for Caledonian intrusions in Ireland. A subset of this data presented as an age distribution histogram (Fig. 8.6) clearly illustrates the original distinction of Read (*op. cit.*) in that a small group of ages cluster at around 475–460 Ma, i.e. during the Grampian Orogeny, and a later suite straddles a much broader time period from 430 to 380 Ma. The older suite (which also includes basic or intermediate intrusions) consists of the syn- to post-tectonic Grampian intrusions of Connemara such as the Cashel–Lough Wheelaun gabbros and the Oughterard Granite, and the plutons of the Tyrone Igneous Complex. These are considered within the context of the Grampian Orogeny in Chapter 4 and are not discussed in detail in this chapter. The younger suite of intrusions spans the final stages of the closure of the Iapetus Ocean in the Silurian, and the Early Devonian Acadian Orogeny, and it continues into post-orogenic times. The term 'Late Caledonian' is used in this chapter to describe this extended period of magmatism.

The Late Caledonian magmatism in Ireland and Britain was volumetrically far greater, and compositionally more granitic, than the older magmatic suite. Three intrusions, those in Leinster, Galway and Donegal, reached batholith proportions. These intrusions, and the Donegal batholith in particular, have served as a superb natural laboratory for generations of geologists, and proved central in one of the great geological debates of the 20th century, the 'Granite Controversy' (e.g. Read, 1957). This is the question of whether granites formed through melting or through granitisation – the *in situ* metasomatic transformation of pre-existing rocks by processes of solid diffusion. Although the determination of the granite system *liquidus* diagram by Tuttle and Bowen (1958) largely resolved the debate in favour of melting, a magmatic origin

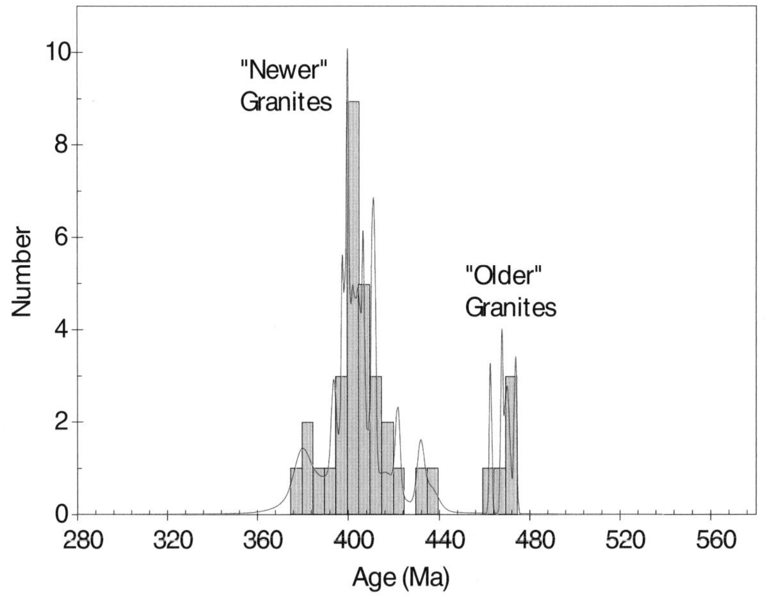

Fig. 8.6 Age distribution diagram of all the Caledonian intrusions in Ireland based on Table 8.1. Where there were duplicate analyses from the same pluton, a subjective decision was made as to the most accurate and precise age. Any demonstrably false Rb–Sr isochrons were rejected.

for granite still presented problems in explaining how the country rocks make room for incoming magma – the so-called 'space problem'.

This problem is particularly acute where we deal with large, batholithic volumes (1×10^5 km^3) of magma. Traditional ('cartoon') models of large-scale diapirs rising through the crust are now thought not to be viable on thermal and mechanical grounds (e.g. Petford *et al.*, 2000), and increasingly granite plutons are viewed as being fed by narrow feeder dykes through which magma at depth rapidly rises through the crust in pulses, feeding fast-growing magma chambers at a high level (Clemens, 2005). Tectonic activity plays an important role in making space in the crust for the injection of magma, and field studies of the Late Caledonian granites in Britain and Ireland have contributed greatly to a clearer understanding of the relationship between tectonics and magma intrusion (e.g. Hutton, 1982). In particular, many of the major Late Caledonian NE–SW trending strike-slip faults and shear zones have played a fundamental role in facilitating both magma ascent in dykes and subsequent emplacement at shallow levels in the crust.

However, the cause for the generation of such large volumes of magma over a protracted time period (430–380 Ma) still remains largely unresolved, and is discussed in detail at the end of the chapter. Before that, the main Late Caledonian granite plutons, and related smaller intrusive bodies, are considered in turn, taking first those that are located north-west of the Iapetus Suture, followed by those on the south-east side. Of course, many details concerning the structural controls on their emplacement, and their subsequent deformation histories, have been presented already in the first part of this chapter.

The granites of Donegal

Of all the major Late Caledonian granite intrusions in Ireland, probably the best known are the granites of Donegal (Fig. 8.7). During the 20th century they were the subject of much research, and of two detailed memoirs (Pitcher and Berger, 1972; Pitcher and Hutton, 2003). The granites comprise a suite of six contiguous and two separate plutons. The terms *Donegal Batholith* and *Donegal Granite* refer to the six contiguous bodies or to the same six plus the nearby Fanad pluton (i.e. not including the more isolated Barnsmore pluton). The individual bodies exhibit a wide range of intrusion mechanisms, including passive emplacement, ballooning, and sheeted emplacement in a shear zone. Compositionally they are dominated by monzogranite, but the overall composition varies from tonalite through to quartz monzodiorite.

The Donegal Batholith was emplaced into a NE–SW trending ductile sinistral shear zone within Dalradian metasediments (Hutton, 1982; 1988; Fig. 8.8a–c). The structural history of the emplacement of the batholith into this shear zone was considered earlier in the chapter, and is only treated briefly here. The shear zone mechanism provides a comprehensive model for the emplacement of the batholith (Hutton, 1982), and resolved the enigma posed by Pitcher and Berger (1972, p.351), who state 'the most remarkable find of our researches is the considerable range of contact effects at the one level in the crust'.

Hutton (1982) envisages that the shear zone first provided an opening for the Thorr and possibly the Fanad magmas (Fig. 8.8d). The Toories and Ardara plutons then ballooned into a compressional tip of the main shear zone (Fig. 8.8e,f). As the shear zone developed the Main Donegal Granite was sucked into an extensional sector as a series of sheets. Meanwhile, in the relatively brittle flanks of the shear zone, the associated extensional stresses opened the fractures controlling both the intrusion of the Trawenagh Bay stack of sheets and the collapse of the cauldron of the Rosses into one such sheet (Fig. 8.8f). The primary reference for the following pluton descriptions is Pitcher and Hutton (2003), while the sequence of intrusion based on cross-cutting relationships is taken from Hutton (1988) and Hutton (1982):

1a The Thorr Granite is a large zoned sheet with steep contacts, emplaced passively by stoping with abundant xenoliths of country rock. In the south of the pluton, a roof zone of quartz diorite preserves abundant xenoliths that trace out a pronounced ghost stratigraphy. The intrusion is surrounded by a broad, predominantly static sillimanite–andalusite–cordierite hornfels aureole which overprints the earlier (Grampian) regional metamorphic assemblages.

1b The Fanad Granite is a large tabular sheet-like body, emplaced by passive stoping. It consists of a petrologically diverse monzodioritic marginal zone and a main body of coarse pink granite (Carey, 1995). The coarse granite exhibits mingling and mixing relationships with microdiorite enclaves, probably derived from the neighbouring Appinite Suite, described below. A broad static aureole is present.

1c The Appinite Suite is a collection of small bosses, sheets and dykes of appinite, and dykes of lampro-

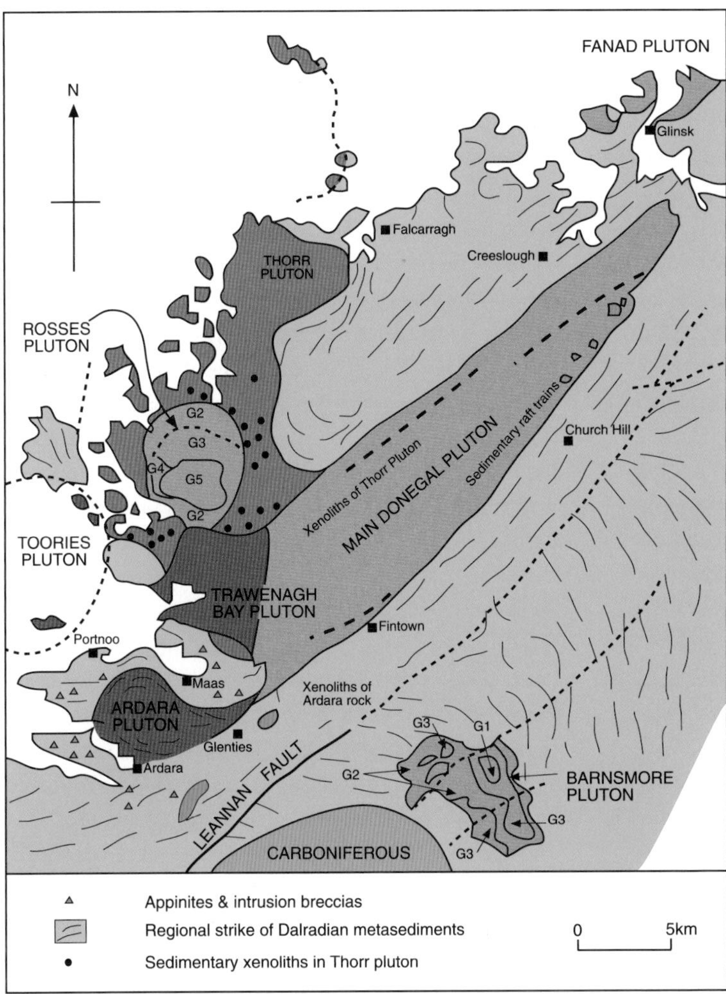

Left Fig. 8.7 The granites of Donegal (*after* Pitcher and Berger, 1972).

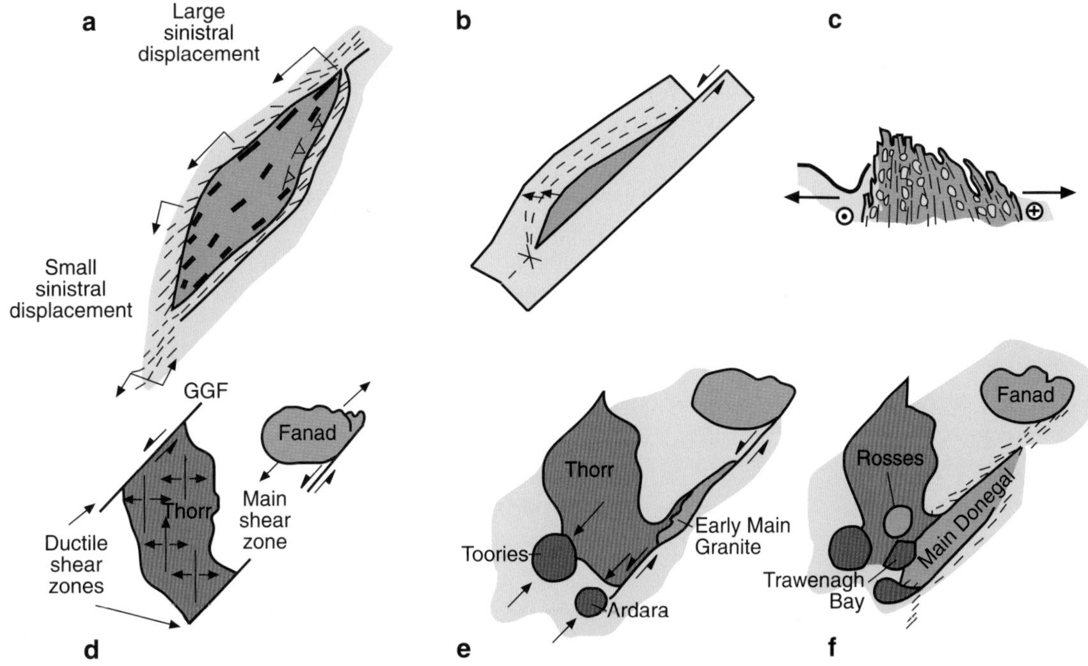

Below Fig. 8.8 Structural controls on the emplacement of the Donegal granites (Hutton, 1988). (a) Strain distribution pattern in Main Donegal Granite. Black lines approximate strain variations in the pluton. (b) Crack opening model for the Main Donegal Granite. (c) NW-SE cross section illustrating aureole deformation and sheeted nature of pluton. (d–f) Evolution of the Donegal batholith with time. GGF = Great Glen Fault. *Reproduced by kind permission of the Royal Society of Edinburgh.*

phyre, closely associated in space and time with the early granites. Many of the appinite bodies are found close to the Ardara pluton (French, 1977), while others occur on a linear trend between the Ardara and the Barnesmore plutons. The appinites are hornblende-rich igneous rocks, usually basic to ultrabasic in chemistry. They are often associated with gas intrusion breccias with tuffisite features formed by the explosive eruption of volatiles (mainly H_2O). The appinites and gas breccias are believed to be the products of mantle-derived basic melts rich in volatiles and potassium (Pitcher, 1997). Their emplacement appears to have been facilitated by deep-seated crustal fractures, possibly shared by their associated granitic intrusions. Their close genetic relationship to their associated granites and their role in the Late Caledonian magmatic episode is discussed later.

2 The Toories Granite is a ballooning pluton with a well-developed marginal foliation, flattened enclaves and deformed country rock, similar to the Ardara Pluton. Only a fragment of this quartz monzodiorite pluton and its contact is exposed.

3 The Ardara Granite is a classic example of a ballooning pluton. Originally interpreted as a diapir by Akaad (1956), a ballooning mechanism was convincingly demonstrated by Holder (1979) who documented the increase in strain towards the pluton margin based on deformed enclaves. The variation in strain was attributed to three consecutive pulses of magma into the expanding pluton. The oldest unit (G1) is a marginal feldspar-phyric quartz diorite, the next youngest unit (G2) is a ring of aphyric granodiorite, and the youngest unit (G3) forms the central core of the pluton and consists of porphyritic granodiorite. Although Paterson and Vernon (1995) and Vernon and Paterson (1993) argue that the Ardara pluton represents a set of nested diapirs emplaced passively by stoping, Molyneux and Hutton (2000) argue from strain data that over 80% of the volume of the pluton was accommodated in the country rocks within 4 km of the pluton, and therefore a ballooning emplacement mechanism is the most plausible model. The aureole is characterised by intense deformation, which involved the tightening of existing folds and the development of new cleavage. The inner aureole contains sillimanite, garnet and cordierite, while the outer aureole contains staurolite, kyanite and andalusite.

4 The Rosses Granite is a centred complex, consisting of an arcuate set of granite sheets cut by four concentric bodies of monzogranite (G1–G4) emplaced by repeated cauldron subsidence. Pitcher (1992) documents knife-sharp, outward-dipping internal contacts between the units, suggesting that each pulse of granite intruded a solid precursor. Hall and Walsh (1971) document the presence of beryl in greisenised granite, in beryl-bearing pegmatite, and in beryl-quartz veins in the Rosses Granite.

5a The Trawenagh Bay Granite is petrographically very similar to the Main Donegal Granite, but is essentially unfoliated. Price and Pitcher (1999) view the Trawenagh Bay Granite as being emplaced as a stack of sheets into the external flank of the shear zone, during the last stages of transcurrent movement.

5b The Main Donegal Granite is a sheeted pluton emplaced syn-kinematically into the main NE–SW trending shear zone; it contains conspicuous raft trains and exhibits strong colour banding and internal foliation. Early sheets were granodiorite in composition, while subsequent sheets were larger and more tonalitic. Porphyritic monzogranite sheets represent the final magmatic episode (Price, 1997). Individual sheets of magma are thought to have been emplaced by the shear zone temporarily locking up on one flank, and subsequently arching or bowing outwards resulting in the 'sucking in' of pulses of magma (Fig. 8.8b). The foliation in the Main Donegal Granite is thought to be tectonic (as opposed to magmatic) in origin, and is sub-parallel to the banding. The origin of the banding is still somewhat enigmatic. Berger (1971) attributes the primary control on the banding to the presence or absence of K-feldspar. The linear raft trains consist of xenoliths of Thorr granite, metadolerite, Dalradian metasediments, appinite and Ardara granite. Typically individual rafts are on the order of a few metres, but locally some rafts are over a kilometre in length. The aureole is intensely deformed (see earlier sections) with growth of cordierite, andalusite, and sillimanite.

The isolated Barnesmore Granite occurs on the southern side of the Leannan Fault. It is thought to have been emplaced by cauldron subsidence. Overall, on the basis of mineralogy the plutonic rocks of the Donegal Batholith are made up mainly of monzogranite and granodiorite, with minor tonalite and quartz diorite (Atherton and Ghani, 2002). They are metaluminous I-type granites characterised by the presence of amphibole and sphene, and the virtual absence of primary muscovite. Their

I-type classification is consistent with their low $^{87}Sr/^{86}Sr_i$ ratios (typically < 0.707).

The Ox Mountains Igneous Complex

The Ox Mountains Igneous Complex (McCaffrey, 1990) comprises the Ox Mountains Granodiorite, the Lough Talt Adamellite and the Easky Lough Adamellite (Fig. 8.9). The Ox Mountains granodiorite crops out as an elongate, NE–SW trending pluton, about 25 km long and 5 km wide. It was intruded into polyphase-deformed metamorphic rocks of the Dalradian Supergroup in the Central Ox Mountains Inlier, and postdates regional amphibolite facies metamorphism that occurred during the Grampian Orogeny. The Late Caledonian structural history of emplacement was referred to earlier, and is briefly described here. It closely resembles that of the Main Donegal Granite, in that magma was emplaced as a series of sheets into an active NE–SW trending sinistral shear zone, and this shear zone was superimposed on rocks that underwent three earlier phases of deformation during the Grampian Orogeny. Similar to the Main Donegal Granite, the Ox Mountains granodiorite exhibits a sub-vertical solid-state foliation, sinistral S–C fabrics, and a strong sub-horizontal mineral lineation.

McCaffrey (1990, 1992) recognised that the Ox Mountains granodiorite consists of four main components. These are (in order of intrusion), (1) small muscovite granite sheets, (2) granodiorite (both equigranular and K-feldspar megacrystic varieties), (3) diorites and (4) tonalites. Of these, the granodiorites (and in particular the equigranular type) are by far the most important volumetrically. McCaffrey (1990, 1992) proposed that the pluton comprises a sequentially-intruded series of sheets. The sheeted nature of the intrusion is immediately apparent in outcrops displaying sheeted contacts with the country rock and in the spectacular sheeted tonalite–granodiorite exposures at Pontoon in the SW corner of the pluton (McCaffrey, 1992, 1994). It is thus possible, or even probable, that the more compositionally homogeneous portions of the intrusion (such as the core, which is composed almost entirely of equigranular granodiorite) were also intruded in a sheet-like manner. The Ox Mountains granodiorite does not have a well-developed thermal aureole, despite the fact that recent U-Pb zircon dating (412.3 ± 0.8 Ma, Chew and Schaltegger, 2005, Table 8.1) confirms that it substantially post-dates the thermal peak associated with Grampian regional metamorphism by approximately

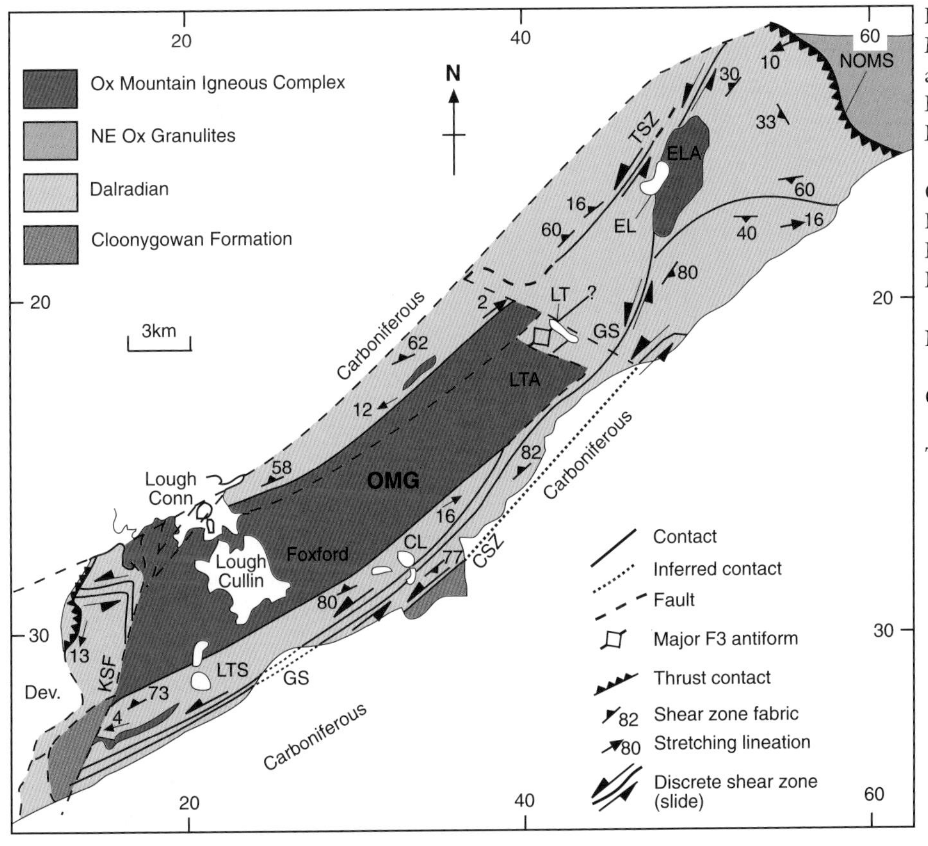

Fig. 8.9 The geology of the Ox Mountains Igneous Complex, after McCaffrey (1994). EL = Easky Lough, ELA = Easky Lough Adamellite, GS = Glennawoo slide, KSF = Knockaskibbole Fault, LT = Lough Talt, LTA = Lough Talt Adamellite, LTS = Lough Talt slide, NOMS = North Ox Mountains slide, OMG = Ox Mountains Granodiorite, TSZ = Tawnaneilleen shear zone.

60 Ma. Perhaps this is because the country rock close to the pluton is dominated by psammite and quartzite, which do not develop mineral assemblages diagnostic of contact metamorphism. The U–Pb zircon date of Chew and Schaltegger (2005) also demonstrates that the Rb–Sr whole rock age of 487 ± 6 Ma of Pankhurst *et al.* (1976) is not recording crystallisation, but is instead most likely an artefact of mixing between magmas from juvenile (probably mantle-derived) and crustal sources. The sheeted nature of the intrusion may have played a role in preventing Sr isotopic homogenisation of the pluton (Chew and Schaltegger, 2005).

The Lough Talt and Easky Lough adamellites are essentially undeformed and exhibit conspicuous aureoles that overprint the regional metamorphism. They were probably emplaced into dilational cavities associated with movement along the main sinistral shear zone (McCaffrey, 1992). These adamellites have been dated at 401 ± 33 Ma (Long *et al.* 1984, Table 8.1).

The Galway Granite

The Galway Granite is a large (approximately 600 km^2) composite Late Caledonian batholith. Its northern margin dips northwards between 55° and 75° (Plant, 1968), and cuts metagabbros and quartz-diorite orthogneisses that form part of the *c.*470 Ma magmatic suite intruded into the Dalradian rocks of Connemara during the Grampian Orogeny. The country rock to the south comprises the Lower Ordovician volcanic and sedimentary rocks of the South Connemara Group (Ryan and Dewey, 2004). The batholith was intruded close to the Skird Rocks Fault (the presumed extension of the Southern Uplands Fault) that juxtaposes the South Connemara Group and the Connemara metagabbro and orthogneiss suite. It post-dates movement on this fault and can thus be considered a terrane 'stitching' pluton. Satellite plutons include the Roundstone Granite, a circular ring of biotite microgranite about 7 km across; the Omey Granite (Fig. 8.2), a circular multiple ring intrusion of biotite adamellite about 6.5 km in diameter emplaced by cauldron subsidence; and the Inish Granite, apparently emplaced by a similar mechanism, which has a marginal zone of aplite and pegmatite containing abundant blocks of country rock (Leake, 1978; Max *et al.*, 1978b; Long *et al.*, 1995; El Desouky *et al.*, 1996).

The batholith is divided into Western, Central and Eastern blocks by two major N–S faults, the Shannawona Fault and the Barna Fault (Fig. 8.10). Gravity studies (Madden, 1987) show that east of the Barna Fault the batholith is 3 to 4 km thinner than to the west. Geobarometric studies based on the total aluminium content of magmatic hornblende (Callaghan, 2005; Leake and Ahmed-Said, 1994) hint that granites in the Central Block crystallised at higher pressures (4–5 kbar) than those in the Western Block (2.5–3 kbar). These data are consistent with the hypothesis that the eastern part was upthrown and then eroded, and now exposes a deeper intrusive level (e.g. Leake, 1978; Leake and Ahmed-Said, 1994).

The following descriptions of the component lithologies of the Western, Central and Eastern blocks are adapted from Callaghan (2005). Western Block lithologies range from granodiorite (Carna Granite) through granite (Errisbeg Townland Granite) to alkali feldspar granite (Murvey Granite) (P. Wright, 1964; Claxton, 1971; Leake, 1974; Max *et al.*, 1978). The Errisbeg Townland and Murvey granites are also present in the Eastern Block (Coats and Wilson, 1971). The Central Block is more varied and contains less evolved lithologies (e.g. diorite), consistent with a deeper intrusive level in a layered magmatic system. Lithologies vary from diorite through granite to alkali feldspar granite (Coats and Wilson, 1971; Leake, 1978; Max *et al.*, 1978; Feely and Madden, 1987; Feely *et al.*, 1991; El Desouky *et al.*, 1996; Crowley and Feely, 1997). The following lithologies are described in order of intrusion and are represented on Figure 8.10 (adapted after Callaghan, 2005).

Three intrusive stages are recognised in the Western and Eastern Blocks:

1 The Errisbeg Townland Granite, characterised by K-feldspar megacrysts, is the most common granite facies within the batholith. It outcrops in both the Western and Eastern blocks. Minor layering is present and the most common lithology is a pink to pale grey adamellite.
2 The Carna Granite intrudes the Errisbeg Townland Granite in the Western Block of the batholith, and is a pink or grey granodiorite with varying proportions of K-feldspar. Layering on scales of 10 cm–300 m has been described by Leake (1974).
3 The Murvey Granite is represented by a suite of late-stage, silica-rich leucogranites, which locally contains garnet. It forms the margin of much of the Western and Eastern blocks (Leake, 1978; Max *et al.*, 1978b).

In the Central Block, several intrusive stages are recognised:

1 The Megacrystic Granite, again characterised by K-feldspar megacrysts, is only present in the Central

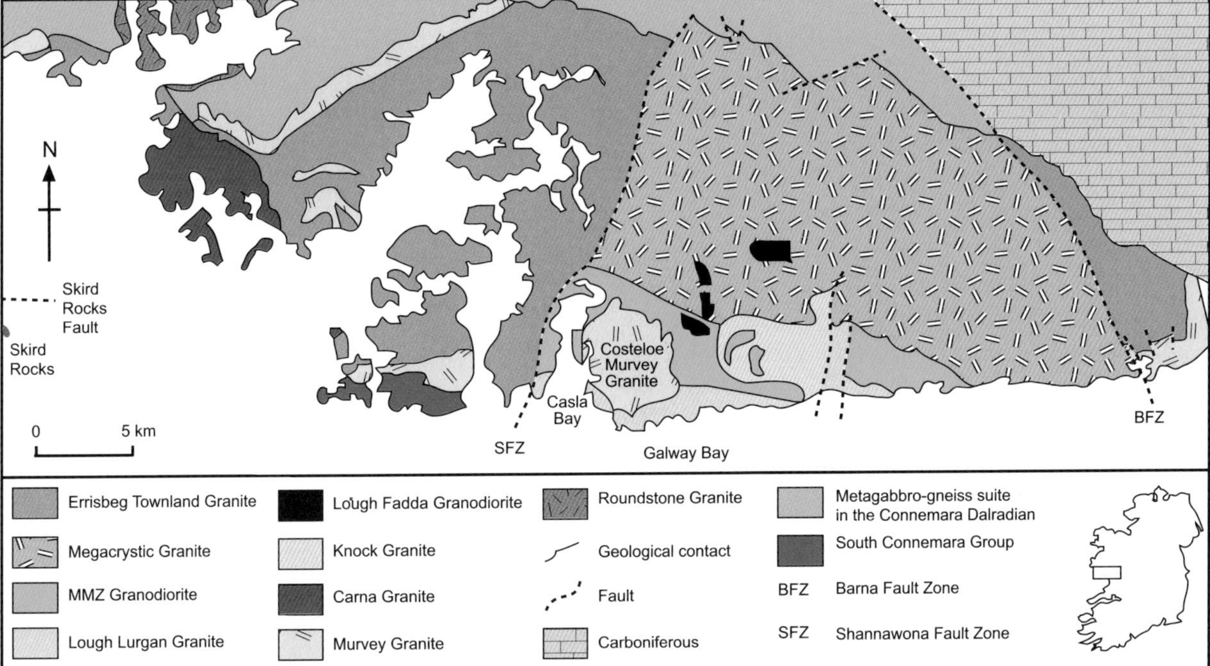

Fig. 8.10 The geology of the Galway Granite batholith, *after* Callaghan (2005).

Block, where it is the dominant lithology. A variety of names have been used for this lithology in different study areas; the term Megacrystic Granite was introduced by El Desouky *et al.* (1996), and is used here. The Megacrystic Granite is poorly exposed, and is distinguished from the Errisbeg Townland Granite by its higher mafic mineral content and generally coarser grain size. It exhibits a steep flow foliation. In the Barna area in the eastern portion of the Central Block, the Megacrystic Granite exhibits evidence for magma mixing on both outcrop and crystal scales (Baxter and Feely, 2002).

2 The Mixing-Mingling Zone Granodiorite (MMZ; El Desouky *et al.*, 1996) incorporates the Banded Zone of Max *et al.* (1978b) and the Composite Zone of Feely (1982). It occurs only in the Central Block where it defines a 4–6 km wide, arcuate, east–west trending mixing zone of foliated host granodiorite and mafic quartz diorite. The diorite is present as distributed groups of mingled mafic blobs which range from ellipsoidal enclaves to severely drawn-out and compressed thin sheets (El Desouky *et al.*, 1996).

3 The Lough Lurgan Granite crosscuts the Mixing–Mingling Zone Granodiorite (Crowley and Feely, 1997).

4a The Knock Granite, a late-stage intrusion restricted to the Central Block, crosscuts the Megacrystic Granite, Mixing–Mingling Zone Granodiorite and Lough Lurgan Granite (El Desouky *et al.*, 1996; Crowley and Feely, 1997).

4b The Lough Fadda Granodiorite intrudes the Megacrystic Granite and the MMZ Granodiorite in the Central Block (Feely, 1982; Crowley and Feely, 1997).

4c The Costelloe-Murvey Granite is unfoliated and crosscuts the Mixing–Mingling Zone Granodiorite and Lough Lurgan Granite east of Casla Bay (Feely, 1982).

In the Western Block, the Murvey Granite is interpreted as a residual liquid, possibly drawn out from the almost-crystallised Errisbeg Townland Granite magma due to block stoping of the batholith margin (Leake, 1974). Re–Os ages from late-stage molybdenite from the Carna and Murvey granites yield ages of 407.3 ± 1.0 Ma and 410.7 ± 1.0 Ma respectively (weighted means of replicate analyses from Selby *et al.*, 2004). Gallagher *et al.* (1992) and O'Reilly *et al.* (1997) demonstrate that the disseminated and vein-hosted Mo-mineralisation hosted by both granites is magmatic in origin, and thus the Re–Os ages reflect granite crystallisation.

In the Central Block, east of the Shannawona Fault, emplacement of the Megacrystic Granite is constrained to *c*.400–395 Ma by single grain U–Pb zircon dating (Megacrystic Granite G1: 394.4 ± 2.2 Ma, Megacrystic

Granite G2: ~402 Ma, Diorite Enclave G3: 397.7 ± 1.1 Ma, Megacrystic Granodiorite G4: 399.5 ± 0.8 Ma; Feely *et al.*, 2003, Table 8.1). The Costelloe–Murvey Granite yields a single grain U–Pb zircon age of 380.1 ± 5.5 Ma. The Central Block provides evidence for a zone of magma mingling and mixing that was active during the emplacement of the batholith. El Desouky *et al.* (1996) infer that an influx of mantle-derived magma triggered melting of lower crustal amphibolite and granulite. Ensuing thorough mixing between the primary and anatectic melts produced a low-viscosity, water-rich dioritic magma which ascended and then led to a second stage of mingling and mixing in the upper part of the newly-established batholith, where granite magma produced from anatexis at more than one crustal level was already in place. Leucogranites (e.g. the Lough Lurgan and Knock granites) were then emplaced into the Mixing–Mingling Zone towards the end of plutonic activity (El Desouky *et al.*, 1996). Following the uplift and uproofing, alkali leucogranite magma rose to a high structural level in the Costelloe sector of the Mixing–Mingling Zone to form the cross-cutting Costelloe–Murvey Granite (Feely and Madden, 1987).

Baxter *et al.* (2005) document a Marginal Deformation Zone in the granite along both the northern contact of the batholith and parallel to internal granite facies boundaries. The intrusion-related fabrics encountered within the Marginal Deformation Zones are regarded as having been produced by co-axial deformation, probably as a result of lateral expansion of successive magma batches in an overall transtensional regime (Baxter *et al.*, 2005). Leake (2006) considers the main phase of emplacement of the central and northern part of the Galway Granite was produced by progressive northward marginal dyke injection and stoping of the Connemara metagabbro-gneiss country rock. The space was provided by synchronous ESE opening of extensional fractures generated by a releasing bend in the sinistrally moving Skird Rocks Fault.

Minor Late Caledonian intrusions in western Ireland

The Galway Granite contains a variety of late-stage silicic sheet intrusions, typically arranged in sets that trend N–NNE perpendicular to the long axis of the batholith. They comprise a series of microphyric dacite dykes that are only found in the western and eastern blocks of the batholith and are absent from the uplifted central block (Mohr, 2003). Mohr suggests that the dacite dykes were produced by mixing of a granitoid magma with a more mafic magma, and that intrusion was facilitated by an east–west stress relaxation of the batholith, coeval with the final plutonic episode which involved the high-level intrusion of alkali leucogranite sills such as the Murvey Granite.

Two small 'permissive' Late Caledonian plutons (stocks or small cauldron intrusions with sharp discordant contacts and static thermal aureoles) have been documented in County Mayo. They are the Blacksod (or Termon Hill) Granite on the Mullet peninsula (381 ± 10 Ma Rb/Sr whole rock, Leutwein, 1970; Table 8.1) and the Corvock Granite in South Mayo (387 ± 12 Ma Rb/Sr whole rock., O'Connor, 1989; Table 8.1) which intrudes Ordovician metasediments of the South Mayo Trough. The Corvock Granite was intruded as a thin sheet roughly 25 km^2 in area but less than 1 km thick (Leake, 1978). It is vertically differentiated, passing from biotite granodiorite upwards into a leucogranite. It has a distinctive biotite hornfels aureole. The granitic sheet was probably intruded into two sets of cracks, which were previously exploited by pre-granite feldspar porphyry dykes. These early dykes are affected by the regional S2 cleavage, and final granite emplacement was post-tectonic (i.e. post local D3), as shown by the static aureole. Associated radial and tangential dyke swarms range in composition from quartz and feldspar porphyries to appinites and lamprophyres (Graham *et al.*, 1989).

A number of minor intrusions (all undated) cut the Lower Palaeozoic successions in South Mayo (Graham *et al.*, 1989). Numerous heavily altered hornblende lamprophyres intrude the Silurian rocks in a belt extending from south of Killary Harbour to Lough Mask. Sills and dykes of microdiorite or microgranodiorite also occur south of Killary Harbour. Later andesite sills and dykes cut both sets of earlier intrusions. Graham *et al.* (1989) attribute the abundance of intrusions in this region to movements on the (buried) terrane-bounding fault zone that juxtaposes the Connemara Dalradian with Ordovician rocks of the South Mayo Trough.

Tuffisite dykes and plugs intrude the Louisburgh–Clare Island Silurian succession on Clare Island (Phillips, 1973). They contain euhedral biotites and olivines (usually serpentinised), and are heavily altered to chlorite, carbonate, quartz and limonite. Xenoliths of quartzite and biotite schist occur, as do talc and chromite. The lack of a thermal aureole and the brecciation of the country rock at the dyke margins suggest fluidisation under high gas pressures at low temperature (Phillips *et al.*, 1970). There is a strong spatial association with major faults such as the Kill and Leck faults. The presence of olivine and chromite suggest derivation from the mantle.

The Newry Igneous Complex

The Newry Igneous Complex intrudes folded greywackes and slates of the Silurian Longford-Down succession. The complex is 40 km long and consists of three overlapping granodiorite plutons aligned NE–SW. A narrow static biotite-cordierite hornfels aureole is developed. A small ultramafic–intermediate intrusion is found at the north-eastern end of the complex (Fig. 8.11). It is cut by two generations of Late Caledonian lamprophyres and at its south-western end by the Tertiary intrusive complex of Slieve Gullion. Meighan and Neeson (1979) report a minimum ^{238}U–^{206}Pb age of 387 Ma with no quoted error for the north-east granodiorite pluton, while a marginal facies of this pluton yields a Rb/Sr whole rock–biotite age of 399 ± 3 Ma. The ultramafic–intermediate complex has yielded a Rb/Sr whole rock–biotite age of 403 Ma. P.J. O'Connor (1975) quotes a Rb/Sr age of 391 ± 21 Ma for the granodiorites in the south-west.

The petrogenesis of this complex provided much discussion in the middle of the 20th century. Reynolds (1944) proposed that the entire complex (including all the ultramafic–intermediate granodioritic rocks) was derived by large-scale *in-situ* metasomatism ('granitisation') of the Silurian sedimentary rocks, while other workers argued for a magmatic origin (e.g. Grant, 1961). Meighan and Neeson (1979) interpret the ultramafic rocks (predominantly biotite pyroxenite and associated diorites) in the ultramafic–intermediate intrusion at the north-eastern part of complex as cumulates from an intermediate magma. The diorites grade into more evolved, leucocratic diorite and monzonite, and the whole ultramafic–intermediate complex is thought by Meighan and Neeson (1979) to be part of a layered intrusion. These authors suggest that the three main granodiorite plutons may be derived by crystal fractionation at depth from an intermediate magma based on geochemical evidence (major and trace element geochemistry and $^{87}Sr/^{86}Sr_i$ ratios). These parental intermediate magmas are interpreted as having similar compositions to the parental intermediate magmas of the ultramafic complex, and

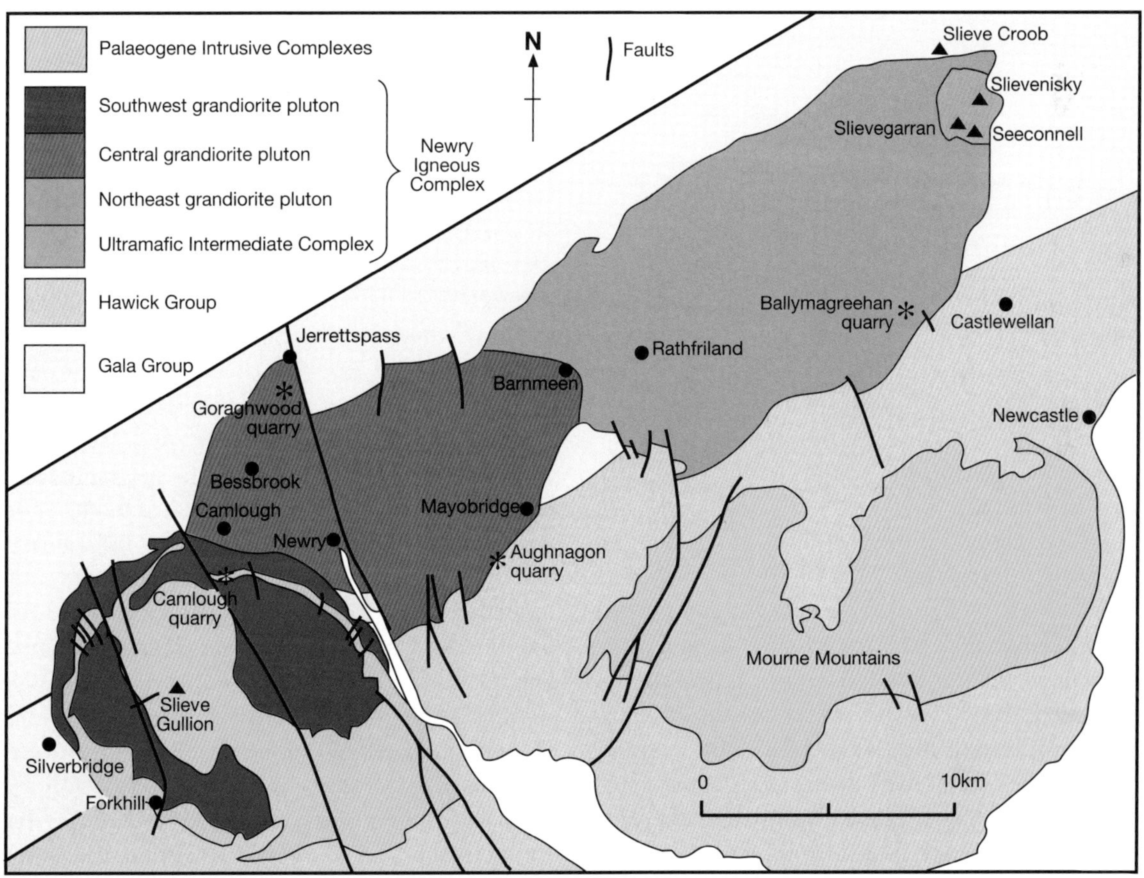

Fig. 8.11 The geology of the Newry Igneous Complex, *after* Cooper and Johnston (2004b).

as having an upper mantle origin. The north-eastern and south-western granodiorite plutons have silica-rich central portions with more mafic margins, while the central pluton exhibits the opposite trend. The marginal zones of each pluton are strongly foliated, while the central portions of each pluton are not. The ultramafic–intermediate complex is veined by granodiorite of the north-east pluton.

Minor Late Caledonian intrusions in eastern Ireland

The Longford-Down terrane is intruded by numerous lamprophyre dykes, which are generally believed to be petrogenetically associated with the Late Caledonian granite suite. Two generations of dykes have been identified on the Ards Peninsula (Reynolds, 1931). The earlier lamprophyres post-date local F1 folds but are folded by F2. The younger dykes post-date folding, but are affected by Acadian strike-slip faulting. Both generations of lamprophyres are intruded parallel to bedding, and hence follow the NE–SW structural grain of the Longford-Down terrane. Anderson and Oliver (1996) quote a K–Ar hornblende age of 415 ± 12 Ma for a xenolith-bearing lamprophyre from the structurally younger suite. Vaughan (1996) suggests that the lamprophyre dyke suite south-west of Clogher Head was coeval with sinistral transtension, and that the magmatism was a result of transient decompression of metasomatised mantle during simple shear. This decompression event triggered the release of a pulse of volatile-rich potassic magma, which ascended up deep-seated strike-slip faults (Vaughan, 1996).

The small (4 km$^{2)}$ Crossdoney Granite has similarities with the Newry Igneous Complex. It consists of melanocratic granodiorite, monzonite and quartz diorite in the northern part of the intrusion, and hornblende-biotite granodiorite with adamellite to the south. It cuts Silurian greywackes with a sometimes sharp, sometimes diffuse contact (Skiba, 1952). Similar diffuse gradational contacts are also seen in the Newry Igneous Complex.

The Leinster Granite Batholith

The Leinster Granite is a composite intrusion more than 1500 km^2 in area, and is the largest body of granite in Ireland and Britain. It is a two-mica peraluminous granite which exhibits a wide range of mineralogical and textural varieties. The granite forms the core of the rounded Dublin and Wicklow mountains, yet in general it is quite poorly exposed due to a thin veneer of upland peat cover. Much of the following account is adapted from McConnell and Philcox (1994).

The five plutons which comprise the Leinster Granite Batholith were intruded apparently simultaneously (Brindley, 1973), in contrast to the significantly smaller Donegal Batholith, which was intruded over a period of approximately 25 Ma (Condon *et al.*, 2004). The five plutons that comprise the Leinster Granite are aligned en echelon, parallel to the NE–SW trending regional foliation in the Lower Palaeozoic country rocks. Septa of country rock are recognised between the plutons, some of which pass into xenolith-rich zones within the granite. Arcuate roof pendants are also seen. These are particularly common in the Lugnaquilla pluton (Brück and Reeves, 1983) indicating that the pluton has been unroofed to a shallow depth. Gravity profiles across the granite demonstrate that it has a thin, sheeted form and that it thins progressively northwards (O'Brien, 1999). A subsequent onshore seismic refraction experiment (LEinster Granite Seismics or LEGS) was carried out across the Leinster Granite in 1999. Two NW–SE profiles and one axial NE–SW profile were shot, and the interpreted data demonstrate that the granite has a maximum thickness in the south of approximately 5 km, and thins northwards towards Dublin Bay where the thickness varies between 1 and 3 km (Hodgson, 2001).

The granite was intruded during the Acadian Orogeny, broadly contemporaneously with the development of D2 shear zones in the country rock, such as the East Carlow Deformation Zone (McArdle and Kennedy, 1985). The Northern pluton has yielded a U–Pb monazite age of 405 ± 2 Ma (O'Connor *et al.*, 1989) and a Rb/Sr whole rock age of 404 ± 24 Ma (O'Connor and Brück, 1978). The granite is cut by a number of major NW–SE trending fractures whose subsequent erosion has formed a series of deeply incised valleys such as Glencullen, Glencree, Glendasan, and Glenmalure. Some fractures, such as the one exploited by the valley of Glencullen, have produced late offsets of both internal and external granite contacts. Other fractures, notably that exploited by the valley of Glenmalure, appear to have controlled intrusion of the granite, with different intrusive bodies either terminating at the fractures or inter-digitating across them. These fractures are also sites of lead–zinc mineralisation at the granite margin.

The country rock consists primarily of Lower Ordovician shales and siltstones of the Ribband Group. Outside of the granite aureole, pelitic rocks of the Ribband Group are sub-greenschist facies. As the granite contact is approached, pelitic rocks become increasingly coarse-grained, with the development of muscovite, biotite and garnet. The inner aureole is characterised by the appearance of andalusite and staurolite (Brück, 1968). The margin of the granite is commonly foliated, and this foliation is parallel to a strong foliation in the country rock. This strong foliation is related to the development of D2 shear zones, which decrease in intensity away from the granite contact. The aureole rocks are also somewhat unusual in that they contain a very distinctive lithology known as *coticule*. Coticule typically consists of thin quartzite beds that contain numerous, tiny manganese garnet crystals (spessartine), and commonly exhibit complex ptygmatic folding patterns. The close association between coticule, tourmalinite and lead and zinc mineralisation in the aureole led Williams and Kennan (1983) to propose that the Lower Ordovician country rocks represent metalliferous sediments deposited on an ocean floor around a hydrothermal vent. Lithium-bearing pegmatites are found on the south-east margin of the granite and include up to 60 wt% spodumene with Li_2O contents over 4 wt% (Luecke, 1981).

From north to south the five plutons are as follows: The Northern (1), Upper Liffey Valley (2), Lugnaquilla (3), Tullow (4), and Blackstairs (5) plutons (Fig. 8.12). The Northern pluton (1), and its satellite pluton, the Upper Liffey Valley pluton (2) (Fig. 8.12) are rounded bodies with broadly concentric internal zonation of granite types (Brück and O'Connor, 1977). Banding and a tectonic foliation are parallel to the contacts, which are subvertical on the western side, and which dip moderately to steeply outwards on the eastern and southern margins. Static thermal metamorphism produced andalusite and staurolite porphyroblasts that overgrow the S1 cleavage in the country rocks. The northern end of the granite is in faulted contact with Carboniferous rocks.

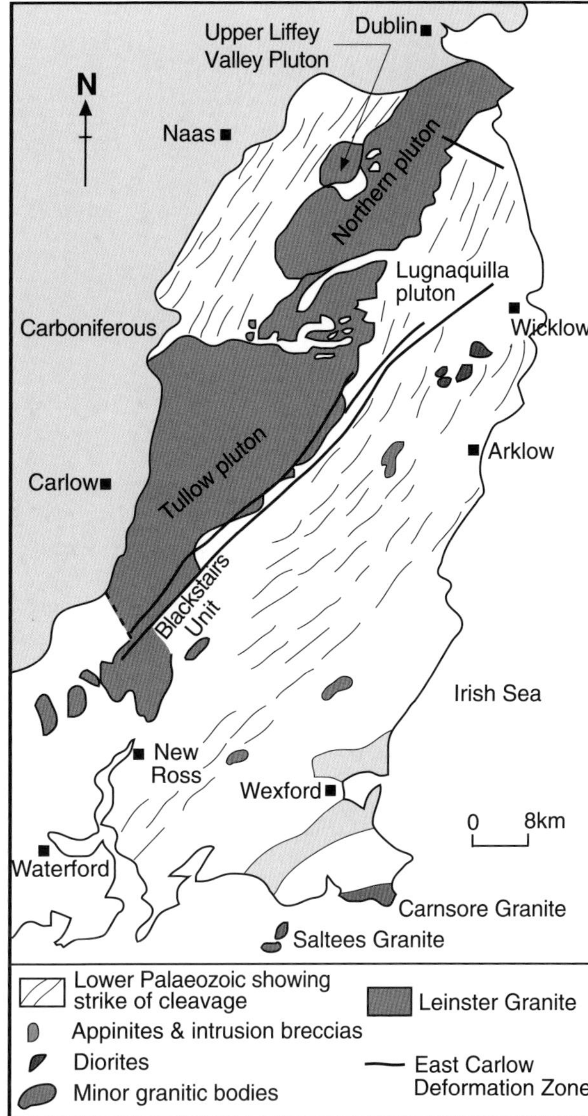

Fig. 8.12 The Leinster Batholith (*after* Brück, 1978 and McConnell and Philcox, 2004).

Six varieties of granite are defined within the Northern and Upper Liffey Valley plutons (Brück and O'Connor, 1977). Type 1 is an aphyric granodiorite which occurs as small bodies both around the margins of the pluton and also internally. It appears to represent an earlier intrusion that was broken up by later intrusive pulses. These small bodies have sharp contacts with the other granite facies. Types 2, 3 and 4 are adamellites that occur in three broadly concentric zones (Fig. 8.13). Type 2 is composed of microcline, plagioclase, muscovite, biotite, and quartz. There are two subdivisions, Type 2e (equigranular) and Type 2p (porphyritic), which has euhedral microcline phenocrysts up to 30 mm long. Types 3 and 4 are similar to Type 2e and Type 2p respectively, but coarse (up to 70 mm) muscovite phenocrysts are present. The coarse muscovite phenocrysts were originally thought to be late-stage hydrothermal products, but detailed petrographic study has demonstrated a magmatic origin (Roycroft, 1991). The contacts between types 2, 3, and 4 are observed in the field, but are thought to be gradational, as intermediate varieties can be found. Types 1–4 form a geochemical continuum, and are interpreted as an evolutionary sequence derived by fractional crystallisation from a single parental magma by Brück and O'Connor (1977). Combined with the concentric zonation pattern, this suggests several intrusive pulses from a magma body evolving at a slightly deeper level. The final intrusive phase was an aplite, which is commonly seen at the margins of the pluton.

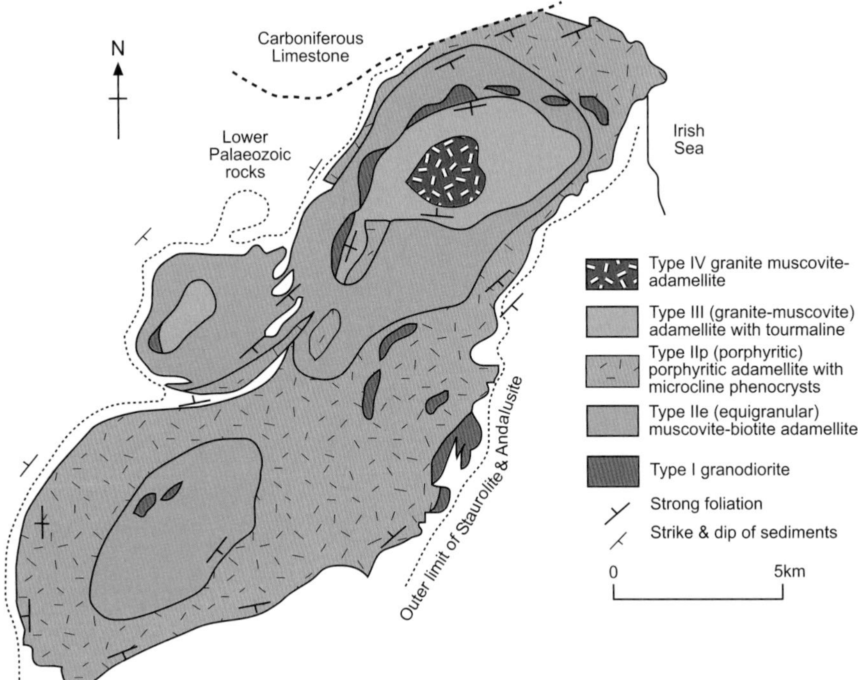

Fig. 8.13 The Northern and Upper Liffey Valley Units of the Leinster Batholith (*after* Brück *et al.*, 1978).

The Lugnaquilla pluton (3) forms the highest topography in Leinster due to the presence of numerous schist bodies within the granite that are substantially more resistant to erosion than their granitic host rock. One such body, a coticule-bearing schist capping the granite on Lugnaquilla Mountain with a shallowly-dipping contact, may represent the roof of the pluton. Five granite varieties have been distinguished (Brück and Reeves, 1983): quartz diorite, granodiorite, adamellite, aplogranite, and aplite. These varieties are not concentrically zoned as in the Northern Pluton, but form separate bodies with irregular, complex boundaries (Fig. 8.13). The quartz diorite, which forms an elongate north-east trending body on the western side of the pluton, is the most basic, with abundant biotite (up to 20% by volume). The granodiorite that forms the central part of the pluton is more leucocratic but still contains conspicuous biotite (*c.*10%). The adamellite forming the north-eastern part is a coarse two-mica adamellite with microcline phenocrysts. The aplogranite forms an irregular body between the granodiorite and the adamellite, and inter-digitates with the country rock schists. It grades southward into the aplite, which contains abundant sub-millimetric almandine-spessartine garnets, and is cut by numerous pegmatite dykes.

The Tullow pluton (4) is in contact with the aplite that defines the southern boundary of the Lugnaquilla pluton (Fig. 8.13). It is poorly exposed, and consists of an adamellite with both equigranular and microcline–porphyritic varieties (as in the Northern pluton), but with a less distinct zonal distribution. It is separated from the southernmost pluton, the Blackstairs pluton, by a structurally complex zone of foliated granite and abundant septa of country rock that forms part of the East Carlow Deformation Zone (McArdle and Kennedy, 1985).

The Blackstairs pluton (5) contains granite varieties similar to those described elsewhere in the batholith. Both porphyritic and equigranular varieties of Types 1 and 2 are present using the terminology established by Brück and O'Connor (1977), Type 2e being the most abundant (Sweetman, 1988). Granite emplacement followed intrusion of appinites, and was controlled by the orientation of the pre-existing regional S1 cleavage, resulting in a sheeted pluton with elongate schist lenses (Sweetman, 1988).

Grogan and Reavy (2002) have described complex magmatic disequilibrium textures in both the Northern (1) and Lugnaquilla (3) plutons. Petrographic textures indicative of disequilibrium include boxy cellular plagioclase, oscillatory zoning, and calcic spike zones within individual plagioclase feldspar crystals. The quartz diorites and K-feldspar megacrystic granites and granodiorites demonstrate the best evidence for disequilibrium and magma mixing and mingling. The textures are interpreted as having formed in a dynamic magmatic environment that facilitated acid–acid mixing on a variety of scales and stages between source and final emplacement. Such mixing is not consistent with the fractional crystallisation petrogenetic scheme proposed by Brück and O'Connor (1977), although there are large areas of the Leinster Granite batholith, mainly in the Northern (1) and Lugnaquilla (3) plutons, where disequilibrium textures are absent and which exhibit broad fractionation trends compatible with fractional crystallisation (Grogan and Reavy, 2002). The source magma of the S-type Leinster granite appears to have been an anatectic melt of upper crustal metasediments. Initial Sr and Nd isotopic ratios are highly variable, but are related to geographic position about the axis of the batholith. This variability is probably a result of a heterogeneous metasedimentary source (Mohr, 1990).

Carnsore and Saltees Granites

At the extreme south-east of Ireland, the Carnsore and Saltees granite plutons intrude the basement rocks of the Rosslare Complex (Chapter 2). The Carnsore granite is a pink, homogeneous porphyritic biotite adamellite with a cordierite–andalusite hornfels aureole, though the actual contact is unexposed (Baker, 1968). Pressure–temperature estimates for mineral assemblages in the aureole rocks yield an emplacement depth of 2–3 kbar (Treloar and Max, 1984). The Carnsore Granite has yielded a Rb–Sr whole rock isochron of 428 ± 11 Ma (O'Connor *et al.*, 1988). The same authors also quote a U–Pb zircon upper intercept age of $409^{+31}/_{-11}$ Ma and a mean $^{207}Pb/^{206}Pb$ age of 432 ± 3 Ma.

The Saltees Granite crops out approximately 10 km to the south-west of Carnsore on the Saltee Islands and on the mainland at Kilmore Quay. The Saltees Granite is a fine–medium grained, quartz-albite-microcline rock which is strongly foliated, particularly on the northern margin at Kilmore Quay (Max *et al.*, 1979). The foliation typically strikes NE–SW and dips vertically. The same authors quote a Rb–Sr whole rock isochron of 436 ± 7 Ma.

Drogheda-Kentstown and Rockabill

In north Leinster, there is evidence for the presence of a major granite pluton at depth, buried beneath Carboniferous limestones. A large negative Bouguer anomaly centred on Kentstown, County Meath, some 10 km east-south-east of Navan and immediately south of the Navan–Silvermines Fault, was identified by T. Murphy in 1952 and interpreted as a buried granite body. Subsequently, two boreholes drilled by Tara Exploration in the centre of this gravity anomaly intercepted granite at depths of 492 m and 662 m beneath a cover of Dinantian limestones. More recently, on the north-eastern margin of the gravity low, shallow (~ 10 m) excavation works east of Drogheda exposed granite subcrop, which was subsequently drilled by the Geological Survey of Ireland in 1998. The main facies of the Drogheda Granite is a hornblende-biotite-quartz monzonite (McConnell and Kennan, 2002). There is no conspicuous mineral alignment, and diorite enclaves are common. The Drogheda Granite has yielded a Rb–Sr whole rock isochron of 379 ± 18 Ma (McConnell and Kennan, 2002). In the Lower Palaeozoic Balbriggan inlier to the east of the gravity low, there is evidence for the presence of granite at depth. Ordovician rocks are hornfelsed and are injected by a suite of minor intrusions, including sills and dykes of porphyritic microgranite, granophyre, and lamprophyre (Murphy, 1987b). The nearest surface exposure to the Drogheda–Kentstown granite is on the small island of Rockabill, east of Skerries (Brindley and Kennan, 1972).

Minor Late Caledonian intrusions in south-eastern Ireland

Appinites and lamprophyres are widespread in the Leinster Batholith. A number of appinite bodies occur west of the Lugnaquilla pluton in the Donard–Baltinglass area, where they predate the aplite phase of granite intrusion. Lamprophyres and appinites are also well developed in the Mount Leinster swarm, which is concentrated in the schist septum that separates the Blackstairs Unit from the Tullow pluton (McArdle, 1974; McArdle and O'Connor, 1987). This septum also coincides with the NE–SW trending East Carlow Deformation Zone (McArdle and Kennedy, 1985) referred to earlier. McArdle and O'Connor (1987) argue on structural grounds that the appinite and lamprophyre bodies were emplaced in the interval between the regional D1 and D2 events. The significance of the appinite–lamprophyre suite, their genetic relationship with deep crustal structures, and their role in the Late Caledonian magmatic episode is discussed later.

Aside from the appinite–lamprophyre suite, there is a diverse suite of minor intrusions in south-eastern Ireland, which were emplaced sometime in the time interval between the Ordovician volcanism (Chapter 6) and the Lower Devonian granite plutonism. These include a suite of alkaline felsitic intrusions, such as the group of small syenite bodies in County Waterford (Stillman *et al.*, 1974), which may be possible correlatives of the peralkaline silicic volcanics at Avoca.

A distinctive swarm of extensively altered dolerite dykes intrudes the Lower Ordovician Aghfarrell and Butter Mountain Formation rocks in the Tallaght Hills, north-west of the Leinster granite. The dykes trend NE–SW, and are generally up to 5 m thick. Locally intrusion of dyke upon dyke was extremely intense, producing a sheeted dyke complex. Dykes exhibit both one-sided and symmetrical chilled margins, with only thin screens of country rock preserved between them (McConnell and Philcox, 1994). The dyke swarm pre-dates the granite, as near the granite margin they are thermally metamorphosed to hornblende schist. Intrusion was probably before or during the regional D1 deformation, and resulted in local thermal metamorphism ('spotting') in the phyllitic country rocks (McConnell and Philcox, 1994). The primary mineralogy consists mainly of plagioclase and augite, and chemically they are classified as tholeiitic to transitional dolerites. The uniformity of the dykes and the common presence of gabbro xenoliths suggest derivation from a high-level basaltic magma chamber.

A series of small, intermediate to basic bodies that have experienced at least some of the regional deformation intrude the Duncannon Group (Chapter 6). The largest of these bodies is the Carrigmore Diorite, which has yielded a 409 ± 17 Ma Rb–Sr isochron age (O'Connor and Reeves, 1980). It exhibits a zonation from diorite at the centre to granodiorite at the margin. This lithological variation has been attributed to fractional crystallisation, but abundant partially resorbed xenoliths demonstrate that substantial crustal assimilation could have been partly responsible. Two other bodies, the Westaston Hill and Rockstown diorites, are petrographically similar (McConnell and Philcox, 1994), while a nepheline-normative pyroxene diorite crops out at Arklow Head (Stillman and Maytham, 1973). McConnell (1987) believes that all these diorite bodies were emplaced in a single magmatic episode, and represent a basic magma derived from enriched sub-continental lithosphere at an active continental (i.e. Andean-type) margin. Chemical and petrographic variations were attributed to variable amounts of assimilation by different magma batches during ascent and crystallisation. The Duncannon Group in the region is also cut by a suite of dolerite intrusions, many of which are concentrated in an intrusion complex on Castletimon Hill (Reeves, 1977). Intrusive relationships suggest that they are syn- to slightly post-diorite in age. They appear to be related to the diorite suite as they overlap both spatially and temporally, and exhibit similar geochemical characteristics (McConnell, 1987).

Petrogenesis of the Late Caledonian Granites

Large volumes of granitic magma were intruded in Ireland and Britain during Late Caledonian times, and the petrogenesis of these granites has preoccupied geologists for many years. Detailed field and geochemical studies and high-resolution geochronology have contributed to a better understanding of their origin, but a single mechanism remains difficult to establish. The origin of this widespread magmatism has been ascribed (e.g. Brown *et al.* 1985) to a combination of circumstances: (a) continued subduction, (b) melting by increased temperatures caused by tectonic thickening of the crust during Late Caledonian times and/or (c) post-collisional exhumation resulting in adiabatic (decompression) melting.

However, there are problems with these models. Most of the Late Caledonian granites in Ireland and Britain are metaluminous, often hornblende-bearing, granodioritic plutons of essentially calc-alkaline 'I' type character (Atherton and Ghani, 2002). Their chemistry is clearly compatible with a subduction-related origin (i.e. an Andean-type arc), but the chief difficulty is the timing of Late Caledonian magmatism relative to the closure of the Iapetus Ocean. Figure 8.6 illustrates that the Late Caledonian magmatism spans the time period from 430 to 380 Ma, while palaeomagnetic and faunal evidence implies that final closure of the Iapetus Ocean in Ireland and Britain had occurred by the Wenlock (~ 425 Ma) (MacNiocaill, 2000). Thickening of the crust causing high temperature melting at depth is another potential mechanism, but the amount of tectonic thickening that accompanied the Late Caledonian magmatism is thought to be small. The collision of Laurentia with Avalonia was strongly oblique (Soper *et al.*, 1992; Dewey and Strachan, 2003), with abundant evidence for sinistral transpression and no substantial crustal thickening.

Decompression melting as a result of rapid post-orogenic exhumation (due to extensional collapse and/or rapid erosion) has also been suggested. This model can potentially account for a portion of the voluminous Late Caledonian granite suite in Scotland, where Lower Devonian conglomerates adjacent to the Highland Boundary Fault contain coarse granite clasts derived from the Grampian terrane. One of these clasts has yielded a

Rb–Sr biotite age (412 ± 4 Ma) within error of the presumed stratigraphic age, implying rapid exhumation (Haughton *et al.*, 1990). However, it seems unlikely that rapid exhumation resulting in decompression melting can account for the genesis of the entire Late Caledonian granite suite, particularly south of the Fair Head–Clew Bay Line where Silurian rocks have commonly experienced sub-greenschist facies metamorphism implying relatively small amounts of exhumation.

It is clear that many of the major Late Caledonian plutons in Ireland (e.g. Leinster, Donegal and the Ox Mountains) were emplaced in dilational cavities associated with major movements on sinistral (predominantly transpressional) shear zones and faults. A good example is the model for the emplacement of the Donegal batholith of Hutton (1982) which was discussed in detail earlier. Additionally, many of the Late Caledonian granites are sited on 'old' (i.e. pre-Caledonian) lineaments (e.g. Jacques and Reavy, 1994; Hutton and Alsop, 1996), and are associated in time and place with clusters of appinite and lamprophyre dykes. These volatile- and potassium-rich magmas in general immediately preceded the emplacement of their associated granite intrusions, but often they overlap in time with the early stages of pluton emplacement, and are also commonly coincident with major Late Caledonian shear zones (e.g. Donegal, Leinster).

Atherton and Ghani (2002) attribute the Late Caledonian granitic magmatism (north of the Iapetus Suture) to slab break-off. Following final closure of the Iapetus Ocean and the suturing of the Laurentian and Avalonian plates, they envisage that the subducting plate ruptured, possibly at the interface between oceanic and continental crust (Fig. 8.14). This 'cavity' in the base of the lithospheric mantle was then filled by an influx of hot asthenospheric mantle, producing substantial basaltic magmatism, which ascends and underplates the base of the crust with lamprophyric magma. Some of this magmatism also produced high-level lamprophyre dykes and appinite complexes in the crust, with magma ascent controlled by movement on deep-seated fracture systems (Fig. 8.14). High temperatures at the base of the lithosphere persisted, causing partial melting of the mafic lower crust and the lamprophyric underplate to form granitic magmas. Their ascent through the crust to their final intrusion level was probably in the form of feeder dykes controlled by the same major fracture systems that facilitated emplacement of the earlier appinite suite.

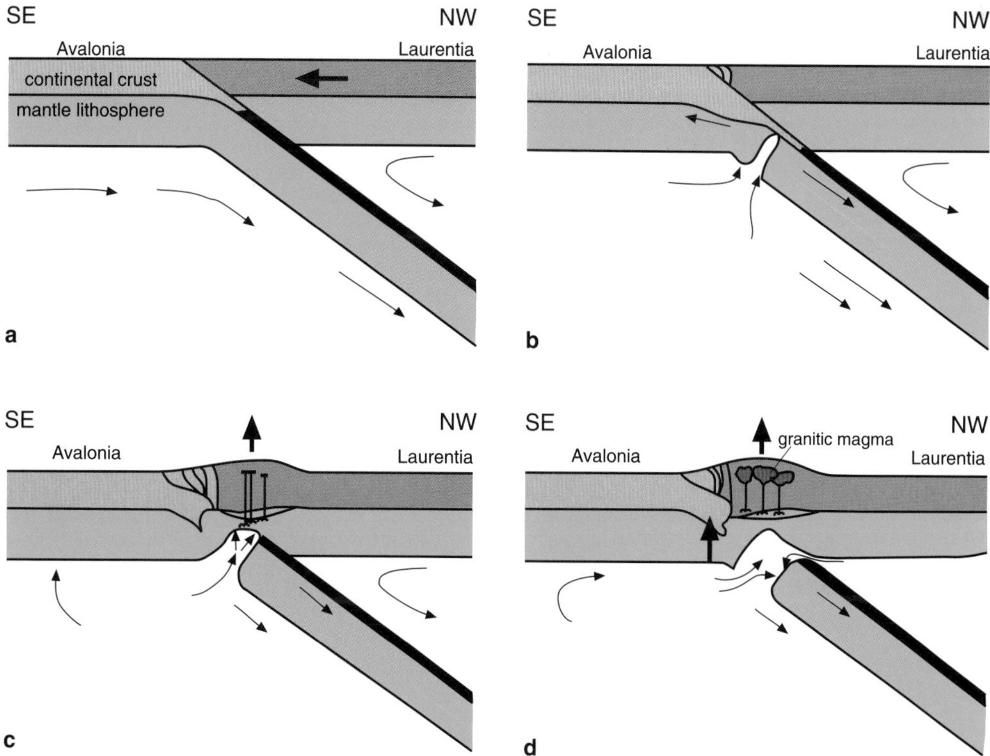

Fig. 8.14 Cartoons showing the sequence of events induced by slab break-off following the collision of Avalonia with Laurentia which led to plutonism north of the Iapetus Suture (modified after Atherton and Ghani, 2002).

9

Devonian

J. R. Graham

During Devonian times the palaeogeography of Ireland and adjacent regions was dominated by the presence of a large landmass (Old Red Continent of House (1968)) that lay mainly to the north of Ireland (Fig. 9.1). This landmass was produced by the final closure of the Iapetus Ocean in the Silurian. Whilst this collision resulted in extensive nappe/thrust style tectonics in Scandinavia, it was a relatively mild event in the British Isles sector, possibly due to earlier collision here when Avalonia (southern British Isles) was accreted onto Laurentia during the late Ordovician (Hutton and Murphy, 1987; Pickering *et al.*, 1988).

Reconstructions of the positions of continental blocks during the Devonian Period have been made using

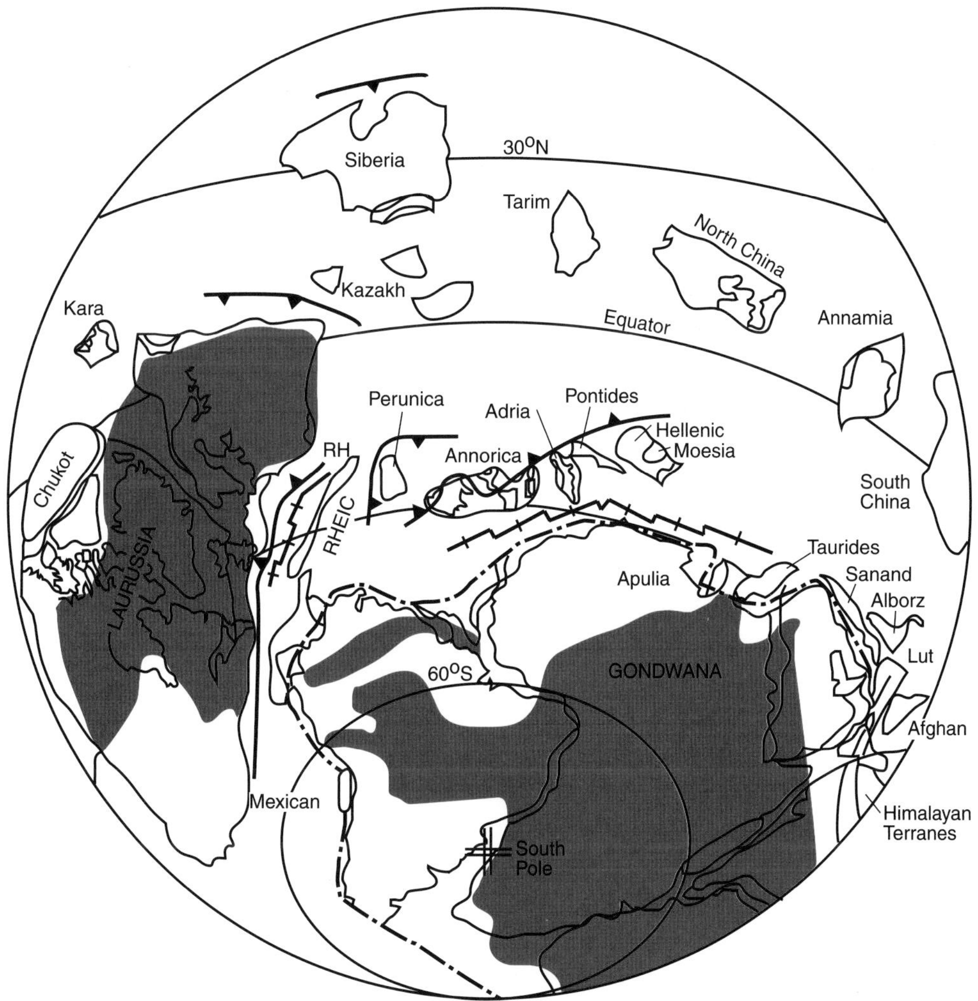

Fig. 9.1 Continental reconstruction at 400 Ma after Torsvik and Cocks (2004). The area labelled Laurussia is roughly equivalent to the Old Red Sandstone landmass of House (1968).

palaeoclimatic, palaeontological, and palaeomagnetic data (Heckel and Witzke, 1978; Tarling, 1985; Scotese *et al.*, 1985; Livermore *et al.*, 1985; McKerrow *et al.*, 2000; Morel and Irving, 1978; McKerrow and Scotese, 1990; Torsvik and Cocks, 2004; Van der Voo, 1988; Witzke, 1990; Witzke and Heckel, 1988) all supplemented by reasoned tectonic interpretations. The synthesis of palaeoclimatically sensitive sediments by Heckel and Witzke (1978) provides an internally consistent pattern in which Ireland was situated about 35°S during the mid-Devonian and moved gradually nearer the equator with time. The palaeomagnetic reconstructions are in agreement with this progressive shift northwards with time, but generally place Ireland some 10–20° nearer the equator at any given time. Figure 9.1 illustrates the likely position of Ireland in the early Devonian based primarily on palaeomagnetic data.

There is a lack of major evaporite accumulations and also of aeolian sediments in rocks of early Devonian age in Ireland. Limited developments of pedogenic carbonate are present in the late Devonian, as are some aeolian deposits, although the latter may be due in part to special topographic effects. Wherever the local water table remained sufficiently high to prevent pervasive oxidation of the sediments, plant fossils are common with no sign of specialised floras. Bioturbation is often intense in late Devonian mudrocks (Graham, 1983) indicating a large indigenous biota. Thus at least seasonal rainfall is indicated throughout the Devonian, a conclusion also reached by Woodrow *et al.* (1973) for Devonian strata at similar palaeolatitudes in eastern North America. These observations are consistent with Ireland's projected position on the eastern, windward side of the Old Red Sandstone landmass during the Devonian (Fig. 9.1).

Devonian rocks in Ireland are mostly of Old Red Sandstone facies. Figure 9.2 shows the distribution of Old Red Sandstone rocks, most of which are likely to be Devonian in age. Marine rocks are limited to a development in the south of the island in the latest Devonian. These marine rocks are related to a major northward marine transgression which affected much of northwest Europe. Biostratigraphical data are limited for much of the Irish Devonian. Dating is based mainly on palynology with subsidiary information derived from fish and macroplants. Correlation of these data with independently dated marine sequences is often imprecise. Other dates are based on isotopic decay systems and often cannot be linked to local biostratigraphical data. Interpretation of what these isotopic ages mean in terms of Devonian stages is by reference to Kaufmann (2006) and House and Gradstein (2004) (Figure 9.3) and these correlations may change as more data become available to refine the geological time scale. Although about nine reliable ages exist for the Devonian where U–Pb TIMS data can be precisely located biostratigraphically, interpolation essentially relies on assumptions concerning sedimentation rates or evolution rates.

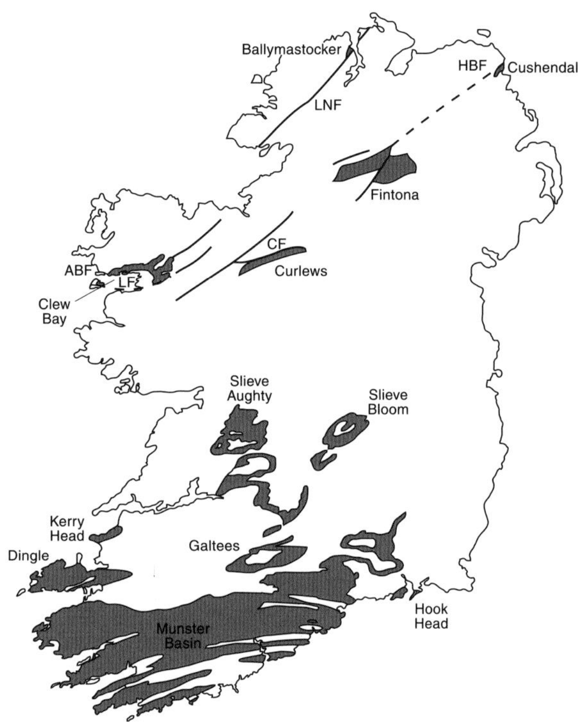

Fig. 9.2 Distribution of Old Red Sandstone rocks in Ireland LNF – Leannan Fault, HBF – Highland Boundary Fault, CF – Curlews Fault, LF – Leck Fault, ABF – Achill Beg Fault.

The Old Red Sandstone rocks display three different tectono-sedimentary styles operating over different time spans:

1 Small basins associated with (mainly strike slip) faults during late stage adjustments at the end of the Caledonian collision event. These basins often show contemporaneous volcanicity of intermediate to acid composition. Some of these basins are or may be Silurian in age, such as the probable Wenlock succession of Louisburgh-Clare Island (Chapter 7), and the Ludlow and younger rocks of the lower part of the Dingle Group.
2 A major extensional basin in the south of Ireland, the Munster Basin, which developed during the late Middle Devonian, and which developed marine facies in its southern part in the latest Devonian.

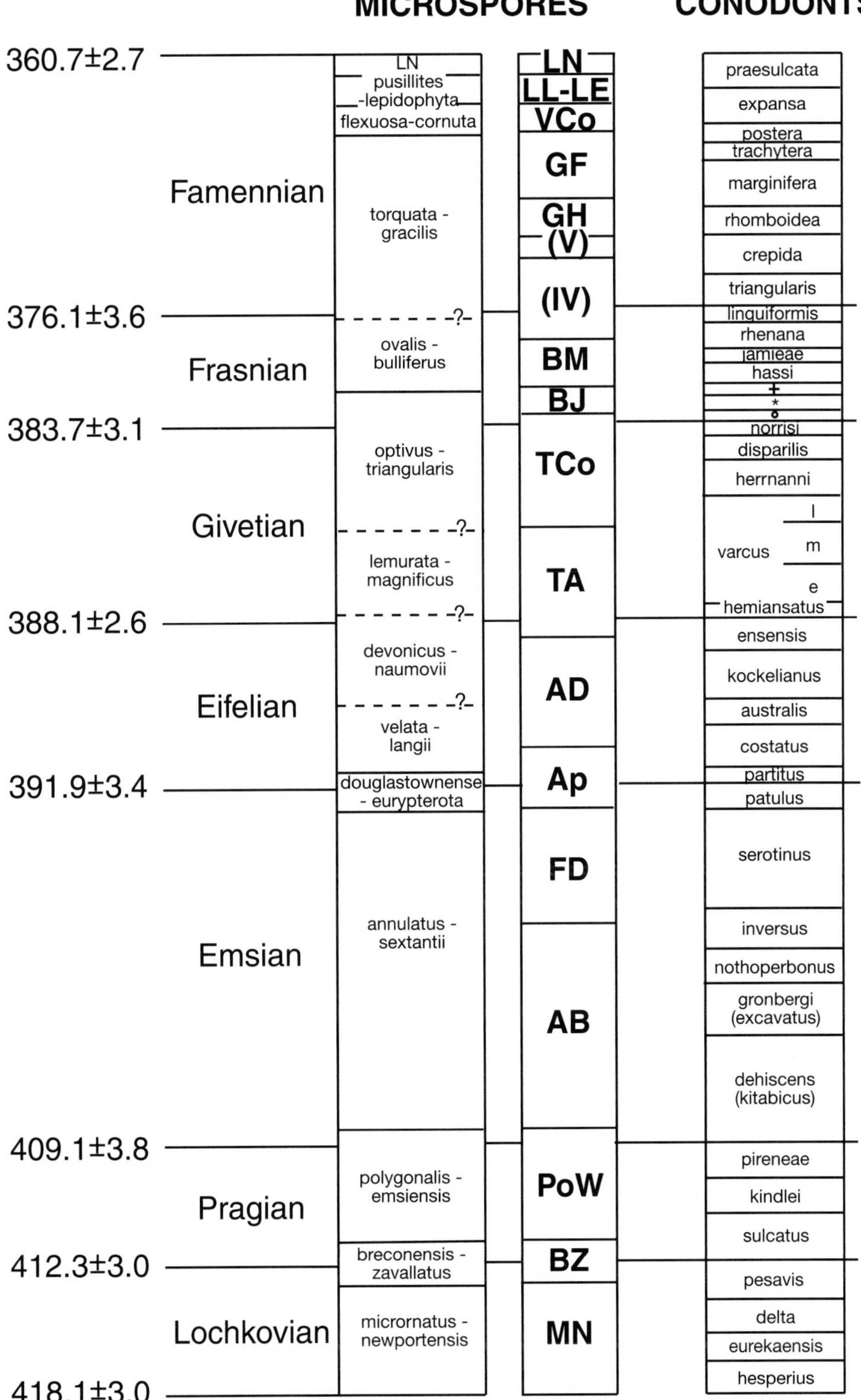

Fig. 9.3 Subdivisions of the Devonian System with conodont biozonation and palynological biozonation of Richardson and McGregor (1986) and Streel *et al.* (1987) (*after* Williams *et al.*, 2000). Best estimate dates of the stage boundaries are from Kaufmann (2006). Note that there are insufficient data to draw the zonal boundaries to scale. Unlabelled conodont biozones: °*falsiovalis (rotundiloba)*; * *transitans*; + *punctata*.

3 Widespread alluviation preceding the Famennian to Viséan transgression which produced generally thin but widespread terrestrial clastics in response to the rising base level. This style continued into the Early Carboniferous.

Late Caledonian basins

Cushendall

In the north-east of Ireland near the coastal town of Cushendall, County Antrim (Figs. 9.4, 9.5) a 1.38 km thick succession of Old Red Sandstone facies has been described in detail by Simon (1981, 1984a), who also summarises earlier work by Wilson (1953) and Sanderson (1970). The strata, which have been termed the Cross Slieve Group, display a simple threefold lithostratigraphy. An oldest Cushendun Formation (620 m) consists of conglomerates dominated by polycyclic quartzite clasts. A middle Ballyagan Formation (160 m) consists mainly of sandstones with large-scale cross strata. The youngest Cushendall Formation (600 m) consists of coarse conglomerates rich in igneous debris.

The basal part of the Cushendun Formation locally displays breccias 5–10 m thick consisting of angular clasts of schist and vein quartz identical to the adjacent Dalradian metamorphic rocks. Simon (*pers. comm.*, 1988) exposed faulted contacts in four stream sections. It appears that a complex series of minor faults is present along this boundary, although it was probably originally an unconformity. Overlying these basal breccias Simon (1981) described up to 320 m of clast-supported conglomerates with minor amounts of pebbly sandstones and mudrocks, the latter displaying desiccation cracks. These conglomerates form laterally extensive beds that are internally massive or horizontally stratified. The associated sandstones are variably parallel laminated or cross-stratified, and some contain imbricated clasts. Simon (1984a) shows variable palaeocurrents with a vector mean towards 172°. Mitchell (2004b) suggested that these conglomerates represent flows to the east-south-east. Sanderson (1970)

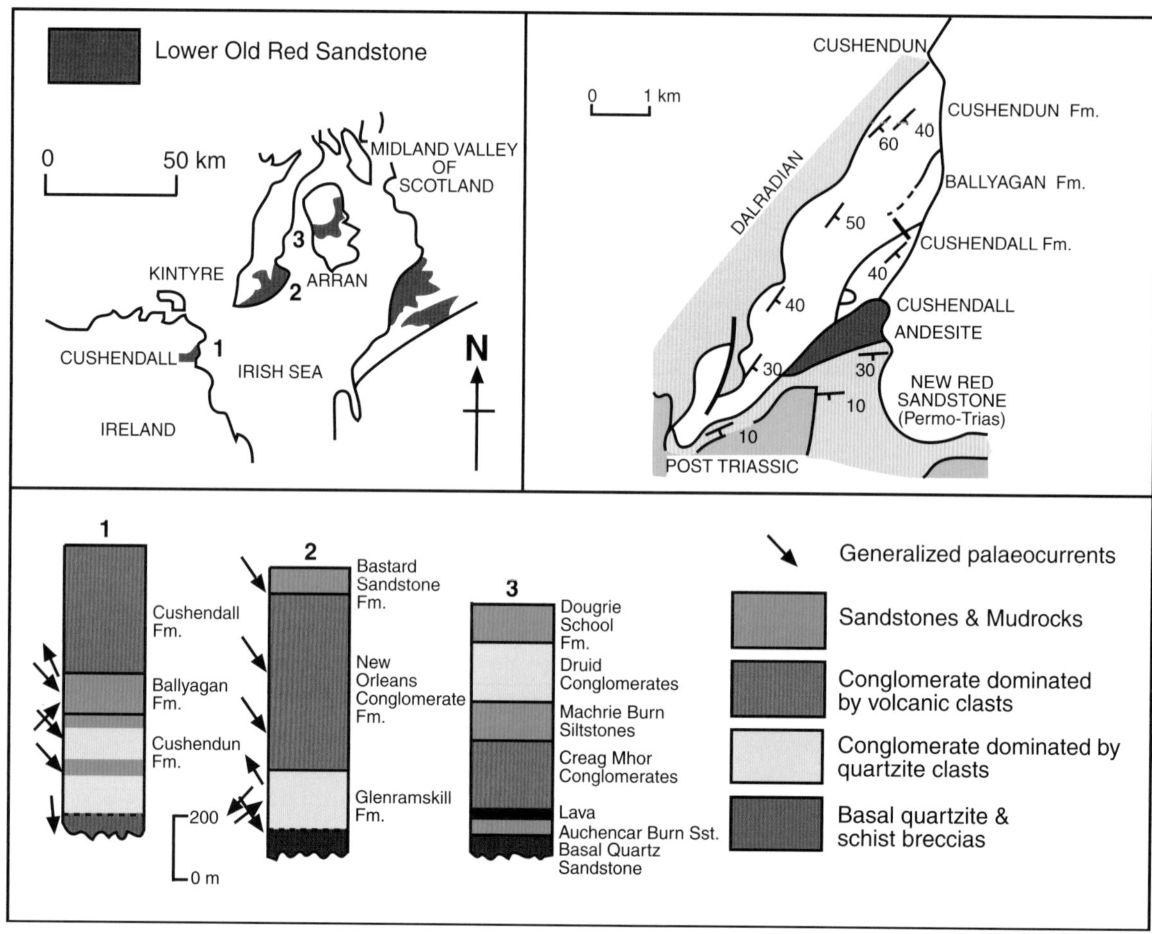

Fig. 9.4. Old Red Sandstone succession of the Cushendall area and comparisons with adjacent successions in SW Scotland.

Fig. 9.5. A, B) Quartzite-rich conglomerates and interbedded sandstones of the Cushendun Formation, Cushendun: Note the large number of well-rounded quartzite clasts of probable polycyclic origin. C) Imbrication in conglomerates of the Cushendun Formation, Cushendun. D) Matrix-supported volcaniclastic conglomerates of the Cushendall Formation, Cushendall.

noted that the conglomerates that are so prominent on the coastal section interdigitate with sandstones when traced along strike to the south-west.

These conglomerates are composed predominantly of well-rounded quartzite clasts (maximum particle size (mps) up to 800 mm) (Fig. 9.5), with 1–2% of sub-angular vein quartz and welded tuff clasts (mps up to 200 mm). The quartzite clasts are interpreted by Simon (1981, 1984a) as being polycyclic because they are all well rounded and include some with broken and smoothed edges. Also, no obvious sources are apparent in the adjacent Dalradian strata. The angular lithic arenite matrix of the conglomerates is interpreted as first cycle detritus and contains fragments of schist, quartzite, and welded tuff. The ultimate source of the polycyclic quartzite clasts is unknown. Similar observations have been made in successions adjacent to the Highland Boundary Fault in Scotland (Bluck, 2000b). The upper part of the Cushendun Formation consists of coarse sandstones, mudrocks, well sorted exotic conglomerates (mps up to 150 mm), and intraformational conglomerates. The conglomerates are generally one clast thick and laterally extensive and are interpreted as lag deposits. The associated sandstones were also deposited in thin laterally extensive sheets which are internally parallel-laminated or planar cross-stratified. Palaeocurrent data presented by Simon (1984a) suggest derivation from the north-west.

The Cushendun Formation has been interpreted as a series of alluvial fan sediments with the conglomerate dominant lower part representing sheetflood deposits on the mid- and proximal fan, and the sandstone dominant upper part representing shallow braided stream deposits on a distal fan (Simon, 1984a).

The overlying Ballyagan Formation is dominated by large scale cross-stratified sandstones, with typical set thicknesses of 1–3 m. The sandstones are lithic arenites containing a high proportion of intermediate igneous rock fragments. Subordinate thin conglomerates contain both quartzite and intermediate igneous debris. Interbedded mudrocks show desiccation cracks. Palaeocurrents differed in the two areas sampled by Simon (1981) showing vector means towards 028° and 121°. Simon and Bluck (1982) interpreted these strata and their palaeocurrents as representing deposits of large bedload rivers displaying an axial flow to the north-east, i.e. parallel to the general Caledonian strike.

The youngest part of this succession, the Cushendall Formation, is dominated by conglomerates containing mainly intermediate igneous debris. Some of these are coarse-grained debris flow deposits (mps up to 2 m; Fig. 9.5D); others are clast-supported tractional flows. Associated sandstones are lithic arenites with a composition that mirrors that of the conglomerates. Palaeocurrents from sandstone cross-strata are bimodal with modes to the NNW and SE (Simon, 1984a). These data are difficult to interpret, particularly as there are possibly some aeolian horizons. The formation has been interpreted as the deposits of a proximal to mid- alluvial fan. Evidence for southerly derivation of material comes from lateral changes in bed thickness and limited imbrication. Although this formation contains abundant evidence for adjacent volcanic strata, which are generally assumed to be contemporaneous, Simon (1984a) claims that interbedded tuffs and lavas are absent, although tuffs were reported by Sanderson (1970). The spatially associated Cushendall Porphyry has been interpreted as both intrusive (Wilson, 1953) and extrusive (Simon, 1984a), and has been described as both dacitic (Wilson, 1953) and andesitic (Simon, 1981) in composition. Despite this uncertainty it is likely to be connected with this Old Red Sandstone sedimentation.

Coastal sections of redbed conglomerates, sandstones, and mudrocks to the south of this porphyry were suggested as forming an 'Upper Old Red Sandstone' by Wilson (1953) with an implied Devonian age. They were described in detail by Simon (1984b) who termed them the Red Arch Formation. However, discovery of halite pseudomorphs in the finer grained strata (Simon, 1986) suggests that they might be better assigned to the New Red Sandstone (Permo-Trias) as suggested by Sanderson (1970). Palaeomagnetic data from Turner *et al.* (2000) suggest acquisition of magnetic signature in the late Carboniferous–early Permian. Remagnetisation of many Devonian rocks is known to have occurred at this time, and thus the data show only that the rocks are at least as old as early Permian, but cannot be constrained beyond this.

This Old Red Sandstone basin clearly lies close to, if not masks, the south-west continuation of the Highland Boundary Fault Zone of Scotland (Max and Riddihough, 1975; Pitcher, 1969). The sequence of coarse-grained alluvial fans is suggestive of contemporaneous fault activity. The Old Red Sandstone shows much steeper dips (60° cf. 30°) near the contact with the Dalradian strata and cracked and sheared clasts are noted in the Cushendun Formation. Sanderson (1970) interpreted the variation in dip as due to progressive basement uplift. The indication of sediment derivation across strike from both NW and SE might suggest that the basin was not particularly wide. Placing limitations on the along-strike length rely largely on correlation with strata exposed in the western part of the Midland Valley of Scotland. The successions in the Cushendall area are similar to those in Kintyre and Arran (Fig. 9.4) in that both contain conglomerates dominated by quartzite clasts that are inferred to be polycyclic and other conglomerates that are dominated by volcanic clasts. The successions in Arran lie just south of or mask the Highland Boundary Fault, whereas those in Kintyre lie to the north of this fault (Friend and McDonald, 1988). Such palaeocurrent data as are available for Scotland are also quite variable, suggesting that there may have been complex basin geometries or subsequent telescoping (Bluck, 2000). In the absence of any evidence of flora or fauna, except for one example of the trace fossil *Diplichnites* (Graham, unpublished), assessing the age of these rocks also relies on possible correlations with Scotland. There are no direct dates from either Arran or Kintyre, and assignment to the Lower Devonian relies on gross lithological similarity to the lower parts of the Strathmore Syncline farther north-east where a Lockhovian date has been obtained (Trewin and Thirlwall, 2002).

Ballymastocker

A small outcrop of Old Red Sandstone facies occurs at Ballymastocker in the northern part of Donegal (Fig. 9.6). The outlier is traceable only 6 km along strike and 0.7 km across strike. A succession of *c.*250 m was briefly described by Pitcher and Berger (1972) and was re-examined by Simon (1981). The lowermost beds are coarse conglomerates containg quartzite and schist clasts that could have been derived from the Dalradian basement to the north-west. Above this are a series of red micaceous sandstones with pebbly and silty seams, which are then succeeded by clast-supported conglomerates. The latter contain fragments of quartzite, schist, metabasite, and porphyry, all relatively well rounded.

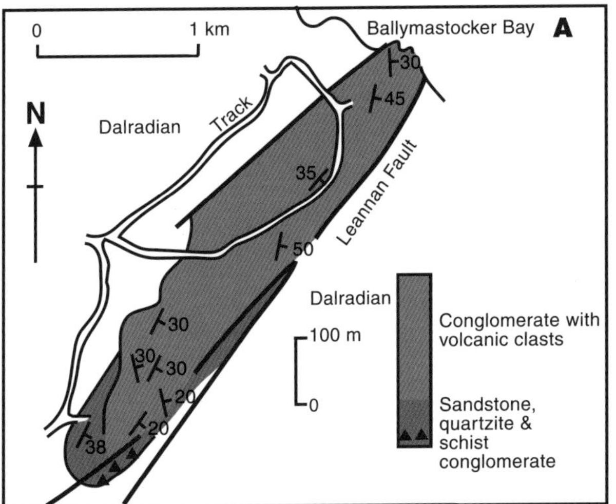

Fig. 9.6 **A)** Summary map of the Old Red Sandstone outcrop at Ballymastocker, Co. Donegal. **B)** Poorly sorted conglomerates rich in porphyry clasts at Ballymastocker Bay.

The south-eastern limit of the outcrop is formed by the Leannan Fault (Pitcher *et al.*, 1964) which is a major Caledonian fault in Donegal and is probably part of the Great Glen Fault System of Scotland (Pitcher, 1969). Some of the Old Red Sandstone strata are obviously affected by subsequent movement along this fault. Clasts of the distinctive quartzite which crops out on the south-east side of this fault are notably absent from the Old Red Sandstone strata, and this fact was used by Pitcher and Berger (1972) to suggest that considerable movement had taken place along the Leannan Fault subsequent to the deposition of the Old Red Sandstone. McSherry *et al.* (2000) suggested that the movement pre-depositional of the ORS succession was sinistral whereas the subsequent deformation had a predominantly dextral component. There is no evidence that the Leannan Fault controlled sedimentation.

The depositional environment is thought to be that of an alluvial fan (Graham and Clayton, 1988; McSherry *et al.*, 2000). Palaeoflow toward the south-east is suggested by McSherry *et al.* (2000) on the basis of a few palaeocurrent measurements. These strata are considered to be Devonian only by general regional comparison and the porphyry clasts are taken to represent penecontemporaneous igneous activity. McSherry *et al.* (2000) suggested that the porphyry clasts may be derived from the Lorne Plateau in Scotland but the large size of the clasts (largest up to 1 m) make this unlikely. This outcrop may well be part of a relatively small basin, similar to others seen along strike in Scotland (Mykura, 1983) but the field data are insufficient to provide reasonable constraints. There is abundant evidence for sinistral strike-slip faults active in the Devonian in this region, e.g. Hutton (1982), and the prominent Devonian granites of Donegal testify to the presence of igneous melts, some of which may have had a surface expression.

Fintona

On most earlier maps the poorly exposed Fintona Block (Fig. 9.2) represents the largest outcrop area of the late Caledonian basins. Assessment of age was largely on the basis of a pteraspid fish fragment recovered by Harper and Hartley (1938) from just north of Lisbellaw. This was re-examined (Forey *in* Simon, 1984c) and a Lochkovian age was thought most likely. However, these rocks have subsequently yielded Carboniferous miospores. Reassessment of this area by Mitchell and Owens (1990) has produced a very considerable revision of the geology, revising earlier work by Simon (1984c) and the Northern Ireland Geological Survey (1978, 1982). Devonian rocks, referred to as the Fintona Group, are much more restricted in distribution than was originally thought, and many of the terrestrial successions have been shown to be Carboniferous in age (see Chapters 10,11).

The Devonian outcrop is both bounded by and traversed by major NE–SW faults that naturally subdivide the block (Fig. 9.7). In the area south of the Tempo–Sixmilecross Fault an estimated 3.5 km of rock is present (Mitchell, 2004b). The lowermost Shanmaghery Sandstone Formation shows thin basal conglomerates containing clasts of vein quartz, quartzite, tuff, jasper, and mica schist that appear to be locally derived from the Tyrone inlier. The overlying Gortfinbar Conglomerate Formation is 2.7 km thick and contains two interbedded

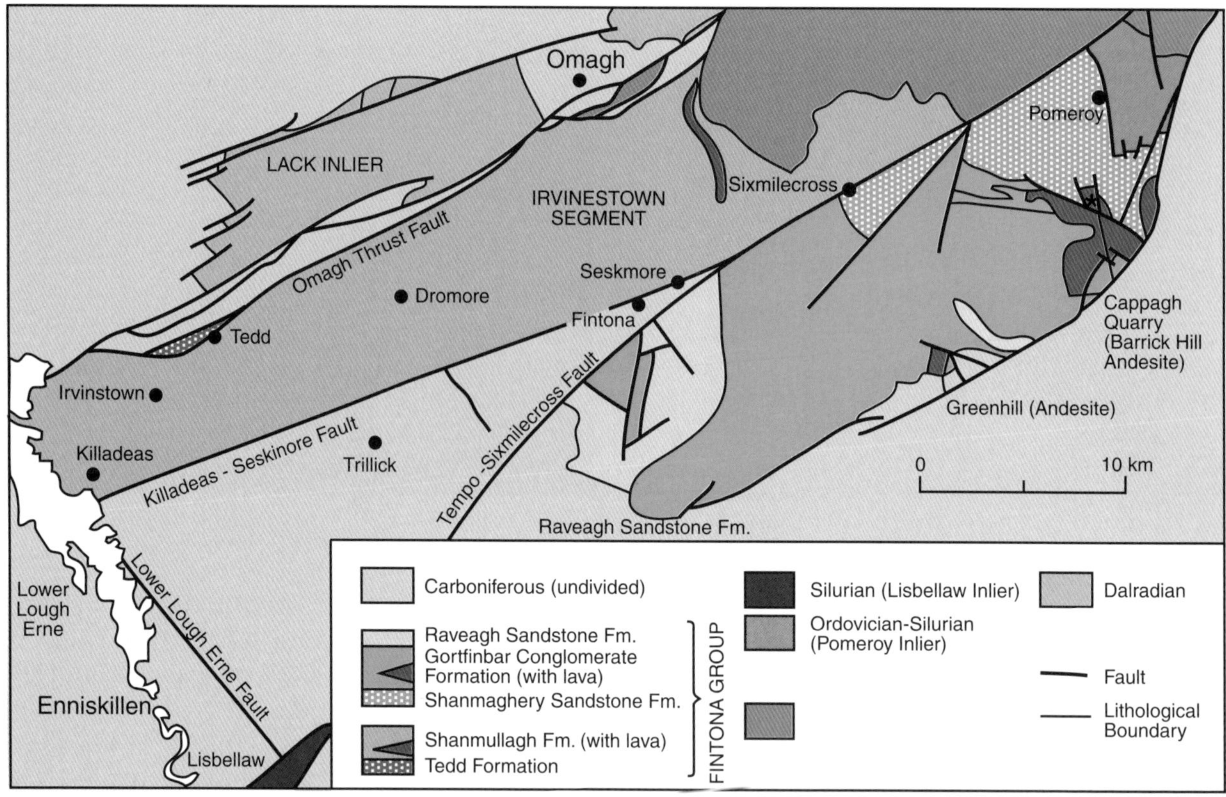

Fig. 9.7 Summary map of the Devonian rocks of the Fintona Block (*after* Mitchell, 2004b).

andesitic lava members. The lower of these, the Barrack Hill Andesite Member, has yielded a K–Ar whole rock date of 376 +/- 12 Ma (Rundle, 1978) and a Rb–Sr age of 437 +/- 6 Ma (Thirlwall, 1988). Four samples dated by Rundle (1988) produced, along with incorporation of the earlier data of Rundle (1978), two distinct groups of ages at 375 +/- 2 Ma and 275 +/- 6 Ma. It is difficult to interpret these data. If the 375 +/- 2 Ma age was the age of extrusion, the age of this succession would be strongly anomalous with respect to adjacent regions. The tentative age assignment in Figure 9.36 is based only on regional arguments. The conglomerates are dominated by volcanic clasts (mps up to 400 mm), and in the southern outcrops vein quartz, quartzite and greywacke clasts are noted. These sediments are interpreted as the products of alluvial fans produced by erosion of contemporaneous volcanics (Simon, 1984c). Sandstone-rich horizons interbedded with the conglomerates consist of sheet-like beds <0.6 m thick which are parallel laminated or contain climbing ripples. Thin (<50 mm) mudstone beds commonly show desiccation cracks. The sand- and mud-dominant Raveagh Sandstone Formation overlies the andesitic conglomerates, and this has been interpreted as the deposits of playas by Simon (1984c).

Devonian strata also crop out between the Omagh Fault to the north-west and the Killadeas–Seskinore and Tempo–Sixmilecross faults to the south-east (Irvinestown Segment). Here the Fintona Group is represented by the Shanmullagh Formation that was assigned a Pragian age (Mitchell and Owens, 1990). This was subsequently revised to a late Devonian age on some published maps (Geological Survey of Northern Ireland, 1995, 1997). Palynomorph assemblages from this formation have been reassessed as indicating an early Lockhovian to early late Emsian age (Stephenson and Mitchell, 2002). This formation comprises fine to medium-grained purple-brown sandstones and thin green siltstones displaying ripples and desiccation cracks. The association probably represents playa deposition (Simon, 1984c). Probable Devonian rocks also crop out north-west of the Omagh Thrust where about 500 m of fine sandstone and mudrock of the Tedd Formation is present. To date this predominantly red bed formation has yielded only non-diagnostic palynomorph assemblages (Mitchell, 2004b).

Curlew Mountains

Devonian rocks crop out over approximately 100 km² in the Curlew Mountains (Figs. 9.2, 9.8) although exposure for the most part is poor. These rocks were described by Charlesworth (1959, 1960b) and Simon (1984c) who both inferred a Lower Devonian age. Charlesworth recognised five lithological units totalling 1.8 km in thickness, whereas reinvestigation by Simon (1984c) suggested a simpler twofold division.

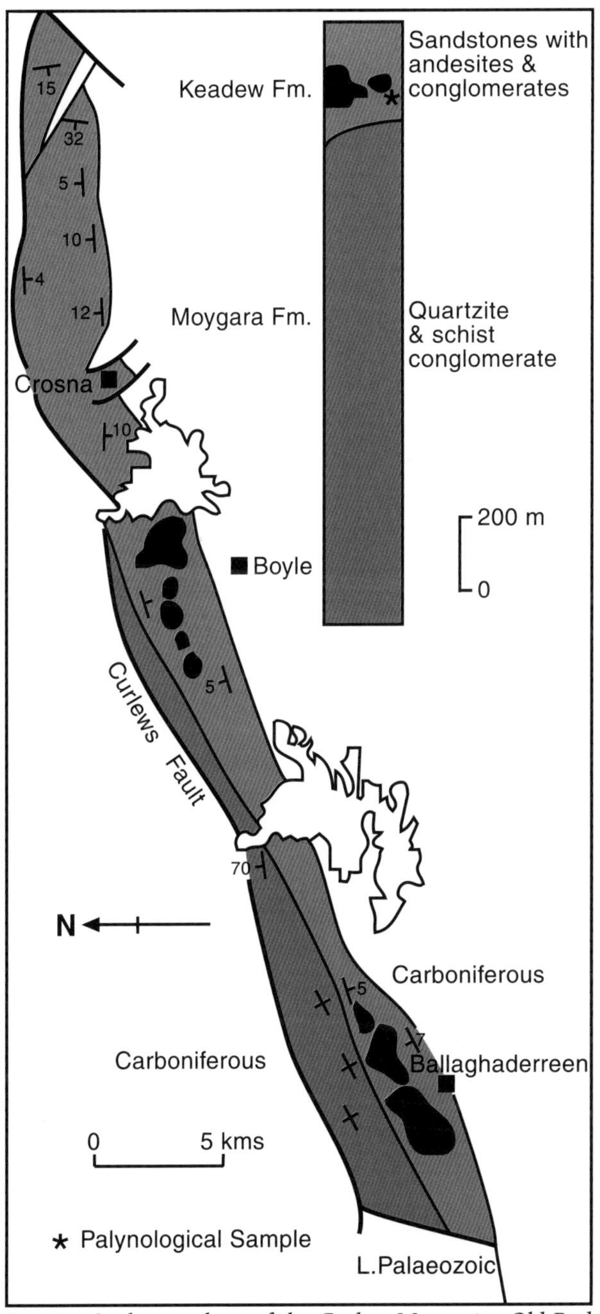

Fig. 9.8 Outline geology of the Curlew Mountains Old Red Sandstone (*after* Graham and Clayton, 1988).

A lower unit of conglomerates (mps up to 480 mm) and pebbly sandstones (Moygara Formation) is 1.5 km thick. Laterally extensive, massive conglomerates have been interpreted as alluvial fan sheetfloods (Simon, 1984c). Associated pebbly sandstones form lenticular beds with cross-stratification indicating derivation from the north, and have been interpreted as braided stream deposits (Simon, 1984c). Locally extensive, interbedded red mudrocks are taken to be playa deposits. The clasts in the conglomerates are mainly flaggy quartzite, vein quartz and jasper with subordinate schist, slate, and sandstone. All could be derived from the regional Proterozoic and Lower Palaeozoic basement.

The overlying Keadew Formation (275 m) consists of sheetlike, massive or parallel laminated quartzose sandstones and desiccated mudrocks, which have been interpreted as distal sheetfloods on an alluvial plain (Simon, 1984c). Locally green, laminated, calcareous mudrocks are preserved in the north-eastern part of the outcrop. There are insufficient palaeocurrent data to estimate dispersal directions. This formation is further distinguished by the presence in some areas of flow-banded andesitic lavas and/or pyroclastic layers. Local volcanogenic conglomerates have maximum particle sizes up to 220 mm. It appears that the dispersed outcrops of volcanic rocks occur at about the same stratigraphical level. At Crosna (Fig. 9.8) (G876076) grey-green mudrocks some 5 m beneath a prominent agglomerate have yielded a poorly preserved miospore assemblage of low diversity. This assemblage, which is dominated by *Retusotriletes spp.* and *Dibolisporites sp.*, is not age diagnostic as these taxa are long-ranging, but the low diversity assemblage is suggestive of a Lower Devonian age (Graham and Clayton, 1988).

The Curlews inlier has a long strike extent parallel to the prominent NE–SW Caledonian faults, most of which show Variscan reactivation. To the south-east the Devonian rocks are unconformably overlain by Carboniferous rocks but the north-west margin is formed by the prominent Curlews Fault. Although this fault has demonstrable early and late Carboniferous movement (Sevastopulo, 1981; Philcox *et al.*, 1989), the marked steepening of dips (up to and beyond vertical) in the Devonian rocks approaching this fault (Fig. 9.8) suggest there is also pre-Carboniferous movement. However, the relatively fine grained nature of the Keadew Formation and the lack of any proximal fan facies in the Moygara Formation suggest that any syn-depositional basin margin faults in the Devonian lay farther to the north-west.

Clew Bay

Devonian rocks of Old Red Sandstone facies crop out between Clew Bay and Lough Conn in County Mayo (Figs. 9.2, 9.9). The outcrop forms a NE–SW belt some 25 km long but less than 10 km across. Two separate conglomerate dominant groups have been recognised which are presently in fault contact, although an original unconformable relationship has been inferred (Graham, 1981b; Graham *et al.*, 1983).

In the south of the outcrop the Islandeady Group (800 m) has been shown to rest unconformably with basal breccias on both Dalradian metamorphic rocks and Ordovician slates and sandstones (Graham, 1981b; Graham and Smith, 1981). A productive palynological sample from 2 m above the basal unconformity with the Dalradian metasediments has yielded a restricted assemblage of palynomorphs indicative of probable Pragian or early Emsian age (Graham *et al.*, 1983). The group is dominated by massive and horizontally bedded clast-supported conglomerates (mps up to 300 mm) and flat to low-angle cross-stratified coarse sandstones (Fig. 9.10A). These have been interpreted as the deposits of mid- distal alluvial fans (Graham, 1981b). Limited palaeocurrent data from clast imbrication suggest a southerly source.

Apart from local debris in the basal conglomerates, the clasts in the Islandeady Group consist of three main types: quartzite, porphyry, and sandstone. The quartzite clasts are rounded and could possibly be polycyclic. The sandstone clasts are lithic arenites that were probably derived from subjacent Lower Palaeozoic rocks. The igneous clasts vary from andesitic to rhyolitic in composition, with the more acid types predominating; some are clearly pyroclastic. The presence, in poorly exposed ground, of one area of contemporaneous rhyolite with well developed shard texture suggests that many of the igneous clasts may represent erosion of Devonian volcanics. However, caution is necessary in this area as there is also evidence for both Ordovician and Silurian acid volcanism.

The Beltra Group, which forms the larger part of the outcrop, comprises three formations and totals 900–1000 m in thickness. The base of this group is only seen in

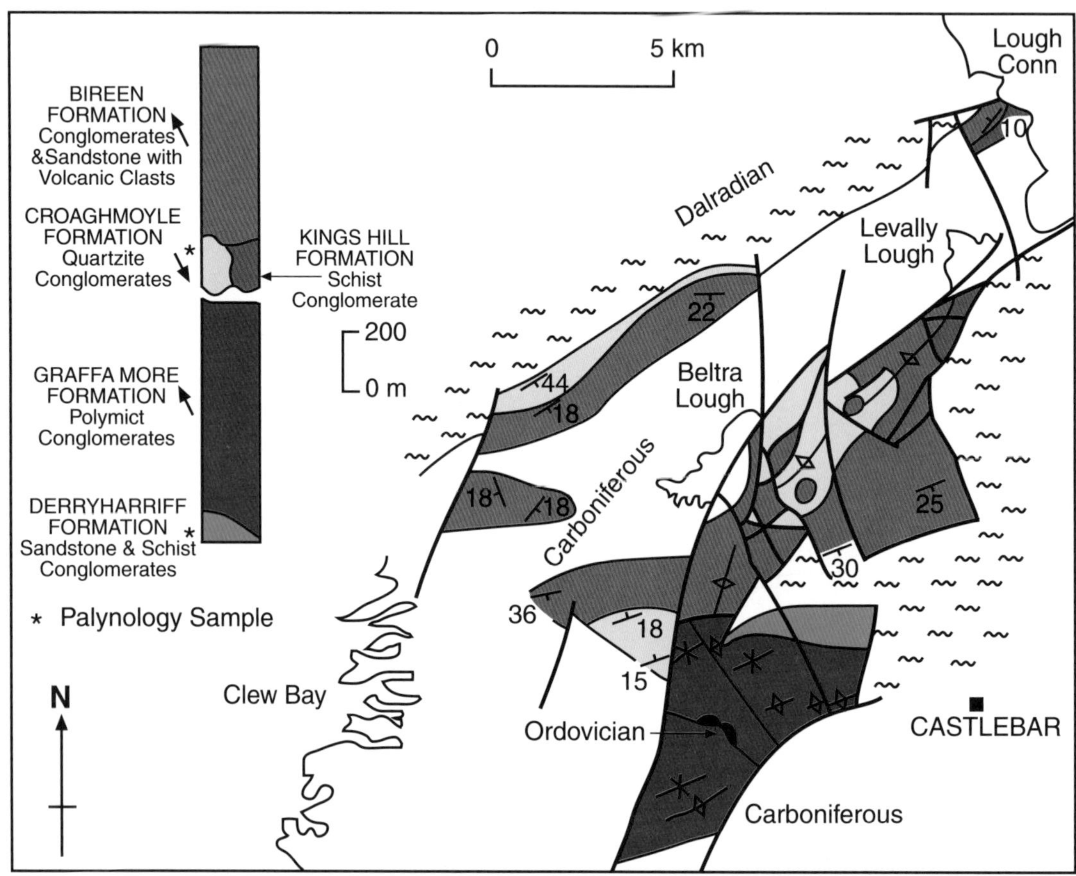

Fig. 9.9 Outline geology of the Devonian rocks between Clew Bay and Lough Conn (*after* Graham, 1981).

Fig. 9.10 **A)** Flat and low-angle cross-bedded conglomerates of the Graffamore Formation, Letter. Strata dip and young to the left. **B)** Burren Mountain (right) and King's Hill (left) from Croaghmoyle. **C)** Poorly sorted debris flow conglomerate, King's Hill Formation, King's Hill. Strata dip and young to the left but bedding is not obvious in these proximal deposits. **D)** Clast-supported, sheet-like conglomerates and pebbly sandstones from the Birreen Formation, Burren Mountain.

the northern part of the outcrop (Fig. 9.9) where it rests unconformably on Dalradian quartzites. In the north and west the lowest formation (Croaghmoyle Formation, 10–230 m) comprises conglomerates formed almost exclusively of quartzite clasts (mps up to 350 mm) along with pebbly sandstones and thin green mudrocks. A productive palynological sample from green- grey mudrocks at the top of the Croaghmoyle Formation has yielded an assemblage of likely early to mid Eifelian age (Graham *et al.*, 1983). Clast imbrication and cross-stratification in the Croaghmoyle Formation indicate transport from the north-west where large potential sources of quartzite exist. In the north-east of the outcrop this formation passes laterally into the King's Hill Formation (100–180 m) which is dominated by coarse conglomerates (mps up to 450 mm) containing mainly semi-pelites, pelites, and psammites with minor quartzites, metabasites, and vein quartz. Clast imbrication indicates transport from east to west, and the clast assemblage closely matches the composition of the Dalradian basement presently exposed to the east. Both of these formations contain matrix-supported conglomerates interpreted as debris flows (Fig. 9.10C) and well structured clast-supported conglomerates produced by dilute tractional flows (Fig. 9.10D). Deposition in proximal to mid-fan areas is suggested (Graham, 1981b) with fans building from both east and north-west into a narrow fault-controlled basin.

Both the Croaghmoyle and King's Hill formations pass gradationally upwards into the Birreen Formation (min. 750 m) which is dominated by flat and low-angle cross-bedded pebble conglomerates and pebbly sandstones rich in volcanic clasts (Fig. 9.10D). This formation has been interpreted as the deposits of mid-distal alluvial fan. Both imbrication and cross-stratification measurements indicate transport from the south-south-east (Graham, 1981b). In detail the clast composition is totally different from that of the Lower Devonian Islandeady Group. The sedimentary clasts in the Birreen Formation are almost all of fine red quartzose sandstone dissimilar to the lithic arenites found in the Islandeady Group. The igneous clasts also show less variety, being mainly porphyritic, of rhyolitic or dacitic composition, and lacking obvious pyroclastics. A source in contemporaneous volcanics is again considered to be most likely but cannot be proven.

These Devonian rocks are unconformably overlain by Lower Carboniferous clastic sediments. The pattern of pre-Carboniferous fold axial traces and faults suggests post-depositional wrench movement in a sinistral NE–SW shear couple (Graham, 1981b; McCaffrey, 1997). Whilst the Devonian sedimentation was probably fault-controlled in a similar wrench regime, the original basin margin faults cannot be recognised in most cases.

Dingle Peninsula

The Dingle Peninsula, along with parts of southern Wales, best demonstrates the change from the Caledonian cycle to the Variscan cycle. The record is preserved almost exclusively in non-marine Old Red Sandstone facies that are by nature poorly fossiliferous. Historically this has led to considerable debate regarding correlation of the Dingle ORS with the thick ORS to the south in Cork and south Kerry. Detailed fieldwork over the last forty years plus the discovery of some dateable horizons in both areas have considerably clarified the geological history (Fig. 9.11). The succession appears to be most complex in the western coastal area of the Dingle Peninsula (Todd *et al.*, 1988a; 1988b) (Fig. 9.12). In places up to four sequences of terrestrial clastics separated by unconformities occur above the marine Silurian rocks of the Dunquin Group. The lowest of these, the Dingle Group, has, in places, a transitional base with Dunquin Group rocks of Ludlow age. This group records the major diachronous regression seen in many parts of the British Isles associated with Caledonian (Acadian) deformation. In terms of basin development there is greater affinity with the late Caledonian basins seen in the northern part of Ireland than with the younger Munster Basin. Younger coarse-grained, conglomerate-bearing sequences which rest unconformably on Dingle Group and older rocks differ between the north-west and south-east of the peninsula, and cannot be related unambiguously to each other except for the regionally extensive upper unit of vein quartz conglomerates and sandstones, the Glengarriff Harbour Group of Horne (1974).

The modern basis for the stratigraphy of the Dingle Group was provided by Horne (1974). Subsequent work based on PhD theses has suggested some modifications to this valuable framework that are summarised in Todd *et al.* (1988a, 1988b); Todd (1989a); Boyd and Sloan (2000);

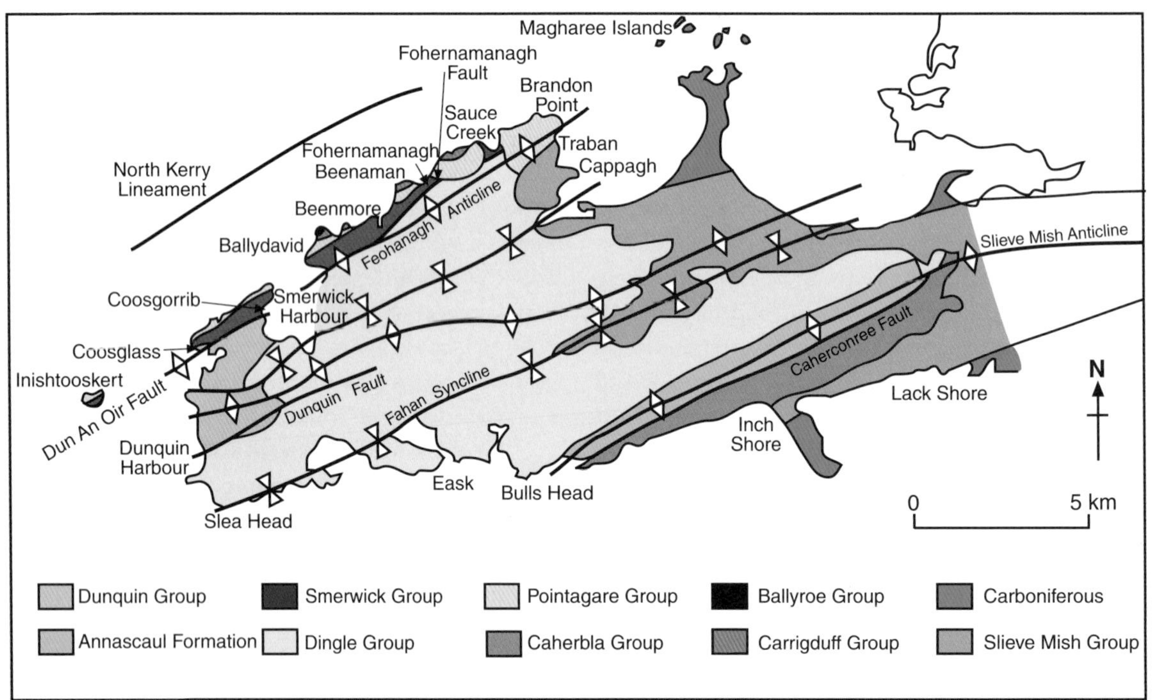

Fig. 9.11 Summary geology of the Dingle Peninsula (*after* Richmond and Williams, 2000).

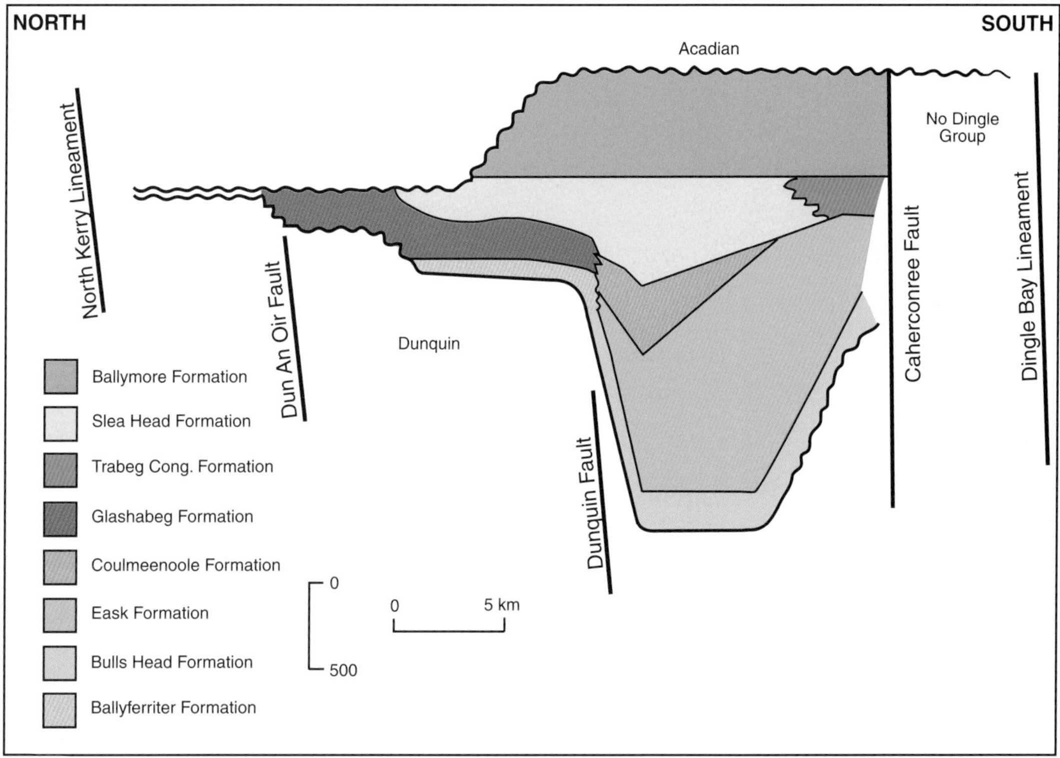

Fig. 9.12 Simplified profile of Dingle Group stratigraphy (*after* Todd, 2000).

Richmond and Williams (2000); and Todd (2000). The Dingle Group is largely confined between the Dun An Oir Fault in the north-west and the Caherconree Fault in the south-east (Figure 9.12). Although most exposed contacts between the Dingle Group and the Silurian Dunquin Group are faulted (Holland, 1987; Boyd and Sloan, 2000), a gradational contact is seen at Glanlick near Dunquin and on the island of Inishnabro (Todd 1991). This fixes the base of the Dingle Group as Ludlow in age. The lower part of the Dingle Group coarsens upward from this basal transition from marine to non-marine environments to a series of conglomerates, which are then overlain in the upper part of the group by an extensive sandstone formation (Figure 9.12).

The lowest part of the Dingle Group is described in detail by Boyd (1983) and Boyd and Sloan (2000). The lowest Bulls Head Formation is dominated by heterolithic facies with wave ripples and desiccation cracks (Fig. 9.13) that are interpreted as the deposits of ephemeral lakes. The overlying Eask Formation contains both coarser grained sandy facies, with sandstones >1 m, and mud-dominant facies with indications of pedogenesis. The sandstones are more common and coarser grained in the south with palaeocurrent data from cross-stratification indicating flow to the north-north-west. The Eask Formation is interpreted as the deposits of a north-draining ephemeral fluvial system. A palynological sample from the middle part of the Eask Formation has been assigned by Higgs (1999) to the W interval of the PoW spore zone of Streel *et al.* (1987) (Figure 9.3) indicating an early late Pragian age.

In the southern part of the peninsula the Eask Formation passes gradationally upwards into the Trabeg Conglomerate Formation, which indicates sediment transport to the north-west (Todd, 1989a,b). Lateral and vertical distribution of clast types within this formation have been used to infer sinistral strike-slip offset of the feeder systems during deposition. Todd (1989a, 2000) interprets the Dingle Bay Fault Zone as being the controlling basin margin structure. He attempts to estimate the drainage basin size and suggests that it occupied at least the present Iveragh Peninsula. Data on clast types (Todd, 2000) show a predominance of quartz wackes and sublithic wackes, some of which had experienced low grade, possibly contact, metamorphism and also demonstrate a pre-depositional cleavage. Particularly important are clasts of limestone, first reported by Jukes and Du Noyer (1863), that contain Silurian fossils best assigned to the Wenlock. Farther north the Eask Formation passes gradationally upwards into a sequence of sandstones, pebbly

Fig. 9.13 Desiccation polygons in the Bull's Head Formation, Coosatorig.

sandstones and conglomerates termed the Coumenoole and Slea Head formations. These formations contain abundant cross-beds that clearly demonstrate transport to the east-north-east, approximately perpendicular to transport in the Trabeg Conglomerate Formation. A wide variety of clast types are reported (Todd, 2000) including quartzites, arenites, wackes, tuffs, basalts, rhyolite, porphyry, and limestone. Todd (2000) suggested a much larger drainage basin than that of the Trabeg Conglomerate Formation. A palynological assemblage from about 200 m below the top of the Slea Head Formation has been described by Higgs (1999) and considered to belong to the *annulatus-sextanti* spore Biozone of Richardson and McGregor (1986) and the AB Biozone of Streel *et al.* (1987) (Figure 9.3). This suggests a probable early Emsian age for this horizon.

Near the Dun An Oir Fault, in the north-western part of the Dingle Group outcrop, another conglomeratic unit, the Glashabeg Formation, can be shown to be laterally equivalent to both the Slea Head and Trabeg Conglomerate Formations (Todd *et al.*, 1988a, b). The Glashabeg Formation comprises conglomerates and sandstones interpreted as channel deposits (Todd and Went, 1991) and calcretised mudrocks interpreted as floodplain deposits. Palaeocurrents indicate derivation from the north-west and Todd (2000) estimates a drainage basin extending some 100 km north of the Dingle Peninsula. Clast types are dominated by volcanics, both crystalline and tuffaceous, but also contain red sandstones, mudrocks, microgranite, jasper and vein quartz and, locally, limestone. The limestone clasts have yielded Wenlock brachiopods. A schematic reconstruction of this conglomerate-rich part of the Dingle Group is shown in Figure 9.14.

The upper part of the Dingle Group in all areas where it is seen is formed by the Ballymore Formation. This sandstone-dominant formation represents a finer grained continuation of the axial fluvial system of the Slea Head Formation. A tuff horizon 380 m above the base of the formation described by Horne (1974) has been termed the Cooscrawn Tuff Bed. This appears to have provided a zircon-based age of 411 +/- 3.3 Ma. However, this age appears only as a personal communication in Higgs (1999) and Richmond and Williams (2000) and as a provisional age in Williams *et al.* (2000a), and the data on which the age is based have not yet been published. It would be difficult to reconcile such a date with the current biostratigraphical data and geological time scale.

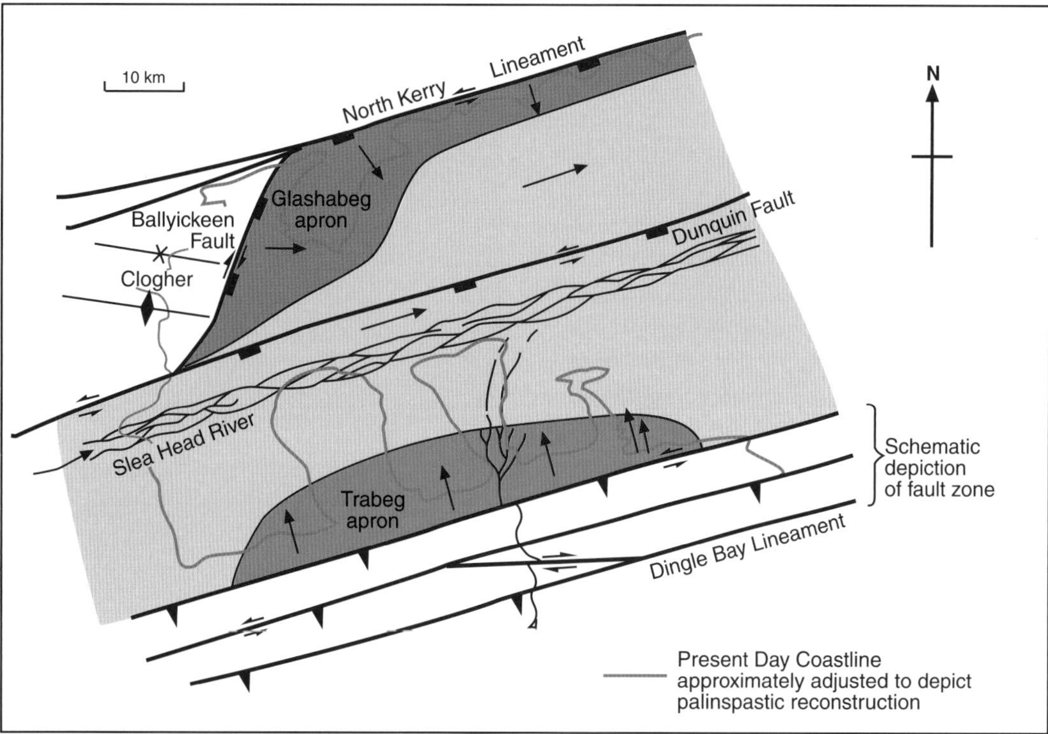

Fig. 9.14 Schematic palaeogeography of the Dingle Group (*after* Todd, 2000).

In the north-west of the Dingle Peninsula near Smerwick, the Smerwick Group (1200 m) rests unconformably on the Dingle Group and comprises northerly derived fluvial sandstones and conglomerates containing quartzite, vein quartz, and sandstone clasts. All three formations in this group have been interpreted in terms of interdigitations of fluvial and aeolian sediments (Todd *et al.*, 1988a, b; Richmond and Williams, 2000; Williams, Sloan and Richmond, 2000). The last authors provide some detailed logs and interpretations. There is nothing to constrain the age of the Smerwick Group other than its bounding unconformities.

Unconformably overlying this (Fig. 9.15) is the 270 m thick Pointagare Group of vein quartz and quartzite conglomerates and aeolian sandstones (Dodd, 1986; Todd *et al.*, 1988a, b). The fluvial conglomerates and sandstones of the Pointagare Group represent bedload rivers flowing to the SE or ESE, whilst the largely overlying aeolian sandstones represent winds directed NNW (Dodd, 1986; Richmond and Williams, 2000). Most workers have considered the unconformity that forms the top of

Fig. 9.15 Unconformity between Smerwick Group (below) and Pointagare Group (above) at Sauce Creek. (*Photo from Lorna Richmond*)

the Pointagare Group to be the base of the local 'Upper Old Red Sandstone', the lowest unit being a widely distributed resistant conglomerate termed the Lough Slat Formation. However, Richmond (1998) and Richmond and Williams (2000) suggested that there are two further lithostratigraphical groups bounded by unconformities, a lower Carrigduff Group and an upper Ballyroe Group, below the Lough Slat Conglomerate Formation. Facies are predominantly fluvial although tidal incursions are stated for the Ballyroe Group; few details are given. The base of the Carrigduff Group has been assigned an early to mid-Frasnian age based on a personal communication by Higgs in Richmond and Williams (2000, p.155). There are obvious similarities between the rocks of north-west Dingle and the fan–fluvial–aeolian complex described from the Galtee Mountains in terms of position and palaeoflow directions. The unconformities in this north-western area are not thought to represent major breaks in deposition but rather punctuated uplift and development of a half-graben basin controlled by a fault lying just north of the Dingle Peninsula (Todd *et al.*, 1988a, b). This small basin is considered to be related to the development of a larger half-graben basin to the south (Munster Basin) as shown by Todd (2000).

In the south-east of the Dingle Peninsula there crops out the 900 m thick Caherbla Group bounded by unconformity on the Silurian Dunquin Group, and by inference on the Dingle Group, below, and by the unconformably overlying Glengarriff Harbour Group above. The rocks comprise alluvial fan conglomerates of the Inch Conglomerate Formation (Capewell, 1951; Horne, 1971) (Fig. 9.16A) and laterally equivalent and overlying aeolian sandstones of the Kilmurry Formation (Horne, 1975; Dodd, 1986) (Fig. 9.16B). The Inch conglomerates (mps – 500 mm) (Fig. 9.16A) show clear evidence of S to N transport, whilst the aeolian transport was generally the reverse (Dodd, 1986; Fig. 9.17). Todd (2000) estimated a relatively small drainage basin for the Inch system, likely to extend <20 km to the south. Clasts in the Inch Conglomerate Formation are mainly metamorphic, many showing mylonitic fabrics, suggesting the uplift and erosion of an earlier shear zone (Todd, 2000). Horne (1975) interpreted these clasts as debris shed northwards from a basement ridge in the area of Dingle Bay. A detailed examination of the aeolian facies by Dodd (1986) suggested deposition in the lee of a fault scarp up to 500 m high in the east but decreasing in relief westwards (Fig. 9.17). The aeolian foresets are steeply dipping (15–30°) and represent the deposits of single, compound, and barchan-transverse dunes; set heights of up to 28 m have been recorded. The direction of dune migration is anomalous when compared to other Devonian aeolian dunefields in the southern British Isles and is thought to be due to the topographic effects of the Dingle Bay fault scarp (Dodd, 1986; Todd *et al.*, 1988b). Richmond and Williams (2000) suggested temporal equivalence of the Caherbla and Pointagare groups based essentially on similar facies assemblages and the presence of first cycle quartz-mica schist and gneiss clasts. In contrast Todd (2000) suggested that the Pointagare Group is younger than the Caherbla Group. Both correlations are possible with the present data base.

Fig. 9.16 A) Inch Conglomerate Formation, Inch Strand. Note the obvious foliation in the large clasts. B) Large-scale cross-strata of aeolian origin, Kilmurry Bay. Person is standing on regional bedding plane.

Summary of Late Caledonian basins

The late Caledonian basins of the northern part of Ireland and the Dingle Peninsula described above display successions ranging from 0.25–3.2 km in thickness. Much

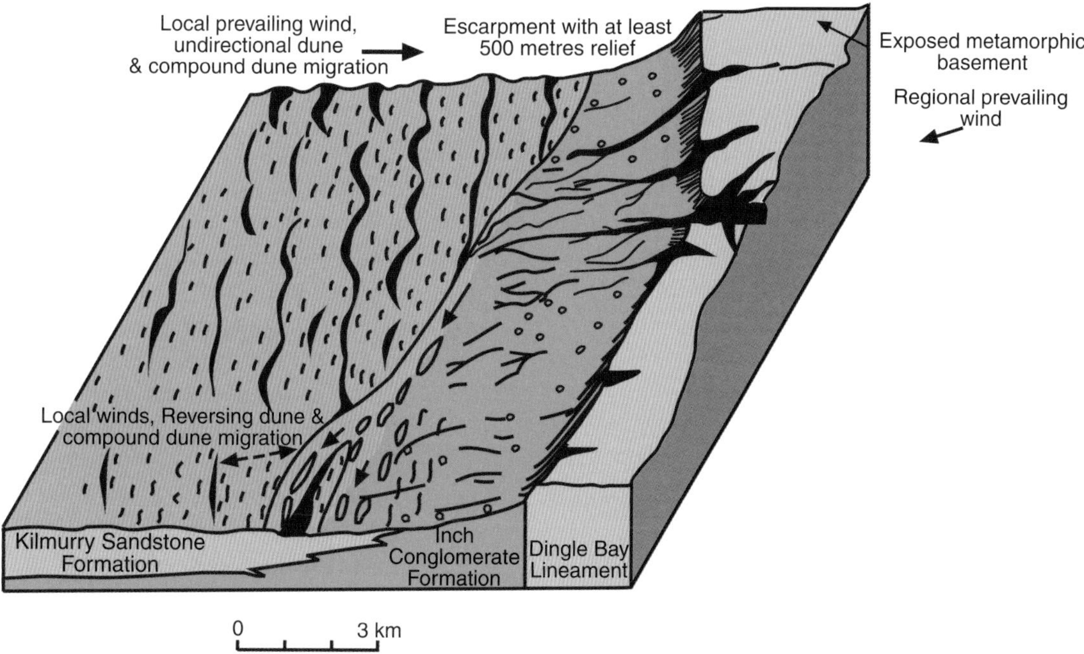

Fig. 9.17 Caherbla Group palaeogeography (*after* Todd *et al.*, 1988a; Todd, 2000).

of this variation may only reflect the amount of later erosion. There is no obvious lithostratigraphical correlation between any of them despite the features that they have in common. Although Devonian is the most likely age for all the sequences described above, in some cases it is only inferred and a pre-Devonian age is possible, as it is for some other Old Red Sandstone basins in Scotland (Thirlwall, 1988; Trewin and Thirlwall, 2002). Indeed it is possible that there is little if any overlap in age among the basins in the north of Ireland. However, despite the lithological and possible age differences among these sequences, all of the late Caledonian basins of Ireland have several features in common: 1) Conglomerates are present and usually predominate. The depositional environments have been interpreted mainly as alluvial fans with associated fluvial and playa deposits. Sedimentation is likely to have been rapid but short-lived. 2) Contemporaneous volcanicity is demonstrable in the sense of interbedded volcanic layers or can be reasonably inferred from conglomerates rich in fresh volcanic clasts. The volcanicity was intermediate to acid in composition. 3) Basin size was generally small with the largest proven across strike extent being less than 50 km. 4) The deposits are spatially associated with major faults which have a typical Caledonian (NE–SW) trend, although most of these faults have been reactivated subsequent to Carboniferous sedimentation. The more E–W trend seen in Dingle reflects a regional Caledonide strike swing seen throughout the western part of Ireland, although there is some debate as to whether this is primary or due to oroclinal bending (MacNiocaill *et al.*, 1998). These major faults form part of a larger suite extending throughout much of Scotland (Pitcher, 1969). Where sense of displacement is known, there is nearly always sinistral strike-slip motion, although downthrow directions are more variable (Soper and Hutton, 1984; Watson 1984). 5) There is a consistent pattern of clast composition in the different basin fills which is essentially an upward change from metamorphic/sedimentary clasts to dominantly volcanogenic debris. It is possible that this sequence occurs in deposits of different ages and may be related to the mode of development or location of the sedimentary basins. The envisaged origin is as local extensional zones within a strike-slip regime

Thirlwall (1981, 1988) has related most of the late Silurian/Devonian volcanism of NW. Britain to the final stages of active subduction of Iapetus oceanic crust. This appears to be at variance with most authors, who suggest that terminal collision was much earlier. Nevertheless the volcanic rocks are calc-alkaline types such as those typically associated with active subduction zones, and there are numerous associated granite plutons as seen in the American Cordillera (Pitcher, 1982). The association of appinites and lamprophyres with many of the granites suggests that these granites are mantle-related and not just a product of recycling of old continental crust. The slab break-off model of Atherton and Ghani (2002) seems to explain the geochemistry, the timing, and the uplift history of the Laurentian block. Moreover, the close spatial and temporal relationships of late Caledonian igneous activity and sinistral strike-slip deformation noted by Francis (1978), Leake (1978), Pitcher (1982), and Watson (1984) would be favoured by the heat input generated from asthenospheric injection at depth. McClay *et al.* (1986) have argued that most Old Red Sandstone basins, including those in the British Isles, were a result of extension following gravitational collapse of a thickened Caledonide orogen, but there is little structural evidence to support this in the Irish/British sector of the fold belt. Thus it seems likely that strike-slip displacement on major faults was the main control on both basin development and preservation of the record of late Caledonian uplift and magmatic history. This tectonic and magmatic history is treated in more detail in Chapter 8.

The Munster basin

The term Munster Basin (Capewell, 1965) has been used to describe a large area of southern Ireland over which Old Red Sandstone facies developed in the later part of the Devonian. The limits of this basin have been arbitrarily defined by either a zone of major thickness change in the Old Red Sandstone sequence, or by the 1 km Old Red Sandstone isopach, the two being approximately coincident (Naylor and Jones, 1967; Clayton *et al.*, 1980; Graham, 1983; Sanderson, 1984) (Fig. 9.18). Only the northern and eastern margins of this basin can be observed, and here the Old Red Sandstone is seen to rest unconformably on Lower Palaeozoic rocks. These margins can be traced from Dingle in the west, north of the Galtee Mountains, and then southwards via the Comeraghs to Ballyvoyle Head on the south coast (Fig. 9.18). A westerly closure can only be tentatively inferred from palaeocurrent trends (Graham, 1983; Fig. 9.18). There is no evidence for southerly derived sediment at any stage in the basin history.

Away from the northern (Fig. 9.19) and eastern margins the base of the Old Red Sandstone is never seen. The various published isopach maps plot known rather than total thickness of Old Red Sandstone and suggest a south-facing half graben structure consistent with early analysis of the regional gravity data of Murphy (1960) (see Naylor and Jones, 1967; Clayton *et al.*, 1980a; Naylor *et al.*, 1980; Matthews *et al.*, 1983; Sanderson, 1984). Williams *et al.* (1989) and Williams (2000) have argued for a more symmetrical basin cross-section, suggesting

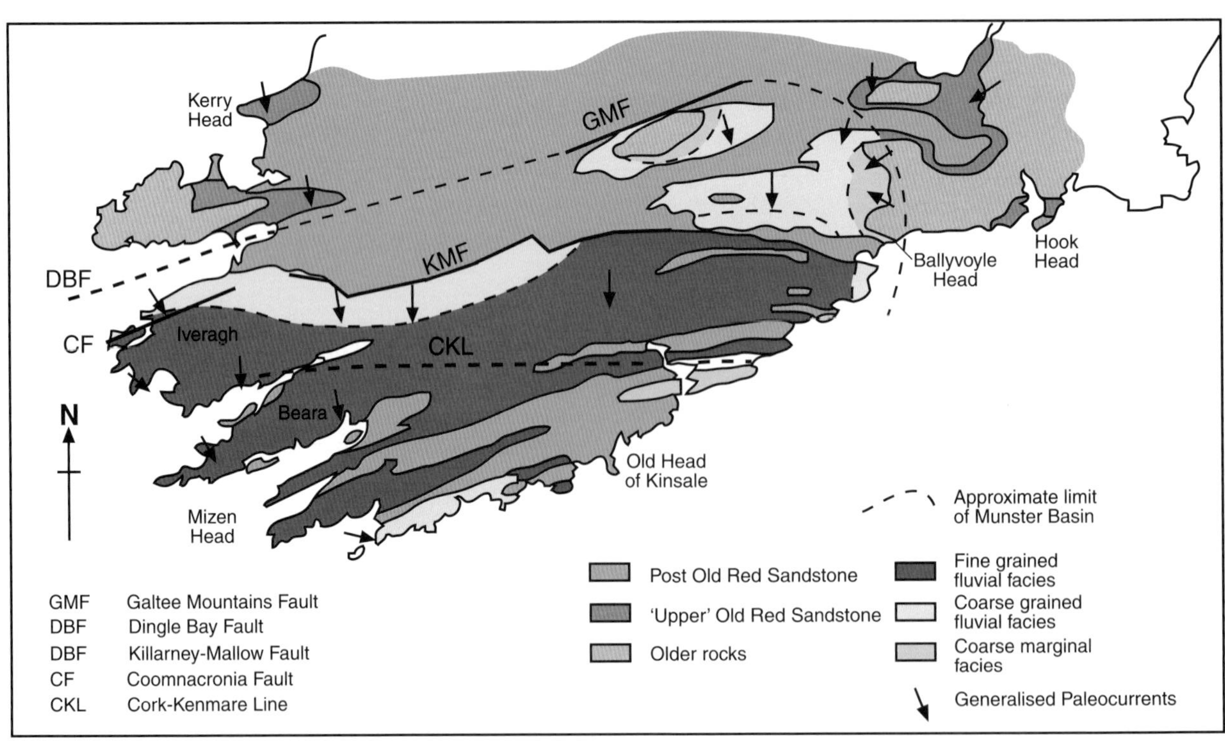

Fig. 9.18 Summary of the Old Red Sandstone of the Munster Basin.

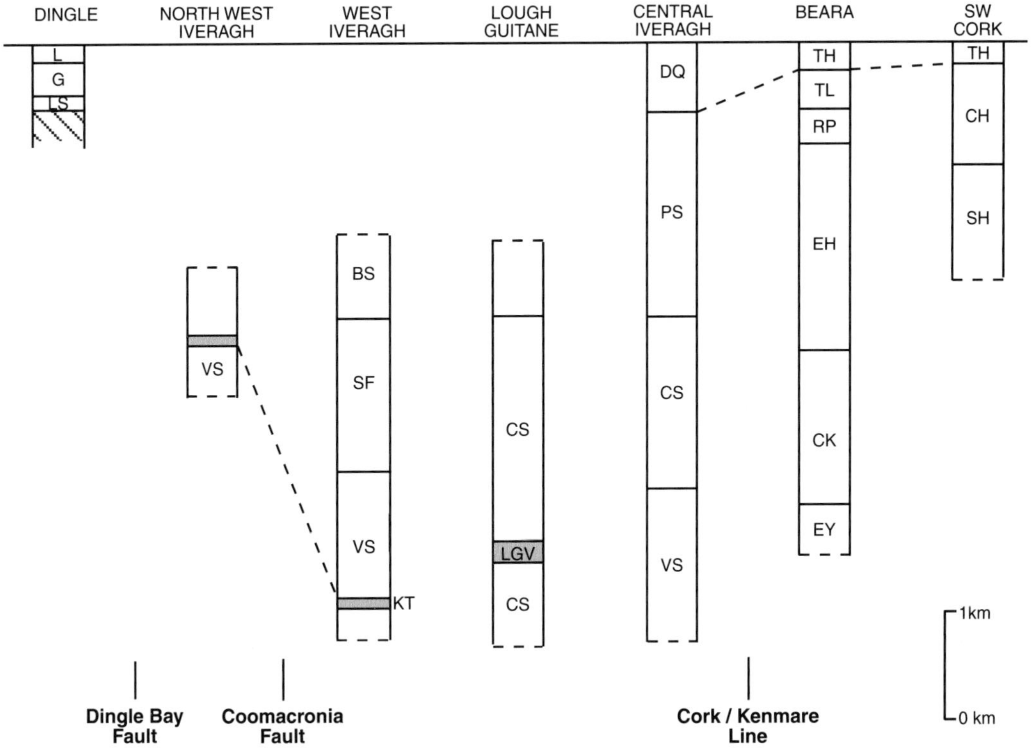

Fig. 9.19 Simplified correlation of local successions in western part of Munster Basin: L – Lack formation, G – Gearhane Sandstone Formation, LS – Lough Slat Formation, VS – Valentia Slate Formation, SF – St. Finan's Sandstone Formation, BS – Ballinskelligs Formation, KT – Keel Tuff Bed, CS – Chloritic Sandstone Formation, LGV – Lough Guitane Volcanics, PS – Purple Sandstone Formation, DQ – Derryquin Formation, EY – Eyeries Formation, CK – Caherkeen Formation, EH – Eagle Hill Formation, RP – Reen Point Formation, TL – Tholane Formation, TH – Toe Head Formation, SH – Sherkin Formation, CH – Castlehaven Formation. Data from Capewell (1975), Graham and Reilly (1972), James and Graham (1995), Russell (1984), Williams *et al.* (2000b).

a Dunmanus–Castletown Fault as being a major north-dipping extensional structure. Whilst the successions south of this putative fault are certainly thinner than those much farther north, the nature of this thinning is difficult to verify. Moreover, the abrupt shift of palaeocurrents that is claimed to occur across this structure is debatable. In fact a westerly component to the palaeocurrent pattern is seen throughout the exposures near the west coast and could be argued to shift progressively from NW to W going southwards (Graham, 1983; Russell, 1984; Graham *et al.*, 1992; James and Graham, 1995). The lack of velocity contrasts makes recognition of such syn-sedimentary structures on seismic data difficult, although they are clearly required to produce the accommodation for the thick Old Red Sandstone succession. There is general agreement that the Killarney-Mallow Fault Zone and the Dingle Bay–Galtees Fault Zone were major syn-depositional structures, along with the Cork–Kenmare Fault Zone, in the latest Devonian and into the Carboniferous. These structures can also be clearly recognised on more recent geophysical data (Vermeulen *et al.*, 1999). Beyond that identification of specific syn-sedimentary faults is possible in only a few cases, e.g. Coomnacronia Fault (Williams *et al.*, 2000b).

Sedimentary facies and provenance

The Old Red Sandstone of the Munster Basin consists predominantly of sandstones and mudrocks with conglomerates locally important near the exposed basin margins. The overall facies distribution has been described by Graham (1983), who recognised four major facies: (a) coarse-grained marginal facies, often with alluvial fans and, more rarely, aeolian sediments; (b) a coarse-grained fluvial facies dominated by 'in channel' deposition; (c) a fine-grained fluvial facies with rare or poorly defined channels; (d) a fluvial coastal plain facies. In gross terms, there is a fining towards the basin centre and also a bulk fining upwards. The facies distribution operative for most of the Upper Devonian has been interpreted as a fluvial distributary system (Graham, 1983).

Figure 9.18 provides a summary of average facies distribution within the Old Red Sandstone. More recent work has refined some of the details presented therein (e.g. Kelly and Olsen, 1993; Kelly and Sadler, 1995; Sadler and Kelly, 1993; Williams, 2000: Williams *et al.*, 1989, 2000b) but has substantiated the basic pattern. Although the marginal sediments of the Munster Basin appear to be locally derived, the huge volume of sediment in the fluvial Munster Basin (>38,000 km^3) necessitates a larger drainage area. The most likely sources are thought to be the metamorphic basement rocks presently exposed in the north and west of Ireland (Graham, 1983; Price, 1986). Attempts at estimating drainage basin parameters by Kelly and Sadler (1995) suggested values of the order of 620 km drainage length, 47,000 km^2 drainage basin area, and 230 m^3 s^{-1} mean annual discharge.

Coarse marginal facies

This facies crops out near the northern and eastern margins of the basin. In the east in the Comeragh Mountains the Coumshingaun Conglomerate Formation (mps – 400 mm) and parts of the Comeragh Conglomerate Formation represent an alluvial fan and proximal braided fluvial assemblage (Penney, 1980b; Capewell, 1957a; Boldy, 1982) and have been interpreted as products of repeated back faulting of the eastern margin (Boldy, 1982). Clast assemblages can be readily matched with the adjacent Lower Palaeozoic strata dominated by metasedimentary clasts in the more northerly Comeragh Mountains (Penney, 1980b), but containing many more igneous clasts from the Waterford volcanic belt (Ordovician) in the more southerly Monavullagh Mountains. (Boldy, 1982). Thus there is clear evidence of easterly provenance (Fig. 9.18) and of no lateral displacement between the conglomerates and their source.

In the Galtee Mountains on the northern margin, coarse-grained conglomerates (mps – 750 mm) are also present (Doran *et al.*, 1973; Carruthers, 1985). Here Carruthers (1985, 1987) has demonstrated the presence of syn-sedimentary basin margin faults and also the presence of aeolian facies (Galtymore Formation) interfingering with alluvial fan deposits (Pigeon Rock Formation). The clasts in the alluvial fan conglomerates are again locally derived, indicating no lateral displacement between source and deposit. The fans built from north to south but the predominant aeolian transport was from south to north. In some sections the fluvial and aeolian deposits are interbedded. The aeolian sands are characterised by large-scale cross-strata (sets up to 15 m but typically <5 m thick) representing deposition in dune and draa bedforms, an interpretation supported by the recognition of hierarchical truncation surfaces (Carruthers, 1987). Grains show a low degree of rounding and petrological maturity suggesting a local source, probably from contemporaneous fluvial deposits to the south. An interpreted palaeogeography showing the location of this Galtymore erg controlled by the faulted northern margin of the Munster Basin, is shown in Figure 9.20. This situation invites direct comparison with the younger parts of the succession seen in north-west Dingle.

The consistent southerly directed palaeocurrents throughout the Upper Devonian succession on the Iveragh Peninsula to the south argue against any persistent ridge in Dingle Bay. Coarse marginal facies occur within the main basin succession at Doulus Head and on Valencia Island, just to the south of Dingle Bay, and can be shown to fine rapidly southwards into fine-grained fluvial facies. The clasts in this local, distal alluvial fan deposit are predominantly metamorphic and probably represent reworking of the (by inference) older Inch Conglomerate Formation (Graham, 1983; Russell, 1984; Todd *et al.*, 1988b). Thus the north-facing scarp interpreted as being present during deposition of the Caherbla Group must have been removed to allow southerly reworking of the Inch Conglomerates, or have been much reduced in relief if the metamorphic clasts were to be interpreted as first cycle detritus.

Coarse-grained fluvial facies

This facies occurs in slightly more basinal areas and/or higher in the Old Red Sandstone succession and is predominantly sandy with some gravelly representatives. The facies is best seen in the Killarney area (Capewell, 1957b; Husain, 1957; Walsh, 1968; Avison, 1984a; Price, 1986) and in the eastern part of the basin (Penney, 1980b; Boldy, 1982; Carruthers, 1985; Capewell, 1957a). In the Chloritic Sandstone Formation (Capewell, 1957b) near Killarney coarse grained, often pebbly, sandstones are dominated by trough cross-bedding with subordinate flat and low angle cross-bedding and planar tabular cross-bedding (Figs. 9.21, 9.22). In general terms these rocks can be interpreted as the deposits of stacked fluvial channels, considered to be part of the feeder zone to the fluvial fan by Kelly and Olsen (1993). When traced south and west (Capewell, 1957b; Russell, 1984; Husain, 1957) there is a clear fining of grain size, increase in interbedded mudrocks, and decrease in the amount of large-scale cross-stratification. There is also an increase

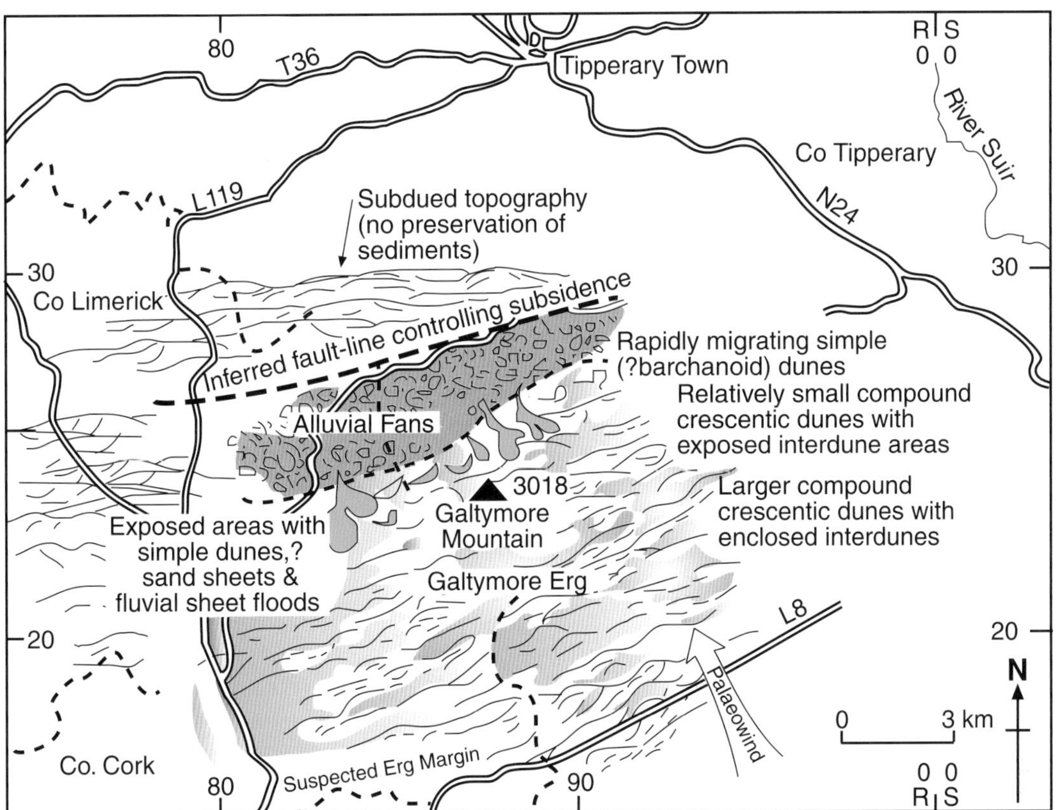

Fig. 9.20 Schematic palaeogeography of the lower part of the succession in the Galtee Mountains (after Carruthers,1987).

in the importance of flat-bedded sandstones relative to cross-bedded sandstones. Individual beds of flat-bedded and cross-bedded sandstone are frequently separated by muddy drapes, each probably representing the final deposits of a flood. This unit was interpreted as the proximal part of a fluvial distributary system (Graham, 1983) in which the rivers decreased in size downstream and the facies passed laterally into the fine-grained fluvial facies described below. Later work south and east of Killarney (Williams *et al.*, 1989; Kelly and Olsen, 1993) has supported and refined this model. Williams *et al.* (1989) suggested that the 'Chloritic Sandstone Fan' was succeeded by a younger fan, which they term the Gun Point Formation. Although somewhat finer grained, this fan occupied approximately the same location as the older Chloritic Sandstone. In a more detailed analysis of sedimentation patterns within the Gun Point Formation, Sadler and Kelly (1993) and Kelly and Sadler (1995) have recognised cyclic patterns of maximum set size, coset size, and sandstone percentage, which they claim can be correlated from Iveragh to Beara. North to south palaeocurrents and downstream fining are clearly demonstrated with facies associations assigned to different parts of the terminal fan. Interdigitated fine grained fluvial facies in areas occupied by the distal areas of these fans suggest two major pulses of coarse grained sediment influx.

In the east of the basin most of the Galtee, Knockmealdown and Monavullagh and part of the Comeragh mountains exhibit successions dominated by coarse-grained fluvial facies. Following the deposition of the coarse-grained marginal fans of the Coumshingaun Conglomerate Formation (Comeraghs) and the Pigeon Rock Formation (Galtees), sandstones and pebble conglomerates of the coarse-grained fluvial facies predominate (Slievenamuck, Ballydavid, Ardea formations in Galtees; Comeragh and Nier formations elsewhere). Throughout this eastern part of the basin the palaeocurrents in this facies are consistently from the north (Penney, 1980b; Boldy, 1982; Carruthers, 1985; Pilling, 1988) and the conglomerates are all dominated by resistant materials, mainly vein quartz and quartzite. Thus the majority of the sediment is likely to have been derived from well outside the Munster Basin and its immediate margins. In addition the above workers have documented consistent fining of grain size and lateral passages from conglomerate-dominant to sandstone-dominant

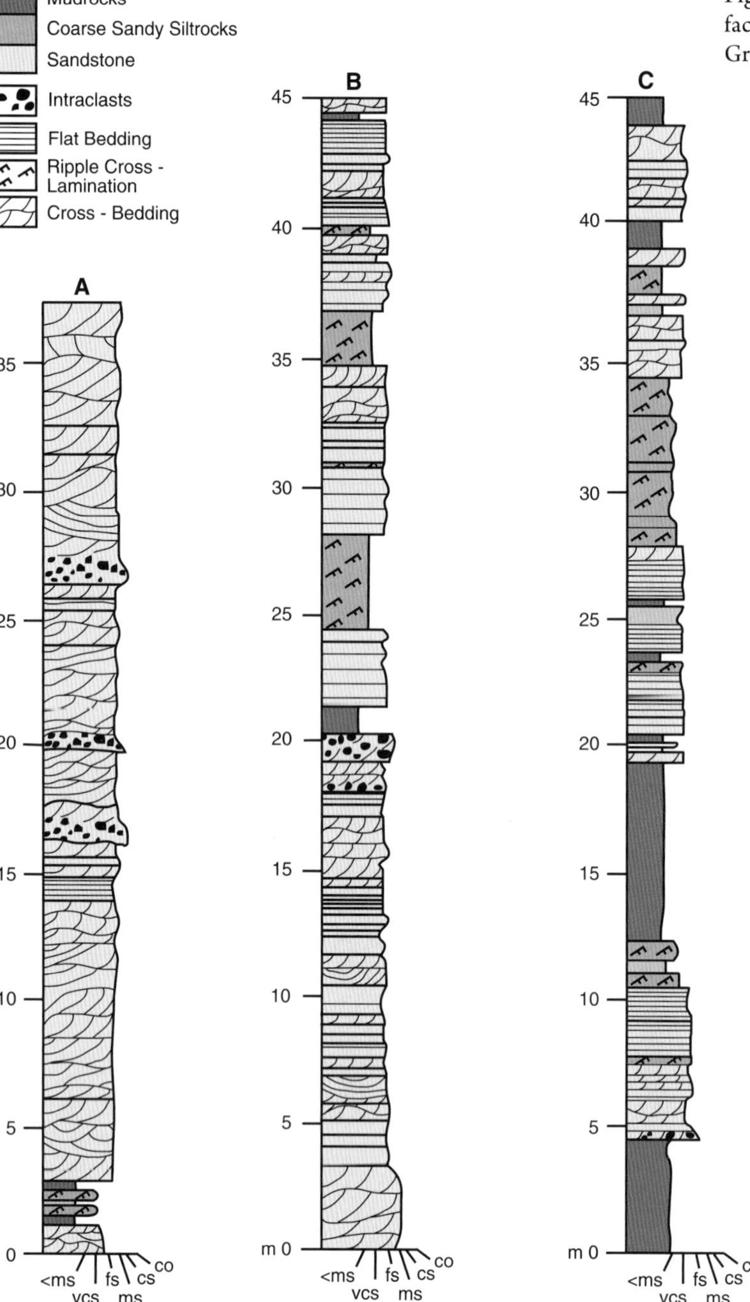

Fig. 9.21 Logs of coarse-grained fluvial facies from the Munster Basin (after Graham, 1983).

sequences to the south and west in the direction of the basin centre. The conglomerates are mainly pebble grade, locally cobble rich in the more proximal areas, and most commonly show planar cross-stratification interpreted as the deposits of transverse bars in bedload rivers. The geometry of these conglomerates varies from sheet-like to wedge-shaped, indicating the presence of relatively wide bedload channels. The associated sandstone-dominated parts of the sequence display both planar and trough cross-stratification as the main sedimentary structures and also show both sheet-like and channel geometries. Intraformational conglomerates are common and desiccation cracks, caliche, and rootlet traces are recorded from the subordinate interbedded mudrocks (Boldy, 1982; Caruthers, 1985). All of these sandstone sequences have been interpreted as the deposits of sandy bedload rivers. The general paucity of floodplain sediments in both the conglomerate- and sandstone-dominant sequences probably reflects extensive lateral migration of these rivers on a braidplain and bypassing of the finer sediment to more

Fig. 9.22 **A, B)** Chloritic Sandstone Formation, Glenflesk, east of Killarney. **C)** Sherkin Formation sandstones, south coast of Cape Clear Island; beds young to the right.

distal floodbasins. A separate braidplain 'fan' is figured by Williams (2000).

The sole representative of this facies in the southern part of the basin is the Sherkin Formation of SW Cork (Graham and Reilly, 1972; Kelly 1992, 1993) (Fig. 9.22C). Here sediment does not exceed medium sand in size, and the palaeocurrents suggest a more westerly provenance, perhaps reflecting the influence of a western basin margin. This formation appears to be interfingering with fine-grained fluvial facies near the eastern (down-current) end of its outcrop, but investigation is hampered by an easterly fold plunge. It is unlikely that equivalent levels in the stratigraphy are reached in areas immediately to the north (Mizen Peninsula), especially given the northern increase in thickness of the total succession.

Fine-grained fluvial facies

This most voluminous facies of the Munster Basin is dominant in the more central parts and constitutes most of the succession in all the major west coast peninsulas, where up to 6 km of sequence is seen (Russell, 1984). The 7 km+ thickness reported for the Beara Peninsula by Coe and Selwood (1968) is likely to be an overestimate, as there is clearly repetition of strata here (James and Graham, 1995). A multiplicity of local lithostratigraphical terms has been applied (see Graham, 1983), mainly of necessity due to the many uncertainties of correlation between areas. However, this has concealed the overall lithological similarity of sediment types over large areas. Subtle changes do occur within sequences dominated by this facies (Graham *et al.*, 1992) but these all occur within the same gross environmental setting. This facies is characterised by very limited development of sandstone bodies displaying flat and cross-bedding and by the predominance of fine sand/silt sediment showing mainly small scale cross- lamination, often in the form of climbing ripples, and parallel lamination (Figs. 9.23, 9.24). Beds composed mainly of clay grade material are also rare. Desiccation cracks and bioturbation are common features. Beds that contain sand are usually 50–500 mm

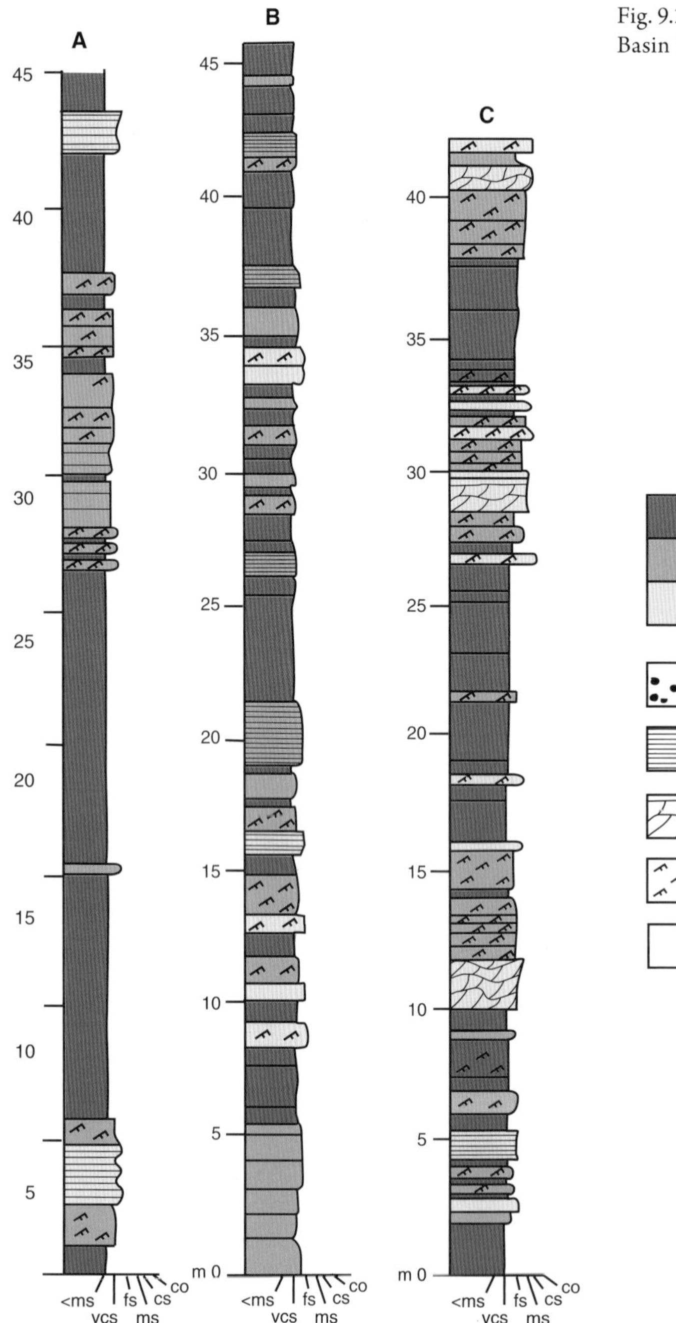

Fig. 9.23 Logs of fine-grained fluvial facies from the Munster Basin (*after* Graham, 1983).

thick and frequently show an upward fining of grain size, an upward decrease in flat lamination, and an increase in climbing ripple drift. Coarser silt beds (usually 5–50 mm) commonly have sharp bases and gradational tops and fine upwards. Finer grained siltrocks show mainly fine lamination and an alternation of coarser and finer grain sizes on the scale of a few centimetres. The great lateral continuity of beds indicates featureless depositional surfaces subject to periodic sheetfloods. This facies has been termed 'background facies' by Williams *et al.* (1989) and 'basinal deposits' by Kelly and Olsen (1993), both of whom give a similar description and interpretation. Thus this facies represents mainly vertical accretion on broad, flat, poorly channelised alluvial plains that show considerable spatial and temporal extent.

Fig. 9.24 **A)** Valentia Slate Formation, Puffin Sound, western Iveragh. **B)** Hungry Hill, Beara Peninsula. **C)** Valentia Slate Formation SW of Portmagee, western Iveragh.

Fluvial coastal plain facies

This facies marks a change to predominantly non-red colours in most areas of the basin and generally is characterised by numerous, often thick, sandstone bodies and the common presence of interbedded sandstones and mudrocks. In areas from Seven Heads on the south coast to Kenmare River on the west coast the term Toe Head Formation is applicable (Graham, 1975a; McAfee, 1987), whilst the term Kiltorcan Formation can be applied in the eastern areas (Capewell, 1957a; Penney, 1980b; Boldy, 1982; Carruthers, 1985).

The Toe Head Formation varies from 185–420 m except for one anomalously thick section at South Dunmanus where Williams *et al.* (1989) record 680 m and where Naylor (1975) estimated 835 m. As the formation has an estimated thickness of 380 m only 9 km along strike to the ENE (Peter Haughton *pers. comm.*, 1988), the possibility of structural repetition here must be considered. The formation is entirely of latest Devonian age having yielded palynological dates from LL, LE, and LN biozones (Higgs *et al.*, 1988). The top is clearly diachronous, progressing from LL to LN in age northwards. At present there are insufficient data to demonstrate conclusively the spatial variation in the age of the base of the formation. The base is characterised by a marked increase in sandstones above the underlying, generally fine-grained purple mudrocks. These sandstones are distinctive in the majority of the sections examined to date in their paucity of typical cross-beds with avalanche foresets. Instead there is a predominance of flat bedding and low-angle cross-bedding. In many cases the latter structures are in the form of swaley and hummocky cross-strata (Cotter and Graham, 1991; Figs. 9.25, 9.26). Although these structures are normally associated with combined flow in marine environments, these Toe Head sandstones are interbedded with desiccated mudrocks and mudrocks containing calcareous or iron rich palaeosols. Moreover, the organic residues show a lack of pyritisation and scolecodonts, both of which typically appear near the base of the overlying formations (Old

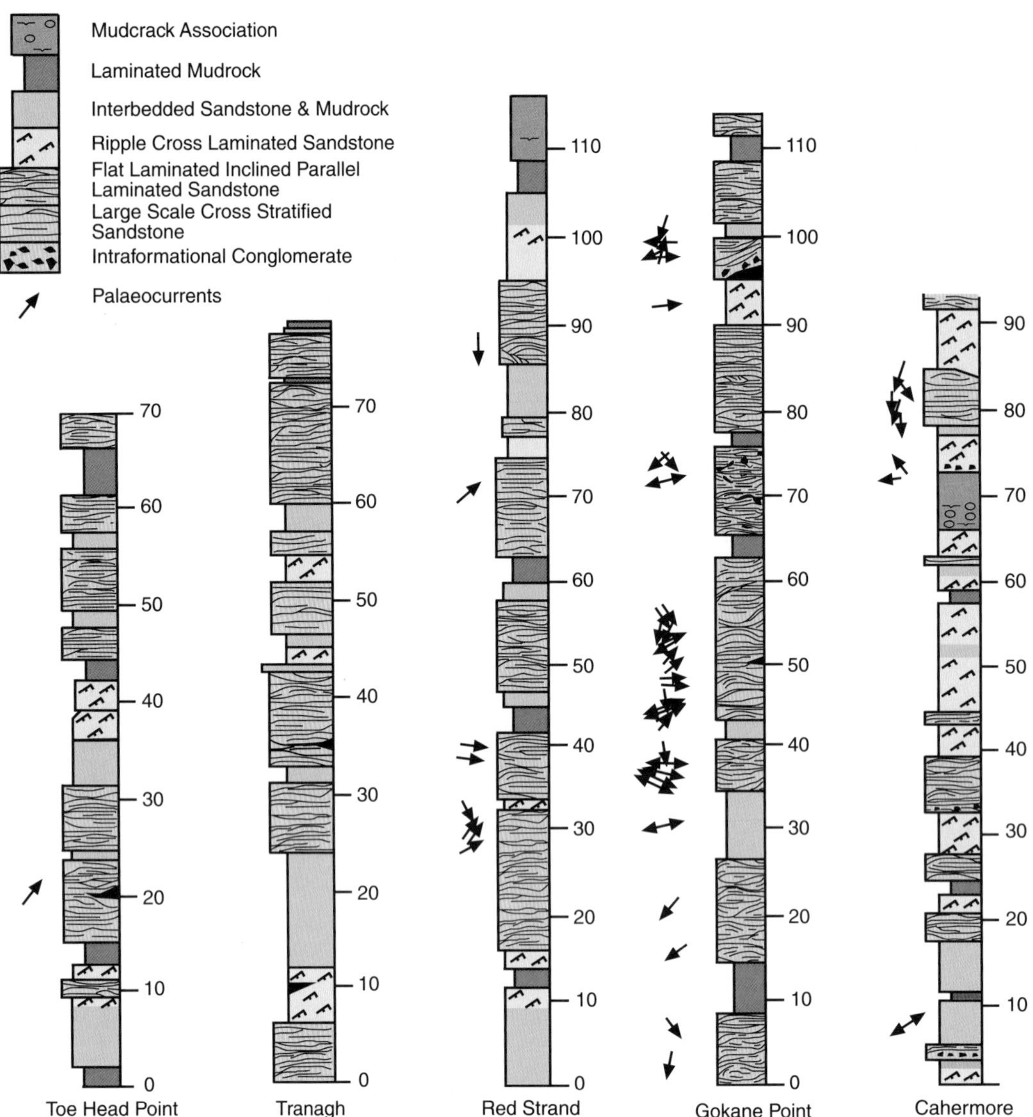

Fig. 9.25 Logs of Toe Head Formation (*after* Cotter and Graham, 1991).

Head Sandstone, Ardaturrish), which have indisputable marine influence. This apparent contradiction of environmental indicators may be due to the importance of fine to very fine sand sizes and possibly the amount of suspended load as controls on the formation of low-angled cross-stratification (Cotter and Graham, 1991). Williams *et al.* (1989) have used the presence of 'bundled bedding', heterolithic lithologies, and evidence for wave action to indicate marine influence during deposition of the Toe Head Formation. However, they and most previous workers (Graham. 1972, 1975a, 1983; Kuijpers, 1975; Naylor, 1975; McAfee, 1987; Reilly and Graham, 1976) suggest that deposition was mainly on a fluvial coastal plain.

The upper part of the Toe Head Formation shows a fining of grain size, decrease in thickness of sandstones and an increase in the presence of heterolithic lithologies, albeit with limited grain size separation (Graham, 1972, 1975a; Reilly and Graham, 1976; Naylor, 1975; Kuijpers, 1975; Williams *et al.*, 1989). It has been suggested by Naylor *et al.* (1977) and Williams *et al.* (1989) that these differences may be sufficiently distinct to allow division into two members. All workers are agreed that the contact with the overlying formations is extremely gradational and generally takes place in a mud-dominant sequence.

Published palaeocurrent data are available only for the sections south and east of Dunmanus Bay and are

Fig. 9.26 **A)** Inclined parallel laminae resembling hummocky cross-stratification, Cahermore, Beara Peninsula (*Photo Ed Cotter*). **B)** Thinly interbedded sandstone and mudrock, Toe Head Point (*Photo Ed Cotter*). **C, D)** Toe Head Formation, Toe Head Point. **E)** Finely laminated grey mudrocks, locally rich in plant debris, eastern side of Toe Head (*Photo Ed Cotter*).

consistently west to east, although bimodal patterns are evident at some localities (Cotter and Graham, 1991). Generalised palaeocurrent arrows presented by Williams *et al.* (1989) from Dunmanus Bay northwards are consistently north to south, although they note bimodal patterns for some localities. It is not yet clear to what extent this palaeocurrent change is simply a variation in space or whether it may be partly time dependent.

There are several possible causes of the increased proportion of sandstones, predominance of non-red colours, and change in sedimentary structures evidenced by the Toe Head Formation. Factors such as higher water tables on what must have been very flat coastal plains, rising base level, and local subsidence patterns may all have been important.

The Kiltorcan Formation exposed in the north and east of the Munster Basin and overstepping its margins was accumulated over a much longer time span that spans the Devonian–Carboniferous boundary (LL to BP biozones). As thicknesses (200–550 m) are similar to those for the Toe Head Formation, subsidence rates were clearly less in this area. This formation is less well exposed than the Toe Head Formation but is likewise characterised by thick, non-red sandstones, often in channel forms, intraformational conglomerates, and both red and non-red mudrocks, some with evidence for exposure of the sediment surface. The area around Kiltorcan is famous for a very well preserved flora deposited in shallow ephemeral lakes on a coastal floodplain (Colthurst, 1978). These deposits have yielded excellent examples of the fern-like *Archaeopteris*, lycopods, and also the first Devonian platyspermic seed (Chaloner *et al.*, 1977). The flora is associated with fossil fish, eurypterids, crustacea and the large freshwater bivalve *Archanodon jukesi*. The formation has been interpreted as the deposits of floodplains traversed by single channel river systems derived from the north and north-east (Colthurst, 1978; Boldy, 1982; Carruthers, 1985). It has been noted by Penney (1980b) that these sandstones are generally very feldspathic, containing abundant microcline, and he suggested that this may be related to unroofing of the Leinster Granite to the north-east.

With the exception of Leflef (1973), descriptions of this fluvial coastal plain facies are much less detailed for areas between the outcrops of the Toe Head and Kiltoran formations. It is apparent, however, that similar non-red strata beneath the overlying marine clastics have only a limited development (Naylor, 1969; MacCarthy, 1974; MacCarthy *et al.*, 1978; Clayton *et al.*, 1982; Wingfield, 1968). Thus it is appropriate that the Kiltorcan and Toe Head formations remain distinct and confined to their present areas of usage. These differences in the development of the fluvial coastal plain facies may be related to the known pulsatory nature of the Tournaisian transgression (Clayton *et al.*, 1986a; MacCarthy, 1987).

Correlation of non-marine successions within the Munster Basin

Erection of even locally viable lithostratigraphies has proven to be a challenge for all workers on the Munster Basin Old Red Sandstone successions. This is due to the relatively thick packages of sedimentary rocks and highly gradational contacts between formations in any one area. The presence of large-scale lateral facies changes has meant that successions traced across large-scale fold structures are subtly different. Coupled with the lack of distinctive marker horizons in most areas, this has led to uncertainties of correlation. Considering some local successions as facies sequences has allowed the elucidation of changes in fluvial style through time, e.g. the superposition of two distinct fluvial fans by Williams *et al.* (1989) and the upward changes from sheetflood to channelised fluvial systems by Graham *et al.* (1992). Although the approach of considering the basin fill in total has greatly aided our understanding in the context of a fluvial distributary system (Graham, 1983; Williams *et al.*, 1989) it does not provide a basis for local correlation.

At present a reasonably consistent although unrefined correlation exists for areas within the influence of the major fluvial fans present south of Killarney (Capewell, 1957a; Husain, 1957; Graham, 1983; Williams *et al.* 1989; Fig. 9.19). However, correlation of this succession with events around the basin margin in Dingle is still uncertain. The obvious similarity of the sequences seen in north-west Dingle and the Galtees has been noted above. It is tempting to suggest that these are time correlatives reflecting tectonic events affecting the whole basin margin, but such a hypothesis is very difficult to test. Matching of the Galtee sequence with that in the basin to the south-east essentially relies on matching influxes of coarse grained sediment. However, there are inherent problems with this approach as the conglomeratic levels are known to pinch out at variable distances southwards and some of the relevant marginal correlatives could become buried during backfaulting of the basin margin. The problem is compounded by the apparent petrographic similarity of these coarse influxes, all being dominated by vein quartz, quartzite, and other resistant clasts. Thus the arrangement shown in Fig. 9.27 is just one possible correlation. The resistant assemblage implies significant transport histories for the surviving sediment.

Further progress in correlation within this large sediment prism can only come from an increased amount of biostratigraphical and chronostratigraphical data. Biostratigraphical data are sparse for much of the Munster Basin. At the western end of the Iveragh Peninsula (Figs. 9.19; 9.36), Russell (1978) has described fish fragments of the genus *Bothriolepis* that were thought to indicate a maximum age of uppermost Middle Devonian.

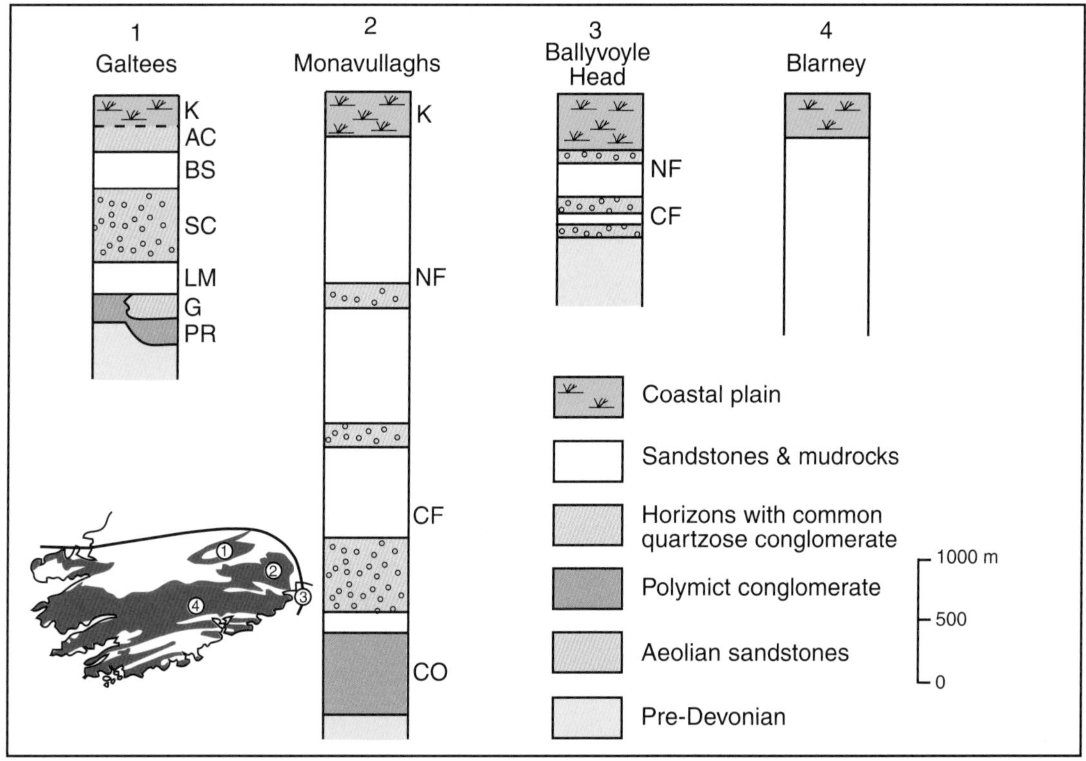

Fig. 9.27 Simplified correlation of Old Red Sandstone in the eastern part of the Munster Basin. Data from Boldy (1982), Carruthers (1985), MacCarthy *et al.* (1978).

Re-examination of this material by Bray (1990) has indicated that it belongs to a new species, *B. russelli*. An age range of Lower Frasnian to mid-Famennian was suggested by comparison with close relatives. Material identified as a *Sauripterus* scale from *c*.1.8 km above the base of the 3.8 km Old Red Sandstone succession here (within the St. Finan's Sandstone Formation) was taken to indicate a Famennian age, but re-examination of this by Bray (1990) suggests it is indeterminate Holoptychian material of little biostratigraphical significance.

A microflora from the Valentia Slate Formation, the oldest exposed formation in western Iveragh, termed the Reenagaveen microflora (Williams *et al.*, 2000b), has been assigned to the TCo spore biozone of Streel *et al.* (1987), which indicates a late Givetian to early Frasnian age. At the eastern end of the Iveragh Peninsula, Higgs and Russell (1981) described a palynological assemblage from low in the succession, here estimated to be >5 km thick (Fig. 9.36), for which they suggest a lower Frasnian age. Re-examination of this microflora suggests a mid-Frasnian age (BM spore zone of Streel *et al.*, 1987) (Williams *et al.*, 2000b). A closely comparable assemblage was previously described by Clayton and Graham (1974) from Clear Island in SW. Cork and assigned a lower Frasnian or upper Givetian age. Re-collection of some of the Cape Clear localities now allows a more refined early to mid-Frasnian age to be suggested, also assigned to the BM spore biozone of Streel *et al.* (1987) (Higgs *et al.*, 2000). In addition associated palynofacies analysis has identified a level with acritarchs and scolecodonts indicating a period of marine incursion and confirming the marine connection necessary for the distribution of the fish faunas.

There is a limited presence of volcanic rocks within the Munster Basin that casts some light on basin origin and also allows the possibility of some chronostratigraphical control. Minor developments of lavas in the east of the basin are of basic composition (Penney, 1978; Boldy, 1982), as are associated intrusive and extrusive rocks at the western end of the Iveragh Peninsula (Capewell, 1975; Russell, 1984; Graham *et al.*, 1995; E.A. Williams *et al.*, 1997) (Figs. 9.28; 9.29). However, volumetrically the most important igneous rocks are the thick dacitic and rhyolitic lavas and associated pyroclastics developed in three separate but closely spaced centres east of Killarney, termed the Lough Guitane Volcanics. Detailed descriptions by Avison (1984a,b) (Fig. 9.30) show ring faults and syn-sedimentary grabens to be associated with

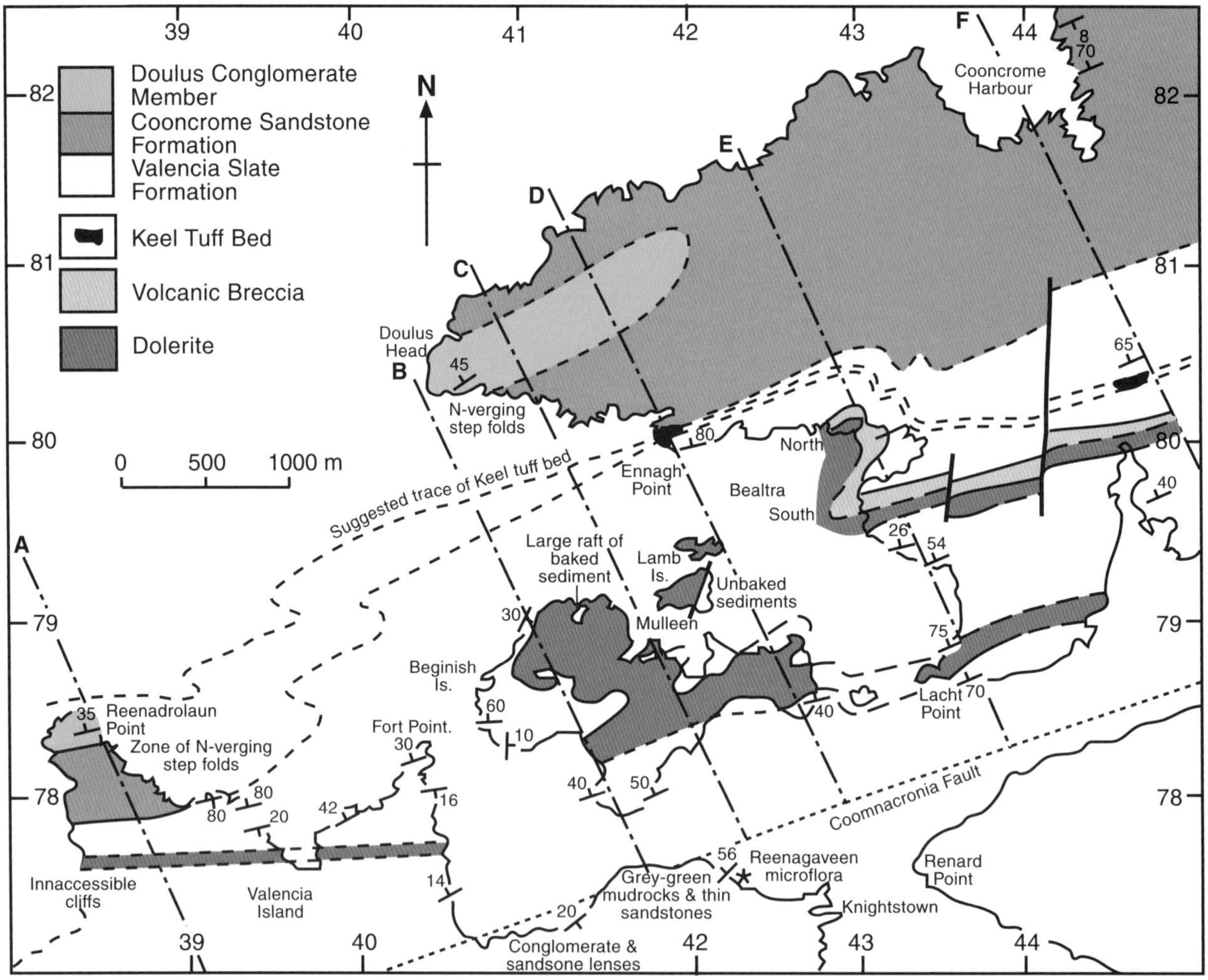

Fig. 9.28 Map of the volcanic rocks of Valentia Harbour (after Graham *et al.*, 1995; Williams *et al.*, 2000).

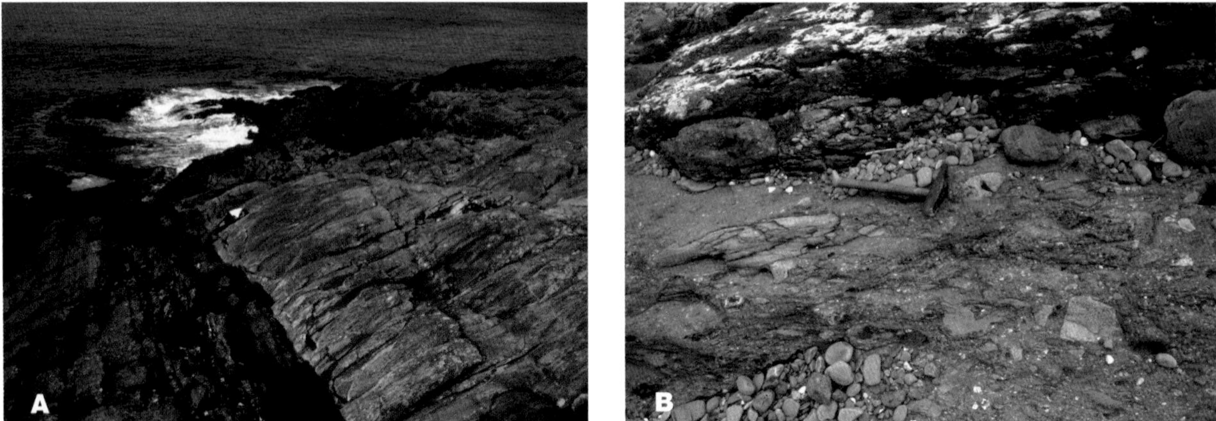

Fig. 9.29 **A)** Keel Tuff Bed at Ennagh Point, which is 9.5 m thick; person for scale near base. The relatively massive tuff is picked out by its distinctive jointing. **B)** Peperite from Bealtra Bay North described by Graham *et al.* (1995).

these centres. The strong petrographic and geochemical similarities of these three centres led Avison (1984a) to propose a major granitic intrusive body at depth, along with control of magma ascent by deep crustal fractures. The available gravity data have also been used to suggest a major acidic intrusion (Howard, 1975; Readman *et al.*,

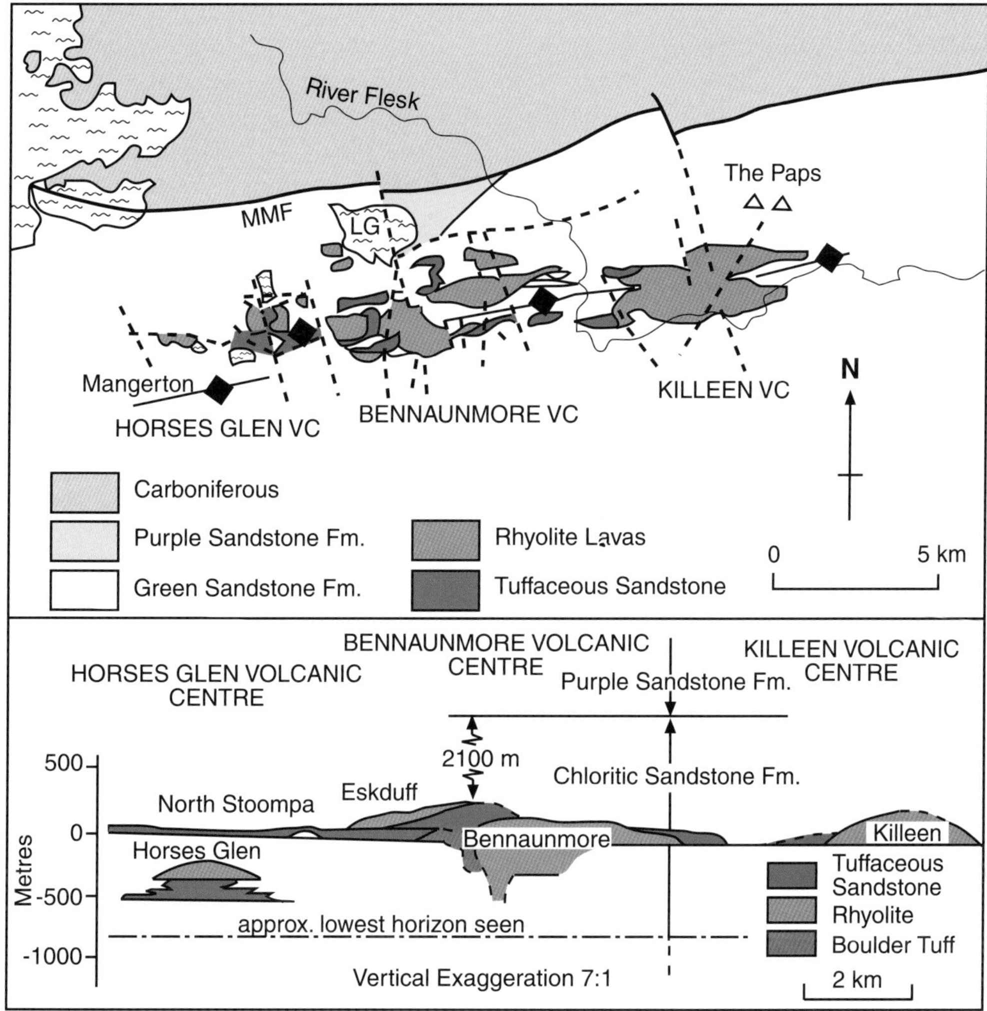

Fig. 9.30 Summary map and profile of the Lough Guitane volcanic centre (after Avison 1984a,b); LG – Lough Guitane.

1997). These acidic rocks have been reinvestigated by Williams *et al.* (2000b). All occur within the Chloritic Sandstone Formation and correlation with areas to the south and west is by tracing through the major fold structures. On this basis, the younger Horse's Glen centre (Fig. 9.30) is considered by Williams *et al.* (2000b) to be at an approximately equivalent horizon to the mid-Frasnian Moll's Gap microflora. The lower tuffs of the Horse's Glen centre contain some inherited zircons, but an aggregate of nine very small needle-like zircons that are interpreted to be magmatic yields a concordant age of 378.5 +/- 0.2 Ma. The Killeen Tuffs, considered to be stratigraphically higher by Avison (1984a, b) (Fig. 9.30) yield discordant, but older ages. Williams *et al.* (2000b) suggested that a weighted mean $^{207}Pb/^{206}Pb$ age of 384.5 +/- 1.0 Ma represents a crystallisation age. Thus, either this discordant age does not represent a crystallisation age, or the correlations shown in Fig. 9.30 are incorrect and there are previously unrecognised faults, the explanation favoured by Williams *et al.* (2000b). This problem is currently unresolved.

A tuff horizon up to 9 m thick, the Keel (Ennagh) Tuff Bed (Fig. 9.29A). described by Graham *et al.* (1995) was shown to occur in markedly different facies across the Portmagee Anticline, a surface distance of some 12 km. The tuff occurs some 800 m above the lowest *Bothriolepis* record. The initial geochemical data of Graham *et al.* (1995) was confirmed and amplified by E.A. Williams *et al.* (1997, 2000b) who also noted differences between this thick tuff and the Lough Guitane volcanics, suggesting derivation from different sources. E.A. Williams *et al.* (1997, 2000b) suggested a crystallisation/eruption age based on a combined $^{207}Pb/^{206}Pb$ mean age of 384.9 +/- 0.4 Ma. Recognition of trackways made by a tetrapod of *c.*1 m length (Stossel 1995) from beneath the dated tuff horizon are ecologically interesting but cannot be used biostratigraphically.

A suite of basic igneous rocks occur in western Beara (Pracht and Kinnaird, 1997) as intrusive sills within the Eagle Hill Formation of James and Graham (1995), approximately 1 km below the top of the Old Red Sandstone succession. However, because there is no clearly related extrusive activity these rocks can only be stated with any certainty to be pre-cleavage in age.

Devonian marine facies

Marine Devonian rocks first appear in southern Ireland in the late Famennian. This marks the start of a transgressive phase that eventually led to marine rocks being deposited in the northernmost parts of Ireland during the Early Carboniferous (George *et al.*, 1976). The common preservation of palynomorphs has allowed erection of a refined biostratigraphical scheme for this interval (Higgs *et al.*, 1988). The simplest interpretation of the available biostratigraphical data is that the shoreline initially progressed northwards, roughly parallel to the present south coast (Clayton and Higgs, 1979; Clayton *et al.*, 1986a; Quin, *in press*; Fig. 9.31).

In essence, the environmental sequence in southernmost Ireland is from fluvial coastal plain through marginal marine clastics to offshore clastics, the last named commencing mainly in earliest Carboniferous times (Clayton *et al.*, 1986a). The bases of the Old Head Sandstone Formation in the south and the Ardaturrish Formation in the west are both defined on the incoming of small-scale structured heterolithic units composed of flaser, wavy, and lenticular bedding (Fig. 9.32). Although isolated examples of these structures occur in the underlying fluvial coastal plain facies, it is only at this level that thick heterolithic units start to dominate the succession. Whilst thin heterolithic beds are unlikely to be environmentally sensitive, thick sequences indicate prolonged, relatively low energy, variable currents typical of wave and tidal current-dominated nearshore environments. This interpretation is supported by the incoming of scolecodonts, which are generally taken to be good indicators of marine waters, in organic residues from this level (McNestry, 1989). In most areas the transition from the underlying coastal plain facies is extremely gradational and occurs in mud-rich sequences (Graham, 1975a, b; Naylor, 1975; Jones, 1974; Naylor *et al.*, 1977). Typically there is an increase in grain size separation between muddy and sandy laminae, a decrease in the amount of climbing ripples, and an increase in features diagnostic of wave action. This may imply extensive coastal mudflats over which there was a gradual landward attenuation of these marine sorting processes.

In the type area of the Old Head Sandstone Formation at the Old Head of Kinsale, 840 m of marine late Devonian rocks are present in which grain size seldom exceeds fine sand. Gradual changes in types and proportions of facies were noted by Naylor (1966) and Kuijpers (1972) and similar changes were shown to occur in time-equivalent strata in areas along strike to the west (Graham, 1975b; Fig.9.33) despite major differences in overall thickness. Essentially these changes are a decrease in heterolithic lithologies upwards and an accompanying increase in flat and low-angle bedded sandstones and uniform mudrocks. In the lower parts sparse faunas of restricted *Modiola*-type occur (Kuijpers, 1972; Naylor, 1966; Naylor *et al.*, 1977), a feature also noted for the Ardaturrish Formation (Jones, 1974). There are also horizons displaying polygonal desiccation cracks and channels filled both by flaser bedded sandstones and by muddier heterolithic facies. Palaeocurrents derived from small-scale ripples typically show a bipolar pattern with modes directed north–south (Kuijpers, 1972) or very variable directions (Van Gelder, 1974). Taken together, these features suggest deposition on tidal flats, although it is not possible to assess what proportion were intertidal and what subtidal. The recognition of coarsening upward, fining upward, coarsening upward–fining upward, and random sequences within the Old Head Sandstone Formation has also been noted in other tidal flat sequences (Reineck, 1969; Terwindt, 1971). Quin (*in press*) suggests that the Old Head Sandstone Formation can be correlated throughout the South Munster Basin and similarly the distinction between the lower and upper parts can also be recognised. He terms these the Daunt and Tower members respectively, based on divisions first recognised in the Roberts Head area by Van Gelder (1974) (Fig. 9.34).

The upper parts of the formation, the Tower Member, typically show indications of higher energy conditions in the sandstones which display parallel lamination and cross-bedding as well as ripple structures. There is also an increase in indicators of wave action (Kuijpers, 1971, 1972; Graham, 1975b; De Raaf and Boersma, 1971; Van Gelder, 1974). These sandstones are interpreted as a complex of submerged and emergent coastal barriers and offshore sands (Graham, 1975b; MacCarthy, 1987; Williams *et al.*, 1989). However, Quin (*in press*) suggests that many of the sharp-based sandstones may

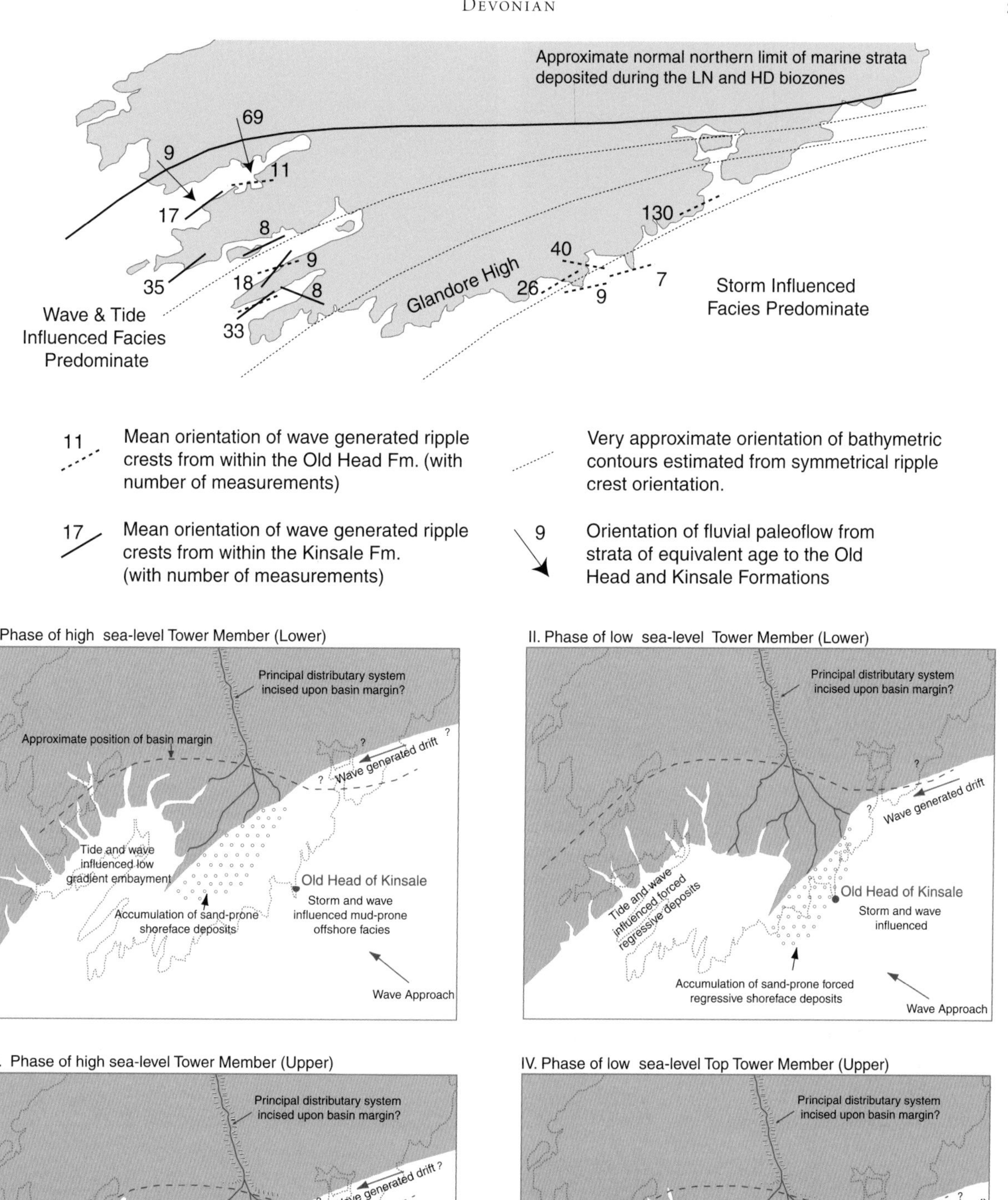

Fig. 9.31 Coastal orientation and bathymetry interpreted from sedimentary structures and stratigraphy (upper part) and interpreted evolution of coastal zone sedimentation during Old Head Sandstone Formation times (*after* Quin, *in press*).

A
B
C
D
E
F
G
H

Fig. 9.32 **A, B)** Longitudinal and cross-sections through a small tidal channel in heterolithic facies, Galley head. **C, D)** Typical heterolithic facies, Galley Head. **E)** Wave-generated ripple structures, Galley Head. **F)** Gutter casts in offshore mudrocks, Seven Heads. **G)** Strongly bioturbated heterolithic facies, Galley Head. **H)** Contact between the Old Head Sandstone Formation (left) and Kinsale Formation (right). The dark colour of the Castle Slate Formation identifies the contact, Dunnycove Bay, Galley Head. **J)** Sheet-like rippled sandstones interpreted as sandy tidal flats, Galley Head.

Below Fig. 9.33 Thickness and facies sequence comparisons between Old Head of Kinsale and successions deposited on the Glandore High.

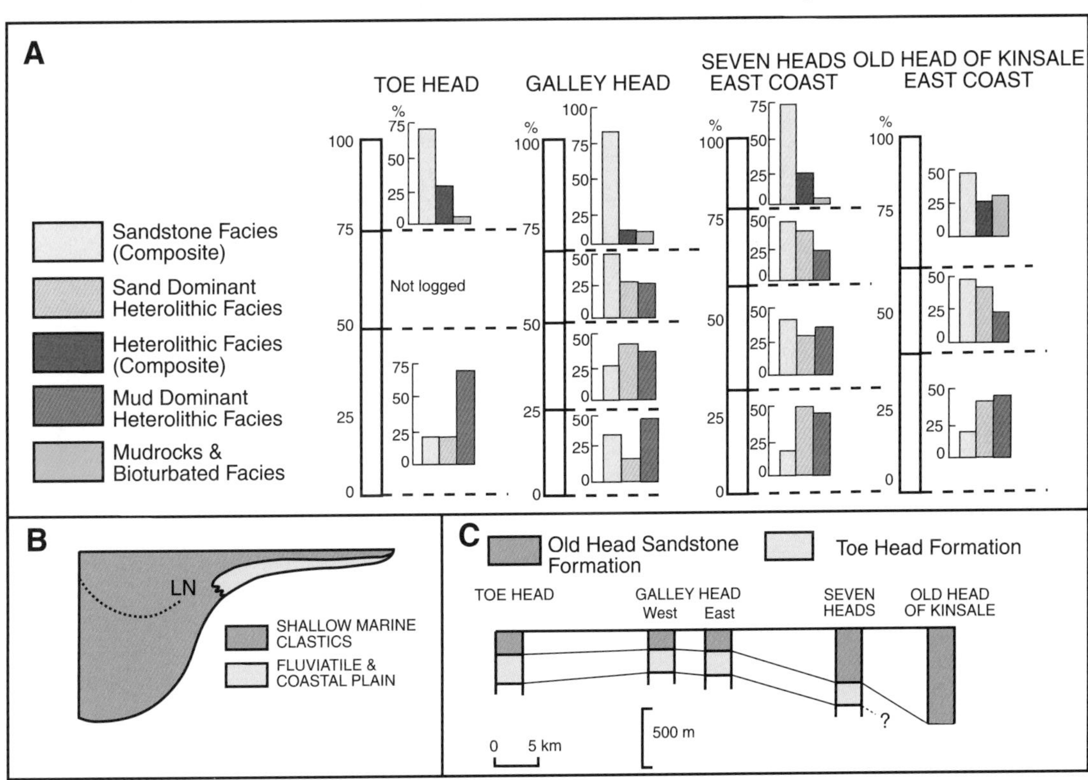

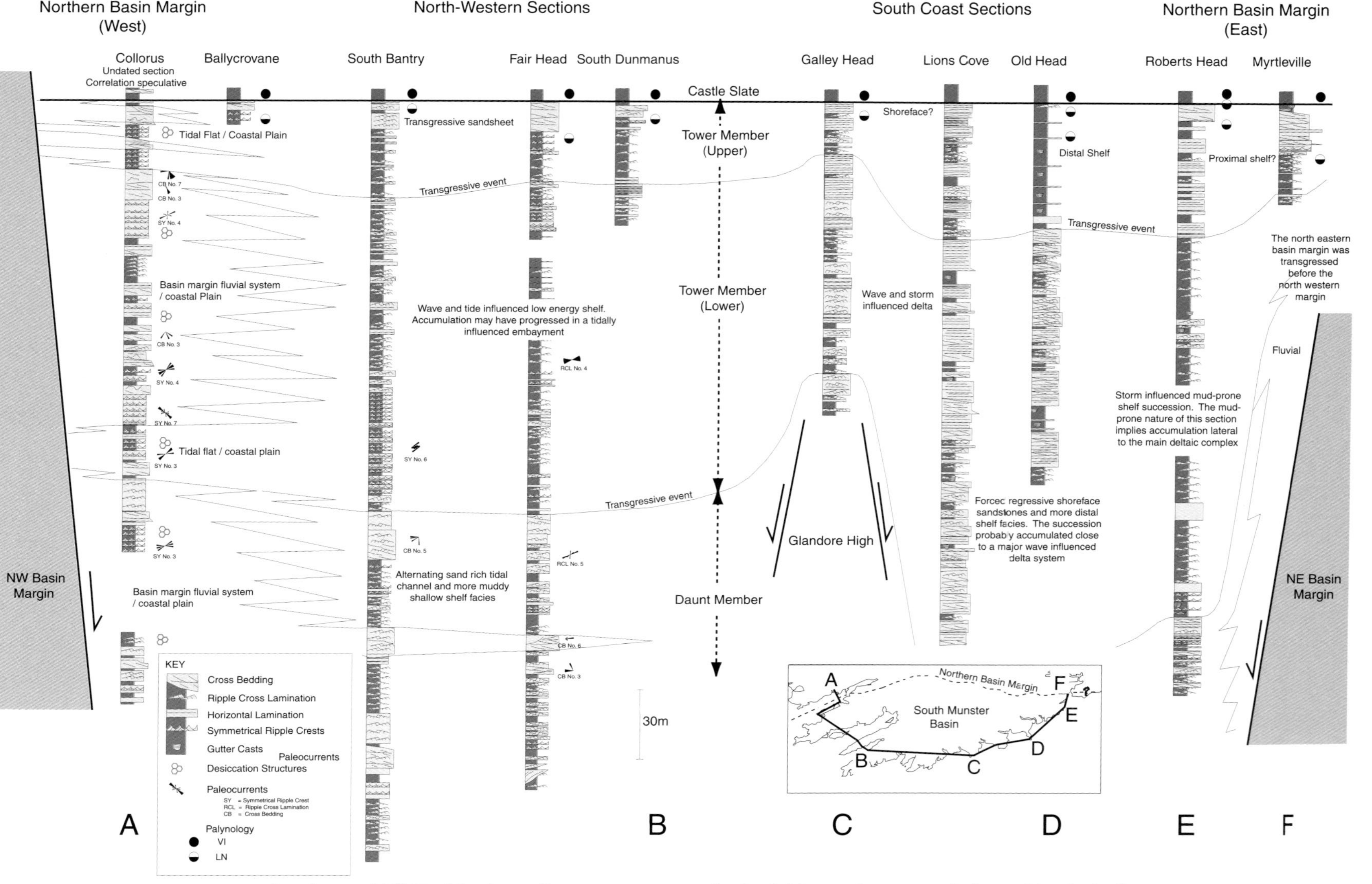

Fig. 9.34 Correlation of Old Head Sandstone Formation sections in the South Munster Basin using the base of the Castle Slate Formation as a regional datum. (*after* Quin, *in press*).

represent lowstand shoreface deposits. Locally bipolar currents and channels indicate the importance of tidal currents (Kuijpers, 1972; Graham and Reilly, 1976), whilst the recognition of hummocky cross-strata in many of the sandstones (Williams *et al.*, 1989) has been used to suggest the dominance of storms. The associated mudrocks which characterise this upper part of the Old Head Sandstone Formation are commonly extensively bioturbated and have mainly been interpreted as lagoonal or interdistributary bay deposits (Kuijpers, 1972; Jones, 1974; Van Gelder, 1974; Graham, 1975b), but also in part as shelf deposits (MacCarthy, 1987; Williams *et al.*, 1989). Quin (2001 *in press*) has recorded numerous gutter casts at this level, which are generally indicative of more open, storm-influenced marine settings. The top part of the Old Head Sandstone Formation is typically the coarsest grained sandstone in the sequence, frequently shows trough cross-bedding, and commonly contains shell debris, particularly crinoids (Graham, 1975b; Naylor *et al.*, 1977; MacCarthy, 1987). In the Dunmanus area this interval is the first significant sandstone above the Daunt Member and is associated with the first evidence, in these sections, of strong storm influence in the form of gutter casts and graded beds in the associated mudrocks. The marked contrast in energy level between the conditions of deposition of this sandstone and that of the overlying early Carboniferous mudrocks (Castle Slate Formation) suggests that there is a small non-sequence at this level produced by shoreface erosion during a transgressive phase (Graham, 1975b; MacCarthy, 1987). MacCarthy (1987) also recognised hummocky cross-stratification within this unit which he interprets as forming in a shallow marine, storm influenced subtidal shelf.

Rocks assigned to the Ardaturrish Formation of the Bantry-Beara area are undoubtedly similar in their lower part to the lower part of the Old Head Sandstone Formation. Although the early Carboniferous Castle Slate has been recognised on the north side of Bantry Bay (Van der Zwan and Van Veen, 1978; MacCarthy, 1987; Higgs *et al.*, 1988), its precise position on the ground relative to the boundaries of the Ardaturrish Formation mapped and described by Jones (1974) is unclear, and this hinders regional correlation; nevertheless it is clearly recognizable at Ballycrovane (Quin, 2001) on the northern side of Beara. These north-western sections show little evidence of strong storm influence and Quin (*in press*) has suggested that these areas may represent a more protected coastline (Fig. 9.31).

Of particular interest is the identification of two major structural elements within the Munster Basin which cannot be proven or do not exist in older strata. During the late Famennian or Strunian differential subsidence across the northern margin of the Munster Basin was reduced and a new region of maximum subsidence, the South Munster Basin, developed, of which the well defined northern margin, the 'Cork–Kenmare Line', appears to have a simple east-west trend, although it is likely to be composed of a series of *en echelon* segments (Naylor *et al.*, 1989). This hinge line persisted for at least 30 Ma, forming the southern margin of the Lower Carboniferous carbonate shelf, and controlling the position of the latest Devonian–earliest Carboniferous shoreline (Naylor and Sevastopulo, 1979; Clayton *et al.*, 1986a). During the late Devonian and early Carboniferous a zone of reduced subsidence has also been demonstrated in south Cork (Graham, 1972, 1975b) termed the Glandore High (Naylor *et al.*, 1974). This high had no surface expression during the late Devonian as identical environmental sequences are present on and off the high (Graham, 1975b) and the shoreline position during the transgression in the LL palynozone is not deflected by this high (Naylor *et al.*, 1983). This suggests that the transgression was caused by either a eustatic rise of sea level or by regional subsidence. A substantial northward movement of the shoreline to the Cork-Kenmare line occurred during the LN palynozone. This top Devonian palynozone is thought to occupy only a short time interval because it is represented by a small thickness of rock everywhere outside the Munster Basin. Nevertheless, more than 250 m of rock was deposited in south Cork during the time represented by this biozone (Clayton *et al.*, 1986a). The late Famennian deposits from southern Ireland are the thickest known in Europe and clearly indicate rapid subsidence here in the latest Devonian that persisted through into the early Carboniferous. To what extent this is a function of regional subsidence control operating on a linear clastic shoreline and to what extent it reflects load-enhanced subsidence due to the presence of a wave- and tide-dominated delta remains debatable. However, it is clear that the succession does not represent a simple and gradual northward transgression but rather a series of pulsed transgressive and regressive events superimposed on this overall pattern (Fig. 9.35). It is the general overall balance between rates of subsidence and sediment supply in the late Devonian and earliest Carboniferous in the South Munster Basin that has led to the accumulation of such exceptionally thick shallow marine sequences.

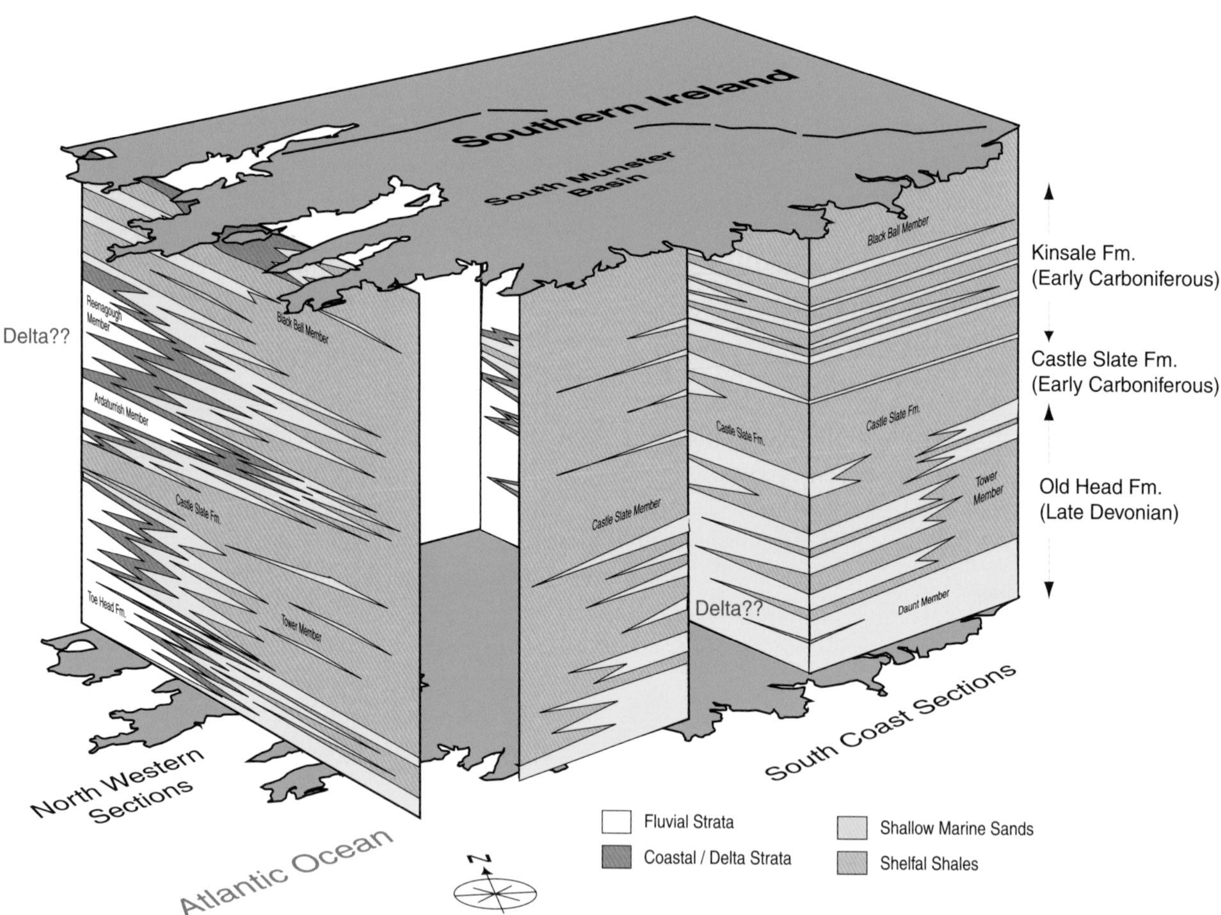

Fig. 9.35 Block diagram of latest Devonian to early Carboniferous sedimentation in the South Munster Basin (*after* Quin, *in press*). Superimposed thickness changes and the effects of Variscan deformation are not included.

Evolution of the Munster Basin

The Munster Basin contains the greatest thickness of non-marine sediment of Upper Devonian age anywhere in Europe. The general asymmetrical profile and present thicknesses of >6 km in places have been demonstrated above. Cooper *et al.* (1984) stated that properly balanced and restored true scale sections removed the need for basin margin faults. However, the direct evidence of marginal faults (Carruthers, 1985) and the indirect evidence from thickness and facies changes seems compelling and is generally accepted (Naylor and Jones, 1967; Clayton *et al.*, 1980a; Graham, 1983; Naylor *et al.*, 1981, 1983; Sanderson, 1984; Todd, 1989; Price, 1989; Williams, 2000; Williams *et al.*, 1989, 2000b). Rapid thickness changes reflecting syn-depositional fault control are most obvious in profiles across Dingle Bay where a long-lived zone of crustal weakness concentrated the extension (Price and Todd, 1988; Todd, 1989; Price, 1989; Figs. 9.17, 9.19). Farther east no single lineament is quite so dominant and there is progressive northward migration of syn-sedimentary faulting during areal expansion of the basin (Boldy, 1982; Carruthers, 1985; Price, 1989). The dominant ENE–WSW syn-depositional faults are parallel to the Caledonian structural grain and probably represent reactivation of basement inhomogeneities. Footwall uplift during extension has been advocated for the Dingle area (Todd, 1989) and can explain the southward distribution of metamorphic clasts into the Doulus Conglomerate on the Iveragh Peninsula, probably by reworking of the Inch Conglomerate (Graham, 1983).

Whilst there is general acceptance that the Munster Basin is a large extensional basin, there are differences in detail in attempts to model basin development (Price, 1989; Sanderson, 1984; Williams, 2000). B-values of between 1.3 and 2.0 have been suggested. Uncertainty still remains as to the total timescale involved, although recent isotopic dates (Williams *et al.*, 2000b) have greatly assisted here. Initial estimates of time-averaged sedimentation rates of 0.4–0.5 mm a^{-1} for the Iveragh Peninsula (Graham, 1983; Price, 1989) that were consistent with the then-available data would now need to be modified downwards. Estimates utilising the recent isotopic data of Williams *et al.* (2000b) would suggest rates of 0.17–0.25 mm a^{-1}, although rates are unlikely to have been constant throughout the basin history. These are similar to the >2mm a^{-1} for the South Munster Basin during the LN biozone estimated by Naylor *et al.* (1989).

The presence of cyclical variation in characters such as sandstone percentage, coset thickness, and maximum set thickness has been recognised by Kelly (1992, 1993), Kelly and Olsen (1993), Kelly and Sadler (1995), and Sadler and Kelly (1993). They recognise two main scales of cyclicity that are spaced in roughly 1:4 thickness ratio. They argue for climatic causes related to the 412,00 and 111,000 year Milankovitch eccentricity cycles. This interpretation was supported by correlation of predicted time-averaged sedimentation rates of spans of 0.38–0.46 mm a^{-1}. However, use of the rates suggested by Williams *et al.* (2000b) would produce timespans that no longer match the Milankovitch bands. Nevertheless, the cyclicity they recognise requires an explanation and the timescale on any estimate appears shorter than that expected for tectonically driven subsidence changes such as those suggested by Graham *et al.* (1992) and Sadler and Kelly (1995).

The southward migration of the main depocentre in latest Devonian times, when the South Munster Basin was initiated, may reflect the locking of the northern basin bounding faults due to a combination of fault rotation and applied sediment load. This depocentre migration transverse to the basin axis is strong support for an origin by orthogonal extension rather than by transtensional pull-apart (Price and Todd, 1988). The overall expansion of the basin in the later stages of its history may reflect the flexural response of the lithosphere to the synrift sediment load (Diemer and Bridge, 1987; Graham and Clayton, 1988). The underlying cause of the extension which produced the Munster Basin remains uncertain but is likely to relate to events farther south in the Variscan realm. An origin by back arc rifting has been suggested (Leeder, 1982).

Late Devonian alluviation

It can be seen from Fig. 9.2 that Old Red Sandstone rocks, including late Devonian representatives, are present over a substantial area to the north and east of the Munster Basin. Details of known Devonian dates are given in Higgs *et al.* (1988). These Old Red Sandstone sequences are everywhere relatively thin (up to a few hundred metres) and of broadly similar type. The most detailed description is from Kerry Head (Fig. 9.2) (Bridge *et al.*, 1980; Bridge and Diemer, 1983; Diemer *et al.*, 1987). Here Diemer *et al.* (1987) divided the exposed non-marine succession of *c.*700 m into three formations on the basis of mean grain size and relative proportion of sand bodies. Mean grain size of the sandstones decreases upwards through the sequence and there is a progressive loss of red coloration. The sandstone bodies are interpreted as the deposits of laterally migrating, southerly flowing river channels with sinuosities of <1.2 and which decreased in slope as the palaeocoastline approached from the south. Bridge and Diemer (1983) analysed well-exposed coastal sections displaying lateral accretion surfaces within the sandstone bodies. These allowed a quantitative estimation of depth, velocity and discharge using empirical equations derived from modern rivers, relating these parameters to channel morphology. They recognised two sizes of channel, one 1–2 m deep with bankfull discharges of 15–30 m^3s^{-1}, and another 3–4 m deep with bankfull discharges of 130–180 m^3s^{-1}. The larger of these would be roughly equivalent in size to one of the larger rivers in the British Isles at the present day. The sandstone–mudrock interbeds are interpreted as the deposits of vegetated floodplains beneath which the water table was progressively rising.

Similar interpretations have been advanced for these late Devonian strata elsewhere, although based on less favourable exposures (Doran, 1971; Emo, 1978; Colthurst, 1978; Feehan, 1982b). Marginal conglomerates and pebbly sandstones formed in less sinuous, bedload dominant rivers, particularly near the fall line. The sequence at Dunmore East and Hook Head in south-east Ireland best displays this feature (Gardiner and Horne, 1981; Ori and Penney, 1982; Sleeman *et al.*, 1983). There is an upward progression from coarse-grained proximal braided stream deposits of latest Devonian age to the deposits of more sinuous single channel river systems. In contrast to the Munster Basin, this area offers close comparisons with the Upper Old Red Sandstone sequence of south-west Wales (Williams *et al.*, 1982).

Overstep of the margins of the Munster Basin might be expected due to lithospheric flexure (Watts *et al.*, 1982), but this is only a partial explanation for the observed distribution of late Devonian Old Red Sandstone facies. The late Devonian Old Red Sandstone extends for a much greater distance away from the Munster Basin than might be expected from a simple flexural response (Fig. 9.2). Also this facies belt migrates gradually northwards ahead of the northerly transgressing sea in the Lower Carboniferous (Sevastopulo and Wyse Jackson, 2001). Thus these sequences can be viewed as primarily being a response to rising base level during the late Devonian. As is well known from Quaternary systems, a rise in sea level promotes alluviation in river valleys and on alluvial plains. This interpretation adequately explains both the widespread nature and the relatively similar thicknesses of the deposits, although some topographic/tectonic factors may be expected to operate locally.

Summary

A simple plot of known biostratigraphical and numerical age determinations against the Devonian time scale emphasises the distinction between those successions related to the Caledonian cycle and those related to the Variscan. Figure 9.36 plots the known age determinations and estimates in an approximate north-south profile. Two features are clear in this plot. Firstly, the age constraints are poorest in the coarser grained 'end Caledonian' Old Red Sandstone successions that were deposited mainly in strike-slip basins. The Dingle Group is proven to extend down into the Silurian and it is possible that some of the other successions also do. Secondly there is a clear switch in the mid-Devonian to Variscan extensional tectonics, and an accompanying geographical shift towards deposition in the southern part of Ireland. The change to areally more extensive deposition is accompanied, in general, by the presence of finer grained sediment. This may reflect both the reduction of topography and the greater size of the drainage systems. The extension of Old Red Sandstone facies northwards and into the early Carboniferous is largely linked to alluviation preceding the Dinantian transgression.

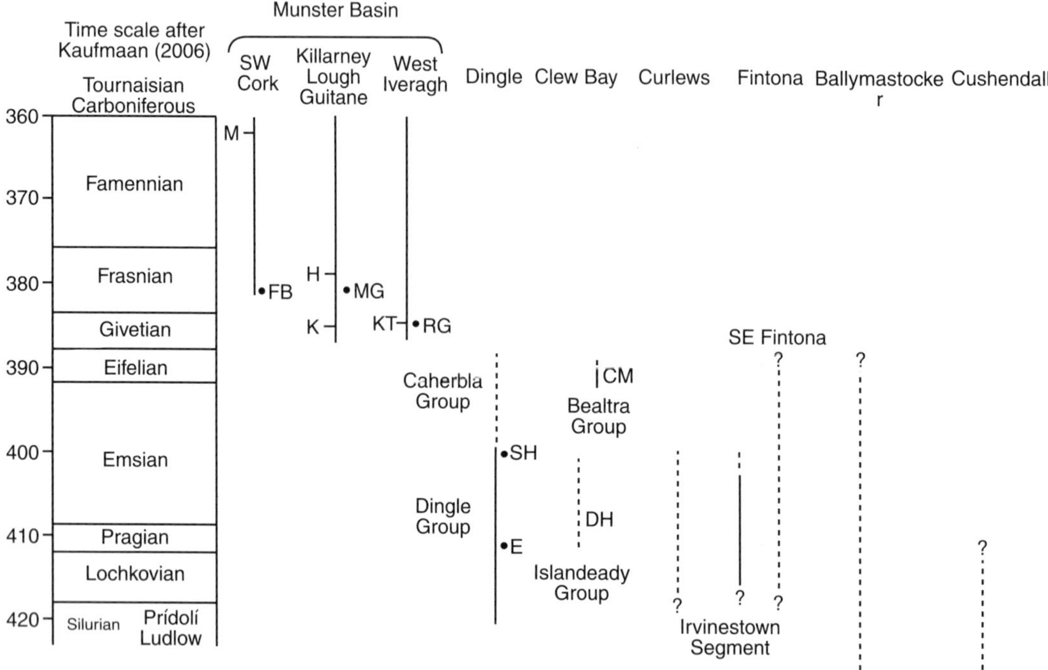

Fig. 9.36 Summary of the age distribution of Devonian successions in Ireland. M – Earliest marine beds, FB – Foilcagh Bay flora, H – Horse's Glen, K – Killeen, MG – Moll's Gap, KT – Keel Tuff, RG – Reenagaveen flora, E – Eask Formation flora, SH – Slea Head flora, DH – Derryharriff flora, CM – Croaghmoyle flora.

10

Carboniferous: Mississippian (Tournaisian and Viséan)

G. D. Sevastopulo and P. N. Wyse Jackson

Rocks of Carboniferous age occur at the surface or beneath Quaternary deposits over nearly half the land area of Ireland. They have been, and still are, of great economic importance: they are quarried extensively for aggregates and industrial minerals, host major base metal deposits, contain coal, and influence water supply and agriculture. Much progress in the study of these rocks has been made since the first version of this book in 1981, to the extent that the broad outlines of the stratigraphy are securely based. Nevertheless, a great deal remains to be done to establish the details of correlation, facies, and palaeogeography.

The Carboniferous System is now divided into the Mississippian and Pennsylvanian Subsystems (Heckel and Clayton, 2006). The Mississippian itself is divided into three stages: Tournaisian, Viséan, and Serpukhovian. The Tournaisian and Viséan stages in Ireland are represented mostly by carbonate rocks and are described in this chapter. The Serpukhovian, in contrast, is represented predominantly by shales and sandstones, sedimentologically more akin to those of the Pennsylvanian; the Serpukhovian and Pennsylvanian are therefore described together in Chapter 11.

Stratigraphical framework

The base of the Carboniferous System has been defined with a GSSP at La Serre, Montagne Noire, France, at a level chosen to coincide with the base of the *Siphonodella sulcata* conodont Biozone. This biostratigraphical datum cannot be recognised in Ireland because the rocks around the Devonian/Carboniferous boundary are of unsuitable facies. However, the boundary between two miospore biozones, the LN and VI Biozones, which has been identified in many sections in the south of the country, correlates very closely, possibly even exactly, with the base of the Carboniferous. The tool for definition of the base of the Viséan stage, it has been agreed, is the first appearance of the foraminiferan *Eoparastaffella simplex* within the evolutionary plexus of *Eoparastaffella*. A GSSP in South China was proposed (Devuyst *et al.*, 2003) and has been ratified. The base of the Serpukhovian (and thus the top of the Viséan) has not yet been specified. For the purposes of this account it is taken at the horizon of the ammonoid *Cravenoceras leion*; it is probable that the GSSP, when it is eventually chosen, will be at a slightly older horizon.

Because the international stages Tournaisian and Viséan provide too coarse a chronostratigraphical framework, a scheme of regional substages derived from Belgium, Britain, and Ireland is used here for the description of the stratigraphy (Table 10.1). The Tournaisian is divided into two substages, Hastarian and Ivorian, named in Belgium (Hance and Poty, 2006; Hance *et al.*, 2006). For the Viséan, the stages (now substages) proposed by George *et al.* (1976) for Britain and Ireland, with the exception of the Chadian, have been adopted in preference to Belgian chronostratigraphical divisions. The oldest division of the Viséan is here given an informal name 'Lower Viséan' because the base of the Chadian Stage is older than the base of the Viséan and, in any case, cannot be identified with confidence away from the stratotype section in Lancashire, England. It should be noted that 'Lower Viséan' in this sense has a more restricted stratigraphical range than the same term formerly in use in Belgium, where the notation V_1 would have included the early part of the Arundian Substage. Some authors have used the term Upper Chadian to signify the Viséan part of the Chadian, but since no stratotype for the Upper Chadian has been proposed, it is in fact a biostratigraphical unit and its use in chronostratigraphy should be discontinued. Now that the GSSP for the base of the Viséan has been ratified, a substage to replace 'Lower Viséan' should be proposed with the same GSSP as the base of the Viséan.

Biostratigraphical correlations have been effected using macrofossils (mostly brachiopods, corals, and ammonoids) and microfossils (palynomorphs, calcareous algae, foraminiferans, and conodonts). Higgs *et al.* (1988a) have made a valuable compilation of Irish Tournaisian palynostratigraphy, which is the source of all the palynological data quoted here whose origins are not acknowledged. Table 10.1 shows the correlation of the various biostratigraphcal schemes and the stages and substages used in this account. While the correlations between many of the boundaries of biozones based on different fossil groups are considered to be accurate, others are less so and there are some unresolved conflicts in correlations based on different biozonal schemes. This is particularly the case at the base of the Brigantian Substage (Cózar *et al.*, 2006; see below p. 261).

The few stratigraphically well-constrained, accurate radiometric ages for the Tournaisian and Viséan stages, together with interpolated ages for the boundaries of the regional substages are shown in Table 10.1.

Table 10.1 Chronostratigraphical divisions of the Tournaisian and Viséan applicable in Ireland; radiometric ages; ammonoid, miospore, conodont, and foraminiferan biozones; and comparison with Belgium. Radiometric ages (from Trapp *et al.*, 2004; Davydov *et al.*, 2005) in bold have good precision and are stratigraphically well constrained; the remainder (from Davydov *et al.*, 2005) are interpolated.

SYSTEM	SUBSYSTEM	STAGE	REGIONAL SUBSTAGE IRELAND	RADIOMETRIC DATE (Ma)	AMMONOID BIOZONE	MIOSPORE BIOZONE	CONODONT BIOZONE	FORAMINIFERAL BIOZONES (MFZ)	REGIONAL SUBSTAGE BELGIUM
CARBONIFEROUS	MISSISSIPPIAN (PART)	VISEAN	BRIGANTIAN	326 **326 ± 0.8**	P_{2c} P_{2b} P_{2a} P_{1d} P_{1c} P_{1b} P_{1a}	NC VF	*bilineatus*	**15** *Janischewskina typica* **14** *Howchinia bradyana*	WARNANTIAN
			ASBIAN	332	B_2 B_1	NM TC	*bilineatus* *commutata*	**13** *Neoarchaediscus*	
			HOLKERIAN	339		TS	*commutata*	**12** *Pojarkovella nibelis*	LIVIAN
			ARUNDIAN	343		TS Pu		**11** *Uralodiscus rotundus* **10** *Planoarchaediscus Ammarchaediscus*	MOLINIACIAN
			LOWER VISEAN	345		Pu	*homopunctatus*	**9** *Eoparastaffella simplex*	
		TOURNAISIAN	IVORIAN	349		Pu CM	*mehli* / *anchoralis* *multistriatus* / *carina*	**8** *Eoparastaffella* M1 **7** *Darjella monilis* **6** *Tetrataxis* **5** *Paraendothyra nalivkini*	IVORIAN
			HASTARIAN	**360.7 ± 0.7**		PC BP HD VI	*inornatus* *spicatus* / *Siphonodella*		HASTARIAN

Regional setting

During the Tournaisian and Viséan, Ireland lay in tropical latitudes (Fig. 10.1) on the southern margin of Laurussia. The configuration and dynamics of the plates that now constitute present-day western Europe at that time were complex and are not fully resolved. Through the late Palaeozoic, ribbon-like terranes, which had rifted from the northern margin of Gondwana, moved northward and docked with the amalgamated Laurentian, Avalonian, and Baltic plates, closing the Rheic ocean, opening the Palaeotethys, and ultimately forming the supercontinent of Pangaea. Ireland was situated in the outer part of the Variscan orogenic belt, which formed as a result of these collisions. It is generally considered to have been in a back arc setting, north of the Ligerian arc, which has been identified in southern Brittany (Cartier and Faure, 2004).

Through the Tournaisian and Viséan, a marine transgression progressed northward across Ireland until by the beginning of the Sepukhovian, most of the country had been covered by the sea. The transgression resulted from both tectonic subsidence generated by crustal extension and, to a lesser extent, eustatic rise of sea level. By the Viséan, Ireland had become the site of a shallow water carbonate shelf, which enclosed local deeper water basins. In the north, mixed terrigenous and carbonate sediments accumulated along the margins of the shrunken remnant of the Old Red Sandstone continent. In the south, a persistent basin, the South Munster Basin, was the site of accumulation of dominantly terrigenous sediments. The carbonate shelf can be traced eastward through Britain, across the North Sea, and through northern France and Belgium into north-western Germany. The South Munster Basin contains rocks similar in some respects to the Culm facies of south-west England and of the Rheinische Schiefergebirge in Germany. It seems unlikely, however, that these three areas were ever part of a single, elongate basin. Tournaisian and Viséan rocks of shelf facies extend across the continental shelf west

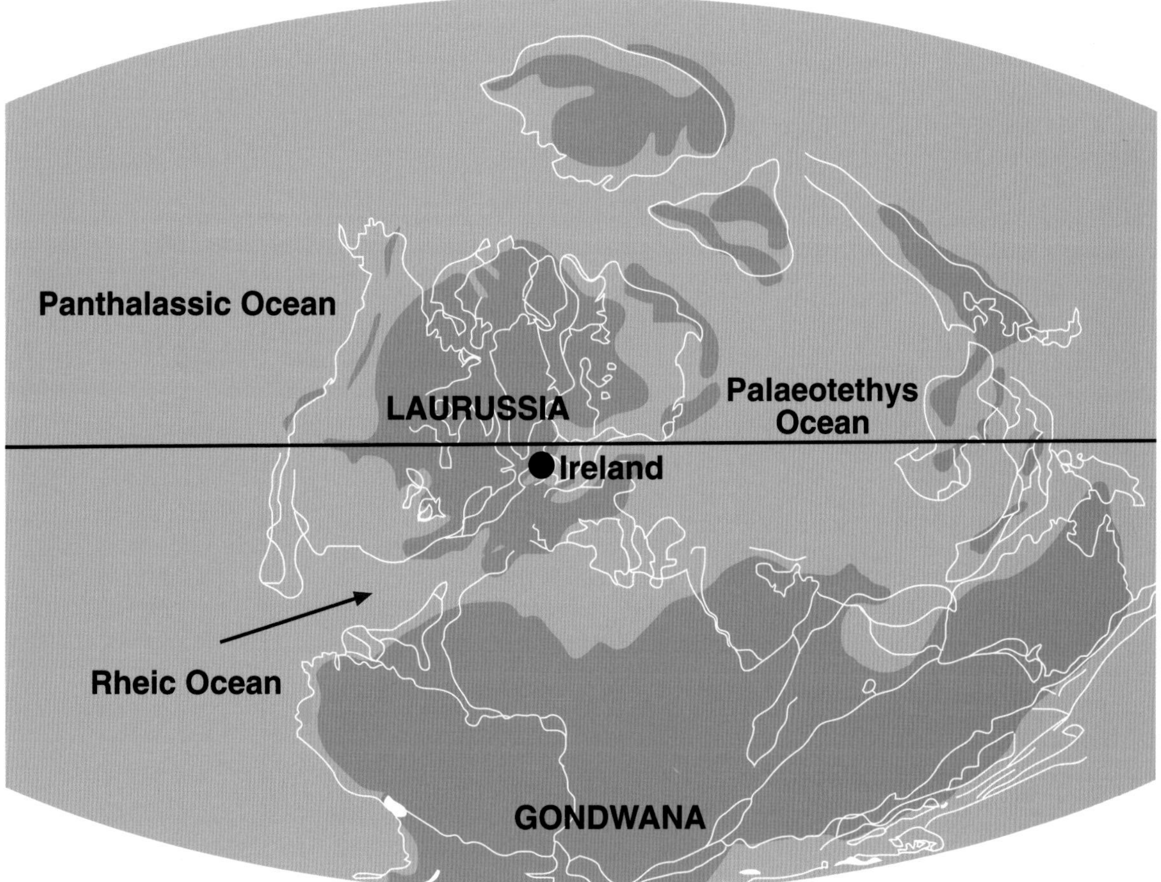

Fig. 10.1 Global palaeogeography in early Carboniferous time (based on the palaeogeographical reconstruction of C.R. Scotese, PALEOMAP Project – www.scotese.com).

of Ireland, at least as far as the Porcupine Basin. They are found next on the west side of the Atlantic in the East Canadian Shelf and in the Maritime Provinces of Canada.

The Tournaisian and Viséan geology of Ireland is most conveniently presented by describing separately the history of the South Munster Basin, followed by that of the regions to the north.

South Munster Basin

Since the earliest mapping in the south of Ireland, it has been recognised that there is an important geological boundary extending westward from Cork Harbour to the Kenmare River (Fig. 10.2). North of this divide, Carboniferous successions are dominated by limestone; to the south, in contrast, the rocks are almost exclusively shale and sandstone. George *et al.* (1976) named this southern region the South Munster Basin in recognition of its distinctive late Devonian and Carboniferous history.

The northern margin of the South Muster Basin can be identified with reasonable accuracy from Cork Harbour to Crookstown, west of Cork City, and in the vicinity of Kenmare, but between these two areas Carboniferous rocks have been removed by erosion. Naylor *et al.* (1989, 1996) and Williams *et al.* (1989) have suggested that the margin is likely to have had a saw-tooth configuration with sectors parallel to and normal to the tectonic strike, rather than having been an east–west line (oblique to tectonic strike) between Cork Harbour and the Kenmare River, as was commonly shown in earlier literature. The southern margin of the basin lay either close to the south coast of County Cork or farther south; the debate about its position is outlined below.

Subsidence and style of sedimentation followed a broadly similar course in different parts of the basin. There are, however, important differences in stratigraphy between the eastern part (Cork Harbour to the Seven Heads) and the western part around Bantry Bay. Naylor *et al.* (1989) referred to these two regions as the Kinsale and Bantry Sub-basins (Fig. 10.2). Between the two there is a region where the succession is extremely attenuated; Naylor *et al.* (1974) named it the Glandore High.

Four significant phases can be recognised in the Carboniferous evolution of the Basin (Naylor *et al.*, 1989). During Phase 1, subsidence was very rapid and the basin fill consisted of sands and muds deposited in fluctuating, but relatively shallow, marine and marginal marine environments. Phase 2 was characterised by a drastic reduction in the influx of sand; but although subsidence also slowed, the basin floor became deeper.

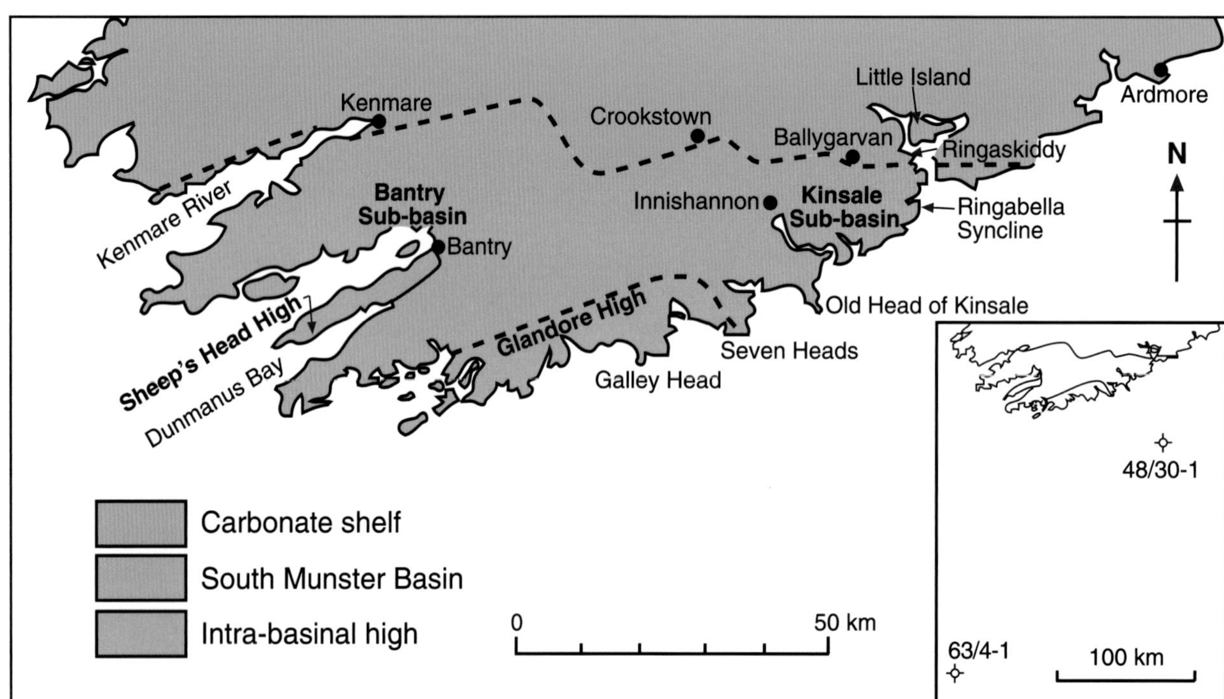

Fig. 10.2 The northern margin of the South Munster Basin and the location of the Glandore and Sheep's Head Highs. The inset map shows the locations of wells 48/30-1 and 63/4-1.

The basin became starved of sediment during Phase 3, and, although subsidence probably was relatively slow, the basin floor remained deep. The renewed input of sand during the Serpukhovian, described in Chapter 11, led to filling of the basin.

Phase 1 (Late Devonian to Hastarian)

Pracht and Sleeman (2002) and Quin (2008) have simplified the stratigraphical nomenclature of the Carboniferous part of this phase; the Castle Slate Formation (formerly the Castle Slate Member of the Kinsale Formation) and the Kinsale Formation are recognized throughout the basin. The facies differences between the Kinsale and the Bantry Sub-basins are highlighted by the definition of different members of the Kinsale Formation in the two areas (Figs. 10.3, 10.4).

The type sections on the Old Head of Kinsale (Naylor, 1966; Naylor *et al.*, 1985) provide a representative succession of the Castle Slate and the Kinsale formations. As is the case over much of the basin, the Castle Slate Formation there abruptly overlies the relatively coarse upper beds of the Old Head Sandstone Formation. This boundary on the west side of the Old Head was chosen by George *et al.* (1976) as the boundary stratotype of the Courceyan Stage (now redundant), the oldest of the regional chronostratigraphical units of the Carboniferous of Britain and Ireland; it coincides with the boundary between the LN and VI miospore Biozones, which is correlated with an horizon close to, or at the base of the Carboniferous.

The Castle Slate consists of dark mudrock with siliceous nodules and a few thin lenses and beds of limestone.

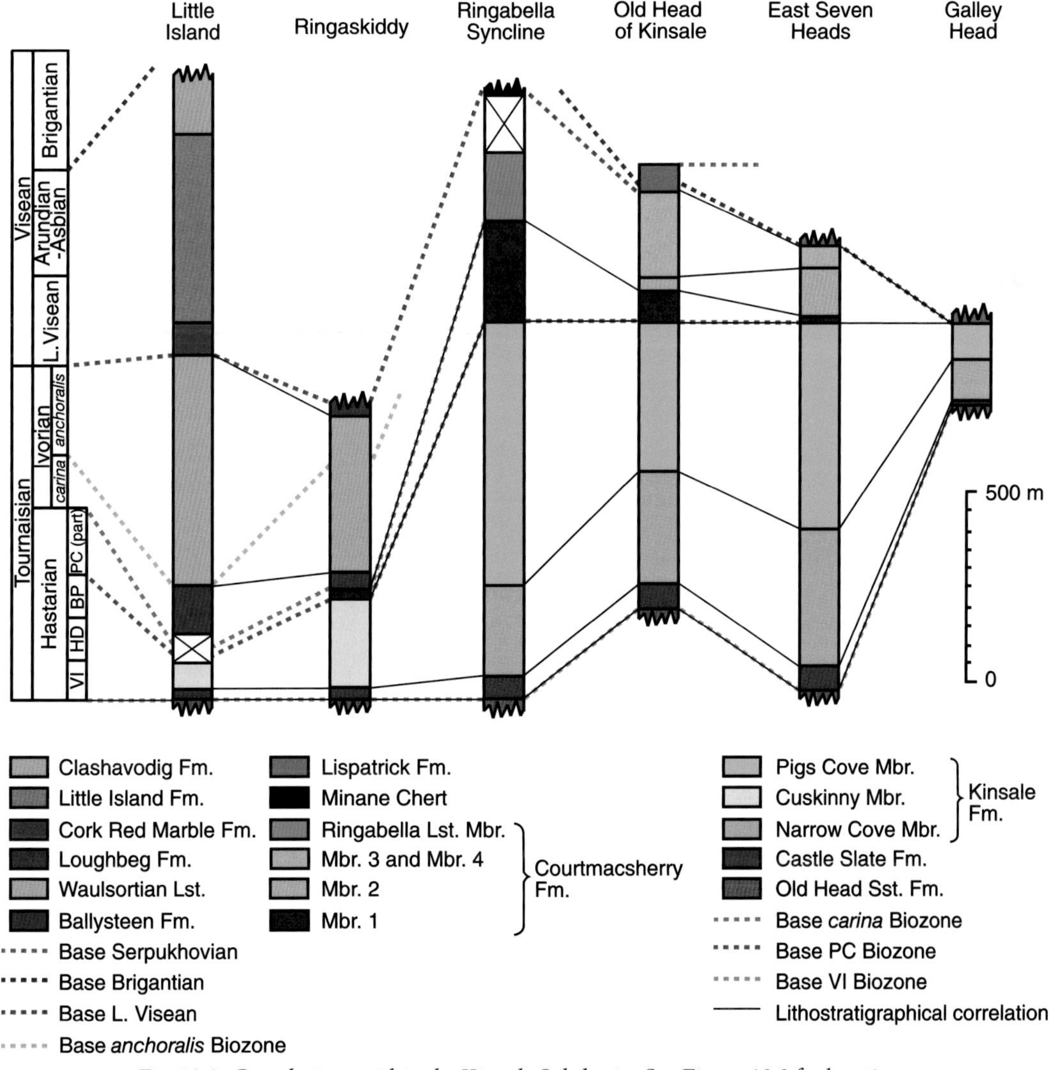

Fig. 10.3 Correlations within the Kinsale Sub-basin. See Figure 10.2 for locations.

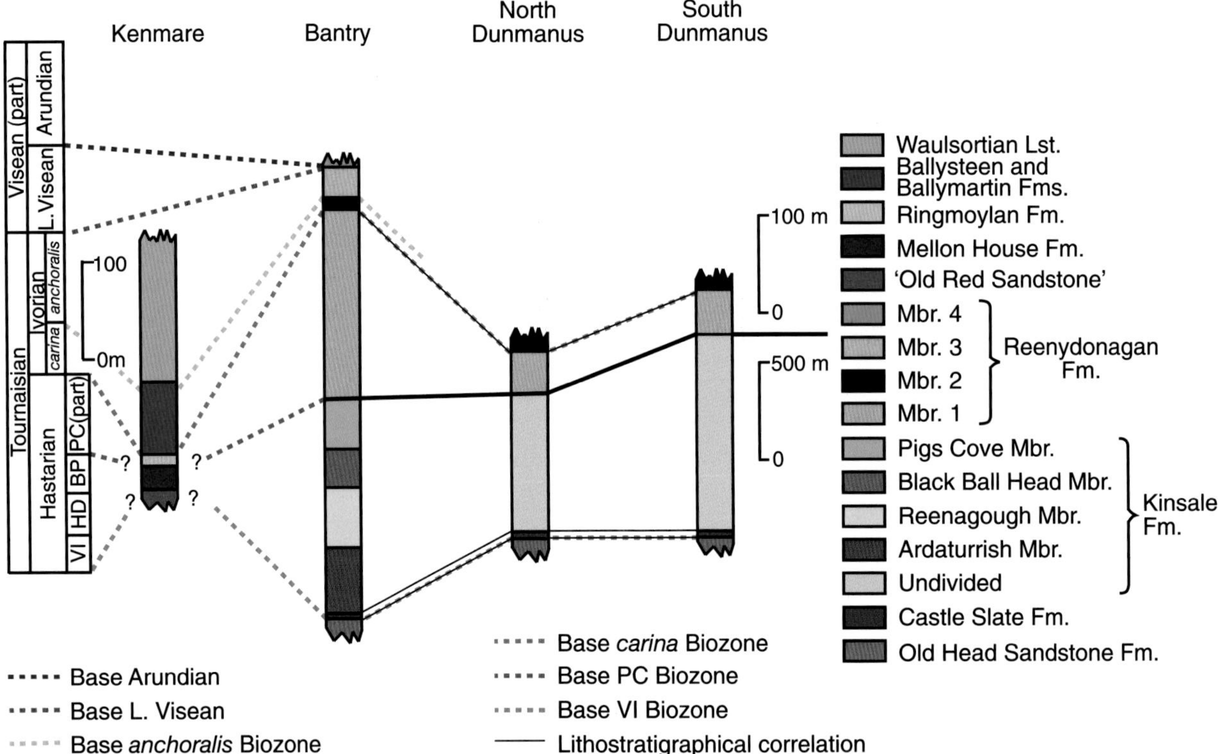

Fig. 10.4 Correlations within the Bantry Sub-basin. Note that the Reenydonagan Formation and all of the Kenmare succession are drawn at a scale different from that of underlying formations. See Figure 10.2 for locations.

On the Old Head of Kinsale and in several other sections, a prominent tuff band occurs within it. The formation occurs throughout the South Munster Basin with its thickest development in the Kinsale Sub-basin between the Old Head and Cork Harbour. While it does not extend north of the basin margin in the west, in the east (Figs. 10.2, 10.6) it is known as far north as Ardmore. It contains a low diversity fauna of delicate crinoids, ostracodes, shallow water conodonts, such as *Patrognathus*, small fish teeth, rare ammonoids (Matthews, 1983), nautiloids, and in a few localities, brachiopods. Although the Castle Slate superficially resembles deep water cephalopod mudstone, a facies widely distributed in the Culm basins of Europe during the Upper Palaeozoic, the ostracodes and conodonts that it contains suggest that it was deposited in relatively shallow water, as does its gradational passage into the overlying Kinsale Formation. The sharp change from the Old Head Sandstone to the Castle Slate has been attributed to a rise in sea level, which altered the base level of the rivers draining the land area to the north; sand, instead of being transported into the shallow sea in County Cork, was trapped in the coastal region farther north. As a result, only fine-grained sediment reached the marine basin. The cause, rate, and magnitude of the rise in sea level have been debated. Clayton *et al.* (1986a) and several other authors have suggested that the rise was eustatic, rapid, but not of great magnitude. MacCarthy (1987) concluded that the transgression took place in two pulses, the first in the late Devonian, and the second in the earliest Carboniferous. Both resulted from slow rises in sea level, which were at least partly eustatic. Glacial deposits of latest Devonian age have been identified in South America (Isaacson *et al.*, 1999), raising the possibility that the sea level changes were glacio-eustatic in origin.

The resumption of the deposition of sand in the basin was gradual, since laminae and very thin beds of sandstone are found sporadically in the upper part of the Castle Slate. The sandstone content rises markedly at the base of the Kinsale Formation, whose lower part over much of the Kinsale Sub-basin is termed the Narrow Cove Member. This member consists of sandstone and dark grey mudrock, interbedded at different scales, ranging from interlamination and flaser and linsen bedding at one extreme, to thick units of sandstone or mudrock at the other. Bedforms include ripples, swales, and hummocks. De Raaf *et al.* (1977) interpreted the mudrock and sandstone as the products of sedimentation on a muddy sea floor that was below wave base in fair weather. The sandstone bodies formed as sandy shoals, oriented parallel to the shore (mostly south-west to north-east), under the influence of waves generated by storms. The Narrow Cove Member contains very few macrofossils, but the VI/HD miospore biozonal boundary falls within it. In the north-east of the Kinsale Sub-basin, the Kinsale Formation above the Castle Slate is more sand-rich than farther south and has been termed the Cuskinny Member (MacCarthy *et al.*, 1978). Cotter (1985) has interpreted sandstone units in the Cuskinny Member, which contain concentrations of granules and small pebbles of quartz, chert, and acid volcanic rocks, as bars on a shallow shelf, whose tops were winnowed during severe storms. The Narrow Cove and equivalent members probably reflect a period of relative stillstand of sea level during which sand prograded southward from the basin margin. Palynological data are too sparse to prove that the Castle Slate/Narrow Cove boundary is diachronous, but it is worth noting that an HD miospore assemblage has been recorded from low in the Narrow Cove Member on the Old Head of Kinsale, whereas VI assemblages appear to range relatively higher in the member near the margin of the basin in the east.

Dark mudrock with subordinate sandstone forms the upper, Pigs Cove Member of the Kinsale Formation on the Old Head of Kinsale and throughout most of the South Munster Basin. The member straddles the HD/BP miospore Biozone boundary. The reduction in the amount of sand perhaps reflects a general deepening of the basin floor, to a depth where it was only occasionally affected by waves generated by storms.

The Castle Slate and Kinsale formations can each be traced westward from the Old Head of Kinsale to the east side of the Seven Heads with little change in thickness or lithology. However, on the west coast of the Seven Heads, at Galley Head, and farther west, their thicknesses are much reduced, reflecting the influence of the Glandore High (Fig. 10.3).

Thick developments of the Kinsale Formation occur above the Castle Slate Formation in the Bantry Sub-basin; the formation there is divided into the Ardaturrish, Reenagough, Black Ball Head, and Pigs Cove members (Fig. 10.4).

The Ardaturrish Member (Jones, 1974) consists of mixed mudstone and sandstone of HD miospore biozone age (O'Liathain, 1992) that Quin (2008) has interpreted as having been deposited in fluctuating coastal plain and shallow water shelf environments. Close to the north-western margin of the basin, the environments represented were dominantly coastal plain fluvial channels and floodplains, whereas in the Bantry and Dunmanus areas farther south, strata that originated in coastal and more distal, wave-dominated, shallow shelf environments alternate, probably as a result of small-scale changes of sea level.

The Reenagough Member (Reenagough Formation of Jones (1974)), also of HD miospore biozone age, is sandstone rich. In the north-west, it represents fluvial channel, floodplain and tidal flat deposits, whereas in the southern part of the sub-basin it consists of stacked small-scale alternations of lower shoreface and storm-influenced distal shelf deposits, interpreted by Quin (2008) to have been produced by cycles of sea level change and consequent coastal migration.

The Black Ball Head Member was introduced by Quin (2008) for the relatively mud-rich strata formerly included in the top of the Reenagough Member. He interpreted the member as reflecting an initial transgression with tidal delta deposits close to the north-west margin of the basin and storm influenced distal shelf deposits farther south, followed by alternations between inner-shelf/shoreface and more distal shelf setting, both showing the influence of storms. The highest beds of the member, where the proportion of sandstone increases, are interpreted to represent a time of regression.

The Pigs Cove Member (synonymous with the Ardnamanagh Formation (Jones, 1974) of earlier literature) is dominantly mudrock with subsidiary sandstone, which decreases in importance upwards. Quin (2008) has interpreted the deposits as predominantly those of a muddy distal shelf, influenced by storms. He suggested that the formation marks a relatively rapid phase of transgression and deepening in comparison to the Black Ball Head Member, and also that the coastline may have overstepped the northern margin of the basin in the west at that time.

The relative rate of subsidence of the basin and shelf during phase 1 is shown graphically by a comparison of the thickness of the Bantry succession and of that at Kenmare (Fig. 10.4). The marine transgression over the coastal plain (Old Red Sandstone) facies reached the latter localities on the shelf at, or just after, the beginning of the Carboniferous. At most, 40 m of marginal marine and shallow marine sandstone and mudrock at Kenmare is the equivalent of the 700 m or more thick succession in Bantry Bay. A similar reduction of thickness has been demonstrated in the east of the Kinsale Sub-basin around Cork Harbour (Sleeman *et al.*, 1978) (Fig. 10.3).

Phase 2 (Hastarian to Ivorian)

Phase 2 is represented by the Courtmacsherry Formation in the Kinsale Sub-basin and by the lowest member of the Reenydonagan Formation in the Bantry Sub-basin.

The Courtmacsherry Formation has been divided into four unnamed members, with their type sections on the Old Head of Kinsale. Its base marks a conspicuous change in depositional style throughout the Kinsale Sub-basin. The upper, generally sandy, beds of the Pigs Cove Member, which almost everywhere appear to be of BP biozonal age, in the North Ringabella section have yielded miospores assigned to the PC Biozone. This suggests that there was a period of non-deposition and, in many places, a small amount of erosion, before the resumption of sedimentation at the base of Member 1 of the Courtmacsherry Formation. The lower part of Member 1 consists of calcareous mudrock, containing thin, lenticular, crinoidal limestones with small solitary corals, rare ammonoids, abundant conodonts, and common granules and small pebbles of phosphate. The limestone and phosphate are concentrated in the lower part of the member, when the sediment of the basin floor was sporadically winnowed by currents. Higher in the member, the grain size and content of carbonate is progressively, but rapidly, reduced. The preservation of delicate lamination in the mudrock suggests that the basin floor was then below wave base. Member 1 has been shown in many sections to be of PC biozonal age. Conodonts from the base are assigned to the *Siphonodella* Biozone (Matthews and Naylor, 1973; Naylor *et al.*, 1988). Correlations with successions of the shelf to the north suggest that Member 1 was deposited during a regional rise in sea level; its base appears to be a flooding surface. Comparison of thickness and facies show that subsidence and water depth were both greater within the basin than on the shelf.

Member 2 of the Courtmacsherry Formation, identified with certainty only between the Old Head of Kinsale and the east side of the Seven Heads (Fig. 10.2), consists of grey mudrock containing thin beds and laminae of sandstone. The sandstones are commonly lenticular and exhibit flaser bedding and cross sections of ripples. They accumulated in an environment that was probably influenced by storms, similar to that envisaged for the Narrow Cove Member of the Kinsale Formation. Naylor *et al.* (1988) recorded miospore assemblages of PC Biozonal age from the member on the Seven Heads. The origin of the sand in Member 2 raises interesting questions. It is unlikely to have been shed from the shelf to the north: at the time that Member 2 was being deposited, the zone of sand accumulation was several tens of kilometres north of the South Munster Basin and the coeval outer-shelf sediments were devoid of siliciclastic sand. It is also unlikely that the sand was derived from erosion of the Kinsale Formation on the Glandore High, because the sand content of Member 2 decreases westwards from the Old Head of Kinsale. It may have been derived by erosion of the Kinsale Formation from actively faulted and tilted blocks at the northern margin of the basin (Sleeman *et al.*, 1986; Naylor *et al.*, 1996), perhaps to the west of Cork Harbour.

Members 3 and 4 are both made up predominantly of mudrock, probably deposited in water as much as a few hundred metres deep. Member 4 also includes thin beds of argillaceous, fine-grained carbonate, which contain rare solitary corals and ammonoids, and, on the east coast of the Seven Heads, conodonts tentatively assigned to the *carina* Biozone (Naylor *et al.*, 1988). Because of the paucity of biostratigraphical data, it is difficult to judge when Phase 2 ended in the Kinsale Sub-basin, but it is likely to have been at an horizon within the Ivorian (*carina* or *anchoralis* biozones) in Member 4. The Courtmacsherry Formation can be traced from the Old Head westward to the east side of the Seven Heads. However, it is markedly attenuated over the Glandore High: on the west side of the Seven Heads it is reduced to less than 17 m of sandy strata, and on Galley Head it is absent (Fig. 10.3).

Close to the northern margin of the Kinsale Sub-basin, rocks of Ivorian age are more carbonate-rich than farther south (Fig. 10.3). The Ringabella Limestone Member of the Courtmacsherry Formation in the south part of Cork Harbour consists of limestones and shales with abundant reworked conodonts of Hastarian age (Matthews and Naylor, 1973; Naylor *et al.*, 1996). The Loughbeg Formation, consisting of chert and limestone of *anchoralis* Biozone age (Sleeman *et al.*, 1986), may represent a slope facies at the margin of the basin. Thinly bedded, cherty mudstone, the Minane Chert, in the core of the Ringabella syncline, is probably a distal equivalent of the Loughbeg Formation.

Just as in the Kinsale Sub-basin, Phase 2 in the Bantry Sub-basin (Fig. 10.4), represented by Member 1 of the Reenydonagan Formation, began at the base of the PC Biozone; but it ended earlier at the beginning of the

Ivorian (Naylor and Sevastopulo, 1993). Member 1 in the Bantry Bay area consists of bioclastic limestone interbedded with dark grey shale and siltstone. The fauna of brachiopods, bryozoans, corals, and crinoids suggests an open shelf environment, and cross-stratification in the limestones near the base attests to current activity. In the upper part of the member, the limestone becomes finer grained, probably as a result of deepening. In the Dunmanus Syncline, Member 1 is drastically reduced in thickness and the carbonates are finer grained.

Phase 3 (Ivorian to Brigantian)

Within the Kinsale Sub-basin, Phase 3 is represented by the Lispatrick Formation and part of Member 4 of the Courtmacsherry Formation. The Lispatrick Formation is characterised by black, richly pyritic, carbonaceous shales, with subordinate beds of fine-grained carbonate, now largely composed of ferroan dolomite, and horizons of nodular phosphate. Biostratigraphical studies have shown that the succession is extremely condensed (Naylor *et al.*, 1985). On the Old Head of Kinsale, the basal bed has yielded conodonts no younger than Arundian, while bedding surfaces 13 m higher are covered with specimens of the thin-shelled bivalve *Posidonia becheri*, indicating a Brigantian age. Several horizons in the upper part of the formation on the Old Head yield pyritised ammonoids, including *Hibernicoceras*, which indicate a P_1d biozonal age (Brigantian). The highest 10 m, which are undated, contain notable amounts of chert and may be of P_2 Biozone or early Serpukhovian age. Farther west, on the west side of the Seven Heads (Naylor *et al.*, 1988) and on Galley Head (Naylor *et al.*, 1985), the Lispatrick Formation is thinner than on the Old Head and rests directly on the Kinsale Formation of Hastarian age. Faunas of Brigantian age are found 3 m from the base of the formation on the Seven Heads and 8 m from the base on Galley Head, pointing to extreme condensation, or, more probably, non-deposition of pre-Brigantian strata over the Glandore High. Naylor *et al.* (1989) have argued that although the Lispatrick Formation represents relatively deep water environments, the rate of subsidence in the basin was very low.

The Bantry Sub-basin (Fig. 10.4) was starved of sediment from the beginning of the Ivorian Stage. Member 2 of the Reenydonagan Formation, 10 m thick in Bantry Bay, consists of black, carbonaceous, phosphatic, and pyritic shale. It represents extremely slow sedimentation in an anoxic or dysoxic environment, below wave base. Member 3 (100 m thick) consists of dolomitised limestone, with subsidiary siltstone and sandstone, interbedded with black pyritic shale. Jones (1974) interpreted the carbonate and sandstone beds as turbidites; bounce and groove casts suggest current flow from the north-east or south-west. The lowest beds of Member 3 have yielded conodonts assigned to the *anchoralis* Biozone and the upper beds are of either late Ivorian or early Viséan age (Naylor and Sevastopulo, 1993). Only a few metres of Member 4 – interbedded chert, dark mudstone, and bioclastic limestone – are seen in the Bantry area. The member is at least as young as Arundian, because one limestone bed contains the foraminiferan *Uralodiscus* sp., as well as other skeletal grains of shallow water origin, and re-worked Tournaisian conodonts (Naylor and Sevastopulo, 1993). Sevastopulo and Naylor (1981) described a varied suite of carbonate cobbles and boulders, ranging in age from Ivorian to Asbian, that have been washed out of the boulder clay on the south shore of Bantry Bay. Some of the boulders are of Waulsortian Limestone of *carina* Biozone age. A borehole has now demonstrated that the limestones are locally derived (Pracht and Sleeman, 2002), which suggests that during the Lower Carboniferous the Sheep's Head peninsula (Fig. 10.2) was a structural high on which shallow water sediments accumulated. Whether the limestones capped a narrow horst or extended farther south over the region from which Ivorian and younger rocks have been stripped is not known.

Facies and thickness relations across the northern margin of the South Munster Basin in the east have been discussed by Sleeman *et al.* (1986) and Naylor *et al.* (1989, 1996). The zone of rapid thinning of the Kinsale Formation approximately coincides with the southern limit of thick successions of carbonate (Fig. 10.2). The basin margin mapped using these criteria (Fig. 10.5) extends nearly east–west across the southern part of the mouth of Cork Harbour as far as just west of Ballygarvan, where it turns north-west. It is identified again at the western end of the Cork Syncline near Crookstown. The margin clearly was controlled by structures, presumably basement faults, some of which trended parallel to, and others across later Variscan folds. Its course westward from Crookstown is speculative because of erosion of Carboniferous rocks. However, it must lie south of the carbonate succession at Kenmare.

Details of the lateral passage from shelf to basin have been established in a few areas. In both the Cloyne Syncline and the Crookstown area (Fig. 10.5), thick accumulations of Tournaisian Waulsortian limestone pass

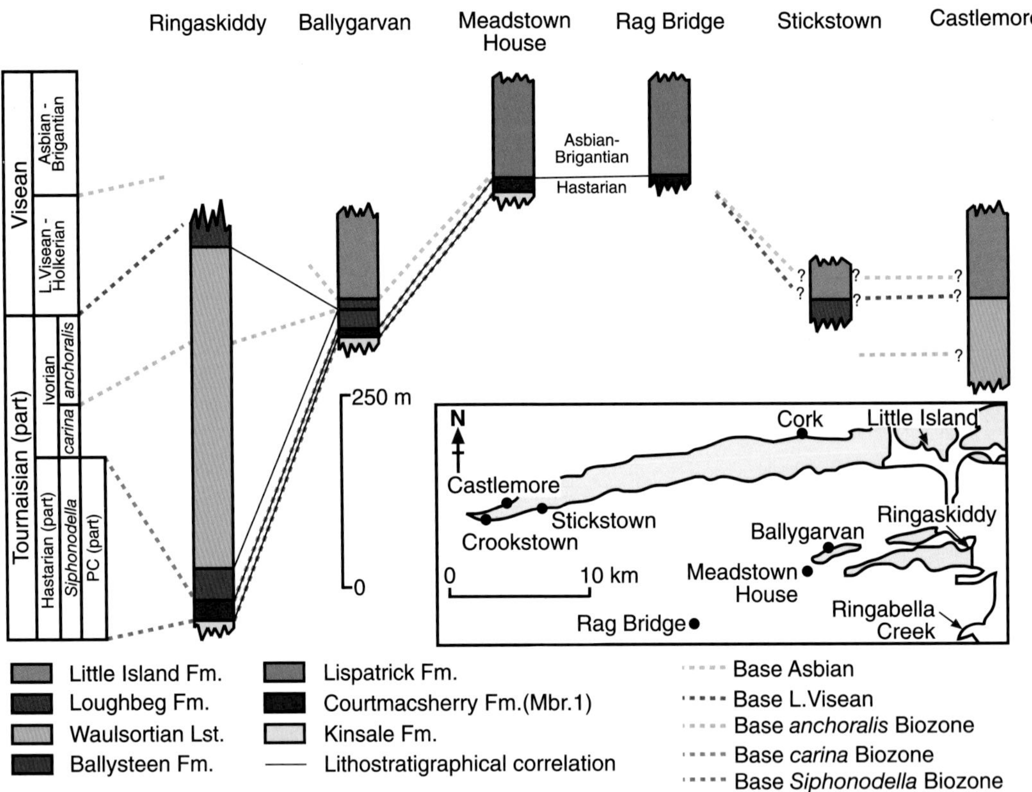

Fig. 10.5 Stratigraphical relations along the northern margin of the South Munster Basin. Distribution of limestone shown in blue.

basinward to thin successions of limestone and chert (Loughbeg Formation). There is evidence of substantial wedges of carbonate derived from the shelf to the north during the Viséan in some areas. Intervening parts of the basin slope with attenuated successions were probably bypassed by sediment. Ballygarvan quarry, on the north flank of the Cloyne Syncline, provides an example of a thick carbonate-dominated Viséan succession conformable on the Loughbeg Formation (Sleeman *et al.*, 1986; Sleeman *et al.*, 1994). The carbonates, most of which are Asbian in age, include both breccias and *in situ* shallow-water facies that may represent slight progradation of the shelf, and thin tongues of black shale comparable in facies to the Lispatrick Formation. An area of sediment bypass has been identified a short distance to the west, near Meadstown House: there, the Lispatrick Formation with tongues of carbonate breccia of Asbian age rests on Member 1 of the Courtmacsherry Formation (Hastarian). Farther west in the vicinity of Rag Bridge, near Innishannon, an analogous contrast has been established. Naylor *et al.* (1996) described a section where the stratigraphical interval between the Hastarian (Member 1 of the Courtmacsherry Formation) and the late Asbian (the lowest dated horizon in the Lispatrick Formation) is represented by at most 10 m of strata. A succession drilled close by and described by Fusciardi (1995) is more like that at Ballygarvan with a much greater thickness of apparently Ivorian and Viséan carbonate-rich strata below black shales of late Asbian or younger age, which are assigned here to the Lispatrick Formation.

There has been discussion regarding the position of the southern margin of the basin. The debate has centred on the nature of the Glandore High (Fig. 10.2). The high is characterised by an attenuated succession but not by major changes of facies. For this reason, and also because Esso Marathon well 48/30–1, some 50 km south of Glandore (Fig. 10.2), intersected a significant thickness of strata interpreted as the Kinsale Formation (Higgs, 1983), it seems likely that the Glandore structure was an intra-basinal high rather being part of the southern margin of the basin as suggested by MacCarthy and Gardiner (1987). Williams *et al.* (1989) have suggested that the Glandore High was positioned near the crest of a rollover anticline generated in the hanging wall of the extensional faults that controlled the northern margin of the basin. This model is difficult to test, but it would require some particular local circumstance to explain the marked east-west change in thickness across the eastern margin of the High on the Seven Heads (Fig. 10.3).

Rocks that are likely to have formed on the platform to the south of the South Munster Basin have been found offshore in the Fastnet Basin (Fig. 10.2). City Services well 63/4–1 intersected Tournaisian shelf limestone (Reeves *et al.*, 1978). Viséan limestones have been reported from the region of the Goban Spur farther south.

Regions north of the South Munster Basin

The evolution of the regions north of the South Munster Basin is most conveniently treated by describing the Tournaisian and the Viséan separately. This division also reflects differences in the patterns of sedimentation. During the Tournaisian, the sea transgressed northward into Ulster; depositional environments became progressively deeper, at least in the later part of the Tournaisian; but, although the thickness of successions varies from place to place, with particularly thick successions around the Shannon estuary (Shannon Basin) and in the North Midlands (Dublin Basin), there was no clear distinction on the basis of lithofacies of shelves and basins until the very latest part of the Tournaisian. During the Viséan, there was a clear differentiation of shallow water shelves and deeper water basins and there was important volcanic activity, which is described in a separate section (p. 265).

Regions north of the South Munster Basin: Tournaisian

Tournaisian stratigraphical successions north of the South Munster Basin differ from region to region (Figs. 10.6–10.9). This lateral variation is conveniently discussed by reference to palaeogeographical provinces, each characterised by a particular stratigraphical succession. The boundaries between provinces are generally gradational.

Limerick Province

The Limerick Province (Philcox, 1984), with a reference section in north-west County Limerick (Shephard-Thorn, 1963; Somerville and Jones, 1985), extends from the northern margin of the South Munster Basin into the Midlands (Fig. 10.7). The succession there divides naturally into three. It is for future research to show to what extent each division can be interpreted as a sedimentary sequence that evolved in response to a major regional, or more widespread, tectonic event or change in sea level. The first division does seem to be such a sequence: it records the marine transgression that brought the sea for the first time into the central and northern part of the Province, and the subsequent southward progradation of the shoreline facies. It is the most laterally variable of the three divisions, and a comprehensive lithostratigraphy for it remains to be established. Eight representative sections give an impression of the lateral and vertical facies changes (Fig. 10.6).

At Ardmore, County Waterford, marine conditions had already been established when the Castle Slate, of VI biozonal age, was deposited. The overlying Crows Point Formation, a more sand-rich equivalent of the Cuskinny Member in Cork Harbour, is dominated by pale sandstone with dispersed pebbles, and is of shallow marine origin. It grades upward to linsen- and flaser-bedded mudstone and sandstone with rare shelly horizons, of BP and PC biozonal age. From these there is a sharp change to well-washed, skeletal limestones with subsidiary shales, which are interpreted as having accumulated on a storm-swept, shallow shelf. The limestones yield shallow-water conodonts assigned to the *Polygnathus spicatus* Biozone. The sequence is capped by mudstone with subordinate thin beds of sandstone; in contrast to the successions farther to the north, there is no evidence of emergence at this horizon. The boundary with the overlying Ringmoylan Shale Formation, which begins the second major sedimentary sequence, is sharp. Successions of this character are typical of the southernmost part of the Limerick Province, but away from the south-east coast the earliest marine deposits are slightly younger.

At Hook Head, County Wexford (Sleeman, 1977; Sleeman *et al.*, 1974, 1983) the equivalent of the Mellon House Formation has been called the Houseland Sandstone Member of the Porters Gate Formation and is of HD to PC miospore Biozone age. The base of the VI Biozone is approximately 55 m below the base of the Houseland Member, within red fluviatile sandstones and shales. The top of the Houseland Member consists of shallow water limestones, including oolite. Sleeman *et al.* (1983) have shown that in the Wexford outlier to the east of Hook Head, sediment first started to accumulate during the Hastarian, when red conglomerates, sandstones, and shales of the Duncormick Conglomerate Formation were formed in fluviatile environments. The Houseland Member there is thinner and coarser grained than at Hook Head.

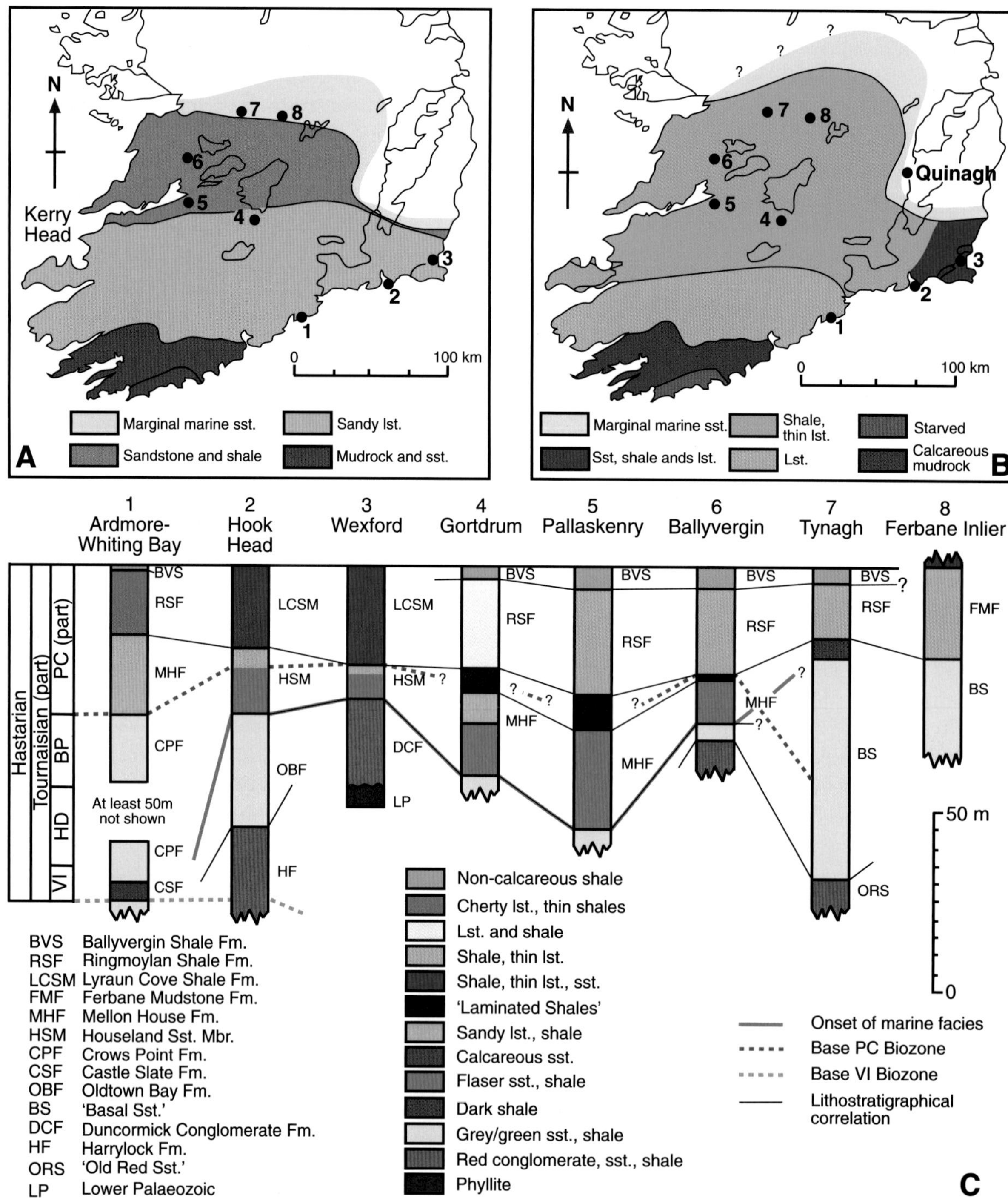

Fig. 10.6 Palaeogeography and correlations of successions of Hastarian age. (A) Palaeogeography during the BP miospore Biozone (*spicatus* conodont Biozone). Outlines of Lower Palaeozoic inliers shown. (B) Palaeogeography during the early part of the PC miospore Biozone (*Siphonodella* conodont Biozone). (C) Representative Hastarian stratigraphical successions.

The next three sections to the north are linked by the common occurrence at their tops of a distinctive unit of laminated, flaser- and linsen-bedded sandstone and mudstone, and subordinate skeletal limestone. This unit was given the name 'Laminated Shales' at Gortdrum, County Tipperary (Steed, 1986). It has been identified widely in the central part of the Limerick Province and is interpreted as having been deposited in tidally influenced environments, which, on the evidence of the occurrence of desiccation cracks, included tidal flats.

At Gortdrum, grey, green, and white sandstone with generally subordinate shale overlies red beds. This facies, which is widespread throughout the Limerick Province, has been assigned by some authors to the Old Red Sandstone but was called by Philcox (1984) the 'Basal Sandstone'. It probably contains the transition from non-marine to marine environments. It is overlain by mixed shale, sandstone, and subordinate limestone of shallow marine origin. This association is followed by sandy, well-washed limestone, some of it oolitic, interpreted here as having accumulated as bars in shallow water. The overlying 'Laminated Shales' formed during the regression which ended the first sequence.

In north-west County Limerick (Pallaskenry section in Fig. 10.6), the mixed shale, sandstone, and subordinate limestone that overlie the 'Basal Sandstone' have been named the Mellon House Formation (Shephard Thorn, 1963; Somerville and Jones, 1985). The 'Laminated Shales' facies occurs at the top, but the unit of high energy limestones that is widespread farther to the south is not developed. Diemer and Bridge (1988) have made a detailed study of the sedimentology of a similar, but more carbonate-rich succession (the Clashmelcon Formation) at Kerry Head, County Kerry (Fig. 10.6). They interpreted it as having been formed in an estuary on a meso- to macro-tidal coast, and recognised within it subtidal, estuarine channel-bar and channel-fill, and tidal flat deposits.

Successions to the north contain less carbonate: at Ballyvergin, County Clare (Hudson and Sevastopulo, 1966), the 'Laminated Shales', there dated as spanning the BP/PC miospore biozonal boundary, directly overlie sandstone, siltstone, and dark mudstone, which contain no skeletal carbonate. The latter facies just above its contact with the underlying Old Red Sandstone is of HD Biozone age. It is not clear whether it is of marine or non-marine origin. Farther north at Tynagh, County Galway (Philcox, 1984), the 'Laminated Shales' are not developed, and the Ringmoylan Shale rests on 'Basal Sandstones', which at their top contain rare horizons with marine fossils. Successions of this character are interpreted as being near the margin of the marine embayment created by the initial transgression of the sea northward into the Midlands (Fig. 10.6).

The second division of the Tournaisian succession within the Limerick Province begins with a transgression and ends with a sharp change from shale-rich to shale-free carbonates, which perhaps reflects shallowing. It comprises three formations through most of the Province: the Ringmoylan Shale, Ballyvergin Shale, and Ballymartin Limestone formations. The first and last of these have type sections in north-west County Limerick (Somerville and Jones, 1985) and the second at Ballyvergin, County Clare (Clayton *et al.*, 1980b; James, 2003).

The Ringmoylan Formation is contained within the PC miospore Biozone and its top is approximately at the top of the *Siphonodella* conodont Biozone (the top of the Hastarian), showing that it is equivalent to Member 1 of the Courtmacsherry Formation of the South Munster Basin. In its type area (Pallaskenry section in Fig. 10.6), the formation consists of 31 m of thinly interbedded, dark shale and subordinate skeletal limestone that is commonly lenticular. As it is traced southward, the proportion of limestone increases and nodular chert is developed, as for example, at Whiting Bay and in north County Cork (Campbell, 1988). To the north-east, the proportion of limestone decreases and the formation is dominated by calcareous shale. The basal contact of the formation is sharp and demonstratively erosive in many places. The lowest limestones contain rolled granules and pebbles of phosphate; thin beds of poorly developed haematitic ironstone occur in many sections. This lower part of the Ringmoylan Shale is interpreted as having been deposited as sea level rose. Subsequently, transport of mud from the north led to the establishment of a muddy inner shelf, and an outer shelf with extensive spreads of skeletal carbonate sand. The fauna was dominated by crinoids, brachiopods, bryozoans, and horn corals and was considerably more diverse than in the underlying Mellon House and equivalent formations. The conodont fauna also reflects a change to more open marine conditions with the incoming of siphonodellid conodonts.

The extent of the transgression by the end of the Hastarian Substage can be traced approximately. In south County Wexford (Fig. 10.6), the Lyraun Cove Member of the Porters Gate Formation (the equivalent of the Ringmoylan Shale) includes beds of sandstone, which near Wexford town contain granules of pink feldspar (Sleeman *et al.*, 1983). The latter were almost certainly derived from the Carnsore Granite, which crops out to the south-east. However, the granite is likely to have formed an offshore island because the regional palaeoslope inferred from the underlying Old Red Sandstone suggests that the coast would have been to the north. How much of the Leinster Lower Palaeozoic and granite inlier was inundated is difficult to assess. Important information comes from a borehole drilled at Quinagh, near Carlow (Fig. 10.6), which shows mudstone and sandstone of shallow marine origin resting directly on the granite (Tietzsch-Tyler and Sleeman, 1994b). The basal beds are probably the equivalent of the Ringmoylan Shale. Granules of feldspar within them show that the granite was still emergent, presumably somewhere to the east. Farther to the west and north, marine mudstone equivalent to the Ringmoylan Shale (the lower part of the Ferbane Mudstone Formation: Philcox, 1984; Gatley *et al.*, 2005) extends into the southernmost part of the North Midlands Province, but farther north and west is replaced by non-marine sandstone.

Throughout much of the Limerick Province the Ringmoylan Shale is overlain abruptly by the Ballyvergin Shale Formation, a distinctive unit, up to 10 m thick, of non-calcareous, grey-green shale with laminae of siltstone and fine sandstone, and a restricted fauna of brachiopods with a notable lack of bryozoans and crinoids. Clayton *et al.* (1980b) and James (2003) showed that the shale contains, in addition to the indigenous microflora of PC biozonal age, a rich assemblage of Lower Palaeozoic palynomorphs, dated as being of Tremadoc/Arenig and mid-Silurian age. It is likely that the sediment that formed the Ballyvergin Shale was derived from the north-west as a result of the rapid erosion of an area of Silurian mudrock (itself containing reworked Tremadoc/Arenig palynomorphs), which up to that time had not been a prominent sediment source. The fine-grained sediment was transported across the shelf and formed a blanket over the different facies of the Ringmoylan Shale. Either deposition was so rapid, or the substrate so soft, that only chonetid and a few other brachiopods were able to colonise it.

The sedimentation of the Ballyvergin Shale ended as abruptly as it had begun. The overlying Ballymartin Limestone Formation in its type area in north-west County Limerick (Fig. 10.7) consists of interbedded limestone and shale with a rich fauna of brachiopods, bryozoans, crinoids, and corals such as *Michelinia* and *Syringopora*. It is, in part at least, of PC biozonal age. Regional facies variation follows the same pattern as in the Ringmoylan Shale, with successions being more lime-rich to the south and generally shalier to the north. In some areas there is little difference in lithology above and below the Ballyvergin Shale. In the north and north-east (Fig. 10.7), where the Ballyvergin Shale is not recognised, the Ballymartin Formation is replaced by the upper part of the Ferbane Mudstone Formation and overlying Cloghan Sandstone Formation (Philcox, 1984), which extend into the North Midlands Province. In the north-west (Fig. 10.6), the formation is known from boreholes around Craughwell, County Galway (Pracht *et al.*, 2002). By this time the shoreline lay north of the Limerick Province.

The third division of the succession in the Limerick Province records overall deepening followed by shallowing, at least in some areas. It comprises the Ballysteen Formation and overlying Waulsortian facies limestone, and in some areas post-Waulsortian strata. The Waulsortian limestone, which occurs widely outside the Limerick Province and in places extends stratigraphically upward above the Tournaisian, is discussed later, as are Tournaisian-aged post-Waulsortian strata.

The type section of the Ballysteen Formation (Somerville and Jones, 1985) is in north-west County Limerick (Fig. 10.7). The lower part of the formation consists of thick-bedded, dark, skeletal limestone, nearly devoid of shale, and contains a fauna of crinoids, brachiopods, bryozoans, and corals. In much of the central part of the Limerick Province, close to the base of the formation, there is a distinctive unit of fine-grained, well-washed grainstone, 3–10 m thick, which is locally oolitic (Philcox, 1984). It has been named the Pallaskenry Member (Somerville and Jones, 1985). Higher in the formation the proportion of shale increases substantially. The highest part, immediately below the Waulsortian facies limestone, commonly contains nodular chert and, where this is the case, has been named the Ballynash Member. The lower part of the formation was deposited in water shallow enough that the sea bed was commonly above storm wave base, but the upper part reflects muddy environments that were below the reach of storm waves. The base of the Ballysteen Formation coincides almost exactly with the base of the *Pseudopolygnathus multistriatus* conodont Biozone. The top of the formation is diachronous, but is generally within the *P. mehli* Biozone. The main lateral facies change within the Ballysteen Formation is the widespread development of oolite at different horizons within the formation in eastern parts of the Province (Fig. 10.7). An areally extensive body of oolite close to the base of the *Polygnathus mehli* Biozone has been named the Lisduff Oolite Member (Archer *et al.*, 1996). In the north-west, a southerly thinning unit of marine calcareous sandstone, the Loughrea Sandstone (Philcox, 1984; Pracht *et al.* 2002) occurs close to the base of the formation. Northward the Ballysteen Formation

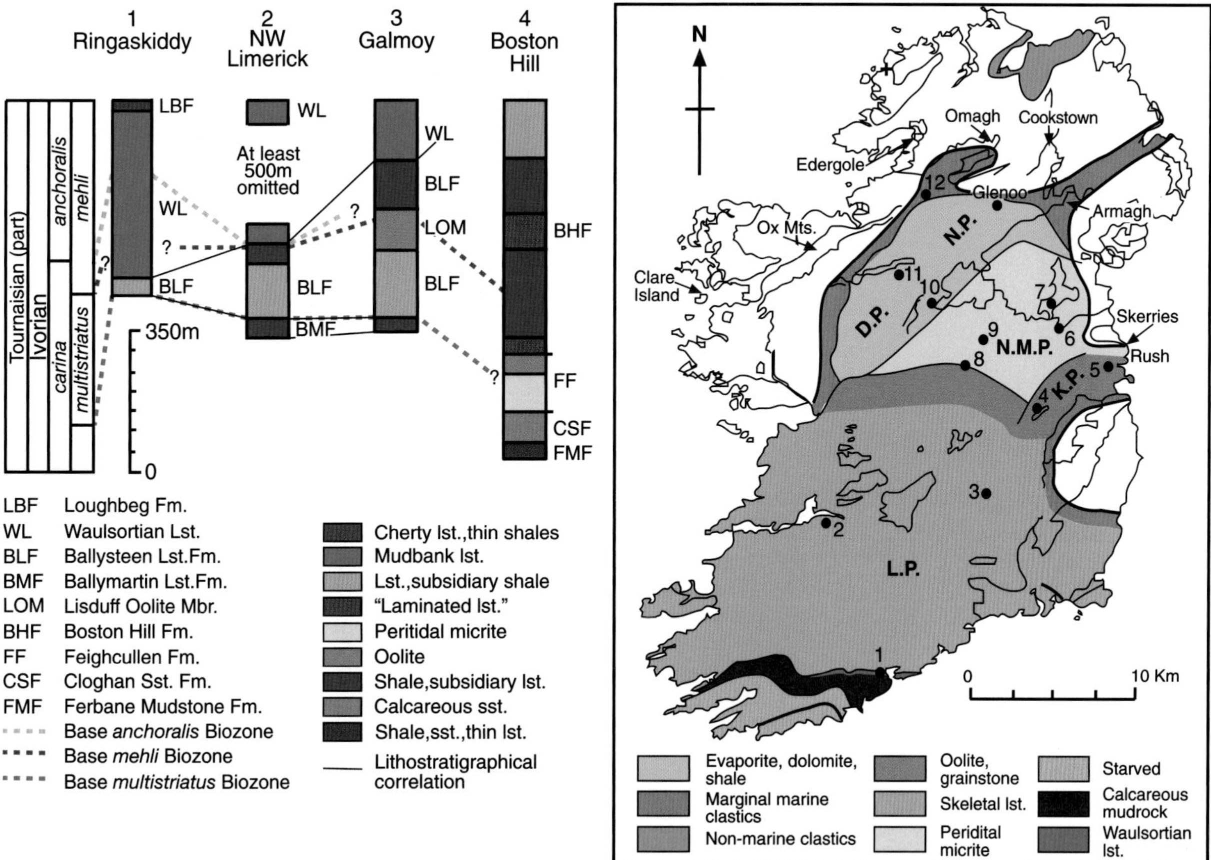

Fig. 10.7 Correlation of successions of Ivorian age in the Limerick Province and speculative palaeogeography for the early part of the *multistriatus* Biozone (early part of the CM miospore Biozone). L.P. – Limerick Province; K.P. – Kildare Province; N.M.P. – North Midlands Province; D.P. – Dunmore Province; N.P. – Northern Province (Adapted from Philcox, 1984).

passes into the Navan and Cruicetown groups of the North Midlands Province.

Kildare Province

Philcox (1984) recognised a Kildare Province and a North Midlands Province (Fig. 10.7) bounding the Limerick Province to the north-east and north respectively. The successions of the Kildare Province contrast with those of the Limerick Province in containing in their lower part developments of peritidal micrite. In this they are similar to the North Midlands Province, but they lack the important units of sandstone seen in the latter. The original description of the stratigraphy of the Kildare Province was based on the succession proved by drilling at Boston Hill, County Kildare and has been revised by McConnell and Philcox (1994). In many respects the succession is similar to that established in the Dublin region, the marine part of which has been termed the Malahide Limestone Formation (Jones *et al.*, 1988).

Throughout the Kildare Province, the marine succession appears to be underlain by red beds, including pebble conglomerates, named the Donabate Formation in the Dublin area (Nolan, 1989). The limited information available suggests that these are generally less than 70 m thick; they have not been dated directly but are likely to be of Carboniferous age. They are succeeded in the Boston Hill area by the Ferbane Mudstone Formation, interlaminated shale, siltstone and thin sandstones, overlain by the Cloghan Sandstone Formation, grey sandstone with subsidiary siltstone, shale and limestone (Fig. 10.7). Both these units are of marine origin and are recognised in the adjoining southern part of the North Midlands Province (Fig. 10.8). The Cloghan Sandstone is of PC miospore biozonal age and is correlated with the Ballymartin Limestone Formation of the Limerick Province; it probably does not extend into the Dublin region. In the Boston Hill area, the sandstones are succeeded by the Feighcullen Formation, which consists of

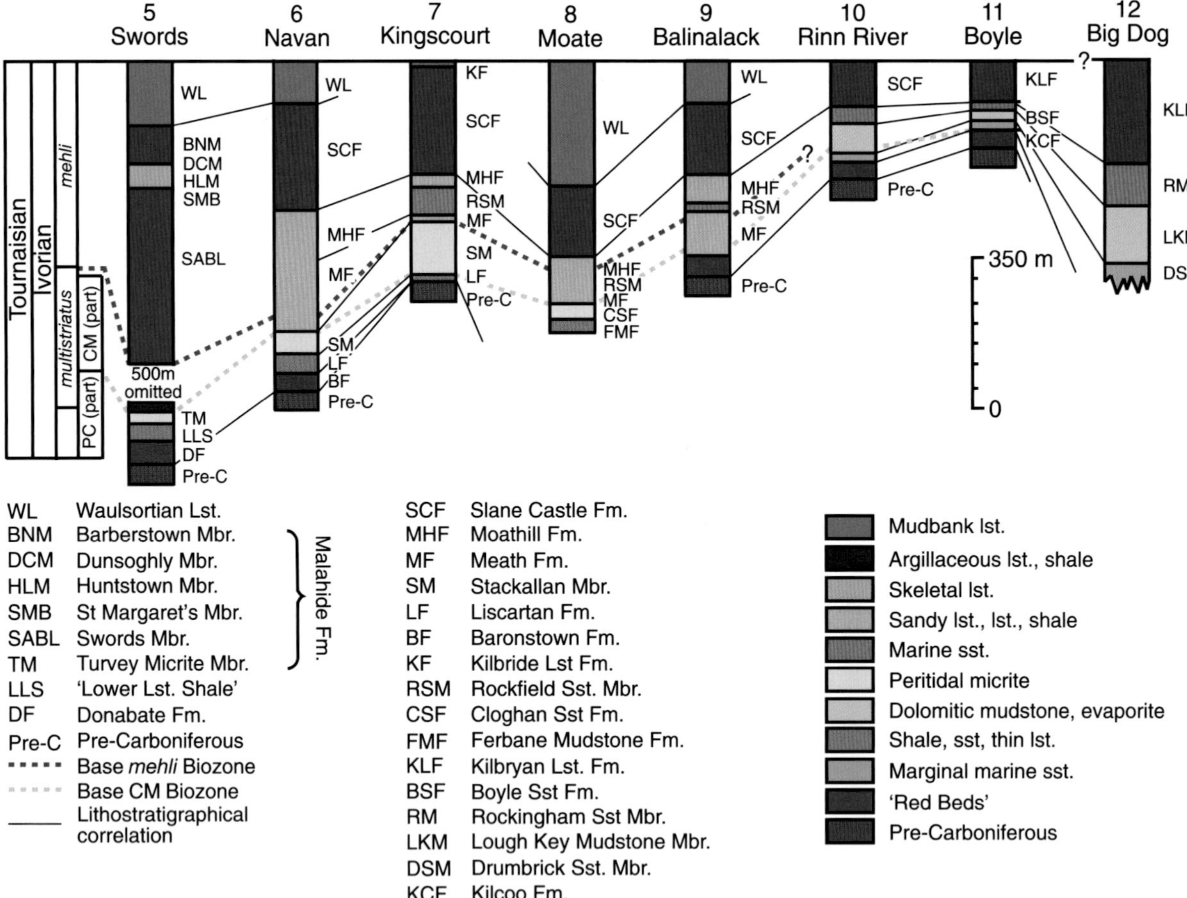

Fig. 10.8 Correlation of successions of Ivorian age in the North Midlands and Northern Regions. See Figure 10.7 for locations.

a varied succession of very shallow water facies, including skeletal grainstone, oolite, and fenestral micrite. Similar developments of peritidal micrite, believed to occur at the same horizon, have been described from close to the Leinster Lower Palaeozoic inlier to the north-east (Brück, 1971b; Strogen and Somerville, 1984). They occur also in the Dublin region, where they make up the Turvey Micrite Member (Fig. 10.8) of the Malahide Formation (Jones *et al.*, 1988). Limited biostratigraphical evidence is consistent with correlation of the peritidal micrites with the base of the Ballysteen Formation of the Limerick Province. Following this episode of shallow-water sedimentation, the environment appears to have become deeper, leading to the deposition in the Boston Hill area of the Boston Hill Formation, which consists mostly of a monotonous succession of argillaceous, skeletal wackestone and packstone, with lesser amounts of interbedded limestone and shale and units of conspicuously laminated silty limestone, a facies characteristic of the Kildare Province. The upper part of the formation reflects shallowing. At Harberton Bridge, 7 km north-east of Boston Hill (Emo, 1986), it includes a cross-stratified grainstone, and at Lyons Hill, 25 km east-north-east of Boston Hill (Strogen and Somerville, 1984), horizons with granules of vein quartz, feldspar, and muscovite, suggesting that part of the Leinster Granite and its Lower Palaeozoic envelope were emergent during the later part of the Tournaisian. At Harberton Bridge, the shallow-water grainstones are succeeded by interbedded shale and limestone, perhaps reflecting deepening, followed by cherty limestone, which underlies the Waulsortian facies limestone.

The Malahide Formation in the Swords area of County Dublin (Fig. 10.8) has been correlated with the succession at Boston Hill (Jones *et al.*, 1988). The Swords Argillaceous Bioclastic Member consists of decimetre-thick units of mud-rich wackestone separated by thinner developments of grainstone and distinctive banded limestone. The latter, which is a lithology typical of the Kildare Province, consists of interleaved laminae or very thin beds of dark silty micrite and fine carbonate grainstone, many with a notable siliciclastic content. Just as in Kildare, the upper part of the succession shows evidence of considerable variation in energy level of the depositional environment, with the Huntstown Laminated and Dunsoghly Massive Crinoidal members representing shallow water environments. Above these, there was a return to lower energy, deeper water environments before the advent of the Waulsortian facies limestone.

The Balbriggan Lower Palaeozoic inlier was emergent during the Tournaisian and the coast lay between Rush and Skerries (Fig. 10.7). In the Lane borehole on the coast south of Skerries (Nolan, 1989), the McGuinness Formation, 60 m of feldspathic sandstone and dolomite with horizons of peritidal micrite, rests on Lower Palaeozoic rocks (Fig. 10.12). The CM/Pu miospore biozonal boundary lies within the lower part of the formation. Most of the shallow water Lane Limestone Formation is of latest Tournaisian age.

North Midlands Province

The North Midlands Province extends westward from the Kildare Province to an arbitrary boundary with the Dunmore Province (Fig. 10.7). Its northern boundary is also taken arbitrarily at the southern margin of the Longford-Down inlier, except in the east where it is drawn to include the Kingscourt outlier.

Because of the large number of exploration boreholes, the Tournaisian stratigraphy is well known; it is conveniently divided into three parts. At the base are variably thick developments of fluviatile red beds. The marine succession comprises the Navan Group (Andrew and Ashton, 1985), which is of shallow water origin and shows considerable geographical variation, and the Cruicetown Group (Rees, 1992), which formed in deeper water and shows less regional variation.

Red beds and Navan Group. The succession in the neighbourhood of Navan, County Meath can be used as a standard with which to compare other sequences (Fig. 10.8). Tournaisian sedimentation began with the deposition of the Baronstown Formation (Rees, 1992), which consists of interbedded red, polymict conglomerates, breccias, sandstone, and siltstone, with nodules of pedogenic carbonate at some horizons. The formation is variable in thickness, up to a maximum of 45 m at Navan (Ashton *et al.*, 1986). Rees (1992) interpreted it as having been deposited by sheetfloods or braided streams with sporadic mudflows. The marked lateral variation in thickness of the formation over short distances suggests local relief of the land surface.

The Liscartan Formation (Strogen *et al.*, 1990), the lowest formation of the Navan Group, is conformable on the Baronstown Formation and records the marine transgression. It consists of a lower Portnaclogh Member and an upper Bishopscourt Member. The Portnaclogh Member, of PC miospore biozonal age (McNestry and Rees, 1992), is made up of bioturbated, flaser- and linsen-bedded sandstone and mudstone, with rare horizons of skeletal carbonate and ooids, and a bed of anhydrite, now mostly replaced by silica. The Bishopscourt Member consists of argillaceous, peloidal and skeletal limestone, with a fauna of brachiopods and corals, and is restricted to the Navan region. McNestry and Rees (1992) have interpreted the Liscartan Formation as representing three cycles, each of which reflects northward migration of offshore barrier sands over intertidal and supratidal, siliciclastic mudflats, followed by regression. The uppermost transgression (Bishopscourt Member) introduced conditions of more normal salinity into the Navan region.

A significant regression occurred at the top of the Liscartan Formation. The marine limestones of the Bishopscourt Member are overlain by the Stackallan Member (Rees, 1992) of the Meath Formation (Strogen *et al.*, 1990). This consists of numerous shallowing-upward cycles, with bioturbated carbonate mudstones and peloidal wackestones and packstones capped by fenestral micrites, in many cases exhibiting vadose cements and geopetal crystal silt. The biota is restricted, being dominated by calcispheres, ostracodes, and *Girvanella*. Rizzi and Braithwaite (1996) have interpreted the cycles as representing shoaling from subtidal lagoons to supratidal carbonate flats. Rees (1992) recorded a gradual change to environments with more normal marine salinity in the upper part of the Stackallan Member in the Slane area, 11 km north-east of Navan. The remainder of the Meath Formation consists of interbedded terrigenous and carbonate sandstones, siltstone, and several beds of shale and wackestone, with faunas of brachiopods, bryozoans, crinoids, and corals, that suggest that the salinity generally was of normal marine values. The carbonate rocks are skeletal, peloidal, and ooid grainstones and packstones; they commonly contain terrigenous sand, and grade into calcareous sandstone. Rizzi and Braithwaite (1996) have interpreted the succession as representing several aggradational cycles, each with argillaceous, burrowed wackestones deposited below fair weather wave base, overlain by carbonate and siliciclastic sandstones, interpreted to have accumulated as shoals on a high energy, shallow water shelf. Some of the basal shaly units of the cycles have been traced over substantial distances. In the Navan area, the Stackallan Member varies from over 60 to less than 10 m in thickness (Andrew and Ashton, 1985); isopachs define a north-west to south-east trending depression, which Rizzi and Braithwaite (1996) have interpreted as a channel incised into the Stackallan Member and filled by the younger part of the Meath Formation, with local developments of granule conglomerates containing quartz clasts. Channels of the same trend and in the same location have been mapped also in the Liscartan Formation suggesting the earliest manifestations of structural control, which become conspicuous during the latest Tournaisian and Viséan in the Navan area.

Little detailed palaeontological work has been published about the Meath Formation in the Navan region. However it is possible to identify some important biostratigraphical boundaries and thus to correlate with successions in the Limerick Province. According to Andrew and Poustie (1986), the PC/CM miospore biozonal boundary occurs above the Stackallan Member at Tatestown, a short distance north-west of Navan, which is consistent with information in McNestry and Rees (1992); the *Pseudopolygnathus multistriatus/Polygnathus mehli* conodont biozonal boundary occurs in the upper part of the Meath Formation, consistent with information from the Slane area reported by Rees (1992). This suggests a broad equivalence of the Meath Formation with part of the Ballysteen Formation of the Limerick Province, the micrites of the Stackallan Member being equated with the lower part of the latter.

The Moathill Formation (Strogen *et al.*, 1990), the youngest formation of the Navan Group, reflects a transition from the shallow water depositional environments of the Meath Formation to deeper water settings represented by the Cruicetown Group described below. It consists of a varied succession of dark, commonly unfossiliferous shale, skeletal, particularly bryozoan-rich, packstone and wackestone, siltstone and sandstone. It accumulated in environments below fair weather wave base, subject to infrequent reworking by storm-generated waves. In the Slane area (Rees, 1992) the upper part of the Moathill Formation is of shallower water facies and includes intertidal and supratidal deposits.

Using the succession in the Navan area as a standard, it is possible to assess variation in the stratigraphy of the Navan Group throughout the North Midlands Province.

South-east of Navan, at the western end of the Balbriggan inlier, Pickard *et al.* (1992) have identified the Baronstown, Liscartan, Meath, and Moathill formations, which do not differ significantly from their equivalents at Navan and Slane, except for the effects of syn-sedimentary movement of an east–west trending fault, which probably became active during deposition of the Meath Formation.

The lower part of the Trim No.1 well, drilled 20 km south-south-west of Navan, was named the Rathmolyon Basal Clastic Formation (Sheridan, 1972b), and is clearly the equivalent of the Navan Group, although it is less rich in terrigenous sand than successions to the north and west.

The most complete succession of the Navan Group for which information has been published in the southern part of the North Midlands Province is at Moyvoughly (Fig. 10.8), north of the Moate inlier (Poustie and Kucha, 1986). There, rocks of Old Red Sandstone facies, perhaps as much as 200 m thick and, in part, almost certainly of Carboniferous age, are overlain by the Ferbane Mudstone and the Cloghan Sandstone, two formations that provide a link with the Limerick Province. The Cloghan Sandstone is overlain by a sequence characterised by peritidal micrite, clearly equivalent to the Stackallan Member of the Meath Formation. Overlying it is a varied succession of siliciclastic siltstones and sandstones and skeletal and ooid grainstones, which were called the Moyvoughly Beds, and which are now assigned to the Meath Formation. They are overlain by a distinctive unit of marine sandstone, which were assigned by Sevastopulo and Wyse Jackson (2001) to the Rockfield Sandstone Member of the Moathill Formation (see below). Younger parts of the Moathill Formation are silty shales, sandstones, and skeletal and peloidal limestones. The PC/CM miospore biozonal boundary occurs within the Stackallan Member, suggesting that the onset of the micrite facies may have been slightly earlier than in the Navan region. The *Pseudopolygnathus multistriatus–Polygnathus mehli* conodont biozonal boundary occurs in the Moathill Formation (Johnston, 1976) above the Rockfield Sandstone Member, again suggesting slight diachronism between Moyvoughly and Navan.

Variation in the Navan Group to the north, west, and south-west of Navan is well known as a result of exploration drilling (Philcox, 1984). The successions in the Kingscourt outlier (Strogen *et al.*, 1995), at Ballinalack, County Westmeath (Bradfer, 1984; Philcox, 1984), and near Rinn River, County Leitrim (Philcox, 1984; Polgar,

1980) are used here to illustrate some of the changes (Fig. 10.8).

In the Kingscourt outlier, the red beds of the Baronstown Formation are only patchily developed and the Liscartan Formation, which has been proved to extend into the northern half of the outlier, in places rests on the Lower Palaeozoic. The Stackallan Member, which also extends into the northern half of the outlier, is thinner than farther south, but forms a proportionately greater part of the Meath Formation. A conspicuous unit of marine sandstone, the Rockfield Sandstone Member, which is not present at Navan but is widespread in the western part of the North Midlands Province (as the Upper Sandstone; Philcox, 1984), occurs at the base of the Moathill Formation (Strogen *et al.*, 1995). Isopachs of the Liscartan Formation and Stackallan Member describe a trough trending north-north-east and closing to the north. Its axis coincides with the trace of the north–south trending Kingscourt Fault, which bounds the west side of the outlier, and Strogen *et al.* (1995) have argued that this is evidence of movement on the fault during the Tournaisian.

At Ballinalack, to the west of Navan, a thin development of red beds resting on Lower Palaeozoic rocks is overlain by flaser-bedded sandstone, comparable to the Liscartan Formation. Overlying peritidal micrites are similar to the Stackallan Member of the Meath Formation at Navan but they are succeeded by interbedded sandstone, shale, and micrite with fenestral fabrics, in many cases containing geopetal fills of crystal silt. This suggests that the fluctuating environments represented by the upper part of the Meath Formation were more frequently inter- or supra-tidal at Ballinalack than at Navan. The Rockfield Sandstone Member of the Moathill Formation, which follows, must have formed a complex of shallow water, generally subtidal, sand bars, at a time of major influx of terrigenous sediment from the west or north-west (Phillips and Sevastopulo, 1986). The upper part of the Moathill Formation at Ballinalack is less sandy than at Navan.

The westward (and northward) trend towards more sandy and more restricted environments within the Navan Group is well illustrated by the succession at Rinn River, County Leitrim (Fig. 10.8). Above a reddened surface of Lower Palaeozoic rocks is a sequence of conglomerate, overlain by red sandstone and mudstone that probably should be assigned to the Fearnaght Formation (Morris *et al.*, 2003), followed, in turn, by grey flaser-bedded sandstone and mudstone. The latter unit contains a miospore assemblage assigned to the PC Biozone, dating the marine transgression in the area. The lower part of the Meath Formation consists of micrite and dolomicrite with fenestral fabrics, overlain by dolomitic siltstone and sandstone with common horizons of anhydrite. The Rockfield Sandstone Member there makes up all of the Moathill Formation and is overlain directly by the Cruicetown Group.

Cruicetown Group. Throughout the North Midlands Province, there is a conspicuous but gradational change from the varied, commonly sandstone-rich succession of the Moathill Formation to argillaceous, skeletal packstone, wackestone, and subsidiary grainstone, interbedded with fossiliferous, calcareous shale named the Slane Castle Formation by Rees (1992). The Geological Survey of Ireland (McConnell *et al.*, 2001; Morris *et al.*, 2003; Gatley *et al.*, 2005) has assigned these rocks to the Ballysteen Formation, but the name Slane Castle Formation is maintained here pending detailed study of both formations. The lower part of the Slane Castle Formation contains small amounts of terrigenous silt and sand, and in the south and cenral part of the Province is particularly argillaceous (the Woodtown Member; Strogen *et al.*, 1996). The fauna is diverse and includes brachiopods, bryozoans, corals, and echinoderms. The limestone beds probably accumulated below fair-weather wave base, although some show evidence of deposition following storms. Through much of the North Midlands Province, the formation is overlain by Waulsortian limestone, discussed below, but in the north, it replaces the Waulsortian limestone and extends into the youngest part of the Tournaisian. In the Kingscourt outlier, a unit of coarse-grained, crinoidal grainstone of latest Tournaisian age, which overlies the Slane Castle Formation, has been named the Kilbride Formation (Strogen *et al.*, 1995).

There is no direct biostratigraphical evidence that the deepening signalled by the passage from the Moathill to the Slane Castle Formation is diachronous across the North Midlands, but the data are few. In several places, the foraminiferan *Tetrataxis*, indicating the MFZ 6 or younger biozone, occurs in the lowest part of the Slane Castle Formation, suggesting that the latter is no older than the latest part (*bouckaerti* Sub-biozone) of the *carina* conodont Biozone and thus the same age as the base of the Waulsortian Limestone in the axis of the Shannon Basin (see below, p. 236).

Dunmore Province

Philcox (1984) distinguished this province from the North Midlands Province to the east principally because it was not possible to recognise formations of the Navan Group below relatively thin developments of the Cruicetown Group and Waulsortian limestone. The succession below the limestone and shale facies is incompletely known but appears to be dominated by sandstone and shale and to contain little carbonate. According to data reported by Higgs *et al.* (1988a), the base of the limestone and shale facies is close to the PC/CM miospore biozonal boundary, which suggests an approximate correlation with the base of the Ballysteen Formation of the Limerick Province. However, Johnston (1976) showed that farther north, near Glenamaddy, County Mayo (Fig. 10.7), the base of a similar unit of limestone and shale could be correlated with the base of the Cruicetown Group of the North Midlands Province. More recent unpublished work strongly suggests that the palynological data should be re-appraised.

Devuyst (2006) has described the Tournaisian succession near Oughterard, County Galway (Fig. 10.9). There the Tonweenroe Formation, 63–120 m thick, rests on the Oughterard Granite. The lower part of the formation consists of repeated sequences of red conglomerates overlain by red sandstone and siltstone, all probably derived largely from the granite. The upper part records the marine transgression and is made up of interbedded grey sandstone and black siltsone and shale, which contain crinoid ossicles towards the top. The overlying Oughterard Formation (*c.*160 m thick) is similar in many respects to the Slane Castle Formation of the North Midland Province and the Kilbryan Formation farther north. It consists of grey to dark grey, variably silty, skeletal carbonates, with grainstones in the lower part and packstones and wackestones in the upper part. A tuff occurs *c.*125 m above the base of the formation. The Oughterard succession is particularly significant because it contains rich faunas of foraminiferans and Devuyst was able to identify Biozones MFZ6–9 (Table 10.1), and in particular to correlate the base of the Viséan accurately to a level just above the tuff horizon. Although it seems likely that the shoreline had moved to the Oughterard area only in the late part of the Ivorian, the history of its migration across the Dunmore Province is uncertain.

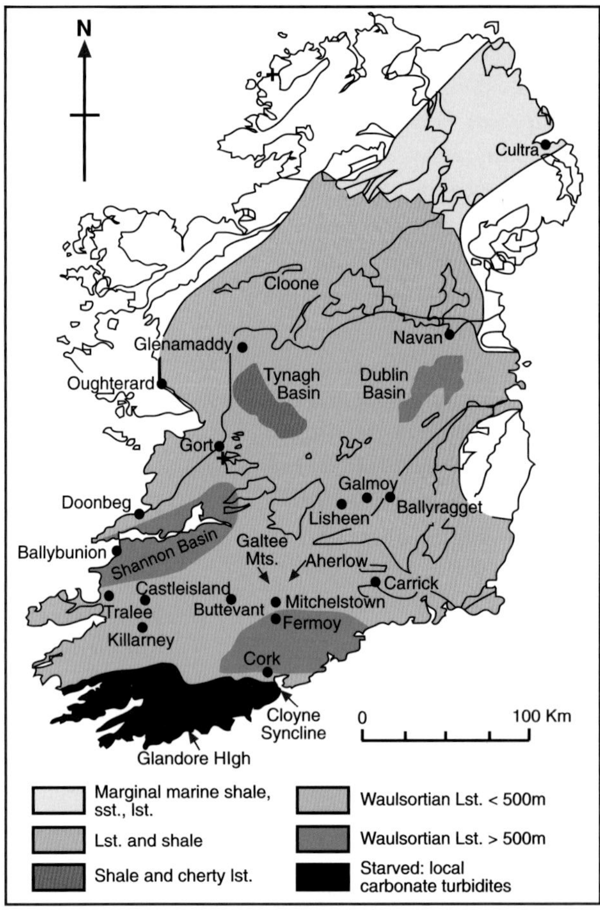

Fig. 10.9 Palaeogeography during the late part of the *anchoralis* Biozone, showing the limit of marine facies and the distribution of the Waulsortian. D.B. - Dublin Basin; S.B. - Shannon Basin; T.B. - Tynagh Basin. Based partly on Andrew (1992).

Waulsortian facies limestones

The Ballysteen Limestone Formation in the Limerick Province and its equivalents in the Kildare, North Midlands, and Dunmore provinces are overlain almost everywhere by a distinctive suite of light grey, fine grained limestones, commonly containing conspicuous sparry masses. Douglas (1909) noted the similarity of these limestones to those termed Waulsortian Phase in Belgium (named for the town of Waulsort), and the Belgian geologist Délépine (1949) endorsed this opinion. The Waulsortian facies in Ireland and elsewhere has been comprehensively reviewed by Lees and Miller (1995). It has now been identified in the south-western and western United States, Ireland, Wales, England, France, and Belgium. However, it is more extensively developed in Ireland, as regards both thickness and geographical extent, than in any other country. The limestones have been given formal stratigraphical names – Limerick Limestone Formation in the south of the country (Somerville *et al.*, 1992a); Feltrim Limestone Formation in the Dublin Basin (Jones *et al.*, 1988) – but because it is common, particularly in the North Midlands, for Waulsortian banks to be completely enclosed in limestone and shale, conventional lithostratigraphical terms are difficult to apply.

The Waulsortian facies, commonly referred to as 'reef limestone' but certainly not analogous to Recent coral/algal reefs, is variable in lithology, but in almost all cases contains a substantial amount of heterogeneous micrite, which can be shown to have formed from several generations of original carbonate mud. A common lithology is skeletal wackestone containing masses of sparry calcite, which originated as cement precipitated in cavities, either primary or, commonly, secondarily enlarged by solution, in the originally mud-rich carbonate sediment. Stromatactis structures – cavities with their greatest dimensions sub-parallel to the stratification, and with flat or stepped floors and digitate tops – are common. Their origin is enigmatic. Other cavities can be seen to have been bounded by skeletal material, such as the valves of brachiopods, or, particularly in some beds, sheet-like or conical zoaria of fenestellid bryozoans. The floor of a cavity is commonly carpeted with geopetal micrite (this produces the flat base of stromatactis cavities). The earliest generation of sparry calcite is typically of elongate, so-called 'fibrous' calcite, which is interpreted as having been originally a high magnesium calcite, marine cement. Other common rock types include skeletal packstone, in some places dominantly crinoidal, skeletal wackestone, and carbonate mudstone, in which original cavities are small or absent. The fauna of the limestones in places is very rich (see, for example, Hudson *et al.*, 1966b) but lacks any organisms with skeletons capable of constructing a rigid framework. The biotic constituents include sponges, diverse brachiopods, fenestellid and other bryozoans, bivalves, gastropods, rostroconchs, nautiloids, ammonoids, crinoids, trilobites, and ostracodes. Corals, which are common and taxonomically moderately diverse in coeval limestone and shale facies, are rare in the Waulsortian, where the most widely reported genus is *Amplexus*. Lees and Miller (1995) have suggested that the carbonate mud characteristic of the Waulsortian was precipitated as a result of the activities of microbial communities that populated some areas of the surface of the banks. Devuyst and Lees (2001) demonstrated, using three localities in the Limerick Province and one in the Dunmore Province, that there is a thin transition zone between the underlying strata and Waulsortian limestones characterised by clotted or peloidal carbonate muds, which they termed precursor muds.

The surveyors of the Geological Survey in the nineteenth century recorded that the Waulsortian facies limestones are commonly massive. Where stratification could be identified, the dips in many places were seen to be quaquaversal. Lees (1964) demonstrated that stromatactis cavities are sub-parallel to the stratification. Mapping the orientation of bedding and of stromatactis allows the geometry of Waulsortian banks to be reconstructed. The morphology of individual banks and aggregates of banks, which grew one on another, is very varied (Lees, 1994). Depositional slopes range from zero to over 40° and depositional relief from zero to as much as 200 m (in an example from Belgium). There is a systematic change in petrography of the limestone from the slope of a bank to its toe and thence into off-bank equivalents (Lees, 1964; Miller, 1986): spar-filled cavities disappear; carbonate mudstone diminishes in importance; and shale, chert, and crinoid ossicles tend to become volumetrically more important.

Probably as a result of failure on the steep depositional slopes, extensional fissures opened in the partly lithified limestone. They were filled contemporaneously, typically with laminated micrite, in which the lamination is commonly subparallel to the fissure walls. The source of the micrite is not clear. Repeated opening of such fissures in places led to the development of spectacular breccias such as those described by Philcox (1963).

Lees *et al.* (1977) showed that in the Dinant region of Belgium, the Waulsortian limestones form composite build-ups. The earliest part of the build-ups there are formed dominantly of banks of cavity-bearing packstone and wackestone with locally high depositional dips, and interbank crinoidal limestone. During the evolution of the build-up, this facies association was largely replaced by the 'biomicrite facies', composed of pale grey biomicrite without sparry masses. A similar replacement of cavity-rich limestones by 'biomicrite facies' has been reported in several places in Ireland (Keeley, 1983; Murphy, 1988; Strogen, 1988) but is not universal.

Lees and Miller (1995) provided a summary of the evidence bearing on the bathymetry of Waulsortian banks. Typically, the constituents of the limestones increase in diversity upward through the build-ups; the composition of the skeletal and non-skeletal components, divided into four phases, A–D, has been interpreted as being related to depth. Phase A, thought to have been deposited in deep, aphotic conditions (perhaps as much as 300 m deep) is characterised by a low diversity assemblage of crinoid ossicles, fenestellid bryozoans, and ostracodes. Phase

B contains hyalosteliid sponge spicules in addition to the components of Phase A; in Phase C, plurilocular foraminiferans first appear. Phase D, interpreted as having accumulated within the photic zone, is characterised by the presence of calcareous algae and cryptalgal coating on grains and cavity walls. Histon and Sevastopulo (1993) showed that many of the nautiloids that are found in the bank limestones probably lived at substantial depths.

The geographical and stratigraphical distribution of the phases in Ireland is very imperfectly known (Lees and Miller, 1985, Sleeman *et al.*, 1986, Somerville *et al.*, 1992c; Dehantschutter, 1995). In the south of the country, the lower part of the Waulsortian appears generally to be of Phase A or B, whereas in the North Midlands, limestones at the base are commonly of Phase C. In general, lower (deeper water) phases are succeeded stratigraphically upward by higher (shallower water) phases but reversals of this trend are known. Abrupt changes of phase laterally over short distances have been recorded, suggesting considerable topographic relief on the bank surfaces.

The age of onset and demise of the Waulsortian facies has been discussed by Sevastopulo (1982). The oldest occurrence is in the Cloyne Syncline on both sides of Cork Harbour (Figs. 10.5, 10.7), at the margin of the South Munster Basin (Sleeman *et al.*, 1986) and probably also on the Sheep's Head High within the basin (see p. 223). In the Cloyne Syncline, a succession estimated to be 420 m thick begins with limestone identified as Phase A, in the early part of the *carina* conodont Biozone. The uppermost part consists of bedded biomicrite and is succeeded by the Loughbeg Formation – cherty mudstone containing thin beds of silicified, skeletal limestone, and nodular, micritic limestone. Westward along the Cloyne Syncline, the Waulsortian limestone passes laterally into a much thinner succession of bedded limestone and part of the Loughbeg Formation. At the western end of the Cork Syncline, near Crookstown (Fig. 10.5), the oldest part of the Waulsortian also is of *carina* conodont Biozone age (Naylor *et al.*, 1996), and, there too, it passes laterally (to the south) into the Loughbeg Formation. In the Cork Syncline around Cork Harbour, however, the Waulsortian was first deposited somewhat later, approximately at the base of the *anchoralis* conodont Biozone. This stratigraphical horizon seems to have been the time of rapid spread northward of the Waulsortian into counties Cork, Kerry and Limerick, but the fact that the oldest known Waulsortian in this region is in the axis of the Shannon Basin (Fig. 10.9), where it is of latest *carina* Biozone (*Dollymae bouckaerti* Sub-biozone) age, suggests that in detail the pattern of advance is likely to have been complex. In the Dublin Basin, the base of the Waulsortian lies substantially above the base of the *anchoralis* Biozone. It is not known whether the northward progradation between the Shannon and Dublin Basins was pulsed or steady, or whether the Waulsortian advanced as a linear front or, as seems more likely, by the development of a series of separate initial growth centres, which later merged.

The age of the demise of the Waulsortian has proved difficult to establish in detail. As a broad generalisation it seems to have persisted longer in the south, where in many places the youngest developments are of early Viséan, and exceptionally in the Tralee Bay area (Fig. 10.9) of Arundian age (Labiaux, 1997). In the Carrick-on-Suir area, Keeley (1983) recorded conodonts from the uppermost beds of the Waulsortian, which suggest that the death of the Waulsortian banks there was nearly coincident with the base of the Viséan. In the Midlands, the youngest Waulsortian is generally of late Tournaisian age, but some lithologically distinctive and generally small, isolated Lower Viséan occurrences are known.

The Waulsortian varies markedly in thickness throughout the country; in general the thickest developments occur in areas characterised by high rates of subsidence earlier in the Tournaisian and later in the Viséan. There appear to have been three main centres: the region between Cork and the Galtee Mountains, and the Shannon and Dublin Basins.

The thickest developments occur in the axial region of the Shannon Basin. Estimates of the thickness in north-west County Limerick range up to as much as 1200 m (Shephard-Thorn 1963), of which some 600 m has been proved by drilling. At Ballybunion (Fig. 10.9), in north County Kerry, a thickness of at least 800 m seems probable. As originally shown by Lees (1961), the thickness decreases markedly away from the Shannon Estuary. Strogen (1988) determined that in the Limerick Syncline the isopachs trend east-north-east and that the thickness is reduced from 400–450 m in the axial region of the syncline to 290 m on the south-east limb. Farther south in the Glen of Aherlow (Carruthers, 1985) only 100–150 m are present. North of the Shannon, the thickness decreases to less than 100 m in the vicinity of Gort, County Galway (Lewis, 1986) and to zero between Gort and Loughrea, County Galway (Pracht *et al.*, 2002). No Waulsortian was encountered in the Doonbeg No. 1 well in west County Clare (Sheridan, 1977), where time-equivalent strata are cherty, fine-grained limestone.

The Galtee inlier (Fig. 10.9) appears to have been a contemporary 'high' because to the south of it the thickness of the Waulsortian is in excess of 350 m in the Carrick-on-Suir and Mitchelstown synclines (Keeley, 1983; Shearley, 1988), perhaps 700 m in the Fermoy Syncline, and 600 m in the Cork Syncline around Cork City. Thicknesses of 300 m or more have been recorded from farther west in the Buttevant area, County Cork (Hudson and Philcox, 1965) and around Castleisland, County Kerry (Hudson *et al.*, 1966a). However, around Killarney, County Kerry only 190 m of Waulsortian is present (Price, 1986).

North-east of a line drawn from Carrick-on-Suir to Gort, the thickness of the Waulsortian varies markedly, both locally and regionally. According to the isopachyte map of Andrew (1992, 1993), the pattern of thickness variation in the North Midlands has a strong north-east grain interpreted as being controlled by basement faults. The thickest developments (over 500 m) occur in the depocentre of the Dublin Basin (Strogen *et al.*, 1996) and around Ballinasloe, County Galway (Gatley *et al.*, 2005). Jones *et al.* (1988) have shown that in north County Dublin, to the east of the depocentre, the Waulsortian is discontinuous and varies in thickness from 200 m to zero over short distances, partly by strong diachronism at its base and top. In County Kildare, Emo (1986) showed the Waulsortian limestone thinning from 300 m to 100 m on the fault-controlled, western flank of the Kildare inlier and extending eastward toward the Leinster Lower Palaeozoic and granite inlier as a series of isolated buildups. The limit farther south is not well controlled. At Lisheen (Hitzman *et al.*, 1992) and Galmoy (Doyle *et al.*, 1992) the Waulsortian appears to be continuous and 180–200 m thick, although it thins to zero farther south. Farther east, Sheridan (1977) recorded off-bank equivalents of the Waulsortian in the Ballyragget well (Fig. 10.9). East of the Leinster Coalfield and in the Wexford outlier, it seems probable that the youngest part of the Ballysteen Limestone is laterally equivalent to the Waulsortian to the west.

The northern margin of the main area of development of the Waulsortian can be traced westward from the Dublin area with some confidence because of the density of exploration drilling. It skirts the southern and western margin of the Balbriggan inlier (Nolan, 1986) and then very approximately parallels the southern margin of the Longford-Down inlier. Close to this margin, there are Waulsortian build-ups that are up to 200 m thick, but show substantial thinning and pass laterally over short distances into bedded limestone and shale of the Slane Castle Formation. They consist of superimposed mounds, separated from each other by interbedded shale and limestone.

The western margin is less clearly defined. Discontinuous, rather thin (generally less than 100 m) Waulsortian build-ups are known, principally from drilling, west of the south-west corner of the Longford-Down inlier, around Glenamaddy, County Mayo, and farther south in east County Galway. Immediately north-east of Slieve Aughty, there is a conspicuous area which is rimmed by Waulsortian (in places very thick) but within which the mudbank limestones are replaced by nodular, cherty limestone and shale (Gatley *et al.*, 2005). This area has been called the Tynagh Basin (Fig. 10.9).

A small number of isolated Waulsortian build-ups are known in the Northern Province north-west of the Longford-Down inlier. These include several concentrated around Cloone, County Cavan (Fig. 10.8), and others, east of Belleek and at Bellanaleck and elsewhere in County Fermanagh (Sheridan, 1972b; Mitchell, 2004c).

Post-Waulsortian strata of late Ivorian age

The task of identifying the base of the Viséan in Ireland is in its infancy. However, it is clear that in the past the base was commonly identified at an horizon substantially older than its true position. Perhaps the most extreme miscorrelation was in the eastern part of the Dublin Basin, where the base, instead of being at the base of the Tober Colleen Formation, has been tentatively identified in the lower part of the Rush Conglomerate Formation by Devuyst (2006) (see below, p. 247). Farther west it may be possible to identify the base by reference to widespread thin grey/green clay bands of volcanic origin, which occur in a variably thick succession of interbedded limestone and shale above Waulsortian limestone. At Oughterard, the base of the Viséan occurs just above a tuff, but the latter's equivalence with clays of volcanic origin, which occur widely across the north Midlands (Philcox, 1984), in some places at more than one horizon, remains to be established.

In this account, strata that may be of late Ivorian age are included in the sections on the Viséan, with appropriate reference to their probable age.

Northern Province

Rocks of Tournaisian age occur widely north of the Longford-Down inlier and in north-west Ireland. It is proposed to refer to this region as the Northern Province, whilst recognising that it contains disparate Tournaisian successions. Mitchell (2004c) has provided a valuable synthesis of the stratigraphy, which is largely followed here, but much of the detailed information on which it is based has not been published.

The stratigraphy of the tract north-west of the Longford-Down inlier, for which information is available from outcrop as well as from the subsurface, is discussed first, starting with successions closest to the inlier. In the poorly exposed ground of Cavan and north County Monaghan (Geraghty, 1997), the following succession has been established from drill core. At the base, red beds of the Fearnaght Formation (probably less than 200 m thick in most places) are overlain by the Cooldaragh Formation, an estimated 125 m of grey mudstones, siltstones, and algal and evaporite-bearing micrites. The overlying Ulster Canal Formation, 30–60 m thick, consists of sandstones, silty and sandy limestones, and shales, with a marine fauna. Succeeding limestones and shales – the Ballysteen Formation of Geraghty (1997) – are here assigned to the Slane Castle Formation. Conglomerates close to the base of the succession at Killashandra, County Cavan, some 30 km north-east of Rinn River (Figs. 10.7, 10.8) are of CM miospore biozonal age. Limestones containing foraminiferans indicating MFZ 6 (Table 10.1), reported by Morris *et al.* (1980) from limestones near Carrigallen, County Leitrim, probably belong to the Ulster Canal Formation. Farther to the north-east along the Longford-Down inlier, the Tournaisian succession has been established in some detail in the Armagh area (Fig. 10.7). Three formations of Tournaisian age have been recognised (McPhilemy, 1988; Somerville *et al.*, 1996b; Mitchell, 2004c). The Killuney Formation (80–100 m) at the base consists of conglomerate, containing clasts of Lower Palaeozoic rocks and sandstone. It has been interpreted as the deposits of ephemeral streams flowing off the Longford-Down inlier. The overlying Retreat Siltstone Formation (75–90 m) is made up of sandstone and mudrock and subordinate carbonate, thought to have been deposited in shallow, brackish water. It has yielded miospore assemblages assigned to the CM Biozone. The Ballynahone Micrite Formation (100 m thick) is composed of interbedded micrite and shale, of peritidal and shallow marine origin. Miospores of the Pu Biozone have been recorded from the base of the formation; the base of the Viséan probably should be correlated with a level in the upper part of the formation from which the foraminiferan *Eoparastaffella simplex* has been reported.

North-east of Armagh, the Carboniferous/Lower Palaeozoic contact is covered by younger rocks, but it emerges in the small inlier of Carboniferous rocks at Cultra (Fig. 10.9) on the south side of Belfast Lough (Griffith and Wilson, 1982; Clayton, 1986; Mitchell, 2004c). The rocks, which may be compared in many respects with the Upper Old Red Sandstone and Cementstone facies in the Midland Valley of Scotland, have been assigned to the Holywood Group. The lower Craigavad Sandstone Formation (approximately 140 m thick) consists of sandstone with subsidiary mudstone and conglomerate containing pebbles of quartz and greywacke. The overlying Ballycultra Formation (of which some 140 m is preserved beneath Permian rocks) is made up of micrite, shale, and thin beds of sandstone. The common occurrence of desiccation cracks, stromatolites, and calcitised evaporites, suggests deposition in hypersaline, peritidal environments; however, a few horizons containing brachiopods indicate that the salinity sometimes had normal marine values. Clayton (1986) has shown that the middle part of the Ballycultra Formation is of CM miospore biozonal age.

Tournaisian successions farther to the north-west of the Longford-Down inlier, such as the 470 m of interbedded grey mudstone, sandstone, micrite and dolomicrite (with horizons of evaporite near the base) in the Glenoo No. 1 well (Sheridan, 1972a) (Fig. 10.7), are much thicker than those adjoining it, confirming that the present inlier was a contemporary high, as suggested by the nature of the basal conglomerates that flank it.

Mitchell (2004c, Fig. 7.8, Table 7.7) distinguished the succession on the south side of the Clogher Valley Fault, typified by the Glenoo No 1. well, from that on the north side. In the latter area (Griffith, 1970; Mitchell and Owens, 1990; Mitchell, 2004c), the lowest formation of the Tyrone Group, the Ballyness Formation, consists of approximately 300 m of red and purple sandstone and conglomerate with pebbles of vein quartz and metamorphic rocks. It has been interpreted as having been deposited in a fluvial environment and as being of Pu miospore Biozone age (Mitchell, 2004c). The overlying Clogher Valley Formation (400 m thick) consists of three parts: the lower formed of grey mudstone, siltstone and sandstone, micrite with fenestral fabric, and evaporites; a median sandstone unit; and an upper part of mudstone, siltstone and sandstone, with thin crinoidal limestone beds. It is possible that younger limestones in this area, assigned by Mitchell (2004c) to the base of the Ballyshannon Limestone Formation, are of latest

Ivorian age, on the evidence of the conodonts reported by Austin and Mitchell (1975). Mitchell (2004d, Fig. 8.4) interpreted a south-east to north-west section from the Longford-Down inlier to the Fintona block as follows: after the initial marine transgression during CM miospore biozonal times along the margin of the inlier, renewed extension in Pu miospore biozonal times led to north-westward overstep across the Clogher Valley Fault and progressive deepening of the sedimentary environment as shown by the succession through the Ballyness and Clogher Valley formations.

Marginal marine conditions were probably established farther north by the end of the Tournaisian. In County Derry (Mitchell, 2004c), the Roe Valley Group extends from the south side of Lough Foyle to 5 km north-west of Draperstown. It consists of the Spincha Burn Conglomerate Formation, which rests on the Dalradian and is made up of 25–100 m of sandstone and coarse conglomerates with pebbles of quartz and quartzite. The overlying Barony Glen Formation, 150–200 m thick, in its lower part is formed of purple, red and green sandstone and mudrock with pedogenic carbonates. The upper part, dated to the CM miospore Biozone (Mitchell, 2004c), contains mudrocks and micrites, with evidence of marginal marine conditions.

Only a short distance to the south-east of the outcrop of the Roe Valley Group, the succession in the Draperstown area appears very different (Mitchell, 2004c) and has been assigned to the Tyrone Group. There the Iniscarn Formation consists of 400 m of boulder conglomerates and red, muddy sandstones, overlain by feldspathic breccias. The younger Altagoan Formation (*c.*550 m thick) consists of sandstone, grey mudstone and micrite, with beds of replaced evaporites. Miospore assemblages from the upper part of the formation (Owens *et al.*, 1977) contain rare specimens of *Lycospora pusilla*, probably indicating the base of the Pu miospore Biozone.

Mitchell (2004c) suggested that the succession in the Draperstown area recorded the same transgression (of Pu miospore biozone age) as seen farther to the south-west, whereas the Roe Valley Group was deposited in an intermontane basin with a marine transgression in CM miospore Biozone times. Further sedimentological and palynological work is required to verify this hypothesis.

The limit of the Tournaisian sea can be identified with more confidence in the west. Carbonates of early Viséan age overlie thin successions of terrigenous rocks on the south-east flank of the Ox Mountains–Ballyshannon High (Lough Derg and Ox Mountains inliers). In several places, the shales near the base have yielded miospore assemblages of Pu biozonal age (Higgs, 1984). In contrast, relatively thick successions of Tournaisian rocks were intersected in deep wells, such as Big Dog (Figs. 10.7, 10.8) in the Lough Allen Basin, not far to the south-east (Sheridan, 1972b, 1977). Philcox *et al.* (1992) have described the evolution of the basin, which developed in the hanging wall of extensional faults parallel to the south-eastern margin of the Ox Mountains. The lowest Kilcoo Formation forms the lower part of a wedge of sandstone which is red and coarsely conglomeratic near the north-western margin of the basin and becomes finer and is interbedded with mudstone containing pedogenic carbonate towards the basin centre. The Kilcoo Formation has been interpreted as the products of alluvial fans, which formed along the faulted basin margin. The overlying Boyle Sandstone Formation, first described by Caldwell (1959), has its type section in the Boyle area (Figs. 10.7, 10.8), where it rests on a thin development of the Kilcoo Formation. It consists of three members, which can be traced through most of the basin. The lower Drumbrick Sandstone Member, probably also of alluvial origin, grades from grey, pebbly sandstone near the north-western margin of the basin, to silty mudstone with sporadic developments of evaporitic sulphate near the basin centre. The middle Lough Key Limestone Member consists of grey and green or dark mudstone, with thin beds of sandstone, micrite, evaporitic sulphate, and minor skeletal carbonates. It marks the initial marine incursion and is of CM miospore biozonal age. Near the north-western margin of the basin the Lough Key Mudstone is absent and skeletal carbonates (the Rosskit Limestone Member) rest directly on the Drumbrick Sandstone Member. Philcox *et al.* (1992) suggested that a narrow, fully marine gulf developed along this margin of the basin. The upper Rockingham Sandstone Member, also of CM biozonal age, varies from coarse, pebbly, non-marine sandstone in the north, to marine sandstone, mudstone, and skeletal limestone farther south. The overlying Kilbryan Limestone Formation consists of limestone and shale of open marine origin. The CM/Pu miospore biozonal boundary occurs low in the formation, confirming that at least some of it is of Tournaisian age. The succession of facies in the formations in the Lough Allen Basin and Boyle area (Kilcoo, Boyle Sandstone, and Kilbryan) is similar to that of the North Midlands Province (Red beds, Navan Group, and Slane Castle Formation).

Tournaisian rocks have been reported to the north-east of the Lough Allen basin in the Omagh Syncline

(Simpson, 1955; Mitchell, 2004c). There the Omagh Sandstone Group consists of up to 600 m of conglomerate, sandstone, and shale. The lower part, which contains red beds and possibly pedogenic carbonate nodules, is most probably of non-marine origin. Marginal marine environments are represented higher in the succession by algal limestones, replaced evaporites, together with bivalves, ostracodes, and 'serpulids'. Mitchell (2004c) has reported that these rocks have yielded miospores of CM Biozone age and has interpreted them as having formed in a gulf extending north to Omagh.

North of Omagh, in the poorly known Newtownstewart outlier, County Tyrone (Mitchell, 2004a), the Owenkillew Sandstone Group, approximately 1500 m thick, includes purple and greenish grey sandstones and mudrocks with rare thin algal micrites. Mitchell (2004c) reported that it contains miospores, interpreted as indicating the lower part of the Pu miospore Biozone.

Farther west in County Donegal (Fig. 10.7), Graham and Clayton (1994) have described an isolated but thick succession of coarse conglomerate and sandstone of CM miospore biozonal age, the Edergole Formation, which was interpreted as having been formed on an alluvial fan during an early phase of extension.

Some of the non-marine sequences to the west of the Ox Mountains–Ballyshannon High are of Tournaisian age. The best information is from the north side of Clew Bay and Clare Island (Fig. 10.7) where Phillips and Clayton (1980) have described the Maam and Capnagower Formations (in total 300 m thick), which Graham (1981b) has interpreted as having been deposited in a fluvial environment by rivers flowing from the north-west. The CM-Pu miospore biozonal boundary lies near the top of the Capnagower Formation.

Regions north of the South Munster Basin: Viséan

In the latest Tournaisian and early Viséan, there was a marked change in the distribution of facies in Ireland, probably related to renewed crustal extension. A transgression extended the area of marine sedimentation, particularly in the west, and relatively deep water basins developed with their depocentres generally close to the sites where maximum subsidence had occurred during the Tournaisian. The basinal sediment was terrigenous clay and silt, interbedded with carbonate transported from the bounding shelves, where varied shallow water carbonate facies developed. The basin margins in some cases were steep, reflecting their control by active faulting.

Viséan stratigraphy north of the South Munster Basin will be described using broad geographical divisions, which correspond, as far as possible, to the distribution of major facies, from south to north.

North Munster Shelf

Between the South Munster Basin and the Shannon and Dublin basins was a broad shelf, which has been referred to as the North Munster Shelf. The Viséan stratigraphy of this shelf has been described in terms of numerous locally defined formations; although in detail there is considerable complexity of facies, there is greater uniformity than the current lithostratigraphical terminology (Fig. 10.10) suggests. In particular, cyclothemic sequences of platform limestone were widespread during the later part of the Asbian and the Brigantian (Gallagher, 1996; Gallagher and Somerville, 1997, 2003). The stratigraphy is described here on a regional basis, starting in the south.

The southern margin of the Viséan carbonate shelf is preserved only in the Cork Syncline and locally along the northern flank of the Cloyne Syncline (discussed above); farther west, in the Kenmare Syncline, Viséan rocks have been eroded. In the Cork Syncline (Heselden, 1991; Sleeman and McConnell, 1995), the Waulsortian limestone is succeeded in the east by the Cork Red Marble Formation (Lower Viséan and possibly Arundian), which consists of 80 m of pale grey limestone, including oolites, with horizons of red limestone breccia and chert (Figs. 10.10, 10.11). The red breccias, known as Cork Red Marble, have been used widely as a decorative stone. Heselden (1991) has interpreted their origin as the product of intensive stylolitisation of originally bedded limestones of shallow water origin. The succeeding Little Island Formation consists of carbonate mudbank facies with an abundant fauna, which was listed by Turner (1937). The upper part of the formation contains foraminiferans (Gallagher and Somerville, 2003) indicating an Asbian (MFZ 13) age; the boundaries of the Arundian and Holkerian stages have not been identified within it. The overlying Clashavodig Formation consists of well-bedded limestones. Heselden (1991) has shown that the succession consists of nine asymmetrical cycles, each recording upward shallowing. The lower of these cycles have been interpreted by Gallagher and Somerville (2003) as being formed in relatively deeper water, but a prominent palaeokarst at approximately the middle of the formation ushers in a sequence of cycles with brachiopod-rich shell beds and oolites at their tops.

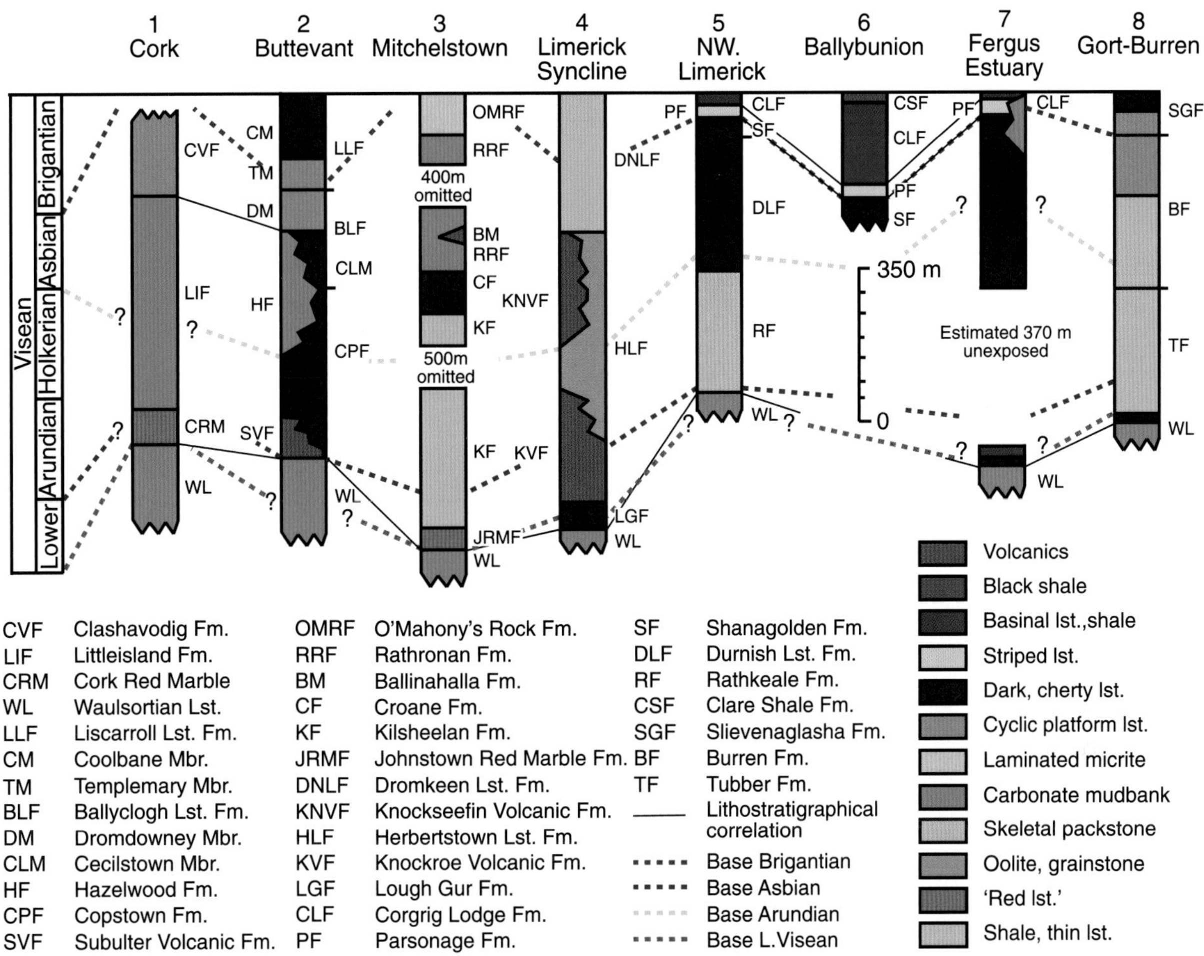

Fig. 10.10 Correlation of successions of Viséan age of the North Munster Shelf, the Shannon Basin, and the southern part of the Galway-Roscommon Shelf. See Figure 10.11 for locations.

These terminate at surfaces interpreted as palaeokarsts, which in some cases are capped by clay-rich palaeosols. The beginning of these high frequency cycles coincides with the first appearance of foraminiferans indicating an Asbian (MFZ 14) age (Gallagher and Somerville, 2003). Rocks of Brigantian age are not preserved within the Cork Syncline.

The evidence from the Cork and Cloyne synclines suggests that the margin of the North Munster carbonate shelf prograded only very slightly to the south during the Viséan. Both in the Cloyne Syncline at Ballygarvan and at the western end of the Cork Syncline near Stickstown (Fig. 10.5), the Little Island Formation, at some levels consisting of limestone breccia, overlies the slope deposits of the Loughbeg Formation (Naylor *et al.*, 1996).

The nearest occurrence of Viséan rocks to the north of the Cork Syncline is in a narrow tract from west of Mallow, County Cork to east of Clonmel, County Tipperary (Sleeman and McConnell, 1995). The stratigraphy within this tract, which follows the Variscan strike, is broadly consistent and is similar to that of the Cork Syncline.

The area around and to the west of Mallow, including Buttevant (Figs. 10.10, 10.11), has been the most intensively studied. The upper part of the Waulsortian limestone there is of early Viséan age, as it contains the conodont *Gnathodus homopunctatus* (Clipstone, 1992). The overlying Copstown Limestone Formation, dark, fine-grained skeletal limestones with chert, ranges in age from Arundian to Asbian. Local developments of volcanic rocks in the lower part of the Copstown Formation, the Subulter Volcanics (Hudson and Philcox, 1965), are discussed below. The Copstown Formation is succeeded by, and probably is also partly laterally equivalent to,

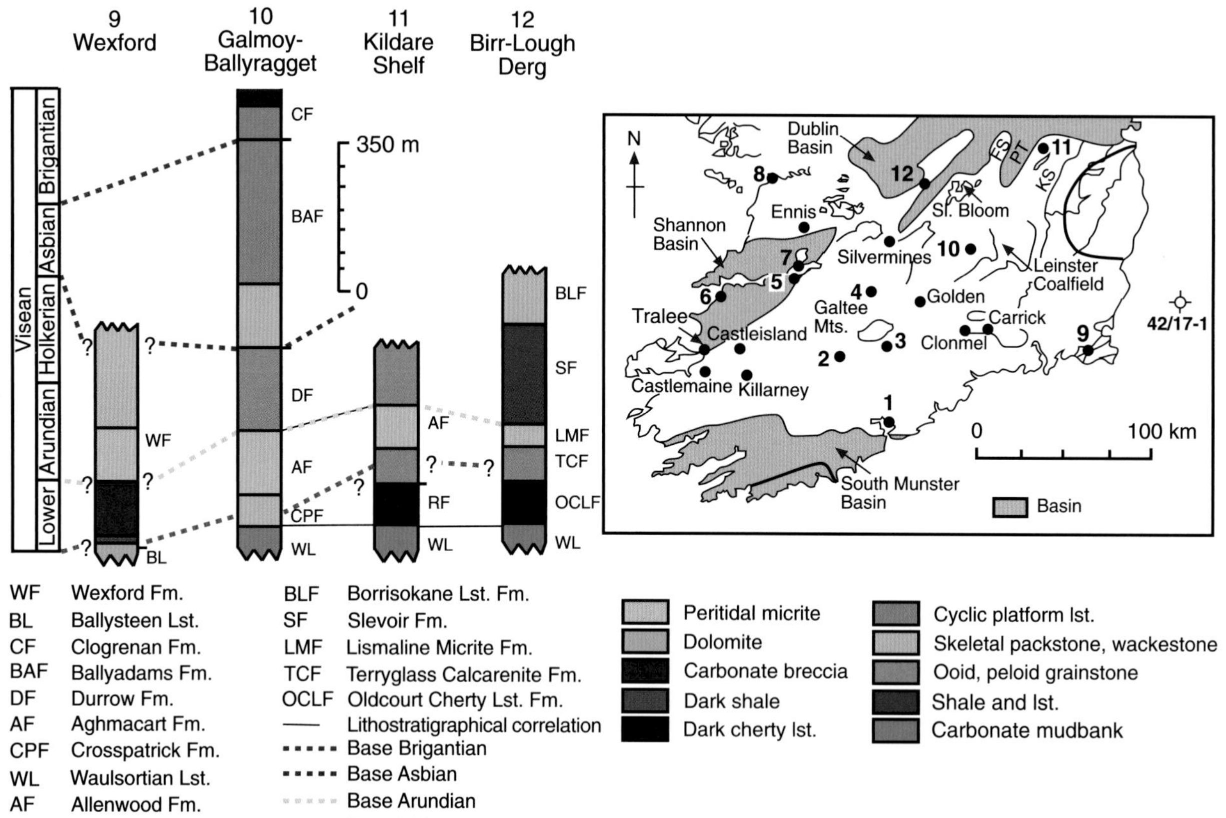

Fig. 10.11 Correlation of successions of Viséan age of the northern and eastern part of the North Munster Shelf. Map shows the South Munster and Shannon Basins and the southern part of the Dublin Basin. KS - Kildare Shelf; PT - Portarlington Trough; ES - Edenderry Shelf.

the Hazelwood Formation, typically fine-grained, grey, mudbank limestone and related limestones of Asbian age. The overlying Ballyclogh Limestone Formation of Asbian age is divided into two members (Gallagher and Somerville, 1997). The lower Cecilstown Member consists of dark, thinly bedded, cherty limestones, interpreted to have been deposited below wave base and to be laterally equivalent to the upper part of the Hazelwood Formation. It contains foraminiferans indicating an Asbian (MFZ 13) age. The upper Dromdowney Member consists of stacked, shallowing-upward cycles, in some cases with evidence of emergence at their tops, broadly similar to, and the same age (MFZ 14) as, those described in the upper part of the Clashavodig Formation of the Cork Syncline. In contrast to the situation in the Cork Syncline, Brigantian strata (the Liscarroll Limestone Formation) are preserved to the west of Mallow. The Liscarroll Formation has been divided into two. The lower Templemary Member consists of cycles of thickly bedded, crinoidal limestone overlain by thinly bedded limestone with coral thickets. The overlying Coolbane Member consists of dark, well-bedded, cherty limestone, interpreted by Gallagher and Somerville (1997, 2003) as having been deposited below fair weather wave base.

Viséan successions generally similar to that of the Mallow area have been described from the Mitchelstown Syncline by Shearley (1988) and from the Carrick-on-Suir Syncline by Keeley (1983) (Figs. 10.10, 10.11). Sleeman and McConnell (1995) have provided a synthesis, simplifying the stratigraphical nomenclature. In the Mitchelstown Syncline, the Waulsortian limestone, locally of early Viséan age at its top, is succeeded by the Johnstown Red Marble Formation, which is lithologically similar to the Cork Red Marble and is also of early Viséan age. In the Carrick-on-Suir Syncline, the Waulsortian limestone is succeeded by cherty limestones of the Silverspring Formation, which varies in thickness from 100 m in the east to less than 10 m in the west. The succeeding Kilsheelan Formation, identified by Sleeman and McConnell (1995) within both synclines, consists of a thick succession of bedded, medium to dark grey, skeletal limestones of shallow water origin. It ranges in age from Lower Viséan to early in the Asbian and is comparable in age to the Copstown Limestone Formation of the Mallow area. Overlying the Kilsheelan Formation is a distinctive unit of thinly bedded, algal-rich, cherty, fine-grained limestone, the Croane Formation. In the Carrick-on-Suir Syncline, this was originally interpreted (Keeley, 1983) as occurring higher in the succession. However, in the Mitchelstown Syncline, Shearley (1988) demonstrated that it forms a transitional facies between the underlying bedded, shelf limestones and the overlying massive, generally fine-grained Rathronan Formation. The latter is predominantly of carbonate mudbank facies and is of Asbian age, broadly comparable in both respects to the Littleisland Formation of the Cork Syncline and the Hazelwood Formation of the Mallow area. It is overlain in the Mitchelstown Syncline by laminated micritic limestone, the O'Mahony's Rock Formation. In the western part of the Carrick-on-Suir Syncline, Keeley (1983) mapped a thick body of oolite, the Ballyglasheen Formation, which was interpreted as being laterally equivalent to the Rathronan Formation. The cyclic platform carbonates overlying the Asbian carbonate mudbank limestones that might have been anticipated from the stratigraphy of the Mallow area have not been recognised within the Mitchelstown and Carrick-on-Suir synclines and were apparently removed by pre-Pennsylvanian erosion.

Viséan successions north of the Galtee Mountains (Carruthers, 1985; Archer *et al.*, 1996) and in the Limerick Syncline (Ashby, 1939; Strogen, 1988; Somerville *et al.*, 1992a; Strogen *et al.*, 1996) differ significantly from those of the Mallow, Mitchelstown, Carrick-on-Suir tract in the greater proportion of very shallow water facies, such as oolite, and in the absence of carbonate mudbanks. In the Limerick Syncline (Figs. 10.10, 10.11), the Waulsortian limestones are overlain by bedded, commonly cherty limestones of the Lough Gur Formation, which ranges in age from late Tournaisian to early Viséan. Similar cherty limestones (Knockordan Formation) also occur in this stratigraphical position immediately north of the Galtee Mountains (Carruthers, 1985). The Lough Gur Formation is succeeded in the Limerick Syncline by the Knockroe Volcanic Formation (Strogen, 1988), which consists of a complex series of basalts and trachytes (see below). The upper boundary of the volcanics is diachronous: early Viséan to early Arundian in age in the east of the Limerick Syncline and Holkerian in the west (Somerville *et al.*, 1992a). The overlying Herbertstown Limestone Formation of the Limerick Syncline is dominated by high energy grainstones, including oolite, of shallow water origin. It extends into the Asbian in the western part of the syncline but to the east its upper, Asbian part is replaced by the Knockseefin Volcanic Formation (see below). In the later part of the Asbian and Brigantian, shallow water platform conditions continued with the deposition of the Dromkeen Limestone Formation.

North-east of the Galtee Mountains (Fig. 10.11), in the so-called 'Golden Gulf' (Archer *et al.*, 1996), dark, cherty, fine-grained limestones and shales, with locally developed volcanics, apparently of early Viséan age, have been named the Athassel Formation (Carruthers, 1985); they have been interpreted by Archer *et al.* (1996) as being of basinal origin. They are overlain by shallow water limestone, including oolite (Suir Formation) of Arundian age. A younger, poorly dated unit of thin-bedded, dark, cherty, fine-grained limestone, the Lagganstown Formation, compared by Archer *et al.* (1996) with the Croane Formation of the Carrick-on-Suir Syncline, may indicate a return to deeper water conditions, but the overlying Hore Abbey Formation of Asbian age is of shallow water facies. It is reported to contain a range of limestone types including skeletal wackestone, packstone, oolitic grainstone, and horizons of algal laminites. Although cycles have not been recorded from the Hore Abbey Formation or the Dromkeen Limestone Formation of the Limerick Syncline, the lithological similarity of the two formations to the Ballyclogh Formation of the Mallow area suggests that cryptic cycles may be present.

During the Viséan, the region between the Lower Palaeozoic envelope of the Leinster Granite and the inliers of Slieve Bloom and Slieve Felim–Devilsbit Mountain, east of Sivermines (Fig. 10.11) was a shallow water carbonate ramp facing west, which evolved to a platform that, at least during the Lower Viséan and in the later part of the Asbian, was periodically emergent. The succession in the Wexford outlier is similar in some respects to that west of the Leinster Granite.

The Viséan succession to the west of the Leinster Coalfield (Fig. 10.11) has been described by Archer *et al.* (1996) and Gatley *et al.* (2005). In this region, the Waulsortian limestone is followed by the Crosspatrick Formation, consisting of bedded, cherty crinoidal packstones. The latter varies in thickness as a result of the topography of the underlying Waulsortian limestone and probably ranges in age from late Tournaisian to Early

Viséan. The succeeding Aghmacart Formation (109 m thick in the Durrow borehole and containing Lower Viséan (MFZ 9) aged foraminiferans) reflects a shallowing of the depositional environment. It is predominantly made up of dark, fine-grained, peloidal limestone, with oolitic and skeletal grainstones and, at some horizons, evidence of evaporites. The overlying Durrow Formation reflects a change to generally more open marine conditions and consists of coral-bearing limestone, fossiliferous shale, cross-stratified oolite, and thin units of fenestral micrite. It is of Arundian age at its base and extends into the early part of the Asbian (Gatley *et al.*, 2005). The succeeding Ballyadams Formation, which occurs to the east of the coalfield as well as to the west, is of Asbian age and consists of skeletal limestones, with horizons rich in corals, including *Lithostrotion* thickets, and brachiopods. The upper part of the formation, with foraminiferans indicating MFZ 14 (Asbian), exhibits well-developed cycles with palaeokarstic surfaces, overlain in some cases by thick palaeosols; these have been described in County Carlow by Cózar and Somerville (2005) and in south County Kilkenny by Gallagher (1996). The youngest Clogrenan Formation (Brigantian) also occurs around the margin of the coalfields and consists of a lower cyclic unit and an upper unit of dark, cherty limestone.

In the very poorly exposed ground east of the Leinster Coalfield, the succession below the Ballyadams Formation encountered in boreholes and formerly named the Milford Formation (Tietzsch-Tyler et al., 1994) has been revised by Nagy et al. (2005a). The Crosspatrick Formation (c.30 m seen) consists of dolomitised crinoidal limestone with chert, similar in facies, apart from more intense dolomitisation, to its occurrences to the west of the coalfield. The Aghamacart Formation, 125 m thick, is also mostly pervasively dolomitised. The lowest part consists of dolomitised, cross-stratified, ooid and peloid grainstones, interpreted as having been deposited in the intertidal zone. They are succeeded by dolomitised, bioturbated wackestone and packstone, which pass upward into dolomitised, peritidal facies with algal stromatolites, fenestral micrite, evaporite pseudomorphs, ostracode-rich micrite with detrital quartz, and rare ooid grainstones. The upper part of the formation is less dolomitised and consists of ooid and peloid grainstone and contains corals indicating a Lower Viséan age. The Durrow Formation (108 m seen) consists of skeletal limestone, indicating open marine conditions but with sporadic occurrence of very shallow water facies. A diverse microbiota shows that it is of Arundian age at the base and extends into the Holkerian. South of the Leinster Coalfield, the poorly known Butlersgrove Formation, dark limestones, in part peloidal (Tietzsch-Tyler and Sleeman, 1994b), occupies a stratigraphical position analogous to that of the Aghmacart Formation.

The succession in the Wexford outlier (George, 1960; Carter and Wilbur, 1986; Clayton *et al.*, 1986b; Tietzsch-Tyler *et al.*, 1994) has been revised by Nagy *et al.* (2005b). The Waulsortian Limestone is not developed and the Wexford Formation (approximately 500 m of micrite, dolomite, evaporite, limestone breccia and skeletal limestone of Lower Viséan to Asbian in age) rests on dolomitised late Ivorian Ballysteen Limestone (Fig. 10.11). Nagy *et al.* (2005b) have interpreted the temporal and geographical facies variations of the Wexford Formation within the outlier as follows. Throughout the Viséan, a sedimentary ramp extended from the Rosslare area (the shallowest water area) south-westward. There was also periodic reactivation of the important Caledonian Wexford Boundary Fault, which forms the south-east margin of the outlier, leading to the generation of debris flows. The lower 225 m of the succession between Rosslare and Wexford town consists of micrite, commonly dolomitised, with a variety of features, such as pseudomorphs after evaporites, evaporite collapse breccias, and a very restricted fauna, indicating hypersaline conditions. The upper part of the succession (100 m of skeletal wackestone and packstone, now mostly dolomite, seen in drill core) suggests deposition under conditions of more normal marine salinity. In the south-west of the outlier, the lowest unit of the Wexford Formation consists of a relatively thin unit of interbedded black, non-calcareous mudrock and grey micrite, a facies not seen in the Rosslare region. This is followed by a unit of limestone breccias (50–120 m thick); the breccias have been interpreted as including debris flows generated on the fault-bounded margin of the half graben to the south-east, and collapse breccias, formed as a result of subsequent solution of deep water evaporites. The succeeding unit, up to 127 m thick, consists of limestones of shallow water facies, including fenestral micrites and oolites, but with some beds of skeletal limestone with microfossils interpreted as indicating an Arundian age. The youngest part of the succession (230 m preserved) consists of skeletal limestones, including oolite, with a fauna of brachiopods and corals, which indicate a Holkerian to Asbian age.

Offshore well 42/17–1, drilled in the St Georges Channel (Fig. 10.11), penetrated Pennsylvanian coal measures overlying Mississippian limestones (Maddox *et al.*, 1995). This is similar to the stratigraphy established in south County Wexford, perhaps suggesting a similar Viséan succession. Seismic data suggest that Mississippian limestones extend northward through the Central Irish Sea: the St George's Land or Leinster–Wales Massif shown on palaeogeographical maps (George, 1958) probably had been drowned by the late Viséan.

The Viséan succession at the eastern end of the southern margin of the Dublin Basin has been summarised by McConnell and Philcox (1994) who, following Hitzman (1995), have interpreted it as being embayed, with a narrow gulf extending south-westward between the Kildare Shelf and the Edenderry Shelf, named by them the Portarlington Trough. On the Kildare Shelf (Fig. 10.11), the Waulsortian limestone is overlain by the Allenwood Formation, although locally there is an intervening unit of dark cherty limestone, the Rickardstown Formation. The Allenwood Formation consists of a lower unit of peloidal and ooid grainstone with an open marine fauna, a middle unit of pelmicrite and micrite, including fenestral micrite, and an upper unit of peloidal grainstone (Emo, 1986). The lower and middle units are of late Ivorian and early Viséan age; the upper unit is of Arundian age. Younger, unnamed, shelf carbonates also occur in this area. On the Edenderry Shelf, the Allenwood Formation consists of oolitic grainstones, many showing cross-stratification; McConnell and Philcox (1994) have named these the Edenderry Oolite Member, part of which is probably late Ivorian in age.

Hitzman (1995) has interpreted the margin of the Dublin Basin to the west of Slieve Bloom as also being embayed, with a narrow basinal gulf extending south-westward to the Silvermines area. This gulf, which may not have been persistent throughout the Viséan, is bounded to the west by a shallow water shelf area, which was described south of Birr by Brück (1985) and Gatley *et al.* (2005). In this area (Fig. 10.11), Waulsortian limestones are overlain by dark, cherty limestones of varying thickness (Oldcourt Cherty Limestone Formation), which may well be of late Tournaisian age. The younger Terryglass Calcarenite Formation in places rests directly on Waulsortian limestone and varies in thickness from 30 m to perhaps as much as 200 m; it is probably of late Tournaisian and early Viséan age. It consists of very well-sorted oolitic, peloidal and skeletal grainstone, which is cross-stratified at some localities. It is overlain by Lower Viséan peritidal micrites, the Lismaline Micrite Formation. The micrites are followed abruptly by dark limestone and shale (the Slevoir Formation) with a fauna indicating normal marine salinity. This drowning of the platform has been dated as coinciding with the base of the Arundian (cf. Gatley *et al.*, 2005, p. 41). Shallow water platform carbonate sedimentation resumed with the deposition of skeletal sands, which formed pale grey packstones of the Borrisokane Calcarenite Formation, of Arundian and perhaps younger age.

Shannon Basin

The region around the Shannon estuary accommodated thick successions of Tournaisian rocks and also subsided rapidly during the Serpukhovian. However, neither the geographical extent of the basin during the Viséan nor its history are well understood. The axis of the basin, at least in the late Viséan, probably extended from Ballybunion, County Kerry to the islands in the Fergus Estuary; there is good evidence that the basin closed to the north-east. Unfortunately, the succession in the axial region is incompletely known because the Arundian–Asbian part is cut out by faulting on the coastal section at Ballybunion and little detail is currently available from the Fergus estuary.

The best-documented succession is in north-west County Limerick (Figs. 10.10, 10.11) (Shephard-Thorn, 1963; Somerville and Strogen, 1992). Overlying the Waulsortian limestone conformably is a thick succession of dark, generally unfossiliferous mudrock with thin beds of mostly fine-grained carbonate, the Rathkeale Formation. Somerville and Strogen (1992, fig. 3) showed the formation as being almost entirely Arundian in age with the base of the stage just above the base of the formation; this suggests that part of the Waulsortian is of early Viséan age. The overlying Durnish Limestone Formation consists of dark, cherty limestone with a fauna of brachiopods and solitary, and rarer colonial, corals. The lower part is probably of Holkerian age; the upper part is Asbian. The overlying Shanagolden Formation, of Asbian age, consists of dark limestones with a distinctive fauna of solitary corals, and also contains horizons of breccia (Sleeman and Pracht, 1999). The succeeding Parsonage Formation is most distinctive and has been recognised widely throughout the basin. It consists of generally unfossiliferous, fine-grained limestone, which contains units of conspicuously striped dark and pale limestone, which may be replaced evaporites (MacDermot and Sevastopulo, 1972). These are succeeded by deep water limestones and shales of the Corgrig Lodge Formation, which in turn are overlain by the Serpukhovian and

Pennsylvanian Clare Shale Formation. Somerville and Strogen (1992) interpreted the succession in terms of a north-westward dipping ramp. They suggested that the Rathkeale Formation was deposited on the deeper, outer part of the ramp and showed that coeval Arundian rocks closer to Limerick city were deposited in shallower water. The succession above the Rathkeale Formation was interpreted as representing a progressive upward shallowing, with the Parsonage Formation reflecting very shallow water environments. An alternative interpretation (Sevastopulo, 1981a) is that the Parsonage and Corgrig Lodge formations are of deep water origin and that the striped limestones represent replaced basinal evaporites.

On the north side of the Shannon (Figs. 10.10, 10.11), the Viséan succession is partially exposed on a number of islands in the Fergus estuary (Tattersall, 1964; Sleeman and Pracht, 1999). The Waulsortian limestone is succeeded by cherty limestones and bedded, fine-grained limestone, from which Austin and Husri (1974) recorded conodonts likely to be of early Viséan age. There is a substantial exposure gap, where beds equivalent to the Rathkeale Formation might be anticipated (limestones of basinal facies, the Finlough Formation (Lewis, 1986; Sleeman and Pracht, 1999) occur at this position in the East Clare Syncline north-east of the estuary). West of the exposure gap, a succession of approximately 400 m of poorly dated limestones, including spectacularly slumped beds, has been established below the Parsonage Formation. Striped limestones within the Parsonage Formation, including slumped and brecciated facies, are very well displayed on the islands. As in the type area in north-west County Limerick, the Corgrig Lodge Formation (Inistubrid Beds of Tattersall) has been shown to contain conodonts of Brigantian age (Austin and Husri, 1974). Although there is an exposure gap, it seems probable that there is no stratigraphical hiatus between the Corgrig Lodge Formation and the Serpukhovian Clare Shale Formation. As the upper beds of the limestone are followed northward towards Ennis, they exhibit pronounced facies changes, the details of which are not fully established. The most striking feature is the development of thick accumulations of mudbank limestones, of both Asbian and Brigantian age, basinward of the cyclic platform limestones that extend southward from the Burren (Figs. 10.10, 10.11). It is probable that basinal facies extended to the north-west under the Serpukhovian cover, because Sheridan (1977) recorded possible striped limestones in the Doonbeg No. 1 well.

The Viséan succession in north Kerry is poorly known except at Ballybunion, where, however, it is incomplete. On the coast there (Figs. 10.10, 10.11), the Waulsortian limestone extends into the Lower Viséan and is overlain by dark, cherty, argillaceous limestone of Arundian age. A fault cuts out most of the Arundian–Asbian part of the succession. Dark, slumped limestones, breccias, and striped limestones described by Kelk (1960) have been assigned by Sleeman and Pracht (1999) to the Parsonage Formation. They are followed by a thick succession of tabular bedded, basinal, fine-grained limestones, which are cherty at some levels and include a few graded calciturbidites. These have been assigned to the Corgrig Lodge Formation. Fossils include conodonts and ammonoids, such as *Hibernicoceras* sp., which confirm the Brigantian age of the formation. Kelk (1960) showed that the youngest part of the Brigantian was represented by ammonoid-bearing black shale, here assigned to the Clare Shale Formation.

Viséan rocks occur around Tralee Bay (Figs. 10.10, 10.11) and eastward to the Castleisland area (Hudson *et al.*, 1966a; Thornton, 1966; Pracht, 1996; Labiaux, 1997). Around Tralee Bay, Waulsortian limestone extends into the Arundian in some places; the carbonate banks of Viséan age tend to be isolated and set in off-bank facies and generally to be mud-rich (Labiaux, 1997). In the Castleisland area, the Waulsortian is overlain by the Rockfield Limestone Formation, dark, fine-grained, in places cherty limestones, with developments of breccias at the base. Fossils recorded by Hudson *et al.* (1966a) suggest that the formation ranges in age from Lower Viséan to Arundian. The succeeding Cloonagh Limestone Formation in the Castleisland area consists of both mudbank (in the north and west) and laterally equivalent, bedded limestone facies (in the south and east). Its base is reported to be of Holkerian age but the bulk of it is Asbian. Farther west, around Tralee Bay, there are carbonate mudmounds of Asbian age, overlain by breccias and striped limestones, which in places are spectacularly slumped. The succession containing the striped limestones can clearly be equated with the Parsonage Formation. Overlying limestones and shales can be assigned to the Corgrig Lodge Formation. The succession above the Cloonagh Limestone in the Castleisland area has been referred to as the Dirtoge Limestone, which is of Brigantian age; records of striped limestones at its base and fine-grained cherty limestones and shale above suggests that it is probably the equivalent of the Parsonage and Corgrig Lodge formations. Both around Tralee Bay and in the Castleisland area, there is reported to be a slight stratigraphical hiatus between the youngest Viséan rocks and the Serpukhovian.

The limestone outcrop may be followed south-westward from the Castleisland area to Castlemaine (Fig. 10.10) and thence eastward to Killarney. Very little is known about the Viséan rocks of this region, which are poorly exposed (Turner, 1962).

Dublin Basin

The Dublin Basin, which almost certainly extended across the Irish Sea to link with the Craven Basin of Lancashire and Yorkshire, covered much of the northern part of the Irish midlands during Viséan time. It has been studied intensively on the coastal sections and in boreholes in north County Dublin and in County Meath; elsewhere, there is less published information, although basinal rocks have been intersected in numerous boreholes drilled for mineral exploration. Where it has been most extensively studied, the basin is fault-bounded. Movement on the marginal faults occurred several times, generating coarse debris flows containing carbonate from the adjacent shelves. Other, intrabasinal, faults were also active. It is not known whether any of them produced intrabasinal highs capped by shallow water limestone; sediment flows associated with these faults appear to contain only older, Tournaisian, limestones and fragments of basement. In a general way, the grain size of the basinal rocks decreases towards the axis of the basin, where thick successions of argillaceous, commonly cherty, spicular, fine-grained packstones and wackestones are common.

In the Dublin region (Fig. 10.12), which is described first, the basin was bounded by two carbonate shelves: the Balbriggan Shelf, with a floor of Lower Palaeozoic rocks cored by Caledonian granite to the north; and a shelf which covered the Leinster Granite and its Lower Palaeozoic envelope, to the south (Nolan, 1989). The latter has been eroded in the area of the map in Figure 10.12 in south County Dublin, where its former presence may be inferred from shelf carbonate detritus transported northward into the basin, but it is preserved farther to the south-west, where it has been called the Kildare Shelf (Fig. 10.12).

Rocks of basinal facies (the Fingal Group; Nolan, 1989) are well exposed on the coast of north County Dublin. There, the oldest rocks are cleaved mudrocks of the Tober Colleen Formation (formerly called the Rush Slates; Matley and Vaughan, 1906). Although these are not seen in contact with underlying formations in their type section on the coast south of Rush, elsewhere, such as Feltrim, they overlie Waulsortian Limestone. Although the Tober Colleen Formation has in the past been regarded as being of Viséan age, preliminary study by Devuyst (2006) has shown that in the Dublin coastal section it is entirely of late Ivorian age. The formation, which varies in thickness between 50 m and more than 250 m, generally contains little carbonate, but in places there are boulder beds, the product of debris flows, in which the clasts are composed of Tournaisian, including Waulsortian, limestone. The mudstones contain a sparse benthonic fauna, including small solitary corals, and, near the top of the formation at Rush, a rich fauna of ammonoids (Smyth, 1951).

On the coast at Rush, the Tober Colleen Formation is overlain by the Rush Formation, which consists of a succession of spectacular conglomerates with pebble- to boulder-sized, commonly moderately rounded clasts of Tournaisian limestone, Ordovician volcanic rocks, vein quartz, and Lower Palaeozoic sandstone and siltstone. The conglomerates are interbedded with limestone, sandstone, and shale. Debris flows occur at several horizons. The formation on the coast is approximately 270 m thick, of which less than 40 m are of latest Ivorian age. The upper 50 m, formerly known as the Carlyan Limestone, consists of resedimented oolitic sands and is of earliest Arundian age (Marchant, 1978). Nolan (1986, 1989) interpreted the Tober Colleen Formation as representing deep water, basinal muds; and the overlying Rush Formation, as turbidites and sediment flow deposits generated as a result of extensional faulting along the margin of a shallow water shelf to the east. The composition of the clasts suggests that the pre-Carboniferous geology of this shelf was similar to that of Lambay Island (Fig. 10.12) at the present day. As the Rush Formation is followed inland and to the south it becomes finer grained.

The Lucan Formation, which extends into the Asbian, overlies the Rush Formation on the coast, but farther south succeeds the Tober Colleen Formation. It consists of dark-coloured, argillaceous, cherty, spicular micrites and shales, with horizons of graded, skeletal limestones, containing ooids and other shallow water grains; it varies in thickness from 300 m to 800 m. It accumulated in relatively deep water as hemipelagic mud and silt and distal turbidites. Limestones from this formation have been called Calp Limestone (Marchant and Sevastopulo, 1980) and were extensively quarried and used for building materials in the Dublin region in the 18th and 19th centuries.

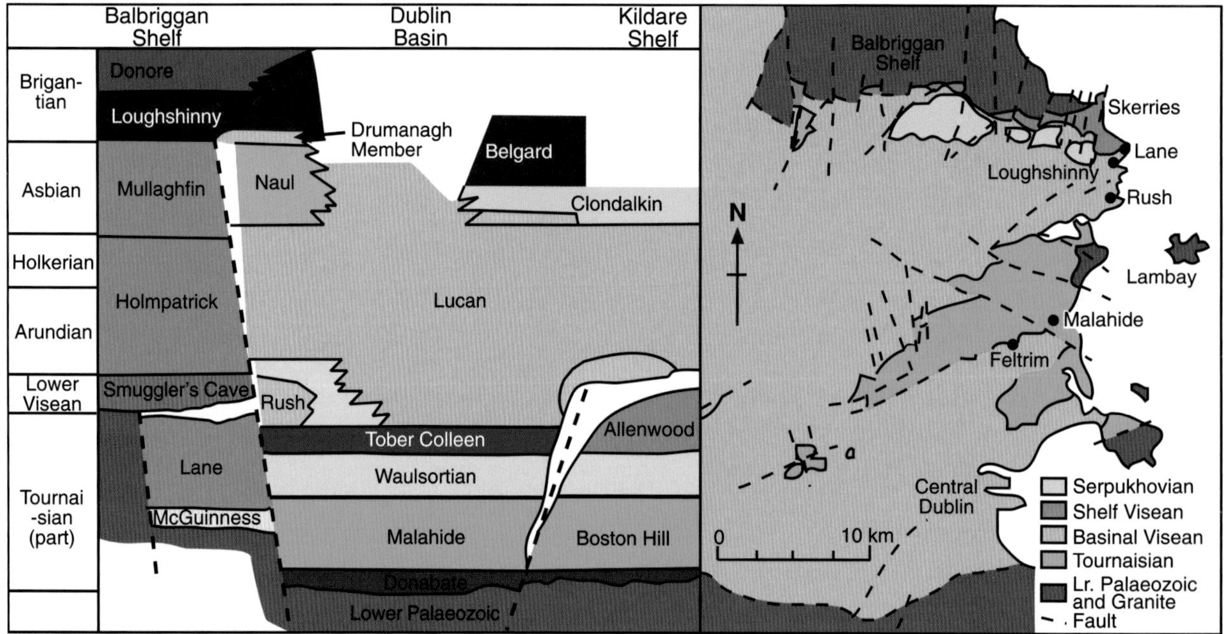

Fig. 10.12 Age, distribution and relationships of formations on the south side of the Balbriggan Shelf, the eastern part of the Dublin Basin, and the northern edge of the Kildare Shelf.

In the north-eastern part of the basin, the Lucan Formation is overlain by the Naul Formation (Asbian) which consists of more than 100 m of grey, relatively shale-free, graded skeletal limestone and calcisiltite deposits derived from the Balbriggan Shelf. It seems probable that it was deposited in shallower water than the Lucan Formation and may reflect a phase of progradation of the shelf. Southward toward the basin centre, the Naul Formation passes laterally into the Lucan Formation.

The Loughshinny Formation (from < 100 m to 150 m thick and of Brigantian age) contains evidence of farther extensional tectonic activity within the basin. The lower part of the formation (Drumanagh Member) consists of coarse limestone breccias formed by debris flows and turbidites that moved down relatively steep slopes created by active faulting. The younger parts of the formation are made up of graded limestones, containing shallow water skeletal material, interbedded with argillaceous limestones and dark shales, with ammonoids at some horizons. At the shelf margin north of Loughshinny (Fig. 10.12), the Loughshinny Formation disconformably overlies debris flows of early Viséan age (see below).

The details of the Viséan succession in south County Dublin (Turner, 1950) are not as well understood as those in the north because of the lack of exposure. The basin margin probably coincided with the present day fault between the Carboniferous and the Leinster Granite and its Lower Palaeozoic envelope. Tectonic activity along this margin occurred during the Arundian, because in the Rathcoole/Newcastle area of south County Dublin (west of the south-west corner of the map in Figure 10.12), coarse conglomerates of Arundian age in places overlie late Tournaisian limestones (Nolan, 1989; Andrew, 1993; Hitzman, 1995). A further episode of tectonic activity in the Asbian led to the northerly transport of carbonate turbidites and debris flows which now make up the Clondalkin Formation, estimated to be up to 250 m thick. A conspicuous feature of these rocks is the occurrence at several localities of angular pebbles and boulders of Leinster Granite and Lower Palaeozoic slate and greywacke. These were derived probably from an active submarine fault scarp. The turbidites contain abundant carbonate grains derived from a shallow water shelf that formerly covered the Leinster Granite. The Belgard Formation consists of interbedded argillaceous fine-grained limestone and shale. An estimated 300 m are preserved.

The northern margin of the basin (Fig. 10.13) can be traced just south of the southern boundary of the Balbriggan Lower Palaeozoic inlier and thence westward to the north-trending Walterstown fault (Pickard *et al.*, 1992, 1994), which controlled the western margin of the Balbriggan-Kentstown Shelf. West of the fault at Skreen, the Lucan Formation is very thick relative to coeval shelf

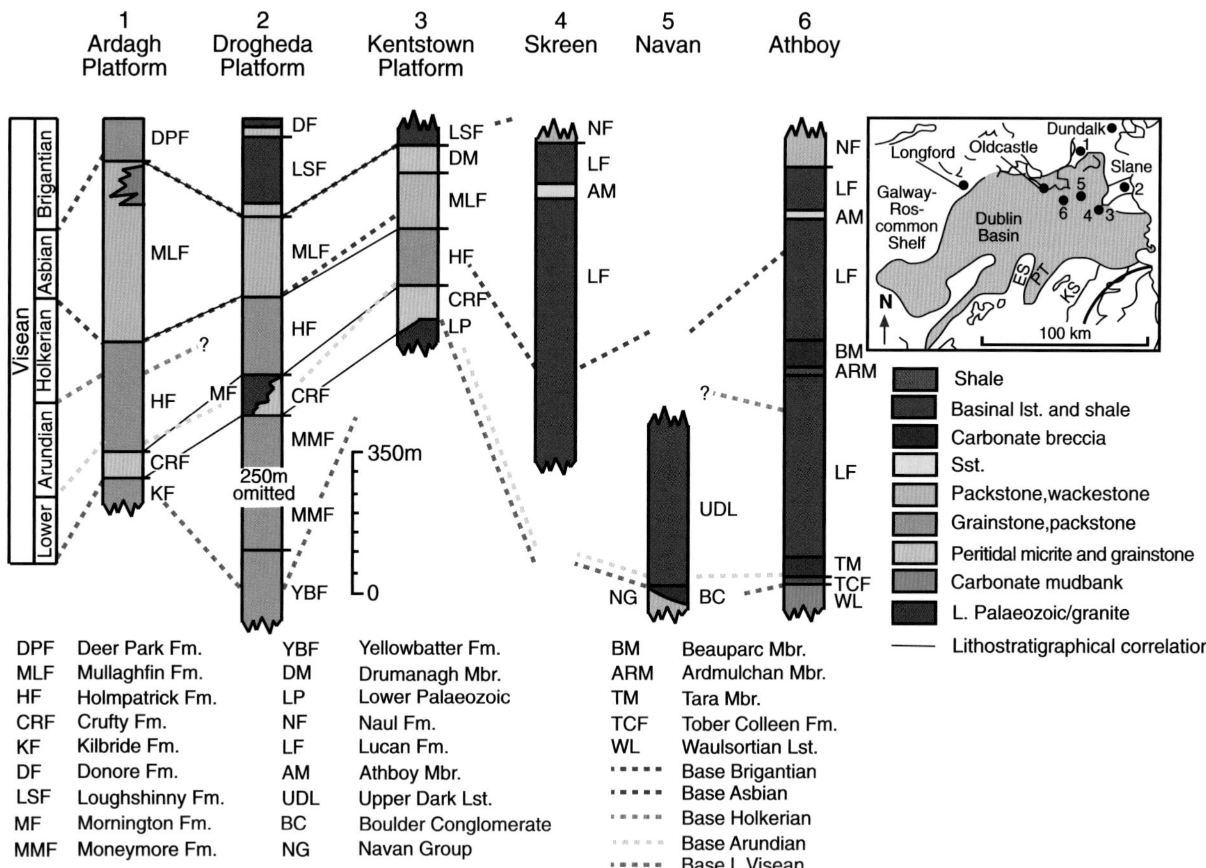

Fig. 10.13 Correlations between basinal sequences of the Dublin Basin and platform sequences to the north. ES - Edenderry Shelf; KS - Kildare Shelf; PT - Portarlington Trough.

limestones east of the fault. Pickard *et al.* (1994) interpreted the relative sparsity of coarse limestone breccias as indicating that the western margin of the Balbriggan Shelf was bypassed by sediment during deposition of the Lucan Formation. As is the case on the coast north of Dublin, the Loughshinny Formation (Brigantian) steps across the former shelf edge: coarse breccias assigned to the Drumanagh Member rest on Asbian shelf limestones east of the Walterstown Fault.

The margin of the basin is inferred to have extended northward from the vicinity of Slane to the southern part of the Kingscourt outlier, where Strogen *et al.* (1995) identified the Tober Colleen, Lucan, and Loughshinny formations; the last, which is Brigantian in age, steps northward onto the former carbonate shelf.

West of Slane, around Navan Mine (Fig. 10.13), the basinal Viséan succession has been extensively drilled (Philcox 1989; Ashton *et al.*, 1992). It rests on a spectacular discontinuity surface, which lies above Waulsortian Limestone to the north-west of the mine, but cuts down into the Lower Palaeozoic to the south-east, where approximately 600 m of the Tournaisian succession has been removed. Lying above the discontinuity is the Boulder Conglomerate, which consists of up to 50 m of debris flow deposits, in which the clasts can be matched with the local Tournaisian and Lower Palaeozoic succession. Some blocks of Waulsortian Limestone in the Boulder Conglomerate are as much as 8 m in maximum dimension. The conglomerates are overlain by limestones of the Lucan Formation, locally referred to as the Upper Dark Limestone. They range in age from Lower Viséan to Asbian and consist largely of fine-grained carbonate, which at some horizons is cherty, and dark, commonly carbonaceous shale. At some horizons there are graded skeletal limestones containing resedimented, shallow water carbonate grains. The erosion surface, and the debris flows that cover it, have been interpreted as having formed as a result of active extension across several east-north-east trending, southerly dipping faults. It is envisaged that large masses of the hanging walls of these faults became detached and moved downslope, leaving a scar over which debris flows derived from farther upslope were deposited. The faults were probably related to re-activated Caledonian structures, but there is no evidence that the latter controlled the shelf margin, which was probably some distance to the north; the shallow water material in the turbidites of the Lucan Formation was probably transported westward from the margin of the Balbriggan shelf. Other erosion surfaces overlain by debris flows of Arundian age have been reported from south of Oldcastle (Brand and Emo, 1986; Hitzman, 1995), associated with an active fault, which may have been close to the basin margin (Fig. 10.13).

A succession farther from the basin margin in this region was penetrated in the Athboy borehole (Strogen *et al.*, 1990). The Tober Colleen (here probably of Lower Viséan age), Lucan (Arundian to Asbian), and Naul (Asbian) formations are represented by a thick succession of basinal limestone and shale (Fig. 10.13). The Tober Colleen Formation is very thin, but the Lucan Formation is 1123 m thick and has been interpreted by Strogen *et al.* (1990) as consisting of four megacycles, each of which can be recognised over a substantial distance in the north of the basin. Each megacycle starts with beds that contain coarse detritus, interpreted as reflecting tectonic activity. The lower two are of Arundian age; the third, whose base corresponds to the contact between the mud-rich Ardmulchan Member and the breccias and resedimented ooid sands of the Beauparc Member (Rees, 1987), is Holkerian; and the fourth starts within the Holkerian and extends into the Asbian. The fourth cycle contains towards its top a development of sandstone and shale (the Athboy Member), which has been interpreted as being deposited at depths within the reach of storm waves. The Naul Formation (Asbian) in this part of the basin and in north County Dublin is similar in character. Successions farther from the basin margin are very fine grained. In the region around Lough Derravaragh, north of Mullingar, the upper part of the Lucan Formation is very cherty, with the development of so-called 'festoon cherts' (Nevill, 1958a)

No Viséan rocks of shelf facies have been recognised between Navan and the Longford area, apart from a north-east to south-west trending promontory east of Lough Sheelin (Morris *et al.*, 2003). From the Longford area, the basin margin trends south-south-west to Athlone (Crowe, 1986; Crean, 1987; Strogen *et al.*, 1996; Morris *et al.*, 2003). South-west of Athlone, Viséan rocks have been eroded. The Tynagh and Lorrha Basins (Fig. 10.13), which contain variable, but in places substantial thicknesses of the Lucan Formation (Brück, 1985; Gatley *et al.*, 2005), may have been connected with the Dublin Basin.

Few details have been published about the southern margin of the basin to the west of Dublin. The identification of narrow tracts of Viséan basinal limestone extending southward into the carbonate shelf have led Hitzman (1995), McConnell and Philcox (1994), and Gatley *et al.* (2005) to recognise the Portarlington and Tullamore Troughs, east and west of Slieve Bloom respectively (Fig. 10.11). It is not known whether these structures persisted throughout the Viséan.

Shelf north of the Dublin Basin

The shelf bounding the Dublin Basin (Fig.10.13) is preserved along the southern and western margin of the Balbriggan Lower Palaeozoic inlier (Nolan, 1986; Pickard *et al.* 1992), at the eastern end of the Boyne Valley (Rees 1987), and in the Kingscourt outlier (Strogen *et al.*, 1995). The shelf sequences have been assigned to the Milverton Group (Nolan, 1989). Poorly known carbonates of shallow water origin and of Viséan age occur around Dundalk farther from the shelf margin. Elsewhere, apart from the north-east to south-west trending promontory east of Lough Sheelin (Morris *et al.*, 2003), which includes coral-bearing limestones of Asbian age, the shelf deposits have been removed by post-Viséan erosion.

On the coastal section in north County Dublin (Fig. 10.12), there is a thin development of shallow water, skeletal limestone, the Lane Formation (Nolan, 1989); inland, the upper part of the formation includes shallow water, carbonate mudbanks that developed along the shelf margin (Somerville *et al.*, 1992b). The Lane Formation preserved on the coast is of late Ivorian age (MFZ 8; Devuyst, 2006) but the mudbank limestones inland are of Lower Viséan age (MFZ9). The Lane Formation was tilted northward and its top was subjected to karstic weathering before the overlying non-marine Smuggler's Cave Formation was deposited on it. The latter is a coarse, fluviatile conglomerate containing sub-rounded boulders of greywacke, which were derived from the north and are closely comparable in lithology to the Lower Palaeozoic rocks of the Balbriggan inlier. Farther north, the Smugglers Cove Formation overlaps the Lane Formation to rest directly on Lower Palaeozoic rocks. Inland, Lower Viséan rocks are absent at the contact with the Balbriggan Lower Palaeozoic inlier, either through non-deposition or through pre-Arundian uplift and erosion (Pickard *et al.*, 1994). A marine transgression at the beginning of the Arundian resulted in ooid sands followed by skeletal sands with corals and brachiopods (the Holmpatrick Formation) being deposited over the Balbriggan Shelf. Similar ooid sands were probably the source of the re-sedimented ooids in the upper part of the Rush Formation within the basin to the south. Nolan (1986) has shown that the Holkerian and Asbian are extremely thin on the Balbriggan Shelf and that the Asbian, termed the Mullaghfin Formation by Pickard *et al.* (1994), contains carbonate mudbanks. The shelf margin is spectacularly exposed on the coast. Boulder conglomerates interbedded with finer grained facies overly a southerly dipping discontinuity surface, which was termed the 'angular fault' by Matley and Vaughan (1908). The conglomerates consist of limestone (including blocks of carbonate mudbank) from the Lane Formation and Lower Palaeozoic material derived from the Smugglers Cave Formation. The boulder conglomerates were formerly considered to be of Asbian age, but are now interpreted to be of Lower Viséan age. Basinal limestones and shales of the Loughshinny Formation (Brigantian) drape the boulder beds. Presumably footwall uplift prior to the Brigantian led to the erosion of any Arundian–Asbian material that had accumulated on the shelf margin, prior to the deposition of the Loughshinny Formation. Because the upper part of the Loughshinny Formation contains little coarse-grained shelf material, it has been suggested that the Balbriggan Shelf became drowned during the Brigantian.

The succession in the eastern part of the Boyne valley around Drogheda (Pickard *et al.*, 1994; Rees, 1987) developed in an asymmetric graben, controlled by the faults that bounded the Longford Down and Balbriggan Lower Palaeozoic inliers (Fig. 10.13). Movement of the northern fault was generally greater, resulting in northerly tilting. The Lower Viséan consists of a thick succession of peloidal and skeletal limestones of shallow water origin (Yellowbatter and Moneymore formations). Drowning of this first phase of platform development in the Drogheda area by dark basinal limestones (the Mornington Member of the Lucan Formation, which continues into the lower part of the Arundian), reflects accelerated subsidence related to movement on the northern boundary fault. Farther south (up ramp), peloidal, commonly sandy, limestones of shallow water origin (Crufty Formation) are time equivalents of the Mornington Member. They prograded northward and throughout most of the Arundian and Holkerian, the region was a shallow water, high energy shelf, floored with skeletal and, at times, ooid sands, which form the Holmpatrick Formation. The succeeding Mullaghfin Formation (Asbian) is lithologically varied: high energy, crinoidal grainstones and locally developed carbonate mudbanks occur around the western margin of the platform, which may have prograded slightly westward at this time; and lower energy, slightly shaly limestones are found away from the margin. Pickard *et al.* (1994) argued that because there are no signs of the aggradational cycles capped by palaeokarsts that characterise the later part of the Asbian in other areas of Ireland and Britain, water depths over the Balbriggan and Drogheda shelves may have been somewhat greater during deposition of the Mullaghfin Formation than over other contemporary shelves. In the Brigantian, basinal limestones and shales of the Loughshinny and Donore Formations stepped across the basin margin and the shelf became progressively drowned.

Shelf limestones of Viséan age occur in the northern part of the Kingscourt outlier (Fig. 10.13), which Strogen *et al.* (1995) have named the Ardagh Platform. The Lower Viséan Crufty Formation is probably disconformable on the late Tournaisian Kilbride Formation and consists of varied rock types of peritidal to shallow subtidal origin, including sandstones, shales, and peloidal limestones and micrites, some of which show fenestral fabrics. The overlying Holmpatrick Formation (early Viséan to Holkerian in age) reflects a change to more open water, high energy conditions, under which coarse-grained, crinoidal sands were deposited. Although the exact position of the shelf margin is difficult to determine because of poor exposure, it seems probable that during deposition of the Mullaghfin Formation (Asbian), the shelf prograded southward. The lithofacies of the Mullaghfin Formation are similar to those of the Holmpatrick Formation but include evidence of emergence in the form of palaeokarsts and, in two places at its top, developments of carbonate buildups. The latter, described by Somerville *et al.* (1996a), consist of several lime mudstone and wackestone bodies, typically containing fossil cyanobacteria, interbedded with coarser skeletal and intraclastic limestones. One of the buildups has been shown to have continued growing during the early part of the Brigantian when an unusual phylloid algal boundstone developed. To the north of the more northerly buildup, deposition of shelf carbonates continued during the early Briagantian: the Deer Park Formation consists of bedded limestones, some of which are dark coloured and many of which are cherty. These are equivalent in age to the lower part of the basinal Loughshinny Formation farther south, which can be seen to overlie the Mullaghfin Formation. Thus it appears likely that the Ardagh Platform was drowned during the Brigantian and was covered by the Loughshinny Formation; in some areas the latter was eroded prior to deposition of the Serpukhovian.

Galway–Roscommon Shelf

The Galway–Roscommon Shelf (Figs. 10.11, 10.13) covers a large area extending from east County Clare and County Galway in the south, to counties Mayo and Roscommon in the north.

North of the Shannon Basin in counties Clare and Galway, the Viséan succession consists almost entirely of limestones of shallow water, shelf facies. The stratigraphy established in the Burren and the Gort lowlands (Fig. 10.10) can be applied, with some modification, to the northern limb of the east Clare Syncline (Sleeman and Pracht, 1999). The Tubber Formation (*c.*300 m thick) is of early Viséan to Holkerian age (Gallagher *et al.*, 2006). It consists predominantly of grey crinoidal packstones, which are cherty at the base where they overlie the Waulsortian limestone; peloidal and oolitic facies occur at some horizons. The overlying Burren Formation also consists of limestones of shallow water origin. Its base has been mapped at the lowest horizon with *Lithostrotion* spp., which, according to Gallagher *et al.* (2006), is within the Holkerian. The formation can be divided into two parts (Gallagher, 1996; Gallagher *et al.*, 2006). The lower part, 215–238 m thick, subdivided into named members (not enumerated here), consists of varied skeletal limestones, which are generally grey but also include dark cherty lithologies, and in the east Clare Syncline, but not commonly in the Burren region, oolitic limestone. The upper part, named the Ailwee Member (152 m thick), consists of cyclic developments of generally thick-bedded, massive skeletal limestone and is characterised by foraminiferans indicating MFZ 14. Gallagher (1996) and Gallagher *et al.* (2006) have interpreted the sequence within each cycle as follows. An initial shallow subtidal phase, during which peloidal limestone was deposited, was followed by slight deepening when crinoids became more prevalent. This was followed in turn by shallowing during which peloidal limestones, containing abundant *Kamaenella* and *Ungdarella*, putative algae, were deposited. Shell beds with productoid brachiopods and coral thickets are generally restricted to the limestones of this shallowing phase. Evidence of emergence, in the form of palaeokarstic surfaces and, more rarely, overlying clay wayboards (interpreted as palaeosols) have been found at the top of most of the cycles. The cycles, which control the scarp and terrace morphology of the Burren and the Aran Islands (Fig. 10.14), have been interpreted as having had a glacio-eustatic origin. The Burren Formation is overlain by the Slievenaglasha Formation, which is Brigantian in age and consists of two developments of crinoidal limestones with a middle unit of intraclastic limestone containing another putative alga, *Fasciella*. The Brigantian succession has been interpreted as representing generally slightly greater water depths than occurred during the Asbian (Gallagher, 1996; Gallagher *et al.*, 2006). The crinoidal limestones form cycles, each of which reflects shoaling; in contrast to the Asbian cycles, the shallowing did not lead to emergence. The *Fasciella*-bearing limestones indicate slight deepening to between fair weather and storm wave base. The surface of the uppermost bed of the formation is a palaeokarst, suggesting sudden shallowing. Within the Slievenaglasha Formation in the east Clare Syncline there is an isolated occurrence of volcanics (Turret Volcanic Member) (see p. 267 below). South of the Burren, a thin unit of dark limestone, the Magowna Formation, occurs between the Slievenaglasha Formation and the Clare Shale Formation. The base of the Serpukhovian occurs within it.

The succession between Galway and Oughterard has been described by MacDermot and Sevastopulo (1972), Conil and Lees (1974), Pracht *et al.* (2002), and Devuyst (2006). Around Oughterard (Fig. 10.9), the upper part (*c.*30 m) of the Oughterard Formation contains foraminiferans assigned to the Lower Viséan MFZ 9. It consists of interbedded pale grey and dark, very argillaceous limestones, interpreted by Devuyst (2006) as having been deposited within the photic zone; towards the top, the environment became shallower. The overlying Lower Viséan Cregg Formation marks an abrupt change to a generally higher energy environment; it consists of packstones and grainstones with granules and small pebbles of silciclastic material, particularly at the base and the top, and commonly developed cross-stratification. It is very variable in thickness from a maximum of over 150 m south of Oughterard to zero on the east shore of Lough Corrib, where its equivalent is the Oakport Formation (see below). The localised influx of coarse silciclastic material probably reflects movement on a fault to the west. The overlying Oldchapel Formation (*c.*64 m thick) consists largely of micrites with a restricted fauna and evidence of emergence at some levels, suggesting the development

Fig. 10.14 Mullach Mór, County Clare. A gentle fold in limestone of the Burren Formation with typical limestone pavement in the foreground. Note the scarp and terrace morphology evident on the left of the photograph.

of a carbonate tidal flat. It is mainly of Lower Viséan (MFZ9) age, although the uppermost part has not been securely dated. The Oldchapel Formation is overlain with a sharp contact by the Illaunagappul Formation (*c.*84 m thick), which consists of brachiopod- and coral-bearing skeletal limestones with a siliciclastic component at some horizons. It was deposited in open marine conditions and contains foraminiferans assigned to MFZ 11, suggesting that there may be a hiatus at its base. The environment appears to have then deepened slightly, and the overlying Ardnasillagh Formation (*c.*66 m thick) consists of crinoidal limestones with significant amounts of fenestellid bryozoans. It is of Arundian and, at its top, Holkerian age. The Aughnanure Oolite Formation (*c.*32 m thick and of Holkerian age) signals shallowing and is followed by brachiopod- and coral-bearing cherty limestone of the Holkerian-aged Corranellistrum Formation, some 100–120 m thick. The youngest Holkerian and Asbian part of the succession has not been investigated in detail but is believed to be similar to that of the Burren.

Northern region

Viséan rocks are widely distributed north of a line from Clew Bay to the western end of the Lower Palaeozoic Longford-Down inlier (Fig. 10.15). It is proposed to refer to this area as the Northern region. Within it there are major changes in facies and thickness, reflecting both fault-controlled differential subsidence and relative proximity to the source of siliciclastic sediment. Its northern part lay close to, or included, the margin of marine deposition from early Viséan times onwards; as a result very little limestone was ever deposited north of Cookstown, County Tyrone and the Omagh and Donegal synclines.

The thickest successions are found in the Lough Allen Basin, between the north-eastern Ox Mountains, which formed part of the Ox Mountains–Ballyshannon High, and an eastern shelf area between Armagh and Cookstown.

Many of the stratigraphical names that are current in the Northern region derive from the Sligo Syncline (Oswald, 1955), which together with south County Donegal and north County Mayo, is described first. All three lie on or to the north-west of the Ox Mountains–Ballyshannon High (Fig. 10.15).

The Viséan succession characteristic of the Sligo Syncline and the Lough Allen Basin has been divided into two groups (Brandon, 1972, 1977; Mitchell, 2004c),

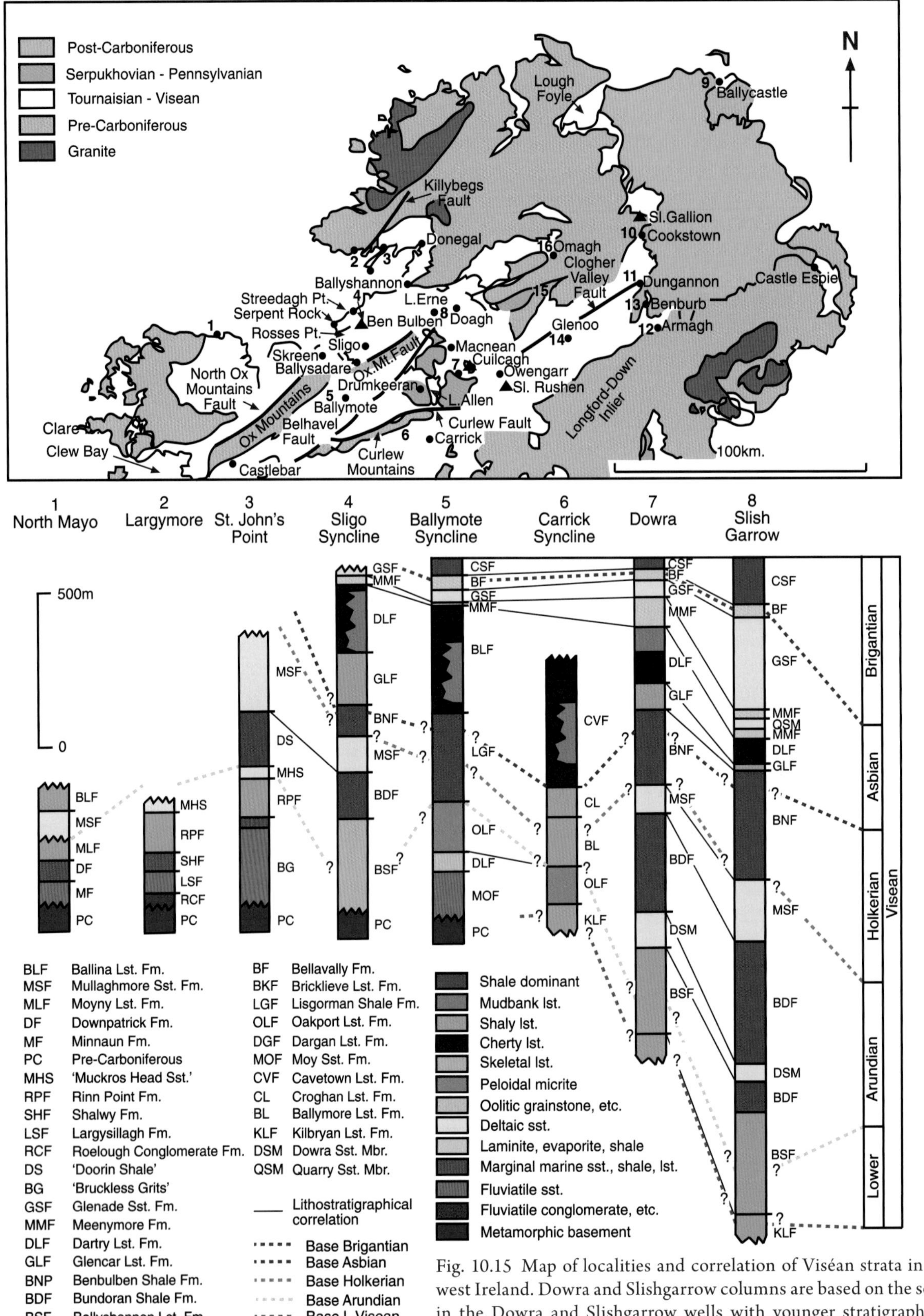

Fig. 10.15 Map of localities and correlation of Viséan strata in northwest Ireland. Dowra and Slishgarrow columns are based on the sections in the Dowra and Slishgarrow wells with younger stratigraphy from nearby areas. Only those faults referred to in the text are shown.

a lower Tyrone Group (Lower Viséan–?Asbian) and an upper Leitrim Group (?Asbian–Serpukhovian).

In the Sligo Syncline (Oswald, 1955), the Ballyshannon Limestone Formation, pale grey, crinoidal limestone of shallow water origin, rests either directly on the metamorphic basement (Fig. 10.15), as is the case east of Ballyshannon and at Rosses Point, or is underlain by a thin succession of pebbly sandstone, argillaceous micrite, skeletal calcarenite and oolite, as at Ballysadare (Hitzman, 1986; MacDermot *et al.*, 1996). Its base is probably of early Viséan age and it extends into the Arundian. The lower part of the Ballyshannon Limestone at Ballysadare contains beds of very coarse sandstone and granules probably derived from the Ox Mountains, and farther west at Skreen there are conspicuous developments of quartz pebble beds within the limestone, suggesting that the North Ox Mountains fault may have been active. Long *et al.* (1992) have recorded conglomerates interpreted as having a similar origin farther to the south-west.

To the west, along the north and south coasts of Donegal Bay (Fig. 10.15), the Ballyshannon Limestone is replaced by a mixture of carbonates and siliciclastic rocks. In the Largymore Syncline on the north coast (George and Oswald, 1957), the succession records a marine transgression (Graham, 1996). The Roelough Conglomerate Formation, which rests on the metamorphic basement, is clast-supported and contains lenses of sandstone and mudrock. The overlying Largysillagh Sandstone Formation is dominated by sandstone with intercalated red and green-coloured mudrocks, and common pedogenic carbonate nodules; the sandstones contain plant debris and the trace fossil *Beaconites*. Both formations are interpreted as having been deposited in a fluviatile environment by small rivers flowing in a generally southerly direction. The succeeding Shalwy Formation, strongly bioturbated mudrocks and sandstones with gastropods and ostracodes, represents marginal marine environments, subject to tidal currents. Some horizons within this formation contain remarkable concentrations of fusain, interpreted as fossil charcoal (Nichols and Jones, 1992). The overlying Rinn Point Formation consists of alternating limestone and shale with a rich, fully marine fauna of brachiopods and corals (Hubbard and Pocock, 1972) and is equivalent to the early Viséan part of the Ballyshannon Limestone farther to the east. A succession similar to, but thicker than, that of the Largymore Syncline is well exposed on the coastal sections of the west limb of the Donegal Syncline. A few kilometres north of the coast, however, the Rinn Point Formation is replaced by a much less calcareous succession capped by the coarse Bannagher Hill Grits (George and Oswald, 1957) whose base coincides with the Lower Viséan/Arundian boundary. The thicker successions of the west limb of the Donegal Syncline have been attributed to more rapid subsidence related to active movement on the Killybegs (Leannan) Fault.

On the coast of north County Mayo (Graham, 1996; Long *et al.*, 1992) the lowest unit above the metamorphic basement, the Minnaun Formation (Fig. 10.15), is dominated by coarse-grained sandstones deposited by small rivers flowing from the north-west. The overlying Downpatrick Formation (Fig. 10.16) is broadly comparable to the Shalwy Formation of Donegal and represents the transition to the fully marine conditions found in the fossiliferous limestone and shales of the Moyny Limestone Formation, which is of early Viséan age. Farther south, on the north side of Clew Bay, dark limestone and shale of the poorly known Rockfleet Bay Formation (Long *et al.*, 1992), which represents marginal marine conditions, lie above the non-marine Capnagower Formation, which is of Pu miospore biozonal age.

The Ballyshannon Limestone in the Sligo Syncline is overlain by the Bundoran Shale Formation, which consists of dark shales that are silty, particularly in their upper part. They are variably fossiliferous; at their base at Bundoran, there is a rich fauna of crinoids, echinoids, brachiopods, bryozoans, and corals. The same unit, albeit probably somewhat older (early Arundian) at its base, has been called the Doorin Shale, and also the Coolmore Shale, in County Donegal (George and Oswald, 1957).

The Bundoran Shale was the prelude to an influx of siliciclastic sand, which formed the Mullaghmore Sandstone Formation. This formation is widespread (Figs. 10.15, 10.19) and has been called by several other names in different parts of north-west Ireland (George and Oswald, 1957; Hubbard 1966a; Connolly, 2003). It consists of grey, feldspathic sandstones, which in some cases are pebble-bearing, grey shales, and rarer limestones, and is interpreted as having been deposited in a deltaic environment. Much of the sandstone, as well as the shale and limestone, is likely to be of marine or marginal marine origin and there is a rich ichnofauna (Buckman, 1992), as well as scattered body fossils. Graham (1996) has identified fluvial channel deposits in coarse-grained, cross-stratified sandstone units associated with rootlet beds and palaeosols. They formed as point bars in river channels up to 200 m wide; cross-strata indicate a

Fig. 10.16 Doonbristy, a sea stack on the coast of north County Mayo, made mostly of the Downpatrick Formation. The lowermost limestone bed of the Moyny Limestone Formation is the prominent light coloured bed at the top of the stack. Photograph by Dr J.R. Graham.

unidirectional transport direction, in general from the north-north-west, although locally on the north Mayo coast, channels within which transport was from the south have been identified. The top of the Mullaghmore Formation is oolitic in several places and Graham (1996) has suggested that it may represent a period of reworking during a rapid rise of sea level. Miospore assemblages from the formation have been assigned to the TS Biozone, which indicates a late Arundian to Holkerian age (Higgs, 1984).

There is generally a sharp contact between the Mullaghmore Formation and overlying limestones and shales that contain a rich, fully marine fauna. In the type area around the Benbulben range (Fig. 10.15), the lowest beds, which are particularly shale-rich, have been called the Benbulben Shale Formation, and they are succeeded by the Glencar Limestone Formation, rhythmic alternations of limestone and shale (Schwarzacher, 1989), which on the coastal sections at Streedagh Point and Serpent Rock, County Sligo contain large, geniculate, solitary corals (*Siphonophyllia*) and colonies of *Siphonodendron* (Hubbard, 1966b), spectacularly displayed on wide bedding surfaces (Fig. 10.17). In north Mayo, facies comparable to the Glencar Limestone rest directly on the Mullaghmore Formation. The base of the Asbian Stage is probably within the upper part of the Benbulben Shale in its type section. In the Donegal Syncline, rocks above the Mullaghmore Formation are not preserved.

The overlying Dartry Limestone Formation (Asbian) includes two distinct facies: well-bedded, fine-grained, typically cherty limestones with lithostrotionids which are commonly silicified (Fig. 10.18); and unbedded, or poorly bedded mudbank limestones, well displayed in the Dartry Mountains, particularly at the head of Gleniff, north-east of Benbulben, where clinoform beds at the margin of a build-up show that it had relief above the seafloor of some 120 m (Schwarzacher, 1961; Warnke and Meischner, 1995; MacDermot *et al.*, 1996).

Fig. 10.17 Concentration of corals, *Siphonophyllia samsonensis*, on bedding surfaces, Glencar Limestone Formation, Streedagh Point, County Sligo. Lens cap is 50 mm in diameter.

Fig. 10.18 Benbulben viewed from the west. The Dartry Limestone Formation forms the cliffs and the upper part of the Glencar Limestone Formation the bedded succession at their base. The slope corresponds to the Benbulben Shale Formation and the lower part of the Glencar Limestone Formation.

The mudbank limestones are similar in general appearance to the Waulsortian facies, with spar-filled cavities including stromatactis, but differ in many of the skeletal components that they contain. They are locally very fossiliferous with diverse productoid brachiopods, rare ammonoids such as *Beyrichoceras*, and, in some thin-bedded flank beds, common blastoids (Waters and Sevastopulo, 1984). Corals are generally rare, in constrast to the rich and stratigraphically important faunas of *Lithostrotion, Clisiophyllum, Dibunophyllum*, and other solitary forms found in the bedded facies. In the Sligo Syncline, the mudbank facies, which was typically initiated at the base of the formation, is found in an elongate zone along the north-west side of the Ox Mountains Fault (Fig. 10.15), suggesting that its distribution may have been controlled by subsidence related to the fault. However, the substantial developments of mudbanks in the Dartry Mountains do not appear to be related to a mapped fault.

Only small erosional remnants of the Leitrim Group (late Asbian and Brigantian) are preserved to the west of the Ox Mountains–Ballyshannon High. They are referred to in the description of the main outcrops of the Leitrim Group below.

The stratigraphy described from the Sligo Syncline can be identified to a large extent in the region between the Ox Mountains–Ballyshannon High and the Cookstown–Armagh shelf, although many authors describing the stratigraphy of particular areas have done so in terms of local stratigraphical names.

Along the south-east flank of the Ox Mountains–Ballyshannon High (Fig. 10.15), the earliest Carboniferous rocks are variably thick (up to 140 m) pebbly conglomerates and red, green, and grey sandstones (the north-westerly derived Moy Sandstone Formation of the Ballymote Syncline, (Dixon, 1972) and Castlebar area (Long *et al.*, 1992) and the lower part of the Twigspark Formation (MacDermot *et al.*, 1996) at the north-east end of the Ox Mountains). They are succeeded by limestones, commonly with admixed siliciclastic material, of shallow water origin, equivalent to the lower part of the Ballyshannon Limestone Formation. These include the Lough Akeel Oolite and Castlebar River Limestone Formations of the Castlebar area, the Dargan Limestone Formation of the Ballymote Syncline, which also contains an oolite member, and the upper part of the Twigspark Formation farther to the north-east. The younger part of the Ballyshannon Limestone Formation (of early Viséan and Arundian age) over the Ox Mountains–Ballyshannon High is of shallow water facies, but in the subsurface of the Lough Allen Basin it consists of dark, fine-grained, argillaceous limestone and subordinate shale and has been interpreted as a deeper water facies (Philcox *et al.*, 1992).

On the south side of the Curlew Mountains in the Carrick Syncline (Fig. 10.15), where fully marine conditions had already been established during the Tournaisian, the earliest Viséan rocks are probably in the upper part of the Kilbryan Limestone Formation. The succeeding Oakport Limestone Formation, which is equivalent to part of the Ballyshannon Limestone, consists of shale-free, pale-coloured limestone with a conspicuous development of peloidal micrite, at some horizons with fenestral fabrics, in the middle of the formation. It clearly represents extremely shallow water, or, at some levels, emergent conditions. Philcox *et al.* (1989) considered that the Oakport Limestone in its type area does not extend above the top of the Lower Viséan, which conflicts with the age implied for the base of the formation in the Ballymote Syncline, where it overlies the Dargan Limestone, from which Dixon (1972) recorded *Siphonodendron*, indicating an Arundian age.

In the Omagh Syncline (Simpson, 1955; Mitchell, 2004c), the Claragh Sandstone Formation is unconformable on the Tournaisian Omagh Sandstone Formation, which it oversteps to rest on the Dalradian. It consists of a thin basal conglomerate, coarse feldspathic sandstone, thin mudstone, and limestone, and is reported to be of Lower Viséan age (Fig. 10.19). It is overlain by Termon River Limestone Formation, skeletal limestone and oolite, which thickens conspicuously from north-east to south-west. The overlying Bin Mountain Sandstone Formation is a wedge-shaped body of calcareous sandstone and oolitic limestone, thickest in the north-east. Its lower part interfingers with the Termon River Limestone Formation and its upper part, which is Arundian in age, with the Ballyshannon Limestone Formation (Pettigo Limestone of Simpson). To the south-west around Lough Erne, the base of the Ballyshannon Formation is of early Viséan age (Legg *et al.*, 1998).

Ballyshannon Formation, also of early Viséan and Arundian age and shallow water facies, overlies the Clogher Valley Formation between the Fintona block and the Longford-Down inlier (Mitchell, 2004c).

Philcox *et al.* (1989, 1992) described evidence for differential subsidence in and around the Lough Allen Basin during the early Viséan and Arundian (Fig. 10.15). The Ballyshannon Limestone is reduced in thickness and the Kilbryan Limestone Formation is not present over the

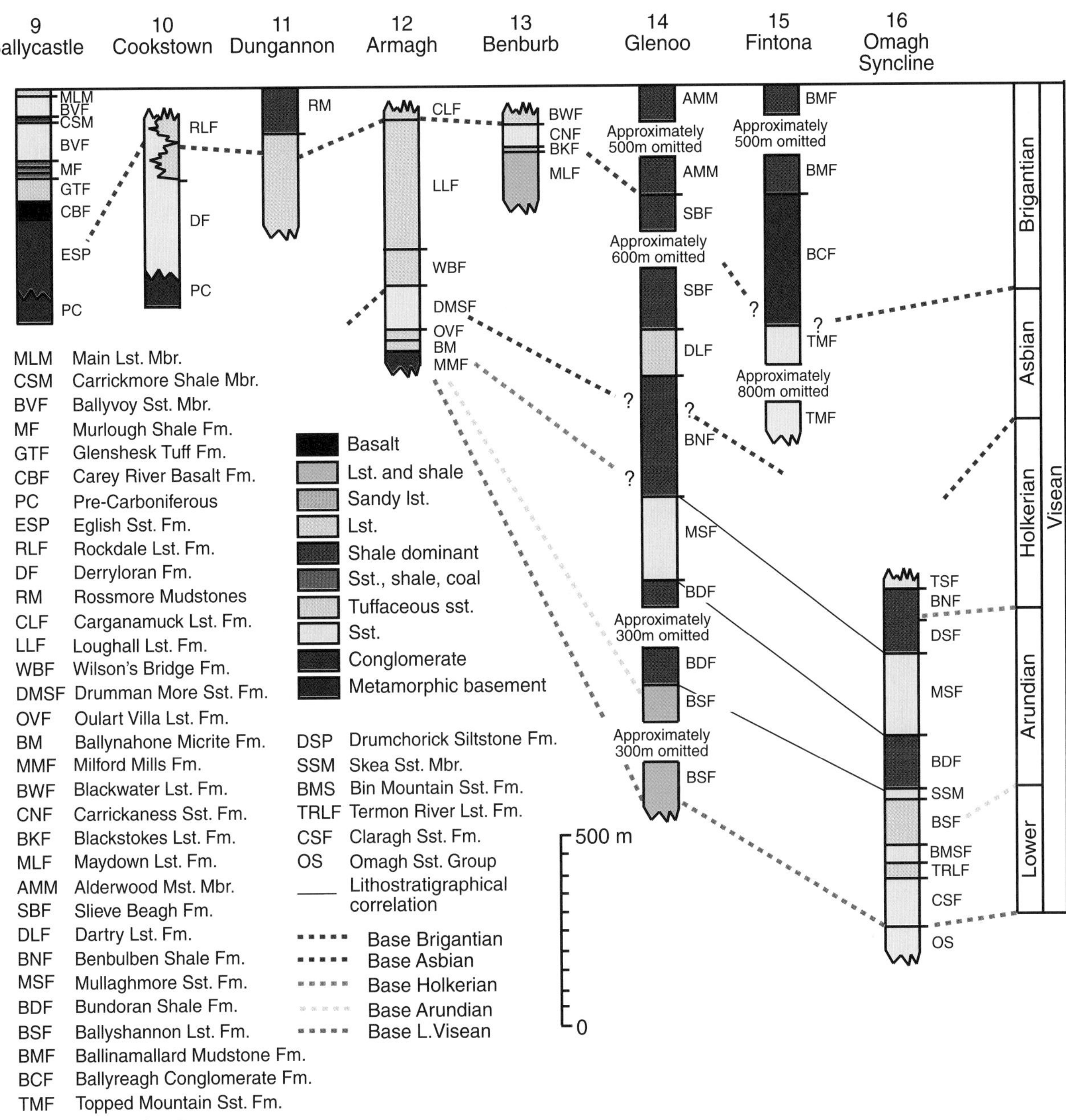

Fig. 10.19 Correlation of Viséan strata in north-east Ireland and the Omagh Syncline. Glenoo column is based on the Glenoo well and younger rocks from the Slieve Beagh area. For localities see Figure 10.15.

intrabasinal Dowra–Macnean High, which trends north-eastward towards Lough Erne (Legg *et al.*, 1998) and is inferred to have been fault-bounded. There was major movement, probably down to the north, on the Curlew Fault close in time to the Lower Viséan/Arundian boundary. The Arundian Greyfield Formation, a heterogeneous unit of breccia composed of sandstone blocks and limestone including carbonate mudbank facies, rests, at different places in the hanging wall of the fault, on Lower Devonian sandstone and on a karstic surface of Oakport Limestone. Kelly and Somerville (1992) have described a development of carbonate mudbanks of similar (Arundian) age at the north-east end of the Curlew inlier.

The influx of siliciclastic sediment represented by the Bundoran Shale and the Mullaghmore Sandstone can be recognised over much of the Northern region (Figs. 10.15, 10.19). The general pattern of distribution of facies and thickness – thick, coarser grained and carbonate-poor in the north, and thinner, finer grained and with more carbonate in the south – strongly suggests transport from a generally northerly direction, in response to regional uplift in the source area.

Within the Lough Allen Basin (Fig. 10.15), the earliest pulse of sand, the Dowra Sandstone Member, occurs at, or close above, the base of the Bundoran Shale. It forms a southerly thinning and fining wedge and may be the same age as the Skea Sandstone Member in the Omagh Syncline (Mitchell, 2004c), which rests disconformably on the Ballyshannon Limestone (Fig. 10.19). The Drumkeeran Sandstone Member, which occurs at a similar horizon in the south of the basin, has been interpreted as having been derived from the south (Philcox *et al.*, 1992).

The Mullaghmore Formation (Arundian and possibly Holkerian) in the Lough Allen Basin also forms a southerly thinning wedge (Fig. 10.15). The south-eastern limit of the sandstone facies lies north-west of the Dowra, Drumkeeran, and Owengarr wells, in which the formation is represented by siltstone and silty mudstone. Within the Ballymote Syncline (Dixon, 1972), the Mullaghmore Sandstone can be identified only in the north-east; to the south-west it is replaced by shales which form part of the Lisgorman Shale Formation (Fig. 10.15). On the south side of the Curlew Mountains, the Ballymore Limestone Formation (Caldwell, 1959) is probably of Arundian and early Holkerian age (Morris *et al.*, 2003) and therefore can be correlated with the combined Bundoran Shale and Mullaghmore Sandstone. It consists of coral- and brachiopod-bearing limestone and subordinate shale and also contains small carbonate mudbanks (Kelly and Somerville, 1992).

In the Omagh Syncline (Fig. 10.19), there is a thick development of siliciclastic rocks represented by the Mullaghmore Sandstone and the overlying Drumchorick Siltstone Formation (Mitchell, 2004c), both reported to be of Arundian age.

On the south-east side of the Fintona inlier, the Glenoo No. 1 well (Sheridan, 1972b) proved a thick succession of shale and argillaceous limestone above equivalents of the Ballyshannon Limestone, with a thick intercalation of sandstone, with subordinate shale and rare limestone. This succession has now been identified as the Bundoran Shale, Mullaghmore Sandstone, and overlying Benbulben Shale Formations. Farther to the south-west, sandstones pinch out and the probably equivalent strata are dark, generally fine-grained limestone, mudstone, and shale, which have been called the Drumgesh Shale Formation (Geraghty, 1997).

In the Sligo Syncline, it seems probable that the boundary between the Mullaghmore Sandstone and the overlying Benbulben Shale represents a significant flooding surface, and this also seems to be the case in the Lough Allen Basin. Published data are insufficient to demonstrate that this is the case elsewhere, but it is likely to be so, and the top of the Mullaghmore Sandstone is, therefore, probably isochronous. There is some uncertainty over the age of this boundary. George *et al.* (1976) considered the Mullaghmore Sandstone to be of Holkerian age. Higgs (1984) and Philcox *et al.* (1992) recorded miospore assemblages assigned to the TS Biozone from outcrop and from the subsurface, suggesting an Arundian or Holkerian age. However, Legg *et al.* (1998) reported an Arundian microfauna from above the Mullaghmore Sandstone south of Lough Erne, and Mitchell (2004c, Table 7.7) shows the Arundian/Holkerian boundary above the base of the Benbulben Shale Formation (see below) throughout the northern region.

The Tyrone Group above the top of the Mullaghmore Sandstone in the Lough Allen Basin is made up of a succession grading from shale (Benbulben Shale Formation), through limestone and shale (Glencar Limestone Formation), to generally shale-free, commonly cherty limestone, or mudbank limestone (Dartry Limestone Formation) (Philcox *et al.*, 1992). The boundaries between these facies are clearly diachronous. The Benbulben Shale and Glencar Limestone contain the Holkerian (and possibly the latest Arundian) and earliest part of the Asbian stages and the Dartry Limestone is of Asbian age. There are conspicuous changes in thickness of individual formations within the basin. Philcox *et al.* (1992) suggested from subsurface data that these probably resulted largely from facies changes rather than local differences in rates of subsidence; however, Kelly (1996), Legg *et al.* (1998), and Mitchell (2004c, d) concluded that during the Asbian there was local tectonic control of subsidence, with the development of ramps. The mudbank facies of the Dartry Limestone is extensively developed over the Dowra-MacNean High (Philcox *et al.*, 1992; Kelly, 1996; Legg *et al.*, 1998) and along the trace of the Belhavel Fault, at the northern margin, and of the Curlew Fault, at the southern margin of the basin (Fig. 10.20) (MacDermot *et al.*, 1996, fig. 19).

In the Ballymote Syncline (Fig. 10.15), strata equivalent to the Benbulben Shale to Dartry Limestone are the upper part of the Lisgorman Shale Formation, which becomes progressively more limestone-rich to the south-west, and the Bricklieve Limestone Formation (Dixon, 1972), which is generally pale coloured with several horizons of coral biostromes made up of *Siphonodendron* spp. Cózar *et al.* (2005b) have identified the Holkerian/Asbian boundary within the Lisgorman Shale Formation; the coral biostromes are of late Asbian age (MFZ 14). A thick development of the Bricklieve Limestone in the Drumkeeran well in the south of the Lough Allen Basin may represent a northward prograding body of carbonate sediment of shallow water origin, which contrasts with coeval thinner, darker, deeper water facies of the Dartry Limestone closer to the basin centre (MacDermot *et al.*, 1996). The Croghan Limestone Formation (Caldwell, 1959) to the south of the Curlew Mountains is a less shale-rich equivalent of the upper part of the Lisgorman Shale, and the overlying Cavetown Limestone Formation is similar to the Bricklieve Limestone farther north.

Only the Benbulben Shale, and a local development of sandstone, the Tubbrid Sandstone Formation (Holkerian) (Fig. 10.19), are preserved above the Mullaghmore Sandstone in the Omagh Syncline (Mitchell, 2004c).

The Glenoo well (Sheridan, 1972) proved thick developments of the Benbulben Shale, Glencar Limestone, and Dartry Limestone formations (Fig. 10.19). To the north-east, around Aughnacloy, County Tyrone (Mitchell, 2004c), the Benbulben Shale is succeeded by the Aughnacloy Sandstone Formation, which contains miospores assigned to the TC miospore Biozone and corals and brachiopods of early Asbian age. Mitchell (2004d) has interpreted the ingress of sandstone as reflecting uplift and erosion of older Carboniferous rocks, originally deposited over the region, between the present day Lisbellaw and Lack inliers. The uplift was caused by local shortening generated at a restraining bend in a zone of dextral strike-slip. The overlying Maydown Limestone Formation contains brachiopods and corals of Asbian age.

The succession around Benburb (Fig. 10.19) was described in detail by Mitchell and Mitchell (1983). Despite its proximity to places where the late Viséan is developed in shelf facies and assigned to the Armagh Group (see below, p. 264), the succession according to Mitchell (2004c) bears a closer resemblance to the Tyrone Group, described above. The Asbian consists of shaly limestone with rich coral and brachiopod faunas (Maydown Limestone Formation), mudstones and bituminous limestones (Blackstones Limestone Formation), sandstones and mudstones (Carrickaness Sandstone Formation), and thin developments of micrite and oolite (lower part of the Blackwater Limestone Formation). The Carrickaness Sandstone, which is at least 60 m thick, is partly of shallow water marine origin, but is believed to have included thin coals, now no longer exposed, reflecting a non-marine environment. The Brigantian sequence is truncated by the pre-Triassic unconformity; it consists of approximately 20 m of limestones, including algal limestones, shales, and thin sandstones (the upper part of the Blackwater Limestone Formation).

The Leitrim Group represents a major change in facies from the underlying Dartry Limestone Formation. There has been considerable discussion of the age of the base of the group. Correlations based on ammonoids suggest that the lower part is of Asbian age. However, Cózar *et al.* (2006) have argued on the evidence of the microfauna that the base is of Brigantian age, and that the previously accepted equation of ammonoid and other biozonal schemes at the Asbian/Brigantian boundary is incorrect. A resolution of this problem may be that Cózar *et al.* (2006) have abandoned the definition of the base of the Brigantian based on its stratotype, for a lower horizon based on biostratigraphical criteria (Cózar and Somerville, 2004). Until there is a formal chronostratigraphical re-definition of the Brigantian Stage, and pending additional work on the biostratigraphy, correlation of the base with the P_1a/P_1b ammonoid biozonal boundary is retained.

The lowermost Meenymore Formation, described by West *et al.* (1968), Brandon (1972, 1977), Brunton and Mason (1979), and Legg *et al.* (1998), is preserved in a small outlier in County Sligo (and probably also in County Mayo: the Rinmore Formation of Long *et al.*, 1992), and east of the Ox Mountains–Ballyshannon High, in the hills around Lough Allen, the Fermanagh highlands, Slieve Rushen, and Slieve Beagh (Fig. 10.15). It is made up predominantly of grey shale, which is mostly unfossiliferous, but at some horizons contains faunas of ammonoids, gastropods, bivalves, ostracodes, crinoids, brachiopods, and corals. Interbedded with the shales are beds of fine-grained, commonly laminated limestone and dolomite, interpreted as stromatolites. In some cases these show desiccation cracks and breccias with clasts of laminite, which were probably dried flakes of stromatolitic carbonate. There is abundant evidence of evaporites, including beds of gypsum proved in the subsurface (Grennan, 1992), limestones with macrocell structure, pseudomorphs after gypsum, and rare pseudomorphs after halite. Sandstones are conspicuous in the north but less so in the south. The base of the Meenymore

Formation is a disconformity and reflects a relative drop in sea level throughout the area. In several places where the underlying Dartry Limestone is developed in carbonate mudbank facies, the base of the Meenymore Formation onlaps the mudbanks; at the top of some mounds the formation is absent and the overlying Glenade Sandstone Formation rests directly on the Dartry Limestone. The depositional environment of the Meenymore Formation has been interpreted as a coastal sabkha and the sandstone bodies as deltaic. The formation varies in thickness, partly because of the topography of the upper surface of the underlying Dartry Limestone, but also because of regional and fault-controlled, local differential subsidence. According to Brandon (1977), the thickness along the western margin of the Lough Allen Basin is approximately 18 m; this increases to 35–40 m in the south of the basin and in the Dowra No. 1 well. In the Doagh outlier in the Fermanagh Highlands, the formation is 18 m thick. An increase to 90 m some 10 km to the west has been attributed to local fault-controlled subsidence (Legg *et al.*, 1998). Mitchell (1992, 2004d) suggested that such thickness changes resulted from dextral movement along re-activated Caledonian faults, which produced local pull-apart basins. South-eastward from the Doagh outlier the thickness increases to 48 m and 100 m on the north and east side of Cuilcagh repectively, and to approximately 240 m on Slieve Rushen, where there are thin, probably turbiditic sandstones in the lower part of the formation (Brandon, 1977). This suggests that the Clogher Valley Fault system actively controlled subsidence during the late Asbian. Further evidence for significant tectonic activity comes from the Slieve Beagh area farther along the Clogher Valley fault to the north-east (Fig. 10.15). There the Meenymore Formation is at least 300 m thick and contains a 75 m thick unit of pebbly sandstone, the Carnmore Sandstone Member (Mitchell *in* Geraghty *et al.*, 1997). It contains proportionately less laminated carbonate than farther east and more sandstone and shale. Padget (1951) recorded several horizons with ammonoids within the lower part of the formation north of Slieve Beagh.

To the north of the Clogher Valley Fault on the Fintona 'Block', the late Viséan facies are in marked contrast to those to the south (Mitchell and Owens, 1990). The Topped Mountain Formation (Fig. 10.19), consisting of over 1000 m of fine- to coarse-grained grey sandstone, interbedded with greyish red and green mudstone, has been dated near its top by palynology as being probably of Asbian age. It is interpreted as being of non-marine origin and reflects uplift of the southern margin of the Fintona Block. Renewed uplift, inferred to have been at the beginning of the Brigantian, gave rise to the succeeding Ballyreagh Conglomerate Formation. This consists of 350 m of conglomerate and pebbly sandstone, interpreted as having been deposited on alluvial fans. It is overlain by the Ballinamallard Formation, approximately 1000 m of greyish-red mudrocks and thin sandstones with ripples and desiccation cracks, which near its top has been dated by palynology as late Brigantian or early Serpukhovian. Farther north, there is another unit of non-marine strata, the Greenan Sandstone Formation, which at its base is probably of late Brigantian age. It consists of a basal breccia resting on Dalradian metamorphic rocks, overlain by at least 500 m of purple-grey sandstones with laminae of mudstone. Mitchell and Owens (1990) interpreted the Greenan Sandstone Formation as being part of the fill of the Omagh-Kesh Basin, now preserved only at its southern margin. The basin was uplifted during the Asbian and renewed uplift to the south in the Brigantian led to non-marine sedimentation over what had been a dominantly marine basin (Fig. 10.20).

The Meenymore Formation of the Lough Allen Basin is succeeded by the cliff-forming Glenade Sandstone Formation, whose type area is around Glenade, south of Benbulben, County Sligo. It forms a southerly thinning wedge, estimated to be 300–350 m thick in the Fermanagh Highlands (Legg *et al.*, 1998), 70–90 m on Cuilcagh, and 4 m at the south end of Lough Allen. The formation consists predominantly of grey, fairly mature, quartzose sandstone. It has generally been referred to as being of deltaic origin, but little detailed sedimentological information has been published. It contains very few marine fossils, but plant material is common. Smith (1996), on the evidence of palynofacies, identified fluvial and lagoonal environments and a short-lived anoxic marine incursion near the base of the formation south-west of Cuilcagh. The vegetation, reflected by the plant miospores, consisted of ferns, tree ferns, and small seed ferns. Legg *et al.* (1998) have demonstrated that on Cuilcagh the lower part of the formation consists of at least four cycles, in which sandstones, which are thin to medium bedded and cross-stratified at the base, become coarser upward, culminating in thick beds of reverse graded, very coarse-grained pebbly sandstone and conglomerate. The upper part of the formation on Cuilcagh consists of sandstones with common trough cross-stratification and channeling. The provenance of the sediment, on the evidence of generally well-rounded pebbles of vein

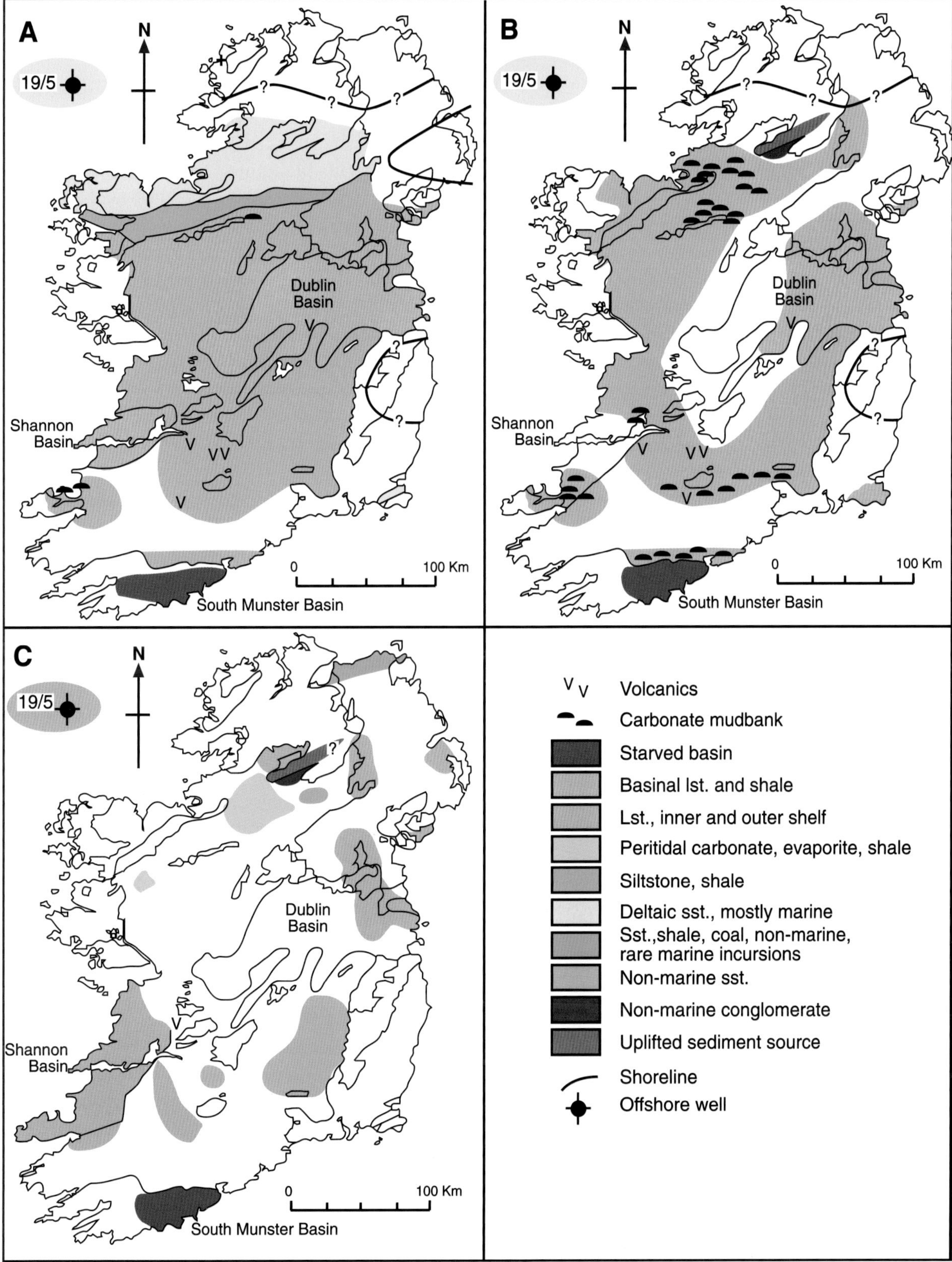

Fig. 10.20 Generalised palaeogeographical maps. **A)** Late part of the Arundian, showing the distribution of the Mullaghmore Sandstone Formation. **B)** Asbian, prior to the deposition of the Leitrim Group of northern Ireland. (C) Brigantian, at the time of the depostion of the Bellavally Formation of north-west Ireland.

quartz, quartzite, and psammite, was probably from metamorphic rocks some distance to the north, which is consistent with the few palaeocurrent measurements published.

The overlying Bellavally Formation (Brandon and Hodson, 1984) consists of an alternation of lithologies (each recognised as a named member) representing shallow marine environments of normal salinity, and intertidal to supratidal environments, which were commonly hypersaline. The marine environments are represented by fossiliferous shale and micrite, with ammonoids, and rarer gastropods, brachiopods and corals; the littoral environments by laminated limestone, commonly associated with evaporites, and by sandstone. In contrast to the Meenymore Formation, which it otherwise resembles, the Bellavally Formation is remarkably consistent across the basin with the majority of the members varying little in thickness. The only persistent sandstone, the Doobally Sandstone Member, has been described by Legg *et al.* (1998) on Cuilcagh, where it consists of two coarsening-up cycles, with pebbly sandstone at the top. The Doobally Sandstone lies close to or at the Asbian/Brigantian boundary, determined by ammonoids. Ammonoid faunas of the B_2 Biozone occur below it and a fauna of the P_1b Biozone occurs immediately above it (Legg *et al.*, 1998).

The Bellavally Formation is succeeded by the Carraun Shale Formation, which consists of dark grey shales and mudstones and five persistent limestone units that have been recognised as named members. Ammonoids at several horizons in the shales and limestones show that the formation ranges from the P_1b/P_1c Biozonal boundary to the top of the P_2 c Biozone. Stromatolites and pseudomorphs after evaporites in the lower two limestone members are evidence of very shallow water environments of deposition; the three younger limestone members probably reflect deeper water conditions. Thin clay bands, interpreted as K-bentonites, occur in the P_1b and P_2c Sub-biozones (Brandon and Hodson, 1984).

The Armagh-Cookstown shelf

Carboniferous successions on the west side of Lough Neagh between Armagh and Cookstown differ substantially from those of County Sligo and the Lough Allen Basin and reflect a shallow water marine shelf, where sedimentation was interrupted by emergence. In the Armagh area (Higgs *et al.*, 1988b; Somerville *et al.*, 1996b; Mitchell, 2004c), the Tournaisian and earliest Viséan Ballynahone Formation is overlain by the Milford Mills Formation (Fig. 10.19), which consists of polymict conglomerate, red and white sandstone, and red and green mudrock with pedogenic carbonate. It has been interpreted as representing mainly non-marine environments. Towards the top of the formation, there are thin crinoidal limestones indicating the incoming of marine conditions. The overlying Oulart Villa Limestone Formation consists of skeletal limestone and dark shale and has been interpreted as having formed in open marine conditions. It contains foraminiferans indicative of MFZ 10 at its base and MFZ 11 higher up; it is noteworthy that the early part of the Arundian, here as in many places in Ireland, coincides with conspicuous deepening. The succeeding Drumman More Sandstone Formation is made up of non-calcareous, red and brown sandstone, with coal streaks, and red, green and yellow mudrocks; it is dominantly of non-marine origin and contains miospores assigned to the TS and TC Biozones. The overlying fossiliferous limestones of Asbian and Brigantian age have been assigned to the Armagh Group, which contains formations deposited on a shallow water, periodically emergent, shelf, extending from Armagh to east of Draperstown, County Derry (Fig. 10.20). The oldest formation assigned to the Group in the Armagh area, the Wilson's Bridge Limestone Formation (Asbian), is made up of skeletal limestones containing *Lithostrotion* and *Siphonodendron* and thin shales. The overlying Loughall Limestone Formation, also of Asbian age, consists of pale, thick-bedded, skeletal and oolitic grainstones, which have been widely used as a building stone. It exhibits evidence of exposure in palaeokarstic surfaces, palaeosols, limestone boulder conglomerates, and thin channel sandstones. The youngest Carganamuck Formation is made up of thin-bedded, dark wackestones and in the upper part of the formation, ooid grainstones. It, too, contains evidence of emergence. The diverse fauna of corals, such as *Actinocyathus*, brachiopods, such as *Gigantoproductus*, and the microbiota (Cózar *et al.*, 2005a) indicate a Brigantian age.

Fowler and Robbie (1961) have described the poorly exposed Viséan rocks of the Dungannon area (Fig. 10.19). The Ballytrea borehole proved approximately 250 m of limestone, shale, subordinate sandstone, and a thin coal and underlying rootlet bed. The limestones contain rich faunas of corals and brachiopods, on the evidence of which the upper *c.*65 m is of Brigantian age. A substantial thickness of Viséan rocks is inferred below the horizon of the base of the borehole. Limestones, shales, and sandstones at outcrop have been inferred to be of Arundian age, but the evidence is tenuous.

The Brigantian limestone is overlain by the Rossmore Mudstone Formation (Leitrim Group), over 100 m of mudstones with limestones, including oolite, towards the base. The fauna, including rare ammonoids, suggests equivalence with the P_2 ammonoid Biozone.

Farther north, in the Cookstown area, the lowest unit of the Armagh Group, the Derryloran Grit Formation, is made up of over 590 m of red sandstone and conglomerate with an intercalation of limestone. It rests unconformably on the metamorphic rocks of the Sperrin Mountains and is of Brigantian age. It passes laterally into shelf limestone of the Rockdale Limestone Formation, which contains corals such as *Lonsdaleia*.

North of Slieve Gallion, the Tournaisian-aged Altagoan Formation is overlain by the Desertmartin Formation (Mitchell, 2004c), which consists of sandy and pebbly limestone, with corals and brachiopods that indicate an Asbian age.

To the north-east of Armagh, along the margin of the Longford-Down inlier, Viséan rocks are either covered by Permo-Triassic sandstones or Tertiary basalts, or are not preserved, as is the case around Belfast Harbour. However, limestones of Brigantian age containing very large specimens of the orthoconic nautiloid *Rayonnoceras* were formerly quarried at Castle Espie near the head of Strangford Lough (Fig. 10.15).

Viséan rocks occur at Ballycastle, County Antrim (Wilson and Robbie, 1966). Mitchell (2004c) has provided a revised stratigraphy (Fig. 10.19), assigning the whole succession to the Ballycastle Group and showing that only Brigantian strata are represented. The lower part of the sequence is known from boreholes. The Eglish Sandstone Formation lies unconformably on the Dalradian. It consists of sandstone, conglomerate, and mudstone, containing miospores. It is overlain by the Carey River Basalt and Glenshesk Tuff Formations, described below (p. 267). The Murlough Shale Formation consists of ten thin coal seams and fireclays, interbedded with dark shales, some with marine fossils. The lower part of the youngest formation, the Ballyvoy Sandstone Formation, consists of red sandstone with mudstone partings, which exhibit desiccation cracks at the base. Overlying shale and argillaceous limestone with a marine fauna (the Carrickmore Marine Band) mark a probably short-lived marine incursion, which was followed by a return to formation of red sandstones, with thin coals at two horizons. The overlying Main Limestone represents the most significant marine incursion in the succession and consists of several metres of richly fossiliferous limestone overlain by fossiliferous shale. It was followed by a return to non-marine conditions represented by sandstones with subordinate shales and thin coals below the Main Coal, which has somewhat arbitrarily been taken to mark the base of the Serpukhovian. Mitchell (2004c) has suggested that the Ballycastle succession was deposited in a fault-bounded basin to the north-west of the Highland Border Ridge as a result of extension during the Brigantian. A borehole at Magilligan, County Derry, (Fig. 10.19) penetrated a succession with thin coals containing late Viséan miospores (Wilson, 1972); it also was probably deposited in the Ballycastle basin.

Offshore west of Ireland

Robeson *et al.* (1988) have summarised the information from two wells in the Porcupine Basin and one in the Erris Trough (see Fig. 11.2, p. 272).

Well 35/15–1, drilled on the eastern margin of the Porcupine Basin, approximately 150 km west of the Shannon estuary, proved interbedded skeletal limestone, shale, and sandstone with horizons of volcanic material, reported to include ignimbrites, overlying a succession of sandstone with subordinate limestone. The drilled section, which is very poorly dated, is approximately 110 m thick. Well 26/26–1 drilled in the northern part of the Porcupine Basin encountered 60 m of poorly dated sandstone, siltstone, and mudstone with evaporite stringers, overlain by 200 m of possibly Lower Viséan or Arundian/Holkerian sandstone and limestone, succeeded in turn by marine mudstone with traces of anhydrite thought to be of late Viséan age. Because of the uncertainty of the age of the two drilled sequences, it is difficult to compare them with the Viséan stratigraphy onshore.

Well 19/5–1, drilled approximately 100 km west of the coast of County Donegal (Fig. 10.20), proved 26 m of grey sandstone, dated as Holkerian; the Asbian (approximately 325 m thick) is represented by sandstone and brown to brick-red siltstone, overlain by skeletal limestone and shale; the Brigantian (430 m thick) consists predominantly of limestone, shale, and sandstone cycles. This succession appears significantly different from that of Donegal Bay except that the lowest grey sandstone unit might be correlated with the Mullaghmore Sandstone Formation.

Tournaisian/Viséan volcanicity

Although rocks of volcanic origin occur through much of the Mississippian, those in the Tournaisian are tuffs that are volumetrically insignificant and have an unknown source. The Viséan examples, by contrast, include large volumes of intrusive and extrusive rocks related to volcanic centres in Munster and the midlands and lesser amounts farther north (Fig.10.21).

Within the Tournaisian of the South Munster Basin, thin tuffs have been identified in the Castle Slate Formation (Hastarian) (Naylor, 1966; Graham and Reilly, 1976) and in the Ringabella Limestone and the Loughbeg Formations (Ivorian).

Despite the large number of boreholes drilled below the Waulsortian of the North Munster Shelf, clay bands of volcanogenic origin have been identified only at Lisheen and elsewhere to the south-east of the Devilsbit Mountains (Hitzman *et al.*, 1992; Shearley *et al.*, 1996), where they occur close to the base of the Waulsortian. Thin clay bands of volcanogenic origin are widely distributed within the upper part of the Waulsortian and just above its top in the midlands and elsewhere (Philcox, 1984); the majority of occurrences are probably of latest Tournaisian age, but some are certainly of early Viséan age.

Of the widespread occurrences of volcanic rocks of Viséan age in Munster, the most important and best known is the Limerick volcanic region, which has been described by Ashby (1939), Strogen (1974a, 1988, 1995), Somerville *et al.* (1992a), and Strogen *et al.* (1996). Substantial volcanic activity began with the deposition of the Knockroe Formation (Lower Viséan and Arundian), which consists of a bimodal suite of slightly alkaline basalts and syenites and trachytes. Six volcanic centres, defining two east-north-east trending lineaments, have been identified on the limbs of the Limerick Syncline. Each centre exhibits a similar volcanic history, although the volcanic activity along the lineaments migrated from west-south-west to north-north-east through time. The earliest phase of eruption was Surtseyan, when highly explosive, initially submarine, basaltic eruptions built up low-angle, vitric tuff rings several kilometres in diameter. These were buried by subaerial emission of large volumes of basaltic lavas that flowed into the sea and spread over substantial areas. There followed phases of intrusion of trachyte dykes and episodes of vent clearing that produced lithic

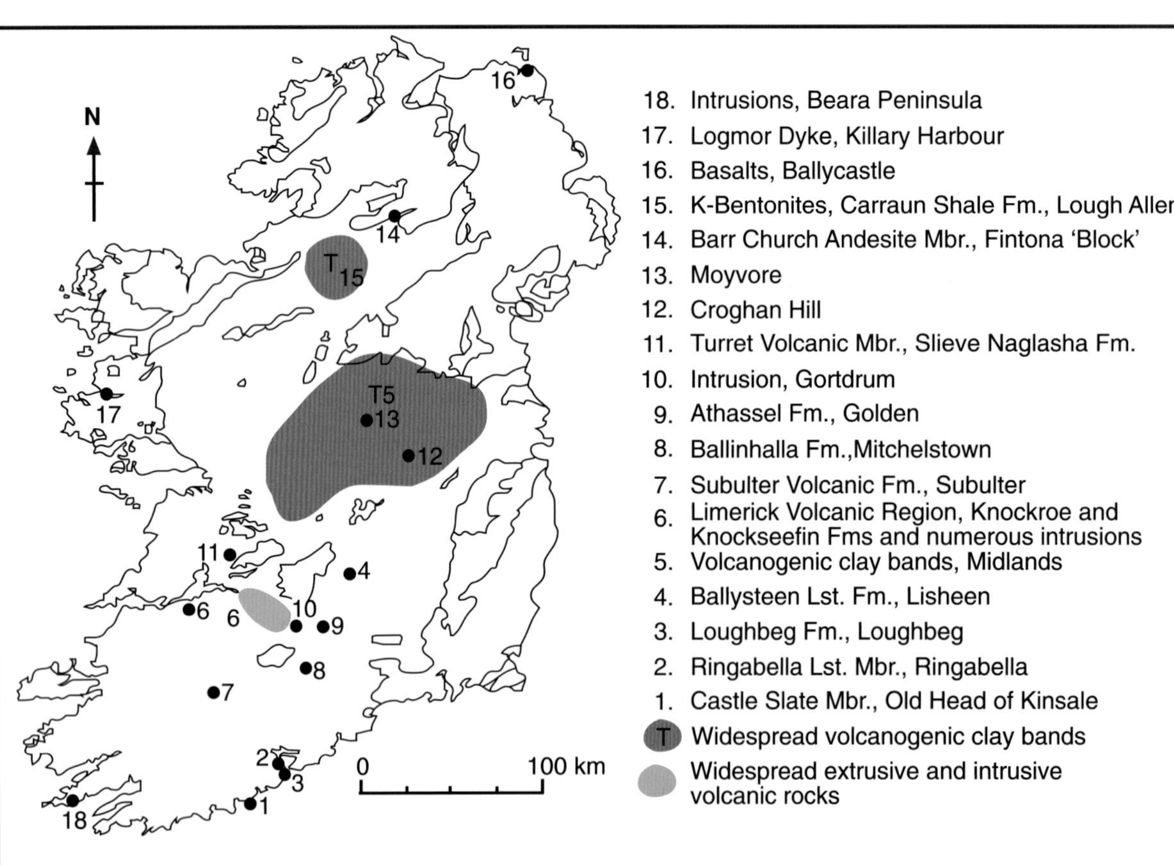

Fig. 10.21 Distribution of Tournaisian and Viséan volcanic rocks in Ireland.

tuffs, rich in fragments of trachyte. Other occurrences of volcanic rocks near the River Shannon and south of the Limerick Syncline also appear to be aligned and, on this evidence, three additional east-north-east trending lineaments have been postulated. Basic tuffs in the Mallow area, named the Subulter Volcanic Formation by Hudson and Philcox (1965) are the same age (Arundian) as the Knockroe Formation. Tuffs have also been reported from within the Athassel Formation (Lower Viséan) around Golden, County Tipperary (Archer *et al.*, 1996).

Volcanic activity resumed in the Asbian, when the Knockseefin Volcanic Formation of the Limerick Syncline formed. It consists of ankaramitic and limburgitic flows with subordinate tuff-breccias. Unlike the Knockroe Formation, there are no associated trachytes. Intrusions of ankaramite at Gortdrum, County Tipperary may be part of this phase of volcanism. Shearley (1988) recorded tuffs of Asbian age in the Mitchelstown Syncline.

The only record in this region of volcanic rocks of Brigantian age is an isolated occurrence of basaltic, coarse tuffs and flows, the Turret Volcanic Member of the Slievenaglasha Formation, south of Tulla, County Clare (Schultz and Sevastopulo, 1965; Lewis, 1986; Sleeman and Pracht, 1999).

The geochemistry and petrology of the volcanic rocks of the Limerick region indicate that they were generated as a result of a low degree of melting of undepleted mantle, suggesting only moderate extension of the crust during the Viséan. Small amounts of basaltic magma rose passively and fractionated extensively at high levels in the crust, eventually exploiting tectonic lineaments as conduits to the surface.

There was substantial volcanic activity within the Dublin Basin during the Viséan, but most of the evidence is in poor surface exposures or in the subsurface. The most striking outcrop is at Croghan Hill, County Offaly, described by Haigh (1914). The hill consists of an accumulation of coarse tuffs and agglomerates and subsidiary flows, which are ankaramites in composition (Strogen *et al.*, 1996; Gatley *et al.*, 2005) and are of early Viséan to Arundian age (McConnell and Philcox, 1994). The agglomerates contain fragments of pre-Carboniferous rock (Strogen, 1974b), including phyllite, interpreted as being of Lower Palaeozoic origin, and gneiss of granulite grade, interpreted as having been derived from a high grade basement. Strogen (1974b) reported similar agglomerate with gneissose xenoliths from two localities in County Westmeath, and Van den Berg *et al.* (2005) described granulite facies xenoliths from the same area. MacCarthy (1990) described tuffisites within Tournaisian micrites near Moyvore that contained clasts of Waulsortian limestone derived from above. The indication from the widely separated occurrences of igneous rocks in this poorly exposed area is that volcanic activity was relatively widespread; this is supported by unpublished evidence from drilling and magnetic surveys (Hitzman, 1995). However, the volume of volcanic rocks in the Dublin Basin was probably not as great as in the Limerick region.

The only other areas where extrusive volcanics of Viséan age have been reported are in Ulster. The Barr Church Andesite Member is a flow-banded andesite within the Topped Mountain Sandstone Formation of Asbian age in the Fintona 'Block'. Thin clay bands, interpreted as K-bentonites, occur within the Brigantian Carraun Shale around Lough Allen (Brandon and Hodson, 1984; Legg *et al.*, 1998). In the Ballycastle coalfield (Wilson and Robbie, 1966; Mitchell, 2004c), volcanic rocks of Brigantian age are well displayed along the coast. The Carey River Basalt Formation consists of some 40 m of subaerial basaltic flows; the overlying Glenshesk Tuff Formation (60 m thick) is composed of vitric and lithic tuffs and sandstone and mudstone.

It has been suggested that several intrusions in pre- or early Carboniferous rocks are of Viséan age. Mitchell and Mohr (1987) have described the north-south trending Logmor dyke, which is intruded into the Lower Palaeozoic rocks of north County Galway and south County Mayo. K–Ar whole rock analyses of the dyke, which has a mildly alkaline basaltic composition, have been interpreted to indicate an age of emplacement of 320 Ma (late Viséan), although a substantialy older date seems more probable.

In the South Munster Basin, there are concentrations of sills, dykes and intrusive pyroclastic pipes within the late Devonian and Tournaisian sandstones and mudrocks. They had been interpreted by Pracht and Kinnaird (1995) to be of late Viséan age. However, both their structural context and more recent radiometric dates suggest that they are younger; they are described in Chapter 11 (p. 274).

Base metal mineralisation

No account of the Carboniferous rocks of Ireland would be complete without at least brief reference to the very large accumulation of metals, principally zinc and lead, that they contain. Details of the mined deposits (including that at Navan – the largest zinc deposit in Europe), the numerous prospects and occurrences of base metal sulphides that have contributed to the perception of an 'Irish Orefield' are described in Andrew *et al.* (1986), Bowden *et al.* (1992), K. Anderson *et al.* (1995), Sangster (1996), Hitzman *et al.* (2002), and Kelly *et al.* (2003).

Some generalisations about the deposits can be made. Almost all of the significant mineralisation is hosted in carbonate rocks of Tournaisian age in the hanging wall of faults, which, in many cases, can be shown to have been active during the late Tournaisian and early Viséan and which were inverted during Variscan shortening. The majority of the deposits contains lead and zinc ores of typically simple mineralogy, galena and sphalerite being the main ore minerals. However, the more southerly deposits also contain, or are principally formed of, copper sulphides, generally at lower stratigraphical horizons than the lead/zinc mineralisation. Stratiform barytes is associated with a few of the deposits. The host rocks are pervasively dolomitised in some deposits, but not in others. The bulk of the sulphides in the Irish Orefield were emplaced during burial of the host carbonate, but there is some evidence for deposition on the sea floor in some of the mined deposits. The metals were probably derived from Lower Palaeozoic basement. The sulphur is dominantly isotopically light and was derived from contemporary seawater sulphate. A small proportion was probably derived from sulphide in the basement. Precipitation of ore occurred where saline brine, derived from seawater, circulated downward and mixed with much lower salinity fluid, carrying metals, which ascended from the basement. Temperatures in the environment of sulphide precipitation in most deposits exceeded 125°C. The plumbing system was largely controlled by steep faults. The age of the mineralisation has been the subject of controversy. Only at Navan, where mineralised clasts of Tournaisian age have clearly been incorporated in the Boulder Conglomerate (Lower Viséan), is there direct stratigraphical evidence of the age. Elsewhere, the evidence is consistent with mineralisation having occurred through a protracted period from the late Tournaisian to at least as young as the Arundian.

11

Carboniferous: Mississippian (Serpukhovian) and Pennsylvanian

G. D. Sevastopulo

The stratigraphical framework within which the Serpukhovian and Pennsylvanian rocks are described is shown in Table 11.1. The Series formerly in use in north-west Europe (Namurian, Westphalian, and Stephanian) have been discarded as globally applicable chronostratigraphical units (Heckel and Clayton, 2006). The Pennsylvanian is now divided into Lower, Middle, and Upper Series and four global Stages that have been adopted from eastern Europe (Table 11.1). GSSPs have not yet been selected for the Serpukhovian, or for any of the Pennsylvanian stages, and in some cases the biostratigraphical tools that will be used to identify the bases of the stages have not been agreed. For these reason correlations shown in Table 11.1 are approximate. Because the global stages do not provide a framework of fine enough resolution to describe the stratigraphy of late Carboniferous rocks in Ireland, this account utilises the Pendleian–Yeadonian stages of Ramsbottom *et al.* (1978) and the Langsettian–Bolsovian stages of Owens *et al.* (1985), reduced in rank to regional substages (Table 11.1). For the younger parts of the Pennsylvanian, the former division 'Westphalian D' is referred to the Asturian regional substage and the Stephanian and Autunian are used as regional stages.

Serpukhovian and Pennsylvanian rocks, particularly the latter, have a restricted distribution on land in Ireland (Fig. 11.1). It is interesting to speculate on their former extent. Studies of the thermal maturation of older Mississippian rocks (p. 307) suggest that the present day occurrences of Serpukhovian and Pennsylvanian strata are the remnants of a thicker and geographically more extensive cover, much of which was removed by erosion following Variscan uplift. The now isolated coalfields of Pennsylvanian age were almost certainly part of a continuous region of deposition, more or less congruent with that of the Mississippian. In those areas (the Shannon and Dublin basins) in which the Pennsylvanian might be expected to have been thickest, structural inversion has resulted in the preservation of only the tiny Crataloe and Kanturk coalfields; the Pennsylvanian strata that have been preserved (the Leinster and Slieve Ardagh coalfields, the Kingscourt and Wexford occurrences and the Tyrone coalfield) were deposited on, or close to Carboniferous depositional highs. Preservation of strata younger than Langsettian is restricted to small areas in Counties Wexford and Tyrone. By comparison with offshore regions, where Serpukhovian and Pennsylvanian rocks occur extensively (Fig. 11.2), and with Britain, it is probable that sedimentation onshore in much of Ireland continued into Asturian times and possibly into the Stephanian. The erosion of Serpukhovian and Pennsylvanian rocks in the Kingscourt region occurred before the Upper Permian.

Throughout Ireland, Carboniferous rocks of Serpukhovian and younger age are almost exclusively terrigenous, the exceptions being rare limestones and thin coals. The switch from carbonate sedimentation to terrigenous muds and sands was not at exactly the same biostratigraphical horizon everywhere, and thus is not precisely coincident with the base of the Pendleian. However, the contrast between the Viséan 'Carboniferous Limestone' and the Serpukhovian 'Millstone Grit' in Ireland and Britain is sufficiently striking to demand a general explanation. Clearly, very large river systems encroached on Britain and Ireland at approximately the beginning of Serpukhovian time. They may have developed as a consequence of a change in the climate during the northward movement of the plate containing Britain and Ireland, suggested by palaeomagnetic data; but the substantial quantity of clastic sediment that was transported into the basins suggests that tectonic uplift of source areas was the principal factor. The southern basins may have been fed by sediment sourced from the south (Fig. 11.3) from the rising Variscan mountains, which are

Table 11.1 Statigraphical divisions of the late Mississippian and Pennsylvanian Subsystems, showing global and regional stages/substages (stages in bold face), Serpukhovian-Langsettian ammonoid biozones, Langsettian non-marine bivalve biozones, Serpukhovian-Langsettian miospore biozones (after Owens *et al.*, 2004), and interpolated radiometric dates for the boundaries of the global stages (after Davydov *et al.*, 2005).

	GLOBAL STAGE	REGIONAL STAGE/ SUBSTAGE	AMMONOID BIOZONE		BIVALVE BIOZONE	MIOSPORE BIOZONE	AGE (Ma)
PENNSYLVANIAN	**GZHELIAN**	**AUTUNIAN**					299.0
	KASIMOVIAN	**STEPHANIAN**					303.9
	MOSCOVIAN	ASTURIAN					306.5
		BOLSOVIAN					
	BASHKIRIAN	DUCKMANTIAN			*Anthroconaia modiolaris*		311.7
		LANGSETTIAN			*Carbonicola communis*	RA	
			G_2	*Gastrioceras listeri*	*Carbonicola lenisulcata*	SS	
				Gastrioceras subcrenatum			
		YEADONIAN	G_{1b}	*Cancelloceras cumbriense*		FR	
			G_{1a}	*Cancelloceras cancellatum*			
		MARSDENIAN	R_{2c}	*Verneuillites sigma*			
				Bilinguites superbilinguis			
			R_{2b}	*Bilinguites metabilinguis*			
				Bilinguites eometabilinguis		KV	
				Bilinguites bilinguis			
			R_{2a}	*Bilinguites gracilis*			
		KINDERSCOUTIAN	R_{1c}	*Reticuloceras coreticulatum*			
				Reticuloceras reticulatum			
			R_{1b}	*Reticuloceras stubblefieldi*			
				Reticuloceras nodosum			
				Reticuloceras eoreticulatum			
			R_{1a}	*Reticuloceras dubium*			
				Reticuloceras todmordense			
				Reticuloceras subreticulatum			
				Reticuloceras circumplicatile			
				Hodsonites magistrorum			
		ALPORTIAN	H_{2c}	*Homoceratoides prereticulatus*			
				Vallites eostriolatus			
			H_{2b}	*Homoceras undulatum*			
			H_{2a}	*Hudsonoceras proteum*		SR	
		CHOKIERIAN	H_{1b}	*Homoceras beyrichianum*			
			H_{1a}	*Isohomoceras subglobosum*			318.1
MISSISSIPPIAN	**SERPUKHOV-IAN**	ARNSBERGIAN	E_{2c}	*Nuculoceras nuculum*		SV	
				Nuculoceras stellarum			
			E_{2b}	*Cravenoceratoides nititoides*			
				Cravenoceratoides nitidus			
				Cravenoceratoides edalensis		TK	
			E_{2a}	*Eumorphoceras yatesae*			
				Cravenoceras gressinghamense			
				Eumorphoceras ferrimontanum			
				Cravenoceras cowlingense			
		PENDLEIAN	E_{1c}	*Cravenoceras malhamense*		Vm	
			E_{1b}	*Tumulites pseudobilinguis*			
				Cravenoceras brandoni			
			E_{1a}	*Cravenoceras leion*			326.4

inferred to have been uplifted progressively northward through time. Basins in the northern half of Ireland, in contrast, received sediment from the north. Evidence from the north of England suggests that the northern source may have been as far away as Scandinavia or Greenland (Drewery *et al.*, 1987).

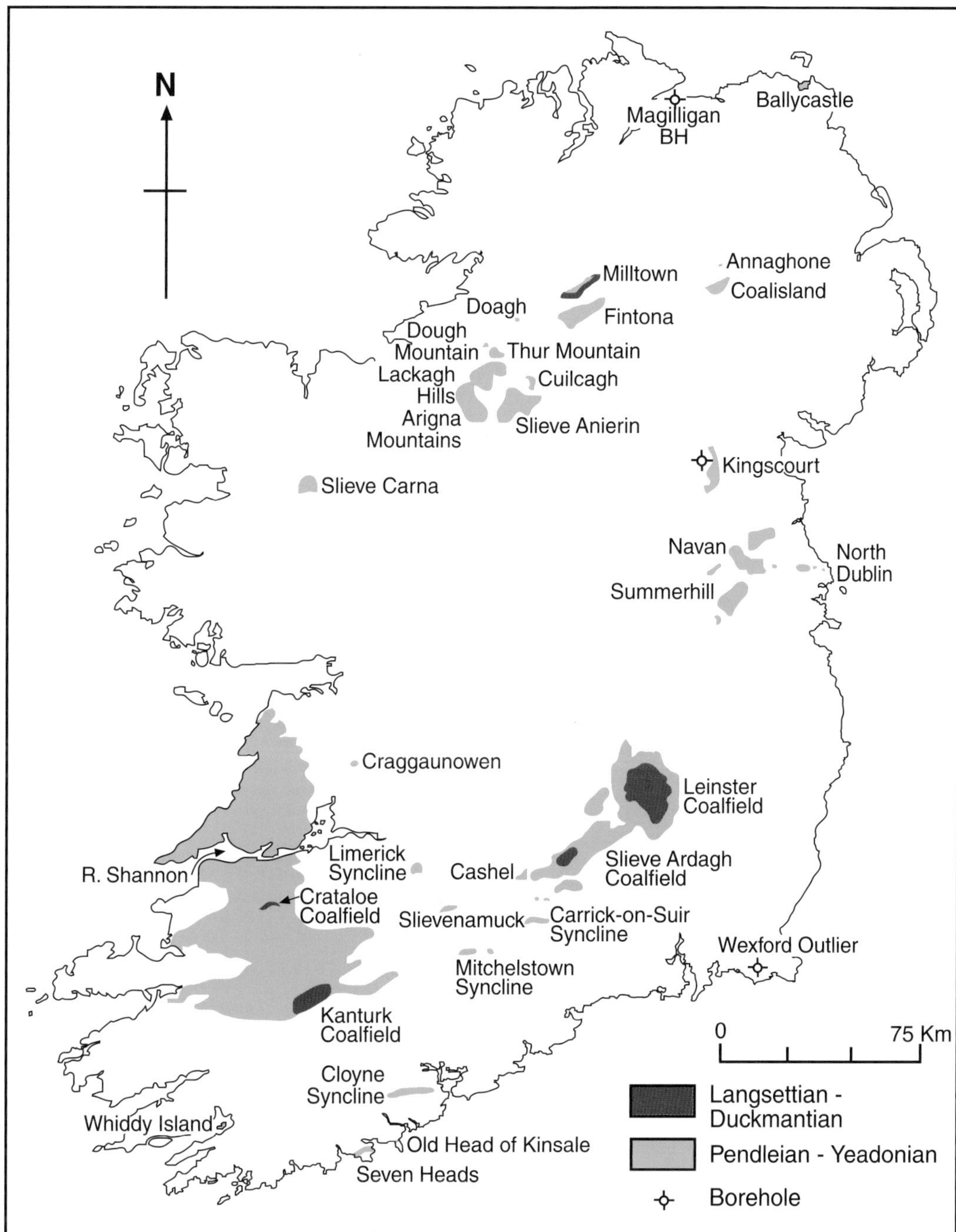

Fig. 11.1 The distribution of Serpukhovian and Pennsylvanian rocks in Ireland.

The changes in sedimentary facies close to the base of the Serpukhovian were paralleled by changes in the biota. During Pendleian–Yeadonian time, facies with brachiopods, corals, and echinoderms were far less common than in the Tournaisian and Viséan, whereas ammonoids and thin-shelled bivalves were widespread and have proved to be valuable biostratigraphical tools. Both of the last groups of fossils are typically found within dark, carbonaceous shale, a facies characteristic of rocks of this age in northwest Europe. They are commonly concentrated in discrete bands that are separated from each other by barren strata. Where fully developed, almost every faunal band contains a distinctive assemblage of ammonoids and thin-shelled bivalves, which have been used to erect Biozones and Sub-biozones that may be correlated throughout Britain and Ireland, and, in many cases, much farther afield. The

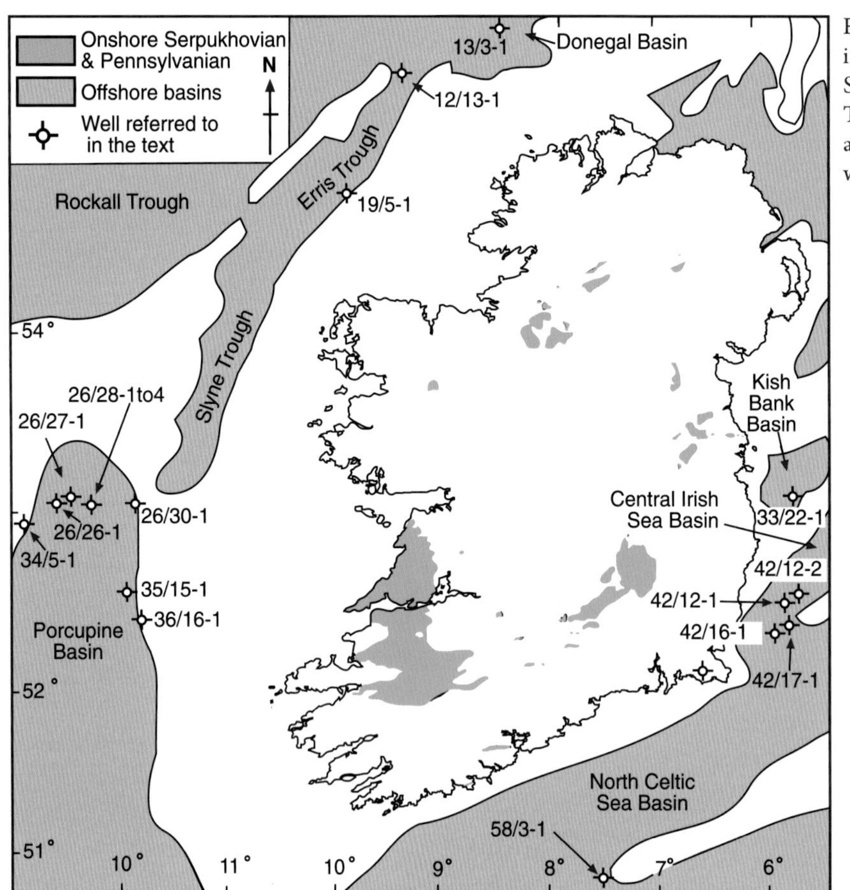

Fig. 11.2 The location of offshore wells in Irish waters that have intersected Serpukhovian and Pennsylvanian rocks. The basins are Mesozoic and Tertiary in age and are not known to be coincident with late Palaeozoic basins.

Below Fig. 11.3 Tentative and generalised palaeogeographical reconstruction of the Serpukhovian and Pennsylvanian in Ireland. **A)** Pendleian, based as far as possible on E_{1c}. **B)** Early Kinderscoutian.

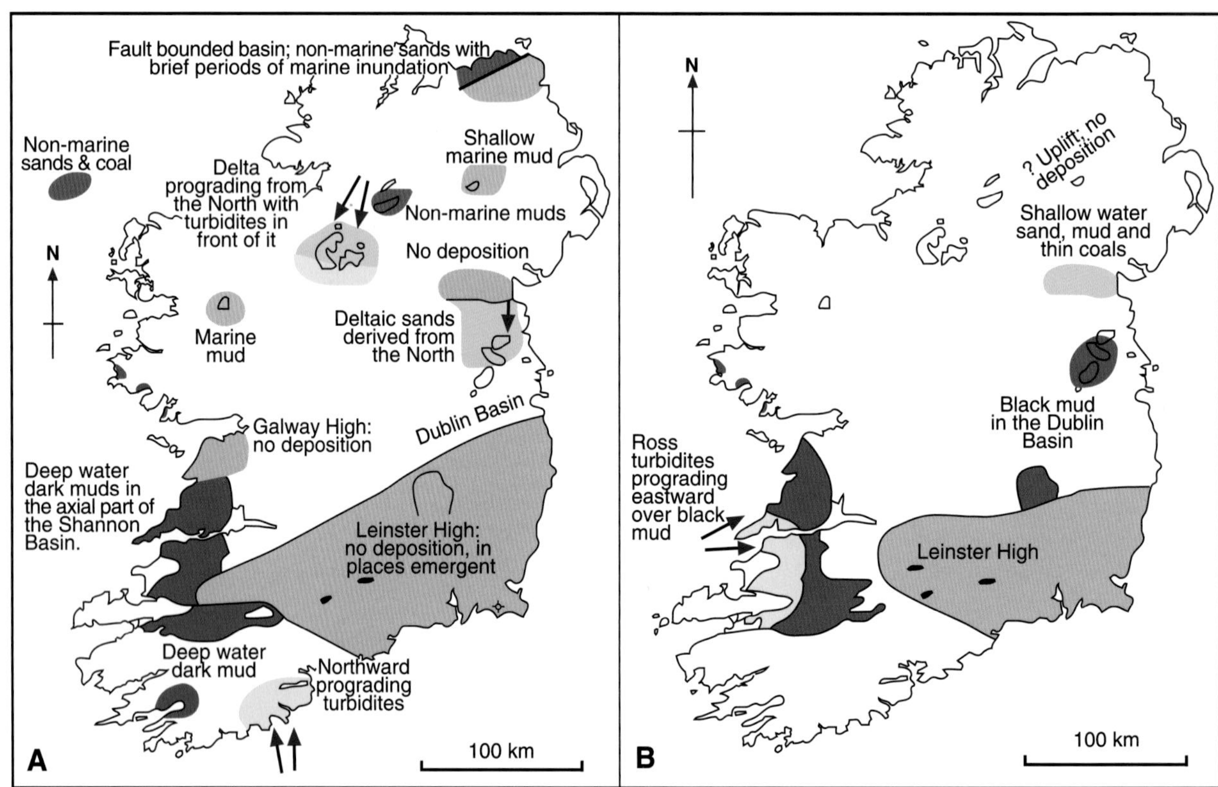

Substages used in this account were defined with reference to sections in the north of England (Ramsbottom *et. al.*, 1978) and are readily identified by the ammonoid Biozones (Table 11.1). From the Langsettian onwards, the successions become progressively less marine and ammonoid-bearing shales less frequent; biostratigraphical correlations are effected through the use of non-marine bivalves and plant spores.

Several authors have discussed the sedimentological setting of black shales of Pendleian–Yeadonian age and the reasons for the restriction of the fauna to widespread, discrete horizons. There is general agreement that during the Serpukhovian and Pennsylvanian there were high frequency, eustatic changes of sea level, controlled by the fluctuations in the extent of ice sheets in Gondwana. Holdsworth and Collinson (1988) suggested that in Britain and Ireland during Pendleian to Yeadonian time, sediment accumulated in several interconnected, small, intracratonic basins, distant from any open ocean. During periods of high stand of sea level, there were substantial connections with the open oceans, salinity was at typical marine values, and normal marine faunas populated the seas. During periods of low stand, individual basins had much more restricted connections with each other and with the open oceans, and the marine waters were rapidly displaced by fresh water delivered by major river systems. As a result, the salinity was reduced and the fully marine fauna was extinguished. This model is based partly on the evidence of the distribution of the fauna within the shales in some sequences. The thick-shelled ammonoids, which are used in biostratigraphical correlation, are thought to have flourished only in water of normal marine salinity; in some cases, the band containing these ammonoids is succeeded by, or succeeds, horizons with thinner shelled ammonoids, such as *Dimorphoceras* and *Anthracoceras*, which have been interpreted as having tolerated lower salinity. Horizons with ammonoid spat or bivalves only, or with *Lingula*, have been interpreted to indicate even lower salinity.

Hampson *et al.* (1997), however, interpreted the ammonoid-bearing black shales as condensed deposits formed during maximum flooding events rather than high stands, and considered that barren beds in some cases were deposited under conditions of normal marine salinity.

Pendleian–Yeadonian rocks are sufficiently widely distributed onshore Ireland to allow attempts at tentative, generalised, palaeogeographical reconstructions (Fig. 11.3).

In what follows, the post-Viséan stratigraphy is described on a regional basis. The successions offshore are referred to by the names of the Mesozoic–Tertiary aged basins (for example, the Porcupine basin) in which they occur. There is no implication in the use of these names that the basins were also Palaeozoic in age.

South Munster Basin

Rocks of Serpukhovian and Pennsylvanian age are preserved in the cores of several of the synclines of the South Munster Basin (Fig. 11.4). The most completely studied section is on Whiddy Island in Bantry Bay (Naylor *et al.*, 1978; Jones and Naylor, 2003). The oldest unit, the East Point Member of the Reenydonagan Formation, consists of at least 171 m of black, carbonaceous, and, at some horizons, pyritic mudrock. Rare thin beds of very fine sandstone also occur. Bullions near the base and in the upper half of the formation have yielded ammonoids indicating an Arnsbergian age (it is possible that the Pendleian is also represented), with the highest beds being assigned to E_{2b}. Black mud deposition was succeeded by an influx of fine sand. The Middle Battery Formation (231 m) consists of very fine-grained sandstone and dark grey mudrock in varying proportions. Sandstones, particularly in the lower two-thirds of the formation, show sedimentary structures, including common sole structures, indicative of turbidites. Flow was from the east-south-east. The youngest Kilmore Formation (110 m preserved) is dominated by fine sandstone with subordinate dark grey mudrock. The lower 75 m of the formation consists of beds of sandstone, interpreted as turbidite sheets, separated by thin beds of mudrock, with an increasing frequency of slump horizons in younger beds. The highest 35 m consist of sandstone beds of medium thickness, amalgamated into thick sheets, separated by mudrock, within which are thin sandstones, some with flute moulds on their bases. The majority of the sandstones in the amalgamated units exhibit parallel lamination but some sheets consist of lenticular, cross-stratified beds with common channeling. Transport was from the east-north-east. The age of the Middle Battery and Kilmore formations is uncertain; poorly preserved microfloras suggest that they are no younger than Kinderscoutian. Jones and Naylor (2003) interpreted the Whiddy Island succession as recording the progradation of a submarine fan (Middle Battery and Kilmore formations) across the basin floor (East Point Formation). The Middle Battery Formation was interpreted as having been deposited

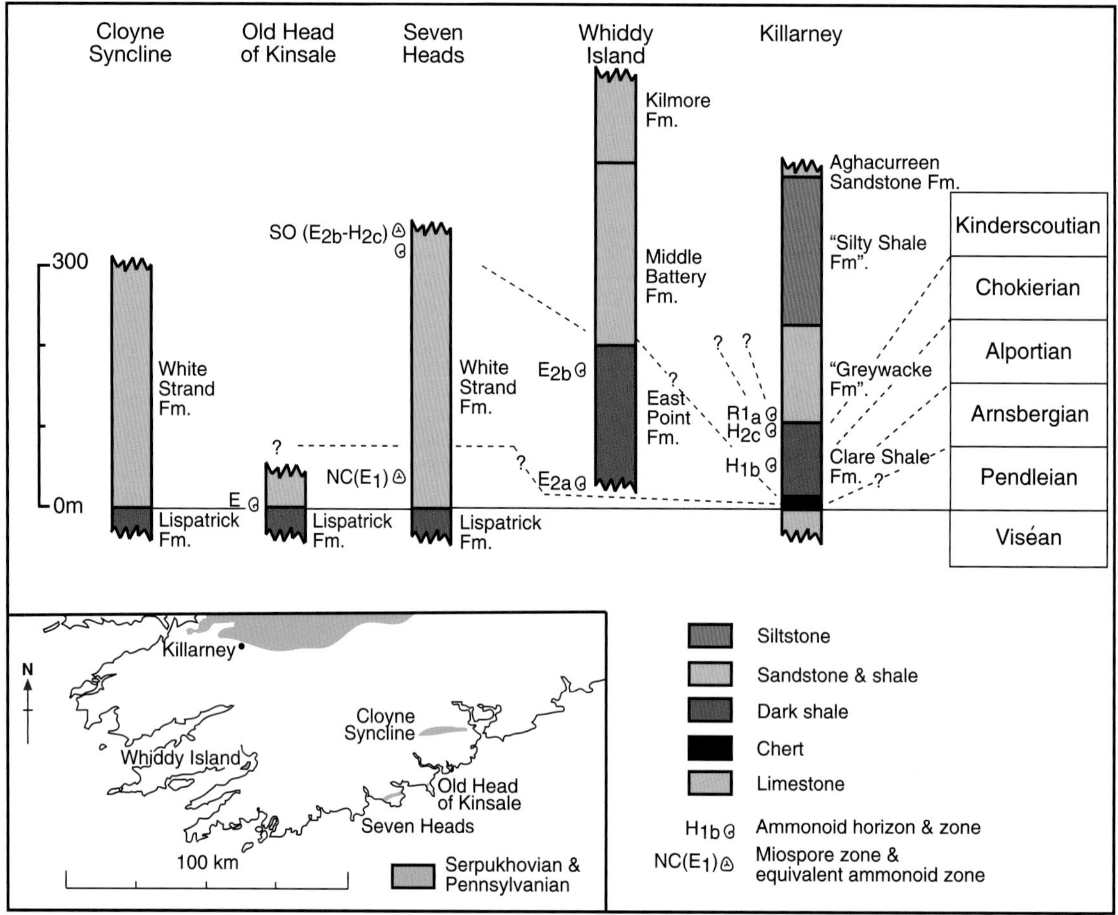

Fig. 11.4 Serpukhovian and Pennsylvanian stratigraphical successions in the South Munster Basin compared with that at Killarney.

on the fringe of a fan, and the Kilmore Formation near channel mouths and in inter-channel environments. The fine-grained character of the post-Viséan succession on Whiddy Island may reflect erosion of an uplifted area of fine-grained sandstone.

It is probable that the high level igneous intrusions on the Beara Peninsula described by Coe (1969) and Pracht and Kinnaird (1995, 1997) are of early Pennsylvanian age (Pracht and Timmerman, 2004; Quinn *et al.*, 2005), only slightly younger than the youngest preserved sedimentary rocks of the Kilmore Formation on Whiddy Island 40 kilometres to the east-south-east. The intrusions occur in two suites, in the axial zone and on the south limb, respectively, of the major anticline that forms the spine of the peninsula. The intrusions of the axial zone are subalkaline basalts, while those of the southern limb exhibit a fractionation trend from alkaline basalt to trachyte. Associated with the southern suite are zoned pipe-like bodies of breccia. The pipes consist of a central core of amphibole, phlogopite, and titaniferous magnetite megacrysts, and nodules of amphibole pyroxenite, which is surrounded by an outer shell consisting mostly of fragments of country rock. The volcanic pipes have been interpreted as having formed from extremely volatile-rich magmas that ascended rapidly from the mantle.

Rocks of Serpukhovian to Alportian age found at several localities in south County Cork between the Seven Heads and Cork Harbour have been assigned to the White Strand Formation. At its type section on the Old Head of Kinsale (Naylor *et al.*, 1985), 44 m of interbedded fine sandstone and mudstone are preserved in the core of the major regional syncline. Many of the sandstone beds are slumped. Close above the base of the formation is a horizon with *Eumorphoceras* sp., which, because of its proximity to the late Viséan levels in the underlying black shale of the Lispatrick Formation, is likely to be Pendleian in age.

Sleeman (1987) has described a poorly exposed but extensive development of the White Strand Formation in the core of the Cloyne syncline (Fig. 11.4). Scattered palynological records and rare occurrences of posidoniellid bivalves suggest that the succession ranges at least as high as the Arnsbergian (E_{2b}). However, the record of *Homoceras* (Ramsbottom *et al.*, 1978, p.43) cannot be substantiated. Interesting elements of the fauna are coelacanth fish, which were discovered in the last century. The thickest succession in south County Cork is on the west side of the Seven Heads (Naylor *et al.*, 1988). The White Strand Formation there (346 m preserved) consists of interbedded sandstone and dark mudrock, generally sandstone-rich but with a mudrock-rich member, 75 m thick, 80 m above the base. Palynological data suggest that the base of the formation is of Pendleian age, which is consistent with the Brigantian age of the top of the underlying Lispatrick Formation. Although a marine band occurs just below the youngest beds preserved, the ammonoids and bivalves it contains have proved to be indeterminate. Palynological evidence indicates that the youngest beds are at least as young as Chokierian and possibly range into the Alportian (Higgs, 1990). Detailed sedimentological studies of the formation have not been published. The sandstones are fine-grained and are commonly slumped. While they do not show many features of classical turbidites, there are rare flute moulds in the lower part of the formation, which indicate derivation from the south. Higgs (1990) recorded the consistent occurrence of reworked elements of early Tournaisian age in the miospore assemblages and suggested that the Kinsale Formation to the south might have been eroded and transported northward by turbidity currents during Serpukhovian and early Pennsylvanian time.

Higgs (1983) recorded miospores of early Serpukhovian age from 260 m of mudstone and siltstone encountered at the bottom of offshore well 58/3-1, approximately 100 km south of the south Cork coast (Fig. 11.2).

The rather meagre data suggest that during the early Pendleian, probably as the result of uplift and erosion of an area of early Tournaisian mudrock and fine-grained sandstone similar to the Kinsale Formation, there was an influx of sand from the south into the region that now forms the south coast of County Cork. It is possible that the sand facies prograded northward over a black mud floor to reach Bantry Bay by the Arnsbergian. Alternatively, the Bantry Sub-basin (p. 218) remained a separate entity and was supplied by fine sand from a distinct source, possibly the Sheep's Head High. (p. 223).

Shannon Basin and adjoining areas, including the Porcupine Basin

The largest area of Serpukhovian and Pennsylvanian rocks in the country (Fig. 11.1) extends from the coast of County Clare across the Shannon to Killarney, County Kerry, and to near Mallow, County Cork. Much of the detailed sedimentological information relating to this area is derived from the magnificent coastal exposures in west County Clare, but the pattern of facies and thickness changes has been established by careful mapping of much of the outcrop belt.

Some of the stratigraphical divisions used by Rider (1974) in west County Clare (Fig. 11.5) are recognisable throughout the area; it is proposed to use a modified version of his stratigraphical nomenclature where possible, even though some of the stratigraphical divisions had been named earlier elsewhere.

In outline, the succession records an overall shallowing-upward trend: from deep water, black shales and laterally equivalent turbiditic sandstones; through mudstone-rich basin floor and slope deposits (the Shannon Group); to shallower water cyclothemic deposits (the Central Clare Group) of deltaic origin. The pattern of subsidence evident in the late Viséan – a platform in north County Clare and a basin with its depocentre in the Shannon region – persisted, at least in terms of the relative thicknesses of the successions, into the Pennsylvanian. Despite the substantial amount of detailed investigation of the rocks of the Shannon Basin, which has also been called the West Clare Basin (Gill, 1979) and the Western Irish Namurian Basin (Martinsen, 1989), there is still debate about such fundamental aspects as its orientation, the location of its margins, and the provenance of the sediments that filled it. Different interpretations of the basin and its history have been given by Wignall and Best (2000, 2002) and Martinsen and Collinson (2002). The description below largely follows the views of Martinsen and Collinson.

The lower part of the Shannon Group (Fig. 11.5) consists of two formations of deep water origin, the Clare Shale Formation and the Ross Sandstone Formation; the latter has been shown by biostratigraphical methods to be laterally equivalent to part of the former.

Below the Ross Sandstone Formation in the west of the Shannon Basin, and replacing it to the east (Figs. 11.5, 11.6), is a unit mostly formed of black, carbonaceous shale, named by Hodson (1954) the Clare Shales (now the Clare Shale Formation). The shale contains discrete horizons with ammonoids and bivalves, but otherwise is generally barren of fossils. In most cases the ammonoids

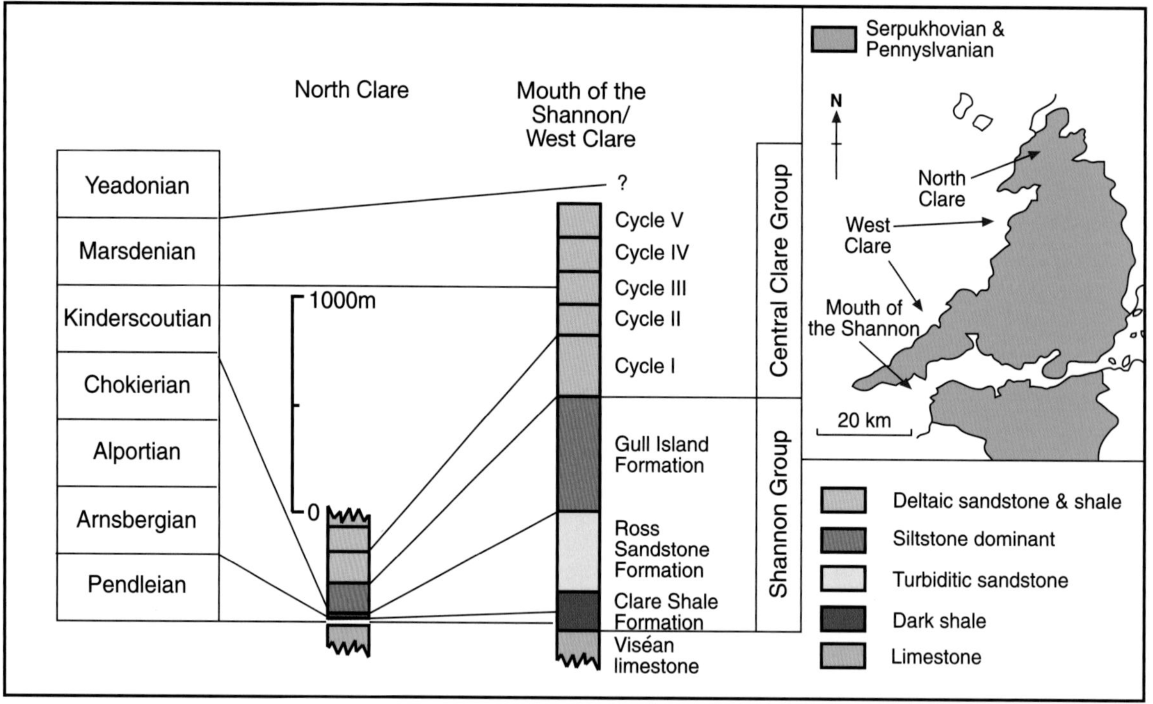

Fig. 11.5 Composite Serpukhovian and Pennsylvanian stratigraphical succession of west County Clare and around the mouth of the Shannon compared to that of north County Clare.

are preserved as flattened moulds, but at some horizons they occur in calcareous bullions, from which beautiful, three-dimensional specimens have been collected. The fauna was probably mostly pelagic; in addition to common thin-shelled bivalves and ammonoids, it includes nautiloids, conodonts, radiolaria, and fish teeth. Biota derived from the land include plant stems, that are interpreted to have floated until they became waterlogged and sank, and an insect (Monaghan, 1995). There is little evidence in the Clare Shale Formation of reduced salinity (Braithwaite, 1993). In addition to black shale, the formation also contains carbonate concretions (bullions), bedded, argillaceous, fine-grained limestone (close to the base), chertified mudrock with abundant sponge spicules (Lewarne, 1963), a facies which occurs in other areas of Ireland in the Arnsbergian, and concentrations of phosphate, particularly where the succession is condensed. As the Clare Shale Formation is traced away from the River Shannon, both to the north and the south, it shows systematic thinning and the development of condensed successions and non-sequences in its lower part. The northerly thinning is well illustrated by comparing the successions between Iniscorker (Hodson and Lewarne, 1961; Braithwaite, 1993), close to the Shannon, and Slieve Elva in north County Clare (Fig. 11.6). The lowest ammonoid horizon recorded at Iniscorker is that of *Tumulites pseudobilinguis* (E_{1b}), but it is probable that the lower, covered, part of the formation contains earlier Serpukhovian faunal horizons, as it does at Ballybunion, County Kerry, which is in an analogous axial position in the basin. At Magowna, the lower part of the Serpukhovian includes argillaceous, fine-grained limestone, containing *Cravenoceras leion* (E_{1a}) and calcareous foraminiferans, which rests on Brigantian platform limestones; the thickness of the Sepukhovian is reduced relative to Iniscorker. The cherty shales of the Arnsbergian thin and are overlapped northward (MacDermot *in* Gill, 1979, p. 10), so that around Lisdoonvarna and Slieve Elva, shales with a basal phosphorite lie on limestone. The oldest ammonoid horizon in this region is that of *Homoceras beyrichianum* (Chokierian; H_{1b}); it occurs less than 3 m above Viséan limestone. Thicker developments of phosphate between the *beyrichianum* horizons and the top of the limestone occur farther south around Lisdoonvarna and Roadford, where they were formerly exploited. They occur as thin, laterally extensive lenses of granular apatite and collophane, which, together with pyrite and carbon, are set in a cement of apatite and carbonate. The phosphates contain phosphatised limestone pebbles, granules of quartz, abundant fish remains and

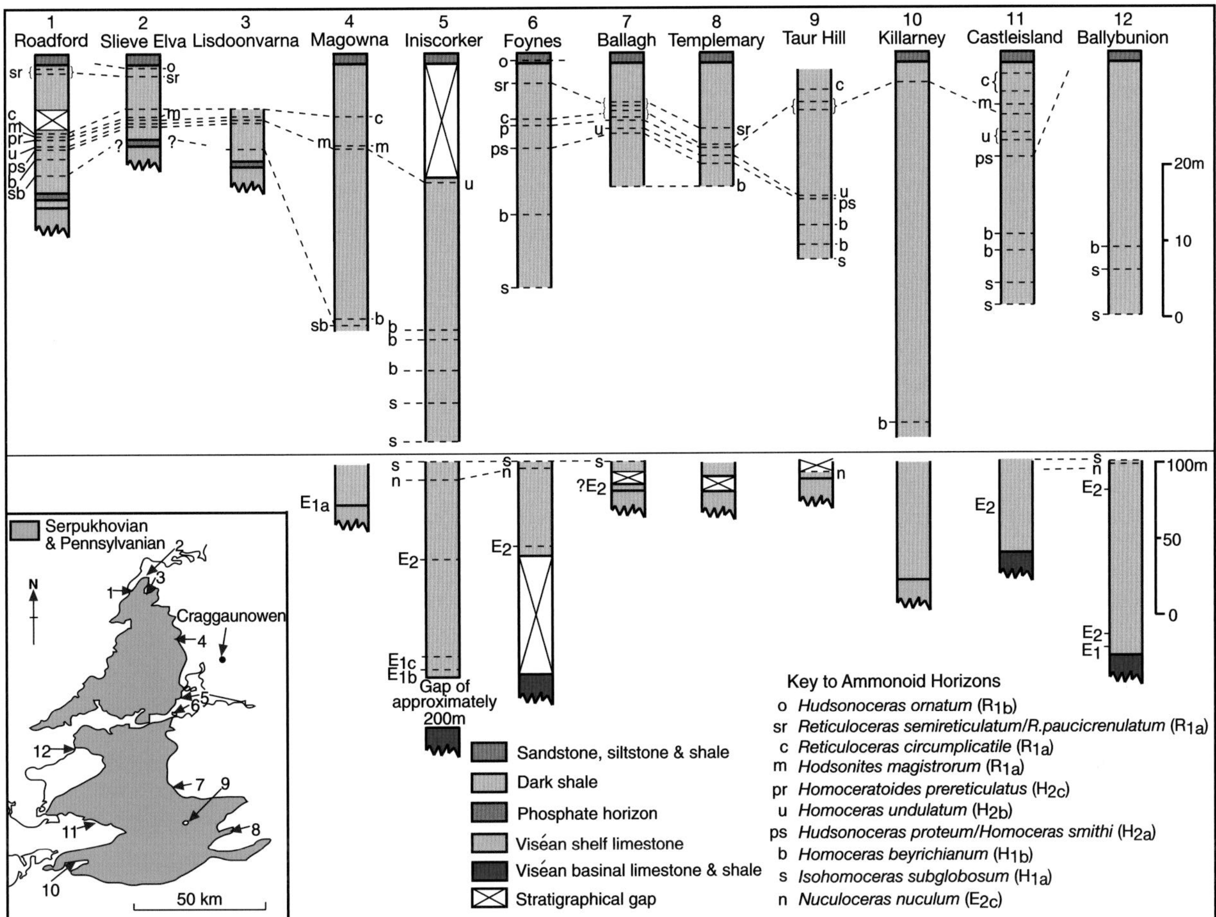

Fig. 11.6 Correlation of sections of the Clare Shale Formation in Counties Clare, Limerick, Kerry and Cork (Serpukhovian and Pennsylvanian outcrop in green). Note the difference of scale in the lower and upper part of the figure.

conodonts, and must reflect very slow sedimentation rates during the Arnsbergian.

The Clare Shale Formation is widely distributed south of the Shannon. In the axial region of the Shannon Basin at Ballybunion (Kelk, 1960) the formation is over 180 m thick and its base is of late Brigantian (P_2 ammonoid Biozone) age. It consists of a lower part (Brigantian–Arnsbergian), characterised by black shale, much of it siliceous and platy; and an upper part (Chokierian), formed of grey shale, with darker horizons containing ammonoid-bearing bullions. Turbiditic sandstone of the Ross Sandstone Formation enters the succession close to the Chokierian/Alportian boundary. As on the north side of the Shannon, the Clare Shale Formation thins away from the axial region of the basin; its top, marked by the incoming of turbiditic sandstone, becomes younger to the east. In the Castleisland area (Brennand, 1965), as at Ballybunion, the lower part of the formation consists of platy, siliceous shale; and the upper part, of black pyritic shale with several ammonoid horizons ranging from youngest Arnsbergian to early Kinderscoutian (R_{1a}: horizon of *Reticuloceras circumplicatile*). The lowest faunal horizon within the siliceous shale is of Arnsbergian (E_{2b}) age, but it is probable that much, if not all, of the Pendleian is represented. East of Castleisland, in the Taur inlier, the Clare Shale Formation has been reported by Morton (1965) to contain late Arnsbergian (E_{2c}) ammonoids close above Viséan limestone. Little is known of the formation farther south, except in the Killarney region (Walsh, 1967), where it is relatively thick and where its upper boundary is just above the *Homoceratoides prereticulatus* marine band.

Along the eastern outcrop of the Serpukhovian and Pennsylvanian south of the Shannon, the Clare Shale Formation is poorly exposed, but it clearly thins southward from Foynes where it is at least 100 m (and possibly as much as 200 m) thick. Near Ballagh, County Limerick, cherty shales that contain sponge spicules,

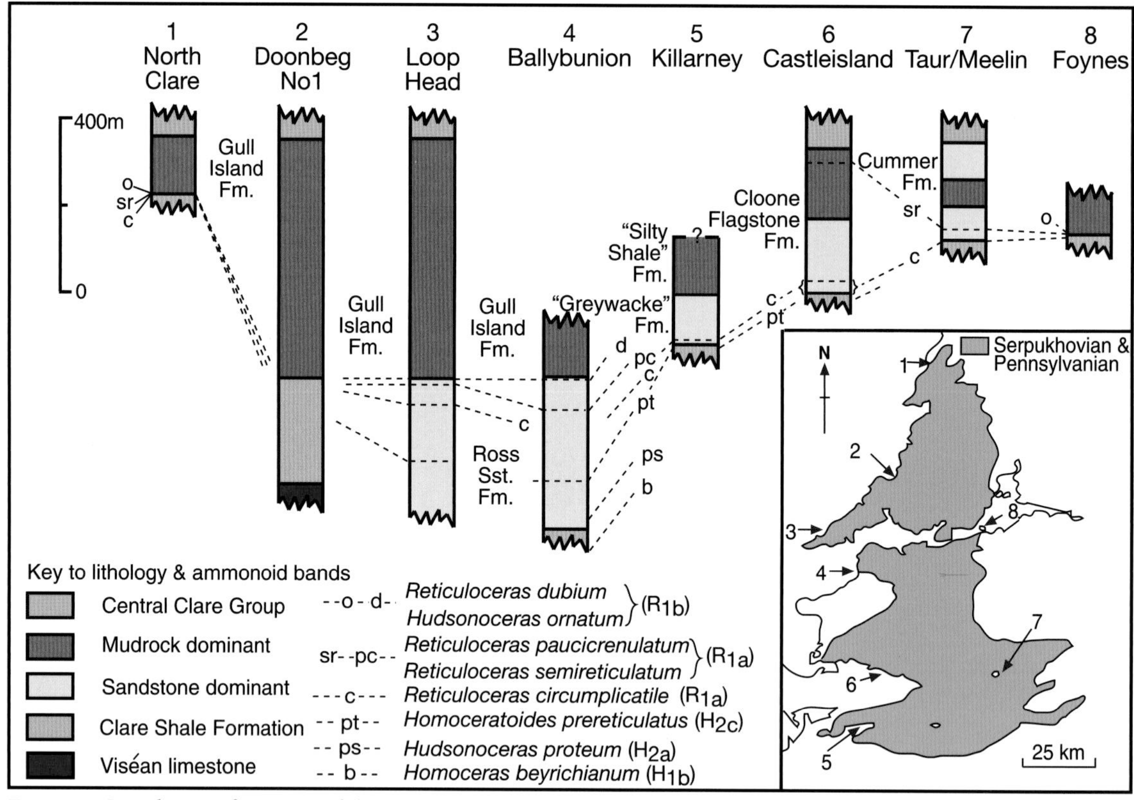

Fig. 11.7 Correlation of sections of the Ross Sandstone and Gull Island Formations and their equivalents south of the River Shannon.

of probable Arnsbergian age (Morton, 1965), rest on Viséan limestone; they are followed by black shales with ammonoid horizons, extending up to the early Kinderscoutian (level of *Reticuloceras subreticulatum*; R_{1a}), the whole succession being less than 40 m thick. According to Morton, the shales pinch out in very poorly exposed ground south of Ballagh, and reappear near Templemary, County Cork (Philcox, 1961). This area may represent the western tip of an extensive region to the east, which appears to have remained emergent until the Kinderscoutian (Fig. 11.3).

The Ross Sandstone Formation (Rider, 1974; Collinson *et al.*, 1991; Chapin *et al.*, 1994; Elliot, 2000 a, b; Lien *et al.*, 2003) has been the subject of intense research because it provides an analogue at outcrop for hydrocarbon reservoirs in some offshore, deep water basins. It was first described from the Loop Head area, where it contains three marine bands (Fig. 11.7), those of *Reticuloceras dubium* (R_{1a}) and *R. paucicrenulatum* (R_{1a}) at its top and *Homoceratoides prereticulatus* (H_{2c}) close above the oldest beds exposed (its base is not seen). The complete formation is exposed north of Ballybunion, where it is approximately 460 m thick and where *Homoceras smithi*, indicating the lower part of the Alportian, occurs close above the base. The lower 170 m at Ballybunion consists of fine-grained turbidites, which exhibit tabular bedding with no channels. In contrast, the upper part of the formation, which is particularly well displayed around Loop Head, consists of turbidites that are commonly either amalgamated to form the fills of channels, which are individually up to 10 m thick and 100 m wide, or arranged in thickening-upward packages. Lien *et al.* (2003) have interpreted the environment of deposition of the upper part of the formation as a deep water setting in which turbidites accumulated on a sea floor traversed by sinuous channel and associated spillover belts. As the channels filled, more and more turbidites spilled farther and farther overbank. Thickening-upward packages developed as a result of channel migration: initially muds and thin-bedded turbidites that were spilled from channels were deposited up to 1 km away; thick-bedded, amalgamated turbidites reflect proximity to the channel margins. Lien *et al.* (2003) suggested that the upper part of the formation represents the more proximal, and the lower part, with no channels, the more distal part of the turbidite system. Palaeoflow directions from the Ross Formation vary substantially, both geographically

Fig. 11.8 Sand volcanoes on top of the Ross Slide, Ross Sandstone Formation, Bridges of Ross, County Clare. Note feet in the background for scale. *Photograph by Dr J. R. Graham.*

and stratigraphically. The vector mean for the formation as a whole is approximately towards the north-east (Lien *et al.*, 2003). Directions derived from the amalgamated, thick-bedded facies show a much higher concentration around the modal direction than do those from the thin-bedded facies.

Slumping occurs at several horizons within the Ross Formation. The Ross Slump (Gill, 1979; Martinsen and Bakken, 1990; Strachan, 2002; Lien *et al.*, 2003) is the most spectacular example. It consists of several metres of siltstone and overlying sandstone, displaying a range of deformational structures, including recumbent folds, sedimentary cleavage, and imbricate thrusts. The sandstone unit in the upper part of the slump is largely dismembered in places. On its upper surface there are well-preserved, large sand volcanoes (Fig. 11.8). Strachan (2002) has shown that the direction of movement of the slump was to the north-east, rather than to the south-south-east as originally suggested by Gill (1979).

As currently interpreted, the Ross Formation is restricted to the western part of the Shannon Basin. It does not appear to be represented in the Doonbeg No.1 well (Sheridan, 1977), only 30 km north-east of Loop Head; nor, according to Rider (1974) and most later authors, is it present along the eastern margin of the Serpukhovian and Pennsylvanian outcrop in County Clare, where it is replaced by the Clare Shale Formation. It should be noted, however, that the definition of the upper limit of the Ross Formation in its type area is essentially biostratigraphical (the boundary with the overlying Gull Island Formation has been taken at the *Reticuloceras dubium* (R_{1a}) marine band). In contrast, along the eastern outcrop of the Clare Shale Formation the oldest sandstones are of R_{1b} age at their base.

The Ross Sandstone Formation is succeeded in south-west County Clare by the mud-rich, extensively slumped Gull Island Formation (Rider, 1974; Martinsen, 1989; Martinsen *et al.*, 2000, 2003). The formation is 550 m thick in the axial region of the basin, but only 130 m thick on the northern margin, where it was previously described by Hodson and Lewarne (1961) as the Ribbed Beds, the Cronagort Sandstone, and the Doonagore Shale. It has been informally divided into an upper and lower part by Martinsen *et al.* (2000, 2003). The lower part consists of beds of silty mudstone, 75% of which are deformed by slumping and sliding, and fine-grained, turbiditic sandstones, which show little syn-sedimentary deformation. Many of the examples of mass movement, including the spectacular Fisherstreet Slump, have been described by Gill (1979), Martinsen (1989), Martinsen and Bakken (1990), and Strachan and Alsop (2006). The slumped mudstones are interpreted as having accumulated on the basin floor at the foot of a slope that faced south-east to east-south-east (except at Fisherstreet where movement was to the north-east), while the turbidites flowed along the basin floor towards the east-north-east. Martinsen *et al.* (2003) have interpreted the thickness (420 m thick near the axis and 50 m on the northern margin) and facies relations (a greater proportion of sandstone near the axis) of the lower part of the formation as indicating progressive filling of the basin and onlap onto the basin margins through time. The upper part of the formation varies much less in thickness than the lower, contains very few sandstone beds, but exhibits progressive coarsening-upward and has a gradational contact with the overlying delta-related deposits of the Central Clare Group. Martinsen *et al.* (2003) interpreted it as reflecting the south-eastward progradation of the earliest delta of the Central Clare Group.

The onset of sandstone deposition in the region south of the Shannon can be demonstrated to have been progressively younger from west to east: close to the base of the Alportian at Ballybunion; within the early Kinderscoutian (horizon of *Reticuloceras circumplicatile*; R_{1a}) around Castleisland; slightly younger in the Kinderscoutian (horizon of *Reticuloceras subreticulatum*; R_{1a}) at Taur Hill; and in the mid Kinderscoutian (marine band containing *Hudsonoceras ornatum*; R_{1b}) at Foynes (Fig. 11.7). However, it is difficult to distinguish from descriptions in the literature, equivalents of the Ross Formation and the Gull Island Formation in the succession between the Clare Shale Formation and the Central Clare Group, except at Ballybunion, where Collinson *et al.* (1991, fig. 8) recorded that approximately 160 m of the Gull Island Formation is preserved (Fig. 11.7).

Walsh (1967) described the succession above the Clare Shale Formation near Killarney as consisting of a greywacke unit, some 120 m thick, with the *Reticuloceras circumplicatile* marine band in its lower part, a relatively thick overlying unit of silty shales, and a thin unit of sandstone. This succession bears some resemblance to that around Castleisland, County Kerry (Brennand, 1965), where the Cloone Flagstone Formation consists of a lower unit (Keam Flagstones) dominated by greywacke, a middle shaly unit (Lackabaun Shale), and an upper, relatively thin unit of turbiditic sandstone (Coombe Sandstone). The Keam Flagstones contain the R. *circumplicatile* marine band and probably equate lithostratigraphically with the Ross Sandstone Formation. Transport directions show a considerable spread with a vector mean between east and north-east. The Lackabaun Shale is similar to the Gull Island Formation, in that it is dominated by mudstone and shale and is extensively slumped. However, it is older than the development of the Gull Island Formation in west County Clare because near its top there is a marine band containing *Reticuloceras subreticulatum*, which is correlated with the *Reticuloceras paucicrenulatum* horizon that occurs close to the top of the Ross Formation in the Loop Head region. The Cummer Formation in north County Cork (Morton, 1965) shows a similar three-part division, with a lower and upper unit of turbidites, separated by a unit of slumped mudstone. However, in this easterly region the marine band with *Reticuloceras subreticulatum* is close to the base of the lower turbidite sequence.

The Central Clare Group in south-west County Clare (Rider, 1974; Pulham, 1989) is made up of five (and part of a sixth) large-scale, repeated sequences of mudstone, siltstone, and sandstone, and, in some cases, coal, which have been referred to as cyclothems (Fig. 11.10). The ideal cyclothem (Rider, 1974) was envisaged as consisting of a black shale with marine fossils, overlain in succession by siltstone, laminated sandstone, channel sandstone, and further small cycles of mudstone, siltstone, and sandstone. It was interpreted as the record of the advance of a delta into the basin. The channel sandstones near the top of the cyclothems were interpreted as representing the fluviatile distributary systems of the prograding deltas. The fluviatile sandstone bodies in the lower two cyclothems were found to be laterally extensive and were named the Tullig and Kilkee Sandstones (Fig. 11.10). The cyclothems of which they were interpreted to be part were given the same name as these prominent sandstones; three younger cycles were referred to as the Doonlicky Cyclothem, and Cyclothems IV and V. However, there has been a radical reinterpretation of the relationship of the laterally extensive fluvial sandstone bodies to the underlying deltaic sequences, which, if correct, necessitates a revision of the stratigraphical nomenclature. Elliott and Pulham (1990) and Davies and Elliott (1996) have proposed that the sandstones fill valleys incised in the underlying deltaic sequences (although this has been questioned by Wignall and Best (2000) in the case of the Tullig Sandstone) and, therefore, occur above major sequence boundaries, rather than below them. If the incised valley model is correct, the Tullig Sandstone would be at the base of the major sequence formerly referred to the Kilkee Cyclothem, and so on. There is a need to establish conventionally defined formations for the Central Clare Group; for present purposes, the major sequences will be identified by Roman numerals I to VI, as was originally done by Rider (1974). Sequence I (that below the Tullig Sandstone) has no marine band associated with it. Sequence II contains the *Reticuloceras* aff. *stubblefieldi* band (Kinderscoutian; R_{1b}) and Sequence VI, the *Bilinguites superbilinguis* band (Marsdenian; R_{2c}).

Pulham (1989), in his study of the magnificent exposures of the Central Clare Group in the almost continuous cliff sections of west County Clare, recognised three main types of depositional sequence within a cyclothem. These are arranged here in the order of deposition suggested by the studies of Elliott and Pulham (1990), Davies and Elliott (1996), and Hampson *et al.* (1997), and are interpreted following those authors' analyses. The first type is dominated by multistorey, channel sandstones, which exhibit unidirectional palaeocurrents and rest on deeply incised erosional surfaces. The sandstones are now interpreted as the fills of valleys, incised during low stands of sea level, that were deposited by rivers as sea level rose. Davies and Elliott (1996) have shown that sandstones in Cyclothem IV (Doonlicky Sandstone) filled palaeovalleys at the same time as palaeosols developed on the interfluves. The abandonment of the channel sandstones is marked in some areas by rootlet beds and thin coals, and in others by intense bioturbation. The trace fossils include *Zoophycos*, which indicates marine flooding of the abandoned fluvial system. Above this initial flooding horizon are coarsening-upward sequences, interpreted by Davies and Elliott (1996) as prograding shelf deltas, deposited as sea level continued to rise. In most of the cyclothems these are very thin, but in Cyclothem II they are several tens of metres thick. They are overlain by carbonaceous shales with ammonoids, the marine bands referred to above, which are condensed and mark the maximum rate of sea level rise. Above the marine bands are siltstone and sandstone sequences interpreted as the products of the progradation of shelf deltas during the period of high stand of sea level. The upper part of these sequences are laterally variable with prominent sandstone bodies, interpreted as mouth-bar sands, and more mud-rich facies that were deposited probably in lateral settings. The sandstones show the dominant influence of fluvial processes in their construction and the subsidiary influence of waves, which impinged on the delta from the east-south-east in each of Cyclothems I to III. The third type of sequence, which occurs normally in the upper parts of the major cycles, is interpreted as the fill of interdistributary bays. The sequences are typically 20 m or less thick and exhibit an upward transition from laminated mudstone, to laminated and cross-laminated siltstone, to sandstone, and rarely to thin coals.

Syn-sedimentary deformation is widespread within the Central Clare Group and is spectacularly displayed along the Atlantic coast of Clare (Rider,1978; Gill, 1979; Pulham, 1989; Wignall and Best, 2004). Large-scale deformation is confined to delta front sequences. In axial and marginal mouth bar settings, where sediment accumulation was rapid, mud diapirs and growth faults are common (Figs. 11.9, 11.11). Wignall and Best (2004) described a complex of at least thirteen growth faults affecting up to 60 m of strata in Cyclothem II exposed for some 3 km in the Cliffs of Moher and the coast to the south. Faulting started at the seaward (north-east) end of the complex; a single fault appeared to have been active at any one time; after it had ceased to move, a new fault was initiated to the south-west, resulting in a landward progression of faulting. Individual faults are listric and sole out into the top of the *R.* aff. *stubblefieldi* marine band, or, in the case of the younger faults, a higher level in the cyclothem. In the final stages of growth-fault movement, erosion of the crests of rollover structures resulted in the highest strata being restricted to the proximity of the fault. Pulham (1989) attributed the widespread occurrence of the syn-sedimentary deformation to the generally fine-grained character of the succession.

Pulham (1989) has discussed the palaeogeography of west County Clare during the deposition of the Central Clare Group. Some of his conclusions require modification in the light of the reinterpretation of the significance of the major fluviatile channel sandstone bodies. Cyclothem I represents the progradation of a substantial, fluvially dominated, lobate delta over the relatively deep-water, slope deposits of the Gull Island Formation. Pulham tentatively inferred a source of sediment in the west or north-west, although progradation of mouth bar sands was toward the east or north-east in sections along the Atlantic Coast. There are some regional differences (Fig. 11.11), which mainly reflect differences in subsidence rates. In the north, around Liscannor Bay, where the Clare Shale Formation is very thin, Cyclothem I is also thin (less than 100 m) and there is little syn-sedimentary deformation; in contrast, sections farther south are thick and syn-sedimentary deformation is common.

Fig. 11.9 Growth fault in Cyclothem IV, Central Clare Group, at Foohagh Point, County Clare. Note the thickening of the pale coloured sandstone unit towards the fault. *Photograph by Dr J. R. Graham.*

The Tullig Sandstone, which Pulham (1989) regarded as genetically linked to the underlying deltaic successions of Cyclothem I, was, according to Hampson *et al.* (1997), deposited in an incised valley as sea level rose in the early part of Cyclothem II by rivers flowing to the north-east. Abandonment facies above the sandstone are marine in the south-west and non-marine in the north-east. The continuing rise of sea level culminated in the deposition of two *Reticuloceras* aff. *stubblefieldi* (R_{1b}) marine bands, similar to many of the other marine bands in the Central Clare Group. They consist of black shale (and, in some places, a thin bed of limestone) containing ammonoids, nautiloids, bivalves, and crinoid stems. The mouth bars in the overlying delta front succession prograded to the east-south-east. Flaggy sandstones in the mouth bar sequences (Liscannor Flags), which have been widely used as paving and facing material, occur in the upper part of Cyclothem II above the Cliffs of Moher. Their surfaces are covered with spectacular examples of the trace fossil *Psammichnites*.

The Kilkee Sandstone, at the base of Cyclothem III, shows south-south-easterly directed palaeocurrents (Pulham, 1989) and is contained according to Hampson *et al.* (1997), in two discrete incised valleys, separated by an interfluve with palaeosols. The deltaic succession above the *Reticuloceras reticulatum* band (R_{1c}) is particularly affected by syndepositional faulting. The direction of progradation of mouth-bar sand bodies is to the south-east and the faults trend north-east.

Cyclothems IV, V, and VI have a restricted geographical distribution in west Clare. Rider (1974) stated that transport vectors measured within them are from directions with a pronounced westerly component.

Studies of the cyclothemic sequences south of the Shannon by Brennand (1965), in the Castleisland area, and by Morton (1965), near Newmarket in north-west

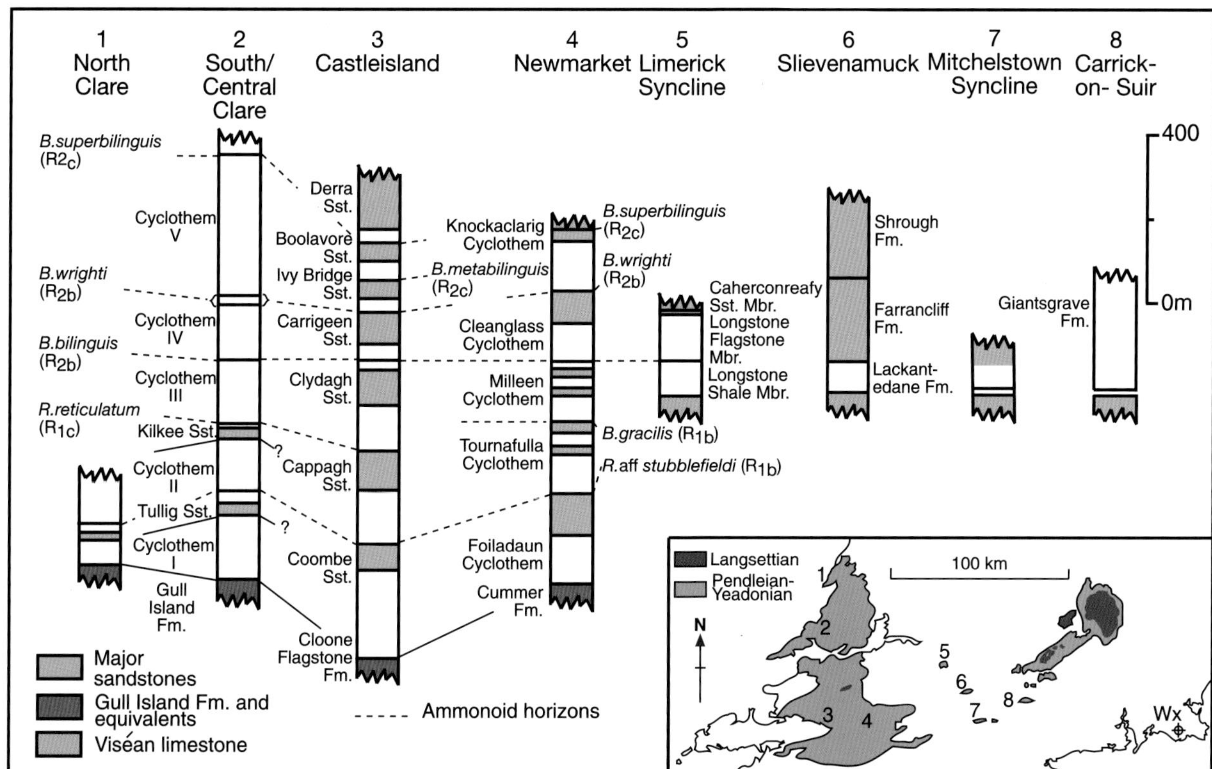

Fig. 11.10 Correlation of cyclothemic successions in Munster and the locality of subsurface Pennsylvanian in the Wexford Outlier (Wx). Shale dominant parts of the section are shown without ornament.

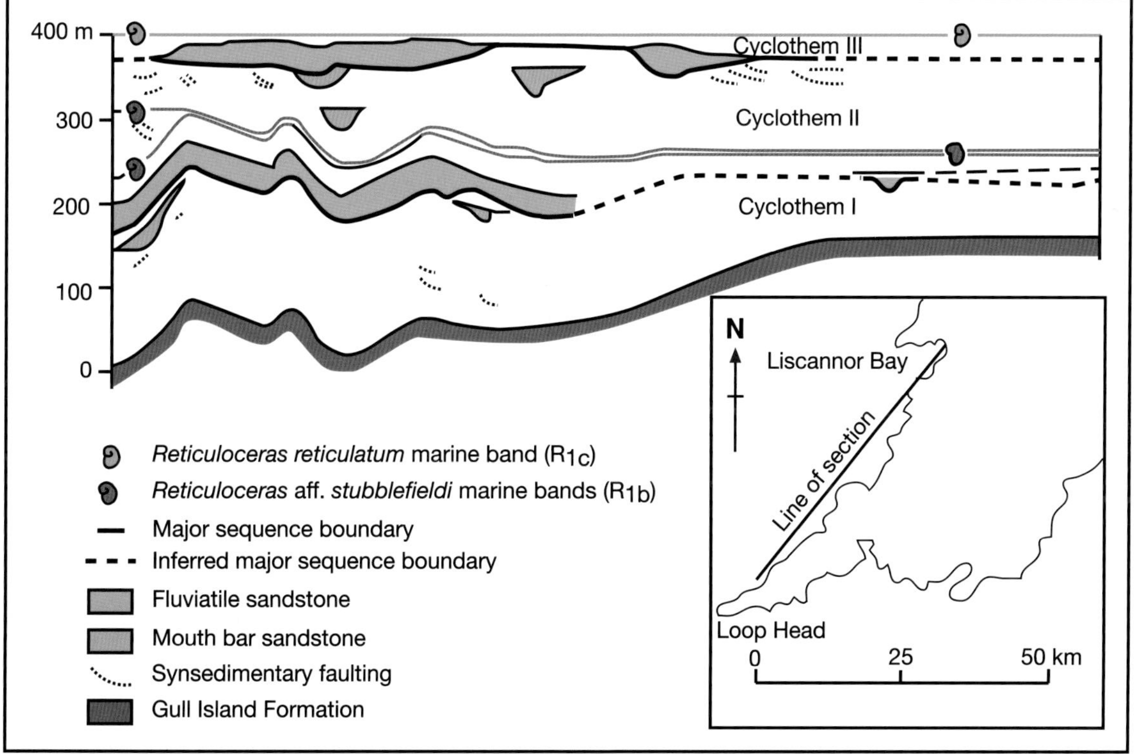

Fig. 11.11 Cross section of cyclothems I and II, Central Clare Group, along the Atlantic coast of County Clare. Adapted from Pulham (1989, fig. 17).

County Cork, have resulted in the correlation of individual cyclothems, through the identification of marine bands, with those in County Clare (Fig. 11.9). The general nature of the cyclothems is similar, but the quality of the exposures does not permit the same level of sophistication of interpretation as north of the Shannon. The following conclusions may be drawn. First, cyclothems up to the *B. superbilinguis* marine band (R_{2c}) are preserved in both areas. Secondly, the thickness of individual cyclothems varies relatively little, suggesting that differential subsidence within the basin had been greatly reduced. Thirdly, there are sandstones that are thick and apparently laterally extensive like the Tullig and Kilkee Sandstones in County Clare. It is not known, however, whether they fill incised valleys. Fourthly, very thin coals are widely developed. The direction of transport of the sediments has not been recorded in detail, but Brennand has shown that palaeoflow was generally to the east-south-east.

The youngest Pennsylvanian rocks in Munster occur in the Crataloe coalfield of north County Limerick and the Kanturk coalfield of County Cork (Figs. 11.1, 11.16). The former consists of a thin remnant extending to just above the level of Ward's Seam of the Leinster coalfield (Nevill, 1958b) (see p. 287). The *Gastrioceras subcrenatum* and *G. listeri* horizons are both represented by ammonoid-bearing shale. A thick sandstone, correlated with the Clay Gall Sandstone of the Leinster and Slieve Ardagh coalfields, occurs above the *G. listeri* horizon; the sand appears to have been transported from the south-west.

The Kanturk coalfield (Figs. 11.1, 11.16) is structurally complex and poorly exposed. Eagar (1974) has revised the correlations proposed by Nevill (1966). Two horizons of shale containing *Planolites opthalmoides* that lie between the Sweet and Rock veins (coals), have tentatively been correlated with the *G. subcrenatum* and *G. listeri* horizons. A thick sandstone that lies above the Rock Vein is probably the correlative of the Clay Gall Sandstone of the Leinster coalfield. A non-marine bivalve fauna from the highest beds proved has been assigned to the upper part of the *lenisulcata* Biozone (Langsettian).

As noted above, Wignall and Best (2000, 2002) presented a substantially different picture of the evolution of the Shannon basin from that outlined above. According to their analysis, the condensed sections of the Clare Shale in north County Clare were a result of that area being under the deepest water in the basin (rather than on a marginal high) during deposition of the younger part of the Clare Shale Formation. The Gull Island Formation was deposited on a north-easterly (rather than south-easterly) facing slope with the deepest water conditions in north County Clare. The progradation of the delta in Cyclothem I was also in a northerly direction. The Tullig Sandstone, rather than being the fill of an incised valley, represents the main distributary channels of the delta of Cyclothem 1, which, through continued avulsion, formed a thick continuous sheet in south County Clare. Their distal (downslope) equivalents in north County Clare were isolated channel sandstones.

The details of the architecture of the Shannon basin and the controls on its location are poorly understood and will remain so until seismic profiles across the basin are acquired. The basin was long-lived – the Shannon-Fergus estuary was the locus of enhanced, perhaps episodic subsidence from at least as early as the Hastarian (see p. 225). Several authors have suggested that subsidence was controlled by crustal extension, perhaps by reactivation of faults that were part of the Iapetus suture zone. Maximum subsidence during Pendleian to Kinderscoutian time was along an axis trending approximately east-north-east from Ballybunion, County Kerry through Killadysert, County Clare, with the greatest thickness of rock accumulating around the mouth of the Shannon. There is disagreement about the positions of the margins of the basin. The south-south-eastern margin is likely to have been south of the main outcrop belt. The south-western extension of the Leinster high (see p. 286) probably divided the eastern part of the basin into two. Most authors have interpreted the thinning of the Clare Shale Formation and Gull Island Formation in north County Clare as a reflection of the proximity of the north-north-western margin. Wignall and Best (2000, 2002), however, interpreted the thin successions as reflecting sediment starvation in the deepest part of the basin; by inference, the margin would have been much farther north.

Some further light on the nature of the north-north-western margin may emerge from inshore seismic surveys in Galway Bay and surrounding areas. Croker (1995) has interpreted existing seismic data as indicating a continuation of the post-Viséan rocks to the west of the coast of County Clare in the so-called 'Clare Basin'. Post-Viséan rocks have been encountered in two wells drilled at the western end of this putative basin, on the margin of the Mesozoic-Tertiary Porcupine Basin (Fig. 11.2), some 150 km west of the coast of County Clare (Robeson *et al.*, 1988). In well 26/26-1, 50 m of pale coloured sandstone, alternating with grey carbonaceous, glauconitic siltstones and black shales, have been tentatively dated as Serpukhovian. This unit rests on mudstones with a marine fauna, and is itself at least partly marine, because it contains acritarchs. In well 36/16-1, 340 m of micaceous, brown grey, silty claystone and subordinate argillaceous, fine sandstone were assigned a Yeadonian age. It is not clear how these rocks correlate with those onshore. It has been suggested that there is a widespread Serpukhovian/Pennsylvanian unconformity, offshore western Ireland. If there is, the effects of the event that caused it are not obvious onshore in County Clare.

The wells in the Porcupine Basin penetrated a full sequence of Langsettian to Stephanian rocks. Strata of Langsettian age have been reported only from the eastern margin of the basin (well 36/16-1), where 230 m of coal-bearing sandstones and shales are conformable on the Yeadonian, in contrast to some of the wells drilled at the north end of the basin, where younger strata are unconformable on pre-Langsettian rocks.

Duckmantian sequences have been identified in several wells. Well 36/16-1 proved 280 m of siltstones with thin coals. At the north end of the basin, the lowest 35 m of the succession drilled in well 26/28-1 consists of pebble and cobble conglomerates with clasts of granite and metamorphic rocks, suggesting that the well terminated close to an unconformity with pre-Carboniferous rocks. The conglomerates have been interpreted as the deposits of braided streams. The succeeding150 m of sandstone with subsidiary siltstone, shale, and thin coals represent the progradation and abandonment of the lobes of an alluvial delta. Two wells in the north of the basin (34/5-1 and 26/27-1b) proved grey measures, with several coal seams in the later part of the Duckmantian.

On the west side of the basin (well 34/5-1), there is a thick succession (458 m) of Bolsovian grey measures, containing 22 coal seams. At the north end of the basin (well 26/27-1b) and on its eastern margin (well 36/16-1), the Bolsovian succession is thinner but of generally the same facies. However, at the north-east end of the basin (well 26/30-1), there is a coal-poor succession that unconformably overlies granite. Sandstones, claystones, and shales, some of them red, predominate; coals, up to a metre thick, occur only close to the boundary with the Asturian.

Asturian sections show an upward transition from paralic coal deposits to red beds. 427 m of grey measures with thin coals have been proved on the eastern margin of the basin (well 36/16-1). On the west side of the basin (well 34/5-1) and in the north-west (well 26/27-1b), where the Asturian successions are truncated by Mesozoic unconformity surfaces, the sections are dominantly grey with thin coals, but are reddened at their tops. At the north end of the basin several wells (for example, 26/28-2) encountered coal-poor successions, with sporadic development of red beds.

Thin successions of Stephanian age have been identified in wells 34/5-1, 26/28-1, and BP 26/28-2. They include grey and red-brown mudstones, sandstones, and traces of coal. Other wells intersected thicker successions of varied shales and sandstones, which may be of Stephanian or Autunian age, but which are poorly dated.

East of the main crop of the Clare Shale Formation in Munster, there are several outliers which provide at least tenuous evidence of palaeogeography. In the core of the East Clare syncline, near Craggaunowen (Fig. 11.6), there is a short section of Clare Shale Formation, probably of Arnsbergian age, resting on Brigantian limestone (Lewis, 1986). This occurrence is interesting because the shales occur close to where the axis of the Shannon Basin might be projected. The suggestion that the earliest Serpukhovian is missing is consistent with closure of the basin in a north-eastward direction.

Pennsylvanian rocks are preserved in the core of the Limerick syncline (Fig. 11.10). They were described by Shelford (1967), whose work has been revised by Strogen (1988). All the rocks have been assigned to the Longstone Formation. The lowermost Longstone Shale Member (75 to 80 m thick) rests on karstified limestone at one locality, and on weathered volcanic material at two others. It consists predominantly of olive, buff-weathering shale and flaggy mudstone, with rare thin beds of sandstone near its top. Ammonoids indicating the *B. bilinguis* Zone (R_{2b}) have been found in two thin black shales high in the member. The overlying Longstone Flagstone Member consists of 105 m of flaggy siltstones with common trace fossils, including *Psammichnites*, similar to those of the Liscannor Flags of County Clare. The uppermost Caherconreafy Sandstone Member consists of 6 m of calcareous sandstone. This outlier is particularly significant in demonstrating that not only was there no deposition until the Marsdenian, but that the underlying Viséan rocks were tilted and subaerially weathered prior to the deposition of clastic sediments. On the north side of Slievenamuck (Fig. 11.10), Shelford (1963) described grey mudstone lying on a surface of cherty limestone with solution hollows, taken to be the base of an approximately 550 m thick succession of mudstone, shale, and sandstone, in which fossils have not been found. Because no black shale similar to the Clare Shales are present, it seems unlikely that rocks of Pendleian to Kinderscoutian age are present. The same situation obtains in two small outliers of presumed Pennsylvanian age in the Mitchelstown syncline (Shearley, 1988) and in the Carrick-on-Suir syncline (Keeley, 1983). Although a stratigraphical contact between the Viséan and the Pennsylvanian is not exposed in either case, grey mudrock crops out close to Viséan limestone, leaving little room for any black shale. The Giantsgrave Formation, north of Clonmel (Keeley, 1983), consists of 300 m of dominantly grey mudstone and siltstone, with yellow sandstone appearing towards the top of the succession.

Although the evidence from south-east and south central Ireland is incomplete, it supports Hodson's (1959) contention that a large part of the region was emergent through the Serpukhovian (Fig. 11.3). Sevastopulo (1981b) referred to this area as the Leinster High.

Leinster and Slieve Ardagh coalfields

The second largest region of Serpukhovian and Pennsylvanian rocks in the country consists of the Leinster and Slieve Ardagh coalfields and neighbouring outliers (Figs. 11.12, 11.16). Because the Leinster coalfield is better known, through the work of Eagar (1974), Higgs (1986), Higgs and O'Connor (2005), and Nevill (1956), it is described first. The oldest rocks assigned to the Serpukhovian are 20 m of thin-bedded, black, spicular cherts and shales (the Coonbeg Chert Member of the Luggacurren Shale Formation). They have been tentatively assigned by Higgs (1986) to the Arnsbergian because of their lithological similarity to Arnsbergian (E_{2b}) cherts in County Limerick. They apparently have a disconformable relationship with underlying limestone of late Viséan age. They are overlain by grey to black shales (the Glenview Shale Member), which have yielded an Arnsbergian fauna of *Dimorphoceras* sp. and bivalves. The third member of the Luggacurren Formation, the Black Trench Limestone Member (approximately 10 m thick), provides excellent biostratigraphical information: thin limestones, interbedded with dark shales, contain the *Nuculoceras nuculum* (E_{2b}), *Isohomoceras subglobosum* (H_{1a}), and *Homoceras beyrichianum* (H_{1b}) horizons; above a few metres of barren shales containing ironstone nodules, younger limestones contain most of the faunal horizons from the *Hudsonoceras proteum* (H_{2a}) to *Reticuloceras circumplicatile* (R_{1a}) bands. The Knockanhonagh Shale Member consists of 22 m of black and dark grey shale with marine bands ranging from *Reticuloceras circumplicatile* (R_{1a}) to *R. stubblefieldi* (R_{1b}) and *R. reticulatum* (R_{1c}). The overlying Killeshin Siltstone Formation (275 m thick) consists mainly of argillaceous siltstone. Its base lies within the Kinderscoutian (R_{1c}) in the south of the coalfield but within the Marsdenian (*R. gracile* horizon; R_{2a}) in the north; its top lies just above the *Cancelloceras cancellatum* (G_{1a}) Marine Band. In the south and particularly in the south-east of the coalfield, at about the horizon of *Bilinguites bilinguis* (R_{2b}), there is a distinctive unit, 30 m thick, of flaggy sandstones with conspicuous *Psammichnites*, called the Carlow Flagstone Member. Northward this changes to sporadic thin sandstone beds within the Killeshin Formation. The Bregaun Flagstone Formation, 50 m of flaggy sandstone

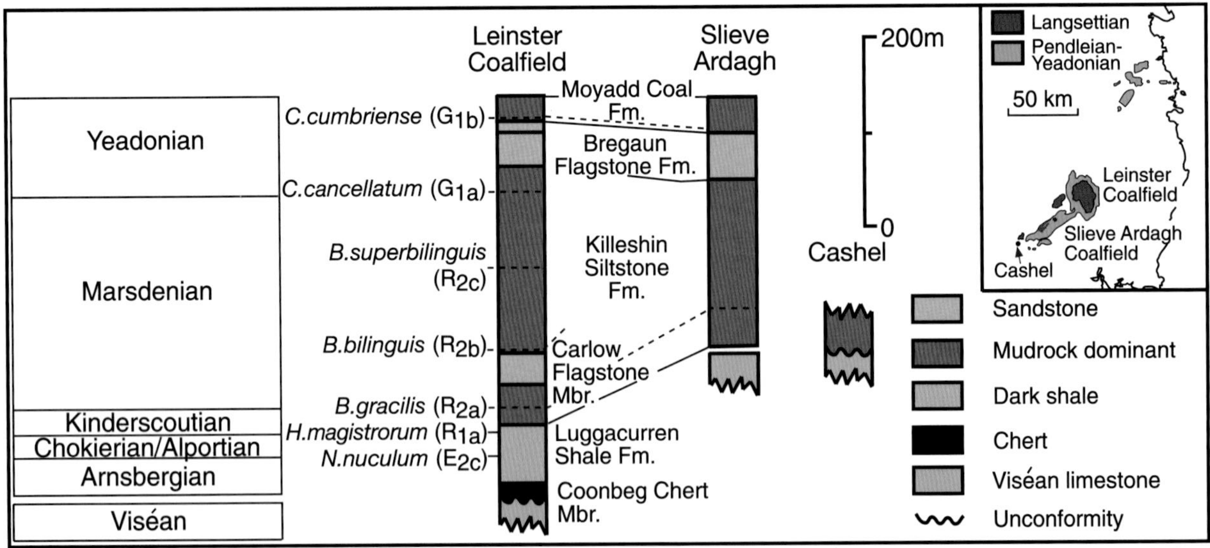

Fig. 11.12 Correlation of Serpukhovian and Pennsylvanian (Chokierian to Yeadonian) successions of the Leinster and Slieve Ardagh coalfields and the Pennsylvanian section at Cashel.

and siltstone with subordinate shale, provides a link with the Slieve Ardagh coalfield to the south, where it was first described.

The overlying Moyadd Coal Formation is divided into two members; the lower Rockafoil Member is 25 m thick and is of Yeadonian age. It has at its base the lowest workable coal of the coalfield, the No.1, or Rockafoil Coal, which is up to 0.25 m thick in the east but thins westward to less than 60 mm. The roof to the coal is black shale containing the *Cancelloceras cumbriense* marine band. Nevill (1956) has documented the north to south faunal changes in this marine band: in the northern half of the coalfield, the fauna is nektonic with ammonoids and pectinoid bivalves, whereas in the south, it contains brachiopods with only rare ammonoids. Above the marine band are siltstones and shales, with commonly a second thin coal, whose roof shale contains non-marine bivalves. The upper part of the Moyadd Coal Formation is of Langsettian age. The Rockafoil Member is overlain by the lithologically distinctive Fleck Rock, a bioturbated, grey, argillaceous siltstone. In the northern part of the coalfield the Fleck Rock contains common ammonoids, including *G. subcrenatum*, whereas farther south the fauna is dominated by the brachiopod *Martinia*; this trend from deeper water environments in the north to shallower water in the south mirrors the situation in the *Cancelloceras cumbriense* marine band, described above.

The marine conditions of the *G. subcrenatum* horizon were succeeded by paralic environments, in which the moderately thick Skehana and Marine Band coals were formed. The youngest widespread marine transgression is represented by dark shale containing *Gastrioceras listeri* and bivalves. The uppermost 15 m of the Moyadd Formation is a distinctive unit of hard black siltstone, named the Black Rock Siltstone Member.

The Moyadd Formation is overlain by the Clay-Gall Sandstone Formation, a sandstone conspicuously coarse-grained and more feldspathic than most others in the Pennsylvanian of southern Ireland. Its name is derived from pebbles of mudstone (clay-galls), derived from the underlying strata, which characterise the lower part of the formation. In the northern part of the Leinster coalfield the sandstone, which is up to 50 m thick, lies well above the *G. listeri* marine band, but farther south it cuts down into rocks of the *G. subcrenatum* cycle. No detailed studies have been made of the sedimentology of the sandstone, which is poorly exposed. Although marine bivalves and productoid brachiopods have been recorded from near its base, it is likely that most of the unit is of non-marine origin. The inclination of foresets in cross-stratified units suggests that it was deposited from currents flowing towards the north, or north-east.

The remainder of the succession (approximately 170 m preserved) has been named the Coolbaun Coal Formation. At its base is Ward's Seam, a shaly coal, in most places 0.2–0.3 m thick. It is overlain by 25–35 m of argillaceous strata that show evidence of fluctuations of the depositional environment, with horizons rich in non-marine bivalves of the *Carbonicola lenisulcata* Biozone, interspersed with marine bands. Of these, the lowest, just above Ward's Seam, has yielded *Lingula*, while the others contain the burrow *Planolites* and estheriid conchostracans. The succeeding Swan Sandstone Member, up to 28 m thick, is lithologically similar to the Clay-Gall Sandstone Formation but lacks the clay galls. Approximately 15 m of shale and siltstone containing non-marine bivalves occur between the Swan Member and the distinctive Double Fireclay Member. The latter consists of two thick beds of grey fireclay, separated by grey mudrock, which contains the Fairy Mount marine band, with ammonoids and chonetid brachiopods (Eagar, 1974). The strata above the Double Fireclay consist of 65 m of dark shale with a *Lingula* band and horizons with non-marine bivalves indicative of the *Carbonicola communis* Biozone near the base, followed by siltstone and several thin sandstones. Higgs and O'Connor (2005) placed the boundary between the *Triquitrites sinani-Cirratriradites saturni* (SS) and the *Radiizonates aligerens* (RA) Miospore Biozones just above the Double Fireclay Member.

The succeeding Jarrow Coal was an economically important, but areally restricted anthracite seam, 0.23–0.31 m thick. In the Jarrow colliery, shales immediately above the coal yielded a diverse fauna of vertebrates, including eight genera of amphibians and several fish. Other elements of the biota included numerous plants, amongst them *Neuropteris* and *Sphenophyllum*, and the small xiphosuran *Belinurus* (also recorded from shales above Ward's Seam). The thickest coal of the whole succession, the Old Three Foot Seam (averaging 0.91 m thick), is separated from the Jarrow Seam by 40–50 m of shale with a few levels containing non-marine bivalves indicating the *communis* Biozone. The uppermost 40 m of the succession are poorly known but include three thin coal seams. The youngest beds have yielded miospores assigned to the RA Miospore Biozone and non-marine bivalves that are older than the *Athroconaia modiolaris* Biozone, which spans the Langsettian/Duckmantian boundary.

The succession in the Slieve Ardagh coalfield (Figs. 11.12, 11.16) is thought to be similar to that of the Leinster coalfield, but biostratigraphical information is sparse and the upper part is poorly known. Nevill (1957a) argued that although the contact with the underlying Viséan limestone is not exposed anywhere around the coalfield, there is insufficient space between the limestone and the oldest siliciclastic beds seen to accommodate the Arnsbergian to Kinderscoutian black shale succession present in the northern end of the Leinster coalfield. A section described by Carruthers (1985) from an outlier just south of Cashel (Fig. 11.12) where grey, silty shales rest directly on limestone of Asbian age, is consistent with this suggestion. Higgs (1986) has proposed that the lowest 244 m of dominantly grey, sandy shale in Slieve Ardagh should be assigned to the Killeshin Formation, and Archer *et al.* (1996) identified the Carlow Flagstone Member through the north-eastern half of the coalfield. Higgs (1986) recorded an ammonoid fauna with *Bilinguites gracilis* (R_{2a}) from a horizon low within the Killeshin Formation, and Archer *et al.* (1996) referred to a record of an R_{1b} marine band in the south of the coalfield. The Bregaun Flagstone Formation, recognised in both coalfields, contains small sand volcanoes and slumps, which indicate a palaeoslope toward the north or north-east (Nevill, 1957a). The equivalent of the Rockafoil Member of the Moyadd Coal Formation in Slieve Ardagh is dominated by grey shale and probably includes two thin coals, the lower capped by the *C. cumbriense* marine band, which contains ammonoids. The equivalent of the Fleck Rock is a siltstone containing only *Lingula*, consistent with the north–south offshore–onshore gradient inferred in the Leinster Coalfield at this horizon. The lowest coal worked in the coalfield, the 0.25 m thick Glengoole Coal, is capped by the *G. listeri* marine band, which contains ammonoids. The Clay-Gall Sandstone Formation, approximately 15 m thick (Nevill, 1957a), is present in the north of the coalfield but apparently absent in the south-east. It is capped by the most important coal of the coalfield, the 0.5 m thick Upper Glengoole coal. The younger strata of the coalfield are poorly known but appear more sandy than the Coolbaun Formation of the Leinster coalfield. The correlations of the younger coals with those of the Leinster coalfield (Fig. 11.16) are tentative. However, bivalves indicative of the *communis* Biozone first occur just above Pat Maher's Seam of Slieve Ardagh.

In summary, the Leinster and Slieve Ardagh coalfields, which lay on the Leinster High, apparently remained emergent until the Arnsbergian (Fig. 11.3), when marine conditions spread as far south as the southern part of the Leinster coalfield (Higgs, 1986), but probably not to the Slieve Ardagh area. Deposition continued in the north of the Leinster coalfield through the Chokierian to Kinderscoutian, but it is not clear whether this was the case farther south. Deposition in the Slieve Ardagh area began in the latest Kinderscoutian or the Marsdenian. There is very little sedimentological interpretation of the Serpukhovian and early Pennsylvanian successions in eastern Ireland apart from Nevill's (1957a) record that cross-stratification suggested that most of the sandstones were transported from the south-west. Higgs and O'Connor (2005) recorded reworked acritarchs, including one Silurian taxon, and late Devonian spores from horizons in the Langsettian of the Leinster coalfield, and concluded that they were derived from the south Midlands area. An alternative hypothesis is that the source of the exotic palynomorphs was uplifted regions south of the South Munster Basin.

Onshore and offshore east of Ireland

Clayton *et al.* (1986b) described concealed Serpukhovian and Pennsylvanian strata from boreholes drilled in south County Wexford (Fig.11.1). There, the Park Formation, of which 32 m of grey mudrock with horizons of fine sandstone were recorded, rests on a karstified surface of limestone of Asbian age. It has been tentatively assigned a Serpukhovian or early Pennsylvanian age. A second borehole penetrated Permo-Triassic conglomerates and sandstones and proved 38 m of sandstone and mudrock, including a thin bituminous coal. The strata were named the Richfield Formation and contained miospores of Asturian age. In addition to the indigenous spores, a small number of spores reworked from rocks of Devonian–early Carboniferous, late Viséan–Serpukhovian, and Langsettian age were identified.

Pennsylvanian rocks have been encountered in wells drilled on the margins of, and within the Central Irish Sea Basin and the Kish Bank Basin (Fig. 11. 2) but, up to the present, rocks of Serpukhovian age have not been discovered. In well 42/17-1, in St George's Channel, Pennsylvanian strata rest directly on Viséan rocks. In well 42/12-1, drilling terminated in Pennsylvanian strata and interpretation of seismic data suggests that the Serpukhovian and earliest Pennsylvanian is likely to be absent (Maddox *et al.*, 1995). Naylor *et al.* (1993) have revised the stratigraphy of well 33/22-1, drilled on the south-eastern rim of the Kish Bank Basin, a few kilometres south-east of Dublin. The well proved 740 m of Duckmantian to Asturian strata resting unconformably on Lower Palaeozoic slates. The succession contains numerous coals, particularly within the Bolsovian. The upper 114 m of the succession contains red shales. Reworked miospores of Langsettian age in a sample of Asturian age are evidence of intra-Pennsylvanian uplift.

The simplest explanation of these records is that during the Serpukhovian there was a landmass in the southern Irish Sea. Alternatively, Serpukhovian rocks were deposited, but stripped during intra-Carboniferous uplift; or marine Serpukhovian strata were deposited in a gulf confined to the deeper (and undrilled) parts of the Central Irish Sea Basin, which was connected to the sea to the north. Pennsylvanian strata probably occur widely in the Irish Sea, but there is, as yet, little published information about their extent and stratigraphy.

Dublin Basin and surrounding areas

North of the Leinster coalfield there are Serpukhovian and earliest Pennsylvanian outliers in north County Dublin, and adjacent parts of County Meath, the remnants of the post-Mississippian fill of the Dublin Basin. Of these, the outlier near Summerhill, County Meath (Nevill, 1957b), while very poorly exposed, contains the youngest Carboniferous rocks preserved in the basin (Fig. 11.13). In contrast to the situation in the Leinster coalfield (the Leinster High), the base of the Pendleian, identified by the occurrence of *Cravenoceras leion*, is within a sequence of basinal limestone and shale, described by Nevill from a quarry near Summerhill. There, beds of turbiditic sandstone occur 4.5 m above the base. No younger sections within the Pendleian are exposed, but Nevill believed that the rocks of this age were probably

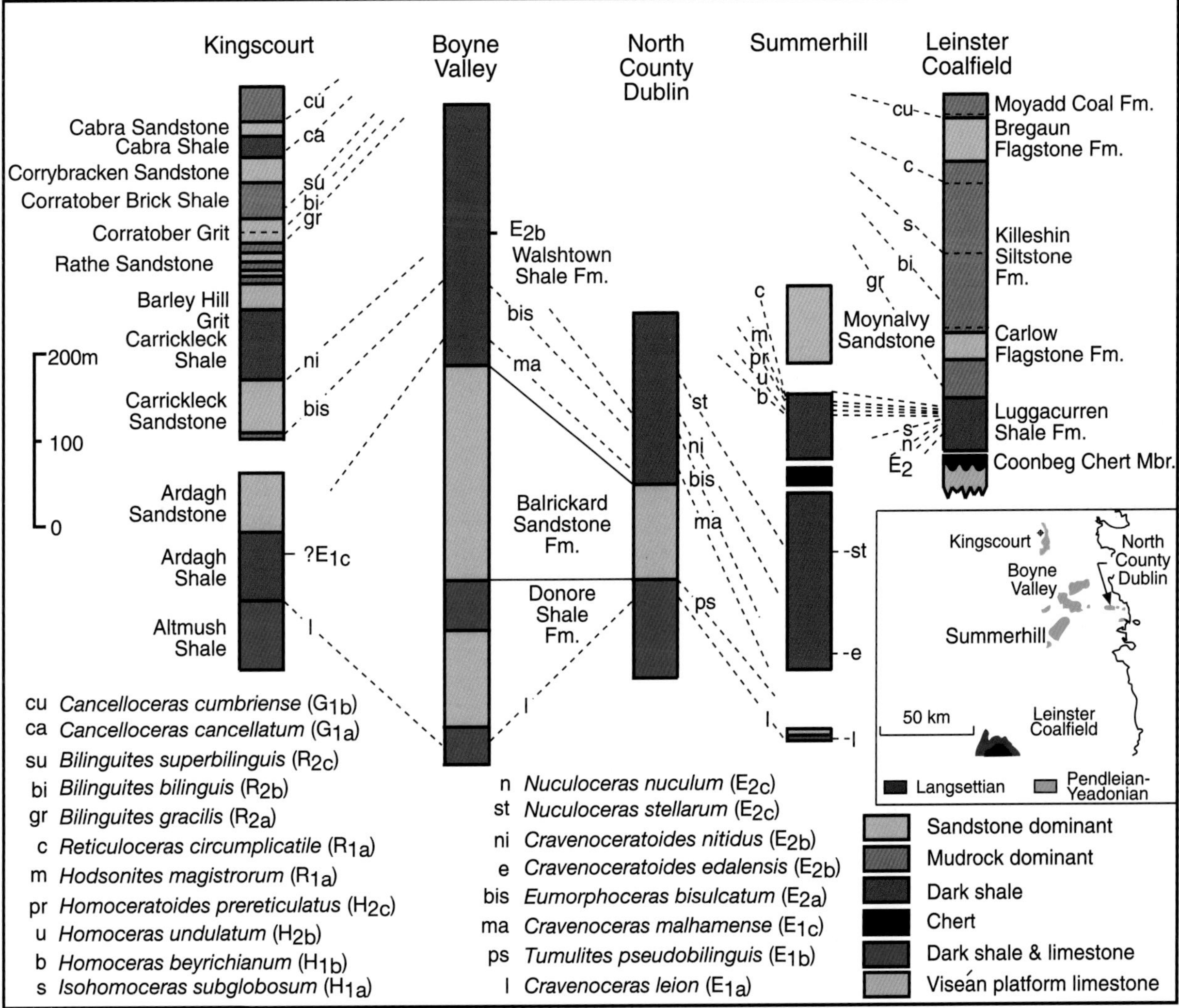

Fig. 11.13 Correlations of Serpukhovian and Pennsylvanian (Chokierian to Yeadonian) successions of the Dublin Basin with those of the north end of the Leinster coalfield and of the Kingscourt outlier.

not very thick. The Arnsbergian (E_{2b}) has been identified in a thick succession (200 m seen) of dark shales, with subordinate thin beds of fine sandstone. Nevill estimated that the combined thickness of the Pendleian and Arnsbergian stages might be as much as 600 m. The next youngest strata are 30 m of black shales with *Anthracoceras* and bivalves, succeeded by black shales containing ironstone nodules, which have yielded many of the ammonoid bands from *Homoceras beyrichianum* (H_{1b}) to *Reticuloceras circumplicatile* (R_{1a}). The youngest Moynalvy Sandstones consist of fine-grained sandstone and is of unknown age.

The succession in north County Dublin (Fig. 11.13) was described by Smyth (1950), and revised by Harrison (1968) and by Nolan (1986, 1989). Harrison also described the outliers in the vicinity of Navan, County Meath, later studied by Rees (1987). The base of the Serpukhovian lies within a succession of basinal shales and thin limestones, the Donore Shale Formation. In all the outliers, there is a prominent unit of sandstone, the Balrickard Sandstone Formation, which Nolan interpreted as being probably deltaic in origin. Smyth (1950) mistakenly believed that the sandstone was of Arnsbergian age and suggested that there was an unconformity in north County Dublin. However, Harrison showed that the sandstone was underlain by a marine band of E_{1b} age. The overlying Walshestown Shale Formation, consisting mainly of shale, ranges in age from late Pendleian (E_{1c}) to Arnsbergian (E_{2b}).

The Serpukhovian and Pennsylvanian of the Kingscourt area, County Cavan (Figs. 11.13, 11.16) were described by Jackson (1965). The Pendleian Ardagh Shale, dark ammonoid-bearing shales with clay-ironstone bands, is conformable on the Viséan in the south of the outlier, but is cut out and overlapped to the north, where the Ardagh Sandstone, a grey micaceous sandstone, rests directly on Viséan limestone. Farther north, the Ardagh Sandstone is also cut out. The younger, pebbly Carrickleck Sandstone, which probably is of fluviatile origin and which, on the evidence of ammonoids below and above it, is of Arnsbergian (E_{2a}) age, comes to rest on the limestone. This progressive northward overlap, which, in conjunction with northward thinning, results in the attenuation of the Pendleian–Yeadonian succession from about 600 m in the south to less than 350 m in the north, reflects the position of the outlier straddling the northern margin of the Dublin Basin. The Carrickleck Sandstone is succeeded by the ammonoid-bearing Carrickleck Shale (E_{2b}). Above the Carrickleck Shale are two units of sandstone, the Lower and Upper Barley Hill Grits, separated by a shale containing *Lingula*. These are followed by a succession, approximately 35 m thick, of shale, some of it with coal streaks, siltstone, and two prominent sandstones, the Rathe Sandstones. Jackson suggested, by comparison with sequences in the north of England, that the Barley Hill Grits may be correlated with the Chokierian and Alportian; and the overlying succession containing the Rathe Sandstones, with the lower part of the Kinderscoutian. Above the Rathe Sandstone are grey and black shales with *Reticuloceras reticulatum* (R_{1c}), the lower part of a cyclothem capped by a sandstone, the Clontrain Grit. The Corratober Grits and Shales consists of two cyclothems together with the overlying *B. superbilinguis* marine band (R_{2c}). The *B. gracilis* (R_{2a}) band lies at the base of the lower cyclothem, while the younger, which contains a seat earth and a thin coal, begins with the *B. bilinguis* (R_{2b}) band. The overlying Corratober Brick-Shales, grey and brown mudstone and siltstone with ironstone, continues the *B. superbilinguis* cyclothem, which is capped by the generally coarse-grained Corrybracken Sandstone, and overlying fireclay. Above this level and below the *C. cancellatum* (G_{1a}) marine band, which marks the base of the Cabra Shale, there are shales with *Lingula* and non-marine bivalves, and thin sandstones. The Cabra Shale and Cabra Sandstone make up the next cyclothem; the latter is overlain by the *C. cumbriense* (G_{1b}) marine band, which is followed by siltstone and fine-grained sandstone. The Langsettian has been proved below the Permian in boreholes (Jackson, 1965; Eagar, 1974). *Gastrioceras subcrenatum* occurs in dark shales with pectinoid bivalves and productoid brachiopods. A higher marine band, containing *Lingula* and *Orbiculoidea*, has been tentatively correlated with the *G. listeri* horizon. The youngest beds proved are dominantly sandstone, with some shale and a few stringers of coal.

North-west Ireland

Serpukhovian and Pennsylvanian rocks occur in several areas in north-west Ireland (Fig. 11.14) and also offshore in the Donegal Basin and Erris Trough (Fig. 11.2).

The most southerly occurrence onshore is the poorly known Slieve Carna outlier (Fig. 11.1), described briefly by Ramsbottom *et al.* (1978). It was reported to consist of Pendleian and Arnsbergian shales capped by Arnsbergian sandstone. Arnsbergian strata are reported to overlap the Pendleian shales to the north-west to rest on Viséan limestone.

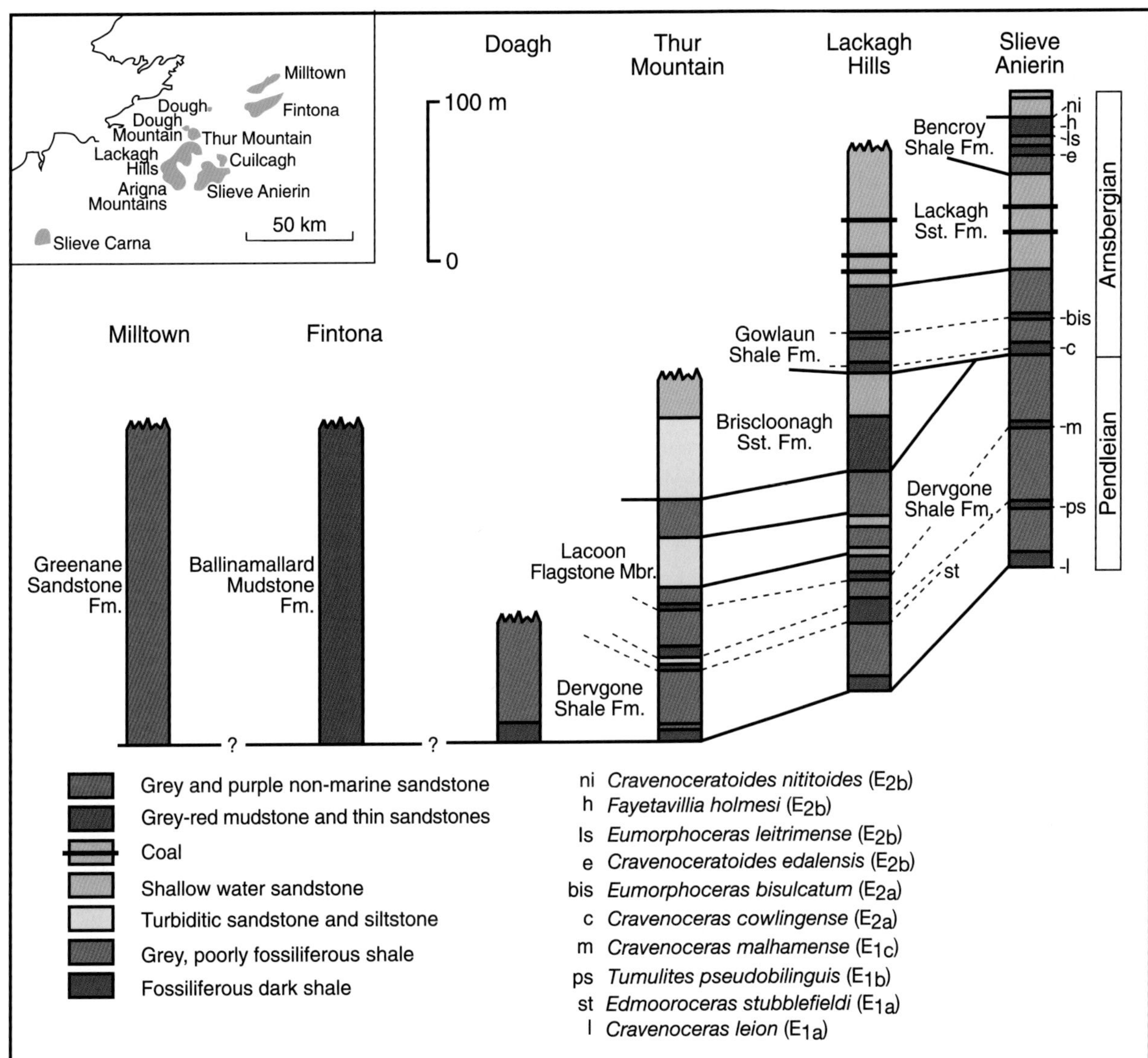

Fig. 11.14 Serpukhovian successions around Lough Allen contrasted with the probable Serpukhovian rocks of the Fintona and Milltown areas. Thicknesses of the last two are very uncertain.

Serpukhovian rocks form the high ground around Lough Allen, and there is a small outlier also at Doagh, in the Fermanagh Highlands, south-west of Lough Erne (Fig. 11.1). These areas have been comprehensively described by Yates (1962), Brandon (1972), and Brandon and Hodson (1984), who give references to earlier literature, Legg *et al.* (1998), and Mitchell, 2004c). The base of the Pendleian is identified at the base of the Dervgone Shale Formation. The underlying Carraun Shale Formation (Brigantian; P_{2c} at its top) is of broadly similar facies, and there is no suggestion of a break in deposition at the boundary. The lower half of the Dervgone Shale, predominantly alternations of dark ammonoid-bearing shales, in some cases with carbonate bullions, and barren or sparsely fossiliferous shales, extends to the top of the Pendleian (E_{1c}). Rare beds of turbiditic sandstone occur on Thur Mountain in the north of the outlier but do not extend southward. The upper part of the formation in the south of the outlier around Slieve Anierin consists of barren grey shale, containing clay ironstone lenses and silty, micaceous shale. North-west of Lough Allen, three thin persistent turbiditic sandstone beds are intercalated with the shales. In the Thur Mountain area, turbidites, the Lacoon Flagstone Member, are 27 m thick. Sole marks indicate transport from the north-north-east. The overlying Briscloonagh Sandstone

Formation consists of a lower unit of turbiditic, fine-grained sandstone, also showing evidence of derivation from the north-north-east, and an upper unit of shallow water, medium- to coarse-grained sandstone, with a seat earth but no coal at its top. The formation thins markedly southward, from approximately 70 m thick at Thur Mountain, to between 6 m and 10 m in the south part of the Lackagh Hills, and pinches out south of Slieve Anierin. Brandon and Hodson interpreted the upper part of the Dervgone Shale and the Briscloonagh Sandstone as the products of the southward progradation of a delta. Marine sedimentation recommenced with the Gowlaun Shale Formation, which close to its base has a marine band of Arnsbergian (E_{2a}) age. The bulk of the formation, which varies little in thickness, consists of dark, sparsely fossiliferous shale, with clay ironstone at some levels. A second marine band near the middle of the formation is also of Arnsbergian (E_{2a}) age. The overlying Lackagh Sandstone Formation is the main scarp-forming unit in the hills around Lough Allen. It consists predominantly of coarse, in places pebbly, thick-bedded sandstone, and contains as many as five thin coal seams (Haughey and McArdle, 1990), which are probably impersistent. The formation thins southward, from 90 m north of Lough Allen to approximately 60 m thick at Slieve Anierin. The youngest Serpukhovian preserved is the Bencroy Shale Formation, which forms the tops of the highest hills east of Lough Allen. It consists of dark ammonoid-bearing shales of Arnsbergian (E_{2b}) age, barren shales with clay ironstones, silty micaceous shales, and thin sandstones.

Serpukhovian and Pennsylvanian rocks occur in two fault-bounded tracts to the east of Lower Lough Erne (Mitchell and Owens, 1990; Mitchell, 2004c, 2004d).

In the area identified as Fintona in Figure 11.1, there is a thick succession of sandstone and conglomerate, the Kilskeery Group, of late Viséan to early Serpukhovian age. Mitchell (2004d, Fig. 8.7) referred to the Kilskeery basin, which was an active graben bounded by the Tempo–Sixmilecross and the Seskinore–Killadea Faults. The older two formations of the group are considered to be of Brigantian age (see p. 262) but the upper part (at least 300 m thick) of the youngest formation, the Ballinamallard Mudstone Formation, has been shown by palynology to be of Pendleian age. The formation consists of 1000 m of greyish-red mudstones with carbonate nodules, and fine-grained sandstones with ripple marks and desiccation cracks. Both these non-marine facies contrast with the episodically fully marine succession in the Lough Allen Basin to the south.

Mitchell and Owens (1990) identified Serpukhovian and Pennsylvanian rocks south of the Lack inlier, between the Castle Archdale fault and the Omagh thrust fault – an area identified as Milltown in Figure 11.1. The Greenan Sandstone Formation (Fig. 11.14) consists of at least 500 m of medium- to coarse-grained, purple-grey feldspathic sandstone, with scattered clasts of metamorphic rocks. At its base there are coarse breccias, which rest on the Dalradian. Miospores found low in the formation suggest a Brigantian or Pendleian age. Rhyolitic volcanic flows occur towards the top of the formation (Mitchell, 2004c). Mitchell and Owens interpreted the upper part of the formation as Serpukhovian in age and concluded that it accumulated in an extensional half graben bounded by the predecessor of the Omagh thrust fault. Following regional uplift, the Slievebane Group of Pennsylvanian age was deposited. The older Tullanaglare Mudstone Formation (late Langsettian–early Duckmantian) consists of 50–80 m of grey-green sandstone and mudstone, capped by a pedogenic limestone. There is a sharp contact with the succeeding Drumlish Conglomerate Formation, which consists of 1000 m of conglomerates, rich in coarse volcanic detritus. The latter were deposited in the proximal part of an alluvial fan, by braided streams flowing from the north (Simon, 1984a). A miospore assemblage from close to the base of the formation is of Duckmantian age. The Tedd Formation, 450 m of predominantly fine-grained red beds of unknown stratigraphical age, structurally isolated from the other formations, is probably the youngest formation within the area. Mitchell (1992) has suggested that deposition of the Slievebane Group took place in a basin formed by rapid subsidence, as a result of transtension caused by dextral movement on the bounding faults in Duckmantian time. Calc-alkaline volcanic material was extruded around the basin margin, eroded, and transported to contribute to the fanglomerates that now comprise the Drumlish Conglomerate Formation. Alluvial fans were followed by the fluvial and lacustrine facies of the Tedd Formation.

Although rocks of the same age as the succession at Coalisland (see below) do not crop out in the Fintona Block, it seems likely that the Kilblane Group was local in occurrence and that it does not reflect a major change in the style of sedimentation throughout the region at the end of Langsettian time.

Late Carboniferous rocks were found in three wells drilled in the Erris Trough and Donegal Basin (Fig.11.2).

Serpukhovian and Pennsylvanian rocks were encountered in well 19/5-1 (Fig.11.2), drilled in the Mesozoic

Erris Trough (Robeson *et al.*, 1988). Interbedded thin crinoidal limestone and sandstone, overlain by sandstone, siltstone and coals, in all totalling 350 m, have been dated as Brigantian/Pendleian. Only 37 m of Arnsbergian sandstone, claystone, and coal is preserved below an unconformity. Above the unconformity are 51 m of sandstone and silty claystone of Langsettian age overlain by 250 m of paralic coal measures of Duckmantian age, including 16 coal seams.

Well 13/3-1, in the Donegal Basin, encountered a thick succession of Pennsylvanian rocks below the Jurassic. It terminated 309 m below the Duckmantian–Bolsovian boundary. The Duckmantian-aged succession consists of grey siltstones and sandstones with eight coal seams; the Bolsovian is represented by at least 475 m of similar facies containing 42 coal seams; the Asturian section (of which 164 m is preserved) consists of coal-poor siltstone and sandstone, with three thick units of volcanic tuff. Well 12/13-1, in the Donegal Basin, terminated in sandy strata of Duckmantian age a short distance below the sub-Permian unconformity.

North-east Ireland

Poorly exposed Serpukhovian and Pennsylvanian rocks occur between Coalisland and Dungannon, County Tyrone (Fig. 11.1). Most information about them is derived from boreholes put down to investigate the several coal seams that were formerly worked in the Coalisland and Annaghone coalfields (Fowler and Robbie, 1961). The base of the Serpukhovian (Fig. 11.15) probably occurs within the Rossmore Mudstones, dark mudstones grading upward to shallow-water, burrowed, laminated siltstone and fine sandstone, but diagnostic fossils are scarce. The overlying 'Millstone Grit' consists of shales with marine fossils, sandstones that are commonly pebble-bearing, seat earths, and several coals. Many of the mudstones are calcareous, but there is only one conspicuous carbonate, the Lurgaboy Dolomite, which is a useful marker in the Dungannon area and has also been found in a borehole at Magilligan, 75 km to the north. The faunas of the Rossmore Mudstones and the 'Millstone Grit' include bivalves, gastropods, and brachiopods, but ammonoids are rare. A nautiloid fauna of Pendleian age has been recognised near the top of the Rossmore Mudstones and above the Lurgaboy Dolomite. Faunas identified as Arnsbergian have been found above the Congo Coal. According to Mitchell (2004c), a major hiatus, spanning the Arnsbergian to Marsdenian, has been identified within the 'Millstone Grit' by palynological means. The Langsettian is represented by a relatively thick paralic succession, with eight coal seams of workable thickness (Fowler and Robbie, 1961; Eagar, 1974). The Coalisland Marine Band contains *G. subcrenatum*, pectinoid bivalves, and productoid brachiopods. Higher marine bands identified by Eagar lack ammonoids. The *lenisulcata/communis* biozonal boundary lies somewhere between the bivalve bands above the Monkey Coal and those below the Shining Coal. Little detail of the lithological succession is available of the measures above the Bone Coal.

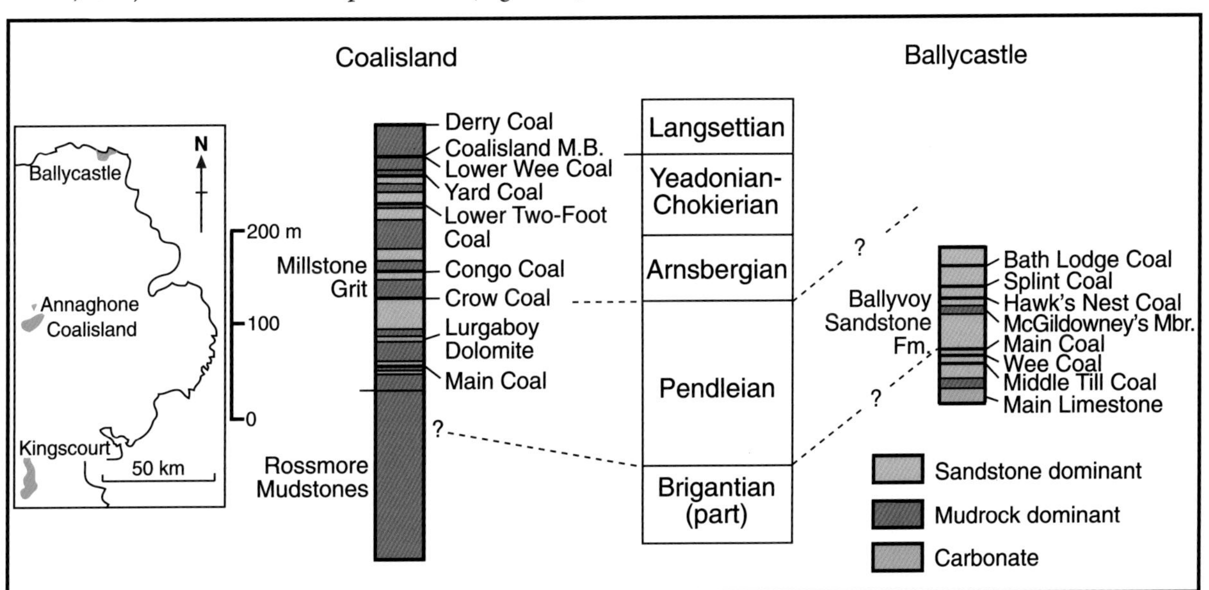

Fig. 11.15 Correlation of Serpukhovian and Pennsylvanian (Chokierian to Yeadonian) succession at Dungannon, County Tyrone with that at Ballycastle, County Antrim.

Highly arenaceous Serpukhovian rocks are preserved in the Ballycastle coalfield (Wilson and Robbie, 1966), which Mitchell (2004b) interpreted as having accumulated in a separate fault-bounded basin to the north of the Dalradian ridge. The base of the Pendleian has been placed at the Main Coal (Fig. 11.16), which is correlated with a similarly named coal in the Macrahanish coalfield across the North Channel in Kintyre, Scotland. Brachiopods from McGildowneys Marine Band are regarded as Arnsbergian in age. The benthonic, as opposed to nektonic, invertebrate faunas of the Dungannon and Ballycastle areas suggest that the water was shallow, even at high stands of sea level.

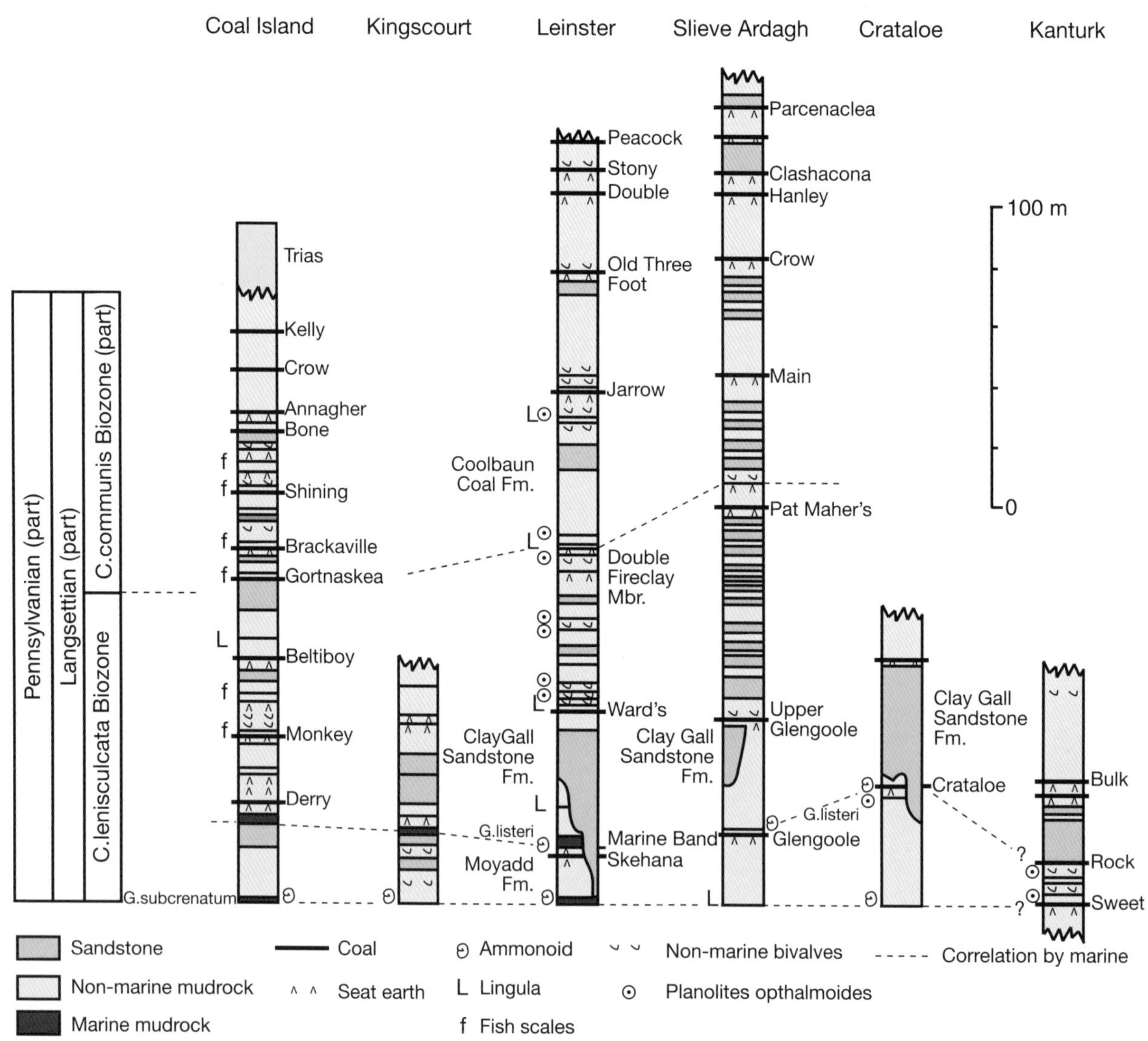

Fig. 11.16 Correlation of the Pennsylvanian (Langsettian) stratigraphy of the Irish coalfields. Compiled from Eagar (1974).

12

Variscan deformation and metamorphism

J. R. Graham

The Variscan orogenic belt can be traced from Russia through western Europe and across the Atlantic to the eastern seaboard of the United States. It resulted from the collision of a large southern continent, Gondwana, with the combined northern landmasses of Laurentia, Baltica and Avalonia that had been assembled into a single large continent during the Caledonian orogenic cycle. In detail, several micro-continental blocks are thought to have existed between the northern and southern continents, complicating the evolution of the Variscan Orogen. The literature is littered with terms for deformation phases at different times and in different places. The European Variscides are usually described in terms of a series of orogen-parallel zones (Figure 12.1). In Ireland the orogeny is manifested by widespread deformation of Devonian and Carboniferous rocks which straddle the northern foreland of the orogen, and the Rheno-Hercynian zone to its south. The terms Hercynian and Armorican are effectively synonymous with Variscan, but for consistency the term Variscan is used throughout this chapter.

In Ireland there are obvious changes in the style of deformation from south to north. Tight folds in the south give way to broad gentle folds and fault reactivation in the

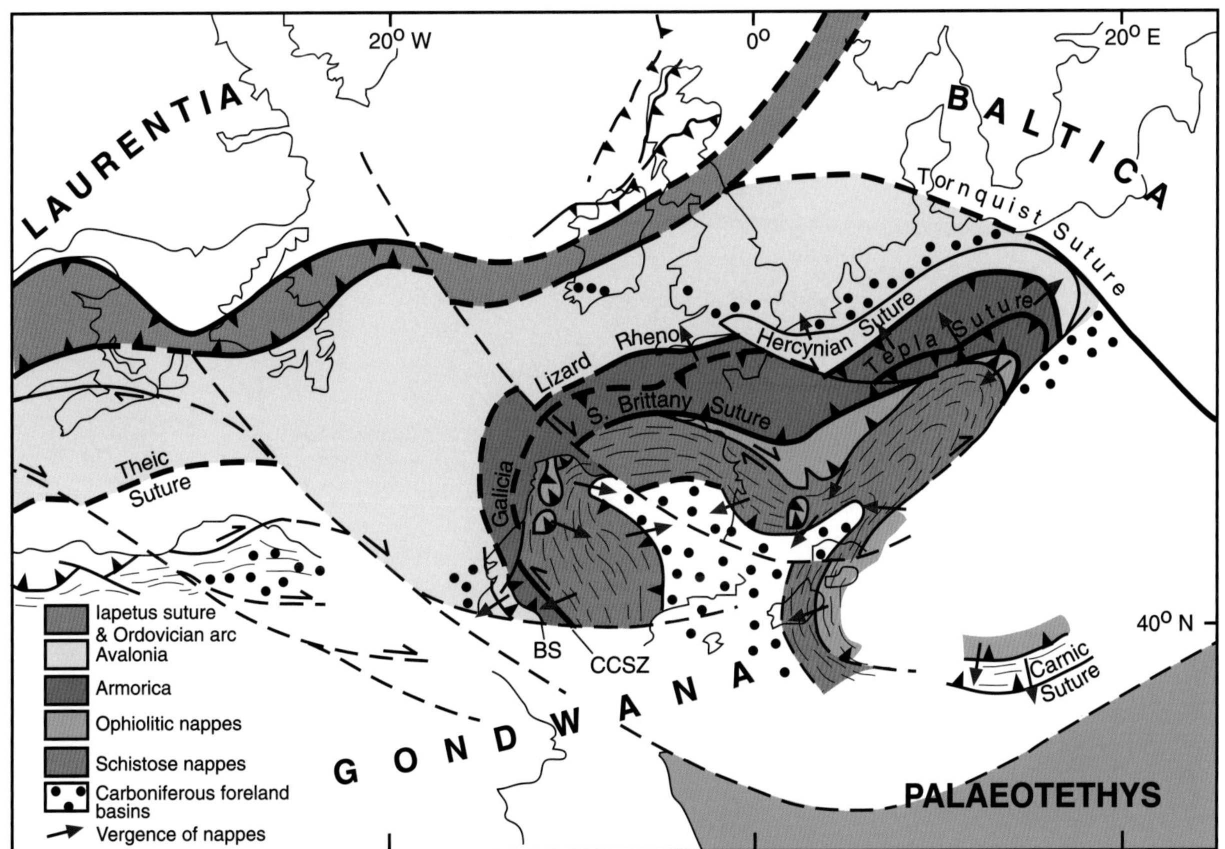

Fig. 12.1 Variscan fold belt of western Europe in Permian times (after Matte, 2001). BS – Beja Suture; CCSZ – Coïmbra–Cordoba Shear Zone.

north (Figure 12.2). Gill (1962) attempted to formalise this transition by delimiting three zones of contrasting tectonic style. His scheme was adapted and modified by Cooper *et al.* (1986). However, the boundaries between the zones are not precise but rather arbitrarily imposed on the gradational changes in structural style from south to north. In this zonal scheme a feature that acquires some significance is the 'Dingle–Dungarvan Line'. On many geological maps of the British Isles and north-west Europe this is labelled as the Variscan (Hercynian) Front (i.e. the boundary between the Rheno-Hercynian zone and the foreland). By implication this is regarded as the outer margin of the fold belt where there has been significant northwards translation over the 'foreland', similar

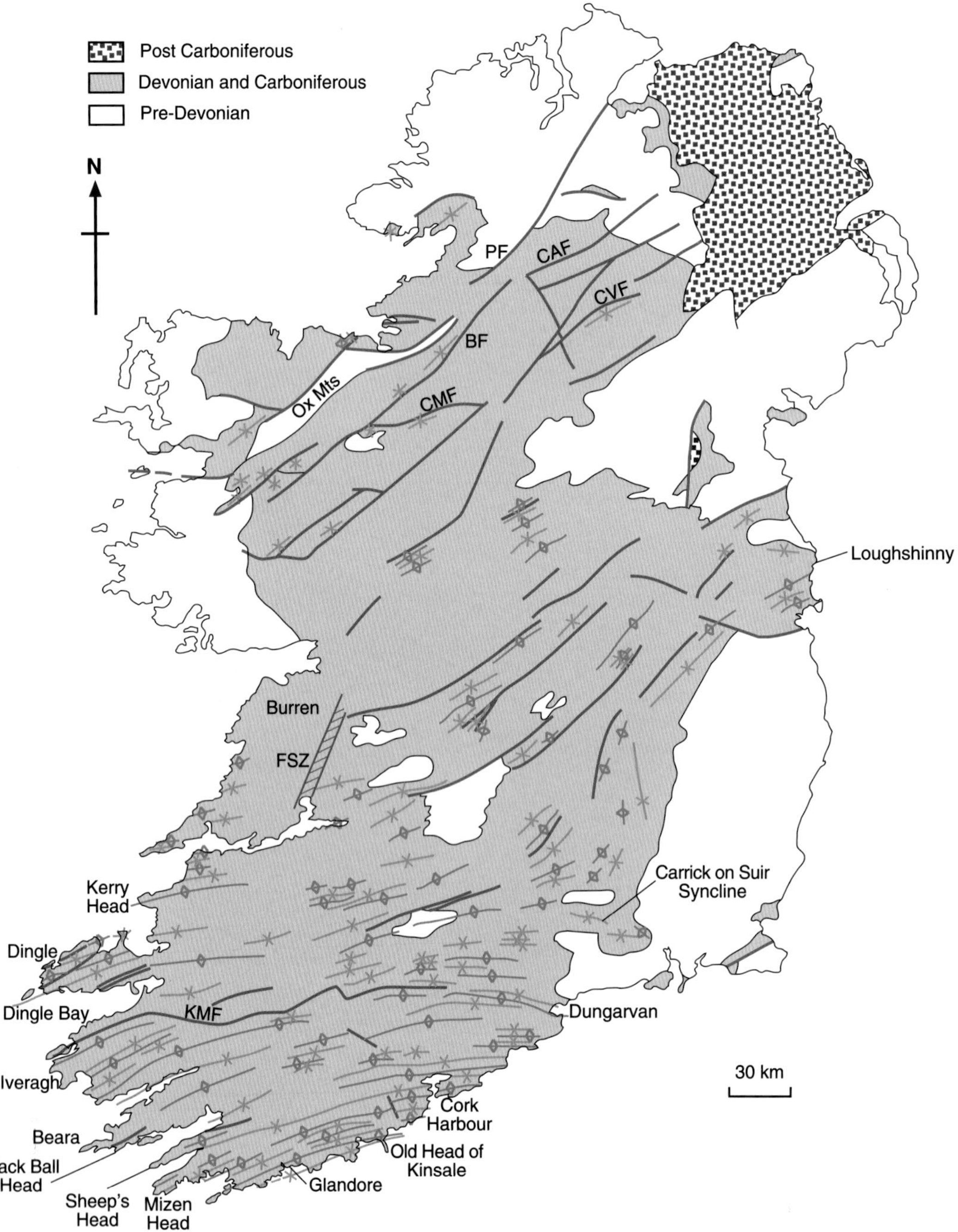

Fig. 12.2 Main Variscan fold and fault traces in Ireland. BF – Belhavel Fault; CAF – Castle Archdale Fault; CMF – Curlew Mountains Fault; CVF – Clogher Valley Fault; FSZ – Fergus Shear Zone; KMF – Killarney Mallow Fault; PF – Pettigo Fault.

to the Faille du Midi in Belgium. However, in Ireland this is clearly not the case, and the Dingle–Dungarvan line demarcates neither an original facies divide nor a sharp strain gradient (Naylor and Sevastopulo, 1979; Bresser, 2000). In places there is not even the certainty of a fault (F.X. Murphy, 1990). Thus, in this chapter the zonal scheme is not employed, and the Variscan structures are treated sequentially in an approximately south to north traverse from the Munster Basin to north-west Ireland. Traces of the main Variscan fold axes and faults are shown in Figure 12.2.

The Munster Basin

The rocks of the Munster Basin and subsequent South Munster Basin (Chapter 9) display ubiquitous ductile deformation in the form of folds and associated cleavage. The overall structural grain is evident on the geological map of Ireland and is also clearly expressed in the topography. The various peninsulas of south-west Ireland correspond to the anticlinal cores of Old Red Sandstone whilst the bays and valleys correspond to complex synclines cored by marine clastics or limestones of latest Devonian or Carboniferous age. A map of the main fold axial traces (Fig. 12.2) shows an arcuate pattern from ENE–WSW trends in the west to more E–W trends in the east. This strike swing may, at least in part, represent inheritance of earlier 'Caledonian' basement structures. The major folds have amplitudes and wavelengths of several kilometres. Many of them can be traced for tens of kilometres along strike and have been individually named.

Published north–south profiles across the Munster Basin show a clear fanning of axial planes (Naylor, 1978). This fanning is also apparent in the southern part of the profile shown in Fig. 12.6. Most of the major folds are upright with near vertical axial planes in the Mount Gabriel area (Reilly and Graham, 1976; Naylor, 1975; Phillips, 1985). To the south of this, towards the south coast, axial planes dip >60° northwards (Reilly and Graham, 1972, 1976). Here parasitic folds are more common on the northern limbs of the anticlines and parasitic folds are mainly absent on the locally overturned southern limbs. North of the zone of vertical axial planes this pattern is reversed, with parasitic folding more common on the southern limbs of the anticlines (Philips, 1985). This simple pattern is modified in the northern part of the basin just south of Dingle Bay. Here an anomalous zone of north-verging folds have gentle, south-dipping axial planes (Capewell, 1975; Price, 1986) (Fig. 12.3). This zone indicates a component of northerly thrusting in this region, probably related to rapidly shallowing basement.

There is considerable variation in the plunge of both major and minor folds throughout the region with the presence of numerous culminations and depressions. These have generally been explained as due to variations in along strike extension (Sanderson, 1984; Cooper *et al.*, 1984, 1986; Ford, 1990). Two aspects of minor folds and fold plunges deserve special mention. In West Cork regional mapping of lithological boundaries suggests that the plunges of the major folds, the Rosscarbery Anticline and the Skibbereen Syncline, are <6° and 3° to the east respectively (Graham and Reilly, 1972; Reilly and Graham, 1972), whereas measured bedding cleavage intersections and minor folds plunge predominantly to the west. A similar situation occurs south-west of Killarney (Walsh, 1968) where the measured minor fold plunges are greater by 15–20° than the overall plunge suggested by the outcrop pattern. Here there may be some local offsetting of minor fold plunges by normal faults. A second notable feature is where minor folds plunge in opposing directions on different limbs of the same major fold (Sevastopulo, 1981c).

Almost all the rocks in the Munster Basin show a well-developed cleavage (Fig. 12.4). In finer-grained lithologies the cleavage fabric is penetrative, but in coarser-grained lithologies it is spaced, the spacing depending on both lithology and whether position is on the

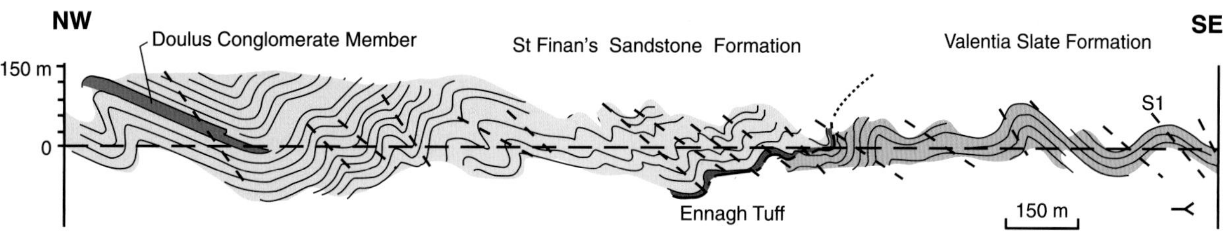

Fig. 12.3 Zone of northerly vergence in northern Iveragh immediately south of Dingle Bay (*after* Price, 1986).

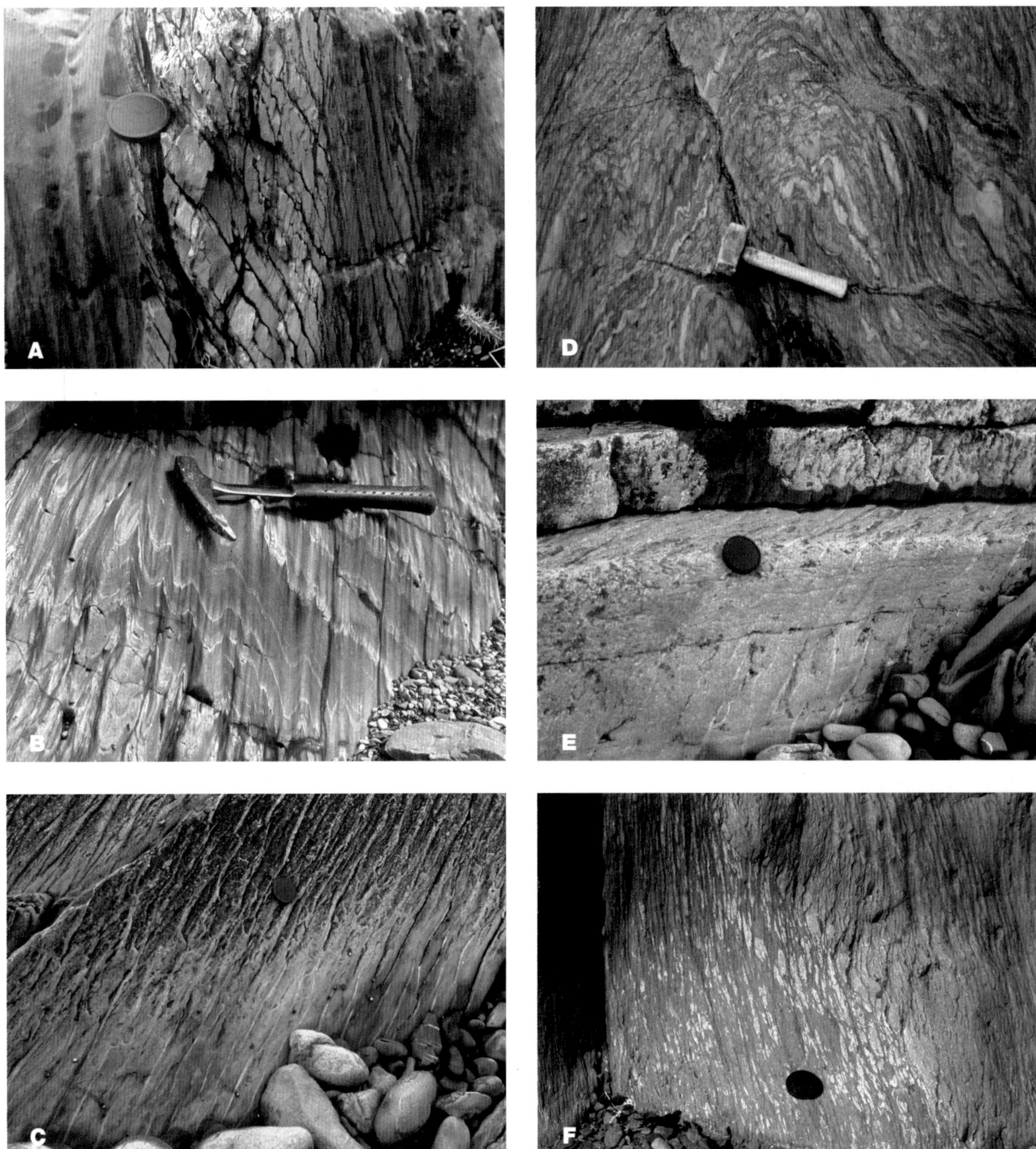

Fig. 12.4 Minor structures from the Munster Basin: **A**) Contrast between slaty cleavage in fine siltrocks and spaced pressure solution cleavage in fine sandstones, Castlehaven Formation, Ballylinchy Tower, SW Cork; **B**) Slaty cleavage and minor folds in fine-grained marine mudrocks at Bollinglanna, Seven Heads; **C**) Plan view of pressure solution cleavage seams in Ballinskelligs Formation, Ballinskelligs, Iveragh; **D**) Small-scale folding in Old Head Sandstone Formation, Toe Head; **E**) Profile view of pressure solution cleavage seams in Ballinskelligs Formation, Ballinskelligs, Iveragh; **F**) Marked cleavage refraction and variation in cleavage style with grain size, Castlehaven Formation, Castlehaven.

limb or in the hinge of a fold. In most cases cleavage is approximately axial planar to the folds, although fanning across the hinge zones of the major folds is common (Capewell, 1957; Walsh, 1968; Reilly and Graham, 1972, 1976; Graham and Reilly, 1976; Naylor, 1978; Naylor *et al.*, 1978; Cooper *et al.*, 1984; 1986; Phillips, 1985; Ford, 1990; Bresser, 2000). In thin section, the cleavage occurs as an anastamosing network of dark seams of clay concentration, in some instances with associated pyrite. Pressure solution controlled by diffusion creep has been suggested as the cause of cleavage formation by Sanderson (1984) and also by Beach (1982), the latter figuring examples from the Munster Basin as typical of this feature. The dominant role of pressure solution in the deformation is also evident from abundant quartz grain overgrowths and pressure fringes. Some cases in which cleavage is not axial planar have been described from the Beara Peninsula (Sanderson, 1984; James, 1989; James and Graham, 1995). In most of these cases cleavage strikes anticlockwise of the major folds. These transecting cleavages are consistent with rotation of folds prior to and during cleavage development and suggest a component of dextral shear. Locally a second cleavage (S2) is present, in places rotated 10–30° relative to S1, which has been interpreted as due to dextral shear during progressive deformation (Bresser, 2000).

Cooper *et al.* (1984) estimated shortening from balanced cross sections as between 33 and 42% with an increase towards the west. However, these balanced sections assume plane strain (i.e. they do not take account of any movement in or out of the section, and this effect may have been considerable). Nor do they take account of volume loss due to pressure solution, which may be up to 35% according to Beach and King (1978).

The amount and orientation of strain within the Munster Basin is consistent with oblate flattening. Phillips (1985) gave strain ratios (X:Z) of 2.1–3, but with local anomalies, based on data from both the Beara and Mizen peninsulas. He concluded that variation in strain did not show a simple regional pattern in this area, but was related to position within the local fold structures. Work by Ford and Ferguson (1985) on Galley Head also shows strongly oblate strains and X:Z ratios within the range 2.13–4 based on estimating the strain ellipse from stretched arsenopyrite crystals. This compares to estimates of 2.07–2.54 based on deformed trace fossils (Bamford and Ford, 1990) and 1.7–2.5 based on flattened ripples (Bamford and Cooper, 1989). More recent work by Bresser and Walter (1999) in a N–S transect gives values ranging from 1.72 in the north to 2.42 in the south, based on measurements of deformed quartz grains in thin section.

Extensional structures such as boudins and veins in the Munster Basin indicate vertical extension in most areas, but many areas also show significant strike-parallel extension (Sanderson, 1984; Phillips, 1985; Cooper *et al.*, 1986; James, 1989; Ford, 1990).

A large number of faults were produced during the Variscan orogeny. Displacement on these is often difficult to determine in the Old Red Sandstone areas due to thick lithostratigraphic units with subtle boundaries (Graham, James and Russell, 1992). Variable cross-cutting relationships between faults of different trends suggest their approximate contemporaneity, and it is likely that the fault system represents the latter part of one phase of major Variscan deformation. The main fault trends recorded from field observations are:

i) near strike faults
ii) E–W and ESE–WNW faults which are a little clockwise of strike
iii) faults at a high angle to strike that vary from NW to NE in trend.

For the near strike-parallel faults, the amount of vertical displacement is always much greater than the amount of horizontal displacement. Many of these have been reinterpreted as thrusts by Cooper *et al.* (1986) irrespective of whether they excise or repeat stratigraphy. However, many of the faults just clockwise of the strike show significant dextral displacements and many of them are mineralised (Reilly and Graham 1972, 1976; Walsh, 1968; Sanderson, 1984; Reilly, 1986; Graham and Reilly, 1976; Capewell, 1957; Phillips, 1985; James and Graham, 1995). The faults at a high angle to the strike generally show both vertical and horizontal displacement. Where two trends are present they appear to form a conjugate set; NNE-trending faults are mainly sinistral and NNW-trending faults are mainly dextral. A recent comprehensive analysis of Landsat imagery by Naylor (2003) is consistent with these field observations, although it also includes lineaments produced by joints and possibly some lithological contacts. This analysis highlights some variation along strike and shows a common NW trend in many places.

The fault with the largest apparent displacement is the Killarney–Mallow Fault (Figs. 12.2; 12.5) (Walsh, 1968; Price, 1986; Gill, 1962) where locally Namurian rocks to the north are juxtaposed with horizons low in the Old Red Sandstone succession. Near Killarney this fault is vertical, with evidence for dextral transpressional movement (Price, 1986), although several thrusts are present in limestones to the north (Wingfield, 1968; Price, 1986; Nex *et al.*, 2003). Further east towards Mallow, Gill (1962) interpreted this fault as a low-angle structure, one of several affecting rocks ranging in age from the Old Red Sandstone to the Namurian. Although this fault acquires

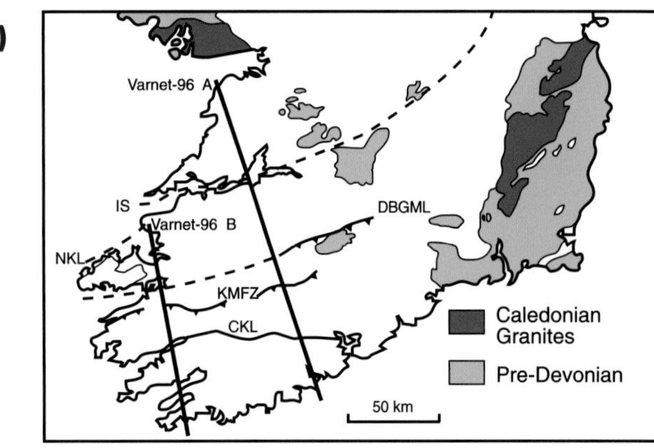

Fig. 12.5 VARNET seismic profiles (*after* Vermeulen *et al.*, 1999).
a) Location of VARNET lines A and B.
b) Interpretation of line A.
c) Interpretation of line B.
Subsurface numbers in **b** and **c** are seismic velocities in km/s.
IS – Iapetus Suture; NKL – North Kerry Lineament; DBGML – Dingle Bay–Galtee Mountains Line; KMFZ – Killarney–Mallow Fault Zone; CKL – Cork–Kenmare Line.

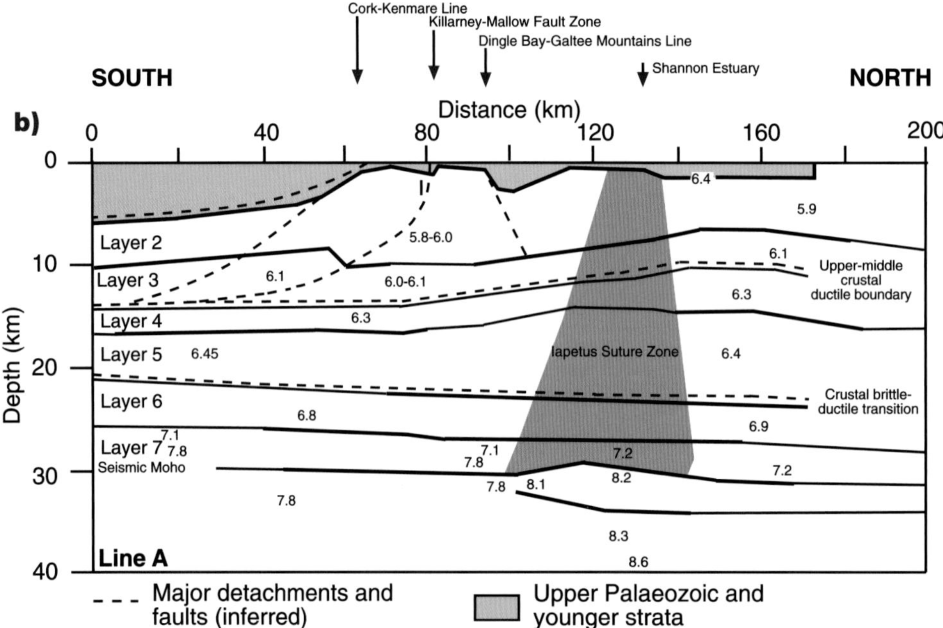

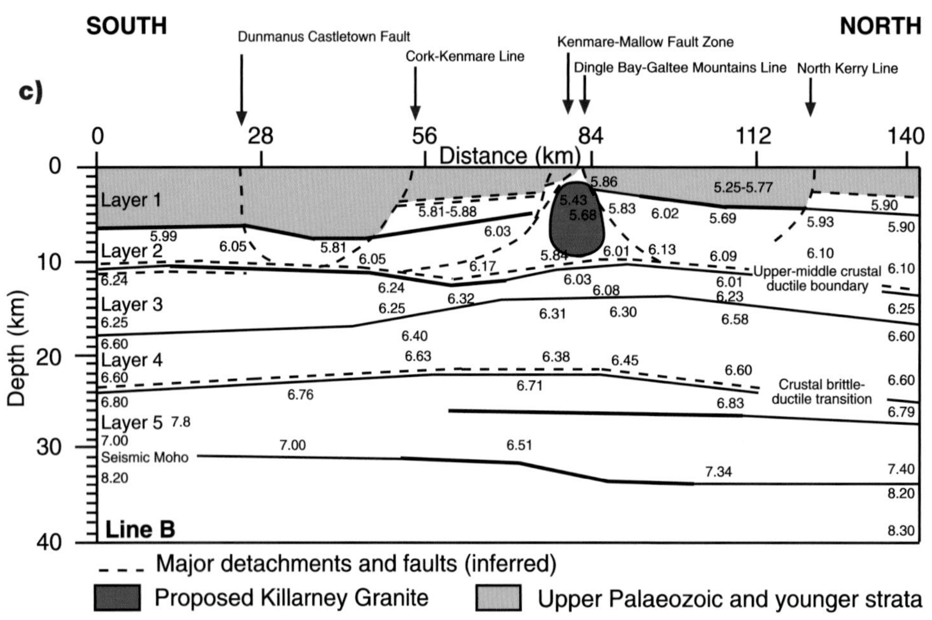

great significance on many regional maps, being considered as the 'Variscan Front', there are few reasons for considering it to be unique (Naylor and Sevastopulo, 1979; Sevastopulo, 1981c; Ford *et al.*, 1992; Bresser and Walter, 1997). Significant thrusting, cleavage, and asymmetrical folds with steep north-younging limbs occur well to the north of this line, and Bresser (2000) records no significant change in internal deformation across this fault. The fault does not coincide with either the northern margin of the Devonian Munster Basin (Chapter 9) or the major facies divide of the younger Cork–Kenmare Line, the northern margin of the South Munster Basin (Chapters 9, 10). Moreover, the fault dies out to both west (Walsh, 1968) and east (F.X. Murphy, 1990) and is not, therefore, a continuous 'line'.

The deformation style in Upper Palaeozoic rocks north of the Killarney–Mallow Fault shows no sharp break from that in rocks to the south. Philcox (1964) shows a series of E–W folds with associated wrench faults and steep reverse faults in the Buttevant area in north County Cork. He also demonstrates a compartmentalisation of deformation which he ascribes to basement control, with structural disharmony between the Dinantian and the Namurian, and inferred décollement in the basal Namurian shales. A similar pattern appears to hold in the Killarney area. Although disharmony is evident, there is no reason for proposing major lateral translation along any décollement horizon within the Namurian.

In general, there is agreement on the nature of the surface structure, but there has been considerable debate on several issues, such as the relative importance of N–S pure shear and E–W simple shear; the explanation of the arcuate fold pattern; the influence of basement structure on deformation; and the nature of the structure at depth. Although there is evidence of dextral transpression, this appears to be of much less importance than N–S pure shear (Naylor, 1978; Price and Todd, 1988). F.X. Murphy (1990) suggested the arcuate nature of the fold belt was due to the development of a surge zone within northerly propagating thrust sheets. However, his suggestion of N–S trending dextral faults in the east and N–S trending sinistral faults in the west is at variance with much of the field data from the west which suggests E–W dextral shear (Sanderson, 1984; Bamford, 1987; James and Graham, 1995). Other possible causes for this pattern are variable amounts of shortening (Cooper *et al.*, 1984) or the configuration of underlying basement blocks.

The precise influence of basement blocks on Variscan deformation of the Munster Basin is difficult to determine. However, the evidence for marked thickness and facies changes during deposition of the Upper Palaeozoic succession are strong evidence for syn-sedimentary normal faulting. This must have produced a very irregular basement topography. Strong evidence for this occurs at the eastern margin of the Glandore High (Fig. 9.33), where there is a zone of steep minor fold plunges (Graham and Reilly, 1976; Bamford and Ford, 1990; Ford, 1990). However, analysis by Naylor (2003) suggests the absolute limit of the Glandore High, and its influence, extends farther east. Stratigraphical and structural evidence suggest a major fault, which may be a high-angle thrust, along the northern side of the Sheep's Head Peninsula, and the presence of a structural high beneath the peninsula (Naylor, 2003). The change in fold style approaching Dingle Bay in western Iveragh (Capewell, 1975; Price, 1986) has been attributed to shallowing basement controlling north-directed thrusting. Basement structure may also have controlled some of the 'compartments' in the Buttevant area (Philcox, 1964), changes in fold orientation in western Beara (Sanderson, 1984; James, 1989; James and Graham, 1995), and the location of igneous intrusions in southern Beara (Matthews *et al.*, 1983).

The nature of the Variscan structure at depth is difficult to test in the absence of boreholes and high-resolution geophysical data. Both thin-skinned thrust sheet models (Cooper *et al.*, 1984, 1986) and thick-skinned models with steep basement faults (Sanderson, 1984; Phillips, 1985) have been proposed. In the thin-skinned model most of the major displacements occur in 'blind' structures. The surface geology does not require any significant lateral displacements although there is clearly disharmony in the folds developed in the Namurian rocks. Thus both models are plausible and each concurs with the structures observed at the surface.

Geophysical evidence includes gravity data (Murphy, 1960; Ford *et al.*, 1991; Readman *et al.*, 1997) and both vertical and wide angle seismic reflection profiling (Le Gall, 1991; Ford *et al.*, 1992; Vermeulen *et al.*, 1999; 2000; Landes *et al.*, 2000, 2005). From analysis of the gravity data, Ford *et al.* (1991) suggest that basement was involved in Variscan deformation with any detachment lying at mid-crustal levels. Readman *et al.* (1997) also suggest that Variscan compressional deformation in the south did not involve the entire crust, and they support a 'thin-skinned' tectonic model. They further propose a south-westward extension of the Leinster Granite at greater depth. The Killarney gravity low (Howard, 1975) is also interpreted as a granitic body at depths similar to this south-westward extension of the Leinster Granite. Previous gravity models (Ford *et al.*, 1991) did not discriminate between the possibilities of granite at depth or a greatly thickened Old Red Sandstone succession in this area. However, if such a granite was to be contemporaneous with the surface volcanic activity (378–385 Ma, Williams *et al.*, 2000b; Chapter 9) it could not be the same age as the Leinster Granite (405 Ma, O'Connor *et al.*, 1989; Chapter 8). Readman *et al.* (1997) also note that typical 'Variscan' gravity trends occur well to the north of Gill's (1962) 'thrust front', thus questioning the validity of a simple northerly margin to strong deformation.

Several interpretations of the deep structure of the Munster Basin have been made on the basis of deep seismic reflection data (WIRE, SWAT and WINCH seismic lines). Many shallow dipping reflectors have been interpreted as Variscan thrusts (BIRPS and ECORS, 1986; Coward and Trudghill, 1989; Le Gall, 1991; Sibuet *et al.*, 1990). However, these are invariably projected on land, both in southern Ireland and in South Wales, to areas where major thrusts are absent, such as the Dungarvan area (F.X. Murphy, 1990). Thus an interpretation of these reflectors as 'Caledonian' structures would appear to be more plausible (Ford *et al.*, 1992). Analysis of data from two major wide-angle seismic profiles (VARNET-96), one from Clare to just west of Cork Harbour and another from Kerry Head to West Cork, is summarised in Figure 12.5. Vermeulen *et al.* (1999) suggest that the Dingle Bay–Galtee Mountains line represents the southern margin of the Dingle Basin only, and not the northern margin of the Munster Basin. The gravity low just south of the Killarney–Mallow Fault Zone is interpreted as a buried granite at only a few kilometres depth. This fault zone is interpreted as the northern syn-rift margin of the Munster Basin, and the Cork-Kenmare line is interpreted as a major syn-depositional fault. Brown and Whelan (1995) provide supporting evidence from magnetotelluric data of steep structures coinciding with these two fault zone traces. These authors interpret the Killarney–Mallow Fault Zone as a north-dipping structure with at least 2 km of Carboniferous sedimentary rocks present in the immediate footwall. They see no evidence for structural linkage between the middle and lower crust, and conclude that the two behaved independently. Based on geophysical data, the greatest vertical thickness of Old Red Sandstone is about 8 km, and is situated beneath the Beara Peninsula. Deformation of the basin floor is not obvious either, because of detachment faulting or reduced seismic resolution.

Thus, despite some unresolved issues, the consensus view of the structure of the Munster Basin is one of dextral transpression with some degree of detachment between the Upper Palaeozoic cover and its basement. This is summarised in Figure 12.6, taken from Bresser (2000).

Dingle Peninsula

Perhaps more than anywhere else in Ireland, the Dingle Peninsula displays the progressive change from 'Caledonian' to Variscan sedimentation and tectonics. The presence of major faults controlling the nature and orientation of the depositional basins is now well documented (Todd *et al.*, 1988a, b; Todd, 1989a) and has been discussed in Chapter 9. The numerous unconformities in the succession testify to successive periods of deformation as discussed in Chapters 8 and 9. Early workers noted that only one prominent cleavage affected all the rocks of the Dingle Peninsula. Shackleton (1940) suggested that the location of the Variscan folds may have been controlled by that of the Caledonian folds, and noted that the imposition of a later cleavage on previously tilted strata produced anomalous bedding–cleavage relationships. Certainly the Variscan fold axial traces are essentially parallel to those seen in the underlying Dingle Group, which is of Ludlow to Emsian age (Chapter 9).

Several workers have now documented 'Caledonian' folds (Parkin, 1976b; Todd *et al.*, 1988a, b; Todd, 1989a, 2000; King, 1989; Chapter 8) that locally contain evidence for sinistral shear, particularly near the margins of the Annascaul inlier. These structures are demonstrably older than the Glengarriff Harbour Group, the youngest package of Old Red Sandstone sediments. King (1989) also concluded that the pre-Variscan deformation did not produce a cleavage, and that the prominent cleavage cuts across the unconformity at the base of the Glengarriff Harbour Group without change in orientation. This unconformity is spectacularly exposed just north of Bull's Head (Figs. 9.11; 12.7) where it was first described by Jukes and Du Noyer (1863), whose drawing was reproduced in Horne (1976). Todd (2000) also ascribed the main regional cleavage to the Variscan, and stated there was no earlier cleavage in the Dingle Beds except locally in fault zones.

These findings are disputed by Meere and Mulchrone (2006), who suggest significant cleavage development due to an Acadian event. They note high strain values associated with a penetrative cleavage fabric in the Dingle Group, and they present data that suggest, at least for the

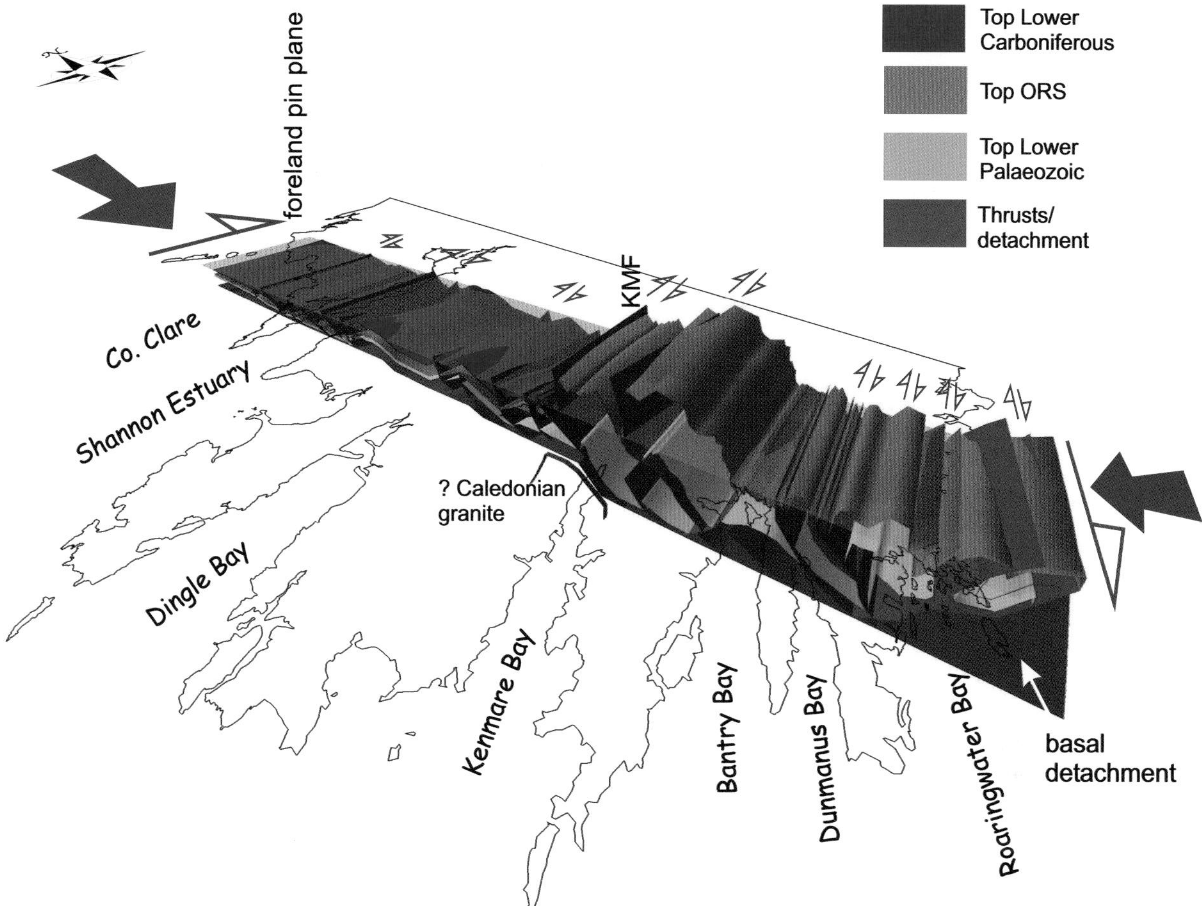

Fig. 12.6 A three-dimensional model of the structure of south-west Ireland (diagram kindly supplied by Georg Bresser).

Bull's Head area, that the fabrics above and below the famous unconformity are discordant. They use Rf/Φ analysis to demonstrate markedly higher strain below the main unconformity, which they attribute to predominantly Acadian deformation (*see* also Chapter 8).

The Variscan cleavage is generally a spaced pressure solution cleavage, although the penetrative fabric in some of the finer lithologies in the Dingle Group has been attributed to the same cleavage-forming event by many authors. Strain has been estimated using reduction spots, conglomerates, desiccation cracks and trace fossils, depending on availability in the various groups on the peninsula (King, 1989). In the rocks of the Glengarriff Harbour Group, shortening has been estimated at 17% from bed length balancing on cross sections and at 36% including volume loss (King, 1989). Unfolding of the Glengarriff Harbour Group demonstrates the nature of the early folds in the Dingle and Dunquin groups, which were subsequently tightened during Variscan deformation. King (1989) estimated 24% pre-Variscan shortening for rocks of the Dingle Group. These early folds are transected by the Variscan cleavage, but there is no significant transection of folds in the Glengarriff Harbour Group. King (1989) suggests a two-stage Variscan deformation with earlier NNW-directed compression, which tightened earlier folds, followed by a progressive increase in dextral shear with some local overthrusting. There is no discrete change in deformation style across the supposed Variscan Front through Dingle Bay.

East Waterford – Wexford

Deformation and metamorphism in the south-east corner of Ireland is markedly different to that seen a short distance to the west in the Munster Basin. Dips are low and the rocks generally lack cleavage. Rare exceptions are at Ballydowane Bay and other coastal localities west of Tramore, where rotated blocks of Old Red Sandstone crop out in cliffs of spectacular vertical beds. Otherwise this area is quite different from the Munster Basin, not

Fig. 12.7 Du Noyer's unconformity at Bull's Head. Steeply dipping Bull's Head Formation (Dingle Group) with steeply plunging folds is overlain in the upper part of the cliff by Upper Old Red Sandstone. A sub-vertical Variscan cleavage cuts through the whole succession.

only in terms of its stratigraphy and structure, but also in its markedly different thermal history, as discussed below. The reasons for these differences are still not fully established.

Central Ireland

The lack of a simple Variscan Front is further shown by the progressive northward decrease in the intensity of deformation. Lower Palaeozoic inliers form the anticlinal cores of folds rimmed by Carbonbiferous limbs. Deformation within the basement rocks seems to be accommodated mainly by movement along pre-existing faults. The orientation of folds within the Carboniferous rocks clearly reflects the underlying Caledonian basement with ENE and NE trends being common (Fig. 12.2). Strongly flattened limestones occur as far north as the Carrick-on-Suir Syncline and Tralee, Co. Kerry. Cleavage development is widespread in the muddier lithologies, particularly in the Dublin Basin and the Shannon Trough where thick Carboniferous successions are present.

Minor folding is strongly controlled by lithology, and there is usually disharmony between folds in the Carboniferous Limestone and those in the Namurian shales and sandstones (Brennand, 1965; Morton, 1965; Dolan, 1984). Folds are typically upright with axial planes that are near vertical or dip steeply to the south.

Folding in the limestone-shale sequences of late Dinantian and early Namurian age of the Dublin Basin and the Shannon Trough is particularly spectacular (Fig. 12.8c). Near Ballybunnion, Dolan (1984) describes E–W trending folds with an axial planar pressure solution cleavage. The style, size and tightness of the folds are lithologically controlled, with close to tight chevron folds developed in the limestone–shale sequences. Thrusts are developed on the limbs but there is no sign of major décollement. However, Le Gall (1991) suggests a major décollement within the Clare Shales of a few kilometres to explain the vertical variations in bulk shortening. Some box folds are present, with cleavage locally absent on the undeformed flats. Box folds are also present in the Namurian rocks north of the Shannon Estuary where the pressure solution cleavage is less well developed and largely confined to fold hinges. Dolan (1984) estimates 11% N–S shortening from section balancing, but did not produce any quantitative strain analysis. Le Gall (1991) estimates shortening at *c.*15% for a section from the Cliffs of Moher to Ballybunnion, although it is not clear what method was used. Analysis of minor folds and faults again indicates a late component of dextral shear. The box folds of south Clare grade northwards into progressively more open folds, commonly widely spaced monoclines (Fig. 12.8a). In the Burren district of County Clare, uncleaved limestones are very gently folded (Fig. 12.8b). Superb exposures here have permitted detailed analysis of the veins and joints affecting the Mississippian limestones. The veins are interpreted as forming during N–S Variscan compression under an overburden estimated to be in the range of 0.6–2.5 km (Gillespie *et al.*, 2001). In contrast, the joints formed during uplift under low differential stress and are not necessarily related to the Variscan. The presence of E–W mineralised faults, such as those at Tynagh, Silvermines, and Navan (Moore, 1975; Coller, 1984; Philcox, 1989) offers further comparison with areas to the south.

Fig. 12.8 (a) South-verging fold from Bridges of Ross, County Clare; (b) Open, gentle folds in the Carboniferous Limestone of the Burren; (c) Chevron folds in the Brigantian Loughshinny Formation, Loughshinny, County Dublin (photo by George Sevastopulo).

Structural information for much of central Ireland is rather sparse, only partly due to the limited exposure. Compilation maps produced by Hitzman (1993) have considerably improved knowledge of the basic structures. It is obvious from these (*see* also Fig. 12.2) that a NE–SW structural trend is predominant, and this is thought to be inherited from the Caledonian basement. Major faults of this trend generally show top to the north-west movement and are interpreted as thrusts.

In the western half of central Ireland, Variscan deformation is characterised by ductile–brittle transcurrent shear zones, major open folds and heterogeneous vertical cleavage. Deformation was the result of N–S compression and E–W to ENE–WSW dextral shear (Coller, 1984). An early period of mid-Dinantian deformation involved E–W dextral shear and was closely associated with base-metal mineralisation (Coller, 1984). The western margin of this zone of deformation has been taken as the NNE trending sinistral Fergus Shear Zone which is 5–15 km wide and has an overall displacement of *c.*10 km. Coller (1984) presents some geophysical evidence to suggest this shear zone is basement controlled. To its east, transpressive deformation is characterised by linear dextral shear zones bounding less deformed tectonic units, such as the Nenagh Block, with its large-scale open folds, lack of cleavage, and poorly developed vein systems. Higher strain and pressure solution cleavage is concentrated in ENE-trending dextral shear zones associated with brittle faults. Coller (1984) suggested deformation took place during transpression with the NNW pure shear component being significantly greater than the strike-parallel dextral shear. This is closely comparable to the situation further south in the Munster Basin (Price and Todd, 1988) and re-emphasises the continuity of structure across the island.

In areas north-west of the main Leinster Lower Palaeozoic inlier, Rothery (1989) showed that fold orientation is also strongly controlled by pre-Variscan structures with steep pressure solution cleavage only locally developed in the limestones. He interpreted obliquity of minor and major folds to the cleavage as due to ESE–WNW compression followed by later extension (relaxation phase) rather than as progressive transpression.

Deformation of late Dinantian rocks is well displayed on the coastal sections north of Dublin. Cleavage is well developed in muddy lithologies such as the Tober Colleen Formation (Rush Slates of earlier usage) and the Loughshinny Formation (Fig. 12.8c). Johnston (1993) described a complex series of *en echelon* folds and vein arrays from the famous sections at Loughshinny Bay. Cleavage is seen to transect the *en echelon* folds that formed under local transpression. The overall pattern is consistent with NNW shortening. Le Gall (1991) considers the structures in the Dublin Basin to be anomalous and related to basement reactivation. Corfield *et al.* (1996) suggest overall shortening of the Dublin Basin of at least 10% with locally up to 30% in the Loughshinny area, although they do not state how these estimates were derived. They also show north-directed thrusts that dip moderately south, but their section is at variance with the detailed and widely accepted work of Nolan (1986, 1989).

North and north-west Ireland

In the northern and north-western part of Ireland the Variscan structures are dominated by major NE–SW faults with continuity over several tens of kilometres. These faults clearly inherit their structural grain from the Caledonian basement, and many were active during Carboniferous extension and sedimentation, as well as later. Worthington (2006) argues, by considering the scale of known Cenozoic movements in NE Ireland, that the majority of movement on these structures can be attributed to the Carboniferous. He shows that several faults present in pre-Carboniferous rocks, that had previously been considered only to show pre-Carboniferous movement, show similar histories and probably represent Carboniferous extensional faults that may have experienced some subsequent Variscan inversion. Most of the major faults exhibit net extension, but the limited control on the amount of early Carboniferous rifting means that the total amount of Variscan inversion and the amount of post-Variscan extension are difficult to estimate. In general the Variscan deformation was strongly heterogeneous, yielding open flexures with no associated cleavage. However, steep dips and local cleavage are seen where deformation is concentrated along these major fault zones.

The Curlew Mountains–Clogher Valley Fault System (Fig. 12.2) shows spectacular syn-sedimentary movement in the Dinantian (Philcox *et al.*, 1989) as well as later movement during Variscan deformation, and a similar history has been inferred for the Ox Mountains–Pettigo Fault System (George and Oswald, 1957). Even more spectacular syn-sedimentary faulting is seen in the Fintona area (Fig. 12.9), e.g. the Belhavel–Castle Archdale Fault System, (Mitchell, 1992; Mitchell and Owens, 1990) with movement in early and late Dinantian and early Westphalian times (Chapters 10, 11) preceding Variscan inversion. Price and Max (1988) present limited seismic evidence to suggest these major faults in north-west Ireland resemble flower structures, while Mitchell (1992) interpreted the faults in the Fintona area as strike-slip faults.

In the area of thickest sediment accumulation, the Lough Allen Basin, a series of NW–SE trending folds are developed which, when they interfere with the basin axial trends, result in dome and basin structures of *c.*8 km diameter. Millar (1990) suggests that this NW–SE trend is also basement inherited. In the western sections around Donegal Bay subsidiary ENE–WSW lineaments are present that are also thought to be basement inherited (Millar, 1990; Ni Bhroin, 1999).

Although the Variscan movement on the major NE–SW faults is largely dip slip, there is also some evidence for dextral shear (Millar, 1990) and the region shows similarities with structures in the Midland Valley of Scotland. Overall the structures in this region represent a continuation of the Variscan structural pattern seen further south, but with a gradual decrease in strain and an increase in the importance of reactivation of basement structures.

Variscan metamorphism

In contrast to most other parts of the European Variscides, metamorphism in Ireland is low grade, and syntectonic granites are absent. Nevertheless, substantial peak palaeotemperatures are indicated for large areas of southern

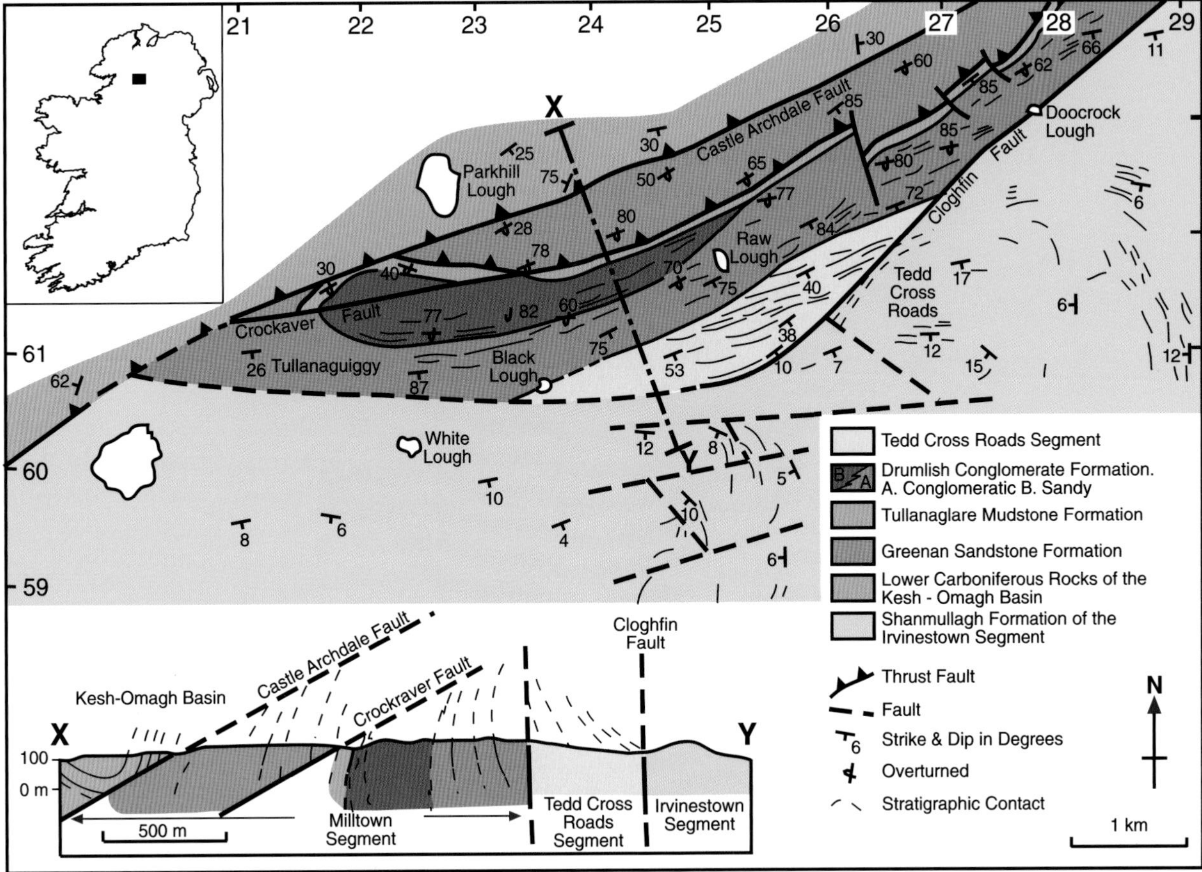

Fig. 12.9 Deformation associated with major faults in Tyrone (*after* Mitchell and Owens, 1990).

Ireland that are difficult to explain. These high temperatures have been inferred from the maturation of organic matter, from conodont alteration indices, and from crystallinity of white mica and chlorite. All of these palaeothermometers are thought to be both temperature and time dependent.

The maturation map of Ireland (Clayton *et al.*, 1989) plots calibrated values from vitrinite and conodont data of rocks at the surface (Fig. 12.10). It shows that much of the south and west of Ireland record high maturation values that appear to be independent of stratigraphic position. This contrasts with the situation in NE Ireland and SE Wexford (Burnett *et al.*, 1990) where maturation levels are much lower and can be explained in terms of reasonably expected overburden and palaeogeothermal gradients.

Blackmore (1995) inferred peak palaeotemperatures of *c.*350 °C for pre-late Bashkirian (Westphalian) rocks in the Munster Basin based on >250 vitrinite measurements. Meere (1995) independently produced similar palaeotemperature estimates (275–325°C) based on illite and chlorite crystallinity. Information from West Clare comes from Serpukhovian–Bashkirian (Namurian) strata in which deformation levels are relatively low, as discussed above. Here E. Fitzgerald *et al.* (1994) record R_{max} values of 6–7.5% from coals and siltstones consistent with CAI (conodont alteration index) values of 5 from conodonts in limestone bullions. Maximum palaeotemperatures in excess of 350°C were estimated. The 'pressure-solution' cleavage fabrics described above are generally considered as typical of temperatures of circa 250–400°C, above which crystal plastic flow processes start to dominate (Rutter, 1976, 1978). Development of this fabric is dependent on both temperature and lithology, and its presence in areas of relatively low strain, such as Clare, is consistent with the high palaeotemperatures indicated by independent metrics. There is also evidence in Clare that coalification, during which the vitrinite values were acquired, predates folding (Clayton and Baily, 1999).

In a profile across the Old Head of Kinsale–Kinsale Harbour area, Clayton (1989) demonstrated that there

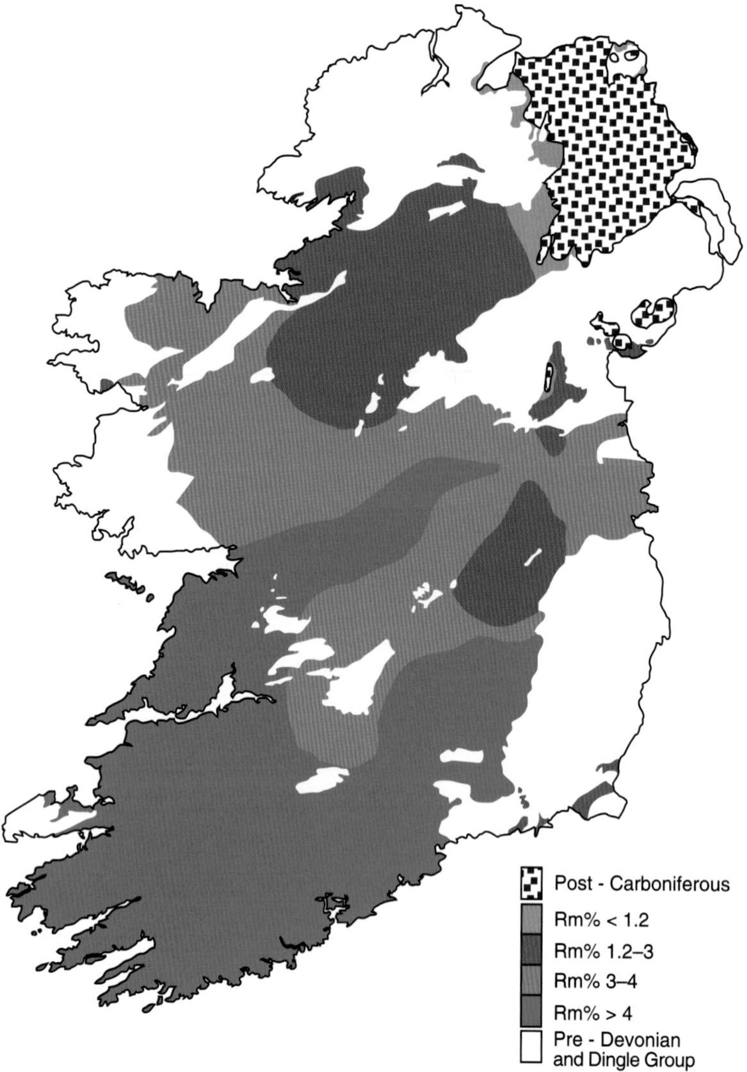

Fig. 12.10 Thermal maturation levels of Devonian and Carboniferous rocks of Ireland (from Clayton *et al.*, 1989).

was little correlation between vitrinite reflectance and stratigraphic position. This suggested that maturation levels were attained or were held at their peak values until the termination of Variscan folding. Blackmore (1995) confirmed the initial work by Clayton (1989) and found little variation of reflectance values with stratigraphic position. Temperature estimates from fluid inclusions in quartz veins in Clare produced during the Variscan deformation give estimates of 220–250°C, significantly lower than the vitrinite peak temperatures. Lower temperatures for vein emplacement were also suggested from work on fluid inclusions in veins from the Munster Basin by Meere (1995). However, whilst it is reasonable to assume that these veins are associated with Variscan deformation, there is little control on their timing.

Estimated thickness of Carboniferous cover appears to be inadequate to generate the inferred palaeotemperatures by simple burial at normal geothermal gradients. As noted above, there is little evidence for substantial crustal thickening by thrust stacking to provide elevated temperatures. Moreover, there is virtually no overall change in maturation between the top and bottom of the >3 km Doonbeg borehole, with a level of reverse gradient in the upper part of the Dinantian carbonates (Fitzgerald *et al.*, 1994; Clayton and Baily, 1999). Later work by Goodhue and Clayton (1999) also encountered reverse gradients in the Slievecallan borehole, but with the reversal taking place in the clastic succession above the limestones. Simple burial metamorphism is again inadequate to explain the data, and fluid advective heating has been suggested as the most likely mechanism (Fitzgerald *et al.*, 1994; Clayton and Baily, 1999).

The maturation map (Clayton *et al.*, 1989) shows that thermal anomalies are widespread in Ireland, and satisfactory explanations require processes that can operate on the required spatial scale. There is potentially much fruitful research to be done in this field to develop and test thermal models, such as the effect of a mantle plume suggested by Keeley (1996).

Timing of deformation and metamorphism

It is difficult to constrain the timing of the Variscan deformation in Ireland with any great precision. The youngest exposed rocks affected by the orogeny are latest Bashkirian (Westphalian B *c.*312 Ma) in age (Chapter 11) although younger Moscovian and Kasimovian (Westphalian and Stephanian *c.*303 Ma) strata are known from the subsurface and offshore. In the southern part of Ireland the orogeny led to a major gap in the depositional record between the Carboniferous and the Jurassic. Jurassic strata are not metamorphosed and have experienced little deformation. Further north in the Kingscourt area, deposition had evidently resumed by the Upper Permian. Thus the most likely timing of the peak Variscan deformation is latest Carboniferous to early Permian.

Critical in assessing the timing of deformation are a suite of felsic and alkaline dykes, sills and pipes in the western part of the Beara Peninsula. These intrusions have been described by Boldy (1955), Coe (1966, 1969), Pracht and Kinnaird (1995, 1997), Pracht and Timmerman (2004) and Quinn *et al.* (2005). The dykes and sills are mainly trachytic whilst the pipes have lamprophyric composition. In most places, the intrusive rocks are strongly cleaved, and thus dating their age of crystallisation can provide a maximum age for the deformation. The breccia pipe at Black Ball Head is particularly important due to the radiometric dates that have been obtained from material in its inner zone, which contains amphibole-pyroxene nodules and kaersutite (Ti-rich hornblende), Ti-magnetite and phlogopite megacrysts. It is suggested that the megacrysts and nodules all formed in the same magma, as they plot in the same PT field. The assemblage suggests crystallisation at *c.*75 km depth (Pracht and Kinnaird, 1995). The megacrysts occur only as individual crystals, which is interpreted as evidence for the expansive nature of the fluidised igneous material during emplacement. Megacrysts are unaltered apart from thin rims of chlorite and clay minerals and minor calcite along cleavage planes. (Pracht and Kinnaird, 2004). There is general agreement that this inner zone of the intrusion represents rapid ascent of magma to prevent re-equilibration of the megacryst assemblage with the magma at lower PT conditions. Dating of two kaersutite crystals by Ar–Ar step heating yielded an estimate for this age of 318 ± 3 Ma (Pracht and Timmerman, 2004). Dating of large phlogopite crystals by $^{40}Ar/^{39}Ar$ from this same inner zone of the Black Ball Head intrusion yielded ages of 296.34 ± 13.03 Ma to 322.2 ± 1.61 Ma, and a weighted mean age of 314.44 ± 1.0 Ma (Quinn *et al.*, 2005). Quinn *et al.* (2005) argue that as the closure temperatures for these minerals is significantly greater than the estimated ambient temperature of the country rocks, dating the megacrysts can be reasonably interpreted as dating the time of emplacement of the pipe, although the depth of emplacement is very difficult to estimate.

The geometry of the large pipe at Black Ball Head, which has near vertical margins with the country rock, has been used to suggest that this pipe post-dates much of the deformation (Quinn *et al.*, 2005). The consistent strike and uniformly steep dip of the cleavage fabric have been taken to indicate that this fabric has not been significantly rotated, and thus largely post-dates folding. Xenoliths of fine-grained country rock contain one cleavage, and the fact that cleavage in the tuff matrix wraps around the xenoliths has been used to suggest syntectonic emplacement (Pracht and Timmerman, 2004). These authors present stereographic data that show parallelism between cleavages in the pipe and those in the country rock. However, a plot of cleavage fabric from the same locality by Quinn *et al.* (2005) suggests the orientation of the cleavage fabric cutting the intrusion is consistently clockwise of the regional cleavage and overprints the marginal breccia. However, the interpretation of these geometries in an area with strong lithological contrasts may not be unambiguous.

The field relations of many of the minor intrusions suggest that dykes and sills were emplaced during both ductile and brittle regimes. Dates from other cleaved intrusions range from 298.08 ± 0.61 Ma to 296.88 ± 0.6 Ma (Quinn *et al.*, 2005), which would represent an earliest Permian age of intrusion. The youngest of these ages comes from the centre of a thin composite dyke that appears to lack fabric, although the outer parts of the dyke do display a fabric. This has been interpreted by Quinn *et al.* (2005) as indicating a post-deformation age.

These dates provide some constraints on the timing of both deformation and metamorphism, although apparent inconsistencies remain. If the younger dates from cleaved intrusions are accepted, then ductile deformation continued until *c.*297 Ma. If the cleaved Black Ball Head intrusion is accepted as being syn-cleavage formation, as argued by Quinn *et al.* (2005) then ductile deformation was already in progress at 314–318 Ma. This gives perhaps an unreasonably long period of ductile deformation. An alternative explanation is that the Black Ball Head intrusion is entirely pre-deformation and that the duration of ductile deformation is much less.

In the context of Ireland, it is possible that deformation and uplift may have been progressive and occurred over a significant time span rather than in one short pulse. There is evidence for major fault movements in the late Bashkirian (Westphalian) in the Fintona area (Chapter 11) that may relate to an overall compressional regime, despite sedimentation in areas of localised extension. Similarly there is evidence for reworking of earlier Carboniferous sediments in some of the Moscovian (Westphalian D) strata of the Kish Basin and Wexford, although these Westphalian strata also show Variscan deformation.

There is also the problem, particularly in the northern part of Ireland, of separating deformation that is truly Variscan from that which is much later, i.e. Mesozoic or Tertiary, reactivation. Thus the major structures marked on Figure 12.2 that affect Carboniferous rocks are considered here to have a component of their history that is Variscan, but probably a later component also. This makes the Variscan component difficult to quantify.

13

Permian and Mesozoic

M. J. Simms

Much of Ireland's surface geology comprises a thin and patchy veneer of, often poorly consolidated, late Pleistocene glacial sediments over a basement of Palaeozoic or older rocks. Only in the north-east of Ireland are intervening strata present across a large area, and even here the outcrop is dominated by Cenozoic rocks, principally the Palaeocene Antrim Lava Group and the Oligocene Lough Neagh Group. Elsewhere Permian and Mesozoic rocks are confined to a few widely scattered, and mostly very small, outcrops. Although Permian and Mesozoic strata occur extensively beneath the Cenozoic cover in Ulster, their actual outcrop constitutes only a tiny proportion, probably less than 1%, of the total sub-Pleistocene outcrop across the whole of Ireland, even though this time period spans some 185 million years, or about 10%, of Ireland's geological rock record.

Within this 185 million year interval the succession is far from complete. For instance, a gap of almost 100 Ma exists between the youngest preserved Jurassic strata (lower Pliensbachian Stage) and the oldest preserved Cretaceous rocks (Cenomanian Stage) in Northern Ireland. Farther south the succession is far less complete than this, with post-Triassic strata represented by only three tiny, and somewhat enigmatic, outcrops. Despite this remarkably sparse record the Permian and Mesozoic succession provides a great deal of information on a significant part of the post-Carboniferous history of Ireland, while also posing many intriguing questions.

Substantially thicker and more complete Permian and Mesozoic successions occur at various locations offshore around Ireland (Chapter 17). These offshore basins are genetically linked to those present onshore, being formed by the same processes at similar times, but they will not be discussed further here.

Ireland in a global context

The geological history of Ireland through the Permian and Mesozoic should not be viewed in isolation. Although the preserved succession inevitably reflects local or regional factors, global factors and events necessarily influence what is formed and preserved in the geological record.

Permian times saw the final coalescence of most of the Earth's continental plates into a single supercontinent, Pangaea, with the creation of its oceanic counterpart Panthalassa. What is now Ireland lay only about 10° North of the Equator and about 1000 km north-west of the Palaeo-Tethys Ocean at the westernmost extremity of Panthalassa. Ireland was separated from the ocean by the Variscan mountain chain, a situation that largely persisted throughout Permian and Triassic times. By late Triassic times the breakup of Pangaea had commenced and, by mid-Jurassic times, the Atlantic Ocean had begun to open. Ireland continued to drift northwards, to about 20° North by the end of the Triassic and about 40° N by late Cretaceous times. Inevitably, such changes in latitude and geographical setting have profoundly influenced the climate and environment that Ireland experienced through the Permian and Mesozoic.

Relative and global sea level have played an important part in the formation of Ireland's geological record. Global sea level was generally low through the Permian and Triassic, as recorded in the predominantly non-marine successions from these periods. Relatively brief marine transgressions are represented by the limestone and evaporites found at several levels. There was a general rise in sea level from the late Triassic into the early Jurassic, and it attained a global maximum in late Cretaceous times, as evidenced by the widespread deposition of pure coccolith chalks. Other global-scale events, such as the eruption of major flood basalts at the end of the Permian and end of the Triassic, and the end-Cretaceous meteorite impact, have left little or no direct evidence in Ireland's stratigraphical column although they must have had a profound influence on the environment here.

Permian and Mesozoic basins

The history of Permian and Mesozoic basins in Ireland is intimately connected with the breakup of Pangaea and the subsequent sequence of events that led, ultimately, to the creation of the North Atlantic Ocean. However, these basins often contain within them 'echoes' of much earlier plate tectonic movements extending far back into the Palaeozoic. Lower Palaeozoic rocks in Ireland may be divided into a series of NE–SW trending tectonic terranes bounded by major faults (Murphy *et al.*, 1991). These sometimes form the bounding faults for much later Mesozoic basins, while Permian and Mesozoic basins developed on each of these terranes may show their own distinctive character (Anderson *et al.*, 1995). Tensional stresses within the European Plate, with Ireland located at its western edge, caused a series of discrete sedimentary basins to develop, some now located onshore with others lying offshore. Although the underlying tensional stress was determined by the overall directions of movement of the European and North American plates, the actual fault-bounded configuration of individual basins remained strongly influenced by pre-existing fractures. Most of these basins are located in the north of Ireland and there is considerable overlap of Mesozoic fills between adjacent basins, such that the outcrop and subcrop of Mesozoic strata is fairly continuous across a large area of north-east Ireland. Several distinct depocentres, containing substantially thicker Permian and Mesozoic fills, help to define individual basins within this broader outcrop/subcrop (Fig. 13.1).

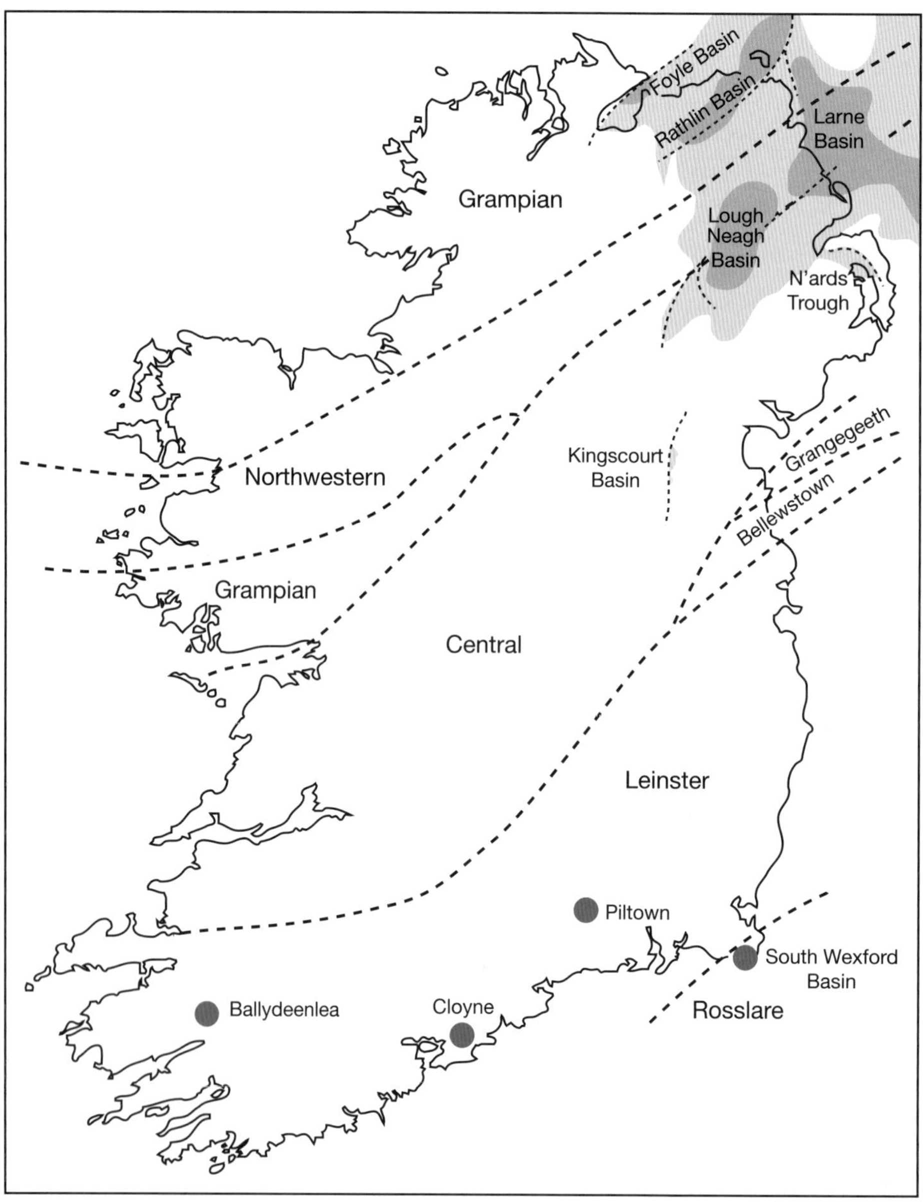

Fig. 13.1 Location of the main depositional basins, and other occurrences of Permian and Mesozoic strata in Ireland, and their relationship to the tectonic terranes of Murphy *et al.* (1991).

Lough Foyle and Rathlin basins

These contiguous transtensional half grabens are bounded respectively by the Lough Foyle Fault to the north-west and the Tow Valley Fault to the south-east, both of which represent reactivated Caledonide strike-slip faults associated with the Grampian Terrane. The obvious manifestation of the Lough Foyle Fault is spectacular, with the Palaeocene basalt scarp of Binevinagh looming over the Mesozoic strata of the Magilligan lowlands, and facing Dalradian metasediments that form the hills of Inishowen to the north-west of the fault. Both basins contain thick Mesozoic successions which in the Rathlin Basin, far the larger of the two, attain a thickness of perhaps 2 km at its depocentre near the Tow Valley Fault. Within the confines of the two main bounding faults deposition was strongly influenced by extension on NNW–SSE normal faults.

Lough Neagh and Larne basins

Like the Lough Foyle and Rathlin basins to the north-west, these two basins have been influenced by several major ENE–WSW Caledonide structures, among them the Highland Boundary and Southern Upland fault zones (T.B. Anderson *et al.*, 1995). However, subsidence was, judging from outcrop patterns, more strongly influenced in post-Palaeozoic times by other normal faults trending NE–SW (Lough Neagh Basin) or NNW–SSE (Larne Basin). The Larne and Lough Neagh basins are located on the Midland Valley Terrane and are less clearly defined and less elongated than basins on other terranes. Across much of the Midland Valley Terrane, both here and in Scotland, the Lower Palaeozoic basement is separated from the Permian and Mesozoic fill by Carboniferous strata that may have 'dampened' the influence of reactivation of the underlying faults (T.B. Anderson *et al.*, 1995).

The Permian to Lower Jurassic fills of these two basins are thicker than correlative successions in many basins in Britain, reaching perhaps 3–4 km in thickness towards their depocentres, but the Cretaceous succession is, in contrast, significantly thinner. The pre-Cenozoic fill of the Larne Basin is patchily exposed at the surface around the margins of the basalt plateau, but in the Lough Neagh Basin is largely concealed by Palaeocene basalts. However, both basins have been penetrated by deep boreholes which show that their Mesozoic successions are fairly similar, although the underlying Permian shows significant differences (Shelton, 1997).

Newtownards Trough

A NW–SE trending half graben, bounded on its NE side by the Newtownards Fault, cuts strikingly across the ENE–WSW 'grain' of the surrounding Lower Palaeozoic mudstones and greywackes of the Longford-Down massif. Located on the Southern Uplands Terrane, it is more similar to Permo-Trias basins in the Southern Uplands of Scotland than to other Mesozoic basins in Northern Ireland. T.B. Anderson *et al.* (1995) suggested that it formed by dextral strike-slip movement along pre-existing N–S or NW–SE Caledonide fractures in response to ENE–WSW crustal stretching. As much as 700 m of Permian and early Triassic strata are preserved in this basin.

Kingscourt Basin

This, the only major outcrop of Permian and Mesozoic strata south of Ulster, is a half graben bounded to the west by the N–S oriented Kingscourt Fault, which downthrows to the east by as much as 2100 m. Permian and Triassic red beds, with economically viable gypsum deposits, reach a maximum thickness of 550 m (Gardner and McArdle, 1992). Like the Newtownards Trough, the Kingscourt Basin lies on the Southern Uplands Terrane and the Kingscourt Fault formed by extensional reactivation of a sinistral strike-slip fault in the Lower Palaeozoic basement beneath (T.B. Anderson *et al.*, 1995).

South Wexford Basin

Drilling in south County Wexford, within the Rosslare Terrane of Murphy *et al.* (1991), unexpectedly revealed, beneath a thick Pleistocene overburden, the presence of a small complex half graben containing presumed Permo-Triassic strata resting unconformably on the Carboniferous (Clayton *et al.*, 1986b). No confirmatory evidence for their age has yet been found, and their assignment to the Permian or Triassic is based solely on their red-bed facies and presence of Westphalian strata beneath the unconformity.

Stratigraphy of the Permo-Triassic

Permian and Triassic strata are best considered together, since they are dominated by red-bed sediments which are often rather poorly dated. Permian strata show particular diversity across the different basins in Ireland and include evidence both for volcanism and for a major marine incursion. In contrast the Triassic successions are more uniform across the country (Fig. 13.2).

SYSTEM		STAGE		Lithostratigraphy (Northern Ireland)	Lithostratigraphy (Southern Ireland)
Palaeocene		**Danian**			
Cretaceous	Late	Maastrichtian	Late		
			Early	Ulster White Limestone Formation	'Ballydeenlea Chalk' *(Campanian)*
		Campanian			
		Santonian		Hibernian Greensands Formation	
		Coniacian			
		Turonian			
		Cenomanian			
	Early	Albian			
		Aptian			
		Barremian			
		Hauterivian			
		Valanginian			
		Berriasian			'Piltown Clay' *(Kimmeridgian-Berriasian)*
Jurassic	Late	Portlandian			
		Kimmeridgian			
		Oxfordian			
	Middle	Callovian			
		Bathonian			'Cloyne Clay' *(late Toarcian-Bathonian)*
		Bajocian			
		Aalenian			
	Early	Toarcian			
		Pliensbachian		Waterloo Mudstone Formation	
		Sinemurian			
		Hettangian			
Triassic	Late	Rhaetian		Penarth Group	
		Norian		Mercia Mudstone Group	
		Carnian			
	Middle	Ladinian			
		Anisian			
	Early (Scythian)	Olenekian		Sherwood Sandstone Group	Kingscourt Sandstone Formation
		Induanian		? — ?	
Permian	Late – Lopingian	Changhsingian		Belfast Group	Kingscourt Gypsum Formation
		Wuchiapingian		? — ?	
	Middle – Guadalupian	Capitanian			
		Wordian			
		Roadian		? — ?	
	Early – Cisuralian	Kungurian		Enler Group	Killag Formation (Wexford Basin) *(? Permo-Trias)*
		Artinskian		? — ?	
		Sakmarian			
		Asselian			
Carboniferous		Stephanian			

Fig. 13.2 Generalised Permian to Cretaceous stratigraphy of Ireland.

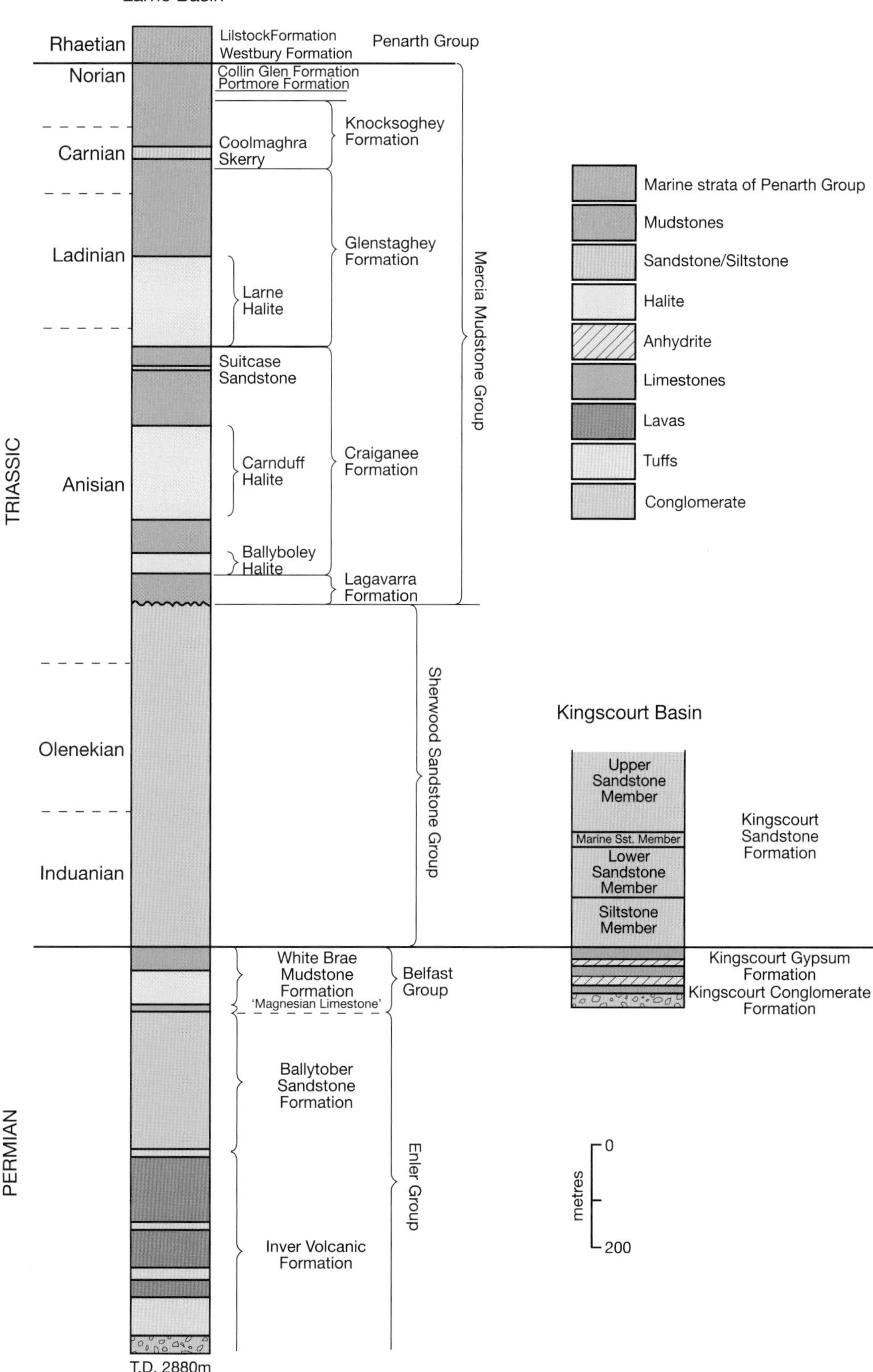

Fig. 13.3 Generalised Permian and Triassic stratigraphy of Northern Ireland and the Kingscourt Basin. The Larne Basin succession is based largely on that encountered in the Larne No. 2 Borehole and differs from that recorded in Larne No. 1 Borehole. The position of System and Stage boundaries is not known precisely.

Fig. 13.4 Red and grey Permian mudstones with gypsum beds at the former Knocknacran Quarry, in the Kingscourt Basin (Grid ref. N 811996). The Upper Gypsum Bed, in the background, and the thicker Lower Gypsum Bed are separated by red mudstones.

The Wexford and Kingscourt basins

In the Kingscourt Basin (Fig. 13.3) Visscher (1971) recognised two conformable formations, corresponding broadly to the Permian and Lower Triassic parts of the succession. The 120 m thick *Kingscourt Gypsum Formation* is dominated by red mudstones with two significant gypsum/anhydrite accumulations (Fig. 13.4); the lower 20–35 m thick and the upper 6–10 m thick. At the base of the formation a conglomerate, up to 18 m thick, unconformably overlies reddened Westphalian strata. It thickens towards the fault and is dominated by Lower Palaeozoic and some Silesian clasts, but Dinantian clasts are absent (Geraghty, 1997). The overlying *Kingscourt Sandstone Formation*, dated on palynological evidence as Lower Triassic, is more than 400 m thick and comprises mostly terrestrial red-bed siltstones and sandstones. However, marine foraminifera and acritarchs in the Marine Sandstone Member, up to 30 m of grey-white sandstone around the middle of the formation, testify to a minor marine incursion (Gardiner and Visscher, 1971).

Little is known about the supposed Permo-Triassic strata in the South Wexford Basin. The assumed age is based on the unconformable relationship to underlying Carboniferous strata and their red-bed facies, and it is possible they could even be as young as Jurassic. Clayton *et al.* (1986b) assigned this post-Westphalian succession to a single lithostratigraphical unit, the *Killag Formation*, comprising about 240 m of sandstone, siltstone and conglomerate resting unconformably on Dinantian to Westphalian rocks. The polymodal and polymictic conglomerates contain clasts, some more than 0.4 m across, derived from local Dinantian limestones, Rosslare Complex gneisses, granites similar to those at Carnsore Point and Rosslare Harbour, siltstone from the Tagoat Formation, and a rhyolite of unknown source. The sandstones and siltstones are predominantly red with a few green-grey patches. A minor development of carbonate nodules possibly represents calcrete.

Permian of the Northern Ireland basins

By far the most extensive development of Permo-Triassic strata in Ireland is found in the Larne (Fig. 13.3) and Lough Neagh basins, although much of this is deeply buried beneath later Mesozoic and Cenozoic strata. Thick Permo-Triassic successions are also present to the north-west, in the Rathlin and Lough Foyle basins, and to the south in the Newtownards Trough. Outcrops, where they occur at all, are narrow and confined to the edges of the vast basalt plateau. The few exposed sections reveal only a rather limited range of lithologies. However, deep boreholes have revealed a greater diversity of facies.

The foreshore exposure at Cultra (Fig. 13.5), on the south shore of Belfast Lough, exemplifies the Permian succession generally found around the margins of the Lough Neagh and Larne basins and comprises three distinct facies. A basal breccia unconformably overlies thinly bedded strata of the Lower Carboniferous Ballycultra Formation but is dominated by clasts of Lower Palaeozoic slatey mudstone and greywacke like that which outcrops in the low hills to the south. The normally dark grey mudstones and dolomites of the Ballycultra Formation are discoloured to mauves and yellows for several metres below the unconformity surface. Formerly called 'Brockram', this breccia now forms part of the *Enler Group*. The Enler Group is barely two metres thick at Cultra, but in various boreholes along the southern edge of the Lagan Valley its thickness varies from a few metres to more than 130 m, and it may be divisible into an upper sandstone and a lower breccia (Smith, 1986). In the adjacent Newtownards Trough the Ballyalton Borehole proved almost 300 m of breccias with subordinate sandstones (Bazley, 1975) that probably can be correlated, at least in part, with the Enler Group. Breccias of uncertain, but possibly Permian, age occur in the south-west of the Lough Neagh Basin. Near Cookstown the clasts are of granite and schist, while around Armagh they are of Carboniferous limestone (Hartley, 1949). These various basin margin breccias probably represent screes and outwash fans deposited by occasional floods within a prevailing arid climate, bringing material down from the Palaeozoic uplands to the south and west.

Fig. 13.5 Seaward-dipping Permian strata exposed on the shore at Cultra, on the south side of Belfast Lough (Grid ref. J413809). Thinly-bedded strata of the Ballycultra Formation (Lower Carboniferous) in the foreground are unconformably overlain by a brown, poorly-bedded breccia of the Enler Group. The basal part of the orange-brown Magnesian Limestone forms a prominent feature around mid-shore, with its top just visible below the red-brown siltstones of the succeeding Connswater Marl Formation which are exposed towards low water. The lower slopes of Knockagh Mountain, on the north side of the lough, are of Triassic Mercia Mudstone Group, overlain unconformably by Ulster White Limestone Formation and the succeeding Palaeocene Antrim Lava Group. *All photographs in this chapter: M. J. Simms.*

On the shore at Cultra the dull brown breccia of the Enler Group is succeeded by more than 8 m of creamy yellow dolomitic limestone, long termed the Magnesian Limestone. The Magnesian Limestone in Northern Ireland is now incorporated in a more inclusive Belfast Harbour Evaporite Formation, more than 30 m thick, which encompasses a thin sandstone and siltstone below the limestone and more than 10 m of anhydrite and evaporitic intraformational breccia above. It has been encountered in various boreholes around the south-western margins of the Larne and Lough Neagh basins, although not in the Newtownards Trough, and is also exposed, albeit poorly, to the west of Lough Neagh near Stewartstown. In the Belfast Harbour Borehole (Smith, 1986) the Magnesian Limestone comprises a series of distinct micrite and oolite units, together more than 15 m thick, while near Stewartstown it is more than 20 m thick. A low diversity fauna of marine fossils, principally casts of bivalves and gastropods, is locally common at some levels. The most abundant is the bivalve *Bakevillia binneyi* but the brachiopod *Horridonia horrida* was recorded by Hull et al. (1871). They establish the Upper Permian age of these beds and their probable correlation with the Lower Magnesian Limestone, EZ1 Cycle, of north-east England (Smith *et al.*, 1986). The Magnesian Limestone and its fossil fauna provide evidence for a brief marine incursion to the region during a major late Permian transgression. Evidence from the more fossiliferous succession in County Tyrone in fact indicates two transgressive–regressive cycles within it (Pattison, 1970). The succeeding evaporites probably represent supratidal or sabkha deposition during the subsequent slow regression. The Belfast Harbour Evaporite Member is succeeded by the Connswater Marl Formation, forming the upper part of the Belfast Group, comprising up to 110 m of brick-red, often gypsiferous or anhydritic, silty mudstones. A Permian age for all or part of this formation is only assumed and the boundary with the succeeding Sherwood Sandstone Group, traditionally considered Triassic in age, is gradational.

Deep boreholes away from the basin margins, such as at Ballymacilroy in the Larne Basin, and the Portmore Borehole in the Rathlin Basin, have proved broadly similar successions. The Belfast Harbour Evaporite Member is widely recognised in these boreholes but other lithostratigraphical units, which in some instances have been afforded Formation or Member status in borehole reports, cannot be widely correlated.

The Larne No. 2 Borehole, close to the depocentre of the Larne Basin, revealed a Permian succession more than 1200 m thick and significantly different and more varied than any other in Ireland. The Belfast Harbour Evaporite Formation here comprises 14 m of Magnesian Limestone capped by 6 m of anhydrite, but 113 m of halite separates the anhydrite from red mudstones, presumed correlatives of the Connswater Marl Formation, which continue up to the base of the Sherwood Sandstone Group. Beneath the Magnesian Limestone the borehole penetrated more than 1000 m of strata, assigned to the Enler Group, without reaching the presumed Carboniferous basement. More than 440 m of reddish- to purplish-brown sands with some pebble beds, the Ballytober Sandstone Formation, overlie more than 550 m of basaltic to trachytic and trachyandesitic lavas and tuffs of the Inver Volcanic Formation. The lowest strata reached comprise about 60 m of coarse clastics, coarsening downwards to a breccia-conglomerate with a tuffaceous matrix and lava clasts. The lavas are heavily altered and yielded a radiometric date of 245 ± 13 Ma (Late Permian to Mid-Triassic) (Penn *et al.*, 1983), although an Early Permian age is probable both on account of its position substantially below the Magnesian Limestone and through comparison with early Permian volcanics in south-west Scotland (Shelton, 1997).

Triassic of the Northern Ireland basins

The Triassic succession in Northern Ireland (Fig. 13.3), as elsewhere in the UK, shows a clear lithostratigraphical division into three distinct Groups. The lower two of these, the Sherwood Sandstone and Mercia Mudstone groups, account for more than 90% of the total thickness in basinal successions and are dominated by various non-marine red-bed facies. At the top lies the Penarth Group, a much thinner succession of quasi-marine mudstones, siltstones and sandstones forming a transition to the fully marine conditions that became established in the early Jurassic.

Sherwood Sandstone

The Sherwood Sandstone Group succeeds the late Permian Belfast Group, although precise lithostratigraphical and chronostratigraphical boundaries are not clearly defined. Palynological evidence suggests that the Permo-Triassic boundary actually lies below the base of the Sherwood Sandstone Group but, for convenience, the System boundary is placed at the lithostratigraphical boundary, which marks a general change from predominantly argillaceous rocks of the Belfast Group to sandstones in which mudstones are subordinate. Fowler and Robbie (1961) and Naylor *et al.* (2003) noted a broad bipartite lithological division of the Sherwood Sandstone Group. This was formalised by the latter authors as an upper Toomebridge Sandstone Formation, dominated by weakly cemented sandstones, and a lower Drumcullen Formation which includes a higher proportion of red mudstones. The lower of these perhaps corresponds to the Lagan Sandstone and Loughside Sandstone formations

Fig. 13.6 Thinly-bedded sandstones, siltstones and subordinate mudstones of the Sherwood Sandstone Group exposed on the shore of Belfast Lough at Cultra (Grid ref. J 403799).

encountered in the Belfast Harbour Borehole (Smith, 1986).

The Sherwood Sandstone Group, almost 650 m thick in the Larne Basin and more than 500 m in the Rathlin Basin, thins dramatically and shows significant facies changes towards the basin margins and over basement highs. Typical basinal facies can be seen at various locations on the northern and southern shores of Belfast Lough (Fig. 13.6), the east shore of Strangford Lough, and in disused quarries inland, notably at Scrabo Hill in north County Down where extensive vertical sections have formed the basis for the most detailed interpretation of this part of the succession in Northern Ireland (Buckman *et al.*, 1998). As its name implies, fine- to medium-grained sandstones predominate, with mudstones forming a minor but significant component. Individual beds may be up to a metre or two thick, with trough and planar cross-bedding commonplace along with smaller sedimentary structures such as ripples, primary current lineation, and desiccation cracks. These sedimentary structures, interpreted as largely fluvial in origin, have been used to define a series of distinct facies. Cross-beds are interpreted as fluvial dunes, and mudstones as overbank deposits, although several lines of evidence point to significant aeolian activity at times (Buckman *et al.*, 1998).

Basinal facies of the Sherwood Sandstone Group contain a sparse, yet surprisingly diverse, assemblage of trace fossils. Burrows, trails, and trackways, variously ascribed to scorpions and other invertebrates, have been described from Scrabo along with rare vertebrate footprints assigned to the ichnogenus *Cheirotherium* (Buckman *et al.*, 1998; Young, 1883). Body fossils are rarer still. In a quarry, now long infilled, at Rhone Hill near Dungannon, intact specimens of a small fish, *Palaeoniscus catopterus*, were found crowded together over a few square metres of a single sandstone bedding plane (Murchison, 1836; Portlock, 1843). Otherwise the only body fossil is the branchiopod ('water flea') *Euestheria minuta,* which has been found at various levels in boreholes and exposures (Fowler and Robbie, 1961). Although rare, these various trace and body fossils indicate that temporary freshwater or brackish lakes developed at times during deposition of the Sherwood Sandstone. None of these fossils has any biostratigraphical significance, but sparse palynomorphs from boreholes indicate an Olenekian age (Lower Triassic, Scythian Stage) for the middle and upper parts of the Sherwood Sandstone Group in Northern Ireland and an Anisian age (Middle Triassic) towards the top (Warrington, 1995). However, it is probable that the top of the Sherwood Sandstone may well be diachronous, particularly towards the basin margin where the Triassic succession onlaps onto Palaeozoic sandstones. Under such circumstances significantly younger parts of the Triassic may be developed in sandy facies through reworking of the underlying sands.

Towards the basin margins the Sherwood Sandstone Group thins markedly and a range of 'marginal' facies are developed as the influence of surrounding upland areas becomes more marked. Conglomerates, predominantly with clasts from adjacent or underlying lithologies, occur at various levels within the Sherwood Sandstone Group, particularly at or towards the base. These marginal conglomerates probably represent alluvial fans derived from uplands at the basin margins. Such an attenuated marginal succession, only 26 m thick and with a conspicuous basal conglomerate, is well exposed at Murlough Bay in County Antrim, on the basement horst of the Highland Border Ridge. It mostly overlies Dalradian metasediments and the conglomerate accordingly is dominated by quartz and schist pebbles. Where it overlaps onto adjacent Carboniferous, it contains occasional pebbles of Carboniferous basalt and sandstone.

Beds of oolite up to a metre thick occur towards the base of the Sherwood Sandstone Group in boreholes and at outcrop, particularly west of Lough Neagh (Mitchell, 2004), and are well exposed in Drapersfield Quarry, near Cookstown. Their occurrence appears to be determined by the adjacency of Carboniferous limestone basement. They may have formed in shallow, wave-agitated lakes fed by carbonate-saturated springs emerging from the limestone (Milroy and Wright, 2000). Even in basinal successions the adjacency of the basin margin to the depocentre may be very evident. In the Foyle Basin stringers and distinct beds of quartzite and schist pebbles, derived from the Dalradian metasediments to the west, are common in the more than 400 m thick development of Sherwood Sandstone Group penetrated by the Magilligan Borehole (Bazley *et al.*, 1997). These probably represent gravel bars deposited in migrating channels of braided streams. Similarly, the Portmore Borehole lies close to the depocentre of the Rathlin Basin, with the Sherwood Sandstone Group more than 500 m thick, yet also is close to the basin margin and the Highland Border Ridge to the east. As in the Magilligan Borehole, stringers and beds of pebbles, mostly of quartz and quartzite, are common within the sandstones and there is a major development of conglomerates, totalling almost 80 m, towards the middle of the group (Wilson and Manning, 1978).

The Red Arch Formation near Cushendall, County Antrim, has in the past been considered as Triassic in age (Wilson, 1953) but palaeomagnetic data indicate a probable Devonian depositional age with remagnetisation in the early Permian. However, the conglomerates are cut by thin (cm-scale) sandstone dykes which significantly postdate this early Permian remagnetisation and probably were remagnetised in the Palaeogene. On the basis of overall Northern Ireland lithostratigraphy, and an appropriate orientation relative to the regional stress field during the Triassic, a Triassic age has been postulated for these dykes (Turner *et al.*, 2000).

Mercia Mudstone

The Mercia Mudstone Group (Fig. 13.3), which conformably succeeds the Sherwood Sandstone Group, is the thickest of the three major lithostratigraphical units in the Triassic of Northern Ireland, and represents substantially the greatest length of geological time. Thicknesses towards basin depocentres are very substantial. In the Larne Basin (Larne No. 1 Borehole) it is more than 1000 m thick although this includes almost 400 m of halite (Manning and Wilson, 1975). It is at least 500 m thick in the Lough Neagh Basin (Ballymacilroy Borehole), more than 600 m in the Rathlin Basin (Port More Borehole), and more than 370 m in the Foyle Basin (Magilligan Borehole). The Mercia Mudstone Group is dominated by reddish to brown calcareous or dolomitic mudstone and siltstone. Sandstones generally are a very minor component except near the base, where there is a gradational transition from the Sherwood Sandstone Group below. Much of what is known of the stratigraphy of the Mercia Mudstone Group in Northern Ireland is derived from borehole data. Six lithostratigraphical divisions have been formally named, based originally on the succession in the Port More Borehole (Wilson and Manning, 1978), but surface exposures generally are so poor and of such limited extent that only the highest and lowest of these divisions can be recognised with any confidence in isolated exposures.

The *Lagavarra Formation*, the lowest of the six, clearly represents a facies transitional from the Sherwood Sandstone Group below and comprises dm- to m-scale alternations of mudstone with sandstone and siltstone, often with ripples, desiccation cracks, or load casts. Foreshore exposures near Greenisland, on the north shore of Belfast Lough, clearly represent this formation. The succeeding Craiganee, Glenstaghey, Knocksoghey, and Port More formations are each dominated by reddish-brown silty mudstones, green-mottled in places and with varying developments of evaporites.

Three major halite units were encountered within the Craiganee and Glenstaghey formations in Larne boreholes No. 1 and No. 2, both drilled close to the depocentre of the Larne Basin. Although less than a kilometre apart, there are very substantial differences in the thicknesses of the halite units between them. In Larne No. 1 borehole the two lower halites (Carnduff Halite and Ballyboley Halite members) are relatively minor (41 m and 26 m thick respectively) and incorporate >50% mudstone, while the Larne Halite Member is >440 m thick, of which >330 m is salt. Intriguingly, correlative thicknesses in Larne No. 2 borehole, <600 m away, are very different, at 40 m, 180 m, and 178 m for the Ballyboley, Carnduff, and Larne Halite members respectively, while in the Newmill No. 1 borehole, 8 km to the SE, four halite zones of (from the base upwards) 74 m, 86 m, 77 m, and 24 m were encountered (McCann, 1990). To what extent these differences represent original depositional thicknesses or are due to halokinesis remains unclear. Halite has not been found in either the Lough Neagh or Foyle/Rathlin basins, though deposits might exist closer to the depocentres of these basins. Halites are, of course, entirely absent at outcrop of the Mercia Mudstone Group, although nodular and fibrous gypsum occurs locally in some of the mudstones, and siltstone pseudomorphs after halite are common at some horizons. Subsurface dissolution of halite, and associated subsidence, may account for locally steep dips at outcrop, as for instance on the shore east of Carrickfergus.

In the massive mudstones of the Knocksoghey Formation there is an almost 10 m thick bed of hard green siltstone and sandstone interbedded with mudstone. Known as the Coolmaghra Skerry Member, its stratigraphical position about 150 m below the top of the Mercia Mudstone Group, and the presence within it of late Triassic (Carnian) palynomorphs, suggests correlation with the Arden Sandstone of the English midlands and, farther afield, the German Schilfsandstein. These widespread sandstones have been taken as one line of evidence for a period of more humid climate during the arid to semi-arid climate that prevailed through much of the Triassic (Simms and Ruffell, 1990).

The uppermost unit of the Mercia Mudstone Group, the *Collin Glen Formation*, is one of the thinnest, at barely 10 m, but also one of the most distinctive, being composed of pale greyish-green silty mudstone lying immediately beneath the much darker, and strikingly different, Penarth Group. It is well exposed on the shore at Waterloo Bay but Palaeocene thermal metamorphism has altered the colour of several tens of metres of the underlying Portmore Formation, from brick-red to grey-green, and hence the two are difficult to separate at this locality.

No macrofossils have been found in the Mercia Mudstone Group, but palynomorphs from a few horizons indicate a mid-Triassic (Anisian) age for the Lagavarra to lower Glenstaghey formations, a mid- to late Triassic (Ladinian to Carnian) age for the top beds of the Glenstaghey Formation and succeeding Knocksoghey Formation, and a late Triassic (Norian) age for the Collin Glen Formation.

Marine strata of the latest Triassic

Although there is some evidence for a marine influence during deposition of the Mercia Mudstone Group halites, and in the later Collin Glen Formation, it is not until very latest Triassic times that unequivocally marine conditions were re-established in Northern Ireland for the first time since the late Permian more than 50 million years earlier. These late Triassic strata comprise the Penarth Group and the lowest few metres of the succeeding, largely Jurassic, Lias Group. They have a very limited outcrop and still more limited exposure. That on the foreshore at Waterloo Bay, on the outskirts of Larne, affords the finest single exposure through the Triassic–Jurassic boundary interval (Fig. 13.7). It extends from the Portmore Formation of the Mercia Mudstone Group (Upper Triassic, Norian Stage), through the entire Penarth Group (Upper Triassic, Rhaetian Stage), into the Lias Group (Lower Jurassic, Hettangian and lowermost Sinemurian Stage). Information from this site is supplemented by additional data from other exposures along the Antrim coast and inland, and from several boreholes.

Fig. 13.7 Upper Triassic and Lower Jurassic mudstones and thin limestones of the Penarth Group and Lias Group exposed on the shore at Waterloo Bay, Larne, County Antrim (Grid ref. D 409036). The conspicuous pale limestones in the foreground form the upper beds of the Langport Member, at the top of the Penarth Group, and are succeeded by dark mudstones of the Lias Group. The first ammonites occur a metre or two this side of the disused sewer pipe.

The Penarth Group

The Penarth Group overlies the Collin Glen Formation with an abrupt facies change, from greenish-grey silty mudstones to almost black pyritic shales and sandstones. In Northern Ireland, as throughout the United Kingdom, the Penarth Group comprises two relatively thin, but distinctive, formations. The lower of these, the *Westbury Formation*, is about 13 m thick and of pyritic black shales and mudstones with subordinate sandstones. The latter are commonly rich in fish debris and a rich assemblage of trace fossils, but in the mudstones fossils are generally confined to specific levels and typically occur as low diversity assemblages of just one or a few species. Bivalves dominate the fauna and include *Rhaetavicula contorta*, probably the most distinctive of all Rhaetian fossils and originally described by Portlock (1843) from material found near Limavady. Unequivocally marine taxa such as ammonites and brachiopods are absent from the Westbury Formation. In general the restricted composition of the fauna, with an often high abundance but low diversity, suggests an environment of restricted salinity.

Another abrupt facies change, to paler lithologies dominated by fine sand and silt, marks the base of the *Lilstock Formation* which, at Larne, is also about 13 m thick. The most striking feature of the lowest 5 m or so is the extreme degree of soft-sediment deformation that it exhibits. This clearly occurred soon after deposition and has been attributed to the effects of a uniquely powerful seismic event that can be traced across the entire British outcrop of the Penarth Group (Simms, 2007). Macrofossils are virtually unknown from this part of the succession in Northern Ireland, although a mixture of marine and non-marine fossils is found in correlative strata elsewhere in Britain. Desiccation cracks a little higher in the succession at Larne attest to a brief period of emergence. Higher beds in the Lilstock Formation are dominated by dark mudstones with thin siltstone laminae and contain a rather sparse fauna of oysters and shallow-burrowing bivalves. The top of the formation is marked by a distinctive series of thin shelly limestones, succeeded by dark mudstones of the Lias Group, which contain an increasingly diverse marine fauna. Ammonites appear at about 6.4 m above the the top of these shelly limestones. At the time of writing a decision had yet to be made as to the precise stratigraphical level of the Triassic–Jurassic boundary globally. Significantly, the Triassic–Jurassic boundary section at Waterloo Bay is one of four candidates for the Global Stratotype Section and Point for the boundary, based on its exceptional thickness and good fossil record through the critical interval.

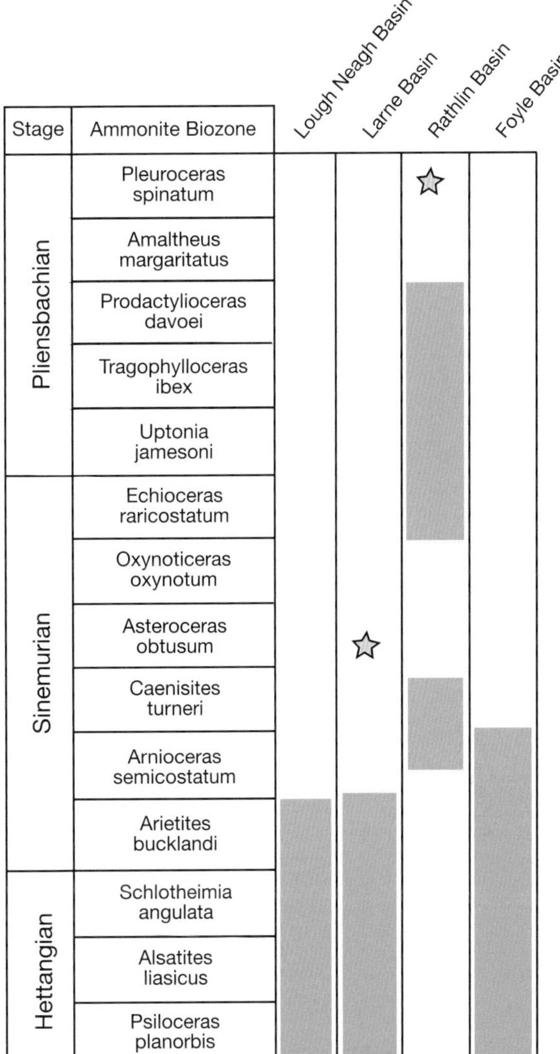

Fig. 13.8 Biostratigraphy of the Lower Jurassic Lias Group (Waterloo Mudstone Formation) in the Northern Ireland basins. Stars indicate ammonite biozones known only from derived material preserved in Cretaceous conglomerates. The entire Hettangian and parts of the Sinemurian remain unproven in the Rathlin Basin, being apparently cut out by major dolerite sills in the Portmore Borehole.

The Jurassic and Cretaceous Systems in Northern Ireland

To an even greater extent than is the case for the Permian and Triassic systems, Irish Jurassic and Cretaceous rocks are confined almost entirely to Northern Ireland. The total known outcrop and subcrop of rocks of this age in the south amounts to probably little more than 1 km^2, although the significance of these particular sites is vastly disproportionate to their actual area. In Northern Ireland Jurassic strata conformably succeed the Triassic strata beneath, but both the Jurassic and Cretaceous successions are far from complete. The two are separated by an unconformity which at a minimum accounts for about 90 Ma of geological time. Even at its fullest development much of the Lower Jurassic, all of the Middle and Upper Jurassic, all of the Lower Cretaceous and several minor parts of the Upper Cretaceous are missing. Inevitably there has long been conjecture as to what extent this is due to original non-deposition, and to what extent it reflects subsequent erosion. This is particularly true for the Mesozoic record in the south. For Northern Ireland at least, the absence of much of the Lower Jurassic is demonstrably due to subsequent erosion, while for the Lower Cretaceous non-deposition seems a likely cause at least in part.

The Jurassic System

In Northern Ireland the Jurassic System is represented solely by the lower part of the Lower Jurassic (Fig. 13.8). The exclusively marine, often richly fossiliferous mudrock-dominated succession here is grouped into a single lithostratigraphical unit within the Lias Group, the *Waterloo Mudstone Formation*. The lowest few metres of the formation lack ammonites and are assigned to the Triassic System pending an international decision on the position of the Triassic–Jurassic boundary. There are very few good exposures of the formation, and most are small and/or disturbed by landslipping. The limited data derived from these surface exposures are supplemented by additional information from boreholes in the various basins.

Both surface exposures and borehole records indicate that the Jurassic succession in the Larne and Lough Neagh basins was eroded to a significantly lower stratigraphic level, prior to deposition of Cretaceous rocks, than was the case in the Rathlin and Foyle basins. At the type locality for the Waterloo Mudstone Formation, at Waterloo Bay on the outskirts of Larne, approximately 100 m of dark grey mudstone with subordinate limestone extends to only a little above the base of the Sinemurian Stage (Ivimey-Cook, 1975). *Ex situ* ammonites from other sites in the Larne Basin indicate that, at most, the succession here extends only a little higher, perhaps into the base of the *semicostatum* Biozone. Limited borehole evidence from the Lough Neagh Basin indicates that the succession there ranges no higher than the *bucklandi* Biozone at the base of the Sinemurian Stage (Cope, Getty, *et al.*, 1980). In both basins the entire Lias Group may be greatly reduced or even absent altogether, removed by erosion prior to deposition of Cretaceous strata. For instance, only 12 m of Lias Group is preserved at Collin Glen, on the outskirts of Belfast, and it is entirely absent north of Garron Point as the northern edge of the basin is approached, with Upper Cretaceous strata resting directly on red mudstones of the Mercia Mudstone Group.

Although only a small proportion of the total Lias Group survives in Northern Ireland, what does remain can be exceptionally thick and stratigraphically complete compared with other basins in the United Kingdom. For instance, at Waterloo Bay, close to the depocentre of the Larne Basin, the *planorbis* Biozone at the base of the Hettangian is more than twice as thick as correlative strata on the Somerset coast, and thicker even than that in the Mochras Borehole (Simms *et al.*, 2004). The rest of the Hettangian at Waterloo Bay also appears to be thicker than equivalent strata on the Somerset coast, and the same is true of the succession in the Lough Neagh Basin penetrated by the Mire House Borehole (Cope, Getty, *et al.*, 1980). These exceptional thicknesses testify to rapid subsidence of these basins. Although no strata younger than basal Sinemurian are known in these basins, a phosphatised fragment of *Asteroceras* in the conglomerate at the base of the local Cretaceous succession north of Larne testifies to the former presence of higher parts of the Sinemurian Stage here.

Surface information on the Lias Group of the Rathlin and Foyle basins is sparse, confined to a handful of coastal and stream sections. These provide little information on the actual thickness of the Lias Group here, which instead has come from the Portmore Borehole, in the Rathlin Basin, and Magilligan Borehole in the Foyle Basin. These have proven critical to providing a broader stratigraphical framework against which to correlate the patchy surface data. Surface exposures of Lias Group mudstones are confined to White Park Bay, Portnakillew, and the classic hornfels locality at Portrush (*see* Chapter 19). They expose strata from the upper part of the Sinemurian (*raricostatum* Biozone) to the lower part of the Pliensbachian (*jamesoni* to *davoei* biozones). Ammonites from the patchy and disturbed exposures on the shore at White Park Bay suggest that, as at Larne, the Lias Group succession is exceptionally complete here. Evidence from the Portmore Borehole confirmed this, penetrating several hundred metres of Lias Group below the sub-Cretaceous unconformity, with the *raricostatum* and *jamesoni* biozones alone 160 m thick. The youngest strata seen *in situ* are from the *ibex* Biozone (*valdani* Sub-biozone) at Portnakillew, but septaria found loose at White Park Bay contain large *Oistoceras* characteristic of the upper part of the *davoei* Biozone (*figulinum* Sub-biozone). Although the Portmore Borehole continued down into Triassic and possibly Permian strata, substantial parts of the Jurassic succession, including the entire Hettangian Stage, appear to have been cut out by major dolerite intrusions and remain unproven in the Rathlin Basin.

Evidence for the presence of higher levels in the Lias Group is equivocal. Upper Pliensbachian ammonites (*Amaltheus* and *Pleuroceras*) have been found in loose blocks of sandstone at Ballycastle, Ballintoy and White Park Bay (Wilson and Robbie, 1966), but their lithology is identical to the Scalpay Sandstone Formation of parts of the Hebrides (Simms *et al.*, 2004) and it remains unclear whether they were brought from there by Pleistocene ice or were derived from offshore outcrops within the Rathlin Basin.

The Lias Group of the Foyle Basin is known from a few stream sections and from the Magilligan Borehole (Bazley *et al.*, 1997). Only the Hettangian and part of the Lower Sinemurian are proven, with progressively older strata cut out from north to south by the sub-Cretaceous unconformity. The succession appears thinner than correlative strata in the Larne Basin, and of a more marginal facies. It includes, in the *semicostatum* Biozone near the top of the proven succession, a 13 m thick sandstone, the Tircrevan Sandstone Member.

That younger parts of the Jurassic were formerly present in the region, but were removed by Cretaceous erosion, is demonstrated by fragments of the ammonite genera *Pleuroceras* and *Dactylioceras* s.l., indicative of the Upper Pliensbachian and Lower Toarcian, in the basal conglomerate of the Cretaceous at Murlough Bay, on the Highland Border Ridge to the east (Hartley, 1933a; Savage, 1963; Versey, 1958).

Beyond the sedimentary basins of Ulster the Jurassic record in Ireland is extremely scant, being known from only two small 'pocket deposits' within the Carboniferous Limestone outcrop in counties Cork and Kilkenny. In the former, dark grey, red and yellow clays, formerly worked from two small pits near Cloyne, were found to contain a rich palynoflora of non-marine Mesozoic taxa. The most stratigraphically restricted of these indicate a late Toarcian to Bathonian age for the deposit (Higgs and Beese, 1986). Near Piltown, County Kilkenny, three boreholes intercepted brown, yellow and fawn clays within and overlying the Carboniferous Limestone beneath Pleistocene till. Palynomorphs from these clays are predominantly non-marine and indicate a late Jurassic (Kimmeridgian) to earliest Cretaceous (Berriasian) age for the deposits (Higgs and Jones, 2000).

The Cretaceous System

Although much of the Permian to Jurassic succession in Northern Ireland is poorly exposed and/or missing altogether, this is less true for the Cretaceous System. Spectacular white limestone cliffs, correlative with the Chalk Group in Britain, form some of the most striking coastal scenery and contrast with dark Palaeocene basalts that overlie them. Inland, numerous quarries have exploited this pure limestone, such that our knowledge of this part of the succession is better than for any other part of the Mesozoic (Fletcher, 1977). Conversely, we know nothing of the early Cretaceous in Ireland; the oldest rocks known anywhere are of Cenomanian age (basal Upper Cretaceous).

Although the hard white limestones of the Chalk Group are the most conspicuous element of the Irish Mesozoic succession, glauconitic sandstones locally are a significant component of the lower part of the Cretaceous succession in Northern Ireland. Accordingly, a bipartite division is recognised for the Cretaceous here (Fig. 13.9), with a lower Hibernian Greensands Formation and an upper Ulster White Limestone Formation. The former is confined largely to part of the Larne Basin in south-east County Antrim, with a very minor development in the Rathlin Basin, but the latter extends far more widely across Ulster.

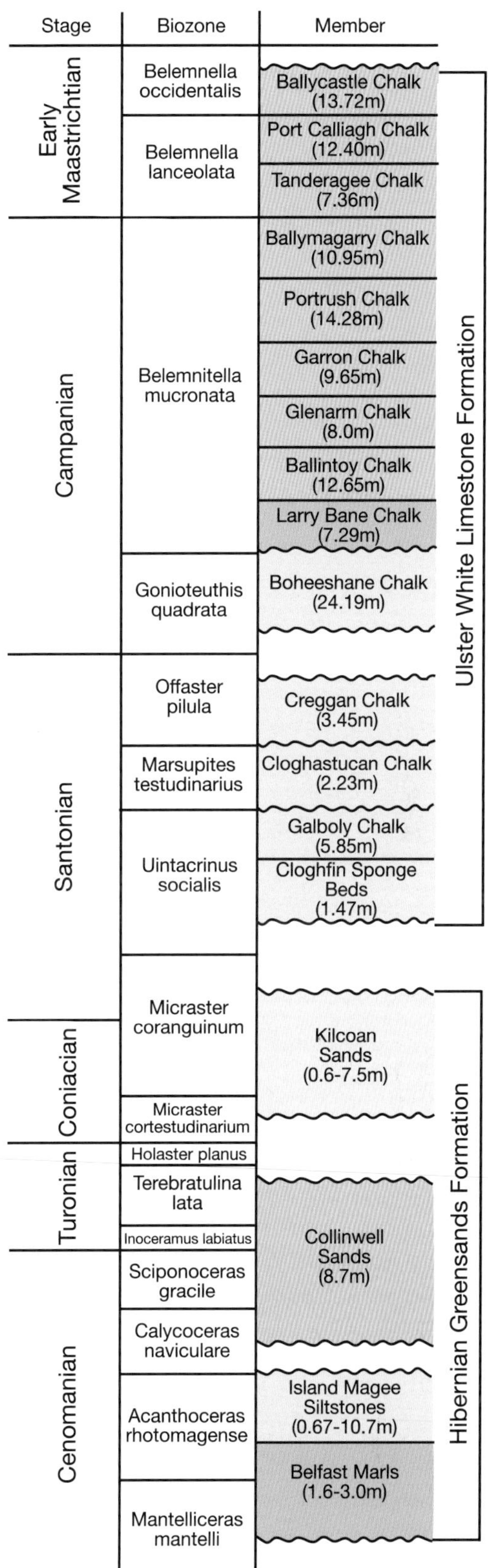

Fig. 13.9 Stratigraphy of the Cretaceous succession in Northern Ireland. Wavy contacts indicate erosion surfaces.

Hibernian Greensands Formation

Although the Hibernian Greensands Formation is superficially similar to the Aptian and Albian (Lower Cretaceous) greensands of southern England, ammonites and other stratigraphically diagnostic fossils demonstrate that they are not chronostratigraphical correlatives. Instead the Hibernian Greensands Formation ranges from Cenomanian to lower Santonian in age, and hence correlates with the lower part of the Chalk Group (Upper Cretaceous) in southern England (Rawson *et al.*, 1978).

In its fullest development, in south-east County Antrim, the Hibernian Greensands Formation comprises four distinct lithostratigraphical units which together reach a composite thickness of ~30 m. Good exposures can be seen on the eastern coast of Islandmagee, particularly at Cloghfin Port and Portmuck. The lowest unit, the Belfast Marls Member, rests with marked and sometimes angular unconformity on much older strata (Mercia Mudstone to Lias Group). Typically a muddy dark-green sandstone, its colour is due to abundant glauconite grains. It is often conspicuously fossiliferous, particularly with the exogyrid bivalve *Amphidonte obliquata*. The ammonite *Schloenbachia* proves a lower Cenomanian age for the base of the member, with the belemnite *Actinocamax planus* indicating a mid-Cenomanian age for its top. There is a gradational contact with the succeeding Island Magee Siltstone Member, a pale grey silty marl weathering to pale yellow brown. Less fossiliferous than the Belfast Marls Member, it has yielded ammonites and inoceramid bivalves that indicate a mid-Cenomanian age.

Above the Island Magee Siltstone Member is the Collinwell Sands Member, comprising two distinct units of largely unconsolidated sand. The lower unit, coarse-grained and glauconitic, is piped down into the Island Magee Siltstone Member beneath by deep burrows, indicating a significant break in deposition. Although fossiliferous, it lacks stratigraphically diagnostic taxa. The distinctive exogyrid bivalve *Rhynchostreon suborbiculatum* indicates an age somewhere between late Cenomanian and mid-Turonian, with rare *Inoceramus* ex grp. *lamarcki* suggesting the later date. The upper unit, a fine- to medium-grained white sand, is unfossiliferous and hence of uncertain age.

The highest unit of the Hibernian Greensands Formation is the Kilcoan Sands Member, which rests unconformably on eroded remnants of the underlying three members and even, locally, on older Mesozoic rocks. The base of the Kilcoan Sands Member is typically conglomeratic but the remainder is predominantly of pale-green fossiliferous and glauconitic sand. Three distinctive bands of inoceramid bivalve fragments occur in the lower part of the member. These, and other fossils, particularly echinoids, indicate an early Coniacian age for the lowest part, below the *Inoceramus* bands, and a Santonian age for the upper part of the member. The rudist bivalve *Durania borealis* has been found at two sites on the Antrim coast, these being among the most northerly occurrence of rudists anywhere in the world.

Ulster White Limestone Formation

The Ulster White Limestone Formation is predominantly a hard white coccolithic micrite limestone. Although a direct correlative of the Chalk Group of southern England, its much greater hardness and lower porosity have long been a source of discussion (Hancock, 1963). The overlying basalt commonly has been cited as a cause of this, either through thermal metamorphism or the effects of pressure (Maliva and Dickson, 1997), but the presence of joint-aligned, 'pre-basalt' palaeokarst features demonstrates unequivocally that pore-filling cementation, and subsequent extensional fracturing, occurred significantly prior to emplacement of the basalts (Simms, 2000b). This early cementation may be a diagenetic effect of relatively slow deposition in relatively shallow water (Jeans, 1980). Compared with chalks in southern England, the Ulster White Limestone Formation is significantly thinner and includes stromatolite-encrusted hardgrounds, clear evidence of deposition at relatively shallow depth within the photic zone.

The limestone is exceptionally pure, with <1% detrital clastics except where reworked material has been incorporated into the basal beds directly from underlying strata. This low level of detrital input has been ascribed to both unusually high sea levels in the late Cretaceous, and to Ireland's palaeogeographical location in an arid climatic belt at this time (Cope, 2006). Flints are the most conspicuous minor component of the Ulster White Limestone Formation. Composed of cryptocrystalline silica derived largely from sponge spicules, flint precipitated out early during diagenesis at locations where sediment porosity and chemistry had been altered by burrows and depositional breaks. The shape, spacing and configuration of some of these flint nodules and bands are characteristic of particular lithostratigraphical units (Fletcher, 1977). A diverse macrofauna is known from the Ulster White Limestone Formation, but preservation is often poor due to the hardness of the limestone. Irregular echinoids and belemnites are the most common and conspicuous elements of the fauna and have long been important for biostratigraphical subdivision of the Chalk Group here as elsewhere (Fletcher, 1977).

At outcrop the Ulster White Limestone Formation reaches a maximum thickness of >120 m in the Ballycastle area of the Rathlin Basin, its top being truncated by the 'sub-basalt' unconformity. However, an unbottomed sequence >150 m thick was reported in the Aughrimderg Borehole of the Lough Neagh Basin (Fowler and Robbie, 1961). Fourteen distinct lithostratigraphical members have been recognised in the Ulster White Limestone Formation at outcrop (Fletcher, 1977), based on the characteristics and spacing of distinctive flint bands and bedding planes, but supported by faunal characteristics. Nearly all of these members can be traced with reasonable confidence from the Larne and Lough Neagh basins into the Rathlin Basin.

The 14 members recognised by Fletcher (1977) in the Ulster White Limestone Formation (Figs. 13.9, 13.10) can be broadly divided into two groups, Post-Larry Bane Chalk and Pre-Larry Bane Chalk, that lie respectively above and below a distinctive unit, the Larry Bane Chalk Member. The base of the formation overlies a marked erosion surface on the underlying Hibernian Greensands Formation or, where this is absent, on pre-Cretaceous rocks (Fig. 13.11). Where the Hibernian Greensands Formation is well developed, as in east Antrim, the lower beds of the Ulster White Limestone Formation are sandy and richly glauconitic due to erosional reworking of the earlier formation. Elsewhere, glauconite is much less abundant or absent altogether from these basal beds but, where the Ulster White Limestone Formation directly overlies pre-Cretaceous strata it typically has a conglomeratic base. Members in the lower part of the succession (Santonian to lower Campanian stages) pinch out or are overstepped towards the basin margins but others, particularly in the higher parts of the succession (middle Campanian to Maastrichtian), can be traced with little change onto the basin margins of the Londonderry Shelf and north County Down. However, later erosion has removed the entire succession across the Highland Border Ridge.

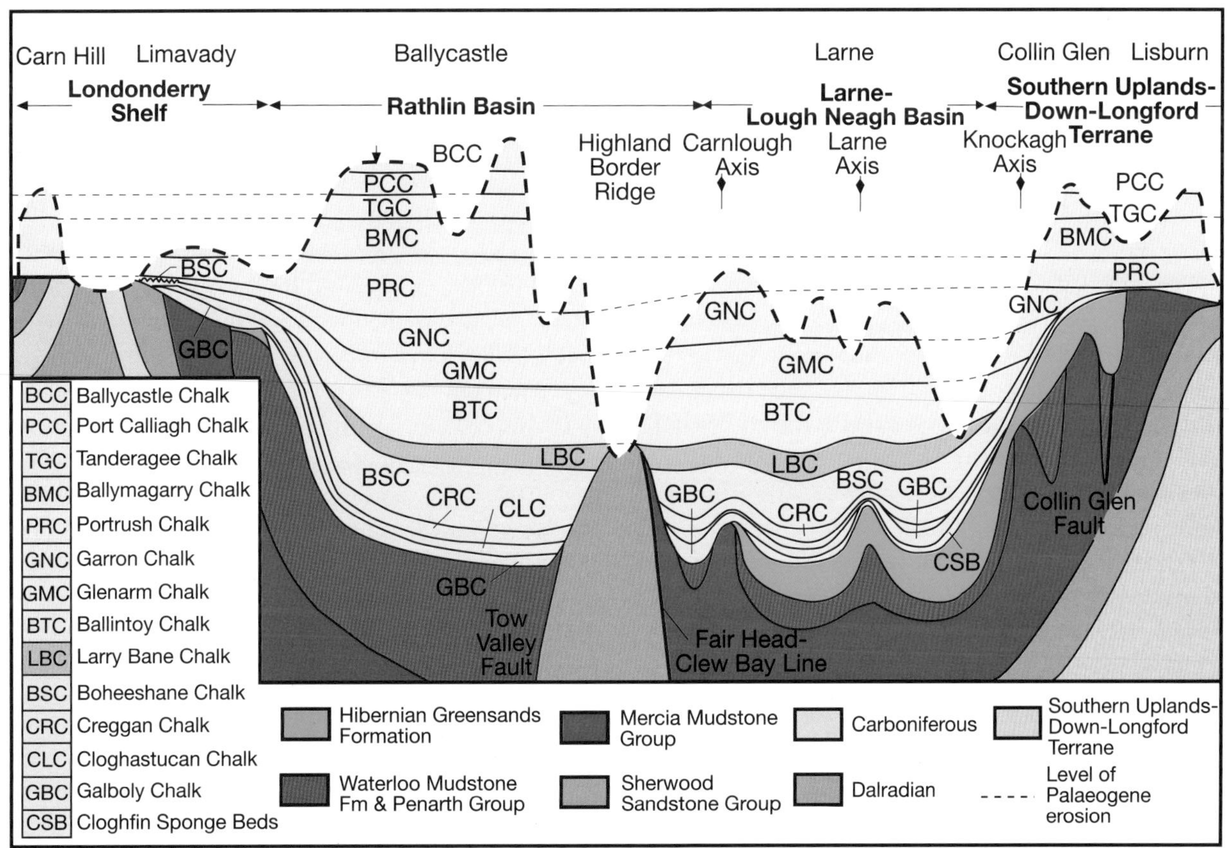

Fig. 13.10 Lateral variation in the Cretaceous succession across Northern Ireland (*after* Mitchell, 2004).

Fig. 13.11 Ulster White Limestone Formation (Cretaceous) unconformably overlying Sherwood Sandstone Group (Triassic) at Murlough Bay, County Antrim (Grid ref. D 194417). The basal Cretaceous conglomerate is visible at the junction.

Pre-Larry Bane Chalk encompasses five members spanning the upper Santonian to lower Campanian interval of the Upper Cretaceous. At outcrop this grouping is thickest in east Antrim, in the Larne Basin, and also is well developed in the Rathlin Basin, but individual members thin out or are overstepped towards the basin margins and are absent across the Highland Border Ridge. The earliest, and easily most distinctive, is the Cloghfin Sponge Beds Member. It is developed only in south-east Antrim, in the same area as the maximum development of the Hibernian Greensands Formation which it overlies unconformably, and similarly is well exposed at coastal sections on Islandmagee, notably at Portmuck and Cloghfin Port. It contains abundant sand and glauconite reworked from below, but its chalky matrix and the unconformity at its base place it firmly within the Ulster White Limestone Formation. Its most conspicuous and distinctive feature is an abundance of fossil hexactinellid sponges. Away from south-east Antrim the lowest member is the Galboly Chalk Member, but it is only well developed where the Cloghfin Sponge Beds are absent and the two are possibly, at least in part, correlatives. The Galboly Chalk Member is well exposed at the base of the white limestone cliff at Murlough Bay (Fig. 13.11), where it is separated from the Triassic Sherwood Sandstone Group beneath by a spectacular basal conglomerate. This conglomerate contains rare derived fossils which provide some of the only evidence for parts of the missing Jurassic succession in Northern Ireland (Hartley, 1933; Savage, 1963; Versey, 1958). The Creggan Chalk Member is a thin, but fairly distinctive, unit characterised by abundant *Inoceramus* shell debris. It is more widely spread than earlier members and commonly forms the basal unit of the formation. Towards the basin margins it commonly is glauconitic, due to reworking of glauconite sands which it oversteps. A marked erosional break occurs at the top of this member and is marked by a distinctive bed containing reworked fossils and glauconite-coated pebbles, the Bendoo Pebble Bed, and locally, in the Larne Basin, by stromatolitic crusts. In parts of east Antrim the stromatolitic surface becomes a major feature of the basal part of the Ulster White Limestone Formation and was referred to by quarrymen as 'mulatto', a term denoting the lowest limit of quarrying.

The Larry Bane Chalk Member (Fig. 13.12) is the most distinctive unit above the Cloghfin Sponge Beds. Although lithologically not especially distinctive, it is sharply demarcated at top and bottom by conspicuous bedding planes, with another about a third of the way from the base. Eight members were recognised above this level by Fletcher (1977), but only some are sufficiently distinctive to be recognised in isolated exposures and the higher members, of Maastrichtian age, have survived 'pre-basalt' denudation in only a few places. The lowest member of the post-Larry Bane Chalk succession, the Ballintoy Chalk Member, is unremarkable except for the uppermost 15–35cm. This topmost bed, known as the Altachuile Breccia, shows disturbed bedding that

Fig. 13.12 Ulster White Limestone Formation on the Antrim Coast Road at Glenarm, County Antrim (Grid ref. D 320155). The Larry Bane Chalk Member can be clearly distinguished by its three sharp bedding planes; the highest of which is about halfway up the cliff. The Boheeshane Chalk Member lies below the Larry Bane Chalk, with the Ballintoy Chalk Member above.

incorporates indurated lumps of chalk and angular flint fragments. It is thickest on Rathlin Island, but can be traced along the scarp from east of Portrush to near Belfast. Ascribed to violent storm activity disrupting the sea bed, it provides clear evidence of the early cementation of the limestone and precipitation of flint at very shallow depth in the sediment. The succeeding Glenarm Chalk Member is most notable for the two North Antrim Hardgrounds, the higher and better developed of which is a valuable lithostratigraphical marker in the Rathlin Basin. In the Larne Basin several hardgrounds are developed in the Larne area, while in the Belfast area the Glenarm Chalk is absent entirely, overstepped by higher parts of the succession. However, a few glauconite-stained pebbles are almost the only evidence of interrupted sedimentation at this level at Glenarm, near the depocentre of the Larne Basin.

Of the three remaining Campanian members of the succession, two are characterised by 'giant' flints. In the Garron Chalk Member these form rings up to several metres across and half a metre thick, while in the Ballymagarry Chalk Member they form paramoudras, barrel-shaped flints a metre or two high, half a metre across, and with a central cavity. The intervening Portrush Chalk Member contains abundant *Inoceramus* fragments but is most notable for the development of hardgrounds on the Londonderry Shelf west of the Lough Neagh Basin, around Moneymore, and especially in the Belfast area where they have been termed the South Antrim Hardgrounds (Fletcher, 1977).

The three highest members, the Tandragee Chalk, Port Calliagh Chalk, and Ballycastle Chalk, contain belemnites indicative of the Maastrichtian Stage. They have a rather limited distribution beneath the sub-basalt unconformity, being notably absent across the Larne Basin. The highest of these, the Ballycastle Chalk Member, is the youngest *in situ* chalk known anywhere onshore in Ireland or Britain but is confined to just two localities, near Portrush and at Ballycastle itself.

The Ballydeenlea Outlier

Despite the extremely poor representation of Mesozoic rocks south of Ulster, it is an outcrop of Upper Cretaceous Chalk that represents one of the most remarkable, and certainly one of the most enigmatic, features of Ireland's geology. At Ballydeenlea, north of Killarney in County Kerry, a tiny outcrop of Campanian Chalk, perhaps no more than a few thousand square metres in area, occurs within a vast tract of Carboniferous rocks. Surrounded by fairly steeply-dipping Namurian shales and sandstones, the Chalk occurs as matrix or discrete pockets within a chaotic breccia of Namurian shale, and occasionally sandstone, blocks. The Chalk itself contains angular fragments of shale, with some even enclosed within flints (Walsh, 1966). Four shallow cored boreholes indicate that the breccia is at least 130 m thick, though barely 100 m across. There is no evidence of any sorting or reworking of any components of the breccia, and it appears that it formed suddenly and catastrophically, and was associated with the downward movement of unlithified chalk mud into the newly formed breccia. The enclosure of shale fragments by both chalk and flint demonstrates how soon after deposition the chalk mud was emplaced within the breccia. Walsh (1966) interpreted the deposit as analogous to a neptunian dyke, and caused by catastrophic collapse into a cavern in the Carboniferous limestone some 150 m below. A more recent interpretation (Evans and Clayton, 1998) envisaged the breccia and chalk deposit as the product of slumping and collapse on a submarine fault scarp, in a manner analogous to the Upper Jurassic fault scarp breccias at Helmsdale in eastern Scotland (Wignall and Pickering, 1993). However, neither the cavern collapse nor submarine fault scarp models adequately explain all of the observed features seen in the Ballydeenlea breccia, and its origin remains one of the great enigmas of Irish geology.

Regardless of its mechanism of emplacement, the presence of Chalk at Ballydeenlea has implications for understanding the late Cretaceous and post-Cretaceous history of south-west Ireland. The absence of any detrital material in the Chalk, other than shale fragments from the immediate vicinity of the deposit, suggests that the Campanian sea floor at this location lay a considerable distance from any land. Such an interpretation conflicts with the present topography, in which uplands less than 20 km away rise hundreds of metres above Ballydeenlea. It suggests that much of this topography may be due to post-Cretaceous differential uplift, unless either the Campanian sea here was more than 1000 m deep or the Ballydeenlea breccia pipe emplaced chalk mud many hundreds of metres below the Campanian sea floor. In this respect it is interesting to note that Evans and Clayton (1998) suggested a burial depth of 1000–1500 m for the Ballydeenlea chalk, based on organic maturation of palynomorphs. A substantial early Cenozoic cover seems unlikely and, although Campanian and Maastrichtian chalks are commonly thicker than earlier parts of the succession (Cope, 2006), this figure still seems excessive unless it is assumed that the original depth of the Ballydeenlea breccia pipe was in excess of a kilometre.

Ireland's Missing Millions

Although Mesozoic strata appear well represented in Northern Ireland, nonetheless very considerable stratigraphical gaps exist compared with southern Britain. In southern Ireland the situation is extreme; Permo-Triassic strata are preserved only in two small basins, with post-Triassic rocks confined to three tiny, and rather enigmatic, 'pocket deposits'. To what extent the absence of particular strata reflects non-deposition or erosion must remain open to conjecture, but several approaches can be taken in an attempt to resolve this issue. The presence of remanié fossils in younger strata is important, and fairly unequivocal, evidence for deposition and subsequent erosion. Consideration of global eustasy, projected subsidence curves, and comparison with the stratigraphy of adjacent regions, is also useful. Indirect methods, such as Apatite Fission Track Analysis (AFTA), vitrinite reflectance and spore colour, also can provide indications of the relative roles of non-deposition and erosion in creating the successions that are preserved today.

Vitrinite reflectance of organic material in Mesozoic rocks now exposed at the surface provides a crude measure of the temperatures experienced since deposition, which broadly equates to depth of burial assuming no additional heat source. Conversely, AFTA data are particularly effective at providing information on times of cooling, which broadly equates to periods of erosion and consequent reduction of overburden thickness. AFTA data for Northern Ireland suggest a cooling period between ~180 Ma and ~150 Ma, with another between ~125 Ma and ~100 Ma (Green *et al.*, 2000). This implies that both the middle Jurassic and early Cretaceous were largely periods of non-deposition and erosion of earlier strata. These AFTA data are broadly in accord with what is seen in the Jurassic and Cretaceous of the Hebrides, in western Scotland, where a thick marine succession from Hettangian to Bajocian (Simms *et al.*, 2004; Cope, Duff, *et al.*, 1980) is succeeded by non-marine Bathonian and then further marine rocks in the Callovian to Kimmeridgian interval. As in Northern Ireland, Lower Cretaceous rocks are entirely absent from western Scotland, and only tiny remnants remain of younger Cretaceous strata. The periods of erosion (AFTA cooling episodes) in Northern Ireland appear to equate with periods of shallowing and emergence in the case of the Middle Jurassic, or an absence of strata in the case of the Lower Cretaceous, in western Scotland. In the south of Ireland, where Mesozoic strata are so poorly represented, the depth to which these small remnants have been buried subsequently is of particular interest. Vitrinite Reflectance data suggest a depth of 1–2 km for the Jurassic clays at Cloyne and the Chalk at Ballydeenlea although, as discussed previously, it is difficult to see how this could be achieved for the latter site solely through burial.

The end of the Mesozoic

Only in north County Antrim are the youngest Mesozoic (late Cretaceous, Campanian and Maastrichtian stages) strata well exposed in a clear stratigraphical context. Often spectacular coastal cliffs present a striking contrast between the Ulster White Limestone Formation and the dark basaltic lavas of the Antrim Lava Group which succeed them. The latter are Palaeocene in age and probably ~60 Ma years old, slightly younger than the initial onset of Palaeogene volcanism in the North Atlantic at ~63 Ma (Pearson, Emeleus, and Kelley, 1996). These basaltic flows rest with marked unconformity on a highly dissected and karstified surface of the Ulster White Limestone beneath, with the two commonly separated by the so-called 'Clay-with-flints', a mixture of unworn flints and volcanic ash (Simms, 2000b). Even the youngest preserved strata range up only into the early Maastrichtian, ~68 Ma ago, and hence there is a time interval of perhaps 8–10 Ma for which we know very little of events in Ireland.

We can only assume that the Ulster White Limestone Formation once extended to the top of the Maastrichtian succession, or possibly into the early Palaeocene. Late Maastrichtian shallowing may have curtailed sedimentation or even initiated removal of the uppermost strata (Hart *et al.*, 2004). The highly dissected and karstified surface of the Ulster White Limestone beneath the basalts testifies to prolonged subaerial exposure prior to emplacement of the Antrim Lava Group, during which time many tens, or possibly even hundreds, of metres of limestone were removed. Consideration of limestone denudation rates suggests that emergence and subsequent denudation of the limestone occurred largely in the Palaeocene, no more than one or two million years before emplacement of the basalts (Simms, 2000b), and probably was a consequence of both eustatic sea level fall and regional uplift.

Major environmental changes occurred at the close of the Cretaceous, 65 Ma ago, notably the eruption of the Deccan flood basalts and, even more spectacularly, the Chicxulub impact. In Denmark, and elsewhere around the world where there is a continuous succession from the Cretaceous into the Palaeocene, the fallout from this impact is clearly recognisable as a thin clay layer containing impact shocked quartz and unusually high levels of the rare metal Iridium (Hart *et al.*, 2004). This impact fallout must have rained down across Ireland at the same time, but any trace of it must have been removed long before the emplacement of the Antrim Lava Group several million years later.

Although the main phase of North Atlantic volcanism clearly is Palaeocene in age, tentative evidence exists for significant volcanism adjacent to Ireland in the late Cretaceous. Glauconite in the Hibernian Greensands Formation and, to a lesser extent, in the Ulster White Limestone Formation, has been ascribed to the alteration of material derived from submarine eruption of basic lavas (Jeans et al., 1982). Basaltic seamounts in the Rockall area, to the north-west of Ireland, have been radiometrically dated to ~70 Ma (O'Connor et al., 2000) and hence volcanic ashfalls may have been significant across Ireland towards the close of the Mesozoic, a precursor of what was to follow.

14

Tertiary igneous activity

J. Preston

The term 'Tertiary' is retained in this chapter although it no longer enjoys formal stratigraphical status. It is retained because it has become deeply embedded in the literature pertaining to magmatic activity in north-eastern Ireland and north-western Scotland – a region of enormous historical importance for igneous petrology.

In the North Atlantic region at the end of the Cretaceous Period the crust was stretched, thinned, and ruptured by plate movement along old and new planes of weakness; many of these ruptures became migration pathways for the more mobile fractions of the Earth's mantle.

In some instances these rock melts were the consequence of deeper-seated mantle processes, and their rise to the Earth's surface was a necessary sequel to the gravitational differentiation of the Earth's interior that energises such major structural changes to its crust. At other times the plate movements themselves affected the underlying mantle, which responded with attendant volcanism.

Widespread emergence of the northern continents from the Mesozoic seas accompanied these mantle changes. Swarms of linear fractures and pipe-like conduits allowed basaltic magma to inundate the surface and establish arrays of central volcanoes over an enormous region known as the Thulean Volcanic Province. Figure 14.1 illustrates the regional extent of this North Atlantic volcanism.

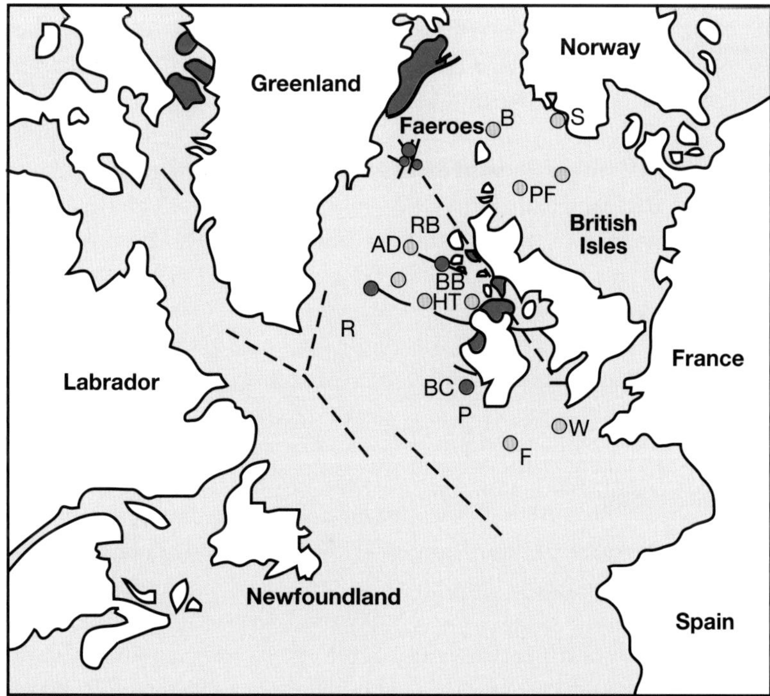

Fig. 14.1 Reconstruction of the relative positions of North America, Greenland, and Europe prior to the opening of the North Atlantic Ocean (*after* Naylor and Shannon, 1982). Mesozoic volcanism – yellow circles; Early Tertiary volcanism – red areas and circles; mid-ocean and incipient rifts – broken lines; dyke trends – solid lines. AD – Anton Dorn seamount; B – Brent; BB – Blackstones Bank; BC – Brendan centre; F – Fastnet; HT – Hebridean Terrace seamount; P – Porcupine/Slyne troughs; PF – Piper Forties centre; R – Rockall; RB – Rosemary Bank; S – Sunnhordland; W – Wolf Rock.

Drifting, rotation, and stretching of continental crustal plates generated graben and half-graben structures and in the stratigraphic sequences found in off-shore basins of sedimentation (Naylor and Shannon, 1982; Upton, 1988) many volcanic events are recorded. Most notable are the Tertiary intrusive and extrusive rocks which cover the Rockall Bank (Roberts, 1969, 1975) and extend marginally into the Rockall and Hatton–Rockall Basins, the Tertiary intrusive centre and associated dyke swarm of the Slyne Trough and Porcupine Basin (Riddihough and Max, 1976), the late Tertiary dolerite sills of the Porcupine Trough (Seemann, 1984), and the Tertiary dolerite sills of the Fastnet Basin (Caston *et al.*, 1981). Similar relationships between plate movements and volcanic activity have been found in older basins of late Mesozoic age, but the culmination of volcanism occurred when the Greenland plate began its movement away from the North European plate some 65 Ma ago.

The basins produced by Tertiary and earlier movements may contain thick sequences of Tertiary marine and lacustrine sediments, but onshore in Ireland the geological record of the Tertiary Period is dominated by volcanic rocks. Though the peak of eruptive activity appears to have been restricted to just a few million years around 58 Ma, the rest of Tertiary time from 65 Ma ago to the beginning of the Quaternary Period, 1.8 Ma ago, is represented by thin *in situ* soils, local lacustrine clays, and the initiation of the landscape as we see it today (see Chapter 15).

In Ireland, flood basalts cover County Antrim and parts of adjoining counties and originally had a more extensive occurrence. Central volcanoes were built in counties Armagh and Louth and the rising magma penetrated upper crustal rocks as dykes, plugs, and sills. Figure 14.9 pictures these and more isolated outposts of this volcanism. A few dykes occur further south as far afield as County Kerry.

Timing and siting

Within the Thulean Volcanic Province certain alignments of the volcanic centres are immediately apparent from Figure 14.1.

One prominent line extends from north-east Ireland up the west coast of Scotland and intersects the west coast of Greenland at the site of the Skaergaard Intrusion. This is a line of central volcanic complexes, yet its significance may be more apparent than real, for no major structural weakness connects the Irish and south Scottish centres with those of north-west Scotland across the older Dalradian rocks of Kintyre, Jura and Islay (George, 1966). Volcanism appears to be related to independent basins of sedimentation; basaltic lavas flooded the lower ground, and central volcanic complexes exploited the marginal weaknesses (Walker, 1979). Swarms of north-west to south-east trending, rift-normal fractures traverse these areas, and the igneous dykes which now occupy them bear witness to the degree and direction of crustal extension that produced them. Rift-parallel fractures follow the main line of rifting between Greenland and northern Europe; such can be found associated with the basalt formations of Greenland and the Faroes. Rift-normal fractures, however, are more characteristic of the North European plate.

The most prominent swarm can be traced from St. Kilda and Boreray (some rift-parallel dykes are found here) in the north-west, across the Outer Hebrides, Mull, and south-western Scotland to north-eastern England. Its trend swings to a more easterly course as it approaches the North Sea Basin. The most intense section of the swarm is centred on the Island of Mull and the Lorn area of Argyll and it may be that the weakened crust around the Mull volcano failed more readily. A second parallel swarm occurs some 200 km to the south connecting Rockall, Donegal, Slieve Gullion, Anglesey, and the English Midlands. This swarm intensifies as it crosses the older, well-jointed, granitic plutons of Donegal and again around the volcanic centres of Slieve Gullion and Carlingford. Similar rift-normal trending dykes can be found in the Faroes and at Lundy Island in the Bristol Channel. Regional doming above a rising mushroom-shaped head of hot material fed by a mantle plume (White, 1988) may have produced both sets of fractures, the thickness of the crust controlling their spacing.

An initial rise of magma into these north-west to south-east fractures across Ireland probably built the extensive lava plateau; later surges exploited lines of weakness and well-jointed, readily-melted granitic rocks to establish more local eruptions.

Radiometric dates offer, as yet, the only stratigraphical correlation across the Thulean Volcanic Province. Various methods of dating have been used; the more reliable results show a remarkable synchronism of rifting, initial plate movement, crustal foundering, and volcanism over the North Atlantic region.

Most of the Irish Tertiary volcanism spanned an interval of 4 Ma from 62 to 58 Ma ago according to Thompson (1985). His use of the ^{40}Ar–^{39}Ar step-wise degassing technique agrees with some of the earlier K–Ar dates and confirms the same sequence of events deduced from geological relationships in the Irish and Hebridean parts of the province. Flood basalt eruption, 61–58 Ma ago (Thompson, *op. cit.*), preceded the establishment of central volcanoes, and though many of the hypabyssal intrusions are younger, dyke and sill intrusion occurred with each phase of activity. Gould (2004) reports a remarkably precise U–Pb zircon age of 60.9 ± 0.5 Ma for a suite of euhedral zircon crystals extracted from a bole horizon sandwiched between the top of the Port na Spaniagh Member and the base of the overlying Causeway Tholeiite Member of the Interbasaltic Formation (see below). Meighan *et al.* (1988) date the Slieve Gullion central complex for the first time at 58–57 Ma; this, when compared with the age of 61–60 Ma for the Carlingford intrusions (Thompson, *op. cit.*), raises difficulties over the contradictory geological evidence. Wilson (1970) used palaeomagnetic data to show that the basalts most probably belonged to one magnetic epoch and could then have been extruded over a much shorter period of 1 Ma.

Macintyre *et al.* (1975) proved a Tertiary age for many outlying examples of plugs and dykes in counties Donegal, Sligo, Mayo, and Galway; these include the arfvedsonitic rhyolite dykes of County Donegal (Whitten, 1954) and the Killala and Inishcrone gabbros and dolerites. Thompson (*op. cit.*) modifies some of the more aberrant K–Ar age dates; the dolerite dykes of the Dingle Peninsula, dated by Horne and Macintyre (1975) at 42 and 25 Ma, are considered to be 59 Ma old. This sheds doubt on the 26 Ma age of the dolerite sills to the west of Ireland, which Seemann (1984) correlated with the Dingle dykes. The Cleggan dyke, dated by Mitchell and Mohr (1986) at 42 Ma, is confined to the interval 60–57 Ma and the Killala Bay intrusions are brought into line with the other activity at 58 Ma ago.

On Skye, Lundy, and St.Kilda granitic plutons postdate the main period of volcanism by 3 Ma and in Ireland the Mourne Granites are also a later phase of intrusion. Gibson *et al.* (1987) give Rb–Sr ages of 56 Ma for the first four ring dykes and offer a tentative age of 52 Ma for the fifth. Thompson *et al.* (1987) put the timing at 55–53 Ma. Later work by Gibson *et al.* (1995) confirmed the age of the G1–G4 granites by $^{40}Ar/^{39}Ar$ determinations but considered 54 Ma as a better estimate for the age of G5.

There is no pattern of migrating volcanism across the Thulean Province in Ireland or elsewhere.

Pre-volcanic conditions

Late Cretaceous times saw the maximum transgression of the Chalk seas over Ireland. Then, many of the Caledonoid (NE–SW) trending land ridges were finally submerged, only to be re-accentuated by early Tertiary uplift. Gentle undulations across the north-eastern corner of Ireland may have controlled both the erosion of the emergent Chalk and the distribution of the first basaltic lavas. Neville George (1967) argued that the Midland Valley (of Scotland) showed only the faintest surface expression across Ireland, and little appears to be known of the role played by the Longford-Down Massif. Cretaceous rocks, if deposited more widely, must have been removed by erosion before the onset of volcanism. Over County Donegal, the present-day drainage pattern originated on a gently southerly-inclined surface which may have been the emergent Cretaceous surface or the basalt plateau itself (Dury, 1959).

Sub-aerial erosion left its mark on this early Tertiary land surface. Simms (2000b) offers a detailed picture of the karst topography over this sub-basaltic surface of Cretaceous Chalk. Small depressions (palaeodolines) are a common feature, but in places spectacular karst features of pinnacles and dolines with a relief of 15 m or more also occur. Cave passages exist and would indicate a well-cemented, low porosity lithology of the Chalk prior to the onset of volcanism. Both Simms (*op. cit.*) and Mitchell (2004) describe the 'clay-with-flints' deposit on top of the Chalk and claim too great an imbalance between flint nodules and red clay to support a 'Terra Rossa' palaeosol origin; they explain the excess of red clay as volcanic ashfall or wind-blown dust.

Over such a surface relatively quiet effusions of basaltic lava rapidly filled the hollows and buried the landscape.

Sub-aerial volcanism

Tomkeieff (1940a), Patterson (1955a, b), and Walker (1959) made the first significant contributions to this facet of Tertiary geology. Lyle (1980, 1985a, 1988), Lyle and Patton (1989), Lyle and Preston (1993) and the Geological Survey of Northern Ireland, notably in the Ballycastle, Causeway Coast, Belfast, Carrickfergus and Bangor memoirs, have added considerably to our appreciation of it.

Tomkeieff, in an address to the British Association in 1952, used data from the deep boreholes around Lough Neagh to demonstrate the far greater thickness of lava that poured into the Lough Neagh basin. In 1964 he published an isopachyte map for the Lower Basalt Formation. The Langford Lodge borehole sited on the south-eastern shore of the lough penetrated the greatest total thickness of basalt flows, which together are known as the *Antrim Lava Group*. This figure of 780.2 m is matched by that from the Ballymacilroy borehole just north of Lough Neagh. Here the total thickness is 770.1 m, but the section is incomplete and much has been removed by erosion.

Tomkeieff (1964) compared this volume of lava with the shield volcanoes building the Hawaiian Islands, but unlike Judd (1874) and Richey (1948) he did not imply the existence of a shield volcano, only stressing the subsidence of the Lough Neagh area which has been a sedimentary basin from Carboniferous times.

Geikie (1897) made comparisons with Icelandic volcanism and argued strongly for fissure eruptions. Past and present volcanicity over Iceland produced a preponderance of fissure-fed plateau basalts, the plateau surface supporting small shield volcanoes built around central vents. Each mode of extrusion is characterised by the geomorphology of the original volcanic construct and its eroded remnants. Plateau basalts show marked terracing in cliff section with interbasaltic beds of tephra and dust; flow thicknesses are typically 5–20 m with ponding of the lava to greater depths. Shield lavas, by contrast, show little or no terracing, cliff sections are more precipitous, and flow thicknesses less than 1 m.

Within the Antrim Lava Group, plateau-building flows are ubiquitous and the sequences of thin pahoehoe flows, which occur locally as on Island Magee or along the coastal cliffs of North Antrim, probably indicate the vicinity of an erupting fissure.

Connections between feeder dyke and lava flow are seldom seen even over the Icelandic lava plateau. An occurrence on Rathlin Island, the Maddygalla dyke near Rue Point, (Wilson and Robbie, 1966; Preston, 1971) is a small-scale example, but other instances along the eastern shore of Brown's Bay, Island Magee, (Patterson, 1950; Walker, 1959) are more probably the infillings of a fissured lava surface by the succeeding flow. Walker (*op. cit.*) has suggested that spatter ramparts may now obscure this vital connection.

Extensive thermal metamorphism is usually absent at the margins of dykes intruding the basalt pile (Patterson, 1950), though many that crop out beyond the confines of the lava plateau, as for example the Blind Rock Dyke (Preston, 1967a), the Fermanagh Dyke (Preston, 1967b), and members of the Killala Bay swarm (Nawaz, 1974) show extensive alteration of their wall rocks, indicating prolonged flow of magma along or up the fissure. Many of the dolerite plugs within and beyond the present outcrop of the basalt lavas show similar features, see Scawt Hill (Tilley and Harwood, 1931), Tievebulliagh (Agrell and Langley, 1958), Tieveragh (Tomkeieff, 1940b), and Slemish (Preston, 1963a), but erosion has removed any erupted rocks. Only the Slemish plug preserves evidence of near-surface activity in the form of a lava lake.

All palaeomagnetic records (Hospers and Charlesworth, 1954; Wilson, 1970) indicate extrusion of the lavas within one reversed magnetic epoch covering a timespan of *c.*1 Ma. Wilson (*op. cit.*) describes the palaeomagnetic stratigraphy of the Lower Basalt Formation at the edge of the Garron Plateau. Successive groups of flows are identified. Within a group the flow intervals are inferred to have been from 10 to 100 years' duration, but the intervals between groups are longer, circa hundreds to a thousand years. This implies building of a lava ridge above a set of fissures followed by migration of the eruptive sites to lower ground, so extending the plateau laterally before eruptions return to the original site and raise the plateau to a higher level.

Lavas are virtually confined to north-eastern Ireland (Figures 14.2 and 14.9) though outliers near Markethill and Poyntz Pass extend almost as far south as the central volcanoes of Slieve Gullion and Carlingford. Open, north-north-west trending fissures fed lava fountains and lava flows; locally, as in North Antrim, an initial explosive phase covered the ground with volcanic ash. Some of the flow surfaces are undulating with tumuli, and on Benbane Head north of Port Moon there exists one fine example (Millar, *pers. comm.*) of what is most probably a lava tube. The massive fill of the tube is now isolated as a sea stack on the tidal rock platform. Other small-scale examples, characterised by radial jointing, were exposed in the basalt quarry at Knockcloghrim, County Londonderry.

The first outpourings of basaltic lava rest on a residual soil (clay with flints) or directly on the eroded surface of the Cretaceous White Limestone (Chalk). Locally there is overstep onto older rocks; north-west of Lough Neagh in the region of Maghera they overlie Carboniferous and Dalradian rocks; south-west of Lough Neagh they rest on Triassic and Carboniferous strata; and south of the Lough they overstep onto Ordovician and Silurian slates. Often these first flows are rubbly and well-weathered as if they encountered wet ground. Here and there they fill in

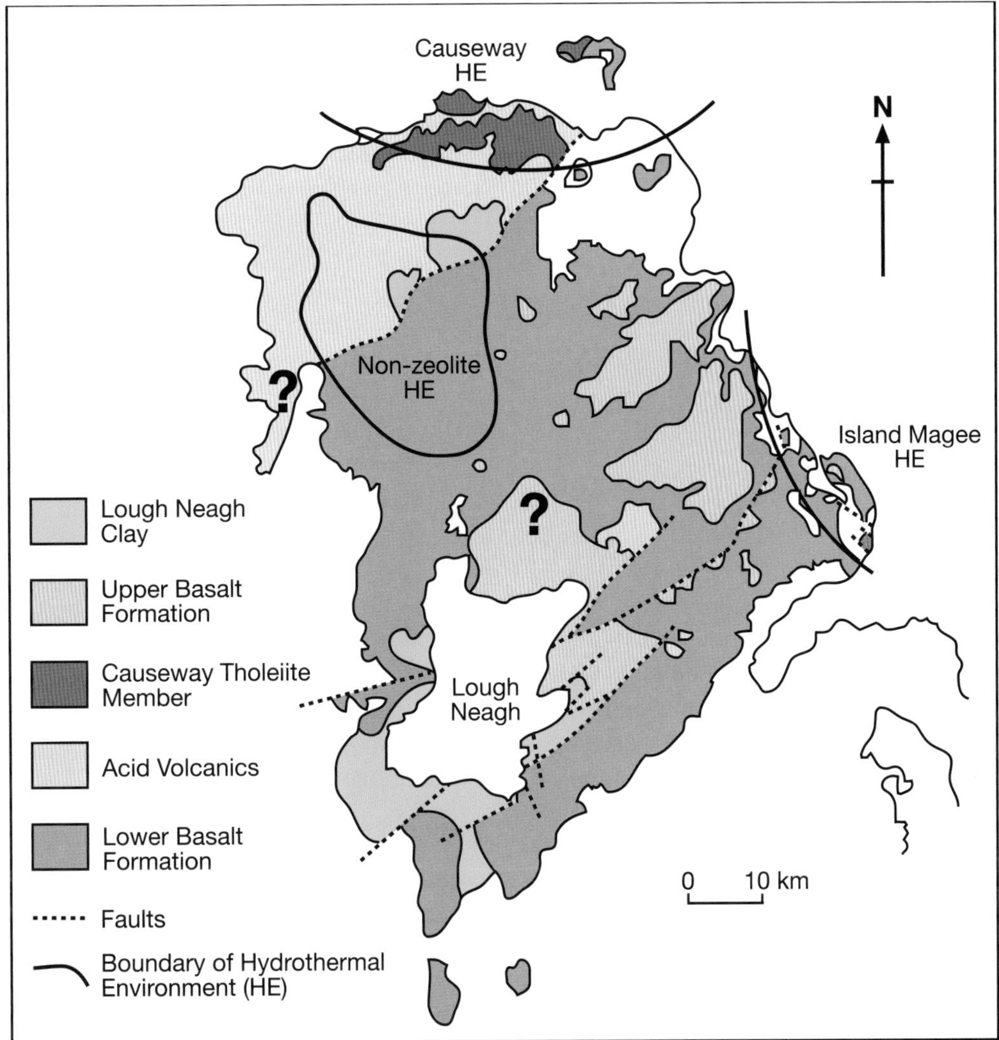

Fig. 14.2. Sketch map of the Tertiary lava formations of the Antrim Plateau (the Antrim Lava Group) based on the maps of the Geological Survey of Northern Ireland. Hydrothermal environments are taken from the work of Walker and of Adair. The distribution of the Upper Basalt Formation (UBF) is in question (marked ?) over two areas, one along the southern extension of the western escarpment and the other to the north of Lough Neagh.

the solution hollows of the Chalk surface, and the rubbly nature of the flow may be mistaken for vent agglomerate. Along lines of weakness, however, the same lines that now define the edge of the plateau or the Antrim coast itself, the volcanism broke through in more spectacular fashion and a series of agglomerate-filled vents were formed. Patterson (1963) has given detailed descriptions of those along the north Antrim coast. Such vents follow the normal faults which define the coastline, throwing down the Cretaceous and Tertiary formations into the Rathlin Trough. It is difficult to determine the relative ages of intrusive sills and extrusive basalts; Patton (1988) has made tentative correlations on a petrochemical basis and it is possible that volumes of sill-magma in the underlying Carboniferous and Jurassic sediments provided the build-up of volcanic gases that drilled conduits through to the Tertiary surface and raised fragments of these deeper-seated rocks to the present level of erosion.

Detailed maps and sections of these vents can be consulted in Patterson (*op. cit.*) and in the Geological Survey Memoirs (Wilson and Robbie, 1966; Wilson and Manning, 1978). A traverse along the edge of the basalt escarpment overlooking Dungiven and Limavady will discover many smaller examples of vents, filled with basalt, Chalk, and flint fragments, which penetrate the Cretaceous and Tertiary rocks.

Stratigraphy of the Antrim Lava Group

Old (1975) revised the nomenclature for the various lava formations described by Tomkeieff and Patterson. His subdivision of the Antrim Lava Group is as follows:

Formation	Members
Upper Basalt Formation (UBF)	Basalt flows
Interbasaltic Formation (IBF)	Ballylaggan Member (Laterite) Causeway Tholeiite Member Port na Spaniagh Member (Laterite)
Lower Basalt Formation (LBF)	Basalt flows

Mitchell *et al.* (1999) would add the Donald's Hill Ignimbrite Formation, DHIF, at the bottom of this listing, but the age of this ignimbrite is open to discussion (see below).

There is unavoidable overlap in this scheme; the topmost flows of the LBF are altered *in situ* to the laterite and lithomarge of the Port na Spaniagh Member. The same observation applies to the top of the Causeway tholeiites; it too is weathered to form the Ballylaggan laterite. There should also be a Postbasaltic Formation where the uppermost flows of the UBF are weathered in similar fashion, though this soil horizon is only seen in borehole sections.

A more detailed stratigraphy is difficult to erect. Flow upon flow successions are conspicuous in quarry and coastal cliff sections but correlation between sequences is often uncertain. Even along the almost continuous cliff sections of the North Antrim Coast any gully that breaks the line of the cliff may conceal a small fault, and correlation across such a feature is uncertain.

Scarp featuring, best developed among the Causeway tholeiites, can be traced with confidence for individual flows, and occasionally a distinctive lithology, as for example the feldsparphyric flow along the edge of the Garron Plateau (Walker, 1959), can be traced for several kilometres.

Groups of flows with a less distinctive lithology are recognised. Thus Doherty (1983) found a lower olivine-rich series of flows capped by a sequence of more feldsparphyric basalts in the LBF along the South Antrim escarpment overlooking the Lagan Valley. The LBF in the Ballymacilroy borehole contains repeated sequences of olivine-rich and olivine-poor basalts. The UBF on Agnew's Hill is capped by a series of finer-grained picritic basalts separated by a conspicuous red bole from the coarser-grained olivine basalts below; this sequence characterises the formation as far north as Troston and as far west as Benevenagh (Lyle, 1985a, 1988). Again, alternating sequences of olivine-rich and olivine-poor basalts occur at the top of the Ballymacilroy borehole. Individual flows of unique mineralogy or chemistry can also be used as marker horizons. A natural gamma radiation detector can identify high potash-bearing flows (see Versey and Singh, 1982; Francis *et al.,* 1986) among the Causeway tholeiites.It is difficult, however, to trace such examples for more than several kilometres.

Wilson (1970) used palaeomagnetic criteria to subdivide the basalt formations of the Garron Plateau into small groups; flows with the same palaeo-declination and -dip were probably extruded at relatively short intervals. Similar work in north-west Iceland (Kristjansson *et al.,* 1975) also identified a succession of small flow groups, each characterised by its own individual petrochemistry. The zirconium content gives the best signature for a group, and analysing for this element together with nickel and chromium in the samples taken from the Ballymacilroy borehole allows a subdivision of the whole succession of flows into some 13 groups.

As yet there are insufficient data to subdivide every region of the Antrim Basalts, but until this is achieved it will be impossible to offer a true picture of the building of the plateau.

Lower Basalt Formation (LBF)

Figure 14.2 shows the areal extent of this formation though there is uncertainty over its exact boundaries. The Ballymacilroy borehole proved 346 m of the UBF below ground level, and this requires some readjustment to the outcrop of the UBF in Mid-Antrim. The basalts along the western escarpment of the Antrim Plateau from Benevenagh to Magherafelt are regarded as UBF flows by Patterson (1955b) and Lyle (1988) and are so illustrated on the Geological Survey's 1:250, 000 geological map of Northern Ireland published in 1997. Earlier maps portrayed an area of LBF flows north of the Tow Valley Fault from Moneymore to Keady Mountain. Cooper (2004) reverts to this distribution of the LBF following the claims of Mitchell *et al.* (1999) for a Donald's Hill Ignimbrite Formation, the DHIF, at the base of the lava pile. This episode of acid volcanism is considered to precede the basaltic outpourings of the LBF. Simms (2000b) however discovered basic dykes (see Figure 14.3) intruding the Cretaceous White Limestone of the Carmean Quarry which are weathered and erosively truncated at the Cretaceous/Tertiary unconformity. These bear witness to an earlier period of basaltic volcanism prior to the lavas resting on the Chalk. Simms (*op. cit.*) also draws attention to Lamplugh *et al.*'s (1904) record of a similarly

Fig. 14.3 An erosively truncated basaltic dyke, one of several, at Carmean Quarry as recorded by Simms. The dyke is weathered well below the erosion surface and overlain by relatively fresh basalt lava. *Photograph courtesy of the quarry manager, D. Donaldson.*

truncated dyke on White Mountain above Lisburn. A comparable hiatus in the extrusion of basaltic lava must have occurred over the south-east sector of the Antrim lava plateau, and both events may be related to the weathered Lower Basalts encountered in the Ballymacilroy borehole where the wide hole caliper indicates soft and incoherent rock. Indeed, if the lava plateau is built of flow groups over migrating fissure feeders, then such intervals must have affected large areas and, given the right conditions notably for groundwater, could have resulted in deep weathering and erosion. There is no proof that such truncated dykes fed lavas to the Tertiary ground surface, but the possibility that any LBF flows that erupted so far west were removed by erosion and the succeeding ignimbrite belongs to such an interval, even the IBF, cannot be ignored.

This Lower Basalt Formation represents a complete cycle of vulcanism (Lyle, 1980) comprised of olivine tholeiitic basalts and, in the waning stages, basaltic andesite and rhyolite flows. In the past the intermediate compositions have been described as quartz trachyte (Patterson, 1951a) and mugearite (Walker, 1960a); they are fine-grained rocks with a platy jointing which, in the field, are difficult to distinguish from basalt. The Geological Survey of Northern Ireland has added considerably to these examples especially in the region of Glenavy, east of Lough Neagh. Lyle and Thompson (1983) have classified, on the basis of their chemistry, all the Tertiary intermediate lavas of north-east Ireland, and in conjunction with Lyle's earlier work on the basalts it is apparent that the whole cycle of volcanism is calc-alkaline in nature and in this respect contrasts with the alkali basalt sequences on Skye and Mull. A few alkali basalts have been recorded; occurrences of nepheline-bearing varieties must be discounted, modal feldspathoid has not been confirmed except in thermally metamorphosed zeolitised basalts; whole rock analyses, however, do in some instances show nepheline in the normative (CIPW) suite of minerals.

In the vicinity of their source the lavas were most probably of pahoehoe (ropy) lava type; well-preserved corded lava crusts full of rounded gas cavities can be found at High Water Mark along the east side of Ferris Bay and to the north of Portmuck on Island Magee, or close to Dunluce Castle on the north Antrim coast. Such flows are usually very thin, less than 1 m, and may represent a succession of flow units related to one episode of eruption. More often the lavas are of aa type with oxidised and scoriaceous surfaces and a thickness of 5–10 m. Flow thickness reflects the viscosity of the lava, and this may be very low close to a vent where the hot melt is highly charged with dissolved gases and vapours. Cooling, degassing, and subsequent crystallisation increased the viscosity and resulted in thicker flows. Ponding of lava in topographic hollows, against earlier flow fronts, or along fault scarps has produced much greater thicknesses.

The recognition of flows and flow units (Walker, 1971) is often difficult and subjective; well-developed lava surfaces can be used with confidence to separate flow from flow, and often the succeeding flow has trapped steam rising from the damp ground below and pipe vesicles, vesicle cylinders, and vesicular zones mark its base. More rarely a hyaloclastite deposit with or without pillow lavas indicates standing water on the older lava surface. Lava fountains must have produced large volumes of basaltic tephra, and the prevailing winds drifted it over the lava plateau. The glassy fragments sifted down onto the cindery surface and were readily oxidised and hydrated to red clay or 'bole'. These red boles resting on top of oxidised aa flows and seemingly veining them have often been mistaken for *in situ* laterite soils, and a longer time interval attributed to them.

The rising magma carried with it well-formed crystals of olivine, plagioclase, and occasionally pyroxene. Olivine formation must have commenced under mantle conditions, for within such crystals occur inclusions of a chrome picotite spinel very close in chemical composition to the mantle spinels themselves (see Griffith and Wilson, 1982). In rare instances such as in a dyke that intersects the marble and granite of the Poison Glen, County

Donegal, the spinel forms phenocrysts in free contact with the rock matrix; usually, however, such spinels are but corroded relicts of the original crystal with a composition much enriched in iron. Not all phenocrystal olivine contains spinel inclusions; some must have nucleated and grown at higher levels in the crust. Translucent green olivine is readily identified in the field; sometimes it is marginally altered to red or irridescent iddingsite, a consequence of high temperature oxidation and hydration, which makes it more conspicuous (Tomkeieff, 1934). Plagioclase can be more evident than olivine, though proof of a deep-seated origin is not forthcoming; single crystals up to 20 cm long have been found; some of the dykes, notably those at St. John's Point and near Green Harbour on the County Down coast, are crowded with labradorite phenocrysts up to 5 cm long but comparable big feldspar basalts are not seen. Prior crystallisation as a cumulus phase in some crustal magma chamber may account for their presence. Small phenocrysts of magnetite or titano-magnetite also occur; early oxidation of the magma against water or carbonate-bearing strata would have enhanced their presence. Pyroxene as calcium-rich or calcium-poor varieties rarely separated at an early stage; they occur in the final matrix of the flow along with interstitial montmorillonite, serpentine, or zeolite. Augite is the typical pyroxene, though as more electron-probe analyses are made on the basaltic minerals more examples of pigeonite and hypersthene are discovered. Zeolite may be a late stage hydrothermal introduction into empty mesostatial cavities; the clay minerals could represent an original glassy mesostasis. Doherty (1983) recorded a brownish glass in some of the south Antrim flows. In Glen Ballyemon in north Antrim the basalt resting directly on the Cretaceous White Limestone contains some 40% of interstitial glass now partially devitrified to dendritic pyroxene and iron oxides; it is curious that this flow exposed in a small disused quarry near the head of the glen should be one of the few examples among the Lower Basalts to show a regular jointing comparable with that of a Causeway tholeiite flow.

Fine-grained rocks have an intergranular texture whereas coarse-grained examples are usually ophitic. Disturbance of the solidifying lava by continued flow may have produced a marked banding by flow differentiation. Early-formed olivine and plagioclase have been pushed aside into more static layers, each with a cumulus or fluxion texture; the intervening liquid, enriched in late crystallising plagioclase and pyroxene, developed a contrasting texture characterised by plates of augite large enough to be mistaken for phenocrysts. Such bands are only several centimetres thick. Similar banding with more contrasting black and red coloration may be due to varying degrees of oxidation in successive pulses of lava. The regional characteristics are listed below.

Along the *South Antrim escarpment* Doherty (1983) made a flow by flow study of the LBF in quarries and stream sections above the Lagan Valley. As noted above, the first flows are olivine-rich members of an incompatible element-depleted, olivine-plagioclase-porphyritic series. The Cave Hill Compound Picrite Member starts the sequence north of Divis Mountain. A higher flow, the Black Hill Compound Member, can be traced along the whole length of the escarpment for 13 km, and attains a thickness of some 60 m on Black Hill. Ponding may account for the excessive thickness and on Collin Mountain colonnade and entablature are well developed as in the ponded lava at the Giant's Causeway. Both these compound flows contain more than 30% modal olivine. An incompatible element enriched, olivine porphyritic series succeeds and occupies the higher ground; feldsparphyric olivine basalts are the highest flows exposed on Divis Mountain, Squire's Hill and Collinward. Locally on Cave Hill two alkali olivine basalt flows occur (CIPW classification).

On *Island Magee* the LBF is exposed in spectacular cliff sections at Black Head (one of the finest arrays of vesicle cylinders occurs just below the lighthouse) and at the Gobbins. Sequences of thin pahoehoe flow units have already been noted. Walker (1959) described a composite flow near Macilroy's Port with an olivine-rich base full of 'pegmatitic' segregations. These latter may be genuine late fractions of the melt or, since they only occur close to the vent wall at Macilroy's Port, thermally metamorphosed amygdales. The sodalite which Walker (*op. cit.*) first identified and the associated nepheline and aegerine-augite could have recrystallised from the undersaturated zeolites and clay minerals introduced into the basalt by hydrothermal solutions.

In *East Antrim*, north of Island Magee the basalt cliffs are set back from the coast, as at Sallach Braes, or tilted back on huge slipped blocks, as under the edge of the Garron Plateau. Walker (*op. cit.*) noted olivine enriched parts of composite flows at Glenarrif, near Cushendall.

In *North Antrim* the most notable feature along this edge of the plateau is the remarkable variation in thickness from a few hundred metres down to one or two metres. The changes are so abrupt that Patterson (1955a) has suggested penecontemporaneous faulting and the banking up of flows against fault scarps. West of the River Bann the stratigraphical position of the basalts is in doubt; they may belong to the UBF (see Lyle, 1985a; Wallace *et al.*, 1994).

In *Mid Antrim*, apart from a few isolated quarries, exposures are very poor and continuous sequences are known only from boreholes. The Ballymacilroy borehole is the best section. A log based on systematic sampling of the rotary drill chips, and on the density and gamma ray Schlumberger logs shows six groups of flows, each defined by specific trace element chemistry. Two of these groups are of olivine-rich basalts and above the lowest of these the samples and hole caliper indicate a clay horizon, another interbasaltic layer of appreciable thickness which is not seen elsewhere in the lava field.

Near the top of the formation more differentiated lavas appear; they vary in thickness from 2 to over 20 m (Lyle and Thompson, 1983) and as at Uppertown can be traced in outcrop for 5 to 6 km. Their presence must indicate a decline in volcanic activity permitting more time for the source magma to differentiate. Longer and longer interflow intervals culminated in the extrusion of acid volcanics that appeared locally, possibly over high level magma chambers, and broke the tranquillity of the volcanic quiet period which followed.

Interbasaltic Formation (IBF)

At least four members can be recognised in this formation. The lowest of them is the *Port na Spaniagh Member. In situ* weathering of the uppermost flows of the LBF produced first lithomarge (Eyles, 1952) as the basaltic minerals altered to clay (kaolinite and meta-halloysite) without destroying the fabric of the rock; magnesia, lime, alkalies and some silica were leached during this stage but iron and titanium remained along with alumina and most of the silica. This is essentially an oxidation and hydration of the basalt and can occur under many climatic regimes. Under tropical conditions with heavy seasonal rains the weathering goes one stage further; silica is leached extensively to leave behind oxides and hydroxides of iron and aluminium. Gibbsite, boehmite, haematite, goethite, magnetite, maghemite, chamosite, anatase, rutile, and a chlorite-type mineral are present in these Antrim basaltic lithomarges, lateritic lithomarges, siliceous laterites, ferruginous laterites and bauxites. Weathering of this nature not only requires a specific climatic regime but also a long period of time. The fossil flora preserved within these fossil soils does not bear witness to a tropical climate and it must be assumed (Kubiena, 1970) that higher groundwater temperatures not only accelerated weathering but simulated those of a hotter climate. Estimates of interflow time intervals based on these interbasaltic beds must be adjusted accordingly.

Acid volcanic rocks weathered in a similar fashion to grey bauxites, but whether laterite or bauxite is the end product there is usually a far greater thickness of underlying lithomarge. Total thicknesses vary from 6 to 12 m up to a maximum of 28 m; some three to four flows of the LBF may have been competely altered in this way.

Acid Volcanics make up the second lowest member. Rhyolitic lava and ashes probably mark the end of the first cycle of volcanism; they appeared after a longer interflow interval and rest on the lateritic surface of the LBF. The greatest volume can be found in mid Antrim at Tardree Mountain and Sandy Braes. A central dome some 60 m high, the Tardree Rhyolite, of columnar jointed porphyritic rock is surrounded by acid lavas and tuffs. At Sandy Braes a similar structure surrounds a vent filled with obsidian fragments and welded tuff. Smaller volumes of rhyolitic rocks occur at many other localities in counties Antrim and Down. At Straid and Irish Hill, near Ballyclare, a grey bauxite is all that remains of this episode. At Cloughwater a beautiful pink flow-banded rhyolite contains small fragments of what is most probably Dalradian schist; the exposure has recently been destroyed by land improvement. At Kirkinriola a grey rhyolite crops out in the churchyard. Near Hillsborough an isolated mass of rhyolite must be an intrusive plug into Lower Palaeozoic rocks; a similar intrusion was proved at Templepatrick, and penetrated by drilling for 430 m (Brandon and Wilson, 1975). A more unusual occurrence was revealed near Ballydugan House, south of Lurgan, in excavations for a hilltop storage reservoir for water. Here, associated with so-called mugearites near the top of the LBF, occurs a porphyritic rhyolite of dark grey colour, which along joint surfaces has been altered to a white rock almost identical in appearance to the Tardree rhyolite. The latter may represent a similar but much more extensive alteration with fresher and darker rock existing at depth. These and other yet hidden centres of acid volcanism may have erupted pyroclastic flows and wind-blown ashfalls. Smaller examples of grey bauxitic clay appear at the base of the UBF escarpment and in quarry sections, notably those at Craigahullier. Such a clay, full of broken quartz crystals, on Agnew's Hill prompted further exploration for ceramic clays and revealed a thick sequence of altered ashes and welded tuffs. The ignimbrite, 1 m thick, described by Mitchell *et al.*, (1999), which can be traced for 30 km along the western margin of the Antrim basalts may well be related to this period of time. Its occurrence conjures up a picture of volcanic domes as the eruptive centres for numerous radiating ash flows.

These acid volcanics were, in turn, weathered to lithomarge and bauxite; pebbles of the harder rocks can be found among the lateritic soils, an indication that local run-off eroded the deep-weathered surface and carried the more durable rocks to other sites.

The third member is the *Causeway Tholeiite Member.* Penecontemporaneous with most of the rhyolites are the fine-grained, columnar-jointed, silica-rich basalt lavas of North Antrim and Rathlin Island. Unlike the rhyolites, however, they appear to be unrelated to the first cycle of eruptive activity. They could represent an independent episode, a possible response to crustal stretching and its attendant decompression melting, or they could be related to the magma batch that fed the second cycle of volcanism (Lyle, *pers. comm.*)

The first outpourings of lava flooded a land surface of iron pan and lithomarge subject to stream erosion. They ponded in the deep valleys excavated through the soft weathered top of the LBF. Rohleder (1929) and later Tomkeieff (1940a) described the 90 m deep valley that drained an offshore region to the north of the Antrim Coast. The Giant's Causeway itself occupies the bottom of a meander that looped southwards. The tranquil cooling conditions at the base of this deep lava pond were ideal for the growth of a regular hexagonal jointing. Smaller examples of exhumed valleys can be seen at Taggart's Quarry, north of Ballybogey, and again at Craigahullier Quarry to the south-east of Portrush. Where the first outpourings of lava encountered surface water, pillow lava and hyaloclastite formed and weathered to a distinctive ocherous colour, see Figure 14.4, as at the Craigahullier quarry.

Indeed, the most spectacular feature of all the lava flows is their jointing. Tomkeieff (1940a) described a tripartite structure of a basal colonnade surmounted by an entablature comprising a central curvi-columnar zone and an upper pseudo-columnar zone. Lyle (2000) studied the eruptive environment of these multi-tiered columnar basalt flows; ponding of the lava in river valleys offered ideal cooling conditions and the disturbed drainage of surface water modified the cooling regime in the upper part of the flow. These features characterise all flows of the Causeway Tholeiite Member but are not unique to it. Fine examples occur locally within the Lower Basalt Formation as in the first flow

Fig. 14.4 Pillow lava/hyaloclastite deposit at the base of the Upper Basalt Formation, above Milltown on the northeastern edge of the Garron Plateau. The ocherous (yellow/orange) colour of these water-quenched rocks is in marked contrast to the red-coloured lateritic rocks of the interbasaltic horizons.

extruded over the Cretaceous White Limestone at the head of Glen Ballyemon or again in the final flow of the UBF on the Rue Peninsula of Rathlin Island, where it builds Haughton's (1852) Little Causeway and the spectacular jointing at Doon Point. Nature imposed a hexagonal close-packed array on the cooling centres that developed at the base of each flow. A hexagonal fracture pattern developed, its dimensions determined by the rate of cooling; because the corners of the hexagons are furthest from the cooling centres, these directions would experience the greatest contraction and fractures would initiate at these points. Since the corners are common to three neighbouring contraction cells, such fractures would appear as three-rayed stars; these can be observed in drying mud, though here the conditions are rarely as uniform as in the base of a thick and often ponded lava flow. Lateral growth of the fractures would complete the hexagonal network and upward growth would extend the columnar jointing into the flow's interior. This growth would be stepwise as the lava cooled below solidus temperature and the tension built up to a value that would propagate the joint. In places, as at the bottom of deep-ponded lava, the columnar joint surfaces are flat and even as in Figure 14.5; elsewhere they tend to be sinuous as if the fracture at each stage was conchoidal; see Figure 14.6. Tomkeieff (1940a) noticed the occurrence of chisel

Fig. 14.5 Detail of the columnar jointing on the west side of the Grand Causeway. The smooth flat surfaces of the columnar joints were probably exposed by small-scale quarrying in times past. Both kinds of cross-joints can be seen; the corner spines (cf. Figure 14.8) have, in most instances, been lost.

Fig. 14.6 Detail of columnar jointing from the cutting through the root of the Grand Causeway. The columnar joints are somewhat sinuous, a feature that can be seen in most exposures.

marks on the joint surfaces normal to the column length. Such features are far better and more extensively developed in a conserved corner of the Craigahullier Quarry, east of Portrush (Figure 14.7). Here the joints are obviously sequential in their growth, but at each stage of development the fractures must have been initiated at the hexagon corners and propagated laterally. Growth of the fracture in this way produced discrepancies in direction between the fractures growing from opposite corners and with the previous, already established, columnar joints. These discrepancies in joint orientation are bridged by curving fractures. One such bridging occurs along the length of each chisel mark, though it is conceivable that more than one could occur if the columnar joint grew upwards at the same time. The conditions that produced this type of joint growth are not obvious.

With further cooling the columns themselves contracted and cross joints formed. These are often ball and socket joints with a flat rim and projecting spines at the corners. They can be best seen at the Giant's Causeway itself and circa 1740 when Susanna Drury painted her

Fig. 14.7 Columnar jointing in the Causeway Tholeiite Member exposed in the north-east corner of Craigahullier Quarry. Chisel marks, 3 cm wide, cross the columnar joint surfaces and mark the sequential growth of the fracture. Each chisel mark represents a laterally propagated fracture that followed an independent course and is connected to neighbouring fractures by curved bridging joints.

'Prospects' of this unique landmark, the column segments were carefully quarried and all details of their structure preserved, as shown in Figure 14.8. Figure 14.5 pictures the last quarried section on the west side of the Grand Causeway, and as noted by Tomkeieff (1940a), there are two individual fractures comprising each cross joint:n earlier curved fracture that originates within the basalt column and a later planar fracture that intersects from the outer surface. This latter usually cuts off the projecting corner spines, and so they are rarely preserved in weathered sections and usually destroyed by normal quarrying practice. Variations in the spacing of these two fractures could account for the differing aspect of the cross joints. As noted by Preston (1930) cooling from above or below could control the 'way-up' of the ball and socket joint.

Fig. 14.8 A detail from Vivares' engraving (1743/4) of Susanna Drury's painting of The East Prospect of the Giant's Causeway (*c*.1740). The three column segments, with their corner spines intact, have been carefully removed from the east side of the Grand Causeway.

At the surface of the flow cooling could not proceed as evenly; disturbed by degassing and by wind and weather a pseudo-columnar zone developed. But the core of the flow marked by the curvi-columnar zone must have had a more complicated cooling regime. This core remained molten and mobile long after the colonnade and pseudo-columnar zones had formed. Drain-away features can be seen in the cliffs on the east side of Port Reostan, where the pseudo-columnar zone rests directly on the top of the colonnade and at the east face of Craigahullier Quarry, before a land-fill site destroyed it; here localised streams, lava tubes, in the core of the flow eroded the top of the colonnade. This heart of the flow must have cooled very slowly, for it is possible to see the original master cooling joints spaced at 1–2 m intervals. The curvi-columnar zone still breaks more readily along these fractures, as witnessed by the fallen blocks on the foreshore east of the Giant's Causeway. But in the course of time, such joints would become secondary cooling surfaces and initiate a secondary jointing pattern. Under undisturbed conditions these secondary joints would grow perpendicular to the isothermal surfaces within the flow and develop the symmetrical arched systems as figured by Spry (1962). These can be seen in the cliffs above Lacada Point at the western end of Port-na-Spaniagh, though the viewing is fraught with danger due to collapse of the cliff path. Seasonal heavy rains (Saemundsson, 1970) would have allowed surface water to penetrate along the master joints into the heart of the flow to produce the curvi-columnar zone that is seen in most exposures. Rapid cooling by this means would have built up the internal stress and triggered its almost explosive release, as envisaged by Tomkeieff (1940a).

Another highly characteristic feature is the fine-grained texture, which persists throughout the flows from chilled selvedge to the interior. Tomkeieff (1940a) very neatly accounted for this by assuming that the lava chilled to a glass that later devitrified; a condition that would have resulted in smooth joint surfaces. Many lavas have been found, however, which show flow alignment of the plagioclase laths, and so crystal nucleation must have occurred before extrusion, as claimed by Francis *et al.* (1986).

Various estimates have been made in the literature of the number of flows that make up the Causeway Tholeiite Member. Patterson (1955a) counted seven, whereas Wilson and Manning (1978) record a total of nine flows, but their count on the ground differs, as does that of Francis *et al.* (1986) for the coastal exposures east of Benbane Head to Port Bradden. Flow thickness falls off towards the top of the succession and the jointing pattern is less spectacular. Olivine rarely occurs unaltered and is present in only some of the flows. Plagioclase is the most abundant constituent, the labradoritic laths often defining a fluxion structure. Augite and pigeonite are next in abundance, followed by iron-titanium oxides. It is the mesostasis that is probably the most interesting component of these rocks; the glass dusted with oxide grains represents a residual liquid composition and shows evidence of immiscibility into iron-rich and silica-rich fractions. Segregation of this residual melt into a tholeiitic andesite flow unit can be found within flow 5 at Dunseverick Slip, east of Benbane Head. This rock, described in detail by Francis *et al.* (*op. cit.*), can be identified by its coarse-grained fasciculate or spherulitic texture.

Six or seven of these lavas crop out on Rathlin Island but here the stratigraphy is a little more complicated. The flow of the LBF immediately below the laterite and lithomarge of the IBF shows spectacular columnar jointing. It was in this flow that Haughton (1852) found and named the Little Causeway, another horizontal section across a colonnade just below High Water Mark on the east coast. Vertical sections of disturbed columns appear in the low cliffs, and at Doon Point to the south a similar jointing array is associated with pillow lava and hyaloclastite rock. This flow has all the chemical characteristics of an olivine tholeiite and belongs to the LBF.

The chemistry of the Causeway Tholeiite Member suggests a differentiation product of a more primitive magma, but the flow types resemble those from fissure eruptions rather than effusions from a shield volcano. No feeder intrusions have so far been identified.

The fourth and final member of the IBF is the *Ballylaggan Member.* Lateritic and bauxitic soils with underlying lithomarge cap the Causeway tholeiites and acid volcanics described above. Elsewhere over the basalt plateau this period of weathering was uninterrupted and only one member of the Interbasaltic Formation exists.

Upper Basaltic Formation (UBF)

A second batch of olivine tholeiitic magma fuelled the second cycle of volcanism (Lyle, 1980). Only basaltic lavas have been found among the effusive products in most areas; the exceptions are the tholeiitic andesites that occur along the north shore of Lough Neagh, though the basalts in this area have been variously mapped as either LBF or UBF. The basalts have a chemical signature distinct from those of the LBF. They are more primitive in their composition and may have risen from a deeper source or from mantle rock already depleted in basaltic components by the first cycle.

The flows are somewhat thicker than those in the first cycle, show fewer interflow beds, and in east Antrim from Agnew's Hill to Troston Mountain show a marked change to finer-grained picritic lavas above a conspicuous red bole (Emeleus and Preston, 1969; Lyle and Patton, 1989). This succession is not apparent in the Ballymacilroy samples, though the seven flow groups do contain two that are olivine-rich.

A complete section through the UBF is found only in the Washing Bay borehole, sited on the south-west shore of Lough Neagh, where Wright (1924) identified a post-basaltic bole some 22 m thick above 213 m of Upper Basalts. As noted above, greater thicknesses have been recorded in other boreholes where this final *in situ* soil is missing. Along the outcrop much thinner sequences are found; that on Agnew's Hill is about 90 m thick, but because all these basalts are extensively zeolitised Walker (1960b) argues that some 460 m of a zeolite-free zone, at least, have been removed by erosion.

Basalt petrochemistry

The pioneering work of Patterson (1951b) and Patterson and Swaine (1955) on the petrochemistry of these basaltic lavas, and the more detailed work of Lyle (1980) and Lyle and Preston (1993), accounted for the compositional variation by a process of fractional crystallisation along a calc-alkaline liquid line of descent culminating in melts of an intermediate and acid nature. Lyle (1985b) argued from the basis of trace element and notably rare earth element determinations that the LBF probably formed by low degrees of partial melting of a spinel lherzolite mantle whereas the UBF required a higher degree of partial melting. The Causeway tholeiites, in contrast (Lyle and Preston, 1993) may have been the consequence of tectonically controlled decompression melting of the mantle. Strontium isotope data (Wallace *et al.*, 1994; Wallace, 1995), however, demonstrates extensive open system behaviour during differentiation of these basalts. The various magmas equilibrated and assimilated basement rock, possibly sub-continental lithospheric mantle for the LBF and UBF and Dalradian metasediment for the CTM; see Barrat and Nesbitt (1996) for further discussion.

Sub-surface volcanism

The upward migration of basaltic magma from its source in the upper mantle had to establish, maintain, and re-establish magmatic pathways through the crustal rocks to the Earth's surface. By its internal pressure, temperature, and gas content it could melt, drill, shoulder aside, or lift the overburden. Always it would fill any cavities it encountered, notably those produced by crustal extension. In the lower crust where a plastic deformation would be the response to applied force, conduit pathways would have been established. Where the magma encountered rocks of a lower melting point, high level magma chambers could develop and where it came in contact with water or carbonate-bearing strata its pressure and chemistry would have changed. In the upper crust brittle fracture would have opened fissures for the magma to invade and modify before erupting at the surface.

Dyke swarms

A glance at Figure 14.9 will reveal the extent of the dykes. These fissure infills are mostly gregarious flash injections that never fed magma to sills or flows. A few are solitary dykes, which may or may not have acted as magmatic feeders.

The main swarm follows a north-west to south-east trend from County Donegal to south Armagh. Where it traverses the older granitic plutons, it is intensified in the well-jointed rock. This feature is well displayed in the Main Donegal and Barnesmore granites. Where the Carboniferous rocks are penetrated, the water-bearing strata led to local brecciation of the wall rock and the charging of the rising magma with xenoliths; tridymite paramorphed by quartz is recorded from such inclusions at Navar Forest in County Fermanagh. Oxidation of the magma in some of these dykes accentuates the change to more tholeiitic rocks; the Blind Rock dyke in North Donegal Bay is one such example (Preston 1967a). Some of those in Devonian and Carboniferous rock are solitary and by their width and multiple use are equivalent to 30 to 40 of the thin dykes encountered in the granites. Such is the Doraville dyke, north of Enniskillen (Preston 1967b); within the 100 m wide intrusion flow differentiation has produced mineral banding, repeatedly scoured by later surges. Gallagher and Elsdon (1990) describe a more remarkable example from the Lower Carboniferous limestones north of Inver in south-west Donegal; a Permo-Carboniferous age cannot be ruled out. This Parkmore dyke is choked with spinel lherzolite and more locally derived xenoliths. The former contain chrome picotite spinel, forsteritic olivine, enstatitic orthopyroxene, and chrome diopside; this mineral assemblage is consistent with a mantle origin, and the chemistry would imply some prior melt withdrawal from the source rock. An unusual feature reported from nearby dykes, east of Ballyshannon, is the presence of segregation vesicles (Sanders, 1986). These are vesicles into which residual interstitial melt was injected at a late stage of crystallisation, presumably driven by the release of steam.

A more north-north-westerly trend characterises the Antrim and Down swarm; this appears to be offset across the Southern Upland Boundary Fault, which follows the line of the Lagan Valley south-west of Belfast. The Antrim swarm peters out towards Scrabo Hill in North Down; the Down swarm appears to start at Lisburn and can be traced to the South Down coast where it is admirably exposed between St. John's Point and Killough; individual dykes are enriched in conspicuous plagioclase phenocrysts, which were a possible cumulus phase in the source region.

A smaller linear swarm with an east-west trend crosses Killala Bay from County Sligo to County Mayo (Emeleus and Preston, 1969; Mohr, 1987). It can be traced for a distance of 60–70 km. Most members of the swarm are simple flash injections, but a few are earlier in age, multiple in use, and effect a far greater thermal metamorphism on the Carboniferous limestone exposed in the wave-cut platform along the east coast of Killala Bay. These could be the frayed out remnants of the gabbro dyke that crops out on the opposite side of the bay.

This Killala Gabbro warrants special mention, for it is probably the only example of a giant dyke in the Irish

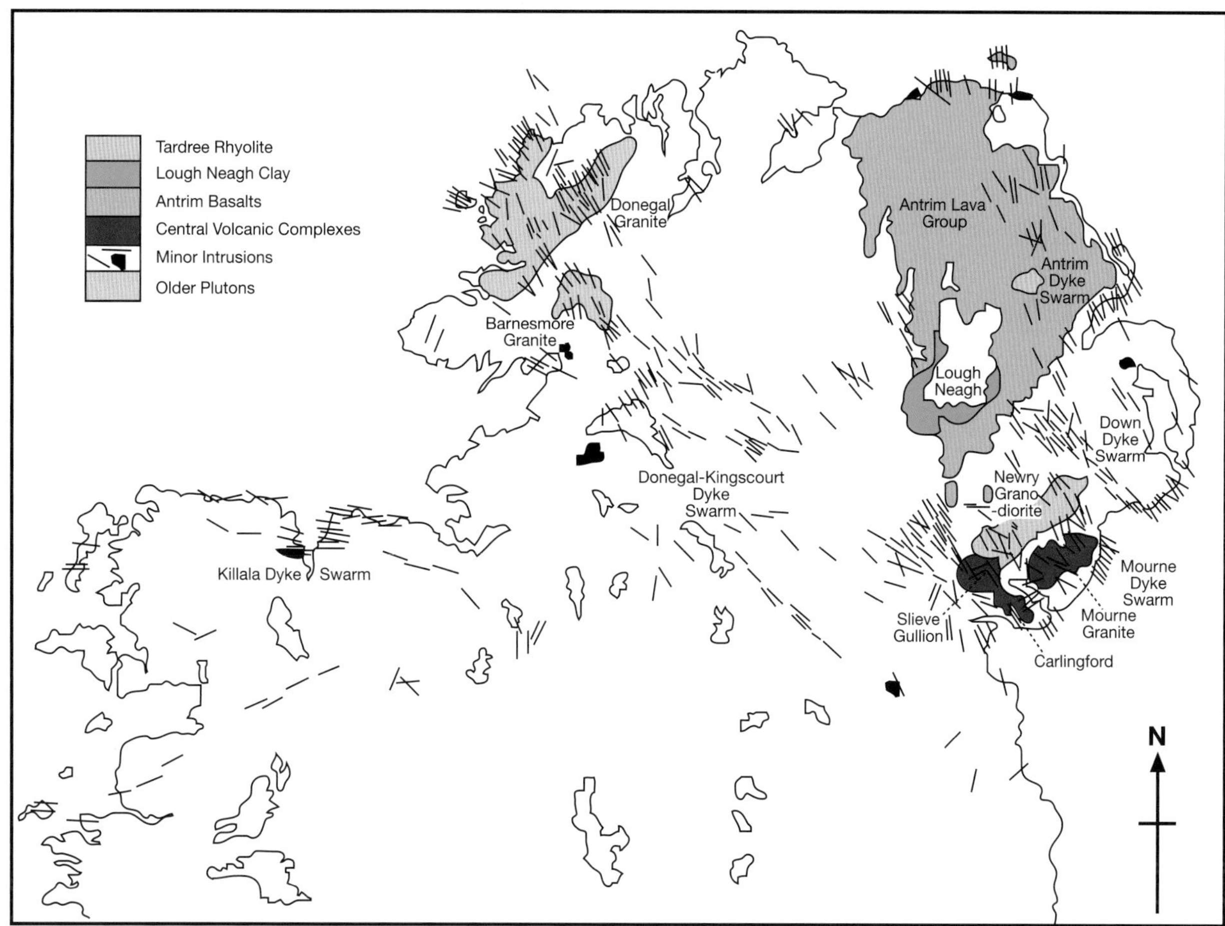

Fig. 14.9 Distribution of Tertiary volcanic rocks over the north of Ireland, including lava flows, dyke swarms, and the sub-volcanic intrusive complexes at Slieve Gullion, Carlingford, and the Mourne Mountains.

Tertiary. Its width is over 400 m in surface outcrop though drilling has proved a funnel-shaped cross section and a much reduced thickness at depth. It is possible that the north wall of the fissure slumped and floated away on the rising tide of denser magma. The intrusion can be traced from the west shore of Killala Bay westwards across the Palmerston River for some 4 km. Within steeply inclined flow-banded margins of dolerite occur near-horizontally laminated gabbros and anorthosites. Trace element variation (see Figure 14.10) exhibits two cycles of layering. The differentiation process culminated in an alkaline facies preserved as an aegerine, arfvedsonite, riebeckite, alkali feldspar, and quartz assemblage lining the cavities of an unusual drusy monzodiorite.

Farther south in counties Mayo and Galway a number of north-east trending dykes appear to radiate from the offshore Brendan centre. Mohr (1982) recognised five systems and drew attention to the chemical gradients established in the stagnant magma which filled this fissure swarm.

A detailed field study of the Donegal–Kingscourt swarm, the Killala swarm and the West Connacht swarm was carried out by Bennett (2006), who demonstrated their emplacement mechanism, and noted a remarkable correlation between their thermal and deformational effects on the country rock. This study also reports (U–Th)/He ages of about 60 Ma for the dykes, in good agreement with their emplacement age and indicative of rapid cooling at a high level.

Local swarms are associated with the Slieve Gullion and Carlingford central complexes and with the Mourne Granites. Some of these are linear but others radiate from the old volcanic hearths.

Dolerite plugs

Conduit pathways were established through the Antrim lava plateau and many can be seen to penetrate the UBF. Walker (1959) has shown that at Donegore and Parkgate they originated as beaded dykes and that many of the isolated examples are elongate in the direction of the Antrim

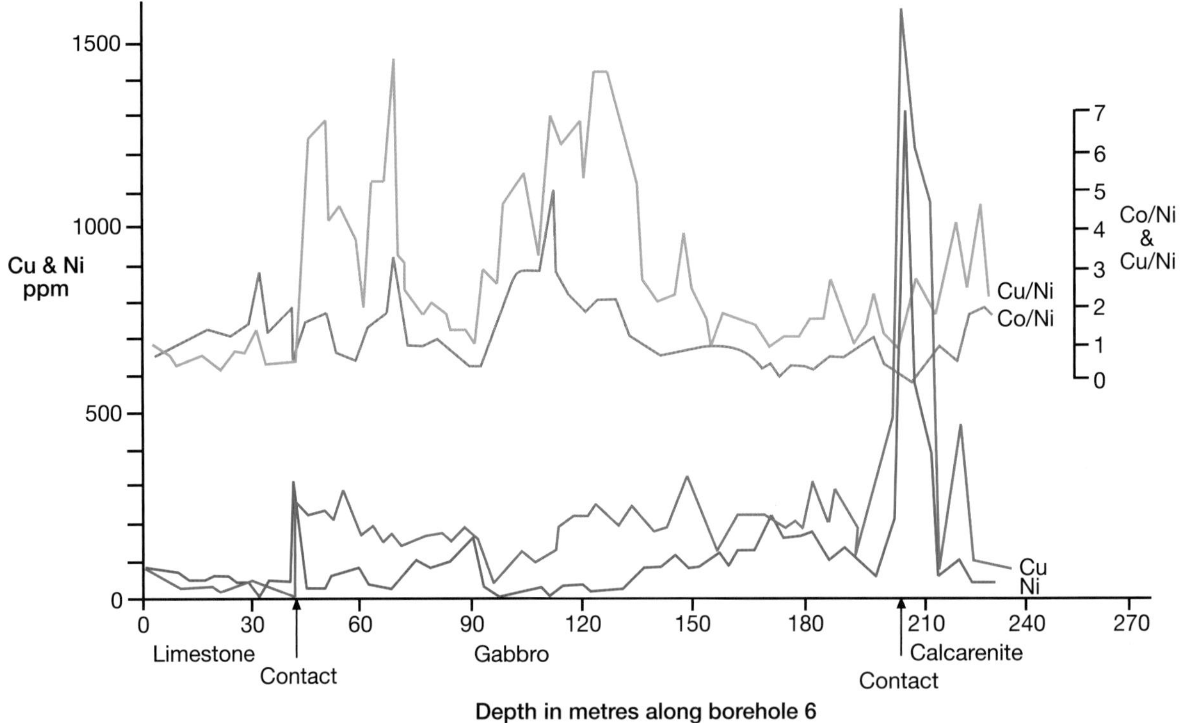

Fig. 14.10 Trace element data for the Killala giant dyke (Cree, *pers. comm.*) from borehole cores drilled by ASARCO. Cu and Ni contents and Co/Ni and Cu/Ni ratios are plotted against depth in inclined borehole 6. Apart from high trace element contents at both wall rock contacts the Ni, and less markedly the Cu, decrease upwards through the layered gabbro. At a depth of 90 m both elements suddenly increase and the pattern is repeated through a higher layered gabbro. Both ratios behave in an inverse fashion.

dyke swarm. Enlargement of a fissure implies continued use of the conduit as a feeder.

The dolerite plug of Scawt Hill is the most outstanding example (Tilley and Harwood, 1931). The pyrometamorphism of the Cretaceous Chalk and flint, and the metasomatic exchange of material across the contact produced many unusual high temperature minerals; larnite and scawtite were discovered here associated with other calc-silicates. Larnite and spurrite developed in the exogenous contact zone; spinel if present gives a dark colour to the rock; rankinite and bredigite can be found close to the plug contact. Flint nodules in the Chalk show progressive metamorphism to larnite-spurrite rock around a core of wollastonite and xonotolite; hydration of these nodules produced a pink-brown gel-like mineral, plombierite. Cavities in these contact rocks yield afwillite, scawtite, portlandite, and hydrocalumite. The endogenous zone shows progressive desilicification of the basic magma to nepheline-titanaugite pyroxenites and ultimately to a melilite-bearing igneous rock. Comparable examples have since been found at Carneal (Sabine and Young, 1975; Griffith and Wilson, 1982), Ballycraigy (McConnell, 1954), and Inishchrone (Emeleus and Preston, 1969). The Blind Rock and the Doonan Rocks in County Donegal have baked the country rock in like manner.

Slemish Mountain in mid-Antrim is built of rock that formed in a conduit pathway close to the Tertiary surface; the pit crater contained a lava lake with a circulation system still defined by the orientation of xenolithic rafts. Figure 14.11 shows some of the larger examples; the great majority are but shreds and wisps of the lava crust. Large olivine xenocrysts, corroded and embayed with spinel reaction products, collected on the floor of this pool (Preston, 1963a).

The hill of Tievebulliagh holds an archaeological as well as a geological interest. Here a large mass of the Interbasaltic lithomarge slid down the foot-wall of an inclined pipe and suffered extreme thermal alteration. Agrell and Langley (1958) detail the mineralogy and development of the mullite, spinel, corundum, cristobalite, haematite porcellanites that early man exploited for his stone axes. The Brockley plug on Rathlin Island yields similar material; its east contact with adhering hackly-jointed metabasalt is shown in Figure 14.12.

Fig. 14.11 Xenolithic rafts in the D3 dolerite at Slemish Mountain. Smaller shreds and wisps are present over the whole exposure at the south end of the plug. A heath fire followed by a winter's rain left bare a bleached surface of the dolerite which revealed every variation in texture and structure of the rock. Weathering and algal growth have since hidden these features.

Fig. 14.12 The east margin of the dolerite plug at Brockley, Rathlin Island. The hackly-jointed, thermally metamorphosed basalt lava still adheres to the plug contact.

Plug intrusions are much less common among the dykes of the west of Ireland; that at Doon Hill in County Galway is known for its contact metamorphism. Possibly fewer magmatic pathways reached the surface in this part of Ireland.

Dolerite sills

Where 'hydrostatic' pressure failed to lift magma to the surface or where a standing column of magma leaked away through the conduit wall, then sills invaded the bedding planes of the sedimentary rocks.

The Fairhead Sill of North Antrim, with its satellitic Binnagapple and Gob sills, is intruded into Carboniferous shales and sandstones. The main sill with its spectacular columnar jointing dips gently to the south and transgresses eastwards up through Triassic sandstones to terminate in the Cretaceous Chalk of Murlough Bay. As noted by Tomkeieff (1939), it contains glomeroporphyritic groups of olivine and plagioclase near the base and complementary pyroxene, plagioclase, magnetite pegmatites in lenses and clots under its roof. At Farragandoo (Figure 14.13) on the west side of the headland its mode of emplacement by fingering along individual bedding planes is clearly visible in cliff section. This accounts for the rafts of porcellanised shales that occur under the roof rock. A possible feeder to the Gob sill is an unusual dyke at Carrickmore; this intrusion and the associated blisters

Fig. 14.13 The cliffs at Farragandoo at the west end of the Fairhead Sill where the dolerite intrusion fingers out into the Carboniferous shales. In the lower cliff section, above the red sandstone, the brown weathered, columnar jointed dolerite interdigitates with the darker coloured shales. Above, the sill is weathered to a grey colour.

(small laccolithic intrusions) have extensively porcellanised the adjacent fireclays. The numerous xenoliths in the first blister show reaction rims of plagioclase, spinel, and corundum in a mullite-cordierite hornfels (Agrell and Langley, 1958).

The Portrush Sill exposed around Ramore Head and in the offshore Skerries invaded Liassic clays. Its association with baked ammonite-rich marine sediments made it the last outpost of Neptunism in Europe; an obvious product of crystallisation from seawater! (Richardson, 1805). Harris (1937) discovered some unique features in the roof zone of this sill, for the invading basic magma mobilised the Liassic shale, the contact shows flow folding, and when the dolerite had cooled and jointed the overburden squeezed the plastic hornfels into the opening cracks. These rheomorphic veins hybridised with still liquid magma below the roof chill.

The main body of the sill is conspicuously banded. On Ramore Head this is a two centimetre scale banding that weathers out on exposed rock but is difficult to appreciate in thin section. At Reviggerly Point and across the foreshore towards the 'Wash Tub' the layers are up to 1 m in thickness and the coarse-grained gabbroic textures show enrichment in olivine and structures associated with disturbed cumulate rocks (Patton, 1988). A later surge of volatile-rich magma most certainly disrupted the olivine gabbro and the small angular fragments are now enclosed in a wide halo of serpentinous material.

Evidence for sub-surface sill intrusions below north Antrim is summarised in McCann (1988).

The Scrabo sill, at one time mapped as an outlier of the basalt plateau in North Down, penetrated and bleached the red Triassic sandstone. The main sill caps the hill of Scrabo and numerous smaller ones follow bedding planes in the floor rock. Small rafts of sandstone near the base were partially melted and crystallised to orbicular tridymite rock (Preston, 1963b). Explosive vents, breccia pipes and tuff dykes (Preston, 1962) are the result of violent degassing of deeper-seated intrusions.

Numerous smaller examples occur around the edge of the plateau and others have been encountered in borehole section. Near Kingscourt in County Cavan, doleritic magma invaded the gypsiferous Permo-Triassic strata. Water of crystallisation released as the gypsum dehydrated converted the base of one sill to loose, friable hyaloclastite material. Morris *et al.* (1981) describe dolerite sills in the Carboniferous rocks near Carrigallen in County Leitrim; olivine is enriched at the base and flow differentiation has produced a layered structure. Similar intrusions were drilled through at Magilligan and Ballymacilroy. Lesser-known ones have been encountered in the off-shore basins (Seemann, 1984).

Sill intrusions at such shallow depth below the Tertiary surface could have been responsible for many of the explosive vents and ash beds which accompanied the effusion of lava.

Central volcanoes

As in the Hebridean regions, though basaltic lavas poured out over the surface of sedimentary basins, the centralisation of activity and the building of shield volcanoes was restricted to the marginal basement rocks. It could be (Meighan and Preston, 1971) that where the linear dyke swarms increase in intensity across the more readily melted rocks, i.e. granite plutons, there is a critical depth where melting of the wall rock between adjacent dykes would lead to the establishment of a high level magma chamber. There are many local examples at the present erosion level where the country rock is partially or completely melted against a Tertiary intrusion, especially where it was completely engulfed in basaltic magma. Kitchen (1981) has re-examined some of the classic localities in Ireland where wall-rock melting has occurred; these include the Crocknageer dyke in the Barnesmore Granite (Walker and Leedal, 1954) and the Tieveragh Plug in the tuffaceous sediments of Old Red Sandstone age (Tomkeieff, 1940b). At the limiting depth where magma can survive the cooling interval between successive dyke intrusions, such rock types would readily melt and a spongy network of acid, basic, and hybrid melts would gradually coalesce into a larger volume. Under such circumstances the Main Donegal Granite would be too shallow. The Barnesmore Granite reaches this limiting depth and some acid dykes are present, whereas the Newry Granite would be ideally situated. As noted by Bailey and McCallien (1956) and Richey and Thomas (1932) the south-west end of the Newry pluton almost exactly coincides with the course of the ring fault which bounds the basal wreck of a central volcano at Slieve Gullion.

The *Slieve Gullion* intrusive complex (Richey and Thomas, 1932) was probably the core of a shield volcano with a summit caldera of subsidence similar to that on Mull. The earliest intrusion of porphyritic felsite followed the southern arc of the ring fault, its mineralogy and texture the consequence of explosive degassing (Emeleus, 1962) as vents were drilled through the overburden. Flow banding defines several foci that mark the channels up which the magma rose. Magma-gas emulsions turned into streams of pumice, glass shards, and broken crystal now compacted and welded into felsitic tuffisite (Bell and Emeleus, 1988). Fragments of Newry Granite and Silurian slate fill these vents, which locally also contain masses of basaltic and trachytic lavas, the only remnants of the superstructure. McDonnell *et al.* (2004) recognise silica-rich margins to this felsite and the later porphyritic granophyre which invaded the same ring fault around the rest of its course, overlapping slightly with the felsite. It congealed more slowly and under a more impervious roof, a feature noted by Richey and Thomas (1930) for successive ring dykes of the Ardnamurchan complex. These variations in rock chemistry suggest a complex, stratified magma chamber with silica-rich granite magma above and tapped first, and a silica-poor melt below.

The central complex postdates the two ring dykes. Reynolds' (1951) 'actualistic interpretation' is one of successive flows, pillow lavas, tuffs, and agglomerates burying the caldera floor. The acid rocks were derived from the older pluton and a ramifying network of acid veins of similar origin penetrated this volcanic pile as incandescent tuffs, and no doubt helped to recrystallise them into rocks of gabbroic and granophyric aspect. Bailey and McCallien (1956) regarded the complex as essentially intrusive, comprising ring dykes and laccolithic tops together with the sill intrusions all authorities agreed on. This view is borne out by subsurface excavation for a pumped storage hydroelectric scheme (Gamble *et al.*, 1976) where the peripheral Lislea Granophyre is seen to be a steeply dipping intrusion and not a horizontal sheet, and could well be contiguous with the summit granophyres. It appears to postdate a layered sequence of granophyre and gabbro (the successive flows of Reynolds, *op. cit.*) whose contacts show every gradation from liquid–liquid to liquid–solid conditions. The chilled basic magma was repeatedly disrupted by violent degassing of the mobile acid melt. Wager and Bailey (1953) first noted this relationship for Slieve Gullion rocks; Elwell (1958) detailed an intriguing example in the base of a thick dolerite sill (Layer 10); here a raft of granophyric material, or possibly the floor of the sill itself, has been remelted and risen as lighter pipes up into the basic magma. Gamble (1982) concludes that partial melting of the older pluton is largely an isochemical process, but that hybrid magmas were produced by mixing of the basic and acid melts.

A second sub-volcanic intrusive centre is located near *Carlingford*, to the east of Slieve Gullion. Evidently the intense intrusive activity within the Slieve Gullion caldera eventually migrated along the linear fissure system, breaching the bounding ring dyke at Anglesey Mountain and Clermont Carn, and establishing another intrusive centre at Carlingford. Le Bas' studies (1960, 1965) detail the subsequent activity that may well have overlapped in time with that of Slieve Gullion.

Carboniferous limestones, domed by volcanic pressures, are covered at Rampark by 100 m of basalt with intercalated hawaiite flows. Early alkali gabbro plugs penetrate this fragment of the volcanic superstructure and a later gabbro lopolith (Le Bas, *op. cit.*) or possibly the laccolithic top of a ring dyke (recent prospecting has proved a deeper margin for the gabbro) was supplied with four pulses of magma. Cryptic layering testifies to successive pulses of high alumina basaltic magma. Explosive vents at Slievenaglogh contain fragments of an earlier granophyre. A later granophyre, a ring dyke with a laccolithic top, crystallised under increasing water-vapour pressure until a reduction in gas tension allowed a more rapid solidification (Le Bas, 1967). High pressure below the complex drove magma up swarms of cone sheets, radial dykes, and linear dyke fractures.

The *Mourne Granites* are a third expression of sub-volcanic magmatism. A high gravity anomaly underlies the Slieve Gullion and Carlingford centres and extends eastwards into Mourne Country, where the deepest level of intrusive activity is now exposed. No features of surface volcanism have been found. Basic xenoliths of cumulate gabbros in early dykes support the geophysical evidence for a large mass of basic rock at depth. Richey (1928) gave the first detailed petrography of the various intrusions and invoked the mechanism of cauldron subsidence on outward-dipping ring-faults to account for their structural relationship. An absence of any foliation and the general lack of country-rock enclaves confirms a passive emplacement of the granite magma. Richey (*op. cit.*) mapped two centres of intrusive activity. The Eastern Mourne Centre contains three ring dykes: an older feldspathic granite G1, a later quartzose granite G2, and a younger aplitic granite G3. Boreholes through the roof of G3 exposed in the Silent Valley penetrated a chilled and drusy contact into what may be a fourth granite for this centre. Nine kilometres to the west-south-west occurs the centre of two more ring-intrusions. Richey (*op. cit.*) described a single pink granite G4 but prompted Emeleus (1955) to map in a later fine-grained, blue-grey granite G5. Only the laccolithic tops of these western ring dykes are exposed. Slight eccentricity among the Eastern Granites

(G1, G2, G3) foreshadowed a westerly migration of the centre of subsidence when the Western Granites (G4, G5), also slightly eccentric, were intruded. Later work has been largely geochemical, but under the guidance of Meighan (Meighan *et al.*, 1984) and the work of Hood (1981), Gibson (1984), and McCormick (1989) this has resulted in a large-scale rearrangement of the geological map for the Eastern Mourne Mountains. Richey's G1 is now restricted to small roof cappings above G2 on Slieve Donard, Slieve Commedagh, and Slieve Corragh, and the steep walls of G1 are incorporated in the geochemically similar G2 intrusion as pulses of the same magma batch. Harry and Richey (1963) found evidence for magmatic pulses in G2, Hood (1981) found further evidence in both the G2 and G3 granites, and Gibson (1984) noted two pulses within each of the Western Mourne granites.

More recently Stevenson *et al.* (2007) have discarded this mechanism of emplacement by cauldren subsidence, and using anisotropic magnetic susceptibility measurements, now propose a laccolithic mode of intrusion by inflation of an initially thin sheet. How such a flat-lying sheet developed in the folded, faulted, and well-jointed Lower Palaeozoic slates is hard to visualise other than by cauldron subsidence. Resurgence of a foundering crustal block could account for the inflation, and the magnetic fabric can only witness the final movement of magma, which in many instances of volcanic activity is a drain-away to other intrusions or drain-back to source.

All the granites except G1 contain drusy cavities where exsolved gases, streaming up the ring dykes, were trapped in the cooler marginal roof zones of an intrusion. It is these cavities that yield a variety of unusual minerals. Haughton (1856) recorded beryl, chrysoberyl, topaz, fluorspar and peridot. Smith (1864) found stilbite, Sollas (1890) zinnwaldite, Seymour (1903) cassiterite, and Greg and Lettsom (1858) and Richey (1928), fayalite. Richey (*op. cit.*) also found a bright green mica, hematite and kaolin.

A single cone sheet appears to surround all centres of subsidence; the composite sheet is best exposed at Glassdrumman Port where the dolerite margins and quartz porphyry core show all the features of two magma mixing. Forgotten exposures on the Mourne foreshore at Green Harbour (Cole, 1894) and further north at Pollderg of a steeply dipping, north trending composite sheet probably follow this conical fracture, which in many of its surface exposures appears to have followed any plane of weakness through the country rock, as the conical fracture failed to penetrate further.

The origin of an acid melt remains one of the outstanding problems of Tertiary igneous activity. It is difficult to envisage a complete absence of crustal melting in the roof zone of a high level magma chamber. The repeated and often simultaneous appearance of two contrasting magmas has prompted the view that the first is mantle derived and basic, and the second is crustal derived and acid. Alternatively the mantle melt could change to more acid and alkaline compositions by fractional crystallisation. Some aphyric intermediate rocks do occur in the Mourne dyke swarm (Tomkeieff and Marshall, 1935) to support this claim.

Meighan *et al.* (1984) use petrochemical data to make a strong case for extreme fractionation with the proviso that Sr isotope evidence indicates some crustal involvement, and Meighan *et al.* (1992) claim crustal contamination of the basic melt.

Nockolds and Richey (1939) found greisen veins in the granite, notably in the north-east quadrant of G2. Robbie (1955) described extensive pneumatolitic alteration of G2 from the Slieve Binnian tunnel; any surface expression of kaolinisation is small-scale and local. McCormick *et al.* (1993) studied the oxygen and hydrogen isotopes of the Mourne complex and found only minor and local interaction with meteoric water.

Waning stages of volcanism

High ground-water temperatures not only influenced the weathering of the rocks but promoted extensive hydrothermal leaching and crystallisation within the lava pile. The silica, alumina, alkali and alkali-earth elements that passed into solution were reconstituted into an assemblage of zeolite, clay, carbonate and silica minerals which lined or filled all mesostatial, vesicular and structural cavities in the lava flows.

The zeolites attracted the attention of mineralogists from the beginning of the 19th century and several new species were discovered at Irish localities, notably gmelinite (Brewster, 1825), garronite (Walker, 1962), and gobbinsite (Nawaz and Malone, 1982). It was Walker (1960b), however, who first appreciated their significance as indicators of temperature/depth zones within the lava pile, each zone being characterised by its own assemblage of amygdale minerals. Thus the higher and cooler portions of the basalt plateau are characterised by a chabazite–thomsonite assemblage of Ca-rich zeolites, whereas in the lower and hotter parts these minerals are joined by Na-rich analcite and natrolite, and locally by stilbite and heulandite.

Walker believed that exothermic reactions between groundwater and basalt were responsible for the higher temperatures in the region of 100 °C. It is a remarkable fact that zeolites do not occur in the underlying rock formations. Rock chemistry influenced this secondary mineralisation; silica-rich zeolites occur in the quartz tholeiite flows. Structural controls allowed a freer circulation of the hydrothermal solutions and possibly gave access to seawater; Island Magee and east Antrim possess a unique Na-rich suite of zeolites. Walker also believed that this kind of activity post-dated all eruptive volcanism.

Against the Slemish and Scawt Hill dolerite plugs (Kitchen, 1985) amygdales in the basalts have been melted or partially melted; a black glass or a pegmatitic crystallisation of plagioclase, pyroxene, and iron oxide now replace the melt. This proves an earlier date for the onset of zeolitisation that may have initiated in the thermal plumes alongside dyke intrusions, as noted by Doherty (1983) for the south Antrim basalts.

Adair's (1987) studies confirmed Walker's zonation and expanded his treatment of the four different environments within the Antrim basalts. The Main Antrim hydrothermal environment is essentially one of calc-alkaline mineralogy throughout both the Upper and Lower Basalt formations. The zeolite zones noted above are present, and a higher temperature zone, near the base of the lava pile, is evidenced by the occurrence of mesolite and laumontite west of Belfast and in the Ballymacilroy borehole. By contrast the Island Magee environment is alkali-rich, gmelinite is the marker zeolite and the unusual chemistry may have been influenced by seepage of seawater or rising brines from the underlying salt deposits. Many of the zeolites present show a significant content of magnesium in the vicinity of dyke intrusions. The Causeway environment shows similar tendencies but it is the leaching of the silica and alkali-rich mesostatial glasses in the quartz tholeiites which gives rise to the distinctive assemblage. A non-zeolite environment corresponds to Walker's plumose calcite/aragonite–quartz zone; it characterises a large tract of the basalt plateau in the north-west. These environments are drawn in Figure 14.2.

Of the 46 accepted species of zeolite, Adair (*op. cit.*) found some 25 within the Antrim basalts; Nawaz (1982) and Nawaz and Foy (1982) have used many of these Irish examples to erect a chemical classification for some of the zeolite groups.

No examples of hot spring or geyser deposits have been described, though such may be encountered in the weathered top of the plateau below the Lough Neagh Clays. Their absence is problematic and may be related to the nature of the water table within the basalt pile.

15

Cenozoic: Tertiary and Quaternary (until 11,700 years before 2000)

P. Coxon and S. G. McCarron

The ages on table 15.1 below are discussed in Bowen and Gibbard (2007) who suggest placing the base of the Quaternary at the Gauss–Matuyama polarity reversal at 2.6 Ma. The onset of the Holocene has recently been assigned an age of 11,554 (Svennson et al., 2005) or 11,784 ± 69 (Rasmussen *et al.*, 2006) ice-core years b2k (before 2000) based on evidence from the NGRIP ice core record. The archived NGRIP core has been suggested as constituting the Global Stratotype Section and Point (GSSP) for the base of the Holocene at 11,700 ± 99 yr b2k (Walker *et al.*, 2008, 2009). This approach is considered to provide more accurate (i.e. calendrical) age estimates than can currently be obtained by calibration of radiocarbon dates. The Quaternary Period is the subject of an excellent recent review by Walker and Lowe (2007).

Dates in the text are given as ka (thousands of years) and Ma (millions of years) BP (before present). The ages of the MIS boundaries on Table 15.2 are taken from Shackleton (1969), Bowen *et al.* (1986), and Martinson *et al.* (1987). Radiocarbon dates are given in '^{14}C ka BP' (thousands of radiocarbon years before present) or, if calibrated, as 'cal ^{14}C ka BP'. These dates are quoted from the original publications in years before 1950 (BP), an accepted standard for radiocarbon dates.

Table 15.2 presents the current knowledge of Irish Pleistocene stratigraphy and key sites are located on Figure 15.1. Formal stratigraphical names have also been proposed (McCabe, 1999) and to facilitate correlation to that scheme the authors have included the marine Oxygen Isotope Stage (MIS – see Figure 15.2), where applicable, in the text.

Table 15.1 The Cenozoic time scale.

Erathem (Era)	System (Period)	Subsystem (Subperiod)	Series (Epoch)		Age of base of unit
			Holocene		11.7 ka
	Quaternary		Pleistocene	Late	127.2 ka
				Middle	780 ka
				Early	2.6 Ma
Cenozoic	Tertiary	Neogene		Pliocene	5.3 Ma
				Miocene	23 Ma
		Palaeogene		Oligocene	33.9 Ma
				Eocene	55.8 Ma
				Palaeocene	65.5 Ma

Where ka = years x 10^3 Ma = years x 10^6

Table 15.2 The subdivision of the Quaternary Period in Ireland. Possible correlations to Marine Oxygen Isotope Stages (MIS) are shown.

Series	MIS	Irish regional stage	age (ka)	Irish regional substage	Comments
Holocene	1	Littletonian			
Pleistocene	2	Midlandian – Late	11.7	Late-glacial: Nahanagan Stadial	Named after glacier activity at Lough Nahanagan in the Wicklow Mountains (Colhoun & Synge 1980). Extensive glaciation has not been recognised in Ireland but many periglacial features and the evidence of small glaciers are found (Coxon 1988; Gray & Coxon 1991; Wilson 1990a, 1990b; Walker *et al.* 1994; Anderson *et al.* 1998).
			12.6	Late-glacial: Woodgrange Interstadial	This complex interstadial (with an early phase of climate amelioration and containing at least one period of erosion and climate deterioration) is recorded in many biogenic sequences from Irish Late-glacial sites (Watts 1977, 1985; Cwynar & Watts 1989; Walker *et al.* 1994). The Woodgrange Interstadial is named after a site in Co. Down (Singh 1970)
			14.7	RIIs KPS LIs CHS CPIs	Distinct events recording millennial scale oscillations of the BIIS during deglaciation of the ISB. Associated with phases of streamlining and moraine formation (McCabe *et al.* 1986; McCabe 1985, 1987, 1993; McCabc & Clark 1998; McCabe, 2005; McCabe *et al*, 2005, McCabe *et al*, 2007b)
			19	Glenavy Stadial	Growth of the BIIS ice sheet to offshore limits at or before 22ka (Bowen et al, 2002). Sediments at Aghnadarragh (McCabe *et al.* 1987b), cosmogenic dates (Bowen et al, 2002), and dates of deglaciation intiation (McCabe *et al*, 2007a, b) constrain this phase of extensive glaciation to the Late/Middle Midlandian.
	3	Midlandian – Middle		Derryvree Cold Phase	Organic silts found between two tills at Derryvree (Colhoun *et al.* 1972) show a treeless, muskeg environment. The mammal remains from Castlepook Cave (Mitchell 1976, 1981; Stuart and van Wijngaarden-Bakker 1985) date from this period (34—35ka). Recent dates for mammal faunas from caves range from 32ka-20ka (Woodman and Monaghan 1993) indicating the possibility of ice free areas in Cork during the Glenavy Stadial.
			c. 40	Hollymount Cold Phase	Organic muds found at Hollymount (McCabe, Mitchell *et al.* 1978), Aghnadarragh (McCabe *et al.* 1987b) and Greenagho (Dardis *et al.* 1985). Fossils suggest cold, open, treeless environments. Possibly a continental climate with high seasonality.
			>48	Aghnadarragh Interstadial	Pollen and beetle evidence from Aghnadarragh (McCabe *et al.* 1987b) suggests cool temperate conditions with woodland, similar to that of Fennoscandia today. Dated to >48ka and tentatively correlated to the Chelford Interstadial (McCabe 1987).
	4	Midlandian – Early		Fermanagh Stadial	Till pre-dating organic beds at Derryvree, Hollymount (McCabe, Mitchell *et al.* 1978) and Aghnadarragh (McCabe, Coope *et al.* 1987) are believed to have covered most of Ulster. Evidence (from the presence of certain tree taxa in the subsequent interstadial) suggests that the glaciation may have been short-lived (Gennard 1986; McCabe 1987).
	5d/5c		*c.* 115	Kilfenora Interstadial	UTD dates place cool temperate organic deposits at Fenit in Co.Kerry early within the Midlandian Glaciation (118,000 years BP -Heijnis 1992; Heijnis *et al.* 1993). The biogenic sediments represent cool conditions during Oxygen Isotope Stage 5a or (more likely) 5c.
	5e	Last Interglacial	*c.* 120		Knocknacran (Co. Monaghan) is a critical site (see text) and may prove to be a Late Pleistocene interglacial of some importance -possibly correlatable to MIS 5e. A reworked ball of organic sediment within the Screen Hills moraine containing a *Carpinus* -rich pollen assemblage (McCabe & Coxon 1993) may also represent a Late Pleistocene warm temperate stage but the evidence from this site is far from conclusive.
	6	Munsterian	*c.* 132ka Minimum date MIS 7: 198ka (or MIS 9: 302ka)		Many deposits of this cold stage are now believed to be Midlandian in age (McCabe 1987; O'Cofaigh 1993; McCabe & O'Cofaigh 1996) but dates on raised beach deposits show that some south coast sequences may indeed belong to MIS 6 (Gallagher & Thorpe 1997)
	7 or 9	Gortian	Maximum date MIS 7: 252ka (or MIS 9: 338ka)	Gn IV Gn IIIb Gn IIIa Gn II Gn I Pre-Gn l-g	Thirteen sites have been described from around Ireland that record part of a characteristic temperate stage deposit with a biostratigraphically similar record. The Gortian is represented by a record of vegetational succession and by a number of fossil assemblages that represent stages which have been described in a number of ways (e.g. by Mitchell 1981; Watts 1985; Coxon 1993, 1996a). Particularly notable aspects of the Gortian are its sudden truncation (Coxon *et al.* 1994) and biogeographically interesting flora (Coxon & Waldren 1995, 1997). Opinion is divided as to the age of the Gortian (Warren 1985; Watts 1985; Coxon 1993, 1996). Biostratigraphically it resembles the Hoxnian of Britain and the Holsteinian of Europe. Amino-acid racemisation results on marine Gortian sediments from Cork Harbour (Scourse *et al.* 1992; Dowling *et al.*, 1998) confirm it is older than MIS 5e and the interglacial may represent MIS 7 or 9 (Dowling *et al.*, 1998; Coxon 1996a). UTD dating of the Gortian has so far proved inconclusive but the method holds promise.
		Pre-Gortian			Prior to the Gortian are sediments of late-glacial aspect, suggesting the temperate stage was preceded by a cool/cold stage.This stage is not represented by long or datable sequences, and the age is unknown.
		Ballyline	date unknown possibly > 428ka		A deposit of laminated, lacustrine, clay over 25 metres thick was discovered in 1979 by the Geological Survey of Ireland filling a solution feature in Carboniferous Limestone below glacial sediments near Ballyline, Co. Kilkenny (Coxon & Flegg 1985). From the evidence available the pollen assemblages can be seen to be typical of Middle Pleistocene sequences in Europe, but a firm correlation to a particular stage is not possible.
Pliocene	103?	Pollnahallia (Pliocene-Pleistocene boundary?)	*c.* 2.6Ma		A complex network of gorges and caves in Carboniferous Limestone at Pollnahallia, Co.Galway, contains lignite deposits —now covered by superficial material including wind-blown silica-rich sands (Tertiary weathering residues) and glacigenic deposits. Palynological results (Coxon & Flegg 1987; Coxon & Coxon 1997) suggest that the lignite infilling the base of the limestone gorge is Pliocene or Early Pleistocene in age. Further work is reported in the text.

Figure 15.1 Location of important sites referred to in the text.

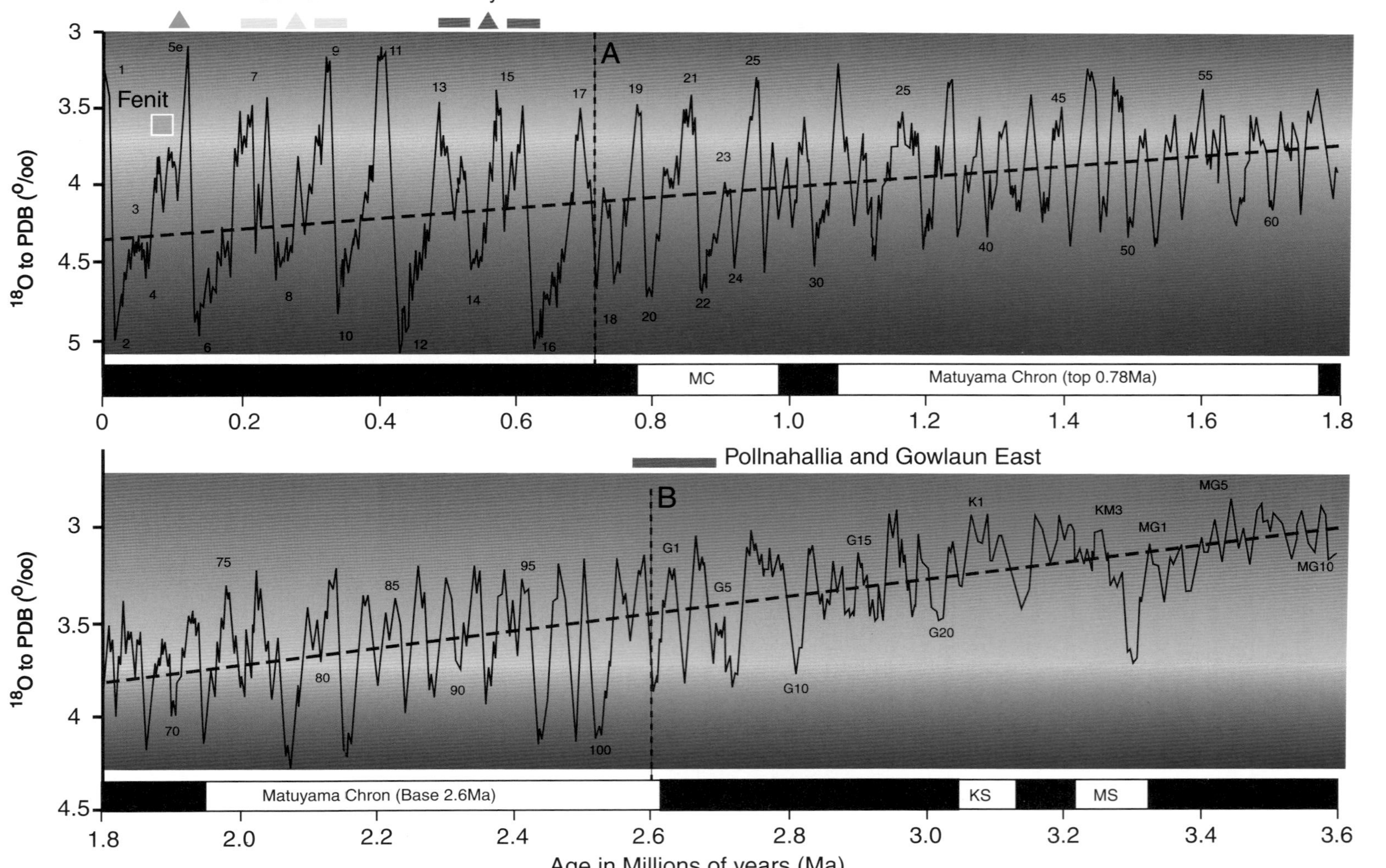

Figure 15.2 A composite oxygen isotope curve for the last 3.5 Ma (*after* Lisiecki and Raymo, 2005; from Walker and Lowe, 2007). The marine isotope stage numbers (MIS) are indicated. The long-term decline in global temperature, the increasing magnitude of climate signals towards the present, the change in climate forcing from a 40,000 year cycle to a 100,000 year cycle (dotted line at A) and the onset of the Quaternary (B) are all apparent. The correlations to the Irish record are tentative only and given to provide an impression of the complexity of the Pleistocene climate record from the ocean cores and the incomplete nature of the terrestrial record. *Reproduced by kind permission of the Geological Society of London.*

Tertiary landscape and vegetation

Relict geomorphological elements of a Tertiary landscape in Ireland were reviewed by Mitchell (1980, 1985) who identified basins, pediments, tectonism, erosion surfaces, limestone solution, and deeply weathered bedrock in many areas of the country. Tertiary rocks in Ireland vary from the extensive (*c.*390,000 ha) basalt plateau and its associated palaeosurfaces (Interbasaltic Beds) found in the north-east, to restricted intrusions and depositional fills or weathering residues. The latter are commonly found within karstic depressions in areas of Carboniferous limestone (Table 15.3 and Figures 15.3 and 15.4) and are associated with palaeosurfaces on other lithologies. The age of the Tertiary deposits, like their geomorphology, is variable, ranging from the Palaeocene of north-eastern Ireland to the less extensive Oligocene sediments found as karstic infills (e.g. Ballymacadam) and the Miocene deposits at Hollymount, County Laois. The latter two examples are well documented, whilst other deposits of possible Tertiary age are less so (see Table 15.3 and Figure 15.4).

Table 15.3 Tertiary sites of note in Ireland.

Site	Nature of evidence/Age	Publication(s)
Interbasaltic Beds, e.g. Giant's Causeway, Craigahulliar and Ballypalady, County Antrim and Washing Bay, County Tyrone	Extensive and numerous organic horizons (some blanket peats or peaty soils) lying on subaerially weathered basalt surfaces. The palaeobotany of these organic sediments tentatively suggests an Early Palaeocene age. Plant fossils include: *Pinus, Tsuga, Cupressus, Araucaria, Alnus* and a number of angiosperms and ferns.	Watts, 1962, 1970 (review); Curry *et al.*, 1978; Mitchell, 1981.
Lough Neagh Clays, southern part of Lough Neagh, e.g. Washing Bay, County Tyrone and Thistleborough, County Antrim.	350m of predominantly lacustrine and swamp sediments deposited in a large subsiding basin. Although the palaeobotany (like that of the Interbasaltic Beds) requires modern research an Oligocene age (Chattian or Rupelian) can be implied. Plant fossils include: *Sequoia, Alnus, Nyssa, Quercus* (?), *Tilia* (?), and Taxodiaceae. The last, and associated taxa, suggest that the vegetation was predominantly that of a lowland swamp.	Watts, 1970; Curry *et al.*, 1978; Boulter, 1980; Wilkinson *et al.*, 1980
Ballymacadam, County Tipperary	A solution pipe in Carboniferous limestone with an infill (*ca.* 10m thick) including clay sediments rich in biogenic material. The pollen content of gymnosperms, '*Quercus*-type', *Engelhardtia, Symplocus*, Ericaceae, and occasional Palmae represents a vegetation with Oligocene affinities, probably similar in age to that of the Lough Neagh Clays	Watts, 1957, 1970; Boulter and Wilkinson, 1977; Curry *et al.*, 1978; Boulter, 1980
Aughinish Island, County Limerick	Deep karstic hollows encountered during site investigations for an aluminium smelting plant. Limited palaeobotanical information implied a Middle Tertiary age.	Clark *et al.*, 1981; Mitchell, 1985
Tynagh, County Galway	Altered sulphide ore lying in karstic hollows developed along faults. Authors suggest that a log of *Cupressus* wood may imply a middle Tertiary age.	Mitchell, 1980; Monaghan and Scannell, 1991
Galmoy and Lisheen, Counties Kilkenny and Tipperary	Deep (40m+) karstification of limestone identified in mineral exploration boreholes. Palynology of organic clays in the depressions suggests a middle Tertiary age.	Unpublished reports
Hollymount, County Laois	20m of quartz sand, lignite and other organic sediments infilling a karst solution hole. Detailed palynology indicates a Miocene or earliest Pliocene age and taxa include *Pinus, Quercus, Corylus, Myrica* , Ericales, *Taxodium* type, *Symplocus, Tsuga, Sciadopitys, Liquidambar* and Palmae type.	Hayes, 1978; Watts, 1985; Boulter, 1980
Ballyegan, County Kerry	Palaeosols containing gibbsite and quartz are exposed in a limestone quarry capped with glacigenic sediments. The authors suggest the weathering is pre-glacial (possibly 'Pliocene?' but noting that they recovered no palynomorphs to prove this conclusion).	Battiau-Queney and Saucerotte, 1985; Battiau-Queney, 1987
Ballygaddy Townland, County Offaly	Karst depression (possibly 40m deep) containing weathered stony clays. A possible Tertiary age was inferred by the authors.	Beese *et al.*, 1983
Listry, County Kerry	A weathered silicious breccia in a matrix of iron-rich sandy clays. A possible Tertiary age is suggested by the author.	Walsh, 1965
Pollnahallia, County Galway	See text	
Gowlan East, County Galway	See text	

Figure 15.3 **A** – Tor exhumed by quarrying in Gowlan East, Connemara. **B** – Residual limestone tower near Fenit, Co. Kerry. **C** – Karstified limestone, Stone Forest, Kunming, China: possible analogue for Ireland's Late Tertiary limestone surfaces (e.g. at Pollnahallia). **D** – Limestone towers, Li river, SW China. **E** – Section at Bothar na Scrathog, Connemara, showing exposed bedrock tor (E1), Neogene palaeosurface and possible Middle Pleistocene palaeosol (E2) separating two tills (see Coxon, 2001, 2005).

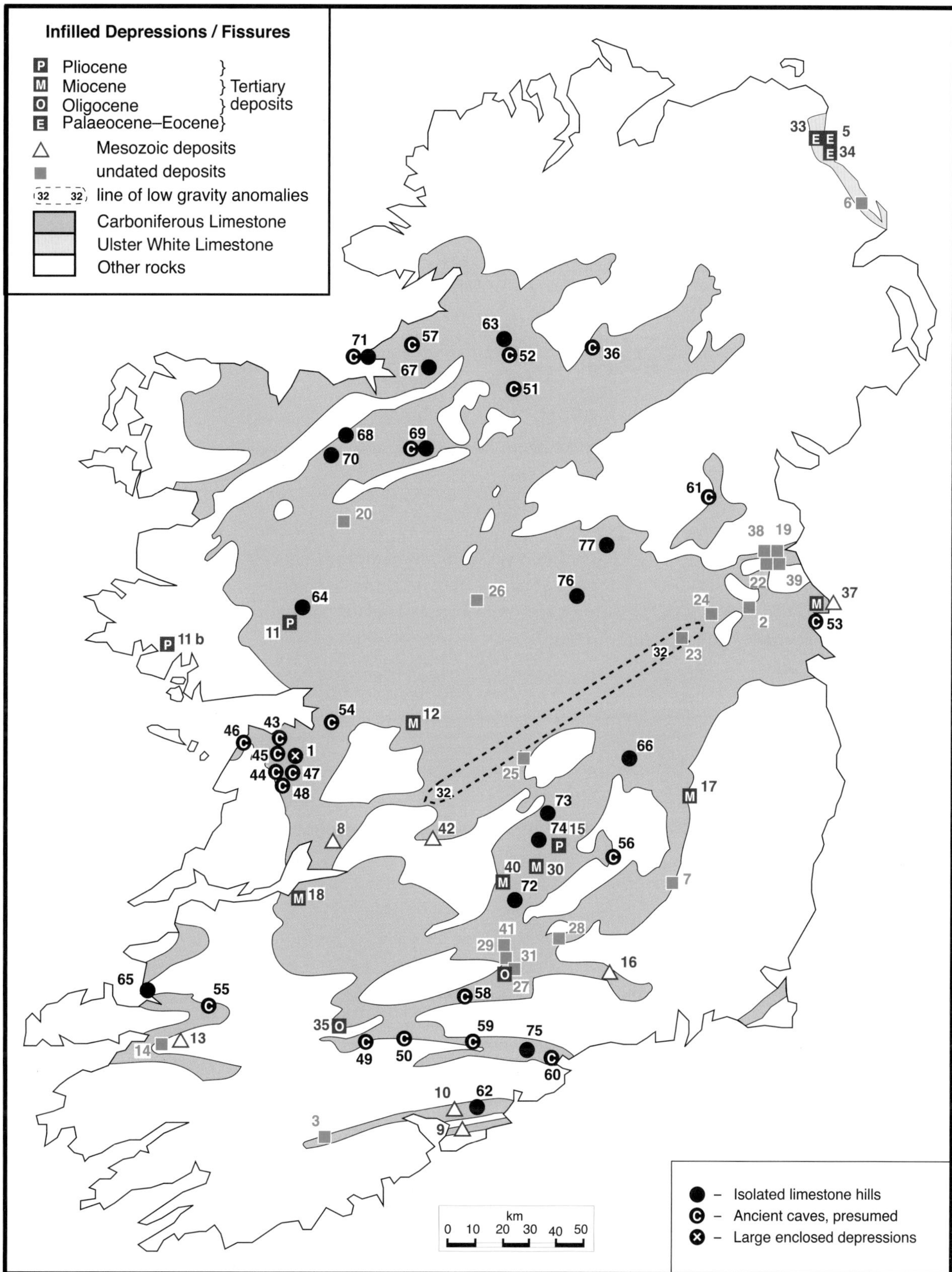

Figure 15.4 Presumed ancient karstic sites and landforms including Pollnahallia (site 11) and Gowlan East (not a karstic feature, site 11b). The figure is modified after Drew and Jones (2000) and the full key to the karstic sites is given in their Table 1.

References in the literature to undated landforms of possible Tertiary age are numerous (Drew and Jones, 2000), as are unpublished reports of deeply weathered limestone and associated infills. Phreatic passages and extensive cave systems now far above (and below) current water table levels have also been widely commented upon (e.g. Simms, 2000a). Some of the landforms recorded are large-scale and thought to be of considerable antiquity, most probably Tertiary, (e.g. the area of The Doons in County Sligo, which may be an area of 'cockpit karst' – Herries Davies and Stephens, 1978; see review by Drew and Jones, 2000 and Coxon, 2006).

The geomorphology and the palaeoecology of the pre-Pliocene Tertiary deposits are summarised in Table 15.3 and Figure 15.4 and the later Tertiary is outlined below.

The Pliocene and the Pliocene-Pleistocene transition

Late Tertiary palaeosurfaces and accompanying palaeosols and sediment fills have been described from limited areas in western Ireland, and they provide a remarkable record of pre-glacial landscape inheritance (reviewed by Coxon, 2006). These relict surfaces have been discovered on limestone, at Pollnahallia in County Galway, and on granite farther west on the flanks of Cnoc Mordáin, near Kilkieran in Connemara.

Gowlan East, Iorras Aintheach, Connemara

Evidence from Connemara to date (Coxon, 2001, 2006) consists of weakly developed palaeosols draping buried tors in the townland of Gowlan East, between Bertraghboy and Kilkieran Bays (the peninsula of *Iorras Aintheach*). This magnificent landscape is scattered with granite tors, some exhumed from the surrounding growan and others still buried (Fig. 15.2). Flanking the northern slopes of Cnoc Mordáin are extensive debris-mantled slopes, and along the toe of one such slope a number of small quarries have exposed buried bedrock tors overlain by tills, soliflucted tills, and palaeosols (location: Fig. 15.1, sections Fig. 15.3). The lower palaeosol lies directly on the buried bedrock and is Neogene (probably Pliocene) in age (Coxon, 2001). The upper palaeosol, probably Middle Pleistocene in age, separates two distinctive tills (Fig. 15.3).

The Pliocene soil contains relatively sparse but well-preserved pollen (see Coxon, 2001, 2006) and the presence of the pollen of *Tsuga, Taxodium, Sequoia, Carya,* and *Castanea* suggests that a diverse forested environment existed here some 2.6 to 5 Ma ago. The assemblage is characteristically Neogene and most likely Pliocene in age (the more exotic taxa of the Miocene being absent). The occurrence of *Quercoidites* (*Tricolpopollenites* cf. *T. microhenrici*) and *Betulaepollenites* type also suggests a pre-Pleistocene age. The discovery of the buried Pliocene soil horizon also defines a palaeosurface and suggests that the granite surface was extensively weathered (the soil is laterally extensive and formed in weathered granite). The existence of tors separated by weathered granite and a palaeosol (now covered by till) suggests that stripping of a weathered mantle had already occurred by the Pliocene and that the landscape was one of exhumed granite tors surrounded by forested soils developed in remnants of the weathering mantle.

The onset of the Pleistocene saw further stripping of the weathered mantle, ice moulding of the bedrock protuberances, and partial (or complete) burial of some bedrock tors by glacigenic or periglacial sediments (Fig. 15.3).

Pollnahallia, County Galway

Far more complete Tertiary sediment sequences are available from the fascinating site at Pollnahallia (location: Figure 15.1). Pollnahallia lies in a geomorphologically complex region on Viséan limestone that includes large shallow depressions (up to 5 m deep), deeper depressions and fractured limestone (up to 15 m deep), deep gorges (up to 20 m deep) and cave passages. The landscape in the immediate vicinity of Pollnahallia (*c*.3.3 km^2 area around the site) is extensively karstified with Pliocene lignites part filling a limestone gorge (Figs. 15.5, 15.6) and even draping the surrounding limestone surface (for a detailed summary see Coxon *et al.*, 2005). The lignite (both inside the gorge and on the limestone surface, see Figure 15.5) was sampled in 1987 and a continuous core was obtained from within a limestone gorge. The latter borehole (Fig. 15.6) was 20.26 m long, and from the base upwards the sediment succession is one of laminated clays, silts, organic detritus and lignites alternating with numerous sandy horizons. Lignite predominates portions of the core whilst the upper part of the core shows an upward transition from silt and clay into laminated silts and clays, into laminated silts, and finally into laminated yellow clays with sand horizons. The loss on ignition data (Fig. 15.6) displays the variable nature of the organic content and shows a marked variability associated with sandier horizons.

The sedimentological interpretation (allied to the geomorphological information) is that the sediments were laid down in an open gorge cut into limestone

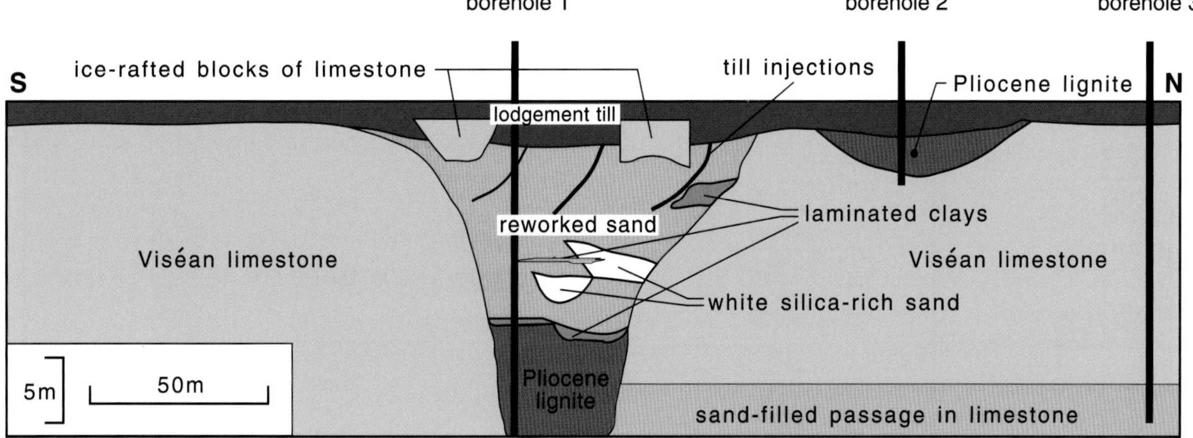

Figure 15.5 A. Schematic cross-section of the Pliocene and Pleistocene deposits at Pollnahallia, County Galway (from Coxon and Coxon, 1997 and Coxon and Flegg, 1987). *Reproduced by kind permission of the Geological Society of London.*

bedrock under conditions with long periods of quiet (or still) water deposition interspersed by occasional flows capable of transporting sand. A considerable volume of organic debris from rotting vegetation accumulated in the narrow gorge. Presumably the latter sediments were derived both allocthonously and autochthonously.

The palaeoecology of the biogenic sediments from Pollnahallia is as fascinating as the geomorphology. The palynology is summarised on two pollen diagrams (Fig. 15.7), and these diagrams can be subdivided into five pollen assemblage biozones (pab).

pab	depth (cm)	assemblage
87/1–5	1050–1100	*Pinus–Betula–Larix*–Gramineae
87/1–4	1100–1275	*Pinus–Corylus*–Ericaceae–*Carya*
87/1–3	1275–1575	*Pinus–Taxus–Taxodium*–Ericaceae
87/1–2	1575–1875	*Pinus–Corylus*–Ericaceae
87/1–1	1875–2012	*Pinus–Taxodium*–Ericaceae–*Sequoia*

The pollen assemblages indicate a vegetation cover dominated by *Taxodium*, ericaceous, cupressaceous and coniferous trees, a diverse assemblage of tree types and assorted shrubs (see Table 15.4 for a summary).

The important biostratigraphical elements of the pollen diagram include the presence of typical late Tertiary taxa; e.g. *Sequoia, Taxodium, Nyssa, Liquidambar, Castanea, Ostrya, Juglans, Sciadopitys, Carya,* and *Pterocarya*. Such taxa are frequently found in Pliocene deposits in the Netherlands (Zagwijn, 1960) and this, the absence of pre-Pliocene marker taxa and the apparent climatic deterioration recorded in the upper part of the sequence, allows a probable correlation to be made to the Reuverian of the Netherlands (Coxon, 1993), possibly Reuverian C.

Pollen zones 87/1–1 to 87/1–4 (i.e. the bulk of the biogenic sediment sequence) are characterised by their diverse assemblages dominated by swamp cypress, cypresses, heathers and pine. The landscape must have been magnificent and the vegetation cover diverse and, in the autumn, very colourful with an affinity to the modern vegetation of the swamps of eastern North America. The assemblages suggest that frosts must have been negligible and the climate (at least in the gorge at Pollnahallia) was warm and wet.

Zone 87/1–4 sees a decrease in the swamp cypress cover associated with a sedimentological change at 12.55 m, and subsequently in pab 87/1–5 there is a marked assemblage change associated with the facies change to laminated clays with sand horizons (at 11.00 m in borehole 87/1). The change involves an increase in taxa indicating climatic deterioration (e.g. the rising values of Ericaceae and *Juniperus*) and the disappearance of a number of the thermophilous taxa. The lithological and palynological changes suggest that the top of the organic sedimentation may represent climatic deterioration at the end of the Pliocene and the beginning of the Pleistocene.

One striking element in the Pollnahallia pollen diagram is the exotic nature of some of the flora. Many of the taxa recorded are no longer native to Europe, and indeed some have a distinctly disjunct modern distribution occurring only in North America and Asia. Table 15.4 partly illustrates this point and the flora, palaeobotanical correlations and a comparison with deposits

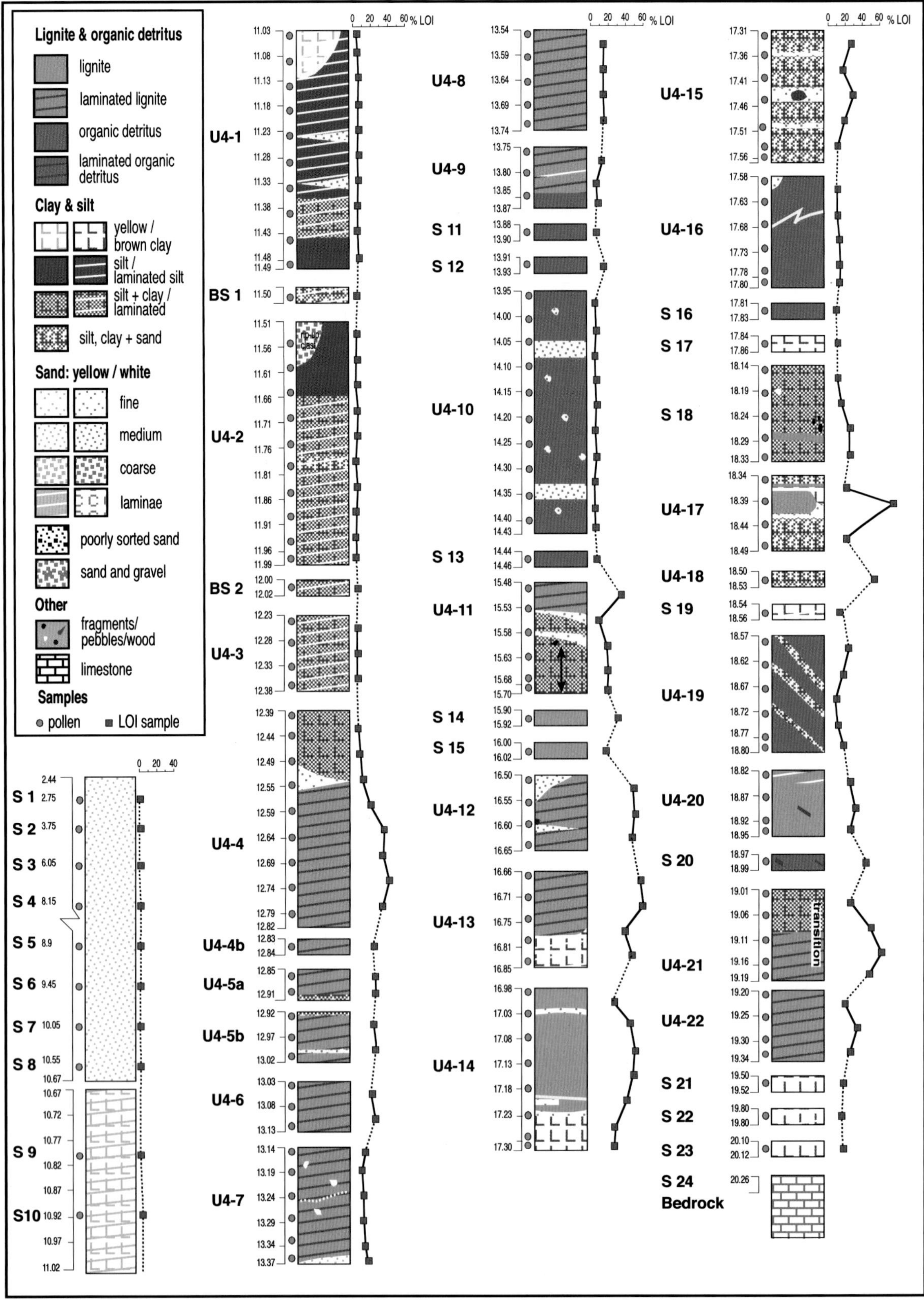

Figure 15.6 The log of borehole 87/1 at Pollnahallia showing the depositional sequence, cores, pollen samples and loss on ignition values (*after* Coxon *et al.*, 2005).

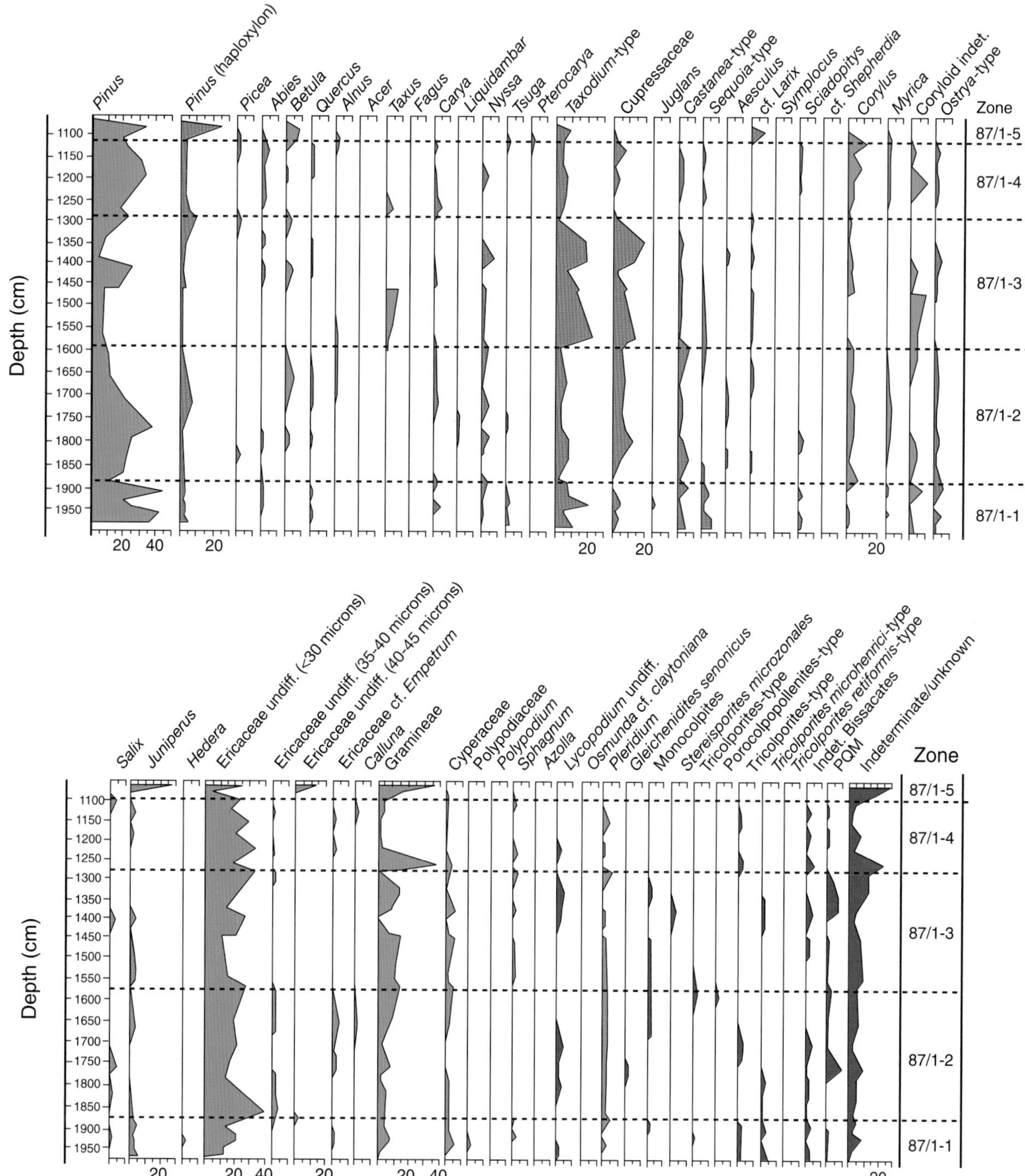

Figure 15.7 Summary pollen diagrams of the organic sediments from borehole 87/1 at Pollnahallia (From Coxon *et al.*, 2005). The taxa are colour coded (a general scheme) as follows: blue – taxa that could be native in the Irish Holocene, orange – taxa found in European Late Tertiary and Early Pleistocene, red – form genera and unknown taxa. *Reproduced by kind permission of the Quaternary Research Association.*

in the UK are made in Coxon and Coxon (1997) whilst Coxon and Waldren (1997) discuss the flora with reference to range retraction during the Pleistocene (Coxon *et al.*, 2005).

The lignite grades upwards into a laminated silt and clay sequence that becomes increasingly sandy, eventually passing into pure silica sand which is at least 9 m thick and fills an elongate basin in the limestone that

Table 15.4 A summary list of the taxa found in the Pliocene biogenic sediments at Pollnahallia (from Coxon *et al.*, 2005). **n.b.** Some of the families and genera have many members (e.g. within the Fagaceae family the genus *Quercus* (oak) contains 600 species) making the identification to species level using palynology difficult/impossible. A further problem is that with deposits of Tertiary age some of the individual species are probably extinct, indeed some authors refer to Tertiary taxa as form genera only (i.e. they describe the pollen morphology rather than attempt an identification to any living taxa e.g. *Tricolpollenites*-type) whilst others might refer to a genus but avoid a direct classification to a living member of that genus, i.e. *Quercoidites*).

botanical name (Family or Genus/species)	common name
Abies	fir
Acer	maple
Aesculus	horse chestnut
Alnus	alder*
Azolla	water fern
Betula	birch*
Calluna	heather (ling)*
Carya	hickory
Castanea	sweet chestnut
Corylus	hazel*
Coryloid indet.	possibly hazel? (or another of numerous triporate pollen taxa)
Cupressaceae	cypress family
Cyperaceae	sedge family*
Empetrum	crowberry*
Ericaceae	heather family*
Fagus	beech
Gleicheniidites senonicus	tropical/warm temperate fern
Gramineae	grass family*
Juglans	walnut
Juniperus	juniper*
Larix	larch
Liriodendron	tulip tree
Liquidambar	sweetgum
Myrica	bog myrtle*
Nyssa	sourgum
Osmunda regalis	royal fern*
Osmunda cf. claytoniana	north American/east Asian fern
Ostrya	hop-hornbeam
Picea	spruce
Pinus (subgenus Diploxylon)	pine*
Pinus (Haploxylon) (Haploxylon pines have a different pollen morphology and represent a subgenus of different species to Diploxylon pines)	pine
Polypodiaceae	ferns*
Pterocarya	wingnut
Quercus	oak*
Salix	willow*
Sciadopitys (cf. *S.verticillata*)	Japanese umbrella pine
Sequoia	giant redwood
Sphagnum	moss*
Taxodium	swamp cypress
Taxus	yew*
Tsuga	hemlock

* denotes representatives native to Ireland in Holocene. (Note many of the taxa listed have been introduced to Ireland but are not native.)

Reproduced by kind permission of the Quaternary Research Association.

slopes downwards from 5 m below surface in the west to a depth of 27 m in the east. The sand is very well sorted and exhibits large-scale foreset bedding, but no smaller scale sedimentary structures were visible in the undisturbed unit. The sands are probably aeolian, but fluvial and glacifluvial reworking has affected some of the material.

The origin of these sands could have been deeply weathered sandstone, quartzite, or even granite, and Mitchell (1980, 1985) has suggested that such weathering, under moist and warm conditions, could have occurred throughout the Tertiary, producing deep residual deposits on the landscape. The onset of the Pleistocene, with the subsequent decline in vegetation cover, would have provided environmental conditions suited to the erosion and transport of such weathered mantles. Several possible source rocks for the sand exist (Coxon and Flegg, 1987) but, given the high purity of the sand and the absence of staining, the most favourable source would be a rock that was virtually monomineralic and also not heavily stained. The quartzites of Connemara are situated 30 km west of Headford and present the most likely source (Coxon and Coxon, 1997).

Glacial action, including that of meltwater, has reworked some of the sand and rafted large sheets of limestone to cover parts of the gorge, as well as depositing an overlying lodgement till (Figs. 15.5, 15.8) that was injected into the sand to a considerable depth. Friction at the base of the ice has transported the rafts of limestone and also moved some of the sand body en bloc (post till injection), causing shearing of the injection structures in the direction of ice movement.

As Davies (1970) pointed out, the information available to us from the Tertiary is scant, but fragments of information from sites such as those above give an indication that Ireland's surface was subjected to warm-temperate or sub-tropical conditions for much of the 63 Ma of the Tertiary and was probably deeply weathered and karstified during this time. At the onset of the Quaternary drastic environmental changes began to modify an extremely old landscape.

Figure 15.8 **A)** The silica sand quarry at Pollnahallia, County Galway. The sand is exposed in the background of the photograph. The main borehole was made in the floor of the pit. **B)** A till injection down into the silica sands at Pollnahallia. **C)** A core of lignite being extruded in the field at Pollnahallia. **D)** Pliocene lignite exposed subaerially at Pollnahallia showing remarkable preservation of a Late Tertiary palaeosurface as the modern soil. **E)** A pollen grain of *Taxodium* (*cf. distichum*) – swamp cypress. The pollen grain is 30 µm in diameter. **F)** Swamp cypress in its native habitat, South Carolina: a possible analogue for lowland Galway in the Pliocene.

The Quaternary: 2.6 million years to 11,700 years before 2000

Introduction

The Global Cenozoic sequence from long marine, lacustrine, and terrestrial records exhibits a characteristic record with a decline in temperature throughout the Miocene culminating in numerous step-like changes in climatic conditions over the last 2.6 Ma (Figure 15.2). Research over the last three decades has confirmed that global climate changes are controlled by regular, orbitally induced, insolation variation (the development of these theories is described in Imbrie and Imbrie, 1979, 1980 and Kutzbach and Webb, 1991). Climate change during the Pleistocene has been noticeably variable in magnitude and frequency, with earlier cyclicity (2.0 Ma–1.4 Ma and 1.4 Ma–0.9 Ma) having a shorter wavelength than later cyclicity. During the period between 0.9 Ma and 1.4 Ma isotopic fluctuations suggest a predominant 41 ka cycle (Ruddiman *et al.*, 1986), whilst from 0.9 Ma to the present the cold stages occur over 100 ka intervals (Shackleton and Opdyke, 1976). This variation is (at least in part) due to the changing relative importance of the three orbital climate-forcing parameters, which have different periodicity (precession 22 ka, tilt 41 ka, and eccentricity 100 ka) (Hays *et al.*, 1976). The climate signal preserved in oxygen isotope records from marine sediments and ice cores displays both the complexity and the regularity of global climate change from the late Tertiary to the end of the Pleistocene. Figure 15.2 shows such a record (oceanic) and allows a comparison to the Irish record as well as putting the latter into a global perspective.

Irish Pleistocene biostratigraphy

Early and Middle Pleistocene

Figure 15.2 shows the paucity of the Irish record from the late Pliocene (c. MIS 103) up until the late Middle Pleistocene (c. MIS 11). Indeed, apart from the uppermost organic sediments at Pollnahallia, there is no documented record of the early Pleistocene in Ireland apart from a glacially reworked fauna of marine shells that originated in the Irish Sea basin (Mitchell, 1981).

Middle and Late Pleistocene

This period of time has been reviewed in detail by Dowling and Coxon (2001) and the bulk of known evidence postdates MIS 11. One interglacial sequence, over 25 m thick, from a solution feature in Ballyline, County Kilkenny (Coxon and Flegg, 1985) is worthy of mention, but the site requires further sampling and subsequent reanalysis. The pollen assemblages from Ballyline can be seen to be typical of Middle Pleistocene sequences in Europe and contain *Abies, Picea, Ulmus, Alnus, Quercus, Carpinus, Pterocarya,* and *Taxus* with areas covered by Gramineae, Ericales, and numerous herbs. The deposits representing this forest/heath/grassland stage are 10 m thick.

Late-Middle Pleistocene: the Gortian Interglacial

Kinahan (1865, 1878) described an interglacial deposit of peat and mud lying below glacial deposits in the banks of the Boleyneendorrish River, near Gort, County Galway. Subsequently the deposit was studied in detail (Jessen *et al.*, 1959) and has been used as the type site of the Gortian Interglacial. We now have the details of thirteen other interglacial sites with Gortian affinities from around Ireland (some are shown on Figure 15.9). The interglacial sites are variable in depositional and geomorphological setting and in the length of the temperate stage that is recorded. The ranges determined from pollen assemblages of the sites are reviewed on Figure 15.10. The Cork Harbour area (Figs. 15.11, 15.12) has provided three long Gortian sequences of estuarine and lacustrine sediments up to 18 m thick that have been the subject of considerable research (Dowling, 1997; Dowling *et al.*, 1998; Dowling and Coxon, 2001). One of the pollen diagrams is reproduced here as Figure 15.13.

The palaeobotany and biogeography of the Gortian exhibits an interesting, diverse flora, and the reader is directed to the original papers for details (Watts, 1985; Coxon and Waldren, 1995, 1997; Coxon, 1996a). Abbreviated pollen diagrams from six of these sites are shown on Figure 15.9 and the vegetational sequences can be seen to follow a general pattern (Watts, 1985). However, the interglacial does not seem to complete a full cycle (*sensu* Iversen, 1958, Turner and West, 1968) but comes to an abrupt end (see figure 15.10). The biostratigraphy of the Gortian is summarised in Table 15.5.

The vegetational record differs slightly between Gortian sites, e.g. *Taxus* is present to a varying degree depending upon local soil conditions. Some of this variation is simply accounted for by the differing substages of the interglacial that are represented at each site. However, as Figures 15.9 and 15.13 show, the principal vegetational succession is distinctive, with *Pinus* and *Betula* remaining important throughout the temperate sequence.

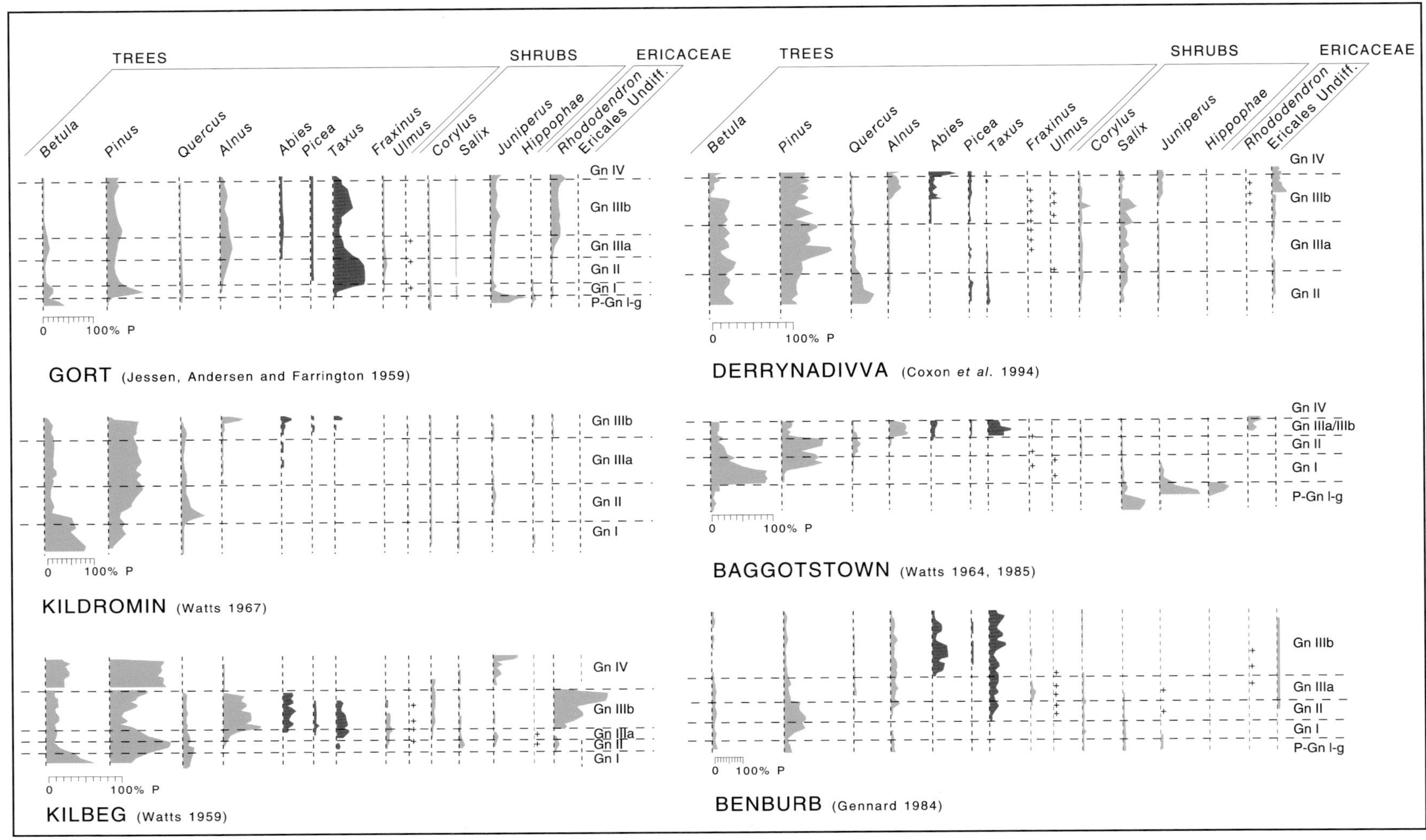

Figure 15.9 Pollen diagrams of selected taxa from the more complete Gortian sequences (*after* Coxon, 1996a).

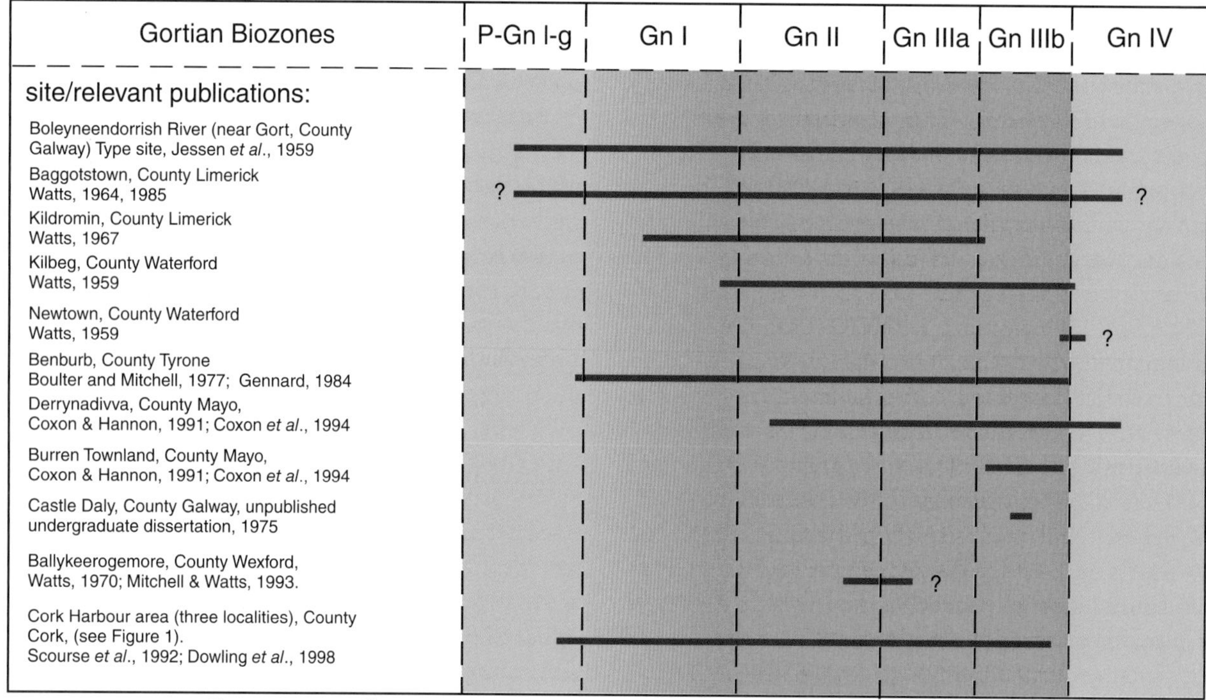

Figure 15.10 Pollen zone ranges for Gortian Interglacial sites (from Coxon, 1996a).

Opposite Figure 15.11 Interglacials. **A** – Drilling Gortian sediments at Douglas, Cork Estuary. 16 metres of interglacial sediment lie just below the surface in a large channel-fill deposit. **B** – Organic sediments interbedded within clays and in a closed-system at Fenit Co.Kerry. U–Th dates obtained from these organic deposits suggest an age of *c*.118,000 years, i.e. towards the end of MIS 5e. **C** – Possible modern analogue for Gortian vegetation. *Rhododendron ponticum* amongst mixed woodland. County Kerry. **D** – Gortian temperate stage peats, Derrynadivva, County Mayo. **E** – A monolith of sediment deposited at the closing stage of the Gortian Interglacial at Derrynadivva, County Mayo. **F** – Organic sediments, possibly last interglacial in age (MIS 5e) below drumlinised till and filling a collapse feature in karstified Permian gypsum, Knocknacran, County Monaghan.

The age of the Gortian temperate stage

Biostratigraphy

The poor Irish Quaternary record, unlike that of Britain (e.g. West, 1980; Jones and Keen, 1993) and continental Europe (e.g. Zagwijn, 1985; Reille and De Beaulieu, 1995; De Beaulieu and Reille, 1995), complicates correlation, making it impossible to place Gortian deposits in the context of a firm Quaternary stratigraphical framework beyond that presented in Table 15.2.

The traditional dating of the Gortian has relied upon biostratigraphical correlation based on the presence/absence of key taxa in assemblages from temperate stage deposits in Britain and continental Europe. This approach has been taken because the age of overlying glacigenic sequences has never been unequivocally established. The crucial biostratigraphical correlation that has linked the Gortian to the Hoxnian of Britain includes the following:

1. 'Relic' taxa, associated with the early Pleistocene in northern Europe, are not recorded (these taxa include *Tsuga* and *Eucommia*) suggesting that the Gortian is late-Middle Pleistocene.
2. Abundant *Hippophae* is present at sites exhibiting the Pre-Gortian late-glacial.
3. Presence of abundant *Abies* in Gn III.
4. Presence of *Azolla filiculoides*.
5. Relative lack of *Carpinus* (cf. last interglacial/MIS 5e records).
6. Presence of Type X (up to 7% ΣP at some sites) throughout stages Gn II –Gn III.
7. Records of *Pterocarya* from Derrynadivva and Cork Harbour.

Nevertheless the correlation has been the subject of some criticism (Warren, 1979, 1985; Coxon, 1993, 1996a) and the Gortian assemblages do imply possibly cooler conditions than those prevailing in the Hoxnian.

If the Gortian is the equivalent of the Hoxnian and Holsteinian temperate stages this suggests that it belongs within MIS 11 (e.g. De Beaulieu and Reille, 1995) that would suggest an age between 352 ka and 428 ka. However,

A
B
C
D
E
F

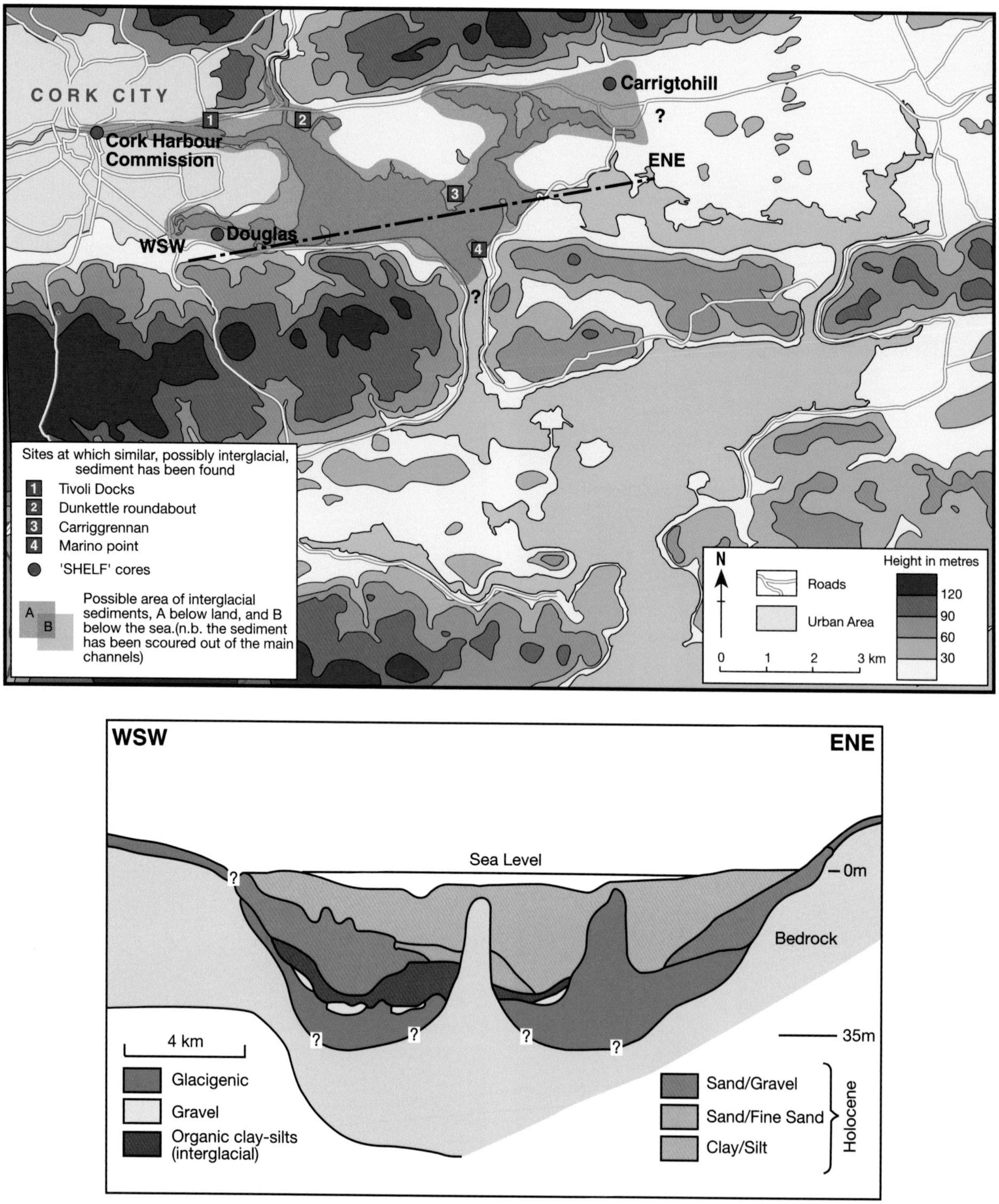

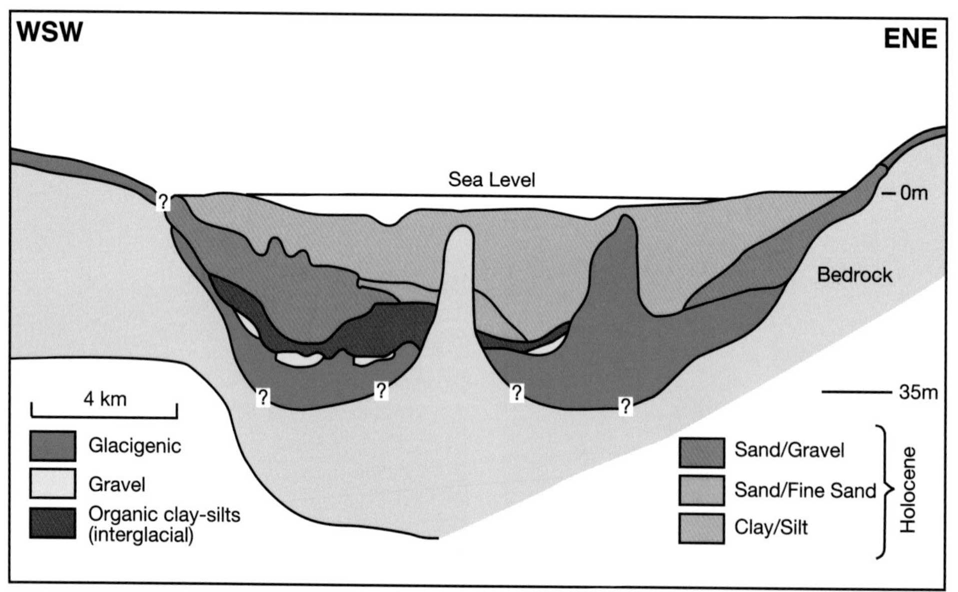

Figure 15.12 The extent of estuarine sediments of Gortian age in the Cork Harbour area and a lithostratigraphic cross-section of the area (*after* Dowling and Coxon, 2001 and Crowley *et al.*, 2005).

if the biostratigraphical correlation is viewed more critically, then it may be realistic to accept the possibility that the rather weakly developed interglacial sequence of the Gortian may represent MIS 9 (302 ka–338 ka) or more likely MIS 7 (195 ka–251 ka).

Table 15.5 The biostratigraphy of the Gortian.

Gortian Substage	Pollen assemblage and vegetation history
Gn IV	**Marked termination of record during *Abies-Picea*-Ericaceae assemblage with notable reworking of thermophilous taxa.** At sites that contain a long record the sequence is abruptly terminated and, although some Gortian deposits are ice-rafted lenses and not *in situ,* there is sufficient evidence of a conformable transition from the wooded, ericaceous heath of GnIII to an open landscape vegetation to conclude that sudden climatic deterioration truncated the interglacial cycle. This record of the closing stages of the Gortian Interglacial at Derrynadivva is a remarkable one as it shows a sudden climatic deterioration evidenced by the extensive reworking of pollen and spores (Coxon *et al.*, 1994). This rapid end to the interglacial may have been due to the proximity of Ireland to the important polar front of the North Atlantic during the onset of the subsequent glaciation – a situation not unlike that proposed to explain marked environmental change during the Nahanagan Stadial.
Gn IIIb	***Pinus–Betula*–with *Alnus–Taxus–Abies–Picea*–Ericaceae (including *Rhododendron ponticum*) assemblage.** The latter part of the Gortian Interglacial is typified by the appearance and/or dominance of *Picea, Abies, Alnus, Taxus,* and a very diverse flora of Ericaceae which, along with sedimentological evidence, indicates that wet conditions and bog development were predominant. These taxa either migrated late into the landscape or were favoured by the wetter conditions and decreasing soil fertility.
Gn IIIa	***Pinus–Betula*–with *Alnus–Taxus–Abies–Picea* assemblage.**
Gn II	***Pinus–Betula-Quercus* assemblage. *Taxus* very important at some sites, with *Hedera* and occasional tree taxa including *Fraxinus*, *Corylus*, and *Ulmus*.** The well-developed and diverse mixed oak forest so characteristic of British temperate stage floras does not develop during the Gortian and *Quercus*, although important at some sites, does not remain a major contributor to the pollen rain after an early peak. This early peak of *Quercus* at Gortian sites (*see* Gort, Kildromin, and Derrynadivva on Figure 15.9) is reminiscent of a similar peak on Hoxnian pollen diagrams from biozone Ho IIa and Ho IIb (e.g. from Marks Tey, Turner 1970). However, in Hoxnian pollen diagrams the diverse woodland contains persistent levels of *Ulmus*, *Corylus*, *Tilia*, and *Fraxinus,* amongst others, from the Early and Late Temperate zones (Ho II and III) whilst the level of these taxa in Gortian sequences is masked by the preponderance of *Pinus* and *Betula*. It is apparent from this that the Gortian temperate stage may be rather poorly developed with the possibility of cooler conditions prevailing. Gortian biozones Gn II and Gn IIIa are also notable for the persistent appearance of charcoal fragments both on the pollen slides and in the macrofossil samples at some sites (Coxon and Hannon, 1991; Coxon *et al.*, 1994; Coxon, 1996a). Fire frequency may have influenced the vegetation and *Pinus*, which can withstand fire events more successfully than the other trees present, may have expanded at the expense of the more fire sensitive taxa.
Gn I	***Betula–Pinus* assemblage.** As soils stabilised and the climate improved the late-glacial pioneering vegetation was shaded out by *Betula* and *Pinus* woodland and many other thermophilous trees and shrubs (e.g. *Quercus, Ulmus, Ilex,* and *Corylus*) were expanding from their glacial refugia and migrating into Ireland.
P-Gn I-g	***Salix–Juniperus–Hippophae* assemblage including a herbaceous component.** A late-glacial vegetation that developed during climatic amelioration and included pioneering plants such as *Salix, Juniperus,* and *Hippophae* as well as a diverse herb flora and *Betula* scrub.

Uranium–thorium dating and amino acid ratios

Attempts have been made to date the Gortian using amino acid racemisation and uranium–thorium disequilibrium dating (UTD). The latter has given ages of 180 ka and 191 ka from Burren Townland and >350 ka (a 'preliminary result') from the type site at Gort (Heijnis, 1992). More recently, UTD results from Cork Harbour suggest an age of 135 ± 35 ka, 120 ± 35 ka, and $150^{+165}/_{-70}$ ka (i.e. MIS 5e or 7) but the dates have very wide errors. Amino acid racemisation results suggest a pre-Eemian (MIS 5e) age (Scourse *et al.*, 1992; Dowling *et al.*, 1998). These data suggest that the Gortian is the equivalent of part of the complex MIS 7 or MIS 9. It is unfortunate that there is still no absolute answer to the problem of the age of the Gortian. The dates quoted here require verification (and a greater degree of between-site consistency) before a firm conclusion can be reached.

Late Pleistocene biostratigraphy

If the Gortian represents marine MIS 7 or 9 that leaves the problem of the missing temperate stages of the Irish Pleistocene. The scale of the problem is apparent on Figure 15.2. In particular, the enigma of the missing last interglacial (MIS 5e) has bedevilled Irish Quaternary studies (see Warren, 1979, 1985; Mitchell, 1981; Watts, 1985; Coxon, 1993). Although well-documented from Britain (as the Ipswichian Interglacial – e.g. West, 1980; Jones and Keen, 1993), from Europe (as the Eemian Interglacial – e.g. Watts, 1988), and from ocean cores as MIS 5e dated to between 132 ka and 122 ka (Fig. 15.2), the last interglacial has not been unambiguously identified in Irish terrestrial sediments (but see Knocknacran below).

Mitchell (1976, 1981) tentatively suggested that the upper part of the sequence at Baggotstown (Watts, 1964) and the estuarine sand at Shortalstown, County Wexford (Colhoun and Mitchell, 1971) might represent the last interglacial. However, the Baggotstown stratigraphy is unclear; the sediments at Shortalstown are glacitectonically disturbed; and the pollen assemblages (Fig. 15.14) may represent Middle (or Early) Pleistocene sediment redeposited from the Irish Sea basin – a scenario made more possible by the occurrence of a seed of *Decodon* sp. and a pollen grain of *Tsuga* (Colhoun and Mitchell, 1971).

A buried podsol recorded from Corraun in County Mayo has been cited as last interglacial in age (Synge, 1968; Finch, 1977) but the deposit has never been analysed in detail and its age remains unknown. The reported

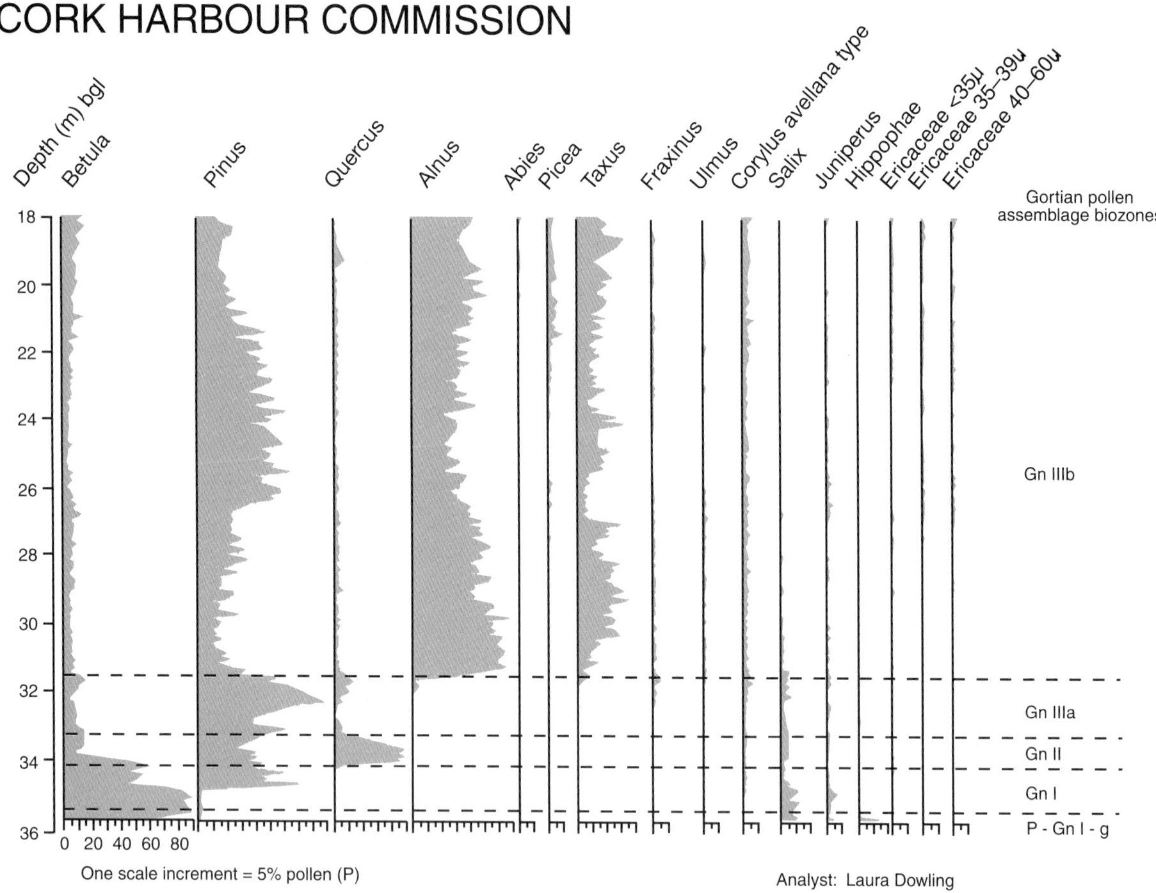

Figure 15.13 A summary pollen diagram (Dowling, 1997) from the remarkable 17.5 m Gortian sequence from Cork Harbour (the depths are 'below ground level', i.e. the top of the Customs House Quay). The inner part of the Cork Harbour estuary is lined with over-consolidated clays, silts and organic sediments up to 18 m thick which belong to the Gortian temperate stage. Interglacial sediments were recognised in 1979 during site investigations for the Eamon de Valera Bridge in Cork City. The widespread nature of the estuary fill and the uniform age has been confirmed in the detailed work of Dowling (1997) and Dowling *et al.* (1997). The pollen diagram has been divided into pollen assemblage biozones as described in the text and the original work defines more local zones.

pollen assemblage is described as follows: '*...Some deciduous tree pollen is present at this site but 30–60 per cent of the total tree pollen is pine. Thick forest cannot have been present, because the bulk of the pollen is from shrubs and grasses...*' (Synge, 1968). With such a description it is impossible to assign a firm correlation, and the pollen assemblage suggests interstadial rather than interglacial conditions.

A redeposited ball of organic sediment within the sands and gravels of the Screen Hills moraine contained a *Carpinus*-rich pollen assemblage (McCabe and Coxon, 1993) with a pollen assemblage that is very similar to those found in Continental Eemian deposits. The assemblage appears to belong to a Late Pleistocene temperate episode.

Knocknacran temperate stage

In 1996 an 18 m wide and 5 m thick organic fill within a solutional feature in gypsum and below drumlinised Midlandian till was recorded from Knocknacran, County Monaghan (Vaughan *et al.*, 2004, Figs. 15.11, 15.14). The solutional fill is a glacitectonised and sheared breccia with an organic matrix. The deposit is very woody in parts (*Quercus* and *Taxus*). Pollen recovered from the deposit include *Quercus, Corylus, Taxus, Betula,* Gramineae, *Pinus, Salix, Ulmus,* Cyperaceae, and *Hedera.* (Vaughan *et al.*, 2004). Interestingly, *Picea* and *Abies* are absent, as are taxa indicative of the late-Middle Pleistocene or earlier. As such the pollen assemblage is unlike any other recorded in Irish deposits (Fig. 15.14) and the site most likely represents the last interglacial (MIS 5e). Infinite

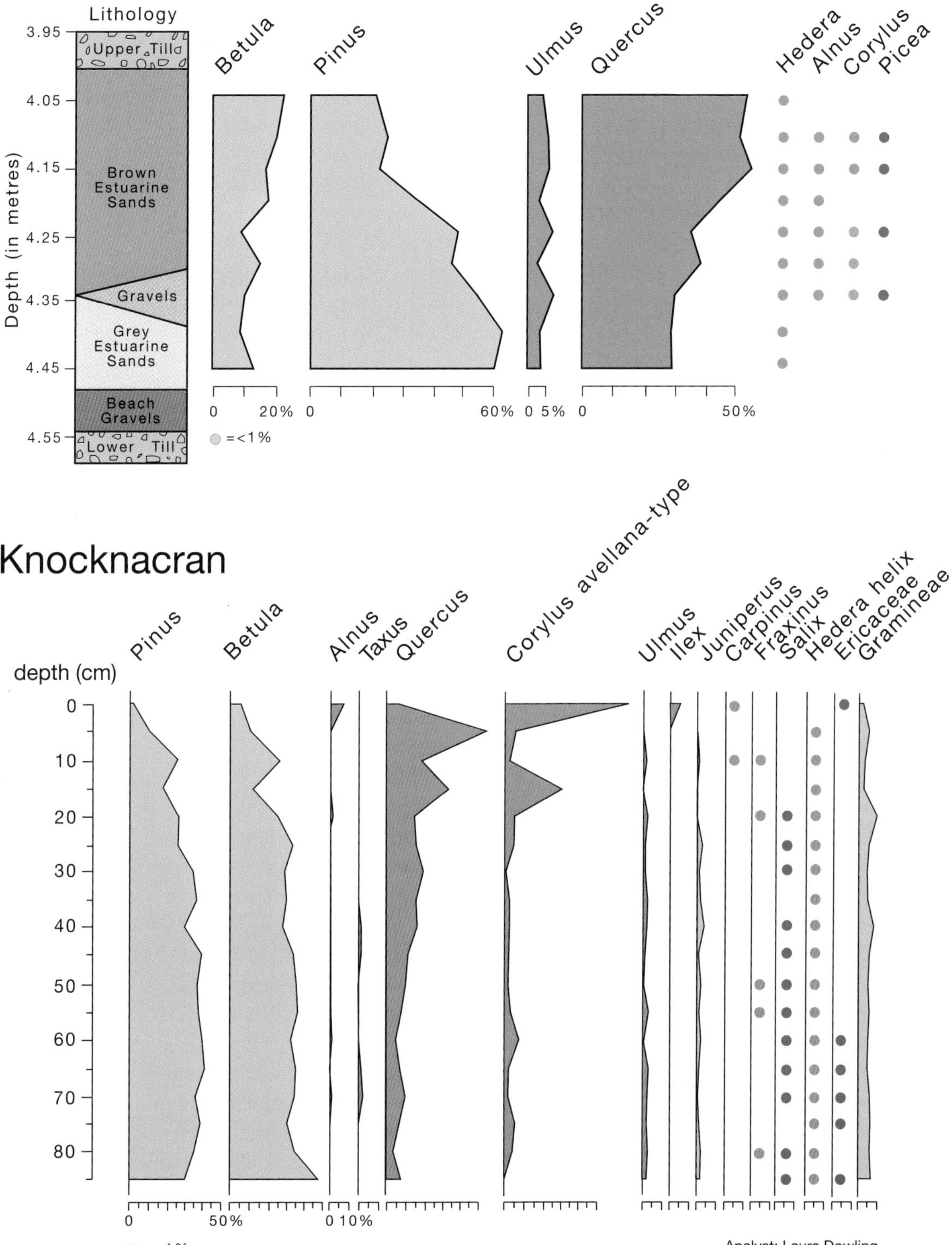

Figure 15.14 **A** – Pollen diagram from the deposit at Shortalstown (*after* Mitchell, 1976 from Colhoun and Mitchell, 1971). **B** – Pollen diagram from the deposit at Knocknacran (*after* Vaughan *et al.*, 2004). This pollen sequence is probably the best palaeoecological evidence of last interglacial deposits in Ireland.

(i.e. beyond the accepted range of radiocarbon dating) dates of >47 ka BP and >48 ka BP show the deposit to predate the Middle Midlandian whilst uranium–thorium dates suggest an age within MIS 4 or 5 (Fig. 15.16).

The Kilfenora Interstadial

The Quaternary deposits in cliff sections between Spa and Fenit along the north shore of Tralee Bay were the subject of an important paper by Mitchell (1970). The sequence of a raised wave-cut bedrock platform overlain by raised beach sediments, biogenic sediments, lower solifluction deposit, glacigenic sediments, and upper solifluction deposit has been summarised by Mitchell (1970, 1981) (see Fig. 15.15).

Research by Ruddock (1990), Heijnis (1992), and Heijnis *et al.* (1993) has shown that the biogenic sediments were deposited in cool temperate conditions. Uranium–thorium disequilibrium (UTD) dating of the biogenics gives an age of between 114 ka and 123 ka. Isochron plots allowed an age estimate of 118 ka (+5 ka/−4 ka) to be made. The lithofacies and their associations suggest that the organic sediments represent deposition in small pools or a lagoon environment frequently inundated with inorganic material. A pollen diagram from one of the organic sites (site F, Ruddock, 1990) is reproduced here on Figure 15.15.

Zone F1 reflects a period when there were open pinewoods in the area, with a field layer composed of Gramineae, Cyperaceae, and other herbs with some Ericaceae. *Pinus* is the dominant tree in Zone F2, but there is also some pollen of *Alnus*. The Ericaceae have increased, indicating that conditions were possibly becoming more oceanic or cool. In the peat layer above this level (Zone F3) *Pinus* pollen disappeared and herb pollen increases. However, *Quercus* pollen is present in greater quantities than below, and at site G *Alnus* is also recorded. Zone F4 at both sites records *Quercus, Alnus, Corylus,* and *Ilex,* indicating a possible amelioration in climate towards the top of the sequence. The pollen diagram is very similar to the data presented by Mitchell (1970) that recorded a thicker sequence and an earlier silt horizon (containing abundant *Pinus*). Mitchell also recorded single grains of *Abies* and the pollen of *Taxus* and *Rhododendron,* whilst Ruddock (1990) recorded a wider range of herb taxa and a single grain of *Picea* from the section that she studied.

The pollen assemblages from these sections are unique in Ireland. Although they contain some taxa found in the Gortian, the assemblage as a whole does not resemble those of the known later parts of that temperate stage (see above and Coxon, 1993, 1996b).

The pollen record of Figure 15.15 seems to be reasonably continuous across lithological boundaries and suggests that the sequence represents cool temperate, sparsely wooded conditions with a predominance of open ground, possibly with increasing thermophilous taxa late in the depositional phase. The pollen sequences may represent the onset of interstadial conditions early in the Midlandian cold stage or a minor amelioration of climate to warmer conditions during the end of a temperate stage.

The pollen assemblages recorded at Fenit cannot be correlated with interstadial floras recorded from the Early Midlandian (e.g. at Aghnadarragh; McCabe, Coope *et al.*, 1987; see below) and for this reason, and the fact that they are dated, they have been named as the Kilfenora Interstadial by Heijnis *et al.* (1993).

The dating of these sediments is crucial, and the dates (between 114 ka and 123 ka) indicate that they represent the latter part of the Eemian (or last) Interglacial or the Early Midlandian. The organic sediments at Fenit appear to belong within MIS 5, possibly towards the end of MIS 5e (132 ka–122 ka) or during 5d or the beginning of 5c. The latter age is suggested by the pollen, indicating a slight amelioration in climate towards the top of the deposit. The age of the interstadial makes it unlikely to be related to the Aghnadarragh Interstadial (below), as it must come before a major expansion of Early Midlandian ice (Table 15.2).

At least some of the raised marine sediments at Fenit are shown to be younger than MIS 5e but the age of the underlying erosional platform remains unknown (for reasons outlined by McCabe, 1987).

The record of the Irish cold stages

Quaternary geological evidence

As previously outlined, the Quaternary has been characterised by large, relatively rapid, periodic changes in climate and environmental conditions. In north-western Europe these climatic oscillations are manifest in a succession of 'glacial' and 'interglacial' stages. Ice core evidence tells us much about the last glaciation and its associated climatic variability – see Figure 15.16.

The long 'glacial' stages are probably better termed *cold stages,* as periods of widespread glaciation (stadials) only made up a small part of their ~100 ka length. In addition to stadials, cold stages also include 'cold phases' and

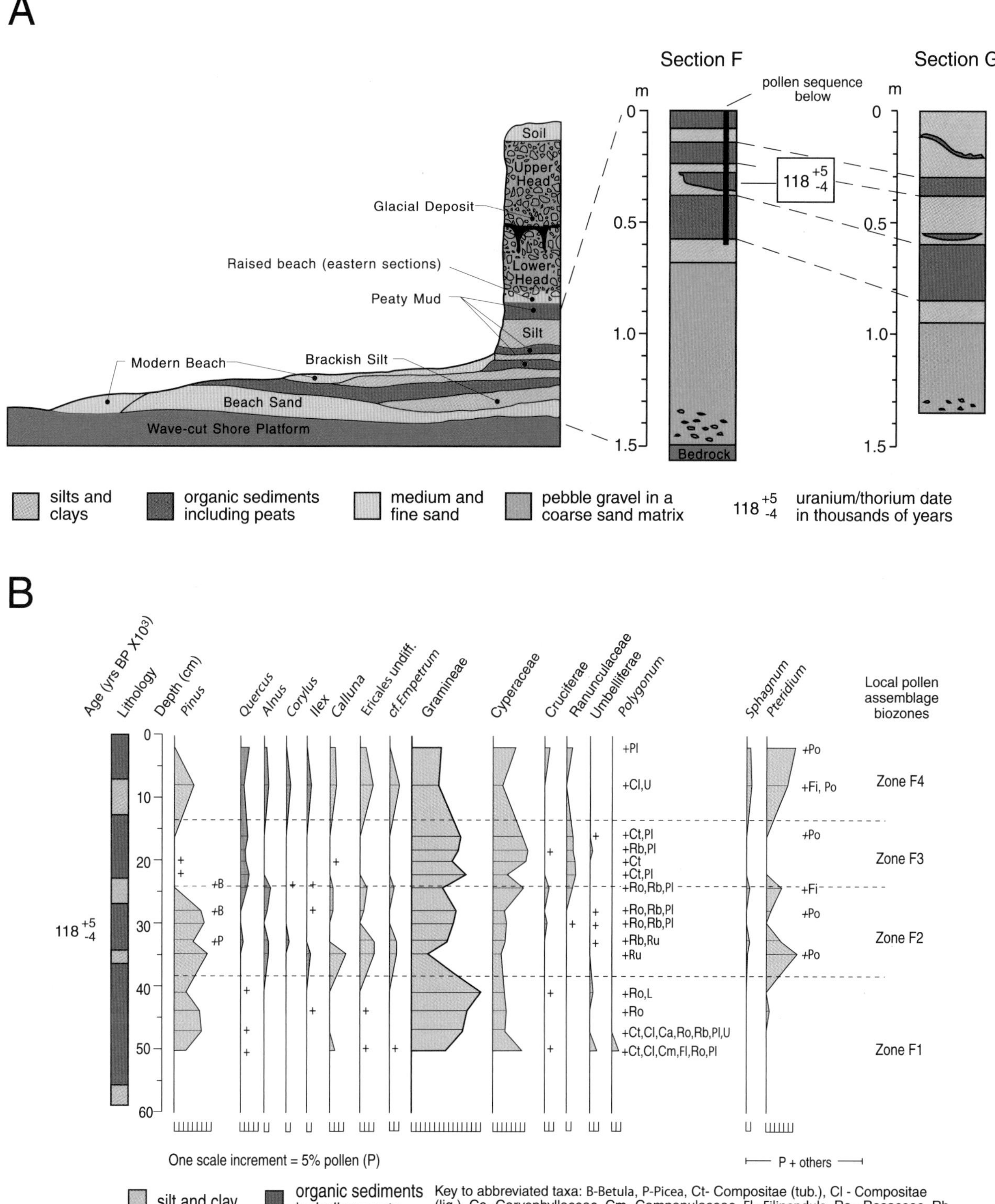

Figure 15.15 Stratigraphy (A) and pollen record (B) from the Late Pleistocene Kilfenora Interstadial, Fenit, County Kerry (*after* Heijnis *et al.*, 1993, from Coxon, 1996b).

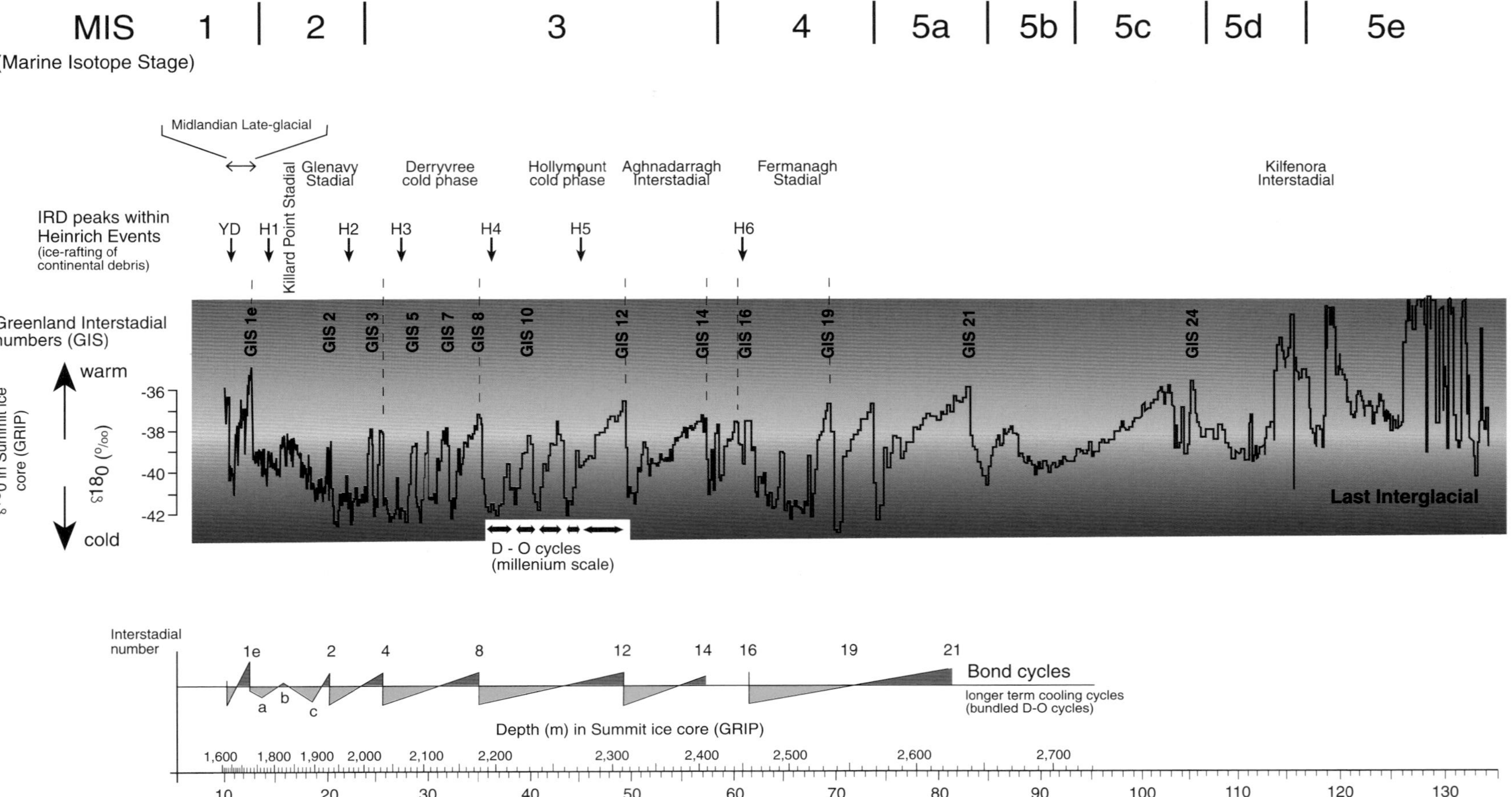

Figure 15.16 The last cold stage showing $\partial^{18}O$ values from the GRIP ice core along with a summary of the long-term cooling cycles (Bond cycles), Dansgaard-Oeschger cycles, and ice-rafted detritus (IRD) peaks within Heinrich events (*after* Bond *et al.*, 1993). The phases of the Midlandian cold stage (Table 2) are also shown.

'interstadial' periods during which the climate ameliorated to some degree, possibly with temperatures almost reaching those of the present day (or warmer) for short periods (i.e. 1–2 ka rather than the 10 ka to 20 ka duration of a full Interglacial stage). Ireland has a physical record of two major cold stages, but may have experienced anything up to five in the last 750 ka (see Figure 15.2). Each one was a long and complex event of ice build up and decay interspersed by cold, periglacial, conditions and by cool (or even warm), interstadial periods.

The terrestrial record of the Irish cold stages is far from complete, and successive glacial events have left only fragmentary evidence, each stadial probably eroding and reworking the products of preceding events. Building on the long tradition of Quaternary geology in Ireland established by workers such as Close, Hull, Dwerryhouse and others, more recently Charlesworth, Orme, Farrington, Mitchell, Synge, McCabe and others have mapped the spatial distribution of Ireland's populous and extensive onshore glacial features (Figure 17). Remote sensing, and continuous all-island digital elevation datasets at decimeter resolution, have provided an additional toolset with which to map glacial features and allowed the recognition of hitherto unrecognised glacial bedforms and ice-flow patterns (e.g. Knight and McCabe, 1997a; Clark and Meehan, 2001). Offshore bathymetric data collection offers the possibility of extending the field constraints of former ice limits on the continental shelf (e.g. King *et al.*, 1998) and near-shore sea level lowstands (e.g. Kelley *et al.*, 2006).

Many good descriptive accounts of Irish glacial features exist, but a definitive stratigraphy recording glacial/non-glacial climatic variability throughout the Quaternary has proved elusive for a number of reasons. Firstly, there is a lack of sufficient dating control. This is true in all areas of the country for glacial events older than the last cold stage (Midlandian), and in many areas older than the last widespread stadial ('Main Phase' of the Late Midlandian/Glenavy Stadial/Last Glacial Maximum[LGM]). Continually tighter dating control of the 'last Termination' (deglaciation from LGM limits) has, however, been proceeding where possible, allowing the context of driving forces (climatic or otherwise) to be considered in the study of Irish glacigenic sediments (McCabe *et al.*, 2005).

Secondly, glacigenic sequences are complex and cannot be analysed simply in terms of advance and retreat cycles (McCabe, 1987). Thus, many 'layer-cake' sections thought to record glacial advance/retreat cycles may have been misinterpreted. Detailed lithofacies analysis has shown that many such sequences in fact comprise lithofacies assemblages formed in generically linked depositional environments (Eyles and McCabe, 1989a).

The nature of past ice sheets

Climatic deteriorations resulted in the repeated growth and decay of ice sheets covering most of the landmass of the British Isles (stadial periods). Whilst generated from independent ice source areas on Ireland and Britain, ice flow patterns indicate that at least during the existence of the last widespread ice sheet, Irish and British sectors were coincident along the axis of the Irish Sea Basin, and can be termed the British Irish Ice Sheet (BIIS). Erratic carriage also evidences ice flows radiating outwards from the Scottish uplands and into the north of Ireland (e.g. forming the Armoy Moraine; Stephens *et al.*, 1975). The distribution of Scottish ice is thought to have been even more widespread throughout Ireland in earlier stadials, transporting erratics from the Firth of Clyde as far west as Donegal, Wexford, and Cork. In contrast, Irish ice dispersal centres formed in upland; e.g. Donegal, Connemara, and lowland areas (as domes); e.g. Omagh Basin, Lough Neagh Basin, north Central Midlands; and Irish ice radiated outwards from these areas, resulting in the vast majority of Irish glacial landforms. Ice sheets were probably comprised of multiple 'sectors' controlled by influences on dynamics such as bedrock topography and their relative proximity to shifting precipitation sources. The presence of marine margins at the limits of Late Midlandian ice sheets was a primary control on deglacial dynamics.

Sediments and products

Ice sheet bedrock erosion and the reworking of pre-existing sediments by repeated glaciation have left a significant imprint on many aspects of the Irish landscape (Fig. 15.18). Ice and meltwater sculpted many distinctive erosional landforms along pre-existing topographic lows and structural weaknesses, e.g. *Glacial valleys*: Glenmacnass and Glendalough, County Wicklow; Glenveagh, County Donegal; Glenahoo, County Kerry; *Fjords*: Killary Harbour, County Clare; Lough Swilly, County Donegal; Carlingford Lough, County Down/Louth. Flowing glacial ice also deposited thick piles of relatively consolidated, impervious, subglacial diamict (till) where conditions within the ice sheet allowed, principally throughout Ireland's lowlands. Subglacial reworking of previous sediment accumulations and gradual

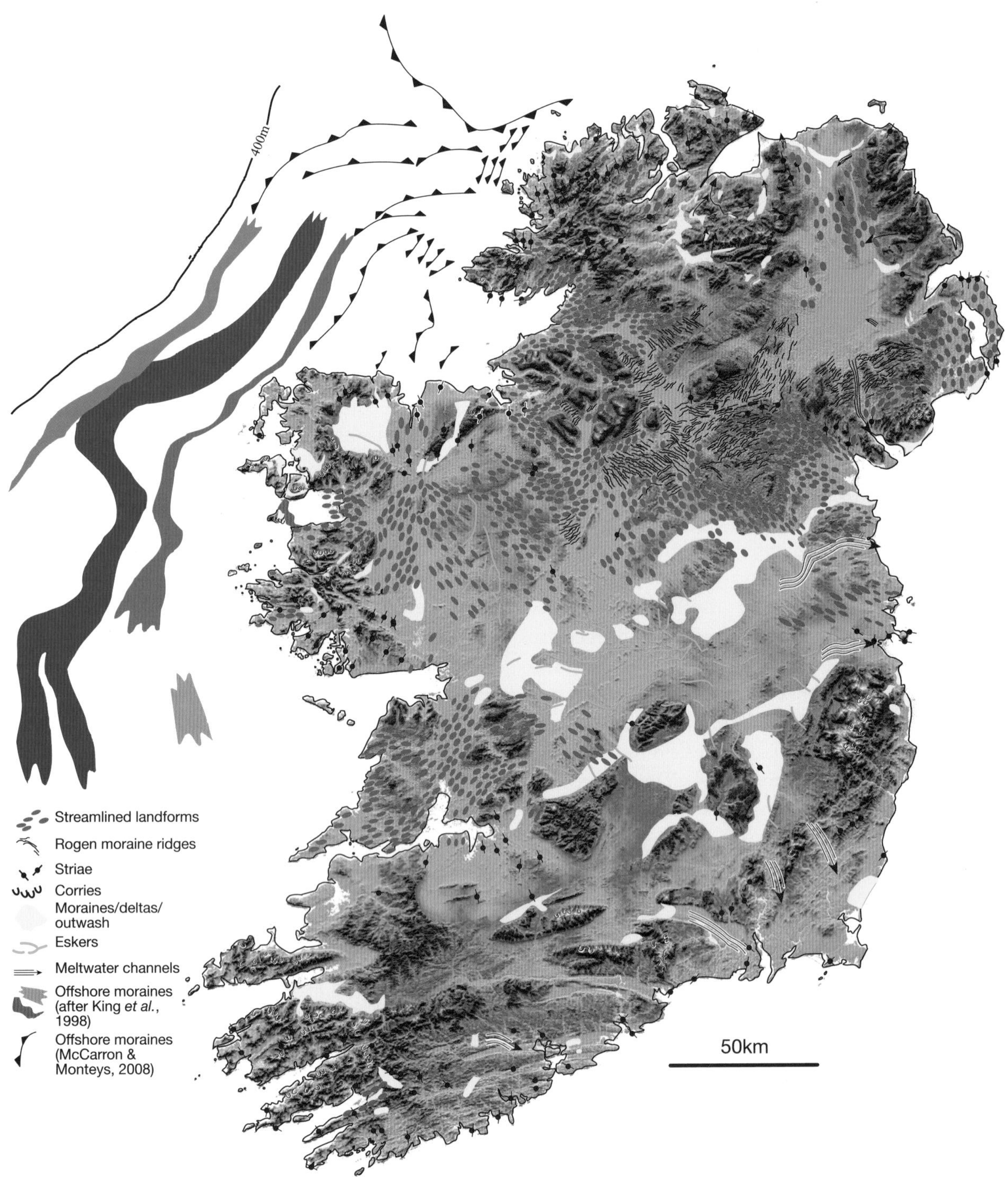

Figure 15.17 The distribution of glacial landforms in Ireland and on the continental shelf (*after* Synge, 1979, McCabe, 1987, 1993; King *et al.*, 1998).

Figure 15.18 A – Lough Nahanagan within Camaderry Mountain corrie, County Wicklow. B – Extensive deltaic foresets, Blessington, County Wicklow. C – Sample site in laminated glaciomarine mud, Corvish, County Donegal. D – Rapidly eroding section in glaciomarine sediment pile, Killiney, County Wicklow. E – Glaciomarine mud, intertidal boulder pavement and subaqueous outwash, Cooley Point, County Louth (*Photo*: Prof. A. M. McCabe, published with permission of the Director of the GSNI). F – Emergent postglacial sequence, Portballintrae, County Antrim.

transfer of sediment towards ice-marginal areas was probably an important process throughout the oscillation between glacial and non-glacial episodes. Some organic material has also survived repeated stadials in isolated pockets, and provides a very fragmentary biostratigraphical record. Preservation of such sediment tends to have occurred in hollows within karstic, uneven parts of the pre-glacial landscape (Coxon and Coxon, 1997; Vaughan *et al.*, 2004; Coxon, 2005a).

The last ice sheet to have covered Ireland decayed rapidly, leaving an extensive legacy of glacial features including the ice-contact, ice-proximal, and ice-distal glaciofluvial and glaciolacustrine deposits that mantle many areas (McCabe, 1987; Coxon and Browne, 1991; Warren and Ashley, 1994). The decay and retreat of ice sheet margins towards centres of ice dispersal left the sand and gravel sediment washed from subglacial sources as eskers, outwash plains, kames and deltas. Some of these stratified accumulations are relatively large, and by the nature of their specific depositional criteria, e.g. the presence of an ice-proximal waterbody or subglacial drainage routes, are spatially concentrated; e.g. the eskers of the central Midlands; near Blessington, north-west County Wicklow; the Sperrin Mountain valleys, County Tyrone; and the Dungiven basin, County Derry. Where they occur, they provide a valuable but diminishing natural resource.

Ice sheet event chronologies

The extent and timing of ice sheet buildup in Ireland was reviewed by McCabe (1987). He noted that, to that date, the sequence of glacial events was known in a relative sense only, predominantly based on morphostratigraphical rather than biostratigraphical or chronometric evidence. Until recently, geochronological constraint of the pattern and timing of ice sheet growth and decay has been as scarce in Ireland as in other regions of the British Isles. The relative paucity of dated ice sheet events, due largely to the relative scarcity of datable organic material on or between glacial sediments (compared to around the margins of the Laurentide ice sheet in North America for example), has prevented confidence in the correlation of glacial advance and/or retreat of sectors of the last BIIS with each other and events recorded elsewhere.

Significant amounts of additional data points, building a geochronological framework constraining ice sheet events (advances/retreats) in the Late Midlandian, have been added since then. This has largely been through AMS ^{14}C dating of *in situ* populations of marine microfauna shells extracted from Irish coastal sediments (e.g. McCabe and Clark, 1998; McCabe and Clark, 2003; Clark *et al.*, 2004; McCabe *et al.*, 2005; McCabe and Clark, 2007a, b). In addition, the quantity of cosmogenically produced ^{10}Be and ^{36}Cl nuclide accretion in 'fresh' rock surfaces (eroded by >2.5 m during the last glacial advance) can provide a minimal date for erosion of that surface (by ice for example) (Bowen *et al.*, 2002; Ballantyne *et al.*, 2006, 2007, 2008). Continued refinement of the application and interpretation of the results of both techniques promises continued advances in bracketing ice events at particular locations into absolute timeframes.

Ice sheet flow patterns

Patterns of ice flow during stadials were not constant and unidirectional, as traditional compilations of ice flow directions reconstructed from grouped ice flow indicators would perhaps suggest (e.g. Mitchell *et al.*, 1973; Warren and Ashley, 1994). Alternatively, ice sheet reorganisation because of internal dynamics and/or climatic forcing produced overprinted bedform sets ('palimpsest' landscapes) in many areas of Ireland, especially throughout the central northern 'drumlinised' lowlands (McCabe *et al.*, 1999; Clark and Meehan, 2001) and around large coastal embayments on the western seaboard (Knight and McCabe, 1997b). Around an isostatically depressed Irish landmass, deglaciation was possibly driven by internal dynamics linked to rapid calving front losses and marine downdraw towards tidewater glacial limits (~7 km/a across a floating terminus 800 m thick and 6 km wide; Jakobshavn Effect, Hughes, 1986). The orientation of streamlined landform long axes in the direction of the major bays of the west of Ireland is unmistakable and gives a strong indication of linkages between marine-influenced ice sheet margins and ice sheet bedform streamlining (Knight and McCabe, 1997b). Large ice volume transfers, high pressure subglacial water, and saturated, deforming sediment layers, would erode and streamline the underlying deposits and deliver large sediment volumes to the marine margins, allowing the reconstruction of significant morainic belts which have field associations with drumlin belts (McCabe *et al.*, 1984).

Deglacial palaeoenvironments around the Irish coastline

In Ireland, a more detailed and age-constrained geological record exists to document the deglaciation of the Irish Sea Basin (ISB) than any other series of glacial events. This is due principally to the presence of tidewater glaciers on the margins of the ISB during deglaciation. Temporary stillstands of the ice front and progressively less extensive ice sheet readvances during overall deglaciation resulted in a generally northward-offlapping sequence of fossiliferous subaqueous sediment wedges along the east coast. Since the publication of Eyles and McCabe (1989b), it has been argued that many of these glacigenic sequences around the ISB are the product of generally glaciomarine conditions (high relative sea levels: HRSLs) at ice margins during deglaciation (Clark *et al.*, 2004; McCabe, 1997; McCabe and Clark, 1998, 2003; McCabe and Haynes, 1996; McCabe *et al.*, 1987a, 1998, 2005, 2007a, b).

Based on published reconstructions of events from detailed sedimentological studies and dated marine fauna, HRSLs in this region were created at a time of a global eustatic minimum (Yokoyama *et al.*, 2000). Isostatic depression resulted in ice sheet destabilisation in rapidly calving tidewater environments and the 'early' onset of deglaciation at about 26 ka BP, preceding the time of the global LGM (McCabe *et al.*, 2007a). The ice thicknesses required to produce enough crustal flexure to result in +30 m HRSLs along the central eastern Irish seaboard (~1 km thickness in the central Irish Sea) follow from the onshore ice sheet thicknesses advocated to explain the pattern of trimline formation in Wicklow (Ballantyne *et al.*, 2006).

Not all authors agree with the model of such levels of isostatic depression and tidewater glaciomarine deposition during deglaciation (e.g. Lambeck, 1996; McCarroll, 2001; Scourse and Furze, 2001; Evans and O'Cofaigh, 2003; Boulton and Hagdorn, 2006). Temporary ice-marginal glaciolacustrine environments are postulated to explain the formation of large subaqueous sediment banks and spreads on the eastern Irish seaboard, south coast, and on the Isle of Man (e.g. Thomas, 1977; G.S.P. Thomas *et al.*, 2004). Deterministic geophysical models combining estimates of earth rheology, ice sheet and substrate configurations (e.g. Shennen *et al.*, 2006; Roberts *et al.*, 2006; Boulton and Hagdorn, 2006), parameterised using Late Glacial and Holocene data, do not produce solutions which agree with age-constrained field evidence for palaeo-sea levels at higher than current elevations around the Irish coastline during the last Termination (McCabe, 1997).

However, revisions of the glaciomarine model are spurred by continuing advances in the production of an internally consistent geochronological framework spanning the Last Termination (see Table 15.6). Dates

Table 15.6 Millennial-scale deglacial events during the Last Termination.

Stage	Approximate date range (^{14}C ka BP)	Description*
Cooley Point Interstadial (CPIs)	≥17 – ≥ 15.0	Along eastern, western and northern Irish seaboards, at Cooley Point (Co. Louth), Cranfield Point and Kilkeel Steps (Co. Down), Fiddauntawnanoneen, Belderg Pier and Glenulra (Co. Mayo) and Corvish (Co. Donegal), dated monospecific samples of Elphidium clavatum from raised in situ massive to laminated marine muds fall in the period from 16,970 ± 190 to 14,705 ± 130 (McCabe *et al.*, 2005). Sediment sequences and dates indicate a period of HRSLs post-dating the global LGM associated with glaciomarine sedimentation along Irish seaboards during deglaciation from offshore limits.
Clogher Head Stadial (CHS)	≤15.0 – ≥ 14.2	At Cooley Point, Co. Louth eastward prograding subaqueous outwash spreads and ice-pushed facies cover laminated to massive mud facies dated to 15,020 ± 110 which underlie an intertidal boulder pavement. At Port, marine muds dated to 15,190 ± 85 are deformed and sheared into overlying subglacial diamict (McCabe *et al.*, 2005; McCabe and Clark, 2007). At Corvish, Co. Donegal deformation of marine muds and formation of an ice-pushed moraine, dated to 15,025 ± 95 (McCabe and Clark, 2003), is correlated with this readvance (McCabe and Clark, 2007).
Linns Interstadial (LIs)	14.2 – ≥13.8	Open marine embayment conditions in Dundalk Bay associated with deglaciation from the Clogher Head Stadial readvance limits and mud formation at both Linns (14,157 ± 69) and Rathcor (14,250 ± 130 14C BP) on the southern and northern bay margins respectively (McCabe *et al.*, 2005).
Killard Point Stadial (KPS)	<14.2 – >13	Marine muds interbedded with stacked, channelised gravels in an ice contact subaqueous morainal bank or apron at Killard Point are dated to 13,995±105 (McCabe *et al.*, 1986). The morainal bank is associated with a regional scale ice sheet reorganisation associated with amphi-North Atlantic climatic events (McCabe and Clark, 1998). Deformation of a diamict at Linns is also associated with KPS readvance to inside CHS limits in Dundalk Bay (McCabe and Clark, 2007) (Figure 16.19).
Rough Island Interstadial (RIIs)	~13.0-	Downdraw of the BIIS during the KPS towards marine calving bays (McCabe and Hynes, 1996; McCabe *et al.*, 1998) and subsequent in situ ablation of the lowered ice sheet profile resulted in rapid Stagnation Zone Retreat and preservation of streamlined bedforms associated with subglacial sediment transfer during KPS. Marine muds comprising part of an inter-drumlin drape are dated to 12,740 ± 95 at Rough Island, Co. Down (McCabe and Clark, 1998). Mud deposition indicates continued HRSLs following drumlinisation and a suppression of isostatic rebound recovery rates relative to the earlier stages of deglaciation (McCabe *et al.*, 2005).

* Dates are ^{14}C yr BP unless stated; Samples corrected for assumed 400yr reservoir effect)

constraining ice sheet events occurring on broadly millennial timescales are derived from AMS ^{14}C dating of *in situ* marine microfauna extracted from fine-grained facies within deglacial sediment sequences (Fig. 15.19). These facies have been shown to contain *in situ* marine biocoenoses, dominated by the foraminiferan *Elphidium clavatum* (89–95%), an opportunistic benthic microfauna species that inhabits low salinity, cold (0.5–2.5°C), turbid water (Hald *et al.*, 1994). The facies are devoid of reworked temperate species forms and therefore provide a robust method to constrain the age of associated glacial events (Haynes *et al.*, 1995).

Geochronological constraint and positioning the sediments within a depositional systems framework (linking fossiliferous sediment deposition and associated landform creation) has allowed the construction of an event history that links ice sheet sector activity across the island and in other sectors of the last BIIS (McCabe *et al.*, 1998; McCabe and Clark, 2003, 2007; McCabe *et al.*, 2005, 2007a, b). Dated events indicate that during the Last Termination largescale events within the last BIIS occurred on broadly millennial timescales, in keeping with other climatic cycles recorded in the ice–ocean–atmosphere system (Bond *et al.*, 1993). The geological evidence indicates an ice sheet that was sensitive to the hemispheric and global scale climatic and glacioeustatic events recorded elsewhere in ice, deep sea sediment, and coral palaeoclimatic archives (McCabe *et al.*, 1998; McCabe and Clark, 1998; Clark *et al.*, 2004). The possible link between drumlinisation in Ireland and Heinrich event 1 (McCabe and Clark, 1998; McCabe *et al.*, 1998) is of considerable importance in understanding ice sheet–climate linkages in the ice–ocean–atmosphere system which have relatively long, crudely defined response-time lags (Ruddiman, 2005). Geological evidence of the response time of the BIIS to other globally significant oceanic events (Clark *et al.*, 2004) demonstrates for the first time the strong links between the behaviour of terrestrial ice masses and the timing of palaeoenvironmental changes in the North Atlantic. The apparent near-synchroneity of events recorded in oceanic and ice cores (Bond and Lotti, 1995) has led to theories of global, or at least circum-North Atlantic, climate change in a coupled ice–ocean–atmosphere system of which the last BIIS was part.

Quaternary chronology

The Munsterian cold stage (MIS 8/6)

Ireland is believed to have been widely glaciated during at least two cold stages: the Munsterian (MIS 8/6) and the Midlandian (MIS 4-2). Given the large number of Quaternary cold stages it is almost certain that pre-Midlandian (>150 ka BP; MIS 5e) glacigenic deposits do exist. However, there is not a single site where such an age can unequivocally be demonstrated for glacigenic sediments. Widespread glacigenic sediments in the southern part of Ireland (Munster) have long been regarded as belonging to an 'older' glaciation on the grounds that they show distinct assemblages of erratics, striae, and glacial limits as well as exhibiting subdued relief, deep weathering profiles, and a lack of 'fresh' glacial landforms (Mitchell *et al.*, 1973; Synge, 1968; Finch and Synge 1966; McCabe 1985, 1987). These older (Munsterian) glacial deposits were thought to have originated from sources both in Scotland, which affected much of the eastern and northern parts of the north of Ireland (Stephens *et al.*, 1975), and from Irish inland centres, in particular the uplands of southern Connemara (Charlesworth, 1929).

The lack of chronometric control has meant that, although the Munsterian deposits exhibit certain unique characteristics, e.g. the widespread distribution of Galway granite recorded by Synge (1979) and Lewis (1974) (but see Warren, 1992) and the recognition of extensive deposits recording glacial advance and retreat (Stephens *et al.*, 1975), the age(s) of the glaciation(s) giving rise to the sediments is unknown. The field relationships of Munsterian and Midlandian glacigenic deposits need to be clarified by detailed mapping; and even then, the inherent complexity of glacial sequences (which hinders lithostratigraphical correlation) and the lack of biostratigraphical markers will continue to make the dating of the Munsterian difficult.

An example of the problem is that research has shown that many sequences of 'Irish sea drift' along the southern and eastern coasts of Ireland traditionally considered to be Munsterian have been shown to be Midlandian in age. This evidence is based on mapping, patterns of crustal deformation, radiocarbon dates, and amino acid ratios (McCabe, 1987). Glaciomarine muds (originally recorded as Munsterian 'shelly tills') in north Mayo have been dated to ~17 ^{14}C ka BP [~19 Cal ka BP] (McCabe *et al.*, 1986; McCabe *et al.*, 2005) and it is likely that many of the subaqueous sedimentary sequences around Ireland's coast also date to the Late Midlandian (<30 ka Cal BP)

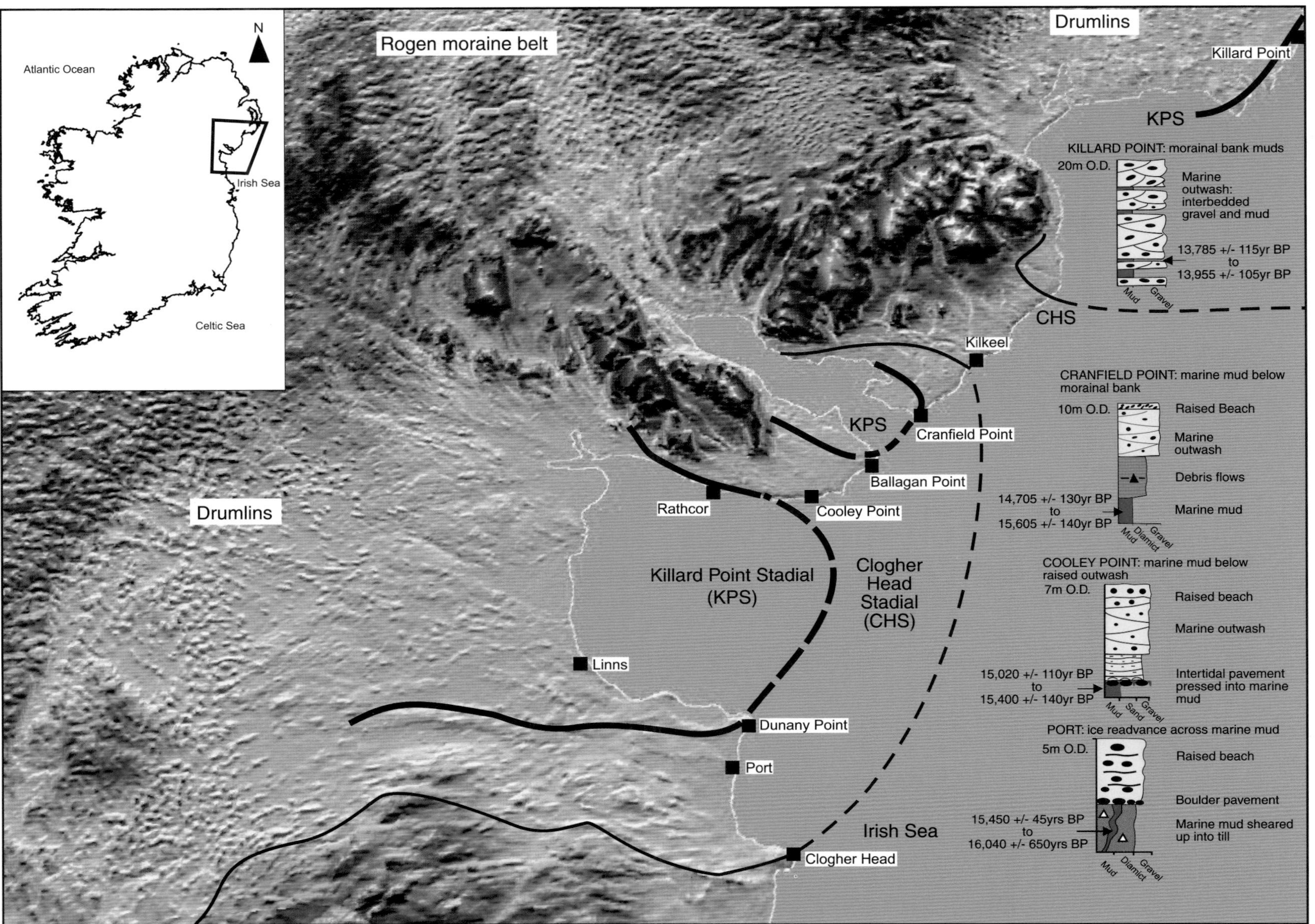

Figure 15.19 Geological evidence of the timing of deglaciation in and around northeast Ireland. (authors' interpretation, and *after* McCabe and Dunlop, 2006).

(e.g. McCabe and Ó'Cofaigh, 1996). Cosmogenic dating of erosion surfaces is redefining the timing and extent of Midlandian glaciations and therefore the probability of finding Munsterian age surficial sediments in the south of Ireland (e.g. Bowen *et al.*, 2002).

Against a background of uncertainty regarding the age of the Gortian Interglacial, and with a lack of recent field mapping and of radiometric dates, the possible presence of pre-Midlandian cold stage deposits remains an enigma. Without geochronological control, the stratigraphical position of deposits assigned to that cold stage remains uncertain.

The Midlandian Cold Stage (MIS 4-2)

Like the Munsterian, the Midlandian has been given no representative type site (Mitchell *et al.*, 1973). Evidence of glaciation and cold climates dating to the last glaciation are very widespread in Ireland, but the diversity of depositional environments and their spatial heterogeneity has meant that the choice of a meaningful type section for the entire cold stage has not been possible.

On a global scale the last cold period (MIS 4, 3, and 2) was an extremely complex event, involving at least two major build-ups of regional scale ice sheets interspersed by a sequence of climatic oscillations that have been recognised from terrestrial evidence (e.g. Coope, 1977), from ocean floor sediments (Bond *et al.*, 1993; Rasmussen *et al.*, 1997), and from the Greenland ice cores (e.g. GRIP, 1993). The latter two have given a detailed picture of millennial-scale climatic oscillations (Dansgaard–Oeschger (D–O) cycles) and longer term cycles (Bond cycles). These have been related to massive effluxes of ice-rafted detritus (IRD) originating from ice sheets around the margins of the North Atlantic ocean (Heinrich layers or Heinrich events; Heinrich, 1988), which were spread across the North Atlantic floor (Gwiazda *et al.*, 1996; Bond *et al.*, 1999; Scourse *et al.*, 2000; Zaragosi *et al.*, 2001). Analyses of deep ocean floor sediments and Greenland ice cores has given a detailed insight into the nature and complexity of ice sheet oscillations during the last cold stage, which can be seen on Figure 15.16.

Evidence of oscillating ice marginal extents on millennial timescales during the last Termination has been identified (McCabe *et al.*, 2005). Interpreting any apparent correlations between ice-sheet and climatic events remains problematic, however, due to the complicated relations between climate and ice sheet behaviour. Continued redefinition of radiocarbon calibration curves and other possible sources of error, including variable ocean water carbon reservoir estimations, all add unquantifiable uncertainty to potential correlations.

However, evidence from biological (pollen, plant macrofossil, and entomological) data, cosmogenic exposure age, and AMS^{14}C dating techniques allows us to reconstruct the nature, extent and timing of Midlandian stadials with more certainty than was previously possible (Table 15.6) (e.g. McCabe *et al.*, 1987b; Bowen *et al.*, 2002; McCabe, 1998; McCabe *et al.*, 2005, 2007a, b). In particular, many dates are becoming available for the build-up, oscillation and retreat of Late Midlandian ice (e.g. McCabe *et al.*, 2007a, b). In addition, studies addressing the sedimentology of Irish glacial deposits have given an insight into the nature, dynamics and timing of the last stadials. The evidence now available from around Ireland allows the stages outlined on Table 15.2 to be identified. The incomplete nature of the Irish sequence is obvious when compared to the complex oxygen isotope record of hemispheric climatic fluctuations.

The Fermanagh Stadial (MIS 4)

Dated organic evidence, which from pollen and insect biocoenoses indicates interstadial rather than interglacial conditions, has led to inferences about a period of glaciation in at least the north of Ireland preceding the lower boundary of MIS 2, but post-dating the last interglacial (MIS 5e) (Fermanagh Stadial in Ireland: McCabe, 1969; Colhoun *et al.*, 1972; Mitchell, 1977, 1981; Bowen *et al.*, 1986b; McCabe *et al.*, 1987b; McCabe, 1991; McCabe, 1999). This period has been named as the Fermanagh Stadial after a number of disparate sites in the county including Derryvree, where *in situ* freshwater organic silts located between two major (regional) till sheets have been dated to 30.5 ± 1.1 ka BP(Colhoun *et al.*, 1972). At Hollymount, also in County Fermanagh, freshwater organic mud overlying a glacial diamict (till) and underlying a streamlined till provided an infinite ^{14}C age of >41.5 ka BP (McCabe *et al.*, 1978). Freshwater organic materials at Aghnadarragh, on the shores of Lough Neagh, County Antrim, are also located between two regionally extensive till sheets, the lower of which contains a distinctive erratic suite from the Tyrone Igneous Complex in the Sperrin Mountains to the west (McCabe, 1987). This ice transport vector (E–W) is not traditionally associated with Late Midlandian (MIS2) ice flows in this region. Wood fragments in the gravels yielded infinite traditional ^{14}C dates of >48.1 ka BP and >47.35 ka BP (McCabe *et al.*, 1987b) and >46.2 ka BP (McCabe,

1999). At Benburb, County Armagh, *in situ* compressed peat and lake mud dated to >46 ka BP and >46.45 ka BP is overlain by drumlinised till (Boulter and Mitchell, 1977) or possibly remobilised till (Dardis, 1980).

Farther north, intact shells of *Turritella communis* taken from gravels at Bovevagh, near Limavady, yielded infinite ^{14}C dates of >45 ka BP (McCabe, 1999). D-aIle/L-Ile ratios (amino-acid racemisation) from marine macrofossils found at the same location are inconclusive, but may indicate an Early Devensian (MIS 4) age (McCabe, 1999). Colhoun (1971) argued that his shelly Bovevagh till from the same region, which contains reworked erratics from a southerly source related to a still earlier glaciation (his Early Sperrin glaciation), was deposited during the onshore movement of ice from the north and north-east which incorporated shells of *Turritella communis*, bivalve shells, crustacea, and foraminiferans as it crossed the North Channel from Scotland. McCabe (1999) argued that the geometry of the gravel beds suggests that the sediments at Bovevagh are *in situ* glaciomarine deposits. At Sistrakeel, just north of Bovevagh on the southern margins of the Foyle Basin, a marine mud containing well-preserved *Arctica islandica* shells and a microfauna dominated by *Elphidium exclavatum* yielded infinite ^{14}C dates of >46,785 BP, >55,100 BP and >55,500 BP (McCabe, 1999). D-aIle/L-Ile ratios on macrofossil fragments indicate they may be Middle Midlandian (MIS 3) in age (McCabe, 1999). The mud stratigraphically overlies a basal diamict containing erratics (chalk, basalt) associated with easterly and north-easterly sources.

When placed in the context of other 'old' dates from the North of Ireland and neighbouring regions, with infinite ^{14}C dates of at least 30 ka BP, glacial diamicts underlying shelly sediments in this area may also possibly be assigned an MIS 4/3 (Early/Middle Midlandian) age (McCabe, 1999). Deposits are also associated with ice expansion of this age in Scotland (Sutherland, 1981) and Scandinavia (Mangerud, 1981). Although the evidence indicates an Early Midlandian (MIS 4) glaciation, no deposits unequivocally attributable to the last interglacial (MIS Substage 5e) have been documented from stratigraphically below the 'older' deposits (cf. Worsley, 1991).

The Aghnadarragh Interstadial (MIS 3)

Organic sediments from between two till sheets at Aghnadarragh, County Antrim gave ages beyond the radiocarbon method (an infinite ^{14}C date of >48,180 BP being the oldest) and have been described in detail by McCabe *et al.* (1987b). The lowest biogenic material in the sequence is a shallow water sequence comprising woody detritus peat containing plant and insect fossils beneath which is a deglacial succession of bottomsets, gravelly muds, and diamicts. This in turn overlies a glacial diamict that represents the Fermanagh Stadial (Table 15.2). The Aghnadarragh Interstadial and the other deposits predating the Glenavy Stadial (MIS 2) probably belong in the complex of MIS 3 (between 59 and 28 Cal ka BP).

The predominant plant fossils include the pollen of *Pinus, Betula*, 'Coryloid' (*Corylus/ Myrica*), *Picea,* and a number of herbs and aquatic plants. The macrofossil plant remains include wood, cones and leaves of *Pinus* and *Picea*, fruits of *Betula* (including *B.* cf. *nana, B.* cf. *pubescens* and *B.*cf. *pendula*), nuts of *Corylus avellana,* seeds of *Taxus baccata* (the latter were possibly derived), and megaspores of *Selaginella selaginoides.*

Abundant insect fossils (Coleoptera) were also recovered from the detrital peat. Although the assemblage was allocthonous, it represents species that might be expected to have been living in available habitats in the locality. The fauna suggest that both acid swampy conditions (with little open water) and drier, partially bare, habitats were present. Two weevils in the fauna, *Rhyncolus strangulatus* and *R.elongatus*, feed on dead or dying *Pinus* or *Abies* trunks and branches, showing that conifers were present at the site.

The insect assemblage includes species that only reach as far north as the southern half of Fennoscandia at the present day and there were no obligate high northern species. The authors concluded that the thermal climate of the Aghnadarragh Interstadial was as follows (and interestingly was just within the present day range):

Mean July Temperature	+15°C to + 18°C
Mean January Temperature	-11°C to + 4°C

Interstadial conditions appear to have been clement enough over a sufficient period to allow the immigration of plants to occur after the preceding stadial. The resultant *Betula–Pinus–Picea* woodland with its adjacent areas of swamp, dry bare ground, and local pools is probably similar in environment to the cool temperate conditions prevailing in Fennoscandian woodlands with similar tree types today.

The wider stratigraphical position of the Aghnadarragh Interstadial is partly based on the correlation of plant and insect assemblages with deposits at Chelford, Cheshire (Simpson and West, 1958; Coope, 1959), where similar woodland species are recorded. One difference between the two sites is the absence of *Ulmus* and *Carpinus* in the Aghnadarragh Interstadial. This was used by McCabe *et al.* (1987b) to suggest that the organic sedimentation at the site did not follow on immediately from an interglacial, and that the *Picea* present was therefore not of interglacial origin.

The absence of *Ulmus* and *Carpinus* and the presence of *Corylus* and *Taxus* (macrofossils and pollen) at Aghnadarragh may raise some doubts concerning the correlation of the interstadial with that at Chelford. A further complication arises with the realisation that there may be many more interstadial periods within the last cold stage than was formerly realised (Fig. 15.16), making such correlations rather tenuous. However, the stratigraphical position of the Aghnadarragh Interstadial seems to place it firmly within the Midlandian.

The Hollymount and Derryvree cold phases (MIS 3)

Three sites in the north of Ireland preserve organic sediments between two (Midlandian) till sheets that contain fossil assemblages with cold climate affinities. These biogenic sediments represent two broad time periods, which together probably cover part of a long, relatively cold period during the Early and Middle Midlandian (MIS 3). The contained fossil assemblages do not suggest very significant climatic deterioration relative to Holocene conditions, and for this reason the term 'cold phase' has been used here in preference to 'interstadial' (Bowen *et al.*, 1986b).

The first of these cold phases is named after the organic muds found at Hollymount, County Fermanagh by McCabe *et al.* (1978), and dated to >41.5 ka BP. Deposits of a similar type have been found at Aghnadarragh (unit 8) and dated to >46.62 ka BP (McCabe *et al.*, 1987b) and to 34.46 ka BP at Greenagho, County Fermanagh (Dardis *et al.*, 1985). Palaeobotanical evidence from Hollymount and Aghnadarragh shows that the landscape during this period was rather barren, with a northern open groundflora dominated by Gramineae and Cyperaceae, along with *Betula* (cf. *B. nana*), Ericaceae (including *Calluna vulgaris*), *Empetrum* sp., *Salix* (including *S. herbacea*), *Artemisia, Thalictrum*, and *Selaginella selaginoides*. At Aghnadarragh it was apparent that some of the pollen was reworked.

The insect assemblages from unit 8 at Aghnadarragh also indicate open ground with sparse vegetation and little open water. There are no southern species and the present day southern limit of some species (e.g. *Diacheila arctica*) is in Arctic Fennoscandia. The authors estimate the thermal climate as follows:

Mean July Temperature	+11°C to + 13°C
Mean January Temperature	-18°C to + -7°C

The later part of this long Midlandian cold episode has been called the Derryvree cold phase, after a site in County Fermanagh (dated to 30.5 ka BP, Colhoun *et al.*, 1972). The fossil content of organic sediments sandwiched between two tills at Derryvree suggested an open tundra landscape. The flora (predominantly Gramineae, Cyperaceae, *Juniperus* and herbs) is diagnostic of an open, tree-less, muskeg environment, and it is associated with a moss flora. The insect assemblages also suggest a muskeg environment with open pools, rich in plant matter and surrounded by mosses. Colhoun *et al.* (1972) concluded that the climate was harsh with cold winters, and the date places the period as later than the Upton Warren Interstadial (MIS 3) of Britain.

A relatively impoverished fauna from Castlepook Cave, County Cork (Mitchell, 1976, 1981; Stuart and Van Wijngaarden-Bakker, 1985; Stuart, 1995) produced two dates, 35 ka BP and 34.3 ka BP, from bones of mammoth (*Mammuthus primigenius* Blumenbach) and spotted hyaena (*Crocuta crocuta* (Erxleben)) respectively. These appear to date the fauna to the Derryvree cold phase. Other mammals found in the cave included wolf (*Canis lupus* L.), brown bear (*Ursus arctos* L.), stoat (*Mustela erminea* L.), red fox or arctic fox (*Vulpes* or *Alopex*), arctic lemming (*Dicrostonyx torquatus* Pallas), arctic or mountain hare (*Lepus timidus* L.) and reindeer (*Rangifer tarandus* L.)(Fig. 15.20).

The Hollymount and Derryvree cold phases cover a long time period that includes the range of the British Upton Warren Interstadial with which they can, in part, be correlated. The Upton Warren Interstadial Complex (Coope, 1975, 1977) started with a warm interval between 43 ka–42 ka, following which the climate became much colder (after ~40 ka), possibly becoming progressively more severe up to the onset of stadial conditions (the Dimlington Stadial in Britain, <26 ka: Rose, 1985; Bowen *et al.*, 1986b). There is no known evidence in Ireland of climatic amelioration equivalent to the onset of the Upton Warren Interstadial. However, the ensuing cold, barren tundra landscape has been likened to conditions in Ireland during the Hollymount and Derryvree cold phases (Lowe and Walker, 1997).

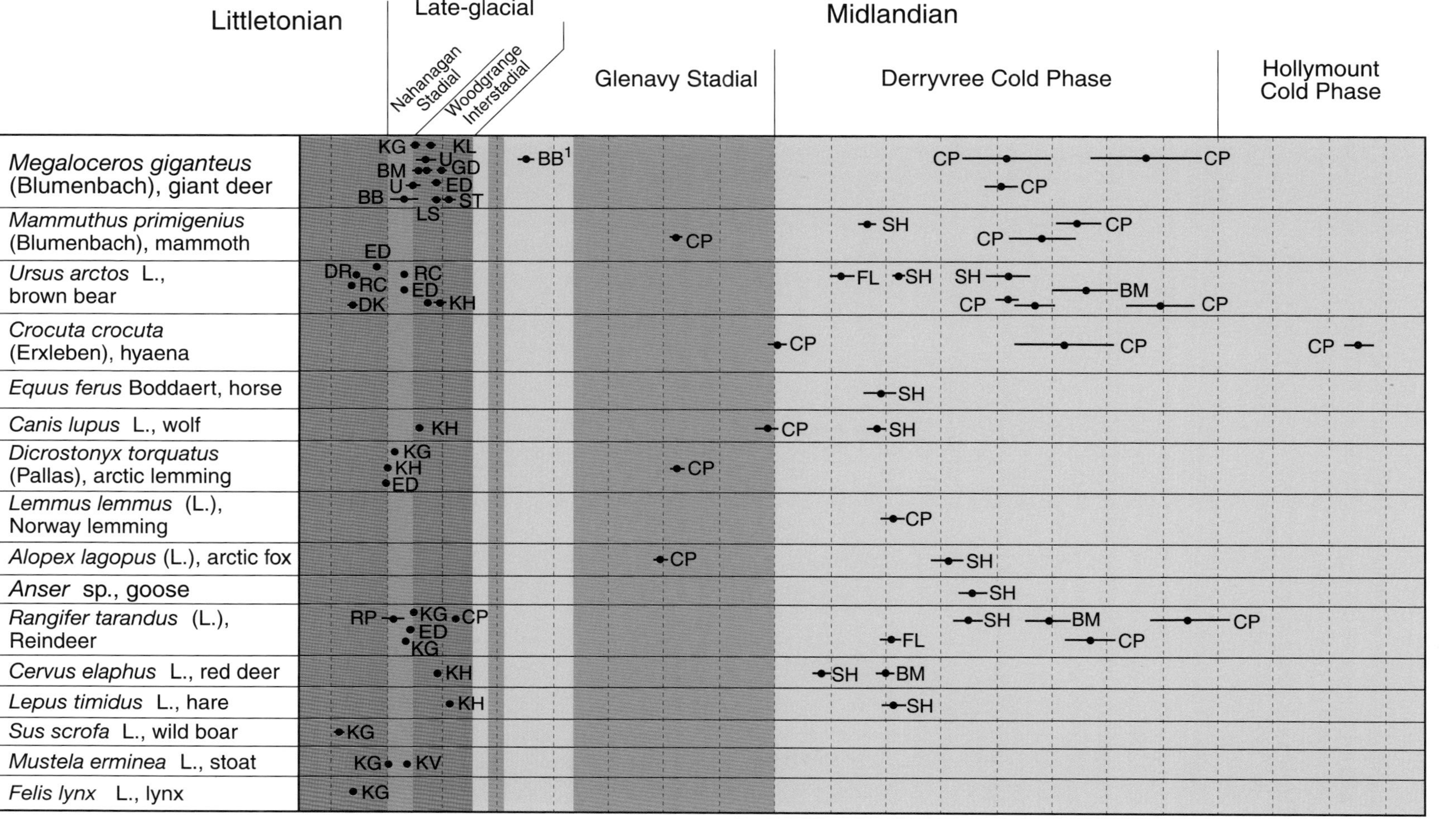

Figure 15.20 Radiocarbon dated mammal remains (showing the one standard deviation estimate of error of the dates) from the Midlandian and early Holocene. Data are from Stuart and Van Wijngaarden-Bakker (1985) and Woodman *et al.* (1997) who quote errors and site details. Key to localities: BB – Ballybetagh Bog, Co. Wicklow (BB[1] – Ballybetagh date possibly too old due to contamination with paraffin preservative), BM – Ballynamintra Cave , Co. Waterford, CP – Castlepook Cave, Co. Cork, DK – Derrykeel Bog, Co. Offaly, DR – Donore Bog, Co. Laois, ED – Edenvale Cave Complex, Co. Clare, FL – Foley Cave, Co. Cork, KG – Kilgreany Cave, Co. Waterford, KH – Kesh Corran Cave Complex) Co. Sligo, KL – Killuragh Cave, Co. Limerick, LS – Liskelly Stream, Co. Cork, RC – Red Cellar Cave, Lough Gur, Co. Limerick, RD – Roddans Port, Co. Down, SH – Shandon Cave, Co. Waterford, ST – Shortalstown, Co. Wexford (date is from associated plant remains) and U – unprovenanced.

The Glenavy Stadial (MIS 2)

The last widespread glaciation of Ireland occurred during the period of generally stable global sea levels and stable polar climate during MIS Stage 2 known as the Last Glacial Maximum (LGM) (23–19 ka Cal. BP; Yokoyama, *et al.*, 2000; Mix *et al.*, 2001). The glaciation in Ireland that encompasses the LGM has been referred to as the Glenavy Stadial, following the proposed Aghnadarragh type-site near Glenavy in County Antrim (McCabe, 1987). Ice-flow indicators (e.g. striae, overlapping belts of streamlined landforms) also indicate the eventual formation and dominance of lowland ice domes that formed the principal ice dispersal centres in the north and north-west of the island. Ice caps also existed on some upland areas e.g. Cork/Kerry (Wright, 1927; Warren, 1977; Herries Davies and Stephens, 1978), Wicklow (Farrington, 1957) and Antrim (Stephens *et al.*, 1975). Glacial ice also extended out of the many steep-sided, bowl-shaped corries that occur on the slopes of Irish upland areas (e.g. Lough Doon and Lough Adoon, Owenmore Valley, Dingle Peninsula, County Kerry; Coumshingaun, Comeragh Mountains; Lough Nahanagan, Camaderry Mountain, County Wicklow (Figure 15.18).

Recent work has indicated ice sheet extents beyond the southern coastline and stadial conditions of much longer duration than previously thought (Bowen *et al.*, 2002). Although MIS Stage 2 expansion of the last BIIS in different sectors was probably slightly diachronous, the main phase of glacial build-up, ice dome migration and ice frontal advances during the Glenavy Stadial have traditionally (e.g. Mitchell *et al.*, 1973) been bracketed between 28 and 22 ka Cal BP and limited to onshore areas north of the South of Ireland End Moraine (Charlesworth, 1928). Erosional features south of this line (McCabe, 1998), south coastal morpho-sedimentary evidence (McCabe and O'Cofaigh, 1996) and the application of cosmogenic dating techniques (Bowen *et al.*, 2002) have all indicated ice extension from Irish lowland sources to just across the southern Irish coastline into the Celtic Sea at this time.

Dating of deglacial sediments in Mayo to ~40–21 ^{14}C ka BP (McCabe *et al.*, 2007a) shows ice sheet advance beyond the western coastline before the traditional timing of the LGM ice sheet maximum extents. Along with cosmogenic exposure age dates from locations around the margins of Ireland (Bowen *et al.*, 2002), and the extent and pattern of Late Midlandian isostatic crustal depression (e.g. McCabe *et al.*, 2007a) the geological data indicate the persistence of ice sheets on Ireland during MIS 2 and into MIS 3. These data seemingly contradict the occurrence of organic materials defining the Hollymount and Derryvree Cold phases late in MIS 3 (above) and evidence from similar paraglacial locations in Scotland (Brown *et al.*, 2007) of open tundra conditions during MIS 3. However, it is conceivable that the sites in Ireland could at least have been outside the limits of an ice sheet localised on the western seaboard uplands at this time, and giving rise to the 'cold phase' (not 'interstadial') conditions in the lowlands farther north. As in the late Midlandian, it is probable that ice sheet sectors responded rapidly to millennial-scale climatic variability in the amphi-North Atlantic. The persistence of the ice masses may also help explain the erroneous extent and patterns of isostatic depression around the Irish coastline derived from deterministic ice sheet models parameterised, calibrated and tested using Holocene data (e.g. Roberts *et al.*, 2006, Boulton and Hagdorn, 2006).

The Last Termination

The Last Termination describes the period from the onset of BIIS deglaciation off the continental shelf through to the onset of the Holocene. The term is used to indicate that the period was one marked not only by general deglaciation but also by distinct events recording millennial-scale oscillations of the BIIS including significant periods of ice mass rejuvenation and ice-marginal readvance during overall deglaciation of the ISB. This period is also associated with phases of streamlining and drumlin formation (McCabe *et al.*, 1986) and is discussed in detail by McCabe, 1993, 1996; McCabe and Clark, 1998; McCabe *et al.*, 1998, 1999; McCabe, 1995; McCabe *et al.*, 2007). The evidence indicates at least two significant temporary reversals in the overall pattern of deglaciation, the Clogher Head Stadial (CHS) and Killard Point Stadial (KPS). Events during the last Termination, as currently understood from age-constrained geological evidence, are summarised in Table 15.5. Key locations in the Dundalk Bay area are shown on Figure 15.19.

The Midlandian Late-glacial (13 ^{14}C ka–10 ^{14}C ka BP)

The last glacial–interglacial transition is an event of considerable complexity that can be recognised on a global scale (e.g. see NASP Members: Executive Group 1994; Walker *et al.*, 1994; Troelstra *et al.*, 1995), and this cold to warm climate transition can be neatly summarised by the comparison of the GRIP ice core and the record of the planktonic foraminiferan *Neogloboquadrina pachyderma* shown on Figure 15.21.

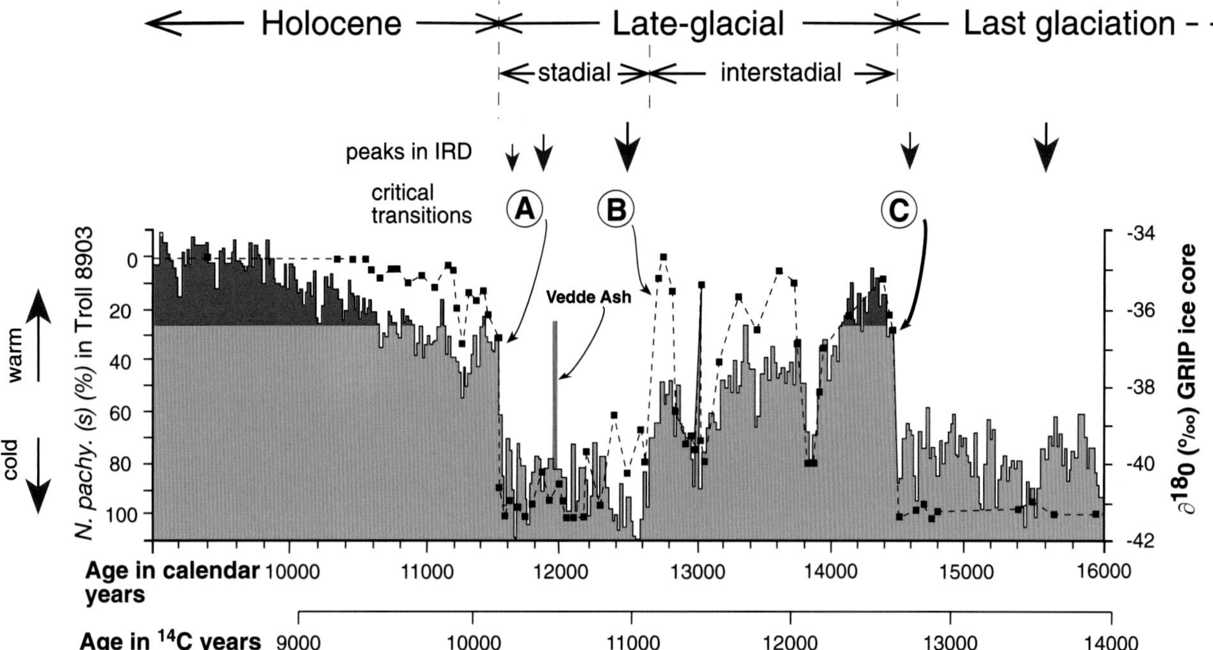

Figure 15.21 The stable isotope record (∂[18]O) from the GRIP ice core (histogram) compared to the record of *N.pachyderma* a planktonic foraminiferan whose presence indicates cold sea temperatures) from ocean sediments (dotted line). High concentrations of IRD from the Troll 8903 core are marked with arrows. *After* Haflidason *et al.* (1995). The transition times for critical lengths of the core were calculated from the sediment accumulation rates by the authors and these gave the following results: Transition A: 9 years; Transition B: 25 years; and Transition C: 7 years. Such rapid transitions have been corroborated from the recent NGRIP ice core data.

In Ireland the period from 13 ^{14}Cka to 10 ^{14}Cka BP is known as the Midlandian Late-glacial and is recorded from many sites, particularly from lake sediments, where extensive palaeoenvironmental information has been obtained (Watts, 1977, 1985; Andrieu *et al.*, 1993; Walker *et al.*, 1994; O'Connell *et al.*, 1999). Indeed, the first study of this period in the British Isles was carried out at Ballybetagh by Jessen and Farrington (1938). The Irish Late-glacial can be subdivided into the Woodgrange Interstadial (*c.*13 ^{14}Cka to 10.9 ^{14}Cka BP) and the Nahanagan Stadial (*c.*10.9 ^{14}Cka to 10 ^{14}Cka BP). Further palaeo-environmental information and more accurate chronologies are likely to be forthcoming from work on speleothems providing uranium–thorium dates and stable isotope proxies for climate (McDermott *et al.*, 2001; McDermott, 2004) as well as from the identification of tephra horizons within sediment sequences (Chambers *et al.*, 2004) as those records are successfully extended into the Late-glacial.

A summary of the palaeoenvironment of the Irish Late-glacial is presented on Figure 15.22. It is important to realise that these subdivisions are simplifications of a complex and global climatic event.

The Woodgrange Interstadial

At 13 ^{14}Cka BP the North Atlantic polar front had retreated from its glacial maximum position. The retreat of the polar front was probably the result of massive meltwater discharges into the North Atlantic and the ensuing collapse of marine-based ice masses (Berger and Jansen, 1995; McCabe and Clark, 1998). The climate of Ireland rapidly ameliorated and we find the first evidence of plant colonisation following the Glenavy Stadial. Initially grasses and herbs flourished, but these were followed by *Rumex* and *Salix herbacea* as biological productivity on the fresh soils increased. About 12.4 ^{14}Cka BP a first peak in *Juniperus* occurs in Irish pollen diagrams associated with organic lacustrine sediments, the period of *Juniperus* domination ends abruptly, sediments become less organic, and an initial phase of soil erosion is suggested by inwashed inorganics at many sites (12 ^{14}Cka to 11.8 ^{14}Cka BP). The vegetation was dominated by grassland and an open herbaceous flora (Figure 15.22).

This steppe-like environment of the Woodgrange Interstadial was home to herds of giant deer (*Megaloceros giganteus*), the fossils of which have been found widely dispersed in Ireland. The male of the species, with its

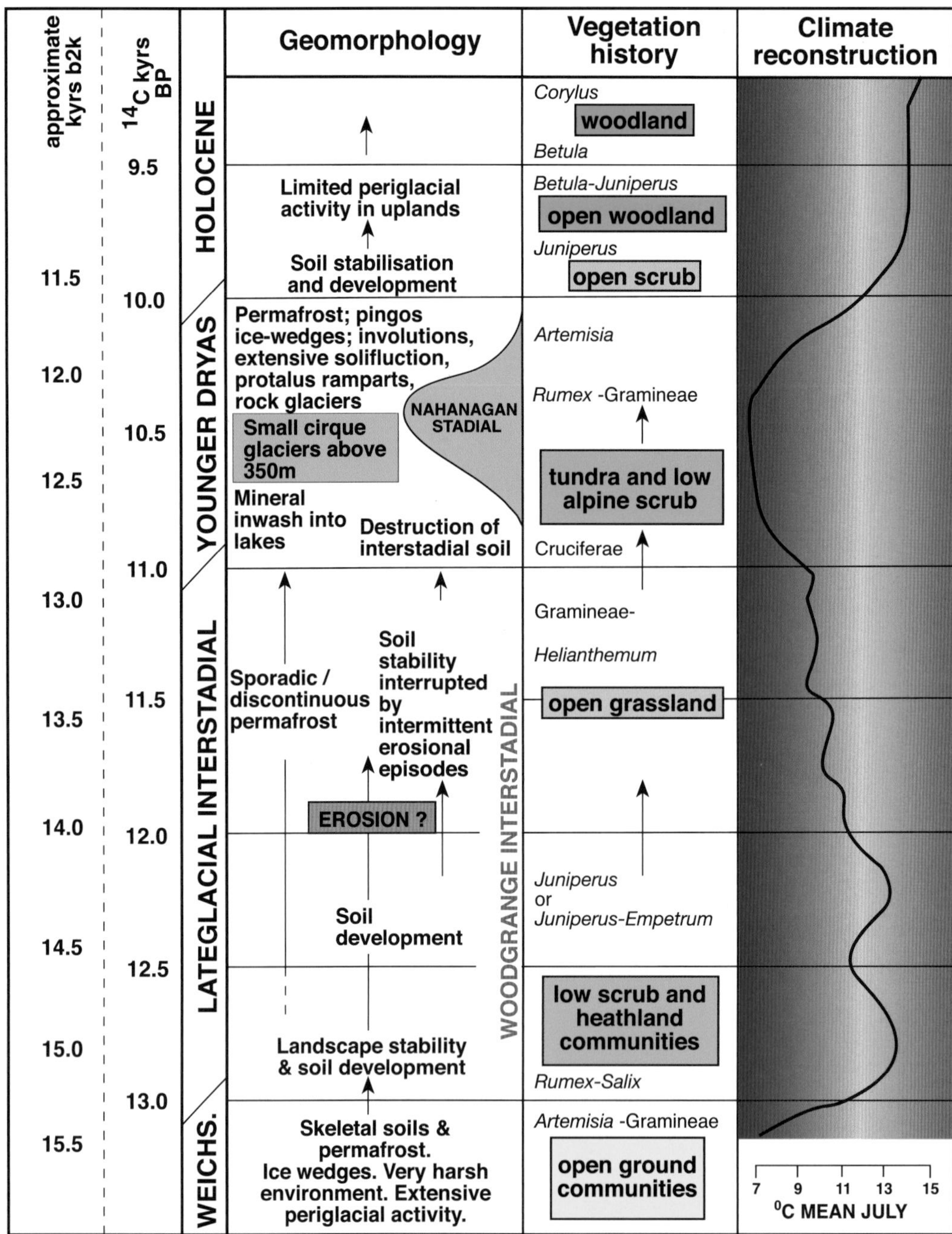

Figure 15.22 Late Midlandian Late-glacial geomorphology, vegetation history and climatic reconstructions (*after* Walker *et al.*, 1994 with modification by the authors).

magnificent antlers, wintered in sheltered sites like Ballybetagh in County Dublin (Jessen and Farrington, 1938), where winterkill led to many younger animals dying (Barnosky, 1985). The giant deer could not survive the ensuing cold and it became extinct at the end of the Woodgrange Interstadial. The ranges of other mammals during the Late-glacial are shown on Figure 15.20.

The Nahanagan Stadial

At 11 ^{14}C ka BP the polar front migrated southwards across the Atlantic and had a dramatic effect on the climate of north-western Europe. The period is known regionally in Europe as the Younger Dryas (YD) and in Britain, where extensive upland glaciation occurred, the cold phase is known as the Loch Lomond Stadial (Gray and Coxon, 1991).

In Ireland the YD has been named informally by Mitchell (1976) as the Nahanagan Stadial. This term stems from work by Colhoun and Synge (1980) at the corrie lake of Lough Nahanagan (Fig. 15.18A) where small moraines formed after the Woodgrange Interstadial. This site does not provide particularly good stratigraphical control; for example it lacks radiocarbon dates bracketing the glacial activity that produced the inner corrie moraines. However, it remains the only site in Ireland that unequivocally shows renewed glaciation in the Late-glacial Stadial through lithostratigraphical analysis, biostratigraphical information, and radiocarbon dating.

The Nahanagan Stadial in the rest of Ireland is well represented lithostratigraphically and biostratigraphically (predominantly from lacustrine sequences; Gray and Coxon 1991) as an inorganic (usually clayey and often containing coarse sand or pebbles) sediment sequence and an *Artemisia* pollen assemblage respectively (Watts, 1977, 1985). Numerous radiocarbon dates are available from this period (e.g. Craig, 1978; Cwynar and Watts, 1989; Browne and Coxon, 1991; Andrieu *et al.*, 1993; Walker *et al.*, 1994; O'Connell *et al.*, 1999) but no published work, other than that detailed above, proves actual glacial activity during the Nahanagan Stadial.

The short cold interval of the Nahanagan Stadial is not only recognised in lake sediments but is also a period during which widespread cold-climate geomorphological processes were active. The environment during this stadial was one of cold tundra with unstable, geliflucting soils and with small glaciers occupying the north-eastern corners of upland cirques. Figure 15.23 is a compilation of periglacial features that have been identified in Ireland and Figure 15.24 illustrates some of these features. Although very few of these landforms are dated, many are considered to have formed in the Nahanagan Stadial (Gray and Coxon, 1991) on the grounds that they are 'fresh' landforms (i.e. they are within areas that underwent severe glaciation during the Glenavy Stadial and hence must postdate it) or that they contain (or are overlain by) material dated to the end of the stadial or the Early Holocene. The problem of dating is apparent when one considers that some protalus ramparts have an Irish name (*Clocha snachta* or snow stones). Wilson (1988) suggested that they could be quite recent, i.e. possibly of a Little Ice Age date (Seventeenth to mid Nineteenth Century).

Examples of the more important features that may be of Nahanagan Stadial age are outlined below:

Feature	Dating evidence	Author(s)
Pingo remnants	Post Glenavy Stadial/ fresh. Features contain 14C dated sediments	Mitchell, 1971, 1973, 1977; Coxon, 1986; Coxon and O'Callaghan, 1987.
Rock glaciers	Post Glenavy Stadial/ fresh Schmidt hammer measurements (see Fig. 23)	Wilson, 1990 a,,b, 1993; Anderson *et al.*, 1998; Harrison and Mighall, 2002.
Protalus ramparts	Post Glenavy Stadial/ fresh Schmidt hammer measurements	Colhoun, 1981; Coxon, 1985; Wilson, 1990 a, b, 1993; Anderson *et al.*, 1998.
Patterned ground Stone stripes	None. Could be Glenavy Stadial and formed on nunataks. May be Nahanagan Stadial or older.	*see* Lewis, 1985; Quinn, 1987 Coxon, 1988; Wilson, 1995.
Landslides	Post Glenavy Stadial/ fresh	Colhoun, 1971; Remmele, 1984; Coxon, 1992.
Solifluction lobes	Post Glenavy Stadial. Ubiquitous and of various ages including Late Holocene	Lewis, 1985; Author's observations; Anderson *et al.*, 1998.
Involutions, Solifluction deposits	Widespread and of various ages but many near surface features are probably Nahanagan Stadial age	Lewis, 1985; Author's observations.

The periglacial landforms for which we have some dating control give both an insight into the geomorphological processes operating in Ireland during the Nahanagan Stadial and the prevailing temperatures at the time. Wilson (1990a) summarises the palaeoclimatic inferences of rock glaciers in detail suggesting a depression of mean annual temperature of c.8.5°C relative to the present. This compares well with estimates made by Colhoun and Synge (1980) of a temperature depression of c.7.2°C. When considering other periglacial evidence as well (pingos and ice wedge casts), Wilson (1990a) suggested that the mean annual temperature of the Nahanagan Stadial was probably lower, with values reaching −2°C to −5°C. Of course the whole Irish YD may not have experienced these low temperatures, and the short-lived stadial may have been a complex event. We shall not know for certain until further geomorphological work is carried out and more dates are available.

One other interesting aspect of the Late-glacial involves the immigration of plants and animals back into Ireland during a rapid rise in sea-level. As Ireland recovered isostatically from the weight of Midlandian ice, relative sea-level fell, but rapid eustatic rise in sea-level is supposed to have isolated Ireland, which may have been linked to Britain by a 'tenuous landbridge'

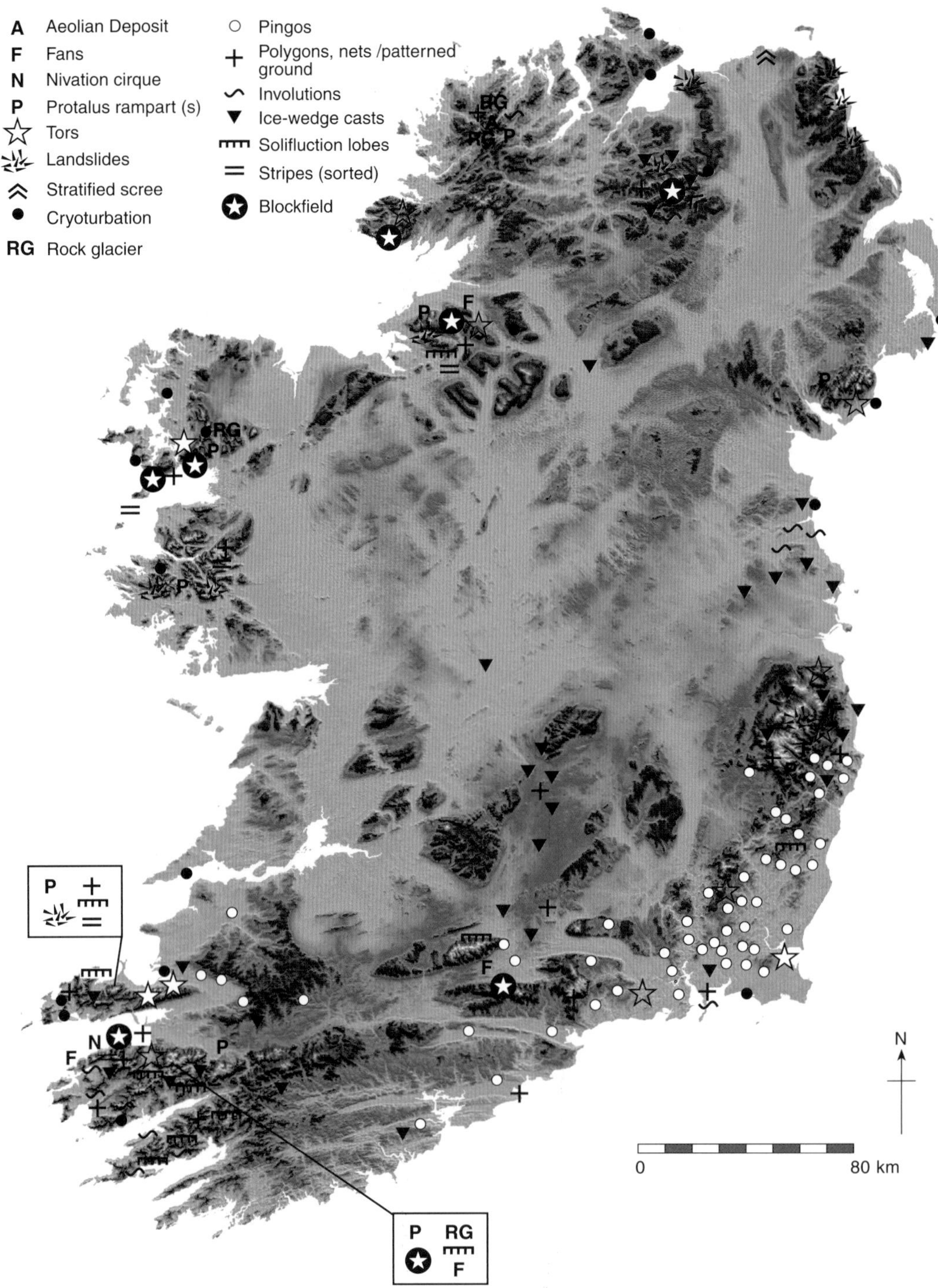

Figure 15.23 Periglacial features in Ireland (*after* Lewis, 1985 with additional information from Wilson, 1990a, 1990b, 1993 and 1995 and personal observations by the authors).

Opposite Figure 15.24 Late Pleistocene. **A** – A pingo rampart containing a flooded depression, Camaross, County Wexford (see Mitchell, 1971). **B** – Core from a pingo scar at Meenskeha, County Cork (Coxon, 1986). B1 – slumped rampart material (latter part of Younger Dryas/Nahanagan Stadial), B2 – Late-glacial/Holocene transition (10,000 ^{14}C a BP, *c*.11,700 yb2k), B3 – Early Holocene laminated muds and marl, B4 – transition into predominantly organic sedimentation ('bulk' date of 9,740 ± 150 ^{14}C BP from between the white lines), **C** – Rock glacier on the northern flank of Errigall, County Donegal (see Wilson, 1990). **D** – Protalus rampart, Gleniff, County Sligo (see Coxon, 1985), **E** – Periglacial patterned ground (sorted nets) on the summit of Truskmore, County Sligo (see Coxon, 1988).

A
B4
B3
B2
B1
2 80
2 90
3 m
B
C
D
E

for a very short period only (between 18 and 14 ^{14}Cka BP according to Lambeck, 1996; or possibly as late as 12.5 ^{14}Cka BP, Brooks *et al.*, 2008) which was a time of cold climate. It is unlikely that many plants and animals survived the maximum of the Midlandian in Ireland or that any other landbridge was possible after 12.5 ^{14}Cka BP. Such a connection to Britain was thus only present in cold-climate conditions, and plants and animals must have migrated into Ireland after it was cut off by the sea in the Midlandian Late-glacial (Woodgrange Interstadial) and in the early Holocene. However, it appears that in the Holocene even terrestrial molluscs managed to cross this barrier with relative ease (Preece *et al.*, 1986).

By 10 ^{14}Cka BP the Polar front had regained its position close to the coast of Greenland and much evidence suggests that the climate around the North Atlantic warmed quickly and dramatically (Troelstra *et al.*, 1995). The climate of Ireland ameliorated rapidly at the onset of the Holocene (some 11,700 years ago) and incredibly the YD/Holocene transition took less than a decade (Rasmussen *et al.*, 2006).

16

The Holocene

Fraser J. G. Mitchell

Introduction

The Holocene extends from the present day back to approximately 11,700 calendar years (Walker et al., 2008, 2009). In many respects it bears close similarity to previous Pleistocene interglacials but is distinguished from them by the blossoming of human culture (Roberts, 1998), the influence of which, combined with climate, have been the significant driving forces in determining the pace and direction of environmental change during the Holocene. These same drivers, and their interaction, are the focus of our current concern over future environmental change, which ensures the immediate relevance of Holocene research. The high proportion of Ireland covered by lakes and peat bogs, coupled with their wide distribution, provide an abundance of well-preserved Holocene sediments ideally suited to meeting this challenge.

Chronology

The Holocene is unrivalled by other epochs in both the range and precision of available dating techniques and this has led to the abandonment of lithostratographical divisions such as the Blytt-Sernander scheme (Birks and Birks, 1980) in favour of absolute dating. The base of the Holocene is clearly demarked in the NGRIP Greenland ice core and has been dated by annual counting to 11,700 calendar years before AD 2000 with a maximum counting error of 99 years (Walker et al., 2008, 2009).

The wide application of radiocarbon dating has made this the principal technique. The isolation of several Icelandic tephras in Irish Holocene lake and peat deposits has also made significant contributions to site chronologies (Hall and Pilcher, 2002). There is, however, some scope for confusion in interpreting radiocarbon chronologies. Publications prior to 2000 almost universally quoted chronologies based on the actual radiocarbon dates. These chronologies adopted an implicit, but incorrect, assumption that the production rate of radiocarbon at its source in the upper atmosphere had remained constant over the duration of the Holocene. Tree ring chronologies, such as the Irish Oak record at Queen's University Belfast, provide a means of calibrating radiocarbon dates to calendar years (Reimer *et al.*, 2004). This development has led to the earlier publication of chronologies based on radiocarbon years and the more recent publication of chronologies of calendar years based on calibrated radiocarbon dates. In this chapter it will be necessary to quote both calibrated and uncalibrated radiocarbon dates. The nomenclature adopted uses the suffix 'BP' to imply uncalibrated radiocarbon dates before present, where present is defined as AD 1950. Calibrated dates are denoted by the suffix 'calbp'. The prefix 'AD' is used to refer to recent calendar ages while the use of the suffix 'BC' is avoided. The degree of deviation between calibrated and uncalibrated radiocarbon dates is minimal for the last 2,000 years but increases significantly before then (see Table 16.1 for details).

Climate

The Holocene climate was relatively stable compared to the dramatic millennial and century scale changes observed during the last glacial cold stage (Chapter 15). The Holocene opens with a dramatic rise in temperature, which in some Greenland ice cores is recorded to have occurred within a decade. In the Irish context, the magnitude of this change was of a tundra climate developing into a temperate climate, not unlike that of the present day, within a decade (Haslett *et al.*, 2006). The most significant deviation from the relatively stable climate record in the Greenland ice cores is the so-called 8.2 ka event (8,200 calbp); high resolution data show the event centred on a 69 year period of significantly reduced temperatures (Thomas *et al.*, 2007). This event has been associated with a collapse in the North American Laurentide ice sheet which resulted in the discharge of meltwater into the North Atlantic and disruption of thermohaline circulation (Clark *et al.*, 2001). It would be expected to be

Table 16.1 Calibration of Holocene radiocarbon dates (*after* Roberts, 1998).

14C Age (BP)	Calendar Age (calbp)	Calendar Age (calbp)	14C Age (BP)
500	520	500	430
1,000	930	1,000	1,120
1,500	1,350	1,500	1,600
2,000	1,940	2,000	2,060
2,500	2,580	2,500	2,450
3,000	3,190	3,000	2,900
3,500	3,790	3,500	3,320
4,000	4,440	4,000	3,670
4,500	5,120	4,500	4,010
5,000	5,730	5,000	4,430
5,500	6,290	5,500	4,790
6,000	6,820	6,000	5,260
6,500	7,380	6,500	5,720
7,000	7,790	7,000	6,130
7,500	8,220	7,500	6,700
8,000	8,780	8,000	7,260
8,500	9,470	8,500	7,760
9,000	9,980	9,000	8,110
9,500	10,510	9,500	8,560
10,000	11,500	10,000	9,040
		10,500	9,420
		11,000	9,800
		11,500	10,000

apparent in the Irish stratigraphy and has been recorded elsewhere in Europe (see references in Head *et al.*, 2007). The apparent absence of this event from the Irish pollen record may relate to its being missed through poor resolution. Head *et al.*, (2007) have found evidence of a wetter and possibly cooler event on the coast of Achill Island around this time, but their chronology is ambiguous and so the association is inconclusive.

The oxygen isotope record from a speleothem in Crag Cave, County Kerry shows remarkable correlation with the GISP2 Greenland ice core record (McDermott *et al.*, 2001). The apparent detection of the 8.2 ka event in this record was subsequently found to be an artifact caused by micro-fracturing of the speleothem calcite during laser analysis (McDermott *et al.*, 2005). Reanalysis using conventional techniques confirmed the absence of a clear 8.2 ka event signal and this was considered to be due to calcite with higher $\delta^{18}O$ being deposited during cooler conditions and therefore compensating for the anticipated lower $\delta^{18}O$ values associated with this event (McDermott *et al.*, 2005). A general circulation model simulation of the 8.2 ka event that explicitly tracked the oxygen isotopic composition of precipitation during and after the putative meltwater pulse also predicted only minor depletions (< 1‰) in the $\delta^{18}O$ of precipitation in the region surrounding the 'British Isles' (LeGrande *et al.*, 2006).

The speleothem calcite $\delta^{18}O$ record can be influenced by several factors including temperature, and so does not necessarily represent a simple climate proxy. Despite this, the close correlation of the rest of the Crag Cave record with the GISP2 $\delta^{18}O$ record is striking. Greatest variability in the record occurs from about 9,000 to 6,000 calbp and the overall trend is for more negative values after 6,000 calbp (Fig. 16.1). These data support the contention that early Holocene climate in NW Europe was more seasonal with warmer, drier summers than at present due to insolation changes (Barber *et al.*, 2004; Wright *et al.*, 1993). This is also supported by the increased charcoal abundance in early Holocene sediments in Kerry, suggesting greater incidence of early Holocene fire (Mitchell and Cooney, 2004). In further support of this, the only finds of *Najas marina* in Ireland are pollen from early Holocene sediments from the Burren (Watts, 1984). This annual aquatic plant is confined to the Baltic today, where it enjoys warmer summer temperatures than are currently available in Ireland.

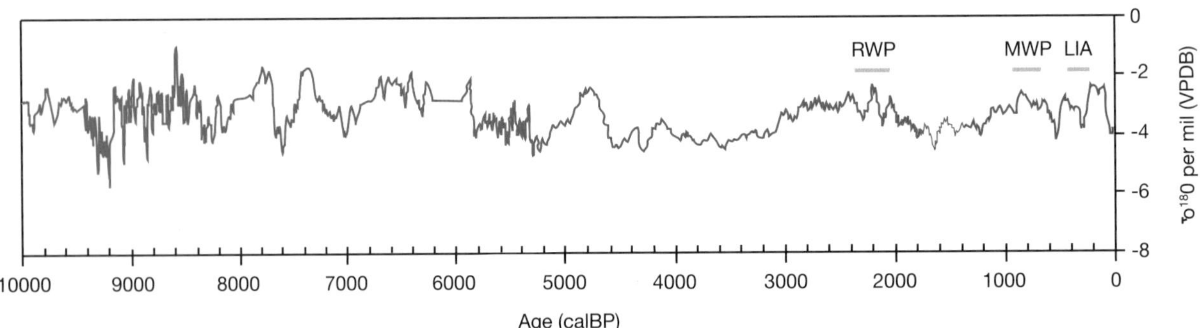

Fig. 16.1 Crag Cave speleothem $\delta^{18}O$ record redrawn from (McDermott *et al.*, 2001) with correction from (McDermott *et al.*, 2005). RWP: Roman Warm Period, MWP: Medieval Warm Period, LIA: the Little Ice Age.

Over the last two millennia the Crag Cave data appear to correlate with historically documented climatic anomalies such as the Roman Warm Period, the Medieval Warm Period and the Little Ice Age (Fig. 16.1). These events are also detected in proxy records of bog surface wetness derived from plant macrofossils, testate amoebae, and peat humification from Ballyduff Bog, County Tipperary, and have been associated with solar forcing (Oksanen *et al.*, 2006). These same proxies have been used at several other Irish peatland sites and serve to illustrate the variability of the Holocene climate and its apparent association with solar forcing. (Barber *et al.*, 2003; Blackford and Chambers, 1995; Hall and Mauquoy, 2005; Plunkett, 2006; Swindles *et al.*, 2007a, 2007b).

Postglacial Migration

The abrupt climate change that heralded the start of the Holocene was followed by rapid environmental change where a treeless tundra landscape was colonised by more thermophilous taxa. The first indication of this is the rapid and universal expansion of juniper scrub. It had been prevalent during the warmer phases of the Lateglacial (Chapter 15) and was presumably close at hand to recolonise. Several hundred years elapsed before the principal tree taxa arrived in Ireland. We can assume that the climate would have supported them, but they were delayed by their migration from refugia to the south. The first tree to arrive was *Betula* (birch) and this expanded very rapidly and led to the dramatic decline in juniper. *Betula* is a pioneer tree that is incapable of regenerating under shade. The subsequent arrival of *Corylus* (hazel), which is more shade-tolerant, led to a replacement of *Betula*. This sequence of replacements took about 500 years and is evident in all early Holocene pollen records. *Corylus* was then restricted (but not replaced) by the arrival of the three principal tree taxa: *Pinus* (pine), *Ulmus* (elm) and *Quercus* (oak). *Corylus* would have become an understorey subordinate to these canopy trees. A full Holocene pollen record from Clara Bog, County Offaly has been selected to illustrate this and subsequent developments (Fig. 16.2).

The migration of trees, and other biota, into Ireland raises the question of how they gained access and whether their migration was facilitated via land bridges from Britain or the Continent. Uncertainties still exist concerning the final isolation of the island of Ireland. The available data suggest that Ireland became isolated as an island several thousand years before the onset of the Holocene, and so there would have been no land bridges when the temperate flora and fauna migrated in during the early Holocene (Mitchell, 2006). Postglacial tree migration into Ireland is not a unique event. The Pleistocene climate had generated these conditions on previous occasions in the past (Chapter 15). It is notable that there are fewer tree genera in Ireland today compared to the mid Pleistocene Gortian interglacial (Coxon, 1996a). Indeed, about one third of the Holocene tree genera in north-west Europe did not gain access to Ireland during

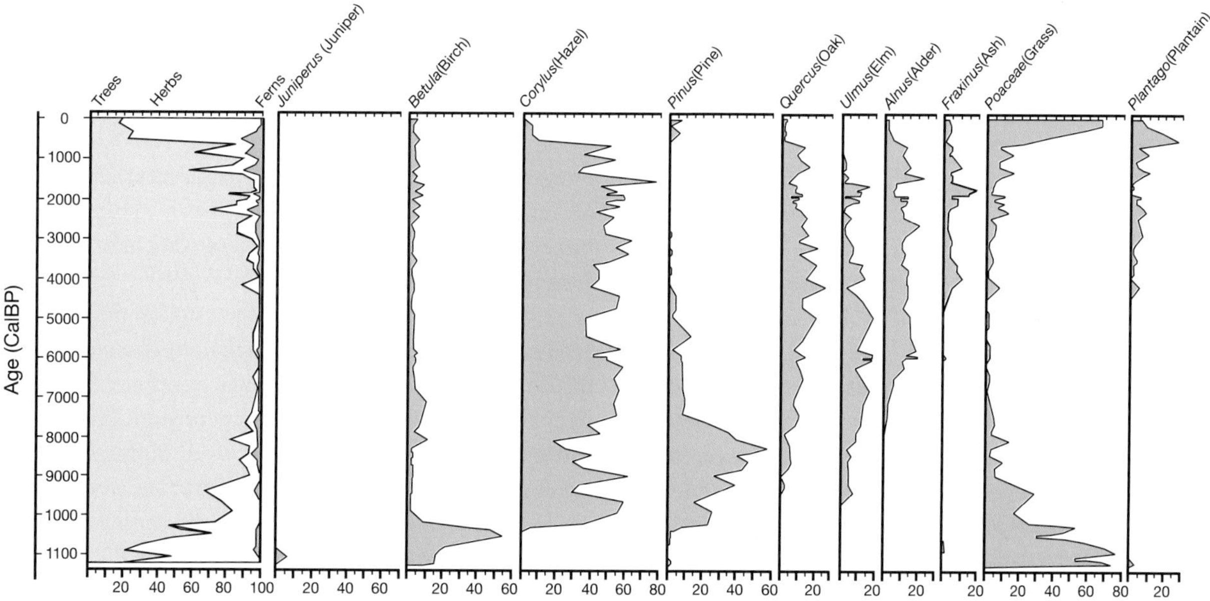

Fig. 16.2 Summary percentage pollen diagram from Clara Bog. Redrawn from Connolly (1999).

the early Holocene although they were all present in Ireland during the Gortian. This suggests that Ireland was a particularly difficult place to gain access to in the early Holocene. Similar parallels have been drawn with fossil faunal data in Britain (Sutcliffe, 1995).

Speculation on the existence and functionality of land bridges may be bypassed by considering the actual data on early Holocene tree migration. This is achieved by mapping the arrival times of trees based on pollen data such as those from Clara Bog. Thirty-two radiocarbon-dated early Holocene pollen records are available throughout Ireland. Abstracting tree arrival times from these records enables the plotting of isochrone maps which permit the patterns of migration rate and direction to be determined (Mitchell, 2006). Juniper and birch spread throughout Ireland so rapidly that they did not facilitate mapping. *Corylus* migrated in a north-westerly direction and had covered almost the entire island by 9,000 BP (Fig. 16.3). Its migration route into Ireland could have been across both the Irish and Celtic Sea basins. *Pinus* had arrived in the south-west by 9,500 BP. This tree migrated in a northerly direction but appeared to migrate faster on the west and east coasts than through the midlands. This may be explained ecologically. *Pinus* is a poor competitor on base-rich soils and had been preceded by *Corylus*, which would have thrived on the base-rich midland soils. *Pinus* would have encountered less competition from *Corylus* on the coasts, which have predominantly more acid soils and greater exposure. *Ulmus* also migrated from the south and moved in a northerly direction. There is an indication that it may have migrated faster in the midlands, and this would be expected for edaphic reasons. The entry point and migration direction for *Quercus* is very similar to that of *Ulmus*. *Quercus* appears to have migrated slightly faster along the coasts than through the midlands. *Quercus* migration lagged *Ulmus* slightly, and so *Quercus* would have met intense competition from *Ulmus* in the midlands; the two species of *Quercus* (*Q. robur* and *Q. petraea*) also have a greater tolerance of acid soils than *Ulmus* (Rackham, 1980). Although the *Quercus* isochrone map depicts only three isolines, there is a strong indication that the migration rate of *Quercus* slowed considerably as it approached the northern coast. Birks (1989) also observed this phenomenon in *Quercus* migration in Scotland.

An isochrone map has not been constructed for *Alnus* (alder). This tree appears late in the Irish pollen record (*c.*7,000 BP) and in the past this has been used as evidence to suggest that it was the late migrant into Ireland (Mitchell and Ryan, 1997). Palynological evidence from Killarney (O'Sullivan, 1990) and Glendalough (unpublished), however, indicates that isolated populations of *Alnus* existed in the early Holocene along with *Pinus* and *Quercus*. A similar situation is evident in Britain, and it appears that this tree migrated into Britain and Ireland during the early Holocene but remained in isolated loci until environmental conditions favoured its expansion after 7,000 BP (Bennett and Birks, 1990).

The migration routes depicted in the isochrone maps suggest that trees migrated into Ireland from the south and did not cross the Irish Sea. Their migration from glacial refugia in southern Europe bypassed Britain and in the case of oak, genetic data support a migration route up the Atlantic coast from an Iberian refugium (Mitchell, 2006; Petit *et al.*, 2002). The migration barrier posed by the Irish Sea is also illustrated by isochrone maps of *Tilia* (lime) and *Fagus* (beech) (Birks, 1989). Both were late immigrants into Britain: *Tilia* arrived on the Welsh coast by 7,000 BP while *Fagus* arrived there by 1,000 BP. Neither tree managed to migrate across the Irish Sea but both thrived in Ireland when introduced.

Migration rates can be derived from these isochrone maps and can further be converted into dispersal in terms of how far seeds must travel, on average, for each generation to achieve the observed migration rate. These data for Ireland indicate that heavy seeds like *Quercus* and *Corylus* had to travel between 4 and 12 km per generation (Mitchell, 2006). This would only have been possible if they were carried by a vector. If a vector is employed to disperse the seeds, then the requirement of a land bridge to facilitate tree immigration becomes obsolete.

Forest Structure and Composition

Several reviews of temporal and spatial variation in Irish Holocene vegetation have already been published and so will not be covered in detail here (Bradshaw, 2001; Mitchell and Ryan, 1997; Mitchell, 1995; Mitchell *et al.*, 1996). The excellent coverage of pollen diagrams in Ireland illustrates that, broadly speaking, base-rich soils of the midlands supported forests dominated by elm, oak, and hazel; oak woods were more prevalent on the more acid soils of the north, east, and southern regions while pine dominated the poorer soils of the western and upland regions. On local scales pollen can provide detailed records of the relative proportions of the tree taxa that formed these forests. There is however, rather more speculation about the structure of these forests.

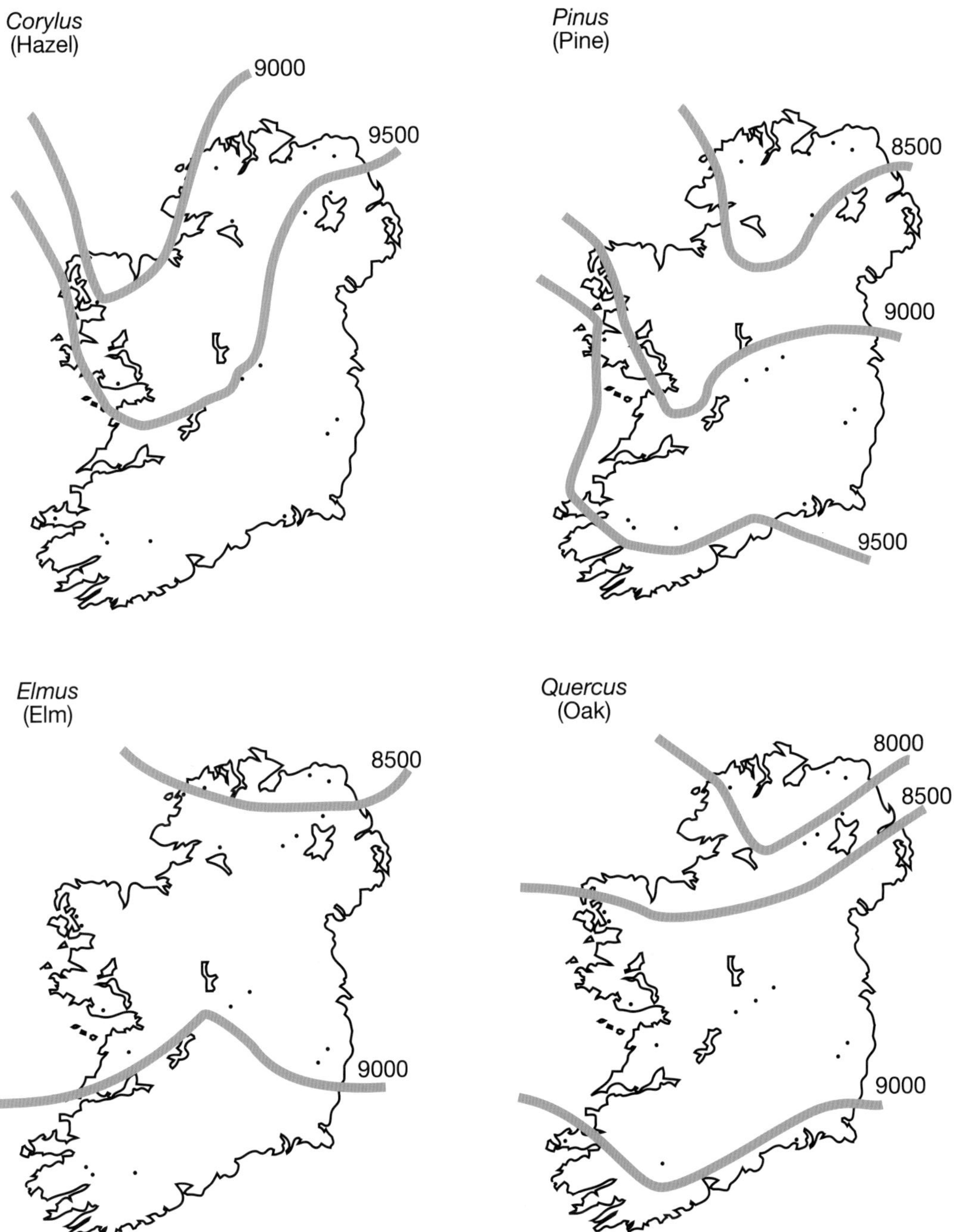

Fig. 16.3 Isochrone maps of Ireland for *Corylus, Pinus, Ulmus,* and *Quercus* redrawn from Mitchell (2006).

The traditional interpretation of western European pollen records has been one of dense primeval forests that covered most of the landscape before they were opened up by the activities of the first farmers some 6,000 years ago. This view has been challenged by Vera (2000) who argued that the activities of large herbivores kept the forest more open so that the landscape was a shifting mosaic of open grasslands, parklands with scattered trees, and small copses of dense woodland. Resolving this debate has some immediate relevance because current and future forest conservation planning in Europe is based on the assumption that natural forest landscapes were dominated by closed canopy woodland rather than parkland (Vera, 2000).

The island of Ireland is an ideal location to test this hypothesis. The large herbivores that were present in Europe during the early Holocene included bison, aurochs, horse, beaver, wild boar, and several species of deer; of these only bison and fallow deer were absent from Britain (Svenning, 2002). This is in stark contrast to the situation in Ireland where very few mammals were present during the early Holocene, especially herbivores. The available evidence suggests that all the above herbivores with the exception of wild boar were absent (Woodman *et al.*, 1997). Red deer is traditionally considered to be native to Ireland, but the earliest evidence is from well after the early Holocene at 4,000 BP and so may represent a Bronze Age introduction (Woodman *et al.*, 1997). Wild boar alone are unlikely to have opened up the Irish woodland canopy; in fact they would be more likely to be agents ensuring its integrity. So we can view Ireland as a primeval grazing exclosure and can compare what was happening in this exclosure with the rest of Europe where the primeval woods were subject to grazing by a suite of large herbivore species. The comparison of key pollen indicators of forest canopy structure reveals no statistically significant difference in Ireland compared to the rest of Europe for the three thousand years that the primeval woodland survived prior to the arrival of the first farmers (Mitchell, 2005). The absence of any appreciable difference here indicates that large grazing animals were not having a significant impact on woodland structure. It is therefore likely that the woodland structure was controlling the density of herbivores by dictating the availability of herbage rather than the herbivores dictating the woodland structure. Evidence for forest canopy openings is apparent later in the Holocene, but these are always associated with human activity (Mitchell, 2005).

Human Impact

The earliest evidence for a human presence in Ireland dates to the early Holocene (Woodman, 1978). These Mesolithic people are considered to have been hunter-gathers and it appears that in Ireland they occupied a landscape dominated by closed canopy forest. With the exception of wild boar, this forest supplied little in the way of game. In Britain, it has been speculated that Mesolithic hunters used fire extensively to modify the landscape to facilitate the hunting of large herbivores; this is unlikely to have happened in Ireland, and so high densities of charcoal in early Holocene deposits may be of natural origin reflecting the impact of climate rather than humans.

The transition to farming in the mid Holocene represents the first major human impact on the Irish landscape. This transition was gradual and straddles the time of the elm decline in Ireland (Woodman, 2000). The highly synchronous decline in *Ulmus* (elm) pollen in the Irish Holocene occurred around 6,000 calbp. The cause was probably disease, the spread of which would have been facilitated by humans (Parker *et al.*, 2002). At many sites in Ireland we have evidence of human activity associated with agriculture before the elm decline. This is not strongly expressed, and this may be due to the fact that isolated activity was taking place in a densely forested landscape and so had poor expression (Edwards, 1993). Following the elm decline, a major forest dominant was removed from the canopy, especially on base-rich soils. This would not only have facilitated agricultural enterprises but also facilitated the expression of these indicators in pollen diagrams. Modelling the impact of the elm decline on a landscape in Westmeath indicated that *Ulmus* covered at least 40% of that landscape, and that following the decline up to 12% of the forest canopy was opened (Caseldine and Fyfe, 2006). The Irish Neolithic is characterised by an initial phase of activity which is clearly registered in the pollen diagrams, especially in the west (e.g. Dodson, 1990; Fossitt, 1994; Molloy and O'Connell, 1991, 1995, 2004; O'Connell *et al.*, 1988). This is followed by an apparent decline in activity (Cooney, 2000), and in some midlands locations such as Clara Bog (Connolly, 1999) there is virtually no human activity recorded during the Neolithic (Fig. 16.2). The establishment of the Céide Fields in North Mayo is dated to this initial phase of the Neolithic. This site includes over 1,000 ha of walled field systems demonstrating a strong reliance of pastoral agriculture (Caulfield, 1978). The blanket bog that subsequently developed to cover and preserve this site became established some 400 years after the apparent abandonment of the site (Molloy and O'Connell, 1995).

Development and spread of much blanket bog in Ireland has been associated with previous human exploitation, much of it dating to the Bronze Age (O'Connell, 1986, 1990). The Bronze Age is characterised by more sustained human impact on the landscape and this is also evident at Clara Bog (Fig. 16.2). One significant cultural development was the use of metal. The earliest known copper mine in western Europe was active on Ross Island, Killarney from 4,400 to 3,900 calbp (O'Brien, 2004). Copper was mined by fire-setting where brushwood was burned against veins of copper ore and then after wetting, to rapidly cool and shatter the rock, it was hammered out with cobbles. The ore was sorted and smelted on site and so must have required prodigious amounts of fuel wood (Mitchell and Cooney, 2004; van Rijn, 2004). Similar, but later, activity has also been investigated on Mount Gabriel, County Cork (O'Brien, 1994).

The transition from Neolithic to Bronze Age also coincided with dramatic declines in pine throughout the country. This was not as synchronous as the elm decline and is probably more closely associated with the expansion of blanket bog rather than direct human impact (Bradshaw and Browne, 1987). It is important to note that both blanket and raised bogs were important habitats for pine when their surfaces dried, and so the decline was also associated with renewed growth in some bogs. Pine, however, did not go extinct in Ireland at this time and is recorded in several pollen sites for at least another 2,000 years in Kerry (Cooney, 1996; Little *et al.*, 1996; Mitchell, 1988) and Clare (Watts, 1984). Some documentary records suggest that pine survived until Medieval times elsewhere in the country (Nelson and Walsh, 1993). Pine is now considered to have been extirpated in Ireland, and all trees in the landscape today are derived from introduced trees. It is, however, tantalising to consider that the last known native pines were present only two or three tree generations ago, and so conceivably some native trees may have survived. Future work on the population genetics of pine in Ireland may help to ascertain whether any native trees did survive to the present day.

One important consequence of the apparent extirpation of pine was the disappearance of beetle species associated with this tree (Whitehouse, 2006). Several other forest beetle species, especially those associated with dead wood, are currently absent from Ireland but are found in the fossil record (Whitehouse, 2006). This is a clear indication of the demise of the Irish forests. The cause of this decline is again associated with human activity. We can assume that most of Ireland was covered by forest in primeval times; unforested areas would have been confined to the upper levels of our higher mountains and wet lowland areas that developed into raised bogs. This would leave at least 80% forest cover. Evidence from blanket bogs suggests that significant areas were permanently deforested during the Bronze Age. Pollen records indicate the case for continued deforestation through early Christian and Medieval times (Hall, 1995) and this is supported by the extirpation of some beetle species indicative of ancient woodland during Medieval times (Whitehouse, 2006). The earliest documentary records for woodland cover in Ireland date from the Civil Survey of Ireland in AD 1654–6. From this Rackham (1995) has estimated that woodland cover was 2.1%. This was further reduced to less that 1% by the end of the nineteenth century (Kelly and Fuller, 1988). The 10% forest cover that Ireland now enjoys is primarily made up of exotic conifer plantations; but how well do the remaining fragments of native woodland match their prehistoric ancestors?

Pollen analysis from small hollows within these woodlands provides some insight (Mitchell, 1988). Data from Derrycunihy wood in the Killarney National Park illustrate that prior to 2,000 BP the wood was composed of a mixture of *Quercus* and *Pinus* (Fig. 16.4). *Pinus* had declined to extinction in most areas around 4,000 BP in association with renewed development of blanket bog (Bradshaw and Browne, 1987; O'Connell, 1990) but this tree survived considerably longer in Derrycunihy Wood and elsewhere in Kerry (Cooney, 1996; Little *et al.*, 1996). The decline in *Pinus* around 2,000 BP is associated with fluctuations in *Betula* and *Quercus* and substantial increases in herbaceous taxa and charcoal. These data are suggestive of human disturbance dating to the Iron Age. The subsequent recovery of the woodland following disturbance does not include *Pinus*, which went extinct at this site. The secondary woodland was more open, as indicated by the higher representation of *Betula*, the understorey shrubs of *Ilex* (holly) and *Corylus* and herbs. Increases in charcoal and disturbances to the woodland canopy towards the top of the diagram relate to human activity in the eighteenth and early nineteenth centuries. This evidence for disturbance coincides with documentary records of charcoal production for iron smelting, timber extraction and grazing in Derrycunihy Wood (Mitchell, 1988). Assessment of the complete pollen records from Derrycunihy Wood and the other sites in Killarney indicate that following disturbance, the botanical and structural diversity of the woods were reduced. The almost monospecific *Quercus petraea* canopy at Derrycunihy Wood today reflects this and the early nineteenth century silvicultural preference for this species. The development of this *Quercus*-dominated canopy is clearly illustrated in the upper levels of the pollen diagram (Fig. 16.4). At no time over the last 5,000 years has *Quercus* had such a dominance in Derrycunihy Wood. The pollen data from Derrycunihy demonstrate that the present composition and structure of the wood bear little resemblance to its primeval counterpart. The complete dominance of *Quercus* in the canopy today has been brought about by repeated human exploitation and

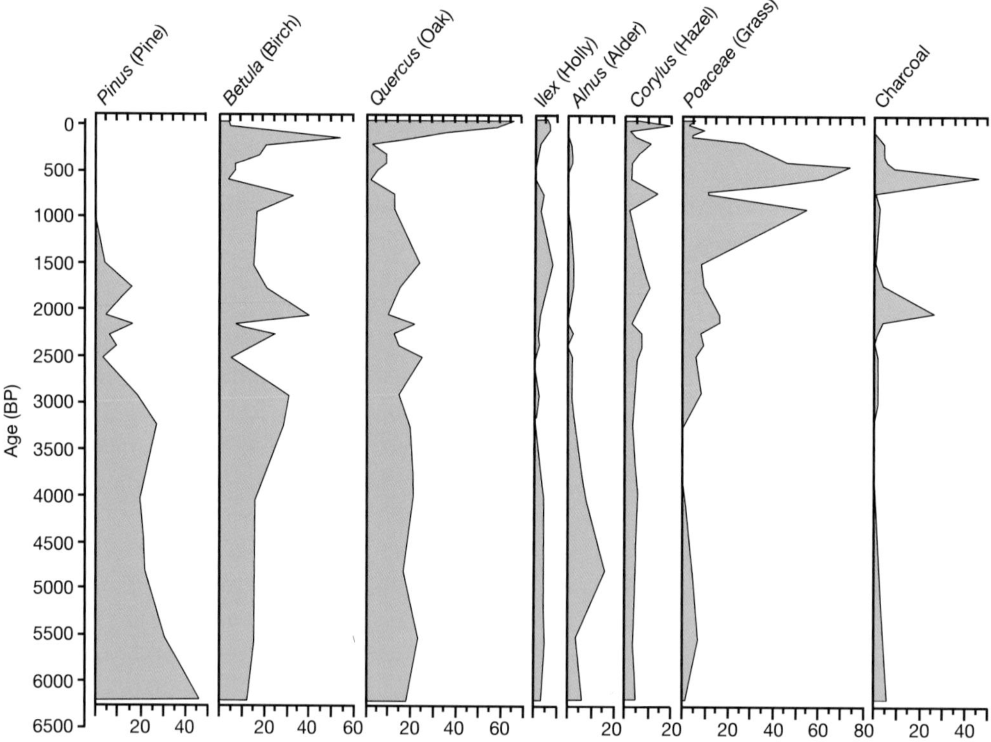

Fig. 16.4 Summary percentage pollen diagram from Derrycunihy Wood. Redrawn from Mitchell (1988).

intervention at the site. The 5,000 year pollen record demonstrates that this composition is unlikely to be sustainable in the long term without substantial intervention.

Relatively recent human-induced changes are also evident at the landscape scale. For example, the building of the North Bull wall in Dublin Bay between 1819 and 1824 facilitated the scouring of the channel that led to Dublin port, but also led to the deposition of sands behind the wall that have formed North Bull Island, which is now over 5 km in length (Jeffrey *et al.*, 1977). In 1981 the island was declared a UNESCO Biosphere. Further inland, the Clara Bog pollen diagram shows a dramatic decline in *Corylus* over recent centuries (Fig. 16.2). This decline is a feature of pollen diagrams across the country. Detailed analyses by Cole and Mitchell (2003) indicated that the decline occurred around AD 1750 and was associated with extensive land use linked to a rapidly rising rural population. Multivariate analysis illustrated that there was clear regional variation between pollen records prior to the decline, but this disappeared subsequently due to intense human manipulation of the landscape (Cole and Mitchell, 2003). This human impact also appears to have overwhelmed any signal from the Medieval Warm Period and Little Ice Age which are apparent from the proxy climate records described above. This aptly illustrates the necessity of separating climate from human impact to facilitate the investigation of their interaction in the past, present and future.

17

Geology of offshore Ireland

D. Naylor and P. M. Shannon

Introduction

The previous chapters have focused upon the details of the geology of the island of Ireland. The geological evolution of the pre-Mesozoic has been pieced together based on a vast amount of scientific study, carried out over more than 150 years and based on many classic geological exposures, supplemented by the results of extensive mineral exploration. However, with the exception of some outcrops in the northeast of the country, relatively little of the younger (post-Carboniferous) history is revealed in onshore outcrops.

This chapter is concerned with the geology of Ireland's offshore regions, which cover an area of approximately nine times that of the onshore. Ireland's offshore area encompasses a large part of the continental shelf beyond the island. It extends eastwards into the Irish Sea, southwards into the Celtic Sea and westwards to the Rockall Bank (Figure 17.1). In contrast to the onshore, the offshore region contains the missing history of the younger geology, especially the Permo-Triassic, Jurassic, Cretaceous and Cenozoic. It also illustrates significant differences in the geological evolution of the onshore and offshore regions during their younger history when, at various times, uplift and denudation of the onshore region provided sedimentary detritus that accumulated in the subsiding offshore basins.

A striking feature of these offshore basins is the presence of a structurally linked system of Mesozoic and Cenozoic sedimentary basins that are separated by Palaeozoic and older rocks. With the exception of the north-east of Ireland, these do not extend onshore (Figure 17.2). The presence of fault-controlled young sedimentary basins had long been recognised in the onshore areas of central and north-west Europe, but the full extend of the ramifying system of basins developed on the Alpine foreland and beyond was not fully appreciated until the 1960s. By that time early offshore surveys carried out by academic institutions, together with the rapidly burgeoning body of data from oil company exploration work in the shallow offshore regions of the North Sea and the preliminary exploration in the Celtic Sea and Atlantic waters, had revealed that the network of basins extended from the land areas out onto the continental shelf of northwest Europe.

Petroleum industry data and research results from academic and government institutions during the past few decades have improved our understanding of the offshore domain considerably. While much of the offshore information results from oil industry exploration, we are concerned in this chapter with the geological information and understanding derived from the data, rather than with the petroleum potential of the offshore. Nonetheless, as these data are largely derived from the search for indigenous hydrocarbons, a brief review of the history of exploration is provided to illustrate the nature, extent and coverage of the data on which the geological understanding is based.

Following the discovery of several large gas fields in the southern part of the North Sea during the 1960s, and the subsequent discovery in the early 1970s of a series of large oil fields in the northern North Sea, the oil industry turned to the virgin areas west of Britain, and to Irish waters. Reviews of oil and gas exploration in and around Ireland are available in Shannon *et al.* (2001a), whilst an assessment of the results and of the future potential of the region can be found in Shannon and Naylor (1998), and Spencer *et al.* (1999).

In 1968 the Irish Government made its first offshore designation of territory and then progressively enlarged the offshore area under its control by a series of designations during the 1970s. The designated area now extends across the deeper waters of the Rockall Trough. The first commercial reflection seismic survey was carried out in the Celtic Sea in 1969. In May 1970 the first well in the Irish offshore (Marathon 48/25-1) was drilled in the North Celtic Sea Basin. The company then drilled another unsuccessful well before returning to drill a

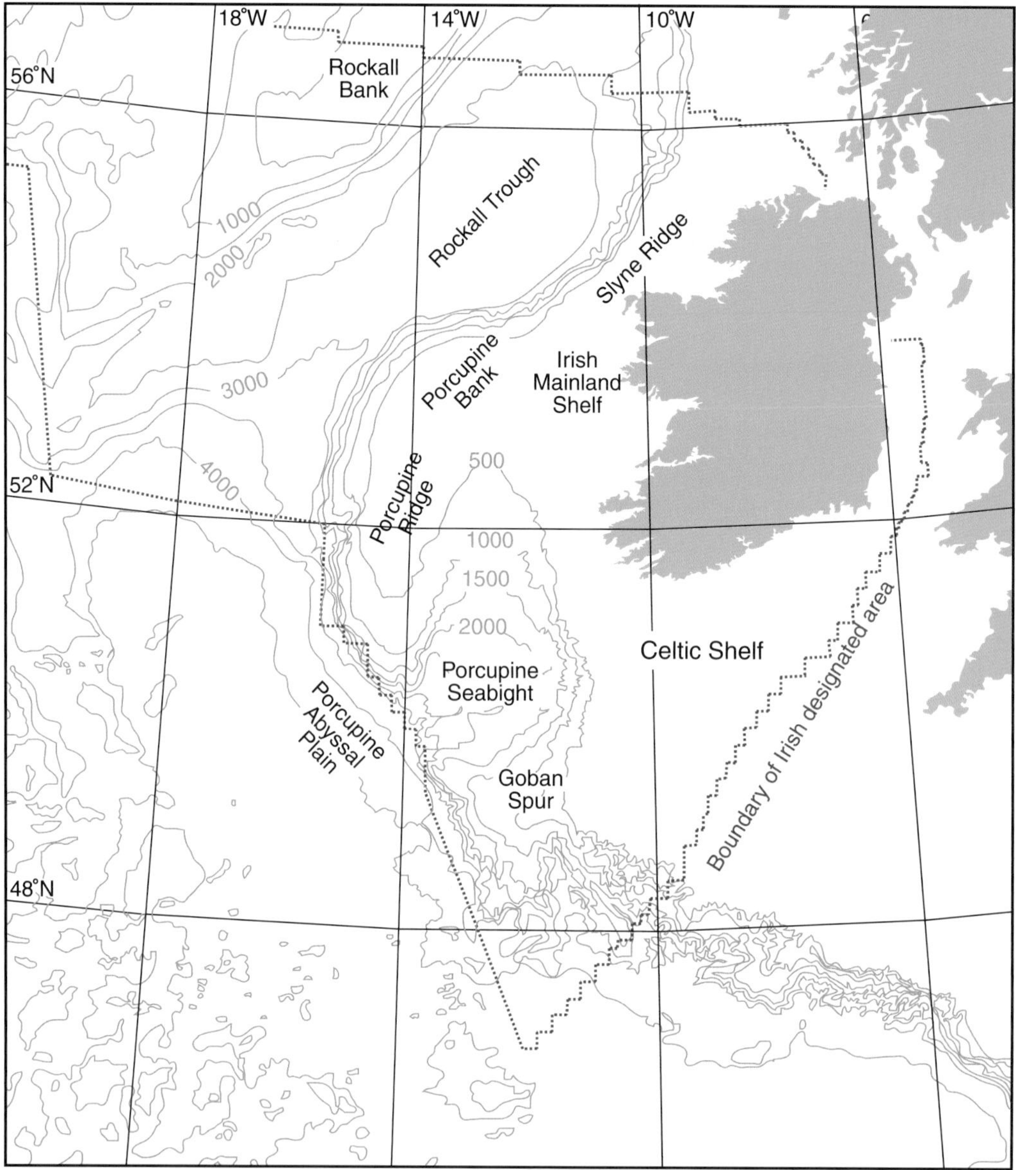

Fig. 17.1 Bathymetric map showing major physiographic features of the Atlantic shelf and margin west of Ireland (*after* Naylor *et al.*, 1999).

second well on block 48/25 in 1971. This was the discovery well of the Kinsale Head Gasfield.

Exploratory drilling in the Celtic Sea region continued throughout the 1970s and in 1976 the first drilling took place in the Fastnet Basin, at the westward end of the Celtic Sea. The following year the first wells were drilled in the Kish Bank Basin east of Dublin, and also in the deeper waters of the Porcupine Basin. Drilling in the Porcupine Basin in the late 1970s encountered many oil shows and several wells flowed oil, but the accumulations proved to be too small or complex for development at that time in this deep-water environment.

The Corrib Gasfield in the Erris Basin was discovered by the Enterprise 18/20-1 well in 1996. The first well (Enterprise 5/22-1) in the Irish sector of the Rockall Basin followed in 2001. Some encouragement was made with a gas condensate discovery in the Enterprise 12/2-1 (Dooish) prospect in the Rockall Basin. There has also been a resurgence of interest, with drilling and small gas and oil discoveries, in the North Celtic Sea Basin, where oil and gas shows were encountered during earlier drilling.

From this brief historical review it will be seen that offshore exploration in Irish waters has differed markedly

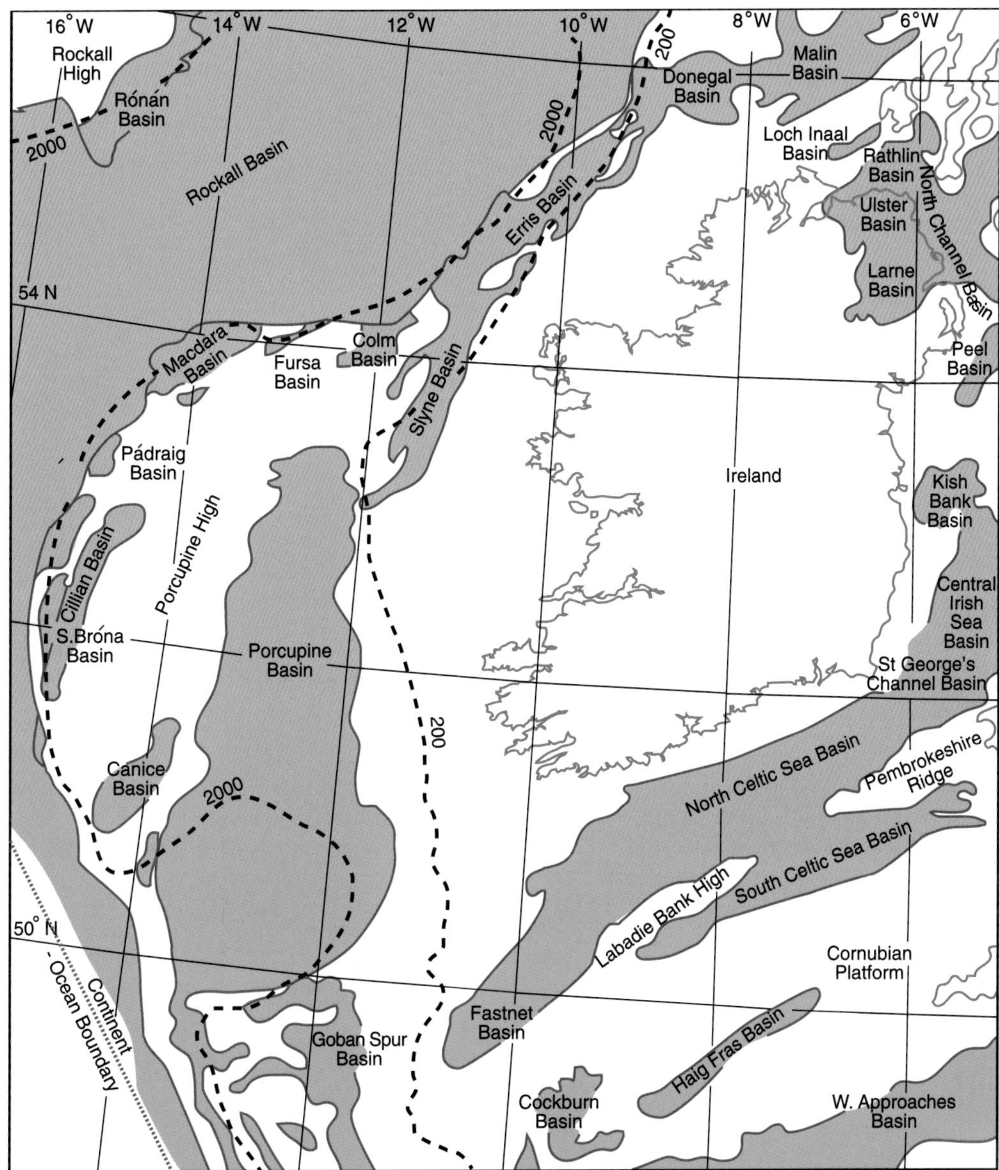

Fig. 17.2 Location of sedimentary basins around Ireland (*after* Naylor *et al.*, 2002). Bathymetric contours at 200 metres and at 2000 metres are shown for reference.

from that of the North Sea. In the North Sea there has been expanding exploration and production as discoveries were made in different geological habitats, to produce a mature production province. Ireland, on the other hand, was distant from the Southern North Sea gas play and the overspill of exploration activity from the North Sea was delayed. Initial successes in the Celtic and Porcupine basins were not sustained. Furthermore, the more promising remaining prospects are in deeper water areas such as the Rockall Basin and the deep southern parts of the Porcupine Basin that are costly to explore, and are only now attracting attention as a result of deep-water successes elsewhere in the world, increased energy demands and higher oil prices, and improved exploration and production technology. For this variety of reasons oil and gas exploration in the Irish offshore region has followed an erratic path, with relatively low annual rates of activity.

At the time of writing a total of 160 exploration and appraisal wells have been drilled in Irish waters at a total estimated cost, in 2008 prices, in excess of €3000 million. In addition, valuable stratigraphic information has been provided through results from the Deep Sea Drilling Projects (DSDP and IODP). The offshore basins have also been covered by numerous reflection seismic surveys with the densest coverage in the Celtic Sea and Porcupine basins. Several hundred thousand kilometres of seismic lines have been shot in Irish waters. Furthermore, major research programmes funded by the Irish Petroleum Infrastructure Programme (PIP) have led to the acquisition of deep seismic profiles in the Rockall and Porcupine basins (Morewood *et al.*, 2005; O'Reilly *et al.*, 2006), deep boreholes and gravity cores on the margins of the Rockall Basin (Haughton *et al.*, 2005; Øvrebø *et al.*, 2005, 2006) and sidescan sonar data on the shelf and slopes of the Rockall Trough (Shannon *et al.*, 2001b; O'Reilly *et al.*, 2005). The Geological Survey of Ireland carried out a comprehensive multibeam survey over the continental shelf within the past few years and this has provided a vast amount of new data on the surface features of the region. The combination of the released well, seismic and other data with published seismic interpretations allows a reasonable picture of the offshore basins to be assembled.

The geology of contiguous British waters is of pertinent and direct interest, and occasions many references in the text. However, no detailed consideration is given the neighbouring basins unless they are a continuation of, or have direct implications for, the adjacent Irish basins. Obvious examples are the Cardigan Bay Basin, the South Celtic Sea Basin, and some basins off the Scottish coast.

The Irish offshore basins owe their existence to the complex tensional regimes that operated through latest Palaeozoic and Mesozoic times. They were later modified by the compressional and inversion events related to the Alpine orogeny and to the far field effects of North Atlantic seafloor spreading rates, plate readjustments and mantle thermal structures resulting from crustal attenuation. The complex nature of this interaction and the role played by older structural units and lineaments is nowhere better illustrated than offshore Ireland. The influence and control exercised by the pre-Mesozoic framework on the development of the younger offshore sedimentary basins is considered in the next section, where the regional tectono-stratigraphic framework is discussed. Later sections describe the stratigraphic evolution of the Irish offshore during each of the main geological periods, with a palaeogeographic summary of each provided. The offshore basins fall conveniently into three main sets: (1) the Northern Ireland–Irish Sea basins where the basins are relatively small and the main period of sedimentation was in the Permo-Triassic, (2) the Celtic Sea basins where the basins are larger and elongate, and where major sedimentation continued through the Jurassic, and (3) the Atlantic Margin basins where the basins are typically very large, underlie bathymetric troughs and contain major Cretaceous and Cenozoic depocentres. A summary of the development of each of the three main groups forms the final sections of the chapter.

Regional geological framework

Ireland lies towards the western edge of the European continent (Figure 17.1), a region where the crust has been affected by multiple orogenic, extensional and inversion episodes over the course of geological time. These events, and their resultant structures, played a major role in influencing the location, orientation and history of the younger sedimentary basins that lie in the offshore. This relationship is complex, although this is at least in part due to the fact that relatively little is still known about the location and boundaries of the various basement blocks and structural terranes in the region. Most of the research to date has, for obvious commercial reasons, focused upon the sediments within the grabenal basins rather than on the non-prospective basement blocks and shelves flanking the basins. In this section the physiography of the extensive continental shelf and the distribution of the offshore sedimentary basins are briefly outlined. There follows a discussion of the nature, structure and likely extent of the basement crust in the region together with its likely control on the location of the sedimentary basins. This provides the background for the stratigraphic development that follows in the succeeding sections of the chapter.

The continental shelf

The waters east and south of Ireland are typically shallow, ranging from approximately 30 m east and north-east of Ireland to little more than 100 m south of Ireland. Unlike the continental shelf west of Ireland there is no evidence from the bathymetry of the presence of the thick sedimentary basins that lie beneath the seabed.

The continental shelf west of Ireland is much more variable in its nature and physiography. It extends out more than 300 km, and considerably more than this in the region of the Rockall Bank (Figure 17.1). In the south-west, the Goban Spur is a remote plateau area on the continental margin some 250 km south-west of Ireland and south of the Porcupine Seabight. Bathymetrically it comprises a smooth platform sloping gently westwards away from the Celtic Shelf to depths of 2000 m. South of the

Goban Spur the continental edge curves eastwards, and the continental slope is cut by a series of deep canyons. North of the Goban Spur the foot of the continental slope swings westwards from Ireland around the Rockall Bank. The Porcupine Seabight and Rockall Trough are deep-water embayments within the shelf that separate a number of higher plateau areas. The Porcupine Seabight is a large north–south trending deep water area which opens south-westwards onto the Porcupine Abyssal Plain. Water depths increase from about 350 m in the north of the Seabight to more than 4 km in the south. The Seabight is bounded on three sides by bathymetrically shallow platforms. The Irish Mainland Shelf lies to the east, the Slyne Ridge to the north, whilst the western boundary is formed by the Porcupine Ridge, extending southwards from the Porcupine Bank. The existence of Mesozoic and older rock units on the Irish Atlantic margin was first indicated by the dredge samples reported by Cole and Crook (1910). The Rockall Trough is bounded to the west by the Rockall Bank, an elongate shallow bank marked at its shallowest point by Rockall islet. Further west lie the Hatton Trough, Hatton Bank and Continental Margin, west of which the slope deepens rapidly into the Iceland oceanic basin where water depths are of the order of 4 km.

Geophysical evidence (gravity, magnetics and seismics) shows that the continent–ocean boundary coincides approximately with the 4 km isobath. It runs to the west of the Goban Spur and the mouth of the Porcupine Seabight, swings east–west, parallel to the Charlie-Gibbs Fracture Zone, to the south of the Rockall Trough and the Rockall Bank, and then turns into a NE–SW orientation to the west of the Hatton Continental Margin. The continental–oceanic boundary is seen on magnetic data (e.g. Srivastava and Verhoef, 1992) along much of its length, although it is somewhat diffuse and unclear in the southwest of the Hatton region, where the boundary may be complicated by the presence of igneous intrusions, seaward-dipping reflections from lava flows, and areas of thin crust. The boundary is seen on deep seismic profiles at the mouth of the Porcupine Seabight (Makris *et al.*, 1988), the mouth of the Rockall Trough (Hauser *et al.*, 1995) and to the west of the Hatton Continental Margin (Vogt *et al.*, 1998). The southern and western continental–oceanic boundaries are different, with the southern boundary, parallel to the Charlie-Gibbs Fracture Zone, reflecting a tectonic (probably transpressional) aspect with a thickened ridge, while the western boundary is marked by a major underplated igneous body at the base of the crust, a feature that is also seen along the contiguous boundary west of the UK and Norway (e.g. White, 1992).

Distribution and development of offshore sedimentary basins

A structurally linked set of Permo-Triassic to Cenozoic sedimentary basins of different sizes, shapes and ages surround Ireland, extending onshore only in the northeast of the country (Figure 17.2). These are part of a regional set of basins extending along and parallel to the European Atlantic margin. The swath of basins extends along strike for several thousand kilometres from Armorica to the Barents Sea, and covers a width of several hundred kilometres from the continental shelf west of Ireland, the UK and Norway, through to the North Sea in the east.

In general, the Irish offshore basins can be grouped into three broad categories, each with distinct size, shape and age profiles. A set of small basins lies between Ireland and the UK in the Irish Sea region. The Lough Indaal, Rathlin and North Channel basins (Figure 17.10) extend between Northern Ireland and Scotland and are generally elongate and run parallel to the Caledonian structural fabric of the region. Farther south in the Irish Sea, the Kish Bank, Central Irish Sea and St George's Channel basins have variable orientations (Figures 17.11 and 17.13). While the Kish Bank Basin is largely circular to slightly elongate, the basins farther south generally follow the Caledonian orientation of the basement rocks in the adjacent Irish and Welsh mainlands. This first group of basins includes the smallest and generally the least elongate of the offshore basins. They are the oldest of the basins, containing a predominant Permo-Triassic basin fill, and typically have a thin to absent Cretaceous and Cenozoic succession.

The second group of basins lies to the south or Ireland (Figure 17.3). These basins are relatively narrow and elongate, and are larger and typically younger than the Irish Sea basins. They contain Triassic strata, with the main phase of basin development being in the Jurassic to Early Cretaceous. They occur in the Celtic Sea and run parallel to the south coast of Ireland and are slightly oblique to the major onshore Variscan structures of Ireland. The Fastnet and North Celtic Sea basins lie to the north of an intermittent basement structure (the Pembrokeshire–Labadie basement high), while the Cockburn and South Celtic Sea basins lie to the south of the ridge. At their eastern end they run, with a major strike swing, into the Irish Sea basins.

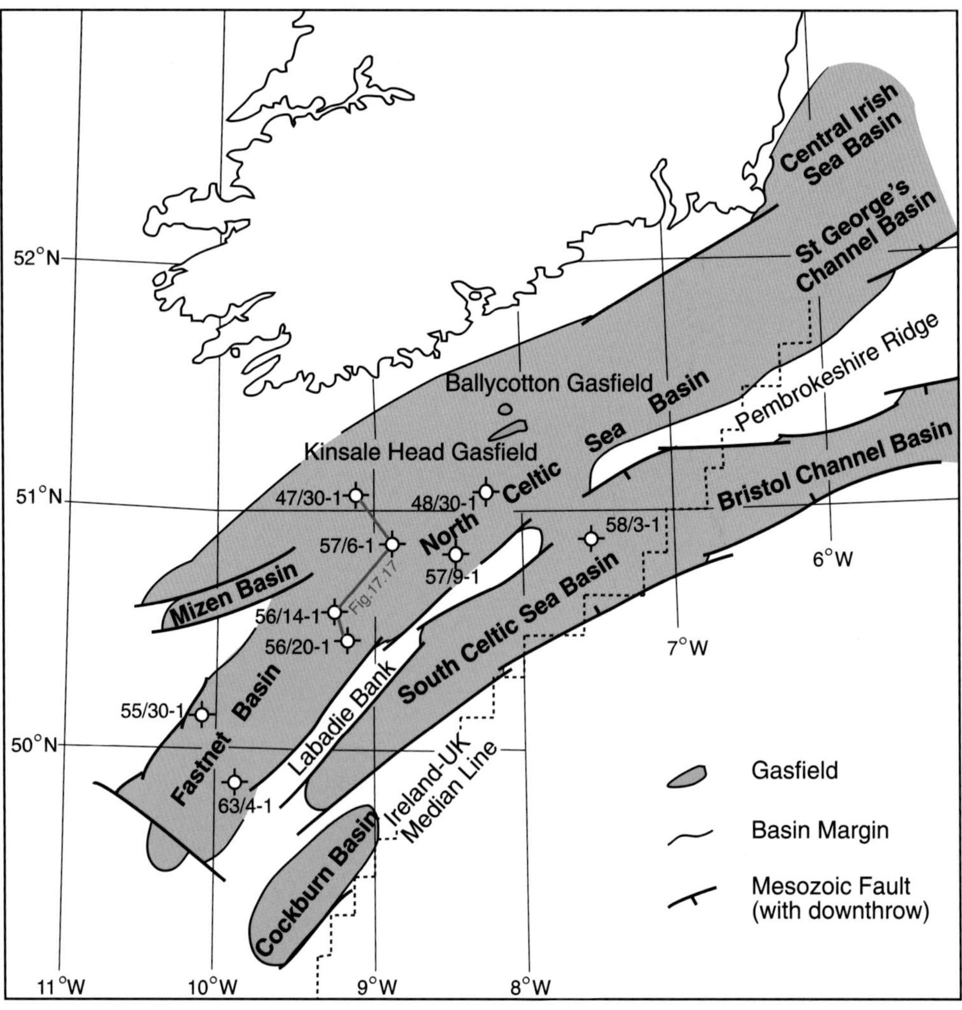

Fig. 17.3 Location of Celtic Sea basins and wells mentioned in the text (after Shannon, 1991b).

The third group of basins is the largest and generally the youngest, and the basins lie in the deep Atlantic waters west of Ireland. Their main phase of sedimentation was in the Cretaceous and especially the Cenozoic. The majority follow a general NNE–SSW trend and form part of a chain of structurally-linked basins that extend from mid Norway to west of Iberia (Doré *et al.*, 1997a, b). Within the Irish sector a band of narrow (inboard) basins includes the Slyne, Erris and Donegal basins. These lie landward of a set of larger (outboard) basins that includes the Porcupine, Rockall and Hatton, basins. A set of small, elongate, probably early Mesozoic basins are located in the footwalls of the main Cenozoic Rockall Basin (Naylor *et al.*, 1999). The notable exceptions to the NNE–SSW basin orientation are the Porcupine Basin that runs north–south, and the Goban Spur Basin, lying south of the Porcupine Basin and west of the Fastnet–Celtic Sea basins system. This third basin type lies exclusively in the Atlantic region and is the farthest west of the three types. Unlike the other basin types, the larger of the basins underlie significant bathymetric depressions, reflecting rapid subsidence and under-sedimentation in mid-Cenozoic times.

The overall progression in size, shape and age of the basins reflects a westward progression of extensional strain towards the area of continental breakup and seafloor spreading. The initial sedimentary basin development took place in a series of localised Permian basins into the foothills of the Variscan mountains and sometimes in depocentres controlled by the gravitational collapse of Variscan orogenic structures. Early to mid-Triassic depocentres were partly infilled topographic lows as well as partly rift induced. Through Late Triassic and into Early Jurassic times, localised fault-controlled depocentres developed, but deposition was largely in broad and shallow thermal subsidence depocentres.

Major rifting occurred during the Middle Jurassic to Early Cretaceous. Rifting in Middle Jurassic time appears to have followed caledonoid-oriented crustal fabrics and structures (e.g. the Celtic Sea and the Slyne–Erris basins). The onset of the north–south elongate rifts in the Porcupine Basin probably developed somewhat later (Late Jurassic). A major Base Cretaceous unconformity is seen in most of the basins and corresponds to a period of plate reorganisation. Residual rifting continued in places (e.g. the Porcupine Basin), with inversion in others (e.g. the Fastnet Basin). A local phase of rifting occurred during the Aptian–Albian and thereafter regional thermal subsidence, accompanied by a eustatic sea level rise, occurred in all the basins, giving rise to major Chalk deposition with very little terrigenous input. The Cenozoic marked a major change with a combination of basin inversion, return to clastic deposition and marked igneous activity especially in the Atlantic margin basins. Various Cenozoic plate reorganisations led to localised major sediment wedges, inversion events and differential basin restructuring. The internal structure of the basins was complicated by igneous activity associated with crustal breakup, and by localised inversion due to plate readjustments. However, the precise influence of older structural fabrics is not well understood or straightforward. In places the basins are parallel to underlying crustal fabrics, while in other places the basins appear to be slightly oblique to the main structures. In yet other places, the younger basins bear no relationship to the orientation of basement structures. Some of these relationships are explored in the next part of the section.

Crustal Structure and Development

Within the past couple of decades a significant amount of industry and academic data, mostly in the form of deep and wide-angle seismics, supplemented by a large volume of gravity, magnetic, seismic reflection and some petroleum exploration boreholes, have provided a general picture of the crustal thickness and broad structure of the Irish offshore, although the location, orientation and influence of the inherited basement fabrics still remain uncertain and loosely constrained. The crust beneath Ireland is approximately 30 km thick, with little evidence of stretching or thinning. Beneath the Irish and Celtic Sea regions it is approximately 25 km thick, with evidence of crustal decoupling to accommodate the slight thinning beneath the basins (O'Reilly *et al.*, 1991). The nature and the thickness of the crust beneath the large outboard Atlantic basins such as the Porcupine and Rockall basins has been the subject of considerable debate during the years (*see* Smythe, 1989; Shannon *et al.*, 1999 for discussion). However, robust geophysical evidence from the RAPIDS (Rockall And Porcupine Irish Deep Seismic) profiles now points to the presence of severely attenuated continental crust beneath the largest of the basins (Figure 17.4), intruded in places by mantle serpentinites or volcanics (Makris *et al.*, 1988, 1991; Hauser *et al.*, 1995; O'Reilly *et al.*, 1996, 2006; Reston *et al.*, 2001). Beneath the centres of the Rockall and Porcupine basins the continental crust is 2–5 km thick and has been modelled to indicate differential stretching, with greater upper and middle crustal extension facilitated by rheologically-controlled detachments at the top of the lower crust. The

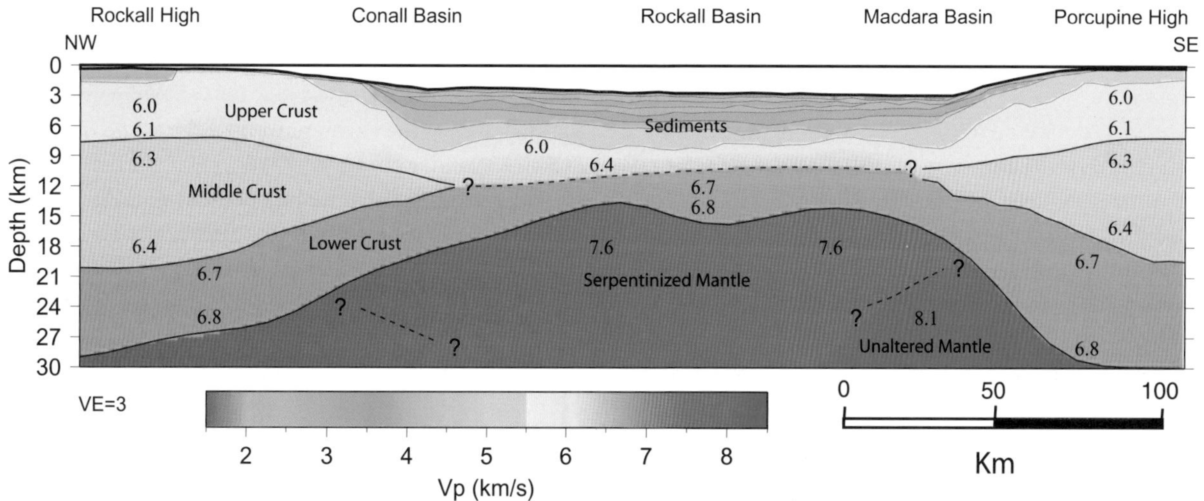

Fig. 17.4 Crustal velocity structure across the Rockall Basin, based on wide-angle seismic reflection studies (*after* Morewood *et al.*, 2005). The location of the section is shown in Figure 17.6, and the geology of the basin fill is shown in Figure 17.25.

uppermost mantle is serpentinised where the stretching is most severe, and this may have provided a degree of added strength to the lithosphere to prevent rupture.

While the thickness of the crust on the continental shelf has been well constrained by deep-seismic profiles (Morewood *et al.*, 2005; O'Reilly *et al.*, 2006), little is known about the composition or detailed nature of the crust or its age. Naylor and Shannon (2005) identified several distinct basement (pre-Permian) provinces in the Porcupine–Rockall region, based on seismic character and on the structure of the overlying basin sediments. The European Atlantic margin was interpreted by Doré *et al.* (1999) as an oblique re-opening of the Caledonian suture and fold system, and thus the present west Rockall margin may follow an earlier controlling lineament. Recent data from the Hatton region (Hitchen, 2004) demonstrate that the area has a thick cover of Cenozoic to possibly Palaeozoic sedimentary rocks. Rockall Bank is thought to belong to the Islay (Rockall Bank basement) terrane, with the boundary against Lewisian rocks lying to the north, in UK waters (Hitchen *et al.*, 1997). Similar rocks may underlie the younger Hatton Bank succession. The Rockall High comprises a distinct province of Precambrian rocks, with no evidence of Palaeozoic cover. This suggests that the extension that formed Rockall Basin took place along an earlier basement structure that had either controlled Palaeozoic deposition, or at the very least controlled the pattern of later denudation. Tyrrell *et al.* (2007) speculated on the nature of basement terranes and terrane boundaries on the Atlantic margin, based upon the geochemistry of the few basement samples taken in the offshore region, integrated with the onshore basement of Ireland, the UK, Greenland and Canada. Their work on Pb isotopic domains suggested the existence of five major domains (an Archaean, two Proterozoic, an Avalonian and a Variscan domain), extending through the offshore region. While some of these are suggested to extend along strike from onshore Ireland and the UK into the Porcupine and Rockall regions, the paucity of *in situ* offshore basement samples places a high degree of uncertainty on their precise boundaries and indeed on whether other unsampled basement terranes may exist in the region.

What is known is that the continental shelf offshore Ireland bears the imprints and structures resulting from Variscan, Caledonian and older orogenic events. Reactivation of some of these structures is likely to have influenced the location, orientation and large-scale structures of the sedimentary basins. Closure of the Iapetus Ocean in late Caledonian times led to the docking and suturing of distinct basement terranes. They have a general north–south to NNE–SSW orientation in Norway, swinging to NE–SW through Scotland and Ireland, and the Mesozoic basins and the continent–ocean margin in these regions are typically parallel to the Caledonian structures. The early Mesozoic basins in the North Channel area follow the NE–SW structural fabric of the Southern Uplands, while the Central Irish Sea Basin has a general NE–SW orientation, parallel to the Caledonian lineaments in the basement rocks of both SE Ireland and Wales/Anglesey. Onshore in Ireland the NE–SW Caledonian fabric in the north of the country changes to an ENE–WSW alignment towards the south coast, so that the Caledonian and Variscan fabrics are sub-parallel. The North Celtic Sea Basin has a more ENE–WSW orientation (Figure 17.3), reflecting either a strike swing in the Caledonian fabric, or a Variscan structure (or possibly a combination with a Caledonian structure that was reactivated in Variscan times). Within the Celtic Sea region, deep seismic profiles suggest that the steep northern margin of the basin is controlled in places by extensional reactivation of a southward-dipping Variscan structure that detaches in the lower crust (Shannon, 1991a). The ENE–WSW fabric is seen again in the eastern part of the Goban Spur province and in the orientation of the Porcupine Fault at its northern margin.

A number of different interpretations have been proposed for the extension of major faults and lineaments westwards from Ireland across the Continental Platform, based in the main on gravity and magnetic data (Young and Bailey, 1973; Riddihough, 1975; Max *et al.*, 1982; Masson *et al.*, 1985). Nevertheless, there is general agreement that the Great Glen and Fairhead–Clew Bay fault systems trend south-westwards towards the north of the Porcupine Basin. Reactivated splays from these systems may have played a role in the segmentation of the Slyne–Erris and North Porcupine basins into discrete sub-basins.

Klemperer (1989) and Klemperer *et al.* (1991) considered that the Great Glen and Clew Bay faults, and possibly the Variscan Front, can be identified on deep seismic profiles. They suggested that north-dipping reflectors west of the Shannon estuary on the BIRPS WIRE 1 profile are structures related to the Iapetus Suture. Any westward extension of the Variscan Front or the Iapetus Suture could be anticipated to cross the Porcupine Basin, but these have not been identified and there are no clear effects of any such deep structures on the orientation or development of the overlying Mesozoic-Cenozoic basin (Figure 17.5).

There is a major swing in strike of the eastern margin of Rockall Basin at approximately 54° N (Figures 17.2 and 17.6). To the north, the margin parallels the NE–SW caledonoid strike of the narrow Slyne–Erris basin system. At 54° N the strike changes almost to E–W. The strike change lies due east of a similar strike swing in the Caledonian of NW Ireland documented by Hutton and Alsop (1996), which they believe to have a deep pre-Caledonian (>600 Ma) structural origin. A possible extension of the important Donegal Lineament of Hutton and Alsop (1996) could have influenced the trend of the Rockall margin at this latitude. The NE–SW faults of the

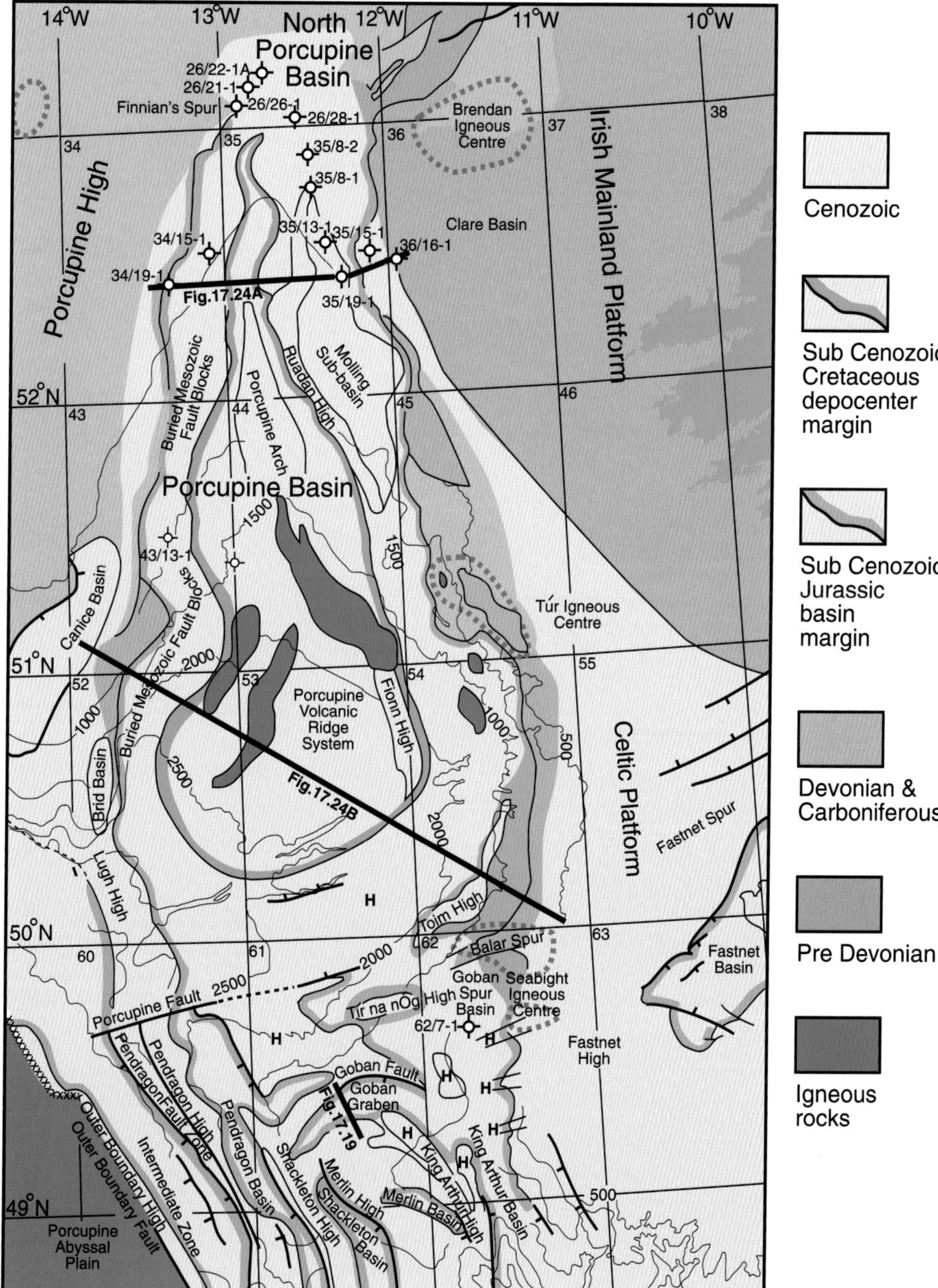

Fig. 17.5 Structural features in the Porcupine Basin and Goban Spur region (after Naylor *et al.*, 2002). Margins of buried early Mesozoic basins and highs beneath the Cenozoic cover are outlined, as is the margin of the main Cretaceous depocentre. Positions of wells mentioned in the text are shown.

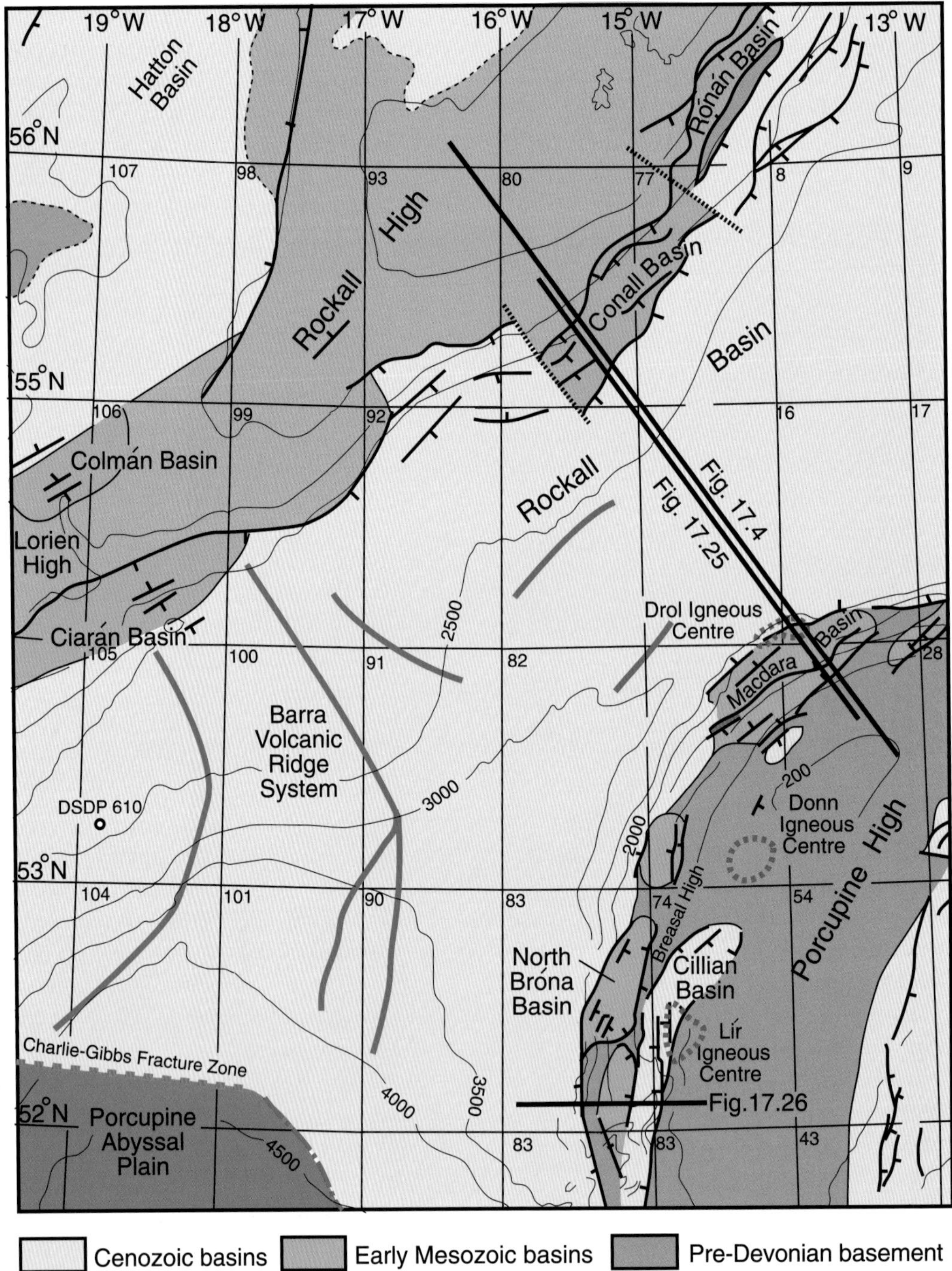

Fig. 17.6 Basins and structural features in the Rockall region (after Naylor *et al.*, 1999).

set of small basins in the footwall of the main Rockall Basin (the Colm, Fursa and Macdara basins of Naylor *et al.*, 1999) along the 54°N sector are oblique to, and intersect, the margin in much the same way as the Leannan and related Caledonian faults are shown as intersecting the older lineaments onshore. Bends in the Rockall Trough in the UK sector were also thought by Musgrove and Michener (1996) to be controlled by the geometry of the Caledonian foreland thrust belt. South of 54°N the eastern Rockall margin is aligned N–S parallel to the axis of the Porcupine Basin. In the Goban province the NW–SE trend is parallel to the main margin faults and to the orientation of the continent–ocean boundary.

Stratigraphy

Palaeozoic Era

Deposition within the basins of the Variscan geosynclinal system of west-central Europe began in Early Devonian time following the Caledonian orogeny. The complex Rheno-Hercynian Basin extended along the southern margin of the East Avalonian plate from south-west Ireland to south-eastern Poland and persisted throughout the Devonian and Carboniferous (*see* Franke, 1989 and Friend *et al.*, 2000 for discussion). The Devonian was a period of extension and transtension, alternating with periods of compression related to the continued docking of Gondwana-derived fragments. The late Viséan saw the onset of Variscan deformation, as the margins of Gondwanaland and Laurussia increasingly impinged on each other. Compressional forces became dominant in the Late Carboniferous and deposition was largely terminated in latest Carboniferous time by the northwards encroaching deformation fronts of the Variscan orogeny.

Offshore distribution of Palaeozoic and older rocks

Much of the evidence regarding the offshore distribution of Palaeozoic rocks in the offshore region is derived from gravity and seismic surveys. Only a scattering of the deep petroleum exploration wells have penetrated the pre-Permian section, and many of these encountered Carboniferous strata. In this chapter we follow Sevastopulo (Chapter 10 of this volume, and Table 10.1) in stratigraphical nomenclatural usage for the post-Viséan Carboniferous section. Specifically the stages (now reduced in rank to regional substages) of Ramsbottom *et al.* (1978) and Owens *et al.* (1985) are employed, together with the Asturian substage, whilst 'Stephanian' and 'Autunian' are retained informally as regional stages. Where possible, the agreed global series and stages are utilised.

The high velocity basement that underlies the Mesozoic basins in the Celtic Sea and forms the dividing platform (Pembrokeshire Ridge–Labadie Bank High) between the parallel set of basins is probably composed of Upper Palaeozoic rocks (Figure 17.3). Within the Celtic Sea basins at least three wells (Marathon 48/30-1, Marathon 58/3-1 and Esso/Marathon 56/20-1) bottomed in deformed Carboniferous shales (Colin *et al.*, 1981; Gardiner and Sheridan, 1981; Higgs, 1983). Palynological studies on a seabed sample from the shelf north of the North Celtic Sea Basin indicate a mid-Tournaisian age (Delanty *et al.*, 1981), while a Tournaisian age was also obtained from the Esso-Marathon 48/30-1 well (Gardiner and Sheridan, 1981). The lithology and age of these samples indicate a southwards extension of the relatively shallow marine conditions and mudrock facies that obtained onshore in southernmost Ireland during the Tournaisian.

Coal Measures are extensively developed along the southern margin of the Palaeozoic massif of St. George's Land. Asturian Coal Measures are also recorded in well Texaco/HGB 103/2-1 within the St George's Channel Basin (Barr *et al.*, 1981) and might therefore be expected to occur in the eastern part of the North Celtic Sea Basin.

In the Fastnet Basin, pre-Permian 'basement' comprises Devonian clastics and Mississippian shelf limestones (Robinson *et al.*, 1981). The Elf Aquitaine 55/30-1 well bottomed in Devonian (possibly Frasnian) continental red beds and tuffs, while Cities Services 63/4-1 penetrated Misssissippian (late Tournaisian) shelf limestones, suggesting that a carbonate shelf existed south of the muddy clastic facies tract at this time.

Upper Palaeozoic rocks have a widespread occurrence in the Atlantic basins west of Ireland (Robeson *et al.* 1988; Tate and Dobson, 1989a, 1989b). In the following descriptions the nomenclature of Naylor *et al.* (1999, 2002) is used for the major tectonic elements in the region. At the western faulted margin of the relatively shallow water Celtic Platform, the subsurface features of the Fastnet High and Fastnet Spur separate the Fastnet Basin from the Goban Spur Basin (Figures 17.3 and 17.5).

The Goban Spur region is composed of a set of fault-bounded basins and structural highs (Naylor *et al.*, 2002). The positive elements appear to be cored by Palaeozoic or older rocks, often with only thin Mesozoic and Cenozoic cover. Three NNW–SSE trending narrow horsts (the Merlin, Shackleton and Pendragon Highs) with intervening narrow basins form the outer part of the Goban Spur, developed in the footwalls of successive down-stepping faults. DSDP Site 548 on the Merlin High (Graciansky *et al.*, 1985) at its base penetrated 20 m of quartzite and black shales, the latter containing Middle to Upper Devonian palynomorphs. On the basis of seismic interpretation, the Shackleton High is a similar feature, with only Cenozoic cover at its crest. The seaward Pendragon High was penetrated at DSDP Site 549 (Graciansky *et al.*, 1985) that terminated in 37 m of foliated sandstones, believed to be Devonian (Old Red Sandstone) in age. It appears likely (Cook, 1987; Naylor *et al.*, 2002) that Upper Palaeozoic rocks are widespread beneath the Variscan unconformity in the Goban region. Dredging on the margin of the Goban Spur (Auffret *et al.*, 1979) had yielded abundant samples of granitic rocks, some clearly *in situ*, with radiometric ages of 275 Ma and 290 Ma, indicating a Variscan intrusive episode. Bioclastic limestones of possible Viséan age were also dredged on the southern margin of the Spur (Auffret *et al.*, 1979), together with fragments of high-grade metamorphic rocks and lithologies of likely Upper Palaeozoic affinity. Three basement elements (Figure 17.5), together with the Porcupine Fault, separate the Porcupine Basin from the Goban Spur Basin – the Balar Spur, the Tír na nÓg High and the Tóim High – all of which, from seismic evidence, are thought to be cored by Upper Palaeozoic sequences (Naylor *et al.*, 2002).

From the evidence presented above it is clear that the pre-Permian Variscan surface of the Celtic Sea and Goban regions is floored mainly by Upper Palaeozoic rocks. Along the Porcupine and Rockall margin, however, positive structural elements with pre-Upper Palaeozoic cores provide a regional framework separating the younger basins. These will be described first, before considering the distribution and stratigraphy of the Upper Palaeozoic sequences.

A series of prominent highs separate the inboard basins along the west coast of Ireland from the Rockall Basin. In the south the Porcupine High is a basement feature that underlies the bathymetric elements of the Porcupine Ridge, Porcupine Bank and part of the Slyne Ridge. The high generally has a thin Cenozoic cover, but the bathymetrically shallower parts may be bald. The pre-Permian geology comprises metasediments of probable Precambrian age, Caledonian granodiorites and Palaeozoic low-grade metamorphic rocks, strongly influenced in their distribution within the Porcupine Bank–Slyne Ridge area by westward extensions of the Great Glen, Leannan and Clew Bay Fault systems (Riddihough and Max, 1976; Bailey *et al.*, 1977; Max, 1978; PESGB, 2005). Cole and Crook (1910) and Auffret *et al.* (1987) recovered Lower Palaeozoic or older metasediments and gneisses by dredging on the Porcupine Ridge and Porcupine Bank.

North of the Porcupine High the positive eastern rim of the Rockall Basin, separating it from the Slyne, Erris and Donegal Basins (Figures 17.7, 17.8), is narrower, less continuous and crossed by major transverse faults. A narrow feature, the Slyne High, extends northwards from the Porcupine High and separates the Colm Basin from the Slyne Basin (Naylor *et al.*, 1999). Further north (Cunningham and Shannon, 1997; Chapman *et al.*, 1999) the Erris High separates the Erris Basin from the main Rockall Basin. The high probably consists of Lower Palaeozoic and/or older rocks and is overstepped by relatively thin Cenozoic sequences. The feature is a fault-bounded elongate horst that plunges southwards, where it probably has a preserved Upper Palaeozoic cover, to form the core of a tilted Mesozoic fault block (Naylor *et al.*, 1999). Dredge samples reported by Cole and Crook (1910) along the margins of the Slyne, Erris and Donegal basins contain a wide range of metamorphic rock types (probably Dalradian) and also Carboniferous-type sandstones, shales and limestones. However, some of this material may have been transported by ice-rafting. The Amoco 12/13-1A well in the northern Erris Basin bottomed in low-grade metasediments of uncertain (possibly Upper Palaeozoic) age. Arkosic Holkerian basal clastics and Brigantian conglomerates with granite fragments (Tate and Dobson, 1989b) in the Amoco 19/5-1 well (Erris Basin) indicate derivation from nearby granitic or metamorphic terrain.

Further west, towards the continental margin, the Rockall High (Figure 17.6) has a thin cover of sediments and Palaeogene lavas resting on acoustic basement. Dredge and drill samples have been dated from the southern part of the Rockall Bank (Miller *et al.*, 1973). It was initially thought that Lewisian (Scourian and Laxfordian), together with Grenvillian metamorphic rocks, were represented in the samples. However, more recent dating techniques (Morton and Taylor, 1991) indicate formation of the Rockall High basement rock sample suite at about 1625 Ma (Late Laxfordian), although the Rockall rocks are considered by Daly *et al.* (1995) to be comparable to those of the Annagh Gneiss Complex of north Mayo, onshore Ireland. This would suggest a somewhat older age of about 1750 Ma for the main phase of crustal growth in Rockall. Also, the southernmost of the sample suite on Rockall Bank is possibly younger (987 ± 5 Ma) and may represent the Grenvillian episode (Miller *et al.*, 1973).

High seismic refraction velocities and reworked granulites and lavas in DSDP boreholes 403 and 404 on the southern margin of Edoras Bank suggest a positive element beneath the Bank. This evidence also indicates that the Bank, together with the adjacent Hatton High, is geologically similar to the Rockall High (Morton, 1984; Morton and Taylor, 1991; Roberts *et al.*, 1979). However, Hitchen (2004), interpreting seismic data acquired across Hatton Bank in UK waters, reported that the High does

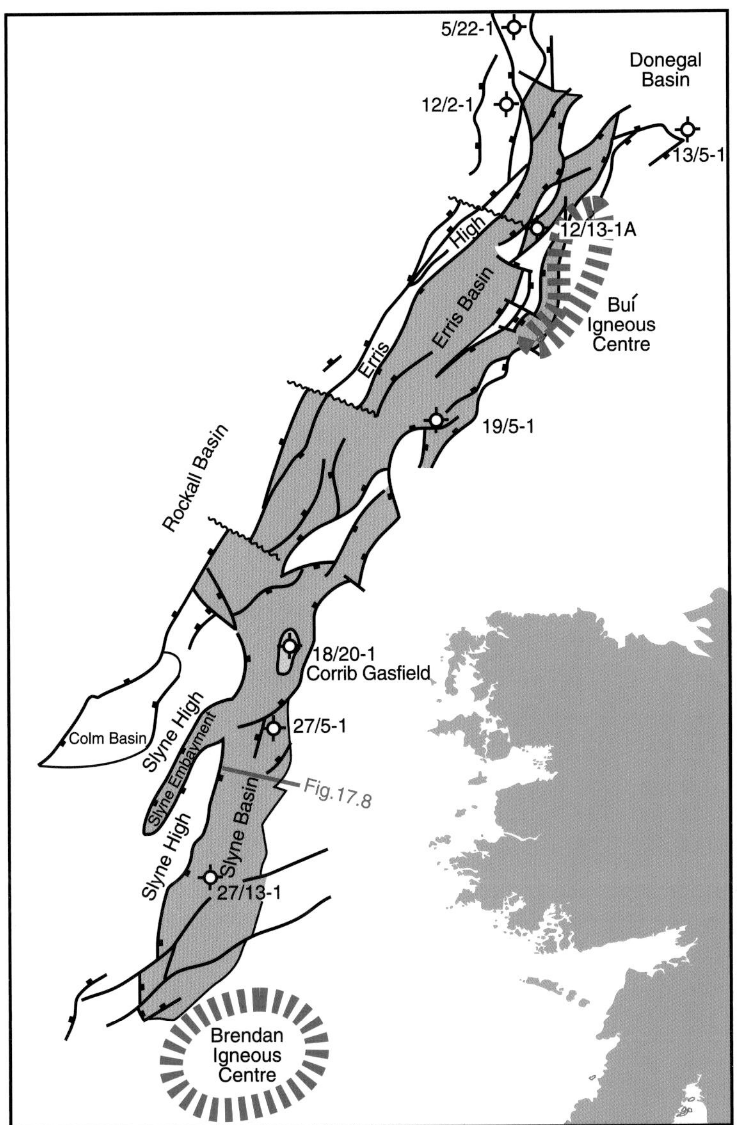

Fig. 17.7 Structural elements in the Slyne and Erris basins (*after* Naylor *et al.*, 1999). The location of the igneous centres is based on gravity and magnetic data.

Below Fig. 17.8 Cross-section through the Slyne Basin (*after* Naylor *et al.*, 1999).

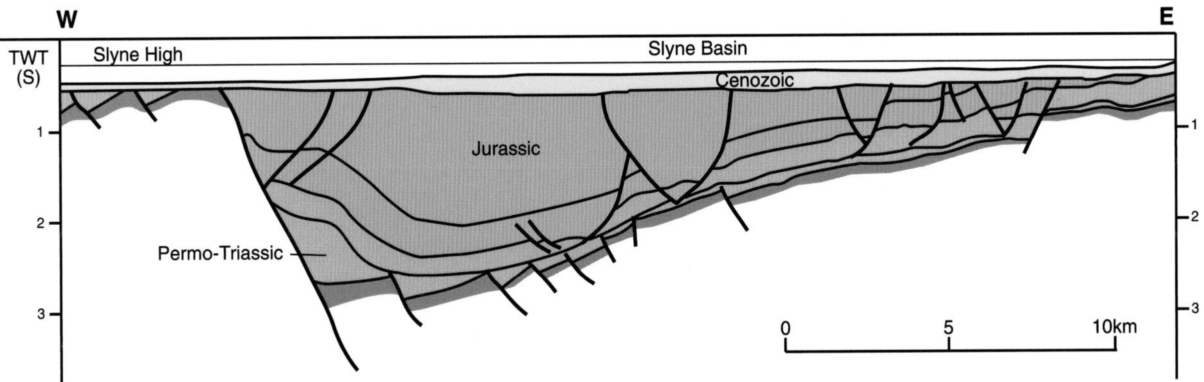

not comprise a simple Precambrian element, but includes discrete faulted basins with probable Mesozoic and possibly older sediments.

Data from offshore petroleum wells (Croker and Shannon, 1987; Robeson *et al.*, 1988; Tate and Dobson, 1989a, b) have added substantially to knowledge of

Carboniferous stratigraphy in the Atlantic basins. Available geophysical data indicate a westwards offshore extension of the onshore Carboniferous West Clare Basin (Gill, 1979) across the shelf towards the Porcupine Basin, and Croker (1995) named this the Clare Basin (Figure 17.5). In the Porcupine Basin, basement is overlain by a thick Carboniferous succession, comparable in the main to that of onshore Ireland. However, Duckmantian to Stephanian rocks appear to be widespread, in contrast to onshore Ireland, where a thicker original sequence was probably eroded following Variscan uplift. Suggested correlations with onshore units have been made by Robeson *et al.* (1988) and Tate and Dobson (1989b). The well data demonstrate that positive elements at the basin margins continued to act as sediment source areas through Carboniferous time. The Shell 26/26-1 well on Finnian's Spur (separating the Porcupine Basin from the North Porcupine Basin; Figure 17.5), encountered terrestrial to marginal-marine Mississippian strata, resting unconformably on probably Dalradian schists and gneisses (Robeson *et al.*, 1988). The nearby Phillips 26/30-1 well is reported to have bottomed in coarse-grained biotite granodiorite, possibly slightly metamorphosed, unconformably overlain by Bolsovian sediments with a basal 'granite wash'. A Duckmantian section in the BP 26/28-1 well contains conglomerates with a range of large clasts, including schist and granite, with a proximal source from a metamorphic and granitic terrain.

Mississippian strata (with a cumulative thickness of more than 1000 m) were penetrated by the Shell 26/26-1 and Phillips 35/15-1 wells in the northern Porcupine and in the Amoco 19/5-1 well in the Erris Basin. Normal marginal marine to fully marine sequences are recorded. In the Erris Basin (Figures 17.7, 17.8) Holkerian continental clastics pass upwards to a marine Asbian succession, generally carbonate-dominated. More varied sub-littoral to marine Brigantian sediments are overlain by Serpukhovian deltaic clastics with coals, terminating in an unconformity. This succession is comparable to the same interval in the north-west of Ireland documented by Higgs (1984). Langsettian beds begin the overlying succession in the Erris Basin and in the main Porcupine Basin (Chevron 36/16-1: Robeson *et al.*, 1988). However, the Shell 26/26-1 well on Finnian's Spur encountered Duckmantian resting on Serpukhovian. Extensive deposition was established across the region (Figure 17.9), with Coal Measures developed in the Porcupine region in the Langsettian, but later (Duckmantian) in the Erris–Donegal basins, and deposition continued into Asturian time (Robeson *et al.*, 1988). There is a gradual upward transition from paralic Coal Measures to red beds. Well Texaco 13/3-1 in the Donegal Basin encountered a thick Duckmantian to Asturian Coal Measure sequence beneath the Jurassic, and terminated in the Duckmantian.

Stephanian rocks are known from possibly six wells in the north and west of the Porcupine Basin (Robeson *et al.*, 1988; Ziegler, 1981; Croker and Shannon, 1987). The overlying Autunian section is slightly anhydritic and is thought to be the product of non-marine playa deposition in arid conditions. Permian strata have not been encountered by wells in the Porcupine area, so that the Triassic rests unconformably on the eroded Carboniferous surface, which is often reddened. In the Erris Basin, however, the uppermost Carboniferous is unconformably overlain by Upper Permian strata.

The offshore basement geology around the north coast of Ireland comprises ridges and platforms extending south-westwards from Scotland, and bounded by major faults (Figure 17.10). North-west of the major Leannan–Loch Gruinart Fault lies the Islay–Donegal basement platform. Northwest of the Foyle Fault, a Dalradian ridge extends from Islay through Middle Bank to the north Irish coast. To the south-east, the Rathlin and Larne basins are separated by a ridge of Dalradian metamorphic rocks, the Highland Border Ridge, that also contains younger rocks. Seismic interpretation by Shelton (1995) shows Carboniferous rocks as underlying much of the North Channel Basin, and resting on Lower Palaeozoic basement.

South of the North Channel Basin a broad syncline of Carboniferous rocks occupies much of the offshore area in the northern Irish Sea (Institute of Geological Sciences 1:250,000 Isle of Man Sheet). Erosion, following upon basin inversion or thermal doming during the Variscan episode, removed the Pennsylvanian sequence beneath the Solway and Peel basins (Figure 17.11). The preserved Carboniferous sequence beneath the unconformity (Newman, 1999) comprises Late Brigantian to Pendleian shallow marine–deltaic sandsones, shales and carbonates in the north-east (Elf 112/19-1: Solway Basin) passing to Holkerian–Asbian shallow-water carbonates in the south-west (Elf 111/29-1: Peel Basin). Pennsylvanian rocks are, however, widespread in the central and southern Irish Sea. The Amoco 33/22-1 well (Figure 17.12) on the margin of the Kish Bank Basin drilled a Duckmantian to Asturian (or possibly Stephanian) section 722 m thick, resting directly on Lower Palaeozoic strata (Jenner, 1981; Naylor *et al.*, 1993). The succession commences with a

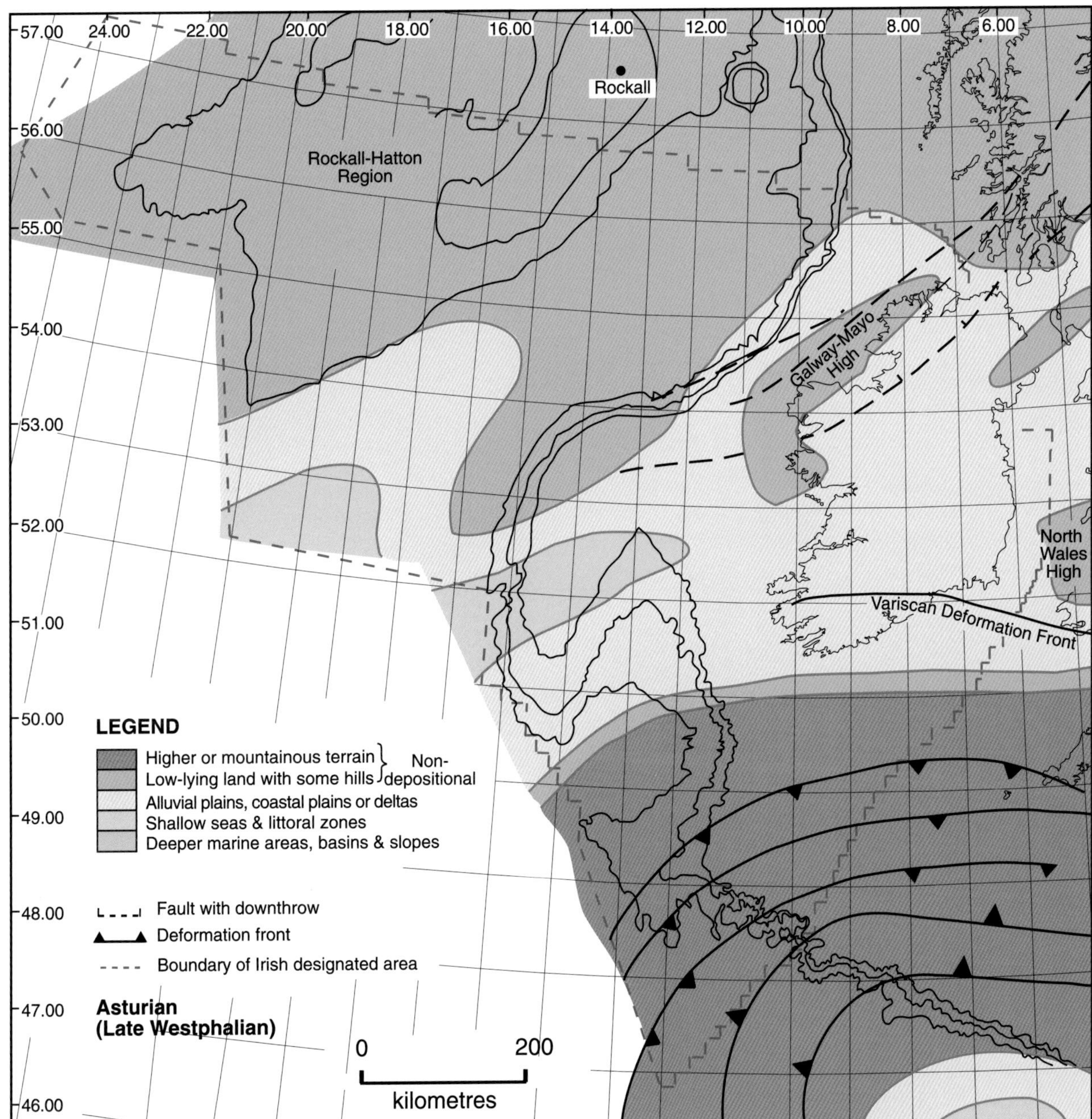

Fig. 17.9 Asturian (late Westphalian) palaeogeography (*after* Naylor, 1998. *Principal sources:* Robeson et al., 1988; Tate and Dobson, 1989b; Ziegler, 1990; Cope *et al.*, 1992b).

basal sandstone horizon overlain by Duckmantian cyclothemic siltstones and sandstones succeeded in turn by Bolsovian fine-grained clastics with coals. The Asturian is represented by 50 m of cyclothemic deposits, topped by reddened siltstones and shales. A reworked miospore assemblage from cuttings at 338 m in the Amoco 33/22-1 well (Asturian) includes late Viséan to late Langsettian taxa (Naylor *et al.*, 1993), suggesting uplift and erosion of older Carboniferous rocks during mid-Pennsylvanian time. Reworked miospores ranging in age from Devonian to Langsettian were also recorded from an onshore Asturian section in County Wexford (Clayton *et al.*, 1986b), suggesting that the erosional episode was relatively widespread.

Wells on the southern margin of the Central Irish Sea Basin (Figure 17.13) have penetrated Duckmantian and Bolsovian deposits overlain by reddened beds of probable Asturian-Stephanian age (Maddox *et al.*, 1995).

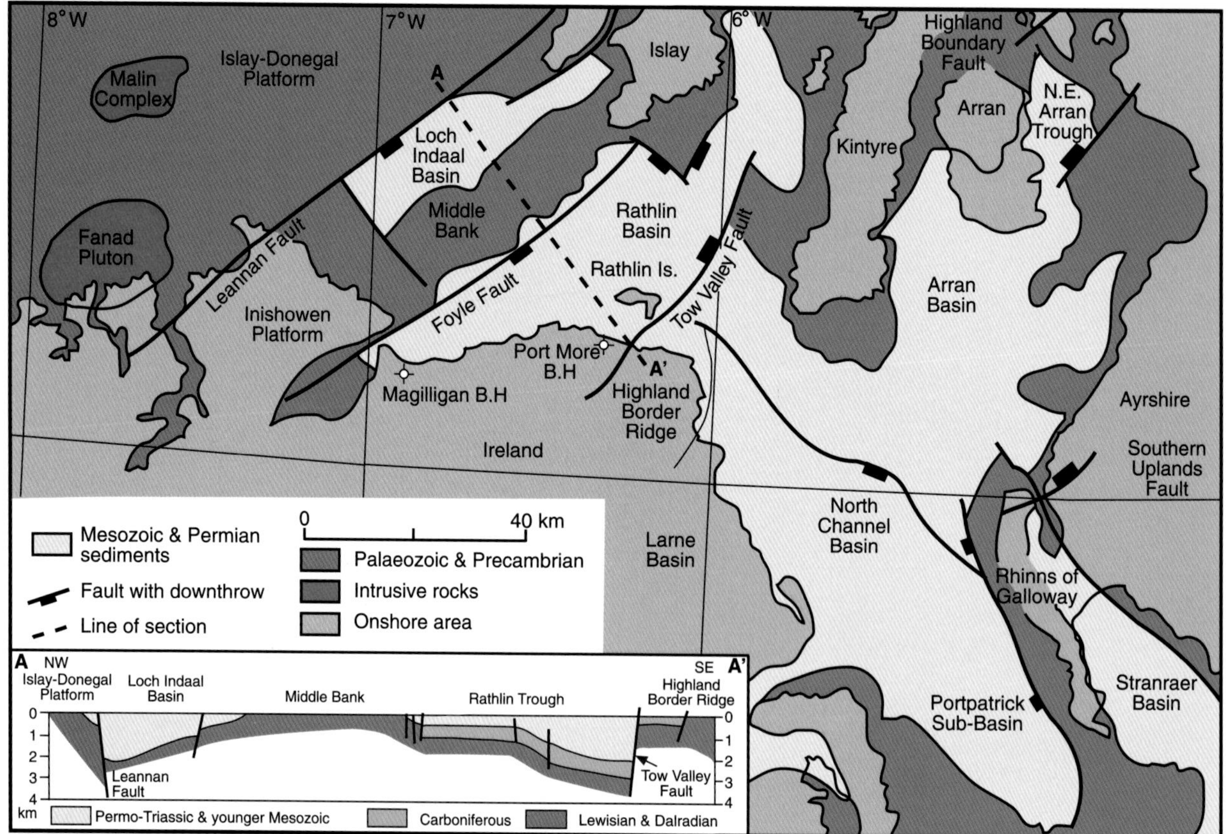

Fig. 17.10 Structural elements between south-west Scotland and north-east Ireland (based on Evans *et al.*, 1980; Naylor and Shannon, 1982; Maddox *et al.*, 1997). *Inset*: Diagrammatic cross-section of the Loch Indall and Rathlin basins (adapted from Evans *et al.*, 1980).

On the southwestwards extension of St Tudwal's Arch (Marathon 42/17-1) Langsettian–Duckmantian deposits are unconformable on Viséan carbonates, the intervening interval being absent through non-deposition or erosion, but possibly represented in the deeper undrilled parts of the basin.

Upper Palaeozoic palaegeography

The data presented above provide evidence for the widespread preservation of Upper Palaeozoic, and particularly Carboniferous, strata in the Irish offshore, with intervening positive elements cored by Caledonian and older rocks. This disposition is largely the result of erosion on the Variscan landscape, although undoubtedly some upstanding elements were further modified by erosion during post-Variscan tectonic phases. A number of structural highs also acted as sediment sources during Upper Palaeozoic deposition, particularly as Variscan deformation encroached into the area. Latest Carboniferous sedimentation probably took place in basins of limited areal extent.

Terrestrial desert conditions prevailed throughout the Devonian over Ireland, and probably also over much of the offshore regions. Deposition was generally in localised alluvial fans, alluvial plains, and playa lake depocentres. Marine conditions had been established in Devon and Cornwall and a shoreline lay somewhere south of Fastnet. The thick (up to 6 km) Old Red Sandstone sequences of the onshore Munster Basin probably range down into the Givetian in age and must extend some distance westwards onto the Atlantic shelf. Much farther south, in the area of the present Goban Spur, DSDP drilling (Sites 548 and 549) encountered sediments of Middle–Upper Devonian age, and it appears likely that similar rocks may be widespread in that region. Events unfolding in the southern ocean, with the northward advance of the Variscan deformation fronts gave rise to a northward marine incursion that reached the southern shoreline of Ireland in Late Famennian time.

Old Red Sandstone facies sediments continued to be deposited across much of onshore Ireland during latest Devonian time, whilst high ground is believed to have

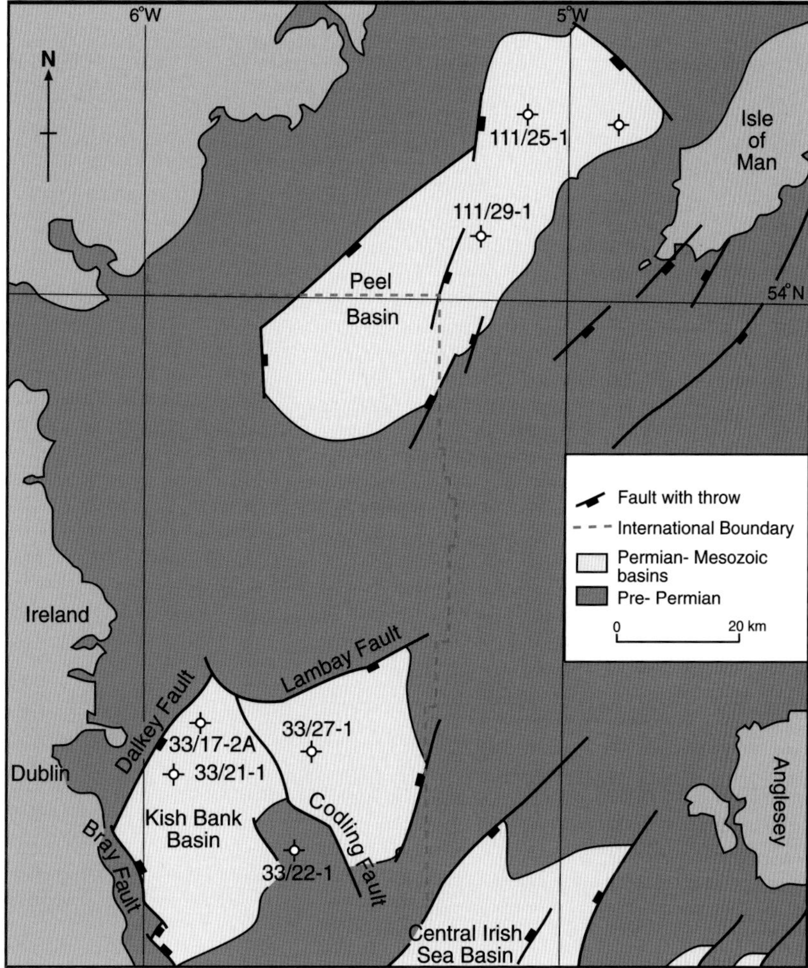

Fig. 17.11 Permian–Mesozoic basins in the northern part of the Irish Sea.

occupied the Galway–Mayo region (Cope *et al.*, 1992a). The general absence of Devonian rocks encountered in offshore drilling in the north Porcupine region (Robeson *et al.*, 1988), suggests that the Galway Mayo High may continue as an upland extension south-westwards into this region.

During the Early Carboniferous the sea continued to advance northwards across the present onshore area of Ireland, and presumably across the adjacent offshore regions. Red bed deposition continued north of the shoreline throughout the period (Clayton and Higgs, 1979). By latest Tournaisian time the sea had transgressed across much of the Irish mainland, although in the north the Galway–Donegal area and the south-west extension of the Southern Uplands were still emergent (Naylor, 1998). The Welsh High (St George's Land) probably also extended across the present Irish Sea into Leinster. It is likely that the region from northern Scotland westwards across the Rockall–Hatton region was largely a continental area of bare plains and low hills, with limited deposition. Carbonate sediments were deposited across much of Ireland with a transition in the south-west of the island into a southern trough with mudrock deposition (South Munster Basin) that drilling suggests extended as far south as the North Celtic Sea Basin. Drilling in the Fastnet Basin indicates that it was part of a carbonate shelf south or south-west of the mudrock-dominant trough.

There was a similar disposition of land and sea areas through Viséan and Serpukhovian times, with no obvious sign of tectonic influence from the southern domain. In the north, the Galway–Mayo and Southern Uplands Highs persisted. On the north Antrim coast, thin coal seams of Brigantian age demonstrate non-marine conditions at that time. During the Serpukhovian, deltaic sandstones prograded south-eastwards from the Caledonian uplands over substantial parts of Ireland. In the south, Serpukhovian and basal Pennsylvanian sediments were

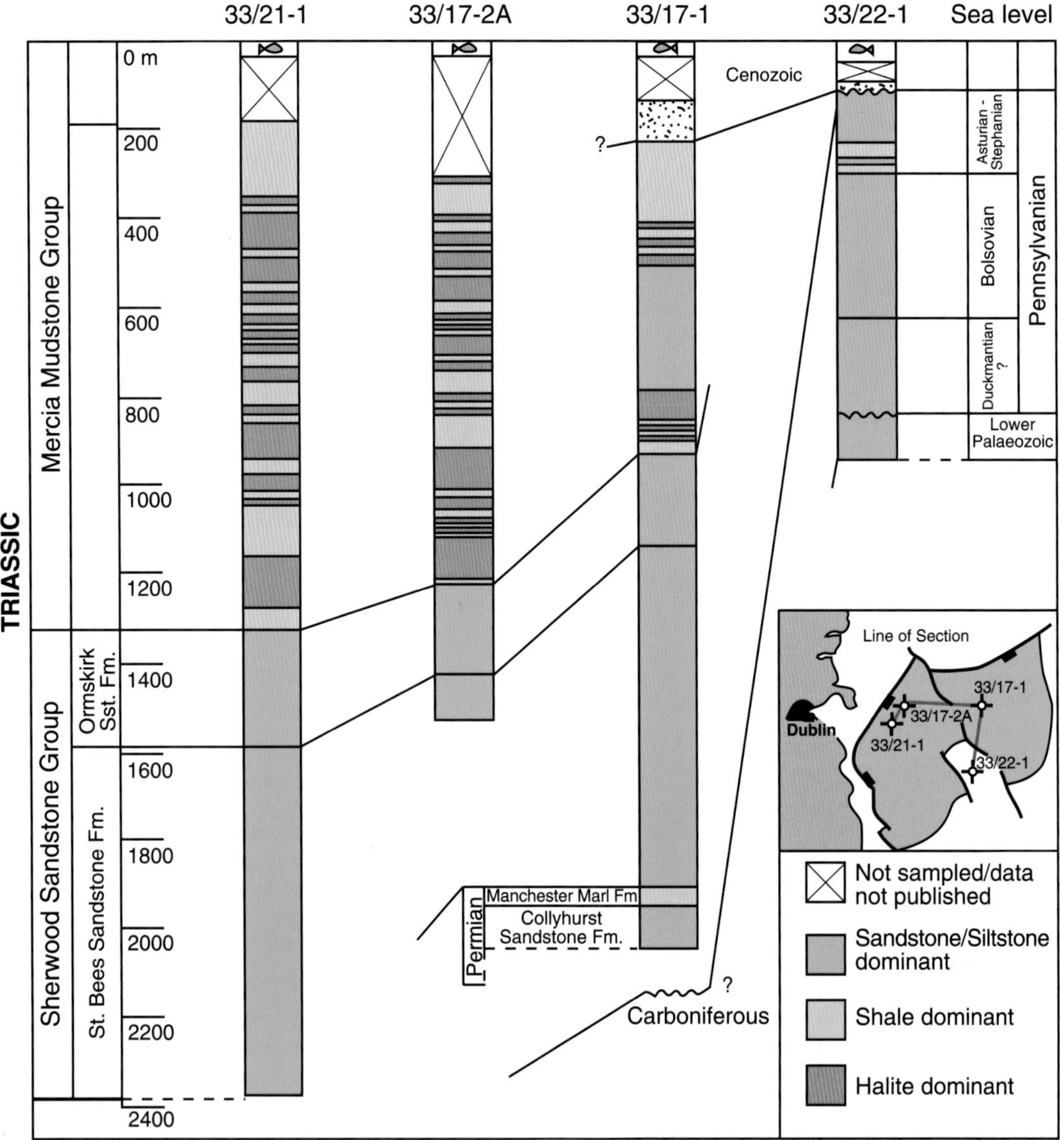

Fig. 17.12 Comparative stratigraphy of the four exploration wells drilled in the Kish Bank Basin (*after* Jenner, 1981; Naylor *et al.*, 1993; Dunford *et al.*, 2001).

probably derived from the developing orogenic areas further south. Evidence from reworking of Carboniferous palynomorphs, both in latest Carboniferous and Triassic strata (Naylor and Clayton, 2000), indicates that a more or less complete Carboniferous sequence was deposited in some parts of the Irish Sea area. As seen earlier, drilling in the Atlantic basins has shown that Serpukhovian and Pennsylvanian deposition extended westwards into the offshore regions. Langsettian to Asturian rocks are widespread offshore, in contrast to the onshore area, and Coal Measure strata are known from the Porcupine Basin northwards to the Donegal Basin. Figure 17.9 shows the palaeogeography of the Asturian.

By Stephanian time the climax of the Variscan tectonic episode had produced fold mountains extending from south-west Ireland across southern Britain and into Central Europe. Stephanian and Autunian rocks are not preserved onshore in Ireland, but have been encountered in drilling off the west coast. Stephanian sediments, apparently conformable upon the Asturian, are well documented in the northern part of the Porcupine Basin (Croker and Shannon, 1987; Robeson *et al.*, 1988). Continental sediments are interbedded with marine intercalations, and pass upwards though marginal marine deposits into anhydritic playa lake deposits. Ziegler (1982) suggested that the marine influences at this time

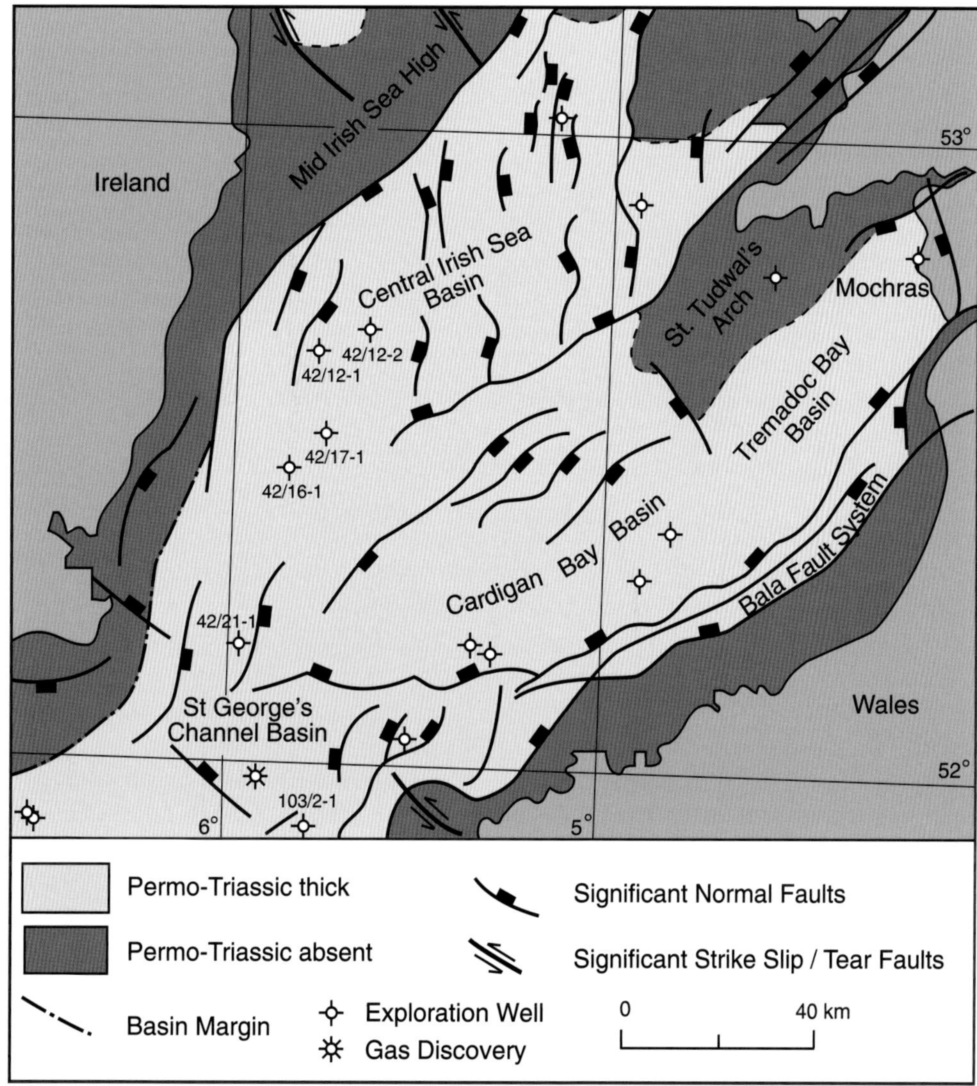

Fig. 17.13 Structural elements of the southern Irish Sea with well locations mentioned in the text (*after* Maddox *et al.*, 1995).

result from marine incursions from the Tethyan domain along the developing Bay of Biscay rift system and into the Porcupine region. This marine access appears to have been closed by the end of the Autunian.

Jackson *et al.* (1997) interpreted up to 4 km of Langsettian to Stephanian strata on seismic data south of the Isle of Man, in the Quadrant 119 Syncline. This, allied to the evidence off the Atlantic coast, suggests a pattern of limited late Carboniferous sedimentation in the axial zones of basins north of the encroaching Variscan deformation front. At the same time, earlier Carboniferous and older sediments were already being eroded at the basin margins.

Permian and Triassic Periods

In Early Permian time, the welding of Laurussia and Gondwanaland along the Variscan (Hercynian) megasuture was complete and the new supercontinent of Pangaea was formed. Continued seafloor spreading in the Pacific was accompanied by rotation of the whole supercontinent (Ziegler, 1982). The resultant stresses, together with the thermal instability that built up rapidly beneath the supercontinent, gave rise to reactivation of earlier fractures and then to the initiation of a set of small and initially isolated rift systems. Although none of the rifts developed into new spreading axes or to the production of new ocean floor, some of the incipient rift systems would later become larger and more elongate and would become important precursors to events later in the Mesozoic.

In the Irish context, deposition on the post-Variscan landscape took place in continental conditions within generally small rift basins developed along reactivated NE–SW Caledonian features (Shannon, 1991a; Johnston *et al.*, 2001). Permo-Triassic rocks were deposited predominantly in non-marine red-bed facies. Seismic evidence points towards rifting in Early and Late Triassic times in different parts of the region (Shannon and MacTiernan, 1993; Musgrove *et al.*, 1995).

Offshore distribution of Permo-Triassic rocks

Unequivocal Permian strata have not been encountered in offshore wells in the Celtic Sea or Fastnet basins (Naylor and Shannon, 1982; Robinson *et al.*, 1981) where a number of wells have penetrated continental Triassic deposits resting unconformably on Palaeozoic basement. The existence of possible small pockets of Permian strata in the South Celtic Sea Basin and in other Celtic Sea basins has been suggested by Petrie *et al.* (1989), Tappin *et al.* (1994), and Shannon (1995), largely from seismic evidence. Onshore along the north coast of the Bristol Channel Upper Triassic rocks rest unconformably on deformed Devonian and Carboniferous metasediments. This is also the situation in the Bristol Channel Basin (Kamerling, 1979). Even if undrilled pockets of Permian strata exist on the shelf south of Ireland, they are clearly of limited extent. The general absence of Permian strata in the region is interpreted as reflecting the intermontane and topographically elevated setting of the region following the Variscan orogeny, with consequent poor sediment preservation potential.

In contrast, Triassic strata are well documented in the Celtic Sea region (Figure 17.3). In the North (NCSB) and South (SCSB) Celtic Sea basins, there is the classic subdivision of the Triassic into a lower fluvial sand-dominant sequence and an upper red-bed mudrock and marly succession with variable amounts of evaporites, the two locally separated by a thin calcareous unit. For example, the Esso 56/20-1 well (Colin *et al.*, 1981) near the northern margin of the NCSB penetrated a 600 m thick Triassic succession including an upper anhydritic 450 m thick Mercia Mudstone sequence (Figure 17.17). The same is also true farther west in the Fastnet Basin (Robinson *et al.*, 1981), where maximum Triassic thicknesses of 600–700 m are indicated from seismic and well data. The Sherwood Sandstone Group is probably restricted to the main topographic or fault-bounded depocentres, whilst the overlying Mercia Mudstone Group extends basinwide. Thick salts are generally absent from the Fastnet and North Celtic Sea basins, except for a narrow zone along the southern margin of the NCSB, but are present in the other basins (Tappin *et al.*, 1994), and are dramatically developed in the South Celtic Sea and St George's Channel basins.

The Mercia Mudstone Group in the Celtic Sea basins passes conformably through a transitional sequence into a fully marine Jurassic succession above. This marine transgression was of Rhaetian to Hettangian age. Non-marine and restricted marine facies were dominant in the Rhaetian (Millson, 1987) and these environments persisted into earliest Jurassic time in marginal and uplifted areas. In the Fastnet Basin (Figure 17.3), Triassic red beds are succeeded by Rhaetian marine carbonates (Figure 17.18B). To the east, the Rhaetian of the Bristol Channel Basin (Kamerling, 1979) is more obviously transitional to marine conditions and comprises dark shales with occasional thin limestones.

Permo-Triassic strata are interpreted as extending throughout both the Goban Spur Basin and Goban Graben depocentres (Figure 17.5; Naylor *et al.*, 2002). The Triassic red beds are succeeded with apparent conformity by Rhaetian marine strata. Rifting of the Variscan basement in the Goban Spur region probably began in mid-Triassic time (Cook, 1987). The overlying succession, interpreted from seismic evidence as marly and evaporitic, is thought to have acted as a detachment zone, where a deformed upper sequence contrasts with the relatively simple faulted horst style of the basement.

Post-Autunian (latest Pennsylvanian) Permian strata have not been encountered by drilling in the Porcupine Basin (Figure 17.5), where locally presumed Triassic strata rest unconformably on the eroded Variscan surface. The Triassic of the basin is little known and poorly dated, with only two of the published wells (BP 26/22-1A and Gulf 26/21-1), both in the north of the basin reportedly penetrating rocks of this age (Figure 17.20). However, the Britoil 35/19-1 well recorded salt of undated age and with unclear stratigraphic relationships. In the BP 26/22-1A well, massive shoreline sandstones are overlain by glauconitic offshore stacked bar sands, and these in turn by evaporitic shales (Croker and Shannon, 1987). These authors suggested, from seismic evidence, that Triassic sediments may be preserved in small rift basins along the south-west and possibly the south-east margins of the Porcupine Basin, and stressed the importance of reactivated Caledonian lineaments in controlling both Triassic and Lower Jurassic sedimentation. Seismic evidence suggests that more than 2 km of Triassic rocks may occur in other parts of the basin (Naylor *et al.*, 2002), while Johnston *et al.* (2001) interpreted Triassic strata as being widespread, if poorly imaged, on seismic sections in the southern part of the Porcupine Basin. Where the Triassic has been drilled in the Porcupine it is conformably succeeded by a limestone–shale succession of Rhaetian age.

Further north along the Atlantic margin (Figure 17.7), the Amoco 19/5-1 well in the Erris Basin penetrated Zechstein-equivalent marine mudstones, dolomites and anhydrites following unconformably on the Carboniferous, while the Amoco 12/13-lA well encountered similar Upper Permian rocks (Tate and Dobson, 1989b) resting on red-bed sandstones and shales of possible Lower Permian age (Murphy and Croker, 1992), although these could be Autunian in age. On seismic evidence, a base Upper Permian–Triassic sequence boundary is recognised throughout the Erris Basin (Chapman *et al.*, 1999). In the Slyne Basin (Dancer *et al.*, 1999) the Enterprise 27/5-1 well drilled a massive evaporite sequence beneath the Triassic Sherwood Sandstone, also assumed to be Zechstein equivalent.

Little is known of the latest Palaeozoic and earliest Mesozoic history of the Rockall region (Figure 17.6). Shannon *et al.* (1995) speculated that Permo-Triassic sedimentation probably extended into the Rockall Basin. Naylor *et al.* (1999), interpreting seismic data, showed a Permo-Triassic sequence within the string of perched Mesozoic basins along the western margin of the Porcupine High. Permo-Triassic to Lower Jurassic sequences, if locally preserved in the basin, are likely to comprise continental to shallow marine sandstones, evaporites and limestones deposited during the initial rift phase marking the breakup of the Pangean supercontinent.

Triassic rocks within the north-west offshore basins were penetrated in the Elf 27/13-1 (Rhaetic only: Scotchman and Thomas, 1995), Enterprise 27/5-1 and Enterprise 18/20-1 wells of the Slyne Basin (Trueblood, 1992; Dancer *et al.*, 1999), and in the Amoco 19/5-1 and Amoco 12/13-lA wells in the Erris Basin (Tate and Dobson, 1989a; Chapman *et al.*, 1999). In the Corrib Gasfield at the northern end of the Slyne Basin (Dancer *et al.*, 2005; Corcoran and Mecklenburg, 2005) a sandstone-dominant Lower Triassic Sherwood Sandstone Group equivalent (400–420 m) is overlain by a Mercia Mudstone Group comprising saliferous and anhydritic mudstones with a thick basal salt section. In general in these basins, limestones become more frequent upwards, suggesting that nearshore sabkha-lagoonal environments were succeeded by deepening conditions as a consequence of the Rhaetian marine transgression. Volcaniclastic beds near the top of the sequence demonstrate early Mesozoic volcanism in the region (Tate and Dobson, 1988).

The Loch Indaal Basin (Figure 17.10) lies off the north coast of Ireland between the positive elements of the Islay–Donegal Platform and Middle Bank. The basin contains a Triassic continental sequence more than 2 km thick, but Permian rocks are unproven. Seabed boreholes in the area (Evans *et al.*, 1980) yielded probable Upper Triassic gypsiferous marls. In the north-eastern portion of the basin there is a transition from the Mercia Mudstone Group to ammonitic black mudstones of the overlying Penarth Group (Westbury Formation; Rhaetian to Early Jurassic), reflecting the ubiquitous latest Triassic to earliest Jurassic marine transgression.

In the onshore portion of the Rathlin Basin, south-east of Middle Bank and the Foyle Fault and near the western basin margin, the Magilligan borehole (Figure 17.10) penetrated a Lower Liassic to Permian sequence (976.3 m) resting on Carboniferous rocks with thin coals. Further east in the same basin, a borehole at Port More was in probable Permian sandstones, beneath a thick Triassic sequence, when terminated at 1864 m. These strata are of continental facies and extend offshore to form the bulk of the 2.4 km thick fill of the offshore portion of the Rathlin Basin. Seismic data indicate that the thickest development is against the Tow Valley bounding fault. Seabed boreholes in the area (Evans *et al.*, 1980) demonstrated gypsiferous marls, presumed to belong to the Mercia Mudstone Group, succeeded by Rhaetic marls and mudstones.

Drilling onshore in the Larne Basin, south of the Highland Border Ridge (*see* Illing and Griffith, 1986 for summary), has also revealed a thick Permo-Triassic succession. The stratigraphy and development of the onshore basins in Ulster are covered elsewhere in this volume. The Larne-2 borehole (Penn, 1981) in the Larne Basin encountered more than 1000 m of Lower Permian strata, including a basal section (617 m thick) of volcanic tuffs, breccias and lava flows, demonstrating an extension of the Scottish volcanic province into this region. Continental red-brown mudstones and major halite horizons more than 200 m thick characterise the Upper Permian. The overlying Triassic sequence in the borehole (1599 m) is divisible into the standard Sherwood Sandstone and Mercia Mudstone groups (Warrington *et al.*, 1980; Penn, 1981; Parnell, 1992), with two thick halite developments within the Mercia Mudstone Group.

The Permo-Triassic sequence penetrated at the coast by the Larne-2 borehole is seen on seismic records to extend seawards into the main North Channel Basin (Figure 17.10), where it is as yet undrilled (Maddox *et al.*, 1997; Shelton, 1997). The Permo-Triassic succession in the offshore North Channel Basin is probably 3 km or more thick, thinning against the Highland Border Ridge to the north and also by overlap across a series of fault blocks towards the eastern basin margin (Shelton, 1995). A succession similar to that of the Larne borehole might be anticipated. The Portpatrick extension of the basin is a half-graben of reversed polarity lying south of the line of the Southern Uplands Fault, with a thick Permian and Triassic section developed in the east against the boundary fault along the Lower Palaeozoic positive element of The Rhinns of Galloway. Possible halokinetic structures are evident on seismic records.

Farther south in the Irish Sea, the Permo-Triassic of the Solway and Peel basins (Figure 17.11) has been penetrated in three wells– Elf 112/19-1, Elf 111/25-1 and Elf 111/29-1 (Newman, 1999). Upper Permian mudrocks and evaporites (up to 162 m) rest on Carboniferous strata, except in well 111/25-1 in which 62 m of Lower Permian sandstones (Collyhurst Sandstone Formation) intervene. The Permian is succeeded by a typical red-bed Lower Triassic Sherwood Sandstone Group sequence that thins south-westwards, from 746 m in the Solway Basin (well Elf 111/25-1) to 585 m thick in the Elf 111/29-1 well of the Peel Basin. The same south-westward thinning is also present in the Upper Triassic Mercia Mudstone Group (724 m to 174 m), in which anhydritic claystones with halites dominate.

In the Kish Bank Basin (Figures 17.11, 17.12), the Carboniferous is overlain unconformably by 2000 m of Permo-Triassic rocks (Jenner, 1981; Naylor *et al.*, 1993; Dunford *et al.*, 2001). Four wells have been drilled within the basin. The Permian Collyhurst Sandstone (100 m drilled) was penetrated only at the base of the Fina 33/17-1 well (Figure 17.12) where it consisted entirely of friable fine to coarse sandstones. The Collyhurst Sandstone is overlain by a thin Upper Permian interval of red-brown claystone, slightly calcareous, with siltstone and sandstone interbeds. The Triassic succession penetrated by the Fina 33/17-1 well (Naylor *et al.*, 1993) consisted of the Sherwood Sandstone Group (1035 m) overlain by the Mercia Mudstone Group (535.2 m).

The lowest unit of the Sherwood Sandstone Group is the St Bees Sandstone Formation, dominated by red, very fine- to fine-grained, moderately well sorted, sandstones. The overlying Ormskirk Sandstone Formation is also predominantly sandy, with claystone intercalations. The sequence is comparable to that of the East Irish Sea Basin (Dunford *et al.*, 2001). The depositional environment of the overlying Mercia Mudstone Group was continental arid to semi-arid, with alternations between fluvial, lacustrine, evaporitic or aeolian conditions. The sequence is dominated by red-brown claystones, occasionally evaporitic or with minor sandstones, separated by halite developments that comprise up to 30% of the sequence in the west of the basin (Dunford *et al.*, 2001). The Mercia Mudstone Group is 535 m thick at the Fina 33/17-1 location but much thicker (1122 m) in the Shell 33/21-1 well, in part due to the presence of thicker halite units. On the basis of palynology, most of the Mercia Mudstone Group is assigned to the Scythian (Naylor *et al.*, 1993).

The Kish Bank Basin is bounded to the south by the Mid-Irish Sea High (Figure 17.13), to the south of which lies the Central Irish Sea Basin (CISB). The CISB is strongly influenced by NNE–SSW and NE–SW faults. The St Tudwal's Arch, a caledonoid-trending extension from the Welsh coastline, in turn separates the CISB from the Cardigan Bay Basin, and both basins merge south-westwards into the St George's Channel Basin (Figure 17.13). Exploration wells within the southern portion of the CISB show that Lower Triassic rocks (Sherwood Sandstone Group) rest unconformably upon deformed and reddened Carboniferous. Permian strata have not been encountered, but may exist in the undrilled depocentre in the north-east of the basin (Maddox *et al.*, 1995). Seabed samples and wells in the CISB, together with drilling on the southern margin of the Cardigan Bay Basin (Barr *et al.*, 1981), have revealed a Triassic succession comparable to that in the East Irish Sea Basin and the English Midlands, with the Sherwood Sandstone Group overlain by saliferous Mercia Mudstone Group strata. The Triassic onlaps the St Tudwal's Arch, a positive fault-controlled Lower Palaeozoic and Precambrian element extending out from the Welsh Massif. The succession on the Arch is thinner and shale-prone. Exploration wells within the southern part of the CISB (Figure 17.14) have penetrated Sherwood Sandstone Group sequences more than 230 m thick, overlain by in excess of 1400 m of Mercia Mudstone Group anhydritic and dolomitic mudstones interbedded with thick halites and minor sandstones (Maddox *et al.*, 1995).

The Rhaetic follows conformably on the Mercia Mudstone Group in the south Irish Sea, and in the

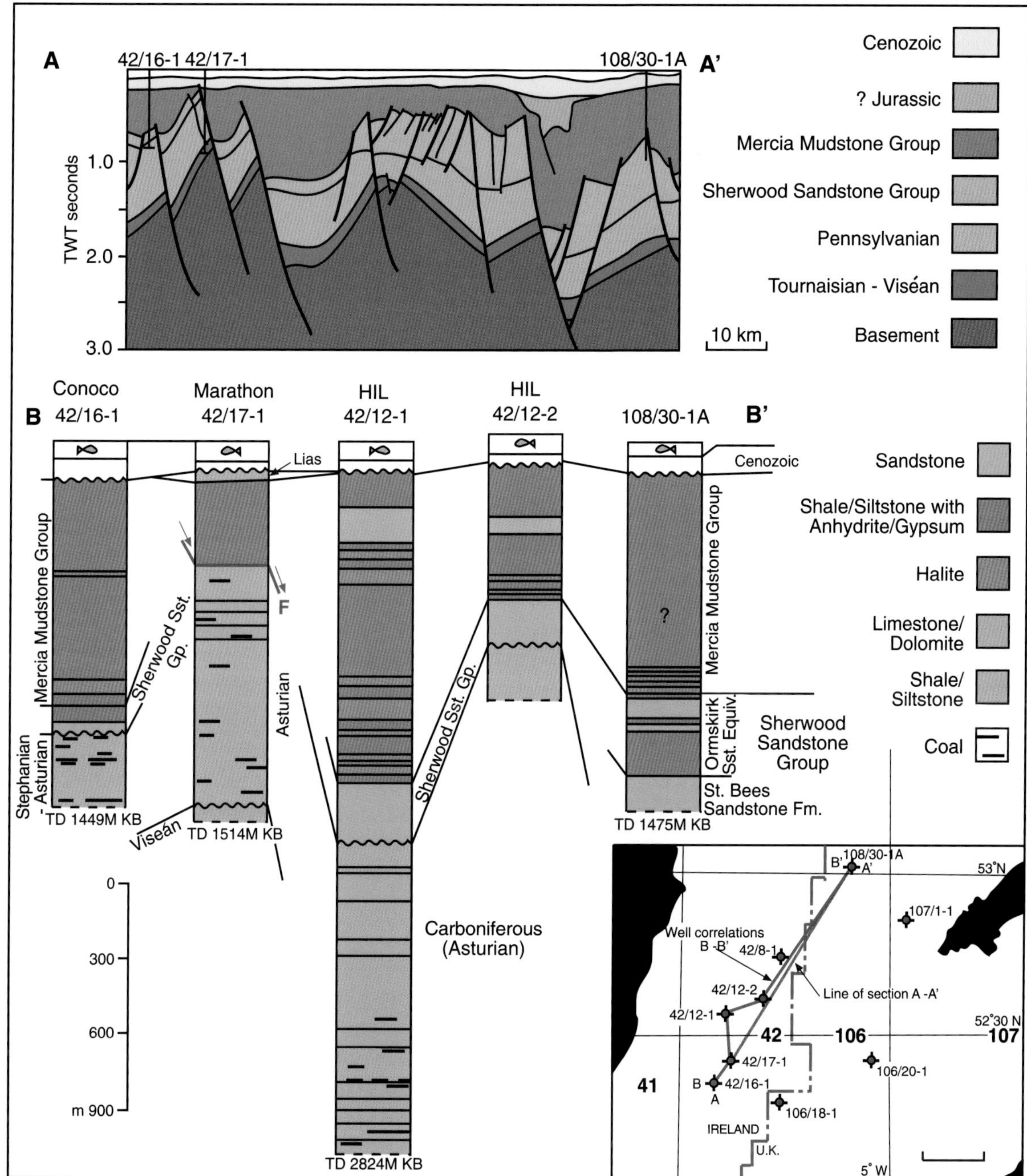

Fig. 17.14 A: Geoseismic section linking wells in the Central Irish Sea Basin (*after* Floodpage *et al.*, 2001). B: Comparative stratigraphy of wells in the Central Irish Sea Basin (based on Green *et al.*, 2001; Floodpage *et al.*, 2001).

Cardigan Bay wells is about 50 m thick and consists of red-brown claystones with occasional sandstones. The Mochras borehole on the Welsh coastline (Woodland, 1971) penetrated 48 m of dolomitic limestones above dolomitic sandstones with red-brown marl bands (Figure 17.13). The lower part of this late Triassic–Rhaetian unit is inferred to be terrestrial in origin, with increasing marine influence in the upper beds.

Offshore wells in the St George's Channel Basin encountered thicknesses of up to 2000 m of Triassic sediments. The Texaco/HGB 103/2-1 well on the southern margin of the basin penetrated 249 m of sandstones containing anhydritic mud partings resting unconformably on Carboniferous Barren Red Measures. The overlying Mercia Mudstone Group (1860 m) is dominated by red-brown mudstones, frequently anhydritic, with a saliferous middle unit. Evidence for salt diapirism is seen on some seismic profiles.

Permo-Triassic palaeogeography

Permian ~ During Permian time Ireland and much of Europe were located in low latitudes. The Variscan uplands still had considerable relief, and erosional products were deposited as fluvial and dune clastics in arid and desert conditions within limited intermontane basins. Ireland was part of a land area that extended westwards from Britain to the Porcupine Bank, with highland areas over much of the present onshore area. Permian strata are unproven on the shelf south of Ireland and, if present, are probably of limited exent. The general absence of Permian strata in the Celtic Sea region is interpreted as reflecting the topographically elevated setting following the Variscan orogeny.

A pattern emerges of subsidence and essentially non-marine deposition during the Permian along the northern and eastern coasts of Ireland to the north of the Variscan highlands. Volcanic activity extended from Scotland across into the Ulster Basin. Non-marine depositional conditions had persisted into latest Carboniferous (Autunian) time in the Porcupine Basin (and possibly also in the Slyne-Erris basins). It is likely that in latest Early Permian time rifting was active along the Greenland–Rockall rift and also the Biscay Rift separating Iberia from southern Europe. These are areas of possible deposition, with a Rockall–Hatton landmass of uncertain geography lying out to the west.

Rifting progressed in the Late Permian (Figure 17.15), and marine conditions extended from the Arctic Boreal Ocean southwards along the Norway–Greenland rift into the Rockall region, and then through the Malin Sea into the depocentres of the Irish Sea. This restricted marine area is known as the 'Bakevellia Sea'. Clastics, evaporites and minor carbonates were deposited within this basin, which had extensions into the East Irish Sea and Cheshire basins, and also into the Larne–Lough Neagh Basin. Intermittent marine flooding is envisaged, and there is evidence of contemporary control on sedimentation by basin margin faults. Wells in the north Erris and Donegal basins have also encountered Zechstein-equivalent mudstone, dolomite and evaporite sequences. The extent of marine transgression southwards along the Rockall rift region is speculative, but might be expected to have reached the latitude of the Porcupine Bank. The Rockall–Hatton landmass was probably an emergent bare area with little deposition, whilst the Irish mainland also remained a positive element, despite continued erosion and deflation of the Variscan landscape.

Triassic ~ During the Triassic Period there was a further extension of the rift system within the Pangaean supercontinent. In the North Atlantic region this resulted in further development of the Boreal Norway–Greenland and Bay of Biscay rift systems, and inception of the Western Approaches, Celtic Sea and Porcupine rifts. The Biscay Rift represented the western extension of a rift system linking back to the Tethyan Ocean in the east.

The Boreal Sea regressed in the Early Triassic, so that in Scythian time the Irish region was entirely an arid continental area with deposition of the Sherwood Sandstone Group. The Variscan mountains had suffered degradation and bare rocky hills covered large areas. Fault-controlled sequences are seen along some margins. The lowland areas were the sites for clastic deposition by seasonal or ephemeral river systems sourced in the upland areas, with further re-deposition of sand in dune fields. Deposits of this type are found throughout the basins of the Irish Sea, Celtic Sea, Porcupine and Slyne–Erris basins. Reworked palynomorphs in the Lower Triassic of the Larne-Lough Neagh Basin and the Central Irish Sea Basin (Naylor and Clayton, 2000) provide evidence of widespread erosion of Carboniferous uplands, with thick sandy deposits shed northwards from the eroding Variscan highlands into major depocentres in the Irish Sea.

In the Celtic Sea the Pembroke Ridge probably formed an upstanding positive element during Triassic time. Triassic stratigraphy shows considerable differences north and south of the Ridge, suggesting that the two basinal areas were separated at that time. Sand input during the Early Triassic from the Pembroke Ridge is likely to have been highly variable along strike, concentrated around fluvial systems, probably located along NW–SE cross-cutting fault zones. The Sherwood Sandstone has long been regarded as the product of a rift episode, with the overlying Mercia Mudstone developed in a thermal sag phase and extending further onto the surrounding basement elements. However, Musgrove *et al.* (1995) questioned this view, and suggested that deposition of the Sherwood Sandstone occurred within palaeogeographic lows whose positions were controlled by faults and resistant Variscan massifs. The Mercia Mudstone they regarded as deposited during a rift phase – *see* also Ruffell and Shelton (1999). Whichever model is correct, the Sherwood Sandstone is of more limited extent than the overlying Mercia Mudstone, and it can be anticipated

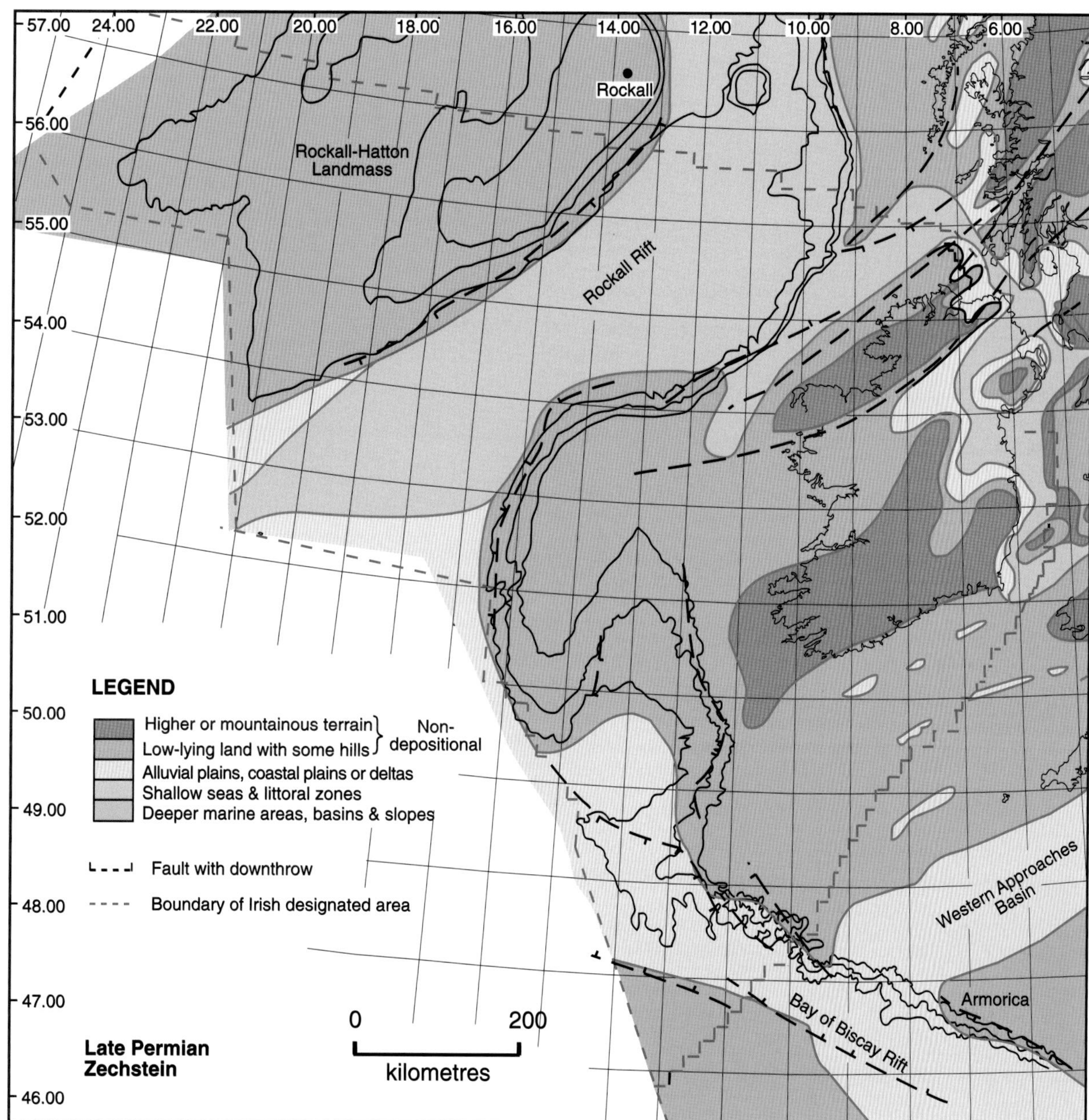

Fig. 17.15 Late Permian: Thuringian (Zechstein) palaeogeography (*after* Naylor, 1998. *Principal sources:* Smith *et al.*, 1992; Ziegler, 1990).

that sandstones would be better developed towards the source areas along the basin margin and specifically towards fluvial input areas. Apatite fission track analysis results suggest that a relatively thick Triassic sequence extended from the Celtic Sea basins northwards over the southern part of the Irish onshore area (Keeley, 1995).

Analysis of the Sherwood Sandstone Group reservoir in the Corrib Gasfield (Slyne Basin: Dancer *et al.*, 2005) indicates deposition within a braided fluvial channel system. Dipmeter and other evidence suggest that drainage flowed from SW to NE along the basin. However, Pb-isotope signatures in detrital feldspar grains (Tyrrell *et al.*, 2007) demonstrate derivation of sediment in Sherwood Sandstone times from the northwest (possibly Greenland) with no indications of either a southerly source or derivation of sediment from the Irish mainland.

Similar arid depositional environments probably also obtained in the Rockall and Hatton basins. The thick Lower Triassic red-bed sequences of the Loch Indall and

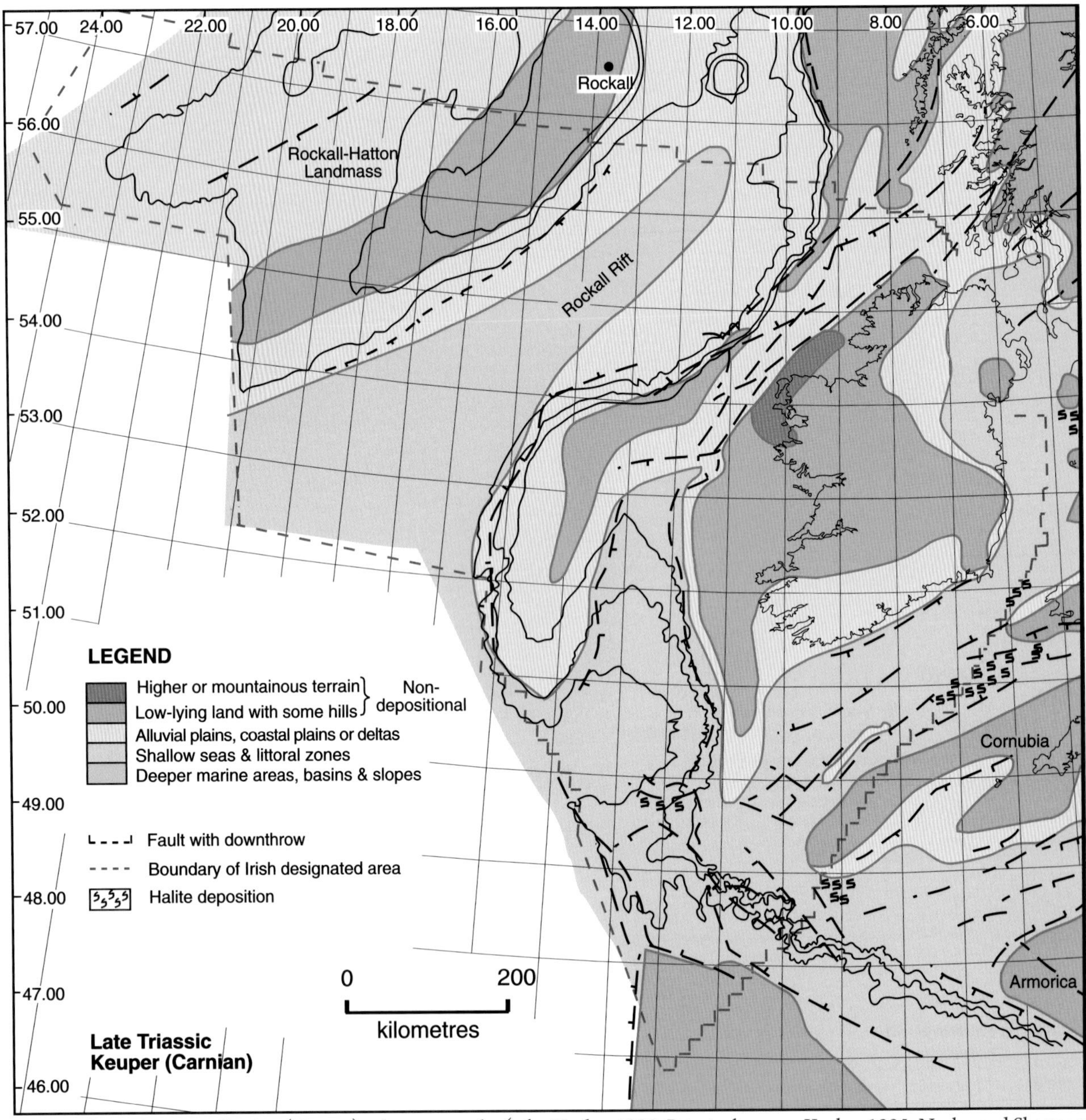

Fig. 17.16 Late Triassic: Carnian (Keuper) palaeogeography (*after* Naylor, 1998. *Principal sources:* Keeley, 1995; Naylor and Shannon, 1982; Naylor, 1992; Warrington and Ivimey-Cook, 1992; Ziegler, 1990).

Rathlin Troughs and the North Channel Basin extend onshore throughout the Ulster basins. The Kingscourt outlier is the only onshore outcrop further south, and the interval there is represented by the siltstones and sandstones of the Kingscourt Sandstone Formation (Visscher, 1971).

The Middle Triassic saw further marine transgression. Shallow marine conditions, with tidal flats, lagoons and sabkhas periodically covered large areas. However, climatic conditions in Ireland remained arid. Marine incursion progressed westwards along the Biscay rift into the Goban–Porcupine–south Rockall region. At the same time the Boreal Sea again advanced southwards towards northern Scotland, although there is no evidence that the two marine tongues became linked in Triassic time.

Higher sea levels in the Late Triassic also gave rise to periodic marine incursions. In consequence there was a widespread development of playa lake and sabkha facies evaporites and marls in the Celtic Sea, Irish Sea and Ulster basins. Figure 17.16, representing the Late

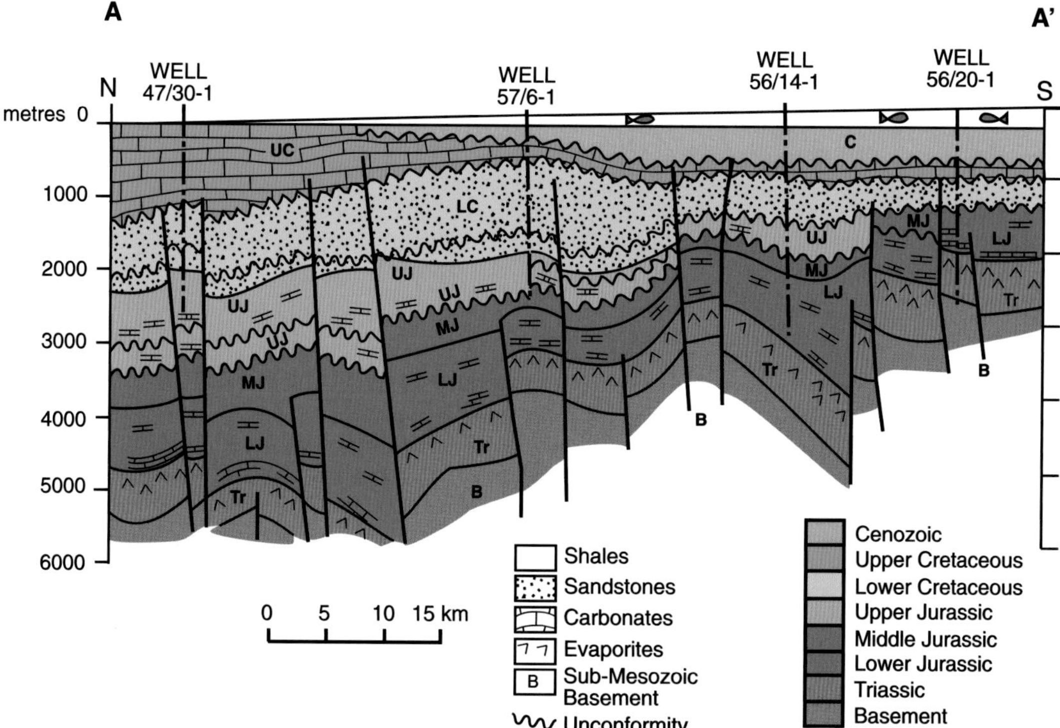

Fig. 17.17 Approximate north–south cross-section through the North Celtic Sea Basin (*after* Colin *et al.*, 1981). The location of the section is shown in Figure 17.3.

Triassic, shows marine seaways established in the main basins around Ireland. The western part of Ireland was part of an emergent area, mostly without significant deposits, that probably extended southwards along the east side of the Porcupine Basin towards the Goban Spur. The land area also linked northwards to the Scottish landmass and through the Slyne–Erris area to the Porcupine Bank. Little is known concerning the Rockall Trough, but the likelihood is that it continued to subside and that marine incursion took place from the south. The Hatton–Rockall region was probably emergent, with alluvial deposition in the lower-lying areas. The Triassic was brought to a close by a widespread marine incursion, terminating the long-lived arid and dominantly non-marine conditions that had existed through Permian and Triassic times.

Jurassic Period

During the Jurassic the chain of rifts extending from Tethys to the Central Atlantic/Caribbean region developed into the main axis along which the Pangaean supercontinent would eventually split. Spreading axes developed along the rifts that eventually gave rise to new ocean floor in the Central Atlantic region during Late Jurassic time, by which time Gondwanaland was again almost separated from Laurasia, except for contact in the Mediterranean region. The onset of tectonism (often referred to as the main Cimmerian event) in earliest Middle Jurassic time provides a natural division of the Jurassic in the Irish offshore basins between Lower and Middle–Upper Jurassic sequences (Figure 17.17), and these are described separately below.

Offshore distribution of Lower Jurassic rocks

Within the Celtic Sea and Fastnet basins there is conformable transition from Triassic red beds to a marine Lower Jurassic succession (Robinson *et al.*, 1981; Millson, 1987). The transgressive sequence begins with a Hettangian shallow marine limestone-dominant sequence in the west. This thins eastwards, interdigitates with, and is overlain by, more argillaceous sediments. Cyclic units with an upward increase in carbonate content are typical of the Hettangian–Sinemurian mudrock sequences. Sinemurian sandstones are found in the Fastnet Basin (Figure 17.18; Robinson *et al.*, 1981), in the eastern portion of the Celtic Sea basins, and intermittently between (Millson, 1987). These are thought by Petrie *et al.* (1989) to have been sourced from exposed Old Red Sandstone in south-west Ireland and from the Leinster massif. Kessler and Sachs (1995) saw this input as related to footwall readjustment at the northern margin of the

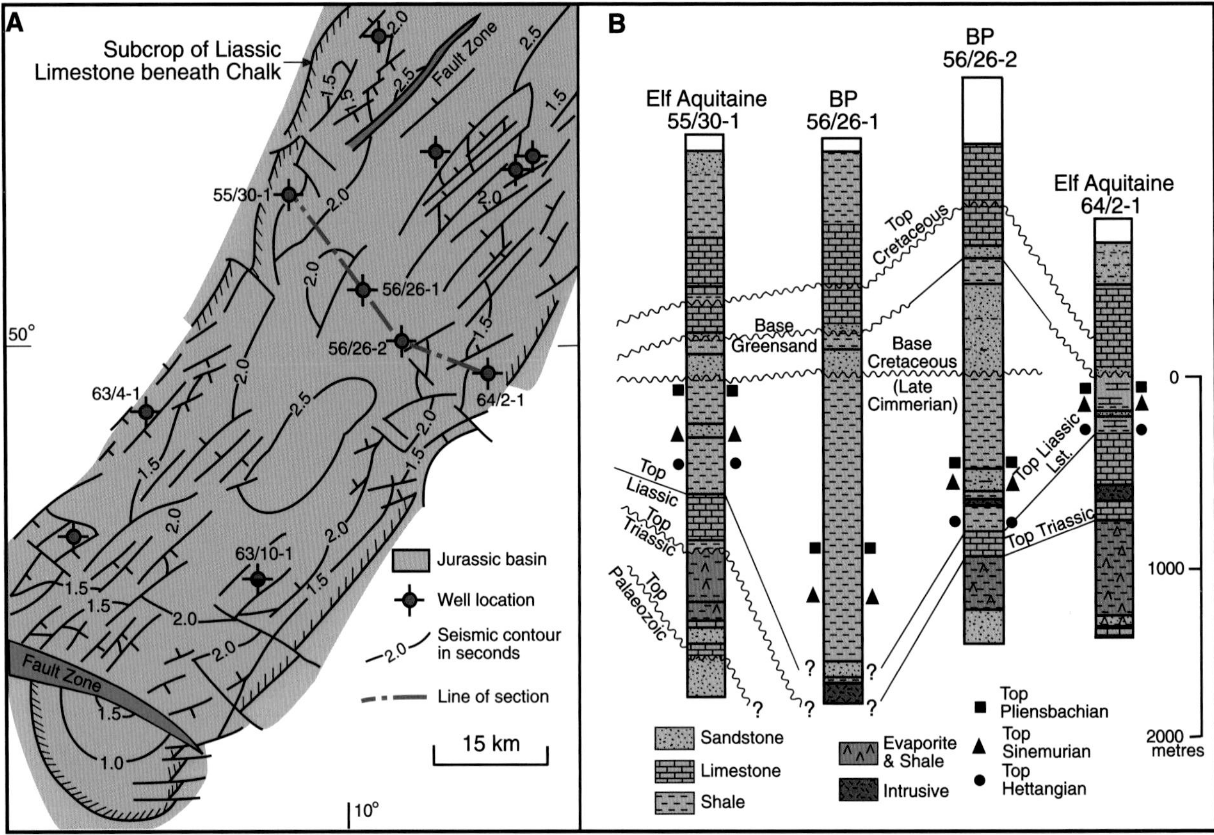

Fig. 17.18 A: Time structure contour map (two-way time) of the Fastnet Basin at Top Triassic Limestone level (*after* Robinson *et al.*, 1981). B: Stratigraphy of exploration wells in the Fastnet Basin. The line of section is shown on the map (*after* Robinson *et al.*, 1981).

North Celtic Sea and St George's Channel basins. These are succeeded by uniform argillaceous and organic-rich Pliensbachian mudstones that in the east, around the Pembroke Ridge extension (Figure 17.3), contain coarser siliciclastics. Shallowing conditions at the end of the Pliensbachian, as indicated by the development of a carbonate-sandstone member, were followed by an abrupt return to deeper conditions in the Toarcian, and deposition of a monotonous mudrock succession. Thin limestone–siltstone units deposited in shallow conditions at the end of the Toarcian brought Early Jurassic deposition in the Celtic Sea area to a close.

Interpretation of reflection seismic data on the Goban Spur, allied with results from the Esso 62/7-1 well (Cook, 1987), suggest that Triassic red beds and evaporites are conformably overlain by Rhaetian to early Sinemurian marine strata. The oldest beds penetrated in the well were early Sinemurian shallow marine claystones and limestones with thin sandstone and siltstone interbeds. These are succeeded by outer neritic late Sinemurian to Toarcian marine claystones. Naylor *et al.* (2002) showed Lower–Middle Jurassic strata extending throughout the Goban Spur Basin (Figure 17.19) and along the margins of the southern part of the Porcupine Basin, based on seismic interpretation.

Liassic rocks in the Porcupine region have been encountered only in the North Porcupine Basin (Figure 17.20; Croker and Shannon, 1987), where they appear to onlap irregular pre-Jurassic topography. Here the succession is a transgressive sequence comprising limestones, shales and intercalated siltstones of Rhaetian to Hettangian age, probably deposited in shallow marine environments. Naylor *et al.* (1999) interpreted Lower-Middle Jurassic strata on geoseismic sections crossing the perched basins on the west flank of the Porcupine High (Figure 17.26) and the east flank of the Rockall High.

The Elf 27/13-1 well in Slyne Trough (Figure 17.7) encountered a full Lower Jurassic sequence. Red beds are succeeded by Sinemurian nearshore limestones and sandstones, and then by Sinemurian to Lower Toarcian marine shales and siltstones. Toarcian shales (84 m) follow and constitute the main petroleum source rock in the basin. A similar sequence in the area of the Corrib

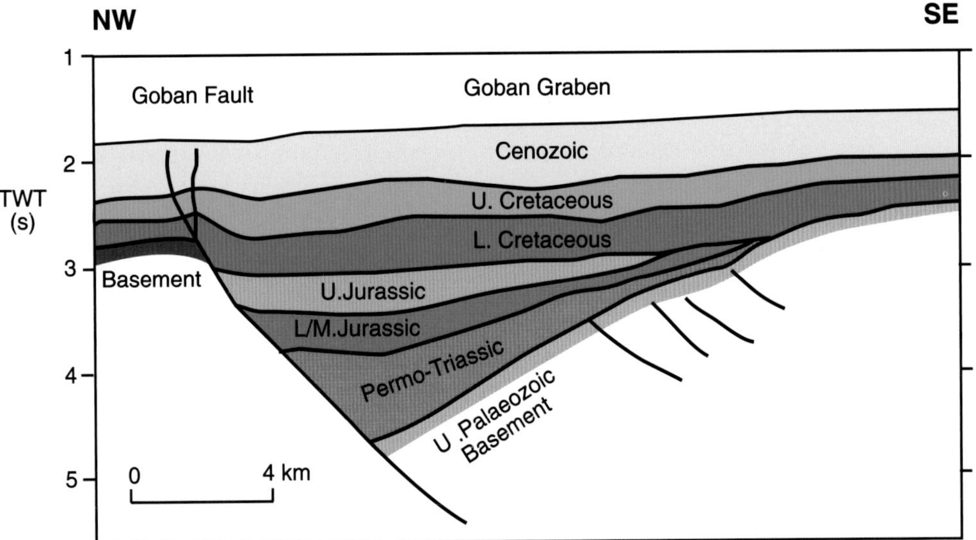

Fig. 17.19 Cross-section of Goban Graben showing growth-faulting in Permo-Triassic to Upper Jurassic successions (*after* Naylor *et al.*, 2002; Naylor and Shannon, 2005).

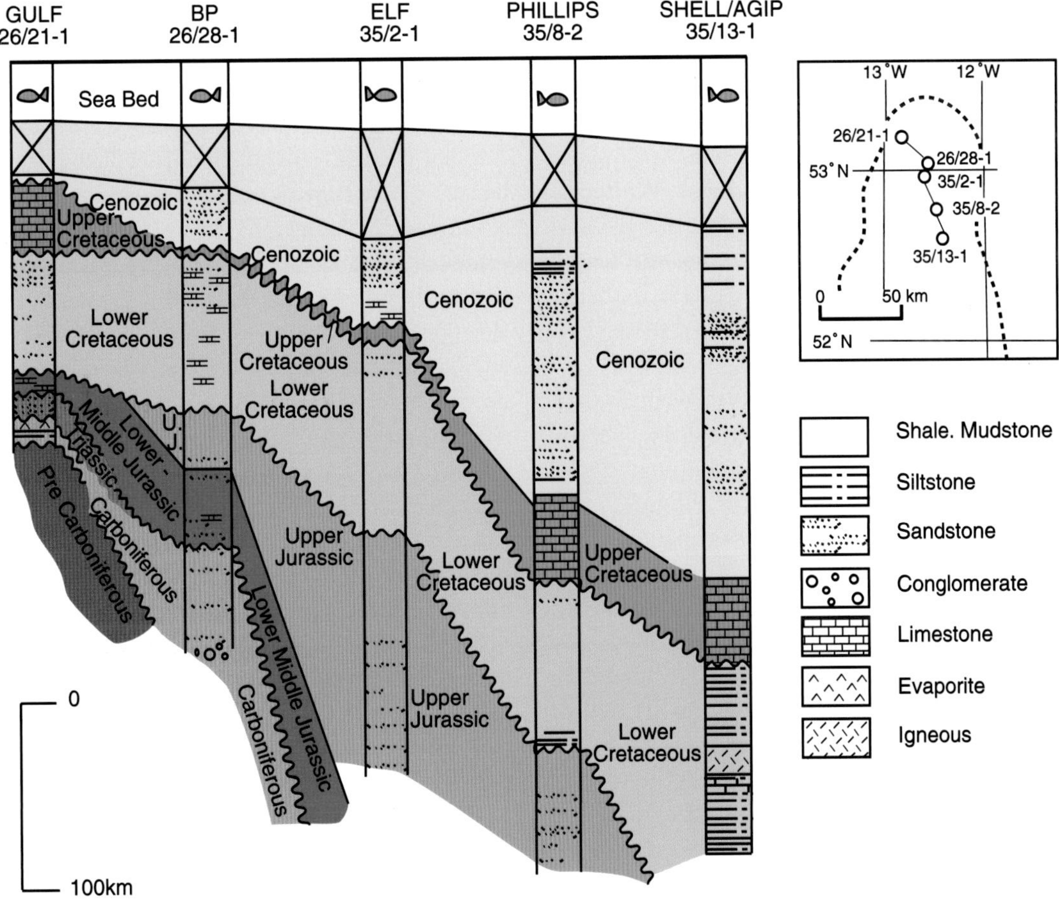

Fig. 17.20 Comparative stratigraphy of exploration wells in the north of the Porcupine Basin and the North Porcupine Basin (based on Croker and Shannon, 1987).

Gasfield (Dancer *et al.*, 2005) is interpreted as a product of open marine deposition with restricted deposits in the deeper parts of the basin (Toarcian anoxic shales). In the Amoco 19/5-1 and Amoco 12/13-1A wells in the Erris Basin (Tate and Dobson, 1989a) the Rhaetian is conformably overlain by Hettangian to early Sinemurian limestones with claystone, siltstone and sandstone interbeds. In contrast to the Goban Spur margin, these were probably the product of freshwater to shallow marine sedimentation. The upper zones of the Liassic are missing in these wells and the sequence is unconformably overlain by Cretaceous or younger sediments. It is likely that the Erris Basin originally had a fuller Jurassic sequence, but that this was removed by latest Jurassic and/or Early Cretaceous erosion (Morton, 1992).

All known onshore Jurassic strata in Northern Ireland belong to the Lower Liassic, although in north Antrim Wilson and Manning (1978) reported that 'derived blocks of shale containing Middle Liassic fossils have been found near Ballintoy and Upper Liassic species have been recovered from the basal conglomerate of the Cretaceous at Murlough Bay'. The extant Lower Liassic deposits generally range from Hettangian to Sinemurian in age, although the sequence (250 m thick) extends up into the Lower Pliensbachian in the Port More borehole. Lower Liassic rocks crop out along the Northern Ireland coastline, as at Magilligan Point, Portrush, Whitepark Bay and Larne, County Antrim. Rocks of this age have also been described from boreholes in the area. Liassic and Rhaetian strata are preserved against the Loch Gruinart boundary fault in the Loch Indaal Basin (Fyfe *et al.*, 1993).

The basal part of the Liassic is preserved locally in parts of the Solway and Peel basins (Newman, 1999). None of the Kish Bank Basin wells encountered Jurassic rocks. However, seismic mapping, seabed sampling and other evidence (Dobson and Whittington, 1979; Jenner, 1981; Broughan *et al.*, 1989) suggest the presence of Jurassic strata along the western and northern margins of the basin (Figure 17.11), with up to 2700 m of possible Liassic rocks preserved against the Bray and Dalkey bounding faults (Broughan *et al.*, 1989). Lithologies similar to those in Northern Ireland or Cardigan Bay are anticipated. Liassic rocks have also been sampled from the sea bed at the northern margin of the basin (Etu-Efeotor, 1976; Dobson, 1977a, b), and Dobson and Whittington (1979) referred to dated Lower Liassic core material composed of dark grey carbonate mudstone and a light grey finely crystalline limestone.

A surprisingly thick section of Liassic sediments within the Cardigan Bay Basin was proved by the Mochras (Llanbedr) borehole on the Welsh coast (Woodland, 1971; Figure 17.13). The 1305 m thick marine Lower Jurassic (Hettangian to Upper Toarcian) is one of the thickest successions known in Europe. The sequence comprises grey, locally silty, calcareous mudstones with some darker mudstone intercalations, particularly in the Lower Liassic. The thin-bedded mudstones alternate with bioturbated silty limestones and hard, massive calcareous siltstones.

Elsewhere in the Irish Sea region a thin succession of Hettangian to Sinemurian marine siltstones and claystones was drilled on the St Tudwal's Arch extension at the southern margin of the Central Irish Sea Basin (Figure 17.14), and seismic evidence points to the possible presence of Jurassic and Cretaceous sediments in the central part of that basin (Maddox *et al.*, 1995). Seabed coring on the flank of the St George's Channel Basin (Whittington *et al.*, 1981) proved Liassic grey laminated shales and limestones.

Early Jurassic palaeogeography

In Western Europe some of the main rift basins established during the Triassic continued to grow and subside. A major marine transgression from Tethys, initiated in latest Triassic time (Rhaetian), progressed in the earliest Jurassic. Passage into the western basins was along established rift systems such as the Bay of Biscay Rift. The Early Jurassic was therefore a period of transgression on the shelf areas west of Britain. There was gradual encroachment across the subdued Rhaetian topography, on which only a few structural intra-basinal highs persisted as areas of non-marine deposition. Land area was reduced to a series of islands with Palaeozoic cores – e.g. Cornubia, part of Wales, western Ireland, Porcupine and Rockall Banks. Deeper marine conditions spread along the main rift axes and by this time there was free communication between Tethys and the Boreal Ocean.

The Liassic of the Irish Sea and Ulster basins is uniformly fine-grained and thin-bedded and is indicative of relatively shallow quiet marine conditions. The basin margin faults show little indication of movement during this period (Woodland, 1971; Cope, 1984; Broughan *et al.*, 1989). Shallow marine and possible non-marine environments extended round to the Erris and Slyne basins, but more varied marine conditions obtained on the Celtic Sea shelf, with deeper, more uniform conditions towards the Goban Spur.

Offshore distribution of Middle–Upper Jurassic rocks

The variation of lithologies and facies in the Middle Jurassic of the Celtic basins is a reflection of the uplift and erosion accompanying Cimmerian tectonism in the region (Naylor and Shannon, 1982; Ziegler, 1990). In the east of the basin there was thick coarse sand input during the Aalenian to early Upper Bajocian interval, whilst calcareous and argillaceous deposition took place in the west. This pattern was interrupted by a transgressive marine pulse that produced widespread mud deposition in early Bajocian time. More widespread marine argillaceous deposition became established in Late Bajocian time. Basic intrusive rocks have been encountered in wells in the central part of the Fastnet Basin, and are imaged on seismic profiles. Caston *et al.* (1981) suggested that they have a widespread occurrence throughout the central region of that basin, associated with a north-east–south-west fault zone. The sills are composed of olivine micro-gabbro and olivine dolerite, and K/Ar dating indicates a Late Bajocian age.

Marine conditions persisted into mid-Bathonian time, when there was a phase of regressive near-coast sand deposition followed by shallow-shelf carbonate sedimentation. Further regression later in the Bathonian produced non-marine red-brown anhydritic mudrocks with local sandstones, probably across much of the basin. Bathonian and older marine rocks are succeeded (Ainsworth *et al.*, 1987) by latest Bathonian freshwater-brackish marine strata and then by presumed Callovian strata, lacking diagnostic fauna, of non-marine continental facies. Millson (1987) suggested that this facies development was related to Central Atlantic rifting between the African and North American plates, and that the transgressive Late Bathonian–Callovian interval that follows coincides with the onset of seafloor spreading in the same region. However, the Callovian is lacking in the west of the region, probably as a result of later erosion. At the eastern end of the North Celtic Sea Basin there was deposition of Callovian to Oxordian non-marine braided fluvial sandstones (Caston, 1995).

The reddened clay deposits recorded onshore adjacent to the south coast of Ireland are probably of Middle Jurassic age (Higgs and Beese, 1986), and represent a thin non-marine deposit. This suggests that thick Jurassic sedimentation was confined to the offshore fault-controlled basins.

A major unconformity near the base of the Upper Jurassic (Figure 17.17) marks a further pulse of Cimmerian tectonism that gave rise to syn-rift sandstone-prone strata of continental to shallow marine origin. A relative rise in sea level produced more shale-prone, and locally limestone-prone, strata during the Late Jurassic throughout the North Celtic Sea Basin. However, there is no evidence of widespread Upper Jurassic strata in the Fastnet and Cockburn basins. Shannon (1995) suggested that the basic intrusions of this age may reflect the presence of a hot spot which caused local doming and prevented the deposition of a thick Upper Jurassic sequence comparable to that in the basins both to the east (NCSB) and to the west (Porcupine Basin).

In the eastern Celtic Sea area, grey and anhydritic mudrocks of Middle–Late Oxfordian age are mainly non-marine and rest unconformably on Middle and even Lower Jurassic strata. North-east of the Kinsale Head Gasfield (Figure 17.3) thin sandstones and limestones have been drilled in the sequence adjacent to the northern boundary of the NCSB (Caston, 1995). A sharply defined marine pulse in the Upper Oxfordian is postulated from microfaunal evidence (Ainsworth *et al.*, 1987) and follows upon an unconformity in the NCSB (Colin *et al.*, 1981). Here the Lower Kimmeridgian is non-marine and contains coarse marginal clastics that grade basinwards into anhydritic mudrocks. Transgression then produced variable Late Kimmeridgian–Portlandian marine facies over much of the region, including the more persistent intra-basinal highs. The SCSB was probably an area of non-deposition throughout the latest Jurassic. The final phase of Portlandian (to earliest Cretaceous) deposition was of argillaceous micrites and carbonates. The varied nature of the Upper Jurassic in the Celtic Sea contrasts with the widespread uniformity of units such as the Kimmeridge Clay in the North Sea, and probably reflects proximity to rifting events in the proto-Atlantic (Millson, 1987).

The Middle Jurassic of the Goban Spur (Cook, 1987) consists of Late Toarcian to Bajocian claystones and siltstones representing a marine middle to outer shelf environment, with rare paralic or marginal sandstone intercalations. The overlying Bathonian to Callovian marine claystones are succeeded by massive stacked beach sandstones. The subarkosic sandstones may have been sourced locally from Devonian rocks to the south-east, similar to those encountered in DSDP drilling (Site 548) in the area (Lefort *et al.*, 1985), and environments are thought to have been non-marine deltaic (Colin *et al.*, 1992). This sequence is succeeded by 215 m of porphyritic basaltic andesite flows that K–Ar dating, reported by Tate and Dobson (1988), indicates are of Valanginian age. In the Esso 62/7-1 well on the Goban Spur the Upper Jurassic is absent probably due to Cimmerian tectonism (Colin *et al.*, 1992). From seismic evidence (Cook, 1987; Naylor *et al.*, 2002), Upper Jurassic rocks may occur elsewhere in the basin, possibly in marginal marine or paralic facies (Figure 17.19).

Middle Jurassic sequences are known from the northern part of the Porcupine Basin, where they rest unconformably on Liassic and older strata (Naylor and Shannon, 1982; Croker and Shannon, 1987). Continental coarse clastics in fining-up cycles were deposited in Bajocian to early Bathonian times by braided stream systems issuing from a northern source and prograding southwards. This sequence is interpreted (Sinclair *et al.*, 1994) as the product of warping prior to the main Cimmerian rifting. The clastics are succeeded in the north of the basin by Bathonian fine-grained sandstones and mudrocks, again dominantly non-marine. The upstanding intrabasinal highs were finally inundated during the Bathonian to Callovian interval, when thin sandstones and mudstones with rare limestones were deposited. The dominantly fluvial deposits in the north of the basin may pass southwards to shallow marine sediments not yet penetrated in exploratory drilling (Croker and Shannon, 1987). The Middle Jurassic in the southern part of the Porcupine Basin is probably comparable to that of the Celtic Sea and Goban Spur. It appears that volcanic activity in the Atlantic basins was relatively limited in comparison with the central part of the North Sea.

Upper Jurassic (Oxfordian to Portlandian) strata have been penetrated in many of the Porcupine Basin wells (Figure 17.20). Sedimentation was apparently continuous through Middle and Late Jurassic time, with no widespread evidence of the intervening unconformity that is a feature of the Celtic Sea region. Nevertheless, late Middle and Late Jurassic time was a period of differential subsidence in the basin, reflected by varied continental sandstone and shale sequences, interrupted by intercalations of shallow marine strata. Oxfordian conglomerates drilled by Phillips 35/8-1, in the north part of the basin, contain quartz and fresh feldspar suggestive of an exposed nearby granite pluton, while Kimmeridgian conglomerates in Shell 34/19-1 contain schistose lithoclasts (Tate and Dobson, 1989b). The overall facies pattern suggests a northward encroachment of transgressive facies (Croker and Shannon, 1987) along the basin axis, with active movement on the marginal bounding faults. The latest Jurassic (Tithonian) exhibits marked changes in thickness and facies, reflecting the build-up to the Late Cimmerian phase of deformation. The BP 26/28-1 well at the northern rim of the basin (MacDonald *et al.*, 1987) demonstrated that subsidence was continuous from Middle Jurassic to Late Jurassic (Bajocian to Tithonian) time. The sequence is a product of braided streams or alluvial fan systems passing upwards to a flood plain environment with meandering rivers. Nearshore marine conditions became dominant in the Kimmeridgian to early Tithonian interval. Smith and Higgs (2001) reported on reworked Carboniferous palynomorphs within late Jurassic sandstones from wells in Quadrants 26 and 35 of the northern part of the Porcupine Basin, with probable derivation from the eastern margin of the basin. From Kimmeridgian time onward local block faulting appears to have exercised increasing influence on sedimentation, with variation in the thickness of stratigraphic units. The whole sequence is truncated by the Late Cimmerian (Base Cretaceous) unconformity surface. As might be expected, the Upper Jurassic section away from the basin margin is less varied (Phillips 35/8-2 well; Croker and Shannon, 1987). Proximal to distal submarine fan clastics and shales are overlain by marine shales (Tithonian).

The nature of Jurassic deposition within the Rockall or Hatton basins is speculative, although Naylor and Shannon (2005) suggested that rocks of this age may be widespread (Figure 17.22). Middle and Upper Jurassic strata are interpreted from seismic data within the small perched basins along both margins of the Rockall Basin (Naylor *et al.*, 1999). Coring on the western flank of the Porcupine High (Haughton *et al.*, 2005) has confirmed the presence of Upper Jurassic strata (early Kimmeridgian to early Portlandian marine sandstones and limestones) in the North Bróna Basin, and possibly in the South Bróna Basin (Figure 17.26).

The Elf 27/13-1 well in the Slyne Basin penetrated an almost complete Middle Jurassic sequence. The Toarcian to Upper Bajocian (320 m) comprises two coarsening-upwards offshore sandbar complexes overlain by 1320 m of Upper Bajocian to Upper Bathonian brackish water strata. In the well, these are overlain by Quaternary sediments, but seismic evidence suggests (Trueblood and Morton, 1991) that Upper Jurassic rocks may exist in the deeper parts of the Slyne Basin (Figure 17.8). Dancer *et al.* (2005) described the Middle Jurassic 1250–1700 m) of the Corrib Gasfield area as overlying a regional unconformity that, although not marked in the field area, becomes more significant towards the basin margins. Upper Jurassic strata (366–506 m thick) are preserved within a collapsed graben on the crest of the field structure, and comprise silty claystones with sandstone, limestone, and coal interbeds.

Middle and Upper Jurassic rocks are unknown onshore in Northern Ireland, or in the offshore basins around the north coast. Rocks of this age could form part of the undrilled sequence in the Kish Basin (Broughan *et al.*, 1989), and in other parts of the Irish Sea, but there is no direct evidence of this. Drilling in the Cardigan Bay–St George's Channel basins (Figure 17.14) has revealed an almost complete Lower Liassic to Lower Purbeckian succession (Barr *et al.*, 1981). The Liassic is succeeded with apparent conformity by 760 m of Middle Jurassic mudstones, sometimes calcareous, with thin limestone interbeds, probably the product of shallow marine neritic and carbonate shelf environments. These are conformably overlain by a complete Upper Jurassic succession of shales, mudstones and calcareous mudstones (1498 m) up to Lower Purbeckian in age. Depositional conditions were probably shallow marine and restricted lagoonal, becoming marginal marine in the uppermost beds. In contrast to the Celtic Sea there is no discordance between Middle and Upper Jurassic strata within the basinal areas (Barr *et al.*, 1981).

Middle and Late Jurassic palaeogeography

The break-up of Pangaea, initiated during Middle Jurassic time, had significant impact in the Irish area. There is evidence that the Boreal and Tethyan faunal provinces were separated during Bajocian–Bathonian times. It is likely therefore, that the northern Rockall and Hatton areas were very shallow and emergent for periods, acting as a barrier. Connection was re-established between the two provinces later in the Middle Jurassic.

The Irish mainland at this time was almost entirely emergent, with land prolongations southwards along both sides of the Porcupine Basin (Figure 17.21). In the Slyne Basin, stacked nearshore sand bars were succeeded by brackish water strata. Middle Jurassic sediments are not preserved in the Erris or Donegal basins, the Northern Ireland basins or the Irish Sea area. Such strata were probably removed by later erosional episodes. A regional pattern emerges in the Middle Jurassic of intermittent subsidence around the southern part of Ireland. Relatively thick sequences accumulated within the basins, from Cardigan Bay out to the Goban Spur, probably with generally westwards-deepening conditions.

With the separation of the European segments of Laurasia from the African portion of Gondwanaland, stresses in Europe related to the Tethyan rift system diminished. The dominant process now was the opening of the Central Atlantic Ocean and the build-up of extensional stresses in the North Atlantic domain (Ziegler, 1987). Rifting increased in the Labrador Sea and within the Greenland-Norway system, and deeper water conditions developed along the main rift axes. From an Irish viewpoint, events henceforth would be dominated by the extensional stresses related to the opening of the North Atlantic.

Upper Jurassic strata along the Atlantic margin reflect deposition in a syn-rift setting, with the development of a range of facies. These range from basin-edge alluvial fans, braided and meandering fluvial systems to submarine fans (Croker and Shannon, 1987, MacDonald *et al.*, 1987; Shannon, 1992, 1993). Seismic evidence suggests that Upper Jurassic sequences are present in the deeper parts of the Goban Spur Basin, possibly in marginal marine facies (Auffret *et al.*, 1979). Elsewhere the sediments are absent due to tectonism and erosion. At the outset of Late Jurassic time in the Porcupine Basin, deltaic non-marine facies were deposited in the north of the basin, with coeval marine shales further south. Gradual trangression resulted in marine shale deposition throughout the basin. The basin bounding faults were active during this period of crustal extension, and submarine fans developed along the margins and in the deeper parts of the basin.

Marine conditions persisted intermittently through Late Jurassic time along the southern seaboard of Ireland. Tectonic influences from the Atlantic domain gave rise to an important unconformity in the Celtic Sea region. In the NCSB tectonic movements produced dominantly non-marine rocks during the Oxfordian and earliest Kimmeridgian. The rise in sea level eventually resulted in Kimmeridgian marine facies over much of the region. A thick sequence of marine and lagoonal shales and mudstones is found in the Cardigan Bay Basin, but is absent further north in the Irish Sea. Shallow seas extended over the areas of the SCSB and the Western Approaches, although no sediments of this age are now preserved. Callovian to mid-Kimmeridgian time was a period of gradually rising sea level, before regression once again set in during the final phases of the Jurassic. Western Ireland was probably emergent and possibly linked, through the Fastnet area, to a land extension from Cornubia. Porcupine Bank, Rockall Bank and Hatton Bank, and residual portions of the Scottish landmass were also probably emergent. Deeper water tongues occupied Rockall Trough and the Bay of Biscay Rift and extended also into the Porcupine Seabight and Hatton Trough.

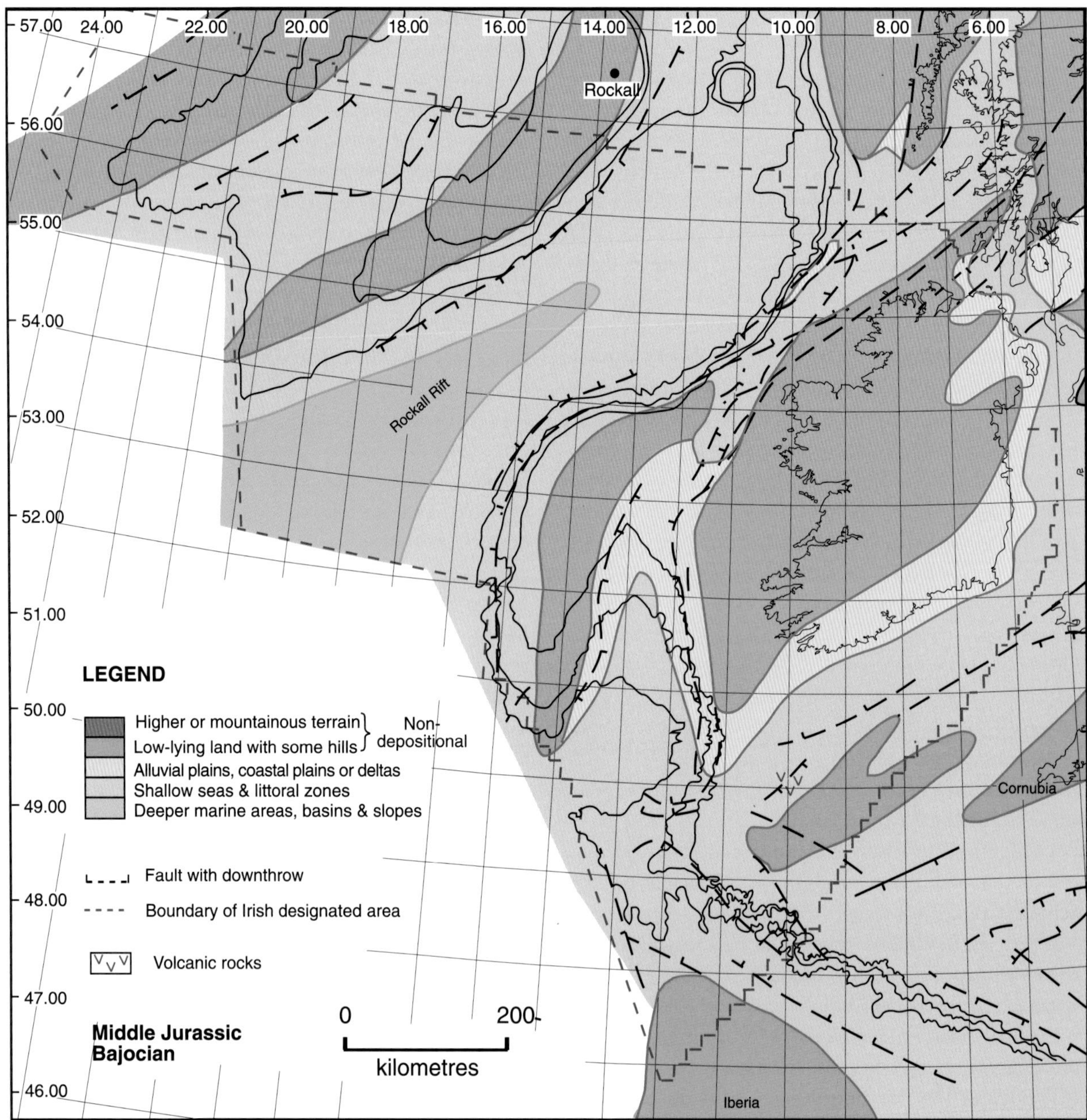

Fig. 17. 21 Middle Jurassic: Bajocian palaeogeography (*after* Naylor, 1998. *Principal sources:* Bradshaw *et al.*, 1992; Keeley, 1996; Naylor and Shannon, 1982; Naylor, 1992; Ziegler, 1990).

The region north and east of Ireland was probably a positive area of thin deposition (or even non-deposition), as in Middle Jurassic time. Thus, despite the widespread marine conditions, Upper Jurassic rocks are not preserved in the Donegal and Erris basins, the Larne–Lough Neagh Basin, the Central Irish Sea basins, or over onshore Ireland, as a result of subsequent Early Cretaceous erosion. An exception to this is the reported occurrence (Higgs and Jones, 2000) of karstic clay deposits with nearshore tidal influences discovered onshore near Carrick-on-Suir, County Kilkenny, containing palynomorphs indicative of a Late Jurassic to Early Cretaceous age. Consideration of regional palaeogeography suggests that a Late Jurassic age is the more likely.

There was a gradual build-up of tectonic activity towards the close of the Late Jurassic, culminating in the Late Cimmerian tectonic episode. Block faulting, tilting and erosion took place, associated with a period of general emergence. This phase is most clearly seen in the Atlantic and Celtic Sea basins, where the time interval is

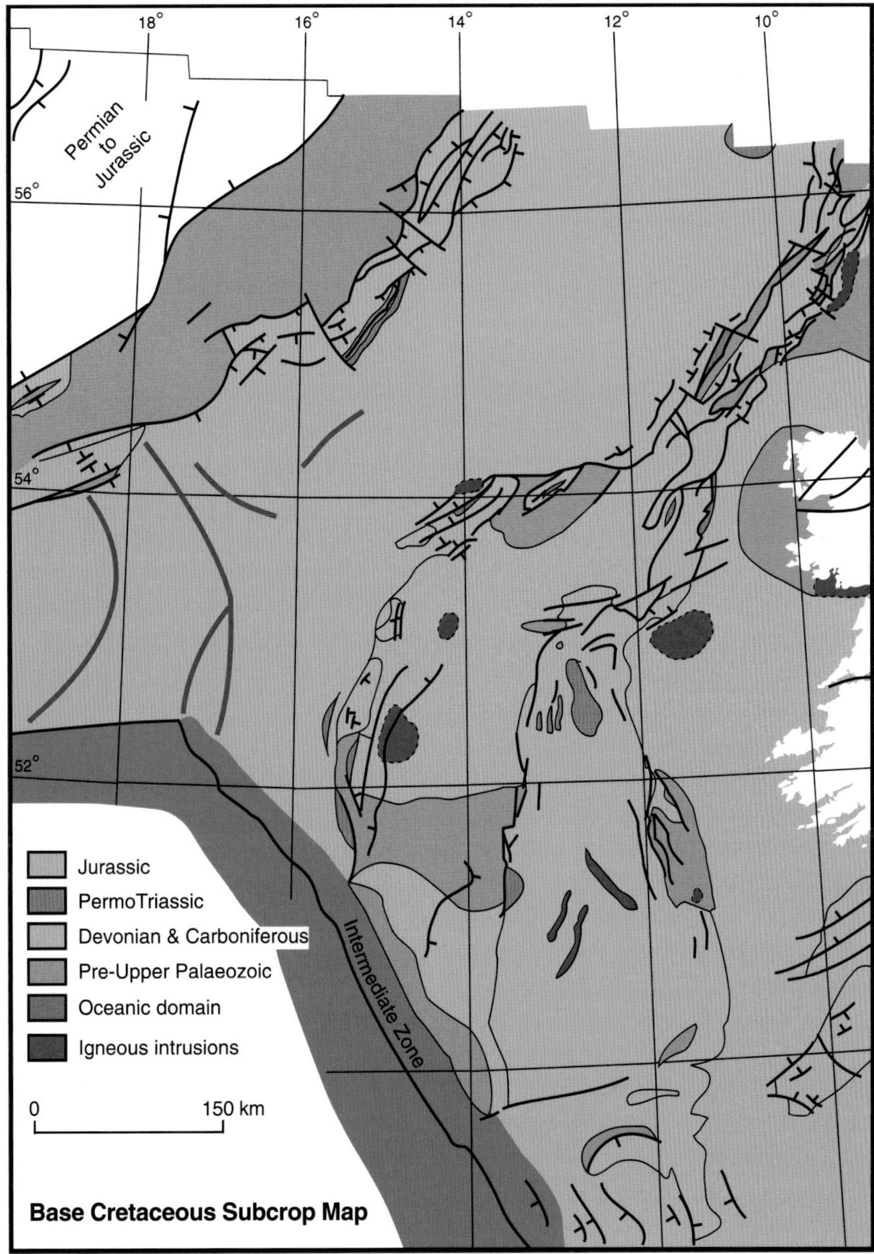

Fig. 17.22 Base Cretaceous subcrop map of the Atlantic margin basins west of Ireland (*after* Naylor and Shannon, 2005)

relatively constrained by the beds above and below the unconformity. In the Malin, Ulster and Irish Sea basins uplift and erosion could have taken place at any time after Early Jurassic, but is most likely to have occurred as part of the Late Cimmerian event. Sea levels began to fall in late Kimmeridgian time and this continued during the Tithonian, with the result that shorelines gradually retreated.

Cretaceous Period

There was an increase in tectonic activity across the Jurassic–Cretaceous boundary in many of the basins. This tectonic phase, referred to as the Late Cimmerian, had a significant impact on Irish offshore geology. Crustal extension across the Arctic–North Atlantic rift system was the driving force throughout much of the Early Cretaceous. Sea levels fell from the high-stand of the Kimmeridgian and reached a low point in late Berriasian–early Valanginian time. The combination of the Late Cimmerian rifting pulse and low sea levels between latest Jurassic to early Valanginian time produced a widespread unconformity (Figures 17.22, 17.24). Large areas of the shelves were exposed and subject to erosion, and deposition was continuous only along the axial zones of the main basins. There was an accompanying change from the marine and carbonate shelf regimes of the Late Jurassic to marginal and non-marine clastic wedges in the Early Cretaceous. Sea level then rose through the remainder of the Early Cretaceous, except for a period of low-stand in the Aptian.

During the Early Cretaceous, rifting propagated rapidly northwards west of Greenland along the Labrador Sea into Baffin Bay. By mid-Aptian there was crustal separation between Iberia and the Goban–Ireland shelf, and seafloor spreading had commenced between Ireland and Newfoundland as the North Atlantic began to open. As extensional stresses were focused along the North Atlantic and Labrador Sea axes, levels of rifting and tectonic activity in the surrounding regions and basins waned.

During the Late Cretaceous the North Atlantic and Arctic spreading systems continued to widen, and rifting continued in the Labrador Sea. Tectonic activity in the other basins was at a minimum. The Late Cretaceous was a time of worldwide sea level rise (Vail *et al.*, 1977). Global sea levels began to rise during the Albian, leading to widespread transgression. Around Ireland the basin margins were progressively overstepped and the Late Cretaceous seas spread onto the surrounding positive massifs.

Offshore distribution of Cretaceous rocks

The sequence in the Celtic Sea basins begins with fluvial, alluvial, deltaic and lacustrine deposition (Robinson *et al.*, 1981; Naylor and Shannon, 1982; Shannon, 1991b) comparable to the Wealden of southern England (Allen, 1981). This non-marine character is in marked contrast to coeval deep marine strata in the Porcupine Basin. Lower Cretaceous strata of pre-Aptian age are generally thin to absent in the South Celtic Sea and Cockburn basins (Smith, 1995), probably due to erosion and non-deposition.

In the North Celtic Sea Basin (NCSB) the Late Cimmerian unconformity is well developed only at the basin margins. Following upon erosion of the Late Cimmerian surface, deposition resumed in Early Cretaceous time. Movement continued on some faults, particularly at the basin margins, and the basin floor topography was only gradually subdued and submerged. Centrally located wells within the NCSB (Esso 47/30-1 and 57/6-1; Figure 17.17) penetrated Berriasian non-marine clastics above Portlandian shales and carbonates, with no obvious unconformity (Colin *et al.*, 1981). Further south within the same basin (Esso-Marathon 56/14-1) Valanginian to Hauterivian beds overstep an Oxfordian–Kimmeridgian alluvial and fluvial sequence. On the flanks of the Pembrokeshire Ridge, Lower Cretaceous beds rest unconformably on the Lower Jurassic (Figure 17.17).

Lower Cretaceous rocks have been intersected by most of the exploration wells drilled in the Celtic Sea basins. The interval is divisible into a lower continental sandy (Wealden) section overlain by a marine sand and shale sequence (Gault and Greensand equivalent). The Wealden section begins with brackish to freshwater, generally alluvial, Berriasian sediments that are succeeded by Valanginian–Hauterivian continental fluvial sandstones and alluvial shales, sourced from the west (Robinson *et al.*, 1981). This western source foundered in Hauterivian time, although the Irish and Welsh areas still provided coarser sediment to the eastern part of the basin. Barremian marine pulses in the western parts of the region produced a complex pattern of fluvial, deltaic and nearshore facies in the centre of the NCSB (Colley *et al.*, 1981).

Marine transgression began in the west during the Aptian and extended across the whole region in Albian time. Ainsworth *et al.* (1987) noted that the Lower Cretaceous of the Fastnet Basin and the southern part of the NCSB shows a stronger marine influence than further north, where marginal marine, deltaic and freshwater conditions prevailed. In the area of the Kinsale Head Gasfield there is a tripartite subdivision of the Albian portion of the Greensand Group section into a Lower Claystone, a middle 'A' Sand and an upper Gault Claystone (Taber *et al.*, 1995). This marks a progression from the estuarine/fluvio-deltaic environments of the Upper Wealden to inner shelf conditions, with sand derivation from the northern basin margin. The 'A' Sand is the main reservoir in the Kinsale Head and Ballycotton gasfields. The Early Cretaceous was a quiescent period for volcanism in the Celtic Sea area, with activity limited to the Cornubian platform, Western Approaches, and Goban Spur margins.

Lower Cretaceous sedimentation in the Fastnet Basin (Robinson *et al.*, 1981) followed along similar lines to that in the NCSB (Figure 17.18). Ainsworth *et al.* (1987) record Lower Cretaceous beds resting unconformably on Aalenian to Toarcian in the centre of the basin, and on lower Toarcian rocks at the basin margins. Wealden equivalents range in age from Berriasian to possible Aptian in age, with provenance areas close to the west. The overlying Greensand-equivalent sandstones and minor claystones range from Barremian and Aptian to Cenomanian in age, and represent deltaic and shallow marine environments.

The Aptian and Albian shallow marine facies comprise an extensive transgressive succession identified in drilling or on seismic sections in all the Celtic Sea basins (Robinson *et al.*, 1981; Taber *et al.*, 1995). The succeeding Gault Clay and Upper Cretaceous Chalk represent a progressive eustatic sea level rise. Chalk (up to 1 km thick) was deposited from Cenomanian time onwards in the depressions of the Celtic Sea, Fastnet and Western Approaches basins. Chalk of Cenomanian to Maastrichtian age crops out over a large area of the sea floor. Turonian to Late Santonian marginal facies along the northern margin of the NCSB suggest that the basin margin at that time may not have been much farther north than the present seabed outcrop limit. Considerable thicknesses of Chalk were later eroded following Cenozoic inversion of the basin. The original thickness has been estimated at *c.*200 m on the basin margin to over 1300 m in the basin centre (Murdoch *et al.*, 1995). Upper Cretaceous rocks, exceeding 1 km thick, are found unconformably above Lower Cretaceous strata in the western part of the Bristol Channel, but thin eastwards and are absent over the eastern part of that basin. The Chalk of the Fastnet Basin (Figure 17.18B) is generally only 120–160 m thick and in the basin centre may be divided into lower and upper units separated by a thin marl (Robinson *et al.*, 1981). The Lower Chalk tends to be slightly arenaceous, being deposited in an inner shelf environment, whereas the Upper Chalk is a product of deeper outer shelf conditions.

Onshore in southern Ireland, Chalk (Upper Campanian) is preserved at only one locality, in a small outlier at Ballydeenlea, near Killarney, County Kerry (Walsh, 1966). This suggests that, with rising sea levels, Chalk deposition overspilled the offshore basin margins and extended onto the Irish massif at least as far north as Killarney. Palynofacies data from the Chalk deposit show a particularly low level of terrestrial influence, supporting a picture of widespread submergence at this time (Evans and Clayton, 1998). Marine transgression also extended to deposit a veneer of Chalk across parts of the Cornubian and Welsh landmasses.

The Esso 62/7-1 well on the Goban Spur intersected a Lower Cretaceous section (580 m thick), onlapping the Cimmerian unconformity surface (Figure 17.19). The Neocomian to Lower Aptian succession consists of sandstones and limestones overlain by limestones and calcareous claystones of shallow shelf origin. Increased seafloor spreading in the Bay of Biscay then produced deeper marine conditions on the Goban Spur, with outer shelf carbonates increasingly represented in the succession. Seismic data in the region indicate the existence of Lower Cretaceous reefs within the Goban Graben (Masson and Roberts, 1981; Cook, 1987), and similar features are known from other localities along this section of the continental margin. The Celtic Shelf margin proximal to the Goban Spur was intruded and covered by Lower Cretaceous (Albian–Aptian) volcanic rocks, identified from geophysical data (Cook, 1987). These are in part likely to be pillow lavas similar to the late Albian basaltic pillow lavas of DSDP site 550 on the southwest flank of the Goban Spur. The Goban Spur succession shows that a major North Atlantic rifting episode began in late Hauterivian or early Barremian time (the age of the oldest syn-rift sediments) and was terminated by sea-floor spreading initiation in the early Albian. The widespread post-rift unconformity here lies between Aptian and Albian strata (Graciansky *et al.*, 1985; Hart and Crittenden, 1984) and extends out to water depths of 2000 m.

The Cretaceous strata throughout most of the Atlantic frontier basins consist of deepening marine shale-prone strata. Shale deposition was interrupted, especially in the Porcupine Basin and locally in the Slyne and Erris basins, by deltaic sandstones that reflect a minor rift episode of Aptian–Albian age. In the Porcupine Basin the thickest development of Lower Cretaceous strata above the Late Cimmerian surface is generally seen along the eastern margin, where the basin margin faults remained active. Parts of the western margin are characterised by progressive onlap across older tilted fault blocks (Figure 17.24). At the centre of the basin the Lower Cretaceous is a thick and uniform sequence, but towards the basin margins it separates into several seismo-stratigraphical units bounded by unconformities and overlain by Cenomanian to Maastrichtian Chalk. The Shell 35/13-1 well drilled 658 m of Cenomanian to Albian calcareous siltstones with sandstones, and this succession thins westwards to the Deminex 34/15-1 well where there are only 60 m of Lower Cretaceous (Hauterivian to Albian) marine claystones. Naylor and Anstey (1987) described

Lower Cretaceous fan deposits related to active fault scarps along the margins of the Porcupine Basin, which in turn are overlain by younger Lower Cretaceous units with less marked internal structure, indicating lower energy environments. Reworked Carboniferous palynomorphs in Barremian and Berriasian sand-prone horizons in Quadrant 35 wells are considered to have been derived from the east flank of the basin (Smith and Higgs, 2001).

In the northern part of the Porcupine Basin (Figure 17.20), the Late Cimmerian unconformity surface is overlain by a northwards-onlapping sequence. The Lower Cretaceous succession has been divided (Moore and Shannon, 1995) into two chronostratigraphic sequences. The lowermost sequence is Valanginian to Late Aptian in age and comprises shallow marine shelf facies sediments around the basin margins and deeper basinal sediments in the basin centre. The upper sequence is Late Aptian–Cenomanian in age and is dominated by marine shelf mudstones and sandstones, with deep-water basinal facies only in the southernmost part of the basin. According to Croker and Shannon (1987) Albian/Aptian rift-generated deltaics prograded from the northern and south-eastern basin margins, and a series of glauconitic offshore bar sandstones rimmed the basin during the Albian. MacDonald *et al.* (1987) recorded a Barremian to early Albian sequence at the northern margin of the basin (Block 26/28) comprising mudstones with thin dolomites. Shallowing in the upper part of the interval was followed by a marked change in sedimentation pattern and the deposition of a widespread 200 m thick blanket of glauconitic calcareous sandstones that pass upwards with apparent conformity into Chalk.

Little is known of the Cretaceous succession in the Rockall Basin (Figure 17.25) due to the paucity of wells in the basin. In the UK sector of the eastern margin of the basin the BP 132/15-1 well drilled a thin section of Aptian/Albian mudstones with interbedded limestones. On seismic profiles (Musgrove and Mitchener, 1996) this thickens basinwards and onlaps a thin wedge of undated, possibly reworked granite resting on crystalline basement. This is overlain by sandstones with interbedded basalts and tuffs of Palaeocene to mid-Eocene age. Various authors (e.g. Corfield *et al.*, 1999; Shannon and Naylor, 1998; Walsh *et al.*, 1999; Shannon *et al.*, 1999) have speculated that marine-dominant Lower Cretaceous strata were deposited in the Rockall Basin. Further west, in the Hatton region, little or no drilling has taken place and the Mesozoic succession is very poorly constrained. However, seismic evidence from the Rockall Basin and the Rockall Bank indicates the presence of tilted fault blocks. These have been interpreted as being of pre-Cenozoic age (e.g. Keser Neish, 1993; Shannon *et al.*, 1994, 1995, 1999). Recent BGS boreholes (99/1 and 99/2A) on the eastern flank of the Rockall Bank drilled up to 30 m of Albian freshwater or lacustrine mudstones and paralic medium- to coarse-grained sandstones and siltstones lying beneath a veneer of Quaternary and Plio-Pleistocene sands and muds (Hitchen, 2004).

Evidence of Lower Cretaceous volcanic activity has been encountered in a number of wells in the Porcupine Basin (Croker and Shannon, 1987; Tate and Dobson, 1988; Tate, 1993). The Phillips 35/8-1 well penetrated 244 m of Hauterivian-Barremian claystones with multiple interlaminated tuff horizons, and 70 m of scoriaceous waterlain airfall deposits, pumice and vitric volcanogenic sediment within the Aptian–Albian interval. Pyroclastics were also recognised at the Barremian–Aptian boundary in the Gulf 26/21-1 well, and at the top of the Albian in the Shell 35/13-1 well. The presence of a Median Volcanic Ridge within the Lower Cretaceous sequence of the Main Porcupine Basin has been postulated from geophysical evidence by a number of authors (Young and Bailey, 1974; Roberts *et al.*, 1981; Ziegler, 1990; Masson and Miles, 1986b). This feature (Figure 17.5) has been further analysed, using multichannel seismic data, by Naylor and Anstey (1987), Tate and Dobson (1988), and Tate (1993). The body appears to comprise two or three discrete flows and is thought (Masson and Miles, 1986b) to be early Aptian or older. This activity, together with related sill intrusion (Naylor and Anstey, 1987), is clearly related to rifting in Early Cretaceous time. Naylor *et al.* (2002) identified additional similar ridges within the basin and named the whole group as the Porcupine Volcanic Ridge System (Figure 17.5). A marked magnetic and gravity anomaly at the south-eastern margin of the Porcupine Basin (49°50' N; 11°30' W) in the region of Balar Spur has been interpreted as an igneous centre (Roberts *et al.*, 1981; Naylor and Shannon, 1982). Later work (Cook, 1987; Tate and Dobson, 1988) showed this to be a penetrative igneous body of possible Aptian–Albian age, which has been named the Seabight Igneous Centre. Within the southern Rockall Trough a number of large curvilinear volcanic ridges (Figure 17.6) – the Barra Volcanic Ridge System – were identified by Scrutton and Bentley (1988). The ridges are 2 km high and 20 km wide and are considered to be of possible mid-Early Cretaceous age, with construction continuing through

Late Cretaceous time. Burial beneath sediment occurred in Eocene time. The Anton Dohrn, Hebrides Terrace and Helens Reef igneous centres in the northern part of the Rockall Trough are all of probable Late Cretaceous age and are the harbingers of the major volcanism that characterised the early Palaeogene of the Atlantic margin basins.

Lower Cretaceous rocks are present in parts of the Slyne and Erris basins and have been identified on published seismic sections (Tate and Dobson, 1989a, 1989b). The sequence in the Corrib Gasfield region at the northern end of the Slyne Basin (Dancer *et al.*, 2005) comprises 50–96 m of Albian claystones and sandstones passing upwards to sandy claystones with occasional limestones, and resting unconformably on Middle to Upper Jurassic strata. In the Amoco 12/13-1A Erris well, Jurassic strata are succeeded by 1.2 km of Cretaceous to Cenozoic rocks (Cunningham and Shannon, 1997). Lower Cretaceous marine claystones and sandstones are overlain by thick submarine-fan sandstones (late Valanginian–Hauterivian) and then by marine claystones grading up to Albian argillaceous limestones (Murphy and Croker, 1992).

Upper Cretaceous strata are widespread on the Irish Atlantic margin. At Goban Spur (Figure 17.19), the Cenomanian to Maastrichtian interval is predominantly Chalk The Turonian is not represented in the Esso 62/7-1 well, nor at DSDP Site 550, possibly due to eustatic uplift (Colin *et al.*, 1992). Upper Maastrichtian strata are also absent from the Esso well, and are probably absent too from the Fastnet and North Celtic Sea basins.

A Cenomanian to Maastrichtian Chalk sequence is found in the Porcupine Basin. About 400 m of Upper Cretaceous is preserved near the margins of the basin, while in excess of 1000 m occurs towards the central axis (Figure 17.24). The succession rests unconformably on an erosional sequence boundary at the basin margins (Moore and Shannon, 1995). The Chalk interval in Block 26/28 at the northern rim of the basin is less than 10 m thick (MacDonald *et al.*, 1987) and is dated as possibly Cenomanian to Maastrichtian. In contrast, possible high-energy calciturbidites are present on the south-eastern and south-western basin margins, probably related to movement on the bounding faults.

A reduced Upper Cretaceous succession is also inferred in parts of the Slyne and Erris basins (Naylor and Shannon, 1982), and it is absent elsewhere due to subsequent erosion. In the area of the Corrib Gasfield (north Slyne Basin) the Upper Cretaceous Chalk, with occasional claystone interbeds, ranges from 79 to 293 m thick (Dancer *et al.*, 2005). Murphy and Croker (1992) reported a typical Upper Cretaceous section comprising micritic limestones and chalks in the Amoco 12/13-1A well in the Erris Basin.

Upper Cretaceous strata are inferred from geophysical evidence in the Rockall Basin (Figure 17.25 and geoseismic sections in Naylor *et al.*, 1999). Marine conditions extended across parts of Ireland and the Celtic Sea and westwards to Rockall. Upper Cretaceous strata in the basins of the Rockall region may be marls rather than pure Chalk (Shannon *et al.*, 1993). In some of the frontier basins (e.g. the Hatton Basin) the Cretaceous succession may be absent owing to basin inversion (Shannon *et al.*, 1995).

Moving to the basins lying north and east of Ireland, Lower Cretaceous strata are again absent from the Malin Sea Basin (Evans *et al.*, 1980), the Ulster Basin and the basins of the Irish Sea. The earliest sediments are the sandy Upper Cretaceous Greensands of the Ulster Basin. By the beginning of the Late Cretaceous a shallow marine embayment extended eastwards, submerging much of the Inner Hebrides, Malin Shelf, Firth of Clyde and Ulster regions. A thin layer of basal sand and Chalk was laid down unconformably across the denuded surface of Jurassic, Triassic and older rocks. Much of this Chalk layer has subsequently been removed by erosion. However, it is preserved over parts of the basins of Western Scotland, but in the Firth of Clyde it survives only as blocks within a volcanic centre on Arran. These younger Mesozoic sequences are not preserved in the basins of the Malin Shelf, but are widely preserved beneath, and around the rim of, the plateau basalt sheet that covers the onshore Ulster Basin. Basal Greensands (Hibernian Greensand Formation) of Cenomanian–Turonian age are followed by widespread Chalk (Ulster White Limestone Formation: up to 110 m) of late Senonian to early Maastrichtian age (Fletcher, 1977).

Cretaceous Palaeogeography

During the Berriasian low-stand episode the Irish massif formed part of a larger landmass which incorporated the Hebridean Platform, Scotland, much of western England, and extended southwards across the Celtic Sea and Western Approaches areas to Armorica (Figure 17.23).

In the Celtic Sea basins the Early Cretaceous deposition began with alluvial and fluvial sediments that gradually extended westwards into the Fastnet Basin. This pattern possibly resulted from thermal uplift in the west

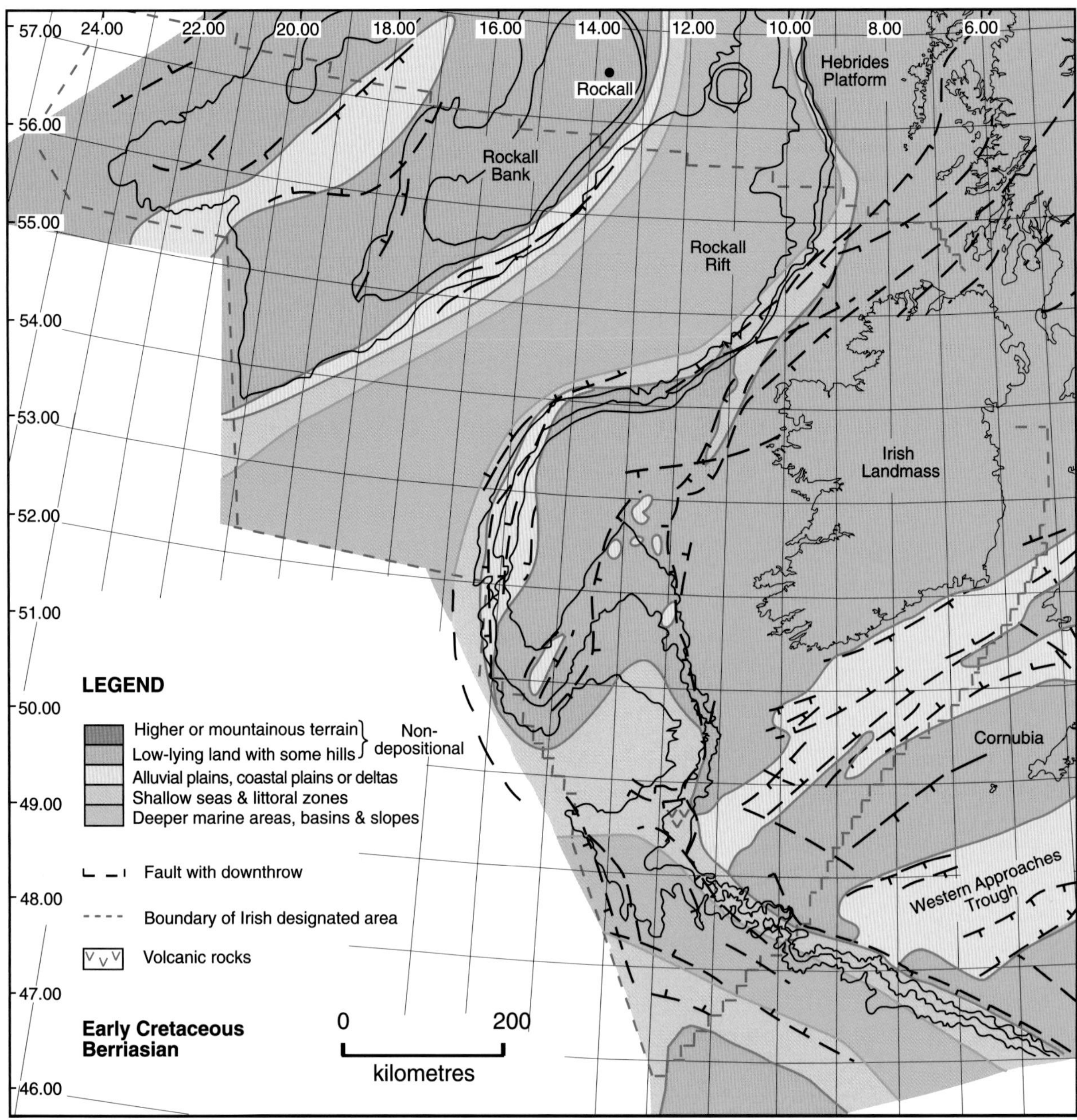

Fig. 17.23 Early Cretaceous: Berriasian palaeogeography (*after* Naylor, 1998. *Principal sources:* Hancock and Rawson, 1992; Naylor and Shannon, 1982; Naylor, 1992; Ziegler, 1990).

associated with rifting. The continental sandy (Wealden) section is overlain by a marine sand and shale sequence. Marine transgression advanced during the Aptian and extended across the whole region in Albian time. In the Bristol Channel–South Celtic Sea basins, Aptian marine sands transgress over eroded Jurassic and older rocks. This kind of complex marginal deposition also occurred in the northern part of the Porcupine Basin and it was only later that fully marine conditions became well established in the southwest regions of the Irish offshore. A marine gulf probably existed along the Rockall Basin, but no details of the succession are available. Non-marine deposition took place in the Hatton Bank region to the west.

The Late Cretaceous was a time of worldwide sea level rise (Vail *et al.*, 1977) in response to major seafloor spreading and plate reorganisation. Global sea levels began to rise during the Albian, leading to widespread transgression. Sea levels in the Campanian may have been 200–300 m above present levels and Chalk deposition was widespread. Around Ireland the basin margins

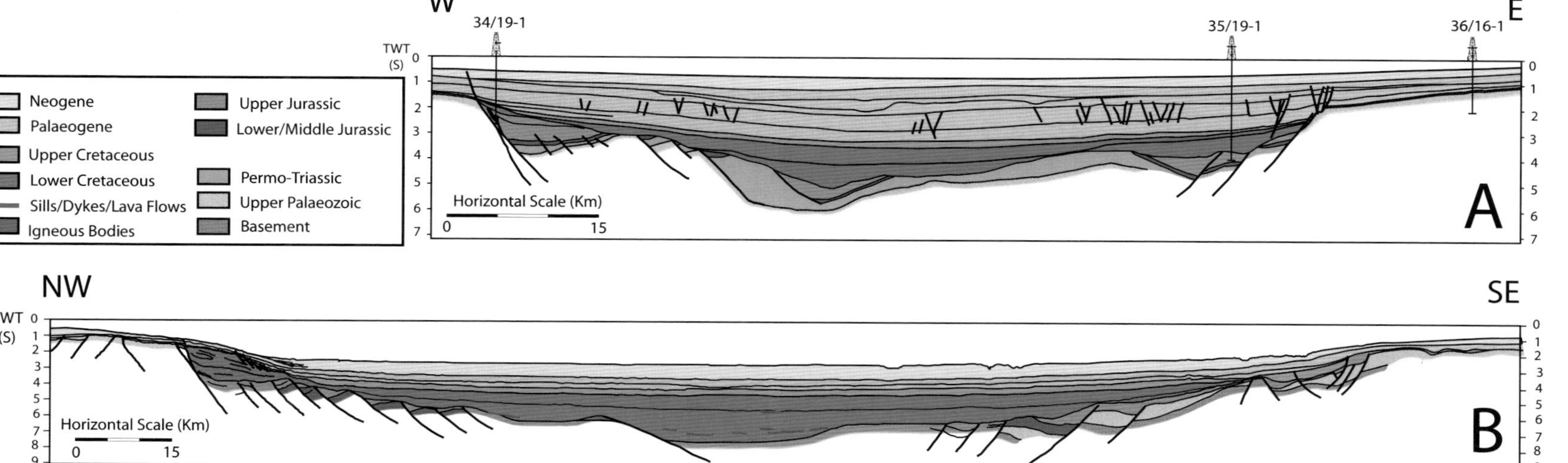

Fig. 17.24 Geoseismic cross-sections across the Porcupine Basin, illustrating Jurassic syn-rift tilted fault blocks beneath a regionally extensive largely thermal subsidence Cretaceous and Cenozoic succession (*after* Naylor *et al.*, 1999). The locations of the sections are shown in Figure 17.5.

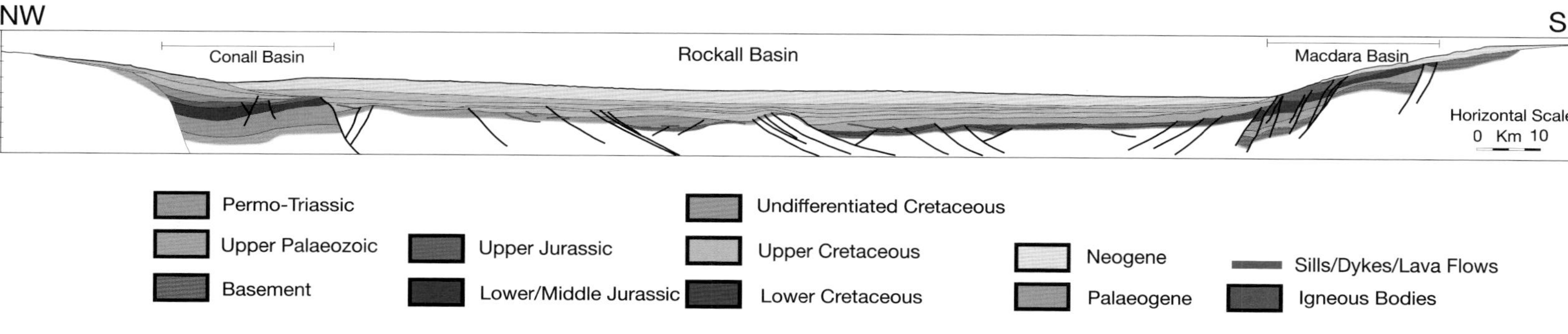

Fig. 17.25 Geoseismic cross-section across the Rockall Basin showing evidence of interpreted Jurassic syn-rift fault blocks beneath an extensive Cretaceous and Cenozoic succession (*after* Naylor *et al.*, 1999). Its location, shown in Figure 17.6, is coincident with the central portion of the crustal velocity cross-section in Figure 17.4.

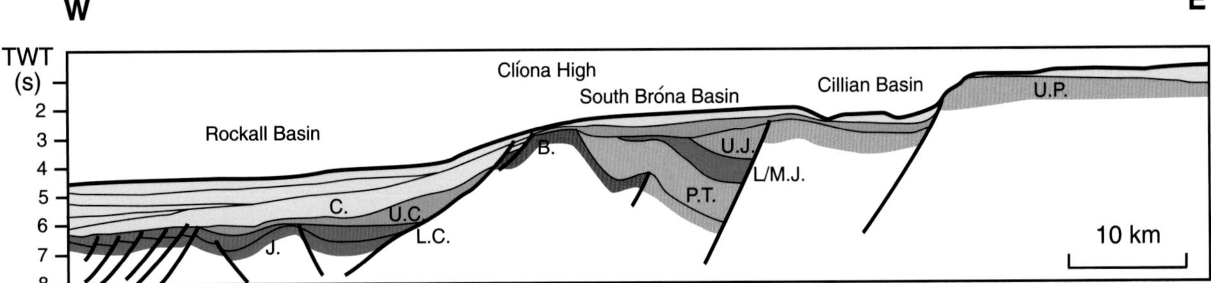

Fig. 17.26 Geoseismic cross-section of the Cillian and South Bróna basins and part of the Rockall Basin showing inferred Permo-Triassic syn-rift strata unconformably overlain by regionally extensive Cretaceous and Cenozoic strata (*after* Naylor *et al.*, 1999; Naylor and Shannon, 2005). U.P. = Upper Palaeozoic; B = Basement; P.T. = Permo-Triassic; L/M.J. = Lower to Middle Jurassic; U.J. = Upper Jurassic; J. = Undifferentiated Jurassic; L.C. = Lower Cretaceous; U.C. = Upper Cretaceous; C. = Cenozoic. The location of the section is shown in Figure 17.6.

were progressively overstepped and the Late Cretaceous seas spread onto the surrounding positive massifs. By Campanian time Ireland was probably totally immersed, with the possible exception of small islands. Marine transgression also extended to deposit a veneer of Chalk across parts of the Cornubian and Welsh landmasses.

Towards the end of the Cretaceous the first major pulses (Laramide) of the Alpine orogeny began to affect the region, causing uplift and erosion. The Chalk Sea retreated and Ireland became emergent, together with large tracts of the surrounding continental shelf.

Cenozoic Era

By earliest Cenozoic (Tertiary) times continental breakup had been completed and seafloor spreading was taking place along the Atlantic margin west of Ireland. This was accompanied by voluminous volcanism and other igneous activity of the North Atlantic Igneous Province (NAIP), spectacularly displayed in the north of Ireland and Scotland. There was a return from the marine Chalk deposition of Late Cretaceous times to shallower water clastic deposition in the Palaeogene. A number of major episodes of Cenozoic epeirogenesis coincided with plate reorganisation. Seafloor spreading in the Labrador Sea ceased in latest Eocene to earliest Oligocene times and resulted in general sagging, differential subsidence and a decrease in coarse clastic deposition. This was coincident with the culmination of the Pyrenean orogenic events.

The early to mid-Miocene period saw the separation of Jan Mayen and Greenland, and this was accompanied by localised compressional inversion and the formation of a regional set of unconformities. It was also coincident with the Late Alpine orogenic effects. The expansion of the deep water system of the North Atlantic is recorded in the build-up of a series of major contourite drift deposits along the Atlantic margin and was coincident with the opening of the Fram Strait between NE Greenland and Svalbart, thus establishing a deep Arctic–NE Atlantic connection and triggering a progressive Northern Hemisphere climatic cooling.

Early Pliocene tectonic movements, associated with late Neogene global plate reorganisation, affected much of the NE Atlantic region and resulted in kilometre-scale domal uplift of onshore areas coeval with offshore subsidence and tilting. Closure of the Central American Seaway by the formation of the Isthmus of Panama resulted in the redirection of warm water masses to northern latitudes. The late Pliocene marked the onset of the widespread northern hemisphere glaciation that prevailed through Pleistocene times.

Offshore distribution of Cenozoic rocks

Cenozoic sedimentary strata are generally thin throughout the NE onshore basins and the basins in the Irish Sea. Typically a few hundred metres at most are preserved. In NE Ireland and the Firth of Clyde, the Cenozoic succession is dominated by the Eocene Plateau basalts which are often in excess of 400 m thick. In the Kish Bank Basin the Fina 33/17-1 well encountered approximately 200 m of Cenozoic to Recent strata, consisting of red-brown to grey sandstones with increasing amounts of claystone with depth. Lignite beds are common in the upper (probably Pliocene to Pleistocene) parts of the succession. A thin Cenozoic succession overlies St Tudwal's Arch, while 600 m of probable Neogene sediments were encountered in the Mochras borehole at the eastern margin of the Cardigan Bay Basin (Figure 17.13).

The Cenozoic succession in the Celtic Sea region is also relatively thin and incomplete, largely due to early Cenozoic tectonism and erosional events. The overall succession thickens and becomes more complete westwards towards the Fastnet Basin and beyond into the Atlantic domain. Within the Celtic Sea region poorly dated Cenozoic strata rest unconformably as a thin blanket on the Pembrokeshire Ridge and in a series of thicker depocentres in the North and South Celtic Sea basins. The succession is cut by two major unconformities, and more than 1 km of strata have been removed by erosion. The first, in early Palaeogene times, resulted in the non-deposition or removal of Palaeocene to early Eocene strata. The second, associated with the Alpine orogenic events, occurred in early Neogene (probably Miocene) times. In the Celtic Sea region, Cenozoic fluvial clastics are overlain by marine strata. Up to 800 m of latest Palaeocene – earliest Eocene to possibly Neogene claystones, thin sandstones and lignites – are documented in the St George's Channel Basin. Farther west, a thicker Cenozoic succession has been encountered by wells, with more than 1 km of strata recorded. Again the succession is attenuated by the two erosional events. It commences with a mid-Eocene to Oligocene shallow marine limestone, of the order of 350 m thick throughout much of the basin but thin or absent in some areas to the southwest. Marine Miocene and Pliocene clays, with some sandy intercalations, overlie the limestones, and the overall succession shows a broad pattern of marine deepening though time.

In the large deep-water basins west of Ireland a thick (up to 4 km) relatively complete Cenozoic succession is preserved. The strata preserved in the Porcupine and Rockall basins provide a comprehensive picture of the sedimentary and oceanographic development of the European passive margin during the period of plate reorganisation following the breakup and separation of the European and North American continents. The earliest Cenozoic strata, of approximately mid-Palaeocene age, rest disconformably on Upper Cretaceous strata, with some of the uppermost Cretaceous succession missing (Figure 17.27). The strata are typically of marly siltstones or limestones, deposited in a marine shelf setting. Where imaged on seismic profiles, these give way to a latest Palaeocene to early Eocene succession of siltstones and sandstones, sometimes showing a bidirectional downlap and indicative of basin floor fan deposits. In the northern part of the Porcupine Basin exploration wells have penetrated a thick succession of deltaic sandstones and southward-prograding clinoforms overlying the basin floor clastics. A number of progradational packages are seen on seismic profiles, each separated by a continuous high-amplitude reflection consistent with a flooding surface (Croker and Shannon, 1987; Moore and Shannon, 1992; Shannon, 1992; Shannon *et al.*, 2005). Each package records sediment input from a northerly quadrant of the basin margin. The preservation of topsets in the deltaic clinoforms points to a relative sea level rise coincident with sediment progradataion. Farther south a coeval succession of pulsed submarine fan mounds has been documented on the margins of the Porcupine Basin (Shannon, 1992). These are especially well displayed on the western margin of the basin, with sediment derived from the Porcupine Bank to the west. The overall pattern of pulsed clastic input in latest Palaeocene to Eocene times has been explained by shelf/flank uplift and basin centre subsidence controlled by regional tectonics (Shannon *et al.*, 2005). The pulsed nature of the succession has previously been interpreted as being due to ridge push (Shannon *et al.*, 1994), while in the North Sea a similar pattern has been attributed to mantle plume tectonics (White and Lovell, 1997).

In addition to the well-known Palaeogene igneous province of Scotland and Northern Ireland, a significant igneous history in the Atlantic offshore basins has been documented by well and seismic data. Major igneous centres have been identified on the basis of seismic, gravity and magnetic mapping (Naylor *et al.*, 1999, 2002). They include the Brendan Igneous Centre in the northeast of the Porcupine Basin (Figure 17.28), which is accompanied by large gravity and magnetic anomalies (Riddihough, 1975; Mohr, 1982; Tate and Dobson, 1988). A number of volcanic cones have been recorded along the eastern margin of the Rockall Basin (Haughton *et al.*, 2001). Although few of the major centres have been dated, their setting is strongly suggestive of an early Palaeogene age. High-amplitude discontinuous seismic reflectors are prevalent through much of the Rockall Basin and the north-eastern part of the Porcupine Basin and have been interpreted as extensive transgressive sills. Seeman (1984) reported tholeiitic sills in the Cenozoic succession of the Shell 35/13-1 well and suggested an Oligocene date. Undated dolerites and other intrusives in other wells are also likely to be of Palaeogene age. Possible feeder dykes to the transgressive sills have been identified on some seismic profiles in the Porcupine Basin. In the north of the basin, Tate and Dobson (1988) identified a series of parallel small igneous plugs, named the Slyne Fissures, and inferred their age to be Palaeogene.

A major regional tectonic and facies change in latest Eocene times is coincident with a major unconformity

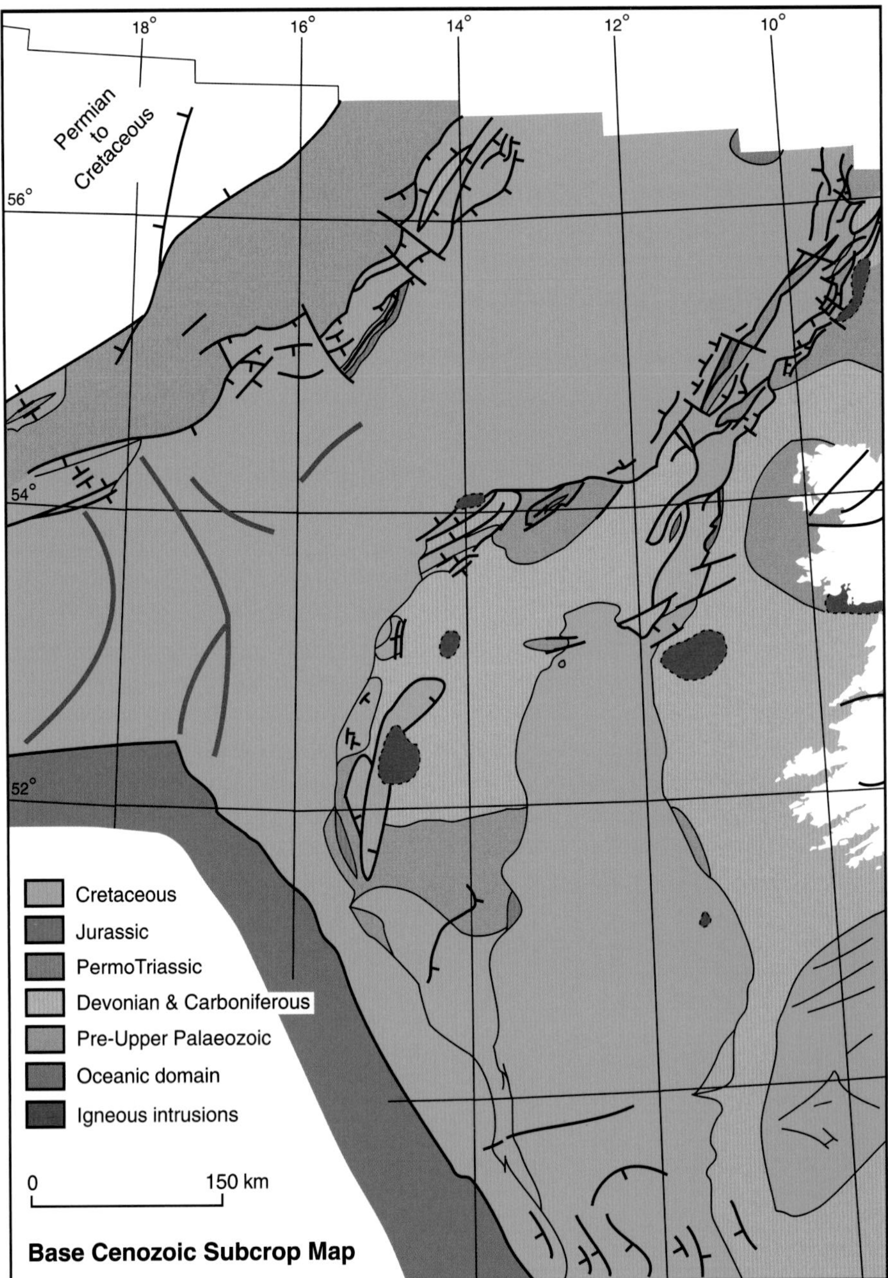

Fig. 17.27 Base Cenozoic subcrop map of the Atlantic margin basins west of Ireland (*after* Naylor and Shannon, 2005).

(C30 of McDonnell and Shannon, 2001; Stoker *et al.*, 2001, 2005b, c). This unconformity locally cuts through several hundred metres of strata in the Porcupine Basin and is also mapped in the Rockall Basin (Shannon *et al.*, 2005). It resulted from a rapid deepening and the onset of vigorous large-scale ocean current circulation. It marks a major change from downslope sediment transportation to mainly alongslope sediment movement. From latest Eocene times basin subsidence outstripped sedimentation in the Atlantic margin basins. This resulted in the development of under-supplied deep-water basins with steep sides.

The early Oligocene sediments are outer shelf and deeper marine marls and siltstones, typically showing very continuous seismic reflections. Praeg *et al.* (2005) suggested a minimum post-Eocene deepening of 700 m in the vicinity of the Anton Dohrn seamount, where mid- to late Eocene nearshore conglomerates crop out on the crest of the seamount. Praeg *et al.* (2005) illustrated the strongly differential nature of the deepening farther

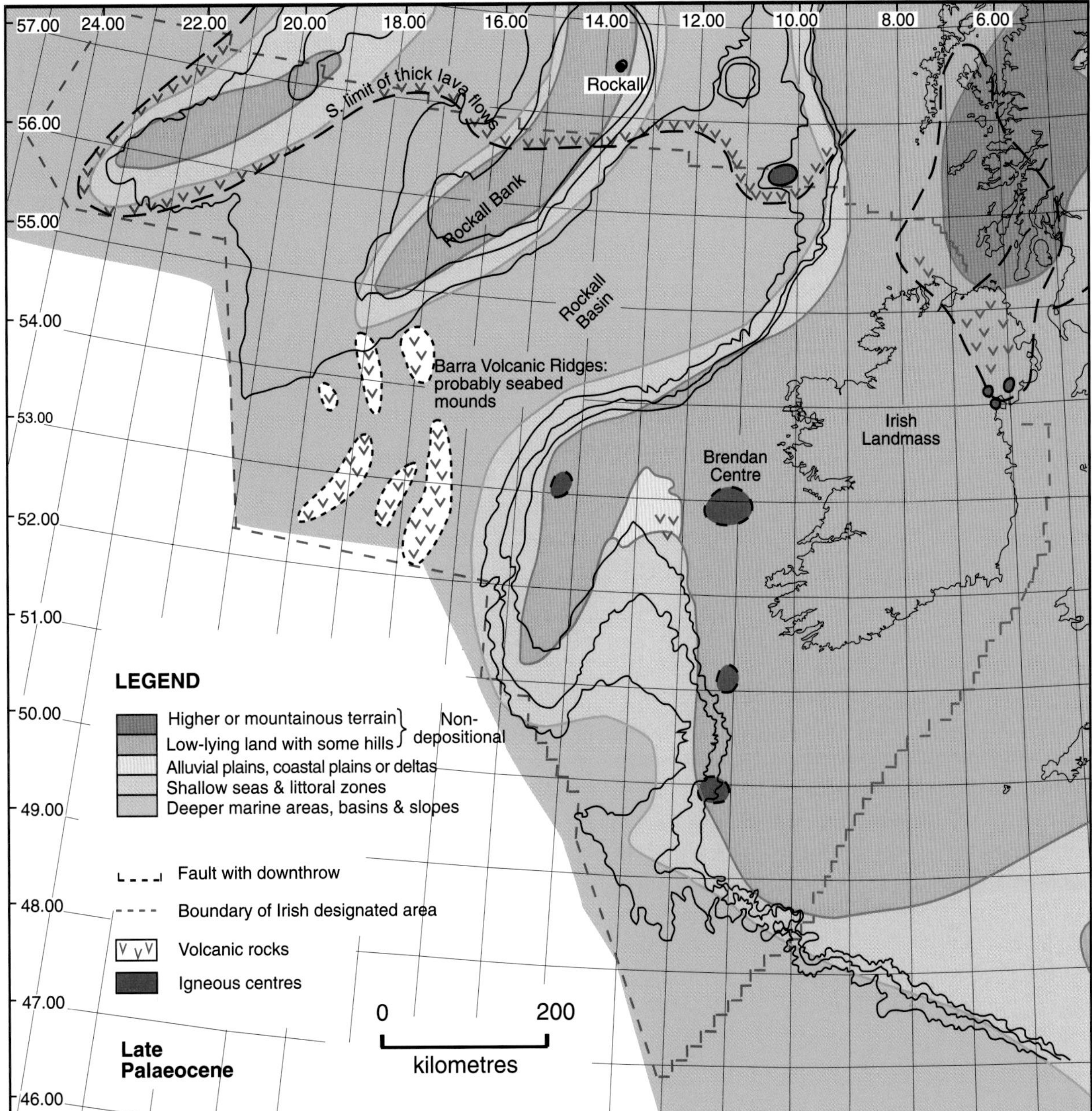

Fig. 17.28 Palaeocene palaeogeography (*after* Naylor, 1998. *Principal sources:* Knott *et al.*, 1993; Murray, 1992; Scrutton and Bentley, 1989; Tate and Dobson, 1988; Ziegler, 1990).

north, where subaerial Palaeocene basalts between Bill Bailey and Lousy banks are overlain by post-Eocene strata and lie in water depths ranging from 500 m to more than 2 km. A similar Oligocene basinward tilt of the Eocene and older strata by several degrees was documented along the NE Rockall margin by other authors (e.g. Cunningham and Shannon, 1997).

The Oligocene and Miocene successions of the Atlantic margin region are characterised by flat-lying parallel reflections within the basins. Where drilled, these are typically mudstones with occasional silt and marl layers. Towards the basin margins the succession typically displays an onlapping, sometimes mounded and locally significant upslope accretion (Stoker *et al.*, 2005c). Muddy and silty contourites have been encountered. The contourites are more pronounced above a set of mid-Miocene unconformities (C20 of McDonnell and Shannon, 2001; Stoker *et al.*, 2001, 2005b, c). The more pronounced of the accretionary contourite bodies are the Porcupine Drift in the Porcupine Basin (McDonnell and Shannon, 2001) and especially the Feni Drift (Stoker *et al.*, 2001) on the western margin of the Rockall Trough. DSDP site 610 sampled the Feni Ridge and encountered lower-middle Miocene to lower Pliocene lower bathyal nannofossil chalk and ooze (Hill, 1987; Stoker *et al.*, 2005c). Broadly similar lithologies were encountered in other boreholes through the same succession along the margin. Along the eastern margin of the Rockall Trough, a thin Eocene to Holocene succession, cut by unconformities and discontinuities, was encountered by drilling (Haughton *et al.*, 2005). This was largely deposited in deep, relatively quiet water conditions. A north to south decrease in the clastic component of the Eocene succession is interpreted as preferential sequestration of clastic sediment into the nearby Porcupine Basin. Slump deposits and mass failures became a feature of the basin margins in the early Miocene (e.g. Moore and Shannon, 1991) and continued periodically through to Holocene time.

The Pliocene to Holocene succession in the Irish Atlantic offshore is generally thin (typically less than 150 m) and lies above a regional unconformity (C10 of Stoker *et al.*, 2001, 2005b, c). The succession is widely distributed as a thin draping sheet of sediment within the Porcupine, Rockall and Hatton basins, with local thickening associated with the Pleistocene Barra Donegal Fan in the north of the region. On the basin floor the strata are typically pelagic foraminiferal oozes. A series of deep-water carbonate mound provinces developed above the Pliocene unconformity (Shannon *et al.*, 2007; Bailey *et al.*, 2003) and appear to be related to a major oceanographic and climatic change, while on the shelf and slope regions a number of dramatic slope failure features have been documented (Shannon *et al.*, 2001b; Unnithan *et al.*, 2001; Elliott *et al.*, 2006). A complex system of canyons, channels and slope failure features are developed on the steep margins of the Rockall Trough, especially in the sediment-starved regions south of the Barra–Donegal Fan region (O'Reilly *et al.*, 2007). Gravity cores indicate the presence of glacial and interglacial successions with characteristic sedimentological patterns (Øvrebø *et al.*, 2006). During full glacial conditions, bottom current activity was weak and multiple episodes of mass wasting took place. During interstadials and interglacials, mud-prone deposition took place along the western margin of the Rockall Trough, while the eastern margin of the basin was characterised by strong bottom currents resulting in periods of erosion and winnowing and the deposition of sandy biogenic (interglacial) and mixed clastic–biogenic (interstadial) contourites.

Cenozoic palaeogeography

Shortly after the commencement of the Cenozoic Era, crustal separation between the European and Greenland plates was complete and new oceanic crust extended northwards through the North Atlantic. The final breakup was accompanied by a change from extensional rifting to passive margin tectonics, and by an increase in igneous activity, which was centred around Scotland and Northern Ireland and adjacent parts of the trailing passive margin (Figure 17.28). This igneous activity possibly reflected mantle circulation cells whose location was influenced by large crustal thickness differences between the major basins and their flanking basement highs (Praeg *et al.*, 2005; Stoker *et al.*, 2005b, c). The overall setting of epeirogenic tilting (coeval subsidence and uplift) caused the basinward-progradation of major sediment wedges deposited as deltas in shallow parts of basins such as the northern part of the Porcupine and basin floor fans in the deeper and more rapidly subsiding parts of the basins. This kilometre-scale epeirogeny saw onshore uplift of up to 1.5 km, coincident with offshore subsidence of up to 600 m in the major sedimentary basins (Praeg *et al.*, 2005). Major domal uplift has also been suggested for the Irish Sea region (Cope, 1994), and the attenuated Palaeogene succession in the Celtic Sea region may be consistent with a period of regional uplift.

The major basin readjustment in latest Eocene–earliest Oligocene times reflects a phase of sagging which was accompanied by the cessation of major coarse clastic deposition, the differential deepening of major basin depocentres, and the onset of large-scale deep-water current systems. The basins became under-supplied and the resulting strata were largely deep-water and fine-grained, with alongslope reworking of sediment to form shelfward accreting silty contourite deposits. The largest of these deposits lies along the western margin of the Rockall Basin, with a smaller contourite drift deposit in the smaller Porcupine Basin. Typified by the pattern observed in the Rockall Basin, currents were largely anticlockwise, with erosion along the eastern margin and deposition on the western margin. This erosion led to the destabilisation of the scoured margin with the formation of canyons, slides and slope failures that provided sediment for reworking on the lower parts of the slope. Local domal inversion in the Miocene resulted in localised topography, erosion and deposition. Such inversion domes (Stoker *et al.*, 2005a) are relatively rare in the Irish offshore, being more prevalent further north in the UK and Norwegian parts of the Rockall and other large basins.

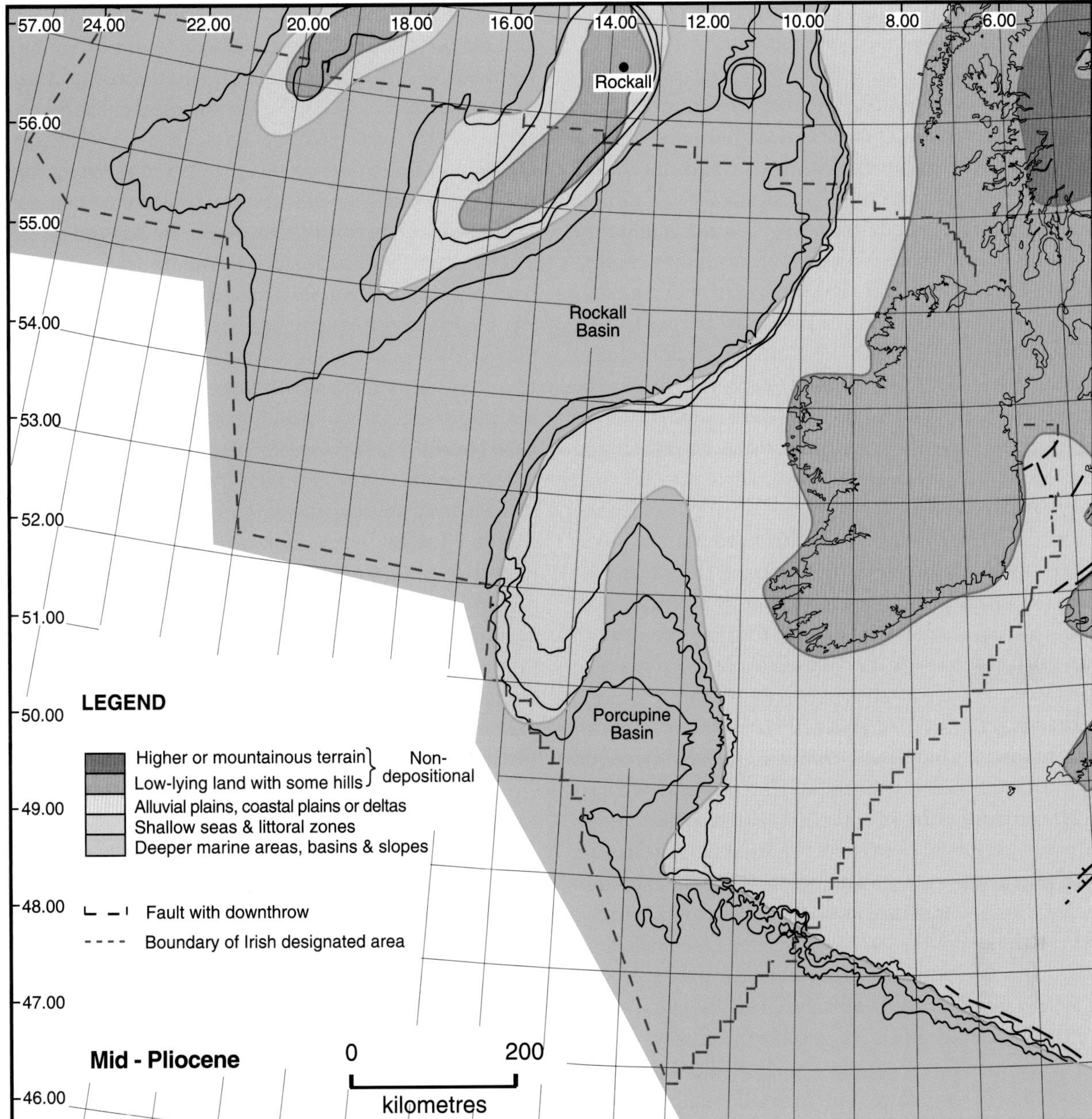

Fig. 17.29 Pliocene palaeogeography (*after* Naylor, 1998. *Principal sources:* Knott *et al.*, 1993; Murray, 1992; Scrutton and Bentley, 1989; Tate and Dobson, 1988; Ziegler, 1990).

The early Pliocene (Figure 17.29) witnessed the third Cenozoic epeirogenic episode in the Atlantic region. Tilting, resulting from plate reorganisation, resulted in a regional unconformity. The Pliocene to Holocene succession was relatively thin throughout most of the Irish offshore, thickening rapidly towards the Irish/UK boundary in the Rockall Trough where the Barra Donegal Fan provided a thick succession of glaciogenic and debris flow clastics shed from the adjacent highlands.

Basin development: Northern Ireland and Irish Sea basins

The basins along the northern and eastern margins of Ireland differ from those of the other offshore regions in that the basin fill largely comprises Permian and Triassic strata (Figure 17.10). This is in direct contrast to the thick Cretaceous and Cenozoic sequences of the Atlantic margin basins, while the basins in the Celtic Sea region contain little Permian, but have a Triassic and a thick Jurassic succession. The succession in the Northern

Ireland and Irish Sea region, discussed earlier in the chapter, may be summarised as follows:

Loch Indaal and Rathlin basins

Both basins contain in excess of 2 km of Permo-Triassic sediments that onlap the intervening Middle Bank positive element. Thin Liassic successions occur against the controlling faults of the half-grabens – against the Leannan Fault in the Loch Indaal Basin, as shown in offshore boreholes, and in the hanging wall of the Tow Valley Fault of the Rathlin Basin, as demonstrated by the Port More borehole (Figure 17.10). Here the Liassic (almost 250 m; Wilson, 1981) is unconformably overlain by the Upper Cretaceous Ulster White Limestone Formation (91.3 m), which is in turn unconformably overlain by the Lower Basalt Formation (Palaeocene).

Larne and North Channel Basins

The onshore Larne Basin extends southwards from the Highland Border Ridge to the Longford-Down Massif and westward from the North Channel to Lough Neagh. It is known from deep boreholes to contain in excess of 2 km of post-Carboniferous strata over much of the onshore area (Thompson, 1979). As in the Rathlin Trough, the Permo-Triassic red beds pass conformably upwards to a thin Liassic sequence that is unconformably overlain by the Hibernian Greensand and Ulster White Limestone Formations. Thick Palaeogene basalts were extruded onto the karstified surface of the limestones. The sedimentary section thickens to 3 km at the coast and extends out into the undrilled North Channel Basin and the Portpatrick Sub-basin. However, there is no seismic evidence (Maddox *et al.*, 1997; Shelton, 1997) for the presence of post-Triassic rocks in the offshore. The Palaeogene lavas are largely absent offshore and were possibly not originally deposited (Maddox *et al.*, 1997).

Solway and Peel Basins

In these basins (Figure 17.11) Permo-Triassic strata rest unconformably on eroded Carboniferous sequences (Holkerian to Pendleian). The Liassic is locally preserved (Newman, 1999) and younger strata are unknown.

Kish Bank Basin

Evidence from drilling (Jenner, 1981) and seismic (Dunford *et al.*, 2001) suggests that Pennylvanian rocks are widespread beneath thick Permian and Triassic successions that comprise much of the basin fill (Figure 17.12). Although the four wells in the basin did not penetrate Jurassic strata, pockets of apparently conformable Liassic strata have been interpreted on seismic data (Jenner, 1981; Naylor *et al.*, 1993; Dunford *et al.*, 2001). Lower Liassic rocks have been sampled from the sea bed at the northern margin of the basin (Etu-Efeotor, 1976; Dobson, 1977 a, b). Broughan *et al.* (1989) interpreted up to 2700 m of Liassic rocks preserved against the Bray and Dalkey bounding faults. Ammonites in glacial deposits on the immediately adjacent coastline, possibly derived from the Kish Bank area, were dated as Lower, Middle and Upper Liassic in age (Broughan *et al.*, 1989).

Central Irish Sea Basin

The Central Irish Sea Basin (CISB) is a NE–SW oriented basin that is separated from the Cardigan Bay and St George's Channel basins by the St Tudwal's Arch and its offshore extension (Figure 17.13). Drilling has shown a Triassic succession resting unconformably on the Carboniferous, and thinning southwards towards the St Tudwal's Arch extension (Figure 17.14). Only Lower Jurassic strata have been proved by drilling, although Maddox *et al.* (1995) suggested that Jurassic and Cretaceous sequences may be preserved in the deeper parts of the CISB.

Cardigan Bay and St George's Channel basins

Exploration drilling in these basins has revealed an almost complete Lower Liassic to Lower Purbeckian succession (Barr *et al.*, 1981). The coastal Mochras borehole (Figure 17.13) penetrated 1305 m of Lower Jurassic (Woodland, 1971), whilst in excess of 3 km of Middle and Upper Jurassic sediments were deposited in the centre of the Cardigan Bay Basin (Maddox *et al.*, 1995).

Discussion

A number of authors have offered observations regarding basin development and the pattern of preserved stratigraphy of the Irish Sea region in terms of phases of exhumation, erosion or non-deposition (e.g. Cope, 1984, 1997; Green *et al.*, 1997, 2001). The following are key observations from the region: (a) Permian and Triassic strata were deposited on the eroded Variscan surface; (b) no Jurassic strata younger than Liassic have been proven north of the Cardigan Bay–St George's Channel basins and (c) no Cretaceous strata are preserved in the region between the onshore Larne Basin and Cardigan Bay Basin.

The marked Caledonian (NE–SW) orientation of some Carboniferous basins, notably the Larne–North

Channel and Solway–Peel basins, points to reactivated Caledonian control on Carboniferous deposition. Late Carboniferous inversion and erosion, partly controlled by the reactivated Caledonian faults, probably resulted in the removal of large thicknesses of Upper Carboniferous strata from the Irish Sea Basins. Corcoran and Clayton (1999) and Quirk *et al.* (1999) emphasised the importance of Early Permian thermal uplift related to rifting in the Irish Sea area. Erosion of the Carboniferous sequences at the basin margins was widespread during the Triassic (Warrington, 1994; Newman, 1999; Naylor and Clayton, 2000). Several authors (Musgrove *et al.*, 1995; Ruffell and Shelton, 1999) have also pointed to the influence of syn-sedimentary faulting during deposition of the Mercia Mudstone Group.

There is general agreement that basin inversion has played a significant role in the region. Apatite fission-track (AFT) analysis, vitrinite reflectance (VR) studies and sonic velocity profile data, together with tectonic modelling, have been used to estimate the timing and the degree of exhumation of the inversion events. The stratigraphic control on exhumation timing, as well as the amount of available and published data, differs from basin to basin. The deep erosion of Mesozoic stratigraphy in the Irish Sea has been interpreted as due to crustal uplift and denudation resulting from plume-related transient dynamic uplift under the Irish Sea during the Palaeocene (Cope, 1994, 1997), or caused by igneous underplating (Brodie and White, 1994; White and Lovell, 1997). Green *et al.* (1997) identified four palaeo-thermal events in the region based on AFT analysis and VR studies. These are pre-Permian (>290 Ma), Late Permian to mid-Triassic (260–220 Ma), Early Cretaceous (140–110 Ma) and Early Palaeogene. Modelling of AFT and VR data by Allen *at al.* (2002) indicated significant Cretaceous denudation onshore in east-central Ireland that poses the possibility that the absence of Cretaceous rocks in the adjacent central Irish Sea may be partly due to Cretaceous non-deposition, rather than solely to erosion during a Palaeocene crustal uplift event.

In the Loch Indaal, Rathlin and onshore Larne basins the remnant stratigraphy places only wide constraints on events. Permian or Triassic rocks rest unconformably on Carboniferous or older strata, indicating late Carboniferous to early Permian inversion and erosion. There is no preserved (or at least proven) stratigraphic record in these basins for the interval between Liassic (Pliensbachian) and the Hibernian Greensand–Ulster White Limestone formations (Cenomanian to early Maastrichtian). The Palaeocene Antrim Lava Group rests unconformably on the karstified limestone surface. However, it is worth noting that deep erosion and removal of the relatively thin limestone formation did not occur. The only substantial deposits younger than the Antrim Lavas are the onshore Lough Neagh Clays (>350 m thick) – a dominantly lacustrine sequence of Oligocene age. Shelton (1997) suggests that accommodation for this sequence was probably due to localised extension around the Lough Neagh depocentre, in contrast to the rest of the Larne Basin that is characterised by uplift.

Stratigraphic evidence in the North Channel Basin (Figure 17.10) affords even less assistance, due to the lack of proven post-Triassic cover, based on seismic interpretation. However, since Liassic and Upper Cretaceous sediments are preserved in the immediate onshore area, it would be surprising if the inversion histories in these basins were substantially different from that of the offshore basins. The lack of Antrim Lava Group basalts, however, may indicate different conditions in Palaeocene time. Shelton (1997), in a study of the Larne and North Channel basins that incorporates geological, well, AFT and sonic/seismic velocity studies, concluded that the late Carboniferous Variscan compressional event was followed by Early Permian and Late Triassic–Early Jurassic extensional episodes. Naylor *et al.* (2003) identified fault reactivation in the Lough Neagh area in Late Permian and possibly Late Triassic time. It is likely that the main phase of uplift and erosion of the Liassic and older section occurred during the Late Cimmerian phase at the end of the Jurassic. The end Cretaceous uplift, although producing a well-defined unconformity surface exhibiting features of exposure and erosion, did not remove significant section. AFT data indicate *c.*450 m of uplift and erosion in the Cenozoic, probably in the Miocene (Shelton, 1997).

The inversion and erosion of the Carboniferous section is clearly seen and documented in the Peel Basin (Newman, 1999). However, the lack of post-Liassic rocks means that Mesozoic and Cenozoic exhumation episodes are largely unconstrained. Using AFT data, Green *et al.* (1997) identified a local cooling event which began between 260 and 220 Ma in the Solway Firth, but this does not appear to have been widespread.

More data are available for the Kish Bank and Central Irish Sea Basins, where AFT, VR and burial modelling studies have been published based on a number of exploration wells (Naylor *et al.*, 1993; Duncan *et al.*, 1998; Corcoran and Clayton, 1999; Dunford *et al.*, 2001; Floodpage *et al.*, 2001; Green *et al.*, 2001). The

AFT–vitrinite studies of Green *et al.* (1997) show that maximum palaeo-temperatures and burial in the East Irish Sea Basin were probably attained in Palaeogene time, prior to uplift. Elsewhere in the Irish Sea, however, the evidence suggests that maximum maturation was attained during earlier periods. Thus Duncan *et al.* (1998) deduced from AFT studies in the CISB that maximum palaeo-temperatures occurred in the Early Cretaceous. However, VR studies led Corcoran and Clayton (1999) to conclude that Carboniferous sediments in the CISB and adjacent areas experienced peak maturation prior to the Variscan orogeny. In the Kish Bank Basin Naylor *et al.* (1993) estimated that a further 1.3 km of cover would have been required to account for maturation levels in the preserved Carboniferous section. They considered that this section was eroded during Late Jurassic uplift, whereas Corcoran and Clayton (1999) saw this as an end Carboniferous event. The balance of more recent data (Floodpage *et al.*, 2001) supports the importance of the Late Cimmerian tectonic episode in the CISB, Kish and Peel basins in initiating cooling from maximum palaeo-temperatures.

The St Tudwall's Arch and its offshore extension serve to separate the CISB from the St George's Channel–Cardigan Bay basins. The latter basins, with extensive Jurassic sequences, are probably best considered within the Celtic Sea domain, rather than with the Permo-Triassic dominated basins of the Irish Sea. The results from the HIL 42/12-1 well at the northern margin of the St George's Channel Basin (Corcoran and Clayton, 1999; Green *et al.*, 1999) suggest that about 1.5 km of section were removed during the main phase of exhumation and that this probably occurred during the Palaeogene, with a further phase in the Late Cenozoic. A more recent comprehensive study of the basin (Williams *et al.*, 2005) concludes that there were at least two major inversion episodes – Late Cretaceous and Neogene, with a minor shortening event in the Eocene. The same authors also presented evidence from the basin margin supporting earlier cooling phases in the Permo-Triassic, following rifting, and in the Mid-Late Jurassic.

Basin development: Celtic Sea basins

The basins of the Celtic Sea region, south of Ireland, lie in relatively shallow (100–200 m) waters and are noticeably elongate (Figure 17.3). The main orientation is ENE–WSW, with some strike swings at the western and eastern ends of the basin system. Two main parallel tranches of basins occur, separated by an intermittent basement high (Pembrokeshire–Labadie Bank). The basins contain up to 9 km of Triassic to Cenozoic strata, resting upon deformed Variscan basement (Shannon and Naylor, 1998). The basins are characterised by having a thick Triassic and Jurassic succession, with a variable thickness of younger strata. The Cretaceous and Cenozoic succession is thinner than in the Atlantic basins while, in contrast to the Irish Sea basins, little or no Permian strata have been encountered, whereas a considerable Jurassic succession is preserved. A number of inversion episodes, notably in the Cenozoic, also characterise the geological development of these basins. A summary of the succession is as follows:

Fastnet Basin

Lying to the north of the Labadie Bank basement high, the Fastnet Basin (Figure 17.18) has a NE–SW orientation. It contains up to 4.5 km of post-Variscan strata. Devonian red beds and Mississippian shelf limestones have been encountered by exploration wells in the basin (Robinson *et al.*, 1981). Permian strata appear to be absent and a widespread Triassic succession, approximately 600 m thick, consists of a lower sandy and an upper marl and mudstone evaporitic succession. This is conformably overlain by a transgressive marine limestone and mudstone Lower Jurassic succession, including a locally thick Sinemurian regressive deltaic sandy succession. Close to 2 km of Lower Jurassic strata have been drilled in the basin. While some Middle Jurassic mudstones and Middle Jurassic igneous rocks have been encountered in drilling (Caston *et al.*, 1981), Upper Jurassic strata are largely absent. The Cretaceous, up to 1 km thick in wells, comprises a sandy and muddy fluvial Wealden succession, a sandy and glauconitic marine Aptian–Albian Greensand and an Upper Chalk succession. Cenozoic sediments are typically up to 1 km thick and comprise a mid-Eocene to Oligocene limestone/chalk succession overlain by Miocene and Pliocene marine claystones and a veneer of Pleistocene sediments.

North Celtic Sea Basin (NCSB)

This is the largest of the Celtic Sea basins and has an ENE–WSW orientation. Deformed Variscan basement is overlain by Triassic and Jurassic strata. It is structurally and stratigraphically linked to the Fastnet Basin. Up to 600 m of Triassic has been encountered in wells on the southern margins of the basin (Esso 56/20-1 and Conoco 57/9-1) and has also been interpreted on seismic profiles (Shannon and MacTiernan, 1993). The Lower and Middle Jurassic are broadly similar to the succession encountered in the adjacent Fastnet Basin, consisting of a basal limestone package overlain by predominantly marine organic-rich mudstones, with occasional sandy units. In contrast to the Fastnet Basin a thick (*c*.1 km) syn-rift Upper Jurassic succession is present in the basin. Fluvial Callovian to Oxfordian sandstones more than 200 m thick (Caston, 1995) are overlain by muddy, silty and calcareous Oxfordian and younger Jurassic strata. The Cretaceous is similar in lithology and stratigraphy, but generally thicker than in the Fastnet Basin. The Cenozoic is thin and only patchily preserved in the basin, and has been eroded following phases of Palaeocene and Oligo-Miocene inversion (Murdoch *et al.*, 1995). The NCSB links eastwards to the St George's Channel and Cardigan Bay basins, which have been addressed, together with the other basins in the Irish Sea region, in the previous section.

South Celtic Sea Basin (SCSB)

This basin is separated from the NCSB by the intermittent Labadie Bank–Pembrokeshire Ridge structural high, a feature that appears to have been active during Mesozoic times, resulting in some differences in the stratigraphic architecture between the two basins. Deformed Variscan basement has been drilled and is overlain by Triassic strata. This tends to be less sandy and contain more dramatic halokinetic structures than the NCSB (Shannon, 1991b, c). Where drilled, the Jurassic typically consists of calcareous and muddy Liassic, with thin to absent Middle and Upper Jurassic strata. The Cretaceous is thinner but broadly similar in stratigraphy and lithology to the Fastnet and North Celtic Sea basins, while the Cenozoic is generally thin but rather more widespread than in the NSCB, reflecting less dramatic late-stage basin inversion.

St George's Channel Basin

This is the eastward extension of the South Celtic Sea Basin into UK waters and contains up to 3 km of post-Palaeozoic sediments. The basin narrows eastwards and shows a strike swing, with a general east–west orientation. The Triassic, with thick salt in the offshore (Kamerling, 1979), onlaps folded Devonian and Carboniferous basement. The Lower and Middle Jurassic are lithologically similar to the other Celtic Sea basins and the locally encountered Upper Jurassic is predominantly muddy. The Cretaceous is generally thin, and again is broadly similar in lithology to the other basins of the region. The Cenozoic is generally very thin or absent.

Cockburn and Haig Fras basins

These lie to the south and south-west respectively of the SCSB. Although undrilled, comparison of seismic profiles with those of the adjacent basins suggests a broadly similar but thinner succession to those encountered in the main Celtic Sea basins (Smith, 1995). A widespread interpreted Triassic succession overlies Variscan basement. In places, especially in the Cockburn Basin, fault-controlled depocentres beneath the Triassic are interpreted as the remains of possible Permian intermontane basins, resulting from orogenic collapse following Variscan mountain building. The Jurassic succession is relatively thin and probably consists of Liassic limestones, marls and organic-rich mudstones. The Cretaceous, resting unconformably on the Liassic, is also relatively thin while the Cenozoic is probably comparable in thickness and lithology to the Fastnet Basin.

Discussion

The Fastnet and North Celtic Sea basins have been the site of close to 100 exploration and development wells and extensive seismic surveys since offshore petroleum exploration commenced in the region in the late 1960s. However, until relatively recently the seismic data quality in the NCSB has been poor, largely due to Cenozoic basin inversion resulting in the exposure of compacted Upper Cretaceous at the seabed. Despite commercial and non-commercial oil and gas discoveries, the deep geology of the basin has therefore been poorly understood. From the large amount of published research literature on the region, the following major points emerge: (a) the Mesozoic sediments rest upon folded and indurated Devonian and Carboniferous (largely Mississippian) metasediments, (b) the basins appear to have an underlying Variscan or older structural control on their location and orientation, (c) although the pre-Jurassic strata are poorly dated, there is very little evidence for the presence of any significant occurrence of Permian strata in the major Celtic Sea basins, (d) thick Upper Jurassic sediments have only been encountered in the NCSB, and (e) the region experienced several phases of basin inversion, most notably in the Cenozoic.

Published seismic profiles have illustrated an underlying structural control on the geometry of the Celtic Sea basins, although the details of that control are still somewhat unclear. A deep seismic profile across the eastern part of the NCSB (McGeary *et al.*, 1987) shows the basin lying in the footwall of a deep crustal structure interpreted as the Vasiscan Front. The north-western segment of the NCSB, somtimes referred to as the Mizen Basin, has a pronounced half-graben morphology and it has been argued (McCann and Shannon, 1993) that the basin shape was controlled by extensional re-activation of a deep Variscan backthrust structure. However, these authors noted that the orientation is somewhat oblique to the dominant Variscan trend, and also suggested the possibility that the reactivated structure is an older Caledonian structural lineament. Overall, the basin orientation is more reminiscent of regional Caledonian rather than Variscan structures. Indeed, it has been argued by Gardiner and Sheridan (1981) that there is no apparent genetic relationship between major Variscan structures and the main offshore structural framework in the Celtic Sea region. While the effects of the Variscan compression and orogeny on the underlying Caledonian and older structures is uncertain, it therefore seems plausible that the major controls are Caledonian and that the strike swing of the Caledonian structures to the south of Ireland was reinforced by the Variscan deformation in this region.

While dating of the extensive early Mesozoic red bed facies in the region is generally poor, regional lithostratigraphic correlations are relatively robust and suggest the regional presence of a Triassic succession, with a lower Sherwood Sandstone and an upper Mercia Mudstone succession. However, it is notable that there is little evidence for Permian strata in the region. A major exception is the half-graben lying beneath the interpreted Triassic in the undrilled Cockburn Basin. This suggests that the Celtic Sea remained an elevated region, shedding sediment into the subsiding Permian basins of the Irish Sea, such as the Kish Bank Basin and the Irish Sea Basin, which lay to the north of the orogenic zone and were generally unaffected by the Variscan orogenesis. Regional early Mesozoic deposition commenced in Triassic times, as discussed in previous sections of this chapter, following crustal cooling, erosion and Variscan orogenic collapse.

With the exception of a narrow belt at the south of the NCSB, thick Triassic salt in the region is confined to the belt of basins lying south of the Labadie Bank–Pembrokeshire Ridge. This suggests that the basement ridge system played a role in the basin development from early Mesozoic times. This is also supported by the general lack of thick Upper Jurassic sediments from the basins south of the ridge, while the effects of Cenozoic inversion are more pronounced north of the ridge system.

Following thermal subsidence, with a regional marine transgression in earliest Jurassic times, minor rift events or localised faulting were manifest by sandy deltaic and shallow marine strata in the western Fastnet and the eastern North Celtic Sea basins. The main rifting in the basins took place in Late Jurassic time but the deposits of this phase are largely confined to the NCSB region. It is unclear if this reflects lack of deposition elsewhere, or removal through later, earliest Cretaceous erosion. The general consensus in the published literature appears to favour the former. The presence of Middle Jurassic volcanics in the Fastnet Basin was interpreted by Shannon (1995) as the possible effect of a thermal anomaly that caused local doming and prevented the deposition of a thick Upper Jurassic sequence in the basin. Little evidence is seen in the SCSB to suggest deposition and later erosion of a significant thickness of Upper Jurassic strata.

The Lower Cretaceous succession in the Celtic Sea basins is of variable thickness. The underlying Upper Jurassic, where present, is dominated by syn-rift deposition. However, the Cretaceous has little evidence of major syn-depositional faulting and exhibits a broad thermal subsidence geometry. In contrast to the Lower Cretaceous of the Atlantic basins the depositional environment in the Celtic Sea was non-marine, dominated by fluvial deposits. Deltaic to marine conditions in the mid Cretaceous and open marine terrigenous-free chalk in the Upper Cretaceous reflected a sea level rise. The non-marine deposition is likely to be due to the effects of the rotation of the Iberian peninsula (Robinson *et al.*, 1981). This imposed compressional stresses on the Celtic Sea region resulting in uplift, erosion and a marked unconformity. However, as seafloor spreading ceased in the Iberian region the Celtic Sea region experienced the delayed thermal subsidence response to the Late Jurassic rifting, resulting in the widespread sag-like geometry of the Cretaceous. The variations in thickness are due partly to the residual rift topography, the effects of the Iberian compression and the effects of later uplift and erosional events.

Evidence is seen for several uplift and erosional episodes in the Celtic Sea region. As mentioned above, the base of the Cretaceous is marked by a pronounced unconformity through much of the region, although the amount of Jurassic strata removed is uncertain. Two Cenozoic inversion and erosional events are recorded in the seismic and well records in the region (Murdoch *et al.*, 1995), and vitrinite reflectance, apatite fission track and seismic velocity analyses provide constraints on the magnitude of erosion. A regional uplift occurred in the Palaeocene, but this was not associated with significant inversion or compression. Up to 500 m of strata were removed during this event. Compressive inversion with doming and reverse fault movement occurred in the late Eocene to early Oligocene. The later event is most pronounced in the northeast of the region, with up to 1.1 km of strata eroded. This event is significant in the region in many regards. It resulted in the exhumation of Upper Cretaceous Chalk to the sea bed, which is the main cause of the poor seismic data quality in the NCSB. It also created a number of elongate inversion structures, some of which provide traps for hydrocarbons. The most obvious of these are the Kinsale Head and the Ballycotton gasfield structures (Figure 17.3).

Basin development: Atlantic margin basins

The sedimentary basins along the Atlantic margin of Ireland wrap around the western shelf and are typically elongate – frequently NE–SW, sometimes N–S and rarely E–W. They typically lie in deep water and the larger basins are overlain by deep-water embayments, indicative of their sediment under-supplied nature in Neogene times. The majority of them, especially the large basins, have a thick Mesozoic and Cenozoic succession, with up to 12 km of sediment interpreted on seismic profiles. However, they have suffered multiple inversion episodes resulting in removal of parts of the late Mesozoic and Cenozoic successions, especially in the shallower basins. A summary of the succession is as follows:

Goban Spur basins

The basins in this region (Figure 17.5) consist of a number of relatively small, elongate sedimentary basins separated by fault-bounded ridges (Naylor *et al.*, 2002). Up to 6 km of Triassic to Cenozoic strata are interpreted from seismic data and from drilling of a single deep exploration well (Esso 62/7-1) in the Goban Basin (Cook, 1987). This well bottomed in Early Sinemurian limestones, having encountered almost 1500 m of Liassic shale-prone marine succession, overlain by Callovian–Bathonian shoreface sandstones and 214 m of Middle Jurassic porphyritic basaltic to andesitic flows. A major unconformity is overlain by Cretaceous sandstones and limestones, followed by almost 1000 m of deep-water Cenozoic calcareous oozes.

Porcupine Basin

This very large, generally north–south oriented basin contains up to approximately 10 km of Upper Palaeozoic to Cenozoic sediments. Approximately 1500 m of Pennsylvanian fluvial to deltaic and brackish sandstones, siltstones, mudstones and coals have been drilled on the eastern margin of the basin (Croker and Shannon, 1987). These are overlain by localised Triassic shallow marine sandstones and evaporitic mudstones, and by extensive Middle and Upper Jurassic fluvial to transgressive shallow marine sandstones, mudstones and thin limestones, with more than 1000 m drilled in places (Croker and Shannon, 1987). The syn-rift Jurassic succession is unconformably overlain by a thick (more than 1000 m drilled in wells) Cretaceous succession of mudstones and local marine and deltaic sandstones, overlain in turn by a thick Chalk succession that onlaps the rifted Jurassic margins of the basin (Figure 17.24). Up to 2 km of Cenozoic mudstones, sandstones and thin limestones have been drilled, with late Palaeocene to Eocene deltaic sandstones in the north of the basin giving way southwards to deep-water equivalents. A basinwide unconformity of latest Eocene to early Oligocene age marks a change in sedimentation style and rate, with the basin becoming dominated by deep-water contouritic siltstones and mudstones, and the products of downslope debris flows. Volcanics (lavas and intrusions) of Cretaceous and Cenozoic age occur in parts of the basin (Naylor *et al.*, 2002).

Rockall and Hatton basins

With the exception of the Neogene succession, the stratigraphy of these basins is poorly constrained by drilling. Interpretation of seismic data suggests that the Rockall Basin (Figure 17.25) contains up to 7 km of Upper Palaeozoic to Holocene rocks (Morewood *et al.*, 2004, 2005), with up to 4.5 km of such strata in the Hatton Basin (Shannon *et al.*, 1995). While Pennsyslvanian to Jurassic strata have been inferred and modelled in these basins, their distribution is still largely conjectural. A series of fault-bounded, 'perched' basins, lying on the slopes of the later Cretaceous to Cenozoic Rockall Basin, have been interpreted as containing thick Permo-Triassic to Jurassic strata (Naylor *et al.*, 1999; Walsh *et al.*, 1999). Upper Jurassic strata were proven at the base of two wells in these 'perched' basins, unconformably overlain by Cretaceous sediments (Haughton *et al.*, 2005). Boreholes on Hatton Bank also drilled Cretaceous (Aptian) shales and sandstones (Hitchen, 2004), while a mud-dominated marine Upper Cretaceous succession overlying Aptian/Albian mudstones with interbedded limestones was penetrated by well BP 132/15-1 on the eastern flank of the Rockall Basin. A Cenozoic succession, up to 2 km thick, has been inferred on the basis of seismic interpretation and comparison with the succession in the nearby Porcupine Basin (McDonnell and Shannon, 2001). The interpretation is supported by evidence from well 132/15-1, where a mud-dominated Palaeocene succession is followed by an Early Eocene sandy succession (McDonnell and Shannon, 2001), overlain in turn by deep marine mudstones above a major early Oligocene unconformity. Eocene to Pleistocene strata have also been drilled on the eastern margin of the Rockall Basin (Haughton *et al.*, 2005), with Neogene deep marine mud-prone strata drilled at various locations within the Rockall Basin (Stoker *et al.*, 2005b, c).

Slyne–Erris–Donegal basins

The Slyne, Erris and Donegal basins (Figure 17.7) are a series of interconnected narrow, half-graben basins lying on the continental shelf east of the Rockall Basin. More than 1500 m of Upper Mississippian and Pennsylvanian fluvial and shallow marine sandstones were encountered in the Amoco 19/5-1 well in the Erris Basin. These are overlain in the Amoco 12/13-1A well by up to 700 m of Permo-Triassic (including Zechstein evaporites), consisting largely of fluvial and locally aeolian sandstones overlain by playa mudstones (Murphy and Croker, 1992; Chapman *et al.*, 1999). Lower and Middle Jurassic mostly marine claystones and carbonates shallowing to deltaic and possibly fluvial sandstones, up to 2 km thick in the Elf 27/13-1 well, have been drilled in the Slyne Basin. The Upper Jurassic and Cretaceous succession is generally thin to absent throughout much of the region, largely due to uplift and erosion along the flanks of the Rockall Basin. However, approximately 600 m of Cretaceous marine sand-prone strata were drilled in the Amoco 12/13-1A well in the Erris Basin. The Cenozoic succession, up to 1 km thick, consists of marine sandstones and mudstones (Trueblood, 1992).

Discussion

The Atlantic margin basins have been the subject of very extensive research and publications in the past two decades, largely fuelled by their petroleum potential and the fact that they are among the last of the lightly explored large frontier exploration provinces of Europe. Publications have focused upon the crustal structure, tectono-sedimentary development and petroleum potential of the region. The following major points may be identified: (a) the continental crust beneath the largest basins (Porcupine and Rockall) is extremely thin (2–5 km) in places, (b) the depositional basins appear to have had different orientations though time, (c) the basins developed through phases of rifting in the Permo-Triassic and Jurassic, (d) the Cretaceous and Cenozoic successions in the large basins are thick and generally devoid of evidence of major fault-related rifting and (e) major igneous activity occurred periodically during the development of the basins.

The crustal structure appears to have played a significant, albeit complex and poorly understood, role in the orientation and development of the basins. The pronounced NE–SW orientation of the Slyne and Erris basins, of the Rockall Basin north of 54° N, and of the set of 'perched' presumed early Mesozoic basins along both margins of the main Rockall Basin (Figure 17.6) is suggestive of an underlying inherited Caledonian structural grain. The east–west alignment of the eastern margin of the Rockall Basin at 54° N, the long-standing east-west Finian's Spur separating the North Porcupine Basin from the main basin to the south at 53° N, the east–west shape of the Carboniferous Clare Basin (Croker, 1995; Naylor *et al.*, 1999) and the pronounced east–west basin and high structures in the Goban Spur region (Naylor *et al.*, 2002) are likely to reflect a different structural control. It is tempting to suggest a Variscan influence on this orientation. However, north of the Goban Spur, the Carboniferous strata encountered in drilling show little if any thermal or structural effects of Variscan deformation. In addition, the putative trend of the Iapetus Suture at the mouth of the Shannon estuary, and the western parts of some major Caledonian structures such as the Clew Bay Fault (e.g. Naylor *et al.*, 1999) show an east–west orientation. These large-scale east–west features may therefore be of pre-Variscan age, with several phases of reactivation to account for the east–west alignment of the Carboniferous Clare Basin, the several phases of Mesozoic reactivation along the Finian's Spur (*see* Naylor *et al.*, 2002), and the cross-cutting relationship of the Cenozoic Rockall Basin with the early Mesozoic 'perched' basins.

The north–south shape of the Porcupine Basin is in sharp contrast with other basin orientations along the Atlantic margin. Croker and Shannon (1987), Shannon (1991a, b) and others have suggested that the early Mesozoic basin development in the region was NE–SW in orientation, along reactivated Caledonian fabrics. This resulted in the extensive Slyne–Erris basin system, with smaller, isolated but similar oriented basins within the Porcupine and Rockall region. The main north–south orientation appears to have developed during Middle and Upper Jurassic times, broadly coincident with the main phase of syn-rift basin development. The Jurassic facies distribution (Croker and Shannon, 1987), with a major basin-edge fan complex seen in the Shell 34/19-1 well, supports this model. However, the general north–south shape of the basin is interrupted by several spaced NW–SE gravity lineaments that correspond to offsets in the basin margins, to changes in the sedimentation style and geometry and to variations in the crustal thickness (Readman *et al.*, 2005). Their regional extent, with some of them extending into the Celtic Sea domain, suggests that they are deep-seated and long-standing inherited crustal structures.

The Atlantic basins show a range of structural and stratigraphic architectures that suggests a composite development through a series of Mesozoic fault-related rift episodes, interspersed with phases of differential non-faulted thermal subsidence and sedimentation. The interplay between these events gave rise to temporal and spatial variations through the region. Relatively little is known about the Late Palaeozoic basin development style, but interpretation of the preserved strata suggests that sedimentation in the Porcupine region took place in an east–west elongate subsiding depocentre (the Clare Basin) in the Porcupine region. A major regional unconformity is recorded between the Pennsylvanian strata and the overlying Mesozoic, and the regional distribution of Mesozoic strata suggests an initial patchy infill of an irregular topography following uplift and erosion of variable amounts of Pennsylvanian strata (Robeson *et al.*, 1988). Permo-Triassic rift strata are patchily preserved in the region, typically in NE–SW oriented fault-controlled basins. The thickest and best-constrained successions are in the Slyne, Erris and North Porcupine basins (Chapman *et al.*, 1999; Dancer *et al.*, 2005), with seismic evidence indicating the preservation of Permo-Triassic strata in the 'perched' basins along the flanks of the Rockall Basin. Their geometry suggests that the original deposition of Permo-Triassic strata was more widespread and that these perched basins simply reflect preserved remains of the early Mesozoic succession (Shannon *et al.*, 1999). Syn-rift Permo-Triassic strata are also inferred in the Goban Spur Basin (Naylor *et al.*, 2002).

The clearest evidence of fault-related rifting and crustal extension is seen in the Jurassic succession in the region. The main rifting appears to be of Middle and especially Late Jurassic age. Tilted fault blocks, with variations in coeval facies from fluvial to basin floor fans, are spectacularly developed on the flanks of the Porcupine Basin (Croker and Shannon, 1987; MacDonald *et al.*, 1987; Naylor and Anstey, 1987; Sinclair *et al.*, 1994; Naylor *et al.*, 2002) where a Late Jurassic age is constrained by drilling. Rifting may have commenced slightly earlier (Middle Jurassic) in the Slyne and Erris basins (Cunningham and Shannon, 1997; Chapman *et al.*, 1999). Late Jurassic syn-rift faulting waned during the earliest Cretaceous times and was followed by rapid marine incursion (Figure 17.22) across the partly peneplaned Late Cimmerian surface (Naylor *et al.*, 2005). Jones *et al.*, (2001) suggested that transient uplift of up to 700 m occurred at this time, followed by subsidence of up to 500 m, coeval with the onset of seafloor spreading along the Goban margin. A regressive episode of deltaic progradation in the north-eastern part of the Porcupine Basin has been interpreted as a minor rift episode that interrupted the regional thermal subsidence following the main Jurassic rifting (Croker and Shannon, 1987; Sinclair *et al.*, 1994). Regional eustatic sea level rise, combined with thermally-driven subsidence of the basin depocentres, resulted in the widespread development of Upper Cretaceous chalk and marls throughout the Atlantic margin basins.

The Early Cenozoic in the region is marked by a sea level regression, a change from carbonate to coarse clastic deposition, widespread igneous activity and another interruption in thermal subsidence (Figure 17.28). A similar pattern is observed in the North Sea, suggesting a regional control. Stoker *et al.* (2005b) and Praeg *et al.* (2005) proposed a deep-seated mantle thermal convection cell control, with differential uplift of the thick crustal regions coeval with subsidence in areas of thinner crust beneath the major sedimentary basins. Similar controls are invoked to explain the flank uplift on the margins of the Rockall Basin, thereby eroding late Mesozoic strata from the Slyne and Erris basins.

The interpreted absence of a thick late Mesozoic succession from the Hatton Basin region has been explained by the uplift and buoyancy provided by an elongate underplated igneous body at the base of the crust adjacent to the continentental–oceanic crustal boundary (Vogt *et al.*, 1998; Shannon *et al.*, 1995, 2007). This represents the terminal phase of rifting and continental crustal extension that drove the Mesozoic rifting in the region, as it migrated westwards through a series of failed rifts until final crustal separation occurred west of the Hatton region. Crustal thinning beneath the nearshore basins (e.g. Slyne and Erris) appears to be slight, while the crust is severely attenuated beneath the large Porcupine and Rockall basins. Wide-angle seismic data from the latter basins (Hauser *et al.*, 1995; Morewood *et al.*, 2005; O'Reilly *et al.*, 2006) show that the continental crust is less than 5 km, and in places less than 2 km thick. It has been suggested that this was achieved, without the creation of oceanic crust and the onset of seafloor separation, by differential stretching facilitated by a lower crustal detachment (Hauser *et al.*, 1995). Upper and middle crustal stretching was more severe than lower crustal and mantle lithospheric stretching. The less attenuated lower crust coupled to less stretched mantle lithosphere resulted in an increase in the overall lithospheric strength, thereby inhibiting crustal breakup. The overall crustal stretching led to serpentinisation of the upper mantle in the Rockall Basin (O'Reilly *et al.*, 1996) and probably in the Porcupine Basin, and also to the major volcanism in Cretaceous and Cenozoic times.

The areas of maximum crustal thinning (Rockall and Porcupine) are coincidentally overlain by bathymetric embayments. These reflect the outstripping of sedimentation by subsidence in Neogene times. A regional unconformity in latest Eocene to earliest Oligocene times resulted in a major change from downslope sediment transport to alongslope reworking in deep-water basins. The onset of contourite currents in turn contributed to the construction, and later localised mass failures, of steep slopes along the basin margins, especially the eastern margin of the Rockall Basin. These were later modified, with subsequent local failures, in Pleistocene times when a combination of localised sediment input through melting glaciers and ice sheets, together with ocean current activity and scouring during glacial times, finally shaped the present dramatic physiography of the deep-water Atlantic margin.

18

Geophysical evidence onshore

T. Murphy,† A. W. B. Jacob† and I. S. Sanders

Geologically Ireland is often considered to be part of a single block along with Scotland, England and Wales, separated by a shallow sea from continental Europe. The geophysical evidence is, however, that it should be considered a distinct unit with fewer links than might at first be thought.

Earthquake evidence

This separateness is well illustrated by the distribution of earthquakes. Historical records go back many centuries. Instrumental records are available within the last 100 years and the density of seismic station coverage has greatly increased in the last thirty years. The evidence is that earthquakes in Britain are about 100 times more frequent than they are in Ireland and the contiguous shallow continental shelf to the west. This suggests some form of decoupling between Ireland and the regional stress field, or that Ireland is behaving as a stable block in which the fault systems are more resistant to stress. At the other end of the scale, there may be movement on faults and strain may be being released by aseismic creep. It is not possible to differentiate between these possibilities with the present data. It has, however, been shown in other parts of the world that major intraplate earthquakes can have a very long return period, i.e., that quiescent periods lasting thousands of years may be followed by isolated large events which would usually have aftershock sequences. This may possibly be the case here.

Historical records since 1700 (Davison, 1924; Jacob, 1993) and the detailed seismic network data gathered since 1978 both agree that the activity in Ireland is currently concentrated (a) along a coastal strip south from Dublin and extending to the region around Cork and (b) in the north-west of the island, mainly in County Donegal (*see* Figure 18.1). There is no well-substantiated evidence for recent events anywhere else in Ireland. Maximum intensities from onshore events seem to be no more than MSK IV, a level at which damage is not expected.

† (Murphy and Jacob are deceased.)

The most significant activity close to Ireland occurs in the Irish Sea. Events may attain magnitudes greater than ML=5. Events of this size may cause quite serious damage if they occur near the surface in populated areas. One quite well-marked trend extends NE–SW through the Irish Sea, from Anglesey to south County Wexford. By Irish standards, this can be quite active. In 1988, for example, more than twenty offshore events were recorded and there was a small earthquake in County Wexford in December.

The largest events within Ireland recorded instrumentally have magnitudes very little more than ML=2. Some such events recorded on the Irish seismic network between 1978 and 1996 were in east County Cork on 24th November 1981 (ML=2.1, maximum intensity MSK IV), in north County Donegal on 24th January

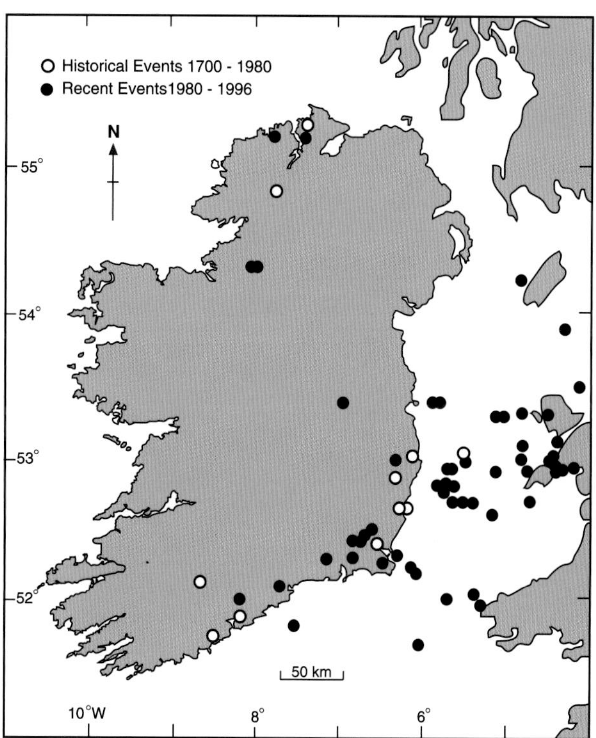

Fig. 18.1 Map of earthquake epicentres in and near Ireland.

1984 (ML=2.1, maximum intensity very low) and two in north County Leitrim on 20th August 1994 (ML=2.2) and 30th November 1994 (ML=2.3). Neither of the Leitrim events were reported as being felt. The intensity of the Donegal event on 24th January 1984 was low because it seems to have been a deep crustal event. Since then, an event near Clonmany in north Donegal was felt with a maximum intensity of MSK IV. The pattern of intensities indicated that this event was a shallow crustal one. In the same period the highest intensity for an offshore earthquake recorded in Ireland reached MSK V at some locations on the east coast of counties Wicklow and Dublin. It was due to an event close to the coast of Wales at about 0757 on 19th July 1984 (Turbitt *et al.*, 1985). The earthquake, with a magnitude ML=5.4, was felt over more than half the area of Ireland and most of Britain, a total felt area of about 250,000 square kilometres. There was some damage, mainly in Wales, but the earthquake's unusual depth, about 22 km, moderated its effects, which might otherwise have been very serious indeed. Three more events with magnitudes over four were recorded in the following month, all of them felt in Ireland, and the aftershock sequence lasted for about ten years. The depth of the events is unusual in that it was below the expected transition in the crust from brittle to ductile rheology.

Evidence from wide-angle reflection seismic studies

Initial indications from surface wave studies on the NE–SW profile across Ireland were that average crustal thickness lies between 25 and 30 km with some evidence of velocities well above 6 km/s in the lower third of the crust. More detail of crustal structure has since been obtained on a number of seismic refraction and wide-angle reflection profiles whose locations are shown on Figure 18.2. The two longest profiles were recorded in the 1980s (Jacob *et al.*, 1985; Lowe, 1988; Lowe and Jacob, 1989; *see* also Cook *et al.*, 1989). The first line ran NE–SW from County Louth to County Clare and was roughly parallel to the Caledonian Suture Zone (the Iapetus Suture). The profile is labelled ICSSP (Irish Caledonian Suture Seismic Project) in Figure 18.2. The second profile extended almost north–south from Donegal Bay to near Ardmore in County Waterford, crossing the main Caledonian strike and the ICSSP line at a high angle. This was called the COOLE project (Celtic Onshore Offshore Lithospheric Experiment) and the profile is again shown in Figure 18.2. The angle of the line made interpretation of the seismic data more difficult but the coverage was good, with sources at reasonable intervals along the line.

Both these profiles lie north of the Variscan Front in a zone affected by the Caledonian orogeny. The general velocity structure of the crust, with sediments and the crystalline basement (about 6 km/s) underlain by an intermediate velocity layer (about 6.4 km/s), which is itself underlain by a lower crustal layer with a higher velocity (between 6.8 and 7.0 km/s), is comparable with that found in northern Britain. A lower crust with these high velocities is commonly found throughout the entire Caledonian–Appalachian orogen, though the total crustal thickness of around 30 km is considerably less than the 45–50 km found in the Caledonides of Norway and in the Appalachian belt on the other side of the Atlantic. The closing of the Iapetus Ocean probably involved a major component of strike-slip movement which would not have resulted in much crustal thickening. Extension in the Mesozoic may have complicated the picture, but there seems to be no evidence of post-Cretaceous regional subsidence that would have resulted from crustal extension and thinning.

A fence diagram of the velocity structure along these two important lines is shown in Figure 18.2. On the ICSSP profile the boundary between the middle and lower crustal zones is at around 20 km depth throughout. The base of the crust, the Mohorovičić discontinuity (or Moho), is slightly shallower at each end than it is in the middle, though its character changes from one end to the other. At the NE end it is a transition zone extending over more than 3 km but, in contrast, the SW end has a well-defined Moho at a slightly shallower depth. A notable feature at the NE end is a pronounced low-velocity body in the upper part of the crust. It is an extended body, which is thickest, and has the greatest depression in velocity (5.5 km/s), just east of Dunany Head close to shotpoint I5 in Figure 18.2.

The COOLE profile, crossing the predominant NE–SW Caledonian strike at a high angle, shows a more heterogeneous structure. The gravity data were used to constrain and improve the seismic model, the main features of which are as follows.

(1) An upper crust approximately 10 km thick where P-wave velocities are variable both laterally and vertically. The lowest near-surface velocities (5.2–5.4 km/s) overlie the large sedimentary basins near the coasts while high near-surface velocities (5.7 km/s) correspond to exposures of Ordovician basic volcanic rocks in the vicinity of shotpoint STR. A large low-velocity body (LVB,

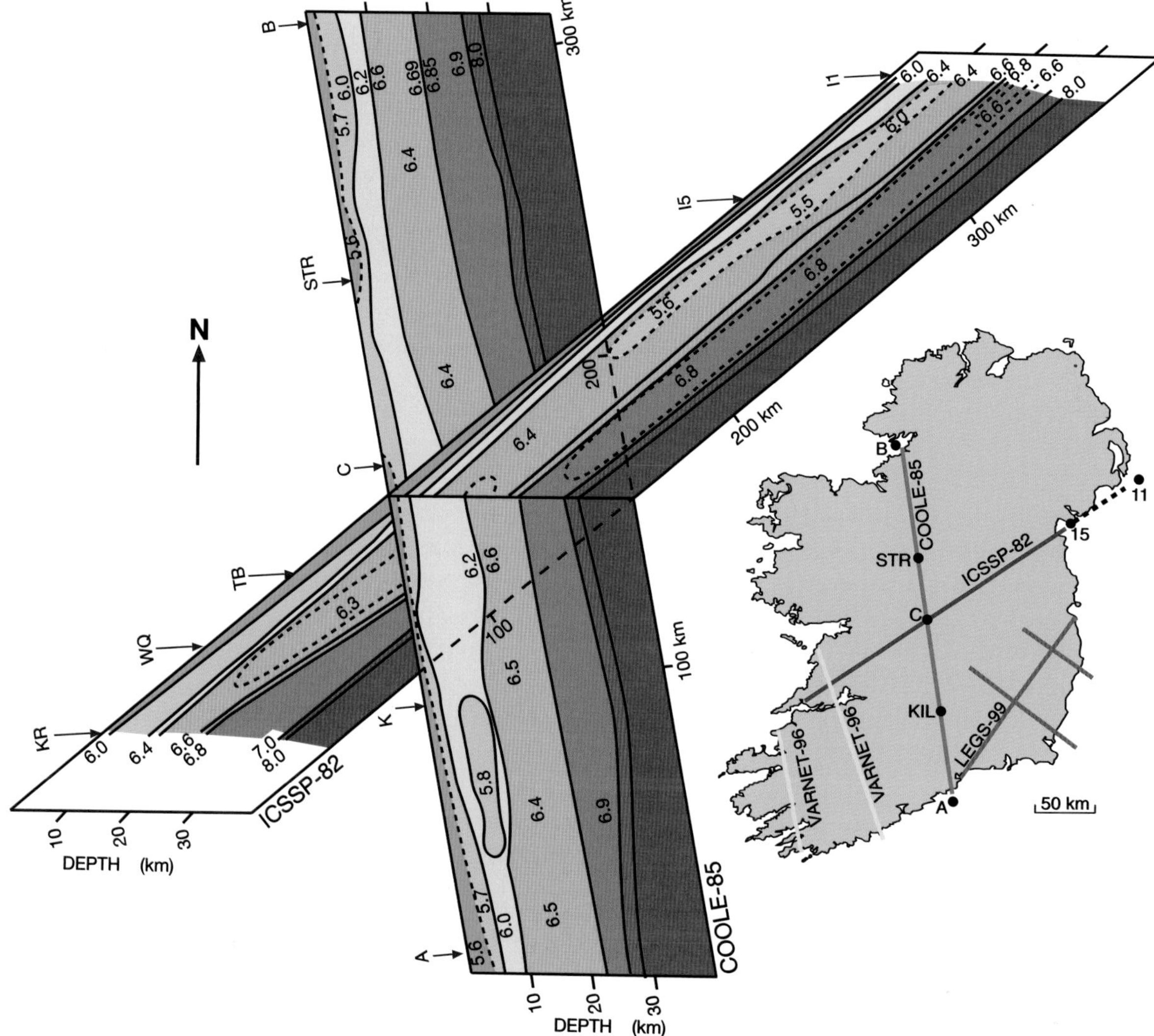

Fig. 18.2 Fence diagram showing crustal velocity structure along the ICCSP and COOLE wide-angle seismic reflection lines. The map shows the location of these two lines, and also of the VARNET (see Chapter 12) and LEGS seismic lines.

mean velocity 5.8 km/s) is observed towards the south of the profile. It corresponds very well with a Bouguer anomaly low and is interpreted as a buried extension of the Leinster batholith.

The southern boundary of the LVB lies approximately 15 km north of the 'Variscan Thrust Front'. This front runs from Dingle to Dungarvan, but evidence of thrusting is confined to the Killarney, Millstreet, and Mallow areas (*see* Chapter 12). The front is considered to mark a transition from a more pervasive style of deformation within the foldbelt to one of pronounced heterogeneous deformation in the foreland. It is quite conceivable, therefore, that the extension of the Caledonian Leinster batholith, as identified by gravity studies and supported by the COOLE profile, had a significant effect on Variscan tectonics in Ireland. Being massive bodies, the plutons comprising the batholith could have acted as resistive blocks impeding further northward migration of the Variscan thrusting regime.

(2) The mid-crust stands out as a distinct layer along the COOLE traverse. It is separated from the upper and lower crust by two laterally continuous velocity discontinuities. The upper discontinuity represents a velocity step from 6.2 km/s to 6.6 km/s. It varies in depth from 7.5 to 13 km. The deeper discontinuity, varying in depth from 17.5 to 22 km, has a velocity increase from 6.7 to 6.85 km/s. The mid-crust shows no velocity increase with depth and may decrease slightly in velocity. Where COOLE crosses the ICSSP profile there is some disagreement in the upper crust. The distribution of sources in COOLE provided better control in the middle of the line for the upper crust. The agreement for the deeper discontinuities is very good.

The depth variation of both reflectors bounding the mid-crust in COOLE occurs mainly across a zone about 60 km wide. Both reflectors shallow significantly by 4 to 5 km northwards, between shotpoints C and STR. This observation is supported by gravity studies. The shallowing of both reflectors, a feature also exhibited by some of the shallower isolines and reflectors in the upper crust, marks a broad zone of crustal change across the Irish Midlands. It was proposed that this zone is the deformation complex associated with the closure of the Iapetus Ocean and the ensuing continent–continent collision during the Late Caledonian orogenic episode (*c.*400 Ma; *see* Chapter 8). Although the Iapetus Suture has traditionally been regarded as a narrow suture zone both onshore and just offshore (*see* Phillips *et al.*, 1976 and Klemperer and Matthews, 1987) there is a growing body of data which supports the view that it is more likely a zone of considerable width. These data include heat-flow studies (Brock, 1989), isotope studies (Pockley, 1961; Greig *et al.*, 1971) and more recent terrane studies. Most of the 'linear sutures' can be accommodated within the proposed deformation complex.

The northerly dipping reflectors identified as the Iapetus Suture on both WINCH (Brewer *et al.*, 1983) and the NEC (Klemperer and Matthews, 1987) deep reflection profiles have been interpreted as indicating a northerly-directed subduction. However, onshore in Ireland, there is little evidence for large-scale uplift and erosion associated with the continent–continent collision, as Late Silurian sediments are still exposed in the Midlands. Consequently, collision tectonics are often interpreted in terms of mainly lateral strike-slip movements rather than 'head-on' collision (Phillips *et al.*, 1976).

(3) The crustal thickness along COOLE is about 30 km. This value is in good agreement with information from the ICSSP data and also with results of seismic refraction profiles in Britain (Bamford *et al.*, 1976; 1978). The base of the crust is defined by the Moho transition zone, which is about 2 km thick. The crust is observed to shallow slightly (1–2 km) towards the coasts, in agreement with gravity observations (*see* below). It is also shallower (about 28.5 km) under part of the Midlands.

The shallowing of the crust under the Midlands cannot be clearly related to the northwards shallowing observed in the mid- and upper-crustal reflectors. However, numerous reflection and refraction studies around Ireland and Britain suggest that the Moho retains no 'memory' of Caledonian or Variscan events, even though large scars remain as a result of younger tectonic movements. This observation led Peddy and Keen (1987) to suggest that the Moho is a relatively young feature, mobile over short time periods. The lower crust varies appreciably in thickness and has an average velocity of 6.9 km/s.

By fortunate coincidence, it has been possible to examine samples of the lower crust from beneath the Midlands. In counties Westmeath and Offaly, not far from the intersection of the COOLE and ICSSP lines, xenoliths were brought to the surface in Carboniferous volcanic vents. A laboratory study of these xenoliths showed that their seismic velocities match closely the measured velocities in the lower crust, when the difference in the pressure and temperature conditions between the laboratory and the present day lower crust are taken into account (Van den Berg *et al.*, 2005). Many of the xenoliths are sedimentary rocks that were metamorphosed at granulite facies (typically around 800 °C), and some appear to be material left behind after partial melting, when the molten fraction of the original sedimentary rock was extracted as granitic magma. U–Pb zircon dating indicates that metamorphism, melting and deformation occurred during the Carboniferous, 340 to 360 Ma. An orthogneiss sample yielded a somewhat earlier Devonian age of 410 Ma, similar to the exposed late Caledonian granite plutons. Inherited zircon cores from this rock yielded ages ranging from the Proterozoic into the early Cambrian. The latter suggest that the protolith material came from Avalonia rather than from Laurentia (Van den Berg, 2005). Thus, the Moho beneath the Midlands may, in effect, be the unconformity between the Iapetus basaltic ocean floor and the overlying Lower Palaeozoic ocean floor sediments, albeit intensely modified by high-temperature metamorphism.

The very high metamorphic temperature reached by these rocks suggests that the geothermal gradient was almost twice as steep during the Late Devonian and Early Carboniferous compared to the present day. Such very high temperatures in the lower crust during the Carboniferous may be connected with the observed flatness of the Moho and the idea that the Moho has been 'renewed' subsequent to the closure of Iapetus.

Wide-angle seismic reflection profiling continued in the 1990s, this time in the south-west of the country traversing the region of the purported Variscan Front (Figure 18.2; Masson *et al.*, 1998; Landes *et al.*, 2000; Landes *et al.*, 2003). The main profiles, referred to as

VARNET lines A and B, are reproduced in Chapter 12 of this book (Figure 12.5). They show a similar pattern to that inferred from the earlier lines, with a multilayered crustal structure. In line A (Landes *et al.*, 2000), for example, which runs from Galway Bay to the Old Head of Kinsale, the upper crust is about 12 km thick, with a velocity of about 6 km/s. However, the top few kilometres are highly variable, with lower velocities (5 to 5.5 km/s) in the sedimentary basins to the south, higher velocities in buried basement ridges (5.8 to 6 km/s), and anomalously high velocities (6.4 km/s) in a wide shallow zone north of the Shannon Estuary. The middle crust has a velocity of about 6.4 km/s and the lower crust between 6.8 and 7.2 km/s, locally up to 7.6 km/s. The total crustal thickness varies from 29 to 32 km.

In south-east Ireland a deep seismic study of the Leinster Granite (LEGS) has been carried out, but remains to be published (Hodgson, 2001). The locations of the seismic profiles are shown in Figure 18.3.

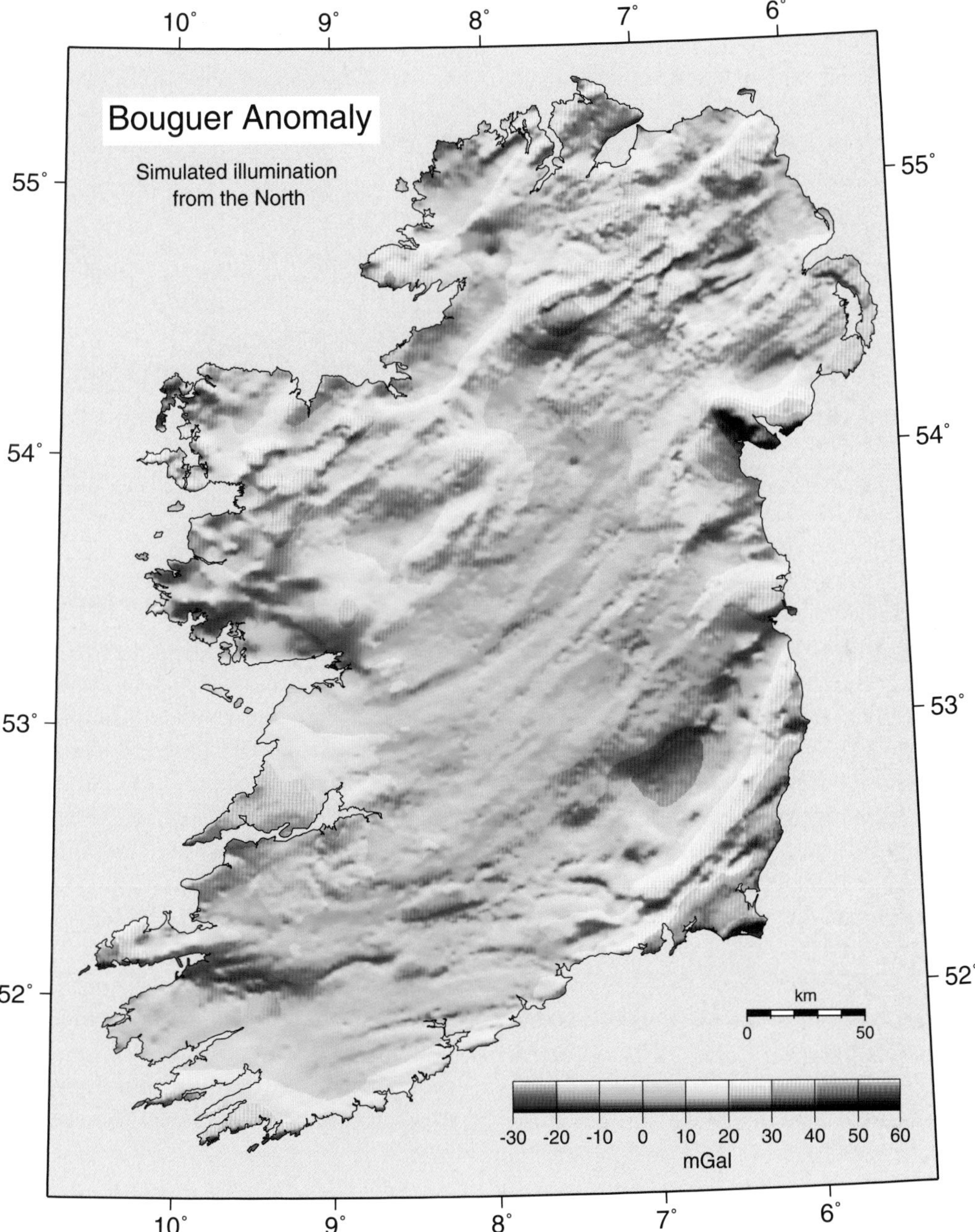

Fig. 18.3 Bouguer anomaly map of Ireland shaded using the first horizontal derivative in the north direction. The shading highlights features that are less obvious in a flat colour contour map. Reproduced from Readman *et al.* (1997) with permission.

Evidence from gravity measurements

This relatively simple picture of the deep-crustal structural setting of Ireland on a global scale is borne out by the results of a gravity survey, illustrated in Figure 18.3. This map shows the Bouguer anomaly, i.e. the departure of the measured gravity at sea level from a theoretical value based on the shape of the earth derived from geodetic measurement. The range of the anomaly is from −25 to +70 mGal with the mean areal value close to zero, which indicates that this part of the world is in isostatic equilibrium and there is no great tendency for the country to rise or sink *vis-a-vis* sea level.

The Bouguer anomaly map of Ireland was originally published by Murphy (1974), but the version of the map in Figure 18.3 was prepared by Readman *et al.* (1997) from the same data by calculating the horizontal first derivative of the change in gravity in a northerly direction. It highlights some remarkable, and even unsuspected, detail in Caledonian and Variscan structural trends. The recalculated map incorporates gravity data supplied by the Geological Survey of Northern Ireland.

The variations in gravity shown on Figure 18.3 are interpreted as being variations in density of parts of the crust with depth, but mainly within 10 km of the surface, that is, down to the seismic–intermediate layer. Because of the limitations of the methods used, the density of the crust below that level is taken to be uniform.

A fundamental feature is the location of high values on the coasts and the small range of values in the interior. This was shown by Morris (1973) and is demonstrated by Figure 18.4. The cause lies probably in the shape of the crust, that is, the thickness is greater in the centre than at the coasts.

However, the most striking feature of Figure 18.3 is the alignment of the contours in a north-east to south-west direction with minor ones in other directions. When the gravity map and geological maps are compared, certain correlations are immediately apparent such as the coincidence of low values over the granite areas of Connemara, Donegal, Leinster, and Down; low values over the Mesozoic in the north and in offshore areas; and high values over the Tertiary volcanic centres under the Carlingford peninsula and nearby Slieve Gullion.

The granite bodies mentioned have fairly uniform densities of about 2,650 kg per metre cube (e.g. Young, 1974) while the mean value for the surface rocks of Ireland is estimated to be 2,725. Although there seems no doubt about the correlation of low gravity areas with granite bodies, the mapped geological boundaries are

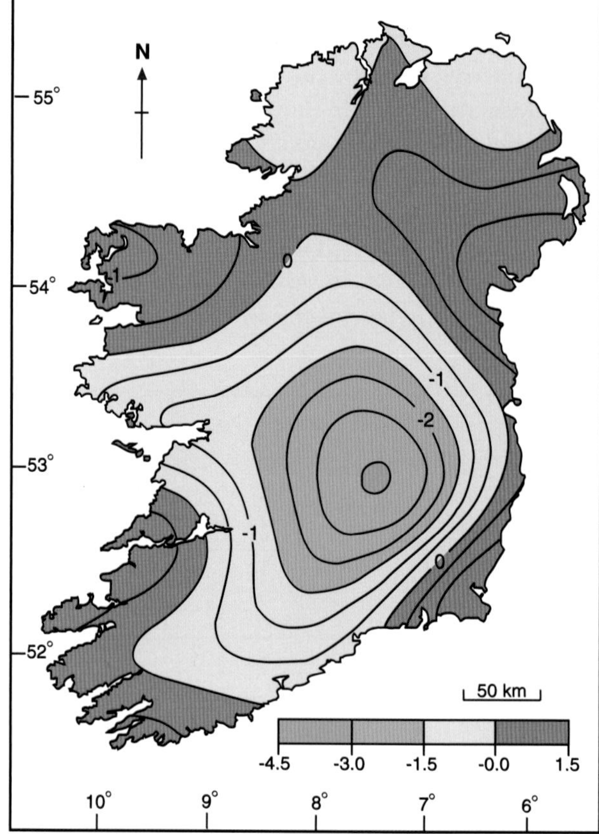

Fig. 18.4 Smoothed gravity map, with the Bouguer anomaly upward continued to 60 km.

only partly in agreement with the gravity pattern. Thus the geophysical evidence points to the buried extension south-eastwards of the Connemara granites under Galway Bay and the coast of Clare, and the continuation of the Leinster granites south-westwards. The suggestion once put forward that the Connemara and Leinster granites formed a chain separated by an east–west fault is not substantiated, nor is the view that the Newry Granite is connected with the Crossdowney pluton.

Detailed studies of the gravity anomalies over the granites have shown that even the large batholithic type are made up of separate bodies with diameters of the order of 10 km, and that some of the smaller plutons, such as those at Crossdowney and Ardara, are more akin to horizontal discs than vertical cylinders.

The Lough Neagh basin, with its low-density infill of perhaps less than 2,300 kg per metre cube, stands out clearly.

In Ireland it has been found that rocks of Lower Palaeozoic age and older, with a few exceptions, have densities which can be related fairly simply to the relative proportions of acidic and basic minerals they contain. The extremes are quartzites and granites (2,650) and gabbro (2,800). The Precambrian, containing great thicknesses of rocks of these types, gives great variations of gravity, particularly over steeply folded formations or where large faults bring together masses of different densities. This is well seen in the north-west and south-east, where the large variations in gravity can be used to delimit various formations and to indicate the lines of large discontinuities, some of which are faults.

The Ordovician is known to contain basic materials such as pillow lavas, as well as acidic extrusives and intrusives, so that it is well-nigh impossible to assign a definite value to it. Silurian formations have a more uniform density about 2,730, differing little from the main mass of the Devonian. The Carboniferous, containing a high proportion of limestone of density 2,700, is marginally less dense. Within any formation there are minor developments of lighter rocks exemplified by porous sandstones in the Devonian and shales in the Namurian, but as these are usually quite thin, less than 300 m, they produce only local effects. Some Namurian shales contain dense minerals such as pyrite, and the density has been measured as high as 3,000; but again the thickness, although unknown, is probably not great. Thus the central plain is thought to consist of a series of basins and swells within the underlying rocks whose density is greater than the overlying Silurian, Devonian and Carboniferous strata. It has already been explained that the crust is probably at its thickest hereabouts.

A map showing offshore gravity has been published by Masson *et al.* (1985), and also by Readman *et al.* (1995), and O'Reilly *et al.* (1996) produced a map which merges the Bouguer anomaly map of Ireland onshore with the marine free-air anomaly derived from satellite altimetry. In the north and at sea there are areas of distinctly low gravity values due to graben structures filled with Mesozoic sediments. The sedimentary basins of the Irish Sea, Celtic Sea, Porcupine, and Rockall result in conspicuous gravity lows.

Evidence from the Earth's magnetic field

The country and adjacent seas have been covered by magnetic surveys at various altitudes and at sea level (Geological Survey of Northern Ireland, 1971; Max and Inamdar, 1983). These surveys have been combined (Morris, 1989) in a composite map, and a smoothed version of this is given in Figure 18.5. The smoothing removed from the original map some conspicuous

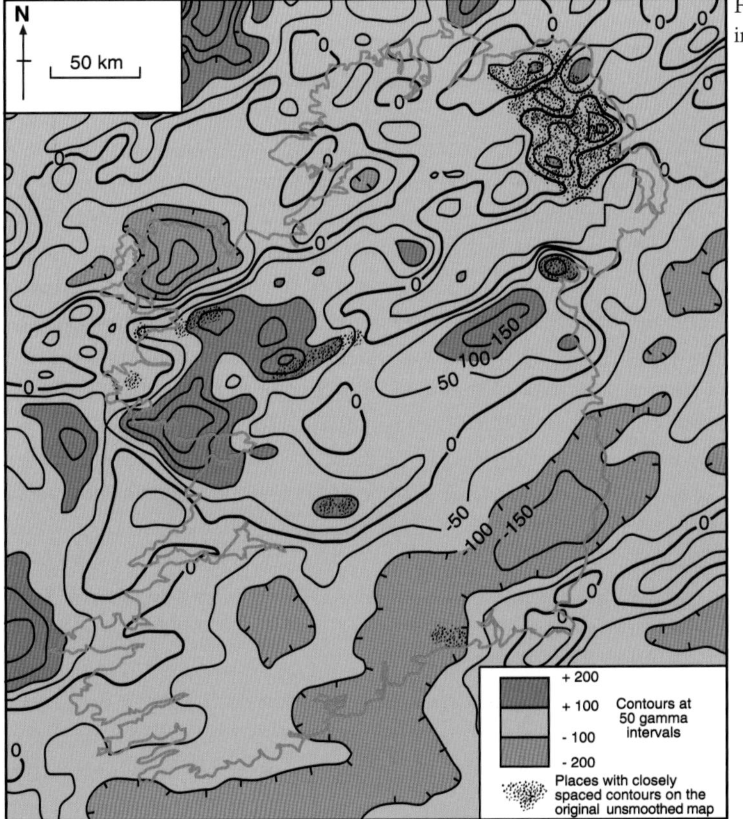

Fig. 18.5 Smoothed magnetic map of Ireland and immediate off-shore area.

anomalies with tightly bunched contours, and the locations of these anomalies have been indicated on Figure 18.5. The magnetic effect produced by rocks is related almost entirely to their content of the mineral, magnetite. In Ireland this varies from virtually zero in the sedimentary rocks to two per cent in the Tertiary gabbros and basalts. Locally high concentrations are present in ore bodies and serpentinites, but the metamorphic rocks, although varying considerably, are in general not very magnetic.

The most notable features on the magnetic map of Ireland are the intense effects due to the Tertiary basalts and gabbros, and to a lesser extent, to the Ordovician, Silurian and Carboniferous basic volcanics, together with the quite characteristically large areas of low value with negligible relief associated with the sedimentary basins. Within the basins small changes in the magnetic field are very useful in ascertaining the depth and character of the underlying basement, even though the difference in magnetic character of the underlying basement may be only slightly different from that of the overlying sediments.

In the north, the magnetic anomaly pattern is dominated by the effect of the basaltic sheets and, when allowance is made for this, a predominantly north-east to south-west trend can be seen. There is also a well-marked system of narrow north-west to south-east anomalies, mainly negative, produced by the Tertiary volcanic dykes north of a line from Sligo to Dublin; but south of this line they are very scarce, and where they do occur it is with a different alignment. In the western part of the country a Tertiary dyke system is evident, which is related to a probable volcanic centre situated at 53° N 11.5° W. An analysis of the magnetic maps has been given (Max and Inamdar, 1983) and should be consulted for details.

It has been pointed out (Morris, 1989) that there is a remarkable difference on the magnetic map between the anomalies north and south of a zone which trends roughly NE to SW across Ireland. North of this line there are numerous large closed anomalies, mainly positive, and narrow linear negative ones due to the Tertiary dykes. In the south the anomalies are few, and these are negative. The cause of this division lies deeper than ten kilometres.

Recently (in 2006) a remarkably detailed magnetic map was produced by the Geological Survey of Northern Ireland (Fig. 18.6). This map of Northern Ireland shows the results of a high resolution, low altitude airborne magnetometer survey carried out in 2005 and 2006. The survey, named the Tellus project, also produced maps of the electrical conductivity and natural (U, Th and K) radioactivity from airborne observations, as well as geochemical maps based on soil and stream sampling on the ground. Most spectacular on Figure 18.6 are the NW–SE trending lines corresponding to Tertiary dykes, particularly those of the Donegal–Kingscourt swarm (*see* Chapter 14). In the west of the province the map reveals a remarkable and previously unknown sinistral offset of the pattern of dykes by several kilometres along a major ENE–WSW-trending wrench fault. The large negative magnetic anomaly over the Antrim basalts reflects the fact that the lava flows were mainly erupted during a period when the Earth's magnetic field was reversed.

When combined, the gravity and magnetic surveys are of value for elucidation of the deep structure. On the whole, however, there is little correspondence between the two maps except for the north-east/south-west alignment. Thus there are several well-marked discontinuities on both maps, and these could possibly be attributed to large faults; but these in general do not coincide with the positions of major faults postulated on geological grounds. These geophysical features do not follow straight lines for any great distance and usually cannot be followed much farther south than the Sligo to Dublin line. The largest, and a somewhat typical one, can be traced on both maps from Fair Head as far as the region of Sligo, but thereafter the prominent features are different on each map. Attempts have been made to trace the continuation of the known large faults of Scotland, in particular the Highland Boundary Fault, but without success.

Over the central part of the country, where the gravity picture exhibits considerable detail, the magnetic map on this scale shows only two large features, both positive and produced by variations in character of the crust at a depth of about ten kilometres. It would seem that, following the major movements of the Caledonian Orogeny, strike faults remained active, permitting basinal structures to develop, and that these continued up to quite recent times.

To the west of the central region the general NE–SW trend on both maps changes direction by roughly 60° clockwise to become WNW–ESE. However, the change is not persistent, and the trend locally reverts back to its former direction, as can be seen, for example, on detailed gravity and magnetic maps of the sea around the Dingle and Iveragh peninsulas. A marked change takes place from east to west near Limerick, the gravity changing from low to high and the magnetic field from high to low. North-westwards from Limerick a distinct correlation

Fig. 18.6 Total Magnetic Intensity image of Northern Ireland. Magnetic data are overlain on a digital terrain model. Topographic data are Crown Copyright and are reproduced with the permission of Land and Property Services under delegated authority from the Controller of Her Magesty's Stationery Office. Crown copyright and database rights PERMIT ID. Magnetic data are provided by the Geological Survey of Northern Ireland.

between the two maps has been commented on by Morris (1989) as marking a line of a major fault which was active in the Caledonian Orogeny, and along which later adjustments have taken place. South-eastwards of the belt of low gravity anomalies the trend is distinctly north-east/south-west with only a locally superimposed hint of the east to west folding of the Variscan Orogeny so well-known geologically in the neighbourhood.

There is a marked belt of high gravity anomaly along the east and south coasts having the same configuration as the low belt attributed to the granite bodies to the north-west and to the low belt of the Mesozoic troughs to the south. On land the highest values occur in the south-east over the Precambrian, so it is thought that a thinning of the crust has taken place in this region, with the subsequent formation of graben structures. On the west coast off Clare the picture is different, and large areas with gentle gravity gradients probably indicate deep Carboniferous basins (Croker, 1995). A remarkable gravity feature occurs at the north-eastern limit of the Leinster granites, which is caused by a deep half-graben, known as the Kish Bank Basin, producing an abrupt change of 35 mGal and gradients up to 10 mGal/km (Readman *et al.*, 1995). There is no corresponding magnetic anomaly.

Evidence from crustal conductivity and heat flow measurements

A magneto-telluric survey (Brown and Whelan, 1995) along a profile south-eastwards from Sligo, and aligned almost perpendicular to the main Caledonian trend, has indicated the presence of two upper crustal conductors, one associated with the Fair Head to Clew Bay lineament and the other with the Virginia magnetic lineament. It was suggested that these effects are due to black shales in sedimentary basins that have been thrust under the continental crusts of Avalonia and Laurentia during the closure of the Iapetus Ocean. Similar anomalies have been observed in the Southern Uplands of Scotland. A more recent magnetotelluric study in the same general region by Rao *et al.* (2007), using broadband and long period measurements capable of imaging structures at greater depth, reveals an undulating highly conductive layer 10 to 15 km thick at mid- to lower-crustal depths which broadly correlates with the middle crust identified by seismic velocity (Lowe and Jacob, 1989). The upper crust is highly resistive to about 10 or 15 km depth, but in two places in particular along the traverse the conductive lower layer bulges upwards to within 5 km of the surface, similar to the conductivity pattern reported by Brown and Whelan (1995). Brown *et al.* (2003) report similar shallow conductivity structures evidently associated with major fault zones in the southwest of Ireland, along the trace of the VARNET seismic line.

Heat flow measurements throughout Ireland were reviewed by Brock (1989). They are higher than might be expected for a terrain which is largely Palaeozoic in age, but they are not unusually high in global terms.

19

A history of Irish geology

G. L. Herries Davies

The pioneers 1770–1820

Modern geology finds its roots in the decades around 1800, and Irish minds were not slow in engaging with the problems then presented by the infant science. Well before the national turmoil of 1798, men armed with hammers had begun the field interrogation of Ireland's rocks. By 1780 Trinity College possessed a geological museum accommodated within the Regent House. There, through hand specimens, undergraduates were introduced to problems of global proportion, while over in London's Soho Square, Sir Joseph Banks (1743-1820), the doyen of British science, made contact with the same specimens through the pages of the catalogue of the College's collection published in 1807. A grant from the Irish Parliament in 1792 allowed the Dublin Society (it assumed the 'Royal' in 1820) to bring from Germany one of the world's premier mineral collections. Exhibited first at the Society's House in Hawkins Street, and then, after 1815, in Leinster House, that collection – the Leskean Cabinet – was one of the Society's most prized possessions.

Irish collectors eager for the expansion of their personal geological cabinets were able to purchase specimens from specialist Dublin dealers. A little later, during the 1830s, Joshua Abell of Eustace Street and Richard Glennon of Suffolk Street seem to have been foremost among those dealers. Members of the Royal Irish Academy (founded in 1785) contributed to international geological debate through the pages of the Academy's *Transactions,* launched in 1787. Richard Kirwan (1733-1812), the President of the Academy from 1799 until his death, contributed to the christening of the new science by, in 1799, publishing a work entitled *Geological Essays.* Another Irishman – William Babington (1756-1833) – in 1807 played a leading role in the foundation of the ground-breaking Geological Society of London, the world's first such society. And when, at its second meeting, that Society named its earliest Honorary Members, five of them possessed Irish addresses: in Ballyshannon, Belfast, Clonfecle, County Tyrone, Cork, and Dublin.

Among the intriguing problems that taxed the votaries of the infant science there was one which loomed particularly large in the eyes of geologists everywhere. What was the origin of basalt? Was it a chemical precipitate, as was claimed in Germany by the influential Abraham Gottlob Werner and his Neptunian followers, or did the truth lie with the Frenchman Nicolas Desmarest and his Vulcanist followers who held the rock to be a consolidated igneous melt? Ireland was well endowed for the shedding of light upon this important issue, and it was the basaltic plateau of Ulster that attracted the particular attention of the pioneers in Irish geology.

Irish opinion in the controversy was divided. The Neptunian interpretation was adopted by Richard Kirwan and by the Rev. William Richardson (1740–1820), a sometime Fellow of Trinity College Dublin who earned international prominence through his claim to have discovered fossil ammonites amidst the olivine-dolerite of the Portrush sill in County Antrim (Fig. 19.1). He had mistaken the fossils within the baked Liassic shales adjacent to the sill for fossils enclosed within the sill itself, but his 'discovery' seemed a confirmation of the Neptunian theory, and in consequence Portrush became for a while one of Europe's most famed geological localities.

The earliest Irish member of the opposing Vulcanist school was Frederick Augustus Hervey (1730–1803), the Bishop of Derry and the fourth Earl of Bristol (Childe-Pemberton, 1925; Fothergill, 1974). By about 1770 he had recognised the volcanic origin of the Giant's Causeway, and for some time he employed an Italian artist to make sketches both there and at other Irish sites of geological interest. A better-known Irish Vulcanist was the Rev. William Hamilton (1755-97), who in 1786 published both an excellent vulcanistic interpretation of north-eastern Ireland and a rudimentary geological map of the Antrim coast between Fair Head and Portrush. This is one of the earliest cartographic representations of Irish

Fig. 19.1 An exposure upon the foreshore at Portrush, Country Antrim, revealing ammonites (*Paltechioceras*) within indurated Liassic shales ('Portrush Rock') adjacent to the Portrush Sill. This olivine-dolerite intrusion was discovered about 1798 by William Richardson, who failed to distinguish between the sill and the adjacent baked shales. Noting the presence of fossils in what he termed 'a silicious basalt', he offered the site as conclusive proof of the Neptunian belief in the sedimentary origin of basalt. His claim to have disproved the Vulcanist interpretation of basalt attracted much attention, and he circulated samples of his 'fossiliferous basalt' to many geologists. In Edinburgh Sir James Hall, John Playfair and Lord Webb Seymour examined some of the specimens and recognised Richardson's mistake. Playfair visited the sill soon after 1802, and in 1816 Berger and Conybeare pronounced a final verdict when they observed that the fossiliferous stratum 'was no other than the slate-clay of the lias formation in an indurated state'. *Photograph: IPR/109-39CT British Geological Survey.*

geology, and in a 1790 edition of the book Hamilton widened his purview by including a map of the whole of north-eastern Ireland, illustrative of the limits of the basaltic plateau. This was certainly the first attempt to represent the extent of one of Ireland's geological formations, and it is sad that Hamilton's murder in County Donegal in 1797 should have cut short so promising a geological career. By that date, however, Hamilton and his contemporaries had made the geology of Antrim so familiar throughout the scientific world that the expression 'a Giant's Causeway' was universally accepted as a technical term applicable to any outcrop of columnar basalt wheresoever it might occur (Fig. 19.2).

In Ireland the debate between the Neptunists and the Vulcanists chiefly involved academics centred upon Trinity College Dublin and upon the Royal Irish Academy, but by the 1780s there existed within the country another group of men possessed of a more practical interest in geology – the members of the Dublin Society. The Society had been founded in 1731 for the encouragement of Irish agriculture and industry, and by the closing decades of the eighteenth century its members were increasingly aware that the study of geology might yield rich economic rewards. England's mineral wealth was beginning to bring her unparalleled prosperity; why should the same not happen in Ireland

Fig. 19.2 As early as the 1690s the Giant's Causeway in Country Antrim was the subject of a series of papers published in the *Philosophical Transactions of the Royal Society of London*, and the site was soon one of the most famed of all geological locations. This engraving (original size 68 x 34 cm) was published on 1 February 1743/4 and is based upon one of a series of paintings of the Causeway by Susanna Drury for which she was in 1740 awarded a premium by the (Royal) Dublin Society. 'Worth seeing? Yes; but not worth going to see,' observed Dr Johnson of the Causeway in 1779, yet by the following decade the members of the Vulcanist School had recognised the location as affording significant evidence of Ireland's fiery past.

if only there could be discovered the mineral basis for an Irish Industrial Revolution? Thus in 1786 the Society commissioned Donald Stewart, a Scotsman, as its 'Itinerant Mineralogist'. His first assignments were to survey County Wicklow and to examine the exposures along the line of the Grand Canal, which was then under construction from Dublin to the Shannon.

The Dublin Society continued to employ Stewart for a quarter of a century, but he was a mineralogist rather than a stratigrapher, and his investigations provided no clear picture of Ireland's geological structure. Such a picture would have emerged had it proved possible to give effect to a scheme that Richard Kirwan placed before the Society in 1802. He recommended the establishment of an Irish Mining Board consisting of 12 expert members, the Board to be responsible both for all Ireland's mines and for the production of a series of county geological maps. Sadly, the scheme was stillborn, but, had it been acted upon, Ireland would have been the first nation in the world to receive the benefit of a detailed official geological survey (Herries Davies, 1983).

Kirwan's scheme may have been premature, but the Dublin Society can certainly take credit for the publication of four early geological maps. Although little known, these four maps, contained in the Society's series of county statistical surveys, are deserving of honourable mention in any international history of geological cartography. In the order of their appearance the maps are County Wicklow, by Robert Fraser (*c.*1760–1831), published in 1801; County Kilkenny, by William Tighe (1766–1816) and County Londonderry, by the Rev. George Vaughan Sampson (1762–1827), both published in 1802; and County Cork, by the Rev. Horatio Townsend (1750–1837), published in 1810 with a second, but by no means improved, edition published in 1815. On the maps by Fraser and Sampson geological symbols or colour washes have been added only to those localities where rock outcrops had received positive identification, and large areas of their sheets are therefore left as geological blanks. The caution of the two authors is nevertheless understandable because theirs was an age when the lack of any secure theoretical basis seemed to render dangerous the extrapolation of geological lines from one outcrop to another.

Tighe and Townsend threw such caution to the winds. Their maps are coloured throughout, and the two sheets thus stand as the earliest comprehensive geological maps of any part of Ireland. Tighe's map of Kilkenny, at a scale of one inch to 2.5 miles (1:158,400) is, for its day, a particularly fine production. Perhaps such rock-categories as 'secondary siliceous schistus and ferruginous argillite' (the Namurian shales) and 'siliceous breccia and red argillite' (the Old Red Sandstone) ring strangely in the ear of the modern geologist, but Tighe's mapping of the various formations was surprisingly good, and in 1802 his sheet was by far the most sophisticated geological map available for any part of either Ireland or Great Britain (Fig. 19.3).

Save for Donald Stewart and Robert Fraser, all the geologists mentioned thus far were Irish-born, but it was during the years around 1800 that Ireland began to receive her first geological visitors from across the water. Arthur Young (1741–1820) toured the country collecting specimens in 1776; John Whitehurst (1713–88), a member of Birmingham's famed Lunar Society, visited County Antrim in 1783; and in 1797 that ardent Wernerian, Robert Jameson (1774–1854), travelled from Edinburgh to Dublin expressly to inspect the geological treasures housed in the museums of the Irish capital. Humphry Davy (1778–1829) made geological excursions through Ireland in 1805 and 1806, accompanied on the latter occasion by George Bellas Greenough (1778–1855) who was shortly to become the first President of the Geological Society of London.

That London Society was soon displaying its own interest in Irish geology, because in 1812 some of its members despatched the Genevan geologist Jean-François Berger (1779–1833) to Ireland to conduct surveys on their behalf (MacArthur, 1987, 1990) and in the following year Greenough himself returned to Ireland where he spent most of his time geologising in western Connaught. Two other visitors from among the ranks of the Geological Society were the Rev. William Buckland (1784–1856) and the Rev. William Daniel Conybeare (1787–1857), who together re-examined some of Berger's ground in northern Ireland. The result of their labours was a major memoir published by the Geological Society in 1816 and accompanied by a geological map depicting the whole of Ireland to the north-east of a line drawn from Dundalk to Derry. This memoir was perhaps the inspiration for a paper on the geology of south-eastern Ireland read to the same Society in 1818 by Thomas Weaver (1773–1855), the English-born and Freiberg-trained manager of the metalliferous mines at Cronbane and Tigroney in County Wicklow. When it was published in 1819 his study, too, was illustrated by an excellent geological map, in this case a sheet showing the whole of south-eastern Ireland as far westwards as the borders of County Limerick.

Fig. 19.3 William Tighe's map of County Kilkenny published in 1802. (Reduced in size.)

Thus, as a result of the debate between the Neptunists and the Vulcanists, as a result of the concern of the Dublin Society for economic geology, and as a result of the activities of her geological visitors, there had emerged by 1820 an outline picture of the geology of at least the eastern half of Ireland. But when, in 1802, the Society of Arts in London offered an award for the completion of a mineralogical map of Ireland, at a scale of at least one inch to ten miles, there was no claimant; nor was there any more enthusiastic a response when, in 1808, the Dublin Society offered a prize of £200 for a mineralogical map of County Dublin. By 1808, however, there was already present in the country the man who was destined to complete the first geological map of Ireland – Richard Griffith, the widely acclaimed 'father of Irish geology'.

Richard Griffith and his map 1809–1839

Richard John Griffith (1784–1878) was the son of a landowner in County Kildare, and after a brief dalliance with the profession of arms, he resolved to apply himself to the study of mining geology (Herries Davies and Mollan, 1980). To this end he in 1802 repaired to London, there to learn the general principles of science, and in 1804 he visited Cornwall in order to familiarise himself at first hand with rocks, minerals, and the latest mining practice. A tour of the other British mineraliferous districts followed, and he concluded his apprenticeship by spending two years in Edinburgh. There he attended the classes of Robert Jameson, and he claims to have assisted Sir James Hall (1761–1832) in some of his famed experimental work on the origin of igneous rocks. In 1808 Griffith returned to reside in Ireland and during the following year the Dublin Society commissioned him to make a detailed survey of the Leinster coalfield, the Society requiring of him a large-scale geological map, horizontal sections of the strata, and an explanatory memoir. While this survey was still in progress, Griffith was in 1812 appointed to the newly-created post of Mining Engineer to the Dublin Society, and this office, together with his considerable private practice as a mining consultant, allowed him to travel the length and breadth of the country laying the foundations of a knowledge of Irish field geology so extensive that it can rarely have been surpassed.

Griffith's memoir on the Leinster coalfield was published in 1814 together with sections and a geological map at a scale of one inch to 1.32 miles (1:83.600). No previous Irish geological map sought to depict a region's structure in such detail, but sadly the map was only a qualified success because Griffith had found difficulty in coping with the multitude of small faults that traverse the area. Failing to appreciate the tectonic complexity, he joined up coal-seams in neighbouring fault-blocks and treated them as continuous, when in reality they are fault-dislocated strata lying at a variety of different stratigraphical levels. As a result his survey was of diminished value for those mineral proprietors whom the Dublin Society must have had in mind when it commissioned the survey in 1809.

With the coalfield survey completed, Griffith turned his attention to a far more ambitious project. In 1811, or perhaps even earlier, G.B. Greenough suggested to Griffith that he should complete a geological map of Ireland as a companion to the sheet for England and Wales that had been engaging Greenough's own attention since 1808. Griffith accepted the proposal, and work upon the new map started in earnest in 1813. An early draft of the map was used to illustrate a course of lectures that Griffith delivered at the Dublin Society in 1814, but a quarter of a century was to elapse before the map was finally ready for publication (Herries Davies, 1983). In the meantime the Society required him to conduct a series of further surveys of mining districts. He surveyed the Connaught coalfield in 1814 and 1815, the Tyrone and Antrim coalfields between 1816 and 1818, and the Munster coalfield between 1818 and 1824, his report upon the Connaught region being published in 1819 (dated 1818) and that upon the Ulster coalfields in 1829. Unfortunately, the Munster survey was never completed, and another of his works which regretfully never reached the stage of publication was a geological section that he and his assistant John Kelly (1791–1869) surveyed across northern Ireland in 1816 from Newcastle, County Down, to the sea at the foot of Benbulben, County Sligo.

The 1820s were a bad period for Irish geology. Perhaps there was some disillusionment with a science that had failed to conjure forth a mineral wealth sufficient to trigger off an Irish Industrial Revolution, but, for whatever reason, the decade yielded singularly few contributions towards a better understanding of the nation's geology. Even Griffith seems to have lost some of his fervour, but in his case two factors obviously contributed to the reduced level of his geological activity.

First, Griffith's efforts to compile a geological map of Ireland were being thwarted by the absence of any accurate base-map upon which to record his observations. Repeatedly he complained of the injury done to his geological lines as a result of his having to plot fieldwork upon misshapen base-maps, and the situation drove him to such despair that in 1819 he resolved to compile his own topographical map of Ireland using triangulation methods. He began his triangulation in County Kerry, but the work was soon abandoned as there dawned upon him the full magnitude of the task he had so naively undertaken.

Secondly, during the 1820s Griffith's time was increasingly taken up in the performance of a variety of onerous public duties. In 1822 he became Engineer of Public Works in the south-west, and in 1825 he received the additional appointment of Director of the General Boundary Survey, in which capacity he had to oversee the location of all Ireland's administrative boundaries so that they might be recorded upon the six-inch maps that the Ordnance Survey was to publish between 1833 and 1846.

These responsibilities, together with his continuing duties at the Royal Dublin Society, left him little time for geological synthesis, but throughout the decade his travels did afford him ample opportunity for adding to his store of geological experience, even if for the moment his interpretations had to remain buried in the pages of pocket notebooks.

The 1830s saw a revival of interest in the geology of Ireland, seemingly stimulated in large measure by economic considerations. Between 1791 and 1831 Ireland's population had almost doubled, but there had been no corresponding expansion of the nation's economy. Population pressure was acute. Now, more than ever before, the obvious palliative seemed to be an Irish Industrial Revolution based upon native mineral resources. One sign of the times was the foundation in 1831 of the Geological Society of Dublin, which initially had hopes of employing surveyors to investigate areas of suspected mineral potential. The Society's finances never proved equal to such a task, although its meetings and journal did greatly assist the national dissemination of geological information (Herries Davies, 1965).

From the very beginning Griffith was prominent in the affairs of the new Society and during the 1830s he certainly found more time for the pursuit of his favourite science. In 1829 he had resigned his post at the Royal Dublin Society, and in 1830, while retaining the Directorship of the Boundary Survey, he assumed the new office of Commissioner of the General Valuation of Rateable Property. In this capacity he was responsible for the valuation of all land and property for the purposes of rate assessment, and for this task he was provided with a staff of around one hundred valuators. The new post was of the greatest importance to Griffith because geology was a legitimate study for valuators seeking to assess soil quality. When the valuation commenced in County Londonderry in 1830, his men were under instruction to visit every quarry, to measure the dip there, and to collect a rock sample. From that moment Griffith was running a rudimentary, if unofficial, geological survey of Ireland. Indeed, there is more than a suspicion that Griffith was guilty of misapplying public funds by employing his valuators upon geological investigations which had no purpose other than the refinement of his geological map. Some of his staff became very proficient field geologists and, as Commissioner, Griffith found that his tours of inspection afforded ample opportunity for geological investigation. He was later to claim that during these years he had frequently travelled from the north to the south of Ireland up to 40 times per year, moving regularly by night in order to avoid the loss of a day in the field.

In 1835 Griffith was the President of the Geological Section when the British Association convened in Dublin for the first of what were eventually to total 12 Irish meetings. He proudly displayed his map of Ireland for the benefit of the visitors and it received universal praise, Adam Sedgwick (1785–1873) considering it 'a thousand pities' that Griffith had failed to publish the map 14 years earlier. But the reason for that failure is clear enough: Griffith was still being haunted by the old base-map problem. The map exhibited in 1835 was founded upon Aaron Arrowsmith's map of Ireland at a scale of about one inch to four miles (1:235,400) and Griffith was at pains to apologise to the visitors for the topographical errors underlying his carefully plotted geological lines. The Ordnance Survey was of course busily publishing its magnificent six-inch maps, but there was still no small-scale Ordnance map depicting the whole of Ireland. Then suddenly the situation changed.

In 1836 Griffith joined a four-man commission appointed to examine Irish railway development, and the commissioners persuaded the Ordnance Survey to compile for their use a quarter-inch topographical map of Ireland. But the new map was to be more than just a topographical sheet. The commissioners decided that it should carry Griffith's geological lines and that a geologically coloured version of the map should be published as a document vital not only to the railway engineer, but also to the railway economist. Were Ireland's limestone regions not both the seat of the country's richest agricultural land and the site of most of the major towns, and were they not therefore the areas that would generate most of the rail traffic? Equally, were the regions of pre-Carboniferous rocks not areas of poor soils, low population densities, and subsistence agriculture? Railways were thus unlikely to be economic if they strayed down the stratigraphical column into systems of Devonian or earlier age.

Following the commissioners' decision, Griffith's map was published in two forms. It appeared first in 1838 in a simplified version at a scale of one inch to 10 miles (1:633,600) and contained in an atlas of six maps accompanying the second report of the commissioners. Then, in May 1839, the full map was published at a scale of one inch to four miles (1:253,440). This large map is a handsome production unequalled in either detail or beauty by any later small-scale geological map of Ireland. True, the map does have its geological weaknesses, and it even bears the trace of hasty compilation, the result, doubtless, of Griffith being caught off balance by the speed with which his long-standing base-map problem had suddenly evaporated. For example, there had not been time to introduce countrywide the tripartite division of the Carboniferous Limestone into Lower Limestone, Calp Beds, and Upper Limestone – a tripartite division that Griffith himself had announced in 1837 – and the earliest issues of the map depicted no Calp Beds whatsoever. Similarly, on the eve of the map's publication Griffith decided that large areas of West Cork and Kerry that he had represented as 'Transition Clay Slate' were in reality underlain by Silurian strata, and he therefore had affixed to the map a small addendum label to that effect. The map is nevertheless a remarkable achievement, not least because one third of Ireland had, to all intents and purposes, never before been represented upon any published geological map. As the first adequate geological map of Ireland, Griffith's masterpiece takes an honoured place alongside William Smith's pioneer map of England and Wales published in 1815 and John Macculloch's equivalent map of Scotland published in 1836.

Ordnance Survey geology 1824–1845

Although Griffith was conducting, albeit unofficially, a geological survey of Ireland from 1830 onwards, his was by no means the first such geological survey to be undertaken. Here we must retrace our steps back to the 1820s because, despite Irish geology then having been temporarily in the doldrums, that decade did witness the inauguration of Ireland's first national geological survey (Herries Davies, 1983). The story began in 1824 when the Ordnance Survey entered Ireland to begin the production of a national coverage of six-inch topographical maps. The man in charge of the operation, Lieutenant-Colonel Thomas Colby (1784–1852), was himself a keen geologist, and he believed that as his men worked their way across Ireland they would be able to collect geological information as a by-product of their normal topographical duties. Indeed, writing in 1826, he looked forward to the completion in Ireland of 'the most minute and accurate geological survey ever published'.

His initial plan was that his surveyors would collect geological samples, mark the collection sites upon maps, and then despatch both samples and maps to the Survey's Dublin headquarters, where a geologist would identify the specimens and interpolate geological lines upon the maps. But the inadequacy of such random methods soon became apparent, and in 1826 Captain John Watson Pringle (*c.*1793–1861), a former Freiburg student, was appointed to reshape the Survey's geological activities. Under his direction the field-parties were themselves instructed in the construction of geological maps, and a few fruits of their geological labours in County Londonderry are still to be seen in the Dublin archives of both the Ordnance and Geological surveys. But Pringle's system was never accorded a fair trial; in September 1828 the economy axe fell. Colby's masters maintained that geology had been interfering with the Survey's more legitimate topographical duties, and he was directed to cease all geological investigations forthwith. Failure though it was, Colby does deserve the credit for having inaugurated one of the earliest of all official geological surveys. His Irish survey pre-dates by almost ten years the geological survey of England and Wales formally established in 1835 under (Sir) Henry De La Beche (1796–1855).

The 1828 prohibition of geology did nothing to extinguish Colby's personal concern for the science, and he was allowed to revivify his geological survey just as soon as Irish interest in geology showed signs of revival in the 1830s. The new survey was placed under the charge of an enthusiastic Lieutenant named Joseph Ellison Portlock (1794–1864), and the earliest achievement of his regime was a geological map and memoir of the parish of Templemore, County Londonderry, published in 1837 as part of the first and only volume of the Ordnance Survey's projected series of parish inventories. Small though the Templemore map may be, it was the first official geological map published for any part of Ireland.

In 1837 Portlock extended his geological activities by opening in Belfast a geological survey office, a museum, and a soils laboratory. His staff was expanded to include eight civilian geologists, among them being men such as Thomas Oldham (1816–78) and George Victor Du Noyer (1817–69) (Croke, 1995), both of whom were destined to make considerable reputations through their Irish geological researches. All seemed to be set fair when suddenly, in February 1840, the activities of Portlock's department were brought to a shuddering halt by a second descent of the economy axe. He was ordered to terminate all his geological activities, to close his Belfast office, and to transfer the contents of the museum from Belfast to Dublin. This last must in itself have proved a formidable task; there had to be removed 1,900 minerals, 4,824 named fossils, 70 boxes of unnamed fossils, 13,580 rock samples, and 16 boxes of soil specimens.

The second suspension of the Ordnance Survey's geological investigation aroused considerable Irish protest and, as a sop to public opinion, it was agreed that Portlock might complete and publish a geological survey of County Londonderry. In making this concession the authorities believed his survey of the county to be virtually finished, but Portlock was a perfectionist who refused to be hurried and his superiors were driven to despair by his repeated failure to meet deadlines. Finally, in February 1843, the report was published as a bulky octavo volume, dealing not only with Londonderry but also with large portions of the neighbouring counties of Fermanagh and Tyrone. Portlock had created a classic of Irish geology – a worthy companion to De La Beche's more famous sister work on south-western England published in 1839 – but commercially it was disappointing and barely a hundred copies were sold in the year of publication. Colby himself was certainly far from satisfied; he found the report too profuse, too pernickety, and too palaeontological. The volume sealed Portlock's fate. He was virtually dismissed from the Ordnance Survey and despatched to Corfu, there to return to the general duties of his corps.

Twice, in 1825 and 1830, Colby had initiated a geological survey of Ireland, and twice, in 1828 and 1840, the survey had come to an abrupt halt. Now Colby resolved to try yet again. His superiors were unenthusiastic, but in July 1844 he secured permission to proceed, and at the head of the revived survey he placed Captain Henry James (1803–77). James had prepared himself for his new duties by spending four months in Britain as a member of De La Beche's field-staff, and late in 1844 he left for County Donegal, there to apply his freshly acquired geological expertise. Accompanying him were six soldiers to be trained as geologists, but from the outset it was a forlorn venture. The Dalradian rocks of Donegal are hardly a suitable training-ground for geological novices, and Colby's third essay in the geological mapping of Ireland had a life of barely three months. For Colby it was the end, but out of the Ordnance Survey's failure there was to develop the modern Geological Survey of Ireland.

The years of the Geological Survey 1845–1924

While James and his men swung their hammers in Donegal, the Government in London was having second thoughts about the geological mapping of Ireland. Ever since its inception in 1835, De La Beche's Geological Survey of England and Wales had been a department of the Ordnance Survey, but De La Beche had long been anxious to escape from Colby's suzerainty. He now found an ally in the Prime Minister, Sir Robert Peel, and the outcome was the foundation on 1 April 1845 of a new Geological Survey of Great Britain and Ireland under the independent control of De La Beche (Herries Davies, 1983, 1995). The staff of the new Survey was to be entirely civilian although in Ireland, as a sweetener for Colby, James was seconded from the Royal Engineers to become the Local Director of the Survey's Irish branch. James was given a staff of five field geologists, and, in addition, it was agreed that De La Beche himself would spend a part of each year working in Ireland, as also would Edward Forbes (1815–54) and Warington Wilkinson Smyth (1817–90), who were respectively the Survey's London-based Palaeontologist and Mining Geologist. With this workforce at his disposal, De La Beche believed that the geological surveying of Ireland could be completed within only ten years at the trifling cost of £15,000.

De La Beche's men of course possessed no monopoly of interest in Irish geology during the decades after 1845. Griffith remained active in Irish geological circles down to the 1860s (indeed, he was aged 89 when he last occupied the chair at a geological assembly), and the members of the Geological Society of Dublin (it became the Royal Geological Society of Ireland in 1864) continued to do sterling work on behalf of their chosen science. Among the amateur geologists one of the most notable was Robert Mallet (1810–81), whose studies of the transmission of shock-waves through various types of rock entitle him to be hailed as one of the founders of modern seismology, that being a term which he himself coined (Cox, 1982).

In the universities, too, geology had taken firm root. John Phillips (1800–74) became the first Professor of Geology in Trinity College Dublin in 1844, and in 1849 geology took its place in the newly-founded Queen's Colleges in Belfast, Cork, and Galway.

Useful research was carried out by the geologists in most of these institutions, and especially noteworthy are the mineralogical papers of Samuel Haughton (1821–97), who held the Trinity College chair from 1851 until 1881, and a group of studies in Irish regional geology by Robert Harkness (1816–78), who taught geology in Queen's College Cork between 1853 and 1878. William King (1809–86), the Professor in Galway from 1849 until 1883, is best remembered for his participation in the international debate over the character of the supposed Precambrian fossil *Eozoön canadense* (Harper, 1988). Only Queen's College, Belfast, failed to make much impact upon the geological scene. There the emphasis was upon natural history rather than geology, and as a result the northern city's chief preceptor of the earth sciences was Ralph Tate (1840–1901), who from 1861 until 1864 was a Belfast-based lecturer for the Department of Science and Art. His local investigations resulted in an important series of papers on Ulster's Mesozoic and Tertiary strata.

Much though these and other geologists accomplished, it was unquestionably the officers of the Geological Survey who dominated the Irish geological world during the second half of the nineteenth century (Herries Davies, 1995). The Survey 'broke ground' at Hook Head, County Wexford, in the summer of 1845, and from there the surveyors advanced northwards, examining every outcrop and producing maps far more detailed than those of which Colby or Griffith had dreamed.

All the field observations were plotted upon the splendid six-inch (1:10,560) Ordnance maps, the existence of which gave the Irish staff a distinct advantage over their British counterparts. In Britain there was, as yet, no six-inch survey, and De La Beche's men perforce were having to map upon the inadequate one-inch scale. But the Irish Survey had its own problems. After only a few weeks of work two members of the staff had to be dismissed, one for incompetence and the other because he was anathema to James. And then in July 1846 James himself resigned because he found the Local Directorship insufficient to satisfy his ego.

James's successor was Thomas Oldham, who is perhaps best remembered in Ireland for his discovery of the fossil *Oldhamia* in the Bray Group of County Wicklow. In 1845 Oldham had succeeded Phillips in the Trinity College chair of geology, but the authorities now agreed to his holding the academic and Survey appointments concurrently, and under his direction the Survey again settled down to its appointed task. But that is not to suggest that all the Survey's problems dissolved with James's departure. Among Oldham's earliest tasks was the delivery of a reprimand to Frederick McCoy (1823–99), one of the field-staff, for his slipshod mapping, and the offender found it prudent to resign rather than face dismissal. This was by no means the end of McCoy's career in science; in 1849 he became Professor of Mineralogy and Geology in Queen's College, Belfast and in 1854 he moved to the chair of Natural Science in the University of Melbourne, where he eventually became the doyen of Australian scientists. In Ireland finance was so tight that one of the two Fossil Collectors had to be discharged, and there was incessant friction between the Survey and the distinguished Irish chemist Sir Robert Kane (1809–90), who in 1845 had established in Dublin a government-financed Museum of Economic Geology (later known as the Museum of Irish Industry). It was unfortunate that the Survey and the museum had to share the same premises at 51 St Stephen's Green East.

Perhaps the most serious problem facing Oldham was, nevertheless, an old and familiar one – that very same base-map problem as had plagued every previous Irish geologist. In Britain the Survey was both mapping and publishing upon the one-inch scale, but in Ireland there was as yet no one-inch map available and the six-inch scale, although admirable in the field, was clearly far too large for publication. Oldham was therefore forced to fall back upon the county index maps that accompanied the six-inch survey, most of the indexes being drawn to a scale of half an inch to the mile (1:126,720). It was absurd to map at the six-inch scale and then to lose all the carefully plotted detail during reduction to the half-inch scale, but Oldham had no alternative. The first county map, that for Wicklow, was published in July 1848, to be followed in 1849 by companion maps of counties Carlow and Kildare. Leaving aside their overly small scale, the three maps may be hailed as accurate and very handsome examples of geological cartography, although the problem of representing both solid and drift geology upon the same sheet had yet satisfactorily to be solved.

In 1850 Oldham resigned both his chair and the Local Directorship in order to become the first Superintendent of the Geological Survey of India, and by 1855 five other members of the Irish Survey's staff had left to join Oldham's team in the east. Oldham's successor in Dublin was Joseph Beete Jukes (1811–69), an officer of the British Survey since 1846 and one of the finest field geologists of his day. In Ireland one of Jukes's earliest tasks was to oversee the publication of two more of the geological county maps, these being for Wexford and Dublin, both of which were published in 1851. The public took little interest in the county maps, however, and by May 1855 only 155 copies of the five sheets had been sold. The geologists themselves were certainly far from satisfied with the maps, and Jukes repeatedly pressed the Ordnance Survey to produce a one-inch (1:63,360) map for geological use. Finally his wish was granted. In 1851 the Ordnance Survey was authorised to construct a one-inch map of Ireland, and the first of the new one-inch geological sheets (the sheets later to be numbered 120, 121, 129, 130, 138, 139, 148, 149) were published in December 1856 with the solid geology represented by the usual hand-applied colour washes and the Pleistocene drifts depicted by a fine stipple.

For thirty years after 1856 the story of the Survey becomes the story of the one-inch geological map. Each year the surveyors examined about a thousand square miles[1] of country, moving first from Leinster into Munster and thence northwards on a broad front, so that by 1870 one-inch maps had been published for the whole of Ireland to the south of a line drawn from Galway Bay to Drogheda. With the passage of time the Survey grew in size. The turbulent George Henry Kinahan (1829–1908) was recruited in 1854 and in 1857 William Hellier Baily (1819–88) was brought over from England to become the Survey's Acting Palaeontologist. It was he who was responsible, year after year, for the identification of the thousands of fossils collected by the field-staff, although to his chagrin he never secured promotion beyond his 'acting' status.

The chief expansion of the Survey occurred in 1867 when the establishment was raised to one Director (Jukes), one District Surveyor, four Senior Geologists, and some ten Assistant Geologists. It was a considerably larger establishment than that vouchsafed to the Scottish Survey, which began its independent existence only in 1867. De La Beche had died in 1855 and the Director-General was now Sir Roderick Murchison (1792–1871), but a single visit of inspection in 1856 was sufficient to convince him first that 'the geology of Ireland is the dullest which I am acquainted with in Europe', and second that the Irish climate was quite unbearable. As a result he scarcely ever returned to Ireland. This threw an additional burden of responsibility upon Jukes (he too found life in Ireland distasteful), and in Ireland the chief legacy of Murchison's reign is the memoirs that accompany each one-inch geological sheet. The preparation of the memoirs was ordered by Murchison in December 1855 and by 1891 every Irish one-inch sheet had its descriptive memoir, a record achieved nowhere else in the British Isles. One noteworthy feature of the Irish memoirs is the fine landscape sketches contained therein, many of them from the artistic pen of the earliest District Surveyor, Du Noyer, whose death from scarlet fever at his Antrim field-station in January 1869 was a serious blow for the Survey.

Du Noyer's was not the only death to strike the Survey that year. During a tour of inspection in July 1864 Jukes had been injured in a fall at Kenmare, County Kerry, and there ensued a long period of ill health which ended with his own death in July 1869. The new Director was Edward Hull (1829–1917), Antrim-born, a Dublin graduate, and a member of the Geological Survey of Great Britain since 1850. It was Hull who in 1870 supervised the transfer of the Survey from St Stephen's Green to 14 Hume Street where the Survey was to have its headquarters until 1984. Sadly the air of Hume Street was soon poisoned by a personality clash between Hull and Kinahan, his newly-promoted District Surveyor. The frequent altercations between the two men became so heated that successive Directors-General had to visit Ireland to oil the troubled waters, and both Hull and Kinahan allowed their personal animosities to cloud their scientific judgment. Field-mapping nonetheless continued, although progress was rather slower than hitherto because the surveyors were now tackling the complex Dalradian districts of the north-west. What was the relationship between the rocks of Ulster and those of Scotland? Was the granite of County Donegal igneous or metamorphic in origin? Did Laurentian rocks exist in Ireland? These were the questions that now taxed the geologists as the primary mapping of Ireland drew to its close.

The survey was completed in 1887 – the task that De La Beche had estimated would take ten years had occupied almost half a century – and the final one-inch sheet was published in 1890. Now the Survey was run down because the authorities believed its task to be completed. Joseph Nolan (1841–1902) was left in charge with a staff of four geologists to answer routine enquiries and to execute minor field-revision. During the primary

1 It has not seemed appropriate to convert to the metric system in this particular chapter.

mapping the Survey had given inadequate attention to the subdivision of what were then described as 'Upper Silurian' and 'Lower Silurian' rocks (the Ordovician System was not so named upon an Irish Survey map until 1913), and the first task awaiting Nolan in the field was a re-examination of the Lower Palaeozoics – a re-examination that resulted in the appearance of sixty-seven revised one-inch sheets between 1900 and 1902.

Around 1900 the Survey's attention was focused very much upon drift geology. In 1897, for instance, James Robinson Kilroe (1848–1927) was instructed to prepare a memoir upon Ireland's agricultural geology, and the resultant work was published in 1907, accompanied by a fine map depicting Ireland's surface geology at a scale of one inch to ten miles (1:633,600). Then, in 1901, it was decided to undertake drift surveys of the vicinities of Ireland's five major cities, and, Nolan having retired, George William Lamplugh (1859–1926), an experienced drift man, was brought over from England to take charge of the project. The outcome was a series of new one-inch, colour-printed drift maps (each with its accompanying memoir) for Dublin (1903, falsely dated 1902), Belfast (1904), Cork (1905), Limerick (1906), and Londonderry (1908).

In April 1905 the link between the Irish Survey and its parent body in Britain was severed, and henceforth the Geological Survey of Ireland was a purely Irish concern, placed at first under the control of the Department of Agriculture and Technical Instruction for Ireland. Lamplugh returned to England and Grenville Arthur James Cole (1859–1924) became the new part-time Director, holding his Survey office alongside the chair of geology in the Royal College of Science for Ireland, that having been his post since 1890 (Wyse Jackson, 1989). Under Cole one of the Survey's chief assignments, between 1907 and 1909, was a re-examination of the interbasaltic beds of counties Antrim and Londonderry to ascertain whether there was sufficient bauxite present to meet some of the needs of a rapidly developing European aluminium industry.

One other Survey project from those pre-1914 years deserves a mention. In 1910 Cole and his one-time student Thomas Crook (1876–1937) published a memoir on the geology of the Irish continental shelf based upon samples dredged by the Department of Agriculture and Technical Instruction's research vessel, the S.S. *Helga*. Although the work is slight, it can now be seen as the Survey's earliest incursion into that realm of offshore geology which in our own day is of such vital significance.

The lean years 1924–1947

Throughout the nineteenth century Ireland occupied a leading position in the geological world. Her geologists were figures of international repute and no foreign student of the science could afford to overlook the geological publications emanating from the Survey, the Royal Dublin Society, the Royal Irish Academy, and the Royal Geological Society of Ireland. Many were the visitors who came to Ireland to inspect localities made famous by Irish geologists or to study such collections as the Leskean mineral cabinet at the Royal Dublin Society, the museum of fossil fish assembled at Florence Court, County Fermanagh, by the third Earl of Enniskillen (1807–86) (James, 1986), or the fine geological exhibit that was opened to the public in August 1890 in the galleries of the Dublin Science and Art Museum (after 1908 the National Museum of Science and Art). But now, in the aftermath of World War I, all this was to change as Ireland sank to the status of being little more than a geological backwater.

A presage of what was in store was seen during the 1880s when the once-flourishing Royal Geological Society of Ireland became moribund before finally dying in 1894. The collapse of most of the increasingly uncompetitive Irish mining industry at about the same time certainly did nothing to revive a flagging public interest in the geological sciences. The run-down of the Geological Survey in 1890, following the completion of the one-inch map, was further evidence of the creeping geological malaise, but it was after World War I that the decay really gathered pace. Two recently recruited young geologists belonging to the Geological Survey had been killed on active service in Flanders. In 1921 William Bourke Wright (1876–1939), the Survey's eminent Senior Geologist, felt disquiet at the political developments taking place around him in Ireland, and he obtained a transfer to England. The government of the newly independent Irish Free State, as an emergency measure, closed the National Museum in June 1922, and in 1924 the geological collections were swept from their gallery in order to create working-space for civil servants. The specimens associated with men such as Griffith, Portlock, Oldham, and Jukes disappeared into the packing-cases which were to remain their home for the next 60 years.

In April 1924 Cole's death removed from the scene the last member of the Irish Survey to be possessed of an international reputation. Two years later his former chair in the Royal College of Science was abolished and the institution itself was extinguished. For a few years the occasional new publication did trickle from the Survey, but a financial strangulation was slowly taking its effect. A proposed national coverage of 16 quarter-inch (1:253.440) colour-printed geological maps was abandoned in 1923 after the publication of only four sheets, and between 1928 and 1943 the emaciated Survey was unable to publish so much as one new map or memoir. In Northern Ireland the situation was even more deplorable because partition of the island had severed cross-border links with Dublin, and for a quarter of a century the Six Counties languished devoid of any established geological survey.

During the 1920s Ireland's political troubles were sufficient to deter several British geologists from crossing the Irish Sea. Earlier in the century, for instance, Sidney Hugh Reynolds (1867–1949) of Bristol and Charles Irving Gardiner (1868–1940) of Cheltenham had published a series of joint papers dealing with the Irish Lower Palaeozoic rocks, but after World War I the pair felt it prudent to abandon their Irish studies and instead turned their attention to Scotland. Thus the birth pangs of an independent Ireland tended to isolate her geologists from scientific intercourse with visitors from overseas.

Even within the Irish universities the flame of geology then burned but dimly. The most renowned Irish earth-scientist of the period was that brilliant polymath John Joly (1857–1933), who held the Trinity College chair of Geology and Mineralogy from 1897 until his death, but his reputation lay in geophysics, and he did little to further an understanding of Ireland's geological evolution (Nudds, 1986). Perhaps the happiest academic event of these troubled years was the decision of Queen's University, Belfast, to re-establish an independent chair of geology to which, in 1921, there was appointed that redoubtable student of the Irish Pleistocene, John Kaye Charlesworth (1889–1972). The decline of Irish geology is nevertheless all too apparent in the Irish scientific periodicals. During the 1880s a total of 190 earth science papers had been published in the pages of those periodicals; during the 1920s a total of only 34 such papers was published (Herries Davies, 1978b).

By the 1930s political stability had been restored in Ireland, and, noting the enfeebled condition of native Irish geology, overseas academic geologists began to move into Ireland in search of exciting research problems. There thus dawned a colonialist phase in the history of Irish geology as foreign geologists – they came overwhelmingly from British universities – staked out claims to pieces of Irish research territory. This episode has been termed the Great Irish Geological Plantation (Herries Davies, 1995), and it eventually reached its climax in the aftermath of World War II.

During the lean years really significant advances were made in only two branches of Irish geology: in our understanding of the igneous rocks of the north-east, and in our decipherment of Ireland's ubiquitous Pleistocene deposits. In each of these branches there are two names deserving of particular mention.

In the realms of igneous petrology the names are those of James Ernest Richey (1886–1968), who investigated the Mourne and Slieve Gullion ring complexes, and Doris Livesey Reynolds (1899–1985), who devoted 30 years to the study of the Newry Granite, Slieve Gullion, and the region's Tertiary dyke swarms.

The two outstanding Pleistocene geologists were the aforementioned Professor Charlesworth, who spent half a century in the study of Ireland's glacial history, and Anthony Farrington (1893–1973), a former Survey officer (he resigned from the Survey in 1928 because of its penurious condition) who became the Resident Secretary at the Royal Irish Academy, and from whose pen there flowed a steady stream of memorable papers on the glacial geology of both Leinster and his native Munster. It was Farrington who in 1934 became Secretary to the newly-constituted Committee for Quaternary Research in Ireland, and as such he played a part that year in bringing the Dane, Knud Jessen (1884–1971), to Ireland to train some selected Irish students in the study of Quaternary deposits. One of those students was George Francis Mitchell (1912–1997), who was destined to become both one of the most eminent authorities upon the Irish Pleistocene and a leading international figure in the field of Quaternary Studies.

The revival since 1947

In April 1947 a branch of the Geological Survey of Great Britain was established in Belfast to commence the revision of the one-inch geological sheets of Northern Ireland. This event may now be seen as the beginning of a renaissance of Irish geology – a renaissance which has allowed the science to regain all of that vigour which it displayed in its Victorian heyday. Three sets of circumstances have been instrumental in rescuing Irish geology from its long years in the doldrums.

First, the closing decades of the twentieth century saw the development of a fresh public awareness of the significance of geology as an underpinning of the national wellbeing. Construction materials for civil engineering; mineral and groundwater supplies for industry; fertile drift deposits to sustain agriculture; fuel resources for an expanding population; safe locations for waste disposal – all are of vital significance within our congested modern world. It was this revived understanding that brought a geological survey back to Northern Ireland in 1947, and from the outset that Survey has been especially concerned with aspects of applied geology. Its publications have included, for example, a memoir on the sources of roadstone aggregate in Northern Ireland, and a map of the engineering geology of the Belfast district.

Within the Republic of Ireland this same revived interest in the nation's resources has led to the remarkable rejuvenation which, over the period since 1967, has transformed the Geological Survey of Ireland. During 1984 the Survey moved into fine new premises at Beggars Bush, Dublin (there is presently a proposal to decentralise the Survey to Cavan), and there the Survey's Customer Centre offers a wide variety of advisory services under titles ranging from Geotechnical and Heritage/Tourism to Bedrock and Seabed. Between 1992 and 2003 the Survey completed a new national Bedrock Map Series at a scale of 1:100,000, each of the coverage of 25 sheets being accompanied by a memoir. Offshore, between 2000 and 2005, there was conducted an Irish National Seabed Survey at a cost of €33 million. During 2006 the Survey published a new 1:500,000 Bedrock Map of the whole of Ireland, compiled from its own 1:100,000 Bedrock Map Series and the Northern Survey's 1:250,000 geological map of Northern Ireland. Presently the Geological Survey of Ireland and the Geological Survey of Northern Ireland are together at work upon a new series of island-wide digital maps at a scale of 1:50,000.

Secondly, the last five decades have witnessed a boom in Irish mining and offshore exploration. Modern methods of geophysical and geochemical prospecting have shown this island to possess that wealth of underground riches of which Griffith and his contemporaries had dreamed, but which their primitive prospecting methods were quite incapable of locating. Since the earliest of the modern mineral strikes – that at Tynagh, County Galway, in 1961 – a number of other internationally significant mineral bodies have been located and the discovery of the offshore Kinsale Head Gasfield – a discovery announced in 1974 – has resulted in one of Ireland's new-found geological resources being piped into innumerable Irish homes. These discoveries have served to stimulate public interest in the earth sciences and to provide employment opportunities for a much-expanded professional geological community.

Thirdly, since 1947 geology has flourished within institutes of higher learning throughout Ireland and Britain, with the result that large numbers of both staff and students have been scouring the maps, the literature, and the ground itself in a quest for worthwhile research problems. For many that hunt has come to a satisfying conclusion amidst the rocks of Ireland. The hundreds of modern researchers who have applied themselves to Irish geological problems are entirely sufficient to gainsay Murchison's hasty 1856 verdict that Ireland was afflicted with the dullest geology to be found anywhere in Europe. A 1990 Directory of research in Irish geology, prepared by the Geological Survey of Ireland, revealed that almost 200 geologists, hailing from many lands, were then engaged in the unravelling of Ireland's geological secrets.

To provide a local publication outlet for geological research conducted in Ireland the country now again possesses its own specialised earth sciences journal. The *Journal of the Earth Sciences Royal Dublin Society* was inaugurated in 1978 as an organ of the Royal Dublin Society, but since 1984 the journal has been published by the Royal Irish Academy under the new title of *Irish Journal of Earth Sciences*. Since the year 2000, 82 different authors have published a total of 51 research papers within the *Irish Journal of Earth Sciences*, and of the 82 authors, 49 have been possessed of an Irish address.

Numerous are the institutions – both at home and abroad – which have sent their geologists to test their intellectual mettle against Ireland's rocks, but four overseas institutions merit particular mention: Imperial College, London; King's College, London; the University of Liverpool; and the University of Glasgow.

Beginning in 1948, members of Imperial College, London, King's College, London, and the University of Liverpool made a detailed study of the finely-exposed granites of County Donegal with the prime intention of shedding light upon both the nature of the granite bodies and the history of their emplacement. The research programme was initiated by Herbert Harold Read (1889–1970), and by 1972 almost 50 other geologists had contributed to the project. Apart from achieving its primary objective of enhancing our understanding of granite, the investigation shed significant fresh light upon the structure and history of the entire Irish Caledonides. A leading figure in the Donegal research was Wallace Spencer Pitcher (1919–2004) of the

University of Liverpool, and he has his final Irish geological memorial in *A Master Class Guide to the Granites of Donegal*, written in collaboration with Donald Herbert William Hutton and published by the Geological Survey of Ireland in 2003.

The University of Glasgow entered Irish geology through the activities of Bernard Elgey Leake, who assumed the chair of geology within that university during 1974. As an undergraduate in the University of Liverpool, in 1951, Leake commenced what was to become a re-investigation of the geology of the entire tract of country lying between Clew Bay and Galway Bay, and the outcome was the publication of two fine one-inch (1:63,360) maps of the region compiled from Leake's own mapping together with that of several other geologists who have deployed their talents in that fascinating and complex district. The two maps have been published by the University of Glasgow, the first – 'Connemara' – in 1981, and the second – 'South Mayo' – in 1985, each map having an accompanying Memoir, that for South Mayo being published by the University of Glasgow in 1989 and that for Connemara by the Royal Irish Academy in 1994.

To cater for the revived professional and public interest in geology there have been established several new societies which now serve amply to fill the void left when the old Royal Geological Society of Ireland disappeared into its grave during 1894. The Belfast Geologists' Society was inaugurated in 1953 'to further the practical study of geology'. The Irish Mining and Quarrying Society was established in 1958 to cater for the interests of those involved in the extractive industries. The Irish Geological Association was launched in 1959 to offer an annual programme of events of interest alike to both the professional and the amateur. The Irish Association for Economic Geology was founded in 1973, following upon discussions held in Athlone, Mullingar, and Port Laoise, and it serves as a forum for all those with a professional involvement in the economic aspects of the earth sciences. The Geotechnical Society of Ireland, a constituent of the Institution of Engineers of Ireland, dates from 1977 and seeks to promote co-operation amongst engineers and scientists in fields such as engineering geology and hydrogeology. The Cork Geological Association was born in 1992 out of the enthusiasm of a group of Cork citizens who had attended an evening diploma course at the local University College. Finally, the Institute of Geologists of Ireland was brought into being in 1999 to promote the geosciences and to represent the interests of the geological profession within Ireland.

For more than 200 years geologists have been probing Ireland's rocks. They have sought to unravel the secrets of Ireland's remote past. They have hoped to discover Irish resources of economic significance. They have tried to learn Irish lessons that might prove to be of universal application. Today the geologists and their allies continue with these tasks, and in their work many have clearly found instruction, guidance, and inspiration within the chapters of the present work founded by Charles Hepworth Holland in 1981 and now here entered upon the third state of its existence.

References

Adair, B.J.I. (1987) *Zeolites and zeolitization of the Tertiary basalts of North Eastern Ireland.* Unpublished Ph.D thesis, Queen's University of Belfast.

Aftalion, M. and Max, M.D. (1987) U–Pb zircon geochronology from the Precambrian Annagh Division gneisses and the Termon Granite, NW County Mayo, Ireland. *Journal of the Geological Society of London,* **144**, 401–406.

Aftalion, M., van Breemen, O., and Bowes, D.R. (1984) Age constraints on basement of the Midland Valley of Scotland. *Transactions of the Royal Society of Edinburgh, Earth Sciences,* **75**, 53–64.

Agrell, S.O. and Langley, J.M. (1958) The dolerite plug at Tievebulliagh, near Cushendall, Co. Antrim. *Proceedings of the Royal Irish Academy,* **59B**, 93–127.

Ainsworth, N.R., O'Neill, M., Rutherford, M.M., Clayton, G., Horton, N.F., and Penny, R.A. (1987) Biostratigraphy of the Lower Cretaceous, Jurassic and Late Triassic of the North Celtic Sea and Fastnet Basins. *In*: Brooks, J. and Glennie, K.W. (eds). *Petroleum Geology of North West Europe, Volume 2.* Graham and Trotman, London, 611–622.

Akaad, M.K. (1956) The Ardara granitic diapir of County Donegal, Ireland. *Quarterly Journal of the Geological Society of London,* **112**, 263–288.

Aldridge, R.J. (1980) Notes on some Silurian conodonts from Ireland. *Journal of Earth Sciences, Royal Dublin Society,* **3**, 127–132.

Allègre, C.J. (2008) *Isotope Geology.* Cambridge University Press. 512pp.

Allen, P. (1981) Pursuit of Wealden models. *Journal of the Geological Society of London,* **138**, 375–405.

Allen, P.A., Bennett, S.D., Cunningham, M.J.M., Carter, A., Gallagher, K., Lazzaretti, E., Galewsky, J., Densmore, A.L., Phillips, W.E.A., Naylor, D., and Solla Hach, C. (2002) The post-Variscan thermal and denudational history of Ireland. *In*: Doré, A.G., Cartwright, J.A., Stoker, M.S., Turner, J.P., and White, N. (eds). *Exhumation of the North Atlantic Margin: Timing, Mechanisms and Implications for Petroleum Exploration.* Geological Society, Special Publications, **196**, 371–399.

Alsop, G.I. (1991) Gravitational collapse and extension along a mid-crustal detachment: the Lough Derg Slide, northwest Ireland. *Geological Magazine,* **128**(4), 345–354.

Alsop, G.I. (1992) Late Caledonian sinistral strike-slip displacement across the Leannan Fault System, Northwest Ireland. *Geological Journal,* **27**(2), 119–125.

Alsop, G.I. (1996) Tectonic analysis of progressive secondary deformation in the hinge of a major Caledonian fold nappe, northwestern Ireland. *Geological Journal,* **31**, 217–233.

Alsop, G.I. and Hutton, D.H.W. (1993) Major southeast-directed Caledonian thrusting and folding in the Dalradian rocks of mid-Ulster: implications for Caledonian tectonics and mid-crustal shear zones. *Geological Magazine,* **130**(2), 233–244.

Alsop, G.I. and Jones, C.S. (1991) A review and correlation of Dalradian stratigraphy in the Southwest and Central Ox Mountains and Southern Donegal, Ireland. *Irish Journal of Earth Sciences,* **11**, 99–112.

Anczkiewicz, R. and Thirlwall, M.F. (2003) Improving precision of Sm–Nd garnet dating by H_2SO_4 leaching: a simple solution to the phosphate inclusion problem. *In*: Vance, D., Müller, W., and Villa, I.M. (eds). *Geochronology: linking the isotopic record with petrology and textures.* Geological Society of London, Special Publication, **220**, 83–91.

Anczkiewicz, R., Platt, J.P., Thirlwall, M.F., and Wakabayashi, J. (2004) Franciscan subduction off to a slow start: evidence from high-precision Lu–Hf garnet ages on high-grade blocks. *Earth and Planetary Science Letters,* **225**, 147–161.

Anczkiewicz, R., Szczepańaksi, J., Mazur, S., Storey, C., Crowley, Q., Villa, I.M., Thirlwall, M.F., and Jeffries, T.E. (2007) Lu–Hf geochronology and trace element distribution in garnet: implications for uplift and exhumation of ultra-high pressure granulites in the Sudetes, SW Poland. *Lithos,* **95**, 363–380.

Anderson, E., Harrison, S., Passmore, D.G., and Mighall, T.M. (1998) Geomorphic evidence of Younger Dryas glaciation in the Macgillycuddy's Reeks, south-west Ireland. *In*: Owen, L.A. (ed.). Mountain Glaciation. *Quaternary Proceedings,* **6**, 75–90. John Wiley and Sons Ltd, Chichester.

Anderson, J.G.C. (1948) The occurrence of Moinian rocks in Ireland. *Quarterly Journal of the Geological Society of London,* **103**, 171–188.

Anderson, J.G.C. (1960) The Wenlock strata of South Mayo. *Geological Magazine,* **97**, 265–275.

Anderson, K., Ashton, J., Earls, G., Hitzman, M., and Tear, S. (eds). (1995) *Irish Carbonate-hosted Zn–Pb Deposits.* Guidebook Series, **21**, Society of Economic Geologists, 1–296.

Anderson, T.B. (1987) The onset and timing of Caledonian sinistral shear in County Down. *Journal of the Geological Society of London,* **144**, 817–825.

Anderson, T.B. (2001) Structural interpretations of the Southern Uplands terrane. *Transactions of the Royal Society of Edinburgh: Earth Sciences,* **91**, 363–373.

Anderson, T.B. (2004) Southern Uplands-Down-Longford terrane. *In*: Mitchell, W.I. (ed.). *The Geology of Northern Ireland- Our Natural Foundation (2nd edition).* Geological Survey of Northern Ireland, Belfast, 9–24.

Anderson, T.B. and Cameron, T.D.J. (1979) A structural profile of Caledonian deformation in Down. *In*: Harris, A.L., Holland, C.H. and Leake, B.E. (eds). *The Caledonides of the British Isles: reviewed.* Geological Society, London, 263–267.

Anderson, T.B. and Oliver, G.J.H. (1986) The Orlock Bridge

Fault; a major late Caledonian sinistral fault in the Southern Uplands terrane, British Isles. *Transactions of the Royal Society of Edinburgh: Earth Sciences*, **77**(3), 203–222.

Anderson, T.B. and Oliver, G.J.H. (1996) Xenoliths of Iapetus suture mylonites in County Down lamprophyres, Northern Ireland. *Journal of the Geological Society of London*, **153**, 403–407.

Anderson, T.B. and Rickards, B. (2000) The stratigraphy and graptolite faunas of the Moffat Shales at Tieveshilly, County Down, Northern Ireland, and their implications for the modeling of the Southern Uplands–Longford Down terrane. *Irish Journal of Earth Sciences*, **18**, 69–88.

Anderson, T.B., Parnell, J., and Ruffell, A.H. (1995) Influence of basement on the geometry of Permo-Triassic basins in the northwest British Isles. *In*: Boldy, S.A.R. (ed.). *Permian and Triassic Rifting in Northwest Europe*. London, Geological Society Special Publication, **91**, 103–122.

Anderton, R. (1982) Dalradian deposition and the late Precambrian–Cambrian history of the N. Atlantic region: a review of the early evolution of the Iapetus ocean. *Journal of the Geological Society of London*, **139**, 421–431.

Anderton, R. (1985) Sedimentation and tectonics in the Scottish Dalradian. *Scottish Journal of Geology*, **21**, 407–436.

Anderton, R. (1988) Dalradian slides and basin development: a radical interpretation of stratigraphy and structure in the SW and Central Highlands of Scotland. *Journal of the Geological Society of London*, **145**, 669–678.

Andrew, C.J. (1992) Basin development chronology of the lowermost Carboniferous strata in the Irish north-Central Midlands. *In*: Bowden, A.A., Earls, G., O'Connor, P.G., and Pyne, J.F. (eds). *The Irish Minerals Industry 1980–1990*. Irish Association for Economic Geology, Dublin, 143–169.

Andrew, C.J. (1993) Mineralization in the Irish Midlands. *In*: Pattrick, R.A.D. and Polya, D.A. (eds). *Mineralization in the British Isles*. Chapman and Hall, London, 208–269.

Andrew, C.J. and Ashton, J.H. (1985) Regional setting, geology and metal distribution patterns of the Navan orebody, Ireland. *Transactions of the Institute of Mining and Metallurgy*, **94B**, 66–93.

Andrew, C.J. and Poustie, A. (1986) Syndiagenetic or epigenetic mineralization: the evidence from the Tatestown Zn–Pb prospect, Co. Meath. *In*: Andrew, C.J., Crowe, R.W.A., Finlay, S., Pennell, W.M., and Pyne, J.F. (eds). *Geology and genesis of mineral deposits in Ireland*. Irish Association for Economic Geology, Dublin, 281–296.

Andrew, C.J., Crowe, R.W.A., Finlay, S., Pennell, W.M., and Pyne, J.F. (eds). (1986) *Geology and genesis of mineral deposits in Ireland*. Irish Association for Economic Geology, Dublin, 709pp.

Andrews, J.R., Phillips, W.E.A., and Molloy, M.A. (1978) The metamorphic rocks of part of the north central Ox Mountains Inlier of Counties Sligo and Mayo. *Journal of Earth Sciences, Royal Dublin Society*, **1**, 173–194.

Andrieu, V., Huang, C.C., O'Connell, M., and Paus, A. (1993) Lateglacial vegetation and environment in Ireland: first results from four western sites. *Quaternary Science Reviews*, **12**, 681–705.

Archer, J.B. (1977) Llanvirn stratigraphy of the Galway–Mayo border area, western Ireland. *Geological Journal*, **12**, 77–98.

Archer, J.B. (1981) The Lower Palaeozoic rocks of the north-western part of the Devilsbit–Keeper Hill inlier and their implications for the postulated course of the Iapetus suture zone in Ireland. *Journal of Earth Sciences, Royal Dublin Society*, **4**, 21–38.

Archer, J.B. (1984) Clastic intrusions in deep-sea fan deposits of the Rosroe Formation, Lower Ordovician, western Ireland. *Journal of Sedimentary Petrology*, **54**, 1197–1205.

Archer, J.B., Sleeman, A.G., and Smith, D.C. (1996) *A geological description of Tipperary and adjoining parts of Laois, Kilkenny, Offaly, Clare and Limerick, to accompany the Bedrock Geology 1:100,000 Scale Map Series, Sheet 18, Tipperary, with contributions by K. Claringbold, G. Stanley (Mineral Resources) and G. Wright (Groundwater Resources)*. Geological Survey of Ireland, 76pp.

Armstrong, H.A. and Owen, A.W. (2001) Terrane evolution of the paratectonic Caledonides of northern Britain. *Journal of the Geological Society of London*, **158**, 475–486.

Armstrong, H.A., Floyd, J.D., Tingqing, L., and Barron, H.F. (2002) Conodont biostratigraphy of the Crawford Group, Southern Uplands, Scotland. *Scottish Journal of Geology*, **38**, 69–82.

Armstrong, H.A., Owen, A.W., Scrutton, C.T., Clarkson, E.N.K., and Taylor, C.M. (1996) Evolution of the Northern Belt, Southern Uplands: implications for the Southern Uplands controversy. *Journal of the Geological Society of London*, **153**, 197–205.

Arndt, N. and Goldstein, S.J. (1987) Use and abuse of crust formation ages. *Geology*, **15**, 893–895.

Arthurs, J.W. (1976a) *The geology and metalliferous mineral potential of the Sperrin Mountains area*. Special report, Geological Survey of Northern Ireland, 119pp.

Arthurs, J.W. (1976b) *The geology and metalliferous mineral potential of northeast Antrim*. Special Report, Geological Survey of Northern Ireland, 52pp.

Ashby, D.F. (1939) The geological succession and petrology of the Carboniferous volcanic area of Co. Limerick. *Proceedings of the Geologists' Association*, **50**, 324–330.

Ashton, J.H., Downing, D.T., and Finlay, S. (1986) The geology of the Navan Pb–Zn orebody. *In*: Andrew, C.J., Crowe, R.W.A., Finlay, S., Pennell, W.M., and Pyne, J.F. (eds). *Geology and genesis of mineral deposits in Ireland*. Irish Association for Economic Geology, Dublin, 243–280.

Ashton, J.H., Black, A., Geraghty, J., Holdstock, M., and Hyland, E. (1992) The geological setting and metal distribution patterns of Zn–Pb–Fe mineralization in the Navan Boulder Conglomerate. *In*: Bowden, A.A., Earls, G., O'Connor, P.G., and Pyne, J.F. (eds). *The Irish Minerals Industry 1980–1990*. Irish Association for Economic Geology, Dublin, 171–210.

Atherton, M.P. and Ghani, A.A. (2002) Slab break-off: a model for Caledonian, late granite syn-collisional magmatism in the orthotectonic (metamorphic) zone of Scotland and Donegal, Ireland. *Lithos*, **62**, 65–85.

Auffret, G.A., Pastouret, L., Cassat, G., De Charpel, O., Gravatte, J., and Guennoc, P. (1979) Dredged rocks from the Armorican and Celtic margins. *In*: Montadert, L., Roberts, D. *et al.* (eds). *Initial Reports on the Deep Sea Drilling Project*, **48**, 995–1013.

Auffret, G.A., Auzende, J.M., Cousin, M., Coutelle, A., Dobson, M., Geoghegan, M., Masson, D., Rolet, J., and Valillant, P. (1987) Géologie des escarpment de Porcupine et de Goben (N.E. Atlantique). Resultats de la campagne de plongéé CYAPORC. Note de groupe CYAPORC. *Comptes Rendus de l'Academie des Sciences, Paris*, **304**, 1003–1008.

Austin, R.L. and Husri, S. (1974) Dinantian conodont faunas of County Clare, County Limerick and County Leitrim: an appendix. *In*: Bouckaert, J. and Streel, M. (eds). *International symposium on Belgian micropaleontological limits from Emsian to Viséan, Namur 1974*. Belgian Geological Survey, **3**,18–64.

Austin, R.L. and Mitchell, M. (1975) Middle Dinantian platform conodonts from County Fermanagh and County Tyrone, Northern Ireland. *Bulletin of the Geological Survey of Great Britain*, **55**, 43–54.

Avison, M. (1984a) The Lough Guitane Volcanic Complex, County Kerry: a preliminary survey. *Irish Journal of Earth Sciences*, **6**, 127–136.

Avison, M. (1984b) Contemporaneous faulting and the eruption and preservation of the Lough Guitane Volcanic Complex, Co. Kerry. *Journal of the Geological Society of London*, **141**, 501–510.

Bachtadse, V. and Briden, J.C. (1990) *Palaeomagnetic constraints on the position of Gondwana during Ordovician to Devonian times*. Geological Society London Memoir, **12**, 43–48.

Bailey, E.B. and McCallien, W.J. (1956) Composite minor intrusions and the Slieve Gullion complex, Ireland. *Liverpool and Manchester Geological Journal*, **1**, 466–501.

Bailey, R.J., Jackson, P.D., and Bennell, J.D. (1977) Marine geology of Slyne Ridge. *Journal of the Geological Society of London*, **133**, 165–172.

Bailey, W., Shannon, P.M., Walsh, J.J., and Unnithan, V. (2003) Spatial distribution of faults and deep sea carbonate mounds in the Porcupine Basin. *Marine and Petroleum Geology*, **20**, 509–522.

Baker, J.W. (1966) The Ordovician and post-Rosslare Series rocks in southeast Co. Wexford. *Geological Journal*, **5**, 1–6.

Baker, J.W. (1968) The petrology of the Carnsore granite intrusion, Co. Wexford. *Proceedings of the Royal Irish Academy*, **67B**, 159–176.

Baker, J.W. (1970) Petrology of the metamorphosed Pre-Cambrian rocks of south-easternmost Co. Wexford. *Proceedings of the Royal Irish Academy*, **69B**, 1–20.

Ballantyne, C.K., McCarroll, D., and Stone, J.O. (2006) Vertical dimensions and age of the Wicklow Mountains ice dome, Eastern Ireland, and implications for the extent of the last Irish Ice Sheet. *Quaternary Science Reviews*, **25**, (17–18), 2048–2058.

Ballantyne, C.K., McCarroll, D., and Stone, J.O. (2007) The Donegal ice dome, northwest Ireland: dimensions and chronology. *Journal of Quaternary Science*, **22**(8), 773–783.

Ballantyne, C.K., Stone, J.O., and McCarroll, D. (2008) Dimensions and chronology of the last ice sheet in western Ireland. *Quaternary Science Reviews*, **27**, 185-200.

Bamford, D., Faber, S., Jacob, B., Kaminski, W., Nunn, K., Prodehl, C., Fuchs, K., King, R., and Willmore, P. (1976) A Lithospheric Seismic Profile in Britain: I. preliminary results. *Geophysical Journal of the Royal Astronomical Society*, **44**, 145–160.

Bamford, D., Nunn, K., Prodehl, C., and Jacob, B. (1978) LISPB–IV. Crustal structure of Northern Britain. *Geophysical Journal of the Royal Astronomical Society*, **54**, 43–60.

Bamford, M.L.F. (1987) Discussion on structural evolution of the Irish Variscides. *Journal of the Geological Society of London*, **144**, 997–998.

Bamford, M.L.F. and Cooper, M.A. (1989) Tectono-sedimentary ripples in Devonian shallow marine siliciclastic sediments, southern Ireland. *Sedimentary Geology*, **63**, 1–20.

Bamford, M.L.F. and Ford, M. (1990) Flexural shear in a periclinal fold from the Irish Variscides. *Journal of Structural Geology*, **12**, 59–67.

Barber, A.J., Max, M.D., and Brück, P.M. (1981) Geologist's Association: Irish Geological Association Field Meeting in Anglesey and southeastern Ireland 4–11 June 1977. *Proceedings of the Geologists' Association*, **92**, 269–291.

Barber, K., Zolitschka, B., Tarasov, P., and Lotter, A.F. (2004) Atlantic to Urals: the Holocene climate record of mid-latitude Europe. *In*: Battarbee, R.W., Gasse, F., and Stickley, C.E. (eds). *Past climate variability through Europe and Africa*. Springer, Dordrecht, 417–442.

Barber, K.E., Chambers, F.M., and Maddy, D. (2003) Holocene palaeoclimates from peat stratigraphy: macrofossil proxy climate records from three oceanic raised bogs in England and Ireland. *Quaternary Science Reviews*, **22**, 521–539.

Barnes, R.P., Anderson, T.B., and McCurry, J.A. (1987) Along-strike variation in the stratigraphical and structural profile of the Southern Uplands Central Belt in Galloway and Down. *Journal of the Geological Society of London*, **144**, 807–816.

Barnosky, A.D. (1985) Taphonomy and herd structure of the extinct Irish Elk, *Megaloceros giganteus*. *Science*, **228**, 340–344.

Barr, K.W., Colter, V.S., and Young, R. (1981) The geology of the Cardigan Bay–St. George's Channel Basin. *In*: Illing, L.V. and Hobson, G.D. (eds). *Petroleum Geology of the Continental Shelf of North-West Europe*. Heyden and Son Ltd., London, 432–443.

Barrat, J.A. and Nesbitt, R.W. (1996) Geochemistry of the Tertiary volcanism of Northern Ireland. *Chemical Geology*, **129**, 15–38.

Barrow, G. (1893) On an intrusion of muscovite-biotite gneiss in the southeast Highlands of Scotland and its accompanying metamorphism. *Quarterly Journal of the Geological Society of London*, **19**, 33–58.

Bartlett, W.H. (1842) *Scenery and Antiquities of Ireland*. London.

Bassett, M.G., Cocks, L.R.M., and Holland, C.H. (1976) The affinities of two endemic Silurian brachiopods from the Dingle Peninsula, Ireland. *Palaeontology*, **19**, 615–625.

Battiau-Queney, Y. (1987) Tertiary inheritance in the present landscape of the British Isles (examples from Wales, the Mendip Hills and south-west Ireland). *In*: Gardiner, V. (ed.). *International Geomorphology* 1986 Part II. J. Wiley and Sons, Chichester, 979–989.

Battiau-Queney, Y. and Saucerotte, M. (1985) Paléosols pré-glaciaires de la carrière de Ballyegan (Co. Kerry, Irlande). *Hommes et Terres du Nord*, **3**, 234–237.

Baxter, S. and Feely, M. (2002) Magma mixing and mingling textures in granitoids: examples from the Galway Granite, Connemara, Ireland. *Mineralogy and Petrology*, **76**(1–2), 63–74.

Baxter, S., Graham, N.T., Feely, M., Reavy, R.J., and Dewey, J.F. (2005) A microstructural and fabric study of the Galway Granite, Connemara, western Ireland. *Geological Magazine*, **142**(1), 81–95.

Bazley, R.A.B. (1975) The Tertiary igneous and Permo-Triassic rocks of the Ballyalton Borehole, County Down. *Bulletin of the Geological Survey of Great Britain*, **50**, 71–101.

Bazley, R.B., Brandon, A., and Arthurs, J.W. (1997) *Geology of the country around Limavady and Londonderry*. Geological Survey of Northern Ireland Technical Report, **GSNI/97/1**, 96pp.

Beach, A. (1982) Spaced cleavage: pressure-solution stripes. *In*: Borradaile, G.J., Bayly, M.B., and Powell, C. McA. (eds). *Atlas of deformational and metamorphic rock fabrics*, 242–243.

Beach, A. and King, M. (1978) Discussion on pressure solution. *Journal of the Geological Society of London*, **135**, 649–651.

Becker, H. (1997) Sm–Nd garnet ages and cooling history of

high-temperature garnet peridotite massifs and high-pressure granulites from lower Austria. *Contributions to Mineralogy and Petrology*, **127**, 224–36.

Beese, A.P., Brück, P.M., Feehan, J., and Murphy, T. (1983) A silica deposit of possible Tertiary age in the Carboniferous Limestone near Birr, County Offaly, Ireland. *Geological Magazine*, **120**, 331–340.

Bell, B.R. and Emeleus, C.H. (1988) A review of silicic pyroclastic rocks of the British Tertiary Volcanic Province. *In*: Morton, A.C. and Parson, L.M. (eds). *Early Tertiary Volcanism and the Opening of the NE Atlantic*. Geological Society Publication, **39**, 365–379.

Bennett, K.D. and Birks, H.J.B. (1990) Postglacial history of alder (*Alnus glutinosa* (L.) Gaertn.) in the British Isles. *Journal of Quaternary Science*, **5**, 123–133.

Bennett, M.C., Dunne, W.M., and Todd, S.P. (1989) Reappraisal of the 'Cullenstown Formation': implications for the Lower Palaeozoic tectonic history of SE Ireland. *Geological Journal*, **24**, 317–329.

Bennett, S.D. (2006) *Tertiary dykes in northwest Ireland: field occurrence, emplacement mechanisms, host rock deformation, and post-emplacement strain and denudations*. Unpublished Ph.D thesis, University of Dublin.

Bentley, M.R. (1988) The Colonsay Group. *In*: Winchester, J.A. (ed.). *Later Proterozoic stratigraphy of the Northern Atlantic Regions*. Blackie, 119–130.

Bentley, M.R., Maltman, A.J. and Fitches, W.R. (1988) Colonsay and Islay: a suspect terrane within the Scottish Caledonides. *Geology*, **16**, 26–28.

Berger, A.R. (1971) The origin of the banding in the Main Donegal Granite, NW Ireland. *Geological Journal*, **12**, 99–112.

Berger, W.H. and Jansen, E. (1995) Younger Dryas episode: ice collapse and super-fjord heat pump. *In*: Troelstra, S.R., van Hinte, J.E., and Ganssen, G.M. (eds). The Younger Dryas. *Koninklijke Nederlandse Akademie van Wetenschappen Verhandelingen, Afd. Natuurkunde, Eerste Reeks, deel*, **44**, 61–105.

Bergstrom, S.M. and Orchard, M.J. (1985) Conodonts of the Cambrian and Ordovician systems from the British Isles. *In*: Higgins, A.C. and Austin, R.L. (eds). *A stratigraphical index of conodonts*, 32–67.

Bergstrom, S.M. and Xu, C. (2007) Ordovician correlation chart compiled by Stig Bergstrom and Chen Xu 2007-7-3 (ISOS) – www.Ordovician.cn

Bickle, M.J., Kidd, R.G.W., and Nisbet, E. (1972) The Silurian of the Croagh Patrick Range, Co. Mayo. *Scientific Proceedings of the Royal Dublin Society*, **A4**, 231–249.

Bingen, B., Demaiffe, D., and Van Breemen, O. (1998) The 616 Ma old Egersund basaltic dike swarm, SW Norway, and Late Neoproterozoic opening of the Iapetus Ocean. *Journal of Geology*, **106**, 565–574.

Bingen, B., Birkeland, A., Nordgulen, Ø., and Sigmond, E.M.O. (2001) Correlation of supracrustal sequences and origin of terranes in the Sveconorwegian orogen of SW Scandinavia: SIMS data on zircon in clastic metasediments. *Precambrian Research*, **108**, 293–318.

Birks, H.J.B. (1989) Holocene isochrone maps and patterns of tree-spreading in the British Isles. *Journal of Biogeography*, **16**, 503–540.

Birks, H.J.B. and Birks, H.H. (1980) *Quaternary Palaeoecology*. Edward Arnold, London.

BIRPS and ECORS (1986) Deep seismic reflection profiling beween England, France and Ireland. *Journal of the Geological Society of London*, **143**, 45–52.

Blackford, J.J. and Chambers, F.M. (1995) Proxy climate record for the last 1000 years from Irish blanket peat and a possible link to solar variability. *Earth And Planetary Science Letters*, **133**, 145–150.

Blackmore, R. (1995) Low-grade metamorphism in the Upper Palaeozoic Munster Basin, southern Ireland. *Irish Journal of Earth Sciences*, **14**, 115–133.

Bluck, B.J. (2000a) Caledonian and related events in Scotland. *Transactions Royal Society of Edinburgh: Earth Sciences*, **91**, 375–404.

Bluck, B.J. (2000b) Old Red Sandstone basins and alluvial systems of Midland Scotland. *In*: Friend, P.F. and Williams, B.P.J. (eds). *New perspectives on the Old Red Sandstone*. Geological Society London Special Publication, **180**, 401–416.

Bluck, B.J. and Dempster, T.J. (1991) Exotic metamorphic terranes in the Caledonides: tectonic history of the Dalradian block, Scotland. *Geology*, **19**, 1133–1136.

Bluck, B.J., Dempster, T.J., and Rogers, G. (1997) Allochthonous metamorphic blocks on the Hebridean passive margin, Scotland. *Journal of the Geological Society of London*, **154**, 921–924.

Bluck, B.J., Dempster, T.J., Aftalion, M., Haughton, P.D.W., and Rogers, G. (2006) Geochronology of a granitoid boulder from the Corsewall Formation (Southern Uplands): implications for the evolution of southern Scotland. *Scottish Journal of Geology* , **42**, 29–35.

Boland, M.E. (1983) *The geology of the Ballyvoyle–Kilmacthomas–Kilfarrasy area, County Waterford with an account of the Lower Palaeozoic geology of County Waterford*. Unpublished M.Sc. thesis, Dublin University.

Boldy, G.D.J. (1955) *The petrology of the igneous rocks near Castletownbere, West Cork, and the tectonic relationships with the Armorican Orogeny in south-west Ireland*. Unpublished M.Sc. thesis, Dublin University.

Boldy, S.A.R. (1982) *The Old Red Sandstone rocks of the eastern Knockmealdown Mountains and Monavullagh Mountains and adjacent areas, Counties Tipperary and Waterford*. Unpublished Ph.D. thesis, University of Dublin.

Bond, G. and Lotti, R. (1995) Iceberg discharges into the North Atlantic on millennial timescales during the last deglaciation. *Science*, **267**, 1005–1010.

Bond, G., Broecker, W., Johnsen, S., McManus, J., Labeyrie, L., Jouzel, J., and Bonani, G. (1993) Correlations between climate records from North Atlantic sediments and Greenland ice. *Nature*, **365**, 143–147.

Bond, G.C., Showers, W., Elliot, M., Evans, M., Lotti, R., Hajdas, I., Bonani, G., and Johnson, S. (1999) The North Atlantic's 1–2 kyr climate rhythm: relation to Heinrich events, Dansgaard/Oeschger cycles and the Little Ice Age. *In*: Clark, P., Webb, R., and Keigwin, L.D. (eds). *Mechanisms of global climate change at millennial time scales*. Geophysical Monograph Series, **112**, 35–58.

Boulter, M.C. (1980) Irish Tertiary plant fossils in a European context. *Journal of Earth Sciences, Royal Dublin Society*, **3**, 1–11.

Boulter, M.C. and Mitchell, I. (1977) Middle Pleistocene (Gortian) deposits from Benburb, Northern Ireland. *Irish Naturalist's Journal*, **19**, 2–3.

Boulter, M.C. and Wilkinson, G.C. (1977) A system of group names for some Tertiary pollen. *Palaeontology*, **20**, 559–79.

Boulton, G. and Hagdorn, M. (2006) Glaciology of the British Isles Ice Sheet during the last glacial cycle: form, flow, streams and lobes. *Quaternary Science Reviews*, **25** (23–24): 3359–3390.

Bowden, A.A., Earls, G., O'Connor, P.G., and Pyne, J.F. (eds). (1992) *The Irish Minerals Industry 1980–1990*. Irish Association for Economic Geology, Dublin, 434pp.

Bowen, D.Q. and Gibbard, P.L. (2007) The Quaternary is here to stay. *Journal of Quaternary Science*, **22**(1), 3-8.

Bowen, D.Q., Rose, J., McCabe, A.M., and Sutherland, D.G. (1986b) Correlation of Quaternary glaciations in England, Ireland, Scotland and Wales. *Quaternary Science Reviews*, **5**, 299–340.

Bowen, D.Q., Phillips, F.M., McCabe, A.M., Knutz, P.C., and Sykes, G.A. (2002) New data for the last glacial maximum in Great Britain and Ireland. *Quaternary Science Reviews*, **21**, 89–101.

Bowen, D.Q., Richmond, G.M., Fullerton, D.S., Sibrava, V., Fulton, R.J., and Velichko, A.A. (1986a) Correlation of Quaternary glaciations in the northern hemisphere. *Quaternary Science Reviews*, **5**, 509–510.

Bowes, D.R. and Hopgood, A.M. (1975) Structure of the gneiss complex of Inistrahull, Co. Donegal. *Proceedings of the Royal Irish Academy*, **75B**, 369–390.

Boyd, J.D. (1983) *Sedimentology of the lower Dingle Group, southern Dingle Peninsula, southwest Ireland*. Unpublished Ph.D thesis, University of Bristol.

Boyd, J.D. and Sloan, R.J. (2000) Initiation and early development of the Dingle Basin, SW Ireland, in the context of the closure of the Iapetus Ocean. *In*: Friend, P.F. and Williams, B.P.J. (eds). *New perspectives on the Old Red Sandstone*. Geological Society London Special Publication, **180**, 123–146.

Bradbury, H.J., Smith, R.A., and Harris, A.L. (1976) 'Older' granites as time-markers in Dalradian evolution. *Journal of the Geological Society of London*, **132**, 677–684.

Bradfer, N. (1984) *Geology of Pb–Zn deposit in Waulsortian (Carboniferous) limestones at Ballinalack, County Westmeath, Ireland*. Unpublished M.Sc. thesis, University of Dublin.

Bradshaw, M.J., Cope, J.W.C., Cripps, D.T., Donovan, D.T., Howarth, M.K., Rawson, P.F., West, M., and Wimbledon, W.A. (1992) Jurassic. *In*: Cope, J.C.W., Ingham, J.K., and Rawson, P.F. (eds). *Atlas of Palaeogeography and Lithofacies*. Geological Society, London, Memoir **13**, 153pp.

Bradshaw, R.H.B. and Browne, P. (1987) Changing patterns in the post-glacial distribution of *Pinus sylvestris* in Ireland. *Journal of Biogeography*, **14**, 237–248.

Bradshaw, R.H.W. (2001) The Littletonial Warm Stage: post 10,000 bp. *In*: Holland, C.H. (ed.). *The geology of Ireland*. Dunedin Academic Press, Edinburgh, 429–442.

Braithwaite, K. (1993) Stratigraphy of a mid-Carboniferous section at Inishcorker, Ireland. *Annales de la Société géologique de Belgique*, **116**, 209–219.

Brand, S.F and Emo, G.T. (1986) A note on Zn–Pb–Ba mineralization near Oldcastle, Co. Meath. *In*: Andrew, C.J., Crowe, R.W.A., Finlay, S., Pennell, W.M., and Pyne, J.F. (eds). *Geology and genesis of mineral deposits in Ireland*. Irish Association for Economic Geology, Dublin, 297–304.

Brandon, A. (1972) The upper Viséan and Namurian shales of the Doagh Outlier, County Fermanagh, Northern Ireland. *Irish Naturalists' Journal*, **17**, 159–170.

Brandon, A. (1977) The Meenymore Formation: an extensive intertidal evaporitic formation in the upper Viséan (B2) of northwest Ireland. *Report of the Institute of Geological Sciences*, **77/23**, 14pp.

Brandon, A. and Hodson, F. (1984) *The stratigraphy and palaeontology of the late Viséan and early Namurian rocks of north-east Connaught*. Geological Survey of Ireland Special Paper, **6**, 1–54.

Brandon, A. and Wilson, H.E. (1975) The geology of the Templepatrick area, County Antrim, Northern Ireland. *Bulletin of the Geological Survey of Great Britain*, **51**, 41–56.

Brasier, M.D., Ingham, J.K., and Rushton, A.W.A. (1992) Cambrian. *In:* Cope, J.C.W., Ingham, J.K., and Rawson, P.F.(eds). *Atlas of Palaeogeography and Lithofacies*. Geological Society, London, Memoir, **13**, 13–18.

Bray, A. A. (1990) *Systematics and palaeobiogeography of some osteolepiform and rhizodentiform crossopterygians*. Unpublished Ph.D thesis, University of Reading.

Brenchley, P.J. and Newall, G. (1984) Late Ordovician environmental changes and their effect on faunas. *In*: Bruton, D.L. (ed.). *Aspects of the Ordovician System*. Palaeontological Contributions University of Oslo, **295**, 65–79.

Brenchley, P.J. and Treagus, J.E. (1970) The stratigraphy and structure of the Ordovician rocks between Courtown and Kilmichael Point, Co. Wexford. *Proceedings of the Royal Irish Academy*, **69B**, 83–102.

Brenchley, P.J., Harper, J.C., and Skevington, D. (1967a) Lower Ordovician shelly and graptolite faunas from southeastern Ireland. *Proceedings of the Royal Irish Academy*, **65B**, 385–390.

Brenchley, P.J., Harper, J.C., Mitchell, W.I., and Romano, M. (1977) A re-appraisal of some Ordovician successions in eastern Ireland. *Proceedings of the Royal Irish Academy*, **77B**, 65–85.

Brenchley, P.J., Harper, J.C., Romano, M., and Skevington, D. (1967b) New Ordovician faunas from Grangegeeth, Co. Meath. *Proceedings of the Royal Irish Academy*, **65B**, 297–304.

Brennand T.P. (1965) The Upper Carboniferous (Namurian) stratigraphy north-east of Castleisland, Co. Kerry, Ireland. *Proceedings of the Royal Irish Academy*, **64**(B), 41–63.

Bresser, G. (2000) *An integrated structural analysis of the SW Irish Variscides*. Aachener Geowissenschaftliche Beiträge, 190pp.

Bresser, G. and Walter, R. (1997) The Irish Variscides and the relation to the Rheinish Massif. *Aardkundige Mededelingen*, **8**, 19–22.

Bresser, G. and Walter, R. (1999) A new structural model for the SW Irish Variscides. The Variscan front of the NW European Rhenohercynian. *Tectonophysics* **309**, 197–209.

Brewer, J.A., Matthews, D.H., Warner, M.R., Hall, J., Smythe, D.K., and Whittington, R.J. (1983) BIRPS deep seismic reflection studies of the British Caledonides. *Nature*, **305**, 206–210.

Brewer, T.S., Storey, C.D., Parrish, R.R., Temperley, S., and Windley, B.F. (2003) Grenvillian age exhumation of eclogites in the Glenelg–Attadale inlier, NW Scotland. *Journal of the Geological Society of London*, **160**, 565–574.

Brewster, D. (1825) Description of gmelinite, a new mineral species. *Edinburgh Journal of Science*, **2**, 262–267.

Briden, J.C., Robertson, D.J., and Smethurst, M.A. (1989) Terrane rotations in the Irish Caledonides. *Journal of the Geological Society of London*, **146**, 909–911.

Bridge, J.S. and Diemer, J.A. (1983) Quantitative interpretation of an evolving ancient river system. *Sedimentology*, **30**, 599–624.

Bridge, J.S., Van Veen, P.M., and Matten, L.C. (1980) Aspects of the sedimentology, palynology and palaeobotany of the Upper Devonian of southern Kerry Head, Co. Kerry, Ireland. *Geological*

Journal, **15**, 143–170.

Brindley, J.C. (1956) The geology of the northern end of the Leinster granite; Part 2: structural features of the country rocks. *Proceedings of the Royal Irish Academy: Section B-Biological Geological and Chemical Science*, **58**, 23–69.

Brindley, J.C. (1973) The structural setting of the Leinster Granite, Ireland: a review. *Scientific Proceedings of the Royal Dublin Society*, **5A**, 27–34.

Brindley, J.C. and Kennan, P.S. (1972) The Rockabill granite, Dublin. *Proceedings of the Royal Irish Academy*, **72B**, 335–346.

Brindley, J.C. and Millan, S. (1973) Variolitic lavas from south County Wicklow, Ireland. *Scientific Proceedings of the Royal Dublin Society*, **A4**, 461–469.

Brindley, J.C., Millan, S., and Schiener, E.J. (1973) Sedimentary features of the Bray Group, Bray Head, Co. Wicklow. *Scientific Proceedings of the Royal Dublin Society*, **A4**, 373–389.

Brock, A. (1989) Heat flow measurements in Ireland. *Tectonophysics*, **164**, 231–236.

Bröcker, M. and Franz, L. (2006) Dating metamorphism and tectonic juxtaposition on Andros Island (Cyclades, Greece): results of a Rb–Sr study. *Geological Magazine*, **143**, 609–620.

Brooks, A.J, Bradley, S.L., Edwards, R.J., Milne, G.A., Horton, B., and Shennan, I. (2008) Post-glacial relative sea-level observations from Ireland and their role in glacial rebound modelling. *Journal of Quaternary Science*, **23**, 175-192.

Brown, C. and Whelan, J.P. (1995) Terrane boundaries in Ireland inferred from the Irish Magnetotelluric Profile and other geophysical data. *Journal of the Geological Society of London*, **152**, 523–534.

Brown, C., Denny, P., Haak, V., Bruton, P., and the VARNET Group. (2003) VARNET MT: interpretation of magnetotelluric data over the Killarney–Mallow Fault Zone in south-west Ireland. *Irish Journal of Earth Sciences*, **21**, 1–17.

Brown, E.J., Rose, J., Coope, R.G., and Lowe, J.J. (2007) An MIS 3 age organic deposit from Balglass Burn, central Scotland: palaeoenvironmental significance and implications for the timing of the onset of the LGM ice sheet in the vicinity of the British Isles. *Journal of Quaternary Science*, **22**(3), 295–308

Brown, G.C., Frances, E.H., Kennan, P.S., and Stillman, C.J. (1985) Caledonian igneous rocks of Britain and Ireland. *In*: Harris, A. L. (ed.). *The nature and timing of orogenic activity in the Caledonian rocks of the British Isles*. Journal of the Geological Society of London, Memoir 9, 1–15.

Browne, P. (1986) *Vegetational history of the Nephin Beg Mountains, County Mayo*. Unpublished PhD. thesis, University of Dublin, Trinity College.

Browne, P. and Coxon, P. (1991) The Nephin Beg Range and the Late-glacial. *In*: P.Coxon (ed.). *Fieldguide to the Quaternary of north Mayo*. Irish Association for Quaternary Studies (IQUA), Dublin, 41–49.

Brück, P.M. (1968) The geology of the Leinster Granite in the Enniskerry–Lough Dan area, Co. Wicklow. *Proceedings of the Royal Irish Academy*, **66B**, 53–70.

Brück, P.M. (1970) Stratigraphy, petrology and structure of the greywacke formations of the Blessington area. *Geological Survey of Ireland Bulletin*, **1**, 31–45.

Brück, P.M. (1971a) Fossil content and age of the greywacke formations west of the Leinster Granite in Counties Dublin, Kildare and Wicklow, Ireland. *Geological Magazine*, **108**, 303–310.

Brück, P.M. (1971b) The Lower Carboniferous rocks of the Naas district, Co. Kildare. *Geological Survey of Ireland Bulletin*, **2**, 211–221.

Brück, P.M. (1972) Stratigraphy and sedimentology of the Lower Palaeozoic greywacke formations in Counties Kildare and west Wicklow. *Proceedings of the Royal Irish Academy*, **72B**, 25–53.

Brück, P.M. (1973) Structure of the Lower Palaeozoic greywacke formations west of the Leinster Granite in Counties Kildare and west Wicklow. *Scientific Proceedings of the Royal Dublin Society*, **A4**, 391–409.

Brück, P.M. (1985) *The geology of the country between Slieve Aughty, the Silvermines–Devilsbit Mountains and Slieve Bloom, Central Ireland*. Department of Geology, University College Cork, Occasional Report Series, RS85/12, 60pp.

Brück, P.M. and Kennan, P.S. (1970) The geology of Shenick's Island, Skerries, Co. Dublin. *Scientific Proceedings of the Royal Dublin Society*, **3A**, 323–333.

Brück, P.M. and O'Connor, P.J. (1977) The Leinster Batholith: geology and geochemistry of the northern units. *Geological Survey of Ireland Bulletin*, **2**, 107–141.

Brück, P.M. and Reeves, T.J. (1976) Stratigraphy, sedimentology and structure of the Bray Group in County Wicklow and south County Dublin. *Proceedings of the Royal Irish Academy: Section B-Biological Geological and Chemical Science*, **76**(4), 53–77.

Brück, P.M. and Reeves, T.J. (1983) The geology of the Lugnaquilla Pluton of the Leinster batholith. *Geological Survey of Ireland Bulletin*, **3**, 97–106.

Brück, P.M. and Stillman, C.J. (1978) Introduction. *In*: *Field guide to the Caledonian and pre-Caledonian rocks of south-east Ireland*. Geological Survey of Ireland, Guide Series, **2**, 5–14.

Brück, P.M. and Vanguestaine, M. (2004) Acritarchs from the Lower Palaeozoic succession on the south County Wexford coast, Ireland: new age constraints for the Cullenstown Formation and the Cahore and Ribband Groups. *Geological Journal*, **39**, 199–224.

Brück, P.M. and Vanguestaine, M. (2005) An Ordovician age for the Muggort's Bay Lower Palaeozoic inlier, County Waterford, Ireland — the southernmost exposure of the Irish Caledonides. *Geological Journal*, **40**, 519–544.

Brück, P.M., Potter, T.L., and Downie, C. (1974) The Lower Palaeozoic stratigraphy of the northern part of the Leinster Massif. *Proceedings of the Royal Irish Academy* **74B**, 75–84.

Brück, P.M., Gardiner, P.R R., Max, M.D., and Stillman, C.J. (1978) *Field guide to the Caledonian and pre-Caledonian rocks of south-east Ireland*. Geological Survey Ireland, Guide Series, **2**, 78pp.

Brück, P.M., Higgs, K.T., Maziane-Serraj, N., and Vanguestaine, M. (2001) New palynological data from the Leinster Lower Palaeozoic massif, Southeastern Ireland. *Transactions of the Royal Society of Edinburgh: Earth Sciences*, **91**, 509–514.

Brück, P.M., Colthurst, J.R.J., Feely, M., Gardiner, P.R.R., Penney, S.R., Reeves, T.J., Shannon, P.M., Smith, D.G., and Vanguestaine, M. (1979) South-east Ireland: Lower Palaeozoic stratigraphy and depositional history. *In*: Harris, A.L., Holland, C.H., and Leake, B.E. (eds). *The Caledonides of the British Isles: reviewed*. Geological Society London Special Publication, **8**, 533–544.

Brunton, C.H.C. and Mason, T.H. (1979) Palaeoenvironments and correlations of the Carboniferous rocks in west Fermanagh. *Bulletin of the British Museum (Natural History), Geology*, **32**, 91–108.

Buckman, J.O. (1992) Palaeoenvironment of a Lower Carboniferous sandstone succession northwest Ireland: ichnological and sedimentological studies. *In*: Parnell, J. (ed.). *Basins on the Atlantic Seaboard: petroleum geology, sedimentology and basin evolution*.

Geological Society Special Publication, **62**, 217–241.

Buckman, J.O., Doughty, P.S., Benton, M.J., and Jeram, A.J. (1998) Palaeoenvironmental interpretation of the Triassic sandstones of Scrabo, County Down, Northern Ireland: ichnological and sedimentological studies indicating a mixed fluviatile–aeolian succession. *Irish Journal of Earth Sciences*, **16**, 85–102.

Burnett, R.D., Clayton, G., Haughey, N., Sevastopulo, G.D., and Sleeman, A.G. (1990) The organic maturation levels of Carboniferous rocks in south County Wexford, Ireland. *Irish Journal of Earth Sciences*, **10**, 145–156.

Burns, V. and Rickards, R.B. (1993) Silurian graptolite faunas of the Balbriggan Inlier, counties Dublin and Meath, and their evolutionary, stratigraphical and structural significance. *Proceedings of the Yorkshire Geological Society*, **49**, 283–291.

Byrne, T. and Hibbard, J. (1987) Landward vergence in accretionary prisms: the role of the backstop and thermal history. *Geology*, **15**(12), 1163–1167.

Cadman, A.C., Heaman, L., Tarney, J., Wardle, R., and Krogh, T.E. (1993) U–Pb geochronology and geochemical variation within two Proterozoic mafic dyke swarms, Labrador. *Canadian Journal of Earth Sciences*, **30**, 1490–1504.

Caldwell, W.G.E. (1959) The Lower Carboniferous rocks of the Carrick-on-Shannon syncline. *Quarterly Journal of the Geological Society of London*, **115**, 163–188.

Callaghan, B. (2005) Locating the Shannawona Fault: field and geobarometric studies from the Galway Batholith, Western Ireland. *Irish Journal of Earth Sciences*, **23**, 85–100.

Cameron, T.D.J. (1981) The history of Caledonian deformation in east Lecale, County Down. *Journal of Earth Sciences, Royal Dublin Society*, **4**, 53–74.

Cameron, T.D.J. and Anderson, T.B. (1980) Silurian metabentonites in County Down, Northern Ireland. *Geological Journal*, **15**, 59–75.

Campbell, K.J.M. (1988) *The geology of the area around Mallow, County Cork*. Unpublished M.Sc. thesis, University of Dublin.

Candela, Y. (2002) Constraints on the age of the Bardahessiagh Formation, Pomeroy, Co. Tyrone. *Scottish Journal of Geology* , **38**, 65–67.

Candela, Y. (2006) Late Ordovician brachiopod faunas from Pomeroy, Northern Ireland: a palaeoenvironmental synthesis. *Transactions of the Royal Society of Edinburgh: Earth Sciences*, **96**, 317–325.

Capewell, J.G. (1951) The Old Red Sandstone of the Inch and Annascaul district, Co. Kerry. *Proceedings of the Royal Irish Academy*, **54B**, 141–168.

Capewell, J.G. (1957a) The stratigraphy, structure and sedimentation of the Old Red Sandstone of the Comeragh Mountains and adjacent areas, County Waterford, Ireland. *Quarterly Journal of the Geological Society of London*, **112**, 393–410.

Capewell, J.G. (1957b) The stratigraphy and structure of the country around Sneem, County Kerry. *Proceedings of the Royal Irish Academy*, **58B**, 167–183.

Capewell, J.G. (1965) The Old Red Sandstone of Slieve Mish, Co. Kerry. *Proceedings of the Royal Irish Academy*, **64B**, 165–174.

Capewell, J.G. (1975) The Old Red Sandstone Group of Iveragh, County Kerry. *Proceedings of the Royal Irish Academy*, **75B**, 155–171.

Carey, M.J. (1995) *The petrology and emplacement of the Fanad Granite, County Donegal*. Unpublished Ph.D. thesis, University of Dublin.

Carlisle, H. (1979) Ordovician stratigraphy of the Tramore area, County Waterford with a revised Ordovician correlation for south-east Ireland. *In*: Harris, A.L., Holland, C.H., and Leake, B.E. (eds). *The Caledonides of the British Isles: reviewed*. Geological Society London Special Publication, **8**, 546–556.

Carruthers, R.A. (1985) *The Upper Palaeozoic geology of the Glen of Aherlow and the Galtee Mountains, counties Limerick and Tipperary*. Unpublished Ph.D. thesis, University of Dublin.

Carruthers, R.A. (1987) Aeolian sedimentation from the Galtymore Formation (Devonian) Ireland. *In*: Frostick, L. and Reid, I. (eds). *Desert sediments: ancient and modern*. Geological Society of London Special Publication, **35**, 251–268.

Carter, J.S. and Wilbur, D.G. (1986) The geological setting of Zn mineralization in the Wexford Permo-Carboniferous outlier. *In*: Andrew, C.J., Crowe, R.W.A., Finlay, S., Pennell, W.M., and Pyne, J.F. (eds). *Geology and genesis of mineral deposits in Ireland*. Irish Association for Economic Geology, Dublin, 471–474.

Cartier, C. and Faure, M. (2004) The Saint-Georges-sur-Loire olistostrome, a key zone to understand the Gondwana–Armorica boundary in the Variscan belt (Southern Brittany, France). *International Journal of Earth Sciences*, **93**, 1437–3254.

Caseldine, C. and Fyfe, R. (2006) A modelling approach to locating and characterising elm decline/landnam landscapes. *Quaternary Science Reviews*, **25**, 632–644.

Caston, V.N.D. (1995) The Helvick oil accumulation, Block 49/9, North Celtic Sea Basin. *In*: Croker, P.F. and Shannon, P.M. (eds). *The Petroleum Geology of Ireland's Offshore Basins*. Geological Society, London, Special Publications, **93**, 209–225.

Caston, V.N.D., Dearnley, R., Harrison, R.K., Rundle, C.C., and Styles, M.T. (1981) Olivine-dolerite intrusions in the Fastnet Basin. *Journal of the Geological Society of London*, **138**, 31–46.

Caulfield, S. (1978) Neolithic fields: the Irish evidence. *In*: Bowen, H.C. and Fowler, P.J. (eds). *Early land allotment in the British Isles*. BAR British Series, Oxford. **48**, 137–143.

Cawood, P.A. and Nemchin, A.A. (2001) Palaeogeographic development of the east Laurentian margin: constraints from U–Pb dating of detrital zircons in the Newfoundland Appalachians. *Geological Society of America Bulletin*, **113**, 1234–1246.

Cawood, P.A., Nemchin, A.A., and Strachan, R.A. (2007b). Provenance record of Laurentian passive margin strata in the northern Caledonides: implications for palaeodrainage and palaeogeography. *Geological Society of America Bulletin*, **119**, 993–1003.

Cawood, P.A., Nemchin, A.A., Smith, M., and Loewy, S. (2003) Source of the Dalradian Supergroup constrained by U–Pb dating of detrital zircon and implications for the East Laurentian margin. *Journal of the Geological Society of London*, **160**, 231–246.

Cawood, P.A., Nemchin, A.A., Strachan, R.A., Kinny, P.D., and Loewy, S. (2004) Laurentian provenance and an intracratonic tectonic setting for the Moine Supergroup, Scotland, constrained by detrital zircons from the Loch Eil and Glen Urquhart successions. *Journal of the Geological Society of London*, **161**, 861–874.

Cawood, P.A., Nemchin, A.A., Strachan, R.A., Prave, T., and Krabbendam, M. (2007a) Sedimentary basin and detrital zircon record along East Laurentia and Baltica during assembly and breakup of Rodinia. *Journal of the Geological Society of London*, **167**, 257–275.

Chaloner, W.G., Hill, A.J., and Lacey, W.S. (1977) First Devonian platyspermic seed and its implications in gymnosperm evolution.

Nature, **256**, 233–235.

Chambers, F.M., Daniell, J.R.G., Hunt, J.B., Molloy, K., and O'Connell, M. (2004) Tephrostratigraphy of An Loch Mór, Inis Oírr, W. Ireland: implications for Holocene tephrochronology in the northeastern Atlantic region. *The Holocene*, **14,** 703–720.

Chapin, M.A., Davies, P., Gibson, J.L., and Petingill, H.S. (1994) Reservoir architecture of turbidite sheet sandstones in laterally extensive outcrops, Ross Formation, western Ireland. *In*: Weimer, P., Bouma, A.H., and Perkins, R.F. (eds). *Submarine fans and turbidite systems, sequence stratigraphy, reservoir architecture and production characteristics*. Gulf of Mexico and International GCSSEPM Foundation 15th Annual Conference, 53–68.

Chapman, T.J., Brooks, T.M., Corcoran, D.V., Duncan, L.A., and Dancer, P.N. (1999) The structural evolution of the Erris Trough, offshore northwest Ireland, and implications for hydrocarbon generation. *In*: Fleet, A.J. and Boldy, S.A.R. (eds). *Petroleum Geology of Northwest Europe: Proceedings of the 5th Conference.* Geological Society, London, 455–469.

Chappell, B.W. and White, A.J.R. (1974) Two contrasting granite types. *Pacific Geology*, **8**, 173–174.

Charlesworth, H.A.K. (1959) *The geology of the Curlew Mountains pericline*. Unpublished Ph.D. thesis, University of Glasgow.

Charlesworth, H.A.K. (1960a) The Lower Palaeozoic inlier of the Curlew Mountains anticline. *Proceedings of the Royal Irish Academy*, **61B**, 37–50.

Charlesworth, H.A.K. (1960b) The Old Red Sandstone of the Curlew Mountains inlier. *Proceedings of the Royal Irish Academy*, **61B**, 51–58.

Charlesworth, J.K. (1928) The glacial retreat from Central and Southern Ireland. *Quarterly Journal of the Geological Society of London*, **84**, 293–345.

Charlesworth, J.K. (1929) The glacial retreat in Iar Connacht. *Proceedings of the Royal Irish Academy*, **B39**, 95–107.

Chew, D.M. (2001) Basement protrusion origin of serpentinite in the Dalradian. *Irish Journal of Earth Sciences*, **19**, 23–25.

Chew, D.M. (2003) Structural and stratigraphic relationships across the continuation of the Highland Boundary Fault in western Ireland. *Geological Magazine*, **140**(1), 73–85.

Chew, D.M. and Schaltegger, U. (2005) Constraining sinistral shearing in NW Ireland: a precise U–Pb zircon crystallisation age for the Ox Mountains Granodiorite. *Irish Journal of Earth Sciences*, **23**, 55–63.

Chew, D.M., Graham, J.R., and Whitehouse, M.J. (2007) U–Pb zircon geochronology of plagiogranites from the Lough Nafooey (= Midland Valley) arc in western Ireland: constraints on the onset of the Grampian orogeny. *Journal of the Geological Society of London*, **164**(4), 747–750.

Chew, D.M., Daly, J.S., Page, L.M., and Kennedy, M.J. (2003) Grampian orogenesis and the development of blueschist-facies metamorphism in western Ireland. *Journal of the Geological Society of London*, **160**, 911–924.

Chew, D.M., Daly, J.S., Flowerdew, M.J., Kennedy, M.J., and Page, L. M. (2004) Crenulation-slip development in a Caledonian shear zone in NW Ireland: evidence for a multi-stage movement history. *In*: Alsop, G.I., Holdsworth, R.E., McCaffrey, K.J.W., and Hand, M. (eds). *Transport and flow processes in shear zones*. Geological Society of London Special Publication, **224**. 337–352.

Chew, D.M., Page, L.M., Daly, J.S., Whitehouse, M.J., and Spikings, R. (2006) Timing of metamorphism in the Clew Bay Complex and Highland Border Complex Ophiolites. *Programme abstracts of the Highlands Workshop of the Tectonic Studies Group*, 4–5.

Chew, D.M., Flowerdew, M.J., Page, L.M., Crowley, Q.G., Daly, J.S., Cooper, M., and Whitehouse, M.J. (2008) The tectonothermal evolution and provenance of the Tyrone Central Inlier, Ireland: Grampian imbrication of an outboard Laurentian microcontinent? *Journal of the Geological Society of London*, **165**, 675–685.

Childe-Pemberton, W.S. (1925) *The Earl Bishop: the life of Frederick Hervey, Bishop of Derry, Earl of Bristol*. London, viii + 330 and 334.

Church, W.R. (1969) Metamorphic rocks of Burlington Peninsula and adjoining areas of Newfoundland, and their bearing on continental drift in North Atlantic. *In*: Kay, M. (ed.). *North Atlantic Geology and Continental Drift: a symposium*. American Association of Petroleum Geologists, Memoir **12**, 212–233.

Clark, C.D. and Meehan, R. (2001) Subglacial bedform geomorphology of the Irish Ice Sheet reveals major configuration changes during growth and decay. *Journal of Quaternary Science*, **16**(5), 483–496.

Clark, P.U., McCabe, A.M., Mix, A.C., and Weaver, A.J. (2004) Rapid sea level rise at 19,000 years ago and its global implications. *Science*, **304**, 1141–1144.

Clark, P.U., Marshall, S.J., Clarke, G.K.C., Hostetler, S.W., Licciardi, J.M., and Teller, J.T. (2001) Freshwater forcing of abrupt climate change during the last glaciation. *Science*, **293**, 283–287.

Clark, R.G., Gutmanis, J.C., Furley, A.E., and Jordan, P.G. (1981) Engineering geology for a major industrial complex at Aughinish Island, Co. Limerick, Ireland. *Quarterly Journal of Engineering Geology*, **14**, 231–239.

Claxton, C.W. (1971) Petrology, chemistry and structure of the Galway Granite in the Rosmuc Area, Co. Galway. *Proceedings of the Royal Irish Academy*, **71B**, 155–169.

Clayton, G. (1986) Late Tournaisian miospores from the Ballycultra Formation at Cultra, County Down, Northern Ireland. *Irish Journal of Earth Sciences*, **8**, 73–79.

Clayton, G. (1989) Vitrinite reflectance data from the Kinsale Harbour–Old Head of Kinsale area, southern Ireland, and its bearing on the interpretation of the Munster Basin. *Journal of the Geological Society of London*, **146**, 611–616.

Clayton, G. and Baily, H. (1999) Organic maturation levels of pre-Westphalian Carboniferous rocks in Ireland and in the Irish offshore. *Neues Jarbuch für Geologie und Paläontologie*, **107**, 25–41.

Clayton, G. and Graham, J.R. (1974) Miospore assemblages from the Devonian Sherkin Formation of Southwest County Cork, Republic of Ireland. *Pollen et Spores*, **16**, 565–588.

Clayton, G. and Higgs, K. (1979) The Tournaisian marine transgression in Ireland. *Journal of Earth Sciences, Royal Dublin Society*, **2**, 1–10.

Clayton, G., Keegan, J.B., and Sevastopulo, G.D. (1982) Palynology and stratigraphy of late Devonian and early Carboniferous Rocks, Ardmore, County Waterford. *Pollen et Spores*, **24**, 511–521.

Clayton, G., Sevastopulo, G.D., and Sleeman, A.G. (1986b) Carboniferous (Dinantian and Silesian) and Permo-Triassic rocks in south County Wexford, Ireland. *Geological Journal*, **21**, 355–374.

Clayton, G., Haughey, N., Sevastopulo, G.D., and Burnett, R. (1989) *Thermal maturation levels in the Devonian and Carboniferous rocks in Ireland*. Geological Survey of Ireland, 36pp.

Clayton, G., Johnston, I.S., Smith, D.G., and Sevastopulo, G.D. (1980b) Micropalaeontology of a Courceyan (Carboniferous) borehole section from Ballyvergin, County Clare, Ireland. *Journal of Earth Sciences of the Royal Dublin Society,* **3**, 81–100.

Clayton, G., Graham, J.R., Higgs, K, Sevastopulo, G.D., and Welsh, A. (1986a) Late Devonian and early Carboniferous paleogeography of southern Ireland and southwest Britain. *Annales de la Société géologique de Belgique,* **109**, 103–111.

Clayton, G., Graham, J.R., Higgs, K., Holland, C.H., and Naylor, D. (1980a) Devonian rocks in Ireland: a review. *Journal of Earth Sciences, Royal Dublin Society,* **2**, 161–183.

Clegg, P. and Holdsworth, R.E. (2005) Complex deformation as a result of strain partitioning in transpression zones: an example from the Leinster Terrane, SE Ireland. *Journal of the Geological Society of London,* **162**, 187–202.

Clemens, J.D. (2005) Granites and granitic magmas: strange phenomena and new perspectives on some old problems. *Proceedings of the Geologists' Association,* **116**, 9–16.

Cliff, R.A. (1985) Isotopic dating in metamorphic belts. *Journal of the Geological Society of London,* **142**, 97–110.

Cliff, R.A. and Rex, D.C. (1989) Evidence for a 'Grenville' event in the Lewisian of the northern Outer Hebrides. *Journal of the Geological Society of London,* **146**, 921–924.

Cliff, R.A., Drewery, S., and Leeder, M.R. (1991) Sourcelands for the Carboniferous Pennine river system: constraints from sedimentary evidence and U–Pb geochronology using zircon and monazite. (1991) *In*: Morton, A.C., Todd, S.P., and Haughton, P.D.W. (eds). *Developments in Sedimentary Provenance Studies.* Geological Society, London, Special Publications, **57**, 137–159.

Cliff, R.A., Yardley, B.W.D., and Bussy, F. (1996) U–Pb and Rb–Sr geochronology of magmatism and metamorphism in the Dalradian of Connemara, western Ireland. *Journal of the Geological Society of London,* **153**, 109–210.

Cliff, R.A., de Jong, K., Rex, D.C., and Guise, P.G. (1997) Evaluation of Rb–Sr hornblende dating of rocks from the Kola Peninsula: an alternative to $^{40}Ar/^{39}Ar$ where excess argon is present. *Abstract Supplement No. 1, Terra Nova,* **9**, 488.

Clift, P.D. and Ryan, P.D. (1994) Geochemical evolution of an Ordovician island arc, South Mayo, Ireland. *Journal of the Geological Society of London,* **151**, 329–342.

Clift, P.D., Draut, A.E., Hannigan, R., Layne, G., and Blusztajn, J. (2003) Trace element and Pb isotopic constraints on the provenance of the Rosroe and Derryveeny formations, South Mayo, Ireland. *Transactions of the Royal Society of Edinburgh: Earth Sciences,* **93**, 101–110.

Clipstone, D. (1992) *Biostratigraphy and lithostratigraphy of the Kilmaclenine area north County Cork.* Unpublished Ph.D. thesis, National University of Ireland.

Coats, J.S. and Wilson, J.R. (1971) The eastern end of the Galway Granite. *Mineralogical Magazine,* **38**, 138–151.

Cobbing, E.J., Manning, P.I., and Griffith, A.E. (1965) Ordovician–Dalradian unconformity in Tyrone. *Nature,* **206**, 1132–1135.

Cocks, L.R.M. and Fortey, R.A. (1982) Faunal evidence for oceanic separations in the Palaeozoic of Britain. *Journal of the Geological Society of London,* **139**, 465–478.

Cocks, L.R.M. and McKerrow, W.S. (1993) A reassessment of the early Ordovician 'Celtic' brachiopod province. *Journal of the Geological Society of London,* **150**, 1039–1042.

Cocks, L.R.M. and Torsvik, T.H. (2002) Earth geography from 500 to 400 million years ago: a faunal and palaeomagnetic review. *Journal of the Geological Society of London,* **159**, 631–644.

Cocks, L.R.M, Holland, C.H., and Rickards, R.B. (1992) *A revised correlation of Silurian rocks in the British Isles.* Geological Society, London, Special Report No.21.

Cocks, L.R.M., Holland, C.H., Rickards, R.B., and Strachan, I. (1971) A correlation of Silurian rocks in the British Isles. *Journal of the Geological Society of London,* **127**, 103–136.

Coe, K. (1966) Intrusive tuffs of West Cork, Ireland. *Quarterly Journal of the Geological Society of London,* **122**, 1–28.

Coe, K. (1969) The geology of the minor intrusions of West Cork, Ireland. *Proceedings of the Geologists' Association,* **80**, 441–457.

Coe, K. and Selwood, E.B. (1968) The Upper Palaeozoic stratigraphy of west Cork and parts of south Kerry. *Proceedings of the Royal Irish Academy,* **66B**, 113–131.

Cohen, A.S., O'Nions, R.K., Siegenthaler, R., and Griffin, W.L. (1988) Chronology of the pressure–temperature history recorded by a granulite terrain. *Contributions to Mineralogy and Petrology,* **98**, 303–311.

Colby, T.F. (1837) *Ordnance Survey of the county of Londonderry. Volume the first. Memoir of the city and north western liberties of Londonderry. Parish of Templemore.* Dublin, 363pp.

Cole, E.E. and Mitchell, F.J.G. (2003) Human impact on the Irish landscape during the late Holocene inferred from palynological studies at three peatland sites. *The Holocene,* **13**, 507–515.

Cole, G.A.J. (1894) On variolites and other tachylites at Dunmore Head, County Down. *Geological Magazine,* **1**, 222.

Cole, G.A.J. (1897) On the geology of Slieve Gallion in the County of Londonderry. *Scientific Transactions of the Royal Dublin Society,* **6**, 213–246.

Cole, G.A.J. (1900) On metamorphic rocks in eastern Tyrone and southern Donegal. *Transactions of the Royal Irish Academy,* **31**, 431–472.

Cole, G.A.J. and Crook, T. (1910) *On rock specimens dredged from the floor of the Atlantic off the coast of Ireland, and their bearing on submarine geology.* Memoirs of the Geological Survey of Ireland, iv + 36.

Colhoun, E.A. (1971) Late Weichselian periglacial phenomena of the Sperrin Mountains, Northern Ireland. *Proceedings of the Royal Irish Academy,* **71B**, 53–71.

Colhoun, E.A. (1981) A protalus rampart from the western Mourne Mountains, Northern Ireland. *Irish Geography,* **14**, 85–90.

Colhoun, E.A. and Mitchell, G.F. (1971) Interglacial marine formation and late-glacial freshwater formation in Shortalstown Townland, Co. Wexford. *Proceedings of the Royal Irish Academy,* **71B**, 211–45.

Colhoun, E.A. and Synge, F.M. (1980) The cirque moraines at Lough Nahanagan, County Wicklow, Ireland. *Proceedings of the Royal Irish Academy,* **80B**, 25–45.

Colhoun. E.A., Dickson, J.H., McCabe, A.M., and Shotton, F.W. (1972) A Middle Midlandian freshwater series at Derryvree, Maguiresbridge, County Fermanagh, Northern Ireland. *Proceedings of the Royal Society,* **B180**, 273–292.

Colin, J.P., Ioannides, N.S., and Vining, B. (1992) Mesozoic stratigraphy of the Goban Spur, offshore south-west Ireland. *Marine and Petroleum Geology,* **9**, 527–241.

Colin, J.P., Lehmann, R.A., and Morgan, B. E. (1981) Cretaceous and Late Jurassic biostratigraphy of the North Celtic Sea Basin. *In*: Neale, J. W. and Brazier, M.D. (eds). *Microfossils from Recent and fossil shelf seas.* Ellis Horwood Ltd., Chichester, 122–155.

Coller, D.W. (1984) Variscan structures in the Upper Palaeozoic rocks of west-central Ireland. *In*: Hutton, D.H.W. and Sanderson, D.J. (eds). *Variscan tectonics of the North Atlantic region.* Special

Publication of the Geological Society, London, **14**, 185–194.

Colley, M.G., McWilliams, A.S.F., and Myers, R.C. (1981) Geology of the Kinsale Head gas field, Celtic Sea, Ireland. *In*: Illing, L.V. and Hobson, G.D., (eds). *Petroleum Geology of the Continental Shelf of North-West Europe*. Heyden and Son Ltd., London, 504–510.

Collinson, J.D., Martinsen, O., Bakken, B., and Kloster, A. (1991) Early fill of the Western Irish Namurian Basin: a complex relationship between turbidites and deltas. *Basin Research*, **3**, 223–242.

Colthurst, J.R.J. (1974) The Lower Palaeozoic rocks of the Slievenamon Inlier, Tipperary. *Scientific Proceedings of the Royal Dublin Society*, **A5**, 265–276.

Colthurst, J.R. (1978) Old Red Sandstone rocks surrounding the Slievenamon inlier, counties Tipperary and Kilkenny. *Journal of Earth Sciences Royal Dublin Society*, **1**, 77–103.

Colthurst, J.R.J and Smith, D.G. (1977) Palaeontological evidence for the age of the Lower Palaeozoic rocks of the Slievenamon inlier, County Tipperary. *Proceedings of the Royal Irish Academy*, **77B**, 143–158.

Compston, W., Williams, I.S., and Meyer, C. (1984) U–Pb geochronology of lunar breccia 73217 using a sensitive high mass-resolution ion microprobe. *Journal of Geophysical Research*, **89B**, 525–534.

Condon, D.J. and Prave, A.R. (2000) Two from Donegal: Neoproterozoic glacial episodes on the northeast margin of Laurentia. *Geology*, **28**, 951–954.

Condon, D.J., Bowring, S.A., Pitcher, W.S., and Hutton, D.W.H. (2004) Rates and tempo of granitic magmatism; a U–Pb geochronological investigation of the Donegal batholith (Ireland) *In*: *Abstracts with Programs: Geological Society of America*. Geological Society of America, Denver, **36**(5), 406.

Conil, R. and Lees, A. (1974) Les transgressions Viséennes dans l'ouest de l'Irlande. Comparaisons avec la Belgique et l'Angleterre. *Annales de la Société Géologique de Belgique*, **97**, 463–484.

Connolly, A. (1999) *The palaeoecology of Clara Bog, Co. Offaly*. Unpublished Ph.D. thesis, University of Dublin.

Connolly, N. (2003) *Sedimentological analysis and correlation of the Mullaghmore Sandstone Formation in the north and north-west of Ireland*. Unpublished Ph.D. thesis, University of Dublin.

Conybeare, W.D. and Berger, J.F. (1816) On the geological features of the north-eastern counties of Ireland. *Transactions of the Geological Society of London*, **3**, 121–222.

Cook, D.R. (1987) The Goban Spur: exploration in a deep-water frontier basin. *In*: Brooks, J. and Glennie, K.W. (eds). *Petroleum Geology of North-West Europe*. Graham and Trotman Ltd, London, 623–632.

Cook, F., Matthews, D.H., and Jacob, A.W.B. (1989) Crustal and upper mantle structure of the Appalachian–Caledonide orogen from seismic results. *In*: Harris, A.L. and Fettes, D. J. (eds). *The Caledonian–Appalachian Orogen*. Geological Society Special Publication, **38**, 21–33.

Cooney, G. (2000) *Landscapes of Neolithic Ireland*. Routledge, London.

Cooney, T. (1996) Vegetation changes associated with Late-Neolithic copper mining in Killarney. *In*: Delaney, C. and Coxon, P. (eds). *Central Kerry. Field Guide No. 20*. Irish Association for Quaternary Studies, Dublin, 28–32.

Coope, G.R. (1959) A Late Pleistocene insect fauna from Chelford, Cheshire. *Proceedings of the Royal Society*, **B151**, 70–86.

Coope, G.R. (1975) Climatic fluctuations in northwest Europe since the Last Interglacial indicated by fossil assemblages of Coleoptera. *In*: Wright, A.E. and Moseley, F. (eds). *Ice ages: Ancient and Modern*, Seel House Press, Liverpool, 153–168.

Coope, G.R. (1977) Fossil coleopteran assemblages as sensitive indicators of climatic change during the Devensian (last) cold stage. *Philosophical Transactions of the Royal Society*, **B, 280**, 313–340.

Cooper, M.A., Collins, D., Ford, M., Murphy, F.X., and Trayner, P.M. (1984) Structural style, shortening estimates and the thrust front of the Irish Variscides. *In*: Hutton, D.H.W. and Sanderson, D.J. (eds). *Variscan tectonics of the North Atlantic region*. Special Publication Geological Society, London, **14**, 167–175.

Cooper, M.A., Collins, D., Ford, M., Murphy, F.X., Trayner, P.M., and O'Sullivan, M. (1986) Structural evolution of the Irish Variscides. *Journal of the Geological Society of London*, **143**, 53–61.

Cooper, M.R. (2004) Palaeogene extrusive igneous rocks. *In*: Mitchell, W.I. (ed.). *The Geology of Northern Ireland*. G.S.N.I. Belfast.

Cooper, M.R. and Johnston, T.P. (2004a) Central Highlands (Grampian) Terrane: metamorphic basement. *In*: Mitchell, W.I. (ed.). *The Geology of Northern Ireland- Our Natural Foundation (2nd edition)*. Geological Survey of Northern Ireland, Belfast, 9–24.

Cooper, M.R. and Johnston, T.P. (2004b) Late Palaeozoic intrusives. *In*: Mitchell, W.I. (ed.). *The Geology of Northern Ireland- Our Natural Foundation (2nd edition)*. Geological Survey of Northern Ireland, Belfast. 61-68.

Cooper, M.R. and Mitchell, W.I. (2004) Midland Valley Terrane. *In*: Mitchell, W.I. (ed.). *The Geology of Northern Ireland: our Natural Foundation (2nd edition)*. Geological Survey of Northern Ireland, Belfast, 25–40.

Cooper, M.R., Crowley, Q.G., and Rushton, A.W.A. (2008) New age constraints for the Ordovician Tyrone Volcanic Group, Northern Ireland. *Journal of the Geological Society of London*, **165**, 333–339.

Cope, J.C.W. (1984) The Mesozoic history of Wales. *Proceedings of the Geologists' Association*, **95**, 373–385.

Cope, J.C.W. (1994) A latest Cretaceous hotspot and the south-easterly tilt of Britain. *Journal of the Geological Society of London*, **151**, 729–731.

Cope, J.C.W. (1997) The Mesozoic and Tertiary history of the Irish Sea. *In*: Meadows, N.S., Trueblood, S.P., Hardman, M., and Cowan., G. (eds). *Petroleum Geology of the Irish Sea and Adjacent Areas*, 73–93. Geological Society, London, Special Publications **124**, 47–59.

Cope, J.C.W. (2006) Upper Cretaceous palaeogeography of the British Isles and adjacent areas. *Proceedings of the Geologists' Association*, **117**, 129–143.

Cope, J.C.W., Ingham, J.K. and Rawson, P.F. (1992a) *Atlas of Palaeogeography and Lithofacies*. Geological Society, London, Memoir **13**, 153pp.

Cope, J.C.W., Guion, P.D., Sevastopulo, G.D., and Swan, A.R.H. (1992b) Carboniferous. *In*: Cope, J.C.W., Ingham, J.K., and Rawson, P.F. (eds). *Atlas of Palaeogeography and Lithofacies*. Geological Society, London, Memoir **13**, 153pp.

Cope, J.C.W., Getty, T.A., Howarth, M.K., Morton, N., and Torrens, H.S. (1980) *A correlation of Jurassic rocks in the British Isles. Part*

1: Introduction and Lower Jurassic. Geological Society, London, Special Report, **14**, 73pp.

Cope, J.C.W., Duff, K.L., Parson, C.F., Torrens, H.S., Wimbledon, W.A., and Wright, J.K. (1980) *A correlation of Jurassic rocks in the British Isles. Part 2: Middle and Upper Jurassic*. Geological Society of London Special Report, **15**, 109pp.

Cope, R.N. (1959) The Silurian rocks of the Devilsbit Mountain district, County Tipperary. *Proceedings of the Royal Irish Academy*, **60B**, 217–242.

Corcoran, D.V. and Clayton, G. (1999) Interpretation of vitrinite reflectance profiles in the Central Irish Sea area: implications for the timing of organic maturation. *Journal of Petroleum Geology*, **22**, 261–286.

Corcoran, D.V. and Clayton, G. (2001) Interpretation of vitrinite reflectance profiles in sedimentary basins, onshore and offshore Ireland. *In*: Shannon, P.M., Haughton, P.D.W., and Corcoran, D.V. (eds). *The Petroleum Exploration of Ireland's Offshore Basins*. Geological Society, London, Special Publications **188**, 61–90.

Corcoran, D.V. and Mecklenburgh, R. (2005) Exhumation of the Corrib Gas Field, Slyne Basin, offshore Ireland. *Petroleum Geoscience*, **11**, 239–256.

Corfield, S., Murphy, N., and Parker, S. (1999) The structural and stratigraphic framework of the Rockall Trough. *In*: Fleet, A.J. and Boldy, S.A.R. (eds). *Petroleum Geology of Northwest Europe: Proceedings of the 5th Conference*. Geological Society, London, 407–420.

Corfield, S.M., Gawthorpe, R.L., Gage, M., Fraser, A.J., and Besly, B.M. (1996) Inversion tectonics of the Variscan foreland of the British Isles. *Journal of the Geological Society of London*, **153**, 17–32.

Corfu, F., Heaman, L.M., and Rogers, G. (1994) Polymetamorphic evolution of the Lewisian Complex, NW Scotland, as recorded by U–Pb isotopic compositions of zircon, titanite and rutile. *Contributions to Mineralogy and Petrology*, **117**, 215–228.

Cotter, E. (1985) Gravel-topped offshore bar sequences in the Lower Carboniferous of southern Ireland. *Sedimentary Geology*, **32**, 195–213.

Cotter, E. and Graham, J.R. (1991) Coastal plain sedimentation in the late Devonian of southern Ireland: hummocky cross-stratification in fluvial deposits? *Sedimentary Geology*, **72**, 201–224.

Coward, M.P. and Trudghill, B. (1989) Basin development and basin structure of the Celtic Sea basins (SW Britain). *Bulletin de la Société Géologique de France*, **3**, 423–436.

Cox, R.C. (ed.). (1982) *Robert Mallet, F.R.S. 1810–1881*. Institution of Engineers of Ireland, vi + 146.

Coxon, P. (1985) Quaternary Geology and Gleniff: protalus rampart. *In:* Thorn, R.H. (ed.). *Sligo and West Leitrim*. Irish Association for Quaternary Studies Field guide. **8**. (1985) IQUA. Dublin, 1–12.

Coxon, P. (1986) A radiocarbon dated early post glacial pollen diagram from a pingo remnant near Millstreet, County Cork. *Irish Journal of Earth Sciences*, **8**, 9–20.

Coxon, P. (1988) Remnant periglacial features on the summit of Truskmore, Counties Sligo and Leitrim, Ireland. *Zeitschrift für Geomorphologie*, **71**, 81–91.

Coxon, P. (1992) Dublin in the grip of an ice age: the Quaternary geology of the Dublin region. *Geographical Viewpoint*, **20**, 35–52.

Coxon, P. (1993) Irish Pleistocene biostratigraphy. *Irish Journal of Earth Sciences*, **12**, 83–105.

Coxon, P. (1996a) The Gortian Temperate Stage. *Quaternary Science Reviews*, **15**, 425–436.

Coxon, P. (1996b) The Quaternary deposits between Fenit and Spa. *In*: Delaney, C. and Coxon, P. (eds.). *Central Kerry*. Irish Association for Quaternary Studies (IQUA), Dublin, 53–65.

Coxon, P. (2001) Understanding Irish landscape evolution: pollen assemblages from Neogene and Pleistocene palaeosurfaces in western Ireland. *Proceedings of the Royal Irish Academy*, **101B** (1–2), 85–97.

Coxon, P. (2005a) The late Tertiary landscapes of western Ireland. *Irish Geography*, **38**(2), 111–127.

Coxon, P. (2005b) Site 3.1. View. Tertiary Planation (?). *In*: Coxon, P. (ed.). *The Quaternary of Central Western Ireland: Field Guide*. Quaternary Research Association, London, 112.

Coxon, P. (2005c) Site 3.2. Quarry 1. Bóthar na Scrathóg. A Neogene palaeosurface. *In*: Coxon, P. (ed.). *The Quaternary of Central Western Ireland: Field Guide*. Quaternary Research Association, London, 113–124.

Coxon, P. and Browne, P. (1991) Glacial deposits and landforms of central and western Ireland. *In*: Ehlers, J., Gibbard, P.L., and Rose, J. (eds). *Glacial Deposits in Britain and Ireland*. Balkema, Rotterdam, 355–365.

Coxon, P. and Coxon, C. (1997) A pre-Pliocene or Pliocene land surface in County Galway, Ireland. *In*: M. Widdowson. (ed). *Palaeosurfaces: Recognition, Reconstruction and Palaeoenvironmental Interpretation*. Geological Society Special Publication, **120**, 37–55.

Coxon, P. and Flegg, A. (1985) A Middle Pleistocene interglacial deposit from Ballyline, Co. Kilkenny. *Proceedings of the Royal Irish Academy*, **85B**, 107–120.

Coxon, P. and Flegg, A.M. (1987) A Late Pliocene/Early Pleistocene deposit at Pollnahallia, near Headford, Co. Galway. *Proceedings of the Royal Irish Academy*, **87B**, 15–42.

Coxon, P. and Hannon, G. (1991) The interglacial deposits at Derrynadivva and Burren Townland. *In*: Coxon, P. (ed.). *Fieldguide to the Quaternary of north Mayo*. Irish Association for Quaternary Studies (IQUA), Dublin, 24-36.

Coxon, P. and O'Callaghan, P. (1987) The distribution and age of pingo remnants in Ireland. *In*: Boardman, J. (ed.). *Periglacial processes and landforms in Britain and Ireland*. Cambridge University Press, 195–202.

Coxon, P. and Waldren, S. (1995) The floristic record of Ireland's Pleistocene temperate stages. *In*: Preece, R.C. (ed.). *Island Britain: a Quaternary Perspective*. Geological Society Special Publication, **96**, 243–268.

Coxon, P. and Waldren, S. (1997) Flora and vegetation of the Quaternary temperate stages of NW Europe: Evidence for large-scale range changes. *In*: Huntley, B, Cramer, W., Morgan, A.V., Prentice, H.C., and Allen, J.R.M. (eds). *Past and Future Rapid Environmental Changes: the Spatial and Evolutionary Responses of Terrestrial Biota. NATO ASI Series. Series I, Global Environment Change*, **47**, Springer Verlag, 103–117.

Coxon, P., Hannon, G., and Foss, P. (1994) Climatic deterioration and the end of the Gortian Interglacial in sediments from Derrynadivva and Burren Townland, near Castlebar, County Mayo, Ireland. *Journal of Quaternary Science*, **9**(1), 33–46.

Coxon, P., Mc Morrow, S., Coxon, C.E., and Nolan, T. (2005) Site 2.3. Pollnahallia: a Tertiary palaeosurface and a glimpse of Ireland's pre-Pleistocene landscape. *In*: Coxon, P. (ed.). *The Quaternary of Central Western Ireland: Field Guide*. Quaternary

Research Association, London, 79–100.

Cózar, P. and Somerville, I.D. (2004) New algal and foraminiferal assemblages and evidence for the recognition of the Asbian/Brigantian boundary in northern England. *Proceedings of the Yorkshire Geological Society*, **55**, 43–65.

Cózar, P. and Somerville, I.D. (2005) Stratigraphy of Upper Viséan carbonate platform rocks in the Carlow area, southeast Ireland. *Geological Journal*, **40**, 35–64.

Cózar, P., Somerville, H.E.A., and Somerville, I.D. (2005a) Foraminifera, calcareous algae and rugose corals in Brigantian (late Viséan) limestones in NE Ireland. *Proceedings of the Yorkshire Geological Society*, **55**, 43–65.

Cózar, P., Somerville, I.D., Aretz, M., and Herbig, H-G. (2005b) Biostratigraphical dating of Upper Viséan limestones (NW Ireland) using foraminifers, calcareous algae and rugose corals. *Irish Journal of Earth Sciences*, **23**, 1–23.

Cózar, P., Somerville, I.D., Mitchell, W.I., and Medina-Varea, P. (2006) Correlation of Mississippian (Upper Viséan) foraminifer, conodont, miospore and ammonoid zonal schemes, and correlation with the Asbian–Brigantian boundary in northwest Ireland. *Geological Journal*, **41**, 221–241.

Craig, A.J. (1978) Pollen percentage and influx analyses in South-east Ireland: a contribution to the ecological history of the late Glacial period. *Journal of Ecology*, **66**, 297–324.

Craig, L.E. (1982) *The Ordovician rocks of North Down*. Unpublished Ph.D. thesis, Queen's University Belfast.

Craig, L.E. (1984) Stratigraphy in an accretionary prism: the Ordovician rocks in North Down, Ireland. *Transactions of the Royal Society of Edinburgh: Earth Sciences*, **74**, 183–191.

Crean, E. (1987) *The Lower Carboniferous geology of south County Longford*. Unpublished M.Sc. thesis, University of Dublin.

Crimes, T.P. (1976) Trace fossils from the Bray Group (Cambrian) at Howth, Co. Dublin. *Geological Survey Ireland Bulletin*, **2**, 54–67.

Crimes, T.P. and Crossley, J.D. (1968) The stratigraphy, sedimentology, ichnology and structure of the Lower Palaeozoic rocks of part of north-eastern County Wexford. *Proceedings of the Royal Irish Academy*, **67B**, 185–215.

Crimes, T.P., Insole, A., and Williams, B.J.P. (1995) A rigid-bodied Ediacaran Biota from Upper Cambrian strata in Co. Wexford, Eire. *Geological Journal*, **30**, 89–109.

Croke, F. (ed.). (1995) *George Victor du Noyer 1817–1869: hidden landscapes*. National Gallery of Ireland, 88pp.

Croker, P.F. (1995) The Clare Basin: a geological and geophysical outline. *In*: Croker, P.F. and Shannon, P.M. (eds). *The Petroleum Geology of Ireland's Offshore Basins*. Geological Society, London, Special Publications **93**, 327–339.

Croker, P.F. and Shannon, P.M. (1987) The evolution and hydrocarbon prospectivity of the Porcupine Basin, offshore Ireland. *In*: Brooks, J. and Glennie, K.W. (eds). *Petroleum Geology of North West Europe*, Graham and Trotman, London, 633–642.

Crow, M.J. (1973) *Geology of metamorphic rocks in part of north-west County Mayo, Ireland*. Unpublished Ph.D. thesis, University of Dublin.

Crow, M.J. and Max, M.D. (1976) The Kinrovar schist. *Scientific Proceedings of the Royal Dublin Society*, **5A**, 429–441.

Crow, M.J., Max, M.D., and Sutton, J.S. (1971) Structure and stratigraphy of the metamorphic rocks in part of north-west County Mayo, Ireland. *Journal of the Geological Society of London*, **127**, 579–584.

Crowe, R.W.A. (1986) The stratigraphic and structural setting of mineralization at Newtown Cashel, Co. Longford. *In*: Andrew, C.J., Crowe, R.W.A., Finlay, S., Pennell, W.M., and Pyne, J.F. (eds). *Geology and genesis of mineral deposits in Ireland*. Irish Association for Economic Geology, Dublin, 331–339.

Crowley, J.S., Devoy, R.J.N., Linehan, D.S., and O'Flanagan, P. (2005) *Atlas of Cork City*. Cork University Press.

Crowley, Q. and Feely, M. (1997) New perspectives on the order and style of granite emplacement in the Galway batholith, western Ireland. *Geological Magazine*, **134**(4), 539–548.

Cummins, W.A. (1954) An Arenig volcanic series near Charlestown, Co. Mayo. *Geological Magazine*, **91**, 102–104.

Cunningham, G.A. and Shannon, P.M. (1997) The Erris Ridge: a major geological feature in the NW Irish Offshore Basins. *Journal of the Geological Society of London*, **154**, 503–508.

Curry, D., Adams, C.G., Boulter, M.C., Dilley, F.C., Eames, F.E., Funnell, B.M., and Wells, M.K. (1978) *A correlation of the Tertiary rocks of the British Isles*. Geological Society of London, Special Report, **12**, 72pp.

Curry, G.B., Bluck, B.J., Burton, C.J., Ingham, J.K., Siveter, D.J., and Williams, A. (1984) Age, evolution, and tectonic history of the Highland Border Complex, Scotland. *Transactions of the Royal Society of Edinburgh: Earth Sciences*, **75**, 113–133.

Cwynar, L.C. and Watts, W.A. (1989) Accelerator-mass spectrometer ages for Late-glacial events at Ballybetagh, Ireland. *Quaternary Research*, **31**, 377–380.

Dale, T.N. (1904) The geology of the Hudson Valley between the Hoosic and the Kinderhook. *United States Geological Survey Bulletin*, **242B**, 1–63.

Dallmeyer, R.D., Strachan, R.A., Rogers, G., Watt, G.R., and Friend, C.R.L. (2001) Dating deformation and cooling in the Caledonian thrust nappes of north Sutherland, Scotland: insights from $^{40}Ar/^{39}Ar$ and Rb–Sr chronology. *Journal of the Geological Society of London*, **158**, 501–512.

Daly, J.S. (1994) Proterozoic geology in western Ireland. *Post-conference excursion guide, "Precambrian Crustal evolution in the North Atlantic Regions", 17th–20th September 1994*, University College Dublin, 22pp.

Daly, J.S. (1996) Pre-Caledonian history of the Annagh Gneiss Complex, north-western Ireland, and correlation with Laurentia-Baltica. *Irish Journal of Earth Sciences*, **15**, 5–18.

Daly, J.S. (2001) The Precambrian. *In*: Holland, C.H. (ed.). *The Geology of Ireland*. Dunedin Academic Press, Edinburgh, 7–45.

Daly, J.S. and Fitzgerald, R.C. (1995) Recognition of Proterozoic terranes in the North Atlantic region. *Terra Abstracts, Abstract supplement No 2 to Terra Nova*, 7, 105.

Daly, J.S. and Flowerdew, M.J. (2005) Grampian and late-Grenville events recorded by mineral geochronology near a basement-cover contact in N. Mayo, Ireland. *Journal of the Geological Society of London*, **162**, 163–174.

Daly, J.S. and Menuge, J.F. (1989) Nd isotopic evidence for the provenance of the Dalradian Supergroup sediments in Ireland. *Terra Abstracts*, **1**, 12.

Daly, J.S. and Van den Berg, R. (2004) Ion microprobe U–Pb zircon evidence from lower crustal xenoliths for Palaeozoic crust formation and crustal development within the Caledonian suture zone in central Ireland. *Geological Society of America Abstracts with Programs*, **36**, no. 5, 48.

Daly, J.S., Flowerdew, M.J., and Whitehouse, M.J. (2004) The Slishwood Division really does have a pre-Grampian high-grade history, but is it exotic to Laurentia? *Irish Journal of Earth Sciences*, **22**, 59.

Daly, J.S., Flowerdew, M. J., and Whitehouse, M.J. (2008b) Grampian high-pressure-granulite facies metamorphism of the Slishwood Division, NW Ireland and its enigmatic eclogite-facies precursor. *Abstracts Volume, 2008 Highlands Workshop, 24th and 25th April*, 20.

Daly, J.S., Muir, R.J., and Cliff, R.A. (1991) A precise U–Pb zircon age for the Inishtrahull syenitic gneiss, Co. Donegal, Ireland. *Journal of the Geological Society of London*, **148**, 639–642.

Daly, J.S., Park, R.G., and Cliff, R.A. (1983) Rb–Sr isotopic equilibrium during Sveconorwegian (=Grenville) deformation and metamorphism of the Orust dykes, SW Sweden. *Lithos*, **16**, 307–318.

Daly, J.S., Balagansky, V.V., Timmerman, M.J., and Whitehouse, M.J. (2006) The Lapland–Kola Orogen: Palaeoproterozoic collision and accretion of the northern Fennoscandian lithosphere. *In*: Gee, D.G. and Stephenson, R. (eds). *European Lithosphere Dynamics*. Geological Society, London Memoir **32**, 579–597.

Daly, J.S., Kirkland, C.L., Lam, R. and Sylvester, P. (2008a) A hafnium isotopic perspective on the provenance and tectonic setting of allochthonous Neoproterozoic sedimentary sequences in the North Atlantic region. *Geochimica et Cosmochimica Acta*, 72, A196.

Daly, J.S., Heaman, L.M., Fitzgerald, R.C., Menuge, J.F., Brewer, T.S., and Morton, A.C. (1995) Age and crustal evolution of crystalline basement in western Ireland and Rockall. *In*: Croker, P.F. and Shannon, P.M. (eds). *The Petroleum Geology of Ireland's Offshore Basins*. Geological Society, London, Special Publications, **93**, 1995, 433–434.

Daly, J.S., Balagansky, V.V., Timmerman, M.J., Whitehouse, M.J., de Jong, K., Guise, P., Bogdanova, S., Gorbatschev, R., and Bridgwater, D. (2001) Ion microprobe U–Pb zircon geochronology and isotopic evidence for a major crustal suture in the Lapland–Kola Orogen, northern Fennoscandian Shield. *Precambrian Research*, **105**, 289–314.

Dalziel, I.W.D. (1994) Precambrian Scotland as a Laurentia-Gondwana link: origin and significance of cratonic promontories. *Geology*, **22**, 598–592.

Dalziel, I.W.D. and Soper, N.J. (2001) Neoproterozoic extension on the Scottish Promontory of Laurentia: paleogeographic and tectonic implications. *Journal of Geology*, **109**(3), 299–317.

Dancer, P.N., Algar, S.T., and Wilson, I.R. (1999) Structural Evolution of the Slyne Trough. *In*: Fleet, A.J. and Boldy, S.A.R. (eds). *Petroleum Geology of Northwest Europe: Proceedings of the 5th Conference*. Geological Society, London, 445–453.

Dancer, P.N., Kenyon-Roberts, S.M., Downey, J.W., Baillie, J.M., Meadows, N.S., and Maguire, K. (2005) The Corrib gas field, offshore west of Ireland. *In*: Doré, A.G. and Vining, B.A. (eds). *Petroleum Geology. North-West Europe and Global Perspectives*. Proceedings of the 6th Petroleum Geology Conference, 1035–1046.

Dardis, G.F. (1980) The Quaternary sediments of central Ulster. *In*: Edwards, K.J. (ed.). *Field Guide no. 3: Co. Tyrone, N. Ireland*. IQUA, Dublin, 5–29.

Dardis, G.F. (1982) *Sedimentological aspects of the Quaternary geology of south-central Ulster, Northern Ireland*. Unpublished Ph.D. thesis, Ulster Polytechnic.

Dardis, G.F., Mitchell, W.I., and Hirons, K.R. (1985) Middle Midlandian interstadial deposits at Greenagho, near Belcoo, County Fermanagh, Northern Ireland. *Irish Journal of Earth Sciences*, **7**, 1–6.

Davies, G., Gledhill, A., and Hawkesworth, C. (1985) Upper crustal recycling in southern Britain: evidence from Nd and Sr isotopes. *Earth and Planetary Science Letters*, **75**, 1–12.

Davies, G.L. (1970) The enigma of the Irish Tertiary. *In*: Stephens, N. and Glasscock, R.E. (eds). *Irish Geographical Studies*, Queen's University, Belfast, 1–16.

Davies, S.J. and Elliott, T. (1996) Spectral gamma ray characterization of high resolution sequence stratigraphy: examples from Upper Carboniferous fluvio-deltaic systems, County Clare, Ireland. *In*: Howell, J.A. and Aitken, J.F. (eds). *High resolution sequence stratigraphy: innovations and applications*. Geological Society Special Publication, **104**, 23–35.

Davison, C. (1924) *A History of British Earthquakes*. Cambridge University Press, 416pp.

Davydov, V., Wardlaw, B.R., and Gradstein, F.M. (2005) The Carboniferous Period. *In*: Gradstein, F.M., Ogg, J.G., and Smith, A.G. (eds). *A geologic time scale 2004*. Cambridge University Press, 222–248.

De Beaulieu, J.L. and Reille, M. (1995) Pollen records from the Velay craters: a review and correlation of the Holsteinian Interglacial with isotopic stage 11. *Mededelingen van de Geologische Dienst*, **52**, 59–70.

De Raaf, J.F.M. and Boersma, J.R. (1971) Tidal deposits and their sedimentary structures. *Geologie en Mijnbouw*, **50**, 479–504.

De Raaf, J.F.M., Boersma, J.R., and Van Gelder, A. (1977) Wave generated structures and sequences from a shallow marine succession, Lower Carboniferous, County Cork, Ireland. *Sedimentology*, **24**, 451–483.

Dean, W.T. (1971—1978) *The trilobites of the Chair of Kildare Limestone (Upper Ordovician) of eastern Ireland*. Palaeontographical Society Monograph.

Dehantschutter, J. (1995) *Aspects of the sedimentology and stratigraphy of the Waulsortian mudbanks of the Dublin Basin, Republic of Ireland*. Unpublished Ph.D. thesis, University of Dublin.

Delanty, L.J., Whittington, R.J., and Dobson, M.R. (1981) The geology of the North Celtic Sea west of 7° Longitude. *Proceedings of the Royal Irish Academy*, **81B**, 37–51.

Délépine, G. (1949) Le Carbonifère d'Irlande et son faciès waulsortiens. *Annales Hébert et Haug*, **7**, 143–159.

Dempster, T.J. (1985) Uplift patterns and orogenic evolution in the Scottish Dalradian. *Journal of the Geological Society of London*, **142**, 111–128.

Dempster, T.J., Hudson, N.F.C., and Rogers, G. (1995) Metamorphism and cooling of the NE Dalradian. *Journal of the Geological Society of London*, **152**, 383–390.

Dempster, T.J., Rogers, G., Tanner, P.W.G., Bluck, B.J., Muir, R.J., Redwood, S.D., Ireland, T.R., and Paterson, B.A. (2002) *Journal of the Geological Society of London*, **159**, 83–94.

DePaolo, D.J. (1981) Neodymium isotopes in the Colorado Front Range and crust–mantle evolution in the Proterozoic. *Nature, (London)*, **291**, 193–196.

Devuyst, F.X. (2006) *The Tournaisian–Visean boundary in Eurasia: definition, biostratigraphy, sedimentology and early evolution of the genus* Eoparastaffella *(foraminifer)*. Unpublished Ph.D dissertation, Université Catholique de Louvain.

Devuyyst, F.X. and Lees, A. (2001) The initiation of Waulsortian buildups in western Ireland. *Sedimentology*, **48**, 1121–1148.

Devuyst, F.X., Hance, L., Hou, H., Wu, X., Tian, S., Coen, M., and

Sevastopulo, G. (2003) A proposed global stratotype section and point for the base of the Viséan Stage (Carboniferous): the Pengchong section, Guangxi, South China. *Episodes*, **26**, 105–115.

Dewey, J.F. (1961) A note concerning the age of the metamorphism of the Dalradian rocks of western Ireland. *Geological Magazine*, **98**, 399–404.

Dewey, J.F. (1962) The provenance and emplacement of Upper Arenigian turbidites in Co. Mayo, Eire. *Geological Magazine*, **99**, 223–234.

Dewey, J.F. (1963) The Lower Palaeozoic stratigraphy of central Murrisk, County Mayo, Ireland, and the evolution of the south Mayo trough. *Quarterly Journal of the Geological Society of London*, **119**, 313–344.

Dewey, J.F. (2005) Orogeny can be very short. *Proceedings National Academy Sciences*, **102**, 15286–15293.

Dewey, J.F. and Mange, M.A. (1999) Petrography of Ordovician and Silurian sediments in the western Irish Caledonides: tracers of a short-lived Ordovician continent–arc collision orogeny and the evolution of the Laurentian Appalachian–Caledonian margin. *In*: MacNiocaill, C. and Ryan, P.D. (eds). *Continental Tectonics*. Geological Society London Special Publication, **164**, 55–107.

Dewey, J.F. and McManus, J. (1964) Superposed folding in the Silurian rocks of Co. Mayo, Eire. *Geological Journal*, **4**, 61–76.

Dewey, J.F. and Ryan, P.D. (1990) The Ordovician evolution of the South Mayo Trough, western Ireland. *Tectonics*, **9**(4), 887–901.

Dewey, J.F. and Shackleton, R.M. (1984) A model for the evolution of the Grampian tract in the early Caledonides and Appalachians. *Nature*, **312**, 115–121.

Dewey, J.F. and Strachan, R.A. (2003) Changing Silurian–Devonian relative plate motion in the Caledonides: sinistral transpression to sinistral transtension. *Journal of the Geological Society of London*, **160**, 219–229.

Dewey, J.F., Rickards, R.B., and Skevington, D.G. (1970) New light on the age of the Dalradian deformation and metamorphism in western Ireland. *Norsk Geologisk Tidsskrift*, **50**, 19–54.

Dhonau, N.B. (1973) The Lower Palaeozoic rocks of the Bannow coast, Co. Wexford. *Geological Survey Ireland Bulletin*, **1**, 231–243.

Dhonau, N.B. and Holland, C.H. (1974) The Cambrian of Ireland. *In*: Holland, C.H. (ed.). *Lower Palaeozoic Rocks of the World, Volume 2, Cambrian of the British Isles, Norden, and Spitsbergen*. John Wiley and Sons, London, 157–176.

Dickin, A.P. (2005) *Radiogenic Isotope Geology*. 2nd Edition. Cambridge University Press, 492pp.

Diemer, J.A. and Bridge, J.S. (1987) Tide-dominated coastal deposits associated with the Tournaisian marine transgression in southwest Ireland. *In*: De Boer, P.L., Van Gelder, A., and Nio, S.D. (eds). *Tide-influenced sedimentary environments and facies*. Reidel Publishing Company.

Diemer, J.A. and Bridge, J.S. (1988) Transition from alluvial plain to tide-dominated coastal deposits associated with the Tournaisian marine transgression in southwest Ireland. *In*: De Boer *et al.* (eds). *Tide-influenced sedimentary environments and facies*. Reidel Publishing Company, 359–388.

Diemer, J.A., Bridge, J.S., and Sanderson, D.J. (1987) Revised geology of Kerry Head, County Kerry. *Irish Journal of Earth Sciences*, **8**, 113–138.

Dixon, O.A. (1972) Lower Carboniferous rocks between the Curlew and Ox Mountains, Northwestern Ireland. *Journal of the Geological Society of London*, **128**, 71-101.

Dobson, M.R. (1977a) The geological structure of the Irish Sea. *In*: Kidson, C. and Tooley, M.J. (eds). *The Quaternary History of the Irish Sea*. Seal House Press, Liverpool, 13–26.

Dobson, M.R. (1977b) The history of the Irish Sea basins. *In*: Kidson, C. and Tooley M.J. (eds). *The Quaternary History of the Irish Sea*, Seal House, Press, Liverpool, 93–98.

Dobson, M.R. and Evans, D. (1974) Geological structure of the Malin Sea. *Journal of the Geological Society of London*, **130**, 475–478.

Dobson, M.R. and Whittington, R.J. (1979) Geology of the Kish Bank Basin. *Journal of Geological Society of London*, **136**, 243–249.

Dodd, C. (1986) *The sedimentology of fluvio-aeolian interactions with geological examples from the British Isles*. Unpublished Ph.D. thesis, Council for National Academic Awards, Plymouth Polytechnic.

Dodson, J.R. (1990) The Holocene vegetation of a prehistorically inhabited valley, Dingle peninsula, Co. Kerry. *Proceedings of the Royal Irish Academy*, **90B**, 151–174.

Dodson, M.H., Compston, W., Williams, I.S., and Wilson, J.F. (1988) A search for ancient detrital zircons in Zimbabwean sediments. *Journal of the Geological Society of London*, **145**, 977–983.

Doherty, D.J.P. (1983) *The Lower Basalt Formation of the South Antrim Escarpment*. Ph.D thesis, Queen's University of Belfast.

Dolan, J.M. (1984) A structural cross-section through the Carboniferous of north-west Kerry. *Irish Journal of Earth Sciences*, **6**, 95–108.

Donovan, S.K. (1986) Pentameric pelmatazoan columns from the Arenig Tourmakeady Limestone, County Mayo. *Irish Journal of Earth Sciences*, **7**, 93–98.

Doran, R.J.P. (1971) *The Palaeozoic rocks between Tipperary town and Milestone, Co. Tipperary*. Unpublished Ph.D. thesis, University of Dublin.

Doran, R.J.P.(1974) The Silurian rocks of the southern part of the Slieve Phelim inlier, Co. Tipperary. *Proceedings of the Royal Irish Academy*, **74B**, 193–202.

Doran, R.J.P., Holland, C.H., and Jackson, A.A. (1973) The sub-Old Red Sandstone surface in southern Ireland. *Proceedings of the Royal Irish Academy*, **73B**, 109–128.

Doré, A.G., Lundin, E.R., Fichler, C., and Olesen, O. (1997a) Patterns of basement structure and reactivation along NE Atlantic margin. *Journal of the Geological Society of London*, **154**, 85–92.

Doré, A.G., Lundin, E.R., Jensen, L.N., Birkeland, Ø, Eliassen, P.E., and Fichler, C. (1999) Principal tectonic events in the evolution of the northwest European Atlantic margin. *In*: Fleet, A.G. and Boldy, S.A.R (eds). *Petroleum Geology of Northwest Europe: Proceedings of the 5th Conference*. Geological Society, London, 41–61.

Doré, A.G., Lundin, E.R., Jensen, L.N., Birkeland, Ø, Eliassen, P.E., and Jensen, L.N. (1997b) The NE Atlantic Margin: implications of late Mesozoic and Cenozoic events for hydrocarbon prospectivity. *Petroleum Geoscience*, **3**, 117–131.

Douglas, J.A. (1909) The Carboniferous Limestone of County Clare (Ireland). *Quarterly Journal of the Geological Society of London*, **65**, 538–586.

Dowling, L.A. (1997) *An investigation of interglacial sediments in Cork Harbour, Southern Ireland*. Unpublished PhD. thesis, University of Dublin.

Dowling, L.A. and Coxon, P. (2001) Current understanding of Pleistocene temperate stages in Ireland. *Quaternary Science Reviews*, **20**, 1631–1642.

Dowling, L.A., Sejrup, H.P., Coxon, P., and Heijnis, H. (1998) Palynology, aminostratigraphy, and U-series dating of marine Gortian Interglacial sediments in Cork Harbour, southern Ireland. *Quaternary Science Reviews*, **17**, 945–962.

Downie, C. (1984) *Acritarchs in British stratigraphy*. Geological Society London, Special Report, **17**, 26pp.

Downie, C. and Tremlett, W.E. (1968) Micropalaeontological evidence on the age of the Clara Group (south-east Ireland). *Geological Magazine*, **105**, 401.

Doyle, E., Bowden, A.A., Jones, G.V., and Stanley, G.A. (1992) The geology of the Galmoy zinc–lead deposit, Co. Kilkenny. *In*: Bowden, A.A., Earls, G., O'Connor, P.G., and Pyne, J.F. (eds). *The Irish minerals industry 1980–1990*. Irish Association for Economic Geology, Dublin, 211–225.

Doyle, E.N. (1994) Storm-dominated sedimentation along a rocky transgressive shoreline in the Silurian (Llandovery) of Western Ireland. *Geological Journal*, **21**, 193–207.

Draut, A.E. and Clift, P.D. (2001) Geochemical evolution of arc magmatism during arc–continent collision, South Mayo, Ireland. *Geology*, **29**(6), 543–546.

Draut, A.E. and Clift, P.D. (2002) The origin and significance of the Delaney Dome Formation, Connemara, Ireland. *Journal of the Geological Society of London*, **159**, 95–103.

Draut, A.E., Clift, P.D., Chew, D.M., Cooper, M.J., Taylor, R.N., and Hannigan, R.E. (2004) Laurentian crustal recycling in the Ordovician Grampian Orogeny: Nd isotopic evidence from western Ireland. *Geological Magazine*, **141**(2), 195–207.

Drew, D.P. and Jones, G.Ll. (2000) Post-Carboniferous pre-Quaternary karstification in Ireland. *Proceedings of the Geologists' Association*, **111**, 345–353.

Drewery, S., Cliff, R.A., and Leeder, M.R. (1987) Provenance of Carboniferous sandstones from U–Pb dating of detrital zircons. *Nature*, **325**, 50–53.

Duncan, W.I., Green, P.F., and Duddy, I.R. (1998) Source rock burial history and seal effectiveness: key facets in understanding hydrocarbon exploration potential in the East and Central Irish Sea basins. *AAPG Bulletin*, **82**, 1401-1415.

Dunford, G.M., Dancer, P.N., and Long, K.D. (2001) Hydrocarbon potential of the Kish Bank Basin: integration within a regional model for the Greater Irish Sea Basin. *In*: Shannon, P.M., Haughton, P.D.W., and Corcoran, D.V. (eds). *The Petroleum Exploration of Ireland's Offshore Basins*. Geological Society, London, Special Publications **188**, 135–154.

Dury, G.H. (1959) A contribution to the geomorphology of Central Donegal. *Proceedings of the Geologists' Association*, **70**, 1–27.

Eagar, R.M.C. (1974) Neuere arbeiten über das Westfal in Irland. *Zentralblatt für Geologie und Paläontologie Stuttgart*. Teil, **1**, 291–308.

Edwards, D., Feehan, J., and Smith D.G. (1983) A late Wenlock flora from Co. Tipperary, Ireland. *Botanical Journal Linnéan Society*, **86**,19–36.

Edwards, K.J. (1993) Models of mid-Holocene forest farming for north-west Europe. *In*: Chambers, F.M. (ed.). *Climate change and human impact on the landscape*. Chapman and Hall, London, 133–146.

Egan, F.W., McHenry, A., and Clark, B. (1899) *In*: *Memoir of the Geological Survey, summary of progress for 1898*, 57–58.

El Desouky, M., Feely, M., and Mohr, P. (1996) Diorite–granite magma mixing along the axis of the Galway granite batholith, Ireland. *Journal of the Geological Society of London*, **153**, 361–374.

Elles, G.L. and Wood, E.M.R. (1910—1918) *A monograph of British graptolites*. Palaeontographical Society Monograph, 530pp.

Elliott, G.M., Shannon, P.M., Haughton, P.D.W., Praeg, D., and O'Reilly, B. (2006) Mid- to late Cenozoic canyon development on the eastern margin of the Rockall Trough, offshore Ireland. *Marine Geology*, **229**, 113–132.

Elliott, T. (2000a) Megaflute erosion surfaces and the initiation of turbidite channels. *Geology*, **28**, 119–122.

Elliott, T. (2000b) Depositional architecture of a sand-rich, channelized turbidite system: the Upper Carboniferous Ross Sandstone Formation, western Ireland. *In*: Weimer, P., Slatt, R.M., Coleman, J., Rosen, N.C., Nelson, H., Bouma, A.H., Styzen, M.J., and Lawrence, D.T. (eds). *Deep-Water Reservoirs of the World*. GCS-SEPM Foundation, 20th Annual Bob F. Perkins Research Conference, 342–373.

Elliott, T. and Pulham, A.J. (1990) Incised valley fills and the sequence stratigraphy of Upper Carboniferous shelf-margin deltaic cyclothems. *American Association of Petroleum Geologists Bulletin*, **75**, 874.

Elwell, R.W.D. (1958) Granophyre and hybrid pipes in a dolerite layer of Slieve Gullion. *Journal of Geology*, **66**, 57–71.

Emeleus, C.H. (1955) The granites of the Western Mourne Mountains, Co. Down. *Scientific Proceedings of the Royal Dublin Society*, **27**, 35–50.

Emeleus, C.H. (1962) The porphyritic felsite of the Tertiary ring complex of Slieve Gullion, County Armagh. *Proceedings of the Royal Irish Academy*, **62B**, 55–76.

Emeleus, C.H. and Preston, J. (1969) *Field Excursion Guide: Tertiary Volcanic Rocks of Ireland*. Belfast.

Emo, G.T. (1978) *The Lower Palaeozoic and Old Red Sandstone rocks of Slieve Aughty, Counties Clare and Galway*. Unpublished Ph.D. thesis, University of Dublin.

Emo, G.T. (1986) Some considerations regarding the styles of mineralisation at Harberton Bridge, County Kildare. *In*: Andrew, C.J., Crowe, R.W.A., Finlay, S., Pennell, W.M., and Pyne, J.F. (eds). *Geology and genesis of mineral deposits in Ireland*. Irish Association for Economic Geology, Dublin, 461–469.

Emo, G.T. and Smith, D.G. (1978) Palynological evidence for the age of the Lower Palaeozoic rocks of Slieve Aughty, counties Clare and Galway. *Proceedings of the Royal Irish Academy*, **78B**, 281–292.

Emslie, R.F. and Hunt, P.A. (1990) Age and petrogenetic significance of igneous mangerite–charnockite suites associated with massif anorthosites, Grenville Province. *Journal of Geology*, **98**, 213–31.

Etu-Efeotor, J.D. (1976) Geology of the Kish Bank Basin. *Journal of the Geological Society of London*, **132**, 708.

Evans, A. and Clayton, G. (1998) The geological history of the Ballydeenlea Chalk Breccia, County Kerry, Ireland. *Marine and Petroleum Geology*, **15**, 299–307.

Evans, D. (1983) *Malin sheet 55°N–08°W 1:250,000 Series*. Institute of Geological Sciences Map.

Evans, D., Kenolty, N., Dobson, M. R., and Whittington, R. J. (1980) *The geology of the Malin Sea*. Institute of Geological Sciences,

Report **79/15**, 44pp.

Evans, D.H. and Graham, J.R. (2001) *Peltoclados clarus* Gen. and Sp. Nov., an Enigmatic Fossil from Clare Island with a Possible Charophyte Relationship. *In*: Graham, J. R. (ed). *New Survey of Clare Island. Volume 2: Geology*. Royal Irish Academy, 49–61.

Evans, D.J.A. and Ó'Cofaigh, C. (2003) Depositional evidence for marginal oscillations of the Irish Sea ice stream in southeast Ireland during the last glaciation. *Boreas*, **32**, 1, 76—101.

Evans, J.A., Fitches, W.R., and Muir, R.J. (1998) Laurentian clasts in a Neoproterozoic tillite from Scotland. *Journal of Geology*, **106**, 361–366.

Eyles, C.H. (1988) Glacially- and tidally-influenced shallow marine sedimentation of the late Precambrian Port Askaig Formation, Scotland. *Palaeogeography, Palaeoclimatology, Palaeoecology*, **68**, 1–25.

Eyles, C.H. and Eyles, N. (1983) Glaciomarine model for upper Precambrian diamictites of the Port Askaig Formation, Scotland. *Geology*, **11**, 692–696.

Eyles, N. and McCabe, A.M. (1989a) The Late Devensian (<22,000 bp) Irish Sea Basin: the sedimentary record of a collapsed ice sheet margin. *Quaternary Science Reviews*, **8**, 307–351.

Eyles, N. and McCabe, A.M. (1989b) Glaciomarine facies within subglacial tunnel valleys: the sedimentary record of glacio-isostatic downwarping in the Irish Sea Basin. *Sedimentology*, **36**, 431–448.

Eyles, V. (1952) *The composition and origin of the Antrim laterites and bauxites*. Memoir of the Geological Survey of Northern Ireland.

Faber, S. and Bamford, D. (1979) Lithospheric structural contrasts across the Caledonides of northern Britain. *Tectonophysics*, **56**(1–2), 17–30.

Fairchild, I.J. (1985) Comment on 'Glaciomarine model for upper Precambrian diamictites of the Precambrian diamictites of the Port Askaig Formation, Scotland.' *Geology*, **13**, 89.

Farrington, A. (1957) Glacial Lake Blessington. *Irish Geography*, **3**, 216–222.

Faure, G. (1986) *Principles of Isotope Geology*. Second Edition. John Wiley, New York, 589pp.

Fearnsides, W.G., Elles, G.L., and Smith, B. (1907) The Lower Palaeozoic rocks of Pomeroy. *Proceedings Royal Irish Academy*, **26B**, 97–128.

Fedo, C.M., Sircombe, K.N., and Rainbird, R.H. (2003) Detrital zircon analysis of the sedimentary record. *In*: Hanchar, J.M. and Hoskin, P.W.O. (eds). Zircon. *Reviews in Mineralogy and Geochemistry*, **53**, 277–303.

Feehan, J. (1982a) The Silurian rocks of the Slieve Bloom mountains, counties Laois and Offaly. *Proceedings of the Royal Irish Academy*, **82B**, 153–167.

Feehan, J. (1982b) The Old Red Sandstone rocks of the Slieve Bloom and northeastern Devilsbit Mountains, counties Laois, Offaly, and Tipperary. *Journal of Earth Sciences, Royal Dublin Society*, **5**, 11–30.

Feely, M. and Madden, J. (1987) The spatial distribution of K, U, Th and surface heat production in the Galway granite, Connemara, western Ireland. *Irish Journal of Earth Sciences*, **8**, 155–164.

Feely, M., McCabe, E., and Kunzendorf, H. (1991) The evolution of REE profiles in the Galway Granite, western Ireland. *Irish Journal of Earth Sciences*, **11**, 71–89.

Feely, M., Coleman, D., Baxter, S., and Miller, B. (2003) U–Pb zircon geochronology of the Galway Granite, Connemara, Ireland: implications for the timing of late Caledonian tectonic and magmatic events and for correlations with Acadian plutonism in New England. *Atlantic Geology*, **39**, 175–184.

Feely, M.M.J. (1982) *Geological, geochemical and geophysical studies on the Galway granite in the Costelloe/Inveran sector, western Ireland*. Unpublished Ph.D. thesis, National University of Ireland.

Ferretti, A., Holland, C.H., and Syba, E. (1993) Problematical microfossils from the Silurian of Ireland and Scotland. *Palaeontology*, **36**, 771–783.

Ffrench, G.D. and Williams, D.M. (1984) The sedimentology of the South Connemara Group, western Ireland: a possible Ordovician trench-fill sequence. *Geological Magazine*, **121**, 505–514.

Finch, T.F. (ed.). (1977) *Guidebook for INQUA excursion C16: Western Ireland*. Geo Abstracts, Norwich, 39pp.

Finch, T.F. and Synge, F.M. (1966) The drifts and soils of west Clare and adjoining parts of counties Kerry and Limerick. *Irish Geography*, **5**, 161–172.

Fitches, W.R., Muir, R.J., Maltman, A.J., and Bentley, M.R. (1990) Is the Colonsay–west Islay block of SW Scotland an alochthonous terrane? Evidence from Dalradian tillite clasts. *Journal of the Geological Society of London*, **147**, 417–420.

Fitches, W.R., Pearce, N.J.G., Evans, J.A., and Muir, R.J. (1996) Provenance of late Proterozoic Dalradian tillite clasts, Inner Hebrides, Scotland. *In*: Brewer, T.S. (ed.). *Precambrian crustal evolution in the North Atlantic Region*. Geological Society, London, Special Publications, **112**, 367–377.

Fitzgerald, E., Feely, M., Johnston, J.D., Clayton, Fitzgerald, L.J., and Sevastopulo, G.D. (1994) The Variscan thermal history of west Clare, Ireland. *Geological Magazine*, **131**, 545–558.

Fitzgerald, R.C., Daly, J.S., Menuge, J.F., and Brewer, T.S. (1993) Palaeoproterozoic evolution of the Annagh Gneiss Complex. *Abstracts, Annual Irish Geological Research Meeting*, 16–17.

Fitzgerald, R.C., Daly, J.S., Menuge, J.F., and Brewer, T.S. (1994) Mayo metabasites: a guide to Dalradian/Basement relationships. *Abstracts, Annual Irish Geology Research Meeting, University of Ulster at Coleraine*, 18.

Fitzgerald, R.C., Daly, J.S., Menuge, J.F., and Brewer, T.S. (1996) Petrogenesis of the Annagh Gneiss Complex, NW County Mayo, Ireland. *Abstracts, Annual Irish Geology Research Meeting*, University College Dublin, 23.

Fletcher, T.P. (1977) *Lithostratigraphy of the Chalk (Ulster White Limestone Formation) in Northern Ireland*. Report of the Institute of Geological Sciences, No. 77/24, 33pp.

Floodpage, J., Newman, P., and White, J. (2001) Hydrocarbon prospectivity in the Irish Sea area: insights from recent exploration in the Central Irish Sea, Solway and Peel basins. *In*: Shannon, P.M., Haughton, P.D.W., and Corocoran, D.V. (eds). *The Petroleum Exploration of Ireland's Offshore Basins*. Geological Society, London, Special Publications **188**, 107–134.

Flowerdew, M.J. (1998) Tonalite bodies and basement-cover relationships in the North-eastern Ox Mountains Inlier, northwestern Ireland. *Irish Journal of Earth Sciences*, **17**, 71–82.

Flowerdew, M.J. (2000) *The thermal history of Proterozoic rocks in the Caledonides of NW Ireland and the response of mineral dating systems to deformation*. Unpublished Ph.D. thesis, University College Dublin.

Flowerdew, M.J. and Daly, J.S. (1997) Nd isotopes and the complicated tectonothermal history of the Slishwood Division, NW Ireland. *Abstracts, Annual Irish Geology Research Meeting*,

Queen's University Belfast, 12.

Flowerdew, M.J. and Daly, J.S. (1999) Indenter tectonics in the Caledonides of Ireland: geochronological and structural evidence. *Journal of Conference Abstracts*, **4**, 83.

Flowerdew, M.J. and Daly, J.S. (2005) Sm–Nd mineral ages and P–T constraints on the pre-Grampian high grade metamorphism of the Slishwood Division, north-west Ireland. *Irish Journal of Earth Sciences*, **23**, 107–123.

Flowerdew, M.J., Daly, J.S., and Rex, D.C. (1995) Timing of metamorphic events in the orthotectonic Caledonides of NW Ireland. *Abstracts, Metamorphic Studies Group Meeting on "Isotopic systematics and metamorphic processes"*, Open University, 9.

Flowerdew, M.J., Daly, J.S., and Rex, D.C. (1997) Comparative isotopic dating systems of metamorphic hornblende: deformation induced resetting of Ar–Ar: fluid resetting of Rb–Sr systems? *Abstract supplement No. 1 to Terra Nova*, **9**, 488.

Flowerdew, M.J., Daly, J.S., and Riley, T.R. (2007) New Rb–Sr mineral ages temporally link plume events with accretion at the margin of Gondwana. *In*: Cooper, A.K., Raymond, C.R., et al. (eds.). A keystone in a changing world: online proceedings of the 10th ISAES. *USGS Open File Report 2007–1047, Short Research Paper 012*, 4p. doi: 10.3133/0f2007–1047.srp012.

Flowerdew, M.J., Daly, J.S., and Whitehouse, M.J. (2005) 470 Ma granitoid magmatism associated with the Grampian Orogeny in the Slishwood Division, NW Ireland. *Journal of the Geological Society of London*, **162**, 563–575.

Flowerdew, M.J., Daly, J.S., Guise, P.G., and Rex, D.C. (2000) Isotopic dating of overthrusting, collapse and related granitoid intrusion in the Grampian orogenic belt, northwestern Ireland. *Geological Magazine*, **137**(4), 419–435.

Flowerdew, M.J., Daly, J.S., Chew, D.M., Millar, I.L., and Horstwood, M.S.A. (2008) *In situ* Hf isotopic measurements of complex zircons from Irish granitoids reveal hidden Palaeoproterozoic and Archaean sources at depth. *Abstracts Volume, 2008 Highland Workshop 24th, 25th April*, 24.

Floyd, J.D. (2001) The Southern Uplands terrane: a stratigraphical review. *Transactions of the Royal Society of Edinburgh: Earth Sciences*, **91**, 349–362.

Forbes, E. (1848) On *Oldhamia*, a new genus of Silurian fossils. *Journal of the Geological Society Dublin*, **4**, 20.

Ford, M. (1990) The stratigraphy and structure of the Galley Head Culmination Zone: an area of enhanced shortening related to basin geometry within the Irish Variscides. *Geological Journal*, **25**, 145–159.

Ford, M. and Ferguson, C.C. (1985) Cleavage strain in the Variscan Fold Belt, County Cork, Ireland, estimated from stretched arsenopyrite rosettes. *Journal of Structural Geology*, **7**, 217–223.

Ford, M., Brown, C., and Readman, P. (1991) Analysis and tectonic interpretation of gravity data over the Variscides of southwest Ireland. *Journal of the Geological Society of London*, **148**, 137–148.

Ford, M., Klemperer, S.L., and Ryan, P.D. (1992) Deep structure of southern Ireland: a new geological synthesis using BIRPS deep reflection profiling. *Journal of the Geological Society of London*, **149**, 915–922.

Fortey, R.A. (1984) Global earlier Ordovician transgressions and regressions and their biological implications. *In*: Bruton, D.L. (ed.). Aspects of the Ordovician System. *Palaeontological Contributions University of Oslo*, **295**, 37–50.

Fortey, R.A. and Cocks, L.R.M. (1988) Arenig to Llandovery faunal distributions in the Caledonides. *In*: Harris, A.L. and Fettes, D.J. (eds). *The Caledonian–Appalachian orogen*. Geological Society London Special Publication, **38**, 233–246.

Fortey, R.A. and Cocks, L.R.M. (2005) Late Ordovician global warming: the Boda Event. *Geology*, **33**, 405–408.

Fossitt, J.A. (1994) Late-glacial and Holocene vegetation history of western Donegal, Ireland. *Biology and Environment: Proceedings of the Royal Irish Academy*, **94B**, 1–31.

Fothergill, B. (1974) *The mitred earl: an eighteenth century eccentric*. London, 254pp.

Fowler, A. and Robbie, J.A. (1961) *Geology of the country around Dungannon*. Memoir of the Geological Survey of Northern Ireland, Her Majesty's Stationery Office, Belfast, **35**, 274pp.

Francis, E.H. (1978) The Midland Valley as a rift, seen in connection with the late Palaeozoic European rift system. *In*: Ramberg, I. and Neumann, E.R. (eds). *Tectonics and geophysics of continental rifts*. Leidal, Holland, 133–147.

Francis, P.E., Lyle, P., and Preston, J. (1986) A tholeiitic andesite flow unit among the Causeway Basalts of North Antrim in Northern Ireland. *Geological Magazine*, **123**, 105–112.

Franke, W. (1989) Tectonostratigraphic units in the Variscan belt of Central Europe. *In*: Dallmeyer, R.D. (ed.). *Terranes in the Circum-Atlantic Palaeozoic Orogens*. Geological Society of America, Special Papers, **230**, 221–228.

Fraser, R. (1801) *General view of the agriculture and mineralogy, present state and circumstances of the county Wicklow, with observations on the means of their improvement*. Dublin, xvi + 289.

French, W.J. (1977) Breccia-pipes associated with Ardara Pluton, County Donegal. *Proceedings of the Royal Irish Academy*. **B**, 77(4), 101–117.

Friedrich, A.M. (1998) *$^{40}Ar/^{39}Ar$ and U–Pb geochronological constraints on the thermal and tectonic evolution of the Connemara Caledonides, Western Ireland*. Unpublished Ph.D. thesis, Massachusetts Institute of Technology.

Friedrich, A.M., Bowring, S.A., and Hodges, K.V. (1997) U–Pb geochronological constraints on the duration of arc magmatism and metamorphism from Connemara, Irish Caledonides. *Abstract supplement No. 1 to Terra Nova*, **9**, 331.

Friedrich, A.M., Bowring, S.A., Martin, M.W., and Hodges, K.V. (1999a) Short-lived continental magmatic arc at Connemara, western Ireland Caledonides: implications for the age of the Grampian orogeny. *Geology*, **27**(1), 27-30.

Friedrich, A.M., Hodges, K.V., Bowring, S.A., and Martin, M.W. (1999b) Geochronological constraints on the magmatic, metamorphic and thermal evolution of the Connemara Caledonides, western Ireland. *Journal of the Geological Society of London*, **156**, 1217–1230.

Friend, C., Kinny, P., Rogers, G., Strachan, R., and Paterson, B.A. (1997) U–Pb zircon geochronological evidence for Neoproterozoic events in the Glenfinnan Group (Moine Supergroup): the formation of the Ardgour granite gneiss, northwest Scotland. *Contributions to Mineralogy and Petrology*, **128**, 101–113.

Friend, C.R.L., Strachan, R.A., Kinny, P.D., and Watt, G.R. (2003) Provenance of the Moine Supergroup of NW Scotland: evidence from geochronology of detrital and inherited zircons from (meta)sedimentary rocks, granites and migmatites. *Journal of the Geological Society of London*, **160**, 247–257.

Friend, P.F. and McDonald, R. (1988) Volcanic sediments, stratigraphy and tectonic background of the Old Red Sandstone of

Kintyre, W Scotland. *Scottish Journal of Geology*, **4**, 265–82.

Friend, P.F., Williams, B.P.J., Ford, M., and Williams, E.A. (2000) Kinematics and dynamics of Old Red Sandstone Basins. *In*: Friend, P.F and Williams, B.P.J. (eds). *New perspectives on the Old Red Sandstone*. Geological Society of London, Special Publications **180**, 29–60.

Fritz,W.J. and Stillman, C.J. (1996) A subaqueous welded tuff from the Ordovician of County Waterford, Ireland. *Journal Volcanology Geothermal Research*, **70**, 91–106.

Frost, C.D. (1993) Nd isotopic evidence for the antiquity of the Wyoming province. *Geology*, **21**, 351–354.

Fusciardi, L.P. (1995) Carboniferous stratigraphy and mineralization in three drillholes from the Innishannon area, South Munster Basin, Co. Cork. *In*: Anderson, K., Ashton, J., Earls, G., Hitzman, M.W., and Tear, S. (eds). *Irish carbonate-hosted Zn–Pb deposits*. Society of Economic Geologists Guidebook Series, **21**, 103–109.

Fyfe, J.A., Long, D., and Evans, D. (1993) *United Kingdom offshore regional report: The geology of the Malin-Hebrides Sea area*. London HMSO for the British Geological Survey.

Gallagher, C. and Thorp, M. (1997) The age of the Pleistocene raised beach near Fethard, County Wexford, using infrared stimulated luminescence (IRSL). *Irish Geography*, **30**(2), 68–89.

Gallagher, S.J. (1996) The stratigraphy and cyclicity of the late Dinantian platform carbonates in parts of southern and western Ireland. *In*: Strogen, P., Somerville, I.D., and Jones, G.Ll. (eds). *Recent advances in Lower Carboniferous geology*. Geological Society Special Publication, **107**, 239–251.

Gallagher, S.J. and Elsdon, R.E. (1990) Spinel lherzolite and other xenoliths from a Tertiary dolerite dyke in southwest Donegal. *Geological Magazine*, **127**, 177–180.

Gallagher, S.J. and Somerville, I.D. (1997) Late Dinantian (Lower Carboniferous) platform carbonate stratigraphy of the Buttevant area, North County Cork, Ireland. *Geological Journal*, **32**, 313–335.

Gallagher, S.J. and Somerville, I.D. (2003) Lower Carboniferous (Late Viséan) platform development and cyclicity in southern Ireland: foraminiferal biofacies and lithofacies evidence. *Rivista Italiana di Palaeontologia e Stratigrafia*, **109**, 159–171.

Gallagher, S.J., MacDermot, C.V., Somerville , I.D., Pracht, M., and Sleeman, A.G. (2006) Biostratigraphy, microfacies and depositional environments of Upper Viséan limestones from the Burren region, County Clare, Ireland. *Geological Journal*, **41**, 61–91.

Gallagher, V., Feely, M., Hogelsberger, H., Jenkin, G.R.T., and Fallick, A.E. (1992) Geological, fluid inclusion and stable isotope studies of Mo mineralization, Galway Granite, Ireland. *Mineralium Deposita*, **27**(4), 314–325.

Gamble, J.A. (1982) The petrology and geochemistry of the Caledonian Newry Granodiorite from the Tertiary igneous complex of Slieve Gullion, North-east Ireland. *Journal of Earth Sciences, Royal Dublin Society*, **5**, 91–105.

Gamble, J.A., Old, R.A., and Preston, J. (1976) *Subsurface exploration in the Tertiary Central Complex of gabbro and granophyre at Slieve Gullion, Co. Armagh, Northern Ireland*. Institute of Geological Sciences, Report No 76/8.

Gardiner, C.I and Reynolds, S.H. (1896) The Kildare inlier. *Quarterly Journal of the Geological Society of London*, **52**, 587–605.

Gardiner, C.I. and Reynolds, S.H. (1897) An account of the Portrane inlier, County Dublin. *Quarterly Journal of the Geological Society of London*, **53**, 520–539.

Gardiner, C.I. and Reynolds, S.H. (1898) The Bala Beds and associated igneous rocks of Lambay Island, County Dublin. *Quarterly Journal of the Geological Society of London*, **54**, 135–148.

Gardiner, C.I. and Reynolds, S.H. (1902) The fossiliferous Silurian beds and associated igneous rocks of the Clogher Head district (Co. Kerry). *Quarterly Journal of the Geological Society of London*, **58**, 226–266.

Gardiner, C.I. and Reynolds, S.H. (1909) On the igneous and associated sedimentary rocks of the Tourmakeady district (County Mayo). *Quarterly Journal of the Geological Society of London*, **65**, 104–154.

Gardiner, C.I. and Reynolds, S.H. (1910) The igneous and associated sedimentary rocks of the Glensaul district (County Galway). *Quarterly Journal of the Geological Society of London*, **66**, 253–280.

Gardiner, C.I. and Reynolds, S.H. (1912) The Ordovician and Silurian rocks of the Kilbride Peninsula (Mayo). *Quarterly Journal of the Geological Society of London*, **68**, 75–102.

Gardiner, C.I. and Reynolds, S.H. (1914) The Ordovician and Silurian rocks of the Lough Nafooey area (Co. Galway). *Quarterly Journal of the Geological Society of London*, **70**, 104–118.

Gardiner, P.R.R. (1978) *Day 2: The Duncannon district: Cambro-Ordovician flysch and Ordovician volcanic sequences*. Geological Survey Ireland Guide Series, **2**, 25–40.

Gardiner, P.R.R. and Horne, R.R. (1981) The stratigraphy of the Upper Devonian and Lower Carboniferous clastic sequences in southwest County Wexford. *Geological Survey of Ireland Bulletin*, **3**, 51–77.

Gardiner, P.R.R. and McArdle, P. (1992) The geological setting of Permian gypsum and anhydrite deposits in the Kingscourt district, counties Cavan, Meath and Monaghan. *In*: Bowden, A.A., Earls, G., O'Connor, P.G., and Pyne, J.F. (eds). *The Irish Mineral Industry 1980–1990*. Irish Association for Economic Geology, 301–316.

Gardiner, P.R.R. and Robinson, K.W. (1970) The geology of Ireland's Eye: the stratigraphy and structure of a part of the Bray Group. *Geological Survey Ireland Bulletin*, **1**, 3–22.

Gardiner, P.R.R. and Sheridan, D.J.R. (1981) Tectonic framework of the Celtic Sea and adjacent areas with special reference to the location of the Variscan Front. *Journal of Structural Geology*, **3**, 317–331.

Gardiner, P.R.R. and Vanguestaine, M. (1971) Cambrian and Ordovician microfossils from south-east Ireland and their implications. *Geological Survey Ireland Bulletin*, **1**, 163–210.

Gardiner, P.R.R. and Visscher, H. (1971) The Permian–Triassic Transition sequence at Kingscourt, Ireland. *Nature*, **299**, 209–210.

Gatley, S., Somerville, I.D., Morris, J.H., Sleeman, A.G., and Emo, G. (2005) *Geology of Galway–Offaly and adjacent parts of Westmeath, Tipperary, Laois, Clare and Roscommon: a geological description to accompany the bedrock geology 1:100,000 scale map series, Sheet 15, Galway–Offaly; with contributions by W. Cox (Minerals), T. Hunter-Williams (Groundwater) and R. van den Berg and E. Sweeney (Carboniferous volcanics); edited by A.G. Sleeman*. Geological Survey of Ireland, vii + 90pp.

Gaudette, H.E., Vitrac-Michard, A., and Allègre, C.J. (1981) North American history recorded in a single sample: high-resolution U–Pb systematics of the Potsdam Sandstone detrital zircons, New York State. *Earth and Planetary Science Letters*, **54**, 248–260.

Geikie, A. (1891) The anniversary address of the president. *Proceedings of the Geological Society. In: Quarterly Journal of the Geological Society of London*, **47**, 48–162.

Geikie, A. (1893) On the pre-Cambrian rocks of the British Isles. *J. Geol.* **1**, 1–14.

Geikie, A. (1897) *The ancient volcanoes of Great Britain, Volumes 1 and 2*, McMillan & Co., London.

Gennard, D.E. (1984) A palaeoecological study of the interglacial deposit at Benburb, Co. Tyrone. *Proceedings of the Royal Irish Academy*, **84B**, 43-56.

Gennard, D.E. (1986) Aghnadarragh. *In*: McCabe, A.M. and Hirons, K.R. (eds). *South-east Ulster Field guide*. Quaternary Research Association, Cambridge, 142–168.

Geological Survey of Northern Ireland (1971) *Magnetic anomaly map of Northern Ireland.*

Geological Survey of Northern Ireland (1995) *1:50,000 solid geology of sheet 33 (Omagh)*. HMSO.

Geological Survey of Northern Ireland (1997) *Northern Ireland. Solid Geology*. (Second Edition). 1:250,000. Keyworth, Nottingham, British Geological Survey.

George, T.N. (1958) Lower Carboniferous palaeogeography of the British Isles. *Proceedings of the Yorkshire Geological Society*, **31**, 227–318.

George, T.N. (1960) The Lower Carboniferous rocks in County Wexford. *Quarterly Journal of the Geological Society of London*, **116**, 349–364.

George, T.N. (1966) Geomorphic evolution in Hebridean Scotland. *Scottish Journal of Geology*, **2**, 1–34.

George, T.N. (1967) Landform and structure in Ulster. *Scottish Journal of Geology*, **3**, 413–418.

George, T.N. and Oswald, D.H. (1957) The Carboniferous rocks of the Donegal Syncline. *Quarterly Journal of the Geological Society of London*, **113**, 137–183.

George, T.N., Johnson, G.A.L., Mitchell, M., Prentice, J.E., Ramsbottom, W.H.C., Sevastopulo, G.D., and Wilson, R.B. (1976) *A correlation of Dinantian rocks in the British Isles*. Geological Society of London Special Report, **7**, 87pp.

Geraghty, M. (1997) *Geology of Monaghan–Carlingford: a geological description to accompany the bedrock geology 1:100,000 scale map series, Sheet 8/9, Monaghan–Carlingford with contributions by I. Farrelly, K. Claringbold (Minerals), C. Jordan, R. Meehan (Quaternary) and M. Hudson (Groundwater)*. Geological Survey of Ireland, 60pp.

Gibbons, W. and Horák, J. (1996) The evolution of the Neoproterozoic Avalonian subduction system: evidence from the British Isles. *In*: Nance, R.D. and Thompson, M.D. (eds). *Avalonian and related peri-Gondwanan terranes of the Circum-North Atlantic*, Geological Society of America, Special Paper 304, Boulder, Colorado, 289-280.

Gibson, D. (1984) *The petrology and geochemistry of the western Mourne granites, Co. Down*. Unpublished Ph.D. thesis, Queen's University of Belfast.

Gibson, D, Lux, D.R., and Meighan, I.G. (1995) New $^{40}Ar/^{39}Ar$ ages for the Mourne Mountains granites, north-east Ireland. *Irish Journal of Earth Sciences*, **14**, 25–35.

Gibson, D., McCormick, A.G., Meighan, I.G., and Halliday, A.N. (1987) The British Tertiary Igneous Province: young Rb–Sr ages for the Mourne Mountains Granites. *Scottish Journal of Geology*, **23**, 221–225.

Gill, R.C.O. (ed.). (1997) *Modern Analytical Geochemistry*. Addison Wesley Longman, 329pp.

Gill, W.D. (1962) The Variscan Fold Belt in Ireland *In*: Coe, K. (ed.). *Some Aspects of the Variscan Fold Belt*. Manchester University Press, 44–64.

Gill, W.D. (1979) *Syndepositional sliding and slumping in the west Clare Namurian basin, Ireland*. Geological Survey of Ireland Special Paper, **4**, 31pp.

Gillespie, P.A., Walsh, J.J., Watterson, J., Bonson, C.G., and Manzocchi, T. (2001) Scaling relationships of joint and vein arrays from The Burren, Co. Clare, Ireland. *Journal of Structural Geology*, **23**, 183–201.

Gillot, J.E. (1951) *Geology of the Slane area, Co. Meath, Ireland*. Unpublished M.Sc. thesis, University of Liverpool.

Gilmore, G. (1992) Scroll coprolites from the Silurian of Ireland and the feeding of early Vertebrates. *Palaeontology*, **35**, 319–333.

Glodny, J., Kühn, A., and Austrheim, H. (2008) Diffusion versus recrystallization processes in Rb–Sr geochronology: isotopic relics in eclogite facies rocks, Western Gneiss Region, Norway. *Geochimica et Cosmochimica Acta*, **72**, 506–525.

Glodny, J., Bingen, B., Austrheim, H., Molna, J.F., and Rusin, A. (2002) Precise eclogitisation ages deduced from Rb/Sr mineral systematics: the Maksyutov complex, Southern Urals, Russia. *Geochimica et Cosmochimica Acta*, **66**, 1221–1235.

Glover, B.W., Key, R.M., May, F., Clark, G.C., Phillips, E.R., and Chacksfield, B.C. (1995) A Neoproterozoic multi-phase rift sequence: the Grampian and Appin groups of the southwestern Monadhliath Mountains of Scotland. *Journal of the Geological Society of London*, **142**, 391–406.

Goldstein, S.J. and Jacobsen, S.B. (1988) Nd and Sr isotopic systematics of river water suspended material: implications for crustal evolution. *Earth and Planetary Science Letters*, **87**, 249–65.

Goodhue, R. and Clayton, G. (1999) Organic maturation levels, thermal history and hydrocarbon source rock potential of the Namurian rocks of the Clare Basin, Ireland. *Marine and Petroleum Geology*, **16**, 667–675.

Gorbatschev, R. and Bogdanova, S. (1993) Frontiers in the Baltic Shield. *Precambrian Research*, **64**, 3–21.

Gould, R.J. (2004) *Antrim Lava Field: flow patterns and provenance of interbasaltic zircons*. Unpublished Ph.D. thesis, University of Dublin.

Gower, C.F. (1996) The evolution of the Grenville Province in eastern Labrador, Canada. *In*: Brewer, T.S. (ed.). *Precambrian crustal evolution in the North Atlantic Region*. Geological Society, London, Special Publications, **112**, 197–218.

Gower, C.F. and Krogh, T.E. (2002) A U–Pb geochronological review of the Proterozoic history of the eastern Grenville Province. *Canadian Journal of Earth Sciences*, **39**, 795–829.

Gower, P.J. (1973) *The middle–upper Dalradian boundary with special reference to the Loch Tay Limestone*. Unpublished Ph.D. thesis, University of Liverpool.

Graciansky, P.C. de, Poag, W.C., Cunningham, R. *et al.* (1985) The Goban Spur transect: geologic evolution of a sediment-starved passive continental margin. *Bulletin of the Geological Society of America*, **96**, 58–76.

Gradstein, F.M., Ogg, J.C., and Smith, A.G. (2004) *A geological timescale 2004*. Cambridge University Press, 589pp.

Graham, J.R. (1972) *The sedimentation of Devonian and Carboniferous rocks in south-west County Cork*. Unpublished Ph.D. thesis,

University of Exeter.

Graham, J.R. (1975a) Deposits of a near-coastal fluvial plain: the Toe Head Formation (Upper Devonian) of southwest County Cork, Eire. *Sedimentary Geology*, **14**, 45–61.

Graham, J.R. (1975b) Analysis of an Upper Palaeozoic transgressive sequence in southwest County Cork, Eire. *Sedimentary Geology*, **13**, 267–290.

Graham, J.R. (1981a) Fluvial sedimentation in the Lower Carboniferous of Clew Bay, County Mayo, Ireland. *Sedimentology*, **30**, 195–211.

Graham, J.R. (1981b) The 'Old Red Sandstone' of County Mayo, northwest Ireland. *Geological Journal*, **16**, 157–173.

Graham, J.R. (1983) Analysis of the Upper Devonian Munster Basin, an example of a fluvial distributary system. *In*: Collinson, J.D. and Lewin, J. (eds). *Modern and ancient fluvial systems*. Special Publication of the International Association of Sedimentologists, **6**, 473–483.

Graham, J.R. (1987) The nature and field relations of the Ordovician Maumtrasna Formation, County Mayo, Ireland. *Geological Journal*, **22**, 347–369.

Graham, J.R. (1996) Dinantian river systems and coastal zone sedimentation in northwest Ireland. *In*: Strogen, P., Somerville, I.D., and Jones, G.Ll. (eds). *Recent advances in Lower Carboniferous geology*. Geological Society Special Publication, **107**, 183–206.

Graham, J.R. (2001a) The Silurian Rocks of Southern Clare Island. *In*: Graham, J.R. (ed.). *New Survey of Clare Island. Volume 2: Geology*. Royal Irish Academy, Dublin, 35–48.

Graham, J.R. (2001b) The geology of Clare Island: perspectives and problems. *In*: Graham, J.R. (ed.). *New Survey of Clare Island. Volume 2: Geology*. Royal Irish Academy, Dublin, 1–17.

Graham, J.R. and Clayton, G. (1988) Devonian rocks in Ireland and their relation to adjacent regions. *In*: McMillan, N.J., Embry, A.F., and Glass, D.J. (eds). *Devonian of the world, Vol. 1*. Canadian Society of Petroleum Geologists, Memoir, **12**, 325–340.

Graham, J.R. and Clayton, G. (1994) Late Tournaisian conglomerates from County Donegal, north-west Ireland: fault-controlled sedimentation and overstep during basin extension. *Irish Journal of Earth Sciences*, **13**, 95–105.

Graham, J.R. and Reilly, T.A. (1972) The Sherkin Formation (Devonian) of south-west County Cork. *Geological Survey of Ireland Bulletin*, **1**, 281–300.

Graham, J.R. and Reilly, T.A. (1976) The stratigraphy of the area around Clonakilty Bay, south County Cork. *Proceedings of the Royal Irish Academy*, **76B**, 379–391.

Graham, J.R. and Smith, D.G. (1981) The age and significance of a small Lower Palaeozoic inlier in County Mayo. *Journal of Earth Sciences, Royal Dublin Society*, **4**, 1–6.

Graham, J.R., James, A., and Russell, K.J. (1992) Basin history deduced from subtle changes in fluvial style: a study of distal alluvium from the Devonian of southwest Ireland. *Transactions of the Royal Society of Edinburgh: Earth Sciences*, **83**, 655–667.

Graham, J.R., Leake, B.E., and Ryan, P.D. (1985) *The geology of South Mayo (map), 1:63,360*. University of Glasgow.

Graham, J.R., Leake, B.E., and Ryan, P.D. (1989) *The geology of South Mayo, western Ireland*. Scottish Academic Press, Edinburgh.

Graham, J.R., Richardson, J.B., and Clayton, G. (1983) Age and significance of the Old Red Sandstone around Clew Bay, NW Ireland. *Transactions of the Royal Society of Edinburgh: Earth Sciences*, **73**, 245–249.

Graham, J. R., Russell, K.J., and Stillman, C.J. (1995) Late Devonian magmatism in West Kerry and its relationship to the development of the Munster Basin. *Irish Journal of Earth Sciences*, **14**, 7–23.

Graham, J.R., Wrafter, J.P., Daly, J.S., and Menuge, J.F. (1991) A local source for the Ordovician Derryveeny Formation, western Ireland: implications for the Connemara Dalradian. *In*: Morton, A.C., Todd, S.P., and Haughton, P.D.W. (eds). *Developments in Sedimentary Provenance Studies*. Geological Society, London, Special Publications. **57**, 199–213.

Grant, N.K. (1961) *The structure and petrology of the north-eastern end of the Newry Complex and the adjacent sediments*. Unpublished Ph.D. thesis, University of Edinburgh.

Gray, J.M. and Coxon, P. (1991) The Loch Lomond Stadial Glaciation in Britain and Ireland. *In*: Ehlers, J., Gibbard, P.L., and Rose, J. (eds). *Glacial Deposits in Britain and Ireland*. Balkema, Rotterdam, 89-105.

Gray, J.R. and Yardley, B.W.D. (1979) A Caledonian blueschist from the Irish Dalradian. *Nature*, **278**, 736–737.

Green, P.F., Duddy, I.R., and Bray, R.J. (1997) Variation in thermal styles around the Irish Sea and adjacent areas: implications for hydrocarbon occurrence and tectonic evolution. *In*: Meadows, N.S., Trueblood, S.P., Hardman, M., and Cowan, G. (eds). *Petroleum Geology of the Irish Sea and Adjacent Areas*. Geological Society, London, Special Publications **124**, 73–93.

Green, P.F., Duddy, I.R, Hegarty, K.A., and Bray, R.J. (1999) Early Tertiary heat flow along the UK Atlantic margin and adjacent areas. *In*: Fleet, A.J. and Boldy, S.A.R. (eds). *Petroleum Geology of Northwest Europe: Proceedings of the 5th Conference*. Geological Society, London, 348–357.

Green, P.F., Duddy, I.R., Bray, R.J., Duncan, W.I., and Corcoran, D.V. (2001) The influence of thermal history on hydrocarbon prospectivity in the Central Irish Sea Basin. *In*: Shannon, P.M., Haughton, P.D.W., and Corcoran, D.V. (eds). *The Petroleum Exploration of Ireland's Offshore Basins*. Geological Society, London, Special Publications **188**, 171–188.

Green, P.F., Duddy, I.R., Hegarty, K.A., Bray, R.J., Sevastopulo, G., Clayton, G., and Johnston, D. (2000) The post-Carboniferous evolution of Ireland: evidence from thermal history reconstruction. *Proceedings of the Geologists' Association*, **111**, 307–320.

Gregg, R.P. and Lettsom, W.G. (1858) *Manual of the mineralogy of Great Britain and Ireland*, London.

Greig, J.A. *et al.* (1971) Lead and sulphur isotopes of the Irish base metal mines in carbonate host rocks. *Society of Mining Geology, Japan, Special Issue*, **2**, 84–92.

Grennan, E. (1992) The Gangevlin gypsum deposit, Co. Cavan. *In*: Bowden, A.A., Earls, G., O'Connor, P.G., and Pyne, J.F. (eds). *The Irish Minerals Industry 1980–1990*. Irish Association for Economic Geology, Dublin, 317–325.

Griffith, A.E. (1970) A review of the Upper Old Red Sandstone and Tournaisian rocks in Northern Ireland. *Comptes Rendus 6ème Congrès de Stratigraphie et de Géologie du Carbonifère, Sheffield 1967*, **2**, 837–841.

Griffith, A.E. and Wilson, H.E. (1982) *Geology of the country around Carrickfergus and Bangor*. Memoir for one inch geological sheet 29, Geological Survey of Northern Ireland.

Griffith, R. (1814) *Geological and mining report on the Leinster Coal District*. Dublin, xxiv + 135.

Griffith, R. (1818) *Geological and mining survey of the Connaught Coal District in Ireland*. Dublin, vii + 108.

Griffith, R. (1829) *Geological and mining surveys of the coal districts of the counties of Tyrone and Antrim in Ireland.* Dublin, ix + 77.

Griffith, R. (1838) *Geological map of Ireland to accompany the report of the Railway Commissioners, 1837. Shewing the different lines laid down under the direction of the Commissioners and those proposed by joint stock companies,* in *Atlas to accompany 2D report of the Railway Commissioners Ireland 1838.*

Griffith, R. (1839) *A general map of Ireland to accompany the Report of the Railway Commissioners shewing the principal physical features and geological structure of the country. Scale One Inch to Four Miles.* Hodges and Smith, Dublin; James Gardner, London.

GRIP (Greenland Ice-core Project Members). (1993) Climate instability during the last interglacial period recorded in the GRIP ice core. *Nature,* **364**, 203–207.

Grogan, S.E. and Reavy, R.J. (2002) Disequilibrium textures in the Leinster Granite Complex, SE Ireland: evidence for acid–acid magma mixing. *Mineralogical Magazine,* **66**(6), 929–939.

Grotzinger, J.P., Bowring, S.A., Saylor, B.Z., and Kaufman, A.J. (1995) Biostratigraphic and geochronologic constraints on early animal evolution, *Science,* **270**, 598–604.

Gwiazda, R.H., Hemming, S.R., and Broecker, W.S. (1996) Tracking the sources of icebergs with lead isotopes: the provenance of ice-rafter debris in Heinrich layer 2. *Palaeoceanography,* **11**, 77–93.

Haflidason, H., Sejrup, H.P., Kristensen, D.K., and Johnsen, S. (1995) Coupled response of the late glacial climatic shifts of northwest Europe reflected in Greenland ice cores: evidence from the northern North Sea. *Geology,* **23**(12), 1059–1062.

Haigh, W.D. (1914) The Carboniferous volcanoes of Philipstown in Kings County. *Proceedings of the Royal Irish Academy,* **22B**, 17–33.

Hald, M., Steinsund, P.J., Dokken, T., Korsun, S., Polyak, L., and Aspeli, R. (1994) *Recent and Late Quaternary distribution of* Elphidium exclavatum *f. cf.* clavatum *in Arctic seas.* Cushman Foundation Special Publication, **32**, 141–155.

Hall, A. and Walsh, J.N. (1971) The beryls of the Rosses district, Donegal. *Mineralogical Magazine,* **38**, 328–334.

Hall, V.A. (1995) Woodland depletion in Ireland over the last millennium. *In*: Pilcher, J.R. and Mac an T-Saoir, S.S. (eds). *Wood, Trees and Forests in Ireland.* Royal Irish Academy, Dublin, 23–33.

Hall, V.A. and Mauquoy, D. (2005) Tephra-dated climate- and human-impact studies during the last 1500 years from a raised bog in central Ireland. *Holocene,* **15**, 1086–1093.

Hall, V.A. and Pilcher, J.R. (2002) Late-Quaternary Icelandic tephras in Ireland and Great Britain: detection, characterization and usefulness. *Holocene,* **12**, 223–230.

Halliday, A.N., Aftalion, M., and Leake, B.E. (1980) A revised age for the Donegal granites. *Nature,* **284**(5756), 542–543.

Halliday, A.N., Graham, C.R., Aftalion, M., and Dymoke, P. (1989) The depositional age of the Dalradian Supergroup: U–Pb and Sm–Nd isotopic studies of the Tayvallich Volcanics, Scotland. *Journal of the Geological Society of London,* **146**, 3–6.

Hamilton, W. (1786) *Letters concerning the northern coast of the county of Antrim. Containing a natural history of its basaltes: with an account of such circumstances as are worthy of notice representing the antiquities, manners and customs of that country.* Dublin, viii + 195.

Hampson, G.J., Elliott, T., and Davies, S.J. (1997) The application of sequence stratigraphy to Upper Carboniferous fluvio-deltaic strata of the onshore UK and Ireland: implications for the southern North Sea. *Journal of the Geological Society of London,* **154**, 719–733.

Hance, L. and Poty, E. (2006) Hastrian. *In*: Dejonghe, L. (ed.). Chronostratigraphic units named from Belgium and adjacent areas. *Geologica Belgica,* **9**, 111–116.

Hance, L., Poty, E., and Devuyst, F-X. (2006) Ivorian. *In*: Dejonghe, L. (ed.). Chronostratigraphic units named from Belgium and adjacent areas. *Geologica Belgica,* **9**, 117–122.

Hanchar, J.M. and Hoskin, P.W.O. (eds.). (2003) Zircon. *Reviews in Mineralogy and Geochemistry,* **53**, 500pp.

Hancock, J.M. (1963) The hardness of the Irish Chalk. *Irish Naturalists' Journal,* **14**, 157–164.

Hancock, J.M. and Rawson, P.F. (1992) Cretaceous. *In*: Cope, J.C.W., Ingham, J.K., and Rawson, P.F. (eds). *Atlas of Palaeogeography and Lithofacies.* Geological Society, London, Memoir **13**, 153pp.

Harkin, J., Williams, D.M., Menuge, J.F., and Daly, J.S. (1996) Turbidites from the Clew Bay Complex, Ireland: provenance based on petrography, geochemistry and crustal residence values. *Geological Journal,* **31**, 379–388.

Harney, S., Long, C.B., and MacDermot, C.V. (1996) *Geology of Sligo–Leitrim. Sheet 7, Bedrock Geology 1:100,000 Map Series.* Geological Survey of Ireland.

Harper, D.A.T. (ed.). (1988) *William King D.Sc. (1809–1886): a palaeontological tribute.* Galway University Press, viii + 90.

Harper, D.A.T. (1992) Ordovician provincial signals from Appalachian–Caledonian terranes. *Terra Nova,* **4**, 204–209.

Harper, D.A.T. and Mitchell, W.I. (1982) Upper Ordovician (Ashgill) brachiopods from the Oriel Brook Formation, County Louth. *Journal of Earth Sciences, Royal Dublin Society,* **5**, 31–35.

Harper, D.A.T. and Parkes, M.A. (1989) Palaeontological constraints on the definition and development of Irish Caledonide terranes. *Journal of the Geological Society of London,* **146**, 413–415.

Harper, D.A.T. and Parkes, M.A. (2000a) Ireland. *In*: Fortey, R.A., Harper, D.A.T., Ingham, J.K., Owen, A.W., Parkes, M.A., Rushton, A.W.A., and Woodcock, N.H. (eds). *A revised correlation of Ordovician rocks in the British Isles.* Geological Society Special Report, **24**, 52–64.

Harper, D.A.T. and Parkes, M.A. (2000b) Late Ordovician fossils from Mweelrea Mountain, South Mayo, western Ireland; Abstracts of the 43rd annual Irish geological research meeting. *Irish Journal of Earth Sciences,* **18**, 134.

Harper, D.A.T., MacNiocaill, C., and Williams, S.H. (1996) The palaeogeography of early Ordovician Iapetus terranes: an integration of faunal and palaeomagnetic constraints. *Palaeogeography, Palaeoclimatology, Palaeoecology,* **121**, 297–312.

Harper, D.A.T., Mitchell, W.I., and Rong, J. (1994) New faunal data from the highest Ordovician rocks at Pomeroy, County Tyrone, Northern Ireland. *Scottish Journal of Geology,* **30**, 187–190.

Harper, D.A.T., Scrutton, C.T., and Williams, D.M. (1995) Mass mortalities on an Irish Silurian seafloor. *Journal of Geological Society of London,* **152**, 917–922.

Harper, D.A.T., Williams, D.M., and Armstrong, H.A. (1989) Stratigraphical correlations adjacent to the Highland Boundary fault in the west of Ireland. *Journal of the Geological Society of London,* **146**, 381–384.

Harper, D.A.T., Graham, J.R., Owen, A.W., and Donovan, S.K. (1988) An Ordovician fauna from Lough Shee, Partry Mountains, Co. Mayo, Ireland. *Geological Journal,* **23**, 293–310.

Harper, D.A.T., Mitchell, W.I., Owen, A.W., and Romano, M. (1985) Upper Ordovician brachiopods and trilobites from the Clashford House Formation, near Herbertstown, Co. Meath, Ireland. *Bulletin British Museum (Natural History) Geology*, **38**, 287–308.

Harper, D.A.T., Parkes, M.A., Hoey, A.N., and Murphy, F.C. (1990) Intra-Iapetus brachiopods from the Ordovician of Eastern Ireland: implications for Caledonide correlation. *Canadian Journal of Earth Sciences*, **27**(12), 1757–1761.

Harper, J.C. (1948) The Ordovician and Silurian rocks of Ireland. *Proceedings Liverpool Geological Society*, **20**, 48–67.

Harper, J.C. (1952) The Ordovician rocks between Collon (Co. Louth) and Grangegeeth (Co. Meath). *Scientific Proceedings of the Royal Dublin Society*, **26**, 85–112.

Harper, J.C. and Brenchley, P.J. (1972) Some points of interest concerning the Silurian inliers of southwest central Ireland in their geosynclinal context: a statement. *Journal of the Geological Society of London*, **128**, 257–262.

Harper, J.C. and Hartley, J.J. (1938) The Silurian inlier of Lisbellaw, County Fermanagh, with a note on the age of the Fintona Beds. *Proceedings of the Royal Irish Academy*, **45B**, 73–87.

Harper, J.C. and Rast, N.J. (1964) The faunal succession and volcanic rocks of the Ordovician near Bellewstown, County Meath. *Proceedings of the Royal Irish Academy*, **64B**, 1–23.

Harper, J.C. and Romano, M. (1967) *Decordinaspis*: a new Caradoc trinucleid trilobite from the Ordovician of Ireland. *Proceedings of the Royal Irish Academy*, **65B**, 305–308.

Harris, A.L. and Pitcher, W.S. (1975) The Dalradian Supergroup. *In*: Harris, A.L., Shackleton, R.M., Watson, J., Downie, C., Harland, W.B and Moorbath, S. (eds). *A correlation of Precambrian rocks in the British Isles*. Geological Society of London, Special Report, **6**, 52–75.

Harris, A.L., Bradbury, H.J., and McGonigal, M.H. (1976) The evolution and transport of the Tay Nappe. *Scottish Journal of Geology*, **12**, 103–113.

Harris, A.L., Baldwin, C.T., Bradbury, H.J., Johnson, H.D., and Smith, R.A. (1978) Ensialic basin sedimentation: the Dalradian Supergroup. *In*: Bowes, D.R. and Leake, B.E. (eds). *Crustal evolution in north-western Britain*, Seel House Press, Liverpool, 115–138.

Harris, A.L., Haselock, P.J., Kennedy, M.J., Mendum, J.R., Long, C.B., Winchester, J.A., and Tanner, P.W.G. (1994) The Dalradian Supergroup in Scotland, Shetland and Ireland. *In*: Gibbons, W. and Harris, A.L. (eds). *A revised correlation of Precambrian rocks in the British Isles*. Geological Society, London, Special Report, **22**, 33–53.

Harris, D.H.M. (1993a) The Caledonian evolution of the Laurentian margin in western Ireland. *Journal of the Geological Society of London*, **150**, 669–672.

Harris, D.H.M. (1993b) *Structure, metamorphism and stratigraphy of Achill Island, Co Mayo, Ireland*. Unpublished Ph.D. thesis, University of Keele.

Harris, D.H.M. (1995) Caledonian transpressional terrane accretion along the Laurentian margin in Co. Mayo, Ireland. *Journal of the Geological Society of London*, **152**, 797–806.

Harris, N. (1937) A petrological study of the Portrush sill and its veins. *Proceedings of the Royal Irish Academy*, **43B**, 95–134.

Harrison, J. (1968) *The stratigraphy of the Namurian outliers adjacent to the Balbriggan massif in Counties Dublin and Meath, Ireland*. Unpublished M.Sc. thesis, University of Dublin.

Harrison, R.K. and Wilson, H.E. (1978) *The granodiorite intrusion of Cushleake Mountain, County Antrim, Northern Ireland*. Institute of Geological Sciences, Report **78/7**, 1–16.

Harrison, S. and Mighall, T.M. (eds). (2002) *The Quaternary of South West Ireland. Field Guide*. Quaternary Research Association, London, 156pp.

Harry, W.T. and Richey, J.E. (1963) Magmatic pulses in the emplacement of plutons. *Liverpool and Manchester Geological Journal*, **3**, 254–268.

Hart, M.B. and Crittenden, S. (1984) Early Cretaceous Ostracoda from the Goban Spur, D.S.D.P. Leg 80, Site 549. *Cretaceous Research*, **6**, 219–233.

Hart, M.B., Feist, S.E., Price, G.D., and Leng, M.J. (2004) Reappraisal of the K–T boundary succession at Stevns Klint, Denmark. *Journal of the Geological Society of London*, **161**, 885–892.

Hartley, J.J. (1933a) Notes on fossils recently obtained from the 'Chloritic' Conglomerate of Murlough Bay, Co. Antrim. *Irish Naturalists' Journal*, **4**, 238–40.

Hartley, J.J. (1933b) The geology of north-eastern Tyrone and the adjacent portions of County Londonderry. *Proceedings of the Royal Irish Academy*, **41B**, 218–285.

Hartley, J.J. (1936) The age of the igneous series, Slieve Gallion, Northern Ireland. *Geological Magazine*, **73**, 226–228.

Hartley, J.J. (1949) Further notes on the Permo-Triassic rocks of Northern Ireland. *Irish Naturalists' Journal*, **9**, 314–316.

Hartz, E.H. and Torsvik, T.H. (2002) Baltica upside down: a new plate tectonic model for Rodinia and the Iapetus Ocean. *Geology*, **30**, 255–258.

Haslett, J., M., S.-T., Wilson, S.P., Bhattacharya, S., Whiley, M., Allen, J.R.M., Huntley, B., and Mitchell, F.J.G. (2006) Bayesian palaeoclimate reconstruction. *Journal of the Royal Statistical Society*, **A169**, 395–483.

Haughey, N. and McArdle, P. (1990) Vitrinite reflectance data from a preliminary study on selected Irish coal seams. *Geological Survey of Ireland Bulletin*, **4**, 201–209.

Haughton, P., Praeg, D., Shannon, P.M., Harrington, G., Higgs, K., Amy, L., Tyrrell, S., and Morrissey, T. (2005) First results from shallow stratigraphic boreholes on the eastern flank of the Rockall Basin, offshore western Ireland. *In*: Doré, A.G. and Vining, B.A. (eds). Petroleum Geology. *North-West Europe and Global Perspectives – Proceedings of the 6th Petroleum Geology Conference*, 1077–1094.

Haughton, P.D.W., Rogers, G., and Halliday, A.N. (1990) Provenance of Lower Old Red Sandstone conglomerates, SE Kincardineshire: evidence for the timing of Caledonian terrane accretion in central Scotland. *Journal of the Geological Society of London*, **147**, 105–120.

Haughton, S. (1852) Notes on the geology of Rathlin Island. *Journal of the Geological Society of Dublin*, **5**, 130–134.

Haughton, S. (1856) Experimental researches on the granites of Ireland. *Proceedings of the Geological Society of London*, **12**, 188–192.

Hauser, F., O'Reilly, B.M., Jacob, A.W.B., Shannon, P.M., Makris, J., and Vogt, U. (1995) The crustal structure of the Rockall Trough: differential stretching without underplating. *Journal of Geophysical Research*, **100**, 4097–4116.

Hawkesworth, C.J. and Kemp, A.I.S. (2006) Using hafnium and ozygen isotopes in zircons to unravel the record of crustal evolution. *Chemical Geology*, **226**, 144–162.

Hayes, F.L. (1978) *Palynological studies in the south-eastern United States, Bermuda and south-east Ireland.* Unpublished M.Sc. thesis, University of Dublin, Trinity College.

Haynes, J.R., McCabe, A.M., and Eyles, N. (1995) Microfaunas from late Devensian glaciomarine deposits in the Irish Sea Basin. *Irish Journal of Earth Sciences,* **14**, 81–103.

Hays, J.D., Imbrie, J., and Shackleton, N.J. (1976) Variations in the Earth's orbit: pacemaker of the ice ages. *Science,* **194**, 1121–1132.

Head, K., Turney, C.S.M., Pilcher, J.R., Palmer, J.G., and Baillie, M.G.L. (2007) Problems with identifying the '8200-year cold event' in terrestrial records of the Atlantic seaboard: a case study from Dooagh, Achill Island, Ireland. *Journal of Quaternary Science,* **22**, 65–75.

Heaman, L. and Ludden, J.N. (eds). (1991) *Applications of Radiogenic Isotopic Systems to Problems in Geology.* Short Course Handbook, **19**. Mineralogical Association of Canada, 498pp.

Heaman, L. and Parrish, R. (1991) U–Pb geochronology of accessory minerals. *In*: Heaman, L. and Ludden, J.N. (eds). *Applications of Radiogenic Isotopic Systems to Problems in Geology.* Short Course Handbook, **19**. Mineralogical Association of Canada, 498pp.

Heaman, L.M. and Smalley, P.C. (1994) A U–Pb study of the Morkheia Complex and associated gneisses, southern Norway: implications for disturbed Rb–Sr systems and for the temporal evolution of Mesoproterozoic magmatism in Laurentia. *Geochimica et Cosmochimica Acta,* **58**, 1899–1911.

Heckel, P.H. and Clayton, G. (2006) The Carboniferous System: use of the new official names for the subsystems, series, and stages. *Geologica Acta,* **4**, 403–407.

Heckel, P.H. and Witzke, B.J. (1978) Devonian world palaeogeography determined from distribution of carbonates and related lithic palaeoclimatic indicators. *In*: House, M.R., Scrutton, C.T., and Bassett, M.G. (eds). *The Devonian System.* Palaeontological Association: Special Papers in Palaeontology, **23**, 99–123.

Heijnis, H. (1992) *Uranium/Thorium dating of Late Pleistocene peat deposits in N.W. Europe.* Published Ph.D. thesis. Wiskunde en Natuurwetenschappen aan de Rijksuniversiteit, Groningen, 149pp.

Heijnis, H., Ruddock, J., and Coxon, P. (1993) A Uranium–Thorium dated Late Eemian or Early Midlandian organic deposit from near Kilfenora between Spa and Fenit, Co. Kerry, Ireland. *Journal of Quaternary Science,* **8**(1) 31–43.

Heinrich, H. (1988) Origin and consequences of cyclic ice rafting in the northeastern Atlantic Ocean during the past 130,000 yrs, *Quaternary Research,* **29**, 142–152.

Hensen, B.J. and Zhou, B. (1995) Retention of isotopic memory in garnets partially broken down during an overprinting granulite-facies metamorphism: implications for the Sm–Nd closure temperature. *Geology,* **23**, 225–228.

Herries Davies, G.L. (1965) The Geological Society of Dublin and the Royal Geological Society of Ireland 1831–1890. *Hermathena,* **100**, 66–76.

Herries Davies, G.L. (1978a) Geology in Ireland before 1812: a bibliographical outline. *Western Naturalist,* **7**, 79–99.

Herries Davies, G.L. (1978b) The earth sciences in Irish serial publications 1787–1977. *Irish Journal of Earth Sciences,* **1**, 1–23.

Herries Davies, G.L. (1983) *Sheets of many colours: the mapping of Ireland's rocks 1750–1890.* Royal Dublin Society, xiv + 242.

Herries Davies, G.L. (1995) *North from the Hook: 150 years of the Geological Survey of Ireland.* Geological Survey of Ireland, xii + 342.

Herries Davies, G.L. and Mollan, R.C. (eds). (1980) *Richard Griffith 1784–1878.* Royal Dublin Society, vi + 221.

Herries Davies, G.L. and Stephens, N. (1978) *Ireland.* Methuen, London, 250pp.

Heselden, R.G.W. (1991) *Sedimentology and stratigraphy of the Courceyan–Asbian limestones (Dinantian, Lower Carboniferous) of the Cork Harbour area, southern Ireland.* Unpublished Ph.D. thesis, National University of Ireland.

Higgs, K. (1983) Palynological evidence for the Carboniferous strata in two wells drilled in the Celtic Sea area. *Bulletin of the Geological Survey of Ireland,* **3**, 107–112.

Higgs, K. (1986) The stratigraphy of the Namurian rocks of the Leinster coalfield. *Geological Survey of Ireland Bulletin,* **3**, 257–276.

Higgs, K. (1990) Upper Viséan and lower Namurian miospore assemblages from County Cork, Ireland. *Irish Journal of Earth Sciences,* **10**, 109–114.

Higgs, K. and Beese, A.P. (1986) A Jurassic microflora from the Colbond Clay of Cloyne, County Cork. *Irish Journal of Earth Sciences,* **7**, 99–109.

Higgs, K. and Jones, G.Ll. (2000) Palynological evidence for Mesozoic karst at Piltown, County Kilkenny. *Proceedings of the Geologists' Association,* **111**, 355–362.

Higgs, K. and Russell, K.J. (1981) Upper Devonian microfloras from Southeast Iveragh, County Kerry, Ireland. *Geological Survey of Ireland Bulletin,* **3**, 17–50.

Higgs, K., Clayton, G., and Keegan, J.B. (1988a) *Stratigraphic and systematic palynology of the Tournaisian rocks of Ireland.* Geological Survey of Ireland, Special Paper, **7**, 93pp.

Higgs, K.T. (1984) Stratigraphic palynology of the Carboniferous rocks in northwest Ireland. *Geological Survey of Ireland Bulletin,* **3**, 171–201.

Higgs, K.T. (1999) Early Devonian spore assemblages from the Dingle Group, County Kerry, Ireland. *Bolletino della Società Palaeontologica Italiana,* **38**, 187–196.

Higgs, K.T. and Jones, G.L. (2000) Palynological evidence for Mesozoic karst at Piltown, Co. Kilkenny. *Proceedings of the Geologists' Association,* **111**, 355–362.

Higgs, K.T. and O'Connor, G. (2005) Stratigraphy and palynology of the Westphalian strata of the Leinster Coalfield. *Irish Journal of Earth Sciences,* **23**, 65–84.

Higgs, K.T. and Williams, D.M. (2001) A Silurian microflora from the Ballytoohy Formation (Upper Clew Bay Complex) of Clare Island. *In*: Graham, J.R (ed.). *New Survey of Clare Island. Volume 2: Geology.* Royal Irish Academy, 19–34.

Higgs, K.T., MacCarthy, I.A.J., and O'Brien, M.M. (2000) A mid-Frasnian marine incursion into the southern part of the Munster Basin: evidence from the Foilcoagh Bay Beds, Sherkin Formation, SW County Cork, Ireland. *In*: Friend, P.F. and Williams, B.P.J. (eds). *New perspectives on the Old Red Sandstone.* Geological Society London Special Publication, **180**, 319–332.

Higgs, K.T., McPhilemy, B., Keegan, J.B., and Clayton, G. (1988b) New data on palynological boundaries within the Irish Dinantian. *Review of Palaeobotany and Palynology,* **56**, 61–68.

Highton, A.J., Hyslop, E.K. and Noble, S.R. (1999) U–Pb geochronology of migmatisation in the Northern Central Highlands: evidence for pre-Caledonian (Neoproterozoic) tectonometamorphism in the Grampian block, Scotland. *Journal of the Geological Society of London*, **156**, 1195–1204.

Hill, P.R. (1987) Characteristics of sediments from Feni and Gardar drifts, sites 610 and 611, Deep Sea Drilling Project Leg 96. *In*: Ruddiman, W.F., Kidd, R.B., Thomas, E., *et al.* (eds.). *Initial Reports of the Deep Sea Drilling Project*, **94**, U.S. Government Printing Office, Washington D.C., 1075–1082.

Histon, K and Sevastopulo, G.D. (1993) Carboniferous nautiloids and the bathymetry of Waulsortian limestones in Ireland. *Proceedings of the Geologists' Association*, **104**, 149–154.

Hitchen, K. (2004) The geology of the UK Hatton–Rockall margin. *Marine and Petroleum Geology*, **21**, 993–1012.

Hitchen, K., Morton, A.C., Mearns, E.W., Whitehouse, M., and Stoker, M.S. (1997) Geochemical implications from geochemical and isotopic studies of Upper Cretaceous and Lower Tertiary igneous rocks around the northern Rockall Trough. *Journal of the Geological Society of London*, **154**, 517–521.

Hitzman, M.W. (1986) Geology of the Abbeytown Mine, Co. Sligo, Ireland. *In*: Andrew, C.J., Crowe, R.W.A., Finlay, S., Pennell, W.M., and Pyne, J.F. (eds). *Geology and genesis of mineral deposits in Ireland*. Irish Association for Economic Geology, Dublin, 341–353.

Hitzman, M.W. (1993) *The geology of the central Midlands of Ireland: 1:100,000 maps 6 sheets*. Geological Survey of Ireland.

Hitzman, M.W. (1995) Geological setting of the Irish Zn–Pb–(Ba–Ag) orefield. *In*: Anderson, K., Ashton, J., Earls, G., Hitzman, M.W., and Tear, S. (eds). *Irish carbonate-hosted Zn–Pb deposits*. Society of Economic Geologists Guidebook Series, **21**, 3–23.

Hitzman, M.W., Beaty, D.W., and Redmond, P. (2002) The carbonate-hosted Lisheen Zn–Pb–Ag deposit, Co. Tipperary, Ireland. Economic Geology, 97, 1627–1655.

Hitzman, M.W., O'Connor, P., Shearley, E., Schaffalitzky, C., Beaty, D.W., Allan, J.R., and Thompson, T. (1992) Discovery and geology of the Lisheen Zn–Pb–Ag prospect, Rathdowney trend, Ireland. *In*: Bowden, A.A., Earls, G., O'Connor, P.G., and Pyne, J.F. (eds). *The Irish Minerals Industry 1980–1990*. Irish Association for Economic Geology, Dublin, 227–246.

Hodgson, J.A. (2001) *A seismic and gravity study of the Leinster Granite: SE Ireland*. Unpublished Ph.D. thesis, National University of Ireland.

Hodson, F. (1954) The beds above the Carboniferous Limestone in north-west County Clare, Eire. *Quarterly Journal of the Geological Society of London*, **109**, 259–283.

Hodson, F. (1959) The palaeogeography of Namurian times in western Europe. *Bulletin de la Société Belge de Géologie, de Paléontologie et d'Hydrologie*, **68**, 134–150.

Hodson, F. and Lewarne, G. (1961) A mid-Carboniferous basin in parts of the counties Limerick and Clare, Ireland. *Quarterly Journal of the Geological Society of London*, **107**, 307–333.

Hoffman, P.F. and Schrag, D.P. (2002) The snowball Earth hypothesis: testing the limits of global change. *Terra Nova*, **14**, 129–155.

Hoffman, P.F., Kaufman, A.J., Halverson G.P., and Schrag, D.P. (1998) A Neoproterozoic Snowball Earth. *Science*, **281**, 1342–1346.

Hofmann, H.J. (1990) Precambrian time units and nomenclature: the geon concept. *Geology*, **18**, 340–341.

Hofmann, H.J. (1999) Geons and geons. *Geology*, **27**, 855–856.

Holder, M.T. (1979) An emplacement mechanism for post-tectonic granites and its implications for their geochemical features. *In*: Atherton, M.P. and Tarney, J. (eds). *Origin of granite batholiths: geochemical evidence*. Shiva, Kent.

Holdsworth, B.K. and Collinson, J.D. (1988) Millstone Grit cyclicity revisited. *In*: Besly, B.M and Kelling, G. (eds). *Sedimentation in a syn-orogenic basin complex: the Upper Carboniferous of North-west Europe*. Blackie, Glasgow and London, 132–152.

Holdsworth, R.E., Strachan, R.A., and Harris, A.L. (1994) Precambrian rocks in northern Scotland east of the Moine Thrust: the Moine Supergroup. *In*: Gibbons, W. and Harris, A.L. (eds). *A revised correlation of Precambrian rocks in the British Isles*. Geological Society, London, Special Report, **22**, 23–32.

Holland, C.H. (1969) Irish counterpart of Silurian of Newfoundland. *In*: Kay, M. (ed.). North Atlantic—Geology and Continental Drift. *Memoir 12, American Association Petroluem Geologists*, 289–308.

Holland, C.H. (1978) Stratigraphical classification and all that. *Lethaia*, **11**, 85–90.

Holland, C.H. (1981) Geology. *In*: McBrierty, V.J. (ed.). *The Howth Peninsula: its history, lore, and legend*. Dublin, 123–128.

Holland, C.H. (1986) Does the golden spike still glitter? *Journal of the Geological Society of London*, **143**, 3–21.

Holland, C.H. (1987) Stratigraphical and structural relationships of the Dingle Group (Silurian), County Kerry, Ireland. *Geological Magazine*, **124**, 33–42.

Holland, C.H. (1988) The fossiliferous Silurian rocks of the Dunquin inlier, Dingle Peninsula, County Kerry, Ireland. *Transactions of the Royal Society of Edinburgh: Earth Sciences*, **79**, 347–360.

Holland, C.H. (2003) *The Irish Landscape – a scenery to celebrate*. Dunedin Academic Press, Edinburgh. 192 pp.

Holland, C.H. and Smith, D.G. (1979) Silurian rocks of the Capard inlier, County Laois. *Proceedings of the Royal Irish Academy*, **79B**, 99–110.

Holland, C.H., Feehan, J., and Williams, E. M. (1988) The Wenlock rocks of the Cratloe Hills, County Clare. *Irish Journal of Earth Sciences*, **9**, 61–69.

Holland, C.H., Lawson, J.D., and Walmsley, V.G. (1963) The Silurian rocks of the Ludlow district, Shropshire. *Bulletin British Museum (Natural History) Geology*, **8**, 93–171.

Hood, D.N. (1981) *Geochemical, petrological and structural studies on the Tertiary granites and associated rocks of the eastern Mourne Mountains, Co. Down, Northern Ireland*. Unpublished Ph.D. thesis, Queen's University Belfast.

Horne, R.R. (1971) Aeolian cross stratification in the Devonian of the Dingle Peninsula, County Kerry, Ireland. *Geological Magazine*, **103**, 151–158.

Horne, R.R. (1974) The lithostratigraphy of the late Silurian to early Carboniferous of the Dingle Peninsula, Co. Kerry. *Geological Survey of Ireland Bulletin*, **1**, 395–428.

Horne, R.R. (1975) The association of alluvial fan, aeolian and fluviatile facies in the Caherbla Group (Devonian), Dingle Peninsula, Ireland. *Journal of Sedimentary Petrology*, **45**, 535–540.

Horne, R.R. (1976) *Geological guide to the Dingle Peninsula*. Geological Survey of Ireland Guide Series, **1**.

Horne, R.R. and Macintyre, R.M. (1975) Apparent age and significance of Tertiary dykes in the Dingle Peninsula, S.W. Ireland. *Scientific Proceedings of the Royal Dublin Society*, **5**, 293–299.

Hospers, J. and Charlesworth, H.A.K. (1954) The natural permanent magnetism of the Lower Basalts of Northern Ireland. *Monthly Notes of the Royal Astronomical Society, Geophysical*

Supplement, **7**, 32–43.

House, M.R. (1968) *Continental drift and the Devonian System*, University of Hull.

House, M.R. and Gradstein, F.M. (2004) The Devonian Period. *In*: Gradstein, F.M., Ogg, J.G., and Smith, A.G. (eds). *A geologic time scale 2004*. Cambridge University Press, 202–221.

Howard, D.W. (1975) Deep-seated igneous intrusions in Co. Kerry. *Proceedings of the Royal Irish Academy*, **75B**, 173–183.

Howarth, R.J. (1971) The Portaskaig Tillite succession (Dalradian) of Co. Donegal. *Proceedings of the Royal Irish Academy*, **71B**, 1–35.

Howell, B.F. (1922) *Oldhamia* in the Lower Cambrian of Massachusetts. *Bulletin Geological Society America*, **33**, 198.

Hubbard, J.A.E.B. (1966a) Facies patterns in the Carrowmoran Sandstone (Viséan) of western Co. Sligo, Ireland. *Proceedings of the Geologists' Association*, **77**, 233–254.

Hubbard, J.A.E.B. (1966b) Population studies in the Ballyshannon Limestone, Ballina Limestone, and Rinn Point Beds (Viséan) of N.W. Ireland. *Palaeontology*, **9**, 252–269.

Hubbard, J.A.E.B. and Pocock, Y.P. (1972) Sediment rejection by Recent scleractinian corals: a key to palaeo-environmental reconstruction. *Geologische Rundschau*, **61**, 598–626.

Hudson, R.G.S. and Philcox, M.E. (1965) The Lower Carboniferous stratigraphy of the Buttevant area, County Cork. *Proceedings of the Royal Irish Academy*, **64**, 65–79.

Hudson, R.G.S. and Sevastopulo, G.D. (1966) A borehole section through the Lower Tournaisian and Upper Old Red Sandstone, Ballyvergin, Co. Clare. *Scientific Proceedings of the Royal Dublin Society*, **A2**, 287–296.

Hudson, R.G.S., Clarke, M.J., and Brennand, T. (1966a) The Lower Carboniferous stratigraphy of the Castleisland area, Co. Kerry, Ireland. *Scientific Proceedings of the Royal Dublin Society*, **A2**, 292–317.

Hudson, R.G.S., Clarke, M.J., and Sevastopulo, G.D. (1966b) A detailed account of the fauna and age of a Waulsortian knoll reef limestone and associated shales, Feltrim, Co. Dublin, Ireland. *Scientific Proceedings of the Royal Dublin Society*, **A2**, 251–272.

Hughes, T. (1986) The Jakobshavns effect. *Geophysics Review Letters*, **13**(1), 46–48.

Hull, E., Nolan, J., Cruise, R.J., and McHenry, A. (1890) *Explanatory note to accompany one-inch sheets 1, 2, 5, 6, and 11 (in part)*. Memoir of the Geological Survey of Ireland, 174pp.

Hull, E.J.L., Warren, W.B., Leonard, W.H., and Baily. (1871) *Explanatory memoir to accompany Sheet 36 of the maps of the Geological Survey of Ireland, etc.* Memoir of the Geological Survey of Ireland.

Husain, S.M. (1957) *The geology of the Kenmare Syncline, Co. Kerry, Ireland*. Unpublished Ph.D. thesis, University of London.

Hutchison, A.R. and Oliver, G.J.H. (1998) Garnet provenance studies, juxtaposition of Laurentian marginal terranes and timing of the Grampian orogeny in Scotland. *Journal of the Geological Society of London*, **155**, 541–550.

Hutton, D.H.W. (1979) Tectonic slides: a review and reappraisal. *Earth Science Reviews*, **15**, 151–172.

Hutton, D.H.W. (1982) A tectonic model for the emplacement of the Main Donegal Granite, NW Ireland. *Journal of the Geological Society of London*, **139**, 615–631.

Hutton, D.H.W. (1987) Strike-slip terranes and a model for the evolution of the British and Irish Caledonides. *Geological Magazine*, **124**(5), 405–425.

Hutton, D.H.W. (1988) Granite emplacement mechanisms and tectonic controls: influences from deformation studies. *Transactions of the Royal Society of Edinburgh: Earth Sciences*, **79**, 245–255.

Hutton, D.H.W. and Alsop, G.I. (1995) Extensional geometries as a result of regional scale thrusting: tectonic slides of the Dunlewy–NW Donegal area, Ireland. *Journal of Structural Geology*, **17**, 1279–1292.

Hutton, D.H.W. and Alsop, G.I. (1996) The Caledonian strike-swing and associated lineaments in NW Ireland and adjacent areas: sedimentation, deformation and igneous intrusion patterns. *Journal of the Geological Society of London*, **153**, 345–360.

Hutton, D.H.W. and Alsop, G.I. (2004) Evidence for a major Neoproterozoic orogenic unconformity within the Dalradian Supergroup of NW Ireland. *Journal of the Geological Society of London*, **161**, 629–640.

Hutton, D.H.W. and Dewey, J.F. (1986) Palaeozoic terrane accretion in the Irish Caledonides. *Tectonics*, **5**(7), 1115–1124.

Hutton, D.H.W. and Holland, C.H. (1992) An Arenig–Llanvirn age for the black shales of Slieve Gallion, County Tyrone. *Irish Journal of Earth Sciences*, **11**, 187–189.

Hutton, D.H.W. and Murphy, F.C. (1987) The Silurian rocks of the Southern Uplands and Ireland as a successor basin to the late Ordovician closure of Iapetus. *Journal of the Geological Society of London*, **144**, 765–772.

Hutton, D.H.W., Aftalion, M., and Halliday, A.N. (1985) An Ordovician ophiolite in County Tyrone, Ireland. *Nature*, **315**, 210–212.

Illing, L.V. and Griffith, A.E. (1986) Gas prospects in the Midland Valley of Northern Ireland. *In*: Brooks, J., Goff, J.C., and Van Hoorn, B. (eds). *Habitat of Palaeozoic Gas in NW Europe*. Geological Society, London, Special Publications **23**, 73–84.

Imbrie, J. and Imbrie, K.P. (1979) *Ice Ages: Solving the mystery*. Macmillan, London and Basingstoke, 224pp.

Imbrie, J. and Imbrie, K.P. (1980) Modelling the climatic response to orbital parameter variations. *Science*, **207**, 943–953.

Institute of Geological Sciences (1975) *Geological Survey of Northern Ireland, Sheet 47, Armagh*.

Institute of Geological Sciences (1979) *Geological Survey of Northern Ireland, sheet 34, Pomeroy*.

Isaacson, P.E., Hladil, J., Shen, J., Kalvoda, J., and Grader, G. (1999) Late Devonian (Famennian) glaciation in South America and marine offlap on other continents: IGCP 421. Special volume. *Abhandlungen der Geologischen Bundesanstalt Wien*, **54**, 239–256.

Iversen, J. (1958) The bearing of glacial and interglacial epochs on the formation and extinction of plant taxa. *Uppsala University Årsskrift*, **6**, 210–215.

Ivimey-Cook, H.C. (1975) The stratigraphy of the Rhaetic and Lower Jurassic in east Antrim. *Bulletin of the Geological Survey of Great Britain*, **50**, 51–69.

Jackson, A.A. (1978) Stratigraphy, sedimentology and palaeontology of the Silurian rocks of the Galty mountain area. *Proceedings of the Royal Irish Academy*, **78B**, 91–112.

Jackson, D.I., Johnson, H., and Smith, N.J.P. (1997) Stratigraphical relationships and a revised lithostratigraphical nomenclature for the Carboniferous, Permian and Triassic rocks of the offshore East Irish Sea Basin. *In*: Meadows, N.S., Trueblood, S.P., Hardman, M., and Cowan, G. (eds). *Petroleum Geology of the Irish Sea and Adjacent Areas*. Geological Society, London, Special Publications **124**, 11–32.

Jackson, J.S. (1965) The Upper Carboniferous (Namurian and Westphalian) of Kingscourt, Ireland. *Scientific Proceedings of the Royal Dublin Society*, **A2**, 131–152.

Jacob, A.W.B. (1993) Seismic hazard in Ireland. *In*: R.K. Maguire (ed.). *The practice of earthquake hazard assessment.* IASPEI and the European Seismological Commission, 150–152.

Jacob, A.W.B., Kaminski, W., Murphy, T., Phillips, W.E.A. and Prodehl, C. (1985) A crustal model for a NE–SW profile through Ireland. *Tectonophysics*, **113**, 75–103.

Jacques, J.M. and Reavy, R.J. (1994) Caledonian plutonism and major lineaments in the SW Scottish Highlands. *Journal of the Geological Society of London*, **151**, 955–969.

Jagger, M.D., Max, M.D., Aftalion, M., and Leake, B.E. (1988) U–Pb zircon ages of basic rocks and gneisses intruded into the Dalradian rocks of Cashel, Connemara, western Ireland. *Journal of the Geological Society of London*, **145**, 645–648.

James, A. (1989) *Stratigraphic and facies analysis of Devonian continental clastics of the Beara Peninsula, SW Ireland.* Unpublished Ph.D. thesis, University of Dublin.

James, A. and Graham, J.R. (1995) Stratigraphy and structure of Devonian fluvial sediments, western Beara Peninsula, southwest Ireland. *Geological Journal*, **30**, 165–182.

James, J.A. (2003) *The provenance of the Mississippian Ballyvergin Shale: an integrated palynological and lithostratigraphical investigation.* Unpublished Ph.D. thesis, University of Dublin.

James, K.W. (1986) *'Damned nonsense!' — the geological career of the third Earl of Enniskillen.* Ulster Museum Publication, **259**, 24pp.

Jeans, C.V. (1980) Early submarine lithification in the Red Chalk and Lower Chalk of eastern England: a bacterial control model and its implications. *Proceedings of the Yorkshire Geological Society*, **43**, 81–157.

Jeans, C.V., Merriman, R.J., Mitchell, J.G., and Bland, D.J. (1982) Volcanic clays in the Cretaceous of southern England and Northern Ireland. *Clay Minerals*, **17**, 105–156.

Jeffrey, D.W., Goodwillie, R.N., Healy, B., Holland, C.H., Jackson, J.S., and Moore, J.J. (1977) *North Bull Island Dublin Bay: a modern coastal natural history.* Royal Dublin Society, 158pp.

Jenner, J. K. (1981) The structure and stratigraphy of the Kish Bank Basin. *In*: Illing, L.V. and Hobson, G.D. (eds). *Petroleum Geology of the Continental Shelf of North-West Europe*, Heyden and Son Ltd., London, 426–431.

Jensen, S., Gehling, J.G., Droser, M.L., and Grant, S.W.F. (2002) A scratch circle origin for the medusoid fossil *Kullingia*. *Lethaia*, **35**, 291–299.

Jessen, K. and Farrington, A. (1938) The bogs at Ballybetagh, near Dublin, with remarks on late-glacial conditions in Ireland. *Proceedings of the Royal Irish Academy*, **44B**, 205–260.

Jessen, K., Andersen, S.T., and Farrington, A. (1959) The interglacial deposit near Gort, Co. Galway, Ireland. *Proceedings of the Royal Irish Academy*, **60B**, 1–77.

Johnson, M.R.W. (1991) Dalradian. *In*: Craig, G.Y. (ed.). *Geology of Scotland.* The Geological Society, 125–160.

Johnston, I.S. (1976) *The conodont biostratigraphy of some Lower Carboniferous (Courceyan Stage) rocks of central Ireland.* Unpublished Ph.D. thesis, University of Dublin.

Johnston, J.D. (1993) Three-dimensional geometries of veins and their relationship to folds: examples from the Carboniferous of eastern Ireland. *Irish Journal of Earth Sciences*, **12**, 47–64.

Johnston, J.D. (1995) Major northwest-directed Caledonian thrusting and folding in Precambrian rocks, northwest Mayo, Ireland. *Geological Magazine*, **132**(1), 91–112.

Johnston, J.D. and Phillips, W.E.A. (1995) Terrane amalgamation in the Clew Bay region, west of Ireland. *Geological Magazine*, **132**, 485–501.

Johnston, S., Doré, A.G., and Spencer, A.M. (2001) The Mesozoic evolution of the southern North Atlantic region and its relationship to basin development in the south Porcupine Basin, offshore Ireland. *In*: Shannon, P.M., Haughton, P.D.W., and Corocoran, D.V. (eds). *The Petroleum Exploration of Ireland's Offshore Basins.* Geological Society, London, Special Publications **188**, 237–263.

Joly, J. (1920) *Reminiscences and Anticipations.* London.

Jones, C.S. (1989) *The structure and kinematics of the Ox Mountains, western Ireland; a mid-crustal transcurrent shear-zone.* Unpublished Ph.D. thesis, University of Durham.

Jones, C.S. and Leat, P.T. (1988) Discussion on trace element geochemical correlation in the reworked Proterozoic Dalradian metavolcanic suites of the western Ox Mountains and NW Mayo Inliers, Ireland. *Journal of the Geological Society of London*, **145**, 1037–1040.

Jones, G.Ll., Somerville, I.S., and Strogen, P. (1988) The Lower Carboniferous (Dinantian) of the Swords area: sedimentation and tectonics in the Dublin Basin, Ireland. *Geological Journal*, **23**, 221–248.

Jones, P.C. (1974) Marine transgression and facies distribution in the Cork Beds (Devonian–Carboniferous) of west Cork and Kerry, Ireland. *Proceedings of the Geologists' Association*, **85**, 159–188.

Jones, P.C. and Naylor, D. (2003) Namurian rocks of Whiddy Island, west Cork: a sedimentological outline and palaeogeographical implications. *Irish Journal of Earth Sciences*, **21**, 115–132.

Jones, R.L. and Keen, D.H. (1993) *Pleistocene Environments in the British Isles.* Routledge, Chapman and Hall, 346pp.

Jones, S.M., White, N., and Lovell, B. (2001) Cenozoic and Cretaceous transient uplift in the Porcupine Basin and its relationship to a mantle plume. *In*: Shannon, P.M., Haughton, P.D.W., and Corocoran, D.V. (eds). *The Petroleum Exploration of Ireland's Offshore Basins.* Geological Society, London, Special Publications **188**, 345–360.

Judd, J.W. (1874) The secondary rocks of Scotland, Second paper. On the ancient volcanoes of the Highlands and the relations of their products to the Mesozoic strata. *Quarterly Journal of the Geological Society of London*, **30**, 220–301.

Jukes J.B. and Du Noyer, G.V. (1858) *Data and descriptions to accompany Quarter Sheet 46 N.W. of the maps of the Geological Survey of Ireland.*

Jukes, J.B. and Du Noyer, G.V. (1863) *Explanation of sheets 160, 161, 171, and part of 172 and of the engraved section, sheet no. 15, of the Geological Survey of Ireland, illustrating part of the County Kerry.* Memoir Geological Survey of Ireland.

Kaljo, D. and Klaamann, E. (1965) The fauna of the Portrane Limestone III: the corals. *Bulletin British Museum (Natural History) Geology*, **10**, 413–434.

Kamerling, P. (1979) The geology and hydrocarbon habitat of the Bristol Channel Basin. *Journal of Petroleum Geology*, **2**, 75–93.

Kamo, S.K., Gower, C.F., and Krogh, T.E. (1989) A birthdate for the Iapetus Ocean? A precise U–Pb zircon and baddeleyite age for the Long Range dikes, southeast Labrador, *Geology*, **17**, 602–605.

Karlstrom, K. E., Bowring, S.A., Dehler, C.M., Knoll, A.H., Porter, S.M., Marais, D.J.D., Weil, A.B., Sharp, Z.D., Geissman, J.W., Elrick, M.B., Timmons, J.M., Crossey, L.J., and Davidek, K.L. (2000) Chuar Group of the Grand Canyon: Record of break-up of Rodinia, associated change in the global carbon cycle, and ecosystem expansion by 740 Ma. *Geology*, **28**, 619–622.

Karson, J. and Dewey, J.F. (1978) Coastal Complex, western Newfoundland: an early Ordovician oceanic fracture zone. *Geological Society of America Bulletin*, **89**, 1037–1049.

Kaufman, A.J., Knoll, A.H., and Narbonne, G.M. (1997) Isotopes, ice ages, and terminal Proterozoic Earth history. *Proceedings of National Academy of Science, USA*, **94**, 6600–6605.

Kaufmann, B. (2006) Calibrating the Devonian time scale: a synthesis of U–Pb ID: TIMS ages and conodont stratigraphy. *Earth Science Reviews*, **76**, 175–190.

Keeley, M.L. (1983) The stratigraphy of the Carrick-on-Suir Syncline, southern Ireland. *Journal of Earth Sciences, Royal Dublin Society*, **5**, 107–120.

Keeley, M.L. (1995) New evidence of Permo-Triassic rifting, onshore southern Ireland, and its implications for Variscan structural inheritance. *In*: Boldy, S.A.R. (ed.). *Permian and Triassic Rifting in Northwest Europe*. Geological Society, London, Special Publications **91**, 239–253.

Keeley, M.L. (1996) The Irish Variscides: problems, perspectives and some solutions. *Terra Nova*, **8**, 259–269.

Keen, C.E., Keen, M.J., Nichols, B., Reid, I., Stockmal, G.S., Colman, S.S.P., O'Brien, S.J., Miller, H., Quinlan, G., Williams, H., and Wright, W.J. (1986) Deep seismic reflection profile across the northern Appalachians. *Geology*, **14**(2), 141–145.

Kelk, B. (1960) *Studies in the Carboniferous stratigraphy of western Eire*. Unpublished Ph.D. thesis, University of Reading.

Kelley, J.T., Coope, R.J.A.G., Jackson, D.W.T., Belknap, D.F., and Quinn, R. (2006) Sea-level change and inner shelf stratigraphy off Northern Ireland. *Marine Geology*, **232**, 1–15.

Kelley, S. and Bluck, B.J. (1989) Detrital mineral ages from the Southern Uplands using ^{40}Ar–^{39}Ar laser probe. *Journal of the Geological Society of London*, **146**, 401–403.

Kelling, G. (1961) The stratigraphy and structure of the Ordovician rocks of the Rhinns of Galloway. *Quarterly Journal of the Geological Society of London*, **117**, 37–75.

Kelling, G. (1962) The petrology and sedimentation of Upper Ordovician rocks in the Rhinns of Galloway. *Transactions Royal Society Edinburgh*, **65**, 107–137.

Kelling, G. (2001) Southern Uplands geology: an historical perspective. *Transactions of the Royal Society of Edinburgh: Earth Sciences*, **91**(3/4), 323–339.

Kelly, D.L. and Fuller, S. (1988) Ancient woodland in central Ireland: does it exist? *In*: Salbitano, F. (ed.). *Human influence on forest ecosystems development in Europe*. Pitagora, Bologna, 363–369.

Kelly, J.G. (1996) Initiation, growth and decline of a tectonically controlled Asbian carbonate ramp: Cuilcagh Mountain area, NW Ireland. *In*: Strogen, P., Somerville, I.D., and Jones, G.Ll. (eds). *Recent advances in Lower Carboniferous geology*. Geological Society Special Publication, **107**, 253—262.

Kelly, J.G. and Somerville, I.D. (1992) Arundian (Dinantian) carbonate mudbanks in north-west Ireland. *Geological Journal*, **27**, 221–241.

Kelly, J.G., Andrew, C.J., Ashton, J.H., Boland, M.B., Earls, G., Fusciardi, L., and Stanley, G. (eds). (2003) *Europe's major base metal deposits*. Irish Association for Economic Geology, Dublin, 552pp.

Kelly, S.B. (1992) Milankovitch cyclicity recorded from Devonian non-marine sediments. *Terra Nova*, **4**, 578–584.

Kelly, S.B. (1993) Cyclical discharge variation recorded in alluvial sediments: an example from the Devonian of southwest Ireland. *In*: North, C.P. and Prosser, D.J. (eds). *Characterisation of fluvial and aeolian reservoirs*. Geological Society London Special Publication, **73**, 157–166.

Kelly, S. B. and Olsen, H. (1993) Terminal fans: a review with reference to Devonian examples. *Sedimentary Geology*, **85**, 339–374.

Kelly, S.B. and Sadler, S.P. (1995) Equilibrium and response to climatic and tectonic forcing: a study of alluvial sequences in the Devonian Munster Basin, Ireland. *In*: House, M.R. and Gale, A.S. (eds). *Orbital forcing timescales and cyclostratigraphy*. Geological Society London Special Publication, **85**, 19–36.

Kelly, T.J. and Max M.D. (1979) The Geology of the Northern Part of the Murrisk Trough. *Proceedings of the Royal Irish Academy*, **79B**(15), 191–206.

Kennan, P.S., Feely, M., and Mohr, P. (1987) The age of the Oughterard Granite, Connemara, Ireland. *Geological Journal*, **22**, 273–280.

Kennan, P.S., Phillips, W.E.A., and Strogen, P. (1979) Pre-Caledonian basement to the paratectonic Caledonides in Ireland. *In*: Harris, A.L., Holland, C.H., and Leake, B.E. (eds). *The Caledonides of the British Isles: reviewed*. Geological Society, London, Special Publications, **8**, 157–164.

Kennedy, M.J. (1969a.) The metamorphic history of north Achill Island, Co. Mayo, and the problem of the origin of albite schists. *Proceedings of the Royal Irish Academy*, **67B**(11), 261–279.

Kennedy, M.J. (1969b.) The structure and stratigraphy of the Dalradian rocks of north Achill Island, County Mayo, Ireland. *Quarterly Journal of the Geological Society of London*, **125**, 47–81.

Kennedy, M.J. (1979) The continuation of the Canadian Appalachians into the Caledonides of Britain and Ireland. *In*: Harris, A.L., Holland, C.H., and Leake, B.E. (eds). *The Caledonides of the British Isles: reviewed*. Geological Society, London, Special Publications, **8**, 33–64.

Kennedy, M.J. (1980) Serpentinite-bearing mélange in the Dalradian of County Mayo and its significance in the development of the Dalradian basin. *Journal of Earth Sciences, Royal Dublin Society*, **3**, 117–126.

Kennedy, M.J. and Menuge, J.F. (1992) The Inishkea Division of northwest Mayo: Dalradian cover rather than pre-Caledonian basement. *Journal of the Geological Society of London*, **149**, 167–170.

Kennedy, M.J., Runnegar, B., Prave, A.R., Hoffmann, K.H. and Arthur, M.A. (1998) Two or four Neoproterozoic glaciations? *Geology*, **26**, 1059–1063.

Keser Neish, J. (1993) Seismic structure of the Hatton–Rockall area: an integrated seismic/modelling study from composite datasets. *In*: Parker, J.R. (ed.). *Petroleum Geology of Northwest Europe: Proceedings of the 4th Conference*. Geological Society, London, 1047–1056.

Kessler, L.G. and Sachs, S.D. (1995) Depositional setting and sequence stratigraphic implications of the Upper Sinemurian (Lower Jurassic) sandstone interval, North Celtic Sea/St George's Channel Basins, offshore Ireland. *In*: Croker, P.F. and Shannon, P.M. (eds). *The Petroleum Geology of Ireland's Offshore*

Basins. Geological Society, London, Special Publications **93**, 171–192.

Key, M.M. Jr., Wyse Jackson, P.N., Patterson, W.P., and Moore, M.D. (2005) Stable isotope evidence for diagenesis of the Ordovician Courtown and Tramore Limestones, south-eastern Ireland. *Irish Journal of Earth Sciences*, **23**, 25–38.

Kilroe, J.R. (1898) *In*: *Summary of progress of the Geological Survey for 1897*.

Kilroe, J.R. (1907) *A description of the soil-geology of Ireland, based upon the Geological Survey maps and records, with notes on climate*. Dublin, 300pp.

Kinahan, G.H. (1865) *Explanation to accompany sheets 115 and 116*. Geological Survey of Ireland, Dublin.

Kinahan, G.H. (1878) *Manual of the Geology of Ireland*. Dublin.

Kinahan, G.H. *et al.* (1860) *Explanations to accompany Sheet 143 of the maps of the Geological Survey of Ireland, illustrating part of the Counties of Clare and Limerick*. Memoirs Geological Survey Ireland.

Kinahan, G.H. *et al.* (1875) Explanation of Sheets 102 and 112. *Geological Survey Ireland Memoir* (2nd Edition).

King, E.L., Haflidason, H., Sejrup, H.P., Austin, W., Duffy, M., Helland, E., Klitgaard-Kristensen, D., and Scourse, J. (1998) End moraines on the Northwest Irish Continental Shelf. *Abstract 3rd ENAM II Workshop, Scotland*.

King, S.C. (1989) *Structural studies in the Dingle Peninsula, Ireland*. Unpublished Ph.D. thesis, Queens University, Belfast.

Kinnaird, T.C., Prave, A.R., Kirkland, C.L., Horstwood, M., Parrish, R., and Batchelor, R.A. (2007) The late Mesoproterozoic–early Neoproterozoic tectonostratigraphic evolution of NW Scotland: the Torridonian revisited. *Journal of the Geological Society of London*, **164**, 541–551.

Kinny P.D. and Maas. R. (2003) Lu–Hf and Sm–Nd isotope systems in zircon. *In*: Hanchar, J.M. and Hoskin, P.W.O. (eds). Zircon. *Reviews in Mineralogy and geochemistry*, **53**, 327–341.

Kirkland, C.L., Daly, J.S., and Whitehouse, M.J. (2005) Early Silurian magmatism and the Scandian evolution of the Kalak Nappe Complex, Finnmark, Arctic Norway. *Journal of the Geological Society of London*, **162**, 985–1003.

Kirkland, C.L., Daly, J.S., and Whitehouse, M.J. (2007a) Provenance and terrane evolution of the Kalak Nappe Complex, Norwegian Caledonides: implications for Neoproterozoic palaeogeography and tectonics. *Journal of Geology*, **115**, 21–41.

Kirkland, C.L., Strachan, R.A., and Prave, A.R. (2008) Detrital zircon signature of the Moine Supergroup, Scotland: contrasts and comparisons with other Neoproterozoic successions within the circum-North Atlantic region. *Precambrian Research*, doi:10.1016/j.precamres.(2008)02.003.

Kirkland, C.L., Daly, J.S., Eide, E.A., and Whitehouse, M.J. (2007b) Tectonic evolution of the Arctic Norwegian Caledonides from a texturally- and structurally-constrained multi-isotopic (Ar–Ar, Rb–Sr, Sm–Nd, U–Pb) study. John Rodgers Volume. *American Journal of Science*, **307**, 459–526.

Kirwan, P.J., Daly, J.S., and Menuge, J.F. (1989) A minimum age for the deposition of the Dalradian Supergroup sediments in Ireland. *Terra Abstracts*, **1**, 16.

Kirwan, R. (1799) *Geological essays*. London, xvi + 502.

Kitchen, D.E. (1981) *A mineralogical and petrochemical study of the partial melting of wall rock against some Tertiary intrusions*. Unpublished Ph.D. thesis, Queen's University of Belfast.

Kitchen, D.E. (1985) The partial melting of basalt and its enclosed mineral-filled cavities at Scawt Hill, Co. Antrim. *Mineralogical Magazine*, **49**, 655–662.

Kjerulf, Th. (1880) *Die Geologie des sudlichen und mittleren Norwege*. Bonn.

Klemperer, S.L. (1989) Seismic reflection evidence for the location of the Iapetus suture west of Ireland. *Journal of the Geological Society of London*, **146**, 409–412.

Klemperer, S.L. and Matthews, D.H. (1987) Iapetus Suture located beneath the North Sea by BIRPS deep reflection profiling. *Geology*, **15**, 195–198.

Klemperer, S.L., Ryan, P.D., and Snyder, D.B. (1991) A deep seismic reflection transect across the Irish Caledonides. *Journal of the Geological Society of London*, **148**, 149–164.

Knight, J. and McCabe, A.M. (1997a) Identification and significance of ice flow-transverse subglacial ridges (Rogen moraines) in north central Ireland. *Journal of Quaternary Science*, **12**, 219–224.

Knight, J. and McCabe, A.M. (1997b) Drumlin evolution and ice sheet oscillations along the NE Atlantic margin, Donegal Bay, western Ireland. *Sedimentary Geology*, **111**, 57–72.

Knott, S.D., Burchell, M.T., Jolley, E.J., and Fraser, A.J. (1993) Mesozoic and Cenozoic plate reconstructions of the North Atlantic and the hydrocarbon plays of the Atlantic margins. *In*: Parker, J.R. (ed.). *Petroleum Geology of Northwest Europe: Proceedings of the 4th Conference*. Geological Society, London, 857–866.

Kokelaar, P. (1988) Tectonic controls of Ordovician arc and marginal basin volcanism in Wales. *Journal of the Geological Society of London*, **145**, 759–775.

Košler, J. and Sylvester, P.J. (2003) Present trends and the future of zircon in geochronology: laser ablation ICPMS. *In*: Hanchar, J.M. and Hoskin, P.W.O. (eds.). (2003) *Zircon*. Reviews in Mineralogy and Geochemistry, **53**, 243–275.

Košler, J., Fonneland, H., Sylvester, P., Tubrett, M., and Pedersen, R.B. (2002) U–Pb dating of detrital zircons for sediment provenance studies: a comparison of laser ablation ICPMS and SIMS techniques. *Chemical Geology*, **182**(2–4), 605–618.

Kristjansson, L., Patsold, R., and Preston, J. (1975) The palaeomagnetism and geology of the Patreksfjordur–Arnarfjordur region of northwest Iceland. *Tectonophysics*, **25**, 201–206.

Krogh, T.E. (1973) A low-contamination method for hydrothermal decomposition of zircon and extraction of U and Pb for isotopic age determinations. *Geochimica et Cosmochimica Acta*, **37**, 485–494.

Krogh, T.E. (1982) Improved accuracy of U–Pb zircon ages by the creation of more concordant systems using the air abrasion technique. *Geochimica et Cosmochimica Acta*, **46**, 637–49.

Kubiena, W.L. (1970) *Micromorphological features of soil geography*. Rutgers University Press.

Kuijpers, E.P. (1971) Transition from fluviatile to tidal marine sediments in the Upper Devonian of Seven Head Peninsula (south County Cork, Ireland). *Geologie en Mijnbouw*, **50**, 443–450.

Kuijpers, E.P. (1972) *Upper Devonian tidal deposits and associated sediments south and southwest of Kinsale, southern Ireland*. Ph.D. thesis, University of Utrecht.

Kuijpers, E.P. (1975) Continental and coastal plain deposits of the uppermost Old Red Sandstone complex of southern Ireland. *Geologie en Mijnbouw*, **54**, 15–22.

Kutzbach, J.E. and Webb, T. (1991) Late Quaternary climatic and vegetational change in eastern North America: concepts, models and data, *In*: Shane, C.K. and Cushing, E.J. (eds). *Quaternary Landscapes*. University of Minnesota Press, 175–217.

Labiaux, S. (1997) Sponges in Waulsortian-type mudmounds at Tralee Bay, Co. Kerry, southwest Ireland. *Facies*, **36**, 253–257.

Lagos, M., Scherer, E.E., Tomaschek, F., Münker, C., Keiter, M., Berndt, J., and Ballhaus, C. (2007) High precision Lu–Hf geochronology of Eocene eclogite-facies rocks from Syros, Cyclades, Greece. *Chemical Geology*, **243**, 16–35.

Laird, M.G. and McKerrow, W.S. (1970) The Wenlock sediments of north-west Galway, Ireland. *Geological Magazine*, **107**, 297–317.

Lambeck, K. (1996) Glaciation and sea-level change for Ireland and the Irish Sea since Late Devensian/Midlandian time. *Journal of the Geological Society of London*, **153**, 853–872.

Lambert, R.St.J. and McKerrow, W.S. (1976) The Grampian Orogeny. *Scottish Journal of Geology*, **12**, 271–292.

Lamont, A. (1941) Irish submarine disturbances. *Quarry Managers' Journal, London*, **24**, 123–127.

Lamplugh, G.W., Kilroe, J.R., McHenry, A., Seymour, B.A., and Wright, W.B. (1903) The geology of the country around Dublin. Explanation of Sheet 112. *Geological Survey Ireland Memoir.*

Lamplugh, G.W., Kilroe, J.R., M'Henry, A., Seymour, H.J., Wright, W.B., and Muff, H.B. (1904) *The Geology of the Country Around Belfast*. Memoirs of the Geological Survey of Ireland.

Landes, M., Ritter, J.R.R., Readman, P.W., and O'Reilly, B.M. (2005) A review of the Irish crustal structure and signatures from the Caledonian and Variscan Orogenies. *Terra Nova*, **17**, 111–120.

Landes, M., Prodehl, C., Hauser, F., Jacob, A.W.B., and Vermeulen, N.J. (2000) VARNET-96: influence of the Variscan and Caledonian orogenies on crustal structure in SW Ireland. *Geophysical Journal International*, **140**(3), 660–676.

Landes, M., O'Reilly, B.M., Readman, P.R., Shannon, P.M. and Prodehl, C. (2003) VARNET-96: three-dimensional upper crustal velocity structure of SW Ireland. *Geophysics Journal International*, **153**, 424–442.

Le Cheminant, A.N. and Heaman, L.M. (1989) Mackenzie igneous events, Canada. Middle Proterozoic hotspot magmatism associated with ocean opening. *Earth and Planetary Science Letters*, **96**, 38–48.

Le Gall, B.(1991) Crustal evolutionary model for the Variscides of Ireland and Wales from SWAT seismic data. *Journal of the Geological Society of London*, **148**, 759–774.

Leake, B.E. (1958) The Cashel–Lough Wheelaun intrusion, Co. Galway, Ireland. *Proceedings of the Royal Irish Academy*, **59B**, 155–203.

Leake, B.E. (1963) The location of the Southern Uplands Fault in central Ireland. *Geological Magazine*, **100**, 420–432

Leake, B.E. (1974) Crystallization history and mechanism of emplacement of western part of Galway Granite, Connemara, Western Ireland. *Mineralogical Magazine*, **39**(305), 498–513.

Leake, B.E. (1978) Granite emplacement: the granites of Ireland and their origin. *In*: Bowes, D.R. and Leake, B.E. (eds). *Crustal evolution in northwest Britain and adjacent regions*. Geological Journal Special Issue, **10**, 221–248.

Leake, B.E. (1989) The metagabbros, orthogneisses and paragneisses of the Connemara complex, western Ireland. *Journal of the Geological Society of London*, **146**, 575–96.

Leake, B.E. (2006) Mechanism of emplacement and crystallisation history of the northern margin and centre of the Galway Granite, western Ireland. *Transactions of the Royal Society of Edinburgh: Earth Sciences*, **97**, 1–23.

Leake, B.E. and Ahmed-Said, Y.A. (1994) Hornblende barometry of the Galway batholith, Ireland: an empirical test. *Mineralogy and Petrology*, **51**(2–4), 243–250.

Leake, B.E. and Singh, D. (1986) The Delaney Dome Formation, Connemara, W. Ireland, and the geochemical distinction of ortho- and para-quartzofeldspathic rocks. *Mineralogical Magazine*, **50**, 205–215.

Leake, B.E. and Tanner, P.W.G. (1994) *The Geology of the Dalradian and Associated Rocks of Connemara, Western Ireland*. Royal Irish Academy. Dublin, 96pp.

Leake, B.E., Tanner, P.W.G., and Senior, A. (1981) *The geology of Connemara: 1:63,360 scale geological map with cross-section, fold traces and metamorphic isograd map*. University of Glasgow.

Leake, B.E., Tanner, P.W.G., Singh, D., and Halliday, A.N. (1983) Major southward thrusting of the Dalradian rocks of Connemara, western Ireland. *Nature*, **305**, 210–213.

LeBas, M.J. (1960) The petrology of the layered basic rocks of the Carlingford complex, County Louth. *Transactions of the Royal Society of Edinburgh*, **64**, 169–200.

LeBas, M.J. (1965) On the occurrence in Ireland of a lava related to Hawaiite. *Proceedings of the Geologists' Association*, **76**, 91–94.

LeBas, M.J. (1967) On the origin of the Tertiary granophyres of the Carlingford complex, Ireland. *Proceedings of the Royal Irish Academy*, **65B**, 325–338.

Leeder, M.R. (1982) Upper Palaeozoic basins of the British Isles: Caledonide inheritance versus Hercynian plate margin processes. *Journal of the Geological Society of London*, **139**, 479–491.

Lees, A. (1961) The Waulsortian 'reefs' of Eire: a carbonate mudbank complex of Lower Carboniferous age. *Journal of Geology*, **69**, 101–109.

Lees, A. (1964) The structure and origin of the Waulsortian (Lower Carboniferous) 'reefs' of west central Eire. *Philosophical Transactions of the Royal Society, Series B*, **247**, 483–531.

Lees, A. (1994) Growth forms of Waulsortian banks: a re-appraisal based on new sections in County Galway. *Irish Journal of Earth Sciences*, **13**, 31–48.

Lees, A. (1997) Biostratigraphy, sedimentology and palaeobathymetry of Waulsortian rocks during the late Tournaisian regression, Dinant area, Belgium. *Geological Journal*, **32**, 1–36.

Lees, A. and Miller, J. (1985) Facies variations in Waulsortian buildups, Part 2: Mid-Dinantian buildups from Europe and North America. *Geological Journal*, **20**, 159–180.

Lees, A. and Miller, J. (1995) Waulsortian Banks. *In*: Monty, C.L.V., Bosence, D.W.J., Bridges, P.H., and Pratt, B.R.(eds). *Carbonate mud-mounds: their origin and evolution*. Special Publication of the International Association of Sedimentologists, Blackwell Science, Oxford, **23**, 191–271.

Lees, A., Noel, B., and Bouw, P. (1977) The Waulsortian 'reefs' of Belgium: a progress report. *Mémoires de l'Institut Géologique de l'Université de Louvain*, **29**, 289–315.

Leflef, D. (1973) A change in rock types associated with the approaching shoreline of the Old Red Continent south of Cork, Ireland. *Geologie en Mijnbouw*, **52**, 335–350.

Lefort, J.P., Peucat, J.J., Deunff, J., and Le Herisse, A. (1985) The Goban Spur Palaeozoic basement. *In*: Bailey, M.G. (ed.). *Initial Reports of the Deep Sea Drilling Project 80*. U.S. Government Printing Office, Washington D.C., 677–679.

Legg, I.C., Johnson, T.P., Mitchell, W.I., and Smith, R.A. (1998)

Geology of the country around Derrygonnelly and Marble Arch. *Memoir of the Geological Survey of Northern Ireland*, Sheets 44, 56 and 43 (Northern Ireland), 82pp.

Leggett, J.K., McKerrow, W.S., and Eales, M.H. (1979) The Southern Uplands of Scotland: a Lower Palaeozoic accretionary prism. *Journal of the Geological Society of London*, **136**, 755–770.

Leggo, P.J., Compston, W., and Leake, B.E. (1966) The geochronology of the Connemara granites and its bearing on the antiquity of the Dalradian Series. *Quarterly Journal of the Geological Society of London*, **122**, 91–118.

LeGrande, A.N., Schmidt, G.A., Shindell, D.T., Field, C.V., Miller, R.L., Koch, D.M., Faluvegi, G., and Hoffmann, G. (2006) Consistent simulations of multiple proxy responses to an abrupt climate change event. *PNAS*, **103**, 837–842.

Lemon, G.G. (1966) Serpentinites of the NE Ox Mountains, Eire. *Geological Magazine*, **103**, 124–137.

Lemon, G.G. (1971) The Pre-Cambrian rocks of the NE Ox Mountains, Eire. *Geological Magazine*, **108**, 193–199.

Lenz, A.C. and Vaughan, A.P.M. (1994) A late Ordovician to middle Wenlockian graptolite sequence from a borehole within the Rathkenny Tract, eastern Ireland and its relation to the palaeogeography of the Iapetus Ocean. *Canadian Journal of Earth Sciences*, **31**, 608–616.

Leutwein, F. (1970) Preliminary remarks on some geochronological analyses of Irish granites and gneisses. *Irish Naturalists' Journal*, **16**, 306–308.

Lewarne, G. (1963) Spongiolites from the Arnsbergian of County Limerick, Ireland. *Geological Magazine*, **100**, 290–298.

Lewis, C.A. (1974) The glaciations of the Dingle Peninsula, County Kerry. *Scientific Proceedings of the Royal Dublin Society*, **5A**, 207–235.

Lewis, C.A. (1985) Periglacial features. *In*: Edwards, K.J. and Warren, W.P. (eds). *The Quaternary History of Ireland*. Academic Press, London, 95–113.

Lewis, D. (1986) *The Carboniferous geology of the East Clare syncline*. Unpublished Ph.D. thesis, University of Dublin.

Lien, T., Walker, R.G., and Martinsen, O.J. (2003) Turbidites in the Upper Carboniferous Ross Formation, western Ireland: reconstruction of a channel and spillover system. *Sedimentology*, **50**, 113–148.

Lindsay, N.G., Haselock, P.J., and Harris, A.L. (1989) The extent of Grampian orogenic activity in the Scottish Highlands. *Journal of the Geological Society of London*, **146**, 733–735.

Lisiecki, L.E. and Raymo, M.E. (2005) A Pliocene–Pleistocene stack of 57 globally distributed benthic ∂18O records. *Palaeoceanography*, **20**, PA1003, doi: 10.1029/2004PA001071.

Little, D.J., Mitchell, F.J.G., von Engelbrechten, S.S., and Farrell, E.P. (1996) Assessment of the impact of past disturbance and prehistoric *Pinus sylvestris* on vegetation dynamics and soil development in Uragh Wood, SW Ireland. *The Holocene*, **6**, 90–99.

Livermore, R.A., Smith, A.G., and Briden, J.C. (1985) Palaeomagnetic constraints on the distribution of continents in the late Silurian and early Devonian. *Philosophical Transactions of the Royal Society, London*, **B309**, 29–56.

Loewy, S.L., Connelly, J.N., Dalziel, I.W.D., and Gower, C.F. (2003) Eastern Laurentia in Rodinia: constraints from whole-rock Pb abd U/Pb geochronology. *Tectonophysics*, **375**, 169–197.

Long, C.B. (1974) A note on the stratigraphy of the probable Dalradian metasediments north and north-east of Castlebar, Co. Mayo. *Geological Survey of Ireland Bulletin*, **1**, 459–469.

Long, C.B. and Max, M.D. (1977) Metamorphic rocks in the SW Ox Mountains Inlier, Ireland: their structural compartmentation and place in the Caledonian orogen. *Journal of the Geological Society of London*, **133**, 413–432.

Long, C.B. and McConnell, B. (1995) *Geology of Connemara. Sheet 10, Bedrock Geology 1:100,000 Map Series*. Geological Survey of Ireland.

Long, C.B. and McConnell, B.J. (1997) *Geology of North Donegal: a geological description to accompany the bedrock geology 1:100,000 scale map series, Sheet 1 and part of Sheet 2, North Donegal*. Geological Survey of Ireland.

Long, C.B. and Yardley, B.W.D. (1979) The distribution of pre-Caledonian basement in the Ox Mountains inlier, Ireland. In: Harris, A.L., Holland, C.H., and Leake, B.E. (eds). *The Caledonides of the British Isles: reviewed*. Geological Society, London, Special Publications, **8**, 153-156.

Long, C.B., Max, M.D., and O'Connor, P.J. (1984) Age of the Lough Talt and Easky adamellites in the central Ox Mountains, NW. Ireland, and their structural significance. *Geological Journal*, **19**, 389–397.

Long, C.B., McConnell, B.J., and Alsop, G.I. (1999) *Bedrock Geology 1:100,000 Scale Map Series, Sheet 3 and part of Sheet 4, South Donegal*. Geological Survey of Ireland.

Long, C.B., McConnell, B.J., and Archer, J.B. (1995) *The Geology of Connemara and South Mayo: a geological description of southwest Mayo and adjoining parts of northwest Galway to accompany the Bedrock Geology 1:100,000 Scale Map Series, Sheet 10, Connemara*. Geological Survey of Ireland.

Long, C.B., McConnell, B.J., and Philcox, M.E. (2004) *Geology of South Mayo: a geological descriptionof South Mayo to accompany the bedrock geology 1:100,000 scale map series, Sheet 11, South Mayo with contributions by W. Cox (Minerals) and U. Leader (Groundwater)*. Geological Survey of Ireland, 50pp.

Long, C.B., MacDermot, C.V., Morris, J.H., Sleeman, A.G., Tietzsch-Tyler, D., Aldwell, C.R., Daly, D.R., Flegg, A.M., McArdle, P.M., and Warren, W.P. (1992) *Geology of North Mayo: a geological description to accompany the bedrock geology 1:100,000 scale map series, Sheet 6, North Mayo*. Geological Survey of Ireland, 50pp.

Loughlin, W.P. (1976) Graptolites of lower Wenlock age from the Slieve Bloom inlier, Kinnitty, Co. Offaly. *Irish Naturalists' Journal*, **18**, 282–283.

Lowe, C. (1988) *A crustal study along a north–south seismic refraction profile in Ireland*. D.I.A.S. Unpublished Ph.D. thesis, Dublin University.

Lowe, C. and Jacob, A.W.B. (1989) A north–south seismic profile across the Caledonian suture zone in Ireland. *Tectonophysics*, **168**(4), 297–318.

Lowe, J.J. and Walker, M.J.C. (1997) *Reconstructing Quaternary Environments*. (2nd Edition). Longman, London and New York, 446pp.

Ludwig, K.R. (1998) On the treatment of concordant uranium–lead ages. *Geochimica et Cosmochimica Acta*, **62**, 665–676.

Ludwig, K.R. (2003) *User's Manual for Isoplot 3.00: A Geochronological Toolkit for Microsoft Excel*. Berkeley Geochronology Center Special Publication No.4.

Luecke, W. (1981) Lithium pegmatites in the Leinster granite (southeast Ireland). *Chemical Geology*, **34**, 195–233.

Lumbers, S.B. and Card, K.D. (1991) Chronometric subdivision of the Archaean. *Geology*, **20**, 56–57.

Lund, K., Aleinikoff, J.N., Evans, K.V., and Fanning, C.M. (2003) SHRIMP U–Pb geochronology of Neoproterozoic Windermere Supergroup, central Idaho: implications for rifting of western Laurentia and synchroneity of Sturtian glacial deposits. *Geological Society of America Bulletin*, **115**, 349–372.

Lyle, P. (1980) A petrological and geochemical study of the Tertiary basaltic rocks of northeast Ireland. *Journal of Earth Sciences, Royal Dublin Society*, **2**, 137–152.

Lyle, P. (1985a) The geochemistry and petrology of Tertiary basalts from the Binevenagh area, County Londonderry. *Irish Journal of Earth Sciences*, **7**, 59–64.

Lyle, P. (1985b) The petrogenesis of the Tertiary basaltic and intermediate lavas of northeast Ireland. *Scottish Journal of Geology*, **21**, 71–84.

Lyle, P. (1988) The geochemistry, petrology and volcanology of the Tertiary lava succession of the Binevenagh–Benbraddagh area of County Londonderry. *Irish Journal of Earth Sciences*, **9**, 141–152.

Lyle, P. (2000) The eruptive environment of multi-tiered columnar basalt lava flows. *Journal of the Geological Society of London*, **157**, 715–722.

Lyle, P. and Patton, D.J.S. (1989) The petrography and geochemistry of the Upper Basalt Formation of the Antrim Lava Group in northeast Ireland. *Irish Journal of Earth Sciences*, **10**, 33–41.

Lyle, P. and Preston, J. (1993) Geochemistry and volcanology of the Tertiary basalts of the Giant's Causeway area, Northern Ireland. *Journal of the Geological Society of London*, **150**, 109–120.

Lyle, P. and Thompson, S.J. (1983) The classification and chemistry of the Tertiary intermediate lavas of Northeast Ireland. *Scottish Journal of Geology*, **19**, 17–27.

MacArthur, C.W.P. (1987) Mineralogical and geological travellers in Donegal 1787–1812. *Donegal Annual*, **39**, 39–57.

MacArthur, C.W.P. (1990) Dr Jean-Francois Berger of Geneva (1779–1833): from the Travelling Fund to the Wollaston Donation. *Archives of Natural History*, **17**(1), 97–119.

MacCarthy, F.J. (1990) *The Lower Carboniferous geology of parts of counties Longford and Westmeath, Central Ireland.* Unpublished Ph.D. thesis, University of Dublin.

MacCarthy, I.A.J. (1974) The Upper Devonian and Lower Carboniferous stratigraphy of the Whitegate area, Co. Cork. *Proceedings of the Royal Irish Academy*, **74B**, 313–330.

MacCarthy, I.A.J. (1987) Transgressive facies in the South Munster Basin, Ireland. *Sedimentology*, **34**, 389–422.

MacCarthy, I.A.J. and Gardiner, P.R.R. (1987) Dinantian cyclicity: a case history from the Munster Basin of southern Ireland. *In*: Miller, J., Adams, A.E., and Wright, V.P. (eds). *European Dinantian Environments*. John Wiley, Chichester, 199–237.

MacCarthy, I.A.J., Gardiner, P.R.R., and Horne, R.R. (1978) The lithostratigraphy of the Devonian–early Carboniferous succession in parts of counties Cork and Waterford, Ireland. *Geological Survey of Ireland Bulletin*, **2**, 265–305.

MacDermot, C.V. and Sevastopulo, G.D. (1972) Upper Devonian and Lower Carboniferous stratigraphical setting of Irish mineralisation. *Geological Survey of Ireland Bulletin*, **1**, 267–280.

MacDermot, C.V., Long, C.B., and Harney, S.J. (1996) *A geological description of Sligo, leitrim, and adjoining parts of Cavan, Fermanagh, Mayo and Roscommon, to accompany the Bedrock Geology 1:100,000 Scale Map Series, sheet 7, Sligo–Leitrim, with contributions by K. Claringbold, D. Daly, R. Meehan and G. Stanley.* Geological Survey of Ireland, 99pp.

MacDonald, H., Allan, P.M., and Lovell, J.P.B. (1987) Geology of oil accumulation in Block 26/28, Porcupine Basin, offshore Ireland. *In*: Brooks, J. and Glennie, K.W. (eds). *Petroleum Geology of North-West Europe*. Graham and Trotman Ltd, London, 643–651.

MacGabhann, B.A., Murray, J., and Nicholas, C. (2007) *Ediacaria booleyi: weeded from the garden of Ediacara? In:* Vickers-Rich, P. and Komarowers, P. (eds). *The Rise and Fall of the Ediacaran Biota*. Geological Society London, Special Publications, **286**, 277–295.

Macintyre, R.M., McMenamin, T., and Preston, J. (1975a) K–Ar results from Western Ireland and their bearing on the timing and siting of Thulean magmatism. *Scottish Journal of Geology*, **11**, 227–249.

Macintyre, R.M., Van Breemen, O., Bowes, D.R., and Hopgood, A.M. (1975b) Isotopic study of the gneiss complex, Inistrahull, Co. Donegal. *Scientific Proceedings of the Royal Dublin Society*, **6**, 301–309.

MacNiocaill, C. (2000) A new Silurian palaeolatitude for eastern Avalonia and evidence for crustal rotations in the Avalonian margin of south-western Ireland. *Geophysical Journal International*, **141**(3), 661–671.

MacNiocaill, C. and Smethurst, M.A. (1994) Palaeozoic palaeogeography of Laurentia and its margins: a reassessment of palaeomagnetic data. *Geophysical Journal International*, **116**, 715–725.

MacNiocaill, C., Smethurst, M.A., and Ryan, P.D. (1998) Oroclinal bending in the Caledonides of western Ireland: a mid-Palaeozoic feature controlled by a pre-existing structural grain. *Tectonophysics*, **299**(1–3), 31–47.

MacNiocaill, C., van der Pluijm, B.A., and van der Voo, R. (1997) Ordovician palaeogeography and the evolution of the Iapetus ocean. *Geology*, **25**, 159–162.

Madden, J.S. (1987) *Gamma-ray spectrometric studies of the Main Galway Granite, Connemara, west of Ireland.* Unpublished Ph.D. thesis, National University of Ireland.

Maddox, S.J., Blow, R., and Hardman, M. (1995) Hydrocarbon prospectivity of the Central Irish Sea Basin with reference to Block 42/12, offshore Ireland. *In*: Croker, P.F. and Shannon, P.M., (eds). *The Petroleum Geology of Ireland's Offshore Basins.* Geological Society, London, Special Publications **93**, 59–77.

Maddox, S.J., Blow, R.A., and O'Brien, S.R. (1997) The geology and hydrocarbon prospectivity of the North Channel Basin. *In*: Meadows, N.S., Trueblood, S.P., Hardman, M., and Cowan, G. (eds). *Petroleum Geology of the Irish Sea and Adjacent Areas.* Geological Society, London, Special Publications **124**, 95–111.

Maguire, C.K. and Graham, J.R. (1996) Sedimentation and palaeogeographical significance of the Silurian rocks of the Louisburgh–Clare Island succession, western Ireland. *Transactions of Royal Society of Edinburgh: Earth Sciences*, **86**, 123–136.

Makris, J., Egloff, R., Jacob, A.W.B., Mohr, P., Murphy, T., and Ryan, P. (1988) Continental crust under the southern Porcupine Seabight west of Ireland. *Earth and Planetary Science Letters*, **89**, 387–397.

Makris, J., Ginzburg, A., Shannon, P.M., Jacob, A.W.B., Bean, C.J., and Vogt, U. (1991) A new look at the Rockall Region. *Marine and Petroleum Geology*, **8**, 410–416.

Maliva, R.G. and Dicksson, J.A.D. (1997) Ulster White Limestone Formation (Upper Cretaceous) of Northern Ireland: effects of basalt loading on chalk diagenesis. *Sedimentology*, **44**, 105–112.

Mange, M.A., Dewey, J.F., and Floyd, J.D. (2005) The origin, evolution and provenance of the Northern Belt (Ordovician) of the Southern Uplands Terrane, Scotland: a heavy mineral perspective. *Proceedings Geologists' Association*, **116**, 251–280.

Mange, M.A., Dewey, J.F., and Wright, D.T. (2003) Heavy minerals solve structural and stratigraphical problems in Ordovician strata of the western Irish Caledonides. *Geological Magazine*, **140**, 25–30.

Mangerud, J. (1981) The Early and Middle Weichselian in Norway: a review. *Boreas*, **10**, 381–393.

Manistre, B.E. (1952) The Ordovician volcanic rocks between Collon (Co. Louth) and Grangegeeth (Co. Meath). *Scientific Proceedings, Royal Dublin Society*, **26**, 113–128.

Manning, P.I. and Wilson, H.E. (1975) The stratigraphy of the Larne Borehole, County Antrim. *Bulletin of the Geological Survey of Great Britain*, **50**, 1–27.

Marcantonio, F., Dickin, A.P., McNutt, R.H., and Heaman, L.M. (1988) A 1800-million-year-old Proterozoic terrane in Islay with implications for the crustal structure and evolution of Britain. *Nature, (London)*, **335**, 62–64.

Marchant, T.R. (1978) *The stratigraphy and micropalaeontology of the Lower Carboniferous (Courceyan to Arundian) of the Dublin Basin, Ireland.* Unpublished Ph.D. thesis, University of Dublin.

Marchant, T.R. and Sevastopulo, G.D. (1980) The Calp of the Dublin district. *Journal of Earth Sciences of the Royal Dublin Society*, **5**, 195–203.

Mariane-Serraj, N., Brück, P.M., Higgs, K.T., and Vanguestaine, M. (2000) Ordovician and Silurian acritarch assemblages from the west Leinster and Slievenamon areas of southeast Ireland. *Review of Palaeobotany and Palynology*, **113**, 57–71.

Martignole, J., Machado, N., and Indares, A. (1994) The Wakeham terrane: a Mesoproterozoic terrestrial rift in the eastern part of the Grenville Province. *Precambrian Research*, **68**, 291–306.

Martinsen, O.J. (1989) Styles of soft-sediment deformation on a Namurian (Carboniferous) delta slope, Western Irish Namurian Basin. *In*: Whately, M.K.G. and Pickering, K.T. (eds). *Deltas: sites and traps for fossil fuels.* Geological Society, London Special Publication, **41**, 167–177.

Martinsen, O.J. and Bakken, B. (1990) Extensional and compressional zones in slumps and slides in the Namurian of County Clare, Ireland. *Journal of the Geological Society of London*, **147**, 153–164.

Martinsen, O.J. and Collinson, J.D. (2002) The Western Irish Namurian Basin reassessed: a discussion. *Basin Research*, **14**, 523–542.

Martinsen, O.J., Walker, R.G., and Lien, T. (2000). Upper Carboniferous deep-water sediments, Western Ireland: analogs for passive margin turbidite plays. *In*: Weimer, P., Slatt, R.M., Coleman, J., Rosen, N.C., Nelson, H., Bouma, A.H., Styzen, M.J., and Lawrence, D.T. (eds). *Deep water reservoirs of the world.* GCSSEPM 20th Annual Research Conference, Houston, TX, 533–555.

Martinsen, O.J., Lien, T., Walker, R.G., and Collinson, J.D. (2003) Facies and sequential organisation of a mudstone-dominated slope and basin floor succession: the Gull Island Formation, Shannon Basin, Western Ireland. *Marine and Petroleum Geology*, **20**, 789–807.

Martinson, D.G., Pisias, N.G., Hays, J.D., Imbrie, J., Moore, T.C., and Shackleton, N.J. (1987) Age dating and the orbital theory of the ice ages: development of a high resolution 0 to 300,000 year chronostratigraphy. *Quaternary Research*, **27**, 1–29.

Masson, D.G. and Miles, P.R. (1986) Structure and development of Porcupine Seabight sedimentary basin, offshore Southwest Ireland. *Bulletin of the American Association of Petroleum Geologists*, **70**, 563–584.

Masson, D.G. and Roberts, D. (1981) Late Jurassic–Early Cretaceous reef trends on the Continental Margin SW of the British Isles. *Journal of the Geological Society of London*, **138**, 437–443.

Masson, D.G., Miles, P.R., Max, M.D., Scrutton, R.A., and Inamdar, D.D. (1985) *A free-air gravity anomaly map of the Irish Continental Margin and a new gravity model across the southern Porcupine Seabight.* Geological Survey of Ireland, Report Series **RS85/4,** 8pp.

Masson, F., Jacob, A.W.B., Prodehl, C., Readman, P.W., Shannon, P.M., Schulze, A. and Enderle, U. (1998) A wide-angle seismic traverse through the Variscan of south-west Ireland. *Geophysics Journal International*, **134**, 689–705.

Matley, C.A. and Vaughan, A. (1906) The Carboniferous rocks at Rush (County Dublin). *Quarterly Journal of the Geological Society of London*, **62**, 275–323.

Matley, C.A. and Vaughan, A. (1908) The Carboniferous rocks at Loughshinny (County Dublin). *Quarterly Journal of the Geological Society of London*, **64**, 413–474.

Matte, P. (2001) The Variscan collage and orogeny (480–290 Ma) and the tectonic definition of the Armorica microplate: a review. *Terra Nova*, **13**, 122–128.

Matthews, S.C. (1983) An occurrence of Lower Carboniferous (*Gattendorfia*-Stufe) ammonoids in southwest Ireland. *Neues Jahrbuch für Mineralogie, Geologie und Paläontologie, Stuttgart. Monatshefte Abt. B*, **H5**, 293–299.

Matthews, S.C. and Naylor, D. (1973) Lower Carboniferous conodont faunas from southwest Ireland. *Palaeontology*, **16**, 355–380.

Matthews, S.C., Naylor, D., and Sevastopulo, G.D. (1983) Palaeozoic sedimentary sequences as a reflection of deep structure in southwest Ireland. *Sedimentary Geology*, **34**, 83–95.

Mattinson, J.M. (2005) Zircon U–Pb chemical abrasion ("CA-TIMS") method: combined annealing and multi-step partial dissolution analysis for improved precision and accuracy of zircon ages. *Chemical Geology*, **220**, 47–66.

Max, M.D. (1973) Caledonian metamorphism in part of northwest Co. Mayo, Ireland. *Geological Journal*, **8**, 375-386.

Max, M.D. (1975) Precambrian rocks of South-east Ireland. *In*: Harris, A.L., Shackleton, R.M., Watson, J., Downie, C., Harland, W.B., and Moorbath, S. (eds). A correlation of Precambrian rocks in the British Isles. *Geological Society of London, Special Report*, **6**, 97–101.

Max, M.D. (1978) Tectonic control of offshore sedimentary basins to the north and west of Ireland. *Journal of Petroleum Geology*, **1**, 103–110.

Max, M.D. (1981) Geology of the sea area adjacent to Malin Head, Ireland. *Progress in Underwater Science*, **6**, 68–69.

Max, M.D. (1983) Sliding and sedimentary structures in the Dalradian along the south side of Clew Bay. *In*: Archer, J.B. and Ryan, P.D. (eds). *Geological guide to the Caledonides of western Ireland.* Geological Survey of Ireland Guide Series, **4**, 42–43.

Max, M.D. (1984) Geology of the metamorphic rocks of the Rosses Point Inlier, County Sligo, Ireland. *Geological Survey of Ireland Bulletin*, **3**, 221–227.

Max, M.D. (1989) The Clew Bay Group: a displaced terrane

of Highland Border Group rocks (Cambro-Ordovician) in Northwest Ireland. *Geological Journal*, **24**, 1–17.

Max, M.D. and Dhonau, N.B. (1971) A new look at the Rosslare Complex. *Scientific Proceedings of the Royal Dublin Society*, **4A**, 103–120.

Max, M.D. and Dhonau, N.B. (1974) The Cullenstown Formation: late Pre-Cambrian sediments in south-east Ireland. *Geological Survey Ireland Bulletin*, **4**, 447–458.

Max, M.D. and Inamdar, D.D. (1983) Detailed compilation magnetic map of Ireland and a summary of its deep geology. *Geological Survey of Ireland*, RS 83/1, 10.

Max, M.D. and Long, C.B. (1985) Pre-Caledonian basement in Ireland and its cover relationships. *Geological Journal*, **20**, 341–366.

Max, M.D. and Riddihough, R.P. (1975) Continuation of the Highland Boundary Fault in Ireland. *Geology*, **3**, 206–210.

Max, M.D. and Roddick, J.C. (1989) Age of metamorphism in the Rosslare Complex, SE Ireland. *Proceeding of the Geologists' Association*, **100**, 113–121.

Max, M.D. and Ryan, P.D. (1975) The Southern Upland Fault and its relation to the metamorphic rocks of Connemara. *Geological Magazine*, **112**, 610–612.

Max, M.D. and Ryan, P.D. (1986) The Tuskar Group of southeastern Ireland: its geochemistry and depositional provenance. *Proceedings of the Geologists' Association*, **97**, 73–79.

Max, M.D., Barber, A.J., and Martinez, J. (1990) Terrane assemblage of the Leinster Massif, SE Ireland, during the Lower Palaeozoic. *Journal of the Geological Society of London*, **147**, 1035–1050.

Max, M.D., Inamdar, D.D., and MacIntyre, T. (1982) *Compilation magnetic map: the Irish Continental Shelf and adjacent areas.* Geological Survey of Ireland Report Series **82/2.**

Max, M.D., Kelly, T.J., and Morris, W.A. (1978a) The Maumtrasna Group problem: possible Devonian rocks in Murrisk, western Ireland. *Journal of Earth Sciences, Royal Dublin Society*, **1**, 115–119.

Max, M.D., Long, C.B., and Geoghegan, M.A. (1978b) The Galway Granite. *Geological Survey of Ireland Bulletin*, 223–233.

Max, M.D., Long, C.B., and MacDermot, C.V. (1992) *Bedrock Geology of North Mayo. Sheet 6, Bedrock Geology 1:100,000 Map Series.* Geological Survey of Ireland.

Max, M.D., Long, C.B., and Sonet, J. (1976) The geological age and setting of the Ox Mountains Granodiorite. *Bulletin of the Geologial Survey of Ireland*, **2**, 27–35.

Max, M.D., O'Connor, P.J., and Long, C.B. (1984) New age data for the pre-Caledonian basement of the north-east Ox Mountains and Lough Derg inliers, Ireland. *Geological Survey of Ireland Bulletin*, **3**, 203-209.

Max, M.D., Ploquin, A., and Sonet, J. (1979) The age of the Saltees granite in the Rosslare Complex. *In*: Harris, A.L., Holland, C.H. and Leake, B.E. (eds). *The Caledonides of the British Isles reviewed.* Geological Society of London Special Publication, **8**. 723–725.

Max, M.D., Rex, D., and Winchester, J.A. (1988) The Farnacht Formation along the south side of the Clew Bay Fault Zone, western Ireland: its chemistry and age of metamorphism. *Geological Journal*, **23**, 249–260.

Max, M.D., Ryan, P.D. and Inamdar, D.D. (1983a) A magnetic deep structural geology interpretation of Ireland. *Tectonics*, **2**, 431–451.

Max, M.D., Treloar, P.J., Winchester, J.A., and Oppenheim, M.J. (1983b) Cr mica from the Precambrian Erris Complex, NW Mayo, Ireland. *Mineralogical Magazine*, **47**, 359–364.

Maziane-Serraj, N., Brück, P.M., Higgs, K.T., and Vanguestaine, M. (2000) Ordovician and Silurian acritarch assemblages from the west Leinster and Slievenamon areas of southeast Ireland. *Review of Palaeobotany and Palynology*, **113**, 57–71.

McAfee, A.M. (1987) Late Devonian coastal plain facies in the Munster Basin of southern Ireland. *Abstracts of Papers, 2nd International Symposium on the Devonian System, Calgary.*

McArdle, P. (1974) A Caledonian lamprophyre swarm in southeast Ireland. *Scientific Proceedings of the Royal Dublin Society*, **5A**, 117–122.

McArdle, P. and Kennedy, M.J. (1985) The East Carlow deformation zone and its regional implications. *Geological Survey of Ireland Bulletin*, **3**, 237–255.

McArdle, P. and O'Connor, P.J. (1987) The distribution, geochemistry and origin of appinites and lamprophyres associated with the East Carlow deformation zone of SE Ireland. *Geological Survey of Ireland Bulletin*, **4**, 77–88.

McArdle, P., Reynolds, N., Schaffalitzky, C., and Bell, A.M. (1986) Controls on mineralisation in the Dalradian of Ireland. *In*: Andrews, C.J., Crowe, R.W.A., Finlay, S., Pennell, W.M., and Pyne, J.F. (eds). *Geology and Genesis of mineral deposits in Ireland.* Irish Association for Economic Geology, 31–43.

McAteer, C.A., Daly, J.S., Flowerdew, M.J., Whitehouse, M.J., and Kirkland, C.L. (2008) Laurentian provenance for the Erris Group and Inishkea Division (=Dalradian Grampian Group?) rocks of NW Mayo, Ireland. *Abstracts Volume, 2008 Highland Workshop 24th, 25th April*, 14.

McAteer, C.A., Daly, J.S., Flowerdew, M.J., Connelly, J.N., Housh, T.B., Whitehouse, M.J., and Menuge, J.F. (2006) Basement–cover relationships in the Scottish and Irish Caledonides: provenance of Neoproterozozic cover sequences on the Inner Hebrides and in North Mayo. *In*: Chew, D.M. (ed.). *Programme abstracts, Tectonic Studies Group Highlands Workshop 12th-15th May, 2006, Achill Island Hotel, Achill Sound*, 17–18.

McCabe, A.M. (1969) The glacial deposits of the Maguiresbridge area, County Fermanagh, Northern Ireland. *Irish Geography*, **6**, 63–77.

McCabe, A.M. (1985) Glacial Geomorphology. *In*: Edwards, K.J. and Warren, W.P. (eds). *The Quaternary History of Ireland.* Academic Press, London, 67–93.

McCabe, A.M. (1986) Glaciomarine facies deposited by retreating tidewater glaciers: an example from the late Pleistocene of Northern Ireland. *Journal of Sedimentary Petrology*, **56**, 880–894.

McCabe, A.M. (1987) Quaternary deposits and glacial stratigraphy in Ireland. *Quaternary Science Reviews*, **6**, 259–299.

McCabe, A.M. (1991) The distribution and stratigraphy of Drumlins in Ireland. *In*: Ehlers, J., Gibbard, P.L., and Rose, J. (eds). *Glacial Deposits in Britain and Ireland.* Balkema, Rotterdam, 421–435.

McCabe, A.M. (1993) The 1992 Farrington Lecture: drumlin bedforms and related marginal depositional systems in Ireland. *Irish Geography*, **26**(1), 22–44.

McCabe, A.M. (1995) Marine molluscan shell dates from two glaciomarine jet efflux deposits, eastern Ireland. *Irish Journal of Earth Sciences*, **14**, 37–45.

McCabe, A.M. (1996) Dating and rhythmicity from the last deglacial cycle in the British Isles. *Journal of the Geological Society of London*, **153**, 499–502.

McCabe, A.M. (1997) Geological constraints on geophysical models of relative sea level change during deglaciation of the western Irish Sea Basin. *Journal of the Geological Society of London*, **153**, 499–502.

McCabe, A.M. (1998) Striae at St. Mullan's cave, County Kilkenny, southern Ireland: their origin and chronological significance. *Geomorphology*, **23**, 91–96.

McCabe, A.M. (1999) Ireland. *In*: Bowen, D.Q. (ed.). *A revised correlation of Quaternary deposits in the British Isles*. Geological Society of London, Special Report, **23**, 115–124.

McCabe, A.M. and Clark, P.U. (1998) Ice-sheet variability around the North Atlantic Ocean during the last deglaciation. *Nature*, **392**, 373–377.

McCabe, A.M. and Clark, P.U. (2003) Deglacial chronology from Co. Donegal, Ireland: implications for deglaciation of the British-Irish ice sheet. *Journal of the Geological Society of London*, *160*, 847—855.

McCabe, A.M. and Coxon, P. (1993) A resedimented interglacial peat ball containing *Carpinus* pollen within a glacial efflux sequence, Blackwater, Co. Wexford: evidence for part of the last interglacial cycle in Ireland? *Proceedings of the Geologists' Association*, **104**, 201–207.

McCabe, A.M. and Dardis, G.F. (1989) Sedimentology and depositional setting of late Pleistocene drumlins, Galway Bay, western Ireland. *Journal of Sedimentary Petrology*, **59**, 944–959.

McCabe, A.M. and Dunlop, P. (2006) *The last glacial termination in the north of Ireland*. GSNI, Belfast, 93pp.

McCabe, A.M. and Eyles, N. (1988) Sedimentology of an ice-contact glaciomarine delta, Carey Valley, Northern Ireland. *Sedimentary Geology*, **59**, 1–14.

McCabe, A.M. and Haynes, J.R. (1996) A Late Pleistocene intertidal boulder pavement from an isostatically depressed emergent coast, Dundalk Bay, eastern Ireland. *Earth Surface Processes and Landforms*, **21**, 555–572.

McCabe, A.M. and Ó'Cofaigh, C. (1994) Sedimentation in a subglacial lake, Enniskerry, eastern Ireland. *Sedimentary Geology*, **91**, 57–95.

McCabe, A.M. and Ó'Cofaigh, C. (1995) Late-Pleistocene morainal bank facies at Greystones, eastern Ireland: an example of sedimentation during ice marginal re-equilibration in an isostatically depressed basin. *Sedimentology*, **42**, 647–663.

McCabe, A.M. and Ó'Cofaigh, C. (1996) Upper Pleistocene facies sequences and relative sea-level trends along the south coast of Ireland. *Journal of Sedimentary Research*. **66**(2), 376–390.

McCabe, A.M., Bowen, D.Q. and Penney, D.N. (1993) Glaciomarine facies from the western sector of the last British ice sheet, Malin Beg, County Donegal, Ireland. *Quaternary Science Reviews*, **12**, 35–45.

McCabe, A.M., Clark, P.U., and Clark, J. (2005) AMS^{14}C dating of deglacial events in the Irish Sea Basin and other sectors of the British-Irish ice sheet. *Quaternary Science Reviews*, **24**, 1673–1690.

McCabe, A.M., Clark, P.U., and Clark, J. (2007a) Radiocarbon constraints on the history of the western Irish ice sheet prior to the Last Glacial Maximum. *Geological Society of America Bulletin*, **35**(2), 147–150.

McCabe, A.M., Dardis, G.F., and Hanvey, P.M. (1984) Sedimentology of a late Pleistocene submarine-moraine complex, County Down, Northern Ireland. *Journal of Sedimentary Petrology*, **56**, 716–730.

McCabe, A.M., Dardis, G.F., and Hanvey, P.M. (1987a) Sedimentation at the margins of a late Pleistocene ice-lobe terminating in shallow marine environments, Dundalk Bay, eastern Ireland. *Sedimentology*, **34**, 473–493.

McCabe, A.M., Haynes, J.R., and Macmillan, N.F. (1986) Late-Pleistocene tidewater glaciers and glaciomarine sequences from north County Mayo, Republic of Ireland. *Journal of Quaternary Science*, **1**, 73–84.

McCabe, A.M., Knight, J., and McCarron, S.G. (1998) Evidence for Heinrich Event 1 in the British Isles. *Journal of Quaternary Science*, **13**(6), 549–568.

McCabe, A.M., Knight, J., and McCarron, S.G. (1999) Ice-flow stages and glacial bedforms in north central Ireland: a record of rapid environmental change during the last glacial termination. *Journal of the Geological Society of London*, **156**, 63–72.

McCabe, A.M., Mitchell, G.F., and Shotton, F.W. (1978) An inter-till fresh-water deposit at Hollymount, Maguiresbridge, Co. Fermanagh. *Proceedings of the Royal Irish Academy*, **78B**, 77–89.

McCabe, A.M., Clark, P.U., Clark, J., and Dunlop, P. (2007b) Radiocarbon constraints on readvances of the British–Irish Ice Sheet in the northern Irish Sea Basin during the last deglaciation. *Quaternary Science Reviews*, **26**, 1204–1211.

McCabe, A.M., Coope, R.J., Gennard, D.E., and Doughty, P. (1987b) Freshwater organic deposits and stratified sediments between Early and Late Midlandian (Devensian) till sheets, at Aghnadarragh, County Antrim, Northern Ireland. *Journal of Quaternary Science*, **2**, 11–33.

McCaffrey, K.J.W. (1990) *The emplacement and deformation of granitic rocks in a transpressional shear zone: the Ox mountains igneous complex*. Unpublished Ph.D. thesis, University of Durham.

McCaffrey, K.J.W. (1992) Igneous emplacement in a transpressive shear zone: Ox Mountains igneous complex. *Journal of the Geological Society of London*, **149**, 221–235.

McCaffrey, K.J.W. (1994) Magmatic and solid state deformation partitioning in the Ox Mountains granodiorite. *Geological Magazine*, **131**(5), 639–652.

McCaffrey, K.J.W. (1997) Controls on reactivation of a major fault zone: the Fair Head–Clew Bay line in Ireland. *Journal of the Geological Society of London*, **154**, 129–133.

McCallien, W.J. (1930) The Gneiss of Inishtrahull, County Donegal. *Geological Magazine*, **67**, 542–549.

McCann, N. (1988) An assessment of the subsurface geology between Magilligan Point and Fair Head, Northern Ireland. *Irish Journal of Earth Sciences*, **9**, 71–78.

McCann, N. (1990) The subsurface geology between Belfast and Larne, Northern Ireland. *Irish Journal of Earth Sciences*, **10**, 157–173.

McCann, T. and Shannon, P.M. (1993) Lower Cretaceous seismic stratigraphy and fault movement in the Celtic Sea Basin, Ireland. *First Break*, **11**, 335–344.

McCarron, S. and Monteys, X. (2008) The pattern and timing of Quaternary ice sheet growth onto the north-western Irish continental shelf. *CGG-01 General contributions to glaciology and glacial geology*, 33rd International Geological Congress, Oslo, Norway.

McCarroll, D. (2001) Deglaciation of the Irish Sea Basin: a critique of the glaciomarine hypothesis. *Journal of Quaternary Science*, **16**(5), 393–404.

McCay, G.A., Prave, A.R., Alsop, G.I., and Fallick, A.E. (2006) Glacial trinity: Neoproterozoic Earth history within the British–Irish Caledonides. *Geology*, **34**, 909–912.

McClay, K.R., Norton, M.G., Coney, P., and Davis, G.H. (1986) Collapse of the Caledonian orogen and the Old Red Sandstone. *Nature*, **323**, 147–149.

McConnell, B. (1987) *Geochemistry of Ordovician peralkaline volcanics at Avoca, Co. Wicklow and their regional associations.* Unpublished Ph.D. thesis, University of Dublin.

McConnell, B. (2000) The Ordovician volcanic arc and marginal basin of Leinster. *Irish Journal of Earth Sciences*, **18**, 41–49.

McConnell, B. and Kennan, P.S. (2002) Petrology and geochemistry of the Drogheda granite. *Irish Journal of Earth Sciences.*

McConnell, B. and Morris, J.H. (1997) Initiation of Iapetus subduction under Irish-Avalonia. *Geological Magazine*, **134**, 213–218.

McConnell, B. and Philcox, M.E. (1994) *Geology of Kildare–Wicklow: a geological description to accompany the bedrock geology 1:100,000 scale map series, Sheet 16, Kildare–Wicklow with contributions by A.G. Sleeman, G. Stanley, A.M. Flegg, E.P. Daly and W.P. Warren.* Geological Survey of Ireland, 70pp.

McConnell, B., Philcox, M.E., and Geraghty, M. (2001) *Geology of Meath: a geological description to accompany the bedrock geology 1:100,000 scale map series, Sheet 13, Meath with contributions by J. Morris, W. Cox (Minerals), G. Wright (Groundwater)and R. Meehan (Quaternary).* Geological Survey of Ireland, 78pp.

McConnell, B.J. and Long, C.B. (1997) *Bedrock Geology 1:100,000 Scale Map Series, Sheet 1 and part of Sheet 2, North Donegal.* Geological Survey of Ireland.

McConnell, B.J., Morris, J.H., and Kennan, P.S. (1999) A comparison of the Ribband Group (southeastern Ireland) to the Manx Group (Isle of Man) and Skiddaw Group (northwestern England). *In*: Woodcock, N.H., Quirk, D.G., Fitches, W.R., and Barnes, R.P. (eds). *In sight of the suture: the geology of the Isle of Man in its Iapetus Ocean context.* Geological Society, London, Special Publications, **160**, 337–343.

McConnell, B.J., Stillman, C.J., and Hertogen, J. (1991) An Ordovician basalt to peralkaline rhyolite fractionation series from Avoca, Ireland. *Journal of the Geological Society of London*, **148**, 71–718.

McConnell, J.D.C. (1954) The hydrated calcium silicates riversideite, crestmoreite, and plombierite. *Mineralogical Magazine*, **30**, 293–305.

McCormick, A.G. (1989) *Isotopic studies on the eastern Mourne centre and other Tertiary acid igneous rocks of north east Ireland.* Unpublished Ph.D thesis, Queen's University, Belfast.

McCormick, A.G., Fallick, A.E., Harmon, R.S., Meighan, I.G., and Gibson, D. (1993) Oxygen and hydrogen isotope geochemistry of the Mourne Mountains Tertiary Granites, Northern Ireland. *Journal of Petrology*, **34**, 1177–1202.

McDermott, F. (2004) Palaeoclimate reconstruction from stable isotope variations in speleothems: a review. *Quaternary Science Reviews*, **28**, 901–918.

McDermott, F., Mattey, D.P., and Hawkesworth, C.J. (2001) Centennial-scale Holocene climate variability revealed by a high resolution speleothem δ18O record from SW Ireland. *Science*, **294**, 1328–1331.

McDermott, F., Mattey, D.P., and Hawkesworth, C.J. (2005) Corrections to 'Centennial-scale Holocene climate variability revealed by a high resolution speleothem δ18O record from SW Ireland'. *Science*, **309**, 1816.

McDonald, S., Troll, V.R., Emeleus, C.H., Meighan, I.G., Brock, D., and Gould, R.J. (2004) Intrusive history of the Slieve Gullion ring dyke, Ireland: implications for the internal structure of silicic sub-caldera magma chambers. *Mineralogical Magazine*, **68**, 725–738.

McDonnell, A. and Shannon, P.M. (2001) Comparative Tertiary basin development in the Porcupine and Rockall basins. *In*: Shannon, P.M., Haughton, P.D.W., and Corcoran, D.V. (eds). *The Petroleum Exploration of Ireland's Offshore Basins.* Geological Society, London, Special Publications **188**, 323–344.

McGeary, S., Cheadle, M.J., Warner, M.R., and Blundell, D.J. (1987) Crustal structure of the continental shelf around Britain derived from BIRPS deep seismic profiling. *In*: Brooks, J. and Glennie, K.W. (eds). *Petroleum Geology of North West Europe.* Graham and Trotman, London, 33–41.

McKee, K. (1976) *The geology of the Lower Palaeozoic rocks around Bellewstown, Co. Meath, Eire.* Unpublished M.Sc. thesis, National University of Ireland.

McKerrow, W.S. (1986) The tectonic setting of the Southern Uplands. *In*: Fettes, D.J. and Harris, A.L. (eds). *Synthesis of the Caledonian rocks of Britain.* Reidel Publishing Company, 207–220.

McKerrow, W.S. and Cocks, L.R.M. (1976) Progressive faunal migration across the Iapetus Ocean. *Nature*, **263**(5575), 304–306.

McKerrow, W.S. and Scotese, C.R. (eds). (1990) *Palaeozoic Palaeogeography and Biogeography*, Geological Society, London, Memoir, **12**, 435pp.

McKerrow, W.S., MacNiocaill, C., and Dewey, J.F. (2000a) The Caledonian Orogeny redefined. *Journal of the Geological Society of London*, **157**(6), 1149–1154.

McKerrow, W.S., MacNiocaill, C., Ahlberg, P.E., Clayton, G., Cleal, C.J., and Eagar, R.M.C. (2000b) The Late Palaeozoic relations between Gondwana and Laurussia. *In*: Franke, W., Haak, V., Oncken, O., and Tanner, D. (eds). *Orogenic processes: quantification and modelling in the Variscan fold belt.* Geological Society London, Special Publication, **179**, 9–20.

McLelland, J., Daly, J.S., and McLelland, J.M. (1996) The Grenville orogenic cycle (*c*.1400—900 Ma): an Adirondack perspective and implications for the Grenville Orogen as a whole. *Tectonophysics*, **265**, 1–28.

McManus, J. (1967) Sedimentology of the Partry Series in the Partry Mountains, County Mayo, Eire. *Geological Magazine*, **104**, 228–231.

McManus, J. (1972) The stratigraphy and structure of the Lower Palaeozoic rocks of eastern Murrisk, Co. Mayo. *Proceedings of the Royal Irish Academy*, **72B**, 307–333.

McNestry, A. (1989) *The palynology of non-marine to marine sequences of Devonian–Carboniferous age from Ireland and south Wales.* Unpublished Ph.D. thesis, University of Dublin.

McNestry, A. and Rees, J. (1992) Environments and palynofacies of a Dinantian (Carboniferous) littoral sequence: the basal part of the Navan Group, Navan, County Meath, Ireland. *Palaeogeography, Palaeoclimatology, Palaeoecology*, **96**, 175–193.

McPhilemy, B. (1988) The value of fluorescence microscopy in routine palynofacies analysis: Lower Carboniferous successions from Counties Armagh and Roscommon. *Review of Palaeobotany and Palynology*, **56**, 345–359.

McSherry, M., Parnell, J., Leslie, A.G., and Haggan, T. (2000) Depositional and structural setting of the (?) Lower Old Red Sandstone sediments of Ballymastocker, Co. Donegal. *In*: Friend, P.F. and Williams, B.P.J. (eds). *New perspectives on the Old Red Sandstone.* Geological Society London Special Publication,

180, 109–122.

Meere, P.A. (1995) Sub-greenschist facies metamorphism from the Variscides of SW Ireland: an early syn-extensional peak thermal event. *Journal of the Geological Society of London*, **152**, 511–521.

Meere, P.A. and Mulchrone, K.F. (2006) Timing of deformation within Old Red Sandstone lithologies from the Dingle Peninsula, SW Ireland. *Journal of the Geological Society of London*, **163**, 461–469.

Meighan, I.G. and Neeson, J.C. (1979) The Newry igneous complex, County Down. *In*: Harris, A.L., Holland, C.H., and Leake, B.E. (eds). *The Caledonides of the British Isles reviewed*. Geological Society of London Special Publication, **8**, 717–722.

Meighan, I.G. and Preston, J. (1971) *Tertiary volcanism in Ireland: U.K. contribution to the Upper Mantle Project*. Royal Society Report, **67**.

Meighan, I.G., Fallick, A.E., and McCormick, A.G. (1992) Anorogenic granite magma genesis: new isotopic data for the southern sector of the British Tertiary Igneous Province. *Transactions of the Royal Society of Edinburgh*, **83**, 227–233.

Meighan, I.G., Gibson, D., and Hood, D.N. (1984) Some aspects of Tertiary acid magmatism in NE Ireland. *Mineralogical Magazine*, **48**, 351–363.

Meighan, I.G., McCormick, A.G., Gibson, D., Gamble, J.A., and Graham, I.J. (1988) Rb–Sr isotopic determinations and the timing of Tertiary central complex magmatism in NE Ireland. *In*: Morton, A.C. and Parson, L.M. (eds). *Early Tertiary Volcanism and the Opening of the NE Atlantic*. Geological Society Special Publication, **39**, 349–360.

Menuge, J.F. and Daly, J.S. (1990) Proterozoic evolution of the Erris Complex, northwest Mayo, Ireland: neodymium isotope evidence. *In*: Gower, C.F., Rivers, T., and Ryan, B. (eds). *Mid-Proterozoic Laurentia-Baltica*. Geological Association of Canada Special Paper, **38**, 41–52.

Menuge, J.F. and Daly, J.S. (1994) The Annagh Gneiss Complex in County Mayo, Ireland. *In*: Gibbons, W. and Harris, A.L. (eds). *A revised correlation of Precambrian rocks in the British Isles*. Geological Society, London, Special Report, **22**, 59–62.

Menuge, J.F., Williams, D.M., and O'Connor, P.D. (1995) Silurian turbidites used to reconstruct a volcanic terrain and its Mesoproterozoic basement in the Irish Caledonides. *Journal of the Geological Society of London*, **152**, 269–278.

Menuge, J.F., Daly, J.S., Chew, D.M., and Scanlon, R.P. (2004) Provenance of the Irish Dalradian(s): Nd isotope evidence. *In*: Daly, J.S. (ed.). *Programme and abstracts for Discussion Meeting 4th May 2004, Tectonics Studies Group Highlands Workshop 2004*, 15.

Mezger, K. (1990) Geochronology in Granulites. *In*: Vielzeuf, D. and Vidal, Ph. (eds). *Granulites and Crustal Evolution*, 451–470.

Mezger, K., Essene, E.J., and Halliday, A.J. (1992) Closure temperatures of the Sm–Nd system in metamorphic garnets. *Earth and Planetary Science Letters*, **113**, 397–409.

Millar, G. (1990) *Fracturing style in the Northwest Carboniferous Basin, Ireland*. Unpublished Ph.D. thesis, Queens Univeristy, Belfast.

Millar, I.L. (1999) Neoproterozoic extensional basic magmatism associated with the West Highland granite gneiss in the Moine Supergroup of NW Scotland. *Journal of the Geological Society of London*, **156**, 1153–1162.

Miller, J. (1986) Facies relationships and diagenesis in Waulsortian mudmounds from the Lower Carboniferous of Ireland and N. England. *In*: Schroeder, J.H., and Purser, B.H. (eds). *Reef diagenesis*. Springer Verlag, Berlin, 311–335.

Miller, J.A., Roberts, D.A., and Dearnley, R. (1973) Precambrian rocks drilled from the Rockall Bank. *Nature*, **244**, 21–23.

Millson, J. A. (1987) The Jurassic evolution of the Celtic Sea basins. *In*: Brooks, J. and Glennie, K.W. (eds). *Petroleum Geology of North West Europe*, Graham and Trotman, London, 599–610.

Milroy, P.G. and Wright, P. (2000) A highstand oolitic sequence and associated facies from a late Triassic lake Basin, south-west England. *Sedimentology*, **47**, 187–209.

Mitchell, A.H.G. and McKerrow, W.S. (1975) Analogous evolution of the Burma orogen and the Scottish Caledonides. *Bulletin of the Geological Society of America*, **86**, 305–315.

Mitchell, F. and Ryan, M. (1997) *Reading the Irish landscape*. Town House, Dublin.

Mitchell, F.J.G. (1988) The vegetational history of the Killarney Oakwoods, S.W. Ireland: evidence from fine spatial resolution pollen analysis. *Journal of Ecology*, **76**, 415–436.

Mitchell, F.J.G. (1995) The dynamics of Irish Post-glacial forests. *In*: Pilcher, J.R. and Mac an T-Saoir, S.S. (eds). *Wood, Trees and Forests in Ireland*. Royal Irish Academy, Dublin, 13–22.

Mitchell, F.J.G. (2005) How open were European primeval forests? Hypothesis testing using palaeoecological data. *Journal of Ecology*, **93**, 168–177.

Mitchell, F.J.G. (2006) Where did Ireland's trees come from? *Biology and Environment: Proceedings of the Royal Irish Academy*, **106B**, 251–259.

Mitchell, F.J.G. and Cooney, T. (2004) Vegetation history in the Killarney Valley. *In*: O'Brien, W. (ed.). *Ross Island mining, metal and society in early Ireland. (Bronze Age Studies number 6)*, Department of Archaeology, National University of Ireland, Galway, 481–493.

Mitchell, F.J.G., Bradshaw, R.H.W., Hannon, G.E., O'Connell, M., Pilcher, J.R., and Watts, W.A. (1996) Ireland. *In*: Berglund, B.E., Birks, H.J.B., Ralska-Jasiewiczowa, M., and Wright, H.E. (eds). *Palaeoecological events during the last 15,000 years: regional syntheses of palaeoecological studies of lakes and mires in Europe*. John Wiley, Chichester, 1–13.

Mitchell, G.F. (1970) The Quaternary deposits between Fenit and Spa on the north shore of Tralee Bay, County Kerry. *Proceedings of the Royal Irish Academy*, **70B**, 141–162.

Mitchell, G.F. (1971) Fossil pingos in the south of Ireland. *Nature*, **230**, 43–44.

Mitchell, G.F. (1973) Fossil pingos in Camaross Townland, Co. Wexford. *Proceedings of the Royal Irish Academy*, **73B**, 269–282.

Mitchell, G.F. (1976) *The Irish Landscape*. Collins, London, 240pp.

Mitchell, G.F. (1977) Periglacial Ireland. *Philosophical Transactions of the Royal Society*, **B**, **280**, 199–209.

Mitchell, G.F. (1980) The search for Tertiary Ireland. *Journal of Earth Sciences, Royal Dublin Society*, **3**, 13–33.

Mitchell, G.F. (1981) The Quaternary: until 10,000 bp. *In*: Holland, C.H. (ed.). *A geology of Ireland*. Scottish Academic Press, Edinburgh, 235–258.

Mitchell, G.F. (1985) The Preglacial landscape. *In*: Edwards, K.J. and Warren, W.P. (eds). *The Quaternary History of Ireland*. Academic Press, London, 17–37.

Mitchell, G.F. and Watts, W.A. (1993) Notes on an interglacial deposit in Ballykeerogemore Townland and an interstadial deposit in Battlestown Townland, both in County Wexford. *Irish Journal of Earth Sciences*, **12**, 107-117.

Mitchell, G.F., Penny, L.F., Shotton, F.W., and West, R.G. (1973) *A correlation of Quaternary deposits in the British Isles*. Geological Society of London, Special Report, **4**, 99pp.

Mitchell, J.G. and Mohr, P. (1986) K–Ar systematics in Tertiary dolerites from West Connacht, Ireland. *Scottish Journal of Geology*, **22**, 225–240.

Mitchell, J.G. and Mohr, P. (1987) Carboniferous dykes of west Connacht, Ireland. *Transactions of the Royal Society of Edinburgh: Earth Sciences*, **78**, 113–151.

Mitchell, W.I. (1977) The Ordovician Brachiopoda from Pomeroy, Co. Tyrone. *Palaeontographical Society Monograph*.

Mitchell, W.I. (1992) The origin of Upper Palaeozoic sedimentary basins in Northern Ireland and relationships with the Canadian Maritime Provinces. *In*: Parnell, J. (ed.). *Basins on the Atlantic Seaboard: petroleum geology, sedimentology and basin evolution*. Geological Society Special Publication, **62**, 191–202.

Mitchell, W.I. (ed.) (2004a) *The Geology of Northern Ireland: Our Natural Foundation*. (2nd edition) Geological Survey of Northern Ireland, Belfast, 133–144.

Mitchell, W.I. (2004b) Devonian. *In*: Mitchell, W.I. (ed.) *The geology of Northern Ireland: our natural foundation* (2nd edition). Geological Survey of Northern Ireland, 69–78.

Mitchell, W.I. (2004c) Carboniferous. *In*: Mitchell, W.I. (ed.). *The geology of Northern Ireland: our natural foundation* (2nd edition). Geological Survey of Northern Ireland, Belfast, 79–116.

Mitchell, W.I. (2004d) Variscan (Hercynian) orogenic cycle. *In*: Mitchell, W.I. (ed.). *The geology of Northern Ireland: our natural foundation* (2nd edition). Geological Survey of Northern Ireland, Belfast, 117–124.

Mitchell, W.I. and Mitchell, M. (1983) *The Lower Carboniferous (Upper Viséan) succession at Benburb, Northern Ireland*. Report of the Institute of Geological Sciences, **82/12**, 9pp.

Mitchell, W.I. and Owens, B. (1990) The geology of the western part of the Fintona Block, Northern Ireland: evolution of Carboniferous basins. *Geological Magazine*, **127**, 407–426.

Mitchell, W.I., Carlisle, H., Hiller, N., and Addison, R. (1972) A correlation of the Ordovician rocks of Courtown (Co. Wexford) and Tramore (Co. Waterford). *Proceedings of the Royal Irish Academy*, **72B**, 83–89.

Mitchell, W.I., Cooper, M.R., Hards, V.L., and Meighan, I.G. (1999) An occurrence of silicic volcanic rocks in the early Palaeogene Antrim Lava Group of Northern Ireland. *Scottish Journal of Geology*, **35**, 179–185.

Mix, A.C., Bard, E., and Schneider, R. (2001) Environmental processes of the ice age: land, oceans, glaciers (EPILOG). *Quaternary Science Reviews*, **20**, 627–658.

Moczydlowska, M. and Crimes, T.P. (1995) Late Cambrian acritarchs and their age constraints on an Ediacaran-type fauna from the Booley Bay Formation, Co. Wexford, Eire. *Geological Journal*, **30**, 111–128.

Mohr, P. (1982) Tertiary dolerite intrusions of west-central Ireland. *Proceedings of the Royal Irish Academy*, **82B**, 53–82.

Mohr, P. (1987) The Cill Ala dike swarm, counties Sligo and Mayo: physical parameters. *Irish Naturalists' Journal*, **22**, 326–333.

Mohr, P. (1990) Cryptic Sr and Nd Isotopic Variation across the Leinster Granite, southeast Ireland. *Geological Magazine*, **128**(3), 251–256.

Mohr, P. (2003) Late magmatism of the Galway Granite batholith: I. dacite dikes. *Irish Journal of Earth Sciences*, **21**, 71–104.

Molloy, K. and O'Connell, M. (1991) Palaeoecological investigations towards the reconstruction of woodland and land-use history at Lough Sheeauns, Connemara, western Ireland. *Review of Palaeobotany and Palynology*, **67**, 75–113.

Molloy, K. and O'Connell, M. (1995) Palaeoecological investigations towards the reconstruction of environment and land-use changes during prehistory at Céide Fields, western Ireland. *Probleme der Küstenforschung im südlichen Nordseegebiet*, **23**, 187–225.

Molloy, K. and O'Connell, M. (2004) Holocene vegetation and land-use dynamics in the karstic environment of Inis Oírr, Aran Islands, western Ireland: pollen analytical evidence evaluated in the light of the archaeological record. *Quaternary International*, **113**, 41–64.

Molloy, M.A. and Sanders, I.S. (1983) The NE Ox Mountains inlier. *In*: Archer, J.B. and Ryan, P.D. (eds). *Geological Guide to the Caledonides of Western Ireland*. Geological Survey of Ireland, Guide Series, **4**, 52–55.

Molyneux, S.J. and Hutton, D.H.W. (2000) Evidence for significant granite space creation by the ballooning mechanism: the example of the Ardara pluton, Ireland. *Geological Society of America Bulletin*, **112**(10), 1543–1558.

Monaghan, N. (1995) Fossil insect from the Carboniferous rocks of Co. Clare. *Irish Naturalists' Journal*, **25**, 155.

Monaghan, N.T. and Scannell, M.J.P. (1991) Fossil cypress wood from Tynagh Mine, Loughrea, Co.Galway. *Irish Naturalists' Journal*, **23**(9) 377–378.

Montadert, L., Roberts, D.G., Auffret, G.A., Bock, W.D., du Peuble, P.A., Hailwood, E.A., Harrison, W.E., Kagami, H., Lumsden, D.N., Muller, C.M., Thompson, R.W., Thompson, T.L., and Timofeev, P.P. (1979) *Initial Reports of the Deep Sea Drilling Project*, Washington, U.S. Government Printing Office, **48**, 1183pp.

Moore, J. McM. (1975) Fault tectonics at Tynagh Mine, Ireland. *Transactions of the Institute of Mining and Metallurgy*, **84B**, 141–145.

Moore, J.G. and Shannon, P.M. (1991) Slump structures in the Late Tertiary of the Porcupine Basin, offshore Ireland. *Marine and Petroleum Geology*, **8**, 184–197.

Moore, J.G. and Shannon, P.M. (1995) The Cretaceous succession in the Porcupine Basin, Offshore Ireland: facies distribution and hydrocarbon potential. *In*: Croker, P.F. and Shannon, P.M., (eds). *The Petroleum Geology of Ireland's Offshore Basins*. Geological Society, London, Special Publications **93**, 345–370.

Morel, P. and Irving, E. (1978) Tentative palaeocontinental maps for the early Phanerozoic and Proterozoic. *Journal of Geology*, **86**, 535–561.

Morewood, N.C., Shannon, P.M., Mackenzie, G.D. (2004) Seismic stratigraphy of the southern Rockall Basin: a comparison between wide-angle seismic and normal incidence reflection data. *Marine and Petroleum Geology*, **21**, 1149–1163.

Morewood, N.C., Mackenzie, G.D., Shannon, P.M., O'Reilly, B.M., Readman, P.W., and Makris, J. (2005) The crustal structure and regional development of the Irish Atlantic margin region. *In*: Doré, A.G. and Vining, B. (eds). *Petroleum Geology: North-West Europe and Global Perspectives – Proceedings of the 6th Petroleum Geology Conference*. Geological Society, London, 1023–1034.

Mørk, M.B. and Mearns, E.W. (1986) Sm–Nd isotopic systematics of a gabbro–eclogite transition. *Lithos*, **19**, 255–267.

Morley, C.T. (1966) *The Geology of South Achill and Achill Beg, Co. Mayo, Ireland*. Unpublished Ph.D. thesis, University of Dublin.

Morris, J.H. (1979) *The geology of the western end of the Lower Palaeozoic Longford-Down inlier, Ireland*. Unpublished Ph.D. thesis, University of Dublin.

Morris, J.H. (1983) The stratigraphy of the Lower Palaeozoic rocks in the western end of the Longford-Down inlier, Ireland. *Journal of Earth Sciences, Royal Dublin Society*, **5**, 201–218.

Morris, J.H. (1984) *The metallic mineral deposits of the Lower Palaeozoic Longford-Down inlier in the Republic of Ireland*. GSI Report Series RS 84/1 (Mineral Resources).

Morris, J.H. (1987) The northern belt of the Longford-Down inlier, Ireland and Southern Uplands, Scotland: an Ordovician back-arc basin. *Journal of the Geological Society of London*, **144**, 773–786.

Morris, J.H., Oliver, G.J.H., and Kassi, A.M. (1988) The petro-tectonic affinity of spilitic volcanics and ferro-magnesian detritus in the Southern Uplands–Longford Down zone, Scotland and Ireland. *Geological Association Canada Annual Meeting, Program with abstracts*, A87.

Morris, J.H., Somerville, I.D., and MacDermot, C.V. (2003) *Geology of Longford, Roscommon, Westmeath, and adjoining parts of Cavan, Laois, Leitrim and Galway to accompany the Bedrock Geology 1:100,000 scale map series, Sheet 12, Longford–Roscommon, with contributions by D.G. Smith, B. McConnell, K. Claringbold, W. Cox, and M. Lee*. Geological Survey of Ireland, ix + 99pp.

Morris, J.H., Howard, D.W., Sevastopulo, G.D., and Williams, C.T. (1980) An occurrence of Tertiary olivine-dolerite sills in Lower Carboniferous (Courceyan) rocks near Carrigallen, County Leitrim. *Journal of Earth Sciences, Royal Dublin Society*, **4**, 39–52.

Morris, J.H., Prendergast, T., Synott, P., Delahunty, R., Crean, E., and O'Brien, C. (1986) The geology of the Monaghan–Castleblayney district, County Monaghan: a provisional summary. *Geological Survey Ireland Bulletin*, **3**, 337–349.

Morris, P. (1973) Density, magnetic and resistivity measurements on Irish rocks. *Dublin Institute for Advanced Studies, Geophysical Bulletin*, **31**, 48.

Morris, P. (1989) A composite magnetic map of Ireland. *Dublin Institute for Advanced Studies, Geophysical Bulletin*, **42**, 7.

Morris, W.A. (1976) Palaeomagnetic results from the Lower Palaeozoic of Ireland. *Canadian Journal of Earth Sciences*, **13**(2), 294–304.

Morton, A.C. (1984) Heavy minerals from Palaeogene sediments. Deep Sea Drilling Project Leg 87: their bearing on stratigraphy, sediment provenance and the evolution of the North Atlantic. *Initial Reports of the Deep Sea Drilling Project*, **81**, 653–661.

Morton, A.C. and Taylor, P.N. (1991) Geochemical and isotopic constraints on the nature and age of basement rocks from the Rockall Bank, NE Atlantic. *Journal of the Geological Society of London*, **148**, 631–634.

Morton, A.C., Todd, S.P., and Haughton, P.D.W. (1991) *Developments in Sedimentary Provenance Studies*. Geological Society, London, Special Publications, **57**, 370pp.

Morton, N. (1992) Late Triassic to Middle Jurassic stratigraphy, palaeogeography and tectonics west of the British Isles. *In*: Parnell, J. (ed.). *Basins on the Atlantic Seaboard: Petroleum Geology, Sedimentology and Basin Evolution*. Geological Society, London, Special Publications, **62**, 53–70.

Morton, W.H. (1965) The Carboniferous stratigraphy of the area north-west of Newmarket, Co. Cork, Ireland. *Scientific Proceedings of the Royal Dublin Society*, **2A**, 47–64.

Muir, R.J., Fitches, W.R., and Maltman, A.J. (1992) Rhinns Complex: a missing link in the Proterozoic basement of the North Atlantic region. *Geology*, **20**, 1043–1046.

Muir, R.J., Fitches, W.R., and Maltman, A.J. (1994a). Rhinns Complex: Proterozoic basement on Islay and Colonsay, Inner Hebrides, Scotland, and on Inishtrahull, NW Ireland. *Transactions of the Royal Society of Edinburgh: Earth Sciences*, **85**, 77–90.

Muir, R.J., Fitches, W.R., Maltman, A.J., and Bentley, M.R. (1994b) Precambrian rocks of the southern Inner Hebrides–Malin Sea region: Colonsay, west Islay, Inishtrahull and Iona. *In*: Harris, A.L. and Gibbons, F.A. (eds). *A Correlation of Precambrian Rocks in the British Isles, 2nd edition* Geological Society, London, Special Report No. **22**, 54–8.

Muir, R.J., Ireland, T.R., Bentley, M.R., Fitches, W.R., and Maltman, A.J. (1997) A Caledonian age for the Kiloran Bay appinite intrusion on Colonsay, Inner Hebrides. *Scottish Journal of Geology*, **33**, 75–83.

Murchison, R.I. (1836) On the recent discovery of fossil fishes (*Palaeoniscus catopterus* Agassiz) in the new red sandstone of Tyrone, Ireland. *Proceedings of the Geological Society*, **2**, 206–207.

Murdoch, L.M., Musgrove, F.W., and Perry, J.S. (1995) Tertiary uplift and inversion history in the North Celtic Sea Basin and its influence on source rock maturity. *In*: Croker, P.F. and Shannon, P.M., (eds). *The Petroleum Geology of Ireland's Offshore Basins*. Geological Society, London, Special Publications **93**, 297–319.

Murphy, F.C. (1984) *The Lower Palaeozoic stratigraphy and structural geology of the Balbriggan inlier, Counties Meath and Dublin*. Unpublished Ph.D. thesis, Dublin University.

Murphy, F.C. (1985) Non-axial planar cleavage and Caledonian sinistral transpression in eastern Ireland. *Geological Journal*, **20**(3), 257–279.

Murphy, F.C. (1987a) Evidence for late Ordovician amalgamation of volcanogenic terranes in the Iapetus suture zone, eastern Ireland. *Transactions Royal Society of Edinburgh: Earth Sciences*, **78**, 153–167.

Murphy, F.C. (1987b) Late Caledonian granitoids and timing of deformation in the Iapetus suture zone of eastern Ireland. *Geological Magazine*, **124**, 251–256.

Murphy, F.C. (1990) Basement–cover relationships of a reactivated Cadomian mylonite zone, Rosslare Complex, SE Ireland. *In*: D' Lemos, R.S., Strachan, R.A., and Topley, C.G. (eds). *The Cadomian orogeny*. Geological Society London Special Publication, **51**, 329–339.

Murphy, F.C. and Hutton, D.H.W. (1986) Is the Southern Uplands of Scotland really an accretionary prism? *Geology*, **14**, 354–357.

Murphy, F.C., Anderson, T.B., Daly, J.S., Gallagher, V., Graham, J.R., Harper, D.A.T., Johnston, J.D., Kennan, P.S., Kennedy, M.J., Long, C.B., Morris, J.H., O'Keefe, W.G., Parkes, M.A., Ryan, P.D., Sloan, R.J., Stillman, C.J., Tietzch-Tyler, D., Todd, S.P., and Wrafter, J.P. (1991) Appraisal of Caledonian suspect terranes in Ireland. *Irish Journal of Earth Sciences*, **11**, 11–41.

Murphy, F.X. (1988) Facies variation within the Waulsortian Limestone Formation of the Dungarvan Syncline, southern Ireland. *Proceedings of the Geologists' Association*, **99**, 205–219.

Murphy, F.X. (1990) The Irish Variscides: a fold belt developed within a major surge zone. *Journal of the Geological Society of*

London, **147**, 451–460.

Murphy, N.J. and Croker, P.F. (1992) Many play concepts seen over wide area in Erris, Slyne troughs off Ireland. *Oil and Gas Journal*, Sept. 14, 92–97.

Murphy, T. (1960) Gravity anomaly map of Ireland, Sheet 5: Southwest. *Dublin Institute of Advanced Studies, Geophysical Bulletin*, **18**.

Murphy, T. (1974) Gravity anomaly map of Ireland. *Dublin Institute for Advanced Studies, Geophysical Bulletin*, **32**.

Murray, J.W. (1992) Palaeogene and Neogene. *In*: Cope, J.C.W., Ingham, J.K., and Rawson, P.F. (eds). *Atlas of Palaeogeography and Lithofacies*. Geological Society, London, Memoir **13**, 153pp.

Musgrove, F.W. and Mitchener, B. (1996) Analysis of the pre-Tertiary rifting history of the Rockall Trough. *Petroleum Geoscience*, **2**, 353–360.

Musgrove, F.W, Murdoch, L.M. and Lenehan, T. (1995) The Variscan fold-thrust belt of southeast Ireland and its control on early Mesozoic extension and deposition: a method to predict the Sherwood Sandstone. *In*: Croker, P.F. and Shannon, P.M. (eds). *The Petroleum Geology of Ireland's Offshore Basins*. Geological Society, London, Special Publications **93**, 81–100.

Mykura, W. (1983) Old Red Sandstone. *In*: Craig, G.Y. (ed.). *Geology of Scotland*, 2nd. Edn., 205–252.

Nagy, Zs.R., Somerville, I.D., Gregg, J.M., Becker, S.P., and Shelton, K.L. (2005a) Lower Carboniferous peritidal carbonates and associated evaporites adjacent to the Leinster Massif, southeast Irish Midlands. *Geological Journal*, **40**, 173–192.

Nagy, Zs.R., Somerville, I.D., Gregg, J.M., Becker, S.P., Shelton, K.L., and Sleeman, A.G. (2005b) Sedimentation in an actively tilting half-graben: sedimentology of the late Tournaisian–Viséan (Mississippian, Lower Carboniferous) carbonate rocks in south County Wexford, Ireland. *Sedimentology*, **52**, 489–512.

NASP Members: Executive Group. (1994) Climatic changes in areas adjacent to the North Atlantic during the last glacial–interglacial transition (14–9 ka bp): a contribution to IGCP-253. *Journal of Quaternary Science*, **9**(2), 185–198.

Nawaz, R. (1974) *Contact metamorphism of some Carboniferous rocks at Killala Bay, Ireland*. Unpublished Ph.D. thesis, Queen's University, Belfast.

Nawaz, R. (1982) A chemical classification scheme for the gismondine group zeolites. *Irish Naturalists' Journal*, **20**, 480–483.

Nawaz, R. and Foy, H.J. (1982) A chemical classification of the chabazite group zeolites. *Irish Naturalists' Journal*, **20**, 435–440.

Nawaz, R. and Malone, J.F. (1982) Gobbinsite, a new mineral from Co. Antrim, Northern Ireland. *Mineralogical Magazine*, **46**, 365–369.

Naylor, D. (1966) The Upper Devonian and Carboniferous geology of the Old Head of Kinsale, Co. Cork. *Scientific Proceedings of the Royal Dublin Society*, **A2**, 229–249.

Naylor, D. (1969) Facies change in Upper Devonian and Lower Carboniferous rocks of southern Ireland. *Geological Journal*, **6**, 307–328.

Naylor, D. (1975) Upper Devonian–Lower Carboniferous stratigraphy along the south coast of Dunmanus Bay, Co. Cork. *Proceedings of the Royal Irish Academy*, **75B**, 317–337.

Naylor, D. (1978) A structural section across the Variscan fold belt, southwest Ireland. *Journal of Earth Sciences, Royal Dublin Society*, **1**, 63–70.

Naylor, D. (1992) The post-Variscan history of Ireland. *In*: Parnell, J. (ed.). *Basins on the Atlantic Seaboard*. Geological Society, London, Special Publications **62**, 255–275.

Naylor, D. (1998) Irish shorelines through geological time. *Royal Dublin Society: Occasional Papers in Irish Science and Technology*, **17**, 20pp.

Naylor, D. (2003) A Landsat TM structural interpretation of the South Munster Basin, Co. Cork. *Irish Journal of Earth Sciences*, **21**, 19–38.

Naylor, D. and Anstey, N.A. (1987) A reflection seismic study of the Porcupine Basin, offshore West Ireland. *Irish Journal of Earth Sciences*, **8**, 187–210.

Naylor, D. and Clayton, G. (2000) Palynological and maturation data and their bearing on Irish post-Variscan palaeogeography. *Irish Journal of Earth Sciences*, **18**, 33–39.

Naylor, D. and Jones, P.C. (1967) Sedimentation and tectonic setting of the Old Red Sandstone of Southwest Ireland. *In*: Oswald, D.H. (ed.). *International Symposium on the Devonian System*, **2**. Alberta Society of Petroleum Geologists, 1089–1099.

Naylor, D. and Reilly, T. (1981) Kinsale Formation, Northeast Seven Heads, County Cork. *Geological Survey of Ireland Bulletin*, **3**, 1–8.

Naylor, D. and Sevastopulo, G.D. (1979) The Hercynian 'Front' in Ireland. *Krystallinikum*, **14**, 77–90.

Naylor, D. and Sevastopulo, G.D. (1993) The Reenydonagan Formation (Dinantian) of the Bantry and Dunmanus synclines, County Cork. *Irish Journal of Earth Sciences*, **12**, 191–203.

Naylor, D. and Shannon, P. M. (1982) *The Geology of Offshore Ireland and West Britain*, Graham and Trotman, London, 161pp.

Naylor, D. and Shannon, P. M. (2005) The structural framework of the Irish Atlantic Margin. *In*: Doré, A.G. and Vining, B.A. (eds). Petroleum Geology: *North-West Europe and Global Perspectives – Proceedings of the 6th Petroleum Geology Conference*, 1009–1021.

Naylor, D., Higgs, K., and Boland, M.A. (1977) Stratigraphy on the north flank of the Dunmanus Syncline, west Cork. *Geological Survey of Ireland Bulletin*, **2**, 143–157.

Naylor, D., Jones, P.C., and Clayton, G. (1978) The Namurian stratigraphy of Whiddy Island, Bantry Bay, West Cork. *Geological Survey of Ireland Bulletin*, **2**, 235–253.

Naylor, D., Jones, P.C., and Matthews, S.C. (1974) Facies relationships in the Upper Devonian–Lower Carboniferous of southwest Ireland and adjacent regions. *Geological Journal*, **9**, 77–96.

Naylor, D., Philcox, M.E., and Clayton, G. (2003) Annaghmore-1 and Ballynamullan-1 wells, Larne–Lough Neagh Basin, Northern Ireland. *Irish Journal of Earth Sciences*, **21**, 47–69.

Naylor, D., Sevastopulo, G.D., and Sleeman, A.G. (1989) Subsidence history of the South Munster Basin, Ireland. *In*: Arthurton, R.S., Gutteridge, P. and Nolan, S.C. (eds). *The role of tectonics in Devonian and Carboniferous sedimentation in the British Isles*. Yorkshire Geological Society, Occasional Publication, **6**, 99–110.

Naylor, D., Sevastopulo, G.D., and Sleeman, A.G. (1996) Contemporaneous erosion and reworking within the Dinantian of the South Munster Basin. *In*: Strogen, P., Somerville, I.D., and Jones, G.Ll. (eds). *Recent advances in Lower Carboniferous geology*. Geological Society, London Special Publications, **107**, 331–343.

Naylor, D., Shannon, P.M., and Murphy, N. (1999) *Irish Rockall Basin region – a standard structural nomenclature system*. Petroleum Affairs Division, Dublin, Special Publication **1/99**, 42pp.

Naylor, D., Shannon, P.M., and Murphy, N. (2002) *Porcupine-Goban region – a standard structural nomenclature system.* Petroleum Affairs Division, Dublin, Special Publication **1/02**, 65pp.

Naylor, D., Haughey, N., Clayton, G., and Graham, J.R. (1993) The Kish Bank Basin, offshore Ireland. *In*: Parker, J.R. (ed.). *Petroleum Geology of North-west Europe: Proceedings of the 4th Conference.* Geological Society, London, 845–855.

Naylor, D., Higgs, K., Reilly, T.A., and Sevastopulo, G.D. (1988) Dinantian and Namurian stratigraphy, Seven Heads Peninsula, County Cork. *Irish Journal of Earth Sciences,* **9**, 1–17.

Naylor, D., Nevill, W.E., Ramsbottom, W.H.C., and Sevastopulo, G.D. (1985) Upper Dinantian stratigraphy and faunas of the Old Head of Kinsale and Galley Head, south County Cork. *Irish Journal of Earth Sciences,* **7**, 47–58.

Naylor, D., Phillips, W.E.A., Sevastopulo, G.D., and Synge, F.M. (1980) *An introduction to the geology of Ireland.* Royal Irish Academy, 49pp.

Naylor, D., Reilly, T.A., Sevastopulo, G.D., and Sleeman, A.G. (1983) Stratigraphy and structure in the Irish Variscides. *In*: Hancock, P.L. (ed.). *Variscan fold belt in the British Isles.* Hilger, 20–46.

Naylor, D., Sevastopulo, G.D., Sleeman, A.G., and Reilly, T.A. (1981) The Variscan fold belt in Ireland. *Geologie en Mijnbouw,* **60**, 49–66.

Nealon, T. (1989) Deep basinal turbidites reinterpreted as distal tempestites: the Silurian Glencraff Formation of north Galway. *Irish Journal of Earth Sciences,* **10**, 55–65.

Nealon, T. and Williams, D.M. (1988) Storm-influenced shelf deposits from the Silurian of western Ireland: a reinterpretation of deep basin sediments. *Geological Journal,* **23**, 311–320.

Nelson, E.C. and Walsh, W.F. (1993) *Trees of Ireland: native and naturalized.* Lilliput Press, Dublin.

Neuman, R.B. (1984) Geology and palaeobiology of islands in the Ordovician Iapetus Ocean. *Bulletin Geological Society America,* **95**, 1188–1201.

Neuman, R.B. (1988) Palaeontological evidence bearing on the Arenig–Caradoc development of the Iapetus Ocean basin. *In*: Harris, A.L. and Fettes, D.J. (eds). *The Caledonian–Appalachian orogen.* Geological Society London Special Publication, **38**, 269–274.

Neuman, R.B. and Bates, D.E.B. (1978) Reassessment of Arenig and Llanvirn age (early Ordovician) brachiopods from Anglesey, north-west Wales. *Palaeontology,* **21**, 571–613.

Neuman, R.B. and Harper, D.A.T. (1992) Palaeogeographic significance of Arenig–Llanvirn Toquima–Table Head and Celtic brachiopod assemblages. *In*: Webby, B.D. and Laurie, D. (eds). *Global perspectives on Ordovician geology.* Balkema, Rotterdam, 241–254.

Nevill, W.E. (1956) The Millstone Grit and Lower Coal Measures of the Leinster coalfield. *Proceedings of the Royal Irish Academy,* **58(B)**, 1–15.

Nevill, W.E. (1957a) Sand volcanoes, sheet slumps and stratigraphy of the Slieveardagh coalfield, County Tipperary. *Scientific Proceedings of the Royal Dublin Society,* **27**, 314–324.

Nevill, W.E. (1957b) The geology of the Summerhill basin, Co. Meath, Ireland. *Proceedings of the Royal Irish Academy,* **58(B)**, 293–303.

Nevill, W.E. (1958a) The Carboniferous knoll-reefs of east-central Ireland. *Proceedings of the Royal Irish Academy,* **58B**, 285–303.

Nevill, W.E. (1958b) A note on the occurrence of Coal Measures in eastern County Limerick, Ireland. *Geological Magazine,* **95**, 20–24.

Nevill, W.E. (1966) The geology of the north Cork (Kanturk) coalfield. *Geological Magazine,* **103**, 423–431.

Newman, P. (1999) The geology and hydrocarbon potential of the Peel and Solway Basins, East Irish Sea. *Journal of Petroleum Geology,* **22**, 305–324.

Nex, P.A.M., Kinnaird, J.A., and Ixer, R.A. (2003) Localised ductile thrusting north of the Variscan Front, Ross Island, south-west Ireland. *Geological Journal,* **38**, 15–29.

Ni Bhroin. C. (1999) *Sedimentology and stratigraphy of the Carboniferous Donegal Basin.* Unpublished Ph.D. thesis, University of Dublin.

Nichols, G. and Jones, T.M. (1992) Fusain in Carboniferous shallow marine sediments, Donegal, Ireland: the sedimentological effects of wildfire. *Sedimentology,* **39**, 487–502.

Noble, S.R., Hyslop, E.K., and Highton, A.J. (1996) High-precision U–Pb monazite geochronology of the *c.*806 Ma Grampian Shear Zone and the implications for the evolution of the Central Highlands of Scotland. *Journal of the Geological Society of London,* **153**, 511–514.

Noble, S.R., Highton, A.J., Hyslop, E.K., and Barreiro, B. (1997) A Rodinian connection for the Scottish Highlands? Evidence from U–Pb geochronology of Grampian Terrane migmatites and pegmatites. *Terra Abstracts, Abstract supplement No 1 to Terra Nova,* **9**, 165.

Nockolds, S.R. and Richey, J.E. (1939) Replacement veins in the Mourne Mountains Granites, N. Ireland. *American Journal of Science,* **237**, 27–47.

Nolan, S.C. (1986) *The Carboniferous geology of the Dublin area.* Unpublished Ph.D. thesis, University of Dublin.

Nolan, S.C. (1989) The style and timing of Dinantian synsedimentary tectonics in the eastern part of the Dublin Basin, Ireland. *In*: Arthurton, R.S., Gutteridge, P., and Nolan, S.C. (eds). *The role of tectonics in Devonian and Carboniferous sedimentation in the British Isles.* Yorkshire Geological Society Occasional Publication, **6**, 83–98.

Northern Ireland Geological Survey. (1978) *Map sheet 34.* HMSO.

Northern Ireland Geological Survey. (1982) *Map sheet 45.* HMSO.

Nudds, J. (1986) The life and work of John Joly (1857–1933). *Irish Journal of Earth Sciences,* **8**, 81–94.

O'Brien, A. (1999) *The shape of the Leinster Granite and some other Caledonian Granites.* Unpublished Ph.D. thesis, University College Dublin.

O'Brien, W. (1994) *Mount Gabriel: Bronze Age mining in Ireland. (Bronze Age Studies number 3).* Galway University Press, Galway.

O'Brien, W. (2004) *Ross Island mining, metal and society in early Ireland. (Bronze Age Studies 6).* Department of Archaeology, National University of Ireland, Galway.

Ó'Cofaigh, C. (1993) *Sedimentology of Late-Pleistocene glacimarine and shallow marine deposits from the south coast of Ireland.* Unpublished MSc. thesis, University of Dublin,.

O'Connell, M. (1986) Reconstruction of local landscapes development in the post-Atlantic based on palaeoecological investigations at Carrownaglogh prehistoric field system, County Mayo, Ireland. *Review of Palaeobotany and Palynology,* **49**, 117–176.

O'Connell, M. (1990) Origins of Irish lowland blanket bog. *In*:

Doyle, G.J. (ed.). *Ecology and Conservation of Irish Peatlands.* Royal Irish Academy, Dublin, 49–71.

O'Connell, M., Molloy, K., and Bowler, M. (1988) Post-glacial landscape evolution in Connemara, western Ireland with particular reference to woodland history. *In*: Birks, H.H., Birks, H.J.B., Kaland, P.E., and Moe, D. (eds). *The cultural landscape: past, present and future.* Cambridge University Press, Cambridge, 487–514.

O'Connell, M., Huang, C.C., and Eicher, U. (1999) Multidisciplinary investigations, including stable-isotope studies, of thick late-glacial sediments from Tory Hill, Co. Limerick, western Ireland. *Palaeogeography, Palaeoclimatology, Palaeoecology*, **147**, 169–208.

O'Connor, E.A. (1975) Lower Palaeozoic rocks of the Shercock–Aghnamullen district, Counties Cavan and Monaghan. *Proceedings of the Royal Irish Academy*, **75B**, 499–530.

O'Connor, J.M., Stoffers, P., Wijbrans, J.R., Shannon, P.M., and Morrissey, T. (2000) Evidence from episodic seamount volcanism for pulsing of the Icelandic plume in the past 70 Myr. *Nature*, **408**, 954–958.

O'Connor, P.G. (1987) *Volcanology, geochemistry and mineralisation in the Charlestown Ordovician inlier, Co. Mayo.* Unpublished Ph.D. thesis, National University of Ireland.

O'Connor, P.G. and Poustie, A. (1986) Geological setting of, and alteration associated with, the Charlestown mineral deposit. *In*: Andrew, C.J., Crowe, R.W.A., Finlay, S., Pennell, W.M., and Pyne, J.F. (eds). *Geology and genesis of mineral deposits in Ireland.* Irish Association Economic Geology, 89–101.

O'Connor, P.J. (1975) Rb–Sr whole rock isochron for the Newry granodiorite, NE. Ireland. *Scientific Proceedings of the Royal Dublin Society*, **5A**, 407–413.

O'Connor, P.J. (1989) Chemistry and Rb–Sr age of the Corvock Granite. *Geological Survey of Ireland Bulletin*, **4**, 99–105.

O'Connor, P.J. and Brück, P.M. (1978) Age and origin of the Leinster Granite. *Journal of Earth Sciences, Royal Dublin Society*, **1**, 105–113.

O'Connor, P.J. and Reeves, T.J. (1980) Age and provenance of the Carrigmore diorite, Co. Wicklow. *Geological Survey of Ireland Bulletin*, **2**, 307–314.

O'Connor, P.J., Aftalion, M., and Kennan, P.S. (1989) Isotopic U–Pb ages of zircon and monazite from the Leinster Granite, southeast Ireland. *Geological Magazine*, **126**(6), 725–728.

O'Connor, P.J., Kennan, P.S., and Aftalion, M. (1988) New Rb–Sr and U–Pb ages for the Carnsore Granite and their bearing on the antiquity of the Rosslare Complex, southeastern Ireland. *Geological Magazine*, **125**(1), 25–29.

O'Connor, P.J., Long, C.B., and Evans, J.A. (1987) Rb–Sr whole-rock isochron studies of the Barnesmore and Fanad plutons, Donegal, Ireland. *Geological Journal*, **22**(1), 11–23.

O'Connor, P.J., Hennessy, J., Bruck, P.M., and Williams, C.T. (1982b) Abundance and distribution of Uranium and Thorium in the northern units of the Leinster Granite, Ireland. *Geological Magazine*, **119**(6), 581–592.

O'Connor, P.J., Long, C.B., Basham, I.R., Swainbank, I.G., and Beddoestephens, B. (1984) Age and geological setting of Uranium mineralization associated with the main Donegal granite, Ireland. *Transactions of the Institution of Mining and Metallurgy: Section B-Applied Earth Science*, **93**(Nov), B190–B194.

O'Connor, P.J., Long, C.B., Kennan, P.S., Halliday, A.N., Max, M.D., and Roddick, J.C. (1982a) Rb–Sr isochron study of the Thorr and Main Donegal Granites, Ireland. *Geological Journal*, **17**(4), 279–295.

O'Keeffe, W. (1986) Age and postulated source rocks for mineralization in central Ireland, as indicated by lead isotopes. *In*: Andrews, C.J., Crowe, R.W.A., Finlay, S., Pennell, W.M. and Pyne, J.F. (eds). *Geology and genesis of mineral deposits in Ireland.* Irish Association for Economic Geology, 617–624.

Oksanen, P.O., Stefanini, B., Mitchell, F.J.G., Blundell, A., and Charman, D. (2006) Multi-proxy high resolution record from Ballyduff Bog, Ireland, reflecting hydrological and climate during the last 4700 years. *Irish Association for Quaternary Studies Newsletter*, **37**, 6.

Old, R.A. (1975) The age and field relationships of the Tardree Tertiary rhyolite complex, Co. Antrim, Northern Ireland. *Bulletin of the Geological Survey of Great Britain*, **51**, 21–40.

Oldham, T. (1844) On the rocks at Bray Head. *Journal of the Geological Society of Dublin*, **3**, 60.

O'Liathain, M. (1992) *The stratigraphic palynology and palynofacies of the Upper Devonian–Lower Carboniferous successions in the South Munster Basin, Ireland and North Devon, England.* Unpublished Ph.D. thesis, University College Cork.

Oliver, G.J.H. (1978) Prehnite-pumpellyite facies metamorphism in County Cavan, Ireland. *Nature*, **274**(5668), 242–243.

Oliver, G.J.H., Chen, F., Buchwaldt, R., and Hegner, E. (2000) Fast tectono-metamorphism and exhumation in the type area of the Barrovian and Buchan zones. *Geology*, **28**, 459–462.

O'Mahony, M.J. (2001) *The structural and metamorphic features of the central and southern portions of the Leinster Granite Complex, SE. Ireland.* Unpublished Ph.D. thesis, University College Cork.

O'Reilly, B.M., Readman, P.W., and Murphy, T. (1996a) The gravity signature of Caledonian and Variscan tectonics in Ireland. *Physics and Chemistry of the Earth*, **21**, 299–304.

O'Reilly, B.M., Readman, P.W., and Shannon, P.M. (2005) Slope failure, mass flow and bottom current processes in the Rockall Trough, offshore Ireland, revealed by deep-tow sidescan sonar. *First Break*, **23**, 45–50.

O'Reilly, B.M., Shannon, P.M., and Readman, P.W. (2007) Shelf to slope sedimentation processes and the impact of Plio-Pleistocene glaciations in the northeast Atlantic, west of Ireland. *Marine Geology*, **238**, 21–44.

O'Reilly, B.M., Hauser, F., Ravaut, C., Shannon, P.M., and Readman, P.W. (2006) Crustal thinning, mantle exhumation and serpentinisation in the Porcupine Basin, offshore Ireland: evidence from wide-angle seismics. *Journal of the Geological Society of London*, **163**, 775–787.

O'Reilly, B.M., Shannon, P.M., and Vogt, U. (1991) Seismic studies in the North Celtic Sea Basin: implications for basin development. *Journal of the Geological Society of London*, **148**, 191–195.

O'Reilly, B.M., Hauser, F., Jacob, A.W.B., and Shannon, P.M. (1996b) The lithosphere below the Rockall Trough: wide-angle seismic evidence for extensive serpentinisation. *Tectonophysics*, **255**, 1–23.

O'Reilly, C., Jenkin, G.R.T., Feely, M., Alderton, D.H.M., and Fallick, A.E. (1997) A fluid inclusion and stable isotope study of 200 Ma of fluid evolution in the Galway Granite, Connemara, Ireland. *Contributions to Mineralogy and Petrology*, **129**(2–3), 120–142.

Ori, G.G. and Penney, S.R. (1982) The stratigraphy and sedimentology of the Old Red Sandstone sequence at Dunmore East,

County Waterford. *Journal of Earth Sciences, Royal Dublin Society*, **5**, 43–59.

Orr, R. (1985) On *Henningsmoenia costa* Orr sp. Nov. *Stereo-Atlas of Ostracod Shells*, **12**, 61–68.

Orr, R. (1987) *Upper Ordovician ostracods from Portrane, E. Ireland*. Unpublished Ph.D. thesis, Queen's University Belfast.

O'Sullivan, A. (1990) *Historical and contemporary effects of fire on the native woodland vegetation of Killarney, S.W. Ireland*. Unpublished Ph.D thesis, University of Dublin.

Oswald, D.H. (1955) The Carboniferous rocks between the Ox Mountains and Donegal Bay. *Quarterly Journal of the Geological Society of London*, **111**, 167–186.

Øvrebø, L.K., Haughton, P.D.W., and Shannon, P.M. (2005) Temporal and spatial variation in Late Quaternary slope sedimentation along the margins of the Rockall Trough, offshore west Ireland. *Norwegian Journal of Geology*, **85**, 279–294.

Øvrebø, L.K., Haughton, P.D.W., and Shannon, P.M. (2006) A record of fluctuating bottom currents on the slopes west of the Porcupine Bank, offshore Ireland – implications for Late Quaternary climate forcing. *Marine Geology*, **225**, 279–309.

Owen, A.W. and Parkes, M.A. (1996) The trilobite *Mucronaspis* in County Wexford: evidence for Ashgill rocks in the Leinster massif. *Irish Journal of Earth Sciences*, **15**, 123–128.

Owen, A.W. and Parkes, M.A. (2000) Trilobite faunas of the Duncannon Group: Caradoc stratigraphy, environments and palaeobiogeography of the Leinster Terrane, Ireland. *Palaeontology*, **43**, 219–269.

Owen, A.W., Harper, D.A.T., and Romano, M. (1992) The Ordovician biogeography of the Grangegeeth terrane and the Iapetus suture zone in eastern Ireland. *Journal of the Geological Society of London*, **149**, 3–6.

Owens, B., Gueinn, K.J., and Cameron, I.B. (1977) A Tournaisian miospre assemblage from the Altagoan Formation (Upper Calciferous Sandstone), Draperstown, Northern Ireland. *Pollen et Spores*, **19**, 313–324.

Owens, B., McLean, D., and Bodman, D. (2004) A revised palynozonation of British Namurian deposits and comparisons with eastern Europe. *Micropaleontology*, **50**, 89–103.

Owens, B., Riley, N.J., and Calver, M.A. (1985) Boundary stratotypes and new stage names for the lower and middle Westphalian sequences in Britain. *Compte Rendu 10ème Congrès International de Stratigraphie et de Géologie Carbonifère, Madrid, 1983*, **4**, 461–472.

Padget, P. (1951) The geology of the Clogher–Slieve Beagh area, Co. Tyrone. *Scientific Proceedings of the Royal Dublin Society*, **26**, 63–83.

Page, L.M., Stephens, M.B., and Wahlgren, C.-H. (1996) $^{40}Ar/^{39}Ar$ geochronological constraints on the tectonothermal evolution of the Eastern Segment of the Sveconorwegian Orogen, south-central Sweden. *In*: Brewer, T.S. (ed.). *Precambrian crustal evolution in the North Atlantic Region*. Geological Society, London, Special Publications, **112**, 315–30.

Palmer, A.R. (1971) The Cambrian of the Appalachian and eastern New England regions, eastern United States. *In*: Holland, C.H. (ed.). *Lower Palaeozoic Rocks of the World, Volume 1, Cambrian of the New World*. Wiley-Interscience, London, 169–217.

Palmer, D. (1970) *Monograptus ludensis* Zone graptolites from the Devilsbit Mountain district, Tipperary. *Scientific Proceedings of the Royal Dublin Society*, **A3**, 335–342.

Palmer, D, Johnston, J.D., Dooley, T., and Maguire, K. (1989) The Silurian of Clew Bay, Ireland: part of the Midland Valley of Scotland? *Journal of the Geological Society of London*, **146**, 385–388.

Pankhurst, R.J., Andrews, J.R., Phillips, W.E.A., Sanders, I.S., and Taylor, W.E.G. (1976) Age and structural setting of the Slieve Gamph Igneous Complex, Co. Mayo, Ireland. *Journal of the Geological Society of London*, **132**, 327–36.

Parker, A.G., Goudie, A.S., Anderson, D.E., Robinson, M.A., and Bonsall, C. (2002) A review of the mid-Holocene elm decline in the British Isles. *Progress in Physical Geography*, **26**, 1–45.

Parkes, M.A. (1993a) Silurian (Telychian) fossil assemblages from the Charlestown inlier, County Mayo. *Irish Journal of Earth Sciences*, **12**, 27–40.

Parkes, M.A. (1993b) Palaeokarst at Portrane, County Dublin: evidence of Hirnantian glaciation. *Irish Journal of Earth Sciences*, **12**, 75–81.

Parkes, M.A. (1994) The brachiopods of the Duncannon Group (Middle–Upper Ordovician) of southeast Ireland. *Bulletin British Museum (Natural History) Geology*, **50**, 105–174.

Parkes, M.A. and Harper, D.A.T. (1996) Ordovician brachiopod biogeography in the Iapetus suture zone of Ireland: provincial dynamics in a changing ocean. *In*: Copper, P.and Jisuo, J. (eds). *Brachiopods*, Balkema, Rotterdam, 197–202.

Parkes, M.A. and Palmer, D.C. (1994) The stratigraphy and palaeontology of the Lower Palaeozoic Kildare Inlier, County Kildare. *Irish Journal of Earth Sciences*, **13**, 65–82.

Parkin, J. (1974) Silurian rocks of Inishvickillane, Blasket Islands, Co. Kerry. *Scientific Proceedings of the Royal Dublin Society*, **A5**, 277–291.

Parkin, J. (1976a) The geology of the Foze Rocks, Co. Kerry: a review. *Irish Naturalists' Journal*, **18**, 308–309.

Parkin, J. (1976b) Silurian rocks of the Bull's Head, Annascaul and Derrymore Glen inliers, Dingle Peninsula, Co. Kerry. *Proceedings of the Royal Irish Academy*, **76B**, 577–606.

Parnell, J. (1992) Hydrocarbon potential of Northern Ireland: 3. Reservoir potential of the Permo-Triassic. *Journal of Petroleum Geology*, **15**, 51–70.

Parrish, R.R. and Noble, S.R. (2003) Zircon U–Th–Pb geochronology by isotope dilution thermal ionization mass spectrometry. *In*: Hanchar, J.M. and Hoskin, P.W.O. (eds.). (2003) Zircon. *Reviews in Mineralogy and Geochemistry*, **53**, 183–213.

Passchier, C.W., Myers, J.S., and Kröner, A. (1990) *Field Geology of High-Grade Gneiss Terrains*. Springer-Verlag, 150pp.

Patchett, P.J. and Arndt, N.T. (1986) Nd isotopes and tectonics of 1.9–1.7 Ga crustal genesis. *Earth and Planetary Science Letters*, **78**, 329–338.

Paterson, S.R. and Vernon, R.H. (1995) Bursting the bubble of ballooning plutons: a return to nested diapirs emplaced by multiple processes. *Geological Society of America Bulletin*, **107**(11), 1356–1380.

Patterson, E.M. (1950) Evidence of fissure eruption in the Tertiary Lava Plateau of north-east Ireland. *Geological Magazine*, **87**, 45–52.

Patterson, E.M. (1951a) An occurrence of quartz-trachyte among the Tertiary basalt lavas of north-east Ireland. *Proceedings of the Royal Irish Academy*, **53B**, 265–287.

Patterson, E.M. (1951b) A petrochemical study of the Tertiary lavas of north-east Ireland. *Geochimica et Cosmochimica Acta*, **2**, 283–299.

Patterson, E.M. (1955a) The Tertiary volcanic succession in the northern part of the Antrim Plateau. *Proceedings of the Royal Irish Academy,* **57B**, 79–122.

Patterson, E.M. (1955b) The Tertiary volcanic succession in the western part of the Antrim Plateau. *Proceedings of the Royal Irish Academy,* **57B**, 155–178.

Patterson, E.M. (1963) Tertiary vents in the northern part of the Antrim Plateau. *Quarterly Journal of the Geological Society of London,* **119**, 419–443.

Patterson, E.M. and Swaine, D.J. (1955) A petrochemical study of Tertiary tholeiitic basalts: the Middle Lavas of the Antrim plateau. *Geochimica et Cosmochimica Acta,* **8**, 173–181.

Pattison, J. (1970) A review of the marine fossils from the Upper Permian rocks of Northern Ireland and north-west England. *Bulletin of the Geological Survey of Great Britain,* **32**, 123–165.

Patton, D.J.S. (1988) *The petrology and petrochemistry of the Tertiary basic hypabyssal intrusions of northeast Ireland.* Unpublished Ph.D. thesis, Queen's University of Belfast.

Peach, B.N. and Horne, J. (1899) *The Silurian Rocks of Britain, Volume 1: Scotland.* HMSO, Glasgow.

Pearson, D.G., Emeleus, C.H., and Kelley, S.P. (1996) Precise $^{40}Ar/^{39}Ar$ age for the initiation of Palaeogene volcanism in the inner Hebrides and its regional significance. *Journal of the Geological Society of London,* **153**, 815–818.

Peddy, C. and Keen, C. (1987) Deep seismic reflection profiling: how far have we come? *The Leading Edge,* **6**, 22–24 and 49.

Penn, I. E. (1981) *Larne No.2 Geological Well Completion Report.* Deep Geology Unit, Institute of Geological Sciences, Report **81/6**.

Penn, I.E, Holliday, D.W., Kirby, G.A., Kubala, M., Soby, R.A., Mitchell, W.I., Harrison, R.K., and Beckinsale, R.D. (1983) The Larne No. 2 Borehole: discovery of a new Permian volcanic centre. *Scottish Journal of Geology,* **19**, 333–346.

Penney, S.R. (1978) Devonian lavas from the Comeragh Mountains, Co. Wexford. *Journal of Earth Sciences, Royal Dublin Society,* **1**, 71–76.

Penney, S.R. (1980a) The stratigraphy, sedimentology and structure of the Lower Palaeozoic rocks of North County Waterford. *Proceedings of the Royal Irish Academy,* **80B**, 305–333.

Penney, S.R. (1980b) A new look at the Old Red Sandstone succession of the Comeragh Mountains, County Waterford. *Journal of Earth Sciences Royal Dublin Society,* **3**, 155–178.

PESGB 2005. *Structural framework of the North Sea and Atlantic margin: 1:750,000 scale map,* 2005 edition. Petroleum Exploration Society of Great Britain, London.

Petford, N., Cruden, A.R., McCaffrey, K.J.W., and Vigneresse, J.L. (2000) Granite magma formation, transport and emplacement in the Earth's crust. *Nature,* **408**(6813), 669–673.

Petit, R.J., Brewer, S., Bordács, S., Burg, K., Cheddadi, C., Coart, E., Cottrell, J., Csaikl, U.M., van Dam, B., Deans, J.D., Espinel, S., Fineschi, S., Finkeldey, R., Glaz, I., Goicoechea, P.G., Jensen, J.S., König, A.O., Lowe, A.J., Flemming Madsen, S., Mátyás, G., Munro, R.C., Popescu, F., Slade, D., Tabbener, H., de Vries, S.G.M., Ziegenhagen, B., de Beaulieu, J.L., and Kremer, A. (2002) Identification of refugia and post-glacial colonisation routes of European white oaks based on chloroplast DNA and fossil pollen evidence. *Forest Ecology and Management,* **156**, 49–74.

Petrie, S. H., Brown, J. R., Granger, P. J., and Lovell, J. P. B. (1989) Mesozoic history of the Celtic Sea Basins. *In*: Tankard, A.L. and Balkwill, H.R. (eds). *Extensional tectonics and stratigraphy of the North Atlantic Margins.* American Association of Petroleum Geologists Memoir **46**, 433–444.

Philcox, M.E. (1961) Namurian shales near Buttevant, north Co. Cork. *Scientific Proceedings of the Royal Dublin Society,* **A1**, 205–209.

Philcox, M.E. (1963) Banded calcite mudstone in the Lower Carboniferous 'reef' knolls of the Dublin Basin, Ireland. *Journal of Sedimentary Petrology,* **33**, 904–913.

Philcox, M.E. (1964) Compartment deformation near Buttevant, County Cork, Ireland, and its relation to the Variscan Thrust Front. *Scientific Proceedings of the Royal Dublin Society,* **2A**, 1–11.

Philcox, M.E. (1984) *Lower Carboniferous lithostratigraphy of the Irish Midlands.* Irish Association for Economic Geology, Dublin, 89pp.

Philcox, M.E. (1989) The mid-Dinantian unconformity at Navan, Ireland. *In*: Arthurton, R.S., Gutteridge, P., and Nolan, S.C. (eds). *The role of tectonics in Devonian and Carboniferous sedimentation in the British Isles.* Yorkshire Geological Society Occasional Publication, **6**, 67–82.

Philcox, M.E., Sevastopulo, G.D., and MacDermot, C.V. (1989) Intra-Dinantian tectonic activity on the Curlew Fault, north-west Ireland. *In*: Arthurton, R.S., Gutteridge, P., and Nolan, S.C. (eds). *The role of tectonics in Devonian and Carboniferous sedimentation in the British Isles.* Yorkshire Geological Society Occasional Publication, **6**, 55–66.

Philcox, M.E., Baily, H., Clayton, G., and Sevastopulo, G.D. (1992) Evolution of the Carboniferous Lough Allen Basin, Northwest Ireland. *In*: Parnell, J. (ed.). *Basins on the Atlantic Seaboard: petroleum geology, sedimentology and basin evolution.* Geological Society Special Publication, **62**, 203–215.

Phillips, E.R., Evans, J.A., Stone, P., Horstwood, M.S.A., Floyd, J.D., Smith, R.A., Akhurst, M.C., and Barron, H.F. (2003) Detrital Avalonian zircons in the Laurentian Southern Uplands terrane, Scotland. *Geology,* **31**, 625–628.

Phillips, S.J.L. (1985) *The stratigraphy and structure of south-west County Cork, Ireland.* Unpublished Ph.D. thesis, Queen's University Belfast.

Phillips, W.E.A. (1973) The pre-Silurian rocks of Clare Island, Co. Mayo, Ireland, and the age of the metamorphism of the Dalradian in Ireland. *Journal of the Geological Society of London,* **129**, 585–606.

Phillips, W.E.A. (1974) The stratigraphy, sedimentary environments and palaeogeography of the Silurian strata of Clare Island, Co. Mayo, Ireland. *Journal of the Geological Society of London,* **130**, 19–41.

Phillips, W.E.A. (1978) The Caledonide Orogen in Ireland. *In*: *I.G.C.P. Project 27, Caledonian Appalachian Orogen of the North Atlantic Region.* Geological Survey of Canada Paper **78–13**, 97–103.

Phillips, W.E.A. (1981a) The pre-Caledonian basement. *In*: Holland, C.H. (ed). *A Geology of Ireland.* Scottish Academic Press, Edinburgh, 7–16.

Phillips, W.E.A. (1981b) The Orthotectonic Caledonides. *In*: Holland, C.H. (ed). *A Geology of Ireland,* Scottish Academic Press, Edinburgh, 17–40.

Phillips, W.E.A. (1983) The Lower Palaeozoic rocks of the Deer Park Complex. *In*: Archer, J.B. and Ryan, P.D. (eds). *Geological Guide to the Caledonides of Western Ireland.* Geological Survey of Ireland Guide Series No. 4, 37–42.

Phillips, W.E.A. (2001) Caledonian deformation. *In*: C.H. Holland (ed.). *The Geology of Ireland.* Dunedin Academic Press, Edinburgh, 179–199.

Phillips, W.E.A. and Clayton, G. (1980) The Dinantian clastic succession of Clare Island, County Mayo. *Journal of Earth Sciences, Royal Dublin Society,* **2**, 115–136.

Phillips, W.E.A. and Sevastopulo, G.D. (1986) The stratigraphic and structural setting of Irish mineral deposits. *In*: Andrew, C.J., Crowe, R.W.A., Finlay, S., Pennell, W.M., and Pyne, J.F. (eds). *Geology and genesis of mineral deposits in Ireland.* Irish Association for Economic Geology, Dublin, 1–30.

Phillips, W.E.A. and Skevington, D. (1968) The Lower Palaeozoic rocks of the Lough Acanon area, Co. Cavan, Ireland. *Scientific Proceedings of the Royal Dublin Society,* **A3**, 141–148.

Phillips, W.E.A., Rickards, R.B., and Dewey, J.F. (1970) The Lower Palaeozoic rocks of the Louisburgh area, Co, Mayo. *Proceedings of the Royal Irish Academy,* **70B**, 195–210.

Phillips, W.E.A., Stillman, C.J., and Murphy, T. (1976) A Caledonian plate tectonic model. *Journal of the Geological Society of London,* **132**, 579–609.

Phillips, W.E.A., Taylor, W.E.G., and Sanders, I.S. (1975) An analysis of the geological history of the Ox Mountains Inlier. *Scientific Proceedings of the Royal Dublin Society,* **5**, 311–329.

Piasecki, M.A.J. (1980) New light on the Moine rocks of the Central Highlands of Scotland. *Journal of the Geological Society of London,* **137**, 41–59.

Piasecki, M.A.J. and van Breemen, O. (1979) A Morarian age for the 'younger Moines' of central and western Scotland. *Nature,* **278**, 734–736.

Piasecki, M.A.J. and van Breemen, O. (1983) Field and isotopic evidence for a *c.*750 Ma tectonothermal event in the Moine rocks in the Central Highland region of the Scottish Caledonides. *Transactions of the Royal Society of Edinburgh: Earth Sciences,* **73**, 119–134.

Pickard, N.A.H., Jones, G.Ll., Rees, J.G., Somerville, I.D., and Strogen, P. (1992) Lower Carboniferous (Dinantian) stratigraphy and structure of the Walterstown–Kentstown area, Co. Meath, Ireland. *Geological Journal,* **27**, 35–58.

Pickard, N.A.H., Rees, J.G., Strogen, P, Somerville, I.D., and Jones, G.Ll. (1994) Controls on the evolution and demise of Lower Carboniferous platforms, northern margin of the Dublin Basin, Ireland. *Geological Journal,* **29**, 93–117.

Pickering, K.T., Bassett, M.G., and Siveter, D.J. (1988) Late Ordovician–early Silurian destruction of the Iapetus Ocean: Newfoundland, British Isles and Scandinavia: a discussion. *Transactions of the Royal Society of Edinburgh: Earth Sciences,* **79**, 361–382.

Pidgeon, R.T. (1969) Zircon U–Pb ages from the Galway granite and the Dalradian, Connemara, Ireland. *Scottish Journal of Geology,* **5**, 375–392.

Pidgeon, R.T. and Compston, W. (1992) A SHRIMP ion microprobe study of inherited and magmatic zircons from four Scottish Caledonian granites. *Transactions of the Royal Society of Edinburgh: Earth Sciences,* **83**, 473–483.

Pilling, D.J. (1988) *The stratigraphy, sedimentology, petrography and structure in parts of Counties Tipperary, Waterford and Cork.* Unpublished Ph.D. thesis, University of Dublin.

Piper, D.J.W. (1967) A new interpretation of the Llandovery sequence of North Connemara, Eire. *Geological Magazine,* **104**, 253–267.

Piper, D.J.W. (1972) Sedimentary environments and palaeogeography of the late Llandovery and earliest Wenlock of North Connemara, Ireland. *Journal of the Geological Society of London,* **128**, 33–51.

Pitcher, W.S. (1969) North-east trending faults of Scotland and Ireland, and chronology of displacements. *In*: Kay, M. (ed.). *North Atlantic: Geology and Continental Drift.* American Association of Petroleum Geologists Memoir, **12**, 724–733.

Pitcher, W.S. (1982) Granite type and tectonic environment. *In*: Hsu, K.J. (ed.). *Mountain building processes.* Academic Press, 19–40.

Pitcher, W.S. (1992) The Rosses multi-pulse pluton: fractures and fractals. *Transactions of the Royal Society of Edinburgh: Earth Sciences,* **83**, 497.

Pitcher, W.S. (1997) *The nature and origin of granite.* Chapman and Hall, London.

Pitcher, W.S. and Berger, A.R. (1972) *The Geology of Donegal: a study of granite emplacement and unroofing.* Wiley-Interscience, New York and London, 439pp.

Pitcher, W.S. and Hutton, D.H.W. (2003) *A master class guide to the granites of Donegal.* Geological Survey of Ireland, Dublin.

Pitcher, W.S., Elwell, R.W.D., Tozer, C.F., and Cambray, F.W. (1964) The Leannan fault. *Quarterly Journal of the Geological Society of London,* **120**, 241–273.

Plant, A.G. (1968) *The geology of the Leam–Shannawona district, Connemara, Ireland.* Unpublished Ph.D. thesis, University of Bristol.

Platt, J.W. (1977) Volcanic mineralization at Avoca, Co. Wicklow, Ireland, and its regional implications. *In*: *Volcanic processes in ore genesis.* The Institute of Mining and Metallurgy and the Geological Society, 163–170.

Plumb, K.A. (1991) New Precambrian timescale. *Episodes,* **14**, 139–140.

Plunkett, G. (2006) Tephra-linked peat humification records from Irish ombrotrophic bogs question nature of solar forcing at 850 cal. yr bc. *Journal of Quaternary Science,* **21**, 9–16.

Pockley, R.P.C. (1961) *Lead isotope and age studies of uranium and lead minerals from the British Isles and France.* Unpublished Ph.D thesis, Oxford University.

Polgar, R. (1980) *The stratigraphy and micropalaeontology of a borehole in the Carboniferous rocks of Knockadrinan Townland, County Leitrim.* Unpublished M.Sc. thesis, University of Dublin.

Portlock, J.E. (1843) *Report of the geology of the county of Londonderry, and parts of Tyrone and Fermanagh.* Dublin: HMSO xxxii + 784.

Poty, E., Devuyst, F.X., and Hance, L. (2006). Upper Devonian and Mississippian foraminiferal and rugose coral zonations of Belgium and Northern France: a tool for Eurasian correlations. *Geological Magazine,* **143**, 829–857.

Poustie, A. and Kucha, H. (1986) The geological setting, style and petrology of zinc–lead mineralization in the Moyvoughly area, Co. Westmeath. *In*: Andrew, C.J., Crowe, R.W.A., Finlay, S., Pennell, W.M., and Pyne, J.F. (eds). *Geology and genesis of mineral deposits in Ireland.* Irish Association for Economic Geology, Dublin, 305–318.

Powell, D. (1965) Comparison of calc-silicate bands from the Moine schists of Inverness-shire with similar bands from Moine-like rocks in Donegal. *Nature,* **206**, 180.

Pracht, M. (1996) *Geology of Dingle Bay: a geological description to accompany the bedrock geology 1:100,000 scale map series, Sheet*

20, Dingle Bay with contributions by G. Wright, P. O'Connor, K. Claringbold, and W.P. Warren. Geological Survey of Ireland, 58pp.

Pracht, M. and Kinnaird, J.A. (1995) Mineral chemistry of megacrysts and ultramafic nodules from an undersaturated pipe at Black Ball Head, County Cork. *Irish Journal of Earth Sciences,* **14**, 47–58.

Pracht, M. and Kinnaird, J.A. (1997) Carboniferous subvolcanic activity on the Beara Peninsula, SW Ireland. *Geological Journal,* **32**, 297–312.

Pracht, M. and Sleeman, A. G. (2002) *A geological description of West Cork and adjacent parts of Kerry to accompany the bedrock geology 1:100, 000 scale map series, Sheet 24, West Cork, with contributions by W. Cox (Minerals) and U. Leader (Groundwater).* Geological Survey of Ireland.

Pracht, M. and Timmerman, M.J. (2004) A late Namurian (318 Ma) $^{40}Ar/^{39}Ar$ age for kaersutite megacrysts from the Black Ball Head diatreme: an age limit for the Variscan deformation in SW Ireland. *Irish Journal of Earth Sciences,* **22**, 33–43.

Pracht, M., Lees, A., Leake, B., Feely, M., Morris, J., and McConnell, B. (2002) *Geology of Galway Bay: a geological description to accompany the bedrock geology 1:100,000 scale map series, Sheet 14, Galway Bay, with contributions by G. Wright (Groundwater) and W. Cox (Minerals).* Geological Survey of Ireland, 76pp.

Pracht, M., Lees, A., Leake, B.E., Feely, M., Long, B., Morris, J.H., and McConnell, B. (2004) *Geology of Galway Bay: a geological description to accompany the bedrock geology 1:100,000 scale map series, Sheet 14, Galway Bay.* Geological Survey Ireland, 76pp.

Praeg, D., Stoker, M.S., Shannon, P.M., Ceramicola, S., Hjelstuen, B.O., and Mathiesen, A. (2005) Episodic Cenozoic tectonism and the development of the NW European 'passive' continental margin. *Marine and Petroleum Geology,* **22**, 1007–1030.

Preece, R.C., Coxon, P., and Robinson, J.E. (1986) New biostratigraphic evidence of the post-glacial colonization of Ireland and for Mesolithic forest disturbance. *Journal of Biogeography,* **13**, 487–509.

Prendergast, B.M. (1972) The Silurian and Devonian inlier of Knockshigowna Hill, County Tipperary, Ireland. *Scientific Proceedings of the Royal Dublin Society,* **A4**, 201–211.

Preston, F.W. (1930) Ball-and-socket jointing in basalt prisms. *Proceedings of the Royal Society, London, Series B,* **106**, 87–93.

Preston, J. (1962) Explosive volcanic activity in the Triassic sandstone of Scrabo Hill, County Down. *Irish Naturalists' Journal,* **14**, 45–51.

Preston, J. (1963a) The dolerite plug at Slemish, County Antrim, Ireland. *Liverpool and Manchester Geological Journal,* **3**, 301–314.

Preston, J. (1963b) An 'orbicular' quartzite at Scrabo Hill. *Irish Naturalists' Journal,* **14**, 149–153.

Preston, J. (1967a) The Blind Rock Dyke, Co. Donegal. *Irish Naturalists' Journal,* **15**, 286–293.

Preston, J. (1967b) A Tertiary feeder dyke in County Fermanagh, Northern Ireland. *Scientific Proceedings of the Royal Dublin Society,* **3**, 1–16.

Preston, J. (1971) The Maddygalla dyke — Rathlin Island. *Irish Naturalists' Journal,* **17**, 88–92.

Price, A.R. (1997) *Multiple sheeting as a mechanism of pluton construction: the Main Donegal Granite, NW Ireland.* Unpublished Ph.D. thesis, University of Durham.

Price, A.R. and Pitcher, W.S. (1999) The Trawenagh Bay Granite: a multipulse, inclined sheet intruded in the flank of a synplutonic shear zone. *Irish Journal of Earth Sciences,* **17**, 51–60.

Price, C. and Max, M.D. (1988) Surface and deep structural control of the NW Carboniferous basin of Ireland: seismic perspectives of aeromagnetic and surface geological interpretation. *Journal of Petroleum Geology,* **11**, 365–388.

Price, C.A. (1986) *The geology of the Iveragh Peninsula, County Kerry, Ireland, incorporating a remote sensing lineament study.* Unpublished Ph.D. thesis, University of Dublin.

Price, C.A. (1989) Some thoughts on the subsidence and evolution of the Munster Basin, southern Ireland. *In*: Arthurton, R.S., Gutteridge, P., and Nolan, S.C. (eds). *The role of tectonics in Devonian and Carboniferous sedimentation in the British Isles.* Yorkshire Geological Society Occasional Publication, **6**, 111–122.

Price, C.A. and Todd, S.P. (1988) A model for the development of the Irish Variscides. *Journal of the Geological Society of London,* **145**, 935–939.

Price, D. (1981) *Tretaspis radialis* Lamont and allied species (Trilobita). *Geological Magazine,* **118**, 289–295.

Prince, C.I., Kosler, J., Vance, D., and Günther, D. (2000) Comparison of laser ablation ICP–MS and isotope dilution REE analyses: implications for Sm–Nd garnet geochronology. *Chemical Geology,* **168**, 255–274.

Pudsey, C.J. (1984a) Ordovician stratigraphy and sedimentology of the South Mayo inlier. *Irish Journal of Earth Sciences,* **6**, 15–45.

Pudsey, C.J. (1984b) Fluvial to marine transition in the Ordovician of Ireland: a humid region fan delta? *Geological Journal,* **19**, 143–172.

Pulham, A.J. (1989) Controls on internal structure and architecture of sandstone bodies within Upper Carboniferous deltas, County Clare, western Ireland. *In*: Whately, M.K.G. and Pickering, K.T. (eds). *Deltas: sites and traps for fossil fuels.* Geological Society of London Special Publication, **41**, 179–203.

Quin, J.G. (2001) *The genesis of exceptionally thick shallow marine sequences exposed within the South Munster Basin of southern Ireland.* Unpublished Ph.D. thesis, University of Dublin.

Quin, J.G. (2008). The evolution of a thick shallow marine succession, the South Munster Basin, Ireland. *Sedimentology,* **55**, 1052-1082.

Quinn, D., Meere, P.A., and Wartho, J. (2005) A chronology of foreland deformation: ultra-violet laser $^{40}Ar/^{39}Ar$ dating of syn/late-orogenic intrusions from the Variscides of southwest Ireland. *Journal of Structural Geology,* **27**, 1413–1425.

Quinn, I.M. (1987) The significance of periglacial features on Knocknadobar, south west Ireland. *In*: Boardman, J. (ed.). *Periglacial processes and landforms in Britain and Ireland.* Cambridge University Press, 287–294.

Quirk, S.G, Roy, S., Knott, I, Redfern, J., and Hill, L. (1999) Petroleum geology and future hydrocarbon potential of the Irish Sea. *Journal of Petroleum Geology,* **22**, 243–260.

Rackham, O. (1980) *Ancient woodland: its history, vegetation and uses in England.* Edward Arnold, London.

Rackham, O. (1995) Looking for ancient woodland in Ireland. *In*: Pilcher, J.R. and Mac an tSaoir, S.S. (eds). *Wood, trees and forests in Ireland.* Royal Irish Academy, Dublin, 1–12.

Rainbird, R. H., Hamilton, M. A., and Young, G. M. (2001) Detrital zircon geochronology and provenance of the Torridonian, NW Scotland. *Journal of the Geological Society of London,* **158**, 15–27.

Ramsay, J.G. and Graham, R.H. (1970) Strain variation in shear belts. *Canadian Journal of Earth Sciences*, **7**, 786–813.

Ramsbottom, W.H.C., Calver, M.A., Eagar, R.M.C., Hodson, F., Holliday, F., Stubblefield, C.J., and Wilson, R.B. (1978) *A correlation of Silesian rocks in the British Isles*. Geological Society, London, Special Report **10**, 81pp.

Rao, C.K., Jones, A.G. and Moorkamp, M. (2007) The geometry of the Iapetus Suture Zone in central Ireland deduced from a magnetotelluric study. *Physics of the Earth and Planetary Interiors*, **161**, 134–141.

Rasmussen, S.O., Andersen, K. K., Svensson, A. M., Steffensen, J. P., Vinther, B.M., Clausen, H.B., Andersen, M.-L.S., Johnsen, S.J., Larsen, L.B., Bigler, M., Röthlisberger, R., Fischer, H., Goto-Azuma, K., Hansson, M.E., and Ruth, U. (2006) A new Greenland ice core chronology for the last glacial termination. *Journal of Geophysical Research*, **111**, D06102, doi:10.1029/2005JD006079, 2006.

Rasmussen, T.L., van Weering, T.C.E., and Labeyrie, L. (1997) Climatic instability, ice sheets and ocean dynamics at high northern latitudes during the last glacial period (58–10ka bp). *Quaternary Science Reviews*, **16**, 71–80.

Rawson, P.F., Curry, D., Dilley, F.C., Hancock, J.M., Kennedy, W.J., Neale, J.W., Wood, C.J., and Worsam, B.C. (1978) *A correlation of Cretaceous rocks in the British Isles*. Geological Society of London Special Report, **9**, 70pp.

Read, H.H. (1923) *The geology of the country around Banff, Huntly and Turriff (Lower Banffshire and North-west Aberdeenshire)*. Memoir of the Geological Survey, Scotland, Sheets 86 and 96 (Scotland).

Read, H.H. (1957) *The Granite Controversy*. Thomas Murby and Co., London.

Read, H.H. (1961) Aspects of the Caledonian magmatism in Britain. *Liverpool and Manchester Geological Journal*, **2**, 653–683.

Readman, P.W., O'Reilly, B.M. and Murphy, T. (1997) Gravity gradients and upper-crustal tectonic fabrics, Ireland. *Journal of the Geological Society of London*, **154**, 817–828.

Readman, P.W., O'Reilly, B.M., Edwards, J.W.F. and Sankey, M.J. (1995) A gravity map of Ireland and surrounding waters. *In*: Croker, P.F. and Shannon, P.M. (eds). *The Petroleum Geology of Ireland's Offshore Basins*. Geological Society Special Publication, **93**, 9–16.

Readman, P.W., O'Reilly, B.M., Shannon, P.M., and Naylor, D. (2005) The deep structure of the Porcupine Basin, offshore Ireland, from gravity and magnetic studies. *In*: Doré, A.G. and Vining, B. (eds). *Petroleum Geology: North-West Europe and Global Perspectives – Proceedings of the 6th Petroleum Geology Conference*. Geological Society, London, 1047–1056.

Rees, J.G. (1987) *The Carboniferous geology of the Boyne Valley area, Ireland*. Unpublished Ph.D. thesis, University of Dublin.

Rees, J.G. (1992) The Courceyan (Lower Carboniferous) succession at Slane, County Meath. *Irish Journal of Earth Sciences*, **11**, 113–129.

Reeves, T.J. (1977) The dolerites in part of eastern Co. Wicklow. *Geological Survey of Ireland Bulletin*, **2**, 159–165.

Reeves, T.J., Robinson, K.W., and Naylor, D. (1978) Ireland's offshore geology. *Irish Offshore Review*, May 1978, 25–28.

Reille, M. and De Beaulieu, J.L. (1995) Long Pleistocene pollen records from the Praclaux crater, south-central France. *Quaternary Research*, **44**, 205–215.

Reilly, T.A. (1986) A review of vein mineralisation in the West Carbery Mining District of County Cork. *In*: Andrew, C.J., Crowe, R.W.A., Finlay, S., Pennell, W.M., and Pyne, J.F. (eds). *Geology and genesis of mineral deposits in Ireland*. Irish Association for Economic Geology, 513–544.

Reilly, T.A. and Graham, J.R. (1972) The historical and geological setting of the Glandore Mines, south-west County Cork. *Geological Survey of Ireland Bulletin*, **1**, 253–265.

Reilly, T.A. and Graham, J.R. (1976) The stratigraphy of the Roaringwater Bay area of south-west County Cork. *Geological Survey of Ireland Bulletin*, **2**, 1–13.

Reimer, P.J., Baillie, M.G.L., Bard, E., Bayliss, A., Beck, J.W., Bertrand, C.J.H., Blackwell, P.G., Buck, C.E., Burr, G.S., Cutler, K.B., Damon, P.E., Edwards, R.L., Fairbanks, R.G., Friedrich, M., Guilderson, T.P., Hogg, A.G., Hughen, K.A., Kromer, B., McCormac, G., Manning, S., Ramsey, C.B., Reimer, R.W., Remmele, S., Southon, J.R., Stuiver, M., Talamo, S., Taylor, F.W., van der Plicht, J., and Weyhenmeyer, C.E. (2004) IntCal04 terrestrial radiocarbon age calibration, 0–26 cal kyr bp. *Radiocarbon*, **46**, 1029–1058.

Reineck, H.E. (1969) Tidal flats. *American Association of Petroleum Geologists Bulletin*, **53**, 737.

Remmele, G. (1984) Massenbewegungen an der Hauptschicht der Benbulben Range. *Untersuchungen zur Morphodynamik und Morphogenese eines Schichtstufenreliefs in Nordwestirland*. Geographischen Instituts der Universität Tübingen. (ISSN 0564–4232).

Reston, T.J., Pennell, J., Stubenrauch, A., Walker, I., and Perez-Gussinye, M. (2001) Detachment faulting, mantle serpentinization, and serpentinite-mud volcanism beneath the Porcupine Basin, southwest of Ireland. *Geology*, **29**, 587–590.

Reynolds, D.L. (1931) The dykes of the Ards Peninsula, Co. Down. *Geological Magazine*, **68**, 97–111.

Reynolds, D.L. (1944) The southwestern end of the Newry igneous complex: a contribution towards the petrogenesis of the granodiorites. *Quarterly Journal of the Geological Society of London*, **395–396**, 205–246.

Reynolds, D.L. (1951) The geology of Slieve Gullion, Foughill, and Carrickarnan: an actualistic interpretation of a Tertiary gabbro–granophyre complex. *Transactions of the Royal Society of Edinburgh*, **62**, 85–143.

Reynolds, N., Mcardle, P., Pyne, J.F., Farrell, L.P.C., and Flegg, A.M. (1990) *Mineral localities in the Dalradian and associated igneous rocks of Connemara, County Galway*. Geological Survey of Ireland Report Series. **RS90/2**, 1–89.

Reynolds, S.H. and Gardiner, C.I. (1896) The Kildare Inlier. *Quarterly Journal of the Geological Society of London*, **52**, 587–604.

Richardson, J.B. and McGregor, D.C. (1986) Silurian and Devonian spore zones of the Old Red Sandstone continent and adjacent regions. *Bulletin of the Geological Survey of Canada*, **364**, 1–79.

Richardson, W. (1803) Inquiry into the consistency of Dr Hutton's Theory of the Earth with the arrangement of the strata, and other phenomena on the basaltic coast of Antrim. *Transactions of the Royal Irish Academy*, **9**, 429–487.

Richardson, W. (1805) Remarks on the basalts of the coast of Antrim. *Transactions of the Royal Society of Edinburgh*, **5**, 15–20.

Richey, J.E. (1928) The structural relations of the Mourne granites Northern Ireland. *Quarterly Journal of the Geological Society of London*, **83**, 653–688.

Richey, J.E. (1948) *British Regional Geology, Scotland: the Tertiary Volcanic Districts.* HMSO, Edinburgh.

Richey, J.E. and Thomas, H.H. (1930) *The geology of Ardnamurchan, North-West Mull and Coll.* Memoir of the Geological Survey of Scotland.

Richey, J.E. and Thomas, H.H. (1932) The Tertiary ring complex of Slieve Gullion, Ireland. *Quarterly Journal of the Geological Society of London,* **88**, 776–849.

Richmond, L.K. (1998) *Fluvial–aeolian interactions and Old Red Sandstone basin evolution, Northwest Dingle Peninsula, Co. Kerry, SW Ireland.* Unpublished Ph.D. thesis, University of Aberdeen.

Richmond, L.K. and Williams, B.P.J. (2000) A new terrane in the Old Red Sandstone of the Dingle Peninsula, SW Ireland. *In*: Friend, P.F. and Williams, B.P.J. (eds). *New perspectives on the Old Red Sandstone.* Geological Society London Special Publication, **180**, 147–184.

Rickards, R.B. (1973) On some highest Llandovery red beds and graptolite assemblages in Britain and Eire. *Geological Magazine,* **110**, 70–72.

Rickards, R.B. and Archer, J.B. (1969) The Lower Palaeozoic rocks near Tomgraney, Co. Clare. *Scientific Proceedings of the Royal Dublin Society,* **A3**, 219–230.

Rickards, R.B. and Smith, W.R. (1968) The Silurian graptolites of Mayo and Galway. *Scientific Proceedings of the Royal Dublin Society,* **A3**, 129–135.

Rickards, R.B., Burns, V., and Archer, J.B. (1973) The Silurian sequence at Balbriggan, Co. Dublin. *Proceedings of the Royal Irish Academy,* **73B**, 303–316.

Riddihough, R.P. (1975) *A magnetic map of the continental margin of west of Ireland involving part of the Rockall Trough and the Faeroe Plateau.* Dublin Institute for Advanced Studies, Geophysical Bulletin **33**.

Riddihough, R.P. and Max, M.D. (1976) A geological framework for the continental margin to the west of Ireland. *Geological Journal,* **11**, 109–120.

Rider, M.H. (1974) The Namurian of west County Clare. *Proceedings of the Royal Irish Academy,* **74B**, 125–142.

Rider, M.H. (1978) Growth faults in the Carboniferous of western Ireland. *Bulletin of the American Association of Petroleum Geologists,* **62**, 2191–2213.

Rivers, T. (1997) Lithotectonic elements of the Grenville Province: review and tectonic implications. *Precambrian Research,* **86**, 117–154.

Rizzi, G. and Braithwaite, C.J.R. (1996) Cyclic emersion surfaces and channels within Dinantian limestones hosting the giant Navan Zn–Pb deposit, Ireland. *In*: Strogen, P., Somerville, I.D., and Jones, G.Ll. (eds). *Recent advances in Lower Carboniferous geology.* Geological Society, London Special Publications, **107**, 207–219.

Robbie, J.A. (1955) The Slieve Binnian tunnel, an aqueduct in the Mourne Mountains, Co. Down. *Bulletin of the Geological Survey of Great Britain,* **8**, 1–20.

Roberts, D.G. (1969) New Tertiary volcanic centre on the Rockall Bank, eastern N. Atlantic. *Nature,* **223**, 819.

Roberts, D.G. (1975) Tectonic and stratigraphic evolution of Rockall Plateau and Trough. *In*: Woodward, A.W. (ed.). *Petroleum and the Continental Shelf of North West Europe,* Vol.1. *Geology.* Applied Science Publishers, London, 77–89.

Roberts, D.G., Montadert, L., and Searle, R.C. (1979) The western Rockall Plateau – stratigraphy and structural evolution. *Initial Reports of the Deep Sea Drilling Project* **48,** U.S.G.P.O., Washington D.C., 1061–1088.

Roberts, D.H., Chiverrell, R.C., Innes, J.B., Horton, B.P., Brooks, A.J., Thomas, G.S.P., Turner, S., and Gonzalez, S. (2006) Holocene sea levels, Last Glacial Maximum glaciomarine environments and geophysical models in the northern Irish Sea Basin, UK. *Marine Geology,* **231**, 113–128.

Roberts, J.L. and Treagus, J.E. (1977) Polyphase generation of nappe structures in the Dalradian rocks of the southwest Highlands of Scotland. *Scottish Journal of Geology,* **13**(2), 237–254.

Roberts, N. (1998) *The Holocene: an environmental history.* (2nd Edition.) Blackwell, Oxford.

Robeson, D., Burnett, R.D., and Clayton, G. (1988) The Upper Palaeozoic geology of the Porcupine, Erris and Donegal Basins, offshore Ireland. *Irish Journal of Earth Sciences,* **9**, 153–175.

Robinson, K.W., Shannon, P.M., and Young, D.G.G. (1981) The Fastnet Basin: an integrated analysis. *In*: Illing, L.G. and Hobson, G.D. (eds). *Petroleum Geology of the Continental Shelf of North- West Europe.* Heyden and Son Ltd, London, 444–454.

Rock, N.M.S. (1985) Value of chemostratigraphic correlation in metamorphic terranes: an illustration from the Colonsay limestone, Inner Hebrides, Scotland. *Transactions of the Royal Society of Edinburgh: Earth Sciences,* **76**, 463–465.

Roddick, J.C. and Max, M.D. (1983) A Laxfordian age from the Inishtrahull Platform, County Donegal, Ireland. *Scottish Journal of Geology,* **19**, 97–102.

Rogers, G. and Pankhurst, R.J. (1993) Unravelling dates through the ages: geochronology of the Scottish metamorphic complexes. *Journal of the Geological Society of London,* **150**, 447–464.

Rogers, G., Dempster, T.J., Bluck, B.J., and Tanner, P.W.G. (1989) A high precision U–Pb age for the Ben Vuirich granite: implications for the evolution of the Scottish Dalradian Supergroup. *Journal of the Geological Society of London,* **146**, 789–798.

Rohleder, H.T.P. (1929) *Geological Guide to the Giant's Causeway district of Northern Ireland and north coast of Antrim between Portrush and Fairhead, Belfast.*

Romano, M. (1970) *The stratigraphy and palaeontology of the Ordovician rocks of eastern central Ireland.* Unpublished Ph.D. thesis, University of Liverpool.

Romano, M. (1980a) The stratigraphy of the Ordovician rocks between Slane (County Meath) and Collon (County Louth), eastern Ireland. *Journal of Earth Sciences, Royal Dublin Society,* **3**, 205–215.

Romano, M. (1980b) The Ordovician rocks around Herbertstown (County Meath) and their correlation with the succession at Balbriggan (County Dublin), Ireland. *Journal of Earth Sciences, Royal Dublin Society,* **3**, 205–215.

Romano, M. and Owen, A.W. (1993) Early Caradoc trilobites of eastern Ireland and their palaeogeographical significance. *Palaeontology,* **36**, 681–720.

Romer, R.L. (1996) Contiguous Laurentia and Baltica before the Grenvillian–Sveconorwegian orogeny. *Terra Nova,* **8**, 173–81.

Rong, J. and Harper, D.A.T. (1988) A global synthesis of the latest Ordovician Hirnantian brachiopod faunas. *Transactions of the Royal Society of Edinburgh: Earth Sciences,* **79**, 383–402.

Rose, J. (1985) The Dimlington Stadial/Dimlington Chronozone: a proposal for naming the main glacial episode of the Late Devensian in Britain. *Boreas,* **14**, 225–230.

Ross, J.R.P. (1966) The fauna of the Portrane Limestone IV: Polyzoa. *Bulletin British Museum (Natural History) Geology,* **12**, 107–135.

Rothery, E. (1989) Transpression in the Variscan foreland: a study in east-central Ireland. *Irish Journal of Earth Sciences*, **10**, 1–12.

Roycroft, P. (1991) Magmatically zoned muscovite from the peraluminous 2-mica granites of the Leinster batholith, southeast Ireland. *Geology*, **19**(5), 437–440.

Ruddiman, W.F. (2005) *Plows, Plagues, and Petroleum: How Humans Took Control of Climate*. Princeton University Press, NJ. 202pp.

Ruddiman, W.F., Raymo. M.E., and McIntyre, A. (1986) Matuyama 41,000-year cycles: North Atlantic Ocean and northern hemisphere ice sheets. *Earth and Planetary Science Letters*, **80**, 117–129.

Ruddock, J. (1990) *An investigation of an interglacial site between Spa and Fenit, Co.Kerry*. Unpublished M.Sc. thesis, University of Dublin, Trinity College.

Ruffell, A. and Shelton, R. (1999) The control of sedimentary facies during phases of crustal extension: examples from the Triassic of onshore and onshore England and Northern Ireland. *Journal of the Geological Society of London*, **156**, 779–89.

Ruffell, A.H. and Coward, M.P. (1992) Basement tectonics and their relationship to Mesozoic megasequences in the Celtic Seas and Bristol Channel area. *In*: Parnell, J. (ed.). *Basins on the Atlantic Seaboard: Petroleum Geology, Sedimentology and Basin Evolution*. Geological Society, London, Special Publication, **62**, 385–394.

Rundle, C.C. (1978) *Preliminary report on dating of samples from the Tyrone Igneous Complex, Northern Ireland*. Institute of Geological Sciences Isotope Geology Unit Internal Report, **78/3** (unpublished).

Rundle, C.C. (1988) *Potassium–argon ages for andesite lavas from the Cappagh quarry, Co. Tyrone, N Ireland*. NERC Isotope Geology Centre, Radiogenic isotopes section Report, **RI/88/2** (unpublished).

Rushton, A.W.A. (1996) *Trichograptus* from the lower Arenig of Kiltrea, County Wexford. *Irish Journal of Earth Sciences*, **15**, 61–70.

Rushton, A.W.A. and Phillips, W.E.A. (1973) A *Protospongia* from the Dalradian of Clare Island, Co. Mayo, Ireland. *Palaeontology*, **16**, 231–237.

Rushton, A.W.A. and Zalasiewicz, J.A. (1999) A distinctive, stratigraphically useful normalograptid graptolite from the Caradoc of eastern Avalonia. *Irish Journal of Earth Sciences*, **17**, 83–90.

Russell, K.J. (1978) Vertebrate fossils from the Iveragh Peninsula and the age of the Old Red Sandstone. *Journal of Earth Sciences, Royal Dublin Society*, **1**, 151–162.

Russell, K.J. (1984) *The sedimentology and palaeogeography of some Devonian sedimentary rocks in Southwest Ireland*. Unpublished Ph.D. thesis, Council for National Academic Awards, Plymouth Polytechnic.

Rutter, E.H. (1976) The kinetics of rock deformation by pressure solution. *Philosophical Transactions of the Royal Society, London*, **A283**, 203–219.

Rutter, E.H. (1978) Discussion on pressure solution. *Journal of the Geological Society of London*, **135**, 135.

Ryan, P.D. and Archer, J.B. (1977) The South Mayo Trough: a possible Ordovician Gulf of California type marginal basin in the west of Ireland. *Canadian Journal of Earth Sciences*, **14**, 2453–2461.

Ryan, P.D. and Archer, J.B. (1978) The Lough Nafooey Fault: a Taconic structure in western Ireland. *Geological Survey Ireland Bulletin*, **2**, 255–264.

Ryan, P.D. and Dewey, J.F. (2004) The South Connemara Group reinterpreted: a subduction–accretion complex in the Caledonides of Galway Bay, western Ireland. *Journal of Geodynamics*, **37**(3–5), 513–529.

Ryan, P.D., Floyd, P.A., and Archer, J.B. (1980) The stratigraphy and petrochemisty of the Lough Nafooey Group (Tremadocian), western Ireland. *Journal of the Geological Society of London*, **137**, 443–458.

Ryan, P.D., Max, M.D., and Kelly, T. (1983a) The petrochemistry of the basic volcanic rocks of the South Connemara Group (Ordovician), western Ireland. *Geological Magazine*, **120**, 141–152.

Ryan, P.D., Sawal, V.K., and Rowland, A.S. (1983b) Ophiolitic mélange separates ortho- and para-tectonic Caledonides in western Ireland. *Nature*, **302**, 50–52.

Ryan, P.D., Soper, N.J., Snyder, D.B., England, R.W., and Hutton, D.H.W. (1995) The Antrim–Galway Line: a resolution of the Highland Border fault enigma of the Caledonides of Britain and Ireland. *Geological Magazine*, **132**(2), 171–184.

Sabine, P.A. and Young, B.R. (1975) Metamorphic processes at high temperature and low pressure: the petrogenesis of the metasomatized and assimilated rocks of Carneal, Co. Antrim. *Philosophical Transactions of the Royal Society*, **A280**, 225–269.

Sadler, S.P. and Kelly, S.B. (1993) Fluvial processes and cyclicity in terminal fan deposits: an example from the Late Devonian of southwest Ireland. *Sedimentary Geology*, **85**, 375–386.

Saemundsson, K. (1970) Interglacial lava flows in the lowlands of Southern Iceland and the problem of the two-tiered columnar jointing. *Jökull*, **20**, 62–77.

Sampson, G.V. (1802) *Statistical survey of the county of Londonderry, with observations on the means of improvement*. Dublin, xxv + 551.

Sanders, I.S. (1979) Observations on eclogite- and granulite-facies rocks in the basement of the Caledonides. *In*: Harris, A.L., Holland, C.H., and Leake, B.E. (eds). *The Caledonides of the British Isles: reviewed*. Geological Society, London, Special Publications, **8**, 97–100.

Sanders, I.S. (1986) Gas filter-pressing origin for segregation vesicles in dykes. *Geological Magazine*, **123**, 67–72.

Sanders, I.S. (1991) Exhumed lower crust in NW Ireland, and a model for crustal conductivity. *Journal of the Geological Society of London*, **148**, 131–135.

Sanders, I.S. (1994) The northeast Ox Mountains inlier, Ireland. *In*: Gibbons, W. and Harris, A.L. (eds). *A revised correlation of Precambrian rocks in the British Isles*. Geological Society, London, Special Report, **22**, 63–64.

Sanders, I.S. and Morris, J.H. (1978) Evidence for Caledonian subduction from greywacke detritus in the Longford-Down inlier. *Journal of Earth Sciences, Royal Dublin Society*, **1**, 53–62.

Sanders, I.S., Daly, J.S., and Davies, G.R. (1987) Late Proterozoic high-pressure granulite facies metamorphism in the north-east Ox inlier, north-west Ireland. *Journal of Metamorphic Geology*, **5**, 69–85.

Sanders, I.S., Van Calsteren, P.W.C., and Hawkesworth, C.J. (1984) A Grenville Sm–Nd age for the Glenelg eclogite in north-west Scotland. *Nature*, **312**, 439–40.

Sanderson, D.J. (1970) The Highland Border Ridge of north-east Ireland. *Geological Magazine*, **107**, 531–538.

Sanderson, D.J. (1984) Structural variations across the northern margin of the Variscides in N.W. Europe. *In*: Hutton, D.H.W. and Sanderson, D.J. (eds). *Variscan Tectonics of the North Atlantic Region*. Special Publication Geological Society London, **14**, 149–165.

Sanderson, D.J., Andrews, J.R., Phillips, W.E.A., and Hutton, D.H.W. (1980) Deformation studies in the Irish Caledonides. *Journal of the Geological Society of London*, **137**, 289–302.

Sangster, D.F. (ed.). (1996) *Carbonate-hosted lead–zinc deposits*. Society of Economic Geologists Special Publication, **4**, 672pp.

Savage, R.J.G. (1963) Upper Lias ammonite from Cretaceous conglomerate of Murlough Bay. *Irish Naturalists' Journal*, **14**, 179–180.

Scherer, E.E., Cameron, K.L., and Blichert-Toft, J. (2000) Lu–Hf garnet geochronology: closure temperature relative to the Sm–Nd system and the effects of trace mineral inclusions. *Geochimica et Cosmochimica Acta*, **64**, 3413–32.

Scherer, E., Münker, C., and Mezger, K. (2001) Calibration of the Lutetium–Hafnium clock. *Science*, **293**, 683–687.

Schiener, E.J. (1974) Syndepositional small-scale intrusions in Ordovician pyroclastics, County Waterford. *Journal of the Geological Society of London*, **130**, 157–161.

Schultz, R.W. and Sevastopulo, G.D. (1965) Lower Carboniferous volcanic rocks near Tulla, Co. Clare, Ireland. *Scientific Proceedings of the Royal Dublin Society*, **A2**, 153–162.

Schwarzacher, W. (1961) Petrology and structure of some Lower Carboniferous reefs in northwestern Ireland. *Bulletin of the American Association of Petroleum Geologists*, **45**, 1481–1503.

Schwarzacher, W. (1989) Milankovich type cycles in the Lower Carboniferous of NW Ireland. *Terra Nova*, **1**, 468–473.

Scotchman, I.C. and Thomas, J.R.W. (1995) Maturity and hydrocarbon generation in the Slyne Trough, northwest Ireland. *In*: Croker, P.F. and Shannon, P.M. (eds). *The Petroleum Geology of Ireland's Offshore Basins*. Geological Society, London, Special Publications **93**, 385–411.

Scotese, C.R. (2001) *Atlas of Earth History, Volume 1, Palaeogeography*, PALEOMAP Project, Arlington, Texas, 52pp.

Scotese, C.R., van der Voo, R., and Barrett, S.F. (1985) Silurian and Devonian base maps. *Philosophical Transactions Royal Society, London*, **B309**, 57–77.

Scourse, J.D. and Furze, M.F.A. (2001) A critical review of the glaciomarine model for Irish sea deglaciation: evidence from southern Britain, the Celtic shelf and adjacent continental slope. *Journal of Quaternary Science*, **16**(5), 419–434.

Scourse, J.D., Hall, I.R., McCave, I.N., Young, J.R., and Sugdon, C. (2000) The origin of Heinrich layers: evidence from H2 for European precursor events. *Earth and Planetary Science Letters*, **182**, 187–195.

Scourse, J.D., Allen, J.R.M., Austin, W.E.N., Coxon, P., Devoy, R.J.N., and Sejrup, H.P. (1992) New evidence on the age and significance of the Gortian Temperate Stage: a preliminary report on the Cork Harbour site. *Proceedings of the Royal Irish Academy*, **92B**, 21–43.

Scrutton, C.T. and Parkes, M.A. (1992) The age and affinities of the coral faunas from the Lower Silurian rocks of the Charlestown inlier, County Mayo, Ireland. *Irish Journal of Earth Sciences*, **11**, 191–196.

Seeman, U. (1984) Tertiary intrusives on the Atlantic continental margin off southwest Ireland. *Irish Journal of Earth Sciences*, **6**, 229–236.

Selby, D., Creaser, R.A., and Feely, M. (2004) Accurate and precise Re–Os molybdenite dates from the Galway Granite, Ireland. Critical comment on 'Disturbance of the Re–Os chronometer of molybdenites from the late-Caledonian Galway Granite, Ireland, by hydrothermal fluid circulation' by Suzuki *et al.*, *Geochemical Journal*, **35**, 29–35, 2001. *Geochemical Journal*, **38**(3), 291–294.

Sevastopulo, G.D. (1981a) Lower Carboniferous. *In*: Holland, C.H. (ed.). *A Geology of Ireland*. Scottish Academic Press, Edinburgh, 147–172.

Sevastopulo, G.D. (1981b) Upper Carboniferous. *In*: Holland, C.H. (ed.). *A Geology of Ireland*. Scottish Academic Press, Edinburgh. 173–187.

Sevastopulo, G.D. (1981c) Hercynian structures. *In*: Holland, C.H. (ed.). *A Geology of Ireland*. Scottish Academic Press, 189–199.

Sevastopulo, G.D. (1982) The age and depositional setting of Waulsortian limestones in Ireland. *In*: Bolton, K., Lane, H.R., and Lemone, D.V. (eds). *Symposium on the environmental setting and distribution of the Waulsortian Facies*. El Paso Geological Society and University of Texas at El Paso, 65–79.

Sevastopulo, G.D. and Naylor, D. (1981) Erratic Carboniferous boulders in Bantry Bay, County Cork. *Geological Survey of Ireland Bulletin*, **3**, 79–84.

Sevastopulo, G.D. and Wyse Jackson, P.N. (2001) Carboniferous (Dinantian). *In*: Holland, C.H. (ed). *The Geology of Ireland*. Dunedin Academic Press, 241—288.

Seymour, H.J. (1903) The occurrence of cassiterite in Tertiary Granite of the Mourne Mountains. *Scientific Proceedings of the Royal Dublin Society*, **9**, 583–4.

Shackleton, N.J. (1969) The Last Interglacial in the marine and terrestrial records. *Proceedings of the Royal Society*, **B174**, 135–154.

Shackleton, N.J. and Opdyke, N.D. (1976) *Oxygen isotope and palaeomagnetic stratigraphy of Equatorial Pacific core V28–239, Late Pliocene to Latest Pleistocene*. Geological Society of America Memoir, **145**, 449–464.

Shackleton, R.M. (1940) The succession of rocks in the Dingle Peninsula, Co. Kerry. *Proceedings of the Royal Irish Academy*, **46B**, 1–12.

Shannon, P.M. (1978a) The stratigraphy and sedimentology of the Lower Palaeozoic rocks of south-east County Wexford. *Proceedings of the Royal Irish Academy*, **78B**, 247–265.

Shannon, P.M. (1978b) The petrology of some Lower Palaeozoic greywackes from southeast Ireland: a clue to the origin of the matrix. *Journal of Sedimentary Petrology*, **48**, 1185–1192.

Shannon, P.M. (1979a) The petrology of the Ordovician volcanic rocks of County Wexford, Ireland. *Journal of Earth Sciences, Royal Dublin Society*, **2**, 41–59.

Shannon, P.M. (1979b) The tectonic evolution of the lower Palaeozoic rocks of extreme SE Ireland. *In*: Harris, A.L., Holland, C.H., and Leake, B.E. (eds). *The Caledonides of the British Isles reviewed*. Geological Society of London Special Publication, **8**, 281–285.

Shannon, P.M. (1991a) The development of the Irish offshore sedimentary basins. *Journal of the Geological Society of London*, **148**, 181–189.

Shannon, P.M. (1991b) Tectonic framework and petroleum potential of the Celtic Sea, Ireland. *First Break*, **9**, 107–122.

Shannon, P.M. (1992) Early Tertiary submarine fan deposits in the Porcupine Basin, offshore Ireland. *In:* Parnell, J. (ed.) *Basins on the Atlantic Seaboard: Petroleum Geology, Sedimentology and Basin Evolution*. Geological Society, London, Special Publications **62**, 351–373.

Shannon, P.M. (1993) Submarine Fan Types in the Porcupine Basin, Ireland. *In:* Spencer, A.M. (ed.). *Generation, Accumulation and Production of Europe's Hydrocarbons. III*. Special Publication of the European Association of Petroleum Geoscientists **3**, Springer-Verlag, Berlin, 111–120.

Shannon, P.M. (1995) Permo-Triassic development of the Celtic Sea region, offshore Ireland. *In*: Boldy, S.A.R. (ed.). *Permian and Triassic Rifting in Northwest Europe*. Geological Society, London, Special Publications **91**, 215–237.

Shannon, P.M. and MacTiernan, B. (1993) Triassic prospectivity in the Celtic Sea, Ireland: a case history. *First Break*, **11**, 47–57.

Shannon, P.M. and Naylor, D. (1998) An assessment of Irish Offshore Basins and petroleum plays. *Journal of Petroleum Geology*, **21**, 125–152.

Shannon, P.M., Corcoran, D.V., and Haughton, P.D.W. (2001a) The petroleum exploration of Ireland's offshore basins: introduction. *In*: Shannon, P.M., Haughton, P.D.W. and Corcoran, D.V. (eds). *The Petroleum Exploration of Ireland's Offshore Basins*. Geological Society, London, Special Publications **188**, 1–8.

Shannon, P.M., McDonnell, A. and Bailey, W.R. (2007) The evolution of the Porcupine and Rockall basins, offshore Ireland: the geological template for carbonate mound development. *International Journal of Earth Sciences*, **96**, 21–35.

Shannon, P.M., Moore, J.G., Jacob, A.W.B., and Makris, J., (1993) Cretaceous and Tertiary basin development west of Ireland. *In:* Parker, J.R. (ed.). *Petroleum Geology of Northwest Europe: Proceedings of the 4th Conference*. Geological Society, London, 1057–1066.

Shannon, P.M., O'Reilly, B.M., Readman, P.W., Jacob, A.W.B., and Kenyon, N. (2001b) Slope failure features on the margin of the Rockall Trough. *In*: Shannon, P.M., Haughton, P.D.W. and Corcoran, D.V. (eds). *The Petroleum Exploration of Ireland's Offshore Basins*. Geological Society, London, Special Publications **188**, 455–464.

Shannon, P.M., Jacob, A.W.B., Makris, J., O'Reilly, B., Hauser, F., and Vogt, U. (1994) Basin evolution in the Rockall region, North Atlantic. *First Break*, **12**, 515–522.

Shannon, P.M., Jacob, A.W.B., Makris, J., O'Reilly, B., Hauser, F., and Vogt, U. (1995) Basin development and petroleum prospectivity of the Rockall and Hatton region. *In*: Croker, P.F. and Shannon, P.M. (eds). *The Petroleum Geology of Ireland's Offshore Basins*. Geological Society, London, Special Publications **93**, 435–457.

Shannon, P.M., Jacob, A.W.B., Makris, J, O'Reilly, B., Hauser, F., Readman, P.W. and Makris, J. (1999) Structural setting, geological development and basin modelling in the Rockall Trough. *In*: Fleet, A.G. and Boldy, S.A.R (eds). *Petroleum Geology of Northwest Europe: Proceedings of the 5th Conference*. Geological Society, London, 421–431.

Sharpe, E.N. (1970) An occurrence of pillow lavas in the Ordovician of County Down. *Irish Naturalists' Journal*, **16**, 299–301.

Shearley, E.P. (1988) *The Carboniferous geology of the Fermoy and Mitchelstown Synclines*. Unpublished Ph.D. thesis, University of Dublin.

Shearley, E., Redmond, P., King, M., and Goodman, R. (1996) Geological controls on mineralization and dolomitization of the Lisheen Zn–Pb–Ag deposit, Co. Tipperary, Ireland. *In*: Strogen, P., Somerville, I.D., and Jones, G.Ll. (eds). *Recent advances in Lower Carboniferous geology*. Geological Society, London Special Publications, **107**, 23–33.

Shelford, P.H. (1963) The structure and relationship of the Namurian outcrop between Duntryleague, Co. Limerick and Dromlin, Co. Tipperary. *Proceedings of the Royal Irish Academy*, **62**(B), 255–266.

Shelford, P.H. (1967) The Namurian and upper Viséan of the Limerick volcanic basin. *Proceedings of the Geologists' Association*, **78**, 121–136.

Shelton, R. (1995) Mesozoic basin evolution of the North Channel: preliminary results. *In:* Croker, P.F. and Shannon, P.M. (eds). *The Petroleum Geology of Ireland's Offshore Basins*. Geological Society Special Publications **93**, 7–20.

Shelton, R. (1997) Tectonic evolution of the Larne Basin. *In*: Meadows, N.S., Trueblood, S.P., Hardman, M., and Cowan, G. (eds). *Petroleum Geology of the Irish Sea and Adjacent Areas*. Geological Society, London, Special Publications **124**, 113–133.

Shennan, I., Bradley, S., Milne, G., Brooks, A.J., Bassett, S., and Hamilton, S. (2006) Relative sea-level changes, glacial isostatic modelling and ice-sheet reconstructions from the British Isles since the Last Glacial Maximum. *Journal of Quaternary Science*, **21**(6), 585–599.

Shephard-Thorn, E.R. (1963) The Carboniferous limestone succession in north-west County Limerick. *Proceedings of the Royal Irish Academy*, **62**, 267–294.

Sheppard, W.A. (1980) The ores and host rock geology of the Avoca Mines, County Wicklow, Ireland. *Norges geologiske Undersokelse*, **360**, 269–283.

Sheridan, D.J.R. (1972a) The stratigraphy of the Trim No. 1 well, Co. Meath and its relationship to Lower Carboniferous outcrop in east-central Ireland.*Geological Survey of Ireland Bulletin*, **1**, 311–334.

Sheridan, D.J.R. (1972b) *Upper Old Red Sandstone and Lower Carboniferous of the Slieve Beagh Syncline and its setting in the northwest Carboniferous basin, Ireland*. Geological Survey of Ireland, Special Paper, **2**, 129pp.

Sheridan, D.J.R. (1977) The hydrocarbons and mineralisation proved in the Carboniferous strata of deep boreholes in Ireland. *In*: Garrard, P. (ed.). *Proceedings of the Forum on Oil and Ore in Sediments*. Geology Department, Imperial College, London, 113–144.

Sibuet, J.C., Dyment, J., Bois, C., Pinet, B., and Ondreas, H. (1990) Crustal structure of the Celtic Sea and Western Approaches from gravity data and deep seismic profiles: constraints on the formation of continental basins. *Journal of Geophysical Research*, **95B**, 10999–11020.

Simms, M.J. (2000a) Quartz-rich cave sediments in the Burren, Co. Clare, Ireland. *Proceedings of the University of Bristol Spelaeological Society*, **22**, 81–98.

Simms, M.J. (2000b) The sub-basaltic surface in northeast Ireland and its significance for interpreting the Tertiary history of the region. *Proceedings of the Geologists' Association*, **111**, 321–336.

Simms, M.J. (2007) Uniquely extensive soft-sediment deformation in the Rhaetian of the UK: evidence for earthquake or impact? *Palaeogeography, Palaeoclimatology, Palaeoecology*, **244**, 407–423.

Simms, M.J. and Ruffell, A.H. (1990) Climatic and biotic change in the late Triassic. *Journal of the Geological Society of London*,

147, 321–327.

Simms, M.J., Chidlaw, N., Morton, N., and Page, K.N. (2004) *British Lower Jurassic Stratigraphy.* Geological Conservation Review Series, **30**, 458pp.

Simon, J.B. (1981) *Old Red Sandstone sedimentation in the northeast of Ireland.* Unpublished Ph.D. thesis, Glasgow University.

Simon, J.B. (1984a) Provenance and depositional history of the Lower Old Red Sandstone of north-east Ireland. *Irish Journal of Earth Sciences*, **6**, 1–13.

Simon, J.B. (1984b) Sedimentation of a small complex alluvial fan of possible Upper Old Red Sandstone age, north-east County Antrim. *Irish Journal of Earth Sciences*, **6**, 109–119.

Simon, J.B. (1984c) Sedimentation and tectonic setting of the Lower Old Red Sandstone of the Fintona and Curlew Mountain districts. *Irish Journal of Earth Sciences*, **6**, 213–228.

Simon, J.B. (1986) Halite pseudomorphs from the Red Arch Formation, Glenariff, County Antrim. *Irish Naturalists' Journal*, **22**, 157–159.

Simon, J.B. and Bluck, B.J. (1982) Palaeodrainage of the southern margin of the Caledonian mountain chain in the northern British Isles. *Transactions of the Royal Society of Edinburgh: Earth Sciences*, **73**, 11–15.

Simpson, I.M. (1955) The Lower Carboniferous stratigraphy of the Omagh Syncline, Northern Ireland. *Quarterly Journal of the Geological Society of London*, **110**, 391–408.

Simpson, I.M. and West, R.G. (1958) On the stratigraphy and palaeobotany of a Late-Pleistocene organic deposit at Chelford, Cheshire. *New Phytologist*, **57**, 239–250.

Simpson, S. (1957) On the trace-fossil *Chondrites*. *Quarterly Journal of the Geological Society of London*, **112**, 475–499.

Sinclair, I.K., Shannon, P.M., Williams, B.P.J., Harker, S.D., and Moore, J.G. (1994) Tectonic control on sedimentary evolution of three North Atlantic borderland Mesozoic basins. *Basin Research*, **6**, 193–218.

Singh, G. (1970) Late-glacial vegetational history of Lecale, County Down. *Proceedings of the Royal Irish Academy*, **69B**, 189–216.

Siveter, D.J. (1989) Silurian trilobites from the Annascaul inlier, Dingle Peninsula, Ireland. *Palaeontology*, **32**, 109–161.

Skevington, D. (1974) Controls influencing the composition and distribution of Ordovician graptolite faunal provinces. *Special papers in Palaeontology*, **13**, 59–73.

Skevington, D.G. (1971) The age and correlation of the Rosroe Grits, northwest County Galway. *Proceedings of the Royal Irish Academy*, **71B**, 75–83.

Skiba, W. (1952) The contact phenomena on the NW side of the Crossdoney Complex, Co. Cavan. *Transactions of the Geological Society of Edinburgh*, **15**, 322–345.

Sleeman, A.G. (1977) Lower Carboniferous trangressive facies of the Porter's Gate Formation at Hook Head, County Wexford. *Proceedings of the Royal Irish Academy*, **77B**, 269–284.

Sleeman, A.G. (1987) The Dinantian and Namurian rocks of the Ringabella and Cloyne synclines, west of Cork Harbour, County Cork. *Geological Survey of Ireland Bulletin*, **4**, 67–76.

Sleeman, A.G. and McConnell, B. (1995) *Geology of East Cork–Waterford and adjoining parts of Tipperary and Limerick to accompany the bedrock geology 1:100,000 scale map series, Sheet 22, East Cork–Waterford, with contributions by K. Claringbold, P. O'Connor, W.P. Warren, and G. Wright.* Geological Survey of Ireland.

Sleeman, A.G. and Pracht, M. (1999) *Geology of the Shannon Estuary: a geological description of the Shannon Estuary region including parts of Clare, Limerick and Kerry, to accompany the bedrock geology 1:100,000 Scale Map Series, Sheet 17, Shannon Estuary, with contributions by K. Claringbold and G. Stanley (Minerals), J. Deakin and G. Wright (Groundwater), and O. Bloetjes and R. Creighton (Quaternary).* Geological Survey of Ireland.

Sleeman, A.G. and Tietzsch-Tyler, D. (1988) *Geological summary and stratigraphical lexicon to accompany the provisional 1:25,000 Geological Map Series: southeast County Wexford map sheets 29/11NW Ferrycarrig, 29/11NE Castlebridge, 29/11SW Johnston Castle, 29/11SE Rosslare, 29/9NW Kilmore Quay, 29/9NE Carnsore Point.* Geological Survey of Ireland Map Report Series **MRS 88/1**, Dublin.

Sleeman, A.G., Higgs, K., and Sevastopulo, G.D. (1983) The stratigraphy of the late Devonian–early Carboniferous rocks of south County Wexford. *Geological Survey of Ireland Bulletin*, **3**, 141–158.

Sleeman, A.G., Reilly, T., and Higgs, K. (1978) Preliminary stratigraphy and palynology of five sections through the Old Head Sandstone and Kinsale Formations (Upper Devonian to Lower Carboniferous) on the west side of Cork Harbour. *Geological Survey of Ireland Bulletin*, **2**, 167–187.

Sleeman, A.G., Thornbury, B., and Sevastopulo, G.D. (1986) The stratigraphy of the Courceyan (Carboniferous: Dinantian) rocks of the Cloyne Syncline, west of Cork Harbour. *Irish Journal of Earth Sciences*, **8**, 21–40.

Sleeman, A.G., Johnston, I.S., Naylor, D., and Sevastopulo, G.D. (1974) The stratigraphy of the Carboniferous rocks at Hook Head, County Wexford. *Proceedings of the Royal Irish Academy*, **74B**, 227–243.

Sleeman, A.G., Pracht, M., Daly, E., Flegg, A.M., O'Connor, P., and Warren, W.P. (1994) *Geology of south Cork: a geological description of south Cork and adjoining parts of Waterford to accompany the bedrock geology 1:100,000 scale map series, Sheet 25, south Cork.* Geological Survey of Ireland, 60pp.

Smethurst, M.A. and Briden, J.C. (1988) Palaeomagnetism of Silurian sediments in W. Ireland: evidence for block rotation in the Caledonides. *Geophysical Journal International*, **95**(2), 327–346.

Smethurst, M.A., MacNiocaill, C., and Ryan, P.D. (1994) Oroclinal bending in the Caledonides of western Ireland. *Journal of the Geological Society of London*, **151**, 315–328.

Smith, C. (1995) Evolution of the Cockburn Basin: implications for the structural development of the Celtic Sea basins. *In:* Croker, P.F. and Shannon, P.M. (eds). *The Petroleum Geology of Ireland's Offshore Basins.* Geological Society, London, Special Publications **93**, 279–295.

Smith, D.B., Harwood, G.M., Pattison, J., and Pettigrew, T.H. (1986) A revised nomenclature for Upper Permian strata in eastern England. *In:* Harwood, G.M. and Smith, D.B. (eds). *The English Zechstein and related topics.* Special Publication of the Geological Society of London, **22**, 9-17.

Smith, D.B., Taylor, J.C.M., Arthurton, R.S., Brookfield, M.E., and Glennie, K.W. (1992) Permian. *In:* Cope, J.C.W., Ingham, J.K., and Rawson, P.F. (eds). *Atlas of Palaeogeography and Lithofacies.* Geological Society, London, Memoir **13**, 153pp.

Smith, D.G. (1975) Wenlock plant spores and tetrads from County Mayo, Ireland. *Geological Magazine*, **112**, 111–114.

Smith, D.G. (1977) Lower Cambrian palynomorphs from Howth, Co. Dublin. *Geological Journal*, **12**, 159–168.

Smith, D.G. (1979a) The distribution of trilete spores in Irish Silurian rocks. *In*: Harris, A.L., Holland, C.H., and Leake, B.E. (eds). *The Caledonides of the British Isles: reviewed.* Geological Society London, Special Publication, **8**, 423–432.

Smith, D.G. (1979b) New evidence for the age of the Ahenny Formation, Slievenamon inlier, County Tipperary. *Journal of Earth Sciences, Royal Dublin Society*, **2**, 61–63.

Smith, D.G. (1981) Progress in Irish Lower Palaeozoic palynology. *Review Palaeobotany Palynology*, **34**, 137–148.

Smith, J. (1996) A palynofacies analysis of the Dinantian (Asbian) Glenade Sandstone Formation of the Leitrim Group, north-west Ireland. *In*: Strogen, P., Somerville, I.D., and Jones, G.Ll. (eds). *Recent advances in Lower Carboniferous geology*. Geological Society, London, Special Publications, **107**, 437–448.

Smith, J. and Higgs, K.T. (2001) Provenance implications or reworked palynomorphs in Mesozoic successions of the Porcupine and North Porcupine basins, offshore Ireland. *In*: Shannon, P.M., Haughton, P.D.W., and Corcoran, D.V. (eds). *The Petroleum Exploration of Ireland's Offshore Basins.* Geological Society, London, Special Publications **188**, 291–300.

Smith, R.A. (1986) *Permo-Triassic and Dinantian rocks of the Belfast Harbour Borehole.* British Geological Survey Report, **18**(6), 13pp.

Smith, W.W. (1864) *Catalogue of mineral collections in the Museum of Practical Geology, London.*

Smyth, L.B. (1950) The Carboniferous System in North County Dublin. *Quarterly Journal of the Geological Society of London*, **105**, 295–326.

Smyth, L.B. (1951) A Viséan cephalopod fauna in the Rush Slates of County Dublin. *Proceedings of the Royal Irish Academy*, **53B**, 289–309.

Smythe, D.K., (1989) Rockall Trough – Cretaceous or Late Palaeozoic. *Scottish Journal of Geology*, **25**, 5–43.

Snyder, D.B., Lucas, S.B., and McBride, J.H. (1996) Crustal and mantle reflectors from Palaeoproterozoic orogens and their relation to arc-continent collisions. *In*: Brewer, T.S. (ed.). *Precambrian crustal evolution in the North Atlantic Region.* Geological Society, London, Special Publications, **112**, 1–23.

Sollas, W.J. (1890) Relation of the Granite to the Gabbo of Barnavave, Carlingford. *Transactions of the Royal Irish Academy*, **30**, 490.

Somerville, I.D. and Jones, G.Ll. (1985) The Courceyan stratigraphy of the Pallaskenry borehole, County Limerick, Ireland. *Geological Journal*, **20**, 377–400.

Somerville, I.D. and Strogen, P. (1992) Ramp sedimentation in the Dinantian limestones of the Shannon Trough, Co. Limerick, Ireland. *Sedimentary Geology*, **79**, 59–75.

Somerville, I.D., Strogen, P., and Jones, G.Ll. (1992a) The biostratigraphy of Dinantian limestones and associated volcanic rocks in the Limerick Syncline, Ireland. *Geological Journal*, **27**, 201–270.

Somerville, I.D., Strogen, P. and Jones, G.Ll. (1992b) Mid-Dinantian Waulsortian buildups in the Dublin Basin, Ireland. *Sedimentary Geology*, **79**, 91–116.

Somerville, I.D., Strogen, P., Jones, G.Ll., and Somerville, H.E.A. (1996a) Late Viséan buildups at Kingscourt, Ireland: possible precursors for Upper Carboniferous bioherms. *In*: Strogen, P., Somerville, I.D., and Jones, G.Ll. (eds). *Recent advances in Lower Carboniferous geology*. Geological Society, London Special Publications, **107**, 127–144.

Somerville, I.D., Strogen, P., Pickard, N.A.H., and Jones, G.Ll. (1992c) Early to mid-Viséan shallow water platform buildups, north Co. Dublin, Ireland. *Geological Journal*, **27**, 151–172.

Somerville, I.D., Strogen, P., Mitchell, W.I., Somerville, H.E.A., and Higgs, K. (1996b) Stratigraphy of Dinantian rocks in WB3 borehole, Co. Armagh. *Irish Journal of Earth Sciences*, **19**, 51–78.

Soper, N.J. (1994a) Was Scotland a Vendian RRR junction?. *Journal of the Geological Society of London*, **151**, 579–582.

Soper, N.J. (1994b) Neoproterozoic sedimentation on the north-east margin of Laurentia and the opening of Iapetus. *Geological Magazine*, **131**, 291–299.

Soper, N.J. and Anderton, R. (1984) Did the Dalradian slides originate as extensional faults? *Nature*, **307**, 357–360.

Soper, N.J. and England, R.W. (1995) Vendian and Riphean rifting in NW Scotland. *Journal of the Geological Society of London*, **152**, 11–14.

Soper, N.J. and Hutton, D.H.W. (1984) Late Caledonian sinistral displacements in Britain: implications for a three plate collision model. *Tectonics*, **3**, 781–794.

Soper, N.J., Ryan, P.D., and Dewey, J.F. (1999) Age of the Grampian orogeny in Scotland and Ireland. *Journal of the Geological Society of London*, **156**, 1231–1236.

Soper, N.J., Webb, B.C., and Woodcock, N.H. (1987) Late Caledonian (Acadian) transpression in north-west England: timing, geometry and geotectonic significance. *Proceedings of the Yorkshire Geological Society*, **46**, 175–192.

Soper, N.J., Strachan, R.A., Holdsworth, R.E., Gayer, R.A., and Greiling, R.O. (1992) Sinistral transpression and the Silurian closure of Iapetus. *Journal of the Geological Society of London*, **149**, 871–880.

Spencer, A.M. (1971) *Late Precambrian glaciation in Scotland.* Geological Society of London, Memoir, **6**, 98pp.

Spencer, A.M., Birkeland, Ø., Knag, G.Ø., and Fredsted, R. (1999) Petroleum systems of the Atlantic margin of northwest Europe. *Petroleum Geology of NW Europe. In*: Fleet, A.J. and Boldy, S.A.R. (eds). *Petroleum Geology of Northwest Europe: Proceedings of the 5th Conference.* Geological Society, London, 231–246.

Spry, A. (1962) The origin of columnar jointing, particularly in basalt flows. *Geological Society of Australia Journal*, **8**, 191–216.

Srivastava, S.P. and Verhoef, J. (1992) Evolution of Mesozoic sedimentary basins around the North Central Atlantic: a preliminary plate kinematic solution. *In*: Parnell, J. (ed.). *Basins on the Atlantic Seaboard: Petroleum Geology, Sedimentology and Basin Evolution.* Geological Society, London, Special Publications **62**, 397–420.

Stacey, J.S. and Kramers, J.D. (1975) Approximation of terrestrial lead isotope evolution by a two-stage model. *Earth and Planetary Science Letters*, **26**, 207–221.

Stanton, W.I. (1960) The Lower Palaeozoic rocks of south-west Murrisk, Ireland. *Quarterly Journal of the Geological Society of London*, **116**, 269–296.

Steed, G.M. (1986) The geology and genesis of the Gortdrum Cu–Ag–Hg orebody. *In*: Andrew, C.J., Crowe, R.W.A., Finlay, S., Pennell, W.M., and Pyne, J.F. (eds). *Geology and genesis of mineral deposits in Ireland.* Irish Association for Economic Geology, Dublin, 481–500.

Steiger, R.H. and Jäger, E. (1977) Subcommission on geochronology: Convention on the use of decay constants in geo- and cosmochronology. *Earth and Planetary Science Letters*, **36**, 359–362.

Stephens, N., Creighton, J.R., and Hannon, M.A. (1975) The Late Pleistocene Period in north-eastern Ireland: an assessment, 1975. *Irish Geography*, **8**, 1–23.

Stephenson, M.H. and Mitchell, W.I. (2002) Definitive new palynological evidence for the early Devonian age of the Fintona Group, Northern Ireland. *Irish Journal of Earth Sciences*, **20**, 41–52.

Stevenson, C.T.E., Owens, W.H., Hutton, D.H.W., Hood, D.N., and Meighan, I.G. (2007) Laccolithic, as opposed to cauldron subsidence, emplacement of the Eastern Mourne pluton, N. Ireland: evidence from anisotropy of magnetic susceptibility. *Journal of the Geological Society of London*, **164**, 99–110.

Stillman, C.J. (1971) Ordovician ash-fall tuffs from Co. Waterford. *Scientific Proceedings of the Royal Dublin Society*, **4A**, 89–101.

Stillman, C.J. and Maytham, D.K. (1973) The Ordovician volcanic rocks of Arklow Head, Co. Wicklow. *Proceedings of the Royal Irish Academy*, **73B**, 61–77.

Stillman, C.J. and Sevastopulo, G.D. (2005) *Leinster. Classic Geology in Europe*, **6**. Terra Publishing, Hertford, 192pp.

Stillman, C.J., Downes, K., and Schiener, E.J. (1974) Caradocian volcanic activity in east and south-east Ireland. *Scientific Proceedings of the Royal Dublin Society*, **5A**, 87–98.

Stoker, M.S., van Weering, T.C.E., and Svaerdborg, T. (2001) A Mid-Late Cenozoic tectonostratigraphic framework for the Rockall Trough. *In:* Shannon, P.M., Haughton, P.D.W., and Corcoran, D. (eds). *The Petroleum Exploration of Ireland's Offshore Basins.* Geological Society, London, Special Publications **188,** 411–438.

Stoker, M.S., Praeg, D., Hjelstuen, B.O., Laberg, J.S., Nielsen, T., and Shannon, P.M. (2005c) Neogene stratigraphy and the sedimentary and oceanographic development of the NW European Atlantic margin. *Marine and Petroleum Geology*, **22,** 977–1005.

Stoker, M.S., Praeg, D., Shannon, P.M., Hjelstuen, B.O., Laberg, J.S., van Weering, T.C.E., Sejrup, H.P., and Evans, D. (2005b) Neogene evolution of the Atlantic continental margin of NW Europe (Lofoten Islands to SW Ireland): anything but passive. *In*: Doré, A.G. and Vining, B. (eds). *Petroleum Geology: North-West Europe and Global Perspectives – Proceedings of the 6th Petroleum Geology Conference.* Geological Society, London, 1057–1076.

Stoker, M.S., Hoult, R.J., Nielsen, T., Hjelstuen, B.O., Laberg, J.S., Shannon, P.M., Praeg, D., Mathiesen, A., van Weering, T.C.E., and McDonnell, A. (2005a) Sedimentary and oceanographic responses to early Neogene compression on the NW European margin. *Marine and Petroleum Geology.* **22**, 1031–1044.

Stone, P. and Merriman, R.J. (2004) Basin thermal history favours an accretionary origin for the Southern Uplands terrane, Scottish Caledonides. *Journal of the Geological Society of London*, **161**, 829–836.

Stone, P., Floyd, J.D., Barnes, R.P., and Lintern, B.C. (1987) A sequential back-arc and foreland basin thrust duplex model for the Southern Uplands of Scotland. *Journal of the Geological Society of London*, **144**, 753–764.

Storey, C.D., Brewer, T.S., and Parrish, R.R. (2004) Late-Proterozoic tectonics in northwest Scotland: one contractional orogeny or several? *Precambrian Research*, **134**, 227–247.

Stossel, I. (1995) The discovery of a new Devonian tetrapod trackway in SW Ireland. *Journal of the Geological Society of London*, **152**, 407–413.

Strachan, L.J. (2002) Slump-initiated and controlled syndepositional sandstone remobilization: an example from the Namurian of County Clare, Ireland. *Sedimentology*, **49**, 25–41.

Strachan, L.J. and Alsop, G.I. (2006) Slump folds as estimators of palaeoslope: a case study from the Fisherstreet Slump of County Clare, Ireland. *Basin Research*, **18**, 451–470.

Strachan, R.A. (2000a) The Grampian Orogeny: Mid-Ordovician arc–continent collision along the Laurentian margin of Iapetus. *In*: Woodcock, N.H and Strachan, R.A. (eds). *Geological history of Britain and Ireland.* Blackwell Science, 88–106.

Strachan, R.A. (2000b) The northern active margin of Iapetus. *In*: Woodcock, N.H. and Strachan, R.A. (eds). *Geological history of Britain and Ireland.* Blackwell Science, 107–123.

Strachan, R.A., Smith, M., Harris, A.L., and Fettes, D.J. (2002) The Northern Highland and Grampian terranes. *In*: Trewin, N.H. (ed.). *The Geology of Scotland.* 4th Edition. The Geological Society, London, 81–147.

Streel, M., Higgs, K.T., Loboziak, S., Riegel, W., and Steemans, P. (1987) Spore stratigraphy and correlation with faunas and floras in the type marine Devonian of the Ardenne–Rhenish regions. *Review of Palaeobotany and Palynology*, **50**, 211–229.

Strogen, P. (1974a) The volcanic rocks of the Carrigogunnel area, County Limerick. *Scientific Proceedings of the Royal Dublin Society*, **5**, 1–26.

Strogen, P. (1974b) The sub-Palaeozoic basement in central Ireland. *Nature*, **250**, 562–563.

Strogen, P. (1988) The Carboniferous lithostratigraphy of south-east County Limerick, Ireland, and the origin of the Shannon Trough. *Geological Journal*, **23**, 121–137.

Strogen, P. (1995) Lower Carboniferous rocks of the Limerick Syncline. *In*: Anderson, K, Ashton, J., Earls, G., Hitzman, M.W., and Tear, S. (eds). *Irish carbonate-hosted Zn–Pb deposits.* Society of Economic Geologists Guidebook Series, **21**, 75–80.

Strogen, P. and Somerville, I.D. (1984) The stratigraphy of the Upper Palaeozoic rocks of the Lyons Hill area, Co. Kildare. *Irish Journal of Earth Sciences*, **6**, 155–173.

Strogen, P., Jones, G.Ll., and Somerville, I.D. (1990) Stratigraphy and sedimentology of Lower Carboniferous (Dinantian) boreholes from west Co. Meath, Ireland. *Geological Journal*, **25**, 103–137.

Strogen, P., Somerville, I.D., Pickard, N.A.H., and Jones, G.Ll. (1995) Lower Carboniferous (Dinantian) stratigraphy and structure in the Kingscourt outlier, Ireland. *Geological Journal*, **30**, 1–23.

Strogen, P., Somerville, I.D., Pickard, N.A.H., Jones, G.Ll., and Fleming, M. (1996) Controls on ramp, platform and basinal sedimentation in the Dinantian of the Dublin Basin and Shannon Trough, Ireland. *In*: Strogen, P., Somerville, I.D., and Jones, G.Ll. (eds). Recent advances in Lower Carboniferous geology. Geological Society, London, Special Publications, **107**, 263–279.

Stuart, A.J. (1995) Insularity and Quaternary vertebrate faunas in Britain and Ireland. *In*: Preece, R.C. (ed). *Island Britain: a Quaternary Perspective.* Geological Society Special Publication, **96**, 111–125.

Stuart, A.J. and van Wijngaarden-Bakker, L.H. (1985) Quaternary Vertebrates. *In*: Edwards, K.J. and Warren, W.P. (eds). *The Quaternary History of Ireland.* Academic Press, London, 221–249.

Sutcliffe, A.J. (1995) Insularity of the British Isles 250,000–30,000 years ago: the mammalian evidence. *In*: Preece, R.C. (ed.). *Island*

Britain: a Quaternary perspective. Geological Society Special Publication, **96**, London, 127–140.

Sutherland, D.G. (1981) The high-level marine shell beds of Scotland and the build up of the last Scottish ice sheet. *Boreas*, **10**, 247–254.

Sutton, J.S. (1971) The stratigraphy and structure of the Moinian and Dalradian metasediments of the Mullet Peninsula, Co. Mayo, Ireland. *Scientific Proceedings of the Royal Dublin Society*, **4A**, 1–13.

Sutton, J.S. (1972) The Pre-Caledonian rocks of the Mullet Peninsula, County Mayo, Ireland. *Scientific Proceedings of the Royal Dublin Society*, **4A**, 121–136.

Svenning, J.-C. (2002) A review of natural vegetation openness in north-west Europe. *Biological Conservation*, **104**, 133–148.

Svenningsen, O.M. (2001) Onset of seafloor spreading in the Iapetus Ocean at 608 Ma: precise age of the Sarek Dyke Swarm, northern Swedish Caledonides. *Precambrian Research*, **110**, 241–254.

Svensson, A., Nielsen, S.W., Kipfstuhl, S., Johnsen, S.J., Steffensen, J.P., Bigler, M., Ruth, U., and Röthlisberger, R. (2005) Visual stratigraphy of the North Greenland Ice Core Project (NorthGRIP) ice core during the last glacial period. *Journal of Geophysical Research*, **110**, D02108, doi:10.1029/2004JD005134.

Swanston, W.S. and Lapworth, C. (1877) On the Silurian rocks of the County Down. *Proceedings of the Belfast Naturalists' Field Club*, Appendix **4**, 1876–77, 107–148.

Sweetman, T.M. (1988) The geology of the Blackstairs unit of the Leinster Granite. *Irish Journal of Earth Sciences*, **9**, 39–59.

Swennen, O. (1984) *The geology of the Greenan–Mount Congreve–Ballyscanlon area, Co. Waterford, Ireland*. Unpublished M.Sc. thesis, University of Dublin.

Swindles, G.T., Plunkett, G., and Roe, H.M. (2007a) A delayed climatic response to solar forcing at 2800 cal. bp: multiproxy evidence from three Irish peatlands. *The Holocene*, **17**, 177–182.

Swindles, G.T., Plunkett, G., and Roe, H.M. (2007b) A multiproxy climate record from a raised bog in County Fermanagh, Northern Ireland: a critical examination of the link between bog surface wetness and solar variability. *Journal of Quaternary Science*, **22**, 667-679.

Synge, F.M. (1968) The glaciation of West Mayo. *Irish Geography*, **5**, 372–386.

Synge, F.M. (1979) Quaternary glaciation in Ireland. *Quaternary Newsletter*, **18**, 1–18.

Taber, D.R., Vickers, M.K., and Winn, R.D. Jr. (1995) The definition of the Albian 'A' Sand reservoir fairway and aspects of associated gas accumulations in the North Celtic Sea Basin. *In*: Croker, P.F. and Shannon, P.M. (eds). *The Petroleum Geology of Ireland's Offshore Basins*. Geological Society, London, Special Publications **93**, 227–244.

Tanner, P.W.G. (1990) Structural age of the Connemara gabbros, western Ireland. *Journal of the Geological Society of London*, **147**, 599–602.

Tanner, P.W.G. (1995) New evidence that the Lower Cambrian Leny Limestone at Callander, Perthshire, belongs to the Dalradian Supergroup, and a reassessment of the exotic status of the Highland Border Complex. *Geological Magazine*, **132**, 473–483.

Tanner, P.W.G. (1996) Significance of the early fabric in the contact metamorphic aureole of the 590 Ma Ben Vuirich Granite, Perthshire, Scotland. *Geological Magazine*, **133**(6), 683–695.

Tanner, P.W.G. (2005) Discussion on evidence for a major Neoproterozoic orogenic unconformity within the Dalradian Supergroup of NW Ireland and reply. Journal, Vol. **161**, (2004) 629–641. *Journal of the Geological Society of London*, **162**, 221–224.

Tanner, P.W.G. and Evans, J.A. (2003) Late Precambrian U–Pb titanite age for peak regional metamorphism and deformation (Knoydartian orogeny) in the western Moine, Scotland. *Journal of the Geological Society of London*, **160**, 555–564.

Tanner, P.W.G. and Leslie, A.G. (1994) A pre-D2 age for the 590 Ma Ben Vuirich Granite in the Dalradian of Scotland. *Journal of the Geological Society of London*, **151**, 209–212.

Tanner, P.W.G. and Shackleton, R.M. (1979) Structure and stratigraphy of the Dalradian rocks of the Bennabeola area, Connemara, Eire. *In*: Harris, A.L., Holland, C.H., and Leake, B.E. (eds). *The Caledonides of the British Isles: reviewed*. Geological Society, London, Special Publication **8**, 243–256.

Tanner, P.W.G. and Sutherland, S. (2007) The Highland Border Complex, Scotland: a paradox resolved. *Journal of the Geological Society London*, **164**, 111–116.

Tanner, P.W.G., Dempster, T.J., and Dickin, A.P. (1989) Time of docking of the Connemara terrane with the Delaney Dome Formation, western Ireland. *Journal of the Geological Society of London*, **146**, 389–392.

Tappin, D.R., Chadwick, R.A., Jackson, A.A., Wingfield, R.T.R., and Smith, N.J.P. (1994) *United Kingdom offshore regional report: The geology of Cardigan Bay and the Bristol Channel*. London HMSO for the British Geological Survey.

Tarling, D.H. (1985) Silurian–Devonian palaeogeographies based on palaeomagnetic observations. *Philosophical Transactions of the Royal Society, London*, **B309**, 81–83.

Tate, M.P. (1993) Structural framework and tectono-stratigraphic evolution of the Porcupine Seabight Basin, offshore western Ireland. *Marine and Petroleum Geology*, **10**, 95–123.

Tate, M.P. and Dobson, M. R. (1988) Syn- and post-rift igneous activity in the Porcupine Seabight basin and adjacent continental margin west of Ireland. *In*: Morton, A.C. and Parson, L.M. (eds). *Early Tertiary volcanism and the opening of the NE Atlantic*. Geological Society, London, Special Publications **39**, 309–334.

Tate, M.P. and Dobson, M.R. (1989a) Late Permian to early Mesozoic rifting and sedimentation offshore NW Ireland. *Marine and Petroleum Geology*, **6**, 49–59.

Tate, M.P. and Dobson, M.R. (1989b) Pre-Mesozoic geology of the western and north-western Irish continental shelf. *Journal of the Geological Society of London*, **146**, 229–240.

Tattersall, J.A. (1964) *The geological succession of the Carboniferous Limestone of south County Clare, Ireland, west of the River Fergus*. Unpublished Ph.D. thesis, University of Southhampton.

Taylor, P.D. and Curry, G.B. (1985) The earliest known fenestrate bryozoan with a short review of Lower Ordovician Bryozoa. *Palaeontology*, **28**, 147–158.

Temperley, S. and Windley, B. F. (1997) Grenvillian extensional tectonics in northwest Scotland. *Geology*, **25**, 53–56.

Terwindt, J.H.J. (1971) Lithofacies of inshore estuaries and tidal inlet deposits. *Geologie en Mijnbouw*, **50**, 515–526.

Theokritoff, G. (1951) Ordovician rocks near Leenane, Ireland. *Proceedings of the Royal Irish Academy*, **54B**, 25–49.

Thirlwall, M.F. (1981) Implications for Caledonian plate tectonic models of chemical data from volcanic rocks of the British Old Red Sandstone. *Journal of the Geological Society of London*, **138**,

123–138.

Thirlwall, M.F. (1988) Geochronology of late Caledonian magmatism in northern Britain. *Journal of the Geological Society of London*, **145**, 951–968.

Thomas, C.W., Graham, C.M., Ellam, R.M., and Fallick, A.E. (2004) $^{87}Sr/^{86}Sr$ chemostratigraphy of Neoproterozoic Dalradian limestones of Scotland and Ireland: constraints on depositional ages and time scales. *Journal of the Geological Society of London*, **161**, 229–242.

Thomas, E.R., Wolff, E.W., Mulvaney, R., Steffensen, J.P., Johnsen, S.J., Arrowsmith, C., White, J.W.C., Vaughn, B., and Popp, T. (2007) The 8.2 ka event from Greenland ice cores. *Quaternary Science Reviews*, **26**, 70.

Thomas, G.S.P. (1977) The Quaternary of the Isle of Man. *In*: Kidson, C. and Tooley, M.J. (eds). *Quaternary History of the Irish Sea*. Seal House Press, Liverpool, 155–178.

Thomas, G.S.P., Chiverrell, R.C., and Huddart, D. (2004) Ice-marginal depositional responses to readvance episodes in the Late Devensian deglaciation of the Isle of Man. *Quaternary Science Reviews*, **23**(1-2), 85–106.

Thompson, P. (1985) *Dating the British Tertiary Igneous Province in Ireland by the ^{40}Ar–^{39}Ar stepwise degassing method*. Unpublished Ph.D. thesis, University of Liverpool.

Thompson, P., Mussett, A.E., and Dagley, P. (1987) Revised ^{40}Ar–^{39}Ar age for granites of the Mourne Mountains, Ireland. *Scottish Journal of Geology*, **23**, 215–220.

Thompson, S.J. (1979) *Preliminary report on the Ballymacilroy No.1 borehole, Ahoghill, Co. Antrim*. Geological Survey of Northern Ireland Open File Report No.63.

Thornton, M.S. (1966) *The Lower Carboniferous limestones of the Tralee Bay area, Co. Kerry, Ireland*. Unpublished Ph.D. thesis, University of Cambridge.

Thorpe, R.S. (1974) Geochemical evidence for the original nature of the Rosslare Complex, S.E. Eire. *Scientific Proceedings of the Royal Dublin Society*, **5A**, 199–206.

Tietzsch-Tyler, D. (1989) The Lower Palaeozoic geology of SE Ireland — a revaluation. *Irish Association of Economic Geology Annual Review*, 112–118.

Tietzsch-Tyler, D. (1996) Precambrian and early Caledonian orogeny in south-east Ireland. *Irish Journal of Earth Sciences*, **15**, 19–39.

Tietzsch-Tyler, D. and Sleeman, A.G. (1994a) *Geology of South Wexford. Sheet 23, Bedrock Geology 1:100,000 Map Series*. Geological Survey of Ireland.

Tietzsch-Tyler, D. and Sleeman, A.G. (1994b) *Geology of Carlow–Wexford: a geological description to accompany the bedrock geology 1:100,000 scale map series, Sheet 19, Carlow–Wexford with contributions by B.J. McConnell, E.P. Daly, A.M. Flegg, P.J. O'Connor and W.P. Warren*. Geological Survey of Ireland, 58pp.

Tietzsch-Tyler, D., Sleeman, A.G., Boland, M.A., Daly, E.P., Flegg, A.M., O'Connor, P.J., and Warren, W.P. (1994) *Geology of South Wexford: a geological description of South Wexford and adjoining parts of Waterford, Kilkenny and Carlow to accompany the bedrock geology 1:100,000 scale map series, Sheet 23, South Wexford*. Geological Survey of Ireland, 62pp.

Tighe, W. (1802) *Statistical observations relative to the county of Kilkenny, made in the years 1800 & 1801*. Dublin, xvi + 763.

Tilley, C.E. and Harwood, H.F. (1931) The dolerite–chalk contact of Scawt Hill, County Antrim. *Mineralogical Magazine*, **22**, 439–468.

Todd, S.P. (1989a) Role of the Dingle Bay Lineament in the evolution of the Old Red Sandstone of south-west Ireland. *In*: Arthurton, R.S., Gutteridge, P., and Nolan, S.C. (eds). *The role of tectonics in Devonian and Carboniferous sedimentation in the British Isles*. Yorkshire Geological Society Occasional Publication, **6**, 35–54.

Todd, S.P. (1989b) Stream-driven, high-density gravelly traction carpets: possible deposits in the Trabeg Conglomerate Formation, SW Ireland and some theoretical considerations of their origin. *Sedimentology*, **36**, 513–530.

Todd, S.P. (1991) The Silurian rocks of Inishnabro, Blasket Islands, County Kerry and their regional significance. *Irish Journal of Earth Sciences*, **11**, 91–98.

Todd, S.P. (2000) Taking the roof off a suture zone: basin setting and provenance of conglomerates in the ORS Dingle Basin of SW Ireland. *In*: Friend, P.F. and Williams, B.P.J. (eds). *New perspectives on the Old Red Sandstone*. Geological Society London Special Publication, **180**, 185–222.

Todd, S.P. and Went, D.J. (1991) Lateral migration of sand-bed rivers: examples from the Devonian Glashabeg Formation, SW Ireland and the Cambrian Alderney Sandstone Formation, Channel Islands. *Sedimentology*, **38**, 997–1020.

Todd, S.P., Boyd, D., and Dodd, C.D. (1988a) Old Red Sandstone sedimentation and basin development in the Dingle Peninsula, southwest Ireland. *In*: McMillan, N.J., Embry, A.F., and Glass, D.J. (eds). *Devonian of the world Vol. 2, Calgary*. Canadian Society of Petroleum Geologists Memoir **12**, 251—68.

Todd, S.P., Murphy, F.C., and Kennan, P.S. (1991) On the trace of the Iapetus Suture in Ireland and Britain. *Journal of the Geological Society of London*, **148**, 869–880.

Todd, S.P., Murphy, F.C., and Kennan, P.S. (1992) Discussion on the trace of the Iapetus suture in Ireland and Britain. *Journal of the Geological Society of London*, **149**, 1049.

Todd, S.P., Williams, B.P.J., and Hancock, P.L. (1988b) Lithostratigraphy and structure of the Old Red Sandstone of the northern Dingle Peninsula, Co. Kerry, southwest Ireland. *Geological Journal*, **23**, 107–120.

Todd, S.P., Connery, C., Higgs, K.T., and Murphy, F.C. (2000) An Early Ordovician age for the Annascaul Formation of the SE Dingle Peninsula, SW Ireland. *Journal of the Geological Society of London*, **157**, 823–833.

Toghill, P. (1968) A new Lower Llandovery graptolite from Coal Pit Bay, County Down. *Geological Magazine*, **105**, 384–386.

Tomkeieff, S.I. (1934) Differentiation in basalt lava, Island Magee, Co. Antrim. *Geological Magazine*, **71**, 501–512.

Tomkeieff, S.I. (1939) Zoned olivines and their petrogenetic significance. *Mineralogical Magazine*, **25**, 229–251.

Tomkeieff, S.I. (1940a) The basalt lavas of the Giant's Causeway district of Northern Ireland. *Bulletin Volcanologique, Ser 2*, **6**, 89–143.

Tomkeieff, S.I. (1940b) The dolerite plugs of Tieveragh and Tievebulliagh near Cushendall, County Antrim, with a note on buchite. *Geological Magazine*, **77**, 54–64.

Tomkeieff, S.I. (1964) Petrochemistry and petrogenesis of the British Tertiary igneous province. *Advancing Frontiers in Geology and Geophysics*, Hyderabad.

Tomkeieff, S.I. and Marshal, C.E. (1935) The Mourne dyke swarm. *Quarterly Journal of the Geological Society of London*, **91**, 251–292.

Torsvik, T.H. (2003) The Rodinia Jigsaw Puzzle. *Science*, **300**,

1379–1381.

Torsvik, T.H. and Cocks, R.M. (2004) Earth geography from 400 to 250 Ma: a palaeomagnetic, faunal and facies review. *Journal of the Geological Society of London*, **161**, 555–572.

Torsvik, T.H. and Trench, A. (1991) The Ordovician history of the Iapetus Ocean in Britain: new palaeomagnetic constraints. *Journal of the Geological Society of London*, **148**, 423–425.

Torsvik, T.H., Trench, A., Svensson, I., and Walderburg, H. (1993) Palaeogeographic significance of mid-Silurian palaeomagnetic results from southern Britain: major revision of the apparent polar wander path for eastern Avalonia. *Geophysical Journal International*, **113**, 651–668.

Torsvik, T.H., Smethurst, M.A., Van der Voo, R., Trench, A., Abrahamsen, N., and Halvorsen, E. (1992) Baltica: a synopsis of Vendian–Permian palaeomagnetic data and their palaeotectonic implications. *Earth Science Reviews* **33**, 133–152.

Townsend, H. (1810) *Statistical survey of the county of Cork, with observations on the means of improvement.* Dublin, xx + 845.

Trapp, E., Kaufmann, B., Mezger, K., Korn, D., and Weyer, D. (2004) Numerical calibration of the Devonian–Carboniferous boundary: two new U–Pb isotope dilution–thermal ionization mass spectrometry single-zircon ages from Hasselbachtal (Sauerland), Germany. *Geology*, **32**, 857–860.

Treloar, P.J. and Max, M.D. (1984) The hornfels and depth of emplacement of the Carnsore Granite, southeast Ireland. *Proceedings of the Geologists' Association*, **5**, 320–324.

Tremlett, W.E. (1959) The Pre-Cambrian rocks of southern County Wicklow, Ireland. *Geological Magazine*, **96**, 58–68.

Trewin, N.H. and Thirlwall, M.F. (2002) Old Red Sandstone. *In*: Trewin, N.H. (ed). *The geology of Scotland*, 4th Edition, 213–250.

Troelstra, S.R., van Hinte, J.E., and Ganssen, G.M. (1995) The Younger Dryas. *Koninklijke Nederlandse Akademie van Wetenschappen Verhandelingen, Afd. Natuurkunde, Eerste Reeks, deel*, **44**. 224pp.

Trueblood, S. (1992) Petroleum geology of the Slyne Trough and adjacent basins. *In*: Parnell, J. (ed.). *Basins of the Atlantic Seaboard: Petroleum Geology, Sedimentology and Basin Evolution.* Geological Society, London, Special Publications **62**, 315–326.

Trueblood, S. and Morton, N. (1991) Comparative sequence stratigraphy and structural styles of the Slyne Trough and Hebrides Basin. *Journal of the Geological Society of London*, **148**, 197–201.

Tucker, R.D. and McKerrow, W.S. (1995) Early Palaeozoic chronology: a review in light of new U–Pb zircon ages from Newfoundland and Britain. *Canadian Journal of Earth Sciences*, **32**, 368–379.

Tunnicliff, S.P. (1982) A revision of late Ordovician bivalves from Pomeroy, Co. Tyrone, Ireland. *Palaeontology*, **25**, 43–88.

Tunnicliffe, S.P. (1983) The oldest known nowakiid (Tentaculitoidea). *Palaeontology*, **26**, 851–854.

Turbitt, T., Barker, E.J., Browitt, C.W.A., Howels, M., Marrow, P.C., Musson, R.M.W., Newmark, R.H., Redmayne, D.W., Walker, A.B., Jacob, A.W.B., Ryan, E., and Ward, V. (1985) The North Wales Earthquake of 19 July 1984. *Journal of the Geological Society of London*, **142**, 567–571.

Turner, C. (1970) The Middle Pleistocene deposits at Marks Tey, Essex. *Philosophical Transactions of the Royal Society*, **B257**, 373–437.

Turner, C. and West, R.G. (1968) The subdivision and zonation of interglacial periods. *Eiszeitalter und Gegenwurt*, **19**, 93–101.

Turner, J.S. (1937) The faunal succession in the Carboniferous Limestone near Cork. *Proceedings of the Royal Irish Academy*, **43B**, 193–209.

Turner, J.S. (1950) The Carboniferous limestone in Co. Dublin south of the River Liffey. *Scientific Proceedings of the Royal Dublin Society*, **25**, 169–192.

Turner, J.S. (1962) The age of reef-limestones of the Maine Valley, Co. Kerry. *Proceedings of the Leeds Philosophical and Literary Society*, **8**, 247–250.

Turner, P., Shelton, R., Ruffell, A., and Pugh, J. (2000) Palaeomagnetic constraints on the age of the Red Arch Formation and associated sandstone dykes (Northern Ireland). *Journal of the Geological Society of London*, **157**, 317–325.

Tuttle, O.F. and Bowen, N.L. (1958) *Origin of granite in the light of experimental studies in the system $NaAlSi_3O_8$– $KAlSi_3O_8$–SiO_2–H_2O.* Geological Society of America, Memoir.

Tyrrell, S., Haughton, P.D.W., and Daly, J.S. (2007) Drainage reorganization during breakup of Pangea revealed by in-situ isotopic analysis of detrital K-feldspar. *Geology*, **35**, 971–974.

Tyrrell, S., Haughton, P.D.W., Daly, J.S., Kokfelt, T.F., and Gagnevin, D. (2006) The use of the common Pb isotope composition of detrital K-feldspar grains as a provenance tool and its application to Upper Carboniferous paleodrainage, northern England. *Journal of Sedimentary Research*, **76**(1–2), 324–345.

Unitt, R.P. (1997) *The Structural and Metamorphic Evolution of the Lough Derg Complex, Counties Donegal and Fermanagh.* Unpublished PhD thesis, National University of Ireland.

Unnithan, V., Shannon, P.M., McGrane, K., Readman, P.W., Jacob, A.W.B., Keary, R., and Kenyon, N.H. (2001) Slope instability and sediment re-distribution in the Rockall Trough: constraints from GLORIA. *In*: Shannon, P.M., Haughton, P.D.W., and Corcoran, D.V. (eds). *The Petroleum Exploration of Ireland's Offshore Basins.* Geological Society, London, Special Publication, **188**, 439–454.

Upton, B.G.J. (1988) History of Tertiary activity in the N. Atlantic borderlands. *In*: Morton, A.C. and Parson, I.M. (eds). *Early Tertiary volcanism and the opening of the NE Atlantic.* Geological Society Special publication. **39**, 429–453.

Vail, P.R., Mitchum, Jr., R.M., Todd, R.G., Widmier, J M., Thompson, S., Sangree, J.B., Bubb, J.N., and Hatfield, W.G. (1977) *Seismic stratigraphy: application to hydrocarbon exploration.* American Association of Petroleum Geologists Memoir **26**, 42–212.

Van Breemen, O., Halliday, A.N., Johnson, M.R.W., and Bowes, D.R. (1978) Crustal additions in late Precambrian times. *In*: Bowes, D.R. and Leake, B.E. (eds). *Crustal Evolution in Northwestern Britain and Adjacent Regions.* Geological Journal Special Issue, **10**, 81–106.

Van den Berg, R. (2005) *Granulite-facies lower crustal xenoliths from central Ireland.* Unpublished Ph.D. thesis, University College, Dublin.

Van den Berg, R., Daly, J.S., and Salisbury, M.H. (2005) Seismic velocities of granulite-facies xenoliths from central Ireland: implications for lower crustal composition and anisotropy. *Tectonophysics*, **407**(1-2), 81–99.

Van der Voo, R. (1988) Palaeozoic paleogeography of North America, Gondwana, and intervening displaced terranes: comparisons of paleomagnetism with paleoclimatology and biogeographical patterns. *Geological Society America, Bulletin* **100**, 311–324

Van der Zwan, C.J. and Van Veen, P.M. (1978) The Devonian–Carboniferous transition sequence in southern Ireland: integration of palaeogeography and palynology. *Palinologia*, **1**, 469–79.

Van Gelder, A. (1974) *Sedimentation in the marine margin of the Old Red Continent south of Cork, Ireland*. Ph.D. thesis, University of Utrecht.

Van Lunsen, H.A. and Max, M.D. (1975) The geology of Howth and Ireland's Eye, Co. Dublin. *Geological Journal*, **10**, 35–58.

Van Rijn, P. (2004) The analysis of charcoal from Ross Island. *In*: O'Brien, W. (ed.). *Ross Island mining, metal and society in early Ireland. (Bronze Age Studies 6)*, Department of Archaeology, National University of Ireland, Galway, 386–401.

Van Staal, C.R., Dewey, J.F., MacNiocaill, C., and McKerrow, W.S. (1998) The Cambrian–Silurian tectonic evolution of the northern Appalachians and British Caledonides: history of a complex, west and southwest Pacific-type segment of Iapetus. *In*: Blundell, D.J. and Scott, A.C. (eds). *Lyell, the Past is the Key to the Present*. Geological Society of London Special Publication, **143**, 199–242.

Vance, D., Müller, W. and Villa, I.M. (2003) Geochronology: linking the isotopic record with petrology and textures: an introduction. *In*: Vance, D., Müller, W., and Villa, I.M. (eds). *Geochronology: linking the isotopic record with petrology and textures*. Geological Society of London, Special Publication **220**, 1–24.

Vance, D., Strachan R.A., and Jones, K.A. (1998) Extensional versus compressional settings for metamorphism: garnet chronometry and pressure–temperature–time histories in the Moine Supergroup, northwest Scotland. *Geology*, **26**(10), 927–930.

Vaughan, A.P.M. (1991) *The Lower Palaeozoic geology of the Iapetus suture zone in eastern Ireland*. Unpublished Ph.D. thesis, Trinity College Dublin.

Vaughan, A.P.M. (1996) A tectonomagmatic model for the genesis and emplacement of Caledonian calc-alkaline lamprophyres. *Journal of the Geological Society of London*, **153**, 613–623.

Vaughan, A.P.M. and Johnston, J.D. (1992) Structural constraints on closure geometry across the Iapetus Suture in eastern Ireland. *Journal of the Geological Society of London*, **149**, 65–74.

Vaughan, A.P.M., McCabe, A.M., Coxon, P., Dowling, L.A., Mitchell, F.J.G., and Lauritzen, S.E. (2004) Depositional and post-depositional history of warm stage deposits at Knocknacran, Co. Monaghan, Ireland: implications for preservation of Irish last interglacial deposits. *Journal of Quaternary Science*, **19**(6), 557–590.

Vera, F.W.M. (2000) *Grazing ecology and forest history*. CABI, Wallingford.

Vermeulen, N.J., Shannon, P.M., Landes, M., Masson, F., and VARNET GROUP. (1999) Seismic evidence for subhorizontal crustal detachments beneath the Irish Variscides. *Irish Journal of Earth Sciences*, **17**, 1–18.

Vermeulen, N.J., Shannon, P.M., Masson, F., and Landes, M. (2000) Wide-angle seismic control on the development of the Munster Basin, SW Ireland. *In*: Friend, P.F. and Williams, B.P.J. (eds). *New perspectives on the Old Red Sandstone*. Geological Society London Special Publications, **180**, 223–238.

Vernon, R.H. and Clarke, G.L. (2008) *Principles of Metamorphic Petrology*. Cambridge University Press, 446pp.

Vernon, R.H. and Paterson, S.R. (1993) The Ardara Pluton, Ireland: deflating an expanded intrusion. *Lithos*, **31**(1–2), 17–32.

Versey, H.C. (1958) Derived ammonites in basal Cretaceous conglomerate. *Geological Magazine*, **95**, 440.

Versey, H.R. and Singh, B.H. (1982) Groundwater in Deccan Basalts of the Betwa Basin, India. *Journal of Hydrology*, **58**, 279–306.

Visscher, H. (1971) *The Permian and Triassic of the Kingscourt outlier, Ireland*. Geological Survey of Ireland, Special Paper, **1**, 1–114.

Vogt, U., Makris, J., O'Reilly, B.M., Hauser, F., Readman, P.W., Jacob, A.W.B., and Shannon, P.M. (1998) The Hatton Basin and continental margin: crustal structure from wide-angle seismic and gravity data. *Journal of Geophysical Research*, **103**, 12545–12566.

Wager, R. and Bailey, E.B. (1953) Basic magma chilled against acid magma. *Nature*, **172**, 68.

Walker, G.P.L. (1959) Some observations on the Antrim basalts and associated dolerite intrusions. *Proceedings of the Geologists' Association*, **70**, 179–205.

Walker, G.P.L. (1960a) An occurrence of mugearite in Antrim. *Geological Magazine*, **97**, 62–64.

Walker, G.P.L. (1960b) The amygdale minerals in the Tertiary lavas of Ireland. III. Regional distribution. *Mineralogical Magazine*, **32**, 503–527.

Walker, G.P.L. (1962) Garronite, a new zeolite, from Ireland and Iceland. *Mineralogical Magazine*, **33**, 173–186.

Walker, G.P.L. (1971) Compound and simple lava flows and flood basalts. *Bulletin Volcanologique*, **35**, 579–590.

Walker, G.P.L. (1975) A new concept of the evolution of the British Tertiary intrusive centres. *Journal of the Geological Society of London*, **131**, 121–141.

Walker, G.P.L. (1979) The environment of Tertiary volcanism in Britain. *Bulletin of the Geological Survey of Great Britain*, Flett volume.

Walker, G.P.L. and Leedal, G.P. (1954) The Barnesmore complex, County Donegal. *Scientific Proceedings of the Royal Dublin Society*, **26**, N.S., 207–300.

Walker, M. and 16 others (2008) The Global Stratotype Section and Point (GSSP) for the base of the Holocene Series/Epoch (Quaternary System/Period) in the NGRIP ice core. *Episodes*, **32**, 264-267.

Walker, M., Johnsen, S., Rasmussen, S.O., Popp, T., Steffensen, J-P., Gibbard, P., Hoek, W., Lowe, J., Andrews, J., Björck, S., Cwynar, L.C., Hughen, K., Kershaw, P., Kromer, B., Litt, T., Lowe, D.J., Nakagawa, T., Newnham, R. and Schwander, J. (2009) Formal definition and dating of the GSSP (Global Stratotype Section and Point) for the base of the Holocene using the Greenland NGRIP ice core, and selected auxiliary records. *Journal of Quaternary Science*, **24**, 3–17.

Walker, M.J.C. and Lowe, J.J. (2007) Quaternary science 2007: a 50-year retrospective. *Journal of the Geological Society of London*, **164**, 1073–1092. doi:10.1144/0016-76492006-195.

Walker, M.J.C., Bohncke, S.J.P., Coope, G.R., O'Connell, M., Usinger, H., and Verbruggen, C. (1994) The Devensian/Weichselian Late-glacial in northwest Europe (Ireland, Britain, north Belgium, The Netherlands, northwest Germany). *Journal of Quaternary Science*, **9**(2), 109–118.

Wallace, J.M. (1995) *Isotopic and geochemical constraints on the petrogenesis of the Tertiary flood basalts of NE Ireland*. Unpublished Ph.D thesis, Queen's University, Belfast.

Wallace, J.M., Ellam, R.M., Meighan, I.G., Lyle, P., and Rogers, N.W. (1994) Sr isotope data for the Tertiary lavas of Northern Ireland: evidence for open system petrogenesis. *Journal of the*

Geological Society of London, **151**, 869–877.

Walsh, A., Knag, G., Morris, H., Quinquis, H., Tricker, P., Bird, C., and Bower, S. (1999) Petroleum geology of the Irish Rockall Trough. *In*: Fleet, A.J. and Boldy, S.A.R. (eds). *Petroleum Geology of Northwest Europe: Proceedings of the 5th Conference*. Geological Society, London, 433–444.

Walsh, P.T. (1965) Possible Tertiary outliers from the Gweestin valley, Co. Kerry. *Irish Naturalists' Journal*, **7**, 100–104.

Walsh, P.T. (1966) Cretaceous outliers in south-west Ireland and their implications for Cretaceous palaeogeography. *Quarterly Journal of the Geological Society of London*, **122**, 63–84.

Walsh, P.T. (1967) Notes on the Namurian stratigraphy north of Killarney, Co. Kerry. *Irish Naturalists' Journal*, **15**, 254–258.

Walsh, P.T. (1968) The Old Red Sandstone west of Killarney, County Kerry, Ireland. *Proceedings of the Royal Irish Academy*, **66B**, 9–26.

Warnke, K. and Meischner, D. (1995) Origin and depositional environment of Lower Carboniferous mud mounds of Northwestern Ireland. *Facies*, **32**, 36–42.

Warren, W.P. (1977) Day 6: North East Iveragh. *In*: Lewis, C.A. (ed.). *Guidebook for INQUA Excursion A15 South and South West Ireland*. Geo Abstracts, Norwich, 37–45.

Warren, W.P. (1979) The stratigraphic position and age of the Gortian Interglacial deposits. *Bulletin of the Geological Survey of Ireland*, **2**, 315–332.

Warren, W.P. (1985) Stratigraphy. *In*: Edwards, K.J. and Warren, W.P. (eds). *The Quaternary history of Ireland*. Academic Press, London, 39-65.

Warren, W.P. (1992) Drumlin orientation and the pattern of glaciation in Ireland. *Sveriges Geologiska Undersökning*, **Ser.Ca 81**, 359–366.

Warren, W.P. and Ashley, G.M. (1994) Origins of the ice-contact stratified ridges (eskers) of Ireland. *Journal of Sedimentary Research*, **A64**(3), 433–449.

Warrington, G. (1994) *Palynology report on the Killary Glebe borehole near Coalisland, County Tyrone*. Geological Survey of Northern Ireland Technical Report GSNI/94-1.

Warrington, G. (1995) *The Permian, Triassic and Jurassic in Northern Ireland: a palynological study with special reference to the hydrocarbon prospectivity of the Larne–Lough Neagh Basin*. Geological Survey of Northern Ireland Technical Report, **GSNI/95/7.**

Warrington, G. and Ivimey-Cook, H.C. (1992) Triassic. *In*: Cope, J.C.W., Ingham, J.K., and Rawson, P.F. (eds). *Atlas of Palaeogeography and Lithofacies*. Geological Society, London, Memoir **13**, 153pp.

Warrington, G., Audrey-Charles, M.G., Elliot, R.E. Evans, W.B., Ivimey-Cook, H.C., Kent, P.E., Robinson, P.L., Shotton, F.W., and Taylor, F.M. (1980) *A correlation of Triassic rocks in the British Isles*. Geological Society, London, Special Report **13**, 78pp.

Waters, J.A. and Sevastopulo, G.D. (1984) The stratigraphical distribution of palaeoecology of Irish Lower Carboniferous blastoids. *Irish Journal of Earth Sciences*, **6**, 137–154.

Watkins, R. (1978) Silurian marine communities west of Dingle, Ireland. *Palaeogeography, Palaeoclimatology, Palaeoecology*, **23**, 79–118.

Watson, J. (1984) The ending of the Caledonian orogeny in Scotland. *Journal of the Geological Society of London*, **141**, 193–214.

Watts, A.B., Karner, G.D., and Steckler, M.S. (1982) Lithospheric flexure and the evolution of sedimentary basins. *Philosophical Transactions of the Royal Society, London*, **A305**, 249–281.

Watts, W.A. (1957) A Tertiary deposit in County Tipperary. *Scientific Proceedings of the Royal Dublin Society*, **27**, 309–311.

Watts, W.A. (1962) Early Tertiary pollen deposits in Ireland. *Nature*, **193**, 600.

Watts, W.A. (1964) Interglacial deposits at Baggotstown, near Bruff, Co. Limerick. *Proceedings of the Royal Irish Academy*, **63B**, 167–89.

Watts, W.A. (1970) Tertiary and interglacial floras in Ireland. *In*: Stephens, N. and Glasscock, R.E. (eds). *Irish Geographical Studies*. Queen's University, Belfast, 17–33.

Watts, W.A. (1977) The Late Devensian vegetation of Ireland. *Philosophical Transactions of the Royal Society*, **B280**, 273–293.

Watts, W.A. (1984) The Holocene vegetation of the Burren, western Ireland. *In*: Haworth, E.Y. and Lund, J.W.G. (eds). *Lake sediments and environmental history*. Leicester University Press, Leicester, 360–376.

Watts, W.A. (1985) Quaternary vegetation cycles. *In*: Edwards, K.J. and Warren, W.P. (eds). *The Quaternary History of Ireland*. Academic Press, London, 155–185.

Watts, W.A. (1988) Europe. *In*: Huntley, B. and Webb III, T. (eds). *Vegetation History*. Academic Publishers, Kluwer, 155–192.

Weaver, T. (1819) Memoir on the geological relations of the east of Ireland. *Transactions of the Geological Society of London*, **5**, 117–304.

Weaver, T. (1838) On the geological relations of the south of Ireland. *Transactions of the Geological Society of London*, series two, **5**, 1–68.

Weir, J.A. (1962) Geology of the Lower Palaeozoic inliers of Slieve Bernagh and the Cratloe Hills, County Clare. *Scientific Proceedings of the Royal Dublin Society*, **A1**, 233–263.

Weir, J.A. (1974) Tectonic style in Lower Palaeozoic rocks: the Clare–south Galway area and the Scottish Southern Uplands contrasted. *Scientific Proceedings of the Royal Dublin Society*, **A5**, 107–112.

Weir, J.A. (1975) Palaeogeographical implications of two Silurian shelly faunas from the Arra Mountains and Cratloe Hills, Ireland. *Palaeontology*, **18**, 343–350.

Wellings, S.A. (1998) Timing of deformation associated with the syn-tectonic Dawros–Currywongaun–Doughruagh Complex, NW Connemara, western Ireland. *Journal of the Geological Society of London*, **155**, 25–37.

West, I., Brandon, A., and Smith, M. (1968) A tidal flat evaporite facies in the Viséan of Ireland. *Journal of Sedimentary Petrology*, **38**, 1079–1093.

West, R.G. (1980) Pleistocene forest history in East Anglia. *New Phytologist*, **85**, 571–622.

Whalen, J.B., Jenner, G.A., Longstaffe, F.J., and Gariepy, C. (1997) Implications of granitoid geochemical and isotopic (Nd, O, Pb) data from the Cambro-Ordovician Notre Dame arc for the evolution of the Central Mobile Belt, Newfoundland Appalachians. *In*: Sinha, K., Whalen, J.B., and Hogan, J. (eds). *Magmatism in the Appalachian Orogen*. Geological Society of America Memoir, **191**, 367–395.

White, N. and Lovell, B. (1997) Measuring the pulse of a plume with the sedimentary record. *Nature*, **387**, 888–891.

White, N., Tate, M.P., and Conroy, J.J. (1992) Lithospheric stretching in the Porcupine Basin, west of Ireland. *In*: Parnell, J. (ed.). *Basins on the Atlantic Seaboard*. Geological Society, London, Special Publications **62**, 327–332.

White, R.S. (1988) A hot-spot model for early Tertiary volcanism in the N. Atlantic, *In*: Morton, A.C. and Parson, L.M. (eds). *Early Tertiary Volcanism and the Opening of the NE Atlantic*. Geological Society Special Publication, **39**, 3–13.

White, R.S. (1992) Crustal structure and magmatism of North Atlantic continental margins. *Journal of the Geological Society of London*, **149**, 84–854.

Whitehouse, N.J. (2006) The Holocene British and Irish ancient forest fossil beetle fauna: implications for forest history, biodiversity and faunal colonisation. *Quaternary Science Reviews*, **25**, 1755.

Whittaker, A. (2001) Karl Ludwig Giesecke: his life, performance and achievements. *Mitteilungen der Osterreichischen Mineralogischen Gesellschaft*, **146**, 451–479.

Whitten, E.T.H. (1954) Two arfvedsonitic rhyolite intrusions from Cloghaneely, Co. Donegal. *Mineralogical Magazine*, **30**, 393.

Whittington, R.J., Croker, P.F., and Dobson, M.R. (1981) Aspects of the geology of the south Irish Sea. *Geological Journal*, **16**, 85–88.

Whittow, J.B. (1974) *Geology and Scenery in Ireland*. Penguin Books, Harmondsworth, 301pp.

Wignall, P.B. and Best, J.L. (2000) The Western Irish Namurian Basin reassessed. *Basin Research*, **12**, 59–78.

Wignall, P.B. and Best, J.L. (2002) Reply to comment on 'The Western Irish Namurian basin reassessed' by O.J. Martinsen and J.D. Collinson. *Basin Research*, **14**, 531–542.

Wignall, P.B. and Best, J.L. (2004) Sedimentology and kinematics of a large, retrogressive growth-fault system in Upper Carboniferous deltaic sediments, western Ireland. *Sedimentology*, **51**, 1343–1365.

Wignall, P.B. and Pickering, K.T. (1993) Palaeoecology and sedimentology across a Jurassic fault scarp, NE Scotland. *Journal of the Geological Society of London*, **150**, 323–340.

Wilkinson, G.C., Bazley, R.A.B., and Boulter, M.C. (1980) The geology and palynology of the Oligocene Lough Neagh clays of Northern Ireland. *Journal of the Geological Society of London*, **137**, 1–11.

Williams, A. (1969) Ordovician faunal provinces with reference to brachiopod distribution. *In*: Wood, A. (ed.). *The Pre-cambrian and Lower Palaeozoic rocks of Wales*, 117–154.

Williams, A. (1972) An Ordovician Whiterock fauna in western Ireland. *Proceedings of the Royal Irish Academy*, **72B**, 209–219.

Williams, A. (1973) Distribution of brachiopod assemblages in relation to Ordovician palaeogeography. *In*: Hughes, N.F. (ed.). Organisms and continents through time. *Special Papers in Palaeontology*, **12**, 241–269.

Williams, A. and Curry, G.B. (1985) Lower Ordovician Brachiopoda from the Tourmakeady Limestone, Co. Mayo, Ireland. *Bulletin of the British Museum, Natural History (Geology)*, **38**, 183–269.

Williams, A., Strachan, I,, Bassett, D.A., Dean, W.T., Ingham, J.K., Wright, A.D., and Whittington, H.B. (1972) *A correlation of the Ordovician rocks in the British Isles*. Special Report of the Geological Society of London, **3**, 74pp.

Williams, B.P.J., Allen, J.R.L., and Marshall, J.D. (1982) Old Red Sandstone facies of the Pembroke peninsula, south of the Ritec Fault. *In*: Bassett, M.G. (ed.). *Geological Excursions in Dyfed, south-west Wales*, 151–174.

Williams, B.P.J., Sloan, R.J., and Richmond, L.K. (2000) Shallow marine to fluvial–aeolian interaction in tectonically active basins, Siluro-Devonian (ORS), Dingle Peninsula, S.W. Ireland. *In*: Graham, J.R. and Ryan, A. (eds). *Field Trip Guidebook, IAS Dublin 2000*. Department of Geology, 71–112.

Williams, D.M. (1980) Evidence for glaciation in the Ordovician rocks of western Ireland. *Geological Magazine*, **117**, 81–86.

Williams, D.M. (1984) The stratigraphy and sedimentology of the Ordovician Partry Group, southeastern Murrisk, Ireland. *Geological Journal*, **19**, 173–186.

Williams, D.M. and Harper, D.A.T. (1991) End-Silurian modifications of Ordovician terranes in western Ireland. *Journal of the Geological Society of London*, **148**, 165–171.

Williams, D.M. and O'Connor, P.D. (1987) Environment of deposition of conglomerates from the Silurian of north Galway, Ireland. *Transactions of the Royal Society of Edinburgh: Earth Sciences*, **78**, 129–132.

Williams, D.M., Armstrong, H.A., and Harper, D.A.T. (1988) The age of the South Connemara Group, Ireland, and its relationship to the Southern Uplands zone of Scotland and Ireland. *Scottish Journal of Geology*, **24**, 279–287.

Williams, D.M., Harkin, J., and Higgs, K.T. (1996) Implications of new microfloral evidence from the Clew Bay Complex for Silurian relationships in the western Irish Caledonides. *Journal of the Geological Society of London*, **153**, 771–777.

Williams, D.M., Harkin, J., and Rice, A.H.N. (1997) Umbers, ocean crust and the Irish Caledonides: terrane transpression and the morphology of the Laurentian margin. *Journal of the Geological Society of London*, **154**, 829–838.

Williams, D.M., Harkin, J., Armstrong, H.A., and Higgs, K.T. (1994) A late Caledonian mélange in Ireland: implications for tectonic models. *Journal of the Geological Society of London*, **151**, 307–314.

Williams, E.A. (2000) Flexural cantilever models of extensional subsidence in the Munster Basin (SW Ireland) and Old Red Sandstone fluvial dispersal systems. *In*: Friend, P.F. and Williams, B.P.J. (eds). *New perspectives on the Old Red Sandstone*. Geological Society London Special Publication, **180**, 239–268.

Williams, E.A., Friend, P.F., and Williams, B.P.J. (2000a) A review of Devonian timescales: databases, construction and new data. *In*: Friend, P.F. and Williams, B.P.J. (eds). *New perspectives on the Old Red Sandstone*. Geological Society London Special Publication, **180**, 1–22.

Williams, E.A., Sergeev, S.A., Stossel, I., and Ford, M. (1997) An Eifelian U–Pb zircon date for the Enagh Tuff Bed from the Old Red Sandstone of the Munster Basin in NW Iveragh, SW Ireland. *Journal of the Geological Society of London*, **154**, 189–193.

Williams, E.A., Sergeev, S.A., Stössel, I., Ford, M., and Higgs, K.T. (2000b) U–Pb zircon geochronology of silicic tuffs and chronostratigraphy of the earliest Old Red Sandstone in the Munster Basin, SW Ireland. *In*: Friend, P.F. and Williams, B.P.J. (eds). *New perspectives on the Old Red Sandstone*. Geological Society London Special Publication **180**, 269–302.

Williams, E.A., Bamford, M.L.F., Cooper, M.A., Edwards, H.E., Ford, M., Grant, C.G., MacCarthy, I.A.J., McAfee, A.M. and O'Sullivan, M.J. (1989) Tectonic controls and sedimentary response in the Devonian–Carboniferous Munster and South Munster basins, south-west Ireland. *In*: Arthurton, R.S., Gutteridge, P., and Nolan, S.C. (eds). *The role of tectonics in Devonian and Carboniferous sedimentation in the British Isles*. Yorkshire Geological Society Occasional Publication, **6**, 123–141.

Williams, F.M. and Kennan, P.S. (1983) Stable isotope studies of sulfide mineralization on the Leinster Granite margin and some observations on its relationship to coticule and tourmalinite rocks in the aureole. *Mineralium Deposita*, **18**(2), 399–410.

Williams, S.H. and Harper, D.A.T. (1994) Late Tremadoc graptolites from the Lough Nafooey Group, South Mayo, western Ireland. *Irish Journal of Earth Sciences*, **13**, 107–111.

Williams, S.H., Harper, D.A.T., Neuman, R.B., Boyce, W.D., and MacNiocaill, C. (1996) *Lower Palaeozoic fossils from Newfoundland and their importance in understanding the history of the Iapetus Ocean*. Geological Survey of Canada Special Paper **41**, 115–126.

Wilson, H.E. (1953) The petrography of the Old Red Sandstone rocks of the north of Ireland. *Proceedings of the Royal Irish Academy*, **55B**, 283–320.

Wilson, H.E. (1972) *Regional geology of Northern Ireland*. Ministry of Commerce, Geological Survey of Northern Ireland. Her Majesty's Stationery Office, Belfast, 115pp.

Wilson, H.E. (1981) Permian and Mesozoic. *In*: Holland, S.H. (ed.). *The Geology of Ireland*. Scottish Academic Press, Edinburgh.

Wilson, H.E. and Manning, P.I. (1978) *Geology of the Causeway Coast*. Memoir of the Geological Survey of Northern Ireland, Sheet 7. Her Majesty's Stationery Office, Belfast, 72pp.

Wilson, H.E. and Robbie, J.A. (1966) *Geology of the country around Ballycastle*. Memoir of the Geological Survey of Northern Ireland, Her Majesty's Stationery Office, Belfast, **8**, 370pp.

Wilson, P. (1988) Early descriptions of pro-talus ramparts. *Journal of Glaciology*, **34**, 141–142.

Wilson, P. (1990a) Characteristics and significance of protalus ramparts and fossil rock glaciers on Errigal Mountain, County Donegal. *Proceedings of the Royal Irish Academy*, **90B**, 1–21.

Wilson, P. (1990b) Morphology, sedimentological characteristics and origin of a fossil rock glacier on Muckish Mountain, north-west Ireland. *Geografiska Annaler*, **72A**, 237–247.

Wilson, P. (1993) Description and origin of some talus-foot debris accumulations, Aghla Mountains, Co.Donegal, Ireland. *Permafrost and Periglacial Processes*, **4**, 231–244.

Wilson, P. (1995) Active patterned ground and cryoturbation on Muckish Mountain, Co.Donegal, Ireland. *Permafrost and Periglacial Processes*, **6**, 15–25.

Wilson, R.L. (1970) Palaeomagnetic stratigraphy of Tertiary lavas from Northern Ireland. *Geophysical Journal of the Royal Astronomical Society*, **20**, 1–9.

Winchester, J.A. and Max, M.D. (1982) The geochemistry and origins of the Precambrian rocks of the Rosslare Complex, S.E. Ireland. *Journal of the Geological Society of London*, **139**, 309–319.

Winchester, J.A. and Max, M.D. (1984) Geochemistry and origins of the Annagh Division of the Precambrian Erris Complex, N.W. Co. Mayo, Ireland. *Precambrian Research*, **25**, 397–414.

Winchester, J.A. and Max, M.D. (1987a) A displaced and metamorphosed peralkaline granite related to the late Proterozoic Labrador and Gardar suites: the Doolough Granite of Co. Mayo, NW Ireland. *Canadian Journal of Earth Sciences*, **24**, 631–642.

Winchester, J.A. and Max, M.D. (1987b) The pre-Caledonian Inishkea Division of north-west Co. Mayo, Ireland: its geochemistry and probable stratigraphic position. *Geological Journal*, **22**, 309–331.

Winchester, J.A. and Max, M.D. (1996) Chemostratigraphic correlation, structure and sedimentary environments in the Dalradian of the NW Co. Mayo inlier, NW Ireland. *Journal of the Geological Society of London*, **153**, 779–801.

Winchester, J.A., Max, M.D., and Long, C.B. (1987) Trace element geochemical correlation in the reworked Proterozoic Dalradian metavolcanic suites of the western Ox Mountains and NW Mayo inliers, Ireland. *In*: Pharoah, T.C., Beckinsale, R.D., and Rickard, D. (eds). *Geochemistry and mineralization of Proterozoic volcanic suites*. Geological Society, London, Special Publications, **33**,489–502.

Wingfield, R.T.R. (1968) *The geology of Kenmare and Killarney*. Unpublished Ph.D. thesis, University of Dublin.

Witzke, B.J. (1990) Palaeoclimatic constraints for Palaeozoic palaeolatitudes of Laurentia and Euramerica. *In*: McKerrow, W.S. and Scotese, C.R. (eds). *Palaeozoic palaeogeography and biogeography*. Geological Society, London, Memoir **12**, 57–73.

Witzke, B.J. and Heckel, P.H. (1988) Palaeoclimatic indicators and inferred Devonian paleolatitudes of Euramerica. *In*: McMillan, N.J., Embry, A.F., and Glass, D.J. (eds). *Devonian of the world, Vol. 1*. Canadian Society of Petroleum Geologists Memoir, **12**, 49–63.

Woodcock, N. and Strachan, R. (2000) *Geological history of Britain and Ireland*. Blackwell Science, Oxford, 421pp.

Woodcock, N.H., Soper, N.J., and Strachan, R.A. (2007) A Rheic cause for the Acadian deformation in Europe. *Journal of the Geological Society of London*, **164**, 1023–1036.

Woodcock, N.H., Awan, M.A., Johnson, T.E., Mackie, A.H., and Smith, R.D.A. (1988) Acadian tectonics in Wales during Avalonia/Laurentia convergence. *Tectonics*, **7**, 483-495.

Woodland, A.W (ed.). (1971) *The Llanbed (Mochras Farm) borehole*. Institute of Geological Sciences, Report **71/18**, 115pp.

Woodman, P., McCarthy, M., and Monaghan, N. (1997) The Irish Quaternary Fauna Project. *Quaternary Science Reviews*, **16**, 129–159.

Woodman, P.C. (1978) *The Mesolithic in Ireland: hunter-gatherers in an insular environment*. BAR British Series no. 58, Oxford.

Woodman, P.C. (2000) Getting back to Basics: transitions to farming in Britain and Ireland. *In*: Price, T.D. (ed.). *Europe's First Farmers*. Cambridge University Press, Cambridge.

Woodman, P.C. and Monaghan, N. (1993) From mice to mammoths: dating Ireland's earliest faunas. *Archaeology Ireland*, **7**(3), 31–33.

Woodrow, D.L., Fletcher, F.W., and Ahrnsbrak, W.F. (1973) Palaeogeography and palaeoclimate at the deposition sites of the Devonian Catskill and Old Red Facies. *Bulletin of the Geological Society of America*, **84**, 3051–3064.

Worsley, P. (1991) Possible Early Devensian glacial deposits in the British Isles. *In*: Ehlers, J., Gibbard, P.L., and Rose, J. (eds). *The Glacial deposits of Great Britain and Ireland*. Balkema, Rotterdam, 47–51.

Worthington, R. (2006) *Geometric and kinematic analysis of Cenozoic faulting onshore and offshore West of Ireland*. Unpublished Ph.D. thesis, University College Dublin.

Wrafter, J.P. and Graham, J.R. (1989) Ophiolitic detritus in the Ordovician sediments of South Mayo, Ireland. *Journal of the Geological Society of London*, **146**, 213–215.

Wright, A.D. (1963) The fauna of the Portrane Limestone, I. The inarticulate brachiopods. *Bulletin British Museum (Natural History) Geology*, **8**, 224–254.

Wright, A.D. (1964) The fauna of the Portrane Limestone II. *Bulletin British Museum (Natural History) Geology*, **9**, 160–256.

Wright, A.D. (1967) A note on the stratigraphy of the Kildare inlier. *Irish Naturalists' Journal*, **15**, 340–343.

Wright, A.D. (1968) A westward extension of the Upper Ashgillian Hirnantia fauna. *Lethaia*, **1**, 352–367.

Wright, A.D. (1970) The stratigraphical distribution of the Ordovician inarticulate brachiopod *Orthisocrania divaricata* (M'Coy) in the British Isles. *Geological Magazine*, **107**, 97–103.

Wright, H.E.J., Kutzbach, J.E., Webb, T.I., Ruddiman, W.F., Street-Perrott, F.A., and Bartlein, P.J. (1993) *Global climates since the last glacial maximum*. University of Minnesota Press, Minneapolis.

Wright, P.C. (1964) The petrology, chemistry and structure of the Galway Granite of the Carna area, Co. Galway. *Proceedings of the Royal Irish Academy*, **63B**, 239–264.

Wright, V.P., Sloan, R.J., Garvie, L.A.J., and Rae, J.E. (1991) A polygenetic palaeosol from the Silurian (Wenlock) of south-west Ireland. *Journal of the Geological Society of London*, **148**, 849–959.

Wright, W.B. (1924) Age and origin of the Lough Neagh Clays. *Quarterly Journal of the Geological Society of London*, **80**, 468–488.

Wright, W.B. (1927) *The geology of Killarney and Kenmare*. Memoir of the Geological Survey of Ireland, Dublin.

Wyse Jackson, P. (1993) *The Building Stones of Dublin*. Town and Country House, Dublin, 67pp.

Wyse Jackson, P.N. (1989) On rocks and bicycles: a bibliography of Grenville Arthur James Cole (1859–1924) fifth director of the Geological Survey of Ireland. *Geological Survey of Ireland Bulletin*, **4**, 151–163.

Wyse Jackson, P.N. (1997) Fluctuations in fortune: three hundred years of Irish geology. *In*: Foster, J.W. and Chesney, H.C.G. *Nature in Ireland: a scientific and cultural* history. Lilliput Press, Dublin, 91–114.

Wyse Jackson, P.N. (2006) The Foze Rocks. *The Kerry Magazine, 2006*, 28–31.

Wyse Jackson, P.N. and Monaghan, N.T. (1994) Frederick M'Coy: an eminent Victorian palaeontologist and his synopses of Irish palaeontology of 1844 and 1846. *Geology Today*, **10**, 231–234.

Wyse Jackson, P.N. and Vaccari, E. (1997) *The Reverend George Graydon (c.1753–1803): cleric and geological traveller*. Royal Irish Academy, 16pp.

Wyse Jackson, P.N., Buttler, C.J., and Key, M.M. Jr. (2002) Palaeoenvironmental interpretation of the Tramore Limestone Formation (Llandeilo, Ordovician) based on bryozoan colony form. *In*: Wyse Jackson, P.N., Buttler, C.J., and Spencer Jones, M.E. (eds). *Bryozoan Studies 2001*, Lisse, A.A. Balkema, 359–365.

Yardley, B.W.D. (1976) Deformation and metamorphism of Dalradian rocks and the evolution of the Connemara cordillera. *Journal of the Geological Society of London*, **132**, 521–542.

Yardley, B.W.D. (1980) Metamorphism and orogeny in the Irish Dalradian. *Journal of the Geological Society of London*, **137**(3), 303–309.

Yardley, B.W.D. and Senior, A. (1982) Basic magmatism in Connemara, Ireland: evidence for a volcanic arc? *Journal of the Geological Society of London*, **139**, 6–70.

Yardley, B.W.D., Barber, J.P., and Gray, J.R. (1987) The metamorphism of the Dalradian rocks of western Ireland and its relation to tectonic setting. *Philosophical transactions of the Royal Society of London*, **A321**, 243–270.

Yardley, B.W.D., Vine, F.J., and Baldwin, C.T. (1982) The plate tectonic setting of NW Britain and Ireland in Late Cambrian and early Ordovician times. *Journal of the Geological Society of London*, **139**, 455–463.

Yates, P.J. (1962) The palaeontology of the Namurian rocks of Slieve Anierin, Co. Leitrim. *Palaeontology*, **5**, 355–443.

Yokoyama Y., Fifield, L.K., Lambeck, K., De Deckker, P., and Johnston, P. (2000) Timing of the Last Glacial Maximum from observed sea-level minima. *Nature*, **406**, 713–716.

Young, D.G.G. (1974) The Donegal Granite: a gravity analysis. *Proceedings of the Royal Irish Academy*, **74B**, 63–73.

Young, D.G.G. and Bailey, R.J. (1973) A reconnaissance magnetic map of the continental margin west of Ireland. *Communication, Dublin Institute for Advanced Studies, Series D. Geophysical Bulletin* **29**.

Young, D.G.G. and Bailey, R.J. (1974) An interpretation of some magnetic data off the west coast of Ireland. *Geological Journal*, **9**, 137–146.

Young, G. (1995) Are Neoproterozoic glacial deposits preserved on the margins of Laurentia related to the fragmentation of two supercontinents? *Geology*, **23**, 153–156.

Young, R. (1883) *Report of the Proceedings of the Belfast Naturalists Field Club, Sec 2, 1881-82*, 116–118.

Zagwijn, W.H. (1960) Aspects of the Pliocene and Early Pleistocene vegetation in the Netherlands. *Mededelingen van de Geologische Stichting*, **CIII** (1), **5**, 1–78.

Zagwijn, W.H. (1985) An outline of the Quaternary stratigraphy of The Netherlands. *Geologie en Mijnbouw*, **64**, 17–24.

Zaragosi S., Auffret, G.A., Turon, J.-L., Garlan, T., Eynaud, F., and Pujol, C. (2001) Initiation of the European deglaciation as recorded in the northwestern Bay of Biscay slope environments (Meriadzek Terrace and Trevelyan Escarpment): a multi-proxy approach. *Earth and Planetary Science Letters*, **188**(3-4): 493–507.

Ziegler, A.M. and McKerrow, W.S. (1975) Silurian marine red beds. *American Journal of Science*, **275**, 31–56.

Ziegler, P.A. (1981) Evolution of Sedimentary Basins in North-West Europe. *In*: Illing, L.V. and Hobson G.D. (eds). *Petroleum Geology of the Continental Shelf of North-West Europe*. Heyden & Son Ltd, London, 3–39.

Ziegler, P.A. (1982) *Geological Atlas of Western and Central Europe. 1st Edition*. Shell Internationale Petroleum Maatschappij, B.V.

Ziegler, P.A. (1987) *Evolution of the Arctic–North Atlantic rift system*. American Association of Petroleum Geologists Memoir, 198pp.

Ziegler, P.A. (1990) *Geological Atlas of Western and Central Europe. 2nd Edition*. Shell Internationale Petroleum Maatschappij B.V.

Index

Page numbers in *italic* denote figures. Page numbers in **bold** denote tables.

C

E

F

G

H

L

N

O